THEORY OF LAMINAR FLOWS

I. Thermodynamics and Physics of Matter. Editor: F. D. Rossini
II. Combustion Processes. Editors: B. Lewis, R. N. Pease, H. S. Taylor
III. Fundamentals of Gas Dynamics. Editor: H. W. Emmons
IV. Theory of Laminar Flows. Editor: F. K. Moore
V. Turbulent Flows and Heat Transfer. Editor: C. C. Lin
VI. General Theory of High Speed Aerodynamics. Editor: W. R. Sears
VII. Aerodynamic Components of Aircraft at High Speeds. Editors: A. F. Donovan, H. R. Lawrence
VIII. High Speed Problems of Aircraft and Experimental Methods. Editors: A. F. Donovan, H. R. Lawrence, F. Goddard, R. R. Gilruth
IX. Physical Measurements in Gas Dynamics and Combustion. Editors: R. W. Ladenburg, B. Lewis, R. N. Pease, H. S. Taylor
X. Aerodynamics of Turbines and Compressors. Editor: W. R. Hawthorne
XI. Design and Performance of Gas Turbine Power Plants. Editors: W. R. Hawthorne, W. T. Olson
XII. Jet Propulsion Engines. Editor: O. E. Lancaster

VOLUME IV

HIGH SPEED AERODYNAMICS

AND JET PROPULSION

THEORY OF LAMINAR FLOWS

EDITOR: F. K. MOORE

PRINCETON, NEW JERSEY

PRINCETON UNIVERSITY PRESS

1964

London: OXFORD UNIVERSITY PRESS

L. C. CARD 62-9129

PRINTED IN THE UNITED STATES OF AMERICA
THE MAPLE PRESS CO., INC., YORK, PENNA.

FOREWORD

On behalf of the Editorial Board, I would like to make an acknowledgement to those branches of our military establishment whose interest and whose financial support were instrumental in the initiation of this publication program. It is noteworthy that this assistance has included all three branches of our Services. The Department of the Air Force through the Air Research and Development Command, the Department of the Army through the Office of the Chief of Ordnance, and the Department of the Navy through the Bureau of Aeronautics, Bureau of Ships, Bureau of Ordnance, and the Office of Naval Research made significant contributions. In particular, the Power Branch of the Office of Naval Research has carried the burden of responsibilities of the contractual administration and processing of all manuscripts from a security standpoint. The administration, operation, and editorial functions of the program have been centered at Princeton University. In addition, the University has contributed financially to the support of the undertaking. It is appropriate that special appreciation be expressed to Princeton University for its important over-all role in this effort.

The Editorial Board is confident that the present series which this support has made possible will have far-reaching beneficial effects on the further development of the aeronautical sciences.

Theodore von Kármán

PREFACE

Rapid advances made during the past decade on problems associated with high speed flight have brought into ever sharper focus the need for a comprehensive and competent treatment of the fundamental aspects of the aerodynamic and propulsion problems of high speed flight, together with a survey of those aspects of the underlying basic sciences cognate to such problems. The need for a treatment of this type has been long felt in research institutions, universities, and private industry and its potential reflected importance in the advanced training of nascent aeronautical scientists has also been an important motivation in this undertaking.

The entire program is the cumulative work of over one hundred scientists and engineers, representing many different branches of engineering and fields of science both in this country and abroad.

The work consists of twelve volumes treating in sequence elements of the properties of gases, liquids, and solids; combustion processes and chemical kinetics; fundamentals of gas dynamics; viscous phenomena; turbulence; heat transfer; theoretical methods in high speed aerodynamics; applications to wings, bodies and complete aircraft; nonsteady aerodynamics; principles of physical measurements; experimental methods in high speed aerodynamics and combustion; aerodynamic problems of turbo machines; the combination of aerodynamic and combustion principles in combustor design; and finally, problems of complete power plants. The intent has been to emphasize the fundamental aspects of jet propulsion and high speed aerodynamics, to develop the theoretical tools for attack on these problems, and to seek to highlight the directions in which research may be potentially most fruitful.

Preliminary discussions, which ultimately led to the foundation of the present program, were held in 1947 and 1948 and, in large measure, by virtue of the enthusiasm, inspiration, and encouragement of Dr. Theodore von Kármán and later the invaluable assistance of Dr. Hugh L. Dryden and Dean Hugh Taylor as members of the Editorial Board, these discussions ultimately saw their fruition in the formal establishment of the Aeronautics Publication Program at Princeton University in the fall of 1949.

The contributing authors and, in particular, the volume editors, have sacrificed generously of their spare time under present-day emergency conditions where continuing demands on their energies have been great. The program is also indebted to the work of Dr. Martin Summerfield who guided the planning work as General Editor from 1949–1952. The cooperation and assistance of the personnel of Princeton University Press and of the staff of this office has been noteworthy. In particular, Mr. H. S. Bailey, Jr., the Director of the Press, and Mr. R. S. Snedeker, who has supervised the project at the Press have been of great help. The figures were prepared by Mr. Zane Anderson. Special mention is also due

Mrs. E. W. Wetterau of this office who has handled the bulk of the detailed editorial work for the program.

Coleman duP. Donaldson
General Editor

PREFACE TO VOLUME IV

Historical research indicates that the present volume of Laminar Flow Theory had its beginnings under the joint editorship of Lester Lees and C. C. Lin. As is often the case with classics of antiquity, much of this early work has been lost. Fortunately, however, the article on basic theory by Prof. Lagerstrom was preserved from the time of its writing, in 1953, to the present; now, as Sec. B, it forms the basis of the present volume, having meanwhile served for years as course notes at California Institute of Technology! In 1957, volume editorship devolved on this writer, and the present group of authors began their work. Thus, except for Sec. B, this book was written in the period 1958–1960.

Though it is regrettable that Prof. Lagerstrom's article has remained unpublished for so long, the hiatus no doubt has had a desirable effect on the reference value of the remainder of the book. The first half of the last decade saw especially vigorous development of boundary layer theory in special categories, such as three-dimensional (A. Mager) unsteady (N. Rott), and hypersonic (F. K. Moore) boundary layers. During the same period, there occurred a surge of interest in body-force flows (S. Ostrach), and the theory of laminar stability (S. F. Shen) was strikingly advanced by consideration of a wider class of disturbances.

Therefore, writing at the time they did, the authors of this book were able to present rather well filled-in pictures of their subjects. These would not require important changes to be brought up to date of publication. It is, of course, true that advances in laminar flow theory have been made in the past few years, but the requisite effort to add discussions of this new material has generally not been made. It is hoped that any deficiencies arising from this cause are outweighed by the merits, for reference use, of such unique reviews as those of Stokes-Oseen flow in Sec. B, and of natural convection in Sec. F, for example.

A separate article on separated flows had originally been planned. This did not materialize, and therefore a rather abrupt discussion of separation and separated flows is included in Sec. A.

Finally, I should like to thank the authors for their great effort and cheerful cooperation; Dr. Donaldson (General Editor) for his invaluable advice and generous assistance; and Princeton University Press for their fine work.

F. K. Moore
Volume Editor

CONTENTS

CONTENTS

Chapter 3. Introductory Discussion of the Navier-Stokes Equations for a Compressible Fluid

Chapter 4. Laminar Boundary Layers in Compressible Fluids

THE FLAT PLATE WITH ZERO PRESSURE GRADIENT

BOUNDARY LAYERS WITH PRESSURE GRADIENTS

Chapter 2. Free Convection. The External Problem

Chapter 3. Thermal Instability

Chapter 4. Natural Convection. The Internal Problem

Chapter 5. Unsteady Flows with Body Forces

Chapter 6. Hydrodynamic Stability

Chapter 3. Extensions and Applications of the Stability Theory of Transverse Wave Disturbances

Chapter 4. Stability Theory Dealing with Longitudinal Vortex-Type Disturbances

Chapter 5. Behavior Subsequent to the Onset of Infinitesimal Disturbances

THEORY OF LAMINAR FLOWS

SECTION A

INTRODUCTION

F. K. MOORE

A,1. Scope of Laminar Flow Theory. The present volume is concerned with fluid motions which are influenced in an important way by viscosity and heat conductivity of the flowing substance. The fluid is regarded as a continuum in which the resistance to shearing motion and the conduction of heat arise by exchange processes occurring on a molecular scale. When viscous and conductive effects arise from this cause, the motion is said to be laminar, as distinct from turbulent motion, in which macroscopic eddies are responsible for high rates of transfer of momentum and energy. These turbulent motions are treated in Vol. V of the present series.

Some of the earliest treatments of viscous effects in fluid motion are provided by Rayleigh [*1*] and Lamb [*2*] in connection with the damping of wave motions. These classical solutions show clearly that if the fluid is one of low viscosity, such as air, viscous effects tend to be confined to the vicinity of solid surfaces. This tendency for a thin "boundary layer" to form in a slightly viscous fluid was exploited by Prandtl [*3*] and others, in a manner particularly appropriate to problems of flight. The Prandtl thin boundary-layer theory applies when the ratio of inertia to viscous effects, expressed as a Reynolds number, is large. Even if viscosity is small, the Reynolds number may be small for slow motions or very small objects, in which case the Stokes-Oseen theory [*4*] applies. Roughly, the high Reynolds-number theory of Prandtl applies to flows about aircraft, whereas the low Reynolds-number theory of Stokes and Oseen has been used, for example, to determine the drag of the exceedingly fine droplets of oil used by Millikan [*5*] to measure the charge of the electron, and to the study of swimming motions of microscopic creatures [*6*].

The basic continuum theory of laminar flows is developed in Sec. B of this volume by Lagerstrom. In that section, chief emphasis is placed on the theory of two-dimensional steady flows of an ideal fluid without body forces, and particular attention is paid to the role of Reynolds number in laminar flows. Lagerstrom's present treatment complements previous treatises on laminar flow theory, among which are those of Schlichting [*7*] and that prepared under the editorship of Howarth [*8*].

Succeeding sections of this volume deal with various elaborations and extensions of the basic theory which have been developed largely in the last decade in response to the appearance of new problems concerning flow technology. Thus, in Sec. C, Mager provides a systematic account of three-dimensional boundary layer theory, emphasizing effects of the "secondary flow" resulting when pressure gradients transverse to the inviscid streamline directions exist. Secondary flow is discussed briefly by Goldstein [*9*] in explanation of the sinuosity developed by water courses. More recently, the development of axial flow compressors has raised many boundary layer questions of a three-dimensional character. Study of the high speed external aerodynamics of slender shapes has involved the analysis of the secondary flows which accompany such configurations at angle of attack.

The viscous damping of wave motions, as we have mentioned, furnishes an early example of unsteady viscous flow, for which a rather general theoretical treatment is provided by Rott in Sec. D. Many dynamical problems in aeronautics involve unsteady boundary layer effects susceptible of theoretical treatment. The stalling flutter of airfoils or compressor blades involve the effect of lift hysteresis, which is intimately connected with the unsteady behavior of boundary layers approaching separation, as do certain problems of panel flutter [*10*]. Also, connections may be drawn between the theory developed in Sec. D with the aerodynamics of flight in a gusty atmosphere.

Modern concern with external hypersonic aerodynamics is reflected in Sec. E which essentially treats laminar boundary layers for which the external flow has very high stagnation temperature. One important consequence of very high stagnation temperature is that, as gas is brought to rest in a boundary layer, chemical reactions between the constituents of air may take place. Such real-gas effects, characteristic of hypersonic flows, are discussed chiefly in terms of flow situations which are otherwise classical. Attention is also paid to boundary layer displacement interactions which are particularly important in hypersonic flow, owing to severe aerodynamic heating in the boundary layer.

In Sec. F, Ostrach has provided an extensive account of the theory of laminar flows with body forces, a class of flows which is especially relevant to modern heat transfer technology. Included are a range of problems, including free convection motions and chimney flows, as well as an introduction to problems of the magnetohydrodynamic boundary layer.

The laminar boundary layer as it proceeds over a surface ends generally in transition to turbulence. Accordingly, it is fitting that this volume terminate with the review provided by Shen of the theory of hydrodynamic stability, which deals with the beginnings of the process of transition. Included is a thorough discussion of the modern work on skew disturbances and compressibility effects.

A,2. The Thin Boundary Layer Assumptions. Much of the theory presented in this volume embodies the Prandtl thin boundary layer assumptions appropriate for high Reynolds number. This matter is discussed quite thoroughly by Lagerstrom in B,8. Under these assumptions, the boundary layer thickness (δ) is shown to be very much less than any characteristic length (L) connected with the surface over which the boundary layer develops; that is, $\delta/L \ll 1$. Under this assumption, as explained in B,8, the differential equations for the motion are simplified. For example, the problem for the stream function of steady, plane, incompressible flow is

$$\left.\begin{aligned} \psi_y\psi_{xy} - \psi_x\psi_{yy} &= -\frac{1}{\rho}p_x + \nu\psi_{yyy} \\ \psi_y(x, \infty) = u_1(x); \qquad \psi(x, 0) &= \psi_y(x, 0) = 0 \end{aligned}\right\} \tag{2-1}$$

where the x coordinate follows the surface and y is normal to the surface. For a thin boundary layer, the pressure does not change across the boundary layer, and is characterized by an inviscid flow which is identical with that which would prevail in the absence of a boundary layer. Thus,

$$\frac{1}{\rho}p_x = -u_1u_{1_x} \tag{2-2}$$

where $u_1(x)$ is the velocity component of the inviscid flow in the surface direction just outside the boundary layer. The boundary condition $\psi_y(x, 0) = 0$ expresses the no-slip condition which is correct provided the fluid is reasonably dense; in other words, when $\delta/l \gg 1$, where l is the molecular mean free path.

When this latter assumption is not met, changes in the theory are necessary to account for slip effects, for example, occurring at the surface. These matters are discussed by Schaaf in another volume of this series (III,H). In addition to the boundary conditions above, it is necessary also to describe the circumstances under which the boundary layer begins its course. For example, if a sharp leading edge is present, as in Fig. A,2, it is appropriate to require that $\psi_y(0, y) = u_1(0)$.

Even though the length of the body (for example, the distance L noted in the idealized sketch shown in Fig. A,2) is large and the boundary layer is generally thin compared with L, it is not assured that the boundary layer assumptions are met everywhere in the flow. Rather, δ must be small relative to any characteristic length which may locally be of interest. For example, near the leading edge in Fig. A,2, the boundary layer at first grows sufficiently fast that over a distance $\nu/u_1(0)$ (a very small distance), the boundary layer assumptions are violated, and a special analysis is required. Reference in this connection may be made to work of Kuo [*11*]. In incompressible flow, this leading edge region is followed by a boundary layer region generally meeting the assumptions required for validity of Eq. 2-1. Now, if the flow proceeds over a shoulder, as sketched, in such a way that the flow is decelerating in a region of increasing pressure, this pressure causes the low-inertia fluid in the boundary layer to reverse its direction and lift away from the surface as sketched. This separation process, like the leading edge process, is one to which the boundary layer equations do not apply, because conditions are changing too rapidly along the surface in comparison with changes across the boundary layer. Nevertheless, as we shall see subsequently, the boundary layer assumptions may, in a certain sense, become valid again in the separated region.

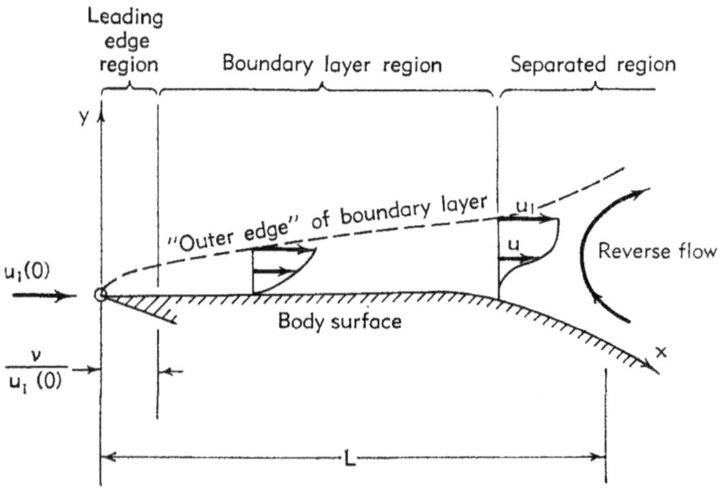

Fig. A,2. **Sketch of boundary layer development from a leading edge through separation.**

From the theoretical standpoint, the work of Kuo [*11*] shows that the deficiencies of boundary layer theory near a leading edge may quite successfully be overcome. Because the leading edge region, if one exists, is so small, it is generally proper to solve boundary-value problems of the type given in Eq. 2-1 for the boundary layer region beginning directly at the leading edge. The separated region, and in particular the point of separation, is much more difficult to analyze, and here the theory is not very firm. We may add that in the boundary layer region, it is quite feasible and, in the case of hypersonic flow, usually necessary to make a first-order correction to boundary layer theory by taking into account deflection of the outer inviscid flow caused by progressive thickening of the boundary layer.

The problem represented by Eq. 2-1 is sufficiently difficult that approximate methods of solution are usually applied. In the remainder of this section we will briefly review certain methods of approximate solution for the boundary layer region, with particular reference to the problem of predicting separation. In addition, we shall consider the application of boundary layer theory to the separated regions associated with wakes and cavities, and arising as a consequence of the impingement of shock waves on thin boundary layers. So far as the leading edge region is concerned, we may simply note that mathematically similar problems concerning boundary layers which, though thin, locally violate the boundary layer assumptions appear in certain three-dimensional cases discussed in Sec. C, an example of which is the boundary layer flowing along a corner, and in unsteady boundary layers such as that of a semi-infinite flat plate impulsively started from rest (Sec. D).

In the discussion to follow, we intend to complement material provided in succeeding sections. Therefore, a reader not already having a certain familiarity with viscous flow theory may wish to put aside the rest of this section until after Sec. B, at least, has been read.

A,3. The Approach to Separation. Before discussing various methods of approximate boundary layer calculation, it is well to inquire into the capacity of the boundary layer Eq. 2-1 to represent the occurrence of separation. Separation is generally thought of as an abrupt thickening of the boundary layer (probably in violation of the thin boundary layer assumption) connected with the vanishing of shearing stress at the surface, as indicated in Fig. A,2, followed by a region of reversed flow. As a matter of general principle, it is not altogether clear whether the abrupt thickening or the vanishing of skin friction provides the more meaningful criterion of separation.

The only exact solution of the thin boundary layer equations which relates to this problem is that of Falkner and Skan [*12*] (B,8), which deals with an external velocity distribution given by

$$u_1 \sim x^m \tag{3-1}$$

where, for negative values of m, the pressure gradient is adverse and separation may be expected. The Falkner-Skan problem is treated in detail in B,8, and we shall mention here only the result for shear stress at the surface as a function of $\beta \equiv 2m/(m+1)$. A fully self-similar solution applies for this problem, which is physically related to the incompressible boundary layer on a downward-inclined ramp. The dimensionless shear stress resulting from the solution, sketched in Fig. A,3, shows that, as pressure gradient is imagined to become gradually more adverse, the shear stress approaches zero with a half-order singularity at $\beta = -0.20$, and reverse flow is found only as a presumably meaningless ambiguity for values of β somewhat greater than this value.

From this problem, one might infer that the thin boundary layer equations are capable of representing separation only by a singularity, recognizing of course that the Falkner-Skan problem is a special and somewhat artificial one. Of course, the appropriate solution of the complete Navier-Stokes equations (B,6) would not show any such singularity. However, our immediate purpose is to understand what sort of description of the separation process may be expected from the thin boundary layer equations. A classical series solution for separation in a linearly decreasing velocity field is due to Howarth [*13*], who found a rather abrupt approach to separation, with a suspicion of a singularity at the separation point, but no definite proof, owing to the fact that the series expansion begins at a point some distance upstream of separation.

A specific attack on the question of a singularity at separation was made by Goldstein [*14*], who treated the vicinity of separation by a Taylor series in x applicable upstream of a place, presumed known, where wall shear is considered to vanish. He found that a singular solution of the boundary layer equations near separation is always a possibility. We may quickly indicate the essential character of his results by the following brief analysis. Eq. 2-1 may be integrated directly to yield the equation

$$u^2 - u_s^2 = 2x(u_1 u_{1_x} + \nu u_{s_{yy}}) + 2\int_0^x u_y \int_0^y u_x dy dx \qquad (3\text{-}2)$$

where x refers to distance measured downstream from the separation point, where the (known) velocity profile $u_s(y)$ has vanishing slope at $x = 0$. Eq. 3-2 clearly suggests that near the surface, as separation is approached, u must approach u_s as $\sqrt{-x}$, provided the right-hand side of Eq. 3-2 is regular in x. In fact, one obtains Goldstein's result if one represents the separation profile (presumed known) by a power series in y of the form

$$u_s = b_2 \frac{y^2}{2!} + b_3 \frac{y^3}{3!} + \cdots \qquad (3\text{-}3)$$

and assumes that at nearby negative values of x the velocity profile has the power series

$$u = a_1 y + a_2 \frac{y^2}{2!} + a_3 \frac{y^3}{3!} + \cdots \qquad (3\text{-}4)$$

where, in Eq. 3-4, the coefficients are presumed to be functions of x. Substitution of Eq. 3-3 and 3-4 into Eq. 3-2 yields the results

$$\left.\begin{aligned} \nu b_2 &= -u_1 u_{1_x}; \qquad \nu b_3 = 0; \qquad a_2 = b_2 \\ a_1 &= \sqrt{-2\nu b_4}\,\sqrt{-x} \end{aligned}\right\} \qquad (3\text{-}5)$$

The coefficient a_1, which represents shearing stress at the surface, is indicated to have a square-root singularity analogous to that sketched for the Falkner-Skan problem in Fig. A,3.

The foregoing argument does not constitute proof of the existence of a singularity, because the unproved assumption is made that the coefficient b_4 of the separation profile does not vanish. However, it seems plausible that the thin boundary layer equations would always indicate a singular abrupt approach to separation; at least, no evidence to the contrary seems to exist. Further, one may infer that a complete Navier-Stokes solution would exhibit an abrupt but nonsingular approach to separation, and that a singular solution is the best representation of the situation of which the thin boundary layer equations are capable.

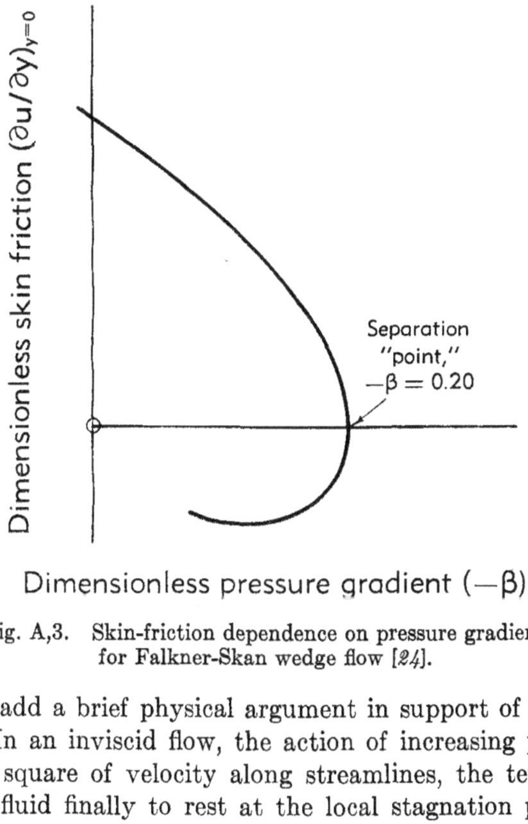

Fig. A,3. Skin-friction dependence on pressure gradient for Falkner-Skan wedge flow [*24*].

One may add a brief physical argument in support of the foregoing conclusions. In an inviscid flow, the action of increasing pressure is to diminish the square of velocity along streamlines, the tendency being to bring the fluid finally to rest at the local stagnation pressure. The tendency of viscous shearing stresses, for the positive profile curvature indicated at separation in Fig. A,2, is to drag fluid elements downstream in opposition to the decelerating effect of adverse pressure gradient. Thus, if the flow adheres to the surface as assumed by thin boundary layer theory, there is no mechanism to produce reverse flow. Accordingly, if shear is to vanish, it must be expected to do so in a singular manner. For solutions of the Navier-Stokes equations, which have a more complete description of viscous stresses, the foregoing argument would, of course, not be correct.

We see, then, that approximate methods of solution of the laminar thin boundary layer equations to be outlined in the next paragraphs must contemplate a very abrupt approach to separation, in order properly to represent the probable behavior of exact solutions, having in mind, of course, that the thin boundary layer equations themselves may not be too accurate near separation.

A,4. Approximate Methods of Boundary Layer Calculation. Eq. 2-1 have been solved exactly only in cases reducible to a single variable, by similarity considerations. This reduction is available only for the class of external flows defined by Eq. 3-1, as explained in B,8. For other cases, approximate methods must be used. In the next paragraphs, we will briefly survey some of the approximate theories, supplementing the discussion in B,17 of the Pohlhausen and Thwaites methods, and emphasizing incompressible flow.

Integral methods. As explained in B,8, the differential Eq. 2-1 may be integrated to give the Kármán-Pohlhausen momentum equation for the boundary layer

$$u_1^2\theta_x + (2 + H)u_1u_{1_x}\theta = \nu(u_y)_{y=0} \tag{4-1}$$

where θ is the momentum thickness, and $H \equiv \delta^*/\theta$, δ^* being the displacement thickness of the boundary layer. In order to achieve an approximate solution based on Eq. 4-1, one begins by assuming a plausible form for the boundary layer profile $u(y)$. If such a profile were completely defined as to shape, and only the y scale were left open (e.g., $u \equiv u(y/\theta)$), then substitution into Eq. 4-1 would yield an ordinary differential equation for θ. This approach is clearly inadequate, because pressure gradient strongly affects profile shape, and, in fact, separation profiles having vanishing gradients are the surface must be represented.

Thus, in the Pohlhausen and Thwaites methods, a free "shape parameter" is admitted in the profile description, with the consequent necessity to provide an additional differential equation. The additional equation is obtained by simply evaluating Eq. 3-1 at the surface

$$u_1u_{1_x} + \nu(u_{yy})_{y=0} = 0 \tag{4-2}$$

Eq. 4-2 is satisfied by requiring that the nondimensional curvature of the profile at the surface is

$$[\partial^2(u/u_1)/\partial(y/\theta)^2]_{y=0} = \theta^2u_{1_x}/\nu \equiv \lambda \tag{4-3}$$

(for the Thwaites method). The quantity λ is the shape parameter of the profile.

Two interpretive remarks should be added here: First, adoption of Eq. 4-3 implies that it is most important to achieve a good approximation to conditions close to the surface; and second, use of the single parameter λ

implies that higher derivatives of $u_1(x)$ are not important. This second point may be clarified by considering the physical meaning of λ. By classical diffusion theory, θ^2/ν represents time required for the effect of some change to diffuse through a region of thickness θ, while u_{1_x} is the reciprocal of a time characterizing how fast the velocity of a fluid element just outside the boundary layer changes. Now, in the boundary layer with pressure gradient, these two effects compete: The exterior velocity changes, and viscous diffusion adjusts the profile accordingly. Thus, if the "diffusion time" is short compared to the "change time," λ is small, and the profile is nearly that for zero pressure gradient. Now, it is necessary to observe that a whole sequence of "change times" may be postulated, including $(u_{1_x})^{-1}$, $(u_1 u_{1_{xx}})^{-\frac{1}{2}}$, $(u_1^2 u_{1_{xxx}})^{-\frac{1}{3}}$, etc. By writing Eq. 4-3, we assume that the first of this group is the most important. This assumption is especially suspect when the velocity is changing rapidly, and, in particular, one should not infer from Eq. 4-3 that the flat plate profile applies where a rapidly changing u_1 passes through a maximum or minimum.

In the Pohlhausen method (elaborated and codified by Holstein and Bohlen (see B,17)), the profile description is completed by requiring the boundary layer velocity to merge smoothly with the outer flow. In the Thwaites method [*15,16*], (B,17), no such full description of the profile is attempted; instead, it is judged more important (especially for adverse pressure gradients leading to separation) to specify a relationship between wall shear stress and λ which is in good agreement with certain series and similarity solutions. Though no single such relationship is "correct," the one proposed by Thwaites (Fig. A,4a) is empirically superior to the one derived by the Pohlhausen method, which generally predicts a too-late separation. For example, the Thwaites relationship agrees rather well with the Falkner-Skan solution sketched in Fig. A,3, including the singularity at separation discussed in Art. 3.

Of course, to apply Fig. A,4a for finding skin friction, it is necessary to know θ. Thwaites suggests a simple quadrature, based on a study of the dependence on λ of various quantities appearing in Eq. 4-1

$$\theta^2 = 0.45\nu u_1^{-6} \int_0^x u_1^5 dx \tag{4-4}$$

We have noted that Eq. 4-2 embodies an assumption that it is important to satisfy the momentum equation precisely at the wall. (The common habit of referring to Eq. 4-2 as a "boundary condition" obscures this point.) Tani [*17*] in effect suggests that such a procedure is inconsistent with an integral method, and is the source of much of the inaccuracy of the Pohlhausen method applied near separation, including a difficulty shared by the Thwaites method, namely, that skin friction should not be a sole function of λ according to various exact and series solutions.

In place of Eq. 4-2, Tani substitutes the kinetic energy equation gotten by multiplying the momentum Eq. 2-1 by u and integrating across the boundary layer. He then adopts exactly the Pohlhausen (quartic) profile, including profile curvature at the wall as the shape parameter (no longer, now, identical to λ as defined in Eq. 4-3). Tani's solution for a family of retarded flows is indicated in Fig. A,4a and agrees beautifully with the more exact solutions (crosses). In Tani's method, one uses Eq. 4-4 to determine θ, and hence λ, and then applies another equation connecting λ with the curvature parameter upon which skin friction

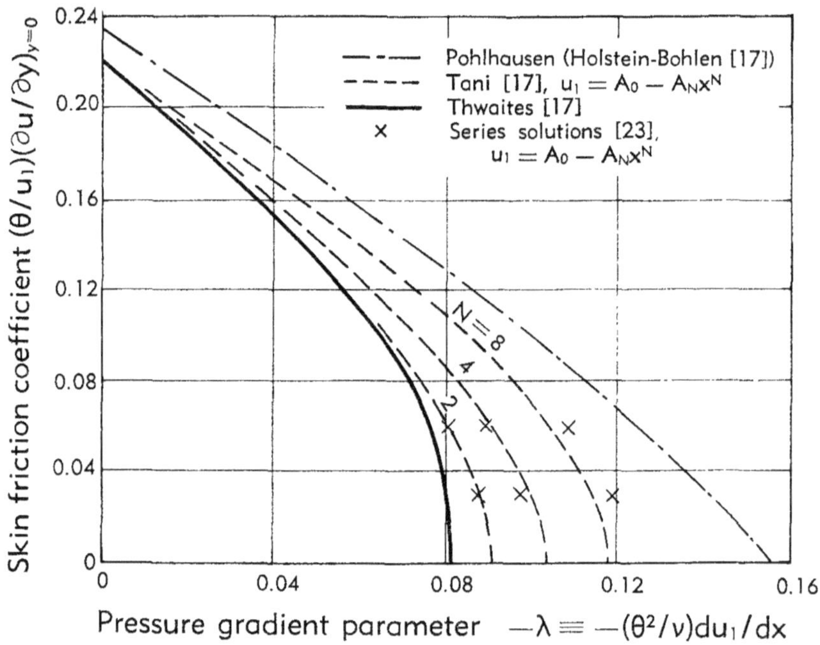

Fig. A,4a. Relation between skin friction and profile shape parameter, for various integral methods.

depends. This latter equation is not especially easy to solve; nevertheless, the accuracy and theoretical directness of his method are remarkable.

As we have remarked earlier, the boundary layer theory may often apply downstream of separation, especially where extensive "dead water" regions are bounded by rather abrupt shear layers. In this connection, we briefly discuss the Crocco-Lees "mixing theory" [*18*], recently re-examined by Glick [*19*]. First, we note that Eq. 4-2 is completely inappropriate for separated flows: Conditions at the surface are certainly not as important as the mass entrainment process occurring in the outer part of the boundary layer. In fact, Glick remarks that the Tani method just described might be adapted to the study of separated flows.

In the Crocco-Lees method, instead of Eq. 4-2 or Tani's energy equation, a continuity equation is adopted, which states that the boundary layer gains mass in proportion to the component of mass flow normal to the outer edge of the boundary layer. The shape parameter is, one may say, invented for the Crocco-Lees method on physical grounds (not derived from the equations, as in the previous methods). In effect, the parameter is a ratio which compares the relative importance of the mass-entrainment process at the outer edge of the boundary layer with the viscous loss of momentum occurring within the boundary layer. In a separated layer, the former effect is more important than in attached flows. The method is finally completed by the specification of certain correlations derived from more exact solutions or from experiment, connecting both skin friction and mass entrainment with the assumed shape parameter. Details of this development are discussed by Glick [*19*] for both attached and separated flows, emphasizing the problem of shock-boundary layer interaction, together with references to other work depending on this ambitious and troublesome theory.

Of course, the integral methods previously discussed have been extended to include compressibility and thermal effects appropriate to high speed flows (see B,17). Tani [*17*] applies his method to compressible flows, and Rott and Crabtree [*16*], as well as Cohen and Reshotko [*20*], employ the Thwaites method for this purpose. Hayes [*21*] deals with the real-gas boundary layer at high temperature by a method related to that of Thwaites.

Series methods. Returning to the original Eq. 2-1, we may observe the possibility of expanding ψ in a suitable power series in x. When the exterior velocity $u_1(x)$ is likewise expanded, and both series are introduced into Eq. 2-1, a sequence of ordinary differential equations results. This classical method is discussed in [*7,8,9*], and will only be mentioned here. If the boundary layer begins at a stagnation point $u_1(0) = 0$, the Blasius-Hiemenz [*22*] form

$$u_1 = \sum_1^\infty A_n x^n; \qquad \psi = \sum_1^\infty f_n(y) x^n \tag{4-5}$$

is appropriate. If the boundary layer begins as a flat plate, the Howarth [*13*] type is applicable:

$$u_1 = \sum_0^\infty A_n x^n; \qquad \psi = \sqrt{x} \sum_0^\infty f_n\left(\frac{y}{\sqrt{x}}\right) x^n \tag{4-6}$$

Various examples have been treated by use of these series, especially, for retarded flows, when

$$u_1 = A_0 - A_N x^N \tag{4-7}$$

and $N = 1$ (Howarth [*13*]), and $N = 2, 4, 8$ (Tani [*23*]). The crosses marked on Fig. A,4a refer to calculations made by this method.

The labor involved in the foregoing series solutions is extreme. For meaningful results approaching separation, 7 or more terms (solutions for the f_n) are required. A generally more economical series method has been proposed by Görtler [*24*], and elaborated for heat conducting flow by Sparrow [*25*]. Görtler's formulation uses insight, gained essentially from the similar solutions of Falkner and Skan for wedge flows (see B,17), to arrange matters so that the leading term of the expansion satisfies the complete boundary conditions for all x. In this way, the role of succeeding terms is limited to the modification of profile shape.

Görtler begins by defining new independent variables

$$\xi \equiv \frac{1}{\nu} \int_0^x u_1(x)dx; \qquad \eta \equiv \frac{u_1 y}{\nu \sqrt{2\xi}} \tag{4-8}$$

and, in place of ψ,

$$F(\xi, \eta) \equiv \frac{\psi}{\nu \sqrt{2\xi}} \tag{4-9}$$

These definitions transform Eq. 2-1 to

$$\left.\begin{aligned} &F_{\eta\eta\eta} + FF_{\eta\eta} + \beta(\xi)(1 - F_\eta^2) = 2\xi(F_\eta F_{\xi\eta} - F_\xi F_{\eta\eta}) \\ &F_\eta(\xi, \infty) = 1; \qquad F(\xi, 0) = F_\eta(\xi, 0) = 0 \end{aligned}\right\} \tag{4-10}$$

where $\beta \equiv 2(u_{1_x}/u_1^2) \int_0^x u_1(x)dx$. Now, from a known function $u_1(x)$, a series

$$\beta = \sum_0^\infty \beta_{n/2}\xi^{n/2}$$

in general may be derived. $F(\xi, \eta)$ may then be expanded in like powers of ξ, each coefficient being a function of η requiring solution. From Eq. 4-10, it is clear that the leading approximation is one of the wedge-flow similar solutions, adjusted in scale to satisfy the condition $\psi_y(x, \infty) = u_1(x)$ for all x.

Görtler [*24*] compares his results, truncated at two terms, with Tani's series solution [*24*] of the form of Eq. 4-6 carried out to 6 or more terms, for various cases of Eq. 4-7, including the case $N = 4$ shown on Fig. A,4b. Apart from a special continuation method used by Tani near separation, the results appear to be in agreement. Despite its efficiency, Görtler's method of course suffers from the usual convergence problems near separation, at least in principle.

Formally, the leading term of Görtler's series (when $\beta_0 = 0$) corresponds closely to the "local similarity" scheme for hypersonic boundary layer calculation introduced by Lees [*26*] and discussed in E,8, though in the latter case the theory is justified by an assumption that β is small.

There are, of course, various other uses to which Taylor series may be put in boundary layer theory (e.g., analysis of hypersonic displacement interactions, E,9) which will appear in the context of succeeding sections.

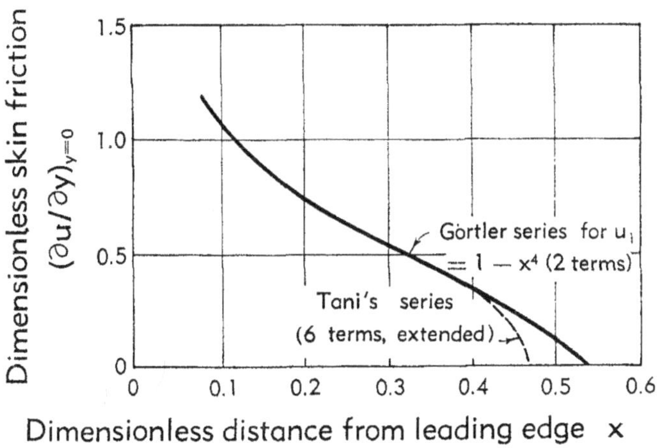

Fig. A,4b. Comparison of Görtler's [*24*] and Tani's [*23*] series solution for skin friction in adverse pressure gradient (from [*24*]).

A,5. Separated Flows.

Flow over a cavity. In previous paragraphs, we have outlined various methods of laminar boundary layer calculation, each of which involves some degree of uncertainty when used to predict separation. In many aerodynamic problems, however, there is no doubt where separation from the surface occurs, and the problem is to analyze the separated flow. Often, especially in hypersonic and supersonic flows, the separated region consists of a cavity of "dead water" connected to the main stream by a thin mixing region, or shear layer. Such a situation is illustrated in Fig. A,5a. The shear layer is, of course, often turbulent, but in many cases may be regarded as a free laminar layer governed by the thin boundary layer equations. An extensive discussion of this class of problems is provided by Chapman, Kuehn, and Larson [*27*].

Fig. A,5a refers to the sudden separation at a rear-facing step, analyzed by Chapman [*28*]. Downstream of the corner, the shear layer grows parabolically under constant pressure and a similar solution of the boundary layer equation applies. The boundary conditions are that $u(x, \infty) = u_1$ (a constant), and $u(x, -\infty) = 0$. Here, we have no actual body on which to require that normal velocity (v) vanish. However, there will be a place in the layer where $v = 0$, defining a "dividing streamline" which proceeds from the corner and separates the fluid entrained out of the main stream from that entrained out of the "dead water" region. The velocity u_d along this dividing streamline is not zero; Chapman finds that $u_d/u_e \cong 0.59$. Chapman [29] has applied this theory to the study of heat transfer in cavity flows.

Reattachment of a mixing layer. It often happens, especially in supersonic flow [27], that a free mixing layer rejoins a surface, as in Fig.

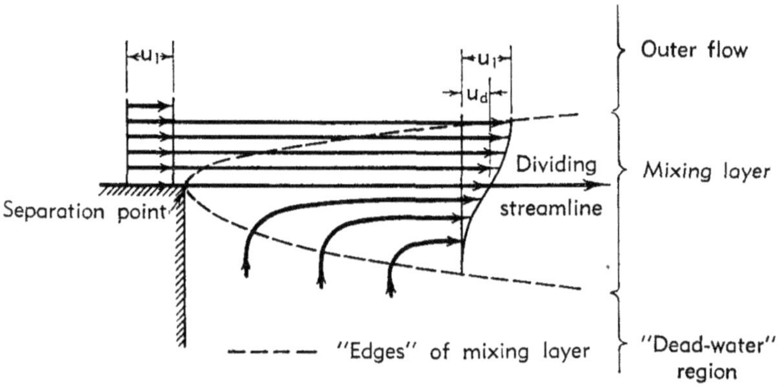

Fig. A,5a. Sketch of separated shear layer, after Chapman [28].

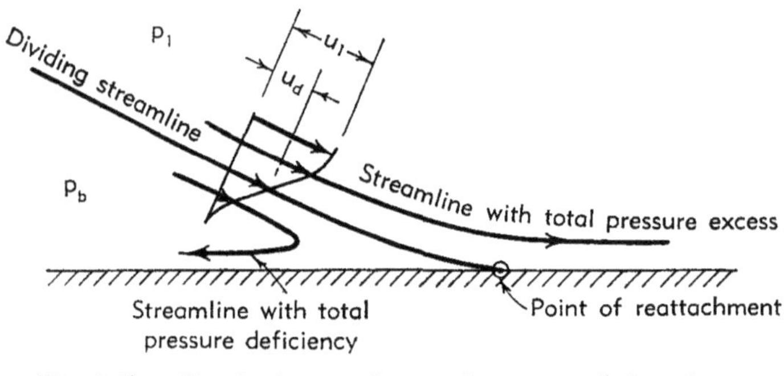

Fig. A,5b. Sketch of reattachment of a separated shear layer, after Chapman, et al. [27].

A,5b. In such a case, a pressure rise occurs during reattachment. When reattachment occurs, the cavity under the mixing layer is thereby closed, and p_b (Fig. A,5b) represents a base pressure which is lower than the pressure p_1 in the outer flow. This base pressure may be estimated [27] by observing that the "dividing streamline" described in Fig. A,5a must be the streamline which bifurcates at reattachment. Now, we know the velocity, and hence the total pressure, of this streamline. This total pressure must equal p_1, so that faster particles negotiate the pressure rise, and slower particles, lacking the necessary total pressure, are turned back. Application of this argument to, for example, incompressible reattachment leads to the result [27]

$$p_b = p_1 - (0.53)\tfrac{1}{2}\rho u_1^2$$

Separation due to "free interaction." In supersonic flow, as a boundary layer thickens and displacement of the outer flow occurs, the turning of the outer flow is accompanied by a pressure increase. This self-induced adverse pressure gradient may cause the boundary layer to separate. Such an action, if it occurs independently of the upstream condition

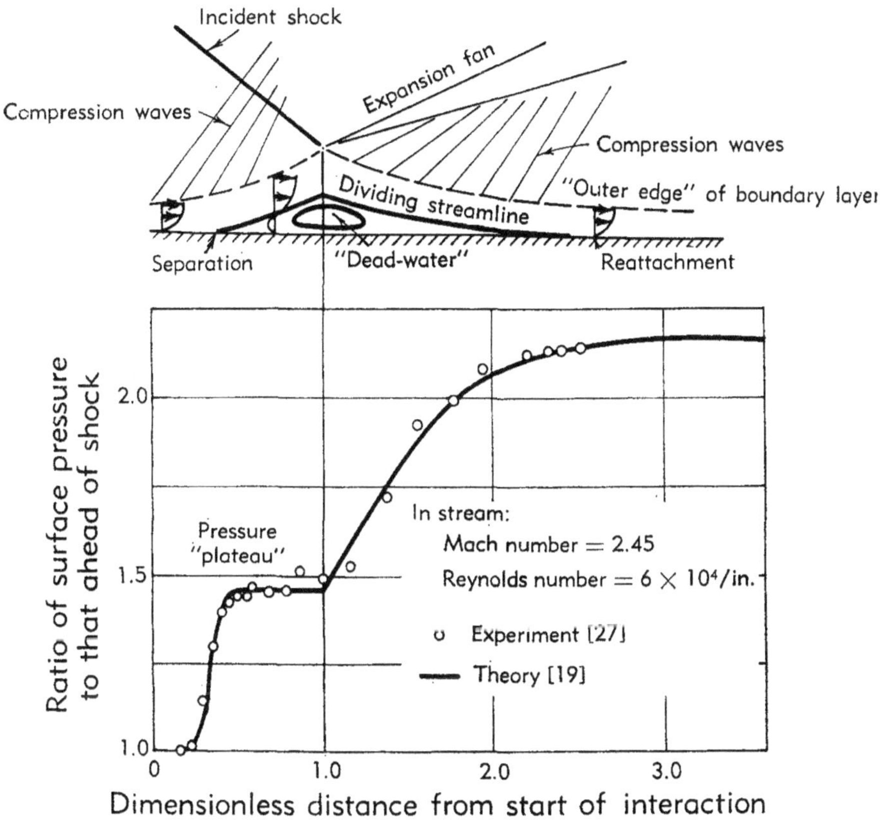

Fig. A,5c. Sketch of a shock-boundary layer interaction process, and an associated surface pressure distribution from [19].

of the boundary layer, is referred to by Chapman, et al [27] as a "free interaction." In such a case, the pressure rise to cause separation is estimated (and verified experimentally) to be proportional to the reciprocal one-fourth root of the Reynolds number at the beginning of interaction.

Shock-boundary layer interaction. The foregoing type of separation often occurs when a shock wave impinges on an otherwise thin boundary layer [*27,30,31,32*]. Fig. A,5c shows the usual picture of this interaction. The shock pressure rise is felt upstream of the actual impingement as the boundary layer thickens, inducing compression waves in the outer flow. Separation may then result, characterized by a pressure plateau. Subsequently, reattachment commonly occurs, and during this process the final shock pressure rise is achieved.

The pressure-rise estimates described in previous paragraphs have application to this problem. However, the development of even an approximate boundary layer theory for this case is very difficult. We may mention that Glick [*19*] has applied the basic ideas of the Crocco-Lees mixing theory to the description of shock-boundary layer interaction. He paid particular attention to the reattachment pressure-rise process, and achieved good agreement (Fig. A,5c) with a particular experiment. He found it necessary to assume the experimental value of plateau pressure in order to fix the necessary correlation functions (Art. 3). Nevertheless, it is argued that these functions, once fixed in relation to a particular experiment, tend to be universal.

In concluding this cursory outline of the theory of separated laminar flow, we should emphasize that the proper application of thin layer theory is of limited scope in such cases, and chief reliance must be placed on experiment, both in formulating and evaluating any theoretical approach.

A,6. Cited References.

1. Lord Rayleigh. *Theory of Sound.* Vols. I and II, Dover, 1945.
2. Lamb, H. *Hydrodynamics.* Dover, 1945.
3. Prandtl, L. Über Flüssigkeitsbewegung bei sehr kleiner Reibung. *Verhandl. 3rd Intern. Math. Kongr.,* Heidelberg 484–491 (1904); reprinted in *Vier Abhandlungen zur Hydrodynamik und Aerodynamik* (L. Prandtl and A. Betz, eds.), 1–8. Göttingen, 1927, Ann Arbor, 1943.
4. Oseen, C. W. *Neuere Methoden und Ergebnisse in der Hydrodynamik.* Akad. Verlagsgesellschaft, Leipzig, 1927.
5. Millikan, R. A. *The Electron.* Univ. Chicago Press, 1927; *Phys. Rev. 29,* 560 (1909).
6. Lighthill, M. J. On the squirming motion of nearly spherical deformable bodies through liquids at very small Reynolds numbers. *Comm. on Pure & Appl. Math. V,* 109–118 (1952).
7. Schlichting, H. *Grenzschicht-Theorie.* G. Braun, Karlsruhe, 1950. English Transl. *Boundary Layer Theory.* McGraw-Hill, 1955.
8. Howarth, L. *Modern Developments in Fluid Dynamics—High Speed Flow.* Vols. I and II, Clarendon Press, Oxford, 1953.
9. Goldstein, S. *Modern Developments in Fluid Dynamics.* Vol. I. Clarendon Press, Oxford, 1938.
10. Miles, J. W. On panel flutter in the presence of a boundary layer. *J. Aeronaut. Sci. 26,* 81–93 (1959).
11. Kuo, Y. H. On the flow of an incompressible viscous fluid past a flat plate at moderate Reynolds numbers. *J. Math. and Phys. 32,* 83–101 (1953).
12. Hartree, D. R. On the equation occurring in Falkner and Skan's approximate treatment of the equations of the boundary layer. *Proc. Cambr. Phil. Soc. 33,* 223–239 (1937).

13. Howarth, L. On the solution of the laminar boundary layer equations. *Proc. Roy. Soc. A164*, 547–579 (1938).
14. Goldstein, S. On laminar boundary-layer flow near a position of separation. *Quart. J. Mech. and Appl. Math. 1*, 43–75 (1948).
15. Thwaites, B. Approximate calculation of the laminar boundary layer. *Aero. Quart. 1*, 245–280 (1949).
16. Rott, N., and Crabtree, L. F. Simplified laminar boundary-layer calculations for bodies of revolution and yawed wings. *J. Aeronaut. Sci. 19*, 553–565 (1952).
17. Tani, I. The approximate solution of the laminar boundary-layer equations. *J. Aeronaut. Sci. 21*, 487–495 (1954).
18. Crocco, L., and Lees, L. A mixing theory for the interaction between dissipative flows and nearly isentropic streams. *J. Aeronaut. Sci. 19*, 649–676 (1952).
19. Glick, H. S. Modified Crocco-Lees mixing theory for supersonic separated and reattaching flows. *Calif. Inst. Technol. Hypersonic Research Project Memo No. 53*, 1960.
20. Cohen, C. B., and Reshotko, E. The compressible laminar boundary layer with heat transfer and arbitrary pressure gradient. *NACA Tech. Note 3326*, 1955.
21. Hayes, W. D. On laminar boundary layers with heat transfer. *Jet Prop. 26*, 270–274 (1956).
22. Hiemenz, K. Die Grenzschicht an einem in den gleichförmigen Flüssigkeitsstrom eigentauchten geraden Kreiszylinder. Dissertation, Göttingen, 1911. *Dinglers Polytechn. Jour.* 326, 321 (1911).
23. Tani, I. On the solution of the laminar boundary layer equations. *J. Phys. Soc., Japan, 4*, 149–154 (1949).
24. Görtler, H. A new series for the calculation of steady laminar boundary layer flows. *J. Math. and Mech. 6*, 1–66 (1957).
25. Sparrow, E. M. Application of Görtler's series method to the boundary layer energy equation. Readers' Forum, *J. Aeronaut. Sci. 25*, 71–72 (1958).
26. Lees, L. Laminar heat transfer over blunt-nosed bodies at hypersonic flight speeds. *Jet Propulsion 26*, 259–269 (1956).
27. Chapman, D. R., Kuehn, D. M., and Larson, H. K. Investigation of separated flows in supersonic and subsonic streams with emphasis on the effect of transition. *NACA Rept* 1356, 421-460 (1958).
28. Chapman, D. R. Laminar mixing of a compressible fluid. *NACA Rept* 958, 1950.
29. Chapman, D. R. A theoretical analysis of heat transfer in regions of separated flow. *NACA Tech. Note 3792*, 1956.
30. Liepmann, H. W. The interaction between boundary layer and shock waves in transonic flow. *J. Aeronaut. Sci. 13*, 623–637 (1946).
31. Ackeret, J., Feldmann, F., and Rott, N. Investigations of compression shocks and boundary layers in gases moving at high speed. *NACA Tech. Mem. 1113*, 1947.
32. Gadd, G. E. Interactions between shock waves and boundary layers. *Boundary Layer Research Symposium Freiburg/Br*, 1957. Edited by H. Görtler, Springer, Berlin, 1958.

SECTION B

LAMINAR FLOW THEORY[1]

P. A. LAGERSTROM

CHAPTER 1. THE NAVIER-STOKES EQUATIONS FOR A VISCOUS HEAT-CONDUCTING COMPRESSIBLE FLUID

B,1. Introduction.

Physical foundations. In this chapter a general system of equations for fluid mechanics is developed. It includes the classical Navier-Stokes equations suitably modified to take into account heat conduction and changes in density and temperature, as well as an energy law and one or more equations of state, taken from equilibrium thermodynamics. This system is referred to simply as the Navier-Stokes equations, and the subsequent theoretical treatment of the present section is based on these equations or on approximations to these equations. It is generally believed that the Navier-Stokes equations are capable of describing most phenomena observed in fluid mechanics. However, it should be remembered that any useful system of equations for fluid mechanics may always be regarded as an approximation to a more accurate system. There are indications that certain special flow problems arising, for example, in the dynamics of strong shock waves or highly rarefied gases, may require more accurate equations for an adequate treatment. For a further discussion of such problems the reader is referred to I,D, I,I, and III,H.

In principle, the range of validity of the Navier-Stokes equations should be apparent from a derivation of these equations. There are two different schemes for deriving equations for fluid mechanics. One

[1] This Section was completed in 1956 and it was not possible to revise it to include newer results and appropriate references to more recent publications. Various people, in particular GALCIT staff members, have rendered invaluable help to the author in preparing this section. H. W. Liepmann worked as co-author during the early stages of writing. Anatol Roshko and Donald Coles are in essence the authors of the articles on experimental results. Saul Kaplun contributed several fundamental ideas. The entire Section was read in great detail by Donald Coles, and his many pertinent suggestions played an essential role in the final revision of the manuscript. I-Dee Chang and many others helped in checking various details.

The author wishes to dedicate this Section to the memory of Saul Kaplun (1924–1964).

is the method of statistical mechanics, in which no a priori assumptions are made about thermodynamics or about the mechanism of heat conduction, the nature of the hydrodynamical stresses, etc. The Boltzmann integro-differential equation is first obtained from general statistical principles with the aid of a simplified model for molecular interactions [*1*]. The hydrodynamical variables are then defined statistically and the equations governing them are derived from the Boltzmann equation by using approximations to the statistical distribution functions [*2*]. Theoretically this method appears to be the ideal one. It fits fluid mechanics into the general framework of molecular physics and exhibits the degree of approximation involved in each system of equations. However, the practical limitations of the statistical method are rather severe. The derivation has so far been carried out in a satisfactory way only for special cases of gases, and even with many simplifying assumptions the derivation is quite complicated. In the case of dense gases and especially of liquids the difficulties are very great. These statistical methods will not be discussed further in the present section; see rather Vol. I and the references given there.

The second and classical method is that of general continuum mechanics. The basic assumption is that a fluid (or solid) consists of continuous matter which at every point has a definite density, velocity, and internal energy. This matter can neither be created nor destroyed. It obeys Newton's classical laws of mechanics and also certain thermodynamical laws. Furthermore, contiguous regions of the medium are assumed to exert forces on each other across their common boundary. With the aid of general arguments from differential geometry the nature of these stresses can to some extent be specified a priori. Another set of relations in continuum mechanics is related to thermodynamics. Classical thermodynamics deals with equilibrium states of very general media. In continuum mechanics the medium is specialized, but on the other hand the notions of thermodynamics are extended to nonequilibrium processes. One assumes that thermodynamic state variables such as pressure, temperature, internal energy, and entropy may be defined even when the medium is not in equilibrium. In the simplest case one also assumes that the relations between the state variables (equations of state) are the same in nonequilibrium as in equilibrium. One also assumes that the law of conservation of energy is valid at any instant for any region of the medium. This law is applied to the total energy, i.e. the internal energy plus the kinetic energy. It is assumed that energy may be transferred between contiguous regions by heat flow through their surface of contact. Various simple possibilities for the heat flow may be discussed by methods similar to those applied to the stresses. A general discussion of the methods of continuum mechanics may be found in [*3*] which also lists further references.

By the methods of continuum mechanics one may obtain various systems of equations of different complexity. Obviously one would like to obtain the simplest possible equations which are consistent with the experimental data for dynamic processes considered. In Art. 2 some very general relations of continuum mechanics will be discussed. In Art. 3 the Navier-Stokes equations will be obtained from these relations by introducing suitable assumptions regarding the fluid stresses and the conduction of heat (some of the derivations are given in more detail in III,A).

Tensor analysis. The mathematical tools necessary for expressing the concepts of continuum mechanics are provided by tensor analysis. Some useful definitions and theorems from tensor analysis are collected below. Only Cartesian coordinates will be used. General curvilinear coordinates will be introduced in Art. 5. For a systematic treatise on the subject the reader is referred to, for example, [*4,5,6*]. Physical illustrations for some of the concepts introduced here will be given in Art. 2 and 3.

Consider vectors attached to a fixed P in space. A linear operator, which by operating on any one of these vectors produces another of them, is called a tensor (of second order) at P. Thus a tensor A associates with each vector $\mathbf{v}$ at P another vector at P denoted by $\mathsf{A}\mathbf{v}$ in such a way that for any two vectors $\mathbf{v}_1$ and $\mathbf{v}_2$ and scalars λ_1 and λ_2

$$\mathsf{A}(\lambda_1\mathbf{v}_1 + \lambda_2\mathbf{v}_2) = \lambda_1\mathsf{A}\mathbf{v}_1 + \lambda_2\mathsf{A}\mathbf{v}_2 \tag{1-1}$$

If a system of orthogonal unit vectors $\mathbf{h}_i$ is chosen as a base, a vector $\mathbf{v}$ may be represented as a row of numbers v_i

$$\mathbf{v} = (v_i), \qquad v_i = \mathbf{v} \cdot \mathbf{h}_i \tag{1-2a}$$

Similarly a tensor may be represented as a matrix (A_{ij})

$$\mathsf{A} = (A_{ij}), \qquad A_{ij} = \mathbf{h}_i \cdot (\mathsf{A}\mathbf{h}_j) \tag{1-2b}$$

The ith component of the vector $\mathsf{A}\mathbf{v}$ is then

$$(\mathsf{A}\mathbf{v})_i = \sum_j A_{ij}v_j \tag{1-2c}$$

Conversely, when a system of coordinates is given, a vector or a tensor may be defined by its components. However, the physical and mathematical significance is shown most clearly when a vector or tensor can be defined invariantly, that is, without any reference to a system of coordinates.

The transpose A^* of a tensor A is defined by

$$A^*_{ji} = A_{ij} \tag{1-3a}$$

or invariantly

$$\mathbf{v} \cdot (\mathsf{A}^*\mathbf{w}) = \mathbf{w} \cdot (\mathsf{A}\mathbf{v}) \tag{1-3b}$$

for all vectors $\mathbf{v}$ and $\mathbf{w}$.

A tensor is called symmetric if $\mathsf{A} = \mathsf{A}^*$ and skew-symmetric if $\mathsf{A} = -\mathsf{A}^*$. Since $(\mathsf{A}^*)^* = \mathsf{A}$, the tensor $\mathsf{A} + \mathsf{A}^*$ is always symmetric and $\mathsf{A} - \mathsf{A}^*$ always skew-symmetric. In a three-dimensional space one may associate a vector $\mathbf{b} = (b_i)$ with a skew-symmetric tensor according to the following rule:

$$\mathbf{b} \to \mathsf{B} \quad \text{if}$$

$$b_1 = -B_{23}, \qquad b_2 = -B_{31}, \qquad b_3 = -B_{12} \tag{1-4a}$$

that is, if

$$\mathsf{B}\mathbf{v} = \mathbf{b} \times \mathbf{v} \quad \text{for any } \mathbf{v} \tag{1-4b}$$

The identity or unit tensor, which operates on any vector $\mathbf{v}$ to reproduce $\mathbf{v}$, will be denoted by I;

$$\mathsf{I}\mathbf{v} = \mathbf{v} \tag{1-5a}$$

The equivalent definition in component form is

$$\mathsf{I} = (\delta_{ij}) \tag{1-5b}$$

where δ_{ij} is the Kronecker delta.

The dyadic product of two vectors $\mathbf{a}$ and $\mathbf{b}$ is a tensor denoted by $\mathbf{a} \circ \mathbf{b}$ and defined by

$$(\mathbf{a} \circ \mathbf{b})\mathbf{v} = \mathbf{a}(\mathbf{b} \cdot \mathbf{v}) \quad \text{for any } \mathbf{v} \tag{1-6a}$$

or

$$(\mathbf{a} \circ \mathbf{b})_{ij} = a_i b_i \tag{1-6b}$$

Then

$$\mathbf{a} \times \mathbf{b} \to (\mathbf{b} \circ \mathbf{a}) - (\mathbf{a} \circ \mathbf{b}) \tag{1-6c}$$

in the sense of Eq. 1-4.

The inner product of two tensors A and B is a scalar

$$\mathsf{A} \cdot \mathsf{B} = \sum_{i,j} A_{ij} B_{ij} \tag{1-7a}$$

This product should be distinguished from the ordinary product AB, which is defined as an application of B followed by an application of A;

$$(\mathsf{AB})\mathbf{v} = \mathsf{A}(\mathsf{B}\mathbf{v}) \quad \text{or} \quad (\mathsf{AB})_{ij} = \sum_k A_{ik} B_{kj} \tag{1-7b}$$

The concepts so far considered belong properly to tensor algebra rather than to tensor analysis since all vectors and tensors are referred to the same point. If in a region V in space a tensor (or a vector) is attached to each point P, one speaks of a tensor field $\mathsf{A}(P)$ (or a vector field $\mathbf{v}(P)$). Tensor analysis deals with the differential calculus of such fields. Whenever components are used in the formulas below it will be assumed that they refer to a Cartesian system of coordinates x_i in V.

The ith unit vector at any point P, $\mathbf{h}_i(P)$, is always taken parallel to the x_i axis.

The gradient of a vector field $\mathbf{v}(P)$ is a tensor field defined by

$$(\text{grad } \mathbf{v})\mathbf{PQ} = \mathbf{v}(Q) - \mathbf{v}(P) \tag{1-8a}$$

where $\mathbf{PQ}$ is the displacement vector from P to the infinitesimally close point Q. The equivalent definition in component form is

$$(\text{grad } \mathbf{v})_{ij} = \frac{\partial v_i}{\partial x_j} \tag{1-8b}$$

The deformation tensor def $\mathbf{v}$ is defined as the symmetric tensor

$$(\text{def } \mathbf{v}) = \text{grad } \mathbf{v} + (\text{grad } \mathbf{v})^* \tag{1-9a}$$

The corresponding skew-symmetric tensor is associated with curl $\mathbf{v}$ if the space is three-dimensional (cf. Eq. 1-4)

$$(\text{curl } \mathbf{v}) \times \mathbf{w} = [\text{grad } \mathbf{v} - (\text{grad } \mathbf{v})^*]\mathbf{w} \tag{1-9b}$$

for any vector $\mathbf{w}$.

Of importance in continuum mechanics are the cases when the field $\mathbf{v}(P)$ consists of displacement vectors or velocity vectors for a continuous medium. Then the tensor grad $\mathbf{v}$ describes the strain or rate of strain respectively for an infinitesimal volume element. If $\mathbf{v}(P)$ is the velocity field of a continuous medium, then $\mathbf{v}$ describes the translational motion of an infinitesimal element of the medium, $\frac{1}{2}$ curl $\mathbf{v}$ is the angular velocity vector of the element, and def $\mathbf{v}$ is a measure of the deformation of the element; if def $\mathbf{v}$ vanishes at a point then the medium moves locally without deformation, i.e., as a rigid body, at this point. The rate of change of volume of the element is div $\mathbf{v}$. For a detailed discussion of the geometrical significance of grad $\mathbf{v}$ and the associated symmetric and skew-symmetric tensors see, for example, [*6*] or [*7*].

The divergence of a tensor field $\mathsf{A}(P)$ is a vector field div A defined by

$$(\text{div } \mathsf{A})_i = \sum_j \frac{\partial A_{ij}}{\partial x_j} \tag{1-10}$$

Let V be a closed region and B its boundary. The infinitesimal volume element $dx_1dx_2dx_3$ of V is denoted by dV. An infinitesimal surface element of B is represented by its outward normal $d\mathbf{n}$ of length equal to the area of the element. The following divergence theorem, the analogue of Gauss' formula in vector analysis, is then valid.

$$\iint_B \mathsf{A}d\mathbf{n} = \iiint_V (\text{div } \mathsf{A})dV \tag{1-11}$$

This formula shows how div A may be defined without reference to any system of coordinates as the flow of the tensor field through the surface of an infinitesimal region divided by the volume of the region.

The following easily proved formulas will also be useful later on. In these formulas B denotes the skew-symmetric tensor associated with the vector $\mathbf{b}$ according to Eq. 1-4. The absolute value of a vector $\mathbf{a}$ is denoted by a.

$$\text{div}\,(\mathbf{a} \circ \mathbf{b}) = (\text{grad}\,\mathbf{a})\mathbf{b} + (\text{div}\,\mathbf{b})\mathbf{a} \tag{1-12a}$$

$$\text{div}\,(\mathsf{A}\mathbf{b}) = (\text{div}\,\mathsf{A}^*) \cdot \mathbf{b} + (\mathsf{A}^* \cdot \text{grad}\,\mathbf{b}) \tag{1-12b}$$

$$\text{div}\,(f\mathsf{I}) = \text{grad}\,f \quad \text{where } f \text{ is a scalar} \tag{1-12c}$$

$$\text{div}\,(\text{grad}\,\mathbf{a})^* = \text{grad div}\,\mathbf{a} \tag{1-12d}$$

$$\text{div}\,\mathsf{B} = -\,\text{curl}\,\mathbf{b}, \qquad \text{div}\,\mathsf{B}^* = \text{curl}\,\mathbf{b} \tag{1-12e}$$

Hence, by Eq. 1-9b,

$$\text{div}\,[\text{grad}\,\mathbf{a} - (\text{grad}\,\mathbf{a})^*] = -\,\text{curl curl}\,\mathbf{a} \tag{1-12f}$$

or, by Eq. 1-12d and defining ∇^2 as div grad,

$$\nabla^2\mathbf{a} = \text{div grad}\,\mathbf{a} = \text{grad div}\,\mathbf{a} - \text{curl curl}\,\mathbf{a} \tag{1-12g}$$

$$\text{div def}\,\mathbf{a} = 2\,\text{grad div}\,\mathbf{a} - \text{curl curl}\,\mathbf{a} = \text{div grad}\,\mathbf{a} + \text{grad div}\,\mathbf{a} \tag{1-12h}$$

$$\text{curl div def}\,\mathbf{a} = -\,\text{curl curl curl}\,\mathbf{a} = \text{div grad curl}\,\mathbf{a} \tag{1-12i}$$

$$\text{grad}\,(\mathbf{a} \cdot \mathbf{b}) = (\text{grad}\,\mathbf{a})^*\mathbf{b} + (\text{grad}\,\mathbf{b})^*\mathbf{a} \tag{1-12j}$$

and hence

$$\begin{aligned}(\text{grad}\,\mathbf{a})\mathbf{a} &= \text{grad}\,\frac{a^2}{2} + [\text{grad}\,\mathbf{a} - (\text{grad}\,\mathbf{a})^*]\mathbf{a} \\ &= \text{grad}\,\frac{a^2}{2} + (\text{curl}\,\mathbf{a}) \times \mathbf{a}\end{aligned} \tag{1-12k}$$

$$\text{curl}\,(\mathbf{a} \times \mathbf{b}) = \text{div}\,[(\mathbf{a} \circ \mathbf{b}) - (\mathbf{b} \circ \mathbf{a})] \tag{1-12l}$$

Thus from Eq. 1-6c and 1-12e

$$\mathsf{B} \cdot \text{grad}\,\mathbf{a} = b \cdot \text{curl}\,\mathbf{a} \tag{1-12m}$$

$$\mathsf{I} \cdot \text{grad}\,\mathbf{a} = \text{div}\,\mathbf{a} \tag{1-12n}$$

The cross product of a vector and a tensor is defined as

$$\mathbf{b} \times \mathsf{A} = \mathsf{B}\mathsf{A} = \Big(\sum_j B_{ij}A_{jk}\Big)$$

or, equivalently,

$$(\mathbf{b} \times \mathsf{A})\mathbf{v} = \mathbf{b} \times (\mathsf{A}\mathbf{v}) \quad \text{for any vector } \mathbf{v} \tag{1-12o}$$

If $\mathbf{r}$ is the radius vector and A is symmetric, then

$$\operatorname{div}(\mathbf{r} \times \mathsf{A}) = \mathbf{r} \times \operatorname{div} \mathsf{A} \tag{1-12p}$$

B,2. General Conservation Laws of Continuum Mechanics.

Integral Relations. The basic integral laws of continuum mechanics express the conservation of mass, momentum, and energy. More precisely, the changes in the last two quantities are accounted for according to certain laws of mechanics and thermodynamics.

In expressing the integral laws, the notation of Eq. 1-11 will be used. The region V and its boundary B will thereby be considered fixed with respect to the system of coordinates chosen, hence in general not fixed with respect to the fluid particles.

The first integral law expresses the assumption that matter is neither created nor destroyed (conservation of mass). It is a relation between the mass density ρ and the velocity $\mathbf{u}$. The vector $\rho\mathbf{u}$ is called the mass flow vector $\mathbf{m}$. (It may also be interpreted as momentum per unit volume.) Thus $\mathbf{m} \cdot d\mathbf{n}$ is the mass flow per unit time through the surface element represented by $d\mathbf{n}$. The principle of conservation of mass may then be written as

$$\iiint\limits_V \frac{\partial \rho}{\partial t} dV + \iint\limits_B \mathbf{m} \cdot d\mathbf{n} = 0, \quad \mathbf{m} = \rho\mathbf{u} \tag{2-1}$$

The second integral law for continuum mechanics is usually referred to as the law for conservation of momentum although it actually describes how the momentum of a part of the continuum changes under the influence of various forces in accordance with Newton's second and third laws. The forces acting on the continuum are volume forces and surface forces. The volume force per unit mass is denoted by $\mathbf{f}$. It is often, sometimes inappropriately, referred to as an external force. It may be due, for instance, to a gravitational field or to mutual electrodynamic interaction between particles in a fluid. The surface forces are expressed with the aid of a stress tensor σ. If a surface element of B is represented by its outward normal $d\mathbf{n}$, $\sigma d\mathbf{n}$ is the force exerted across the surface element by the matter outside V on the matter inside V. From the requirement that the force per unit volume due to surface forces be finite, one proves that the force is a linear function of $d\mathbf{n}$, that is, σ is actually a tensor (cf. Art. 1). By a similar argument about the moment exerted by the stresses one proves that σ is a symmetric tensor. These proofs may be found in standard textbooks (see, for example, [8]). The momentum law then takes the form

$$\iiint\limits_V \frac{\partial(\rho\mathbf{u})}{\partial t} dV = -\iint\limits_B (\rho\mathbf{u})(\mathbf{u} \cdot d\mathbf{n}) + \iiint\limits_V \rho\mathbf{f} dV + \iint\limits_B \sigma d\mathbf{n} \tag{2-2a}$$

It is convenient to introduce the flow of momentum tensor, defined as the dyadic product $\rho\mathbf{u} \circ \mathbf{u}$. Then the first term on the right-hand side of Eq. 2-2a may be written

$$\iint_V (\rho\mathbf{u})(\mathbf{u} \cdot d\mathbf{n}) = \iint_B (\rho\mathbf{u} \circ \mathbf{u})d\mathbf{n} \tag{2-2b}$$

The third integral law is the law of conservation of energy and corresponds to the first law of thermodynamics. The total energy per unit mass of the medium consists of the kinetic energy, $\mathbf{u}^2/2$, and the internal energy, e_q. (For a discussion of the concepts of internal energy, see e.g. [9] or Vol. I.) This law describes how the total energy in a region changes due to transport, work done by the forces, conduction of heat through the boundary, and external heat sources. The external heat added per unit mass is denoted by Q.[2] By an argument similar to that used for the stress function it may be shown that the heat conducted per unit time out of V through a surface element, represented by its outward normal $d\mathbf{n}$, is a linear function of $d\mathbf{n}$. It may hence be represented as $\mathbf{q} \cdot d\mathbf{n}$ where $\mathbf{q}$ is called the heat conduction vector. The law of conservation of energy may then be written as

$$\iiint_V \frac{\partial}{\partial t}\left(\rho\frac{\mathbf{u}^2}{2} + \rho e\right)dV + \iint_B \left[\left(\rho\frac{\mathbf{u}^2}{2} + \rho e\right)\mathbf{u} + \mathbf{q}\right]\cdot d\mathbf{n}$$
$$= \iiint_V \rho\mathbf{f}\cdot\mathbf{u}dV + \iint_B (\sigma\mathbf{u})\cdot d\mathbf{n} + \iiint_V \rho Q dV \tag{2-3}$$

Eq. 2-1, 2-2, and 2-3 are the fundamental integral relations of continuum mechanics. Additional relations may be derived from these as shown by Eq. 2-9b and 2-10′ below.

Differential Relations. By letting the volume of the region V tend to zero, one may derive two types of relations. The first type occurs when V encloses a surface of discontinuity. One then obtains jump conditions such as the shock wave relations in the theory of nonviscous fluids. If, on the other hand, one assumes that the functions occurring in the integral relations have sufficiently many derivatives in V one obtains differential equations. These are usually derived by applying the divergence theorem, Eq. 1-11.

It is generally assumed that discontinuities do not occur in a viscous heat-conducting fluid. Hence only the second type of equations will be

[2] An example of a process where external heat is added is in the combustion of fuel in air. If the mass of the fuel and of the combustion products is neglected relative to that of the air, then the heat released by the combustion may be considered as external heat added to the energy of the air. Another example is that of a medium which loses heat through radiation; in this case Q is negative.

considered. The relations Eq. 2-1, 2-2, and 2-3 then yield the following differential equations:

Continuity equation:

$$\frac{\partial \rho}{\partial t} + \operatorname{div}(\rho \mathbf{u}) = 0 \tag{2-4}$$

Momentum equation:

$$\frac{\partial(\rho \mathbf{u})}{\partial t} + \operatorname{div}(\rho \mathbf{u} \circ \mathbf{u}) = \rho \mathbf{f} + \operatorname{div} \sigma \tag{2-5}$$

Energy equation:

$$\frac{\partial}{\partial t}\left(\rho \frac{|\mathbf{u}|^2}{2} + \rho e\right) + \operatorname{div}\left[\left(\rho \frac{|\mathbf{u}|^2}{2} + \rho e\right)\mathbf{u}\right] + \operatorname{div} \mathbf{q} = \rho \mathbf{f} \cdot \mathbf{u} + \operatorname{div}(\sigma \mathbf{u}) + \rho Q \tag{2-6}$$

The above equations are the basic differential equations of general continuum mechanics. We now derive some consequences of and alternative forms of these equations. It follows from the continuity equation that the density times the substantial derivative of any scalar function g may be written

$$\rho \frac{Dg}{Dt} = \frac{\partial(\rho g)}{\partial t} + \operatorname{div}(\rho g \mathbf{u}) \tag{2-7a}$$

where the substantial derivative is defined by

$$\frac{Dg}{Dt} = \frac{\partial g}{\partial t} + (\operatorname{grad} g) \cdot \mathbf{u}$$

Similarly, by multiplying the continuity equation by a vector $\mathbf{a}$ one proves that (cf. Eq. 1-12a)

$$\rho \frac{D\mathbf{a}}{Dt} = \frac{\partial(\rho \mathbf{a})}{\partial t} + \operatorname{div}(\mathbf{a} \circ \rho \mathbf{u}) \tag{2-7b}$$

where

$$\frac{D\mathbf{a}}{Dt} = \frac{\partial \mathbf{a}}{\partial t} + (\operatorname{grad} \mathbf{a})\mathbf{u}$$

Using these relations and Eq. 1-12k one finds the following alternate forms of the momentum and energy equations:

$$\rho \frac{D\mathbf{u}}{Dt} = \rho \frac{\partial \mathbf{u}}{\partial t} + \rho \operatorname{grad} \frac{|\mathbf{u}|^2}{2} + \rho(\operatorname{curl} \mathbf{u}) \times \mathbf{u} = \rho \mathbf{f} + \operatorname{div} \sigma \tag{2-5'}$$

$$\rho \frac{D}{Dt}\left(\frac{|\mathbf{u}|^2}{2} + e\right) + \operatorname{div} \mathbf{q} = \rho \mathbf{f} \cdot \mathbf{u} + \operatorname{div}(\sigma \mathbf{u}) + \rho Q \tag{2-6'}$$

If the external force $\mathbf{f}$ is conservative and time-independent, that is, $\mathbf{f} = -\operatorname{grad} \varphi$ and $\partial\varphi/\partial t = 0$, then the energy equation may also be written

$$\rho \frac{D}{Dt}\left(\frac{|\mathbf{u}|^2}{2} + e + \varphi\right) + \operatorname{div} \mathbf{q} = \operatorname{div}(\sigma \mathbf{u}) + \rho Q \tag{2-6''}$$

If Eq. 2-5′ is multiplied by $\mathbf{q}$ one obtains

$$\rho \frac{D(|\mathbf{u}|^2/2)}{Dt} = \rho \mathbf{f} \cdot \mathbf{u} + (\operatorname{div} \sigma) \cdot \mathbf{u} \tag{2-8}$$

This relation describes the change in kinetic energy only and is derived from Eq. 2-4 and 2-5 without the use of Eq. 2-6. It is thus a purely mechanical energy law; no concepts from thermodynamics are used in this equation. Subtracting Eq. 2-8 from Eq. 2-6 and using the fact that σ is symmetric one obtains (cf. Eq. 1-12b) the equation for the change in internal energy,

$$\rho \frac{De}{Dt} + \operatorname{div} \mathbf{q} = (\sigma \cdot \operatorname{grad} \mathbf{u}) + \rho Q \tag{2-6'''}$$

The last two equations may be interpreted as follows: The work done per unit time by the surface stresses on the boundary of the region V is $\iint_B \mathbf{u} \cdot (\sigma d\mathbf{n})$. From the symmetry of the stress tensor and from Eq. 1-11 and 1-12b it follows that

$$\iint_B \mathbf{u} \cdot (\sigma d\mathbf{n}) = \iint_B (\sigma \mathbf{u}) \cdot d\mathbf{n} = \iiint_V \operatorname{div} (\sigma \mathbf{u}) dV = \iiint_V [(\operatorname{div} \sigma) \cdot \mathbf{u} + \sigma \cdot \operatorname{grad} \mathbf{u}] dV \tag{2-9a}$$

The work done per unit time per unit volume by stresses on an infinitesimal volume element is thus seen to be [div σ) $\cdot \mathbf{u} + \sigma \cdot$ grad $\mathbf{u}$]. The first term is the work done per unit time by div σ, the net force on the volume element. This work is converted into kinetic energy as shown by Eq. 2-8. The second term, $\sigma \cdot$ grad $\mathbf{u}$, represents according to Eq. 2-6′′′ a source strength per unit time and unit volume for the internal energy. Hence this part of the work is converted into internal energy. Its relation to the notion of dissipation will be discussed in Art. 3. Note that the quantities "work done," "velocity," and "kinetic energy" are not invariant under a Galilean transformation.[3] The second term, however, represents the part of the work that is Galilean invariant since it involves only the derivatives of the velocity. Correspondingly, the internal energy is the part of the total energy which is Galilean invariant.

A combination of Eq. 2-8 and 2-9a gives the following integral law for the kinetic energy:

$$\iint_B (\sigma \mathbf{u}) \cdot d\mathbf{n} = \iiint_V \left[\rho \frac{D|\mathbf{u}|^2/2}{Dt} - \rho \mathbf{f} \cdot \mathbf{u} + \sigma \cdot \operatorname{grad} \mathbf{u}\right] dV \tag{2-9b}$$

[3] A Galilean transformation is a coordinate transformation from x_i, t to x_i', t' of the form $x_i' = x_i + A_i t$, $t' = t$, $A_i = \text{const.}$

A formula for the moment of momentum may be derived from Eq. 2-5. Let $\mathbf{r}$ be the radius vector from a fixed origin to a variable point. The moment of momentum per unit volume is then $\mathbf{r} \times \rho\mathbf{u}$. If Eq. 2-5 is multiplied vectorially by $\mathbf{r}$, an application of Eq. 1-12p shows that

$$\frac{\partial(\mathbf{r} \times \rho\mathbf{u})}{\partial t} + \operatorname{div}\left[\mathbf{r} \times (\rho\mathbf{u} \circ \mathbf{u})\right] = \mathbf{r} \times \rho\mathbf{f} + \operatorname{div}\left(\mathbf{r} \times \sigma\right) \quad (2\text{-}10)$$

The corresponding integral relation is

$$\iiint_V \frac{\partial(\mathbf{r} \times \rho\mathbf{u})}{\partial t}\, dV + \iint_B (\mathbf{r} \times \rho\, \mathbf{u})(\mathbf{u} \cdot d\mathbf{n}) = \iiint_V (\mathbf{r} \times \rho\mathbf{f})\, dV + \iint_B \mathbf{r} \times (\sigma d\mathbf{n}) \quad (2\text{-}10')$$

which expresses the conservation of moment of momentum, or rather how the moment of momentum changes due to the torque of the forces acting on the continuum.

KINEMATICAL RELATIONS. The kinematical relations for a fluid are concerned only with the space-time geometry of the motion. They are independent of the dynamics and the thermodynamics of the continuum, and are based on Eq. 2-1 (or, equivalently, Eq. 2-4).

Streamlines. By integrating the field of the velocity vectors at a fixed time t, one obtains the instantaneous streamlines. They have the property that the instantaneous velocity vector $\mathbf{u}$, and hence also the mass flow vector $\mathbf{m}$, at any point is tangent to the streamline through that point. A surface made up entirely of streamlines is called a stream surface, or, when it has the appropriate shape, a stream tube. If one follows a particle, that is, an infinitesimal region of the continuum, one obtains a position (x_1, x_2, x_3) for each time t, that is, a curve in x_i, t space. This curve represents the time history of the particle, and is sometimes referred to as a generalized streamline. If one considers t as a parameter, the corresponding curve in x_i space is called a particle path. In stationary flow the particle paths coincide with the instantaneous streamlines; in general this is not true for nonstationary flow.

Stream functions. Consider now stationary flow. The continuity equation states that $\rho\mathbf{u} = \mathbf{m}$ is a solenoidal vector field. From general vector analysis it then follows that one may find two scalar functions ψ_1 and ψ_2, called the stream functions, such that (cf [*10*])

$$\mathbf{m} = \rho\mathbf{u} = \operatorname{grad} \psi_1 \times \operatorname{grad} \psi_2 \quad (2\text{-}11)$$

Since then $\mathbf{u} \cdot \operatorname{grad} \psi_i = 0$ $(i = 1, 2)$, ψ_1 and ψ_2 are constant along each streamline. Each surface $\psi_1 = \text{const}$ or $\psi_2 = \text{const}$ is then a stream surface. Each streamline is an intersection of two such stream surfaces.

More generally, a vector potential may be found such that **m** is the curl of this potential. Eq. 2-11 is a special case of this since it may be written

$$\mathbf{m} = \text{curl } (\psi_1 \text{ grad } \psi_2) \equiv - \text{curl } (\psi_2 \text{ grad } \psi_1) \tag{2-11$'$}$$

This formula shows that the mass flow per unit time across any cross section of a stream tube is the line integral of ψ_1 grad ψ_2 along any closed line encircling the stream tube once. Consider in particular the special stream tube whose cross section is shown in Fig. B,2. The mass flow through the area S, in the direction which forms a right-handed system with the curved arrow shown, is

$$\iint_S \mathbf{m} d\mathbf{n} = (a_2 - a_1)(b_2 - b_1) \tag{2-12}$$

This formula shows the intuitive meaning of the stream functions.

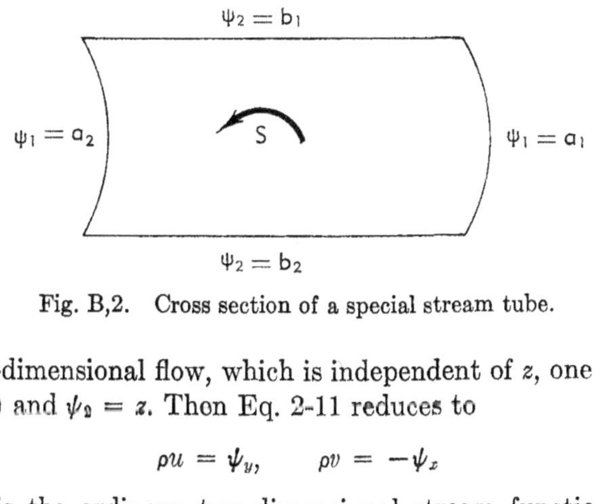

Fig. B,2. Cross section of a special stream tube.

For two-dimensional flow, which is independent of z, one may choose $\psi_1 = \psi(x, y)$ and $\psi_2 = z$. Then Eq. 2-11 reduces to

$$\rho u = \psi_y, \qquad \rho v = -\psi_x \tag{2-13}$$

so that ψ_1 is the ordinary two-dimensional stream function. Eq. 2-12 then corresponds to the familiar interpretation of ψ. If $\rho = \rho_\infty$ and $\mathbf{u} = U\mathbf{i}$ at upstream infinity, ψ has there the form $\rho_\infty Uy +$ const. Similarly, for flow which is axially symmetric about the line $y = z = 0$ one may choose $\psi_1 = \psi(r, x)$, $\psi_2 = -\theta$ where $r^2 = y^2 + z^2$, $\theta = \tan^{-1} y/z$. Denoting the axial and radial velocity components by u and v, Eq. 2-11 reduces to

$$\rho r u = \psi_r, \qquad \rho r v = -\psi_x \tag{2-14}$$

so that in this case ψ_1 is Stokes' stream function which at upstream infinity has the value $\frac{1}{2}\rho U_\infty r^2$.

There is obviously great freedom in choosing ψ_1 and ψ_2 for a given mass flow field. Let **u** be U**i** at $x = -\infty$ (upstream infinity). One may then take any double set of coordinate lines in the plane $x = -\infty$, and assign ψ_1 values to one set and ψ_2 values to the other set in such a way that Eq. 2-12 is satisfied. If then ψ_1 and ψ_2 are taken to be constant along streamlines Eq. 2-11 will be satisfied everywhere because of the continuity equation.

Vorticity. For a given velocity field $\mathbf{u}$, the vorticity field $\mathbf{\Omega} = (\Omega_1, \Omega_2, \Omega_3)$ is defined by

$$\mathbf{\Omega} = \text{curl } \mathbf{u} \tag{2-15a}$$

When two-dimensional flow is considered, the rectangular coordinate axes in general are chosen so that the flow is parallel to the x, y plane. Hence $\mathbf{\Omega} = \Omega_3\mathbf{k}$. We then write Ω instead of Ω_3. If $\mathbf{u} = u\mathbf{i} + v\mathbf{j}$ then

$$\Omega = v_x - u_y \tag{2-15b}$$

Some integral laws for vorticity, whose validity follows from pure vector geometry or pure kinematics, are discussed below. A first example is the formula for the total vorticity in a region V whose boundary B is one of several closed surfaces. According to the divergence theorem (Eq. 1-11) and Eq. 1-4b and 1-12e,

$$\iiint_V \mathbf{\Omega} dV = -\iint_B \mathbf{u} \times d\mathbf{n} \tag{2-16a}$$

The corresponding two-dimensional formula is

$$\iint_V \Omega dV = \int_B \mathbf{u} \cdot d\mathbf{s}, \qquad d\mathbf{s} = \mathbf{k} \times d\mathbf{n} \tag{2-16b}$$

The right-hand side of Eq. 2-16b represents the negative circulation[4] around the curve B. The integral in the right-hand side of Eq. 2-16a may then be regarded as circulation over a surface B.

It follows in particular that if the velocity vanishes at the boundary of a closed region V, then the total vorticity is zero. Consider now a solid moving in an unbounded fluid. Let Σ be a surface enclosing the body and V the region between Σ and the body surface S. Assume further that the no-slip condition is valid, that is, that the velocity of the fluid at a point P on the solid equals the velocity of P. The contribution of S to the surface integral in Eq. 2-16a may then be evaluated as follows. The motion of a solid may be decomposed into a translation and a rotation. The former is described by a velocity vector $\mathbf{U}$, the latter by an angular velocity vector $\boldsymbol{\omega}$, directed along the axis of the rotation. If a center O is chosen on the axis of rotation and $\mathbf{r}$ is the radius vector $\mathbf{OP}$, the velocity of a point P in the solid is then $\mathbf{U} + \boldsymbol{\omega} \times \mathbf{r}$. The vortic-

[4] The line integral in Eq. 2-16b is taken in the positive counterclockwise direction of B, that is, the sense of rotation of B that forms a right-handed system with the z axis. It is customary to define the circulation Γ as the clockwise line integral of the velocity.

ity of the velocity field at a point in the solid is then const $= 2\omega$. The integral of $\mathbf{u} \times d\mathbf{n}$ over the boundary of the solid may be obtained by applying Eq. 2-16a to the solid. Using this result one finds that Eq. 2-16a applied to the region V yields

$$\iiint\limits_V \mathbf{\Omega} dV = -\iint\limits_\Sigma \mathbf{u} \times d\mathbf{n} - 2\omega \text{ (volume of solid)} \tag{2-17}$$

The corresponding formulas for the rate of change of vorticity may be obtained by taking the derivative of Eq. 2-16a, or, alternatively, by the following method: Let $\mathbf{a}$ be the acceleration of a fluid particle. Then (cf. Eq. 2-5′)

$$\mathbf{a} = \frac{D\mathbf{u}}{Dt} = \frac{\partial \mathbf{u}}{\partial t} + \operatorname{grad} \frac{|\mathbf{u}|^2}{2} + \mathbf{\Omega} \times \mathbf{u} \tag{2-18a}$$

and hence by Eq. 1-12l

$$\operatorname{curl} \mathbf{a} = \frac{\partial \mathbf{\Omega}}{\partial t} + \operatorname{div} [(\mathbf{\Omega} \cdot \mathbf{u}) - (\mathbf{u} \cdot \mathbf{\Omega})] \tag{2-18b}$$

By integrating over a volume V whose boundary B is fixed in time one obtains

$$\frac{\partial}{\partial t} \iiint\limits_V \mathbf{\Omega} dV = -\iint\limits_B \mathbf{a} \times d\mathbf{n} + \iint\limits_B \mathbf{u}(\mathbf{\Omega} \cdot d\mathbf{n}) - \iint\limits_B \mathbf{\Omega}(\mathbf{u} \cdot d\mathbf{n}) \tag{2-19}$$

The last term represents material transport of vorticity through the boundary. If one considers the change of vorticity of a part of a fluid rather than in a region in space, the boundary B moves with the fluid and the last term is absent. Using the notation D/Dt for the substantial time derivative one may write

$$\frac{D}{Dt} \iiint\limits_V \mathbf{\Omega} dV = -\iint\limits_B \mathbf{a} \times d\mathbf{n} + \iint\limits_B \mathbf{u}(\mathbf{\Omega} \cdot d\mathbf{n}) \tag{2-19′}$$

The term $-\iint\limits_B \mathbf{a} \times d\mathbf{n}$ represents "conduction" of vorticity through the boundary. To evaluate it, further dynamical equations are needed which relate the fluid stresses to the velocity field (cf. the discussion of free vorticity in Art. 6). Note that only the rotational part of the acceleration contributes to this integral. The last term in Eq. 2-19′ is zero in two-dimensional flow.

It is a theorem of vector geometry that the vector field $\mathbf{u}$ may be decomposed into a solenoidal part $\mathbf{u}_1$ and an irrotational part $\mathbf{u}_2$, where the former may be obtained from the vorticity by the law of Biot and Savart. Thus we may write

$$\mathbf{u} = \mathbf{u}_1 + \mathbf{u}_2 \tag{2-20a}$$

The first component satisfies the equations, whenever the integral converges

$$\mathbf{u}_1 = \frac{1}{4\pi} \operatorname{curl} \iiint \frac{\mathbf{\Omega}(Q)}{r(P, Q)}\, dQ, \qquad \operatorname{div} \mathbf{u}_1 = 0 \tag{2-20b}$$

Here $r(P, Q)$ is the distance between the points P and Q and dQ is the volume element at Q. The second component has a potential ϕ

$$\mathbf{u}_2 = \operatorname{grad} \phi, \qquad \operatorname{curl} \mathbf{u}_2 = 0 \tag{2-20c}$$

The above equations furnish the general solution of Eq. 2-15a regarded as a differential equation for $\mathbf{u}$ with $\mathbf{\Omega}$ prescribed. If the fluid is incompressible, that is, of constant density, the continuity equation and Eq. 2-20c imply that ϕ must be harmonic, i.e., obey Laplace's equation $\nabla^2\phi = 0$. If the fluid is unlimited and the velocity vanishes at infinity, ϕ must be a constant which may be normalized to zero. If on the other hand boundaries are present, one must in general add a potential field grad ϕ to satisfy the boundary conditions at the solid.

B,3. The Navier-Stokes Equations.

Expressions for the Stress Tensor and the Heat Conduction Vector. Relation to Thermodynamics. The three fundamental equations, Eq. 2-4, 2-5, and 2-6, describe the mechanics of continuous matter under very general conditions. However, they are obviously underdetermined. To complete the system of equations one introduces special equations that express σ and $\mathbf{q}$ in terms of the other variables. The form of these additional equations varies with the nature of the medium and the type of flow considered. The special assumptions which lead to the Navier-Stokes equations for a viscous heat-conducting compressible fluid will now be presented. For some of the ideas used here compare the relevant sections of Vol. I and also [*2*,*3*].

First a general remark should be made about the distinction between thermodynamics and continuum mechanics. Classical thermodynamics deals with systems of a very general nature. On the other hand, it is restricted to the study of equilibrium states. The relation between an initial and a final equilibrium state of a system is subject to the laws of thermodynamics, even if the transition between the two states is irreversible. However, statements regarding intermediate nonequilibrium states are outside the scope of thermodynamics. Fluid dynamics and theory of elasticity go beyond thermodynamics in that they attempt a detailed description of the time-dependent processes. The thermodynamic state variables must be generalized so that they have meaning in nonequilibrium states, and concepts must be introduced (stress tensor, heat conduction vector), which have no analogue in equilibrium thermodynamics.

In this sense continuum mechanics is an extension and generalization of thermodynamics. However, there is a sacrifice of generality in continuum mechanics in that the nature of the systems considered is more restricted than in thermodynamics.[5]

It should be noted that thermodynamic concepts have already been used to a limited extent. In the energy equation (Eq. 2-6) it was implicitly assumed that the state variable e is defined in the nonequilibrium state. This energy equation must be considered as an extension rather than a consequence of the first law of thermodynamics.

To obtain the Navier-Stokes equations the following assumptions are made about the stress tensor σ: The stress tensor at a point P depends on the values of ρ, e, and $\mathbf{u}$ in the fluid, more specifically on the local values of these variables and their derivatives at the point P. Actually it cannot depend on $\mathbf{u}$ itself, only on the derivatives of $\mathbf{u}$, if it is to be invariant under a Galilean transformation. It will be assumed that each component of σ is a linear combination of the first derivatives of $\mathbf{u}$ with respect to the spatial variables x_i. Thus σ_{ij} is of the form $A_{ij} + \sum_{k,l} B_{ijkl}\partial \mathbf{u}_k/\partial x_l$. Furthermore it will be assumed that A_{ij} and B_{ijkl} depend only on the local and instantaneous values of ρ and e but not on their derivatives. As pointed out earlier σ is a symmetric tensor. Finally it will be assumed that there are no intrinsically preferred directions in space (isotropy). By geometry one may then prove that σ has the form (for notation see Eq. 1-5 and 1-9a) $A_1\mathsf{I} + A_2(\text{div }\mathbf{u})\mathsf{I} + A_3(\text{def }\mathbf{u})$, where A_i may depend on ρ and e. A fluid defined by these properties, especially the property that the stresses depend linearly on the first derivatives of the velocity, is usually called a Newtonian fluid. On considering the case of the fluid at rest, one sees that the negative A_1 is identical with the thermodynamical state variable called pressure. A_1 is therefore denoted by $-p$, and it is assumed that p has the same dependence on ρ and e as when the fluid is in thermodynamic equilibrium. It should be remembered, however, that A_1 is conceptually of a nature similar to that of A_2 and A_3. The latter quantities are called second and first viscosity and denoted by λ and μ, respectively. The final expression for the stress tensor is thus

$$\sigma = -p\mathsf{I} + \tau \tag{3-1a}$$

where

$$\tau = \lambda(\text{div }\mathbf{u})\mathsf{I} + \mu(\text{def }\mathbf{u}) \tag{3-1b}$$

τ is called the viscous stress tensor.

An expression is also needed for the heat conduction vector. In equilibrium thermodynamics two systems in contact with each other are

[5] For attempts to develop a theory of general systems in nonequilibrium, based on statistical mechanics and extensions of classical thermodynamics, the reader is referred to I,B.

said to have the same temperature T if no exchange of energy takes place without one system doing work on the other. We assume that even in a nonequilibrium state a temperature T with this property is defined and that the equation of state for T as a function of ρ and e is the same as in equilibrium. Furthermore, we assume that $\mathbf{q}$ depends linearly and isotropically on the spatial first derivative of T and also on the local values of ρ and e. The heat conduction vector must then have the form

$$\mathbf{q} = -k \text{ grad } T \tag{3-2}$$

where k is called the heat conductivity and is at most a function of ρ and e.

The assumptions made so far may be loosely summarized as follows: The concepts of density, internal energy, and velocity in continuous media are defined phenomenologically without reference to molecular structure. Two concepts from equilibrium thermodynamics, namely pressure and temperature, are assumed to be defined for nonequilibrium states by their dependence on density and internal energy as observed in the equilibrium state. The stress tensor and heat conduction vector are assumed to be highly specialized linear isotropic functions of velocity gradient and temperature gradient respectively. These assumptions are all of a physical nature, and underlie any mathematical statements about the behavior of the hypothetical fluids treated in this section. In other words, these assumptions are always part of any theoretical analysis based on the Navier-Stokes equations, whether or not the assumptions in question are valid for the real fluids encountered in nature. It may be possible that the Navier-Stokes equations are inadequate under certain extreme conditions, for example in flows at very low density or in flows with very rapid changes of state. However, the limits of validity of these equations can only be established experimentally, and few decisive experiments are known.

The effect of the assumptions just summarized has been to introduce three parameters λ, μ, and k. These are called transport parameters and determine the nonequilibrium behavior of a Newtonian fluid. To complete the Navier-Stokes equations, the specific formulas which relate p, T, and the transport parameters λ, μ, and k to ρ and e have to be given. Because of the special assumption that p and T obey the equilibrium laws, it is not necessary to take ρ and e as basic state variables. The equations for the state variables and the transport parameters will be discussed later.

THE DIFFERENTIAL EQUATIONS. By combining the assumptions discussed above with Eq. 2-4, 2-5′, and 2-6‴ one obtains the general Navier-Stokes equations for a viscous compressible heat-conducting fluid:

Continuity equation:

$$\frac{\partial \rho}{\partial t} + \operatorname{div}(\rho \mathbf{u}) = 0 \tag{3-3}$$

Momentum equation:

$$\rho \frac{D\mathbf{u}}{Dt} + \operatorname{grad} p = \operatorname{grad}(\lambda \operatorname{div} \mathbf{u}) + \operatorname{div}[\mu(\operatorname{def} \mathbf{u})] + \rho \mathbf{f} \tag{3-4}$$

Energy equation:

$$\rho \frac{De}{Dt} = \operatorname{div}(k \operatorname{grad} T) - p \operatorname{div} \mathbf{u} + \Phi + \rho Q \tag{3-5a}$$

The quantity Φ is the dissipation, more properly the dissipation of energy per unit volume and unit time, and is defined by

$$\begin{aligned} \Phi &= (\operatorname{grad} \mathbf{u}) \cdot \tau \\ &= \lambda(\operatorname{div} \mathbf{u})^2 + \mu[(\operatorname{grad} \mathbf{u}) \cdot (\operatorname{def} \mathbf{u})] \\ &= \lambda(\operatorname{div} \mathbf{u})^2 + \frac{\mu}{2}[(\operatorname{def} \mathbf{u}) \cdot (\operatorname{def} \mathbf{u})] \\ &= \lambda\left(\sum_i \frac{\partial \mathbf{u}_i}{\partial x_i}\right)^2 + \frac{\mu}{2}\sum_{i,j}\left(\frac{\partial \mathbf{u}_i}{\partial x_j} + \frac{\partial \mathbf{u}_j}{\partial x_i}\right)^2 \end{aligned} \tag{3-5b}$$

The last equality is valid only in a Cartesian system of coordinates. The form of the Navier-Stokes equations in a general curvilinear system of coordinates will be discussed in Art. 5.

To complete the system of equations one must add the thermodynamic equations of state and the equations for the dependence of the transport coefficients on the thermodynamic state variables. This gives five additional equations of the form

$$p, e, \mu, \lambda, k = \text{functions of } \rho \text{ and } T \tag{3-6}$$

These equations will be discussed later in this article, under the heading "Relations between State Variables and Transport Parameters."

Alternative forms of Eq. 3-4 and 3-5 may be found as in Art. 2. It is often convenient to rewrite the energy equation in terms of the enthalpy or the entropy. For nonequilibrium states these quantities will be defined by equations taken from equilibrium thermodynamics. The enthalpy per unit mass, h, is thus defined by

$$h = e + \frac{p}{\rho} \tag{3-7a}$$

The total, or stagnation enthalpy H is defined by

$$H = h + \frac{|\mathbf{u}|^2}{2} \tag{3-7b}$$

The entropy per unit mass s is a state variable whose dependence on the other state variables is such that

$$ds = \frac{1}{T}\left[de + pd\left(\frac{1}{\rho}\right)\right] = \frac{1}{T}\left[dh - \frac{1}{\rho}\,dp\right] \tag{3-7c}$$

The energy equation (Eq. 3-5a) may now be rewritten as

$$\rho\frac{Dh}{Dt} - \frac{Dp}{Dt} = -\operatorname{div}\mathbf{q} + \Phi + \rho Q \tag{3-8a}$$

or as

$$\rho\frac{DH}{Dt} - \frac{\partial p}{\partial t} = -\operatorname{div}\mathbf{q} + \operatorname{div}(\tau\mathbf{u}) + \rho\mathbf{f}\cdot\mathbf{u} + \rho Q \tag{3-8b}$$

or as

$$\begin{aligned}\rho\frac{Ds}{Dt} &= -\frac{\operatorname{div}\mathbf{q}}{T} + \frac{\Phi}{T} + \frac{\rho Q}{T}\\ &= -\operatorname{div}\left(\frac{\mathbf{q}}{T}\right) - \frac{1}{T^2}(\mathbf{q}\cdot\operatorname{grad}T) + \frac{\Phi}{T} + \frac{\rho Q}{T}\end{aligned} \tag{3-8c}$$

These equations are independent of the special form assumed for τ, if Φ is defined by the first equality of Eq. 3-5b.

The role of the quantity $\sigma\cdot(\operatorname{grad}\mathbf{u})$ was discussed after Eq. 2-9a. This analysis may now be pursued further. Using the notation of Eq. 3-1a we write

$$\sigma\cdot(\operatorname{grad}\mathbf{u}) = -p\operatorname{div}\mathbf{u} + \tau\cdot(\operatorname{grad}\mathbf{u}) \tag{3-9}$$

The first term on the right-hand side represents a change of internal energy according to Eq. 3-5a. This change is reversible, i.e., it does not affect the entropy (Eq. 3-8c). The second term, which is the dissipation Φ, represents an irreversible change; under very general assumptions on τ one may prove that it represents an increase in entropy. These remarks will now be elaborated by a comparison of the entropy equation (Eq. 3-8c) with the second law of thermodynamics.

The quantities Φ and $-1/T(\mathbf{q}\cdot\operatorname{grad}T)$ in Eq. 3-8c enter formally in the same way as ρQ. These quantities may thus be thought of as heat added per unit volume per unit time. When divided by T, the corresponding terms represent entropy sources. The vector $-\mathbf{q}/T$, on the other hand, represents the flow of entropy through the boundary of the system.

Now consider a closed adiabatic system; that is, a fluid element for which $\mathbf{q}\cdot\mathbf{n}$ vanishes on the boundaries and Q vanishes in the interior. Suppose that $\mathbf{q}$ has the special form given by Eq. 3-2, $\mathbf{q} = -k\operatorname{grad}T$. Then $-(1/T)(\mathbf{q}\cdot\operatorname{grad}T)$ is non-negative provided only that $k \geqq 0$. Furthermore, if τ has the special form given by Eq. 3-1b, then the dissipation Φ is non-negative according to Eq. 3-5b provided only that $\mu \geqq 0$, $3\lambda + 2\mu \geqq 0$. This may be seen by studying the minimum of Φ for arbitrary values of the velocity derivatives. With these restrictions on the values of the transport parameters λ, μ, and k it follows that the entropy of a closed system cannot decrease. Eq. 3-8c may then be considered as an extension of the second law of thermodynamics to Newtonian fluids in nonequilibrium states.

Boundary conditions. Various initial and boundary conditions may be specified for the Navier-Stokes equations. The ones occurring most frequently are the following:

A complete set of initial conditions is obtained if $\mathbf{u}$, p, and T are specified at the initial time. Obviously p and T may be replaced by any other pair of independent state variables.

At the boundary of a solid, $\mathbf{u}$ may be specified together with some condition on the temperature. Usually the value of $\mathbf{u}$ is determined by the no-slip condition, that the local velocity of the fluid is equal to the local velocity of the point of the solid with which it is in contact, i.e.[6]

$$\mathbf{u}_{\text{boundary}} = \mathbf{u}_{\text{surface}} \tag{3-10}$$

The condition for the temperature is usually either that the temperature or the heat conduction is prescribed at the surface of the body. It is sometimes assumed that for certain extreme conditions, usually involving low densities, discontinuities in velocity or temperature may occur at the boundary between a viscous fluid and a solid or at an interface. The condition (Eq. 3-10) is then replaced by a relation between $\mathbf{u}$ and its derivatives at the boundary. It is possible that under these conditions equations still more general than the Navier-Stokes equations are required. There are, however, indications that slip conditions together with the Navier-Stokes equations may give physically significant results. These questions are discussed in III,H.

For a body moving in an infinite fluid, the velocity and the thermodynamic state of the fluid at infinity must also be defined.

Some special cases of Eq. 3-3, 3-4, 3-5, 3-6, and 3-7 will now be described in order to clarify the terminology. These are the equations for an incompressible fluid and the equations for a nonviscous fluid. An example of equations for mean flow quantities will also be given.

Incompressible fluid. The classical Navier-Stokes equations for an incompressible fluid are obtained from Eq. 3-3 and 3-4 by making the assumption that ρ and μ are constants. Eq. 3-5a is not considered at all and Eq. 3-6 is replaced by specifying the constant values of ρ and μ. The quantities e, T, λ, k, and Q do not occur in these equations:

$$\operatorname{div} \mathbf{u} = 0 \tag{3-11a}$$

$$\rho \frac{D\mathbf{u}}{Dt} + \operatorname{grad} p = \mu \operatorname{div} (\operatorname{def} \mathbf{u}) + \rho \mathbf{f} \tag{3-11b}$$

[6] This condition is also assumed to be valid at an interface between two fluids. The "body" may be, say, a liquid drop in an air stream. At a point of contact the velocity of the liquid and the gas are then the same. The application of condition (Eq. 3-10) in this case becomes involved since the velocities at the surface of the liquid drop depend on the internal currents of the liquid.

where

$$\rho = \text{const} = \rho_\infty, \qquad \mu = \text{const} = \mu_\infty$$

At the surface of a solid, no-slip condition is retained. At infinity $\mathbf{u}$ and p are specified. As initial values $\mathbf{u}$ and p have to be given. The initial distribution of $\mathbf{u}$ must satisfy Eq. 3-11a.

Nonviscous case. Euler equations. In this case λ and μ are assumed to vanish. Generally, one then also assumes k to be zero. The no-slip condition has then to be weakened to the corresponding condition for the velocity component normal to the solid. The tangential component at the solid is left unspecified. There are no conditions on the temperature at the solid. The resulting equations are called the Euler equations for a compressible fluid. If in addition one assumes that the vorticity is zero, the flow is called potential since $\mathbf{u}$ is then the gradient of a potential ϕ. The potential ϕ obeys the Laplace equation $\nabla^2\phi = 0$ if the flow is also incompressible.

Equations for mean flow quantities. For certain types of nonstationary flow one may form a time average (mean value) of the flow quantities which does not vary with time.[7] It will be denoted by the horizontal bar.

$$\bar{\mathbf{u}} = \frac{1}{2T}\int_{-T}^{+T} \mathbf{u}\,dt, \qquad \bar{p} = \frac{1}{2T}\int_{-T}^{+T} p\,dt, \quad \text{etc.} \tag{3-12}$$

The two important cases in which average values are considered are those of periodic flow and stationary turbulent flow. In the former case $2T$ may be put equal to the shortest period, in the latter case $2T$ should be much longer than the correlation time. The differences between the actual and the mean flow quantities are called fluctuations and denoted by the subscript $_f$:

$$\mathbf{u}(x_1, x_2, x_3, t) = \bar{\mathbf{u}}(x_1, x_2, x_3) + \mathbf{u}_f(x_1, x_2, x_3, t) \tag{3-13a}$$

$$p(x_1, x_2, x_3, t) = \bar{p}(x_1, x_2, x_3) + p_f(x_1, x_2, x_3, t) \tag{3-13b}$$

etc.

If $\mathbf{u}$, p, etc. satisfy the Navier-Stokes equations, the average values will satisfy certain equations which are different from the Navier-Stokes equations but which still must fall under the general scheme of continuum mechanics as described in Art. 2. The equations for the mean values in fluctuating incompressible flow without volume forces are

$$\text{div}\,\bar{\mathbf{u}} = 0 \tag{3-14a}$$

$$\text{div}\,(\rho\bar{\mathbf{u}}\cdot\bar{\mathbf{u}}) + \text{grad}\,\bar{p} = \text{div}\,(\mu\,\text{def}\,\bar{\mathbf{u}} + \sigma_R) \tag{3-14b}$$

where the Reynolds stress tensor σ_R is defined by

$$\sigma_R = -\overline{(\rho\mathbf{u}_f\cdot\mathbf{u}_f)} \tag{3-14c}$$

[7] The time average may also be interpreted as a stochastic average. The notion of stochastic average is especially important when one considers mean values which vary with time.

The components of this stress tensor are thus the correlations of the fluctuations of the velocity components times the density. The equations corresponding to Eq. 3-14 for the compressible case become quite unwieldy, although they are in principle easy to write down.

Eq. 3-14 are evidently a special case of Eq. 2-4 and 2-5. To make them determinate one needs a relation, corresponding to Eq. 3-1, which expresses σ_R in terms of $\bar{\mathbf{u}}$, $\bar{p}$, and ρ. This problem is as yet unsolved. For further details about turbulent flow the reader is referred to V, B and V, C.

Relations Between State Variables and Transport Parameters. Specific equations will now be given for the functional dependence of the state variables p, e and the transport parameters μ, λ, k on the state variables ρ and T (cf. Eq. 3-6). We are concerned here primarily with gases, in particular with air at moderate values of density and temperature.

Equations for state variables. As pointed out in Art. 3, the relations between the state variables of a given medium in nonequilibrium are assumed to be the same as in equilibrium. Relations between state variables are often called equations of state. Sometimes this name is reserved specifically for the equations that express pressure as a function of density (or volume) and temperature.

For the equation of state for the pressure of a gas one often uses the general form

$$\frac{p}{\rho \mathcal{R} T} = 1 + B(T)\rho + C(T)\rho^2 + \cdots \tag{3-15}$$

where $\mathcal{R}$ is a constant[8] and B, C, . . . are called the second, third, etc. virial coefficients. For a fixed temperature, any fluid behaves more like

[8] The quantity $\mathcal{R}$ is the gas constant per unit mass, with the same physical units as $p/\rho T$ or specific heat. The product $\mathfrak{M}\mathcal{R}$, where $\mathfrak{M}$ is the molecular weight of the gas, is the gas constant per mole and is a universal constant for all real gases at sufficiently low density. Its dimension is properly speaking energy per mass mole; it may also be expressed as energy per weight mole if the weight is measured under standard conditions. Its value in the engineering system, where the pound and slug are the units of force and mass respectively, is $\mathfrak{M}\mathcal{R} = 1545$ ft lb/lb mole°R or 1.985 BTU/lb mole°R. In the metric system, where the dyne and gram are the units of force and mass, $\mathfrak{M}\mathcal{R} = 8.314 \times 10^7$ erg/gm-mole°K or 1.985 cal/g mole°K. It follows that for gases at sufficiently low density the quantity $\mathfrak{M}/\rho$ with dimension of specific volume is a universal function of pressure over temperature. In the metric system the value of $\mathfrak{M}/\rho$ is 22,415 cm³/g mole at 0°C and 1.013 dyne/cm².

For mixtures of gases $\mathcal{R}$ may be determined from the universal gas constant with the aid of Dalton's law of partial pressures. Taking the apparent molecular weight of air as 28.97, then $\mathcal{R}$ is 1716 (lb/ft²)/(slug/ft³ °R) or (ft lb)/(slug °R).

The NACA standard atmosphere tables [*11*] specify for air at sea level: p = 14.70 lb/in.² = 2116 lb/ft², ρ = 0.002377 slug/ft³, and T = 518.7°R = 59°F = 15°C. If the density is multiplied by the standard acceleration of gravity, g = 32.18 ft/sec², then ρg is the specific weight at standard conditions, 0.07648 lb/ft³, and $\mathcal{R}/g$ becomes 53.35 ft lb/(lb °R). In the metric system the corresponding values are $p = 1.013 \times 10^6$ dyne/cm², $\rho = 1.225 \times 10^{-3}$ g/cm³, and T = 288.2°K, so that $\mathcal{R} = 2.870 \times 10^6$ (dyne/cm²)/(g/cm³)(°K) or (erg)/(g °K).

a thermally perfect gas,

$$p = \rho \mathcal{R} T \tag{3-16}$$

as the density is reduced indefinitely. In all subsequent work in this section Eq. 3-16 will be used whenever a specific relation between p, ρ and T is needed. The results thus obtained are consequently valid only for perfect gases.

From the point of view of statistical mechanics, the internal energy is the average of the random kinetic energy of the molecules plus the potential energy of intermolecular forces. In general one may therefore expect e, the internal energy per unit mass, to depend on both temperature and density. However, for a fluid that satisfies Eq. 3-16 or even the slightly more general equation,

$$\frac{p}{\rho \mathcal{R} T} = f(\rho)$$

one may prove by thermodynamics that e is a function of T only.

The enthalpy h defined by Eq. 3-7a as $e + (p/\rho)$, is also in general a function of the density and the temperature. For a thermally perfect gas, however, h is evidently equal to $e + \mathcal{R}T$ and hence a function of temperature only.

The specific heats at constant volume, c_v, and constant pressure, c_p, may be defined by

$$c_v = \left(\frac{\partial e}{\partial T}\right)_{\rho=\text{const}} \tag{3-17a}$$

$$c_p = \left(\frac{\partial h}{\partial T}\right)_{p=\text{const}} \tag{3-17b}$$

The difference $c_p - c_v$ then satisfies the relation

$$c_p - c_v = -\frac{T}{\rho^2}\left(\frac{\partial p}{\partial T}\right)_\rho \left(\frac{\partial \rho}{\partial T}\right)_p \tag{3-18a}$$

For a thermally perfect gas this reduces to

$$c_p - c_v = \mathcal{R} \tag{3-18b}$$

c_p and c_v for gases are in general slowly variable functions of the temperature T and it is often possible to neglect this temperature variation completely. The ratio of the specific heats,

$$\gamma = \frac{c_p}{c_v}$$

is an important parameter in compressible flow processes. A fluid for which c_p and c_v, and hence γ, are constant is said to be calorically perfect.

Experimental determination of the ratio of the specific heats is usually carried out by spectroscopic methods or by measuring the velocity of sound a at low frequency and low pressure. Assuming an isentropic or reversible adiabatic process for a perfect gas,

$$a^2 = \left(\frac{\partial p}{\partial \rho}\right)_{s=\text{const}} = \gamma \frac{p}{\rho} \tag{3-19}$$

where s is the entropy defined by Eq. 3-7c. For a gas with n classical degrees of freedom and equipartition of energy, the parameter γ is given by

$$\gamma = \frac{n+2}{n} \tag{3-20}$$

and must therefore lie between unity and $\frac{5}{3}$.

For most real gases it is found that rotational and vibrational degrees of freedom are not fully excited at low temperatures. For example, the vibrational energy of O_2 and N_2 is negligible at room temperature and below, but its effect on γ may not be small at higher temperatures. Numerical values may be found in [*12*] and the references given there; see also Vol. I. The values quoted are usually obtained by spectroscopic methods.

Equations for transport parameters. Most experiments to determine the values of the transport parameters μ and k assume a Newtonian fluid; i.e., a stress tensor and a heat conduction vector having the form prescribed by Eq. 3-1 and 3-2. Consistency of measurements under various experimental conditions is then an a posteriori justification for this assumption (see e.g. [*13,14*]).

In the case of gases, kinetic theory supports the general form of Eq. 3-1 and 3-2 and predicts moreover that μ and k should depend only on temperature. For example, assuming central intermolecular forces proportional to r^{-n}, where r is the distance between centers, μ and k vary as

$$\mu,\, k \sim T^{\left(\frac{1}{2}+\frac{2}{n-1}\right)} \tag{3-21}$$

However, to compute the transport parameters theoretically for any more realistic collision process is a formidable task. The calculations are at present practicable only for gases at moderate density, although there is an extensive literature on the subject (see e.g. [*15*] and I,D).

In the case of liquids, the viscosity μ and the heat conductivity k depend on both pressure and temperature, and the temperature dependence is in general opposite to that observed for gases. For experimental results for liquids, reference is made to handbooks of physical chemistry, especially [*10*]. The kinetic theory of dense gases and liquids is still in a relatively primitive state, but is useful in establishing certain general trends of the parameters and in making it possible to conduct and represent experiments in a rational way.

Compilations of numerical data for μ and k as functions of temperature for several common gases are in progress at the Massachusetts Institute of Technology (see [*17*]) and at the National Bureau of Standards (see [*18,19*]). In each case new measurements have been undertaken where information is absent or of doubtful value. The tabulated values are usually calculated from an interpolation formula fitted to the measured data. This formula may be the Sutherland two-parameter

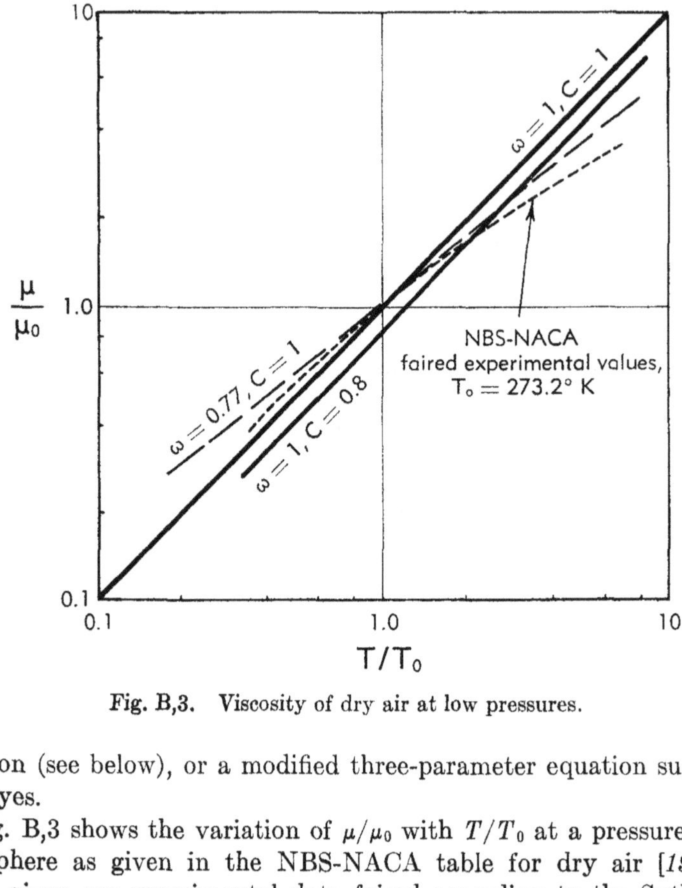

Fig. B,3. Viscosity of dry air at low pressures.

equation (see below), or a modified three-parameter equation suggested by Keyes.

Fig. B,3 shows the variation of μ/μ_0 with T/T_0 at a pressure of one atmosphere as given in the NBS-NACA table for dry air [*19*]. The values given are experimental data faired according to the Sutherland equation,

$$\frac{\mu}{\mu_0} = \left(\frac{T}{T_0}\right)^{\frac{3}{2}}\left(\frac{T_0 + S}{T + S}\right) \tag{3-22}$$

with $S = 110.4°K$ at $T_0 = 273.1°K$. The reference viscosity μ_0 is taken in the original table as 1716×10^{-7} (g/cm³)(cm/sec)(cm), or 3.58×10^{-7} (slugs/ft³)(ft/sec)(ft) in the engineering system.

Two approximate formulas frequently used in aerodynamics are special cases of the empirical equation

$$\log \frac{\mu}{\mu_0} = A \log \frac{T}{T_0} + B \tag{3-23}$$

where A and B are constants. If $A = \omega$ and $B = 0$ then $\mu/\mu_0 = (T/T_0)^\omega$. If $A = 1$ and $B = \log C$ then $\mu/\mu_0 = C(T/T_0)$. It is clear that ω and C are slope and intercept in the logarithmic coordinates of Fig. B,3 and that the two-parameter representation of Eq. 3-23 allows the viscosity to be correctly given for two points, for example, stagnation and free stream conditions. It may be noted that choice of a reference temperature other than 0°C would simply shift the experimental curve without changing its shape. The figure also exhibits the practical limitations of present knowledge at low temperatures, where condensation occurs, and at high temperatures, where dissociation cannot be neglected (cf. [*20*]).

In discussing the heat conductivity k it is usually more convenient to consider the Prandtl number $Pr = \mu c_p/k$, which is a measure of the relative importance of dissipation and heat conduction and is therefore an important parameter in problems of aerodynamic heating and heat transfer. If the Prandtl number is large, thermal energy tends to accumulate in dissipative regions because heat is produced from an adiabatic velocity field by friction faster than it is carried away by conduction. Most liquids fall in this category. Gases, on the other hand, usually have Prandtl numbers in the neighborhood of 0.7, and molten metals such as mercury may have very small Prandtl numbers of the order of 0.01. The kinetic theory of gases predicts that the Prandtl number should be independent of temperature, and the experimental variation is found to be small for permanent gases. However, since c_p, μ, and k must be measured separately, a typical experimental uncertainty of one per cent means that no more than two significant figures in the Prandtl number are dependable for aerodynamic calculations.

The second viscosity λ. The kinetic theory of gases yields for point molecules with central force fields (cf. Eq. 3-21) a relation between λ and μ, namely

$$\lambda + \tfrac{2}{3}\mu = 0 \tag{3-24}$$

The quantity $\lambda + \frac{2}{3}\mu$ is usually called volume or bulk viscosity, while μ and λ are called ordinary or shear viscosity and second viscosity respectively. The relation (Eq. 3-24), first proposed and later repudiated by Stokes, is sometimes assumed to be universally valid. In fact, however, the Stokes relation can be expected to hold for monatomic thermally perfect gases, but not for polyatomic gases or for liquids. As pointed out after Eq. 3-9 the assumption that the dissipation Φ is always non-negative implies that the bulk viscosity is also non-negative.

The symbol p as in this section denotes the part of the stress tensor $\sigma = -p\mathsf{I} + \lambda(\text{div}\,\mathbf{u})\mathsf{I} + \mu(\text{def}\,\mathbf{u})$ which remains finite when the fluid is at rest or in uniform motion, and hence does not depend explicitly on the velocity field. On the other hand, the quantity most accessible experimentally is usually some normal component of the stress tensor σ. For this reason the pressure is sometimes defined as the mean of the normal stresses at a point, i.e., $p - (\lambda + \frac{2}{3}\mu)(\text{div } \mathbf{u})$. If this quantity is to be a state variable in compressible flow the Stokes relation must hold. Although this statement is sometimes presented as a proof of the Stokes relation, the premise of the proof is, actually, an arbitrary postulate. A related spurious proof of the Stokes relation is based on the arbitrary assumption that the dissipation in uniform dilatation is zero.

The value of λ is obviously immaterial when div $\mathbf{u}$ is small. Since the continuity equation implies $\rho \text{ div } \mathbf{u} = -\partial\rho/\partial t - \mathbf{u} \text{ grad } \rho$ it follows that flow processes in which there are sufficiently strong density gradients in the direction of motion, or sufficiently rapid density changes at a fixed point, may be influenced by the second viscosity λ and may therefore provide an opportunity for experimental investigation. Examples of such processes are strong shock waves or intense high frequency sound waves.

Finally, there is some evidence that phenomena associated with finite volume viscosity in gases resemble to some extent phenomena treated from the molecular point of view as relaxation effects. This question has been discussed by Chapman [*21*].

B,4. Forces on a Solid.

Local Viscous Stress. If an infinitesimal surface element of a solid boundary S is represented by its outward normal $d\mathbf{n}$, the force exerted by the fluid on this element is $\sigma d\mathbf{n}$. Under the assumption (Eq. 3-1) this force may be written

$$\sigma d\mathbf{n} = -p d\mathbf{n} + \tau d\mathbf{n} = (-p + \lambda \text{ div } \mathbf{u})d\mathbf{n} + \mu(\text{def } \mathbf{u})d\mathbf{n}$$

This expression may be simplified in special cases, in particular for stationary flow when $\mathbf{u} = 0$ on the solid.

The following lemma is needed:

$$\text{If } \frac{\partial}{\partial t} = 0 \text{ and } \mathbf{u} = 0 \text{ on } S, \text{ then div } \mathbf{u} = 0 \text{ on } S \tag{4-1}$$

Proof: The continuity equation for stationary flow has the form $\mathbf{u} \cdot \text{grad } \rho + \rho \text{ div } \mathbf{u} = 0$. Hence, assuming $\rho \neq 0$, div $\mathbf{u} = 0$ whenever $\mathbf{u} = 0$.

An immediate corollary is that $\tau = \mu \text{ def } \mathbf{u}$ at the solid. This formula may also be written

$$\tau d\mathbf{n} = \mu(\text{grad } \mathbf{u})d\mathbf{n} = \mu \text{ curl } \mathbf{u} \times d\mathbf{n} \tag{4-2}$$

Proof: Choose a system of Cartesian coordinates with origin in the surface element and the x axis directed along $d\mathbf{n}$. At the origin $\partial\mathbf{u}/\partial y = \partial\mathbf{u}/\partial z = 0$ since $\mathbf{u} = 0$ on S. These relations together with Eq. 4-1 imply $\partial u/\partial x = 0$. The component representations of (grad $\mathbf{u}$)* and $d\mathbf{n}$ then show that (grad $\mathbf{u}$)*$d\mathbf{n} = 0$. But this relation is written in invariant form and hence is independent of the particular choice of coordinates. Hence $\tau = \mu[\text{grad } \mathbf{u} + (\text{grad } \mathbf{u})^*]$ may as well be written $\mu[\text{grad } \mathbf{u} - (\text{grad } \mathbf{u})^*]$. If use is made of Eq. 1-9b the proof is thus concluded.

A corollary of Eq. 4-2 is that under the assumptions stated the viscous stress at a point of the solid is tangential to the solid. This is, of course, not true in general. Consider, for example, one-dimensional piston motion along the x axis. Then there is a viscous stress normal to the piston of magnitude $(\lambda + 2\mu)\partial u/\partial x$. Other examples are given in the discussion of potential solutions in Art. 6.

Integrated Forces. The total force exerted by the fluid on a solid (drag, lift, side force) may of course be expressed as an integral of the local force over the surface of the solid. However, it is also possible to relate the total force to integrals over surfaces at a distance from the body and also to volume integrals over the fluid. Since the state of the fluid depends on the time history of the motion, some theorems are in general useful only for stationary flow or flows that are stationary in the mean. Of special interest are formulas which express the body force as an integral over a plane downstream of the solid and perpendicular to the flow direction; such formulas form the theoretical basis for determination of drag from wake measurements.

To simplify the writing, we shall consider only the case in which the volume forces and heat sources ($\mathbf{f}$ and Q) are zero (see IV,F). The formulas derived below may be easily modified for the more general case. Except where the contrary is explicitly stated, it will also be assumed that the flow is stationary and that $\mathbf{u}$ is $U\mathbf{i}$ at infinity and zero on the body surface.

Relations based on continuum mechanics. To begin with, only the general equations of Art. 2 will be used. Let Σ be an arbitrary closed surface fixed in space and enclosing the body surface S and denote by V the regions between S and Σ. One sometimes refers to Σ as a control surface. The normals $d\mathbf{n}$ to S and Σ are taken with the directions indicated in Fig. B,4a. Note that the outward normal with respect to V of a surface element of S is then $-d\mathbf{n}$.

Total Force and Torque. The total fluid force $\mathbf{F}$ on the body is

$$\mathbf{F} = \iint_S \sigma d\mathbf{n} \tag{4-3a}$$

and $\mathbf{M}$, the total moment of the forces (torque with respect to a point is

$$\mathbf{M} = \iint_S \mathbf{r} \times (\sigma d\mathbf{n}) \tag{4-3b}$$

where $\mathbf{r}$ is the radius vector from a fixed center to a variable point on S.

By applying the momentum law (Eq. 2-2a) to the region V one obtains

$$\mathbf{F} = \iint_{\Sigma} \sigma d\mathbf{n} - \iint_{\Sigma} (\rho \mathbf{u} \circ \mathbf{u}) d\mathbf{n} \tag{4-4a}$$

The only boundary condition necessary to derive this formula is that the normal component of $\mathbf{u}$ vanish at the boundary.

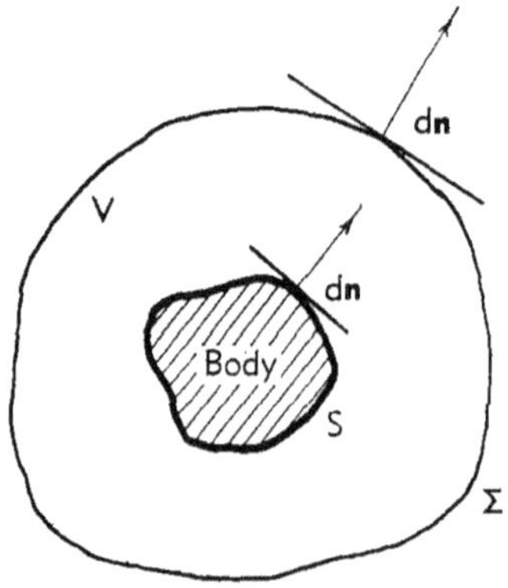

Fig. B,4a.

Similarly Eq. 2-10′ implies

$$\mathbf{M} = \iint_{\Sigma} \mathbf{r} \times (\sigma d\mathbf{n}) - \iint_{\Sigma} \mathbf{r} \times [(\rho \mathbf{u} \circ \mathbf{u}) d\mathbf{n}] \tag{4-4b}$$

It is often convenient to introduce perturbation quantities into the integral laws. The perturbation velocity $\mathbf{u}'$ is defined as

$$\mathbf{u}' = \mathbf{u} - U\mathbf{i} = (u', v', w') \tag{4-5a}$$

where

$$u' = u - U, \qquad v' = v, \qquad w' = w$$

Similarly,

$$\sigma' = \sigma - \sigma_{\infty}, \qquad p' = p - p_{\infty} \quad \text{etc.} \tag{4-5b}$$

Here σ_{∞} is the constant tensor $-p_{\infty}\mathsf{I}$. No linearization of the equation is intended and in general it will not be assumed that, for example, u' is small compared to U. On the contrary, at the solid, $u' = -U$ if no slip is assumed. At large distances from the body, however, the perturbation quantities tend to zero. Using the conservation of mass, one may then modify Eq. 4-4a to

$$\mathbf{F} = \iint_{\Sigma} \sigma d\mathbf{n} - \iint_{\Sigma} (\rho \mathbf{u}' \circ \mathbf{u}) d\mathbf{n} \tag{4-4a'}$$

Here, as in all formulas below, σ may be replaced by σ' since σ_{∞} is constant. Similarly

$$\mathbf{M} = \iint_{\Sigma} \mathbf{r} \times (\sigma d\mathbf{n}) - \iint_{\Sigma} \mathbf{r} \times [(\rho \mathbf{u}' \circ \mathbf{u}) d\mathbf{n}] \tag{4-4b'}$$

Drag. The drag D is the x component of $\mathbf{F}$ and hence

$$D = \mathbf{F} \cdot \mathbf{i} = \mathbf{i} \cdot \iint_{\Sigma} \sigma d\mathbf{n} - \iint_{\Sigma} \rho u \mathbf{u} \cdot d\mathbf{n}$$

$$= \mathbf{i} \cdot \iint_{\Sigma} \sigma d\mathbf{n} - \iint_{\Sigma} \rho u' \mathbf{u} \cdot d\mathbf{n} \tag{4-6}$$

The integral law (Eq. 2-9b) applied to V gives

$$\iint_{\Sigma} \rho \frac{|\mathbf{u}|^2}{2} \mathbf{u} \cdot d\mathbf{n} = \iint_{\Sigma} (\sigma \mathbf{u}) \cdot d\mathbf{n} - \iiint_{V} \sigma \cdot \operatorname{grad} \mathbf{u} dV \tag{4-7}$$

In deriving Eq. 4-7 it is not sufficient to assume that $\mathbf{u} \cdot d\mathbf{n}$ is zero on S; in addition $\sigma\mathbf{u} \cdot d\mathbf{n}$ must vanish on S. For a nonviscous fluid the second assumption follows from the first since then $\sigma d\mathbf{n}$ is parallel to $d\mathbf{n}$. Thus in this case it is sufficient to assume that $\mathbf{u}$ is tangent to S at S. For a viscous fluid, on the other hand, one needs the assumption that $\mathbf{u} = 0$ on S. Thus Eq. 4-7 is not in general valid for stationary viscous flow when the solid is rotating or when the fluid is allowed to slip along the boundary. In these cases $\iint_{S} \sigma\mathbf{u} \cdot d\mathbf{n}$ has to be added to the left-hand side of Eq. 4-7. This term represents the work done on the fluid by the shearing stresses at the solid.

Combining Eq. 4-6 and 4-7 one obtains

$$DU = \iint_{\Sigma} \rho \frac{|\mathbf{u}'|^2}{2} \mathbf{u} \cdot d\mathbf{n} - \iint_{\Sigma} (\sigma \mathbf{u}') \cdot d\mathbf{n} + \iiint_{V} \sigma \cdot \operatorname{grad} \mathbf{u} dV \tag{4-8}$$

An alternative derivation of Eq. 4-8 shows more clearly its physical meaning. In the system of coordinates used so far the work done by the fluid forces at the solid is zero since $\mathbf{u} = 0$ (or, at least, $\mathbf{u} \cdot \sigma d\mathbf{n} = 0$) on S. On the other hand, if the body is considered moving with velocity $-U\mathbf{i}$ the work done on the fluid per unit time is UD. Formally, this amounts to introducing a new system of coordinates by the Galilean transformation $x' = x - Ut$, $y' = y$, $z' = z$, $t' = t$. The velocity $\mathbf{u}'$ at a point with respect to the new system of coordinates is the same as the perturbation velocity in the old system. The notation is thus consistent with Eq. 4-5. The fluid is at rest at infinity, $\mathbf{u}'(\infty) = 0$, and the no-slip condition implies $\mathbf{u}' = -U\mathbf{i}$ on S. In the new system the flow is not stationary. Since $\partial/\partial t = 0$ in the old system, $\partial/\partial t = U\partial/\partial x'$. In this new system of coordinates one may now apply the integral law for the kinetic energy (Eq. 2-9b) to the region between Σ and S. (These surfaces are now regarded as fixed with respect to the new system of coordinates; at the value of t for which the integral is applied, S coincides with the body surface.) Making use of the fact that $\mathbf{u}' = -U\mathbf{i}$ on S, one obtains

$$DU = \iint_{\Sigma} \rho \frac{|\mathbf{u}'|^2}{2} U\mathbf{i} \cdot d\mathbf{n} + \iint_{\Sigma} \rho \frac{|\mathbf{u}'|^2}{2} \mathbf{u}' \cdot d\mathbf{n} - \iint_{\Sigma} (\sigma\mathbf{u}') \cdot d\mathbf{n} + \iiint_{V} \sigma \cdot \operatorname{grad} \mathbf{u}' dV \quad (4\text{-}8')$$

Since $\mathbf{u} = U\mathbf{i} + \mathbf{u}'$, Eq. 4-8′ and 4-8 are identical laws. In the primed system of coordinates Eq. 4-8′ has a very obvious physical interpretation (cf. the discussion after Eq. 2-9a): DU is the work done per unit time by the solid on the fluid in V. The first term on the right-hand side represents the rate of increase of the total kinetic energy in V. It has the form of a surface integral because all changes with time inside V are due to a fixed flow pattern moving past V with constant velocity $-U\mathbf{i}$. The second term represents the flow of kinetic energy through Σ and the third term the work done per unit time on the fluid in V by the stresses on Σ. The fourth term is due to the fact that all the work done on the fluid in V does not reappear as kinetic energy. Part of the work done causes internal changes in the state of the fluid; it is partly stored as compression and partly dissipated (Eq. 3-9).

Energy. An integral law involving the total energy may be obtained by integrating Eq. 2-6‴ over V and adding the resulting formula to Eq. 4-8:

$$DU + Q_w = \iint_{\Sigma} \rho \left(\frac{|\mathbf{u}'|^2}{2} + e \right) \mathbf{u} \cdot d\mathbf{n} + \iint_{\Sigma} \mathbf{q} \cdot d\mathbf{n} - \iint_{\Sigma} (\sigma\mathbf{u}') \cdot d\mathbf{n} \quad (4\text{-}9)$$

where $Q_w = \iint_S \mathbf{q} \cdot d\mathbf{n}$ = heat conducted from the solid into the fluid.

This law may be easily interpreted physically in the same way as Eq. 4-8.

Similarly, one may derive other integral laws corresponding to the various energy equations. As a last example the integral law for the stagnation enthalpy $H = h + (|\mathbf{u}|^2/2)$ will be considered. If it is assumed that σ is of the form $\sigma = -p\mathsf{I} + \tau$, Eq. 3-8b applies. Integrating this equation in the original unprimed system of coordinates one finds

$$\iint_{\Sigma} \rho H \mathbf{u} \cdot d\mathbf{n} = Q_w - \iint_{\Sigma} \mathbf{q} \cdot d\mathbf{n} + \iint_{\Sigma} (\tau\mathbf{u}) \cdot d\mathbf{n} \quad (4\text{-}10)$$

Using the continuity equation one sees that H may be replaced by the perturbation quantity $H' = H - H_\infty$.

Relations based on the Navier-Stokes equations. The integral laws derived above were based on the general equations of continuum mechanics, and the shape of the control surface was arbitrary. From now on it will be assumed that we are dealing with a fluid obeying the Navier-Stokes equation discussed in Art. 3. Furthermore, the control surface will be given certain special shapes. Considerable simplification of the formulas then results. Formulas for lift and drag are derived below. For similar derivations, cf. [*10,22,23*].

In the drag formula, a convenient choice of Σ is an infinite plane downstream of the solid and perpendicular to the free stream direction. This is obtained as a limiting case in the following manner. We first assume Σ to be a parallelopiped bounded by the planes $x = -x_2$, $x = x_1$, $y = \pm y_1$, $z = \pm z_1$. The cross section in the plane $z = 0$ is shown in

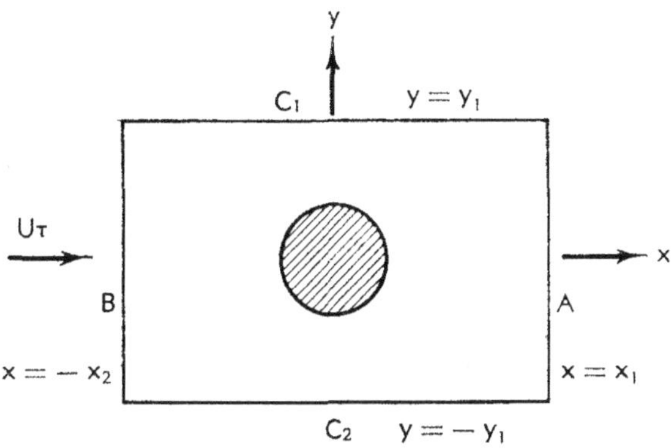

Fig. B,4b. Intersection of special control surface with plane $z = 0$.

Fig. B,4b. The limiting case where Σ is the infinite plane $x = x_1$ is then obtained by letting x_2, y_1 and z_1 tend to infinity. In the two-dimensional case Σ is the rectangle AC_1BC_2.

Assumptions about flow at large distances. In order to evaluate the integrals of the general formulas for this limiting case it is necessary to make assumptions about the magnitude of various flow quantities at large distances from the solid. It is not possible to prove these assumptions conclusively. They will, however, be made plausible by certain arguments in the following chapters (cf. also [*23*]). The underlying assumption is that at large distances an Oseen linearization is valid. (This statement requires some qualifications for supersonic flow.) A more detailed discussion of this assumption and of the behavior of Oseen flow will be given in Art. 7 (for incompressible flow) and in Art. 13 (for compressible flow); cf. also the discussion of laminar wakes in Art. 8. Anticipating these results we shall state below certain properties of the flow field at large distances from a finite body.

For reasons to be explained in the next chapter there exists a region downstream of the body called the wake, where the flow field due to the action of viscosity is of a different nature than in other regions of the fluid. It is not necessary to define the wake region exactly. The important fact is only that it is a region downstream of the body whose cross-sectional area increases parabolically, that is, as $r^{(m-1)/2}$ in the downstream direction. Here r denotes the distance of an arbitrary point in the fluid to a fixed point in the body, and m denotes the dimensions of the space ($m = 2$ or 3).

Consider first the *region outside the wake.* The flow is here essentially nonviscous at large distances. The effect of viscosity is indirect. The strongly viscous regions close to the body and in the wake give the body an apparent shape which is that of a half-open body qualitatively similar to a paraboloid (or parabola in two dimensions). The magnitude of the perturbation quantities at large distances is then at most equal to that of the perturbation flow due to a source and a lifting element. Hence we make the following assumptions for the region outside the wake:

1. Perturbation quantities (of velocity components, pressure, etc.) decrease at least as fast as $r^{(1-m)}$. The viscous stresses, being proportional to velocity derivatives, then decrease as r^{-m}.

Since the flow is essentially nonviscous, a linearized Bernoulli's law is valid.

2. $p' \sim -\rho_\infty U u' \quad \text{as} \quad r \to \infty$

Consider now a surface whose area increases as r^{m-1} as the surface is moved to infinity. Assumption 1 above implies the following two statements for integrals over such a surface:

3. In the limit $r \to \infty$ any integral of the viscous stresses vanishes.

4. Integrals of linear expressions in the perturbation quantities may remain finite in the limit. Integrals of quadratic and higher order terms vanish.

Secondly consider the *wake region.* Here (2) is not valid and (1) has to be replaced by the weaker assumption:

5. Perturbation quantities in the wake decrease at least as fast as $r^{(1-m)/2}$.

The plane (or line) $x = x_1$ intersects the wake in a region of area $x_1^{(m-1)/2}$. In this region perturbation quantities decrease as described by (5) as x_1 tends to infinity. However, the area of this region increases at a correspondingly slow rate and one may hence conclude:

6. Integrals of flow quantities over the plane $x = x_1$ behave as described by (4) as x_1 tends to infinity.

The essential difference between an integral over the plane $x = x_1$ and one over the plane $x = -x_2$ is thus that in the limit $r \to \infty$ Bernoulli's law (2) may be assumed only for the latter integral.

Drag. Consider now the expression for the drag, Eq. 4-6, where σ

is replaced by $\sigma' = -p'\mathsf{I} + \boldsymbol{\tau}$. In the special control surface shown in Fig. B,4b, let all sides except the plane $x = x_1$ tend to infinity. Assumption 3 above shows that, except for the surface $x = x_1$, the integrals involving $\boldsymbol{\tau}$ vanish in the limit. The integral $\mathbf{i} \cdot \iint (-p'\mathsf{I})d\mathbf{n}$ is identically zero on all planes parallel to the free stream direction. On the plane $x = -x_2$ it is $\iint p'dydz$ which in the limit may be replaced by $-\iint \rho_\infty Uu'dydz$ by assumption 2. The integral $-\iint \rho u'\mathbf{u} \cdot d\mathbf{n}$ involves only quadratic and cubic terms of the perturbation velocities on the sides parallel to the free stream and hence vanishes by assumption. By the same assumption it tends to $\iint \rho_\infty u'Udydz$ on the plane $x = -x_2$. This integral will not vanish by itself as x_2 tends to infinity; it actually represents half the strength of the apparent source. However, as seen from the formula above, it cancels with the integral of the perturbation pressure on the same surface. Hence, in the limit, only the integral over $x = x_1$ remains and

$$D = \iint_{x=x_1} [\tau_{xx} - p' - \rho u'(U + u')]dydz \tag{4-11}$$

This formula is valid for any plane $x = x_1$ at a finite distance downstream of the body.

In incompressible flow τ_{xx} is given by

$$\tau_{xx} = 2\mu \frac{\partial u}{\partial x} = -2\mu \left(\frac{\partial v}{\partial y} + \frac{\partial w}{\partial z}\right), \qquad \mu = \text{const}$$

hence its integral over $x = x_1$ is zero.

Thus

$$D = \iint_{x=x_1} [-p' - \rho u'(U + u')]dydz \tag{4-12}$$

if $\rho = \text{const}$, $\mu = \text{const}$.

In viscous flow the drag is not zero. Hence the integrals in Eq. 4-11 and 4-12 remain equal to $D \neq 0$ even as x_1 tends to infinity. By assumption 6 both may be simplified to[9]

$$D = \iint_{x=\infty} [-p' - \rho_\infty Uu']dydz \tag{4-13}$$

[9] It must be kept in mind that we consider only viscous flow past a finite object. If either of these restrictions is removed, terms that are nonlinear in the perturbation quantities may still contribute to the drag even at infinitely distant control surfaces. Consider, for example, a lifting three-dimensional wing in an incompressible non-viscous fluid. The drag may be computed by an integral over the plane $x = x_1$ as x_1 tends to infinity (Trefftz plane). The perturbation velocities connected with the trailing vortices are then not weakened by viscosity, and their magnitude remains such that quadratic expressions must be retained. Another example is flow towards a parabola. In this case the upstream disturbance is so strong that the nonlinear terms on the plane $x = -x_2$ contribute to the drag even as x_2 tends to infinity (this is true independent of viscosity).

The corresponding integral was assumed to vanish for $x = -\infty$. The difference in behavior is that the linearized Bernoulli's law is valid near upstream infinity but not near downstream infinity. As will be seen in the discussion of the Oseen equations, this is due to the occurrence of transversal waves in the wake. The failure of Bernoulli's law may also be illustrated as follows. For convenience only the incompressible case is considered. The stagnation pressure or total head is then defined as $p^0 = p + \frac{1}{2}\rho|\mathbf{u}|^2$. Bernoulli's law for nonviscous flow asserts that p^0 is constant along a streamline. Actually, there is a loss of total head due to viscous stresses. These are negligible upstream but not in the wake whose streamlines have passed close to the body and hence have been influenced by strong viscous stresses. The loss in total head as compared with upstream infinity is

$$(p^0)_\infty - p^0 = -p' - \rho_\infty u'U - \tfrac{1}{2}\rho(u'^2 + v^2 + w^2)$$

Within nonlinear terms, this expression is equal to the integrand of Eq. 4-12 and 4-13. The latter equation then expresses the fact that the drag is equal to the integral of the loss of total head over a plane infinitely far downstream.

Eq. 4-12 may be generalized to the case of fluctuating incompressible flow which is stationary in the mean, as described by Eq. 3-14. Using the experimental fact that the Reynold stresses are negligible, except in the boundary layer and in the wake, one finds by the method used in deriving Eq. 4-12

$$D = \iint_{x=x_1} [-\bar{p}' - \rho u'(U + \bar{u}') - \rho\bar{u}_f^2]dydz \tag{4-14}$$

The last term in the integrand is $(\sigma_R)_{xx}$ (Eq. 3-14c).

The drag formulas (Eq. 4-11, 4-12, and 4-14) are of practical interest since they furnish the basis for experimental determination of the drag by wake measurements. Drag formulas based on Eq. 4-8 have rather a theoretical interest since the volume integral cannot be evaluated experimentally. This equation may be given a simple form if the Navier-Stoke equations are used and all sides of the control surface are removed to infinity. If σ is replaced by $\sigma' = \sigma - \sigma_\infty$, all surface integrals in Eq. 4-8 involve only terms which are at least quadratic in the perturbation quantities. They will then vanish in the limit and one obtains

$$UD = \iiint \sigma' \cdot \text{grad } \mathbf{u} dV = \iiint (p' \text{ div } \mathbf{u} + \Phi)dV \tag{4-15}$$

where the region of integration is the entire fluid. Thus in the case of an incompressible fluid the work done by the drag (cf. discussion of Eq. 4-8) is equal to the total viscous dissipation in the fluid.

Lift. A formula for the force in the y direction will now be derived. This force is arbitrarily called the lift. Actually, the free stream direction is the only distinguished direction. Hence, depending on the choice of coordinates, the lift formula may be interpreted as a formula for the side force or the force in any direction normal to the body.

Take a control surface Σ which has the cross section shown in Fig. B,4b in any plane z = const. It is thus a cylinder with generators parallel to the z axis. From Eq. 4-4a, with σ replaced by σ', one obtains

$$L = \mathbf{F} \cdot \mathbf{j} = \iint_A [\tau_{xy} - \rho v(U + u')]dydz + \iint_{C_2} [p' - \tau_{yy} + \rho v^2]dxdz$$
$$+ \iint_{B_1} [-\tau_{xy} + \rho v(U + u')]dydz + \iint_{C_1} [-p' + \tau_{yy} - \rho v^2]dxdz$$

If now y_1, z_1, x_1, and x_2 tend to infinity the viscous stress terms and the terms at least quadratic in the perturbation velocities may be neglected. Furthermore, p' may be replaced by $-\rho_\infty Uu'$. This would not be allowed on the integral over A; however, its integrand does not contain p'. In the limit then

$$L = \rho_\infty U \left[\iint_A (-v)dydz + \iint_{C_2} (-u')dxdz + \iint_B (v)dydz + \iint_{C_1} (u')dxdz\right]$$

Hence

$$L = \mathbf{F} \cdot \mathbf{j} = \rho_\infty U \int_{z=-\infty}^{z=\infty} \Gamma(z)dz \tag{4-16}$$

where $\Gamma(z)$ is the circulation around a very large contour in the plane z = const. Note that for a viscous fluid the Joukowski formula, Eq. 4-16, is true only in the limit when the circulation is computed for a contour infinitely far away. Near the body the circulation tends to zero because of the no-slip condition. For two-dimensional viscous flow, Eq. 4-16 and 2-16b imply

$$L = -\rho_\infty U\Omega_T \tag{4-17a}$$

where

$$\Omega_T = \iint \Omega dxdy = \text{total vorticity in fluid}$$

The corresponding formula in three dimensions which gives the total force normal to the free stream direction is

$$\mathbf{F} - D\mathbf{i} = \rho_\infty U\mathbf{i} \times \mathbf{\Omega}_T = -\rho_\infty U\mathbf{i} \times \iint_\Sigma \mathbf{u} \times d\mathbf{n} \tag{4-17b}$$

where $\mathbf{\Omega}_T$ is the total vorticity in the fluid and the surface Σ enclosing the body is moved to infinity.

B,5. The Navier-Stokes Equations in General Coordinates. In dealing with the Navier-Stokes equations it is often advantageous to use non-Cartesian systems of coordinates. In many cases one uses some special system adapted to the symmetry properties of a problem; an example is the use of cylindrical coordinates in the study of flows with rotational symmetry. For certain general theoretical investigations it is often convenient to use general curvilinear coordinates; an example is provided by three-dimensional boundary layer theory.

In order to formulate the Navier-Stokes equations (Eq. 3-3, 3-4, and 3-5) for a curvilinear system of coordinates, it is necessary to make use of general tensor analysis. For a discussion of this subject the reader is referred to the standard literature, for example [*4*,*5*,*6*]. However, a few brief comments are made below in order to emphasize certain points which are important in the present context.

In dealing with vector and tensor fields in curvilinear coordinates one may use two different methods. The first method is that of general tensor calculus (or Riemannian geometry). It is applicable to any system of curvilinear coordinates. In this method a distinction is made between contravariant and covariant components of vectors and tensors; also between absolute scalars and scalar densities, between absolute vectors and vector densities, etc. The second method is that of vector geometry and is applicable only to orthogonal coordinates. Except for this restriction it differs from the first method in matters of conventions and definitions rather than in mathematical content. Only one kind of component of vectors and tensors is used, this component being the geometrical mean of the contravariant and covariant components of general tensor calculus. Furthermore, densities are not used. The difference between the two methods is discussed in [*4*, pp. 144ff.].

The Navier-Stokes equations are written below in the notation of general tensor calculus (Eq. 5-8, 5-10, and 5-13). The method of vector geometry is then discussed, and general expressions are given for the operators div and grad. With the aid of these expressions the formulation of the Navier-Stokes equation in the notation of vector geometry is immediate. Finally, examples are given of the application of the general formulas to special systems of coordinates.

Method of general tensor calculus. Let x_i be Cartesian coordinates in space and let ξ_i be a second system of coordinates. The geometry of the second system is determined by the metric tensor

$$g_{ij} = \sum_k \frac{\partial x_k}{\partial \xi_i}\frac{\partial x_k}{\partial \xi_j} \tag{5-1}$$

The functional determinant Δ of the coordinate transformation is denoted by

$$\Delta = \frac{\partial(x_1, x_2, x_3)}{\partial(\xi_1, \xi_2, \xi_3)} = \text{determinant of} \left(\frac{\partial x_i}{\partial \xi_j}\right) \tag{5-2a}$$

The determinant g of the matrix (g_{ij}) is then

$$g = \text{determinant of } (g_{ij}) = \Delta^2 \tag{5-2b}$$

The Christoffel symbols are defined by

$$\Gamma^h_{ik} = \frac{g^{hl}}{2}\left(\frac{\partial g_{il}}{\partial \xi_k} + \frac{\partial g_{kl}}{\partial \xi_i} - \frac{\partial g_{ik}}{\partial \xi_l}\right) \tag{5-3}$$

The right-hand side of the last equation is to be summed over all values of l. This is a special case of the summation convention which is used in general tensor calculus: whenever an upper index is the same as a lower index, the formula is to be summed over all values of this index.

The gradient of a scalar f is a vector whose covariant components are denoted by $f_{,i}$, where

$$f_{,i} = \frac{\partial f}{\partial \xi_i} \tag{5-4a}$$

The gradient of a vector with contravariant components v^i is a tensor whose mixed contravariant-covariant components are (cf. Eq. 1-8a)

$$v^i_{,j} = \frac{\partial v^i}{\partial \xi_j} + \Gamma^i_{jk} v^k \tag{5-4b}$$

If f is an absolute scalar, then $\sqrt{g}\, f$ is a scalar density. Similarly if (v^i) are the contravariant components of a vector then $(\sqrt{g}\, v^i)$ are the contravariant components of the corresponding vector density.

The divergence of a vector density (d^i) or a vector (v^i) may be expressed as

$$\text{div } (d^i) = \frac{\partial u^i}{\partial \xi_i} = \text{scalar density, } (d^i) = \text{vector density} \tag{5-5a}$$

or

$$\text{div } (v^i) = v^i_{,i} = \text{absolute scalar, } (v^i) = \text{absolute vector} \tag{5-5b}$$

It follows that if v^i are the contravariant components of a vector, then

$$\frac{1}{\sqrt{g}} \frac{\partial(\sqrt{g}\, v^i)}{\partial \xi^i} = v^i_{,i} = \frac{\partial v^i}{\partial \xi_i} + \Gamma^i_{ik} v^k \tag{5-5c}$$

The divergence of a contravariant tensor density (D^{ij}) is a vector density whose ith contravariant component is (cf. Eq. 1-10)

$$[\text{div } (D^{ij})]^i = \frac{\partial D^{ij}}{\partial \xi_j} + \Gamma^i_{jk} D^{jk} \tag{5-5d}$$

The identity tensor I (Eq. 1-5) has its simplest form if mixed contravariant-covariant components are used:

$$\mathsf{I} = (\delta^i_j),\ \delta^i_j = 0 \text{ for } i \neq j,\ \delta^i_i = 1 \quad \text{(no summation)} \tag{5-6}$$

If f is an absolute scalar, then the divergence of the tensor density $f\sqrt{g}\,(\delta^i_j)$ is equal to $\sqrt{g}$ times the gradient of f (cf. Eq. 1-12c). Expressed in covariant components this statement reads

$$\frac{\partial}{\partial \xi_i}(f\sqrt{g}\,\delta_j^i) - \Gamma_{ij}^h f\sqrt{g}\,\delta_h^i = \sqrt{g}\,\frac{\partial f}{\partial \xi_j} \tag{5-7}$$

The Navier-Stokes equations will now be written in the notation of general tensor calculus. The physical concept of density is treated as a density in the sense of tensor calculus. If (u^i) are the contravariant components of the velocity vector, then (ρu^i) are the contravariant components of a vector density, namely the mass flow vector. The continuity equation (Eq. 3-3) may then be written (cf. Eq. 5-5a)

$$\frac{\partial \rho}{\partial t} + \frac{\partial(\rho u^i)}{\partial \xi_i} = 0 \tag{5-8}$$

Note that formally one may treat ρ as an absolute scalar, the value of which is independent of the system of coordinates. In the above formula one must then replace ρ by $\sqrt{g}\,\rho$.

In the momentum equation we shall express the flow of momentum and the stress as tensor densities, and consider μ, λ, and p as absolute scalars. The flow of momentum has contravariant density components (M^{ij}) defined by

$$M^{ij} = \rho u^i u^j \tag{5-9a}$$

The contravariant density components of the stress tensor are

$$H^{ij} + \sqrt{g}\,(K - p)g^{il}\delta_l^j \tag{5-9b}$$

where the first term is μ def $\mathbf{u}$ expressed as a tensor density

$$H^{ij} = \mu\sqrt{g}\,(g^{ik}u^j_{,k} + g^{jl}u^i_{,l}) \tag{5-9c}$$

and K is the scalar λ div $\mathbf{u}$ (cf. Eq. 5-5c)

$$K = \frac{\lambda}{\sqrt{g}}\frac{\partial}{\partial \xi_l}(\sqrt{g}\,u^l) = \lambda u^l_{,l} \tag{5-9d}$$

Using the above notation and also Eq. 5-7, one may write the momentum equation (Eq. 3-4) as

$$\frac{\partial(\rho u_i)}{\partial t} + \frac{\partial M^{ij}}{\partial \xi_j} + \Gamma^i_{jk}M^{jk} = \frac{\partial H^{ij}}{\partial \xi_j} + \Gamma^i_{jk}H^{jk} + \sqrt{g}\,g^{ij}\frac{\partial(K - p)}{\partial \xi_j} + \rho F^i \tag{5-10}$$

With the aid of the continuity equation the left-hand side of Eq. 5-10 may be rewritten as a substantial derivative (cf. Eq. 2-7b) defined by

$$\rho\frac{Du^i}{Dt} = \rho\left(\frac{\partial u^i}{\partial t} + u^i_{,j}u^j\right) \tag{5-11}$$

In the energy equation e, T, and k are considered absolute scalars. The contravariant vector densities expressing the transport of internal energy and conduction of heat are then $\sqrt{g}\, e\mathbf{u}^i$ and $\sqrt{g}\, kg^{ij}T_{,j}$, respectively. If the dissipation Φ is considered as a density it may be written (cf. Eq. 3-5b)

$$\Phi = \mu\sqrt{g}\, u^j_{,i}(g_{jk}g^{il}u^k_{,l} + u^i_{,j}) + \lambda(u^i_{,i})(u^j_{,j}) \tag{5-12}$$

The energy equation (Eq. 3-5) may now be written

$$\frac{\partial(\rho e)}{\partial t} + \frac{\partial(\rho e u_i)}{\partial \xi_i} = \frac{\partial}{\partial \xi_i}(\sqrt{g}\, kg^{ij}T_{,j}) - p\frac{\partial}{\partial \xi_i}(\sqrt{g}\, u^i) + \Phi + \rho Q \tag{5-13}$$

The left-hand side may be rewritten as a substantial derivative, defined by

$$\rho\frac{De}{Dt} \equiv \rho\left(\frac{\partial e}{\partial t} + u^i e_{,i}\right) \tag{5-14}$$

Method of vector geometry. In many applications the coordinates ξ_i are orthogonal, in which case the matrix of the (g_{ij}) is diagonal. The following notation is then often used.

$$h_i^2 = g_{ii} = \frac{1}{g^{ii}}, \qquad \sqrt{g} = h_1h_2h_3 \tag{5-15}$$

Covariant and contravariant components of a vector $\mathbf{v}$ are then related by the simple formula $v_i = h_i^2 v^i$ (no summation over i). A new type of component, V_i, is defined by

$$V_i = h_i v^i = \frac{v_i}{h_i} = \sqrt{v_i v^i} \quad \text{(no summation)} \tag{5-16}$$

In vector geometry one uses only the components V_i which are referred to as the components of $\mathbf{v}$ in the sense of vector geometry. Note that the magnitude of V_i is equal to the length of the orthogonal projection of $\mathbf{v}$ on the ξ_i axis, so that the square of the length of $\mathbf{v}$ is

$$|\mathbf{v}|^2 = \sum_i V_i^2 = \sum_i v^i v_i \tag{5-17a}$$

In particular, if $\mathbf{P} = (\xi_i)$ and $\mathbf{Q} = (\xi_i + d\xi_i)$, the contravariant components of the infinitesimal displacement vector $\mathbf{PQ}$ are $d\xi_i$ whereas the components in the sense of vector geometry are $h_i d\xi_i$. The square of the length ds of $\mathbf{PQ}$ is then

$$(ds)^2 = \sum_i (h_i d\xi_i)^2 \tag{5-17b}$$

This formula shows the geometrical meaning of the h_i.

Similarly, in vector geometry the components A_{ij} of a tensor A are obtained from the corresponding contravariant components by multiplying by $h_i h_j$. With the above conventions the equation $A\mathbf{v} = \mathbf{w}$ may be written in component form as $V_i = \sum_j A_{ij} W_j$.

Furthermore, densities are not used explicitly in vector geometry. Note that if (V_i) are the components of $\mathbf{v}$, then the components of the corresponding vector density in the sense of general tensor calculus are $\sqrt{g}\,(V_i/h_i)$, i.e. $h_2h_3V_1$, $h_1h_3V_2$, and $h_1h_2V_3$.

The last remark may be used for deriving the expression for the divergence of a vector $\mathbf{v} = (V_i)$ in vector geometry. One finds from Eq. 5-5a

$$\operatorname{div} \mathbf{v} = \sum_i \frac{1}{\sqrt{g}} \frac{\partial}{\partial \xi_i}\left(\frac{\sqrt{g}\, V_i}{h_i}\right)$$

$$= \frac{1}{h_1h_2h_3}\left[\frac{\partial(h_2h_3V_1)}{\partial \xi_1} + \frac{\partial(h_1h_3V_2)}{\partial \xi_2} + \frac{\partial(h_1h_2V_3)}{\partial \xi_3}\right] \tag{5-18a}$$

Similarly one finds from Eq. 5-5d that the divergence of a tensor A is a vector with components

$$(\operatorname{div} \mathsf{A})_i = \frac{h_i}{\sqrt{g}} \sum_j \frac{\partial}{\partial \xi_j}\left(\frac{A_{ij}\sqrt{g}}{h_i h_j}\right) + \sum_j \frac{\partial h_i}{h_j \partial \xi_j} \frac{A_{ij} + A_{ji}}{h_i} - \sum_j \frac{\partial h_j}{h_i \partial \xi_i} \frac{A_{jj}}{h_j} \tag{5-18b}$$

If f is a scalar, the components of grad f are, according to Eq. 5-4a and 5-16,

$$\operatorname{grad} f = \left(\frac{1}{h_i}\frac{\partial f}{\partial \xi_i}\right) \tag{5-19a}$$

If $\mathbf{v} = (V_i)$, one derives from Eq. 5-4b that the tensor grad $\mathbf{v}$ has components $V_{i,j}$ where

$$V_{i,i} = \frac{1}{h_i}\frac{\partial V_i}{\partial \xi_i} + \sum_{k \neq i} \frac{1}{h_k}\frac{\partial h_i}{\partial \xi_k}\frac{V_k}{h_i} \tag{5-19b}$$

$$V_{i,j} = \frac{1}{h_j}\frac{\partial V_i}{\partial \xi_j} - \frac{1}{h_i}\frac{\partial h_j}{\partial \xi_i}\frac{V_j}{h_j} \quad \text{for } i \neq j \tag{5-19c}$$

The operators div and grad have been expressed above in the notation of vector geometry. Hence one obtains immediately from Eq. 3-3, 3-4, and 3-5 the Navier-Stokes equations in general orthogonal coordinates. The resulting equations will not be written out explicitly here (cf. III,A, and for incompressible flow [*24*, pp. 101ff.]).

Examples. The most frequently occurring case of curvilinear coordinates is that of *cylindrical coordinates,*

$$\xi_1 = r = \sqrt{x_1^2 + x_2^2}$$
$$\xi_2 = \theta = \tan^{-1}\frac{x_2}{x_1} \tag{5-20}$$
$$\xi_3 = z = x_3$$

These coordinates are orthogonal so that (g_{ij}) is diagonal. One finds

$$g_{11} = g_{33} = 1, \qquad g_{22} = \xi_1^2 = r^2, \qquad \sqrt{g} = \xi_1 = r \tag{5-21a}$$

$$\Gamma^1_{22} = -\xi_1 = -r \qquad \Gamma^2_{21} = \Gamma^2_{12} = \frac{1}{\xi_1} = \frac{1}{r} \tag{5-21b}$$

All other g_{ij} and Γ^i_{jk} are zero.

With the aid of these formulas, Eq. 5-8, 5-10, and 5-13 are easily specialized to the case of cylindrical coordinates.

Since the coordinate system is orthogonal one may also use the notation of vector geometry. Then

$$h_1 = h_3 = 1, \qquad h_2 = r \tag{5-22}$$

If f is a scalar, then (Eq. 5-19a) is

$$\operatorname{grad} f = \left(\frac{\partial f}{\partial r}, \frac{1}{r}\frac{\partial f}{\partial \theta}, \frac{\partial f}{\partial z}\right) \tag{5-23}$$

If $\mathbf{v} = (V_i)$ is a vector, then (Eq. 5-19b and 5-19c) are

$$\operatorname{grad} \mathbf{v} = \begin{pmatrix} \dfrac{\partial V_1}{\partial r}; & \dfrac{1}{r}\dfrac{\partial V_1}{\partial \theta} - \dfrac{V_2}{r}; & \dfrac{\partial V_1}{\partial z} \\ \dfrac{\partial V_2}{\partial r}; & \dfrac{1}{r}\dfrac{\partial V_2}{\partial \theta} + \dfrac{V_1}{r}; & \dfrac{\partial V_2}{\partial z} \\ \dfrac{\partial V_3}{\partial r}; & \dfrac{1}{r}\dfrac{\partial V_3}{\partial \theta}; & \dfrac{\partial V_3}{\partial z} \end{pmatrix} \tag{5-24}$$

By summing the diagonal of this matrix, or by using Eq. 5-18a, one finds

$$\operatorname{div} \mathbf{v} = \frac{\partial V_1}{\partial r} + \frac{1}{r}\frac{\partial V_2}{\partial \theta} + \frac{V_1}{r} + \frac{\partial V_3}{\partial z} \tag{5-25}$$

From the skew-symmetric part of the above matrix one finds

$$\operatorname{curl} \mathbf{v} = \begin{pmatrix} \dfrac{1}{r}\dfrac{\partial V_3}{\partial \theta} - \dfrac{\partial V_2}{\partial z} \\ \dfrac{\partial V_1}{\partial z} - \dfrac{\partial V_3}{\partial r} \\ \dfrac{\partial V_2}{\partial r} - \dfrac{1}{r}\dfrac{\partial V_1}{\partial \theta} + \dfrac{V_2}{r} \end{pmatrix} \tag{5-26}$$

Finally, if $A = (A_{ij})$ is a tensor, Eq. 5-18b shows that

$$\operatorname{div} A = \begin{pmatrix} \dfrac{\partial A_{11}}{\partial r} + \dfrac{A_{11}}{r} + \dfrac{1}{r}\dfrac{\partial A_{12}}{\partial \theta} - \dfrac{A_{22}}{r} + \dfrac{\partial A_{13}}{\partial z} \\ \dfrac{\partial A_{21}}{\partial r} + \dfrac{A_{21}}{r} + \dfrac{1}{r}\dfrac{\partial A_{22}}{\partial \theta} + \dfrac{A_{12}}{r} + \dfrac{\partial A_{23}}{\partial z} \\ \dfrac{\partial A_{31}}{\partial r} + \dfrac{A_{31}}{r} + \dfrac{1}{r}\dfrac{\partial A_{32}}{\partial \theta} + \dfrac{\partial A_{33}}{\partial z} \end{pmatrix} \tag{5-27}$$

With these formulas the Navier-Stokes equations in cylindrical coordinates are obtained directly from Eq. 3-3, 3-4, and 3-5 (cf. also III,A,15). The explicit form of the equations is given for the case of incompressible fluids in Art. 6.

In two dimensions *conformal coordinates* are an especially simple case of orthogonal coordinates. ξ_1 and ξ_2 are called conformal coordinates if

$$\zeta = \text{analytic function of } z$$

where

$$z = x_1 + ix_2, \qquad \zeta = \xi_1 + i\xi_2 \tag{5-28}$$

We shall define the Jacobian J as $|d\zeta/dz|^2$. Then (cf. Eq. 5-1 and 5-2b)

$$J = \left|\frac{d\zeta}{dz}\right|^2 = (\Delta)^{-1} = \left|\frac{dz}{d\zeta}\right|^{-2} \tag{5-29a}$$

$$g^{11} = g^{22} = \frac{1}{g_{11}} = \frac{1}{g_{22}} = J \tag{5-29b}$$

$$\sqrt{g} = J^{-1} \tag{5-29c}$$

Because of the last two equations, the covariant components of a vector are equal to the contravariant density components. The equations for incompressible flow are comparatively simple in conformal coordinates (cf. Eq. 6-10 and 6-11) if the notation of general tensor calculus is used.

CHAPTER 2. REVIEW OF VISCOUS INCOMPRESSIBLE FLUIDS

The purpose of the present chapter is to review some of the important properties of viscous incompressible flow in order to provide a background for the discussion of compressible flow. For a more extensive treatment of some of the questions discussed here, the reader is referred to [*7,8,24,25,26*].

The question of when a fluid may be considered as incompressible will be discussed in Art. 12. In the present chapter the assumptions of incompressibility will be made without a discussion of their validity.

GENERAL PROPERTIES OF VISCOUS WAVES. EXACT SOLUTIONS

B,6. The Navier-Stokes Equations for Incompressible Fluids. When the Navier-Stokes equations, as given in Art. 3, are specialized to the case of an incompressible fluid, the following assumptions are made: (1) The density of the fluid is constant. This is, properly speaking, the assumption of incompressibility. However, in addition one usually also assumes that (2) the viscosity is constant, and (3) the thermodynamics of the fluid is irrelevant.

Assumption (3) implies in particular that the energy equation (Eq. 3-5) and the equations of state (Eq. 3-6) are not used. Temperature and internal energy are not considered, and consequently the heat conductivity does not appear in the equations. Pressure enters solely as a dynamical variable. The resulting equations were given in Chap. 1 (Eq. 3-11a and 3-11b). Assuming that the external force **f** is zero and using the relations (Eq. 1-12) one may write the continuity and momentum equations in the following form:

$$\operatorname{div} \mathbf{u} = 0 \tag{6-1a}$$

$$\frac{D\mathbf{u}}{Dt} + \frac{1}{\rho}\operatorname{grad} p = \nu\nabla^2\mathbf{u} \tag{6-1b}$$

where ρ and ν are constant. Alternative forms may be obtained from the relations (cf. Eq. 1-12)

$$\begin{aligned}\frac{D\mathbf{u}}{Dt} &= \frac{\partial\mathbf{u}}{\partial t} + (\operatorname{grad}\mathbf{u})\mathbf{u}\\ &= \frac{\partial\mathbf{u}}{\partial t} + \operatorname{div}(\mathbf{u}\cdot\mathbf{u})\\ &= \frac{\partial\mathbf{u}}{\partial t} + \operatorname{grad}\left(\frac{|\mathbf{u}|^2}{2}\right) + (\operatorname{curl}\mathbf{u})\times\mathbf{u}\end{aligned} \tag{6-1c}$$

and

$$\nabla^2\mathbf{u} = \operatorname{div}(\operatorname{def}\mathbf{u}) = \operatorname{div}(\operatorname{grad}\mathbf{u}) = -\operatorname{curl}(\operatorname{curl}\mathbf{u}) \tag{6-1d}$$

The no-slip condition (Eq. 3-10) is imposed as a boundary condition: the velocity of the fluid at a solid boundary is equal to the local velocity of the solid at a point of contact. For flow past a solid at rest this means

$$\mathbf{u} = 0 \qquad \text{at the surface of the solid} \tag{6-2a}$$

When the domain is infinite the following conditions have to be added for stationary flow past a finite body:

$$\mathbf{u}(\infty) = \text{const}; \qquad p(\infty) = \text{const} = p_\infty \tag{6-2b}$$

Eq. 6-2b is the appropriate condition for flow past a finite body. Boundary conditions for other cases will be given in connection with specific examples (cf. the discussion of Falkner-Skan flow in Art. 8).

As initial conditions in a nonstationary problem, $\mathbf{u}$ and p may be prescribed over the entire domain. However, the initial velocity field must satisfy the relation div $\mathbf{u} = 0$. This last condition has no analogue for compressible fluids because of the occurrence of the term $\partial\rho/\partial t$ in the continuity equation in that case.

Unless otherwise specified, rectangular coordinates are used below. These are denoted by x_1, x_2, x_3 or x, y, z. The corresponding velocity components are denoted by u_1, u_2, u_3 or u, v, w. The free stream direction is taken to coincide with the positive x direction so that $\mathbf{u}(\infty)$ may be written $U\mathbf{i}$, $U =$ const.

According to assumption (3) above, there is no law expressing the conservation of energy. However, the law for the rate of change of kinetic energy (Eq. 2-8) is of course still valid since it is a consequence of the momentum equation. Under the assumptions of incompressibility it becomes

$$\frac{D}{Dt}\left(\rho \frac{|\mathbf{u}|^2}{2}\right) + \mathbf{u} \cdot \operatorname{grad} p = \operatorname{div}(\boldsymbol{\tau}\mathbf{u}) - \Phi \tag{6-3}$$

where the dissipation per unit volume per unit time is

$$\begin{aligned}\Phi &= \boldsymbol{\tau} \cdot (\operatorname{grad} \mathbf{u}) \\ &= \frac{\mu}{2} (\operatorname{def} \mathbf{u}) \cdot (\operatorname{def} \mathbf{u}) \\ &= \frac{\mu}{2} \sum_{i,j} \left(\frac{\partial u_i}{\partial x_j} + \frac{\partial u_j}{\partial x_i}\right)^2\end{aligned}$$

Stream functions may be introduced as in Eq. 2-11 or Eq. 2-13. Since the density is constant it may be omitted from the definitions. Thus in the present chapter stream functions are defined by

$$\mathbf{u} = (\operatorname{grad} \psi_1) \times (\operatorname{grad} \psi_2) \tag{6-4a}$$

and in two dimensions

$$u = \psi_y, \qquad v = -\psi_x \tag{6-4b}$$

Vorticity. By applying the operation "curl" to Eq. 6-1b and using Eq. 6-1c, 6-1d, and 1-121 one obtains the law for the propagation of vorticity $\boldsymbol{\Omega} = \operatorname{curl} \mathbf{u} = (\Omega_1, \Omega_2, \Omega_3)$

$$\frac{\partial \boldsymbol{\Omega}}{\partial t} + (\operatorname{grad} \boldsymbol{\Omega})\mathbf{u} = \frac{D\boldsymbol{\Omega}}{Dt} = (\operatorname{grad} \mathbf{u})\boldsymbol{\Omega} + \nu\nabla^2\boldsymbol{\Omega} \tag{6-5a}$$

where (cf. Eq. 1-12a)

$$(\operatorname{grad} \boldsymbol{\Omega})\mathbf{u} = \operatorname{div}(\boldsymbol{\Omega} \cdot \mathbf{u}) \tag{6-5b}$$

$$(\operatorname{grad} \mathbf{u})\boldsymbol{\Omega} = \operatorname{div}(\mathbf{u} \cdot \boldsymbol{\Omega}) \tag{6-5c}$$

and (cf. Eq. 1-12g and 1-12h)

$$\nabla^2 \mathbf{\Omega} = -\text{curl}\ (\text{curl}\ \mathbf{\Omega}) = \text{div}\ (\text{def}\ \mathbf{\Omega}) = \text{div}\ (\text{grad}\ \mathbf{\Omega}) \tag{6-5d}$$

The last relations show that Eq. 6-5a may be written as a divergence relation

$$\frac{\partial \mathbf{\Omega}}{\partial t} = \text{div}\ [(\mathbf{u} \cdot \mathbf{\Omega}) - (\mathbf{\Omega} \cdot \mathbf{u}) + \nu\ \text{grad}\ \mathbf{\Omega}] \tag{6-5$'$}$$

It follows from Eq. 6-1b and 6-1d that the acceleration vector $\mathbf{a}$ has the value $\mathbf{a} = -(1/\rho)$ grad $p - \nu$ curl $\mathbf{\Omega}$. Hence Eq. 6-5′ is a special case of Eq. 2-18b.

For two-dimensional flow parallel to the x, y plane the vorticity vector reduces to $\Omega_3\mathbf{k}$. Writing Ω for Ω_3 one obtains

$$\frac{D\Omega}{Dt} = \Omega_t + u\Omega_x + v\Omega_y = \nu\nabla^2\Omega \tag{6-6a}$$

where

$$\Omega = v_x - u_y = -\nabla^2\psi \tag{6-6b}$$

or

$$\Omega_t + \text{div}\ (\Omega\mathbf{u}) = \nu\ \text{div}\ (\text{grad}\ \Omega) \tag{6-6c}$$

In the two-dimensional case, Eq. 6-6a together with the definitions of u, v, and Ω in terms of ψ form a system of equations equivalent to Eq. 6-1. Eliminating u, v, and Ω one finds a single equivalent equation

$$\nabla^2\psi_t + \psi_y\nabla^2\psi_x - \psi_x\nabla^2\psi_y = \nu\nabla^2\nabla^2\psi \tag{6-7}$$

Note that in a physically relevant solution $\mathbf{u}$ and p must be single-valued functions whereas ψ may be multivalued. The function $\psi = A\theta + Br^2$ is a formal solution of Eq. 6-7 with a single-valued velocity field. However, the pressure field is single-valued only when A or B is zero.

Curvilinear coordinates. Eq. 6-1 may be expressed in general coordinates by specializing the relevant equations of Art. 5.

As a first example we shall give the equations in cylindrical coordinates. Let u, v, w be the velocity components in the sense of vector geometry in the directions of r, θ, and z respectively. It then follows from the geometrical relations (Eq. 5-22ff.) that the Navier-Stokes equations for an incompressible fluid are

$$\frac{\partial u}{\partial t} + u\frac{\partial u}{\partial r} + \frac{v}{r}\frac{\partial u}{\partial \theta} + w\frac{\partial u}{\partial z} - \frac{v^2}{r} + \frac{1}{\rho}\frac{\partial p}{\partial r} = \nu\left[\nabla^2 u - \frac{u}{r^2} - \frac{2}{r^2}\frac{\partial v}{\partial \theta}\right] \tag{6-8a}$$

$$\frac{\partial v}{\partial t} + u\frac{\partial v}{\partial r} + \frac{v}{r}\frac{\partial v}{\partial \theta} + w\frac{\partial v}{\partial z} + \frac{uv}{r} + \frac{1}{\rho r}\frac{\partial p}{\partial \theta} = \nu\left[\nabla^2 v + \frac{2}{r^2}\frac{\partial u}{\partial \theta} - \frac{v}{r^2}\right] \tag{6-8b}$$

$$\frac{\partial w}{\partial t} + u\frac{\partial w}{\partial r} + \frac{v}{r}\frac{\partial w}{\partial \theta} + w\frac{\partial w}{\partial z} + \frac{1}{\rho}\frac{\partial p}{\partial z} = \nu\nabla^2 w \tag{6-8c}$$

$$\frac{\partial (ru)}{\partial r} + \frac{\partial v}{\partial \theta} + \frac{\partial (rw)}{\partial z} = 0 \tag{6-8d}$$

where

$$\nabla^2 = \frac{\partial^2}{\partial r^2} + \frac{1}{r}\frac{\partial}{\partial r} + \frac{1}{r^2}\frac{\partial^2}{\partial \theta^2} + \frac{\partial^2}{\partial z^2} \tag{6-8e}$$

For two-dimensional flow the general form of Eq. 6-6 and 6-4b (cf. Art. 5) is

$$\sqrt{|g|}\,\frac{\partial \Omega}{\partial t} + \left(\frac{\partial \psi}{\partial \xi_2}\frac{\partial}{\partial \xi_1} - \frac{\partial \psi}{\partial \xi_1}\frac{\partial}{\partial \xi_2}\right)\Omega = \nu \frac{\partial}{\partial \xi_i}\left(\sqrt{|g|}\, g^{ij}\,\frac{\partial \Omega}{\partial \xi_j}\right) \tag{6-9a}$$

$$\Omega = -\frac{1}{\sqrt{|g|}}\frac{\partial}{\partial \xi_k}\left(\sqrt{|g|}\, g^{kl}\,\frac{\partial \psi}{\partial \xi_l}\right) \tag{6-9b}$$

$$u_1 = \frac{\partial \psi}{\partial \xi_2}, \qquad u_2 = -\frac{\partial \psi}{\partial \xi_1} \tag{6-9c}$$

Here Ω and ψ are treated as absolute scalars, and u_1 and u_2 are the components of the velocity when expressed as a contravariant vector density.

The two-dimensional equations are especially simple if conformal coordinates are used. The notation of Eq. 5-28ff. is used. ψ and Ω are considered absolute scalars. The contravariant components of the velocity density $u_1 = \partial\psi/\partial\xi_2$, $u_2 = -\partial\psi/\partial\xi_1$ are in this special case also the covariant components of the absolute velocity vector. The equations are then

$$\frac{\partial u_i}{\partial t} + J \sum_k \frac{\partial u_i}{\partial \xi_k} u_k + \frac{1}{2}|\mathbf{u}|^2 \frac{\partial J}{\partial \xi_i} + \frac{1}{\rho}\frac{\partial p}{\partial \xi_i} = \nu\left[J\nabla^2 u_i + \sum_k \left(\frac{\partial u_i}{\partial \xi_k} - \frac{\partial u_k}{\partial \xi_i}\right)\frac{\partial J}{\partial \xi_k}\right] \tag{6-10a}$$

$$\frac{\partial u_1}{\partial \xi_1} + \frac{\partial u_2}{\partial \xi_2} = 0 \tag{6-10b}$$

where

$$\nabla^2 = \frac{\partial^2}{\partial \xi_1^2} + \frac{\partial^2}{\partial \xi_2^2}$$

and

$$J = \left|\frac{d\zeta}{dz}\right|^2, \qquad \zeta = \xi_1 + i\xi_2, \qquad z = x_1 + ix_2$$

Equivalent equations are

$$\frac{1}{J}\frac{\partial \Omega}{\partial t} + \sum_i u_i \frac{\partial \Omega}{\partial \xi_i} = \nu\nabla^2\Omega \tag{6-11a}$$

$$\Omega = -J\nabla^2\psi = J\left(\frac{\partial u_2}{\partial \xi_1} - \frac{\partial u_1}{\partial \xi_2}\right) \tag{6-11b}$$

Thus, in the stationary case, Eq. 6-11a is invariant under conformal mapping whereas Eq. 6-11b changes with the factor J.

Further examples of the Navier-Stokes equations in orthogonal coordinates may be found in [*24*] and in III,A.

B,7. Qualitative Discussion of Viscous Waves. Simple Examples. This article is devoted to a preliminary qualitative discussion of viscous flow, based on simple exact solutions and on general theorems. The qualitative discussion will be continued in subsequent articles.

An inspection of the Navier-Stokes equations for incompressible fluids reveals the following important properties.

1. The Navier-Stokes equations contain derivatives of one order higher than the corresponding equations for nonviscous flow (the Euler equations). As a consequence the former equations require one more boundary condition than the latter. The condition usually imposed is the no-slip condition. It will be shown later that this fact is responsible for the existence of boundary layers. Due to the presence of the highest order terms the Navier-Stokes equations have a formal resemblance to the heat equation (diffusion equation) in that they are parabolic in the nonstationary case, that is the surfaces t = const are characteristic surfaces in the sense described in [*27*].
2. Viscosity effects are intimately connected with vorticity. Whenever vorticity vanishes, the viscous force on a fluid element, although not necessarily the viscous dissipation, is zero. In particular, Eq. 6-1a and 6-1b show that an irrotational solution of the Euler equations (potential flow) is also a solution of the Navier-Stokes equations.
3. The propagation law for vorticity has an obvious intuitive interpretation, especially for two-dimensional flow. If one puts $\nu = 0$ in Eq. 6-6 one obtains the Euler equation for vorticity. This equation simply states that the vorticity of a fluid particle is constant. In stationary flow, vorticity is thus constant along any streamline. The right side of Eq. 6-6a states that, in addition to being transported with the flow, vorticity diffuses like heat. The simplicity of the diffusion equation for vorticity is, however, only apparent, unless the flow field is assumed to be known. A nonlinear interaction takes place since the vorticity field in its turn determines the velocity field according to Eq. 6-6b. In three-dimensional flow the term (grad $\mathbf{u}$) $\mathbf{\Omega}$ of Eq. 6-5a leads to some additional complications. The significance of this term is discussed in the standard textbooks on perfect fluids.

A serious obstacle in deriving solutions of the flow equations is the lack of methods for treating the nonlinear interaction of the various modes of propagation of flow quantities. Solutions of the Navier-Stokes equations are in general only known for cases where this nonlinear interaction is absent or is very simple due to very special boundary conditions. None of these solutions represents flow past a finite body. Approximate methods of solution usually are based on simplifying assumptions about the transport process (Stokes' and Oseen's methods) or about viscous diffusion and the pressure field (Prandtl's boundary layer theory). The introduction of such simplifying assumptions will of course have to be preceded by an estimate of the relative importance of the various modes of propagation.

Simple examples of viscous diffusion. We shall first discuss viscous diffusion. Its role may be clearly seen in an example, originally studied by Stokes [28], namely the *transversal wave due to an infinite flat plate moving parallel to itself.* Let the fluid occupy the region $y > 0$ and let the x axis coincide with the flat plate. Furthermore, assume that for $t < 0$ the fluid is at rest, $\mathbf{u} = 0$ and $p = p_\infty$. At $t = 0$ the flat plate starts moving parallel to itself with velocity $f(t)$. It is easily seen that v remains zero and p remains constant so that Eq. 6-1b and 6-6a reduce to

$$u_t = \nu u_{yy} \tag{7-1a}$$

$$\Omega_t = \nu \Omega_{yy} \tag{7-1b}$$

where

$$\Omega = -u_y$$

In this case both velocity and vorticity obey the simplest form of the diffusion (or heat) equation. Transport and pressure waves are absent and velocity and vorticity spread like temperature in a homogeneous solid whereby ν corresponds to the thermometric conductivity κ.

In the theory of the heat equation $\partial T/\partial t = \kappa\, \partial^2 T/\partial x^2$, an important role is played by the fundamental solution $\frac{1}{2}(1/\sqrt{\pi\kappa t})e^{-x^2/4\kappa t}$ (for the following, cf. [29]). The fundamental solution represents the one-dimensional temperature wave resulting from instantaneously introducing heat of amount ρc at the point $x = 0$ at $t = 0$ (ρ = density, c = specific heat). The negative x derivative of the fundamental solution is called an instantaneous dipole or doublet.

The problem of the sliding flat plate is of the radiation type. At the plate u is made to assume the value $f(t)$, and the velocity then radiates into the fluid according to Eq. 7-1a. It is known from the theory of the heat equation that the solution may be obtained by a doublet distribution along the plate of strength $f(t)$ per unit length. Thus

$$u(y, t) = \int_0^t \frac{f(\tau)}{2\sqrt{\pi\nu}} \frac{y}{(t-\tau)^{\frac{3}{2}}} e^{-\frac{y^2}{4\nu(t-\tau)}}\, d\tau \tag{7-2a}$$

The vorticity may be obtained by differentiation of the above formula. It may also be found directly as follows. On the plate its normal derivative $\partial\Omega/\partial y = -\partial^2 u/\partial y^2$ is prescribed to be $-(1/\nu)\partial f/\partial t$. The resulting vorticity wave may then be found by a distribution of source strength along the plate equal to twice the normal derivative of Ω. This gives

$$\Omega = \int_0^t \frac{f'(\tau)}{\sqrt{\pi\nu(t-\tau)}} e^{-\frac{y^2}{4\nu(t-\tau)}} d\tau + \frac{f(0)}{\sqrt{\pi\nu t}} e^{-\frac{y^2}{4\nu t}} \tag{7-2b}$$

The skin friction at the plate, τ_w, is directly proportional to Ω;

$$\tau_w = \mu \left(\frac{\partial u}{\partial y}\right)_{y=0} = -\mu\Omega(0, t) \tag{7-2c}$$

The special case when $f(t)$ is a step function which is zero for $t < 0$ and const $= U$ for $t > 0$ is often called *Rayleigh's problem*. In this case the solution for $t > 0$ is

$$u = U \operatorname{erfc} \frac{y}{\sqrt{4\nu t}} \tag{7-3a}$$

$$\Omega = \frac{U}{\sqrt{\pi\nu t}} e^{-\frac{y^2}{4\nu t}} \tag{7-3b}$$

$$\tau_w = -U\sqrt{\frac{\mu\rho}{\pi t}} \tag{7-3c}$$

Here the error function is defined as $\operatorname{erf} z = (2/\sqrt{\pi}) \int_0^z e^{-\zeta^2} d\zeta$ and the complementary error function as $\operatorname{erfc} z = (1 - \operatorname{erf} z)$ (cf. Appendix of [*29*] or [*30*]).

According to the above, the source distribution for vorticity is simply $\delta(t)\delta(y)$ which accounts for the fact that the vorticity wave according to Eq. 7-3b is that of a simple vorticity source of strength $2U$. (Here δ denotes the Dirac delta function.) This could have been seen directly by giving a different physical interpretation to the same boundary-value problem. Let the fluid occupy the entire plane and the initial velocities at $t = 0$ be $u = U_1$ for $y > 0$ and $u = U_2$ for $y < 0$. The initial velocity field then represents an infinite slipstream and the physical problem concerns the smearing out of the initial slipstream discontinuity through the action of viscosity. By choosing a suitable moving system of coordinates one may assume that $U_1 = U$ and $U_2 = -U$. Eq. 7-1 are still valid and the solution is

$$u = U \operatorname{erf} \frac{y}{\sqrt{4\nu t}} \tag{7-4a}$$

$$\Omega = -\frac{U}{\sqrt{\pi\nu t}} e^{-\frac{y^2}{4\nu t}} \tag{7-4b}$$

which differs from Eq. 7-3 only by constants. In this case the initial vorticity distribution prescribed is $-2U\delta(y)$. Hence at $t = 0$ and $y = 0$ a vorticity pulse of strength $-2U$ is introduced. Accordingly, the solution for Ω is $-2U$ times the fundamental solution.

In the examples given there is no wave front. The signal velocity is infinite, since for any $t > 0$, no matter how small, and for any $y > 0$, no matter how large, u is different from zero. Infinite signal velocity is a property of solutions of parabolic equations. It always characterizes non-stationary viscous waves, even in the compressible case. On the other hand, for a fixed t, u decreases exponentially with y so that one may say that effectively the disturbance has spread only a certain distance. The definition of this effective distance δ is to a certain degree arbitrary. One way of defining it is to replace the graph of u versus y with a step function which is U for $0 \leqq y < \delta$ and zero for $y \geqq \delta$ and such that the areas under the two curves are the same. Applied to Eq. 7-3 this yields

$$\delta = \frac{1}{U} \int_0^\infty u dy = \frac{2}{\sqrt{\pi}} \sqrt{\nu t} \tag{7-5}$$

Thus the region of effective influence increases parabolically with t, and the effective velocity of the disturbance is $\sqrt{\nu/\pi t}$. In particular the speed of propagation is proportional to $\sqrt{\nu}$ but varies with time.

A solution of Eq. 7-1 may also be regarded as a superposition, containing various frequencies, of plane harmonic waves of the form

$$u = e^{i(\omega t - ky)} e^{-\beta y} \tag{7-6}$$

where

$$\omega = \text{frequency}$$

$$k = \text{wave number}$$

$$\beta = \sqrt{\frac{\omega}{2\nu}} = \text{attenuation factor}$$

$$\frac{\omega}{k} = \sqrt{2\nu\omega} = \text{phase velocity}$$

The phase velocity, which should be carefully distinguished from the other velocities discussed above, thus depends on the frequency. This accounts for the dispersion of a wave packet, that is the tendency of a wave to become flatter and more uniform. In particular, an initial discontinuity in u (wave front) is instantaneously smoothed out as shown by Eq. 7-4.

The simplicity of the solutions discussed above is due to the fact that the streamlines are straight lines and that velocities and pressure do not change in the x direction. A related set of solutions represents flows *where the streamlines are concentric circles* and velocities do not change in directions tangential to the circles. The new feature exhibited by this example is the effect of streamline curvature.

The notation of Eq. 6-8 is used, so that in particular v denotes the tangential velocity. Because of the symmetries of the problem, the Navier-Stokes equations reduce to

$$\rho \frac{v^2}{r} = \frac{\partial p}{\partial r} \tag{7-7a}$$

$$v_t = \nu \left(v_{rr} + \frac{1}{r} v_r - \frac{v}{r^2} \right) \tag{7-7b}$$

$$\Omega = v_r + \frac{v}{r} \tag{7-7c}$$

$$\Omega_t = \nu \left(\Omega_{rr} + \frac{\Omega_r}{r} \right) \tag{7-7d}$$

The viscous stress tensor τ has components

$$\tau_{11} = \tau_{12} = 0, \qquad \tau_{12} = \tau_{21} = \mu \left(\frac{\partial v}{\partial r} - \frac{v}{r} \right) \tag{7-7e}$$

Note that in this case vorticity still diffuses like heat, whereas in the equation for v, the curvature effect leads to an additional term in the stress tensor.

A simple solution of Eq. 7-7 represents the *decay of a potential vortex.* Assume that at $t = 0$ the velocity field is that of a potential vortex, that is $u = 0$, $v = \Gamma/2\pi r$. This example is similar to that of the infinite slipstream in that the vorticity density initially is a delta function. Vorticity is initially zero everywhere except at the origin where an amount

$$\oint_{r=\text{const}} \frac{\Gamma}{2\pi r} r d\theta = \Gamma$$

is concentrated. The subsequent spreading of vorticity must then be described by Γ times the fundamental solution of the heat equation in the plane. The decay of the vortex is thus described by

$$v = \frac{\Gamma}{2\pi r} \left(1 - e^{-\frac{r^2}{4\nu t}} \right) \tag{7-8a}$$

$$\Omega = \frac{\Gamma}{4\nu t\pi} e^{-\frac{r^2}{4\nu t}} \tag{7-8b}$$

Other solutions of Eq. 7-7 represent flow generated by a rotating cylinder or by two concentric cylinders. For example, consider the flow outside a cylinder of radius a which at time $t = 0$ starts rotating with an angular velocity V/a. This is the analogue for circular flow of Rayleigh's problem. The solutions for v and Ω may be constructed from source and doublet distributions as was done for the sliding flat plate. Since the equations for v are more complicated than the heat equation, the method of the Laplace transform is more convenient for constructing v. This method gives

$$\frac{v}{V}=\frac{1}{2\pi i}\int_{-i\infty}^{i\infty}\frac{K_1\left(r\sqrt{\frac{\sigma}{\nu}}\right)}{K_1\left(a\sqrt{\frac{\sigma}{\nu}}\right)}e^{\sigma t}\frac{d\sigma}{\sigma} \tag{7-9a}$$

$$\Omega=\frac{Va}{\nu t}e^{-\frac{r^2+a^2}{4\nu t}}I_0\left(\frac{ra}{2\nu t}\right) \tag{7-9b}$$

The functions K_1 and I_0 are discussed in [*30*].

The formula for the vorticity may be obtained by assuming an instantaneous source distribution uniformly distributed on the surface of the cylinder (cf. [*31*]).

The skin friction τ_w at the surface of the cylinder is, according to Eq. 7-7e (subscript $_w$ indicates the quantity is evaluated at the wall, i.e. at $r = a$),

$$\begin{aligned}\tau_w &= \mu\left(v_r-\frac{v}{a}\right)_w\\ &= \mu\left(\Omega-2\frac{v}{a}\right)\\ &= \frac{\rho Va}{t}e^{-\frac{a^2}{2\nu t}}I_0\left(\frac{a^2}{2\nu t}\right)-2\mu\frac{V}{a}\end{aligned} \tag{7-9c}$$

As t tends to infinity the solution approaches a potential vortex whose strength is such that $v = V$ for $r = a$. In general the time-independent solutions of Eq. 7-7 are independent of the magnitude of the coefficient of viscosity. The general time-independent solution is

$$v = Ar+\frac{B}{r} \tag{7-10a}$$

$$\Omega = 2A \tag{7-10b}$$

The solution thus represents a superposition of a potential vortex and a flow field where the fluid rotates as a solid. For *stationary flow between two rotating cylinders* (*Couette flow*), A and B are chosen so as to fit the boundary conditions at the cylinder. For flow outside a single cylinder, A must be zero if the solution is the limit of nonstationary motion started from rest. Thus in this case only a potential vortex remains. For flow inside a single cylinder, B is zero since otherwise the vorticity singularity at the origin would diffuse. Thus in stationary flow inside a cylinder the fluid rotates like a solid.

Another generalization of the problem of the sliding flat plate is that of a *sliding circular cylinder*. We shall generalize this problem further by assuming the cylinder to be *expanding radially*. At $t = 0$ the circular cylinder starts sliding along its axis with velocity U. At the same time its radius R expands as t^n $(0 \leqq n \leqq 1)$. A detailed discussion of this problem is given in [*32*]. For $n = 0$ and $n = \frac{1}{2}$ exact solutions exist which are given below.

The axis of the cylinder is taken as the x axis and the radial distance from this axis denoted by r. The velocity components in the x and r direction are denoted by u and v respectively. Because of the symmetries of the problem the equation for the x momentum and the continuity equation reduce to

$$u_t + vu_r = \nu\left(u_{rr} + \frac{1}{r}u_r\right) \tag{7-11a}$$

$$(rv)_r = 0 \tag{7-11b}$$

The boundary and initial conditions are

$$u = U,\ v = \frac{dR}{dt} \text{ for } t \geqq 0 \text{ and } r = R \tag{7-12a}$$

$$u = v = 0 \text{ for } t < 0 \text{ or } r = \infty \tag{7-12b}$$

The continuity equation can be immediately integrated to

$$v = \frac{1}{2r}\frac{d(R^2)}{dt} \tag{7-13}$$

For $n = 0$ we put $R = L = \text{const.}$ The radial velocity is then zero and Eq. 7-11a reduces to the heat equation in the plane with circular symmetry. The solution of this problem is well known (cf. [*29*] and also [*33*]). It may be written

$$\frac{u(r,t)}{U} = \frac{1}{2\pi i}\int_{-i\infty}^{+i\infty} e^{\sigma t}\,\frac{K_0\left(\sqrt{\frac{\sigma}{\nu}}\,r\right)}{K_0\left(\sqrt{\frac{\sigma}{\nu}}\,L\right)}\,\frac{d\sigma}{\sigma} \tag{7-14}$$

where the contour is taken to the right of the imaginary axis. K_0 is the modified Bessel function of the second kind. (See. e.g. [*30*].)

Approximations to the exact solution may be conveniently expressed with the aid of the coordinates

$$\xi = \frac{r}{2\sqrt{\nu t}}, \qquad \eta = \frac{R}{2\sqrt{\nu t}} \tag{7-15}$$

According to the previous discussion of Rayleigh's problem $\sqrt{\nu t}$ is a viscous length that measures the effective spreading of the viscous wave. For η large, i.e. for a viscous layer thin compared to R, an approximation to Eq. 7-14 is

$$\frac{u}{U} = \text{erfc}\,\xi - \frac{1}{\sqrt{Re}}\frac{\bar{r}}{2}\,\text{erfc}\,\bar{\xi} + O\left(\frac{1}{Re}\right) \tag{7-16a}$$

where

$$\bar{r} = \frac{(r - R)\sqrt{U}}{\sqrt{\nu L}}, \qquad \bar{\xi} = \xi - \eta \tag{7-16b}$$

and

$$Re = \frac{UL}{\nu} = \text{Reynolds number} \tag{7-16c}$$

For η small an approximate solution is

$$\frac{u}{U} = \epsilon\frac{E_1(\xi^2)}{2} + O(\epsilon^2) \tag{7-17a}$$

where

$$\epsilon = -\frac{1}{\ln\sqrt{\gamma_0\,\eta}}; \qquad \gamma_0 = e^{\gamma} = 1.781; \qquad \gamma = \text{Euler's const} = 0.577 \tag{7-17b}$$

$$E_1(z) = -E_1(-z) = \int_z^\infty \frac{e^{-s}}{s}\,ds \tag{7-17c}$$

For $n = \frac{1}{2}$ we define a length L from the relation $R = \sqrt{t} \times \text{const}$ in such a way that

$$R = L\sqrt{\frac{Ut}{L}} \tag{7-18}$$

With this value of R one finds that η, as defined by Eq. 7-15, is independent of time:

$$\eta = \frac{\sqrt{Re}}{2} \tag{7-19}$$

By similarity it may be shown that r and t enter only in the combination $rt^{-\frac{1}{2}}$. Hence if ξ (Eq. 7-15) is introduced into Eq. 7-11a one obtains an ordinary differential equation which may be integrated to

$$\frac{u}{U} = \frac{g(\xi, \eta)}{g(\eta, \eta)} \tag{7-20a}$$

where

$$g(\xi, \eta) = \int_\xi^\infty e^{-\sigma^2}\sigma^{(2\eta^2-1)}d\sigma = \tfrac{1}{2}\Gamma(\eta^2, \xi^2) \tag{7-20b}$$

The function Γ is the incomplete gamma function (cf. [*30*], Vol. 2, Chap. 9). Whenever $2\eta^2$ is a positive integer the integral in Eq. 7-20b may be evaluated in terms of elementary functions (error functions and exponentials).

For η large an approximate solution is

$$\frac{u}{U} = \operatorname{erfc}(\sqrt{2}\,\xi) + \frac{1}{\sqrt{Re}}\frac{2}{3}\sqrt{\frac{2}{\pi}}\left[\operatorname{erfc}(\sqrt{2}\,\xi) - (1-\xi^2)e^{-2\xi^2} + O\left(\frac{1}{Re}\right)\right] \tag{7-21}$$

For η small

$$\frac{u}{U} = \frac{E_1(\xi^2)}{E_1(\eta^2)} + O(\eta^2) \tag{7-22a}$$

which also may be written

$$\frac{u}{U} = \epsilon\frac{E_1(\xi^2)}{2} + O(\eta^2) \tag{7-22b}$$

where ϵ is defined by Eq. 7-17b and 7-19.

For $n \neq 1$, $n \neq \frac{1}{2}$ exact solutions have not been found. However, expansions for η small or large may be derived easily. The significance of these expansions will be seen later when approximations valid for small or large Reynolds numbers are discussed.

For other generalizations of Rayleigh's problem, see [*34,35,36*].

Free vorticity. The preceding examples illustrated how viscous waves are generated by the motion of bodies. The general principle is that a change of velocity introduces vorticity at the boundary. This vorticity is then spread into the fluid by diffusion and transport subject to the boundary conditions at the solid. Except for very simple solids it has not been possible to find the exact solutions.

It is of interest also to consider the development of vorticity in an infinite fluid. The results would apply qualitatively to a wake region sufficiently far downstream so that the boundary conditions at the solid have very little influence. Two examples of free vorticity were given previously, namely the decay of a slipstream and the decay of a potential vortex. More generally, consider an infinite fluid where velocity and vorticity tend to zero at infinity. Several theorems for nonviscous fluids then remain valid for viscous fluids.

We first observe that if the vorticity is given at a certain instant the velocity field associated with the vorticity $\boldsymbol{\Omega}$ may be computed by the Biot-Savart law. As pointed out in Chap. 1 (cf. Eq. 2-20 and the discussion thereafter) this is a purely kinematic relation for incompressible matter independent of the special dynamic laws for viscous fluids. The vorticity field will then in general change with time. There is a superposition of two effects, the nonviscous transport of the vortices and the viscous diffusion. At each stage, however, the relation between the vortex field and the velocity field remains the same as in a nonviscous fluid.

In a nonviscous fluid the total vorticity is conserved. Since the viscous diffusion does not change the magnitude of the total vorticity, it is reasonable to assume that a similar statement is true in a viscous fluid. This assumption may indeed be verified from Eq. 2-19 which shows that if vorticity vanishes sufficiently strongly at infinity then

$$\frac{\partial}{\partial t} \iiint \mathbf{\Omega} dV = 0 \tag{7-23}$$

where the domain of integration is the entire space.

Similarly, it may be shown that the moment of vorticity does not vary with time, so that in particular the center of vorticity remains fixed. This theorem is true for nonviscous fluids, and symmetry considerations lead one to expect that viscous diffusion will not change the moment. One may indeed prove the following result under the assumption that velocity is zero at infinity and that vorticity vanishes sufficiently strongly at infinity

$$\frac{\partial}{\partial t} \iiint \mathbf{r} \times \mathbf{\Omega} dV = 0 \tag{7-24a}$$

and similarly in two dimensions

$$\frac{\partial}{\partial t} \iint x\Omega dS = \frac{\partial}{\partial t} \iint y\Omega dS = 0 \tag{7-24b}$$

In nonviscous flow the moment of inertia of vorticity is also an invariant of vortex motion. However, if vorticity is diffused by viscosity the moment of inertia must in general change. The simplest example of this is the decaying potential vortex. The corresponding theorem for two-dimensional flow of viscous fluids is

$$\frac{\partial}{\partial t} \iint (x^2 + y^2)\Omega dS = 4\nu \iint \Omega dS \tag{7-25}$$

For proofs of these theorems the reader is referred to the original paper by Poincaré [*37*] as well as to several papers by Truesdell (see [*38*] and references given there). The latter contains an exhaustive discussion of integral laws for vorticity.

Potential solutions. Dissipation. The examples studied above mainly illustrated viscous diffusion. The solutions were easily obtained due primarily to the fact that the boundary conditions led to a great simplification of the transport and pressure terms in the equations. It is then natural to ask for solutions where roles are reversed, that is where viscous diffusion plays a minor role. As pointed out earlier, any potential solution of the equations for nonviscous flow is a solution of the Navier-Stokes equations. The viscous terms in the latter are automatically zero since the viscous force on a fluid element is proportional to the vorticity. It is, however, important to realize that viscous effects, such as skin friction and dissipation, still may exist in irrotational flow. The following example serves to illustrate this.

As seen above, the viscous flow field generated by a cylinder of radius a and angular velocity V/a tends toward the stationary solution $v = Va/r$. This is an irrotational solution of the Navier-Stokes equation. The net viscous force on an infinitesimal volume element then vanishes since it depends on the rotation, that is, the skew-symmetric part of the gradient of the velocity field. On the other hand the viscous shear on a line and the dissipation depend on the deformation, that is, the symmetric part of the velocity gradient, which does not vanish.

The viscous stress on a circular streamline of radius r according to Eq. 7-7e is equal to $\mu[(\partial v/\partial r - v/r)] = -2\mu Va/r^2$. (This is the force per unit length in the positive θ direction exerted by the fluid outside the circle on the fluid inside.) The skin friction is thus $2\mu V/a$. Note that the viscous stresses vanish as $1/r^2$ at infinity. The viscous torque exerted on a circle has then the magnitude $2\pi\mu V$ independent of the radius of the circle; in particular it does not vanish as the radius of the circle tends to infinity. That this is necessary may be seen by applying the law of conservation of moment of momentum to a region between two circles. There is no flow of moment of momentum out of this region. Hence the torques on the two circles must balance (cf. Eq. 4-4b).

The total work done per unit time by the viscous forces on the cylinder is

$$\int_0^{2\pi} \left(2\mu \frac{V}{a}\right) Vad\theta = 4\pi\mu V^2 \qquad (7\text{-}26a)$$

According to the energy laws (cf. Eq. 2-9) this must equal the total dissipation in the fluid. This is easily checked since the dissipation per time and unit volume is

$$\Phi = \frac{\mu}{2} (\text{def } \mathbf{u} \cdot \text{def } \mathbf{u}) = 4\mu \frac{V^2a^2}{r^4} \qquad (7\text{-}26b)$$

Similar relations may be derived for potential flow past a solid (cf. [39]). The potential solution for flow past a solid is also a solution of the Navier-Stokes equations. It must thereby be assumed that each part of the surface of the solid slides with an appropriate velocity. Since the velocity does not vanish at the surface of the solid, Eq. 4-2 is no longer valid for the skin friction. Except for uniform flow, skin friction will exist even in the absence of vorticity. The work done by this skin friction per unit time equals the total dissipation per unit time in the fluid.

To obtain a general relation between skin friction, vorticity, and dissipation, one may start from the identity valid for any symmetric tensor $\boldsymbol{\tau}$

$$\text{div } (\boldsymbol{\tau}\mathbf{u}) = (\text{div } \boldsymbol{\tau}) \cdot \mathbf{u} + \boldsymbol{\tau} \cdot \text{grad } \mathbf{u}$$

The significance of the various terms in this relation has previously been discussed in connection with Eq. 2-9a. If now $\tau = \mu$ def $\mathbf{u}$ and div $\mathbf{u} = 0$ one may rewrite the equation above as

$$\text{div}\,(\tau\mathbf{u}) = -\mu(\text{curl}\,\mathbf{\Omega}) \cdot \mathbf{u} + \Phi \tag{7-27}$$

If the fluid has a solid boundary B one obtains by integration (cf. Eq. 2-9a),

$$\iint\limits_B (\tau d\mathbf{n}) \cdot \mathbf{u} = \iiint\limits_V \Phi dV - \mu \iiint\limits_V (\text{curl}\,\mathbf{\Omega}) \cdot \mathbf{u} dV \tag{7-28}$$

(Note that this equation is valid even for nonstationary flow.)

For irrotational flow the last term is zero so that the dissipation is balanced by the work done by the skin friction at the solid. Since Φ is positive, its integral is zero only if Φ is everywhere zero. In this case the deformation def $\mathbf{u}$ is everywhere zero and the whole fluid moves like a solid. Eq. 7-10a with B equal to zero represents, within an arbitrary translation of the whole fluid, all viscous flows with zero dissipation.[10] In this case all terms in Eq. 7-27 vanish identically.

For flow past a solid such that $\mathbf{u} = 0$ at the surface of the solid, the first term in Eq. 7-28 vanishes. Since, as explained above, $\Phi \neq 0$, it follows that such flows cannot be irrotational.

B,8. Further Exact Solutions. The exact solutions given above were introduced to illustrate certain qualitative aspects of viscous flow. There are many other exact solutions of the Navier-Stokes equations. Unfortunately most of these are not solutions of physically significant initial or boundary-value problems. A large number of the solutions are obtained by specifying in advance that they should have a certain simple mathematical form, and the interpretation of the solution is done a posteriori. As a result, most of the exact solutions contribute very little to the understanding of practical flow problems. None of them represent flow past a finite solid. For surveys of existing exact solutions the reader is referred to [*40*; *41*, pp. 147–156]. Some exact solutions that either have a direct practical application or at least an instructive value are listed below.

Poiseuille flow. A very simple solution exists for stationary flow in a straight circular pipe. Since this problem is thoroughly discussed in many standard reference books on viscous fluids (see in particular

[10] From purely geometrical considerations it follows that if the deformation is zero everywhere in the fluid, then the velocity field must have the form $\mathbf{u} = \mathbf{\Omega}/2 \times \mathbf{r} + \mathbf{u}_0(t)$ where $\mathbf{u}_0$ and the vorticity $\mathbf{\Omega}$ are independent of the space coordinates. The dynamical equation (Eq. 6-5) then shows that $\mathbf{\Omega}$ is also constant in time. A suitable moving system of coordinates may be chosen such that $\mathbf{u}_0 = 0$. (Note that the equations of motion are invariant under an arbitrary translation if the pressure is suitably redefined.)

[*24*]) only the solution is given here:

$$u = \left(\frac{dp}{dx}\right)\frac{1}{4\mu}(a^2 - r^2) \tag{8-1a}$$

$$\frac{dp}{dx} = \text{const} \tag{8-1b}$$

Here a is the radius of the pipe, r the radial distance of any point from the center line, x the distance along the axis of the pipe, and u the longitudinal velocity component. All other velocity components vanish.

The two-dimensional analogue of this problem has the solution

$$u = \left(\frac{dp}{dx}\right)\frac{1}{2\mu}(a^2 - y^2) \tag{8-2a}$$

$$\frac{dp}{dx} = \text{const} \tag{8-2b}$$

Flow towards a plane. A class of solutions of the Navier-Stokes equations represent flow towards a plane, that is an infinite flat plate perpendicular to the main flow direction. At large distances from the plane the flow is required to approach the corresponding potential flow.

Let the plane be $y = 0$ and the fluid occupy the region $y > 0$. Two cases are considered: (1) Two-dimensional flow independent of z and symmetric about $x = 0$ and (2) three-dimensional flow axially symmetric about the y axis. In the first case u and v denote the x and y components of the velocity; r is used instead of x. In the second case, cylindrical coordinates $r = \sqrt{x^2 + z^2}$, θ ,and y are used. The radial and axial velocities are denoted by u and v. The θ component of the velocity is zero by assumption. The dimension of the flow is denoted by m. The corresponding potential flow, denoted by the subscript e, is then described by the formulas

$$\psi_e = \frac{2k}{m-1} r^{(m-1)} y \tag{8-3a}$$

$$\phi_e = k\left(\frac{r^2}{m-1} - y^2\right) \tag{8-3b}$$

$$u_e = \frac{2k}{m-1} r \tag{8-3c}$$

$$v_e = -2ky \tag{8-3d}$$

$$\frac{p_e}{\rho} = A - 2k^2\left[\frac{r^2}{(m-1)^2} + y^2\right] \tag{8-3e}$$

where k is a constant scale factor of dimension (time)$^{-1}$ and A is another constant. For flow towards the plate, v is negative and hence k positive.

The corresponding solutions of the Navier-Stokes equations will now be discussed.[11] Since v_e and p_e become infinite at upstream infinity ($y = \infty$) one may not require that $\mathbf{u}$ and p tend to constant values at infinity. Instead it will be required that the viscous solution tend to the corresponding potential solution at infinity in the following sense:

$$\frac{\psi}{\psi_e} \to 1 \text{ as } y \to \infty \tag{8-4a}$$

$$\frac{p}{p_e} \to 1 \text{ as } y \to \infty \tag{8-4b}$$

In addition the customary boundary conditions are imposed at the plate

$$\psi(x, 0) = 0 \tag{8-4c}$$

$$\psi_y(x, 0) = 0 \tag{8-4d}$$

One method of solution is to assume that the flow field has the following form:

$$\psi = \frac{2r^{(m-1)}}{m-1}\sqrt{k\nu}\, f(\bar{y}), \qquad \bar{y} = y\sqrt{\frac{k}{\nu}} \tag{8-5a}$$

$$u = \frac{2}{m-1} rk f'(\bar{y}) \tag{8-5b}$$

$$v = -2\sqrt{k\nu}\, f(\bar{y}) \tag{8-5c}$$

$$\frac{p}{\rho} = A - 2k^2\left\{\left(\frac{r}{m-1}\right)^2 + \frac{\nu}{K} g(\bar{y})\right\} \tag{8-5d}$$

where k is assumed to be positive. It will be seen later in the discussion of boundary layer theory that certain similarity considerations actually lead in a natural way to the assumption that the solution have the above form.

By inserting Eq. 8-5 into the Navier-Stokes equations one finds that f and g must satisfy the equations

$$f''' + 2ff'' + \frac{2}{m-1}(1 - f'^2) = 0 \tag{8-6a}$$

$$g' = 2ff' + f'' \tag{8-6b}$$

with boundary conditions

$$f(0) = f'(0) = 0 \tag{8-6c}$$

$$f'(\infty) = 1 \tag{8-6d}$$

$$\frac{g(\bar{y})}{\bar{y}^2} \to 1 \text{ as } \bar{y} \to \infty \tag{8-6e}$$

[11] The first numerical solution for the two-dimensional case was carried out by Hiemenz [42]. The three-dimensional case was first solved by Homann [43]. More recent work on these problems is discussed in Art. 13.

Eq. 8-6a with boundary conditions (Eq. 8-6c and 8-6d) is a special case of the Falkner-Skan equation. Its solution will be discussed in Art. 13 (cf. Eq. 14-32ff.). From the solution for the velocity field one obtains the solution for the pressure: By integrating Eq. 8-6b one finds

$$g = f^2 + f' + c \tag{8-7}$$

If one requires that p tends to p_e as ν tends to zero, the constant of integration c in Eq. 8-7 must be zero.

It may be shown that in the case of negative k, i.e., flow away from the plate, there is no solution for viscous flow that approaches potential flow at $y = \infty$. The intuitive reason for this is that the vorticity generated by the plate is transported with the flow away from the plate. The flow can therefore not be potential at large distances.

Radial diffuser flow. An idealized case of flow in a channel with nonparallel walls will now be discussed. The channel considered is a sector bounded by the two lines $\theta = \pm\theta_0$, $\theta_0 > 0$. The flow is assumed to be stationary, two-dimensional and purely radial, that is only the radial velocity component u is different from zero (the notation used is the same as in Eq. 6-8). This problem was originally studied by Hamel [*44*]. Further discussion is found in [*45,46*]. The most exhaustive analysis of the problem has been given by Rosenhead [*47*].

It follows from the continuity equation that ψ is a function of θ only and that u is inversely proportional to r. The origin is then a singularity which is either a sink or a source. The velocity u is replaced by the nondimensional variable

$$h(\theta) = \frac{1}{\nu}\frac{\partial\psi}{\partial\theta} = \frac{ru}{\nu} \tag{8-8a}$$

The radial momentum equation is then (cf. Eq. 6-8a)

$$2hh' = -(4h' + h''') \tag{8-8b}$$

The no-slip condition at the wall leads to the boundary condition

$$h(\theta_0) = h(-\theta_0) = 0 \tag{8-8c}$$

As a third boundary condition the total flux ψ_0 is prescribed

$$\psi_0 = \int_{-\theta_0}^{\theta_0} ur d\theta = \psi(\theta_0) - \psi(-\theta_0) \tag{8-8d}$$

The nondimensional quantity ψ_0/ν is denoted by F and may be taken as the Reynolds number of the problem. The last boundary condition may thus be written

$$\int_{-\theta_0}^{\theta_0} h d\theta = \frac{\psi_0}{\nu} = F \tag{8-8d'}$$

If F is positive the net outflux is positive and the origin is a source. Similarly a negative F corresponds to a sink at the origin.

Eq. 8-8 may be solved in terms of elliptic functions although the special nature of the boundary conditions introduces complications. For each pair of values θ_0 and F infinitely many solutions exist. (This has not been proved rigorously but appears plausible.) To avoid some complications occurring in the low Reynolds number case we shall restrict ourselves to the case when $\theta_0 \leqq \pi/2$. For negative flux (a sink at the origin) there exists a unique solution with the property that u is everywhere negative (pure inflow). If ν tends to zero for fixed ψ_0, and hence F tends to minus infinity, the solution approaches that of a potential (nonviscous) sink (except for a discontinuity at $\theta = \pm\theta_0$ where u is always zero). On the other hand, pure outflow is not possible for large positive values of F. There occur regions of inflow (backflow, which alternate with the outflow regions. The number of backflow regions increases indefinitely. Thus if ψ_0 and θ_0 are fixed the velocity profile (u versus θ for fixed r) shows an increasing number of oscillations. A limiting value as ν tends to zero does not exist. The solution for very small values of ν is radically different from the nonviscous source solution. There exists thus a fundamental difference between viscous flow in converging and diverging channels, which has no analogue in the theory of nonviscous flow.

FLOW AT LOW REYNOLDS NUMBERS. LINEARIZING APPROXIMATION

The next three articles will be devoted to a discussion of the Stokes equations and the Oseen equations. These equations may be regarded from two points of view: (1) They may be considered as equations which are approximately valid at low Reynolds numbers and (2) they may be formally obtained by linearizing the Navier-Stokes equations. The Stokes equations are obtained by neglecting terms quadratic in $\mathbf{u}$; the Oseen equations are obtained by neglecting terms quadratic in $(\mathbf{u} - U\mathbf{i})$ where $U\mathbf{i}$ = free stream velocity.

The first point of view is probably the more significant one (cf. the discussion of analogous equations for compressible fluids in Art. 20). However, the Oseen linearization is justified for an arbitrary Reynolds number, provided one considers only the flow at large distances from the body. This fact has certain important applications. It was actually used implicitly in the discussion of the momentum theorems in Art. 4 (cf. also [*23*]). Other applications are found in [*48*] and in the discussion of wakes and jets (Art. 13).

Finally the Oseen equations may be used as a mathematical model for the full Navier-Stokes equation. Some other linear model equations are briefly mentioned in Art. 12. If due caution is exercised, solutions of the model equations may be used as a guide to obtain qualitative ideas regarding the solutions of the exact equations.

As an introduction we shall consider various nondimensional forms of the Navier-Stokes equations in Art. 9. These equations will then be used in Art. 10 and 11 for deriving the Stokes and Oseen equations respectively. In Art. 10 the simplest Stokes solutions will be given and the limitations of the Stokes equations will be discussed. In the discussion of Oseen solutions, special attention will be paid to the fundamental solution. This solution will be used to verify various statements made in Art. 4 about the flow field at large distances from a solid. It will also be used in solving boundary-value problems for low Reynolds numbers. The significance of the Stokes and Oseen approximations and their relation to each other will be discussed briefly in the following articles and more fully in Art. 17.

Experimental verifications of the Stokes and Oseen solutions will be discussed in Art. 17.

B,9. Nondimensional Form of the Navier-Stokes Equations. We shall consider flow past a solid which has a characteristic length L. The free stream velocity is assumed to be $U\mathbf{i}$. The parameters of the problem are then U, L, ν, ρ, and p_∞. By dimensional analysis the number of parameters may be reduced to one, namely the Reynolds number Re, defined by

$$Re = \frac{LU}{\nu} \tag{9-1a}$$

To show this we shall write the Navier-Stokes equations in nondimensional form. This may be done in many ways. However, there are two special methods which are of importance for the study of flow at low Reynolds numbers. In addition to the geometrical length L the parameters also determine a viscous length ν/U. There are then two "natural" ways of defining nondimensional independent variables. The following notation is used

Stokes variables. $$t^* = \frac{Ut}{L}, \quad x_i^* = \frac{x_i}{L} \tag{9-1b}$$

Oseen variables. $$\tilde{t} = \frac{U^2t}{\nu} = Ret^*, \qquad \tilde{x}_i = \frac{Ux_i}{\nu} = Rex_i^* \tag{9-1c}$$

The justification of the names will be apparent later. Since both ρU^2 and $\mu U/L$ have the dimensions force per unit area, a nondimensional pressure may be defined by

$$p^+ = \frac{L(p - p_\infty)}{\mu U} \tag{9-1d}$$

or

$$p^* = \frac{p - p_\infty}{\rho U^2} = \frac{p^+}{Re} \tag{9-1e}$$

The nondimensional velocity is defined by

$$\mathbf{u}^* = \frac{\mathbf{u}}{U}, \text{ i.e.} \qquad u^* = \frac{u}{U} \text{ etc.} \tag{9-1f}$$

The Navier-Stokes equations may then be written in the following nondimensional form using Stokes variables:

$$Re \frac{D\mathbf{u}^*}{Dt^*} + \text{grad } p^+ = \nabla^2 \mathbf{u}^* \tag{9-2a}$$

$$\text{div } \mathbf{u}^* = 0 \tag{9-2b}$$

In these equations the vector operations grad, div, etc. should be formed with respect to Stokes variables x_i^*: div $\mathbf{u}^* = \sum_i \partial u_i^* / \partial x_i^*$ etc.

An alternate form of the equations, using Oseen variables, is

$$\frac{D\mathbf{u}^*}{D\tilde{t}} + \text{grad } p^* = \nabla^2 \mathbf{u}^* \tag{9-3a}$$

$$\text{div } \mathbf{u}^* = 0 \tag{9-3b}$$

where grad, div, etc. are formed with respect to the Oseen variables $\tilde{x}_i$.

Eq. 9-2 contain only one parameter, namely *Re*. In Eq. 9-3 all parameters have been eliminated. However, *Re* will reappear in the boundary conditions; for example, if the solid is a sphere of diameter L, then the boundary of the solid is given in Oseen coordinates by $\tilde{R} = Re/2$ where $\tilde{R}^2 = \tilde{x}_1^2 + \tilde{x}_2^2 + \tilde{x}_3^2$.

Using the nondimensional forms of the equations we shall now find equations which are approximately valid for small values of the parameter *Re*.

B,10. The Stokes Equations.

Derivation of equations. If one lets Reynolds number tend to zero in the equations written in Stokes variables (Eq. 9-2), one obtains the Stokes equations[12]

$$\text{grad } p^+ = \nabla^2 \mathbf{u}^* \tag{10-1a}$$

$$\text{div } \mathbf{u}^* = 0 \tag{10-1b}$$

[12] Note that if p^* had been used in Eq. 9-2 the term *Re* grad p^* would have formally disappeared in the limit. The resulting equations would then have been overdetermined. This shows that p^+ is the correct variable to use, in other words that p^* is of the order $1/Re$ for small *Re*.

In dimensional form the Stokes equations are

$$\operatorname{grad} p = \mu \nabla^2 \mathbf{u} \tag{10-2a}$$

$$\operatorname{div} \mathbf{u} = 0 \tag{10-2b}$$

In the stationary case[13] these equations may be obtained by linearizing Eq. 6-1, i.e. by omitting terms that are quadratic in $\mathbf{u}$.

The corresponding Stokes equations for Ω and (two-dimensional case) ψ are

$$\nabla^2 \boldsymbol{\Omega} = 0 \tag{10-3a}$$

$$\nabla^2 \nabla^2 \psi = 0 \tag{10-3b}$$

The Cartesian components of the vorticity vector are thus harmonic and the stream function is biharmonic.

It is seen from Eq. 10-1a or 10-2a that the characteristic property of the Stokes equations is that the transport effect is neglected compared to viscous diffusion. An equivalent statement is that the inertial forces are neglected compared to the viscous and pressure forces. From the dimensional analysis it follows that Stokes flow occurs when the flow is very slow, the object is very small, or when the viscosity coefficient is very large. As an example consider the case when U and L are fixed, but ν becomes very large. It is then clear that for sufficiently large ν the viscous effects must outweigh the transport effect so that the Stokes equations are approximately valid. As it will be seen later, the transport effects are not uniformly small but will dominate at sufficiently large distances, even for very large values of ν.

Similarity. An immediate consequence of the Stokes approximation is that flow past similar objects is similar, independent of the Reynolds number: $\mathbf{u}^*$ and p^+ are the same at corresponding points with the same nondimensional coordinates. Hence if U and L are fixed, $(p - p_\infty)$ is directly proportional to the coefficient of viscosity μ, an important fact in lubrication theory. As the flow is kinematically similar, the viscous stress is also proportional to μ. In nondimensional terms this means that the drag coefficient (force divided by $\frac{1}{2}\rho U^2$ times characteristic area) varies inversely as the Reynolds number. The dimensional drag force itself is proportional to free stream velocity. This result, however, is not true for flow past two-dimensional bodies for reasons discussed later.

Solutions for flow past solids. One would expect the boundary conditions for the Stokes equations to be the same as those for the full Navier-Stokes equations. It was noticed, however, by Stokes himself that solutions do not exist for stationary two-dimensional flow past a

[13] One may also form a nonstationary Stokes equation by adding the term $\partial \mathbf{u}/\partial t$ to the left-hand side of Eq. 10-2a. Its significance can be studied with the methods used in the present article and in Art. 17.

solid which satisfy the boundary conditions at the solid as well as at infinity (cf. [46,49,50]). The significance of this fact will be discussed later in Art. 11 and 17.

For three-dimensional flow, on the other hand, solutions of the Stokes equations may be found which satisfy all the required boundary conditions. Various solutions are discussed in [7,8,24] and in particular in [25] where further references are given. For the case of a sphere with radius a and center at the origin, and with free stream velocity $U\mathbf{i}$, the solution may be written as the sum of an irrotational part $\mathbf{u}_1$

$$\mathbf{u}_1 = \frac{U}{4}\,\mathrm{grad}\left[a^3\frac{\partial}{\partial x}\left(\frac{1}{R}\right) + 3a\left(\frac{x}{R}\right) + 4x\right] \tag{10-4a}$$

and a rotational part $\mathbf{u}_2$ which has only an x component

$$u_2 = -\frac{3}{2}\frac{Ua}{R}, \qquad v_2 = 0, \qquad w_2 = 0 \tag{10-4b}$$

Here

$$R^2 = x^2 + y^2 + z^2$$

The pressure is

$$p - p_\infty = -\frac{3}{2}\frac{\mu U a x}{R^3} \tag{10-4c}$$

As pointed out in Art. 4, only the rotational part contributes to the viscous stress. Because of symmetry the forces may be computed in the plane $z = 0$. If polar coordinates are introduced in this plane by

$$r^2 = x^2 + y^2, \qquad \cos\theta = \frac{x}{r}, \qquad \sin\theta = \frac{y}{r}$$

the viscous stress is (cf. Art. 4)

$$\mu(\mathrm{curl}\,\mathbf{u}) \times \mathbf{n} = -\mu\frac{\partial u}{\partial y}\mathbf{k} \times (\mathbf{i}\cos\theta + \mathbf{j}\sin\theta)$$

$$= \mu\frac{\partial u}{\partial y}(\mathbf{i}\sin\theta - \mathbf{j}\cos\theta)$$

where

$$\frac{\partial u}{\partial y} = \frac{3}{2}\frac{Uy}{a^2} = \frac{3}{2}\frac{U}{a}\sin\theta$$

The corresponding pressure force per unit area is $\frac{3}{2}\mu U(\cos\theta/a)(\mathbf{i}\cos\theta + \mathbf{j}\sin\theta)$. The total force is then $\frac{3}{2}(\mu U/a)\mathbf{i}$. Hence the force per unit area is always directed along the x axis and independent of position. The total drag D of the sphere is then given by the Stokes formula (cf. [51])

$$D = \frac{3}{2}\frac{\mu U}{a}4\pi a^2 = 6\pi\mu U a \tag{10-5a}$$

and the drag coefficient based on the frontal area πa^2 is

$$C_D = \frac{D}{\frac{1}{2}\rho U^2 \pi a^2} = \frac{24}{Re}, \qquad Re = \frac{2Ua}{\nu} \tag{10-5b}$$

Applications. As will be seen in Art. 17 the Stokes formula for the sphere has been verified by experiments for sufficiently small Reynolds numbers. It has found several applications to various physical problems.

An obvious application is to use the Stokes formula for determining the coefficient of viscosity of fluids by measuring the drag of small spheres. This is usually done by dropping spheres in the liquid to be investigated and measuring the time of fall after uniform velocity has been attained. This method is useful for large viscosities, for example, oils, etc. The effects of the walls of the vessel in which the spheres are dropped is important, and several experimental and theoretical investigations have been concerned with this effect. For a detailed analysis reference is made to [*52*].

Another application was made by Millikan in his classical experiments for determining the charge of the electron.

Stokes' formula for the drag of the sphere was also used by Einstein in his theory of Brownian motion of particles suspended in a liquid. Even when these particles are not assumed to be spheres, Einstein's theory gives good agreement with experiments. This is due to the fact that in Stokes flow the drag coefficient is relatively insensitive to the shape of the body. For example, the drag for a circular disk of radius a placed perpendicular in the stream is, according to the computation of Oberbeck, (cf. [*25*])

$$D = 16\mu Ua \tag{10-6a}$$

If the same disk is placed edge-wise into the flow the result is

$$D = \frac{32}{3}\mu Ua \tag{10-6b}$$

Hence even in these two extreme cases the drag differs from that of a sphere only by the ratio of $6\pi:16$ and $6\pi:\frac{32}{3}$ respectively.

Limitations of the Stokes equations. For three-dimensional flow one may thus obtain useful solutions of the Stokes equations. For sufficiently low Reynolds number the agreement with experiments is good at least as far as drag is concerned. There remains, however, the problem of obtaining two-dimensional solutions for low Reynolds numbers. Furthermore one would like to extend the solutions to higher Reynolds numbers. Attempts in this direction have shown that the Stokes solution is not a suitable starting point for an iteration process. Finally, one notices that the Stokes solution for the sphere shows no wake; the streamline pattern is symmetrical upstream and downstream. It thus fails to describe a very important qualitative feature of flow past an object.

These various difficulties led Oseen to reconsider the assumptions of the Stokes equations. His reasoning was essentially as follows. The basic assumption in deriving the Stokes equations was that for Re small, the term $Re D\mathbf{u}^*/Dt^*$ is negligible compared to $\nabla^2\mathbf{u}^*$. However, in the Stokes solution for the sphere, discussed above, the viscosity terms like $\nabla^2 u^*$ are of the order $1/r^{*3}$, and the transport terms like $Re u^* \partial u^*/\partial x^*$ are of the order Re/r^{*2}. The ratio of $Re u^* \partial u^*/\partial x^*$ to $\nabla^2 u^*$ is of the order $Re r^*$. Thus as Re tends to zero, the transport term $Re D\mathbf{u}^*/Dt^*$ becomes negligible compared to the viscous term $\nabla^2\mathbf{u}^*$ but not uniformly in space. Within a given radius r_0, the first term may be made an arbitrarily small fraction of the second by choosing Re sufficiently small. But for this same Reynolds number, the transport terms may be made much larger than the viscous terms by choosing $r \gg r_0$. This indicates that for points which are at a large distance from the sphere the Stokes solution is not valid as an approximation. The Stokes approximation is inconsistent in the sense that the solution does not satisfy the assumptions on which the approximation is based. To remedy the difficulties Oseen proposed a new set of equations which are now discussed.

B,11. The Oseen Equations.

Linearization about the Free Stream Velocity. As pointed out above, the Stokes equations may be derived formally by linearizing the Navier-Stokes equations about the value $\mathbf{u} = 0$, i.e., about the value of $\mathbf{u}$ at the solid. It was thereby understood that the velocity of the solid was zero. Only the stationary case was considered. The Oseen equations may be obtained by linearizing about the free stream velocity $\mathbf{u} = U\mathbf{i}$, i.e. by neglecting terms which are quadratic in $\mathbf{u}' = \mathbf{u} - U\mathbf{i}$. The non-stationary case is included. The resulting equations are then[14]

$$\frac{\partial \mathbf{u}}{\partial t} + U\frac{\partial \mathbf{u}}{\partial x} + \frac{1}{\rho}\operatorname{grad} p = \nu\nabla^2\mathbf{u} \tag{11-1a}$$

$$\operatorname{div} \mathbf{u} = 0 \tag{11-1b}$$

The corresponding equations for $\mathbf{\Omega}$ and ψ are

$$\frac{\partial \mathbf{\Omega}}{\partial t} + U\frac{\partial \mathbf{\Omega}}{\partial x} = \nu\nabla^2\mathbf{\Omega} \tag{11-2a}$$

$$\frac{\partial}{\partial t}(\nabla^2\psi) + U\frac{\partial}{\partial x}(\nabla^2\psi) = \nu\nabla^2\nabla^2\psi \tag{11-2b}$$

Since only its derivatives appear, $\mathbf{u}$ may be interpreted either as the perturbation velocity $\mathbf{u}'$ or as the full velocity $\mathbf{u}' + U\mathbf{i}$. In the former case the boundary conditions for the velocity at the surface of a

[14] These equations are not invariant under a Galilean transformation since the value of U depends on the system of reference chosen. However, in the stationary case there is a unique system of reference such that $\partial/\partial t = 0$.

solid at rest are

$$\mathbf{u} = -U\mathbf{i} \quad \text{at solid} \tag{11-3a}$$

$$\mathbf{u} \to 0 \quad \text{at infinity} \tag{11-3b}$$

In the latter interpretation

$$\mathbf{u} = 0 \quad \text{at solid} \tag{11-4a}$$

$$\mathbf{u} \to U\mathbf{i} \quad \text{at infinity} \tag{11-4b}$$

In general the former interpretation is used in the following.

VALIDITY OF THE OSEEN EQUATIONS. In the stationary case the Oseen equations differ from the Stokes equations through the inclusion of the linearized transport term $U\partial\mathbf{u}/\partial x$. It is to be expected that the former equations are approximately valid at large distances from the solid since there $\mathbf{u}'$ is small relative to $U\mathbf{i}$. Near the solid the linearization is not justified for an arbitrary Reynolds number. However, for very small Reynolds numbers the viscous effects dominate near the solid as pointed out in Art. 10. Hence for a description of the flow near the solid it is irrelevant whether one includes the transport term $U\partial\mathbf{u}/\partial x$ or omits it. On the other hand the inclusion of this term leads to a better description of the flow at large distances from the solid. It will actually be seen later that the Oseen equations are an improvement on the Stokes equations and that they may be used to derive approximations which are uniformly valid over the entire flow field for low Reynolds numbers.

The ideas discussed above are essentially due to Oseen [*25*]. A somewhat different approach to the Oseen equations has been given by S. Kaplun, whose ideas will be briefly indicated below. (A fuller discussion is given in Art. 17.) Let the velocity $\mathbf{u}^*$ be written as a function of the Oseen variables $\tilde{x}_i$ (Eq. 9-1c) and assumed that it has an expansion in powers of ϵ where $\epsilon = \epsilon(Re)$ is some suitable function of Re which tends to zero with Re,

$$\mathbf{u}^* = \mathbf{u}_0(\tilde{x}_i) + \epsilon\mathbf{u}_1(\tilde{x}_i) + \epsilon^2 \cdots \tag{11-5}$$

The first term is then the limit of $\mathbf{u}^*$ as Re tends to zero with $\tilde{x}_i$ fixed. To fix the ideas we shall let L tend to zero while U and ν are fixed. The dimensional coordinates x_i are then also fixed. In this limit a finite body shrinks to a point which cannot cause a finite disturbance in the fluid. Hence the value of $\mathbf{u}^*$ at any fixed point will tend to the free stream value $\mathbf{i}$. The first term $\mathbf{u}_0$ of the expansion (Eq. 11-5) is then equal to $\mathbf{i}$. If one then inserts this expansion into the Navier-Stokes equations as written in Oseen variables (Eq. 9-3) and retains only terms of order ϵ it is found that $\mathbf{u}_1(\tilde{x}_i)$ satisfies the nondimensional form of the Oseen equations (Eq. 11-1).

An expansion of the type given by Eq. 11-5 is actually fundamental for the study of low Reynolds number flow, as will be discussed further in Art. 17.

An interesting consequence of the method of Kaplun is the following. Let L be the radius of curvature at the nose of a parabolic cylinder or a paraboloid of revolution. As L tends to zero the former tends to a semi-infinite plane and the latter to a semi-infinite line ("needle"). A line cannot cause a finite disturbance. Hence the Oseen solution for the paraboloid, but not the one for the parabola, has physical meaning as an approximation for low Reynolds numbers.

The remarks just made refer to the validity of the Oseen approximation over the entire fluid for low Reynolds numbers. The validity of the Oseen equations at large distance from a solid follows from the standard arguments about linearization. A systematic investigation of viscous flow at large distances from a finite body has been made by Imai [*23*].

Propagation of Vorticity. According to the Stokes equations vorticity spreads only by diffusion, i.e. vorticity is conducted like heat in a solid, independent of the motion of the medium. In the Oseen approximation it is in addition assumed that vorticity is transported with the streamlines of the undisturbed flow (the lines y = const). Thus in the neighborhood of a solid the vorticity is transported through the solid rather than with the true streamlines (cf. Art. 6). This paradoxical result shows that the Oseen equations can be approximately valid near a solid only if the Reynolds number is very low. In this case the diffusion of vorticity is the dominant mode of propagation and a physically incorrect assumption regarding the transport of vorticity does not introduce a large error.

It was pointed out previously that a Stokes solution shows no wake, since the transport of vorticity is entirely neglected. However, according to the Oseen equation the vorticity generated at a solid will be transported downstream. For this reason a wake is obtained from the Oseen solutions. Thus an important aspect of real flow can be described at least qualitatively with the aid of the Oseen equations.

Longitudinal and Transversal Waves. Before exhibiting solutions of the Oseen equations, we shall investigate a fundamental property of these equations, namely the possibility of splitting a general solution into two distinct types of waves, longitudinal and transversal.

A solution of Eq. 11-1 is said to be a *longitudinal wave* if the velocity field is irrotational: curl $\mathbf{u} = 0$. Hence the viscous term $\nu\nabla^2\mathbf{u} = -\nu$ curl curl $\mathbf{u}$ drops out of the momentum equation. The equations then reduce to

$$\mathbf{u}_t + U\mathbf{u}_x + \frac{1}{\rho}\,\text{grad}\; p = 0 \tag{11-6a}$$

$$\text{div}\;\mathbf{u} = 0 \tag{11-6b}$$

$$\text{curl}\;\mathbf{u} = 0 \tag{11-6c}$$

The last equation implies that there exists a potential ϕ such that

$$\mathbf{u} = \operatorname{grad} \phi \tag{11-7a}$$

Hence one obtains from Eq. 11-6b

$$\nabla^2\phi = 0 \tag{11-7b}$$

and from Eq. 11-6a

$$p' = p - p_\infty = -\rho U u \tag{11-7c}$$

Eq. 11-6 and 11-7 are equivalent. The velocity components of a longitudinal wave are simply those of a potential flow field. However, the pressure is evaluated from a linearized Bernoulli's law. This result of linearized theory should be compared with the earlier one that a potential flow field satisfies the full Navier-Stokes equations.

A solution of the Oseen equations is called a *transversal* wave if its pressure field is identically zero, that is $p' \equiv 0$, and its velocity field is solenoidal, that is $\operatorname{div} \mathbf{u} = 0$. The equations of a transversal wave are thus:

$$\mathbf{u}_t + U\mathbf{u}_x = \nu\nabla^2\mathbf{u} \tag{11-8a}$$

$$\operatorname{div} \mathbf{u} = 0 \tag{11-8b}$$

One may then prove the following theorem of splitting (cf. [*8*,*53*]): Any solution ($\mathbf{u}$, p) of the Oseen equations (Eq. 11-1) may be decomposed into a longitudinal wave ($\mathbf{u}_L$, p') and a transversal wave ($\mathbf{u}_T$, 0) where ($\mathbf{u}_L$, p') satisfies Eq. 11-6, ($\mathbf{u}_T$, 0) satisfies Eq. 11-8, and

$$\mathbf{u} = \mathbf{u}_L + \mathbf{u}_T \tag{11-9}$$

This splitting is unique if conditions at infinity (say $\mathbf{u} = 0$, $p = p_\infty$) are specified.

Similar theorems of splitting hold for a much wider class of equations (cf. [*53*] and references given there). For certain types of Oseen flows an additional splitting of the transversal wave is possible. Let χ be a scalar function, satisfying

$$U\chi_x = \nu\nabla^2\chi \tag{11-10a}$$

and vanishing at infinity. Then it is easily seen that the following vector field is a transversal wave:

$$\mathbf{u}_T = -\frac{\nu}{U}\operatorname{grad}\chi + \chi\mathbf{i} \tag{11-10b}$$

Conversely, let a transversal wave $\mathbf{u}_T$ be given such that $\int_{-\infty}^{+\infty} v_T dy = 0$. Then one may write

$$\mathbf{u}_T = \mathbf{u}_1 + \mathbf{u}_2 \tag{11-11a}$$

where

$$\mathbf{u}_1 = -\frac{\nu}{U}\operatorname{grad}\chi, \qquad \mathbf{u}_2 = \chi\mathbf{i} \tag{11-11b}$$

and χ is determined from $\mathbf{u}_T$ by

$$\chi = \frac{U}{\nu} \int_y^\infty v_T dy \tag{11-11c}$$

The χ so defined satisfies Eq. 11-10a. Note that neither $\mathbf{u}_1$ nor $\mathbf{u}_2$ is a transversal wave by itself since in general div $\mathbf{u}_1$ and div $\mathbf{u}_2$ are not zero. The second splitting has the advantage of exhibiting an irrotational component of the transversal wave. If $\mathbf{u} = \mathbf{u}_L + \mathbf{u}_T = \mathbf{u}_L + \mathbf{u}_1 + \mathbf{u}_2$, the only rotational component is $\mathbf{u}_2$:

$$\Omega_1 = 0, \qquad \Omega_2 = \frac{\partial u_2}{\partial z}, \qquad \Omega_3 = -\frac{\partial u_2}{\partial y} \tag{11-11d}$$

FUNDAMENTAL SOLUTIONS. A very useful concept in the study of linear differential equations is that of the fundamental solution. For the general theory, see [*27*, Vol. 2, Chap. 4; *25*; *53*]. The fundamental solution of the stationary Oseen equations may be defined as the response to a singular force, that is a solution of

$$U\mathbf{u}_x + \frac{1}{\rho}\,\text{grad}\, p - \nu\nabla^2\mathbf{u} = \mathbf{f} \tag{11-12a}$$

$$\text{div}\, \mathbf{u} = 0 \tag{11-12b}$$

when no external boundaries are present and the force per unit mass is a delta function. More precisely, let $\mathbf{f}$ be concentrated at a point Q and equal to $\delta(P, Q)\mathbf{a}$ where $\mathbf{a}$ is a given vector and $\delta(P, Q)$ (delta function) is a function of the point P which is zero for $P \neq Q$, ∞ for $P = Q$ and whose integral over any domain including Q is unity. Then the resultant velocity field at P, in the absence of boundaries, is determined from the fundamental tensor $\Gamma(P, Q)$ by

$$\mathbf{u}(P) = \Gamma(P, Q)\mathbf{a} \tag{11-13a}$$

and the perturbation pressure $p' = p - p_\infty$ from the fundamental vector $\boldsymbol{\pi}(P, Q)$ by

$$p'(P) = \boldsymbol{\pi}(P, Q) \cdot \mathbf{a} \tag{11-13b}$$

One may consider either the two-dimensional or the three-dimensional case. Similarly, the fundamental solution of the nonstationary Oseen equations describes the response to a singular impulse, i.e. a force distribution that is concentrated in both space and time.

From the linearity of the Oseen equations it follows easily that Γ and $\boldsymbol{\pi}$ are linear functions of $\mathbf{a}$, i.e. actually a tensor and a vector respectively. If a system of coordinates is chosen Γ may then be represented as a matrix (Γ_{ij}). In Cartesian coordinates $(\Gamma_{11}, \Gamma_{21}, \Gamma_{31})$ is the velocity field due to a unit force directed along the x axis ($\mathbf{a} = \mathbf{i}$).

Furthermore, due to the linearity of Eq. 11-12, superposition of fundamental solutions can be used to determine the effect of distributed forces. If **f** is an arbitrary function of space, the solution of Eq. 11-12 for an unbounded domain is

$$\mathbf{u}(P) = \int \Gamma(P, Q)\mathbf{f}(Q)dQ \tag{11-14a}$$

and

$$p'(P) = \int \boldsymbol{\pi}(P, Q) \cdot \mathbf{f}(Q)dQ \tag{11-14b}$$

Here dQ is the infinitesimal volume or area element around Q and the integral sign stands for a triple or double integral.

Since the Oseen equations have constant coefficients, Γ and $\boldsymbol{\pi}$ depend only on the difference of the coordinates of P and Q (homogeneity in space). It is thus sufficient to consider the case when the coordinates of Q are zero.

The fundamental solutions of the Oseen equations are derived and discussed in [*25,53*]. Only the stationary case will be discussed here.

Two-dimensional Case. The results for the two-dimensional stationary case are as follows: The point Q is taken as the origin, and $\mathbf{u}$ denotes the perturbation velocity. Then the fundamental tensor for the velocity is

$$\Gamma = \begin{pmatrix} \Gamma_{11} & \Gamma_{12} \\ \Gamma_{21} & \Gamma_{22} \end{pmatrix} = \begin{pmatrix} \dfrac{\partial A}{\partial x} & \dfrac{\partial A}{\partial y} \\ \dfrac{\partial A}{\partial y} & -\dfrac{\partial A}{\partial x} \end{pmatrix} + \frac{1}{2\pi\nu} e^{\lambda x} K_0(\lambda r) \begin{pmatrix} 1 & 0 \\ 0 & 0 \end{pmatrix} \tag{11-15a}$$

where

$$\lambda = \frac{U}{2\nu}, \qquad r^2 = x^2 + y^2, \qquad A = -\frac{1}{2\pi U}[\ln r + e^{\lambda x}K_0(\lambda r)]$$

and the fundamental vector for p' is

$$\boldsymbol{\pi} = \begin{pmatrix} \pi_1 \\ \pi_2 \end{pmatrix} = \frac{\rho}{2\pi} \operatorname{grad} (\ln r) \tag{11-15b}$$

The corresponding three-dimensional formulas will be given later. First the significance of Eq. 11-15 will be discussed.

Singular flat plate. Consider now the special case when **f** is of magnitude $(1/\rho)\delta(x, y)$ and directed along the negative x axis. Since **f** denotes force per unit mass the total force exerted on the fluid is then of magnitude unity and this force is concentrated at the origin. The corresponding flow field may be interpreted as follows. Consider a flat plate at zero angle of attack near the origin such that $u = -U$ at the plate and the total drag is unity. The total force exerted by the plate on the fluid is then unity and directed along the negative x axis. If one imagines that the length of the plate shrinks to zero while its retarding effect (i.e. $-u$ at the plate) increases in such a way that the drag remains unity, the resulting limiting flow field is identical with the flow field due to the singular force $-\mathbf{i}$ located at the origin. This limiting case of the flat plate is referred to as a singular flat plate. From the above formula for the fundamental solution one finds that the flow field is

$$\mathbf{u} = -\frac{1}{\rho}(\Gamma_{11}, \Gamma_{21}) = \frac{1}{2\pi\mu}\left[\frac{\nu}{U}\operatorname{grad}\{\ln r + e^{\lambda x}K_0(\lambda r)\} - e^{\lambda x}K_0(\lambda r)\mathbf{i}\right] \tag{11-16a}$$

where $\mathbf{u}$ has the dimension velocity per unit force. The pressure field is

$$p' = -\frac{\pi_1}{\rho} = -\frac{x}{2\pi r^2} \tag{11-16b}$$

Note that this flow field is split according to Eq. 7-18 and 7-20. The three terms of the right-hand side of Eq. 11-16a are in order $\mathbf{u}_L$, $\mathbf{u}_1$ and $\mathbf{u}_2$ where $\mathbf{u}_T = \mathbf{u}_1 + \mathbf{u}_2$.

The longitudinal wave represents simply potential source flow. The total outflow from the source is $1/\rho U$. The transversal wave has a sink of the same strength so that the net flow through any closed contour is zero.

Near the origin, $r \ll 1$, the total flow field is approximately

$$u(x, y) \approx \frac{1}{4\pi\mu}\left(\ln\frac{\gamma_0 U r}{4\nu} - \frac{x^2}{r^2}\right) \tag{11-17a}$$

$$v(x, y) \approx -\frac{1}{4\pi\mu}\left(\frac{xy}{r^2}\right) \tag{11-17b}$$

where $\gamma_0 = e^\gamma = 1.781$, $\gamma =$ Euler's const $= 0.577$.

For large values of r the transversal flow field may be given asymptotically as

$$u_T(x, y) \cong -\frac{1}{4\rho}\sqrt{\frac{1}{\pi\nu U r}}\left(1 + \frac{x}{r}\right)e^{-\lambda(r-x)} \tag{11-18a}$$

$$v_T(x, y) \cong -\frac{1}{4\rho}\sqrt{\frac{1}{\pi\nu U r}}\left(\frac{y}{r}\right)e^{-\lambda(r-x)} \tag{11-18b}$$

The value of $e^{-\lambda(r-x)}$ is constant on each curve $\lambda(r - x) = \text{const} = c$. This is a parabola with focus at the origin, and opening downstream. Outside such a parabola with a sufficiently high value of c the transversal wave is then negligible. Consider, for example, the value of u on the x axis. Upstream ($x = -r$) it dies off exponentially whereas downstream of the plate it decreases only as $1/\sqrt{x}$. Thus the tranversal wave, and hence also the vorticity, are negligible except in a parabolic wake.

The existence of the transversal wave is directly due to the retarding effect of the plate. Vorticity generated at the plate diffuses like heat and is at the same time transported downstream with the streamlines of the undisturbed flow. This is the reason why the vorticity is very weak upstream and comparatively strong downstream. The longitudinal wave may be thought of as due to a displacement effect (cf. Art. 1). The slowing down of the fluid at the plate gives the plate an apparent shape. Since the longitudinal wave is a simple source, the apparent shape is that of a semi-infinite body whose thickness downstream approaches a finite value. The effective (or apparent) thickness may be defined as $\int_{-\infty}^{+\infty} u_T dy$ for a fixed value of x. Roughly speaking u_T decreases as $1/\sqrt{x}$ inside the wake and is negligible outside. On the other hand the thickness of the wake increases as $\sqrt{x}$. Hence the apparent thickness due to the transversal wave is of the order $(1/\sqrt{x})\sqrt{x}$ = unity. A more rigorous evaluation of the integral is given below (cf. Eq. 11-20).

The momentum formula for the drag will now be verified for the singular flat plate. According to Eq. 4-13 the drag D is

$$D = \lim_{x_1 \to \infty} \int_{x=x_1} (-p' - \rho U u) dy \tag{11-19}$$

where p' and u are perturbation quantities. The longitudinal component of u cancels the pressure according to Eq. 11-7c. Hence

$$D = \lim_{x_1 \to \infty} \int_{x=x_1} (-\rho U u_T) dy \tag{11-20}$$

Since the corresponding integral over an upstream plane $x = -x_2$, $x_2 \to \infty$ must vanish by Eq. 11-18 it follows from Eq. 11-20 that $D/\rho U$ is the total sink strength of the transversal component. By the continuity equation this is equal to the total source strength of the longitudinal component which is $1/\rho U$ (cf. above). Hence $D = 1$ which checks with the original assumption on the force exerted by the plate. Note that actually the integral in Eq. 11-20 is the same for any $x_1 > 0$. This depends on the fact that we consider linearized equations.

Consider now real incompressible flow past a finite solid of drag D and zero lift. The Navier-Stokes equations are then the governing equations. At large distances the Oseen approximation is valid. We then assume that the real flow field at large distances is the same as Oseen flow past a singular flat plate of drag D. The assumptions about the flow field at large distances made in Art. 4 can then be justified by a comparison with various formulas given above. Note that the loss of total head, discussed in Art. 4, is measured by the strength of the transversal component of the Oseen flow. Strictly speaking, one should actually consider nonstationary Oseen flow since real flow is generally not stable. This refinement, however, is omitted here. If the body also has a lift L one may compute the flow field at large distances by finding the Oseen flow due to an element of lift L. This will now be done. (For further questions regarding torque and higher order approximations, see [*23*].)

Singular lifting element. We consider a singular force located at the origin, of magnitude unity and directed along the negative y axis. The two-dimensional Oseen flow due to such an element is, according to Eq. 11-15,

$$\mathbf{u} = -\frac{1}{\rho}\,[\Gamma_{12}, \Gamma_{22}] = \left(\frac{\partial A}{\partial y}, -\frac{\partial A}{\partial x}\right) \tag{11-21}$$

where

$$A = \frac{1}{2\pi\rho U}\,[\ln r + e^{\lambda x}K_0(\lambda r)]$$

The first term represents the longitudinal wave. It is the flow field due to a potential vortex whose clockwise circulation Γ is $1/\rho U$. The transversal wave at large distances from the origin is

$$u_{\mathrm{T}} \sim -\frac{1}{4\rho}\sqrt{\frac{1}{\pi U\nu r}}\left(\frac{y}{r}\right)e^{-\lambda(r-x)} \tag{11-22a}$$

$$v_{\mathrm{T}} \sim -\frac{1}{4\rho}\sqrt{\frac{1}{\pi U\nu r}}\left(1 - \frac{\nu}{4Ur} - \frac{x}{r} - \frac{3}{4}\frac{\nu x}{Ur^2}\right)e^{-\lambda(r-x)} \tag{11-22b}$$

As before it may be seen that the transversal wave is exponentially negligible outside the wake. Furthermore, the contribution of the transversal wave to the circulation around a contour at large distances from the origin is negligible. Hence if Γ is the clockwise circulation of the total flow field around a very distant contour enclosing the origin, then $\rho U\Gamma = 1 =$ lift. This checks with the formula for the lift obtained by momentum methods (Eq. 4-16).

THREE-DIMENSIONAL CASE. The fundamental solution for three-dimensional stationary Oseen flow is derived in [*25*, p. 34] (for methods of derivation cf. also [*53*]). The result is given below in the more intuitive terminology adopted for the two-dimensional case. There are two cases to be considered, namely the flow field due to a singular force parallel to the free stream direction, and that due to a force normal to the free stream direction.

Singular needle. The first case, force directed along the x axis, may be envisaged as the flow field caused by a "singular needle" located at the origin and of drag unity. The resulting velocity and pressure field is then

$$\mathbf{u} = -\frac{1}{4\pi U\rho}\,\mathrm{grad}\left(\frac{1}{R} - \frac{e^{-\lambda(R-x)}}{R}\right) - \frac{1}{4\pi\mu}\frac{e^{-\lambda(R-x)}}{R}\,\mathbf{i} \tag{11-23a}$$

$$p' = \frac{x}{4\pi R^3} \tag{11-23b}$$

where $\lambda = U/2\nu$ and $R^2 = x^2 + y^2 + z^2$. This flow field may be analyzed in the same way as that due to a singular flat plate of unit drag (Eq. 11-16). The velocity field is split, according to Eq. 11-9 and 11-11,

$\mathbf{u} = \mathbf{u}_L + \mathbf{u}_T$, $\mathbf{u}_T = \mathbf{u}_1 + \mathbf{u}_2$. The longitudinal wave $\mathbf{u}_L$ represents potential perturbation flow due to a source of strength $1/U\rho$. The transversal wave $\mathbf{u}_T$ vanishes exponentially upstream. In the downstream wake, of cross-sectional area x, it decreases in magnitude as $1/x$. The transversal wave has a sink at the origin of the same strength as the longitudinal source. The inflow through a plane $x = x_1$, is zero if x_1 is negative and equal to $1/\rho U$ if x_1 is positive. Hence ρU times the transversal inflow in the wake is unity, i.e. equal to the drag. This checks the momentum formula for the drag.

Singular lifting element. In the second case the force has a direction normal to the free stream. Because of the rotational symmetry of the Oseen equation, it is sufficient to consider any one of these directions. We shall therefore only consider the flow field due to a singular lifting element of unit lift located at the origin. The force exerted on the fluid is then of absolute value unity and is directed along the negative y axis. The solution is

$$\mathbf{u} = \frac{1}{4\pi\rho U} \operatorname{grad}\left[(1 - e^{-\lambda(R-x)}) \frac{\partial}{\partial y} \ln (R - x)\right] - \frac{e^{-\lambda(R-x)}}{4\pi\mu R} \mathbf{j} \qquad (11\text{-}24a)$$

$$p' = -\rho U u_L = \frac{1}{4\pi} \frac{\partial}{\partial y}\left(\frac{1}{R}\right) \qquad (11\text{-}24b)$$

where

$$\lambda = \frac{U}{2\nu}, \qquad R^2 = x^2 + y^2 + z^2$$

The longitudinal wave is identical with the potential flow field due to the infinitesimal horseshoe vortex of unit lift. All the vorticity is contained in the last term of the transversal wave and is

$$\Omega_1 = \frac{\partial v_2}{\partial z}, \qquad \Omega_2 = 0, \qquad \Omega_3 = \frac{\partial v_2}{\partial x} \qquad (11\text{-}25a)$$

where

$$v_2 = -\frac{e^{-\lambda(R-x)}}{4\pi\mu R} \qquad (11\text{-}25b)$$

This function, which is the coefficient of $\mathbf{j}$ in the last term in Eq. 11-24a, satisfies the following equation at any point except the origin

$$\operatorname{div} (\operatorname{grad} v_2 - 2\lambda v_2 \mathbf{i}) \equiv \nabla^2 v_2 - 2\lambda \frac{\partial v_2}{\partial x} = 0 \qquad (11\text{-}26a)$$

The singularity at the origin of the left-hand expression is equal to that of $-(1/4\pi\mu)\nabla^2(1/R)$. By integrating Eq. 11-26a, one then obtains the following equation

$$\iiint_V \frac{\partial v_2}{\partial x} dV = \frac{1}{2\lambda} \iiint_V \operatorname{div} \operatorname{grad} v_2 dV - \frac{1}{\rho U} \qquad (11\text{-}26b)$$

where V is any region containing the origin. If B is the boundary of V and V increases indefinitely in all directions, then

$$\iint_B (\text{grad } v_2) \cdot d\mathbf{n} \to 0 \tag{11-26c}$$

A comparison of Eq. 11-25a, 11-26b, and 11-26c shows that

$$\iiint \Omega_3 dV = -\frac{1}{\rho U} \tag{11-27}$$

if the integral is taken over the entire space. Since the lift is supposed to be unity, this equation checks the lift formula obtained by the momentum method (Eq. 4-17a).

Use of Fundamental Solutions in Boundary-Value Problems. The fundamental solution was used above to describe the flow field at large distances from a solid. This may be done for an arbitrary Reynolds number. The fundamental solution may also be used to advantage in finding approximate solutions of boundary-value problems for Oseen flow past a solid. In this case the entire flow field is determined. However, the solution obtained is a valid approximation for real flow only if the Reynolds number is low. Three examples of such solutions are now given, corresponding to flow past a finite flat plate, a circular cylinder, and a sphere, respectively.

Finite flat plate. Consider Oseen flow past a flat plate of zero angle of attack and zero thickness, located on the x axis between $x = -a$ and $x = a$. The Oseen equations then have to be solved with the following boundary conditions:

$$u = -U, \qquad v = 0 \text{ for } -a \leqq x \leqq a, \qquad y = 0 \tag{11-28a}$$

$$u = v = 0 \quad \text{at infinity} \tag{11-28b}$$

It is natural to try to represent the finite plate as a superposition of singular flat plates with appropriate strength factors. The flow field is then represented according to the two-dimensional form of Eq. 11-14. The y component of the force distribution $\mathbf{f}$ is zero and the x component is $\delta(y)g(x)$ where $g = 0$ for $|x| > a$. The boundary condition $u = -U$ at the plate then gives the following integral equation for g:

$$-U = \int_{-a}^{+a} \Gamma_{11}(x - \xi, 0)g(\xi)d\xi, \quad -a \leqq x \leqq a \tag{11-29}$$

An approximate solution for low Reynolds numbers is now given. The Reynolds number of the plate is defined as

$$Re = \frac{2aU}{\nu} \tag{11-30}$$

In the integral equation the maximum value of the nondimensional quantity $(x - \xi)U/\nu$ is Re. Hence if the latter is small one may use an expression for Γ_{11} valid for small values of $(x - \xi)U/\nu$, such as given by Eq. 11-17a. One then obtains the approximate equation

$$U = \frac{1}{4\pi\nu} \int_{-a}^{a} (\ln|x - \xi| + c)g(\xi)d\xi, \quad -a \leqq x \leqq a \tag{11-31}$$

where

$$c = \ln\left(\frac{\gamma_0 U}{4\nu}\right) - 1$$

The solution of this integral equation is (cf. e.g. [*54*, p. 143])

$$g(x) = \frac{4U\nu}{\ln\left(\frac{\gamma_0 Re}{16}\right) - 1} \frac{1}{\sqrt{a^2 - x^2}} \tag{11-32a}$$

Since the force distribution on the plate is now determined, one finds the drag D by integrating:

$$C_D = \frac{D}{\frac{1}{2}\rho U^2 \cdot 2a} = \frac{8\pi}{Re} \frac{1}{1 - \ln\left(\frac{\gamma_0 Re}{16}\right)} \tag{11-32b}$$

The complete flow field may now be obtained from Eq. 11-14. In this integral one may use the force distribution as determined above. However, except in the immediate neighborhood of the plate, the functions Γ_{11}, Γ_{12} appearing explicitly in Eq. 11-14a may not be approximated by the values given by Eq. 11-17. At large distances Eq. 11-18 should be used. It is easily checked that at large distances the dominant part of the flow field created by the finite flat plate is the same as the flow field due a singular flat plate of drag D. Near the plate the solution given satisfies the Stokes equation (cf. below).

Circular cylinder. We now discuss low Reynolds number flow past a circular cylinder of radius a and center at the origin. The diameter $2a$ is used as the characteristic length. The Reynolds number and the Stokes variables (Eq. 9-1) are then defined by

$$Re = \frac{2aU}{\nu}, \qquad x^* = \frac{x}{2a} \quad \text{etc.} \tag{11-33}$$

Nondimensional variables are used. $\mathbf{u}$ denotes the *complete velocity vector*, i.e. $U\mathbf{i}$ plus perturbation velocity.

The boundary conditions are then

$$\mathbf{u}^* = \mathbf{i} \quad \text{at infinity} \tag{11-34a}$$

$$\mathbf{u}^* = 0 \quad \text{at the cylinder, i.e. for } \tilde{r} = \frac{Re}{2}, \; r^* = \frac{1}{2} \tag{11-34b}$$

An approximate solution for low Reynolds number is

$$\mathbf{u}^* = \mathbf{i} + 2\epsilon\mathbf{u}^{(1)} - \frac{\epsilon Re^2}{8}\mathbf{u}^{(2)} \tag{11-35a}$$

where

$$\mathbf{u}^{(1)} = \text{grad}\left[\ln\tilde{r} + e^{\frac{\tilde{x}}{2}}K_0\left(\frac{\tilde{r}}{2}\right)\right] - e^{\frac{\tilde{x}}{2}}K_0\left(\frac{\tilde{r}}{2}\right)\mathbf{i} \tag{11-35b}$$

$$\mathbf{u}^{(2)} = \text{grad}\left[\frac{\partial}{\partial\tilde{x}}(\ln\tilde{r})\right] \tag{11-35c}$$

$$\epsilon = \frac{1}{\dfrac{1}{2} - \ln\left(\dfrac{\gamma_0 Re}{8}\right)} \tag{11-35d}$$

Here in accordance with a previous convention grad denotes $(\partial/\partial\tilde{x}, \partial/\partial\tilde{y})$.

The flow field corresponding to $2\epsilon\mathbf{u}^{(1)}$ is the perturbation field due to a singular flat plate of drag $2\epsilon \cdot 2\pi\mu U$ as seen by comparison with Eq. 11-16. The function $\mathbf{u}^{(2)}$ represents a potential flow field, namely the perturbation field of a dipole directed along the x axis. Its contribution to the drag is zero, as seen e.g. from momentum considerations (Art. 5).

The remarks above imply that $\mathbf{u}^*$ satisfies the Oseen equation and the boundary conditions at infinity. We now investigate to what degree the condition at the cylinder is satisfied. Since $\tilde{r} = Re/2$ at the cylinder and Re is assumed small, we develop $\mathbf{u}^*$ for small values of $\tilde{r}$. Since $\tilde{r} = Re r^*$, this is equivalent to rewriting Eq. 11-35 in Stokes variables and expanding in orders of ϵ. One finds

$$\mathbf{u}^* = \epsilon\left[\left(\ln 2r^* + \frac{1}{2} - \frac{1}{8r^{*2}}\right)\mathbf{i} + \left(-\frac{x^*}{r^*} + \frac{x^*}{4r^{*3}}\right)\text{grad}\, r^*\right] + O(Re) \tag{11-36}$$

The boundary condition at $r^* = \frac{1}{2}$ (Eq. 11-34) is thus satisfied to order $Re\epsilon$.

It can be easily verified that $\mathbf{i} + 2\epsilon\mathbf{u}^{(1)}$ satisfies Eq. 11-34b to order unity whenever $\epsilon = 1/(c - \ln Re)$, C = arbitrary const. If one chooses $C = \frac{1}{2} - \ln(\gamma_0/8)$ and adds a suitable multiple of a potential dipole Eq. 11-34b will be satisfied to order $Re\epsilon$, as just seen.

It was remarked above that the total drag D is $4\pi\mu U\epsilon$. Hence

$$C_D = \frac{D}{\frac{1}{2}\rho U^2 2a} = \frac{8\pi}{Re}[\epsilon + O(Re\epsilon)] \tag{11-37}$$

This drag formula is due to Lamb (see e.g. [*51*, p. 615]). The connection with the fundamental solution was pointed out by Oseen [*25*, p. 177ff].

The general nature of the flow field is easily found from previous comments on $\mathbf{u}^{(1)}$. A viscous wake exists which far downstream is that of a singular flat plate of drag D. Outside the wake the flow field is essentially potential at large distances from the cylinder. More specifically, it is the flow field due to potential source and potential dipole of appropriate strengths.

In [55] results of numerical computations for Oseen flow past a circular cylinder are given. The computations show two standing eddies on the downstream part of the cylinder. For increasing Re these eddies become more and more elongated in the downstream direction. It is also proved that within the Oseen approximation the pressure drag is exactly equal to the drag due to viscous shear.

Sphere. We now discuss low Reynolds number flow past a sphere of radius a. A solution analogous to Eq. 11-35 has been given by Oseen [25, p. 166ff.]. It is

$$\mathbf{u}^* = \mathbf{i} + \tfrac{3}{4}(1 + \tfrac{3}{16}Re)Re\mathbf{u}^{(1)} - \tfrac{1}{32}\left(1 + \frac{3Re}{16}\right)Re^3\mathbf{u}^{(2)} \tag{11-38a}$$

where

$$\mathbf{u}^{(1)} = \operatorname{grad}\left[\frac{e^{-\frac{\tilde{R}-\tilde{x}}{2}} - 1}{\tilde{R}}\right] - \frac{e^{-\frac{\tilde{R}-\tilde{x}}{2}}}{\tilde{R}}\mathbf{i} \tag{11-38b}$$

$$\mathbf{u}^{(2)} = -\operatorname{grad}\left[\frac{\partial}{\partial\tilde{x}}\left(\frac{1}{\tilde{R}}\right)\right] \tag{11-38c}$$

As before, $R^2 = x^2 + y^2 + z^2$, $Re = 2aU/\nu$, $\tilde{R} = UR/\nu$, etc.

This solution is obviously analogous to that previously given for the circular cylinder, and the discussion of the present case is therefore very brief. However, some special comments are needed about the boundary condition at the sphere ($\mathbf{u}^* = 0$ for $\tilde{R} = Re/2$). The term $Re\mathbf{u}^{(1)}$, which is the flow field due to a singular needle of drag $8\pi\mu Ua$ (cf. Eq. 11-23), may be expanded for small values of $\tilde{R}$ as (cf. Eq. 11-36)

$$Re\mathbf{u}^{(1)} \cong \left[-\frac{x^*}{2R^{*2}} + Re\left(\frac{1}{8} - \frac{x^{*2}}{8R^{*2}}\right)\right]\operatorname{grad} R^* + \left[-\frac{1}{2R^*} + Re\left(\frac{1}{4} - \frac{x^*}{4R^*}\right)\right]\mathbf{i} \tag{11-39a}$$

where $x^* = x/2a$ etc. $Re^3\mathbf{u}^{(2)}$, which represents a potential dipole, is, rewritten in Stokes variables

$$Re^3\mathbf{u}^{(2)} = \frac{1}{R^{*3}}\mathbf{i} - \frac{3x^*}{R^{*4}}\operatorname{grad} R^* \tag{11-39b}$$

It is then seen that if the term $3Re/16$ is omitted in the factor $1 + (3Re/16)$ in Eq. 11-38a then $\mathbf{u}^* = O(Re)$ at $R^* = \frac{1}{2}$ so that the boundary conditions are satisfied to order unity. If the full factor $1 + (3Re/16)$ is retained all terms of order unity and most terms of order Re cancel out at $R^* = \frac{1}{2}$. The uncancelled terms have the property that the x components of the velocity are odd functions of x whereas the y and z components are even. It follows by symmetry that the Oseen solution which may be added to cancel these terms contributes to the drag only with terms of order Re^2 [*25*, p. 173ff.]. (This symmetry argument is based on the fact that near the cylinder the unsymmetrical Oseen equation may be replaced by the symmetrical Stokes equation as explained below.) Hence the solution given by Eq. 11-38 satisfies the boundary conditions only to order unity but gives the drag correctly to order Re. The drag comes only from the $\mathbf{u}^{(1)}$ term. It is then

$$C_D = \frac{D}{\frac{1}{2}\rho U^2 \pi a^2} = \frac{24}{Re}\left(1 + \frac{3Re}{16} + O(Re^2)\right) \tag{11-40}$$

For further discussion of Oseen flow past a sphere see [*55*].

Relation between Oseen Flow and Stokes Flow. Some general statements made earlier in this article regarding the Stokes and Oseen approximations may now be verified and exemplified. The Oseen solutions found above all satisfy the Stokes equations near the solid. This may be seen from the fact that the approximations to the fundamental solutions used are solutions of the Stokes equations. In order to obtain a more precise statement consider the case of the sphere. If one rewrites the Oseen solution given by Eq. 11-38 in Stokes coordinates (cf. Eq. 11-39) and then expands it in powers of Re it is seen that the first term agrees with the previously given Stokes solution (Eq. 10-4). In particular the Stokes solution gives correctly the first term of the drag formula (cf. Eq. 10-5 and 11-40).

Consider now instead the circular cylinder. If the Oseen solution is rewritten in Stokes coordinates and expanded in powers of ϵ the first term is given by Eq. 11-36. This function satisfies the Stokes equation and may hence properly be called the Stokes solution for the circular cylinder. At infinity $\mathbf{u}^*$ becomes infinite because of the term $\ln 2r^*$. Hence the boundary condition at infinity must be given up. The solution is then not uniquely determined by the boundary condition at the cylinder and the Stokes equation. However, if it is required that the value at infinity is of as low an order as possible one finds that the solution is that given by Eq. 11-36 except that ϵ is an as yet undetermined multiplicative constant. This constant may then be found by a matching procedure. One first writes Eq. 11-36 in Oseen coordinates and then expands in orders of Reynolds number. If it is then required that the dominant term be the correct value at infinity, namely $\mathbf{i}$, one finds that ϵ must be

$$\frac{-1}{\log Re} + O\left(\frac{1}{\log Re}\right)$$

For further discussion see Art. 17 and [*32*].

B,12. Other Linearizations. Burgers' Equations. The linearizing methods used may be generalized. One may choose any basic flow $\mathbf{u}_0$, $\mathbf{\Omega}_0$ etc. and write

$$\mathbf{u} = \mathbf{u}_0 + \mathbf{u}' \tag{12-1a}$$

$$\mathbf{\Omega} = \mathbf{\Omega}_0 + \mathbf{\Omega}' \tag{12-1b}$$

etc.

If one neglects quadratic terms in the perturbation quantities, linear equations are obtained. Each quadratic term gives rise in general to two terms in the linear equations, for example,

$$(\text{grad } \mathbf{u})\mathbf{\Omega} \rightarrow (\text{grad } \mathbf{u}_0)\mathbf{\Omega}' + (\text{grad } \mathbf{u}')\mathbf{\Omega}_0 \tag{12-2}$$

The linear equations are then in general no longer conservation equations. If for example $\mathbf{\Omega}_0$ is different from zero the vorticity of the perturbation flow is not conserved. While $(\text{grad } \mathbf{u}_0)\mathbf{\Omega}'$ expresses the transport of vorticity with the basic flow field $\mathbf{u}_0$ the term $(\text{grad } \mathbf{u}')\mathbf{\Omega}_0$ plays formally the role of a sink or source for the vorticity $\mathbf{\Omega}'$. Many examples of such linearized equations occur in the theory of stability of laminar flow and are discussed in Sec. G.

One may of course also obtain linearized equations by another method which may be illustrated by

$$(\text{grad } \mathbf{u})\mathbf{\Omega} \rightarrow (\text{grad } \mathbf{u}_0)\mathbf{\Omega}_1 \tag{12-3}$$

While the first method is a perturbation procedure, the second method may be viewed as an iteration procedure. If $\mathbf{u}_0$ is a certain solution of a flow problem one may construct iteration equations by assuming that the vorticity $\mathbf{\Omega}_1$ of the first iteration $\mathbf{u}_1$ is transported by the velocity of the known solution $\mathbf{u}_2$. If $\mathbf{u}_1$ is found one may then assume that the vorticity $\mathbf{\Omega}_2$ of the second iteration is transported with $\mathbf{u}_1$ etc. The method of Weyl for solving the Blasius equation (cf. [*56*] and Art. 13) is actually such an iteration scheme. The linear equations obtained by the second method are conservation equations. This is made plausible by the transport interpretation and proved formally from the equation (cf. Eq. 1-12a)

$$(\text{grad } \mathbf{u}_0)\mathbf{\Omega}_1 - (\text{grad } \mathbf{\Omega}_0)\mathbf{u}_1 = \text{div } (\mathbf{u}_0 \circ \mathbf{\Omega}_1 - \mathbf{\Omega}_0 \circ \mathbf{\Omega}_1) \tag{12-4}$$

If the basic flow $\mathbf{u}_0$ is irrotational the two methods of linearization are equivalent. This method of linearization was proposed by Burgers as a generalization of the Oseen equations (see [*57*] and also Appendix 1 of [*25*]). Only the case of two-dimensional stationary flow will be considered. Let the basic flow be a potential flow field (u_p, v_p), with stream function ψ_p and potential ϕ_p. The linearized equations are then in the vorticity form

$$u_p \frac{\partial \Omega}{\partial x} + v_p \frac{\partial \Omega}{\partial y} = \nu \nabla^2 \Omega \tag{12-5a}$$

$$\Omega = -\nabla^2 \psi \tag{12-5b}$$

If conformal coordinates

$$\xi = \phi_p$$

$$\eta = \psi_p$$

are introduced, the equations are reduced to

$$\frac{\partial \Omega}{\partial \xi} = \nu \nabla_\zeta^2 \Omega \tag{12-6a}$$

$$\Omega = -J \nabla_\zeta^2 \psi \tag{12-6b}$$

where

$$z = x + iy, \qquad \zeta = \xi + i\eta, \qquad J = \left|\frac{d\zeta}{dz}\right|^2, \qquad \nabla_\zeta^2 = \frac{\partial^2}{\partial \xi^2} + \frac{\partial^2}{\partial \eta^2}$$

These equations are derived in the same manner as Eq. 6-9. If $u_p = 1$, $v_p = 0$, then $z = \zeta$ and the equations reduce to the Oseen equations.

The fundamental solution of the Eq. 12-5 may be found as a double integral by the same method as was applied to the Stokes equations (cf. Eq. 11-28 and 11-29). Eq. 12-6 may be easily solved with an appropriate singular forcing function. Inserting the value into Eq. 12-6b reduces the problem of solving Poisson equations. It appears, however, that the resulting double integral is difficult to evaluate even for relatively simple forms of J.

FLOW AT HIGH REYNOLDS NUMBERS. BOUNDARY LAYERS, JETS, AND WAKES

In preceding articles approximations suitable for flow at very low Reynolds numbers were discussed. The next three articles discuss flow at very high Reynolds numbers. Art. 13 and 14 deal with boundary layers. Art. 15 deals with jets, wakes, and viscous layers on slender bodies. These latter flow phenomena are normally classed together with boundary layers. However, they have some characteristics in common with low Reynolds number flow.

B,13. Basic Notions and Equations of Boundary Layer Theory.

Regions of vorticity. Certain qualitative aspects of viscous layers may be explained from considerations of the spreading of vorticity and the relative importance of transport of vorticity and diffusion of vorticity. The discussion of the sliding plate and related examples (Art. 6) brought out the following facts about diffusion of vorticity. If a solid is set in motion, vorticity is generated at the boundary. This vorticity spreads with infinite signal velocity into the fluid, so that immediately after the motion has started, vorticity exists in the entire fluid. However, except for a region near the body or the wake, vorticity is practically negligible at any finite time after the motion started. In the examples studied, vorticity was effectively confined to a region of thickness $\sim \sqrt{\nu t}$ so that the effective velocity of the viscous wave was $\sim \sqrt{\nu/t}$. Consider now stationary flow past a solid and let x be the distance along its surface measured from the stagnation point or the leading edge. At upstream infinity the flow is irrotational. As a fluid element passes downstream it will eventually receive some of the vorticity generated at the body surface. This process may be thought of as follows: The shear at the wall imparts vorticity to the fluid elements in a stream tube passing along the body. The vorticity thus generated at the wall will diffuse outward with an effective velocity $\sim \sqrt{\nu/t}$ and be transported downstream with a velocity of the order of the free stream velocity U. It is then expected that there will be an effective region of influence of vorticity similar to the region of influence of disturbances in nonviscous supersonic flow. In this analogy the effective velocity of diffusion corresponds to the speed of sound. At the station x_0 the diffusion velocity is $\sim \sqrt{\nu U/x_0}$ since the time of transport of the fluid element from the station $x = 0$ to $x = x_0$ is $\sim x_0/U$. Let $\delta(x)$ be the effective distance away from the body of the boundary of the vorticity region. Then

$$\delta(x) \sim \int_0^x \sqrt{\frac{\nu U}{x}} \frac{dx}{U} \sim \sqrt{\frac{\nu x}{U}} \tag{13-1a}$$

The important part of this estimate is that

$$\delta(x) \sim \sqrt{\nu} \tag{13-1a'}$$

The parabolic dependence on x is, of course, in general oversimplified since the transport velocity is a function of position rather than a constant U. Eq. 13-1a may be written

$$\frac{\delta(x)}{x} \sim \sqrt{\frac{1}{Re_x}} \tag{13-1b}$$

where Re_x is the local Reynolds number Ux/ν. If the body has a characteristic length, then

$$\frac{\delta(x)}{L} \sim \sqrt{\frac{x}{L}} \sqrt{\frac{1}{Re}} \tag{13-1c}$$

where Re is the over-all Reynolds number UL/ν. The function $\sqrt{x/L}$ may have to be replaced by a more complicated function of x/L, the value of which, however, is still independent of viscosity. Eq. 13-1c then shows that for large Reynolds numbers the vorticity layer is very thin relative to the body dimension. Such a layer is called a *boundary layer*.

An important premise of the argument just given is that vorticity is washed away downstream. If backflow occurs this premise is no longer valid. In particular, there may exist regions close to the body in which the streamlines are closed. The vorticity in such separated regions is not carried downstream by transport. It cannot be expected that the thickness of such a region is of order $\sim \sqrt{\nu}$: it is actually finite even in the limit of vanishing viscosity. The phenomenon of separated regions is an example of a complicated nonlinear interaction between velocity and vorticity (cf. the discussion of propagation of vorticity at the beginning of Art. 7). Vorticity is transported with the streamlines. The reasoning leading to Eq. 13-1 is valid if the flow near the body is directed from the front part towards the back part. However, the presence of vorticity may profoundly alter the velocity field so that flow reversal occurs.

The vorticity generated in unseparated regions is transported downstream and diffused at the same time. Using the same argument as above we see that downstream of the body the vorticity is essentially confined to a region called the wake, which far downstream at a distance d from the body has a width of order $\sqrt{\nu d/U}$. At large distances the wake is actually qualitatively independent of the Reynolds number of the body (cf. Art. 15).

The flow picture sketched above is further complicated by certain instabilities which will be discussed later.

Outside the wake and the vorticity region near the boundary the flow is essentially irrotational. The viscous forces on each fluid element are then negligible and the flow field is thus potential. However, it is not accurately described by the solution for potential flow past the solid. The vorticity regions described above give the body an apparent shape. Since the boundary layer proper has a thickness of order $\sqrt{\nu}$ its effect on the potential flow outside is of the same order. This is the effect of displacement thickness. On the other hand, separated regions and the wake have an effect which to first order is independent of viscosity. The important parameter for this effect is not the coefficient of viscosity, but rather some shape parameter. A body is called streamlined if the separated regions have a small effect. When the shape is such that separation has a large effect, the body is called bluff.

Perturbation problem for small values of ν. A different and more mathematical point of view is now adopted. In order to obtain approximate solutions for small ν (or small values of some suitable equivalent nondimensional parameter such as $1/Re$) it is natural to attempt to find a perturbation of the solution for $\nu = 0$. The situation is, however, considerably more complicated than in many other perturbation problems of mathematical physics. The first fundamental difficulty is the following: If ν is put equal to zero in the Navier-Stokes equations, one obtains the Euler equations (cf. Art. 3). There are several solutions of the Euler equations for the same solid and for the same conditions at upstream infinity. The potential solution with zero viscosity represents only one possible solution (cf. the end of Art. 14). Among these possible solutions the relevant solution to be perturbed would be the one that is obtained as a limit of the solutions of the Navier-Stokes equations as viscosity tends to zero. But unless one can guess it directly, finding the relevant solution to be perturbed would presuppose a knowledge of the solution of the Navier-Stokes equation for small ν, that is, exactly the solution to be found by the perturbation procedure. The whole problem is really the same as the problem of separation mentioned above. If the vorticity layer is of thickness $\sqrt{\nu}$, this layer disappears as ν tends to zero. All streamlines then come from upstream infinity and hence carry zero vorticity. The relevant solution for $\nu = 0$ is then the potential solution, which in principle may be assumed to be known. If on the other hand there is a separated region as described above, the limiting solution as $\nu \to 0$ may have vorticity in this region or at its boundary. It is plausible that the separated region grows indefinitely in the downstream direction as ν tends to zero. The limiting shape of this region is called the (nonviscous) wake. Outside this region the flow is then potential since all streamlines there come from upstream infinity. However, this flow is not potential flow past the original body but is flow past a semi-infinite object consisting of the body plus the wake. There are at present, however, no methods for determining the limiting flow. Furthermore, there are additional difficulties due to instability of the flow at high Reynolds number.

For the present we shall ignore these difficulties and assume that the limiting flow is known. An additional difficulty then appears: In the limit the viscous solution will tend nonuniformly to a discontinuous solution. Due to the no-slip condition the velocity must be zero at the boundary, whereas the potential solution or any solution of the Euler equations in general has nonzero velocity at the boundary. This special behavior is due to the fact that the small parameter ν multiplies the derivatives of highest order so that for ν equal to zero, the order of the equations is lowered and one has to give up a boundary condition. For ν small the discontinuity is replaced by a thin layer of rapid transition. This is the boundary layer. The perturbation procedure aims at finding the structure of this transition layer. Perturbation problems that exhibit this peculiarity of nonuniform convergence are called singular and have only recently been studied systematically (cf. e.g. [*58,59,60*]). In a regular perturbation where $f(P, \epsilon)$ converges uniformly to $f_0(P)$ as ϵ tends to zero, one may in general find the first perturbation $f_1(P)$ by assuming a power series expansion $f(P) = f_0(P) + \epsilon f_1(P) + \epsilon^2 f_2(P) + \cdots$. This method fails for singular perturbation problems. In the boundary layer case Prandtl devised the method to be described below. This method has actually great mathematical generality and applies to many other singular perturbation problems.

Prandtl's boundary layer equation. Consider now stationary two-dimensional flow past a solid. Choose a system of orthogonal coordinates x, y such that $y = 0$ on the boundary, x measures the distance along the boundary from some point, for example the stagnation point, and y measures the distance normal to the boundary within an infinitesimal neighborhood of the boundary. Let u and v be the x and y components of the velocity in the sense of vector geometry. The limiting nonviscous flow past the same solid is referred to as the external flow and its velocity components and the pressure is denoted by u_e, v_e, and p_e. Actually the boundary layer can be treated only in unseparated regions. We therefore exclude the wake region and assume that u_e, v_e, and p_e are potential in the regions considered. Note that $v_e(x, 0)$ is equal to zero, whereas $u_e(x, 0)$ is in general different from zero. According to the ideas set forth in the beginning of this article, there exists for viscous flow at high Reynolds numbers a thin boundary layer of thickness $\delta \sim \sqrt{\nu}$ adjacent to the solid. At the solid ($y = 0$) u must be zero according to the no-slip condition. At the outer edge of the boundary layer ($y = \delta$) u has the value $u_e(x, 0)$ plus terms of the order $\sqrt{\nu}$. This is so because the flow outside the boundary layer is essentially potential and the modification of the effective shape of the solid is of order $\sqrt{\nu}$. Hence, considering only terms of order unity, u varies from zero to $u_e(x, 0)$ within a distance $\delta \sim \sqrt{\nu}$. The rapid variation within the boundary layer may thus be expressed by the estimate $\partial/\partial y \sim 1/\sqrt{\nu}$. On the other hand, the variation in the x direction, that is, parallel to the boundary is to first order independent of ν so that $\partial/\partial x \sim 1$. The normal velocity component v is also zero for $y = 0$: and since $v_e(x, 0) = 0$, v is of order $\sqrt{\nu}$ at the edge of the boundary layer. Hence, $v \sim \sqrt{\nu}$ and $\partial v/\partial y \sim 1$. The last equation also follows from the continuity equation. Finally, there is no rapid variation of pressure since no additional boundary condition is imposed on the pressure for viscous flow.

These estimates may then be used to simplify the Navier-Stokes equations. If these equations are rewritten in the orthogonal coordinate system x, y and only the terms of lowest order in $\sqrt{\nu}$ are retained, the following Prandtl boundary layer equations are obtained. (The details of the derivation will be discussed in connection with the general case of three-dimensional compressible flow in Art. 20.)

$$uu_x + vu_y + \frac{1}{\rho} p_x = \nu u_{yy} \tag{13-2a}$$

$$p_y = 0 \tag{13-2b}$$

$$u_x + v_y = 0 \tag{13-2c}$$

The boundary conditions at the wall are

$$u(x, 0) = v(x, 0) = 0 \tag{13-3a}$$

At the edge of the boundary layer ($y = \delta$) u should be approximately equal to $u_e(x, 0)$. It can be shown (cf. Art. 17) that the correct boundary condition is that u assumes the value u_e at $y = \infty$. A similar statement is true for the pressure. Hence

$$u(x, \infty) = u_e(x, 0) = u_{ew} \tag{13-3b}$$

$$p(x, \infty) = p_e(x, 0) = p_{ew} \tag{13-3c}$$

The subscript $_w$ denotes conditions at the wall, i.e. $y = 0$. It is in general not possible to specify $v(x, \infty)$ in advance.

In addition a condition of the form

$$u(0, y) = u_e(0, 0) \tag{13-3d}$$

is needed. It is seen below that this condition is an initial condition necessitated by the parabolic nature of Eq. 13-2.

According to Eq. 13-3b and 13-3c, one has formally to consider the boundary layer solution for very large distances from the solid. This does not imply that the solution of the boundary layer equations has physical significance outside the boundary layer. According to the derivation of the boundary layer equations, they represent a valid approximation for terms of the order unity within a boundary layer of thickness $\sqrt{\nu}$, in the case of unseparated flow. The question of how to join the boundary layer solution with the outer flow solution, and how to obtain higher order approximations will be discussed later, especially in Art. 17.

The structure of Eq. 13-2 will now be discussed. The coordinates x, y as defined above are curvilinear when the body is curved. If the full Navier-Stokes equations are written out in these coordinates, curvature terms appear in addition to the Cartesian terms (cf. Art. 5 and 6). In the boundary layer approximation the curvature terms disappear if the special system described above is chosen. Hence the influence of centrifugal forces, as well as other curvature effects, are neglected. This statement will be proved in Art. 21 where a more systematic derivation of the boundary layer equations will be presented. It will then also be seen that for a three-dimensional body some curvature terms may not be neglected. In two dimensions the effect of curvature on the boundary layer flow enters only indirectly through the boundary conditions. The external flow quantities $u_e(x, 0)$ and $p_e(x, 0)$ are of course dependent on the body shape. Eq. 13-2b and 13-3c imply that $p(x, y) = p_e(x, 0) = p(x, \infty)$ so that the term $(1/\rho)\partial p/\partial x$ enters into Eq. 13-2a as a known forcing term equal to $(1/\rho)dp_e/dx$ or, by Bernoulli's law, equal to $-\frac{1}{2}[d(u_e)^2/dx]$. Within the boundary layer approximation, the flow near the wall is thus the same whether the pressure gradient is due to the shape of the solid or whether the boundary layer occurs along a flat plate with a pressure gradient induced by a neighboring solid.

The most radical simplification of the x momentum equation occurs in the viscous stress term. Not only the curvature terms but also the term u_{xx} are neglected. This means, intuitively speaking, that viscous diffusion does not take place in all directions, but only in the y direction. An equivalent statement is that in the two-dimensional boundary layer approximation the viscous stress tensor τ has components $\tau_{11} = \tau_{22} = 0$, $\tau_{12} = \tau_{21} = \mu \partial u/\partial y$. In this case the symbol τ is often used to denote simply $\mu \partial u/\partial y$. In the language of differential equations (cf. [*27*]) Eq. 13-2 are parabolic and the lines $x = \text{const}$ are characteristic curves. There is a formal resemblance to the one-dimensional heat equation where x corresponds to the time coordinate and y to the space coordinate of the heat equation. For the heat equation the lines $t = \text{const}$ are characteristics in the x, t plane which implies that the present cannot influence the past. The corresponding statement for Eq. 13-2 is that conditions at $x = x_0$ cannot influence conditions for $x < x_0$. Diffusion takes place only in the y direction and transport only in the downstream direction, that is toward increasing x. (It is hereby assumed that $u \geqq 0$. The case of backflow, $u < 0$, falls outside the scope of boundary layer theory.) This parabolic structure of the boundary layer problem is the reason why an initial condition such as Eq. 13-3d is needed. The full Navier-Stokes equations, on the other hand, are elliptic in the case of stationary flow, which means that every point of the fluid is within the domain of influence of any other point. The above analogy between nonstationary one-dimensional heat flow and stationary two-dimensional boundary layer flow should be carefully distinguished from the analogy between the nonstationary Navier-Stokes equations and the heat equation which was discussed in Art. 6. The first analogy is consistent with the intuitive ideas used at the beginning of the present article.

Boundary layer approximation as a limit of the exact solution. The derivation of Eq. 13-3 is essentially the one employed by Prandtl in his original paper [*61*]. For many purposes it is convenient to rephrase this definition in terms of limit processes: The boundary layer solution is defined as a special limit of the exact solution of the Navier-Stokes equations. Only some preliminary remarks are made here. A more exhaustive use of the concepts introduced below will be made in Art. 17 where we shall give a somewhat more mathematical discussion of the relation of the boundary layer solution to the exact solution and to asymptotic expansions for large Reynolds numbers.

The external flow was defined above as the limit of the exact solution as ν tends to zero. It was thereby tacitly understood that in taking the limit the position of each point $P = (x, y)$ was fixed. This limit is called the outer or first limit and is denoted by $\lim_e$. Thus if f is any flow quantity, such as a velocity component or pressure, which depends on P and ν, we define

$$\lim_e f(P, \nu) = \lim_{\substack{\nu \to 0 \\ P \text{ fixed}}} f(P, \nu) \tag{13-4}$$

Other parameters such as U, L etc. are also considered fixed and are not explicitly exhibited in the formulas. Their role can be studied by standard dimensional analysis (cf. Art. 9).

Using the notation of Eq. 13-4 we have

$$\psi_e = \lim_e \psi_{NS} \tag{13-5a}$$

where

$$\psi_e = \text{the stream function of external flow} \tag{13-5b}$$

$$\psi_{NS} = \text{the stream function of the exact solution of the Navier-Stokes equations} \tag{13-5c}$$

Similarly we may define $\mathbf{u}_e$, p_e etc. These quantities as well as ψ_e satisfy the Euler equations which are obtained by equating ν to zero, i.e. applying the first limit to the Navier-Stokes equations.

The boundary layer equations and their solution may also be obtained by a special limit process. Let $\zeta = (\xi, \eta)$ be a coordinate system such that η vanishes at the boundary of the solid and let u and v be velocity components in the ξ and η directions respectively. To derive the boundary layer equations with respect to the ζ system one may first rewrite the Navier-Stokes equations (Eq. 6-1) in this system. Then one replaces η and v by $\sqrt{\nu}\,\bar{\eta}$ and $\sqrt{\nu}\,\bar{v}$ where

$$\bar{\eta} = \frac{\eta}{\sqrt{\nu}} \tag{13-6a}$$

$$\bar{v} = \frac{v}{\sqrt{\nu}} \tag{13-6b}$$

Then ν is put equal to zero, wherever it appears explicitly in the equations. Finally η and v are reintroduced according to Eq. 13-6. The resulting equations are the boundary layer equations. The form of these equations depends on the coordinate system used. If ξ and η are the special coordinates defined in the discussion preceding Eq. 13-2 (or more generally any orthogonal system such that h_1 and h_2 as defined by Eq. 5-15 are unity at the body surface) the equations derived by the method just stated will be formally identical with Eq. 13-2.

The boundary layer equations are thus obtained by a special limit which will be called the second or inner limit. In this limit $\eta/\sqrt{\nu}$ is kept fixed as ν tends to zero. The following notation is used

$$\lim_\zeta f = \lim_{\substack{\nu\to 0\\ \xi,\bar{\eta}\text{ fixed}}} f \tag{13-7}$$

Note that $\bar{\eta}$ fixed implies that the point ξ, η moves toward the boundary as ν tends to zero. The above derivation of the boundary layer equations is equivalent to the definitions

$$u_{\text{BL}} = \lim_\zeta u_{\text{NS}} \tag{13-8a}$$

$$v_{\text{BL}} = \sqrt{\nu}\lim_\zeta \bar{v}_{\text{NS}} \tag{13-8b}$$

where subscripts $_{\text{BL}}$ and $_{\text{NS}}$ stand for "boundary layer" and "Navier-Stokes," respectively.

Boundary layer equations in conformal coordinates. The form of the boundary layer equations as well as the actual flow field described by their solution in general depends on the choice of coordinates. For this reason the coordinate system ζ is exhibited in the notation $\lim_\zeta$. The question of the role of the coordinate system in boundary layer theory will be discussed in Art. 17. Here it is remarked only that the skin friction is independent of the choice of coordinates.

As an example of the method discussed above we shall derive the boundary layer equations in conformal coordinates from Eq. 6-9. Let ξ and η now be conformal coordinates. It can be seen that it is consistent with Eq. 13-8 to require

$$\psi_{\text{BL}} = \sqrt{\nu}\lim_\zeta \bar{\psi}_{\text{NS}}; \qquad \bar{\psi}_{\text{NS}} = \frac{\psi_{\text{NS}}}{\sqrt{\nu}} \tag{13-9a}$$

$$\Omega_{\text{BL}} = \frac{1}{\sqrt{\nu}}\lim_\zeta \bar{\Omega}_{\text{NS}}; \qquad \bar{\Omega} = \sqrt{\nu}\,\Omega \tag{13-9b}$$

Only the stationary case is considered and we write ψ_η, $-\psi_\xi$ instead of u_1 and u_2 respectively. Introducing $\bar{\eta}$, $\bar{\psi}$, and $\bar{\Omega}$ into Eq. 6-11, then putting $\nu = 0$, and finally reintroducing η, ψ, and Ω yields the boundary layer equations in conformal coordinates

$$\psi_\eta\Omega_\xi - \psi_\xi\Omega_\eta = \nu\Omega_{\eta\eta} \tag{13-10a}$$

$$\Omega = -J_w\psi_{\eta\eta} \tag{13-10b}$$

where

$$J = \left|\frac{d\zeta}{dz}\right|^2; \qquad \zeta = \xi + i\eta, \qquad z = x + iy; \qquad J_w = J(\xi, 0) \tag{13-10c}$$

Similarity of boundary layer solutions. In general u_{NS} is a function of ξ, η, and ν. From the definition of the limit processes it follows that u_e is a function of ξ and η only, whereas u_{BL} is a function of ξ and $\eta/\sqrt{\nu}$.

This last statement is equivalent to stating that u varies very rapidly over a region of thickness $\sqrt{\nu}$ and that $\partial/\partial\eta \sim 1/\sqrt{\nu}$. To see this let

$$u_{\mathrm{BL}} = f\left(\xi, \frac{\eta}{\sqrt{\nu}}\right) \tag{13-11}$$

According to Eq. 13-3a and 13-3b, $f(\xi, 0) = 0$, $f(\xi, \infty) = u_{ew}$. Now let $\bar{\eta}_0 = \eta_0/\sqrt{\nu}$ be a number large enough so that $f(\xi, \bar{\eta}_0)$ is almost equal to u_{ew}. This means that in the interval $0 \leqq \eta \leqq \sqrt{\nu}\,\bar{\eta}_0$ the value of u_{BL} varies from zero to almost u_{ew}. Since $\bar{\eta}_0$ may be chosen independently of ν the transition to the free stream value u_e thus takes place in a region of thickness $\sim\sqrt{\nu}$. Furthermore $(\partial u_{\mathrm{BL}}/\partial\eta)_w = (\partial f/\partial\bar{\eta})_{\bar{\eta}=0} 1/\sqrt{\nu}$, where again $(\partial f/\partial\bar{\eta})_{\bar{\eta}=0}$ is independent of ν. Similar remarks apply to ψ and v. This illustrates how the present formulation of the boundary layer approximation is equivalent to the original Prandtl formulation.

A further consequence of the functional form of u_{BL}, v_{BL} etc. is that if the boundary layer solution for flow past a specific body has been found for one value of ν, it is known for all values of ν. (The validity of the solution as an approximation decreases of course with increasing ν.) This statement is explained in nondimensional terms:

Let x, y be the special coordinate system introduced in connection with Eq. 13-2. The notation $x^* = x/L$, $y^* = y/L$, $u^* = u/U$ etc. is used as in Eq. 9-1. We also define

$$\overline{y^*} = y^* \sqrt{Re} \tag{13-12a}$$

$$\overline{v^*} = v^* \sqrt{Re} \tag{13-12b}$$

If now the variables u^*, $\overline{v^*}$, p^*, x^*, and $\overline{y^*}$ are used, the parameter Re does not appear explicitly in either the boundary layer equations or the boundary conditions. Hence

$$u^* = \mathrm{fn}\ (x^*, y^* \sqrt{Re}) \tag{13-13a}$$

$$v^* = \frac{1}{\sqrt{Re}}\,\mathrm{fn}\ (x^*, y^* \sqrt{Re}) \tag{13-13b}$$

where "fn" denotes "function of." This means that within the boundary layer approximation all flows past a body of fixed shape are similar: If the solution is known for one value of Re it is known for all values of Re. This may be compared with the discussion in Art. 10 where it was pointed out that the Stokes approximation leads to a certain type of similarity. It was shown by dimensional analysis in Art. 9 that, if the full Navier-Stokes equations are used, then

$$u^* = \mathrm{fn}\ (x^*, y^*, Re) \tag{13-14}$$

However, the parameter Re cannot be eliminated in this case. Thus the similarity expressed by Eq. 13-13 goes beyond the similarity obtained by dimensional analysis.

Later on we shall consider boundary layers on conical bodies, i.e. bodies without a characteristic length L and hence without an over-all Reynolds number Re. In this case the similarity ideas used above lead to a reduction of the boundary layer equations to an ordinary differential equation (cf. Eq. 13-30).

Perturbation boundary layer. It is sometimes convenient to consider the perturbation boundary layer defined as the result of applying the boundary layer limit to $\psi - \psi_e$ rather than to ψ. (No linearization is implied by the word "perturbation.") Since ψ_e vanishes at the wall $\eta = 0$, ψ_e has an expansion

$$u_{ew}\eta + \left(\frac{\partial^2\psi_e}{\partial\eta^2}\right)_w \frac{\eta^2}{2} + \cdots$$

Hence (cf. Eq. 13-7, 13-9a)

$$\sqrt{\nu}\,\lim_\zeta \frac{\psi_e}{\sqrt{\nu}} = u_{ew}\eta \tag{13-15}$$

It then follows that the perturbation boundary layer has the following stream functions and velocity components:

$$\psi'_{BL} = \psi_{BL} - u_{ew}\eta \tag{13-16a}$$

$$u'_{BL} = u_{BL} - u_{ew} \tag{13-16b}$$

$$v'_{BL} = v_{BL} - \frac{du_{ew}}{d\xi}\eta \tag{13-16c}$$

Displacement effects. In general neither v_{BL} nor v'_{BL} vanish at infinity. Since the perturbation boundary layer obeys the continuity equation we find from Eq. 13-16c

$$(v'_{BL})_\infty = \frac{d}{dx}\int_0^\infty (u_{ew} - u_{BL})dy \tag{13-17}$$

(The special coordinates x, y defined in connection with Eq. 13-2 are used in the following.) The integral on the right-hand side is positive since the flow in the boundary layer is slower than the external flow at the wall. Except for very strong favorable (negative) pressure gradients the region of slow flow thickens with increasing x. Hence $(v'_{BL})_\infty$ is generally positive. Since u_{ew} is independent of y and ν, and u_{BL} depends on these variables only in the combination $y/\sqrt{\nu}$ (Eq. 13-11), it follows that

$$(v'_{BL})_\infty = \sqrt{\nu}\,\text{fn}\,(x) \tag{13-18}$$

The significance of the situation just described may be explained in a somewhat vague intuitive manner as follows: We define an approximate edge of the boundary layer as a line $y = \delta(x) \sim \sqrt{\nu}$ outside of which the viscous stresses, which govern the boundary flow, are practically negligible. For $y > \delta(x)$ the integrand of Eq. 13-17 is then essentially zero. For $y < \delta(x)$ we assume that the boundary layer approximation describes the main features of the flow. Eq. 13-17 then states that the slowing down of the fluid in the boundary layer induces an outflow of order $\sqrt{\nu}$ from the boundary layer; the normal velocity component of this outflow is practically $(v'_{\mathrm{BL}})_\infty$ at $y = \delta(x)$. As a result an additional potential flow field of order $\sqrt{\nu}$ is generated; this flow field is called the flow due to displacement thickness, for reasons that will be explained later.

The above ideas will be made more precise in Art. 17 where in particular the following statements will be justified. We define the first external perturbation with stream function ψ'_e by a repeated application of the first limit process (cf. Eq. 13-4 and 13-5) as

$$\psi'_e = \sqrt{\nu}\, \lim{}_e \frac{\psi - \psi_e}{\sqrt{\nu}} \tag{13-19}$$

ψ'_e then represents the potential flow and is determined by the following boundary condition at the wall:

$$-\psi'_{ew} = \int_0^\infty (u_{ew} - u_{\mathrm{BL}})dy \tag{13-20a}$$

or, equivalently, cf. Eq. 13-17,

$$v'_{ew} = -\frac{\partial \psi'_{ew}}{\partial x} = (v'_{\mathrm{BL}})_\infty \tag{13-20b}$$

In this interpretation the outflow which determines the first external perturbation is placed at the wall, $y = 0$, instead of at $y = \delta(x)$. It will be shown in Art. 17 that it is not necessary to patch the external flow and the first external perturbation with the boundary layer flow along some line $\delta(x)$. Instead one may consider ψ_e and ψ'_e defined everywhere. To the potential flow field $\psi_e + \sqrt{\nu}\,\psi'_e$ one may then add a flow field with stream function $\sqrt{\nu}\,\lim_\xi (1/\sqrt{\nu})[\psi_{\mathrm{NS}} - (\psi_e + \sqrt{\nu}\,\psi'_e)]$. The latter flow represents a perturbation boundary layer similar to the one defined by Eq. 13-16 except that now the flow due to displacement thickness has also been subtracted out. Alternatively, one can find special coordinates with the property that the boundary layer approximation with respect to those coordinates automatically includes both the external flow and the flow due to displacement thickness.

Displacement and momentum thickness. von Kármán's integral relation. The displacement thickness δ^* is defined by

$$u_{ew}\delta^* = \int_0^\infty (u_{ew} - u_{\mathrm{BL}})dy \tag{13-21}$$

Eq. 13-20b may then be written

$$v'_{ew} = (v'_{BL})_\infty = \frac{d(u_{ew}\delta^*)}{dx} \tag{13-22}$$

The first external perturbation may hence be thought of as the perturbation flow resulting when the thickness of the original body is increased by an amount δ^* in the direction normal to the wall. The apparent thickness is due to the fact that the real streamlines, as compared with the corresponding potential streamlines, are displaced away from the wall. This displacement is required by the continuity equation and the fact that the flow in the boundary layer is comparatively slow. In two-dimensional incompressible flow it is natural and convenient to measure this displacement with the aid of an actual thickness δ^*. However, in a more general case this interpretation can be somewhat forced. On the other hand Eq. 13-20, which interprets the displacement as due to an apparent source strength at the wall, can easily be generalized (cf. Art. 34).

The concept of the displacement thickness δ^* plays an important role in a somewhat different connection. We first define a second length, the momentum thickness δ^{**}, by

$$u_{ew}^2\delta^{**} = \int_0^\infty u_{ew}(u_{ew} - u_{BL})dy \tag{13-23}$$

By integrating the momentum equation (Eq. 13-2a) and using the continuity equation (Eq. 13-2c) and the boundary conditions (Eq. 13-3) one then obtains

$$\rho\frac{d(u_{ew}^2\delta^{**})}{dx} + \rho u_{ew}\delta^*\frac{du_{ew}}{dx} = \tau_w \tag{13-24}$$

where

$$\tau_w = \mu\left(\frac{\partial u_{BL}}{\partial y}\right)_{y=0}$$

This equation is called von Kármán's integral relation. In Chap. 4 it will be generalized in several respects and its importance for numerical calculations will be discussed.

B,14. Methods of Solution of the Boundary Layer Equations. Very many methods for solving the boundary layer equations have been developed. In this article only a few of the methods and results are discussed. For further details the reader is referred to standard treatises on this subject, as for example [*24,26*].

Similarity Solutions (Falkner-Skan). For special forms of the external velocity field, the boundary layer problem exhibits certain similarity characteristics with the aid of which one may reduce the equations to an ordinary differential equation. Goldstein [*62*] and Mangler

[*63*] have made systematic studies of the cases when this is possible. The most important case is that of the boundary layer on a semi-infinite wedge at zero angle of attack, in which case the boundary layer equations reduce to the Falkner-Skan equation. The simplest form of the Falkner-Skan equation is the Blasius equation, which corresponds to the case of zero-wedge angle (semi-infinite flat plate). Another special case is that when the total wedge angle is π (flow towards a flat plate normal to the stream). This is the only case where the boundary layer solution is at the same time a solution of the complete Navier-Stokes equations (cf. Eq. 8-6).

Consider an infinite wedge of opening angle $\pi\beta$, where $0 \leqq \beta \leqq 2$ with apex at the origin and center line on the positive x axis (Fig. B,14a).

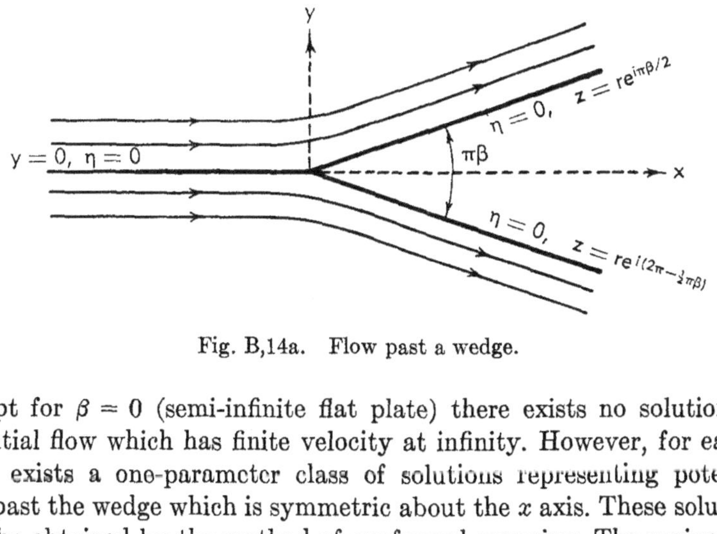

Fig. B,14a. Flow past a wedge.

Except for $\beta = 0$ (semi-infinite flat plate) there exists no solution for potential flow which has finite velocity at infinity. However, for each β there exists a one-parameter class of solutions representing potential flow past the wedge which is symmetric about the x axis. These solutions may be obtained by the method of conformal mapping. The region outside the wedge may be mapped conformally on the ξ, η plane, slit along the positive ξ axis, according to the formula (cf. Fig. B,14a),

$$z = \frac{2}{A_1(2-\beta)} e^{\frac{i\pi\beta}{2}} \zeta^{\frac{2-\beta}{2}} \tag{14-1}$$

where $z = x + iy$ is the complex coordinate in the physical plane, $\zeta = \xi + i\eta$, and A_1 is a real scale factor which is positive for flow towards the wedge. The coordinate system ξ, η is called the streamline coordinate since ζ, regarded as a function of z, is the complex stream function for potential flow past the wedge. The velocity components of the potential flow field in the original Cartesian system are given by

$$u - iv = \frac{d\zeta}{dz} = A_1 e^{-\frac{i\pi\beta}{2}} \zeta^{\frac{\beta}{2}} = A_1 e^{-\frac{i\pi\beta}{2}} \left[A_1\left(1 - \frac{\beta}{2}\right) e^{i\left(\theta - \frac{\pi}{2}\beta\right)}\right]^{\frac{\beta}{2-\beta}} \tag{14-2}$$

where $z = x + iy = re^{i\theta}$. The velocity on the negative x axis ($\theta = \pi$) is then $u = A_1\{A_1 r[1 - (\beta/2)]\}^{\beta/(2-\beta)}$, $v = 0$. For $\beta = 0$ the wedge reduces to a semi-infinite flat plate. Then $\zeta = z$, and the flow field is constant, $u - iv = A_1$.

We shall now consider viscous flow past the wedge which at large distances upstream tends to potential flow past the wedge (in the sense of Eq. 8-4). The boundary layer equations are most conveniently derived with the aid of the conformal coordinates (cf. Eq. 13-10 and [*56*]). The Jacobian of the conformal mapping is $J = |d\zeta/dz|^2 = A_1^2(\xi^2 + \eta^2)^{\beta/2}$ and hence $J_0 = J(\xi, 0) = A_1^2\xi^\beta$. The boundary layer equations are then

$$\psi_\eta\Omega_\xi - \psi_\xi\Omega_\eta = \nu\Omega_{\eta\eta} \tag{14-3a}$$

$$\Omega = -A_1^2\xi^\beta\psi_{\eta\eta} \tag{14-3b}$$

The corresponding boundary and initial conditions are

$$\psi(\xi, 0) = \psi_\eta(\xi, 0) = 0 \quad \text{for} \quad \xi \geqq 0 \tag{14-3c}$$

$$\psi_\eta(\xi, \infty) = 1, \qquad \psi_\eta(0, \eta) = 1 \tag{14-3d}$$

This system of equations has the following similarity property: If a is an arbitrary constant the system is invariant under the mapping

$$\eta \to a\eta, \qquad \xi \to a^2\xi, \qquad \psi \to a\psi, \qquad \Omega \to a^{\beta-1}\Omega \tag{14-4}$$

Hence $\psi/\sqrt{\xi}$ must be a function of $\eta/\sqrt{\xi}$ only.[15] If one then puts

$$\psi = \sqrt{2\nu\xi}\,f(\theta), \qquad \theta = \frac{\eta}{\sqrt{2\nu\xi}} \tag{14-5}$$

then Eq. 14-27a and 14-27b reduce to a fourth order ordinary differential equation which after one integration becomes

$$f''' + ff'' + \beta[1 - (f')^2] = 0 \tag{14-6a}$$

with boundary conditions

$$f(0) = f'(0) = 0 \tag{14-6b}$$

$$f'(\infty) = 1 \tag{14-6c}$$

Eq. 14-6 is the Falkner-Skan equation [*64*]. For $\beta = 0$ it reduces to the Blasius equation (Eq. 14-19a below). For $\beta = 1$, one obtains Eq. 8-6a with $m = 2$. Thus for two-dimensional flow towards a flat plate the boundary layer solution is at the same time a solution of the full Navier-Stokes equations ("exact" solution). For $\beta = \frac{1}{2}$, Eq. 14-6a is identical

[15] The similarity expressed by Eq. 13-14 provides a comparison between flow past two similar solids which differ only, say, by the value of L. For a conical object, such as a wedge, no geometrical length may be defined. Changing the length scale then provides a comparison of the flow quantities at two different points for the same object. Another example of the similarity theory used here is Busemann's theory of conical flow.

with Eq. 8-6a with $m = 3$. Thus the exact solution for axially symmetric flow towards a plane is formally identical with the boundary layer solution for a rectangular wedge.

An alternative derivation of the boundary layer equations for wedge flow is now considered. Instead of the conformal system of coordinates (ξ, η) a system of coordinates X, Y is used, which has the property that at the wedge surface X is the distance from the leading edge and Y the distance from the surface (cf. the discussion preceding Eq. 13-2). It is obviously sufficient to consider only the upper surface of the wedge. Hence, a system of coordinates of the required type may be obtained by a rigid rotation of the x, y system. In the notation of complex variables (cf. Eq. 14-1)

$$Z = X + iY = ze^{-\frac{i\pi\beta}{2}} = \frac{2}{A_1(2 - \beta)} \zeta^{\frac{2-\beta}{2}} \tag{14-7}$$

Since the Z system of coordinates is Cartesian the covariant velocity components will be the same as the velocity components in the sense of vector geometry. From now on u will denote the X component and v the Y component of the velocity. The subscript $_e$ refers to the potential (external flow solution). Since ζ is the complex stream function for the potential flow

$$u_e - iv_e = \frac{d\zeta}{dZ} = \frac{2}{2 - \beta}\left(\frac{2 - \beta}{2A_1}\right)^{\frac{2}{2-\beta}} Z^{\frac{\beta}{2-\beta}} \tag{14-8}$$

On the upper surface of the wedge $v_e = 0$ and

$$u_e(x, 0) = Ax^l \tag{14-9a}$$

where

$$A = \frac{2}{2 - \beta}\left(\frac{2 - \beta}{2A_1}\right)^{\frac{2}{2-\beta}} \tag{14-9b}$$

and

$$l = \frac{\beta}{2 - \beta}, \qquad \beta = \frac{2l}{l + 1} \tag{14-9c}$$

When the external velocity distribution is given by Eq. 14-9c, the boundary layer equations (13-2) become (from now on we shall use x, y in the sense of Eq. 13-2, i.e. they are actually the coordinates that above were denoted by X, Y):

$$uu_x + vu_y - lA^2x^{2l-1} = \nu u_{yy} \tag{14-10a}$$

$$u_x + v_y = 0 \tag{14-10b}$$

$$u(x, 0) = v(x, 0) = 0 \tag{14-10c}$$

$$u(x, \infty) = Ax^l \tag{14-10d}$$

In the present system of coordinates the similarity expressed by Eq. 14-4 now takes the following form: Eq. 14-10a, 14-10b, 14-10c, and

14-10d invariant under the mapping

$$x \to a^2 x, \qquad y \to a^{1-l} y, \qquad u \to a^{2l} u, \qquad v \to a^{l-1} v$$

From this similarity it follows that $\psi x^{-(l+1)/2}$ is a function of $yx^{(l-1)/2}$ only. We then put

$$\psi = \sqrt{\frac{2A\nu}{(l+1)}}\, x^{\frac{l+1}{2}} f(\theta) = \sqrt{\frac{2\nu u_{ew} x}{(l+1)}}\, f(\theta) \tag{14-11a}$$

where

$$u_{ew} = u_e(x, 0) = Ax^l \tag{14-11b}$$

and

$$\theta = \sqrt{\frac{(l+1)A}{2\nu}}\, x^{\frac{l-1}{2}} y = \sqrt{\frac{(l+1)u_{ew}}{2\nu x}}\, y \tag{14-11c}$$

If the above value for ψ is used in Eq. 14-10 one again finds the Falkner-Skan equation (Eq. 14-6a) with the boundary conditions (Eq. 14-6b and 14-6c).

Physical meaning of the Falkner-Skan equation. In the original interpretation β was restricted to the range $0 \leqq \beta < 2$. The corresponding values of l lie between zero and infinity. It can be shown that the Falkner-Skan equation has a unique solution in this range. In addition, this equation may be solved in a range of negative values of β: $\beta_0 \leqq \beta \leqq 0$ where $\beta_0 \cong -0.198838$. The corresponding value of l (Eq. 14-9c) vary between $l_0 \cong -0.090429$ and zero. The velocity thus varies as a negative power of x and hence decreases with x. The pressure gradient is then positive, i.e. unfavorable. The Falkner-Skan solutions may then be interpreted as boundary layers along a flat surface where the external velocity field has been generated in some manner, say by other objects in the flow field. (It must of course be borne in mind that all Falkner-Skan solutions correspond to highly idealized physical situations.)

In all the preceding cases it was tacitly assumed that the external velocity field has the same direction as the positive x direction. Solutions of the Falkner-Skan equation also exist when $\beta > 2$. In this case $l < -1$, with the limiting case $\beta = \infty$ corresponding to $l = -1$. However, Eq. 14-11a then shows that these solutions correspond to real flow fields only if u_{ew} is negative, i.e. the external flow field has the direction opposite to the x axis. The pressure then decreases in the flow direction, so that these are cases of a favorable pressure gradient. It will be shown later that the limiting case $l = -1$ corresponds to the boundary layer in a radial diffuser with sink flow (cf. Eq. 8-8ff. and Eq. 14-33 below). This class of Falkner-Skan flows should again be interpreted as boundary layers on a wall with an artificially created pressure gradient rather than as boundary layers on a wedge.

Computation of Falkner-Skan solutions. Eq. 14-6 with a general value of β was first investigated by Falkner and Skan [*64*]. Further work was done by Hartree [65]. A survey of some of the results is given in the books by Goldstein [24] and Schlichting [26]. Recently very accurate computations have been made by A. M. O. Smith [66]. Weyl studied Eq. 14-6 with $\beta \geqq 0$ from a mathematical point of view in [56]. Some of the methods developed by Weyl lead to very rapid approximate calculations for $\beta = 0$ or $\beta = \frac{1}{2}$. A brief discussion of some of the results obtained by various investigators is now given.

We first observe that the functions $f(\theta) = f(\theta; \beta)$ are characterized by two important constants (or rather functions of β), namely

$$\alpha_1(\beta) = f''(0; \beta) \tag{14-12a}$$

and

$$\alpha_2(\beta) = \lim_{\theta \to \infty} [\theta - f(\theta; \beta)] \tag{14-12b}$$

The physical meaning of these constants is given by

$$\text{Skin friction} = \tau_w = \rho u_{ew}^2 \sqrt{\frac{(l+1)\nu}{2u_{ew}x}}\, \alpha_1 \tag{14-13a}$$

$$\begin{aligned} \text{Displacement thickness} = \delta^* &= \int_0^\infty \left(1 - \frac{u}{u_{ew}}\right) dy \\ &= \sqrt{\frac{2\nu x}{(l+1)u_{ew}}}\, \alpha_2 \end{aligned} \tag{14-13b}$$

The momentum thickness δ^{**} is also simply related to these constants. From Eq. 13-23 and 14-11b we find

$$\delta^{**} = \sqrt{\frac{2\nu x}{(l+1)u_{ew}}} \int_0^\infty f'(1 - f')d\theta \tag{14-14}$$

By integrating Eq. 14-6 from $\theta = 0$ to $\theta = \infty$ and using integration by parts one finds that

$$\int_0^\infty f'(1 - f')d\theta = \int_0^\infty ff''d\theta = \frac{\alpha_1 - \beta\alpha_2}{1 + \beta} \tag{14-15}$$

The Kármán integral relation (13-24) is a differential equation involving τ_w, δ^*, δ^{**}. In a similarity case, when the x dependence is known in advance, it must reduce to an algebraic equation, which in the present case is in fact Eq. 14-15, as may easily be checked.

The mathematical meaning of α_1 and α_2 is the following. Since Eq. 14-6 is of third order and $f(0)$ and $f'(0)$ are known it should be possible to find an expansion of f near $\theta = 0$ in which the only unknown quantity is $\alpha_1 = f''(0)$. One finds in fact

$$f(\theta) = \alpha_1 \frac{\theta^2}{2} - \beta \frac{\theta^3}{6} + (2\beta - 1)\alpha_1^2 \frac{\theta^5}{120} + \cdots \tag{14-16}$$

On the other hand, at $\theta = \infty$ only f' is known so that the expansion near infinity involves two unknown parameters. One of these is α_2; the other one, which does not seem to have a simple physical meaning, is denoted by α_3. One finds

$$f(\theta) = \theta - \alpha_2 + \frac{\alpha_3}{(\theta - \alpha_2)^{2(1+\beta)}} e^{\frac{-(\theta-\alpha_2)^2}{2}} + \cdots \tag{14-17}$$

Semi-infinite flat plate. When $\beta = 0$, $u_e = U =$ const and Eq. 14-11a and 14-11c reduce to

$$\psi = \sqrt{2\nu U x}\, f(\theta) \tag{14-18a}$$

$$\theta = y \sqrt{\frac{U}{2\nu x}} \tag{14-18b}$$

The Falkner-Skan equation reduces to the Blasius equation

$$f''' + ff'' = 0 \tag{14-19a}$$

with boundary conditions

$$f(0) = f'(0) = 0 \tag{14-19b}$$

$$f'(\infty) = 1 \tag{14-19c}$$

The solution of these equations is called the Blasius function and denoted by $B(\theta)$.

The flat plate problem was originally studied by Prandtl [*61*] and Blasius [*67*]; the most complete numerical investigations have been given by Howarth [*68*] and by A. M. O. Smith [*66*]. Details of methods of solution and numerical results are given in [*24*, p. 135ff.] and [*26*, p. 98ff.]. In using the numerical tables in the last two references, it should be kept in mind that different normalizations are used. The Blasius function B as defined above is connected with the corresponding function f of [*24*] by the relation $f(\theta) = \sqrt{2}\, B(\sqrt{2}\, \theta)$; the function f of [*26*] on the other hand satisfies the relation $f(\theta) = \sqrt{2}\, B(\theta/\sqrt{2})$. A highly interesting discussion of the Blasius equation is given by Weyl in [*56,69*]. The method of Weyl not only shows existence and uniqueness of the solution but also leads to a rapidly converging iteration process suitable for numerical calculations (cf. Eq. 14-24ff. below).

For small values of θ Eq. 14-19a and 14-19b lead to the power-series expansion of the Blasius function (cf. Eq. 14-16)

$$B(\theta) = \frac{\alpha_1 \theta^2}{2!} - \frac{\alpha_1^2 \theta^5}{5!} + \frac{11\alpha_1^3 \theta^8}{8!} - \frac{375\alpha_1^4 \theta^{11}}{11!} + \cdots \tag{14-20}$$

where $\alpha_1 = B''(0)$. It was noticed by Weyl [*69*] that the power series for $B(\theta)$ has a finite radius of convergence, somewhere between 3.37 and 5.03.

By replacing f in the first factor of ff'' in Eq. 14-19a by its approximate value at infinity, namely $(\theta - \alpha_2)$, one finds an improved approximation at infinity

$$f \cong \theta - \alpha_2 + \alpha_3 \int_{\theta-\beta}^{\infty} (\theta - \alpha_2 - \zeta) e^{-\frac{\zeta^2}{2}} d\zeta \tag{14-21}$$

Asymptotic expansion of the last integral gives Eq. 14-17 with $\beta = 0$.

One method of solution of the Blasius equation is to determine the constants α_1, α_2, and α_3 by matching the approximate solutions for small and large θ (Eq. 14-20 and 14-21, respectively) at some intermediate θ. This method is in principle somewhat dangerous since the series in Eq. 14-20 has a small radius of convergence and the expansion is an asymptotic expansion valid for only very large θ.

Numerical integration of the Falkner-Skan equation is in general complicated by the fact that one boundary condition is imposed at infinity. It was pointed out, however, by Toepfer [70] that for $\beta = 0$ the problem may be converted into an initial value problem, i.e. all boundary conditions imposed at $\theta = 0$. If $f(\theta)$ is a solution of Eq. 14-19a then $kf(k\theta)$ is another solution for any constant k. Now choose f to satisfy the initial values

$$f(0) = f'(0) = 0, \qquad f''(0) = 1 \tag{14-22}$$

Then it is easily verified that the function $B(\theta)$

$$B(\theta) = kf(k\theta), \qquad k = [f'(\infty)]^{-\frac{1}{2}} \tag{14-23}$$

satisfies the correct boundary conditions, Eq. 14-19b and 14-19c. Thus the solution of an initial value problem gives us the correct Blasius function $B(\theta)$. It is also seen that

$$B''(0) = \alpha_1 = k^3 f''(0) = [f'(\infty)]^{-\frac{3}{2}} \tag{14-24}$$

with f and k defined as above.

For further discussion of various methods of solution the reader is referred to the references cited above. Here we shall only derive an approximate solution using the method of Weyl. The initial value problem (Eq. 14-22) is chosen, although this is not essential for the method. The method of Weyl consists of finding successive approximations f_0, f_1, . . . defined by

$$f'''_{n+1} + f_n f''_{n+1} = 0 \tag{14-25}$$

and the boundary conditions (14-22.) Weyl proves that if one starts with $f_0(\theta) = 0$, the iteration procedure converges to the unique solution, actually quite rapidly. One finds that $f_1 = \theta^2/2$ and hence that f_2 satisfies the equation

$$f_2''' + \frac{\theta^2}{2} f_2'' = 0 \tag{14-26}$$

This yields

$$f_2'' = e^{-\theta^3/6}$$

and hence

$$f_2'(\infty) = \int_0^\infty e^{-\theta^3/6}d\theta = (6)^{\frac{1}{3}}\Gamma(\tfrac{4}{3})$$

Hence,

$$\alpha_1 \cong \frac{1}{\sqrt{6}}[\Gamma(\tfrac{4}{3})]^{-\frac{3}{2}} \cong 0.4838 \tag{14-27}$$

The iteration scheme of Weyl means that the vorticity of the $(n + 1)$th approximation is transported with the flow of the nth approximation. In particular, $f_1 = \theta^2/2$ corresponds to $u_1 = U\theta$ and a corresponding v computed from the continuity equation. The approximate value for u_1 corresponds to replacing the correct velocity profile by its tangent at $y = 0$. Comparison with the more accurate value given below shows that the error in Weyl's second approximation is less than 3.1 per cent. A fair accuracy is thus obtained in spite of the fact that $u_1(\infty) = \infty$. If on the other hand one uses U as transport velocity, as in the Oseen approximation, a much larger error results (Eq. 14-45). This shows that it is more important to choose the transport terms correctly near the wall than at the outer edge of the boundary layer. This fact will be utilized in the study of heat transfer in Chap. 4.

From Eq. 14-13 and A. M. O. Smith's computations of α_1 and α_2 [*66*] (cf. Table B,14) one obtains the following formulas for the skin friction and displacement thickness of the Blasius boundary layer:

$$\tau_w = \rho U^2 \sqrt{\frac{\nu}{2U_x}}\,\alpha_1, \quad \alpha_1 = B''(0) = 0.46960 \tag{14-28a}$$

$$\delta^* = \sqrt{2\nu x}\,\alpha_2, \quad \alpha_2 = 1.21678 \tag{14-28b}$$

$\beta = \frac{1}{2}$; $l = \frac{1}{3}$. A solution of the Falkner-Skan equation with $\beta = \frac{1}{2}$ gives the boundary layer on a wedge of total opening angle $\pi/2$. The same solution is also the exact solution of the Navier-Stokes equations for axially symmetric flow towards a flat plate normal to the stream. This case is almost as simple as the Blasius case due to the fact that a differentiation of the Falkner-Skan equation gives the comparatively simple equation,

$$f'''' + ff''' = 0 \tag{14-29a}$$

Putting $\theta = 0$ in the original Falkner-Skan equation gives the added boundary condition needed for a fourth order equation, namely

$$f'''(0) = -\beta = -\tfrac{1}{2} \tag{14-29b}$$

The comments made about the Blasius equation apply also, with obvious modifications, to Eq. 14-29a. In particular the method of Weyl may be applied in the present case also. The second approximation gives

$$\alpha_1 = \frac{1}{2}\int_0^\infty e^{-\alpha\frac{\theta^3}{6}}\,d\theta = \frac{\alpha^{-\frac{1}{3}}}{2}\,6^{\frac{1}{3}}\Gamma(\tfrac{4}{3}) \tag{14-30a}$$

i.e.,

$$\alpha_1 = \left(\frac{3}{4}\right)^{\frac{1}{3}} \left[\Gamma\left(\frac{4}{3}\right)\right]^{\frac{2}{3}} \cong 0.855 \tag{14-30b}$$

According to more exact calculations,

$$\alpha_1 = 0.92768 \tag{14-31a}$$

and

$$\alpha_2 = 0.80455 \tag{14-31b}$$

Large β. For large values of β the Falkner-Skan equation may be solved by an expansion method. The leading term of this expansion (Eq. 14-38) is easily obtained; it gives values for skin friction and displacement thickness that are reasonably accurate even for rather small values of β, say $\beta = 1$.

As β tends to infinity, Eq. 14-6a becomes

$$1 - (f')^2 = 0 \tag{14-32}$$

The only relevant solution of this equation is $f' = 1$. Hence as β tends to infinity, f' tends to unity everywhere except at $\theta = 0$. The convergence is thus nonuniform at $\theta = 0$ and the problem of obtaining solutions for β large appears to be a singular perturbation problem. This depends of course on the fact that Eq. 14-6a is of higher order than its reduced equation (Eq. 14-32). Following the technique of solving singular perturbation problems used in Art. 13, we introduce new variables $\bar{\theta} = \beta^a\theta$ and $g = \beta^b f$ where a and b are constants, and consider the limits as β tends to infinity while $\bar{\theta}$ and g are fixed. It can easily be verified that the proper choice of exponents is $a = b = \frac{1}{2}$. Furthermore the perturbation problem turns out to be regular in the new variables; for large β the solution of the Falkner-Skan equation may be written as a power series[16] in $1/\beta$

$$g(\bar{\theta}; \beta) = g_0(\bar{\theta}) + \frac{1}{\beta} g_1(\bar{\theta}) + \frac{1}{\beta^2} g_2(\bar{\theta}) + \cdots \tag{14-33a}$$

where

$$\bar{\theta} = \beta^{\frac{1}{2}}\theta = \sqrt{\frac{\beta\mu_{ew}}{(2-\beta)\nu x}}\, y \tag{14-33b}$$

and

$$g(\bar{\theta}; \beta) = \beta^{\frac{1}{2}} f = \sqrt{\frac{\beta}{(2-\beta)\nu u_{ew} x}}\, \psi \tag{14-33c}$$

Note that for large β the new variables are physically more natural than θ and f since $\bar{\theta}$ and g are of the same order of magnitude as y and ψ,

[16] The corresponding problem for compressible flow (cf. Eq. 30-4) remains a singular perturbation problem even after the introduction of new variables. The expansion (Eq. 14-33a) must then be replaced by a composite expansion in the sense explained in Art. 17. The limit with θ fixed then plays the role of an outer limit whereas $\bar{\theta}$ is fixed in the inner limit.

respectively, as β tends to infinity. The Falkner-Skan equation in the new variable is (prime denotes derivative with respect to $\bar{\theta}$)

$$g''' + \frac{gg''}{\beta} + [1 - (g')^2] = 0 \tag{14-34a}$$

with boundary conditions

$$g(0) = g'(0) = 0; \qquad g'(\infty) = 1 \tag{14-34b}$$

The leading term of the expansion (Eq. 14-33c) then obeys the equation

$$g_0''' + [1 - (g_0')^2] = 0 \tag{14-35a}$$

with boundary conditions

$$g_0(0) = g_0'(0) = 0; \qquad g'(\infty) = 1 \tag{14-35b}$$

The boundary conditions for g_0 are the same as those for g. Subsequent g_i obey linear nonhomogeneous equations with variable coefficients and homogeneous boundary conditions.

The function g_0 was introduced as an approximation to g for β large and for $\beta = \infty$ should be exactly equal to g. Now $\beta = \infty$ corresponds to $l = -1$ and hence $u_{ew} = A/x$, $A < 1$ (Eq. 14-9). The Falkner-Skan solution thus describes the boundary layer of viscous sink flow.[17] This may be verified directly. If one derives the boundary layer approximation to Eq. 8-8 by standard method one obtains an equation which, with suitable normalization of the variables, is identical with Eq. 14-36.

To solve Eq. 14-33c we multiply by g_0 and integrate. Using the condition $g_0'(0) = 0$ and $g_0'(\infty) = 1$, one finds

$$(g_0'')^2 + 2g_0' - \frac{2(g_0')^3}{3} = \text{const} = [g_0''(0)]^2 = \frac{4}{3} \tag{14-36}$$

Hence for very large β we have approximately

$$\alpha_1(\beta) = f''(0;\beta) \cong \sqrt{\frac{4\beta}{3}} \tag{14-37a}$$

or (Eq. 14-14)

$$\tau_w \cong \rho u_{ew}^2 \sqrt{\frac{l\nu}{u_{ew}x}}\sqrt{\frac{4}{3}},\ l = \frac{\beta}{2-\beta} \tag{14-37b}$$

By integrating Eq. 14-36 one finds

$$\beta^{\frac{1}{2}} f \cong g_0(\bar{\theta}) = \bar{\theta} - 3\sqrt{2}\tanh\left(\frac{\bar{\theta}}{\sqrt{2}} + \tanh^{-1}\sqrt{\frac{2}{3}}\right) - 2\sqrt{3} \tag{14-38a}$$

[17] As mentioned earlier, the Falkner-Skan solution does not represent a real flow field for $\beta > 2$ when the flow direction and the x direction coincide. In the special case, $\beta = \infty$, this is related to the fact that viscous source flow in a radial diffuser does not tend to a limit as the Reynolds number tends to infinity.

and hence

$$\frac{df}{d\theta} \cong \frac{dg_0}{d\bar{\theta}} = -2 + 3 \tanh^2\left(\frac{\bar{\theta}}{\sqrt{2}} + \tanh^{-1}\sqrt{\frac{2}{3}}\right) \tag{14-38b}$$

The first of these equations gives

$$\alpha_2(\beta) = \beta^{-\frac{1}{2}}(3\sqrt{2} - 2\sqrt{3}) \tag{14-39}$$

or (cf. Eq. 14-13b and 14-9c)

$$\delta^* = \sqrt{\frac{\nu x}{l u_{ew}}}(3\sqrt{2} - 2\sqrt{3}) \tag{14-40}$$

The accuracy of the solution is discussed below.

Numerical values. Some results of numerical calculations are presented below in Table B,14 and in Fig. B,14b and B,14c. The values

Table B,14. Skin friction and displacement thickness in Falkner-Skan flow (cf. Eq. 14-12).

β	$\alpha_1(\beta)$	$\alpha_2(\beta)$	β	$\alpha_1(\beta)$	$\alpha_2(\beta)$
−0.198838	0	2.35885	0.20	0.686708	0.98416
−0.195	0.055172	2.11705	0.30	0.774755	0.91099
−0.190	0.085700	2.00676	0.40	0.854421	0.85263
−0 18	0.128636	1.87157	0.50	0.927680	0.80455
−0.16	0.190780	1.70665	0.60	0.995836	0.76397
−0.14	0.239736	1.59590	0.80	1.120268	0.69868
−0.10	0.319270	1.44270	1.00	1.232588	0.64790
−0.05	0.400323	1.31236	1.20	1.335722	0.60690
0	0.469600	1.21678	1.60	1.521514	0.54402
0.05	0.531130	1.14174	2.00	1.687218	0.49743
0 10	0.587035	1.08032	....	....	....

are taken from [*66*] to which the reader is referred for more complete tables of numerical values and for graphs of velocity distributions.

The leading term of the expansion of $\alpha_1(\beta)$ for β very large was given by Eq. 14-37. It is plotted in Fig. B,14b together with the more accurate curve. The similarity between the two curves is apparent. It is consistent

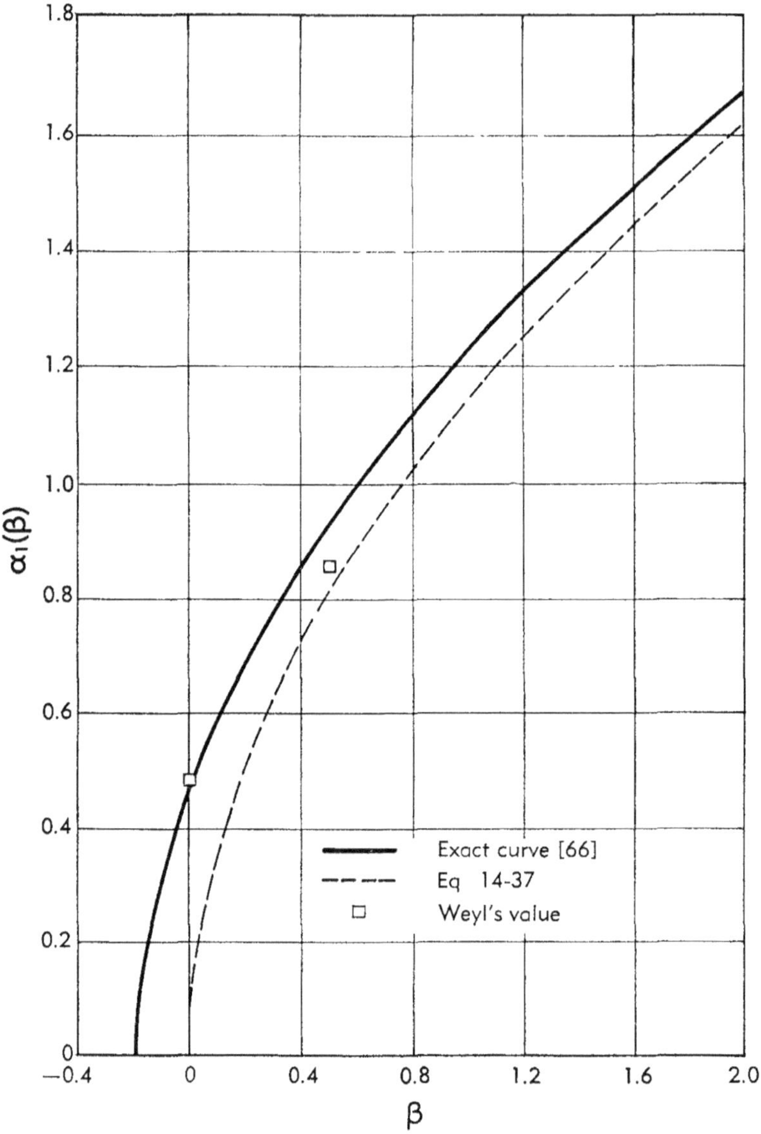

Fig. B,14b. Variation of skin friction with wedge angle.

with the asymptotic method used in deriving Eq. 14-37 to replace its right-hand side by $\sqrt{\frac{4}{3}(\beta + A)}$, A = const. By the suitable choice of A, fair numerical agreement with the accurate values may be obtained for the entire range of β. One may, for instance, choose A so as to give the correct value for $\beta = 0$. Similar comments apply to the asymptotic formula for $\alpha_2(\beta)$, given by Eq. 14-39 and plotted in Fig. B,14c.

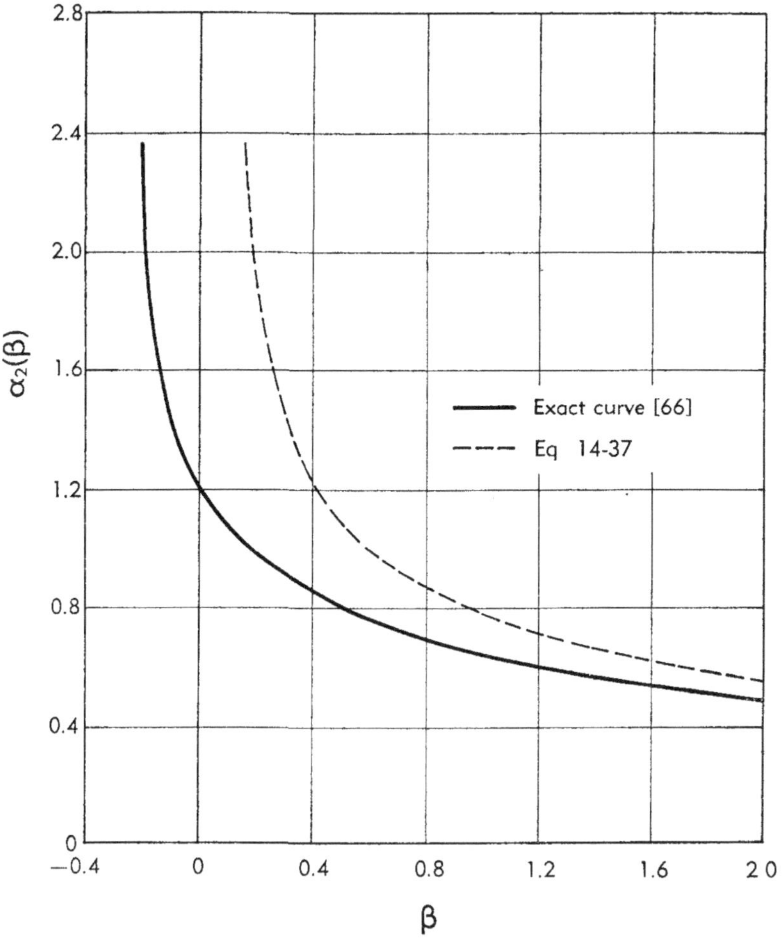

Fig. B,14c. Variation of displacement thickness with wedge angle.

Further Discussion of the Boundary Layer Approximation.

Methods for solving general boundary layer equations. Many methods have been developed for solving the boundary layer equations for flow with an arbitrary pressure gradient. The most useful of these methods is probably the momentum method which will be discussed in Art. 31. Space does not permit a discussion of the other methods; for these the reader is referred to [*24*,*26*] and the references given there.

Boundary layer theory of Burgers' linearized equations. The boundary layer approximation to Eq. 12-5 may be obtained by replacing

$$u\frac{\partial u}{\partial x} + v\frac{\partial u}{\partial y}$$

in Eq. 13-2a by

$$(u_{ew})\frac{\partial u}{\partial x} - \left(y\frac{\partial u_{ew}}{\partial x}\right)\frac{\partial u}{\partial y}$$

Thus the assumed transport velocity in the x direction is the external velocity rather than the actual local value of u. The transport velocity in the y direction is obtained from u_{ew} with the aid of the continuity equation. Since the assumed transport is stronger than the real one, a solution of these linearized boundary layer equations is expected to overestimate the skin friction.

The linearized equations may be used as a mathematical model to obtain qualitative information about real boundary layers. One may easily find the general solution of the linearized boundary layer equations with an arbitrary pressure gradient by transforming these equations into the heat equation (for a different application of such a transformation, cf. [*24*, p. 609]). This transformation may be achieved in many ways. The method used below is essentially due to Burgers [*57*].

We write the linearized boundary layer equations in the form corresponding to Eq. 13-10a, thereby choosing as conformal coordinates φ_e and ψ_e, i.e. the potential and the stream function of the external flow. The equations then are

$$U\Omega_\xi = \nu\Omega_{\eta\eta} \tag{14-41a}$$

$$\Omega = -J_w\psi_{\eta\eta} \tag{14-41b}$$

where

$$U(\xi + i\eta) = \varphi_e + i\psi_e \tag{14-41c}$$

$$J_w = \frac{2}{\rho U^2}(p^0 - p) \tag{14-41d}$$

Here p^0 is the constant stagnation pressure of the external flow: p is the static pressure at the wall and is a function of ξ.

Eq. 14-41 may be interpreted as a conservation law which states that, in the ξ, η plane, vorticity is transported in the ξ direction with velocity U and is conducted in the η direction with a coefficient of conductivity ν. The amount A being transported through the line ξ = const is then (cf. Eq. 14-41b)

$$A(\xi) = \int_0^\infty U\Omega d\eta = -U^2J_w(\xi) \tag{14-42}$$

This vorticity must come from the wall downstream of ξ. At the point of impingement of the free stream at the body ($\xi = 0$) there is then a vorticity source of strength

$$A(0) = -U^2J_0(0) = -\frac{2}{\rho}[p^0 - p(0)] \tag{14-43a}$$

This value is zero only if $\xi = 0$ is a stagnation point. For $\xi > 0$ there are distributed sources along the wall of strength per unit length (measured in ξ) equal to

$$\frac{dA}{d\xi} = -U^2\frac{dJ_0}{d\xi} = \frac{2}{\rho}\frac{dp}{d\xi} \tag{14-43b}$$

Since Eq. 14-41a is formally identical with the heat equation, its solution may be expressed as an integral involving the fundamental solution and the source strength (cf. Eq. 7-2b). We note here only the result for the skin friction τ_w,

$$\tau_w = 2\sqrt{\frac{\nu}{\pi U\xi}}\,[p^0 - p(0)] - 2\sqrt{\frac{\nu}{\pi U}}\int_0^\xi \frac{dp}{d\xi}\frac{d\xi_1}{\sqrt{\xi - \xi_1}} \tag{14-44}$$

For the semi-infinite flat plate at zero angle of attack the Burgers equations reduce to the Oseen equation. The corresponding skin friction according to boundary layer theory is then

$$\tau_w = \rho U^2\sqrt{\frac{\nu}{\pi U\xi}} \tag{14-45}$$

The ratio of this value to the Blasius value (cf. Eq. 14-28a) is

$$\sqrt{\frac{2}{\pi}}\,\frac{1}{\alpha_1(0)}$$

or approximately 1.7.

Separation. A boundary layer is said to separate at a point where the skin friction at the wall $\tau_w = 0$. It is thereby implied that at this point τ_w changes from a positive to a negative value. When τ_w is negative, $\partial u/\partial y$ is negative at the wall, hence u is negative near the wall. Thus downstream of a point where $\tau_w = 0$ backflow occurs.

There are two important theoretical results regarding separation. Firstly, according to a result of Goldstein [*71*] the boundary layer equations have in general no real solution downstream of a point of separation if the external velocity is directed forward.

Secondly, a boundary layer can support only a very weak adverse pressure gradient without separating. In particular, if the external flow has a rear stagnation point, the boundary layer must separate before this point. The following argument due to S. Kaplun supports this statement. Consider first Burgers' linearized boundary layer equations. The total vorticity supplied at the wall up to the rear stagnation point is zero (Eq. 14-43a). An analogy with heat flow then makes it intuitively plausible that τ_w becomes negative somewhere before the rear stagnation point. This can also be seen formally from Eq. 14-44. Note now that the linearized equations overestimate the skin friction. Hence whenever the linearized boundary layer equations show that $\tau_w = 0$ at $\xi = \xi_1$, the

τ_w computed from the corresponding nonlinear boundary layer equations must be zero at some $\xi_2 < \xi_1$.

Consider now a finite body, the trailing edge of which is not a cusp. Potential flow past this body then has a rear stagnation point. If this flow were the external flow in the sense of Eq. 13-5a, then it should be possible to solve the boundary layer equations up to the rear stagnation point with the given potential pressure gradient. However, as shown above, τ_w vanishes before the rear stagnation point and the result of Goldstein makes it implausible that the boundary layer solution can be continued beyond the point where $\tau_w = 0$. Now, assuming that boundary layer theory is the correct perturbation method, it follows that the wrong external flow has been assumed.

Nature of external flow. Streamlined bodies. On the basis of the above results about boundary layer separation we shall now reconsider the

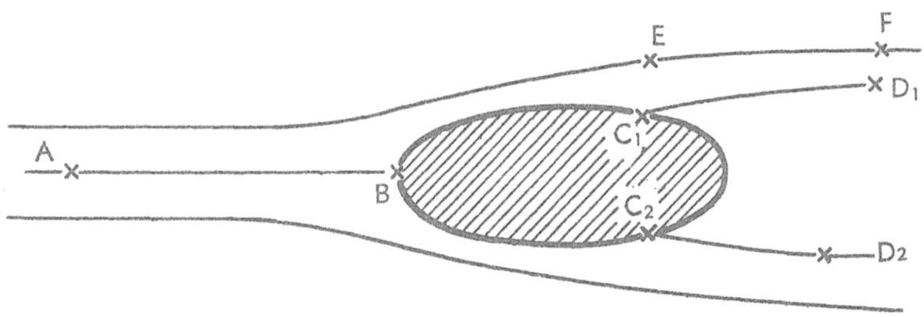

Fig. B,14d. External flow past a blunt body.

external flow. As ν tends to zero the flow pattern past a blunt body tends to a limiting configuration which is qualitatively shown in Fig. B,14d.

The stream function of the external flow ψ_e satisfies the Euler equation. The Euler equations for flow past a body have many solutions, one being that for potential flow. Among these possible solutions ψ_e is the one that is the limit of a solution of the Navier-Stokes equations. As pointed out, ψ_e is not the potential solution.

The qualitative picture of the streamlines of the external flow is then as shown in Fig. B,14d. There is a streamline AB which comes from upstream infinity and at the stagnation point B splits into two streamlines BC_1D_1 and BC_2D_2. Consider the upper one of these two streamlines. It leaves the body surface at a point C_1 which is called the separation point of the external flow. A corresponding separation point C_2 occurs on the lower streamline. The region between the streamlines C_1D_1 and C_2D_2 is called the wake of the external flow. Any streamline outside the wake comes from upstream infinity where the vorticity is zero. Since vorticity is constant along every streamline in Euler flow it follows then that the entire region outside the wake is irrotational. On the other hand it can be shown that vorticity must occur in the wake or at least at the boundary of the wake.

The point of separation of the external flow (C_1 or C_2) is in principle defined in a different way than the point of separation according to boundary layer theory ($\tau_w = 0$). Goldstein's result shows that τ_w cannot become zero before C_1. It is sometimes asserted that the two kinds of separation points coalesce, i.e. that τ_w approaches zero at C_1. While this assertion may be correct, as yet it has not been demonstrated in a satisfactory manner.

If the external flow is known, boundary layer theory may be used to find the flow for small values of ν near the surface along BC_1 and BC_2. In addition some perturbation method is needed to find the flow in the wake and in the immediate neighborhoods of C_1 and C_2. Such perturbation methods are not developed as yet.

The problem of finding ψ_e has not been solved in a satisfactory manner. It is often asserted that ψ_e is Kirchhoff flow past the body, i.e. the velocity in the wake is zero and C_1D_1 and C_2D_2 are free streamlines with constant pressure, a discontinuity in velocity, and hence an infinite vorticity. However, the arguments brought forward in favor of this assertion are at the best very incomplete.

It might then seem that boundary layer theory is of very limited applicability since it presupposes a knowledge of ψ_e. Fortunately there exist body shapes for which ψ_e differs only slightly from the corresponding potential stream function for the body. Such bodies are called streamlined. For practical purposes highly streamlined body shapes may be found. The theoretical question whether or not perfectly streamlined finite bodies exist is not settled; at any rate it seems certain that such a body must have a cusped trailing edge.

Furthermore, even for blunt bodies, boundary layer theory may be used to compute the skin friction from an experimentally determined pressure gradient along BC_1 and BC_2.

Time-dependent flows. Above, it was implicitly assumed that the flow was stationary. The situation is, however, further complicated by the instability of a real fluid. Various types of instabilities appear in the wake at relatively low Reynolds numbers. At higher Reynolds numbers the boundary layer becomes turbulent, especially under the influence of an adverse pressure gradient. A discussion of turbulent flow is beyond the scope of this volume. Some experimental results regarding flow in the wake are given in Art. 16, and the theory of laminar stability is discussed in Sec. G.

Boundary layers may also be nonstationary owing to time dependence of the external flow (or the motion of the body). Examples include the early calculations of the development of the boundary layer on a body whose motion starts from rest ([*7*] and, especially, [*26*, p. 168ff.]); Stewartson's more recent analysis [*72*] of the semi-infinite flat plate impulsively started from rest; and the effects on a boundary layer of slight oscillations of the external flow [*73*]. These, and other problems of the time-dependent boundary layer are discussed in Sec. D.

B,15. Wakes. Jets. Viscous Layers on Slender Bodies.

The flow phenomena to be discussed below have certain characteristics in common with boundary layer flow. At the same time most of them show important similarities with Oseen flow. As everywhere else in this section, only laminar flow will be considered. It must then be kept in mind that the direct practical applicability of the results will be limited. The flow regions to be considered will very often be turbulent in real flow.

Wakes. It was pointed out in Art. 11 that perturbation flow in the laminar wake far downstream of a finite body is described by the fundamental solution multiplied by the drag of the body. In two dimensions the dominant part of the fundamental solution far downstream is given by Eq. 11-18. This equation may be further simplified by assuming that $y \ll x$. This assumption is justified by the parabolic shape of the wake. Thus we may write for the velocity profile far downstream of a two-dimensional body of drag D

$$u = U + u' = U - \frac{D}{2\rho}\sqrt{\frac{1}{\pi\nu Ux}}\, e^{-\frac{y^2}{4\nu x}} \tag{15-1}$$

The corresponding value of v is obtained from the continuity equation or Eq. 11-18b.

For the wake behind a three-dimensional body of drag D we obtain from Eq. 11-23a

$$u = U + u' = U - \frac{D}{4\pi\mu x}\, e^{-\frac{r^2}{4\nu x}} \tag{15-2}$$

These formulas may be obtained directly as follows. Far downstream the wake has certain boundary layer characteristics: The pressure variation p_x and the viscous volume force νu_{xx} in the x direction may be neglected in the first momentum equation. At the same time this equation may be linearized as in Oseen flow since far downstream the perturbation u' is small compared to U. The equation and the boundary conditions for u' are then

$$Uu'_x = \nu u'_{yy} \tag{15-3a}$$

$$u'(0, y) = u'(x, \infty) = 0 \tag{15-3b}$$

$$\left(\frac{\partial u'}{\partial y}\right) = 0 \quad \text{at} \quad y = 0 \tag{15-3c}$$

Since the above equations are homogeneous the solution is determined only within a multiplicative constant. This constant may be determined with the aid of the momentum theorem for drag (Eq. 4-13)[18] and the solution obtained is Eq. 15-1.

In three dimensions Eq. 15-3 is replaced by

$$Uu'_x = \nu\left(u'_{rr} + \frac{1}{r}u'_r\right) \tag{15-4a}$$

$$u'(0, r) = u'(x, \infty) = 0 \tag{15-4b}$$

$$\left(\frac{\partial u'}{\partial r}\right) = 0 \quad \text{at} \quad r = 0 \tag{15-4c}$$

These equations are solved by Eq. 15-2.

Note that even if the original body does not have rotational symmetry about the x axis the wake far downstream tends to become rotationally symmetric. The effective axis of symmetry cannot at present be determined in advance by theoretical methods if the body is of irregular shape. The interpretation of Eq. 15-2 is then that this equation is valid for a suitable chosen axis. In two-dimensional flow the center axis of the wake may also be displaced by downwash due to lift. Similar remarks apply about the location of $x = 0$. In other words, let us assume that the value of u is measured experimentally far downstream of a body and that the flow is approximately steady there. It should then be possible to find an effective origin, $x = 0$ and $y = 0$ (or $r = 0$) such that the measured data agree with Eq. 15-1 or 15-2.

Jets. Solutions for two-dimensional and three-dimensional jets are treated in detail [*24*,*26*]. Only a few comments about the meaning of the solution are made here. To fix the ideas consider the two-dimensional case. Let a jet be emitted through an opening of width L with average velocity U at the same pressure and density as the surrounding air, which is assumed to be at rest except for the action of the jet. Let x measure the distance downstream from the exit. We consider the region

$$\frac{x}{L}, \frac{1}{L}\sqrt{\frac{\nu x}{U}} \gg 1 \tag{15-5}$$

In this region pressure variations and viscous diffusion in the x direction are negligible. The equations of motion are then

$$uu_x + vu_y = \nu u_{yy} \tag{15-6a}$$

$$u_x + v_y = 0 \tag{15-6b}$$

[18] In this connection it should be remembered that the dominant velocity perturbation in the wake is a transversal wave. In Eq. 4-13 one may then put $p' = 0$ (cf. discussion of Eq. 11-19 and 11-20).

Eq. 15-4b and 15-4c are used as boundary conditions. The wake problem is a linearized jet problem. In both cases the flow of momentum through a plane x = const is constant. In the present case we put

$$\int_{-\infty}^{\infty} u^2 dy = A = \text{const} \tag{15-7}$$

A is no longer determined by the drag, but from the outflow at the jet orifice. If there are pressure variations near the orifice A must be determined further downstream.

The solution of the jet problem is

$$u = \frac{2}{3}\alpha^2 x^{-\frac{1}{3}}\left[1 - \tanh^2\left(\frac{\alpha y}{3\sqrt{\nu}\, x^{\frac{2}{3}}}\right)\right] \tag{15-8}$$

where

$$\alpha = \left(\frac{9A}{16\sqrt{\nu}}\right)^{\frac{1}{3}}$$

Denoting the mass flow through any station x = const by B one finds

$$B \cong 3.30(A\nu x)^{\frac{1}{3}} \tag{15-9}$$

Formally the solution (Eq. 15-8) describes a jet that issues from a singular point $x = 0$, $y = 0$. The comments made about the effective (or apparent) origin of the wake apply also to the jet. The solution is of course only valid far downstream of both the real and the apparent orifice of the jet. One may also define an apparent exit velocity U_e from the momentum flow far downstream by

$$A = LU_e^2 \tag{15-10}$$

Eq. 15-6a and 15-6b are the same as the equations for the boundary layer on a flat plate. The interpretation of their validity is, however, different in the two cases. The domain of validity of the boundary layer equations is increased as ν tends to zero, other parameters being fixed. However, the opposite is true for the wake equations. As an example, the mass flow as given by Eq. 15-9 would tend to zero with ν which is physically impossible. As shown by Eq. 15-5 x must be increasingly larger as ν decreases. For ν small, or rather the Reynolds number of the orifice UL/ν large, there are two separate shear layers at each edge of the jet. The thickness of these layers grows as $\sqrt{\nu x/U}$. (This flow is actually of the boundary layer type.) For the jet equations to be valid these two shear layers must be completely merged, which requires that their thickness be much larger than the width of the jet orifice, i.e. the parameter $(1/L)\sqrt{\nu x/U}$ must be small.

The three-dimensional jet can be discussed in a similar manner. For details, see the references given above.

If the jet is initially surrounded by a uniform stream of velocity U_1, the difference between the velocity u in the jet and U becomes eventually small compared to U_1. The momentum equation can be linearized and the wake equation (Eq. 15-3 or 15-4) is obtained. Linearization is not possible only when $U_1 = 0$, in which case there is no reference velocity to gauge the smallness of u.

Slender bodies in viscous flow. We shall first consider the infinite circular cylinder of constant radius R which at $t = 0$ starts sliding along its axis with velocity U. The exact solution is given by Eq. 7-14. An approximation valid for small values of time is given by Eq. 7-16a. The first term of this expansion is the solution according to boundary layer theory. It may be obtained by the formal procedure discussed earlier (cf. Eq. 13-4ff.). The coordinate $\bar{r}$ (Eq. 7-16b) is introduced into Eq. 7-11a whereupon one lets ν tend to zero. The term $(\nu/r)u_r$ then vanishes (note that v is zero since $R =$ const) and one finds the boundary layer equation

$$u_t = \nu u_{rr} \tag{15-11}$$

The equations and the boundary conditions are then formally identical with Rayleigh's problem for the flat plate and the first term of Eq. 7-16a is indeed the Rayleigh solution as given by Eq. 7-3.

The term $(\nu/r)u_r$ describes the part of the viscous force that is due to the transverse curvature of the cylinder. The intuitive reason for neglecting this term is that initially the thickness of the viscous layer, which is $\sim \sqrt{\nu t}$ is much smaller than the radius of curvature, i.e. parameter $\eta = R/2\sqrt{\nu t}$ is large. The consecutive terms of the expansion (Eq. 7-16) represent corrections for the finiteness of R. On the other hand as t becomes very large this expansion fails completely and an expansion for small η is needed (Eq. 7-17a).

The situation has been analyzed in detail in [*32*] for the more general case when $R \sim t^n$, $0 \leqq n \leqq 1$. The boundary layer solution is a valid approximation whenever η is large (this means small values of t for $n < \frac{1}{2}$ and large values of t for $n > \frac{1}{2}$). More generally, an expansion of the high Reynolds number type can be found when η is large. On the other hand, when η is small the flow is of the Oseen type; in particular one may then neglect the nonlinear transport term vu_r but the term $(\nu/r)u_r$ must be retained. The parameter η, which measures the ratio of the instantaneous radius to the viscous length $2\sqrt{\nu t}$, thus plays the role of a Reynolds number. For $n \neq \frac{1}{2}$, η takes on all values between zero and infinity in the interval $0 \leqq t < \infty$. Neither of the two expansions is then uniformly valid in time. For $n = \frac{1}{2}$, $\eta = \sqrt{Re/2} =$ const (Eq. 7-19) and the expansions for η large and small respectively are uniformly valid.

The qualitative ideas introduced above may be applied to stationary flow past a semi-infinite body. In this case the parameter

$$\eta = \frac{L(x)}{\sqrt{\frac{\nu x}{U}}} \tag{15-12}$$

is of fundamental importance. $L(x)$ is a characteristic length of the cross-sectional shape of the body in a plane x = const. The denominator is a viscous length. If L increases with x faster than $\sqrt{x}$, boundary layer theory is applicable far downstream. An example of this is a body with conical afterbody. In the opposite case η tends to zero far downstream and the flow is then of the Oseen type. The second type of viscous flow is often referred to as a boundary layer on a slender body. The term "boundary layer" is, however, somewhat misleading. It has the connotation of a viscous layer of thickness $\sim \sqrt{\nu}$ which is small compared to a relevant geometrical length. However, the thickness of the viscous layer is not of order $\sqrt{\nu}$ and it is not small compared to $L(x)$ as defined above.

As an example, let the afterbody be a circular cylinder of constant radius R whose axis coincides with the free stream direction. As in the case of a boundary layer the pressure gradient and viscous diffusion in the x direction may be neglected. On the other hand the effect of the curvature may not be neglected. The transport term may be linearized as in Oseen flow. Thus the approximate equation for u is

$$Uu_x = \nu\left(u_{rr} + \frac{1}{r}u_r\right) \tag{15-13}$$

This equation is formally identical with Eq. 7-11a for $v = 0$, R = const. The solution is then found from Eq. 7-17a taking the slight change of boundary conditions into account:

$$\frac{u}{U} = 1 - \epsilon\frac{E_1(\xi^2)}{2} + O(\epsilon^2) \tag{15-14a}$$

where now

$$\xi = \frac{r}{2}\sqrt{\frac{U}{\nu x}}, \qquad \eta = \frac{R}{2}\sqrt{\frac{U}{\nu x}} \tag{15-14b}$$

$$\epsilon = -\frac{1}{\log\sqrt{\gamma_0}\,\eta} \tag{15-14c}$$

The skin friction τ_w is given by

$$\frac{R\tau_w}{\mu U} = \epsilon \tag{15-15}$$

For further discussion of this problem, see [*74,75*] and Sec. C. Corrections to boundary layer theory for the upstream part of the cylinder are discussed in [*76,77*]. The methods introduced in [*32*] may also be used for analyzing more general problems.

B,16. Experimental Observations of Viscous Flow past Solid Bodies. A brief discussion of some experimental observations will now be given. It is intended to bring out some characteristic phenomena of incompressible viscous flow. For this purpose a few specific examples such as flow past a cylinder, a sphere, and a flat plate are discussed in some detail. It will be seen that the theory of low Reynolds number flow and also boundary layer theory has been verified experimentally. However, in addition, experiments demonstrate the existence of very many flow phenomena for which no adequate theory exists; this is in particular true for the various types of instabilities occurring in the flow of real fluids.

Before taking up a specific case it is useful to emphasize some very typical and general characteristics of viscous flow past solids. Consider a flow problem characterized by a length L and a velocity U so that a Reynolds number $Re = UL/\nu$ can be assigned. Let the boundary conditions be steady in time. If observations are made for increasing values of Reynolds number one generally encounters three regimes of flow which, depending on the specific problem, are more or less distinct: (1) steady flow, (2) a regime in which regular time-dependent fluctuations occur in the flow, and (3) a regime in which part of the flow is turbulent, some or all of the fluctuations are random in time and space, and the local transfer rates of momentum and matter are increased.

Regimes 2 and 3 have in common the fact that the flow is not steady, even though the boundary conditions are nominally time-independent. They differ in that 2 corresponds to a strictly laminar problem which belongs to the realm of classical viscous flow, while regime 3 (turbulence) does not. Furthermore, the time-dependent fluctuations in 2 can be and are often predominantly two-dimensional disturbances, while turbulence is a strictly three-dimensional motion. A typical example of state 2 is the vortex street behind a cylinder at moderately small values of the Reynolds number.

Flow past a circular cylinder. We first consider the flow past bodies which may be described, roughly, by the term "bluff." For convenience, we start with a specific shape, the circular cylinder. For this case a rather large amount of information is available, both experimental and theoretical (the discussion below owes much to the recent work of Roshko [*78,79,80*]). This example displays most of the important characteristics of the flow, and provides a convenient reference for the later discussion of flows past other cylinder shapes, as well as past three-dimensional bodies.

For the circular cylinder, the different regimes of flow noted above are very distinct. Furthermore the third regime splits into two distinct subregimes. The approximate Reynolds numbers (based on the diameter) which divide these four regimes are indicated by A, B, C in Fig. B,16a. The lower graph of this figure shows the well-known plot of drag coefficient versus Reynolds number, for a two-dimensional circular cylinder. The upper part shows the dimensionless shedding frequency or Strouhal number, to be defined below. The various regimes are now discussed separately.

The lowest Reynolds number for which drag measurements are available is approximately $Re \cong 0.06$ [*82*]. However, there is no reason to suspect that the flow will be of a different nature at still lower Reynolds

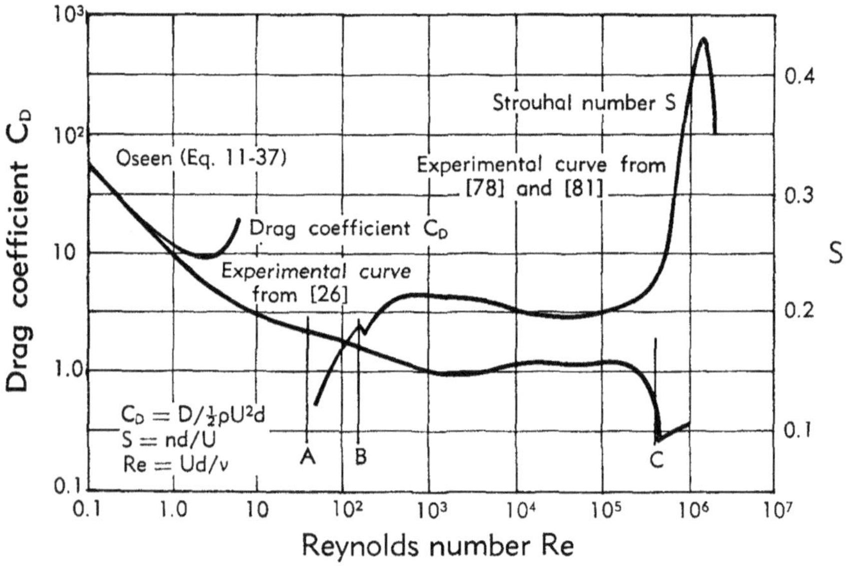

Fig. B,16a. Drag coefficient and Strouhal number of smooth cylinders.

numbers; this idea is supported by measurements on spheres at values as low as $Re \cong 10^{-5}$. The limitation in this direction appears to be only the requirement that the fluid may be considered incompressible (cf. Art. 19).

At the lowest Reynolds numbers measured, the agreement between Oseen theory and the measurements is already good, and it may be expected (again from experience with spheres) that this will be the same at lower Reynolds numbers, and that the drag coefficient will rise rapidly with decreasing Re, as predicted by the theory. Photographs of the flow [*24*] show that behind the cylinder a pair of stationary vortices is formed, symmetrical about the center line. These are very small at first, but increase in size and become elongated downstream as the Reynolds number increases [cf. *55*].

The Oseen theory predicts further, as shown in [*55*], that the total drag of the circular cylinder is divided equally into pressure and friction drags (C_{D_p} and C_{D_f}). Measurements by Thom ([*83*]; cf. also [*24*, p. 425]) in the range $10 < Re < 10^4$ indicate that C_{D_f} is approximately $4/\sqrt{Re}$. In the range $10 < Re < 40$, C_{D_p} has a similar variation but is slightly larger than C_{D_f}. At $Re = 40$ the ratio C_{D_p}/C_{D_f} is approximately 1.5. This ratio decreases with decreasing Reynolds number, and may be assumed to tend to unity for very low values of *Re*.

Summing up, the main characteristics of the flow in this first regime are: (1) the effect of viscosity is most important in the wake but also extends a large distance upstream, (2) the flow is steady, (3) the friction drag and pressure drag are of equal importance. These features are found from the lowest values of *Re* to about $Re = 40$.

The second regime starts at about $Re = 40$ (*A* in Fig. B,16a), where a new feature, the well-known phenomenon of vortex shedding, appears. The eddies, which at slightly lower values of *Re* had become large and elongated, break away from the vicinity of the cylinder and flow downstream in the wake. The breaking away occurs periodically and alternately from either side, so that downstream the vortices arrange themselves in the familiar pattern of the Kármán vortex street. The shedding frequency, *n*, with which the vortices break away from either side may be expressed in dimensionless form by the Strouhal number *S* which is shown in Fig. B,16a. *S* is defined by

$$\text{Strouhal number} = S = \frac{nd}{U} \tag{16-1}$$

where *n* is the shedding frequency and *d* the diameter of the cylinder. In the range between *A* and *B*, *S* increases rapidly with increasing *Re*. The drag coefficient is still decreasing, the effect being due mainly to the decrease of the friction component, C_{D_f}. On the other hand Thom's measurements referred to above show that C_{D_p} no longer decreases with increasing *Re*.

The flow in this range differs from the previous one principally in its nonstationary feature, the alternate breaking away of the vortices. It should also be noted that the wake is unsymmetrical. It is similar to the previous range in that the flow is still nonturbulent and viscous stresses are large. Contrary to what is often stated, the flow does not become turbulent, that is, no transition occurs, even far downstream in the wake; the vortices simply decay by viscous diffusion. This range extends over only a short interval of *Re* from about 40 to 150. It might be pointed out that most of the impressive pictures of vortex streets are made in this range. Recent experiments [*78*] show that the vortices are not necessarily two-dimensional, but that a periodic phase variation may exist parallel to the cylinder axis.

The third regime mentioned in the introduction starts at about

$Re = 150$ when turbulence appears for the first time in the wake. It is intermittent at first; some of the vortices which are shed are purely viscous, while others contain turbulent fluid. By about $Re = 300$ the wake turbulence is well established. The power spectrum of velocity fluctuations in the wake shows a dominant frequency n on a continuous background. This dominant frequency is then the frequency of the vortex shedding, and the Strouhal number can be defined as before [*78*]. At about $Re = 300$ it also becomes possible to consider that the viscous effects on the front of the cylinder are confined to a small region near the surface, and that the flow outside this boundary layer and outside the wake is potential. The boundary layer separates on the front of the cylinder, at about 80° from the stagnation point, and forms free shear layers, one on each side of the wake. A little further downstream these "roll up" and are shed as vortices, alternately and periodically on each side. The actual details of the vortex formation, and the nonstationary flow patterns involved, are not well understood. However, it is known that the transition to turbulence occurs in the free shear layers before the vortices are formed and shed, so that these are composed of turbulent fluid when they pass into the wake. They are consequently diffused more rapidly than was the case in the viscous flow at lower Re, and all trace of the periodicity disappears by 40 or 50 diameters downstream. From there on the wake is fully turbulent, the velocity fluctuations being measurable for over 1000 diameters downstream.

It is sometimes believed that the vortices, when first formed, are approximately potential, with small viscous cores which grow after detachment. Actually, the scale of the motion involved in the rolling up of the free shear layers is large, about the same as the cylinder diameter, and there appears to be considerable mixing of the rotational fluid from the opposite shear layers. The vortices which detach themselves from the mixing region have cores of diameter about equal to the cylinder diameter, and are composed of turbulent fluid.

The characteristics of the vortex formation and shedding in the third regime are essentially unchanged up to a Reynolds number of about 10^5. Fig. B,16a shows that the variations in drag coefficient and Strouhal number are small. C_D varies only from about 0.9 to 1.2, so that the drag is nearly proportional to the square of the velocity. The greater part of the drag is due to pressure. According to the semiempirical formula by Thom given above, which is fairly accurate in this range, the friction drag at $Re = 300$ is still about 20 per cent of the total, but at $Re = 10^5$ it has decreased to only about 1 per cent. An interesting detail is that at about $Re = 10^4$ there is a rather rapid increase in C_D from 0.9 to 1.2. This increasing drag is due principally to an increase in the pressure drag component, and corresponds to an upstream movement of the transition point in the free shear layer [*84*]. When the transition reaches the separation point on the cylinder, the drag coefficient reaches the value 1.2, both transition point and drag coefficient then remaining stationary to about $Re = 10^5$, which is the beginning of the next range. The forward movement of the transition point and increase in drag coefficient can be made to occur more abruptly and at a lower Reynolds number by introducing turbulence into the wind stream, roughening the cylinder, etc.

A distinct subregime starts at $Re \cong 10^5$, the well-known critical Reynolds number, where a rapid decrease of C_D to about the value 0.3 occurs. The point of turbulence transition, which for $10^4 < Re < 10^5$ occurred in the shear layer just downstream of the separation point, now moves ahead of the separation point. The boundary layer becomes turbulent and is able to negotiate a larger adverse pressure gradient before separating. A new equilibrium is established, with the separation point on the back of the cylinder at 120° to 140° from the stagnation point. The resulting improved pressure recovery accounts for the low value of C_D. The drag is still mainly pressure drag, for although the contribution from friction is now a little larger than below the critical *Re*, it is still only about 5 per cent of the total drag.

With separation on the back of the cylinder the free shear layers are concave away from the center line, instead of toward it, and they are much closer together than with separation on the front. Fundamentally, however, the situation is the same, with two shear layers of opposite vorticity, facing each other. They roll up and are shed periodically. Because of the smaller distance between them, the shedding frequency is higher than with separation in front, the Strouhal number being roughly twice as high (Fig. B,16a). No measurements are available for *Re* greater than 2×10^6, but it seems probable that the Strouhal number will decrease, and C_D will increase, as the separation point again moves forward.

Further experiments are found in [*78,79,80*] and references given there.

Other cylinder shapes. The discussion for the circular cylinder is still qualitatively correct for other smooth bluff cylinders. The actual values of the Reynolds number for transition from one range to another are slightly different for different cylinder shapes. Also, the numerical values of the drag coefficients and Strouhal numbers depend on the shape, being related to the "bluffness" of the cylinder [*80*].

The case of a cylinder with a sharp corner or edge, for example a cylinder of triangular cross section (wedge), requires special consideration. At very low Reynolds numbers, that is, when the viscous effects extend far from the cylinder, an edge produces no important change in the flow, as compared to the case of the circular cylinder. This is certainly true at the very low Reynolds numbers in the Stokes range, and appears to be true at higher values, even up to the early stages of the vortex shedding. It is only when the Reynolds number is high enough for the flow to exhibit boundary layer character that the edge becomes important through its effect on the boundary layer separation. On the circular cylinder, the boundary layer separates from a continuous surface, a development which is possible only in an adverse pressure gradient. At an edge, on the other hand, the boundary layer may separate in a favorable gradient,[19] and separation always occurs there, if it has not occurred earlier.

The simplest example of a bluff cylinder with edges is the flat plate perpendicular to the flow. The measurements of C_D and S for this case are not as extensive as those for the circular cylinder, but they bear out the above statements. There are no essential differences in the flow, as compared to the circular cylinder, until a Reynolds number of about 10^6 is reached, where there is no phenomenon corresponding to the critical Reynolds number for the circular cylinder, that is, there is no sudden decrease in the drag. There are two reasons for this: first, it is more difficult for the transition point to move ahead of the separation point than in the case of the circular cylinder; and second, even if the boundary layer on the front of the plate does become turbulent, there can be little effect on the *position* of the separation point, which must still be at the edge of the plate. It seems certain that at a high enough Reynolds number the boundary layer on the front of the plate becomes turbulent, and this may indeed have some effect on the drag coefficient and Strouhal number, but one can hardly expect as abrupt a change as for the circular cylinder. There appear to be no measurements at high enough Reynolds numbers to settle this point.

Three-dimensional bluff bodies. Bluff bodies in three dimensions exhibit the same regimes of flow as two-dimensional bodies. The best-known example is flow past a sphere for which measurements exist over a Reynolds number range from about 10^{-5} to 10^6. Fig. B,16b shows the drag coefficient of spheres versus Reynolds number. The similarity with the cylinder drag is apparent and the various regimes of flow can be equally well identified. At the lowest Reynolds number the drag is very well represented by the formula of Stokes, $C_D = 24/Re$ (Eq. 10-5b), which agrees with the measurements in the range from $Re = 10^{-5}$ up to Reynolds number around unity, a range comparable to a range of validity of Lamb's solution for the circular cylinder. If the Oseen approximation is used to compute the drag of the sphere, the result shows agreement with experiments up to Reynolds numbers between 5 and 10. In general it may be said that, in the whole regime of steady flow, there is no essential difference between flow past a cylinder and flow past spheres or other three-dimensional bodies.

The nonsteady regime, that is regime 2 of the general discussion, starts at a Reynolds number somewhere between 100 and 200. In this regime there exists a significant difference between the flow pattern of a

[19]Admittedly the term "favorable pressure gradient" at a sharp corner may be a matter of definition.

cylinder and a sphere. For the latter the three-dimensionality of the flow becomes important in the shedding of vortices: It is impossible to shed simply alternating vortices from a sphere. The fluctuations in the wake here are due to "vortex loops," inclined with respect to the direction of the motion, which are shed periodically. The detailed motion is far more complicated than for a cylinder and the available experiments are less extensive. In general, the difference between the sphere and the cylinder is due to the distinctly different behavior of vortices in two and three dimensions.

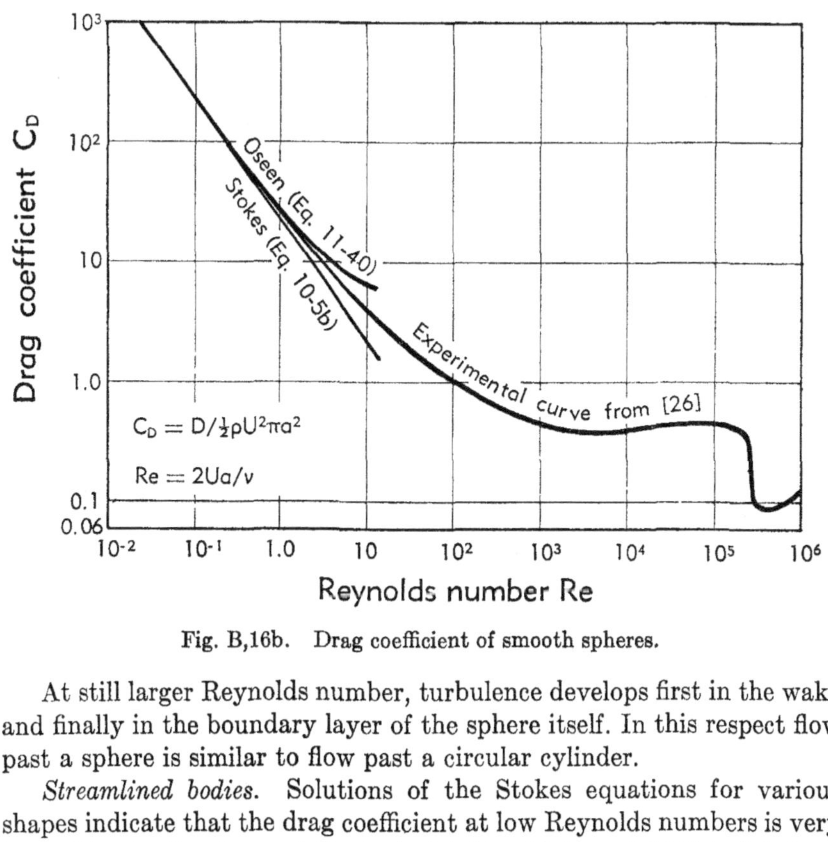

Fig. B,16b. Drag coefficient of smooth spheres.

At still larger Reynolds number, turbulence develops first in the wake and finally in the boundary layer of the sphere itself. In this respect flow past a sphere is similar to flow past a circular cylinder.

Streamlined bodies. Solutions of the Stokes equations for various shapes indicate that the drag coefficient at low Reynolds numbers is very insensitive to the shape of the body (cf. Eq. 10-6a and 10-6b). Solutions of the Oseen equations yield similar results. Experimentally, not much seems to have been done to demonstrate this point, but there is little doubt that it is correct. It appears that when the shape of the body is varied so that, for example, the pressure drag decreases, the skin friction then increases so that the total drag does not vary significantly.

At Reynolds numbers large enough for a boundary layer to exist the situation is different. Boundary layer theory shows that the skin friction decreases with increasing Reynolds number. On the other hand, pressure drag at high Reynolds numbers is nearly independent of Reynolds numbers (except for the significant decrease when the boundary layer becomes turbulent) and is very sensitive to a change in shape. It is very much larger for bluff than for slender bodies (cf. end of Art. 14). Since the variation in pressure drag with shape is not offset by a correspondingly large variation in skin friction, the total drag depends very much on the shape. For example, consider the drag coefficients of a circular disk placed normal (C_{D_1}) and parallel (C_{D_2}) to the stream respectively. The ratio C_{D_1}/C_{D_2} for a Reynolds number near unity is 1.5 while at a Reynolds number of 10^6 it becomes almost 1000.

Laminar separation. Theoretical calculations indicate that even very small positive pressure gradients tend to make a laminar boundary layer separate. This is borne out by experiments. Indeed it is difficult to produce experimentally a positive pressure gradient small enough for a laminar boundary layer to remain unseparated. For practical purposes one may thus say that the laminar boundary layer separates shortly after a point where a minimum occurs in the pressure. The main difficulty in predicting laminar separation lies in the determination of the pressure distribution, which is not known a priori. As an example, if the potential pressure distribution is used, calculations indicate that a laminar boundary layer on a circular cylinder separates at 100° from the stagnation point (cf. [*26*, p. 130]). Measurements indicate that laminar separation takes place at about 81°, a result which can be reproduced by calculations if the measured pressure distribution is used [*42*].

Experiments have therefore been made to test the accuracy of methods of predicting the separation point for a given, that is, measured, pressure distribution; the well-known experiments of Schubauer [*85*] on an elliptic cylinder belong here. The complete problem of laminar separation is actually intimately tied up with the flow past bluff bodies in general. The possibility of sensible and useful measurements has been by no means exhausted. (For a discussion and further references, see [*79,86*]).

A problem in separation which has become increasingly more important in recent times is the problem of local separation with reattachment. In regimes of steep pressure gradients, as near the leading edge of airfoils on sharp edges etc., so-called "separation bubbles" are observed. That is, backflow occurs in a region of small streamwise extent, after which the boundary layer-type flow continues. Detailed experimental studies of this phenomenon are still lacking and it is not even definitely established whether or not transition to turbulent flow has to occur in order to make reattachment possible. Very curious three-dimensional phenomena of this type are observed on swept wings (see, for example [*86*]).

Flow along a flat plate. Local measurements of boundary layer flow. The problem of flow past a flat plate of zero thickness, zero angle of attack, finite chord, and infinite span has been treated at length in the theoretical literature of viscous fluids. As in the case of the circular cylinder or sphere, a theoretical solution is known for sufficiently small Reynolds numbers. Furthermore, the plate is a perfectly streamlined body; i.e. separation does not occur. An accurate boundary layer solution can therefore be obtained for sufficiently large Reynolds numbers without recourse to empirical information about the pressure distribution over the body. Several experimental investigations made on plates whose geometry is close to that assumed in the theory are mentioned below. In addition to over-all drag measurements at relatively small Reynolds numbers, careful studies of local skin friction and velocity profile have been made at Reynolds numbers for which the boundary layer theory might be expected to apply.

For the plate, as for bluff bodies, three general Reynolds number regimes may be distinguished, depending on the stability of the flow. The first regime, that of steady laminar flow, extends to $Re = 10^5$, approximately. The corresponding Reynolds number based on displacement thickness (Re_{δ^*}) is about 500. The second regime is characterized by nonsteady laminar oscillations in the boundary layer. This regime, which may be observed in the range $500 < Re_{\delta^*} < 3000$, was postulated on theoretical grounds by Tollmien and others and was first observed experimentally by Schubauer and Skramstad *[87]*. The third regime, that of turbulent flow, begins at some value of Re_{δ^*} between 500 and 3000 depending on surface roughness, the free stream turbulence level, and other factors. Only the steady flow regime is discussed here since laminar stability, transition, and turbulence are treated in other sections of this volume.

Before discussing the measurements a summary of some theoretical results is given for the flow past a flat plate. A solution valid in the limit of vanishing Reynolds number was derived in Art. 11 (Eq. 11-28ff.). The method involved approximating to the Oseen fundamental solution by terms which dominate near the origin. The corresponding flow field of the fundamental solution is then symmetric about the line $x = \xi$ where $(\xi, 0)$ is the location of the singularity. As a result, this method of solution gives a flow field which is symmetric about the center of the plate. In particular, the skin friction is symmetric; near the edges it varies as $1/\sqrt{x}$ where x is the distance from the leading or the trailing edge. The symmetry is ultimately due to the fact that the effect of transport is entirely neglected compared to the effect of viscous diffusion. The solution in question is therefore a Stokes solution (cf. Art. 17). For increasing Reynolds numbers the local skin friction becomes increasingly asymmetrical; its minimum shifts toward the trailing edge of the plate. This shift, which can be seen in the solutions of the Oseen equations obtained by Tomotika and Aoi *[88]* is a consequence of the asymmetry which appears in the fundamental solution when transport is taken into account. Numerical solutions of the Navier-Stokes equations by Janssen [*89*] for $Re = 0.1$, 1.0, and 10.0 show the same trend. For large Reynolds numbers the minimum of the skin friction must occur very close to the trailing edge. According to boundary layer theory the skin friction decreases as $1/\sqrt{x}$ everywhere, where x is the distance from the leading edge. No singularity is predicted at the trailing edge inasmuch as there is no mechanism in boundary layer theory for upstream spreading of a disturbance. Hence the skin friction on a finite plate is found to be the same as that on the corresponding portion of a semi-infinite plate. However, if correction terms to the boundary layer solution (cf. Art. 17) are taken into consideration, it is seen that the skin friction is infinite at the trailing edge

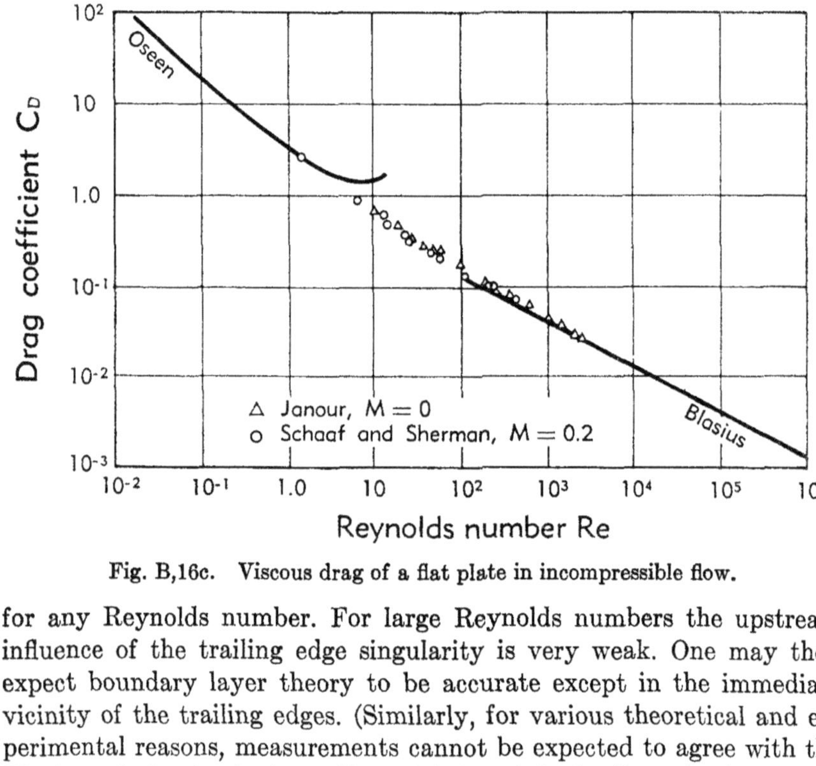

Fig. B,16c. Viscous drag of a flat plate in incompressible flow.

for any Reynolds number. For large Reynolds numbers the upstream influence of the trailing edge singularity is very weak. One may then expect boundary layer theory to be accurate except in the immediate vicinity of the trailing edges. (Similarly, for various theoretical and experimental reasons, measurements cannot be expected to agree with the Blasius solution in the immediate vicinity of the leading edge.)

The over-all drag coefficient C_F for one side of a finite plate is plotted in Fig. B,16c. The theoretical value of C_F for small Reynolds numbers is one-half of the value given by Eq. 11-32b. The theoretical curve for large Reynolds numbers is obtained by integrating the Blasius expression (Eq. 14-28a) for local skin friction over one side of the plate. The over-all drag coefficient is found to be $C_F = 1.328/\sqrt{Re}$. Included in Fig. B,16c are some measurements by Janour [*90*] of viscous drag in the intermediate range $10 < Re < 2000$. These data were obtained in a water channel and corrected for end effects. Fig. B,16c also shows some drag measurements by Schaaf and Sherman [*91*] in air at subsonic speeds, using finite plates of aspect ratio 1, 2, and 5. Comparison with Janour's data does not suggest that the effect of slip, of finite aspect ratio, or of departure from continuum flow is important for the conditions of these experiments in air.

The available drag measurements confirm the supposition that the Oseen formula should apply for $Re < 1$. In addition, it appears that the boundary layer calculation of total drag becomes accurate for finite plates at Reynolds numbers of perhaps 10^4. For $Re = 10^3$ the contribution of the trailing edge singularity to the total drag is appreciable according to

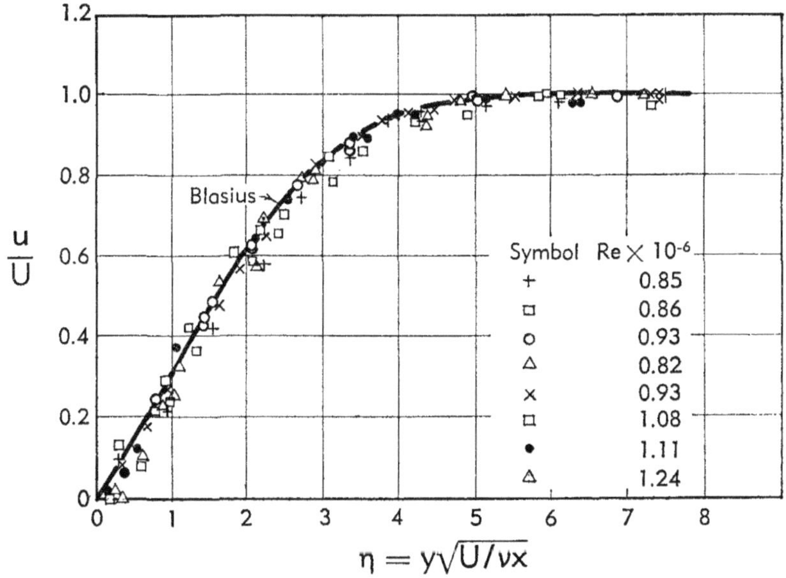

Fig. B,16d. Experimental verification of the Blasius profile [*93*].

the measurements cited in Fig. B,16c. It is possible, however, that the velocity profile, and hence the local skin friction, are given correctly by the Blasius calculation at Reynolds numbers appreciably smaller than 10^4, provided that the station in question is not near the trailing edge.

For Reynolds numbers in the range of boundary layer theory the velocity profile calculated from the Blasius equation has been verified experimentally. Poor agreement in early experiments established the importance of maintaining constant pressure along the plate, and of minimizing effects of finite thickness. Beginning with measurements by Dryden [*92*] and Liepmann [*93,94*], the Blasius profile has been confirmed by several observers. Fig. B,16d shows some typical data from [*93*].

The local friction coefficient $c_f = 2\tau_w/\rho U^2$ for the flat plate boundary layer is half of the mean coefficient C_F. Liepmann and Dhawan [*95*] have verified the theoretical values of c_f by two independent methods. One method involves the differentiation of the observed velocity profile to obtain $\tau_w = \mu_w(\partial u/\partial y)_w$. The other involves measurement of τ_w by the floating-element technique in which the tangential force of a small isolated surface element is observed directly. Fig. B,16e shows the agreement between calculated and measured values of c_f in the range $60{,}000 < Re < 600{,}000$.

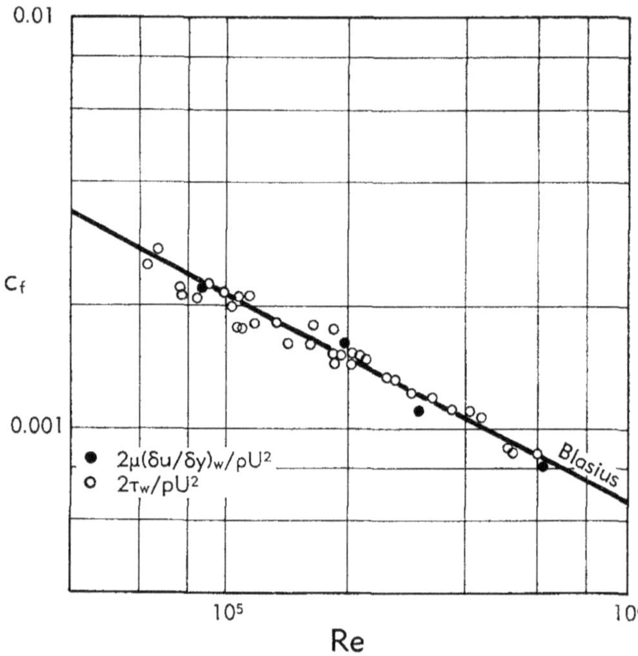

Fig. B,16e. Local skin friction on flat plate in incompressible flow [95].

B,17. Asymptotic Expansions of the Solutions of the Navier-Stokes Equations. The mathematical theory of the Navier-Stokes equations is at present in a highly unsatisfactory state. The principal difficulty is of course due to the nonlinearity of the equations. The situation is further aggravated by the fact that the equations are of fourth order. Furthermore, in many interesting applications they have to be solved for an infinite domain. Not only is the number of exact solutions very small but our knowledge of the general nature of the solutions is very limited. Even such basic questions as those of existence, uniqueness, and boundedness of the solutions have been settled only for very special cases.

We shall only discuss one general mathematical problem, namely the nature of the asymptotic expansions of the solutions for very large or very small Reynolds numbers. It must of course be remembered that many mathematical assertions to be made below cannot be proved rigorously but can only be made plausible on the basis of various physical and mathematical considerations.

The case of high Reynolds numbers is discussed first. The methods used there will then, with suitable modifications, be applied to the case of low Reynolds numbers. In each case two limit processes, an outer and an inner limit, play a fundamental role. The methods used belong to the theory of singular perturbation problems (see in particular [*60*]). Some of the ideas used in the low Reynolds number case are due to Kaplun. For further examples of the use of limit processes, see e.g. [*32,96*].

In general we shall deal only with stationary flow past a solid. The notation and the concepts introduced in Art. 9 are used.

Expansions for High Reynolds Numbers.

General remarks. In Art. 13 we mentioned several basic difficulties connected with the problem of finding approximate solutions of the Navier-Stokes equations for small values of ν. These difficulties are summarized below.

1. The perturbation problem is singular. This follows from the fact that the reduced equations, i.e. the equations obtained by putting $\nu = 0$ are the Euler equations which are of lower order than the Navier-Stokes equations. As a consequence, not all the boundary conditions imposed on the latter equations can be imposed on the former.
2. The solutions of the Euler equations are not unique, and in general it is difficult to select the solution which is the limit of a solution of the Navier-Stokes equations. In a boundary value problem, for the Euler equations one must give up the no-slip condition in order not to make the problem overdetermined. However, the problem then becomes underdetermined, and some condition (or conditions) must be found to make the solution unique (cf. the use of the Kutta condition in wing theory). This condition must be chosen so that the resulting solution is the limit of an exact solution for ν tending to zero.
3. For ν sufficiently small the flow becomes unstable. Hence the solution is time-dependent even for steady boundary conditions.

To begin with, we shall disregard the last two difficulties and consider the form of the expansion for the case when the external flow is the potential flow past the solid. In order to illustrate the ideas we shall discuss a mathematical model, namely the solution of the Oseen equations for flow past a parabola.

Boundary layer solution of the Oseen equations. For high Reynolds numbers the Oseen equations are not a valid approximation to the Navier-Stokes equations. However, one may formally consider solutions of the former equations and their expansions for high Reynolds numbers. The results may be used as a mathematical model in the study of the latter equations, provided that one keeps the differences between the two cases duly in mind. The following difference is of particular importance: In Oseen flow the boundary layer is of thickness ν, rather than $\sqrt{\nu}$, whenever the body is not tangent to the free stream direction.

The mathematical basis for the statement just made was discovered by Latta [*60*]. The full explanation cannot be given here. However, the criterion is the following: Consider any system of equations occurring in fluid dynamics which involves viscosity. Let the solution have a discontinuity along a line (or surface) L as ν tends to zero. For ν small but greater than zero this discontinuity is replaced by a rapid transition layer. If L is a characteristic line of the corresponding reduced equations (obtained by putting $\nu = 0$), the transition layer is of thickness $\sqrt{\nu}$; at any place where L does not have a characteristic direction the transition layer is of thickness ν. The streamlines are characteristics of the reduced Navier-Stokes equations, i.e. of the Euler equations. In this case boundary layers and shear layers are of thickness $\sqrt{\nu}$. On the other hand, shock discontinuities in compressible fluids occur along lines which are not characteristic lines. Accordingly, as discussed in the following chapter, the corresponding shock discontinuities are of thickness ν. The reduced Oseen equations are the linearized Euler equations whose characteristics are the streamlines of the undisturbed flow. As ν tends to zero, discontinuities occur along a surface of the solid. It then follows that the Oseen boundary layer on a flat plate of zero thickness and zero angle of attack is of thickness $\sqrt{\nu}$. On the other hand, for a body whose boundary is not tangent to the free stream direction, such as a parabola, the boundary layer is of thickness ν. Intuitively speaking, according to the Oseen equations, the vorticity generated at the boundary of, say, a parabola is transported into the body rather than along the body (cf. Art. 13). This is the reason why the boundary layer is thinner, i.e. $\sim \nu$ rather than $\sim \sqrt{\nu}$, in this case.

Incidentally, the fact that the Oseen boundary layer in the cases mentioned is of thickness ν shows clearly that the Oseen equations are not valid as a high Reynolds number approximation to the Navier-Stokes equations. This was not realized by Oseen (cf. [*25*]).

Oseen flow past a parabola. We shall now discuss a concrete example, namely the Oseen solution for flow past a parabola, and its high Reynolds number expansion. The advantage of this example is that a comparatively simple exact solution[20] exists.

Let ξ and η be parabolic coordinates defined by

$$z = \zeta^2; \qquad z = x + iy; \qquad \zeta = \xi + i\eta \tag{17-1}$$

If η_0 is a constant then the curve $2\eta_0^2 = \sqrt{x^2 + y^2} - x$ is a parabola; we

[20] This solution was communicated to the author by S. Kaplun. It has, however, been found independently by various other investigators.

shall consider Oseen flow past this parabola. The constant η_0^2 is used as the characteristic length; the Reynolds number is then

$$Re = \frac{U\eta_0^2}{\nu} \tag{17-2}$$

Nondimensional coordinates are defined by

$$\xi^* = \frac{\xi}{\eta_0}, \qquad \eta^* = \frac{\eta}{\eta_0}, \qquad \sigma^* = \frac{\eta - \eta_0}{\eta_0}, \qquad \bar{\sigma} = \sigma^* Re \tag{17-3}$$

As discussed in Art. 11 the Oseen solution may be decomposed into a longitudinal and transversal wave with stream functions ψ_L and ψ_T respectively. We denote the nondimensional stream functions by a star,

$$\psi^* = \frac{\psi}{2U\eta_0^2} \quad \text{etc.} \tag{17-4}$$

The decomposition theorem then states that

$$\psi^* = \psi_L^* + \psi_T^* \tag{17-5}$$

For Oseen flow past the parabola $\eta = \eta_0$ the solution for ψ_L^* and ψ_T^* is

$$\psi_L^* = \xi^*\left(\eta^* - \frac{e^{-Re}}{\sqrt{\pi\, Re}\,\operatorname{erfc}\sqrt{Re}}\right) \cong \xi^*(\eta^* - 1) - \frac{1}{Re}\frac{\xi^*}{2} + \left(\frac{1}{Re}\right)^2\frac{\xi^*}{2} + \cdots \tag{17-6a}$$

$$\psi_T^* = \xi^*\frac{\frac{1}{\sqrt{\pi}}e^{-Re\eta^{*2}} - \sqrt{Re}\,\eta^*\operatorname{erfc}(\sqrt{Re}\,\eta^*)}{\sqrt{Re}\,\operatorname{erfc}\sqrt{Re}}$$
$$\cong \xi^*\left[\frac{1}{Re}\frac{e^{-2\bar{\sigma}}}{2} - \left(\frac{1}{Re}\right)^2\frac{e^{-2\bar{\sigma}}}{2}(\bar{\sigma} - 1)^2 + \left(\frac{1}{Re}\right)^3(\bar{\sigma}^4 - 2\bar{\sigma}^3 + 8\bar{\sigma}^2 + 10\bar{\sigma} + 5)\frac{e^{-2\bar{\sigma}}}{4}\right] + \cdots \tag{17-6b}$$

The asymptotic expansions for large Re are given above; they are found from the asymptotic expansion of the complementary error function (cf. [*29,30*]).

Combining the expansions for ψ_L^* and ψ_T^* one thus sees that ψ^* has a composite asymptotic expansion of the form

$$\psi^* \cong \sum_{j=0}^{\infty}\left(\frac{1}{Re}\right)^j\left[f_j(\xi^*, \eta^*) + \frac{1}{Re}g_j(\xi^*, \bar{\sigma})\right] \tag{17-7}$$

It can be verified that the nth partial sum ψ_n^* has the property

$$\left|\frac{\partial\psi^*}{\partial x_i^*} - \frac{\partial\psi_n^*}{\partial x_i^*}\right| < M_n(Re)^{-(n+1)} \tag{17-8}$$

where $x_i^* = \xi^*$ or η^*, and M_n is independent of ξ^* and η^*. Thus the velocity field associated with ψ_n^* is a uniformly valid approximation of order Re^{-n}.

As in Art. 13 (Eq. 13-4 and 13-7) we define an outer (or first) limit for points off the boundary by

$$\lim_e f = \lim_{Re\to\infty;\,\xi^*,\eta^*\text{fixed}} f \tag{17-9a}$$

and an inner (or second) limit by

$$\lim_\chi f = \lim_{Re\to\infty;\,\xi^*,\bar{\sigma}\text{fixed}} f \tag{17-9b}$$

Here χ stands for the system of coordinates (ξ^*, σ^*).

The transversal component is transcendentally small for any point off the boundary in the sense that

$$\lim_e (Re^n \psi_T^*) = 0, \qquad \eta^* > 1, \quad n \text{ arbitrary} \tag{17-10}$$

Thus the outer expansion of ψ^*, obtained by a repeated application of the first limit process for $\eta^* > 1$, is identical with expansion (Eq. 17-6a) of ψ_L^*, i.e.

$$f_0 = \lim_e \psi^* \tag{17-11a}$$

$$f_1 = \lim_e Re\,(\psi^* - f_0) \quad \text{etc.} \tag{17-11b}$$

The outer expansion of ψ^* is not a uniformly valid approximation to ψ^* near the boundary. For this reason the correction terms $(1/Re)g_j$ are needed in the expansion (Eq. 17-7). These are obtained by a repeated application of the second limit process to $\psi^* - \psi_L^* = \psi_T^*$. It is thereby not necessary to subtract out the entire ψ_L^* but only an appropriate part of its expansion

$$\lim_\chi Re\left[\psi^* - \left(f_0 + \frac{1}{Re} f_1\right)\right] = g_0 \tag{17-12a}$$

$$\lim_\chi Re^2\left[\left(\psi^* - f_0 - \frac{1}{Re} f_1 - \frac{1}{Re} f_2\right) - \frac{1}{Re} g_0\right] = g_1 \quad \text{etc.} \tag{17-12b}$$

If one instead applies the second limit process directly to ψ^* one finds

$$\lim_\chi \psi^* = 0 \tag{17-13a}$$

$$\lim_\chi Re\,\psi^* = \xi^*\bar{\sigma} - \frac{\xi^*}{2} + \frac{e^{-2\bar{\sigma}}}{2} = Re\,f_0 + f_1 + g_0 = h_1 \tag{17-13b}$$

$$\lim_\chi Re^2\left(\psi^* - \frac{1}{Re} h_1\right) = \frac{\xi^*}{2} - \frac{e^{-2\bar{\sigma}}}{2} (\bar{\sigma}^2 - 1)^2 = f_2 + g_1 = h_2 \tag{17-13c}$$

Repeated application of the inner limit to ψ^* gives the inner expansion

$$\psi^* = \sum_{j=1}^{\infty} \left(\frac{1}{Re}\right)^j h_j(\xi, \bar{\sigma}) \tag{17-14}$$

where h_1 is defined above and $h_j = f_j + g_{j-1}$ for $j > 1$. In the present case the inner expansion is thus identical with a simple regrouping of the composite expansion Eq. (17-7). This is due to the special coordinate system used in forming the inner limit (cf. the discussion after Eq. 17-21).

In order to determine the f_j and g_j by solving the appropriate equations rather than by using the exact solution one may substitute the expansion (Eq. 17-7) into the Oseen equations. From the assumed properties of the expansion one finds successively the equations and the boundary conditions for the f_j and the g_j. It is then seen that f_0, $(1/Re)f_1$, and h_1 represent external flow, flow due to displacement thickness, and boundary layer flow, respectively. Some details of this procedure are shown for the nonlinear case which follows.

Expansions of the solutions of the Navier-Stokes equations. Potential external flow. We now return to the problem of expansion of the solution of the Navier-Stokes equations for the case in which the external flow is potential. The basic assumption is made that the stream function ψ has a composite asymptotic expansion[21] analogous to that given by Eq. 17-7, namely

$$\psi \cong \psi_L + \psi_T \tag{17-15a}$$

$$\psi_L \cong \sum_{j=0}^{\infty} (\sqrt{\nu})^j f_j(\lambda, \sigma) \tag{17-15b}$$

$$\psi_T \cong \sum_{j=0}^{\infty} (\sqrt{\nu})^{j+1} g_j(\lambda, \bar{\sigma}) \tag{17-15c}$$

Here $\chi = (\lambda, \sigma)$ is a system of coordinates such that σ vanishes on the boundary, and $\bar{\sigma} = \sigma/\sqrt{\nu}$. The choice of such a system will be discussed later. The equations are written in dimensional form and parameters such as U and L are not exhibited explicitly. They are assumed to be constant in the limit processes to be discussed below; by dimensional analysis it is seen that this convention does not restrict generality.

There are two significant differences between the expansions (Eq. 17-7 and 17-15). (1) In Eq. 17-15, ν has been replaced by $\sqrt{\nu}$. The reason for this was explained above. (2) The splitting theorem (Eq. 11-9) is not valid for the nonlinear Navier-Stokes equation. Hence one cannot expect

[21] The form of this expansion is a slight modification of a form proposed by Latta in [*60*].

that both ψ_L and ψ_T of Eq. 17-15 satisfy the Navier-Stokes equations separately. Actually, if f_0 is potential as assumed, the f_j will also be potential for a large class of problems. In this case ψ_L satisfies the Laplace equation and hence also the Navier-Stokes equations. However, ψ_T is not asymptotic to a solution of the Navier-Stokes equations. It is still convenient to refer to ψ_L and ψ_T as the longitudinal and transversal components of ψ.

It is assumed that, with the exception of the differences mentioned, the general nature of the expansions (Eq. 17-7 and 17-15) is the same. Thus we assume that the components of ψ_T are transcendentally small off the boundary. This implies in particular that each g_j will vanish at $\bar{\sigma} = \infty$. A repeated application of the first limit process (cf. Eq. 13-4, 17-9a, and 17-11) then yields the f_j:

$$f_0 = \lim_e \psi \tag{17-16a}$$

$$f_1 = \lim_e \frac{\psi - f_0}{\sqrt{\nu}} \tag{17-16b}$$

Thus the stream function of the external flow is $\psi_e = f_0$ and that of flow due to displacement thickness (cf. Eq. 13-19) is $\psi_e' = \sqrt{\nu}\, f_1$.

We now discuss the recursive method for successively finding the f_j and the g_j. By inserting the expansion of ψ into the Navier-Stokes equations and applying the first limit one finds that f_0 satisfies the Euler equations. By assumption f_0 is also irrotational and can hence be found in the customary way by imposing all boundary conditions except the no-slip condition.

Below we use the notation

$$u = \frac{\partial \psi}{\partial \sigma}, \qquad v = -\frac{\partial \psi}{\partial \lambda}, \qquad u_e = \frac{\partial \psi_e}{\partial \sigma} \quad \text{etc.} \tag{17-17}$$

A subscript $_w$ indicates values at $\sigma = 0$ and a subscript $_\infty$ values at $\sigma = \infty$. It should be remembered that u and v are now contravariant vector densities rather than components in the sense of vector geometry.

The next step after finding $\psi_e = f_0$ is to find the boundary layer solution ψ_{BL}, and then g, from ψ_{BL}. The boundary layer equations are obtained by applying the second limit to the Navier-Stokes equations ($\bar{\sigma}$ fixed, cf. Eq. 13-7 and Eq. 17-9b). At the wall ψ_{BL} and its first derivatives should vanish. The boundary condition at infinity is found from the relation of ψ_{BL} to the expansion of ψ. By applying the second limit to Eq. 17-15 one finds that

$$\psi_{BL} = \sqrt{\nu} \lim_\chi \frac{\psi}{\sqrt{\nu}} = \sqrt{\nu}\, u_{ew}\bar{\sigma} + \sqrt{\nu}\, f_1(\lambda, 0) + \sqrt{\nu}\, g_0(\lambda, \bar{\sigma}) \tag{17-18a}$$

and hence

$$u_{\mathrm{BL}} = u_{\mathrm{ew}} + \frac{\partial}{\partial \bar{\sigma}} g_0(\lambda, \bar{\sigma}) \tag{17-18b}$$

Eq. 17-18 shows the relation between ψ_{BL} and g_0. From the assumption that g_0 vanishes at $\bar{\sigma} = \infty$ one then finds

$$u_{\mathrm{BL}}(\lambda, \infty) = u_{\mathrm{e}}(\lambda, 0) \tag{17-18c}$$

This boundary condition was previously given as an intuitively plausible condition (Eq. 13-3b). It is now seen to be an exact consequence of the assumed form of the expansion for ψ.

After ψ_{e} and ψ_{BL} have been determined one can find $\psi_{\mathrm{e}}' = f_1$ as follows. By insertion of the expansion (Eq. 17-15) into the Navier-Stokes equations one finds that f_1 satisfies a linearized Euler equation where the transporting velocity is the flow field associated with f_0. Hence vorticity of f_1 is constant along the lines $f_0 = $ const. These lines all come from upstream infinity where the flow is irrotational. Hence f_1 is irrotational and its equation reduces to Laplace's equation.

To find a first boundary condition on f_1 we rewrite Eq. 17-18a as

$$\sqrt{\nu}\, f_1(\lambda, 0) = \int_0^{\sigma} (u_{\mathrm{BL}} - u_{\mathrm{ew}})d\sigma - \sqrt{\nu}\, g_0(\lambda, \bar{\sigma}) \tag{17-18'}$$

Letting $\bar{\sigma}$ tend to infinity, one then finds

$$\psi_{\mathrm{ew}}' = \sqrt{\nu}\, f_1(\lambda, 0) = \int_0^{\infty} (u_{\mathrm{BL}} - u_{\mathrm{ew}})d\sigma \tag{17-19}$$

It is thus seen that Eq. 13-20a is an exact consequence of the assumed form of the expansion of ψ.

The second boundary condition on f_1 should be imposed at infinity. Since f_0 gives the correct flow field at infinity and the flow field associated with g_0 vanishes there, we require that the flow field of f_1 also vanish at infinity. This, however, does not imply that $\partial f_1/\partial\lambda$ and $\partial f_1/\partial\sigma$ vanish at infinity. These derivatives are contravariant vector densities and the metric must be considered. We require only that $|\mathrm{grad}\, f|^2$ tend to zero at infinity or, equivalently, that $\partial f_1/\partial x$ and $\partial f_1/\partial y$ vanish, where x and y are Cartesian coordinates. (Note that in the Oseen solution given by Eq. 17-6a $\partial f_1/\partial\xi^* = -\frac{1}{2} = $ const.)

The equation and the boundary conditions for f_1 have thus been found and in principle f_1 can be determined. Once ψ_{BL}, ψ_{e}, and ψ_{e}' have been found one may determine g_0 from Eq. 17-18a. In this way one determines the first part of the expansion (Eq. 17-15), namely

$$\psi_1 = f_0 + \sqrt{\nu}\, f_1 + \sqrt{\nu}\, g_0 \tag{17-20}$$

In principle one may determine the higher order terms by analogous methods. This is not pursued further here; it is only remarked that it can be made plausible that all the f_i satisfy Laplace's equation (cf. [*32*, p. 871]).

The relation of ψ_1 to ψ_{BL} depends on the choice of the system of coordinates χ, a problem which now will be discussed.

Role of coordinate system in boundary layer theory. Let $\zeta = (\xi, \eta)$ and $\chi = (\lambda, \sigma)$ be two systems of coordinates such that η and σ vanish on the boundary. The second limit process may then be defined with respect to either system. It is denoted by $\lim_\zeta$ and $\lim_\chi$, respectively. The boundary layer approximations obtained in these two systems may be described by two stream functions ψ_ζ and ψ_χ defined by (cf. Eq. 13-9a)

$$\bar{\psi}_\zeta = \lim_\zeta \bar{\psi}_{NS} \tag{17-21a}$$

$$\bar{\psi}_\chi = \lim_\chi \bar{\psi}_{NS} \tag{17-21b}$$

where $\psi_\zeta = \sqrt{\nu}\,\bar{\psi}_\zeta$, $\psi_{NS} = \sqrt{\nu}\,\bar{\psi}_{NS}$ etc.

In general ψ_ζ and ψ_χ are not only formally different, but describe different flow fields. The boundary layer solution for a given flow problem is thus not unique, but depends on the system of coordinates used in the approximation process. The relation between the various solutions needs investigation; one would also like to know what is common to all solutions and whether one solution is in some sense superior to the others. These questions have to a large extent been settled by Kaplun [*96*]; some of his results are now discussed.

The relation between the two stream functions is given by Kaplun's correlation theorem: If

$$\bar{\psi}_\chi = f(\lambda, \bar{\sigma}) \tag{17-22a}$$

then

$$\bar{\psi}_\zeta = f(\lambda_\zeta, \bar{\sigma}_\zeta) \tag{17-22b}$$

where

$$\lambda_\zeta = \lim_\zeta \lambda = \lambda(\xi, 0) \tag{17-22c}$$

$$\bar{\sigma}_\zeta = \lim_\zeta \bar{\sigma} = \left(\frac{\partial\sigma}{\partial\eta}\right)_{\eta=0} \bar{\eta} \tag{17-22d}$$

This correlation formula may be proved by assuming certain continuity properties of the functions concerned, without considering the differential equations. It may also be checked from the boundary layer equations in χ and ζ coordinates respectively.

As an example of the application of the formula, let $\zeta = \xi + i\eta$ be parabolic coordinates with respect to the conformal coordinates $\chi = \lambda + i\sigma$, i.e. $\zeta^2 = \chi$. Then

$$\lambda_\zeta = \xi^2, \qquad \bar{\sigma}_\zeta = 2\xi\bar{\eta} \tag{17-23}$$

Now consider the boundary layer on the semi-infinite flat plate and let $\chi = z = x + iy$ be Cartesian coordinates. According to Eq. 14-18ff. the boundary layer solution in Cartesian coordinates is

$$\bar{\psi}_x = \sqrt{2Ux}\, B\left(\bar{y}\sqrt{\frac{U}{2x}}\right) \tag{17-24a}$$

where B is the Blasius function. From Eq. 17-22b one then finds that the boundary layer solution with respect to parabolic coordinates is (cf. [*96*, p. 131])

$$\bar{\psi}_\zeta = \xi\sqrt{2U}\, B(\sqrt{2U}\,\bar{\eta}) \tag{17-24b}$$

The correlation theorem implies that ψ_ζ and ψ_χ represent the same flow field only if ζ and χ are connected by the relations

$$\lambda(\xi, \eta) = \lambda(\xi, 0) \tag{17-25a}$$

$$\sigma(\xi, \eta) = \left(\frac{\partial\sigma}{\partial\eta}\right)_{\eta=0}\eta \tag{17-25b}$$

or, equivalently,

$$\lambda = f_1(\xi) \tag{17-26a}$$

$$\sigma = f_2(\xi)\eta \tag{17-26b}$$

where f_1 and f_2 are arbitrary functions. It is immediately seen that these relations are not satisfied if ζ is parabolic relative to χ. In particular Eq. 17-24a and 17-24b describe different flow fields.

Invariants. Previously it was pointed out that the flow field obtained by a boundary layer approximation depends on the choice of coordinates. There are, however, many quantities which are invariant. The most important of these is the skin friction. Let τ_w be the skin friction according to the exact solution of the Navier-Stokes equations. The skin friction according to the boundary layer approximation in the ζ system may be defined as (cf. Eq. 13-9)

$$(\tau_w)_\chi = \sqrt{\nu}\, \lim_\chi\left(\frac{\tau_w}{\sqrt{\nu}}\right) \tag{17-27}$$

However, at the wall $\sigma = 0$ and hence keeping σ constant as $\nu \to 0$ is equivalent to keeping $\bar{\sigma} = \sigma/\sqrt{\nu}$ constant. The second limit process is then identical with the first process, which is independent of the choice of coordinates. Hence the skin friction is invariant,

$$(\tau_w)_\zeta = (\tau_w)_\chi \tag{17-28}$$

An incidental result of the derivation of Eq. 17-19 is that the integral on its right-hand side is an invariant. This may also be proved directly from the correlation theorem (Eq. 17-22). This second method of proving the invariance is very general and may immediately be applied whenever the integrand is a homogeneous function of u_{ew} and u and the range of integration is the same as above. This implies (cf. Eq. 13-21 and Eq.

13-23) that the value of the displacement thickness and the momentum thickness is independent of the choice of coordinates.

So far the velocity vector has been represented by its contravariant density components u and v. Let, now, ζ be an orthogonal system and U denote the ξ component in the sense of vector geometry. Then $h_2(\xi, \eta)U = u$. Applying the second limit process and denoting the boundary layer approximation by a subscript ζ gives

$$h_2(\xi, 0)U_\zeta = u_\zeta \tag{17-29a}$$

If $U_e = \lim_e U$ and $U_{ew} = U_e(\xi, 0)$, then

$$h_2(\xi, 0)U_{ew} = u_{ew} \tag{17-29b}$$

Hence

$$\int (u_{ew} - u_\zeta)d\eta = \int (U_{ew} - U_\zeta)d(h_2(\xi, 0)\bar{\eta}) \tag{17-30}$$

which shows how to form invariant integrals from the components in the sense of vector geometry. Note that if η measures true distance in an infinitesimal neighborhood of the wall (cf. the discussion preceding Eq. 13-2) then $h_2(\xi, 0) = 1$.

Optimal coordinates. It has been shown that the boundary layer approximations with respect to various coordinate systems have several important invariants in common. They may also agree approximately within a region of thickness $\sqrt{\nu}$ at the wall. However, if the formally obtained results are taken literally at a distance from the boundary, almost any flow field may be obtained. The result depends on the choice of coordinates rather than on the original physical problem. For this reason one normally uses the solution of the boundary layer equations up to some rather arbitrarily defined edge of the boundary layer, where the solution is patched onto an external flow solution, possibly including the flow due to displacement thickness. There are, however, several conceptual and computational difficulties associated with such a patching procedure, especially when the result is being used as a starting point for higher order approximations. The problem of how and where to patch may, however, be entirely bypassed. One method is to use the expansions (Eq. 17-15). Another method is to carry out the boundary layer approximation in a special system of coordinates. This method is now shown, following the work of Kaplun [*96*].

A coordinate system ζ is called optimal if the corresponding boundary layer approximation includes the external flow and the flow due to displacement thickness in the sense that (cf. Eq. 13-5a and 13-19)

$$\lim_e \psi_\zeta = \psi_e \tag{17-31a}$$

$$\sqrt{\nu} \lim_e \frac{\psi_\zeta - \psi_e}{\sqrt{\nu}} = \psi_e' \tag{17-31b}$$

An equivalent definition is that $\psi_\zeta = \psi_1$ where ψ_1 is defined by Eq. 17-20. It can easily be seen that one optimal system is the following:

$$\xi = \psi'_e, \qquad \eta = \psi_e \tag{17-32}$$

Any other system χ that is optimal is related to ζ by Eq. 17-25 and 17-26. Thus to find ψ_ζ when ζ is optimal one first computes ψ_χ in any system of coordinates χ, not necessarily optimal. From ψ_χ one finds ψ'_e as described earlier. Then one determines ζ from Eq. 17-32. Finally ψ_ζ is found from ψ_χ according to Eq. 17-22.

Eq. 17-24b represents a solution for the semi-infinite flat plate that is optimal in the sense defined above. More generally, the streamline coordinates used in discussing the Falkner-Skan flow (Eq. 14-1) are not optimal whereas coordinates which are parabolic with respect to these are optimal. The former system of coordinates collapses the wedge (Fig. B,14a) onto a semi-infinite line parallel to the flow. Optimal coordinates, on the other hand, open up the wedge to a line perpendicular to the flow. If the wedge angle is π, the optimal coordinates are the original Cartesian coordinates. The boundary layer solution in these coordinates happens to agree with the exact solution of the Navier-Stokes equations (cf. the discussion of flow towards a plane in Art. 8). Similarly the Oseen boundary layer solution for the semi-infinite flat plate gives the exact solution of the Oseen equation if optimal coordinates, i.e. parabolic coordinates or coordinates equivalent thereto in the sense of Eq. 17-25, are used. The special cases mentioned above are discussed in detail in [*96*, p. 126ff.].

Uniform validity of the asymptotic expansion. We now return to the question of the validity of the expansion 17-15. As emphasized earlier the discussion is limited to bodies for which the external flow is potential. For a large class of such bodies it is reasonable to hope that, with a suitable choice of coordinates, the expansion is uniformly valid in the sense of Eq. 17-8 (where, of course, Re has to be replaced by $\sqrt{Re}$). It is, however, dubious whether the expansion can be uniformly valid in the neighborhood of sharp corners or sharp leading edges. Consider, e.g. Falkner-Skan flow. If streamline coordinates are used (Eq. 14-1ff.) the solution has a singularity along the line $\xi = 0$. In particular v is infinite along this line. The use of optimal coordinates alleviates the situation somewhat. However, unless the boundary layer solution happens to be the exact solution, it cannot be uniformly valid. This follows from the fact that no length can be defined from the geometry of the wedge (cf. *32*, Chap. 6.1).

For a three-dimensional semi-infinite body one may define an expansion analogous to Eq. 17-15. It follows, however, from the discussion in Art. 15 that a necessary condition for this expansion to be uniformly valid is that the parameter η as defined by Eq. 15-12 be uniformly large when the over-all Reynolds number is large.

Pressure. From the expansion for the stream function (Eq. 17-15) one finds the corresponding expansions for the velocity components by differentiation. For the pressure, one may assume a similar expansion

$$p = p_{\mathrm{L}} + p_{\mathrm{T}} \tag{17-33a}$$

$$p_{\mathrm{L}} = \rho \sum_{j=0}^{\infty} (\sqrt{\nu})^j \, a_j(\lambda, \sigma) \tag{17-33b}$$

$$p_{\mathrm{T}} = \rho \sum_{j=1}^{\infty} (\sqrt{\nu})^j \, b_j(\lambda, \bar{\sigma}) \tag{17-33c}$$

The longitudinal component p_{L} may be computed from ψ_{L} by Bernoulli's law. It is then seen that a_0 is the pressure of the external flow; $\sqrt{\nu}\, a_1$ is a correction term which one finds from the flow due to displacement thickness. However, in general the transversal wave also contributes a pressure term of order $\sqrt{\nu}$, namely $\sqrt{\nu}\, b_1$. This term may be computed as follows. For convenience we assume that λ and σ are conformal coordinates. As before, we assume that σ vanishes on the boundary. From the second momentum equation one then finds (cf. Eq. 6-10)

$$b_1 = -\frac{1}{2}\left(\frac{dJ}{d\sigma}\right)_{\mathrm{w}} \int_{\bar{\sigma}}^{\infty} (u_{\mathrm{ew}}^2 - u_{\mathrm{BL}}^2)\, d\bar{\sigma} \tag{17-34}$$

The velocity components are here those of a contravariant vector density and J is defined as in Eq. 13-10c. The result is independent of the choice of coordinates. The curvature κ at the wall is

$$\kappa = \left(-\frac{1}{2\sqrt{J}}\frac{dJ}{d\eta}\right)_{\mathrm{w}} \tag{17-35}$$

This quantity is negative for a concave wall. If one chooses a system of coordinates x, y such that $h_1 = h_2 = 1$ at the wall, Eq. 17-34 may be written

$$b_1 = \kappa \int_{\bar{y}}^{\infty} (u_{\mathrm{ew}}^2 - u_{\mathrm{BL}}^2)\, d\bar{y} \tag{17-36}$$

Separation and instability. Regarding the last two difficulties mentioned at the beginning of this subarticle, very few positive statements can be made. Only a few comments are made here beyond the discussion in Art. 14. Assume for the sake of argument that as ν tends to zero the flow past a solid tends to flow of the Kirchhoff type. In order to find the solution for small ν one needs, in addition to boundary layer theory, a technique for studying the flow near the point of separation and the wake flow. At first it seems plausible that for small values of ν the free streamlines bounding the wake may be replaced by simple shear layers. Since, however, the Kirchhoff wake spreads parabolically downstream, the two shear layers, which have thickness $\sim \sqrt{\nu x}$, do not merge for ν sufficiently small. This would mean that the streamlines C_1D_1 and C_2D_2 in Fig. B,14d are modified for small values of ν, but that they actually do not meet. However, this contradicts the assumption, made in Art. 4 and elsewhere, that the velocity far downstream tends to its free stream value. Thus a special perturbation technique is needed which shows how the Kirchhoff wake collapses for $\nu > 0$.

The instability of the flow at high Reynolds numbers was mentioned in Art. 14 and 16. This fact, however, does not make the study of stationary solutions irrelevant. As an example, flow past a circular cylinder is stable only for $Re < 40$. However, from the point of view of expansion methods a Reynolds number such as, say, 30 is a high Reynolds number, and it should be possible to find the corresponding flow from high Reynolds number solutions. As it will be seen later, the usefulness of the low Reynolds number expansions is restricted to a very small range of Reynolds numbers. Furthermore, even in the unstable range, it is of interest to find the stationary solution as a first step in understanding the instability of the flow.

Expansions for Low Reynolds Numbers. Solutions which are approximately valid for low Reynolds numbers were discussed in Art. 9, 10, 11, and 12. We now study the relation of these approximate solutions to asymptotic expansions of the exact solutions. Many of the basic ideas are due to Kaplun. For a deeper analysis of the problems discussed below, reference may be made to [*104,105,52*]. Further applications of the methods used are found in [*31*].

Two limit processes play a fundamental role in the mathematical study of low Reynolds number flow, namely the Stokes limit (inner limit) and the Oseen limit (outer limit). Let l be any nondimensional flow quantity, such as $\mathbf{u}^*$, u^*, p^*, etc. The Stokes limit is denoted by $\lim_s$ and defined by

$$\lim_s l = \lim_{Re\to 0, x_i^* \text{fixed}} l \tag{17-37}$$

Here the x_i^* are the Stokes variables x_i/L (Eq. 9-1b). Similarly, the Oseen limit is denoted by $\lim_O$ and defined by

$$\lim_O l = \lim_{Re\to 0, \tilde{x}_i \text{fixed}} l \tag{17-38}$$

Here the $\tilde{x}_i$ are the Oseen variables x_iU/ν (Eq. 9-1c). In general the outer limit is applied only at points which are not on the boundary of the body.

The Stokes limit may be thought of as the limit where ν tends to infinity while the other parameters are kept fixed. From this it is evident that in the Stokes limit the influence of the boundary is strong. For this reason the Stokes limit is called an inner limit. The value of $\mathbf{u}^*$ at the wall, i.e. zero, will have an increasingly large influence on any fixed point not on the boundary as ν tends to infinity. By analogy with the heat equation it is then seen that

$$\lim_s \mathbf{u}^* = 0 \tag{17-39}$$

in two-dimensional flow. This limit is, however, nonuniform at infinity. For any given ν one can find points sufficiently far away from the body such that $\mathbf{u}^*$ there is arbitrarily close to $\mathbf{i}$. The analogy with the heat equation also shows that Eq. 17-39 is not valid in three-dimensional flow.

The Oseen limit may be thought of as the limit where L tends to zero while the other parameters and the x_i are fixed. The body then tends to a conical body, i.e. a body for which no length can be defined. The limiting conical body may in particular be a point, a line, or a semi-infinite line. Such objects have no arresting power, i.e. they cannot cause a finite disturbance in the fluid. Then (cf. the discussion of Eq. 11-5)

$$\lim_O \mathbf{u}^* = \mathbf{i} \tag{17-40}$$

Thus in the Oseen limit, conditions at infinity determine the value of the limit. For this reason the Oseen limit is called an outer limit. On the other hand, if the limiting conical body has a nonzero surface area inside a finite domain it causes a nonzero flow disturbance. Eq. 17-40 is then no longer valid. This case is not considered here. The discussion below is mainly concerned with finite two-dimensional or three-dimensional bodies. In the former case the limiting cone is a two-dimensional point, i.e. an infinite line in three-dimensional space. In the latter case the limiting cone is a point.

Note that the Oseen limit is nonuniform at the body. No matter how small L is, points sufficiently close to the body may be found where $\mathbf{u}^*$ is arbitrarily small.

Let now $\epsilon_j(Re)$ be the sequence of functions of Re only, such that

$$\epsilon_0(Re) = 1, \lim_{Re\to 0} \frac{\epsilon_{j+1}(Re)}{\epsilon_j(Re)} = 0 \tag{17-41}$$

The ϵ_j are thus of successively smaller order of magnitude. One may then define a repeated application of the Stokes limit to a flow quantity l as follows

$$l_0 = \lim_s l, \qquad l_1 = \lim_s \frac{l - l_0}{\epsilon_1}, \qquad l_2 = \lim_s \frac{l - l_0 - \epsilon_1 l_1}{\epsilon_2} \tag{17-42}$$

If these limits all exist one may associate an asymptotic expansion with l

$$l \to \sum_{j=0}^{\infty} \epsilon_j(Re) l_j(x_1^*, x_2^*, x_3^*) \tag{17-43}$$

This expansion is called the Stokes expansion of l.

Similarly one obtains the Oseen expansion of l by a repeated application of the Oseen limit.

We now indicate how these limit processes may be used in constructing expansions which are uniformly valid for low Reynolds numbers.

Two-dimensional flow. In this case we choose

$$\epsilon_j(Re) = [\epsilon(Re)]^j, \qquad \epsilon = \frac{1}{c - \ln Re} \tag{17-44}$$

where c is a suitably chosen constant. Note that, since $Re = e^{c-(1/\epsilon)}$, Re is transcendentally small, i.e.

$$Re = 0(\epsilon^n) \quad \text{for all } n \tag{17-45}$$

The following form of the asymptotic expansion has been proposed by Kaplun (cf. expansions (Eq. 17-7 and 17-15))

$$\mathbf{u}^* \cong \sum_{j=0}^{\infty} \epsilon_j(Re)[\mathbf{f}_j(\tilde{x}_i) + \mathbf{g}_j(x_i^*)] \tag{17-46}$$

As in expansion (Eq. 17-15) the $\mathbf{f}_j$ are obtained from $\mathbf{u}^*$ by a repeated application of the outer (Oseen) limit. The g_j are then transcendentally small except at the body. They are corrections terms needed to make the expansion uniformly valid in the neighborhood of the body. Formally the g_j are obtained by a repeated application of the Stokes limit to $\mathbf{u}^* - \sum_{j=0}^{n} \mathbf{f}_j(\tilde{x}_i)$ where n is a sufficiently large number (cf. Eq. 17-12).

Consider now flow past the circular cylinder as an example. From Eq. 11-35 we obtain

$$\mathbf{f}_0 = \lim{}_0\, \mathbf{u}^* = \mathbf{i} \tag{17-47a}$$

$$\mathbf{f}_1 = \lim{}_0 \frac{\mathbf{u}^* - \mathbf{f}_0}{\epsilon} = 2\mathbf{u}^{(1)} \tag{17-47b}$$

$$\mathbf{g}_0 = 0 \tag{17-47c}$$

$$\mathbf{g}_1 = \lim{}_s \frac{\mathbf{u}^* - \mathbf{f}_0 - \epsilon \mathbf{f}_1}{\epsilon} = -\frac{1}{8} \operatorname{grad} \frac{\partial}{\partial x^*} (\ln r^*) \tag{17-47d}$$

The operator grad in the last equation stands for $(\partial/\partial x)_i^*$.

Assume for the moment that the Oseen equations are the basic equations. Then $\mathbf{f}_j = \mathbf{g}_j = 0$ for $j > 1$ and the expansion given is complete up to terms of transcendentally small order. (This is true only with the very special choice of ϵ given by Eq. 11-35.) It is then seen that a pure Oseen expansion, i.e. $\mathbf{f}_0 + \epsilon \mathbf{f}_1$, is not uniformly valid near the body. The term $\epsilon \mathbf{g}_1$, which is transcendentally small off the boundary, is of order unity at the boundary.

The expansion as given is still correct for the Navier-Stokes equations up to order ϵ^2. In this case $\mathbf{f}_j$ and $\mathbf{g}_j$ are not zero for $j > 1$. The equations for these terms may be found by inserting the expansion (Eq. 17-46) into the Navier-Stokes equations. Certain conditions for matching the inner and the outer flow have to be used as boundary conditions. These matching conditions are somewhat more complicated than in the high Reynolds number case (cf. Eq. 17-19 and 17-20). An example of the matching procedure is now given (cf. [*32*, p. 873 ff.]).

By repeated application of the Stokes limit to $\mathbf{u}^*$ one obtains a pure Stokes expansion

$$\mathbf{u}^* \to \sum_{j=0}^{\infty} \epsilon^j \mathbf{h}_j(x_i^*) \tag{17-48}$$

Since Re is transcendentally small it follows that all the $\mathbf{h}_j$ obey the homogeneous Stokes equation. For reasons explained above $\mathbf{h}_0 = 0$. To determine $\mathbf{h}_1$ we first find a solution of the Stokes equations which vanishes at the surface of the body and whose divergence at infinity is of as low order a magnitude as possible. This solution is determined within a multiplicative constant, it is in fact given by Eq. 11-36 with ϵ as yet undetermined. To determine ϵ we use the following matching condition:

$$\lim\nolimits_O (\epsilon \mathbf{h}_1) = \lim \mathbf{u}_0^* = \mathbf{i} \tag{17-49}$$

From this it follows easily that

$$\epsilon = -\frac{1}{\ln Re} + O\left[\frac{1}{(\ln Re)^2}\right] \tag{17-50}$$

By this method, which is essentially due to Kaplun, one can thus solve the Stokes equations for flow past a two-dimensional body.

Since $\mathbf{h}_1$ diverges at infinity it is clear that a pure Stokes expansion of $\mathbf{u}^*$ is not uniformly valid. For the same reason $\mathbf{h}_1$ does not occur in Eq. 17-46 but only the part $\mathbf{g}_1$ which is not contained in $\mathbf{f}_0 + \epsilon\mathbf{f}_1$. By this it is meant that

$$\mathbf{g}_1 = \mathbf{h}_1 - \lim\nolimits_s \frac{\mathbf{f}_0 + \epsilon \mathbf{f}_1}{\epsilon} \tag{17-51}$$

Thus $\mathbf{h}_1$ corresponds to the boundary layer solution in the high Reynolds number case while $\mathbf{g}_1$ corresponds to a perturbation boundary layer obtained by the inner limit after part of the external solution has been subtracted out (cf. Eq. 13-15 or, for a closer analogy, Eq. 17-18a).

The higher order terms of the expansion (17-46) may, in principle, be computed by straightforward methods; some very tedious numerical evaluations of integrals seem, however, unavoidable. The $\mathbf{f}_j$ satisfy nonhomogeneous Oseen equations for $j > 1$, whereas the $\mathbf{h}_j$ satisfy the homogeneous Stokes equation. The matching conditions between the Stokes and Oseen expansions are similar to the matching condition described above.

Three-dimensional case. In this case Eq. 11-38 suggests that $\epsilon_j(Re)$ might be defined as Re^j. However, the existence of intermediate orders such as $Re^2 \ln Re$ has been shown in [*158*]. A pure Oseen expansion may be used. Note, however, that while the term $Re^3\,\mathbf{u}^{(2)}$ in Eq. 11-38 is no longer transcendentally small off the boundary, it is of high order, namely Re^3, off the boundary and of order unity on the boundary. It is thus analogous to the Stokes correction term in the two-dimensional case. Actually both the Stokes and Oseen limits may be used in discussing the expansion even in the three-dimensional case. Instead of a pure Oseen expansion one may also conveniently use a composite expansion similar to Eq. 17-46.

Use of the low Reynolds number expansions. Even if the higher order terms of the low Reynolds expansions are found, the usefulness of the expansion is restricted to quite low Reynolds numbers. The parameter ϵ defined by Eq. 11-35d becomes infinite for $Re \cong 7.4$ and is equal to unity for $Re \cong 2.7$. Thus, the expansion of the solution for the circular cylinder is of dubious value, say, for $Re > 2$.

CHAPTER 3. INTRODUCTORY DISCUSSION OF THE NAVIER-STOKES EQUATIONS FOR A COMPRESSIBLE FLUID

The Navier-Stokes equations for a viscous heat-conducting compressible fluid were derived in Chap. 1. The present chapter contains some introductory remarks about these equations; in particular some special solutions are studied, and the meaning of various approximating equations is discussed.

In Art. 18 the Navier-Stokes equations are written in nondimensional form in order to bring out significant parameters such as Mach number, Prandtl number, etc. These parameters play an important role in subsequent discussions.

It was pointed out in Chap. 2 that exact solutions of the Navier-Stokes equations for incompressible fluids may be found only for very specialized types of flow. Very few of these solutions have been extended to take compressibility into account. The most important example of such an extension is the solution for compressible plane Couette flow. In addition, there exist a few exact solutions of the Navier-Stokes equations which have no analogue in the incompressible case. The most important among these is the solution for a one-dimensional stationary shock wave (or shock layer). These solutions are discussed in Art. 19.

The direct applicability of the exact solutions is very limited; hence the study of approximate equations becomes even more important than in the incompressible case. In Art. 20 equations for limiting values of various parameters are derived. The first example studied is that of flow at zero Mach number. In connection with this it is discussed when and in what sense a fluid may be treated as incompressible. Low Reynolds number flow is also discussed; the equations obtained are a generalization of the Stokes equations to the case of compressible fluids. For high values of the Reynolds number the boundary layer equations for compressible fluids are obtained. The shock layer equations are also a limiting case for high Reynolds numbers.

In Art. 21 various simplified equations are studied which may be regarded as mathematical models for the Navier-Stokes equations. The important feature of these equations is that various exact solutions may be obtained. By analyzing these solutions one may obtain qualitative information about certain aspects of the solutions of the full Navier-Stokes equations. Under certain limited conditions these equations may also be regarded as approximations to the Navier-Stokes equations. The linearized equations, for example, may be applied to certain problems for sound waves.

B,18. Nondimensional Form of the Navier-Stokes Equations. Important Parameters.

In analytical discussions of the Navier-Stokes equations dimensional analysis is of great importance. The use of nondimensional variables lends greater generality to theoretical (as well as to experimental) results that are obtained. Furthermore, the validity of approximate equations is dependent on the values of certain nondimensional parameters. There are in general many ways in which a dependent variable such as pressure or an independent variable such as position may be made nondimensional. In many cases there is a natural way of choosing the nondimensional variables and parameters. In other cases the method is less obvious and the choice of the appropriate nondimensional variable is of great importance. The above statements have been previously illustrated in Chap. 2.

In general, the following quantities are given by the boundary conditions: At infinity a free stream velocity U is prescribed and also the thermodynamic state of the fluid, that is ρ_∞, p_∞, and T_∞. This implies that the transport coefficients μ_∞ etc. are also known at infinity. The geometry of the body determines a length L (except for conical shapes). A condition on the temperature or the heat transfer is also given at the body, for example $T = T_w$ at the surface of the body. With the aid of these dimensional constants the variables occurring in the equations may be made nondimensional. The state variables and transport coefficients may be made nondimensional by dividing by the corresponding values at infinity. The result is denoted by $\sim$: $\tilde{\rho} = \rho/\rho_\infty$, $\tilde{\nu} = \nu/\nu_\infty$ etc. The space and time coordinates may be made nondimensional as in the incompressible case, and the notation of Eq. 9-1 is used. The choice of the proper nondimensional pressure presents some difficulties. In the momentum equation pressure enters as a dynamical variable. The choice $p^* = (p - p_\infty)/\rho_\infty U^2$ is then natural. However, pressure is also a thermodynamic variable. When it enters in this capacity, as in the equation of state, the choice $\tilde{p} = p/p_\infty$ may be considered more natural. In the discussion of the Stokes approximation in Art. 10 it was shown that for very low Reynolds number flow the parameter $p^+ = Re\, p^*$ is significant. The coefficient of heat conductivity may be made nondimensional by referring to its value at infinity. It is, however, usually more convenient to use the nondimensional Prandtl number $Pr = \mu c_p/k$.

The following is a list of some important nondimensional variables:

$$x_i^* = \frac{x}{L}; \quad t^* = \frac{tU}{L}; \quad \mathbf{u}^* = \frac{\mathbf{u}}{U}; \quad \Phi^* = \frac{L^2}{U^2}\frac{\Phi}{\mu_\infty}$$

$$\tau^* = \frac{L}{U\mu_\infty}\tau; \quad p^* = \frac{p - p_\infty}{\rho_\infty U^2}; \quad p^+ = \frac{L(p - p_\infty)}{U\mu_\infty}; \quad \tilde{p} = \frac{p}{p_\infty}$$

$$\tilde{\rho} = \frac{\rho}{\rho_\infty}; \quad \tilde{T} = \frac{T}{T_\infty}; \quad \tilde{h} = \frac{h}{h_\infty}; \quad \tilde{\mu} = \frac{\mu}{\mu_\infty}; \quad \tilde{\nu} = \frac{\nu}{\nu_\infty}; \quad \tilde{\lambda} = \frac{\lambda}{\mu_\infty} \tag{18-1a}$$

Here Φ is the dissipation per unit time and volume as defined by Eq. 3-5b and τ is the viscous stress tensor as defined by Eq. 3-1b.

In addition one may form nondimensional parameters from the constants occurring in the boundary conditions. The isentropic speed of sound a is defined as $(\partial p/\partial \rho)^{\frac{1}{2}}$, where the derivative is evaluated at constant entropy. The Mach number may then be defined as $M = U/a_\infty$. In a few cases the local Mach number will be considered. The Mach number at infinity will then be distinguished by the subscript $_\infty$. The Reynolds number may be defined as in the incompressible case if the value of viscosity at infinity is used. The Prandtl number at infinity, Pr_∞, is determined by the thermodynamic state of the medium. Unlike Mach number and Reynolds number, it is independent of the speed of the flow and the dimension of the object. It is in general assumed that the Prandtl number depends on the nature of the fluid only, and is independent of the values of the state variables (cf. Art. 3). The subscript $_\infty$ is then omitted. Another property of the fluid itself is the ratio between the specific heats $\gamma = c_p/c_v$. It is often assumed that this ratio does not vary with temperature, pressure, etc. The following is then a list of the nondimensional parameters used:

$$M_\infty = \frac{U}{a_\infty}; \quad Re = \frac{UL}{\nu_\infty}; \quad Pr_\infty = \left(\frac{\mu c_p}{k}\right)_\infty; \quad \gamma = \frac{c_p}{c_v} \tag{18-1b}$$

In addition, the boundary condition for the temperature at the wall may lead to an additional independent parameter, for instance $\tilde{T}_w - 1 = (T_w - T_\infty)/T_\infty$.

With the aid of the nondimensional variables and parameters listed in Eq. 18-1, the Navier-Stokes equations may easily be put in nondimensional form. However, the results are given only for the case in which the following special assumptions are made about the state variables and transport coefficients:

$$p = \rho \mathcal{R} T, \ \mathcal{R} = \text{cons} \tag{18-2a}$$

$$h = c_p T; \ c_p = \text{const} \tag{18-2b}$$

$$Pr = \text{const} \tag{18-2c}$$

$$\gamma = \text{const} \tag{18-2d}$$

In addition it is assumed that the external heat added and the external forces are zero,

$$\mathbf{f} = 0; \qquad Q = 0 \tag{18-2e}$$

The validity of the assumptions in Eq. 18-2 is discussed in Art. 3. These assumptions imply in particular the following:

$$a^2 = \gamma \mathcal{R} T \tag{18-3a}$$

$$\frac{U^2}{h_\infty} = (\gamma - 1)M^2 \tag{18-3b}$$

$$\tilde{p} = \gamma M^2 p^* + 1 = \gamma \frac{M^2}{Re} p^+ + 1 \tag{18-3c}$$

$$\tilde{h} = \tilde{T} \tag{18-3d}$$

If now Eq. 3-3, 3-4, 3-8a, and 18-2a are multiplied by $L/\rho_\infty U$, $L/\rho_\infty U^2$, $L/Uh_\infty\rho_\infty$, and $1/p_\infty$ respectively, one obtains a *nondimensional form of the Navier-Stokes equations under the special assumptions stated in Eq. 18-2:*

$$\frac{\partial \tilde{\rho}}{\partial t^*} + \text{div}\,(\tilde{\rho}\mathbf{u}^*) = 0 \tag{18-4a}$$

$$\tilde{\rho}\frac{D\mathbf{u}^*}{Dt^*} + \text{grad}\, p^* = \frac{1}{Re}\,\text{div}\,\boldsymbol{\tau}^* \tag{18-4b}$$

$$\tilde{\rho}\frac{D\tilde{h}}{Dt^*} - \frac{\gamma - 1}{\gamma}\frac{D\tilde{p}}{Dt^*} = \frac{1}{Pr\,Re}\,\text{div}\,(\tilde{\mu}\,\text{grad}\,\tilde{h}) + (\gamma - 1)\frac{M^2}{Re}\Phi^* \tag{18-4c}$$

$$\tilde{p} = \tilde{\rho}\tilde{h} \tag{18-4d}$$

Here

$$\text{div}\,\boldsymbol{\tau}^* = \text{div}\,(\tilde{\mu}\,\text{def}\,\mathbf{u}^*) + \text{grad}\,(\tilde{\lambda}\,\text{div}\,\mathbf{u}^*) \tag{18-4e}$$

In these equations the operators div, grad, etc. are to be formed with the nondimensional derivatives $\partial/\partial x^*$, etc. Different pressure parameters may be introduced with the aid of Eq. 18-3c.

B,19. Exact Solutions.

PLANE COUETTE FLOW. Consider two-dimensional stationary flow between two parallel plane walls of infinite extent located at $y = 0$ and $y = L$. The lower wall ($y = 0$) is assumed to be at rest, while the upper wall moves with velocity U in the x direction. The temperature T, or, equivalently, the enthalpy h, has a constant prescribed value at each wall. Conditions at the upper wall are taken as reference conditions ("free stream" conditions) and the values of the state variables and transport coefficients at this wall are denoted by the subscript $_\infty$. The corresponding values at the lower wall are denoted by the subscript $_w$. The "free stream" pressure, that is, the pressure at the upper wall, is also prescribed. The boundary conditions are then

$$u = 0, \qquad v = 0, \qquad h = h_w \quad \text{at} \quad y = 0 \tag{19-1a}$$

$$u = U, \qquad v = 0, \qquad h = h_\infty, \qquad p = p_\infty \quad \text{at} \quad y = L \tag{19-1b}$$

where h_w, h_∞, and p_∞ are constants. The Reynolds number and the free stream Mach number are defined by

$$Re = \frac{UL}{\nu_\infty}, \qquad M^2 = \frac{U^2}{(\gamma - 1)h_\infty} \tag{19-1c}$$

For incompressible fluids the solution is simply

$$u = U\frac{y}{L}, \qquad v = 0, \qquad \tau = \mu\frac{\partial u}{\partial y} = \frac{\mu}{L}U \tag{19-2}$$

In a compressible fluid the heat conducted from the walls and the heat added by dissipation introduce temperature variations in the flow. This temperature variation changes the values of the density and of the viscosity coefficient. Because of this fact the solution given by Eq. 19-2 can no longer be expected to satisfy the dynamical equations. A modified solution, valid for the compressible case, may be obtained as follows.

Since the boundary conditions are independent of x it follows that $\partial/\partial x = 0$. Hence the continuity equation reduces to $\partial(\rho v)/\partial y = 0$. This equation, together with the boundary conditions on v, implies that v is everywhere zero. Hence, all transport terms are zero. The equation for the y momentum then reduces to $\partial p/\partial y = 0$. Since $\partial p/\partial x$ is also zero, the pressure is everywhere constant. Thus

$$v = 0, \qquad p = p_\infty \tag{19-3}$$

The remaining equations reduce to

$$\frac{d}{dy}\left(\mu\frac{du}{dy}\right) = 0 \tag{19-4a}$$

$$\frac{1}{Pr}\frac{d}{dy}\left(\mu\frac{dh}{dy}\right) + \mu\left(\frac{du}{dy}\right)^2 = 0 \tag{19-4b}$$

With the aid of the first equation, the second equation may be written

$$\frac{d}{dy}\left[\mu\left(\frac{1}{Pr}\frac{dh}{dy} + u\frac{du}{dy}\right)\right] = 0 \tag{19-4b'}$$

The nondimensional equations corresponding to Eq. 19-4a and 19-4b′ are (cf. Eq. 18-4)

$$\frac{\partial}{\partial y^*}\left(\tilde{\mu}\frac{\partial u^*}{\partial y^*}\right) = 0 \tag{19-5a}$$

$$\frac{\partial}{\partial y^*}\left[\tilde{\mu}\left(\frac{1}{Pr}\frac{d\tilde{h}}{dy^*} + (\gamma - 1)M^2u^*\frac{du^*}{dy^*}\right)\right] = 0 \tag{19-5b}$$

Note that the Reynolds number does not enter these equations explicitly. Hence flows at different Reynolds numbers are similar in the sense that u^* and $\tilde{h}$ depend on y^* and on the parameters M, Pr, and $\tilde{h}_w$ but not on the parameter Re.

Eq. 19-4 may be reduced to a simpler form by a transformation employed by Crocco in the study of compressible boundary layers (cf. Art. 22; the other transformations described there may also be used). The velocity u is used as independent variable instead of y. As a dependent variable u is replaced by τ where

$$\tau = \mu\frac{du}{dy} \tag{19-6a}$$

The heat transferred per unit time and area from the lower wall into the fluid is then

$$q_w = -\left(k\frac{dT}{dy}\right)_w = -\frac{1}{Pr}\left(\mu\frac{dh}{dy}\right)_w = -\frac{1}{Pr}\left(\tau\frac{dh}{du}\right)_w \tag{19-6b}$$

The following equations are easily obtained by integrating Eq. 19-4a and 19-4b:

$$\tau \doteq \text{const} = \tau_w \tag{19-7}$$

$$\frac{Pr\,q_w}{\tau_w} = \frac{1}{U}\left(h_w - h_\infty - Pr\frac{U^2}{2}\right) \tag{19-8}$$

$$h_w - h = \left(h_w - h_\infty - Pr\frac{U^2}{2}\right)\frac{u}{U} + Pr\frac{u^2}{2} \tag{19-9}$$

The nondimensional form of the last equation is

$$\tilde{h} = \tilde{h}_w + \left[\frac{\gamma - 1}{2}M^2\,Pr + (1 - \tilde{h}_w)\right]u^* - \frac{\gamma - 1}{2}M^2\,Pr\,u^{*2} \tag{19-9'}$$

In principle, the problem is now solved. From Eq. 19-9 one determines h as a function of u. Since μ is a given function of h one may consider μ as a function of u. By integrating Eq. 19-6a one then finds y as a function of u

$$y = \frac{1}{\tau_w} \int_0^u \mu du \tag{19-10a}$$

where τ_w is determined from the condition that $y = L$ for $u = U$,

$$\tau_w = \frac{1}{L} \int_0^U \mu du \tag{19-10b}$$

Thus h, u, μ, etc. are given as implicit functions of y.

It is often convenient to introduce the recovery temperature T_r as the temperature of an insulated wall, i.e.

$$T_r = \text{value of } T_w \text{ for which } q_w = 0 \tag{19-11}$$

The recovery enthalpy h_r is defined in the same way. Eq. 19-8 may then be written

$$\frac{q_w}{\tau_w U} = \frac{\tilde{T}_w - \tilde{T}_r}{(\gamma - 1)M^2 Pr} \tag{19-12a}$$

where

$$\tilde{T}_r - 1 = \frac{\gamma - 1}{2} M^2 Pr \tag{19-12b}$$

(Note that $\tilde{h} = \tilde{T}$ since c_p is assumed to be constant.) Similarly Eq. 19-9′ may be written

$$\tilde{h} = \tilde{h}_w + (\tilde{h}_r - \tilde{h}_w)u^* - \frac{\gamma - 1}{2} M^2 Pr\, u^{*2} \tag{19-13}$$

Some special cases will now be discussed.

$M = 0$. When $q_w \neq 0$ this case may be described as low speed heat transfer. Eq. 18-4c or 19-4b show that in this case no heat is added by dissipation. Furthermore, by Eq. 19-12b the recovery temperature is equal to T_∞. Hence $\tilde{T}_r$ and $\tilde{h}_r$ may be replaced by unity in Eq. 19-12a and 19-13. The velocity profile and the skin friction may be obtained once the dependence of μ on T is known.

If in addition $T_w = T_\infty$, then $q_w = 0$. Eq. 19-13 then shows that T and hence μ is everywhere constant. The solution given then reduces to the solution for the incompressible case (Eq. 19-2). Thus the fluid may be considered incompressible when the heat added by dissipation and by conduction from the walls is negligible.

$M \geqq 0$. Special viscosity law: $\tilde{\mu} = \tilde{T}$. This viscosity law, while not satisfied strictly by any gas, represents an interesting special case and turns out to be very useful in boundary layer theory. For this viscosity

law, Eq. 19-13 and 19-10 can be integrated easily and yield

$$y^* = \frac{\left[1 + \frac{\gamma - 1}{2} M^2 Pr\left(1 - \frac{u^{*2}}{3}\right) + (\tilde{T}_w - \tilde{T}_r)\left(1 - \frac{u^*}{2}\right)\right] u^*}{1 + \frac{\gamma - 1}{3} M^2 Pr + \frac{1}{2}(\tilde{T}_w - \tilde{T}_r)} \tag{19-14a}$$

and

$$\frac{\tau_w}{\rho_\infty U^2} = \frac{1}{Re}\left[1 + \frac{\gamma - 1}{3} M^2 Pr + \frac{1}{2}(\tilde{T}_w - \tilde{T}_r)\right] \tag{19-14b}$$

In principle there is no difficulty in using more accurate laws for μ/μ_∞. The qualitative behavior of the flow pattern remains the same as long as the viscosity increases with temperature: the skin friction increases with increasing Mach number and with heat transfer to the fluid. The effects are simple because of the fixed dimensions of the shear layer. In boundary layer theory it is seen that the temperature effects are not equally simple. Both the viscosity μ and the density ρ vary with temperature and this leads, in boundary layer flow, to two opposing effects. One effect is the increase in μ with temperature; this increases the shear as in the present case. The other effect is a decrease in ρ and hence an increase in boundary layer thickness; this decreases the shearing stress at the wall.

$M \to \infty$, $\tilde{\mu} = \tilde{T}$. The limiting case $M \to 0$ was discussed above. It is of interest also to study the "hypersonic" limit $M \to \infty$. When $q_w = 0$, i.e. $T_w = T_r$, Eq. 19-14a and 19-14b reduce to

$$y^* = \frac{3}{2} u^*\left(1 - \frac{u^{*2}}{3}\right) + O\left(\frac{1}{M^2}\right) \tag{19-15a}$$

$$\frac{\tau_w}{\rho_\infty U^2 M^2} = \frac{\gamma - 1}{3}\frac{Pr}{Re} + O\left(\frac{1}{M^2}\right) \tag{19-15b}$$

According to Eq. 19-9′ and 19-12b h and $\tilde{h}_w$ increase as M^2. Hence $h^* = \tilde{h}/(\gamma - 1)M^2 = h/U^2$ may be regarded as a suitable nondimensional enthalpy variable. h^*_∞ is then negligible. This is consistent with the fact that the solutions given by Eq. 19-15 are independent of the temperature at the upper wall. The heat transferred into the fluid from the upper wall is negligible compared to the heat added by dissipation.

The case when $\tilde{h}_w$ = const may be analyzed in similar terms. The significant variable is again $h^* = h/U^2$.

Velocity and temperature profiles are plotted in Fig. B,19a for various values of the parameters $\tilde{h}_w$ and M.

The compressible Couette flow was first discussed by Illingworth who made a survey of the exact solutions of the Navier-Stokes equations for compressible flow [97]. Other cases considered by Illingworth are flow past an infinite porous plate in a steady stream parallel to the plate and with steady, constant suction applied at the plate; and also circular flow round a circular cylinder with suction at the surface and simple shearing motion between concentric circular cylinders (circular Couette flow).

One-Dimensional Stationary Shock Layer. In plane Couette flow the viscous stresses are caused by velocity gradients perpendicular to the direction of motion. The pressure gradient is everywhere zero. We now discuss another exact solution of the Navier-Stokes equations, namely the solution for a one-dimensional stationary shock layer.[22] In this case the viscous stresses are associated with velocity gradients in the

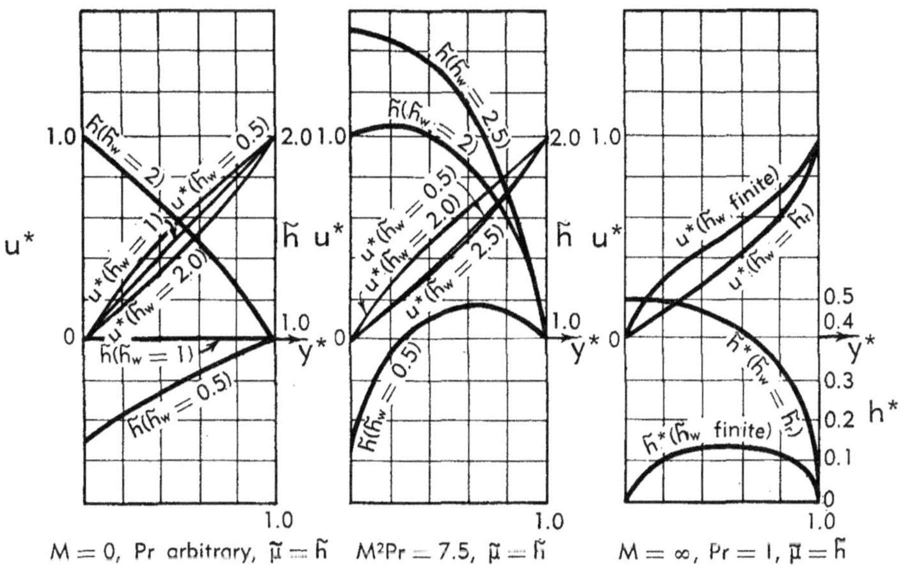

Fig. B,19a. Velocity and temperature profiles for plane Couette flow.

direction of motion, i.e. with compression of the fluid, and the pressure therefore plays an essential role. Loosely speaking, plane Couette flow may be called a purely transversal flow phenomenon, while the shock layer is purely longitudinal.

It is assumed below that all flow quantities are independent of time and depend only on the x coordinate, and that the velocity vectors are parallel to the x direction. According to the theory of nonviscous compressible fluids, the only one-dimensional flow phenomena of this type are contact discontinuities and discontinuous shock waves (cf. e.g. [*98*]). It is easily seen that viscosity and heat conductivity eventually completely obliterate a contact discontinuity, so that the corresponding stationary solution is that of uniform flow. On the other hand, the effect of

[22] In Art. 20 it is shown that the same solution is also approximately valid for general three-dimensional nonstationary shock layers.

viscosity and heat conductivity on a shock wave is to replace the shock discontinuity by a continuous transition between initial and final states. The region of continuous transition is called a shock layer; it is in many respects analogous to a boundary layer. Under normal conditions the greater part of the transition takes place in a very narrow region; however, strictly speaking, the shock layer is infinitely wide and the conditions ahead and behind the shock will have to be replaced by conditions at upstream and downstream infinity respectively. With this reservation the same "jump conditions" should result whether viscosity and heat conduction are considered or not. This may be seen as follows: Assume that in the nonviscous limit the shock would occur at $x = 0$. Consider a region between $x = -x_0$ and $x = +x_0$ ($x_0 > 0$) and apply the integral theorems for conservation of mass, momentum, and energy to this region. These give relations between conditions at $x = \pm x_0$ if no mass, momentum, or energy is added inside the region by external sources for mass or energy or by external forces. In the nonviscous case these relations are the jump conditions and x_0 may be taken arbitrarily small since the flow is uniform upstream and downstream of the shock. In the viscous heat-conducting case the mass relation is the same but the other two conservation laws involve additional terms due to viscous stresses and heat flux. Within the framework of the Navier-Stokes equations these additional terms are proportional to the derivatives du/dx and dT/dx respectively. However, if we let x_0 tend to infinity these derivatives are expected to vanish. Hence the relations for changes through a shock wave are the same in the viscous and in the nonviscous case. In the former case the shock wave is considered to start at $-\infty$ and end at $+\infty$; in the latter case it is infinitely thin (cf. the discussion below of the concept of the thickness of a viscous shock wave). In order to distinguish between the two cases, the continuous shock transition is referred to as a shock layer.

The Navier-Stokes equations, which were obtained by limiting procedures from the integral conservation relations, should of course also lead to the same relations connecting initial and final states of a shock transition. In addition, they should also imply that only compression shocks exist; rarefaction shocks are excluded. In the nonviscous case this does not follow directly from the conservation laws but is proved by entropy considerations, using the second law of thermodynamics. This law refers, not to the conservation of energy, but to the conversion of energy from one form into another. However, the Navier-Stokes equations, due to the specific assumptions about the stress tensor and the heat flux vector, contain implicitly the rules for energy conversions (cf. Art. 3). Thus it should be possible to exclude rarefaction shocks without specifically assuming the second law of classical thermodynamics. Finally, the equations should lead to a detailed description of the structure of the shock layer.

Shock layer equations. Neglecting all terms in Eq. 3-3, 3-4, and 3-8a which contain $\partial/\partial t$, $\partial/\partial y$, $\partial/\partial z$, v, and w, one obtains the following equations for a stationary one-dimensional shock layer.

$$\rho u \frac{du}{dx} + \frac{dp}{dx} = \frac{d}{dx}\left(\mu'' \frac{du}{dx}\right) \tag{19-16a}$$

$$\frac{d(\rho u)}{dx} = 0 \tag{19-16b}$$

$$\rho u \frac{dh}{dx} - u \frac{dp}{dx} = \frac{d}{dx}\left(\frac{\mu''}{Pr''} \frac{dh}{dx}\right) + \mu'' \left(\frac{du}{dx}\right)^2 \tag{19-16c}$$

If the Stokes relation (Eq. 3-24) is assumed, then μ'' is equal to $\frac{4}{3}\mu$. In the more general case μ'' is $(2\mu + \lambda)$. The value of μ'' is irrelevant for the following discussion. However, it must be kept in mind that the Prandtl number is formed relative to μ'' rather than to μ; it is therefore denoted by Pr''.

If the term udp/dx in Eq. 19-16c is eliminated with the aid of Eq. 19-16a (cf. Eq. 3-8b) the shock layer equations may be integrated once with respect to x,

$$mu + p - \mu'' \frac{du}{dx} = mU_i + p_i = mU_f + p_f \tag{19-17a}$$

$$\rho u = m = \rho_i U_i = \rho_f U_f \tag{19-17b}$$

$$m\left(\frac{u^2}{2} + h\right) - \frac{\mu''}{Pr''}\frac{dh}{dx} - \mu'' \frac{d(u^2/2)}{dx} = m\left(\frac{U_i^2}{2} + h_i\right) = m\left(\frac{U_f^2}{2} + h_f\right) \tag{19-17c}$$

Here m denotes the constant mass flow and subscripts $_i$ and $_f$ denote the initial and the final state respectively, i.e. conditions at upstream and downstream infinity. The constancy of the left-hand side of these equations follows from the integration. If in addition it is assumed that u and T tend to finite values at $\pm\infty$, it follows that du/dx and dT/dx are zero there. Then the constants may be evaluated as shown above. In addition there exist mathematically possible but physically irrelevant solutions with exponential behavior at infinity. Discarding these physically irrelevant solutions one sees that Eq. 19-17 contain the integral relations of nonviscous shock theory. The law that entropy increases through the shock layer is, however, not exhibited explicitly. Eq. 19-17 also shows that as boundary conditions for Eq. 19-16 one may prescribe the same quantities as in the theory of the discontinuous shock wave. As an example, one may prescribe U_i, p_i, and h_i at $x = -\infty$.

Alternative form of the shock layer equations for a perfect gas. It will now be assumed that the gas is perfect, i.e.

$$h = c_pT, \text{ where } c_p = \text{const} \tag{19-18a}$$

$$p = \mathcal{R}\rho T = \beta\rho h, \ \beta = \frac{\gamma - 1}{\gamma} = \text{const} \tag{19-18b}$$

Using this assumption we write the shock layer equations in a form suitable for the study of weak shock layers and of shock layers with $Pr'' = 1$.

As shown by Eq. 19-17c, the total enthalpy has the same value, H^0, at upstream and downstream infinity. Inside the shock layer H may differ from H^0. The difference $H - H^0$ is denoted by l. Thus

$$H = h + \frac{u^2}{2} = H^0 + l, \qquad H^0 = h_i + \frac{U_i^2}{2} = h_f + \frac{U_f^2}{2} \tag{19-19}$$

Multiplying the equation of state by u and using Eq. 19-17b one then finds

$$pu = \beta mh = \beta m\left(H^0 + l - \frac{u^2}{2}\right) \tag{19-20}$$

Inserting this relation into the momentum equation (Eq. 19-17a) multiplied by u, one finds

$$\mu'' u \frac{du}{dx} = -u\left(mU_i + \frac{\beta m h_i}{U_i}\right) + u^2 m\left(1 - \frac{\beta}{2}\right) + \beta m\left(h_i + \frac{U_i^2}{2}\right) + \beta ml \tag{a}$$

For $x = \pm\infty$ the left-hand side is zero and l is also zero. The right-hand side, except for the term βml, is therefore a quadratic polynomial in u which vanishes for $u = U_i$ and for $u = U_f$. Hence it must have the form $A(U_i - u)(u - U_f)$ where $-A$ is the coefficient of u^2. One may thus write (a) as

$$\mu'' u \frac{du}{dx} = A(U_i - u)(u - U_f) + \beta ml \tag{19-21a}$$

where

$$A = -m\frac{\gamma + 1}{2\gamma}$$

Using the notation of Eq. 19-19 one may write the energy equation (Eq. 19-17c) as

$$ml - \frac{\mu''}{Pr''}\frac{dl}{dx} + \mu''\left(\frac{1}{Pr''} - 1\right)\frac{d(u^2/2)}{dx} = 0 \tag{b}$$

Eliminating the derivative of $u^2/2$ with the aid of Eq. 19-21a one finds

$$\frac{\mu''}{Pr''}\frac{dl}{dx} = ml\left[1 + \beta\left(\frac{1}{Pr''} - 1\right)\right] + A\left(\frac{1}{Pr''} - 1\right)(U_i - u)(u - U_f) \tag{19-21b}$$

The case $Pr'' = 1$. As is easily seen from either Eq. 19-17c or Eq. 19-21b the total enthalpy is constant for $Pr'' = 1$, i.e. the following energy integral is obtained

$$H = h + \frac{u^2}{2} = \text{const} = h_i + \frac{U_i^2}{2} = h_f + \frac{U_f^2}{2}, \; l = 0 \tag{19-22}$$

The momentum equation (Eq. 19-21a) then reduces to

$$\tau = \mu'' \frac{du}{dx} = \frac{A(U_i - u)(u - U_f)}{u} \tag{19-23a}$$

The corresponding equation for the heat flux q is obtained by differentiating Eq. 19-19. Again using the assumption $Pr'' = 1$ one finds

$$q = -k\frac{dT}{dx} = -\frac{\mu''}{Pr''}\frac{dh}{dx} = u\tau = A(U_i - u)(u - U_f) \quad (19\text{-}23b)$$

Eq. 19-23a and 19-23b can be used to show that u is decreasing and T is increasing. Thus the case of a stationary rarefaction shock is excluded. A special assumption about the entropy increase is not necessary to prove this, but only the assumption that μ'' and k are positive. The law about the entropy increase can then be derived independently of the special assumptions (19-18) by considering the one-dimensional form of Eq. 3-8c. In the discussion following this equation it was pointed out that the entropy sources are positive if μ'', $k > 0$. If Eq. 3-8c is integrated from $x = -\infty$ to $x = +\infty$ the total effect of entropy conduction is zero and the net increase in entropy is positive. Locally, however, entropy conduction is not negligible. Hence it has not been proved that the entropy gradient is everywhere positive.

We now return to the case $Pr'' = 1$. As pointed out above, both τ and q are everywhere negative. They have certain minimum values which are denoted by τ_m and q_m, respectively. ($|\tau_m|$ and $|q_m|$ are maxima of $|\tau|$ and $|q|$, respectively.) We now show that τ_m occurs when u has its critical value u_* and that q_m occurs when u has the value $U_a = (U_i + U_f)/2$. The critical point is defined as the point where the velocity equals the speed of sound, i.e. the point where $u = a = u_* = a_*$. Since the total enthalpy is constant (Eq. 19-22), u_* has the same value as in nonviscous isentropic flow, i.e.

$$u_*^2 = \frac{\gamma - 1}{\gamma + 1}U_i^2 + \frac{2a_i^2}{\gamma + 1}$$

The value of the density at the critical point is $\rho_* = m/u_*$ which is not in general the isentropic value.

Logarithmic differentiation of Eq. 19-23a shows that τ is stationary for $du/dx = 0$ and for $u^2 = U_iU_f$. However, Prandtl's relation for a shock wave in a perfect gas states that $U_iU_f = a_*^2$ (cf. [*63*]). Thus, since τ is negative it must reach its minimum value at the critical point where $u = u_* = a_*$. This value is

$$\begin{aligned} -\tau_m &= \frac{\gamma + 1}{2\gamma}\rho_*(U_i - a_*)(a_* - U_f) \\ &= \frac{\gamma + 1}{\gamma}m(U_a - a_*) \end{aligned} \quad (19\text{-}24a)$$

where $U_a = (U_i + U_f)/2$.

Similarly it follows from Eq. 19-23b that q has its minimum at $u = U_a$

and that

$$-q_m = \frac{\gamma + 1}{8\gamma} m(U_i - U_f)^2 \qquad = \frac{\gamma + 1}{2\gamma} m(U_a^2 - a_*^2) \tag{19-24b}$$

In Eq. 19-23 τ and q were determined as functions of u. The solution has thus been given in Crocco's form (for this terminology, cf. Art. 22 and the discussion of Eq. 19-6). In principle the shock layer structure may then be computed, i.e. u, h, etc. found as functions of x. One way to proceed is first to introduce a distorted coordinate ξ (cf. Howarth's transformation in boundary layer theory, Art. 22) defined by

$$\xi = \int_0^x \frac{dx}{\mu''} \qquad \frac{d}{d\xi} = \mu'' \frac{d}{dx} \tag{19-25}$$

The left-hand side of Eq. 19-23a then becomes $du/d\xi$. Integration gives ξ as an elementary function of u, within an arbitrary additive constant. This constant of integration is not determined by the problem, because the equations and the boundary conditions are invariant under a translation of the x axis. In principle, the functional relation between ξ and u can be inverted to determine u as a function of ξ. From Eq. 19-22 one then finds h as a function of ξ.

In order to eliminate ξ in favor of x we observe that Eq. 19-25 implies[23]

$$x = \int_0^{\xi} \mu'' d\xi = \int_{u(\xi=0)}^{\mu(\xi)} \mu'' \frac{d\xi}{du} du \tag{19-26}$$

If it is assumed that μ'' depends on h only and that Eq. 19-22 is valid, then the second integrand is a function of u only. Integration then gives x as a function of u. By inverting this relation one finds u, and hence h, τ, etc. as functions of x.

Some details of the scheme just proposed are now carried out. First, however, we write Eq. 19-23a in a form suitable for the study of weak shock layers. We introduce the notation

$$U_a = \frac{U_i + U_f}{2}, \qquad V = \frac{U_i - U_f}{2}, \qquad u = U_a + w$$
$$\epsilon = \frac{V}{U_a}, \qquad \bar{\xi} = A\epsilon\xi, \qquad \bar{w} = \frac{w}{V} \tag{19-27}$$

where A is the constant defined following Eq. 19-21a. With this notation one may write Eq. 19-23a as

$$\frac{dw}{d\xi} = A \frac{V^2 - w^2}{U_a + w} \tag{19-28}$$

[23] Throughout this discussion ξ and x may be replaced by ξ + const and x + const, respectively.

or

$$\frac{d\bar{w}}{d\xi} = \frac{1 - \bar{w}^2}{1 + \epsilon\bar{w}} \tag{19-28'}$$

At upstream infinity, that is $x = -\infty$, $\xi = -\infty$, the value of $\bar{w}$ is 1. At downstream infinity $\bar{w} = -1$. By integrating Eq. 19-28′ one obtains

$$\xi = \tanh^{-1} \bar{w} - \frac{\epsilon}{2} \ln (1 - \bar{w}^2) \tag{19-29}$$

This solution is normalized in such a way that $u = U$, i.e. $\bar{w} = 0$, when $\xi = 0$ and thus when $\xi = 0$ and $x = 0$.

From Eq. 19-29 or directly from Eq. 19-28′ it is seen that the solution *for the velocity has the following form:*

$$\bar{w} = f(\epsilon, \xi) \tag{19-30}$$

or

$$u - U_a = Vf\left(\frac{V}{U_a}, \frac{AV\xi}{U_a}\right) \tag{19-30'}$$

In principle, the function f can be found by inverting Eq. 19-29. This is now carried out for weak shock layers.

Weak shock layers. Pr″ arbitrary. The strength of a shock may be defined in various ways (cf. the references given for the theory of non-viscous discontinuous shock waves). In the present context it is most convenient to use the parameter ϵ (Eq. 19-27) as a measure of the shock strength.

We now discuss the structure of the shock layer for a weak shock (ϵ small). The case of $Pr'' = 1$ will be considered first. Afterwards it is shown that, with minor modifications, the results are valid for any value of Pr''.

The discussion of the exact solution for $Pr'' = 1$ shows that the velocity and pressure may be expanded as follows for small values of ϵ (cf. Eq. 19-27 and 19-30 for notation):

$$\bar{w} = f(\epsilon, \xi) = w_0(\xi) + \epsilon w_1(\xi) + \cdots \tag{19-31a}$$

$$\bar{p} = \frac{p - p_a}{mU_a\epsilon} = p_0(\xi) + \epsilon p_1(\xi) + \cdots \tag{19-31b}$$

where $p_a = (p_1 + p_f)/2$.

For $Pr'' = 1$ we then find immediately from Eq. 19-28′ or Eq. 19-29 that

$$w_0 = \tanh \xi \tag{19-32}$$

i.e.

$$u - U_a \cong V \tanh\left[-\frac{Vm(\gamma + 1)}{2\gamma U_a}\,\xi\right] \tag{19-32'}$$

This formula may be obtained directly from Eq. 19-23a if u in the denominator of the right-hand side is replaced by U_a. Eq. 19-28′ also gives the following equation for the first correction to the velocity distribution:

$$\frac{dw_1}{d\bar{\xi}} = -2w_0w_1 - w_0(1 - w_0^2) \tag{19-33}$$

This equation is easily integrated to

$$w_1 = -\frac{\ln(\cosh\bar{\xi})}{(\cosh\bar{\xi})^2} \tag{19-34}$$

Now let Pr'' be arbitrary. We assume that the expansions of velocity and pressure for small values of ϵ have the same form as for $Pr'' = 1$ (Eq. 19-31). It is characteristic of these expansions that ξ occurs only in the combination $\bar{\xi} = \epsilon\xi$. This implies that the thickness of a weak shock layer is inversely proportional to ϵ (cf. the discussion of shock layer thickness below). It also follows that $d/d\xi$, and hence also d/dx, is of order ϵ. Thus the derivatives in Eq. 19-17a and Eq. 19-17c are of higher order than the other terms. In particular, the viscous force may be neglected to first order in the momentum equation (Eq. 19-17a). To first order $mu + p$ is then constant in the shock layer. Formally one obtains by inserting expansions (Eq. 19-31a and 19-31b) into Eq. 19-17a

$$w_0 + p_0 = 0 \tag{19-35a}$$

i.e.

$$p - p_a \cong m(U_a - u) \tag{19-35b}$$

or

$$p - p_i \cong \rho_i U_i(U_i - u) \tag{19-35c}$$

Thus, to the first order, the pressure is found from the velocity with the aid of a linearized Bernoulli's law, familiar from the linearized theory of inviscid flow.

From Eq. 19-17a one also finds

$$\frac{dw_0}{d\bar{\xi}} = -\frac{\gamma+1}{2\gamma}(w_1 + p_1) \tag{19-36}$$

The equations just given relate the pressure to the velocity, but do not give the velocity distribution. For $Pr'' = 1$ the velocity distribution was determined with the aid of the energy integral $H = \text{const}$. This equation is no longer valid for $Pr'' \neq 1$. However, the assumption $d/dx \sim \epsilon$ implies that the change in H, i.e. l (Eq. 19-19), is of the order ϵ^2. This may be seen either from Eq. 19-21a or Eq. 19-21b. If terms of order ϵ^3 are neglected, Eq. 19-21b then reduces to the algebraic equation

$$0 = ml\left[1 + \beta\left(\frac{1}{Pr''} - 1\right)\right] + A\left(\frac{1}{Pr''} - 1\right)(U_i - u)(u - U_f) \tag{19-37}$$

Inserting this value for l into Eq. 19-21a one finds

$$\mu'' U_a \frac{du}{dx} = A'(U_i - u)(u - U_f) + O(\epsilon^3) \tag{19-38}$$

where

$$A' = \frac{A}{1 + \beta\left(\frac{1}{Pr''} - 1\right)} = \frac{-m(\gamma + 1)}{2\gamma + 2(\gamma - 1)\left(\frac{1}{Pr''} - 1\right)}$$

Eq. 19-38 differs from the corresponding equation for $Pr'' = 1$ only by the value of the multiplicative constant. One thus finds from Eq. 19-32 or 19-32′ that the velocity profile for a weak shock layer at an arbitrary value of Pr'' is given by

$$w = \tanh\left(\frac{A'}{A}\xi\right) \tag{19-39}$$

or

$$u - U_a \cong V \tanh\left[\frac{-Vm(\gamma + 1)\xi/U_a}{2\gamma + 2(\gamma - 1)\left(\frac{1}{Pr''} - 1\right)}\right] \tag{19-39′}$$

Since l is of order ϵ^2 one may evaluate the first order changes in enthalpy by taking the total enthalpy as constant. One finds

$$h - h_a \cong U_a(U_a - u) \tag{19-40}$$

From this equation it follows that, within terms of order ϵ^3, the heat flux q is given by Eq. 19-23b with A replaced by A'. Similarly, if the right-hand sides of Eq. 19-24a and Eq. 19-24b are multiplied by A'/A one obtains the expressions for τ_m and q_m, respectively, within terms of order ϵ^3.

Finally the entropy equation (Eq. 3-8c) shows that the total entropy increase $s_f - s_i$ is of the order ϵ^3. Inside the shock layer, however, the contribution of the term $-\operatorname{div}(q/T)$ must be considered. It follows that inside the shock layer $s - s_i$ is of order ϵ^2. In a sufficiently weak shock layer, i.e. $\epsilon^3 \ll \epsilon^2$, the entropy increases to a maximum s_m such that $s_m - s_i = O(\epsilon^2)$ and then decreases to its final value s_f where $s_f - s_i = O(\epsilon^3)$.

General Prandtl number and shock strength. When the Prandtl number Pr'' is different from 1 and the shock strength is not small, Eq. 19-17 may still be successfully attacked using modern theory of systems of ordinary differential equations, even if one in general does not obtain results in closed form. This problem has been discussed by von Mises [*99*] and Weyl [*100*]. The most complete discussion is contained in two papers by Gilbarg [*101,102*] who gives a complete mathematical solution of the shock layer problem for very general fluids. The method of Gilbarg also leads to comparatively simple numerical methods. The reader is referred in particular to [*102*] for details and also for further references.

If $Pr'' \neq 1$ the total enthalpy varies within the shock layer. This variation was given above for the case of very weak shocks. For a very small or a very large Prandtl number the heat flux thickness of the shock layer differs appreciably from the stress thickness (cf. the discussion below of the thickness of a shock wave). In general, however, the case of $Pr'' = 1$ shows the correct qualitative features of a shock layer.

Thickness of a shock layer. Strictly speaking, a shock layer has infinite width since the limiting values U_i, U_f, etc. are assumed to be reached only at $x = \pm\infty$. However, one may define an effective thickness of the shock layer (cf. the similar problem for transversal shear layers, discussed in connection with Eq. 7-5 and 13-21). This effective thickness can be (and has been) defined in many ways. Physically significant and simple definitions are the following (cf. Eq. 19-24)

$$\delta_\tau = \int_{-\infty}^{+\infty} \frac{\tau}{\tau_\text{m}}\, dx \tag{19-41a}$$

$$\delta_q = \int_{-\infty}^{+\infty} \frac{q}{q_\text{m}}\, dx \tag{19-41b}$$

The quantities δ_τ and δ_q are then measures for the extent of the region where transfer processes (viscous stress or heat flux) are important. A more convenient form of these relations is

$$\delta_\tau = \frac{\mu''_*}{|\tau_\text{m}|} \int_{U_\text{f}}^{U_\text{i}} \frac{\mu''}{\mu''_*}\, du \tag{19-42a}$$

$$\delta_q = \frac{k_*}{|q_\text{m}|} \int_{T_\text{i}}^{T_\text{f}} \frac{k}{k_*}\, dT \tag{19-42b}$$

As above, subscript "$_*$" denotes critical values, i.e. values at the point where $u = a$.

For a weak shock we may neglect the variation of μ'' and k with temperature and find (cf. Eq. 19-24 and the discussion after Eq. 19-40)

$$\delta_\tau \cong \frac{\mu''_*}{|\tau_\text{m}|}(U_\text{i} - U_\text{f}) \cong \frac{32\left[\gamma + (\gamma - 1)\left(\dfrac{1}{Pr''} - 1\right)\right]\nu'_*}{3(\gamma + 1)(U_\text{i} - U_\text{f})} \tag{19-43a}$$

$$\delta_q \cong \frac{k_*}{|q_\text{m}|}(T_\text{f} - T_\text{i}) \cong \frac{32\left[\gamma + (\gamma - 1)\left(\dfrac{1}{Pr''} - 1\right)\right]\nu'_*}{3\, Pr''\, (\gamma + 1)(U_\text{i} - U_\text{f})} \tag{19-43b}$$

where $\nu''_* = \mu''_*/\rho_*$. These relations show that the width of the shock layer is inversely proportional to its strength, a fact which was commented on earlier. Note that $\delta_\tau = \delta_q$ for $Pr'' = 1$.

For very strong shocks the thickness of the shock layer depends very much on the assumed viscosity-temperature relation. In the early literature on shock layers it was often stated that the computed thickness of a strong shock is much smaller than the mean free path and hence that the Navier-Stokes equations are inadequate for the description of a shock layer. However, a more careful examination of the problem shows that the early evaluations of the shock thickness were not completely realistic. An especially valuable discussion of this problem has been given by Gilbarg and Paolucci [*102*]. It is clear that for very large Mach numbers various effects may render the equations used in the present article invalid. However, the range of validity of these equations is probably greater than was formerly believed (cf. the discussion of experimental results below).

Note that the thickness of a boundary layer is of the order $\sqrt{\mu}$ (Art. 8); the thickness of a shock layer, on the other hand, is of the order μ''.[24] This is shown by the formulas for δ_τ and δ_q given above. It can also be seen from the fact that μ'' may be eliminated from Eq. 19-17 with the aid of the transformation (Eq. 19-25). Note that it must be possible to eliminate μ'' in some way since the relations between flow quantities at upstream and downstream infinity are independent of μ. As an example the entropy gradient s_x depends on the dissipation $\mu'' u_x^2$. If δ is the thickness of the shock layer, then gradients are of the order $1/\delta$. It then follows that $\delta \sim \mu''$ in order for the total entropy rise to be independent of μ.

Recent experiments on shock layers. Some remarkable measurements of shock layer structure have recently been carried out by Sherman [*106*]. The experimental technique was to measure the equilibrium (recovery) temperature T of a fine wire mounted parallel to a plane stationary shock layer. Interpretation of the data was based on the fact that the wire diameter was small compared to both the shock thickness and the mean free path. Under these conditions the wire temperature is closely related [*107,108,109*] to the molecular distribution function at the wire location.

Sherman's somewhat cautious conclusion is that "the Navier-Stokes equations appear satisfactory to predict the shock transition up to a Mach number of 2" in the sense that "it has been possible to find a distribution function which yields the Navier-Stokes equations when substituted into the Maxwell transport equations for mass, momentum, and energy, and which, when used in the calculation of the free molecule cylinder temperature, yields results in good agreement with experiment up to a Mach number of 2."

Fig. B,19b and B,19c show data obtained for initial Mach numbers

[24] A general mathematical principle explaining the difference between the two cases were discussed by Latta [*60*] (cf. the discussion of the Oseen boundary layer in Art. 17).

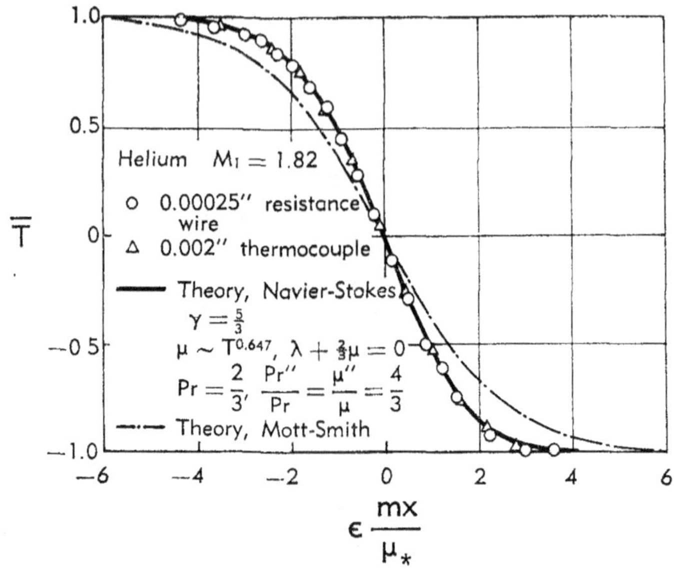

Fig. B,19b. Comparison of theory and experiment for shock layer [*106*]. Monatomic gas.

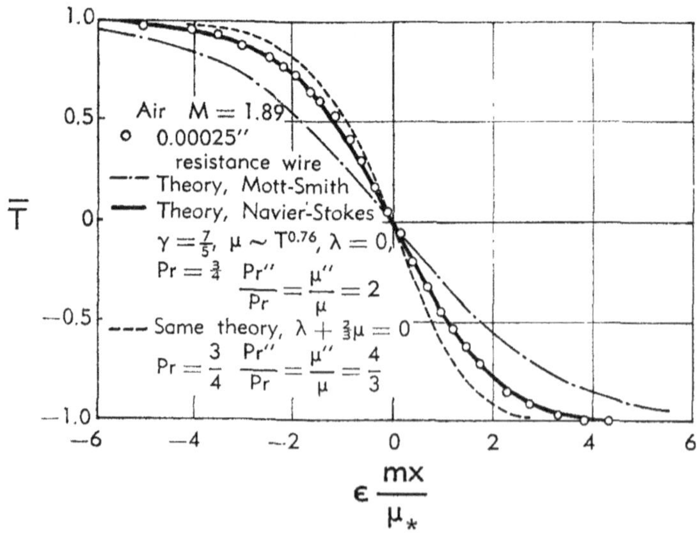

Fig. B,19c. Comparison of theory and experiment for shock layer [*106*]. Diatomic gas.

of 1.82 in helium and 1.89 in air. The ordinate, $\bar{T}$, is the wire recovery temperature T_r normalized in the same way as the velocity function $\bar{w}$ (Eq. 19-27); that is,

$$\bar{T} = \frac{2(T_r - T_{r_f})}{T_{r_i} - T_{r_f}} \tag{19-44}$$

The abscissa, $\epsilon mx/\mu_*$ in the present notation, is a Reynolds number based on the viscosity at sonic conditions. Also shown are profiles computed from the Navier-Stokes equations using values for γ, $Pr = c_p\mu/k$, and μ''/μ as noted in the figures. The solution is found by the method developed by Gilbarg and Paolucci [*101,102*]. Finally there is shown a shock layer profile, computed from the Boltzmann equation by the method of Mott-Smith [*110*].

Sherman concludes that the bulk viscosity $\lambda + 2\mu/3$ in helium is very nearly zero, while that in air is very nearly $2\mu/3$ (i.e. $\lambda = 0$ or $\mu''/\mu = 2$).[25] Thus Sherman's experiments indicate that physically realistic values for the conditions of these experiments are $Pr'' = \frac{8}{9}$ for helium and $Pr'' = \frac{3}{2}$ for air.

B,20. Limiting Cases for Extreme Values of Parameters. The nondimensional form of the Navier-Stokes equations, as presented in Art. 18, contains the nondimensional parameters Re, Pr, M, and γ; in addition, a parameter such as $\tilde{h}_w$ may appear in the boundary conditions. In many cases of practical interest the parameters M and Re may take on extreme values, either very large or very small. The value $\tilde{h}_w = 1$ may also be considered an extreme value since in that case the temperature difference between the solid and the flow at infinity is zero. Occasionally it is also of interest to consider values of Pr which are either very large or very small. The present article discusses various simplified equations that are approximately valid when one or more of the parameters mentioned above assumes an extreme value.[26] In particular, the equations for incompressible fluids discussed in Chap. 2 turn out to be a limiting case for special values of the parameters.

Similar problems were studied in Chap. 2 within the framework of the theory of incompressible fluids, namely flows at very large or very small Reynolds numbers.

[25] The real issue of these experiments is of course not the validity of the Stokes relation, Eq. 3-24, but the adequacy of continuum mechanics for describing phenomena within a shock layer. It is interesting to note that Sherman's testimony in favor of the Navier-Stokes equations rests on an experimental technique which itself ejects the idea of a continuum.

[26] The important case of flow at very large Mach numbers is not considered here (see Sec. E). It is hoped, however, that the perturbation methods used in the present article, and elsewhere in this section, may be profitably employed in the study of this rather difficult problem.

Compressible Flow at Zero Mach Number. Incompressible Flow. Some experience with this problem has been gained from the study of plane Couette flow (Art. 19). However, in that special case, pressure is constant whereas, as it will be seen, special attention must be paid to the pressure term in any more general case.

It seems reasonable to assume that the equations for $M = 0$ can be obtained by equating M to zero in the nondimensional form of the full equations (Eq. 18-4). The result depends, however, on what nondimensional variables are used. In particular a proper nondimensional pressure has to be found. If $\tilde{p} = p/p_\infty$ is used, the term grad $\tilde{p}/\gamma M^2$ dominates the momentum equation. This fact suggests the use of $p^* = (p - p_\infty)/\rho_\infty U^2$, especially since this is a suitable pressure variable for incompressible flow. Introducing this variable into Eq. 18-4 and then equating M to zero, one obtains the following equations:

$$\frac{\partial \tilde{\rho}}{\partial t^*} + \operatorname{div}(\tilde{\rho}\mathbf{u}^*) = 0 \tag{20-1a}$$

$$\tilde{\rho}\frac{D\mathbf{u}^*}{Dt^*} + \operatorname{grad} p^* = \frac{1}{Re}\operatorname{div}\tau^* \tag{20-1b}$$

$$\tilde{\rho}\frac{D\tilde{h}}{Dt^*} = \frac{1}{Pr\,Re}\operatorname{div}(\tilde{\mu}\operatorname{grad}\tilde{h}) \tag{20-1c}$$

$$\tilde{\rho}\tilde{h} = 1 \tag{20-1d}$$

These equations show that at $M = 0$ the heat added by dissipation may be neglected. Furthermore, pressure enters only in its dynamic role, not in its thermodynamic role. In the energy equation the term $D\tilde{p}/Dt^* = \gamma M^2 Dp^*/Dt^*$ is neglected. The full equation of state is $\tilde{\rho}\tilde{h} = 1 + \gamma M^2 p^*$, and here again the pressure term is neglected.

If in addition $\tilde{h}_w = 1$ then Eq. 20-1c and 20-1d, as well as the boundary condition on $\tilde{h}$, are identically satisfied by

$$\tilde{\rho} = 1 \tag{20-2a}$$

$$\tilde{h} = 1 \tag{20-2b}$$

and Eq. 20-1a and 20-1b reduce to the equations for incompressible flow in Chap. 2. Note that if $M = 0$ the assumption $\tilde{h}_w = 1$ is equivalent to the assumption $q_w = 0$. A viscous fluid may thus be considered incompressible if the heat added by dissipation and conduction is negligible, and if furthermore the pressure variation causes negligible enthalpy changes. In a nonviscous and nonheat-conducting fluid, only the last effect is important. In this case μ and k are equated to zero, that is, Re is infinite. The heat added by dissipation is then zero even for $M > 0$. The energy equation reduces to $\tilde{\rho}D\tilde{h}/Dt^* = (\gamma - 1)M^2 Dp^*/Dt^*$. If the right-hand side is neglected, $\tilde{h}$ does not change along streamlines and is hence equal to unity if h is constant at upstream infinity. Similarly, the term $\gamma M^2 p^*$ is neglected in the equation of state for M small so that $\tilde{\rho}$ is also everywhere unity.

For an incompressible fluid, viscous or nonviscous, one has the apparent paradox that pressure may vary with space and time, while density and temperature remain constant. The meaning of this statement may be investigated by further analyzing the limiting procedure that led to Eq. 20-1. Consider solutions for flow past a given object for a fixed value of Re but for various values of $\tilde{h}_w$ and M, in particular for M ranging between zero and a small value. If M is equated to zero, the type of the equations does not change and no boundary conditions have to be given up. One therefore assumes that the various flow quantities show a regular dependence on M. In particular we assume that $\tilde{p}$, $\tilde{\rho}$, and $\tilde{h}$ are power series in M^2 for small values of M:

$$\tilde{p} = p^{(0)} + M^2 p^{(1)} + M^4 p^{(2)} + \cdots \tag{20-3a}$$

$$\tilde{\rho} = \rho^{(0)} + M^2 \rho^{(1)} + M^4 \rho^{(2)} + \cdots \tag{20-3b}$$

$$\tilde{h} = h^{(0)} + M^2 h^{(1)} + M^4 h^{(2)} + \cdots \tag{20-3c}$$

where $p^{(0)}$, $p^{(1)}$, $\rho^{(0)}$ etc. are functions of x_i^* and t^* with $\tilde{h}_w$ as parameter. The first equation implies that

$$p^* = \frac{1}{\gamma M^2}(p^{(0)} - 1) + \frac{p^{(1)}}{\gamma} + \frac{M^2}{\gamma} p^{(2)} + \cdots \tag{20-3d}$$

If we require that grad p^* remain finite as M tends to zero, then $p^{(0)}$ is constant and its value is unity since $\tilde{p}_\infty = 1$. The equation of state is then

$$\tilde{\rho}\tilde{h} = \rho^{(0)}h^{(0)} + M^2(\rho^{(1)}h^{(0)} + \rho^{(0)}h^{(1)}) + \cdots = 1 + M^2 p^{(1)} + \cdots \tag{20-4}$$

Because the zeroeth order term of $\tilde{p}$ is constant, the dominant term of grad p^* is independent of M^2 and equal to $(1/\gamma)$ grad $p^{(1)}$. Thus $p^{(1)}$ occurs in the momentum equation. On the other hand, only the zeroeth order terms of $\tilde{\rho}$ and $\tilde{h}$ need to be considered for M very small. These terms are independent of $p^{(1)}$ as shown by Eq. 20-4. In particular, if $\tilde{h}_w = 1$ or $q_w = 0$ enthalpy and density are constant for very low Mach numbers in the sense that $\rho^{(0)}$ and $h^{(0)}$ are equal to unity. Pressure is still variable in the sense that $p^{(1)}$ varies. These statements do not contradict the equation of state as shown by Eq. 20-4. If $\tilde{h}_w - 1$ is small, $\rho^{(0)}$ and $h^{(0)}$ may be developed in power series of this parameter with leading terms equal to unity. More generally, the expansions (Eq. 20-3) may be replaced by a double power series in M^2 and $\tilde{h}_w - 1$.

Eq. 20-1 may be regarded as the equations for low speed laminar heat transfer. In this connection "low speed" means that the speed is small compared to the speed of sound; the dissipation is negligible. The case when the speed is low in the sense that the Reynolds numbers is low is discussed below.

The remarks above, as well as the discussion in Chap. 2 of incompressible flow at low or high Reynolds numbers, bring out the fact that in deriving approximate equations for small values of a parameter ϵ one has to make assumptions regarding the dependence of various flow quantities on ϵ near $\epsilon = 0$. In the present case it is assumed that the nondimensional quantities used are power series in M^2 and $(\tilde{h}_w - 1)$ with coefficients that are functions of x_i^* and t^*. This means that the perturbation problem is assumed to be regular in the sense that the resulting limiting equations have solutions which satisfy the same boundary conditions as the full equations and which are free of discontinuities. In the special case of Couette flow, the power series expansion can actually be constructed. Equations for higher order approximations may be easily found by inserting the power series expansions into the full equations. Their validity is limited by the radius of convergence of the power series (in terms of M^2 or $h_w - 1$). On the other hand, in the high Reynolds number case studied in Chap. 2, discontinuities appear as $1/Re$ tends to zero, as shown by the fact that the no-slip condition has to be abandoned for $1/Re = 0$. This has the consequence that a power series expansion in $1/Re$ is not possible.

COMPRESSIBLE FLOW AT LOW REYNOLDS NUMBERS. Low Reynolds number flow of an incompressible fluid was studied in Art. 9, 10, and 11, where the Stokes and Oseen equations were derived and discussed. The limiting procedures leading to these equations were studied in Art. 17. The analogous equations for a compressible fluid are now discussed. We consider limiting cases of Eq. 18-4 for very small values of the Reynolds number. The inner equations, analogous to the Stokes equations, are derived first. The outer equations, analogous to the Oseen equations, are then briefly mentioned; these latter equations are studied in greater detail in Art. 21.

The inner (Stokes) equations. We consider the limit of Eq. 18-4 as Re tends to zero. If the ratio of M^2 to Re is left unspecified, one obtains the following equations

$$\frac{\partial \tilde{\rho}}{\partial t^*} + \operatorname{div}(\tilde{\rho}\mathbf{u}^*) = 0 \tag{20-5a}$$

$$\operatorname{grad} p^+ = \operatorname{div} \tau^* \tag{20-5b}$$

$$0 = \frac{1}{Pr} \operatorname{div}(\tilde{\mu} \operatorname{grad} \tilde{h}) + (\gamma - 1)M^2\Phi^* \tag{20-5c}$$

$$\tilde{\rho}\tilde{h} = 1 + \gamma \frac{M^2}{Re} p^+ \tag{20-5d}$$

These equations differ from the full equations (Eq. 18-4) only by the omission of the transport terms in the momentum and energy equations. If the variation of M^2 with Re is specified, various additional simplifications of Eq. 20-5 result, as seen below. In each special case it may be verified that the limit applied to Eq. 20-5 gives the same result applied to the full Eq. 18-4. In this sense Eq. 20-5 are equations for low Re which are uniformly valid for any value of M.

First consider the case in which M^2 tends to zero faster than Re. Eq. 20-5 then reduce to

$$\frac{\partial \tilde{\rho}}{\partial t^*} + \operatorname{div}(\tilde{\rho}\mathbf{u}^*) = 0 \tag{20-6a}$$

$$\operatorname{grad} p^+ = \operatorname{div} \tau^* \tag{20-6b}$$

$$\operatorname{div}(\tilde{\mu} \operatorname{grad} \tilde{h}) = 0 \tag{20-6c}$$

$$\tilde{\rho}\tilde{h} = 1 \tag{20-6d}$$

If in addition $\tilde{h}_w = 1$, or equivalently $q_w = 0$, then $\tilde{\rho} = \tilde{h} = \tilde{\mu} = 1$ and Eq. 20-6 reduce to the Stokes equations for an incompressible fluid (Eq. 10-1 or 10-2). It was pointed out in Art. 10 that the Stokes equations for incompressible flow may be obtained either by linearization or by letting Re tend to zero. That these two procedures give the same result in the incompressible case is fortuitous; for compressible fluids the low Reynolds number equations are nonlinear as shown by Eq. 20-5 and 20-6.

In spite of their nonlinearity, Eq. 20-6 show certain simple features. Since Re does not enter these equations explicitly, the solutions have the same similarity as the Stokes equations. In particular, for an object of a given shape and size and for fixed $\tilde{h}_w$, $p - p_\infty$ at a fixed point is proportional to μ_∞ and to U. Since $\tilde{\mu}$ is assumed to depend on $\tilde{h}$ only, Eq. 20-6c may be solved independently of the other equations. The diffusion of heat takes place as if the medium were at rest. Transport of heat is neglected; furthermore, no heat is generated by dissipation. The equation of state (Eq. 20-6d) is identical with the previously derived equation of state for $M = 0$ (cf. the discussion of Eq. 20-1d). The role of pressure is thus purely dynamic; it does not occur as a thermodynamic variable. Formally, this follows from the fact that p^+ rather than $\tilde{p}$, is the proper nondimensional pressure variable in the case considered. In solving the equations, one first finds $\tilde{h}$ from the energy equation and the $\tilde{\rho}$ from the equation of state. The continuity and momentum equations then become linear equations whose variable coefficients involve the known functions $\tilde{\rho}$ and $\tilde{h}$.

Consider next the case in which M^2 is of the same order as Re, i.e. M^2/Re tends to a nonzero finite constant in the limit. The full equation of state must now be retained. As pressure variable one may choose either p^+ or $\tilde{p}$, the two being proportional to each other. The dissipation term $(\gamma - 1)M^2\Phi^*$ is negligible in the limit.

Finally, there is the case in which Re tends to zero faster than M^2. This case presents some special difficulties and will not be considered here.

The outer (Oseen) equations. In Art. 10, 11, and 17 it was shown that two limit processes play a fundamental role in the study of low Reynolds number flow, namely the inner (Stokes) limit and the outer (Oseen) limit. In the discussion above it was implicitly assumed that the Stokes limit was used. However, in the two-dimensional case it is necessary also to use the Oseen limit. The ideas used in the incompressible case carry over to the present case. The Oseen limit may be thought of as the limit obtained by letting the characteristic length L shrink to zero. In many cases (cf. Art. 17) the body then shrinks to a point or a line without arresting power. Hence the corresponding equations are obtained by linearizing about free stream conditions. The resulting Oseen equations for compressible flow are discussed in Art. 21. Matching between the inner and outer solutions can be carried out with the aid of the general methods developed by Kaplun (see Art. 17).

Note on the validity of the Navier-Stokes equations. It may be objected that the Navier-Stokes equations are no longer valid in the limits considered above, especially not in the limit where M^2/Re tends to infinity. It must certainly be true that the Navier-Stokes equations cease to be valid for extreme values of certain parameters. However, a similar criticism would apply, say, to the classical case of high Reynolds number and zero Mach number. As the Mach number tends to zero, at constant free stream velocity, the velocity of sound, and hence the temperature, tends to infinity. At sufficiently high temperatures any real gas certainly has properties which are not accounted for in the Navier-Stokes equations.

The answer to the objection stated above is that letting a parameter tend to zero, say, is a mathematical device for obtaining approximate solutions for small values of the parameter (more generally one should consider complete expansions of Navier-Stokes solutions for small values of this parameter as was done in Art. 17). The method is physically significant if there are values of the parameter which are sufficiently large for the Navier-Stokes equations to be adequate, and at the same time sufficiently small for the mathematical method used to be approximately valid. As yet, it has not been investigated carefully whether or not such values of the parameters exist in the cases considered above. In order to settle this question it is first necessary to carry out the methods of solutions indicated above. The results must then be compared, either with solutions of more accurate equations, or, preferably, with experiments.

Flow at Large Reynolds Numbers. Boundary Layer and Shock Layer Equations.

Boundary layers. The boundary layer equations for high Reynolds number flow of an incompressible fluid were discussed in Art. 13. The basic qualitative ideas set forth there may be carried over to the case of a compressible fluid.

As before, when the viscosity ν is equal to zero, the no-slip condition for the velocity can no longer be satisfied without admitting discontinuities. For small values of ν there is then a thin layer near the body in which the velocity changes rapidly from zero at the surface to a value close to that predicted by the inviscid solution. In addition, for the compressible case the coefficient of heat conduction k must also be considered. When k is equal to zero, the boundary condition involving the temperature at the solid cannot be satisfied. For small values of k the temperature can be expected to vary rapidly in a layer next to the solid. The compressible case may thus have a temperature boundary layer as well as a velocity boundary layer. For the fluids considered here these boundary layers will nearly coincide, since the assumption of nearly constant Prandtl number of order unity implies that k is approximately proportional to μ. More precisely, it was shown in Art. 13 by an intuitive argument that, when the Reynolds number $Re = UL/\nu_\infty$ is large, a thin velocity boundary layer exists having a thickness δ which is of order $LRe^{-\frac{1}{2}}$. The argument, which was based on the fact that viscosity causes vorticity, or velocity, to diffuse like heat, is still qualitatively correct in the case of compressible flow. This argument may of course be repeated, a fortiori, for the diffusion of heat itself. The parameter corresponding to the Reynolds number is the Péclet number $Pe = UL\rho c_p/k$. For large Péclet numbers a thin temperature boundary layer exists whose thickness δ is of the order $LPe^{-\frac{1}{2}}$. However, $Pe = PrRe$ so that, since Pr is of order unity, the nondimensional thickness of both boundary layers is of order $Re^{-\frac{1}{2}}$ (cf. the nondimensional form of the Navier-Stokes equations given by Eq. 18-4).

Thus boundary layer equations, including both velocity and temperature boundary layers, may be derived by applying the limit procedure discussed in Art. 13 to the Navier-Stokes equations for a compressible fluid. This derivation is discussed below. It is carried out for the general case of nonstationary three-dimensional flow. The three-dimensionality of the flow does not require any new principles of derivation. The result of the derivation, however, shows that the certain curvature terms in the full equations may not be neglected in the general three-dimensional case. In a two-dimensional boundary layer the effect of all curvature terms is negligible. For further discussion of the derivation of the equations see also the recent work of Hayes [*111*] and Howarth [*112*].

Derivation of equations. To obtain boundary layer equations by the limit process, a set of curvilinear coordinates x_i is introduced such that, on the part of the boundary where the boundary layer occurs, x_3 is equal to zero. This condition is the only essential requirement on the system of coordinates. For simplicity we consider only orthogonal coordinates. For a given body surface there are in general many ways of constructing three families of mutually orthogonal coordinate surfaces which contain the body surface as one of the coordinate surfaces. Furthermore, even when the coordinate surfaces are prescribed the metric may be chosen in many ways. In specific examples use is made of this freedom of choice in order to obtain especially simple forms of the boundary layer equations. The metrical properties of the coordinate system are determined by the coefficients e_i (cf. h_i of Art. 5). Different metrics are associated with different ways of labeling the surfaces. If, for instance, two infinitesimally close surfaces of const x_i are labeled $x_i = a$ and $x_i = b$ respectively, and the orthogonal distance between the surfaces is ds at the point considered, then e_i is determined by $ds = e_i|b - a|$. If the two surfaces are relabeled $x_i = a_1$ and $x_i = b_1$ respectively, then e_i changes since ds remains the same. In general the h_i are functions of all x_i.

The velocity component in the x_i direction is denoted by u_i and taken in the sense of vector geometry (cf. Art. 5), that is, u_i is the algebraic length of the projection of $\mathbf{u}$ on the positive direction of the x_i axis, and the square of the magnitude of the velocity is then $u^2 = u_1^2 + u_2^2 + u_3^2$.

In order to carry out the limit procedure, one first writes the full Navier-Stokes in the form corresponding to the coordinate system chosen. Then the new variables

$$\bar{x}_3 = \frac{x_3}{\sqrt{\nu}}, \qquad \bar{u}_3 = \frac{u_3}{\sqrt{\nu}} \tag{20-7}$$

are introduced. Finally ν is equated to zero. The equations then obtained are the boundary layer equations. Finally these equations may be expressed in terms of the original variables by reintroducing x_3 and u_3 according to Eq. 20-7.

As an example of the resulting simplification of the equations consider the acceleration term in the first momentum equation

$$\begin{aligned}\frac{Du_1}{Dt} \equiv \frac{\partial u_1}{\partial t} + \frac{u_1}{e_1}\frac{\partial u_1}{\partial x_1} + \frac{u_2}{e_2}\frac{\partial u_1}{\partial x_2} + \frac{u_3}{e_3}\frac{\partial u_1}{\partial x_3} \\ + \frac{u_1 u_2}{e_1 e_2}\frac{\partial e_1}{\partial x_2} - \frac{u_2^2}{e_1 e_2}\frac{\partial e_2}{\partial x_1} + \frac{u_1 u_3}{e_1 e_3}\frac{\partial e_1}{\partial x_3} - \frac{u_3^2}{e_1 e_3}\frac{\partial e_3}{\partial x_1}\end{aligned} \tag{20-8}$$

In the limit procedure $e_i(x_1, x_2, x_3)$ is replaced by $e_i(x_1, x_2, 0)$, that is, the metric is evaluated at the boundary. Formally e_i should be written $e_i(x_1, x_2, \sqrt{\nu}\,\bar{x}_3)$ and as ν tends to zero the last argument tends to zero. Actually this means that the coordinate system is chosen independently of ν so that, for small values of ν, the e_i vary slowly as x_3 varies within the boundary layer. Also, the e_i derivative $\partial e_1/\partial x_3$ is then of the order unity in the boundary layer since

$$\frac{\partial e_1(x_1, x_2, x_3)}{\partial x_3} = \frac{1}{\sqrt{\nu}} \frac{\partial e_1(x_1, x_2, \sqrt{\nu}\,\bar{x}_3)}{\partial \bar{x}_3} = e_1'(x_1, x_2, \sqrt{\nu}\,\bar{x}_3) \to e_1'(x_1, x_2, 0)$$

(prime denotes derivative with respect to the last argument). Hence in the limit the term $(u_1u_3/e_1e_3)\partial e_1/\partial x_3$ is negligible since u_3 is of the order $\sqrt{\nu}$. The last term containing u_3^2 also vanishes in the limit. All other terms are of the order unity and hence remain in the limit. It is clear that terms containing neither u_3 nor x_3 explicitly are of the order unity. The term $(u_3/e_3)\partial u_1/\partial x_3$ is also of the order unity, since the factor u_3, of the order $\sqrt{\nu}$, is balanced by the factor $\partial u_1/\partial x_3$ which, due to the rapid variation of u_1 with x_3, is of the order $1/\sqrt{\nu}$.

As another example, consider the viscous stress tensor $\boldsymbol{\tau}$. The boundary layer approximation to $\boldsymbol{\tau}$ is $\sqrt{\mu}\lim \boldsymbol{\tau}/\sqrt{\mu}$ where lim is understood to mean the second limit process (cf. Eq. 13-7). Since div $\mathbf{u}$ is of the order unity and λ of the order μ it follows that the limit of the compressive part of the stress tensor λ div $\mathbf{u}\mathbf{I}$ is zero. In the remaining part of $\boldsymbol{\tau}$, namely μ def $\mathbf{u}$, only the terms $\partial u_1/\partial x_3$ and $\partial u_2/\partial x_3$ give a contribution in the limit, in particular the curvature terms all vanish. Hence the boundary layer approximation $\boldsymbol{\tau}_{\mathrm{BL}}$ to the stress tensor $\boldsymbol{\tau}$ is

$$\boldsymbol{\tau}_{\mathrm{BL}} = \mu \begin{pmatrix} 0 & 0 & \dfrac{1}{e_3}\dfrac{\partial u_1}{\partial x_3} \\ 0 & 0 & \dfrac{1}{e_3}\dfrac{\partial u_2}{\partial x_3} \\ \dfrac{1}{e_3}\dfrac{\partial u_1}{\partial x_3} & \dfrac{1}{e_3}\dfrac{\partial u_2}{\partial x_3} & 0 \end{pmatrix} \tag{20-8a}$$

The matrix in the above equation represents the boundary layer approximation to def $\mathbf{u}$, i.e. the symmetric part of the tensor grad $\mathbf{u}$. The antisymmetric part of the same tensor is related to curl $\mathbf{u} = \boldsymbol{\Omega}$ explained in Art. 1. The reduction of $\boldsymbol{\Omega}$, corresponding to Eq. 20-8a, is

$$\boldsymbol{\Omega}_{\mathrm{BL}} = \left(-\frac{1}{e_3}\frac{\partial u_2}{\partial x_3}, \frac{1}{e_3}\frac{\partial u_1}{\partial x_3}, 0\right) \tag{20-8b}$$

If similar reductions are made for all the terms in the Navier-Stokes equations the general boundary layer equations for a compressible fluid are obtained:

$$\frac{\partial \rho}{\partial t} + \frac{1}{e_1e_2e_3}\left[\frac{\partial(e_2e_3\rho u_1)}{\partial x_1} + \frac{\partial(e_1e_3\rho u_2)}{\partial x_2} + \frac{\partial(e_1e_2\rho u_3)}{\partial x_3}\right] = 0 \tag{20-9a}$$

$$\frac{\partial u_1}{\partial t} + \frac{u_1}{e_1}\frac{\partial u_1}{\partial x_1} + \frac{u_2}{e_2}\frac{\partial u_1}{\partial x_2} + \frac{u_3}{e_3}\frac{\partial u_1}{\partial x_3} + \frac{1}{\rho e_1}\frac{\partial p}{\partial x_1} + \frac{u_1 u_2}{e_1 e_2}\frac{\partial e_1}{\partial x_2} - \frac{u_2^2}{e_1 e_2}\frac{\partial e_2}{\partial x_1} = \frac{1}{\rho e_3}\frac{\partial}{\partial x_3}\left(\frac{\mu}{e_3}\frac{\partial u_1}{\partial x_3}\right) + f_1 \quad \text{(20-9b)}$$

$$\frac{\partial u_2}{\partial t} + \frac{u_1}{e_1}\frac{\partial u_2}{\partial x_1} + \frac{u_2}{e_2}\frac{\partial u_2}{\partial x_2} + \frac{u_3}{e_3}\frac{\partial u_2}{\partial x_3} + \frac{1}{\rho e_2}\frac{\partial p}{\partial x_2} + \frac{u_1 u_2}{e_1 e_2}\frac{\partial e_2}{\partial x_1} - \frac{u_1^2}{e_1 e_2}\frac{\partial e_1}{\partial x_2} = \frac{1}{\rho e_3}\frac{\partial}{\partial x_3}\left(\frac{\mu}{e_3}\frac{\partial u_2}{\partial x_3}\right) + f_2 \quad \text{(20-9c)}$$

$$\frac{1}{\rho}\frac{\partial p}{\partial x_3} = 0 \quad \text{(20-9d)}$$

$$\rho\left(\frac{\partial h}{\partial t} + \frac{u_1}{e_1}\frac{\partial h}{\partial x_1} + \frac{u_2}{e_2}\frac{\partial h}{\partial x_2} + \frac{u_3}{e_3}\frac{\partial h}{\partial x_3}\right) - \left(\frac{\partial p}{\partial t} + \frac{u_1}{e_1}\frac{\partial p}{\partial x_1} + \frac{u_2}{e_2}\frac{\partial p}{\partial x_2}\right) = \frac{1}{e_3}\frac{\partial}{\partial x_3}\left(\frac{\mu}{Pr}\frac{1}{e_3}\frac{\partial h}{\partial x_3}\right) + \mu\left[\left(\frac{1}{e_3}\frac{\partial u_1}{\partial x_3}\right)^2 + \left(\frac{1}{e_3}\frac{\partial u_2}{\partial x_3}\right)^2\right] + \rho Q \quad \text{(20-9e)}$$

In these equations e_1, e_2, e_3, f_1, f_2, and Q should be evaluated at the surface of the body, that is, for $x_3 = 0$.

The relations between the various thermodynamic state variables and between these and the transport coefficients are unaffected by the limit procedure. Typical boundary conditions for boundary layer flow obeying Eq. 20-9 are the following (as in Art. 13, subscript $_e$ denotes the corresponding external nonviscous flow past the body):

$$u_i(x_1, x_2, 0) = 0, \; i = 1, 2, 3 \quad \text{(20-10a)}$$

$$u_1(x_1, x_2, \infty) = u_{1e}(x_1, x_2, 0) \quad \text{(20-10b)}$$

$$u_2(x_1, x_2, \infty) = u_{2e}(x_1, x_2, 0) \quad \text{(20-10c)}$$

$$T(x_1, x_2, 0) = \text{given function of } x_1 \text{ and } x_2 \quad \text{(20-10d)}$$

$$T(x_1, x_2, \infty) = T_e(x_1, x_2, 0) \quad \text{(20-10e)}$$

$$p(x_1, x_2, \infty) = p_e(x_1, x_2, 0) \quad \text{(20-10f)}$$

Instead of Eq. 20-10d one often imposes a condition on the heat conduction at the wall. For an insulated wall

$$\frac{\partial T}{\partial x_3} = 0 \quad \text{at} \quad x_3 = 0 \quad \text{(20-10d}')$$

In other cases some relation between heat conduction and temperature may be prescribed.

As pointed out in Art. 14 and 17, it is often difficult to determine the correct external flow. These difficulties will of course remain in the compressible case, where the situation may be further aggravated by such phenomena as shock wave-boundary layer interaction.

Cylindrical flow. Two-dimensional flow. Solutions of the boundary layer equations for special cases will be discussed in Chap. 4. Examples are given there which show how the e_1 may be chosen appropriately (cf. in particular Art. 33). Only one special example of a metric is discussed here, namely a metric which may be used in discussing the boundary layer on a cylinder. (The two-dimensional metric used in Eq. 13-2 is included as a special case.) Let the body be a cylinder, that is, its surface made up of generators which are parallel infinite lines. The planes perpendicular to these generators are taken as the planes x_2 = const. In the two-dimensional case the flow is perpendicular to the generators, that is, $u_2 = 0$ and $\partial/\partial x_2 = 0$. We consider here the more general case of a swept-back cylinder. The flow velocity at infinity may have a component perpendicular to the generators and a component parallel to the generators, so that $u_2 \neq 0$. However, since the boundary conditions and the equa-

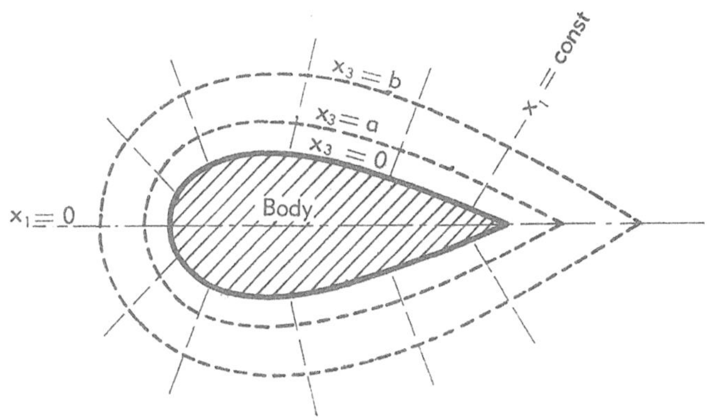

Fig. B,20. Coordinates for cylindrical flow.

tions are invariant under a translation in the x_2 direction, it is assumed that the solution has the same symmetry, that is, $\partial/\partial x_2 = 0$. (As pointed out in Art. 17, this need not always be true for unstable flow of a real fluid.) We choose $e_2 = 1$ which means that the perpendicular distance between the planes $x_2 = a$ and $x_2 = b$ is $|b - a|$. A coordinate system is now chosen in the plane $x_2 = 0$ (cf. Fig. B,20). A set of curves enclosing the curve $x_3 = 0$ are taken as the curves x_3 = const. The following restriction can be made on these curves: The curves $x_3 = a$ which are infinitesimally close to the body surface have a distance ds to the surface which is equal to a. In other words, if $\partial/\partial n$ denotes the derivative in the outward normal direction, then $(\partial x_3/\partial n)_{x_3=0} = 1$ or $e_3 = 1$ at the body surface. The orthogonal trajectories to the curves x_3 = const are taken as the curves x_1 = const. The constants are so chosen that the distance along the body between $x_1 = a$ and $x_1 = b$ is $|b - a|$. Hence $e_1 = 1$ at the body surface. The stagnation point may be defined as the point on the body where the streamlines constructed from the x_1 and x_3 components of the external flow have their forward stagnation point. At this point x_1 is given the value zero. From these coordinate lines in the plane $x_2 = 0$ one constructs coordinate surfaces in the entire space by assuming that x_1 and x_3 do not vary in a direction perpendicular to this plane. The metric so constructed has the following properties:

$$e_1 = fn(x_1, x_3); \qquad e_1 = 1 \quad \text{at} \quad x_3 = 0 \tag{20-11a}$$

$$e_2 = 1 \tag{20-11b}$$

$$e_3 = f_n(x_1, x_3); \qquad e_3 = 1 \quad \text{at} \quad x_3 = 0 \tag{20-11c}$$

In the boundary layer equations (Eq. 20-9) the e_i should be given their value at the body surface, which in this case is $e_i = 1$. Hence, the curvature terms involving $\partial e_i/\partial x_j$ are all zero. Furthermore $\partial/\partial x_2 = 0$ so that the boundary layer equations for cylindrical flow simplify to

$$\frac{\partial \rho}{\partial t} + \frac{\partial(\rho u_1)}{\partial x_1} + \frac{\partial(\rho u_3)}{\partial x_3} = 0 \tag{20-12a}$$

$$\rho\left(\frac{\partial u_1}{\partial t} + u_1\frac{\partial u_1}{\partial x_1} + u_3\frac{\partial u_1}{\partial x_3}\right) + \frac{\partial p}{\partial x_1} = \frac{\partial}{\partial x_3}\left(\mu\frac{\partial u_1}{\partial x_3}\right) + \rho f_1 \tag{20-12b}$$

$$\rho\left(\frac{\partial u_2}{\partial t} + u_1\frac{\partial u_2}{\partial x_1} + u_3\frac{\partial u_2}{\partial x_3}\right) = \frac{\partial}{\partial x_3}\left(\mu\frac{\partial u_2}{\partial x_3}\right) + \rho f_2 \tag{20-12c}$$

$$\frac{\partial p}{\partial x_3} = 0 \tag{20-12d}$$

$$\rho\left(\frac{\partial h}{\partial t} + u_1\frac{\partial h}{\partial x_1} + u_3\frac{\partial h}{\partial x_3}\right) - \left(\frac{\partial p}{\partial t} + u_1\frac{\partial p}{\partial x_1}\right) = \frac{\partial}{\partial x_3}\left(\frac{\mu}{Pr}\frac{\partial h}{\partial x_3}\right) + \mu\left[\left(\frac{\partial u_1}{\partial x_3}\right)^2 + \left(\frac{\partial u_2}{\partial x_3}\right)^2\right] \tag{20-12e}$$

The boundary layer equations for two-dimensional flow are a special case of Eq. 20-12 with $u_2 = 0$.

Free shear layers. The boundary layer may be regarded as a region of rapid transition which in the limit of zero viscosity is replaced by a discontinuity along the streamlines at the body surface. Discontinuities may also occur at free streamlines or stream surfaces in nonviscous flow. Across such a discontinuity, temperature and velocity may jump but pressure must be continuous. Such a discontinuity is sometimes called a contact discontinuity (cf. the discussion in [*98*, p. 126]). A discontinuity in velocity implies that the vorticity is infinite. For $\nu > 0$ this vorticity concentration diffuses in the same manner as in a boundary layer except that the velocity is not zero at the original place of discontinuity. From usual arguments about diffusion of vorticity it is found that the velocity discontinuity in nonviscous flow is replaced by a shear layer of thickness $\sim \sqrt{\nu}$ in which the velocity varies rapidly but continuously. The temperature discontinuity is replaced by the same type of continuous temperature variation. Eq. 20-10 may then be used as approximate equations for the flow in a free shear layer. A very simple example of a free shear layer was given in Art. 7 in connection with Rayleigh's problem. In supersonic flow, contact discontinuities may arise as a result of shock intersections (cf. [*98*, Chap. IV,D]).

Shock layers. A feature of compressible flow with no analogue in incompressible flow is the occurrence of shock waves. In the nonviscous theory these waves take the form of surfaces, across which the velocity and the state variables are discontinuous. These discontinuities do not occur on stream surfaces but demand a normal component of velocity which is supersonic relative to the wave. It is to be expected that, for small but nonzero values of the coefficients of viscosity, viscous and thermal diffusion smooth out the shock waves into zones of rapid but continuous change. The continuous transition layer corresponding to a shock wave is called a shock layer. Shock layers and boundary layers are thus similar, in that viscous and thermal diffusion prevents the occurrence of discontinuities. A more detailed consideration, however, shows significant differences between these two types of transition layers.

In Art. 19 it was shown that the full Navier-Stokes equations may be solved for the special case of a one-dimensional stationary shock layer. As ν tends to zero the solution for the velocity u tends to a discontinuous function which is equal to U_i for $x < 0$ and U_2 for $x > 0$. Other flow quantities such as pressure, temperature, etc. exhibit similar discontinuities in the limit. These discontinuities satisfy the jump conditions for a nonviscous shock wave. Guided by this exact solution for a shock layer one may try to find approximate solutions for shock layers in other cases.

Starting with the simplest case, consider first one-dimensional flow generated by the motion of a piston into a column of gas at rest. It should be noted that the boundary conditions and the initial conditions on velocity and pressure are the same whether the fluid is considered viscous or not. In the former case a continuous solution is possible. However, in general, discontinuities must be admitted in the latter case in order to make a solution possible. Consider now a particular discontinuous shock wave produced by the motion of the piston. The problem is to obtain approximate equations for the corresponding shock layer. Following the method for obtaining the boundary layer approximation one may first replace the original x, t coordinates by x_1, t_1 such that the position of the shock wave is always $x_1 = 0$. In Art. 19 the important fact was emphasized that the shock layer has a thickness $\sim\nu$ whereas the boundary layer has a thickness $\sim \sqrt{\nu}$. In shock layer theory it is hence assumed that $\partial/\partial x_1 \sim 1/\nu$. If then the largest terms are retained in the equations of motion, expressed in the system, the approximate shock layer equations are obtained. An equivalent procedure is to introduce the new variable

$$\bar{x}_1 = \frac{x_1}{\nu_\infty} \tag{20-13}$$

where ν_∞ = kinematic viscosity far away from the shock, into the equations of motion and then let ν tend to zero, keeping $\bar{x}_1$ fixed.

The following example illustrates the procedure. Let the position of a one-dimensional shock wave be $x = f(t)$ according to nonviscous theory. Introduce coordinates (x_1, t_1) which move with the shock wave:

$$x_1 = x - f(t) \tag{20-14a}$$

$$t_1 = t \tag{20-14b}$$

$$u_1 = u - f'(t) \tag{20-14c}$$

where u_1 is the velocity of the gas relative to the shock wave. In this system of coordinates the position of the shock wave is $x_1 = 0$. According to Eq. 20-14

$$\frac{\partial}{\partial t} = \frac{\partial}{\partial t_1} - f'(t_1)\frac{\partial}{\partial x_1} \tag{20-15a}$$

$$\frac{\partial}{\partial x} = \frac{\partial}{\partial x_1} \tag{20-15b}$$

and hence

$$\frac{\partial u}{\partial x} = \frac{\partial u_1}{\partial x_1} \tag{20-15c}$$

$$\frac{\partial}{\partial t} + u\frac{\partial}{\partial x} = \frac{\partial}{\partial t_1} + u_1\frac{\partial}{\partial x_1} \tag{20-15d}$$

The continuity equation is then

$$\frac{\partial \rho}{\partial t_1} + \frac{\partial(\rho u_1)}{\partial x_1} = 0$$

Since $\partial\rho/\partial t \sim 1$ and $\partial(\rho u_1)/\partial x_1 \sim 1/\nu$ the shock layer approximation to the continuity equation is then $\partial(\rho u_1)/\partial x_1 = 0$. In the momentum equation the acceleration

$$\rho\left(\frac{\partial u}{\partial t} + u\frac{\partial u}{\partial x}\right)$$

is equal to

$$\rho\left(\frac{\partial u_1}{\partial t_1} + f''(t) + u_1\frac{\partial u_1}{\partial x_1}\right)$$

The first two terms are of order unity so that they can be neglected with respect to $u_1\partial u_1/\partial x_1$ which is of order $1/\nu$. The pressure and viscous stress terms in the momentum equation are of order $1/\nu$. The energy equation may be reduced in a similar fashion. This procedure yields the following shock layer equations:

$$\rho u_1 \frac{\partial u_1}{\partial x_1} + \frac{\partial p}{\partial x_1} = \frac{\partial}{\partial x_1}\left(\mu'' \frac{\partial u_1}{\partial x_1}\right) \tag{20-16a}$$

$$\frac{\partial(\rho u_1)}{\partial x_1} = 0 \tag{20-16b}$$

$$\rho u_1 \frac{\partial h}{\partial x_1} - u_1 \frac{\partial p}{\partial x_1} = \frac{\partial}{\partial x_1}\left(k \frac{\partial T}{\partial x_1}\right) + \mu'' \left(\frac{\partial u_1}{\partial x_1}\right)^2 \tag{20-16c}$$

where μ'' is defined as in Eq. 19-16.

In these equations the time dependence does not appear and they are in fact equivalent to Eq. 19-16 which are the equations for a stationary shock layer.

The boundary conditions for Eq. 20-16 are that the values of the flow quantities at $x_1 = -\infty$ are the same as the values of the corresponding flow quantities in nonviscous flow at the point $x_1 = -0$, that is, their limiting values at $x_1 = -\epsilon$, $\epsilon > 0$, as ϵ tends to zero. A similar statement holds for the values at $x_1 = +\infty$. This boundary condition is the analogue of the requirement in boundary layer theory that velocity and pressure at infinity in the boundary layer agree with the external flow at the corresponding points on the body surface.

The method outlined above may be applied to the general three-dimensional nonstationary case. Coordinates (x_i, t) are introduced such that the position of the shock wave is $x_1 = 0$ for all values of time. The assumption $\partial/\partial x_1 \sim 1/\nu$ again leads to the shock layer equations (Eq. 20-16). Thus the shock layer equations are always one-dimensional, stationary, and independent of the special flow problem studied. The space variables other than x_1, and the time variable, enter only through the boundary conditions. These will always be of the type used in Art. 19. Thus there is essentially one shock layer problem and this problem may be solved by the methods discussed in Art. 19.

The method of deriving the shock layer equations by a special limit process was explicitly stated by Weyl [*100*]. For further details, see also the work by Ludford [*113*]. Its connection with the general theory of singular perturbations has been investigated by Latta [*60*].

B,21. Linearized Equations. A Quasi-Linear One-Dimensional Equation. We now discuss solutions of two types of simplified equations. The first part of this article deals with linearized equations of motion which are the extension of the Oseen equations to the case of compressible flow. The linearized equations are presented with some comments on their general properties and some simple solutions are given. Their relation to the Navier-Stokes equations is then discussed. Their role as approximate equations is similar to that of the Oseen equations in incompressible flow. However, the main use of the linearized equations so far has been in the study of the effects of viscosity and heat conduction on acoustic waves. The linearized equations may also be considered as a qualitative mathematical model for the full Navier-Stokes equations. Solutions of the linearized equations give a shock layer thickness of order $\sqrt{\nu}$ rather than ν. This limits their usefulness as approximations and as mathematical models.

A simple quasi-linear parabolic equation is studied later in this article. The retention of a nonlinear transport term leads to a correct value ($\sim\nu$) of the shock layer thickness. This equation was introduced by Cole [*53*] for the purpose of deriving approximate equations for nonstationary one-dimensional shock layers. It had previously been used by Burgers [*114*] as a mathematical model. Its general solution is given and a special solution discussed in detail.

LINEARIZED (OSEEN) EQUATIONS FOR COMPRESSIBLE FLOW. The equations studied in this subarticle are obtained by linearizing the Navier-Stokes equations about free stream conditions. It is assumed that the velocity at infinity is normalized to be zero with the aid of a Galilean transformation: $\mathbf{u}_\infty = 0$. Nondimensional perturbation quantities s, $\tilde{\omega}$, and θ of density, pressure, and temperature are introduced as follows[27]

$$\rho = \rho_\infty(1 + s) \tag{21-1a}$$

$$p = p_\infty(1 + \tilde{\omega}) \tag{21-1b}$$

$$T = T_\infty(1 + \theta) \tag{21-1c}$$

Neglecting terms which are nonlinear in $\mathbf{u}$, s, $\tilde{\omega}$, θ, $\nu - \nu_\infty$, and $k - k_\infty$ in the Navier-Stokes equation (Eq. 3-3ff.), one obtains the following equations which generalize the Oseen equations (Eq. 11-1) to the case of a compressible fluid.[28]

$$s_t + \operatorname{div} \mathbf{u} = 0 \tag{21-2a}$$

$$\mathbf{u}_t + \frac{a_\infty^2}{\gamma} \operatorname{grad} \tilde{\omega} = \frac{4}{3} \nu_\infty \operatorname{grad} \operatorname{div} \mathbf{u} - \nu_\infty \operatorname{curl} \operatorname{curl} \mathbf{u} + \mathbf{f} \tag{21-2b}$$

$$\theta_t - k_\infty \operatorname{grad} \theta = (\gamma - 1)s_t + Q \tag{21-2c}$$

$$\tilde{\omega} = s + \theta \tag{21-2d}$$

where

$a_\infty = \sqrt{\gamma \mathcal{R} T_\infty}$ = isentropic speed of sound at infinity
$\mathbf{f}$ = force per unit mass
Q = heat added per unit mass per unit time

[27] The symbol s denotes here a nondimensional density perturbation, as is customary in studies of linearized compressible flow. In all other articles of this section s stands for entropy per unit mass.

[28] The Stokes relation (Eq. 3-24) is assumed for simplicity.

In the following the subscript ∞ is dropped from the symbols a_∞, ν_∞, and k_∞.

A general discussion of the linearized equations has been given in [*115,116*] and, for the case of zero heat conductivity, in [*53*]. Some of the results obtained are summarized below.

Theorem of splitting. The proof of the theorem of splitting as given in [*53*] may easily be extended to the case of nonzero heat conduction and is stated without proof (cf. also Eq. 11-9).

A solution of Eq. 21-2 ($\mathbf{u}$, $\tilde{\omega}$, s, θ; $\mathbf{f}$, Q) may always be expressed as a sum of a longitudinal wave ($\mathbf{u}_{\text{L}}$, $\tilde{\omega}$, s, θ; $\mathbf{f}_{\text{L}}$, Q) and a transversal wave ($\mathbf{u}_{\text{T}}$, 0, 0, 0; $\mathbf{f}_{\text{T}}$, 0). The longitudinal wave satisfies the equations

$$\operatorname{curl} \mathbf{u}_{\text{L}} = 0, \qquad \operatorname{curl} \mathbf{f}_{\text{L}} = 0 \tag{21-3a}$$

$$s_t + \operatorname{div} \mathbf{u}_{\text{L}} = 0 \tag{21-3b}$$

$$(\mathbf{u}_{\text{L}})_t + \frac{a^2}{\gamma} \operatorname{grad} \tilde{\omega} = \tfrac{4}{3}\nu \operatorname{grad} \operatorname{div} \mathbf{u}_{\text{L}} + \mathbf{f}_{\text{L}} \tag{21-3c}$$

$$\theta_t - k \operatorname{grad} \theta = (\gamma - 1)s_t + Q \tag{21-3d}$$

$$\tilde{\omega} = s + \theta \tag{21-3e}$$

The transversal wave satisfies the equations,

$$\operatorname{div} \mathbf{u}_{\text{T}} = 0, \qquad \operatorname{div} \mathbf{f}_{\text{T}} = 0, \qquad Q = 0 \tag{21-4a}$$

$$\tilde{\omega} = s = \theta = 0 \tag{21-4b}$$

$$(\mathbf{u}_{\text{T}})_t = \nu \operatorname{curl} \operatorname{curl} \mathbf{u}_{\text{T}} + \mathbf{f}_{\text{T}} = \nu\nabla^2\mathbf{u}_{\text{T}} + \mathbf{f}_{\text{T}} \tag{21-4c}$$

Furthermore, with a given splitting of the force field into $\mathbf{f}_{\text{L}} + \mathbf{f}_{\text{T}}$, the decomposition $\mathbf{u} = \mathbf{u}_{\text{L}} + \mathbf{u}_{\text{T}}$ is unique (assuming only that $\mathbf{u}$ is given as $t \to -\infty$).

The longitudinal waves satisfy the usual equations of linearized viscous heat-conducting acoustics. All of the pressure, density, and temperature disturbances of the complete flow are associated with the irrotational field of the longitudinal waves. A velocity potential and a force potential exist for the longitudinal waves so that the system of equations (Eq. 21-3) may be replaced by equations for scalar quantities. The coupling of the thermal and pressure fields is easily seen from the equations. The transversal waves are the same as in incompressible flow. All of the vorticity is included in this field and there is no pressure, temperature, or density disturbance.

Longitudinal and transversal waves are separate solutions to the complete system and with suitably simple geometrical configurations each type can be produced. For example, an infinite plane moved parallel to itself produces the purely transversal wave discussed in Art. 6 (cf. Eq.

6-11ff.). An expanding sphere produces a purely longitudinal wave. Other examples are given below. In general, however, the longitudinal and transversal waves cannot be computed separately but are coupled through the boundary conditions.

Heat pulse. If heat is added to a compressible fluid it spreads partly according to the classical laws of heat conduction in solids. However, in addition, changes in density and pressure occur. The increase in pressure and the expansion of the fluid corresponding to the decrease in density cause a pressure wave. This effect is similar to one aspect of the displacement effect of a boundary layer in a compressible fluid. This is especially evident in Rayleigh's problem for compressible fluids as discussed by Van Dyke in [*117*]. If the heat addition is thought of as an explosion, the accompanying pressure wave corresponds to the sound caused by the explosion. For a real fluid a qualitative idea of the two processes mentioned may be obtained from a quantitative solution of the linearized equation. Since the superposition principle is valid in the linearized case, it is sufficient to study the effect of a heat pulse, that is, a unit amount of heat added instantaneously at a point. The resulting flow field and temperature field is then described by a fundamental solution of Eq. 21-2, namely the solution with zero initial conditions, $\mathbf{f} = 0$ and Q a delta function in space and time. This solution may be obtained by the standard methods of Laplace transformation and the result expressed as a contour integral. Approximate values of this integral may be found for small values of time and for large values of time. The results for the one-dimensional case are described briefly below. We assume that $Pr = \frac{3}{4}$. The simplification resulting from this assumption may be illustrated as follows: If Eq. 21-2 are specialized to the one-dimensional case and s and u eliminated, one obtains

$$\left(\frac{\partial}{\partial t} - \frac{\kappa}{\gamma}\frac{\partial^2}{\partial x^2}\right)\theta + \frac{1-\gamma}{\gamma}\,\tilde{\omega}_t = \frac{Q}{\gamma} \tag{21-5a}$$

$$\left(\frac{\partial}{\partial t} - \frac{4}{3}\,\nu\,\frac{\partial^2}{\partial x^2}\right)(\theta_t - \tilde{\omega}_t) + \frac{a^2}{\gamma}\,\tilde{\omega}_{xx} = f \tag{21-5b}$$

where $\kappa = k/\rho c_v$ = thermometric coefficient of heat conductivity. If θ is eliminated one obtains in general a fifth order equation for $\tilde{\omega}$. However, since $Pr = \mu c_p/k = \nu\gamma/\kappa$, $\frac{4}{3}\nu$ is equal to $\frac{4}{3}\,Pr\,\kappa/\gamma$ and hence equal to κ/γ for $Pr = \frac{3}{4}$. In this case the pressure obeys the third order equation,

$$\left(1 + \frac{\kappa}{a^2}\frac{\partial}{\partial t}\right)\frac{\partial^2\tilde{\omega}}{\partial x^2} - \frac{1}{a^2}\frac{\partial^2\tilde{\omega}}{\partial t^2} = \frac{\gamma}{a^2}f - \frac{Q_t}{a^2} \tag{21-6}$$

The solution of the one-dimensional form of Eq. 21-2 with zero initial conditions at $t = 0$, $f = 0$ and $Q = \delta(x)\delta(t)$ is then, for small values of

(a^2t/κ),

$$u = \frac{a^2(\text{sign } x)}{2\kappa(\gamma - 1)}\left(\text{erfc}\sqrt{\frac{x^2}{4\kappa t}} - \text{erfc}\sqrt{\frac{\gamma x^2}{4\kappa t}}\right)\left[1 + O\left(\frac{a^2t}{\kappa}\right)\right] \tag{21-7a}$$

$$\tilde{\omega} = \theta = \frac{1}{\sqrt{4\pi\kappa t}}\, e^{-\frac{x^2}{4\kappa t}}\left[1 + O\left(\frac{a^2t}{\kappa}\right)\right] \tag{21-7b}$$

and for large values of a^2t/κ,

$$u = \left(-\frac{x}{2\gamma\sqrt{4\pi\frac{\kappa}{\gamma}t^3}}\, e^{-\frac{\gamma x^2}{4\kappa t}} + \frac{a}{2\gamma}(\text{sign } x)\, F(x, t)\right)\left[1 + O\left(\frac{\kappa}{a^2t}\right)\right] \tag{21-8a}$$

$$\tilde{\omega} = \tfrac{1}{2}F(x, t)\left[1 + O\left(\frac{\kappa}{a^2t}\right)\right] \tag{21-8b}$$

$$\theta = \left(\frac{1}{\gamma}\frac{1}{\sqrt{4\pi\frac{\kappa}{\gamma}t}}\, e^{-\frac{\gamma x^2}{4\kappa t}} + \frac{\gamma - 1}{2\gamma}F(x, t)\right)\left[1 + O\left(\frac{\kappa}{a^2t}\right)\right] \tag{21-8c}$$

where

$$F(x, t) = \frac{1}{\sqrt{2\pi\kappa t}}\left[e^{-\frac{(x-at)^2}{2\kappa t}} + e^{-\frac{(x+at)^2}{2\kappa t}}\right] \tag{21-8d}$$

and sign $x = x/|x|$. Thus for small values of (a^2t/κ) temperature diffuses according to the classical laws of heat diffusion in solids. The coefficient of diffusivity is κ. Pressure, density, and velocity diffuse in a similar manner. No wave fronts or approximate wave fronts exist. For large values of time, two modes of spreading exist. The first mode corresponds to the terms containing $F(x, t)$. In the wave described by this function two wave centers exist, moving to the right and left respectively, with speed a; in addition, diffusion with an effective coefficient of diffusivity $\kappa/2$ takes place symmetrically about each center. For any κ each wave eventually flattens out as t tends to ∞. Thus $F(x, t)$ describes an acoustic wave which is diffusing about its wave center. Eq. 21-8 show that for large time the velocity, the pressure, and the temperature spread partly by means of such a wave. The second mode of spreading corresponds to the first term in the expressions for θ and u respectively. Thus temperature diffuses about the origin in the same manner as for small time except that the coefficient of diffusivity is now κ/γ. It can be seen by integrating Eq. 21-2a and 21-2c that $\int_{-\infty}^{+\infty} \theta(x, t)\, dt = 1$. This is the conservation law for internal energy. For large times, the distribution is such that a fraction $1/\gamma$ of the original energy added is left to diffuse about the origin, while the remainder $1 - (1/\gamma)$ is contained in the diffused acoustic waves.

Other types of fundamental solutions of Eq. 21-2, such as the solution for a singular flat plate (cf. Art. 11), have also been found. For a discussion of fundamental solutions and their applications the reader is referred to [*53,115,116,118,119*].

Linearized shock layer structure. The linearized equations give a qualitatively incorrect shock layer structure. In a real shock layer the viscous terms have a diffusive effect while the nonlinear transport terms have a steepening effect. When these two opposite effects balance one obtains a stationary shock layer of thickness $\sim \nu$. If the nonlinear terms are omitted, there is no mechanism for counteracting the diffusion and one-dimensional stationary shock layers are not possible. If Eq. 21-2 is solved for flow past a body in supersonic flow one obtains stationary shock layers. However, these are of the thickness $\sim \sqrt{\nu}$ and the thickness also increases with distance from the body due to the diffusive effect (cf. [*53*])[29]. In this connection it should be recalled that the perturbation method used for the study of weak shock layers in Art. 19 led to equations where certain nonlinear terms were retained.

Use of the linearized equations. The applicability of the linearized equations is limited by the fact that they predict a qualitatively incorrect shock layer structure. As a consequence, the remarks made below apply directly only to subsonic stationary flow. In other cases corrections for shock layer structure are needed, a problem discussed later in this article.

As pointed out in Art. 20 the outer solution or low Reynolds number flow obeys the linearized (Oseen) equations, and the inner solution obeys the generalized Stokes equations (Eq. 20-5, or special cases thereof). In incompressible flow the Oseen equations are actually uniformly valid, since the inner limit of these equations gives the classical Stokes equations. This is a fortuitous coincidence, connected with the fact that the Stokes equations for incompressible flow are linear. The Oseen equations for compressible flow are not uniformly valid near the body. In solving a problem of low Reynolds number compressible flow past a solid, it is not sufficient to use the Oseen solution only; the outer solution must be matched with an inner solution, i.e., a solution of the nonlinear Stokes equations. The method of matching is in principle the same as in the case of incompressible flow (cf. Art. 17) although the computational difficulties are of course considerably greater in the case of compressible flow.[30]

The linearized equations also describe the behavior of the flow at

[29] The mathematical reason for this was pointed out by Latta (cf. the discussion of Oseen flow past a parabola in Art. 17). According to the linearized equations for inviscid flow discontinuities occur across Mach waves, i.e. across characteristic lines or surfaces. The corresponding nonlinear equations, on the other hand, predict shock discontinuities only across surfaces that are not characteristic.

[30] In order to obtain a crude estimate of the solution one may solve the Oseen equations as if they were uniformly valid. This method was applied to the study of flow past a hot wire in [*119*].

large distances from a solid even if the Reynolds number of the solid is not small. This fact is of importance in the discussion of integral laws such as the momentum method of evaluating drag (cf. Art. 4).

The main field of application of the linearized equations has been in acoustics, which may be defined as the study of longitudinal waves of small amplitude. Viscosity and heat conduction are important for the understanding of dispersion and absorption of sound. However, these problems are beyond the scope of this section. A recent survey of this field is given in [*120*].

Finally, it should be mentioned that once a linearized solution is obtained, nonlinear effects may be partially accounted for by some iteration process. This method has been applied to Rayleigh's problem for compressible fluids by Howarth [*121*].

A QUASI-LINEAR ONE-DIMENSIONAL EQUATION. The quasi-linear parabolic equation

$$u_t + uu_x = \nu u_{xx} \tag{21-9}$$

has been studied by several authors for various purposes. It was used by Burgers [*114*] as a model in the study of turbulence. Cole [*53*] introduced it as an actual approximation. He showed that for certain nonstationary weak shock layers an approximation to the Navier-Stokes equations may be used which, after suitable normalization of the variables, is identical with Eq. 21-9. In this derivation Cole used the fact that in a weak shock layer the entropy is approximately constant and a linearized Bernoulli's law is approximately valid.

It was noticed by Cole [*53,122*], and independently by Hopf [*123*], that Eq. 21-9 may be reduced to the heat equation by a transformation. If we put

$$u(x, t) = -2\nu \frac{\theta_x}{\theta} \tag{21-10a}$$

then Eq. 21-9 reduces to

$$\theta_t = \nu \theta_{xx} \tag{21-10b}$$

This transformation leads in particular to a complete solution of the initial value problem.

If ν is put equal to zero the resulting equation can be easily solved by the method of characteristics. In certain cases shock discontinuities occur as in a nonviscous compressible fluid. For $\nu \neq 0$ these discontinuities are replaced by shock layers. Using the method described in Art. 20 one sees that the equation for such a shock layer is the stationary form of Eq. 21-9 where the velocity u is measured relative to the velocity of the shock.

As an example, consider the following "piston problem."

$$u(0, t) = \text{const} = u_0 > 0 \tag{21-11a}$$

$$u(x, 0) = 0 \tag{21-11b}$$

With the aid of the transformation (Eq. 21-10) it has been shown by Cole that the exact solution is

$$\frac{u}{u_0} = \frac{\operatorname{erfc}\left(\frac{x - u_0 t}{\sqrt{4\nu t}}\right)}{e^{\frac{u_0}{2\nu}\left(x - \frac{u_0}{2}t\right)} \operatorname{erf} \frac{x}{\sqrt{4\nu t}} + \operatorname{erfc}\left(\frac{x - u_0 t}{\sqrt{4\nu t}}\right)} \tag{21-12}$$

The following approximation

$$\bar{u} = \frac{u_0}{2}\left[1 - \tanh \frac{u_0}{4\nu}\left(x - \frac{u_0}{2}t - \frac{2\nu}{u_0}\ln 2\right)\right] \tag{21-13}$$

has the property that if u is the exact solution given by Eq. 21-12, then

$$\lim_{\nu \to 0} \frac{|u - \bar{u}|}{\nu^n} = 0 \tag{21-14}$$

uniformly in the x, t plane if an arbitrarily small but fixed neighborhood of the origin is excluded.

Following the methods proposed in Art. 20 one may try to obtain an approximate solution as follows: First one finds the solution of Eq. 21-9 with $\nu = 0$. This leads to a shock discontinuity moving with velocity $u_0/2$. Then this discontinuity is replaced by a shock layer with the center at $x = (u_0/2)t$. This gives the solution (Eq. 21-13) except for the "phase shift" $(2\nu/u_0) \ln 2$. However, this term is necessary for Eq. 21-14 to be uniformly valid.

It seems plausible that a similar situation arises for the piston problem for the Navier-Stokes equations. The shock layer solution gives a uniformly valid approximation to the full solution if the correct phase shift is found. At present there are, however, no methods for determining this phase shift a priori. The existence of a phase shift was pointed out to the author by L. N. Howard.

The solution just given illustrates how Eq. 21-9 can be used as a model for the Navier-Stokes equations. Many other instructive examples can be given, which, however, are not discussed here. Following the method of Cole [*53*] one may also obtain approximate solutions of the Navier-Stokes equations from exact solutions of Eq. 21-9.

CHAPTER 4. LAMINAR BOUNDARY LAYERS IN COMPRESSIBLE FLUIDS

B,22. Introduction. The theory of laminar boundary layers occupies a very special place in the study of viscous compressible flow. It is the only case where a well-developed theory exists which has led to many results of great practical importance. The main problem for which boundary layer theory was originally developed is that of computing the skin friction. A related problem is also of great importance, namely that of computing the heat transfer between a solid boundary and a fluid. For high speed flow there is the additional problem of computing the temperature rise at an insulated wall due to viscous dissipation inside the boundary layer. These are the central practical problems of boundary layer theory.

In many problems it is possible to arrive at estimates of skin friction and temperature or heat transfer at a solid boundary without detailed consideration of the flow field. The few relatively complete solutions which have been obtained have, however, considerable significance, both theoretical and practical. They may be used for verifying results obtained by more approximate methods. Furthermore, it is obviously of great importance to be able eventually to deal with transition to turbulence, and knowledge of the laminar velocity profile is essential in any study of stability problems. Finally, detailed solutions of the boundary layer equations are also needed for the computation of corrections to boundary layer theory. The necessity for such correction is in general greater for very high Mach numbers than for low speed flow.

The general boundary layer equations for a compressible fluid were derived in Art. 20. It was pointed out there that for cylindrical flow the curvature terms are negligible within the boundary layer approximation. Two-dimensional flow is the special case of cylindrical flow where the cross flow u_2 vanishes, that is, it is the case of zero sweepback. The corresponding equations are then obtained from Eq. 20-12 by putting $u_2 = 0$. Furthermore we replace the symbols x_1, x_3, u_1, and u_3 by x, y, u, and v respectively. The coordinate x then measures the distance along the boundary of the solid, and y the distance normal to the boundary. With this notation, the boundary layer equations for two-dimensional flow are

$$\rho \frac{Du}{Dt} + \frac{\partial p}{\partial x} = \frac{\partial}{\partial y}\left(\mu \frac{\partial u}{\partial y}\right) + \rho f_1 \tag{22-1a}$$

$$\frac{\partial p}{\partial y} = 0 \tag{22-1b}$$

$$\frac{\partial \rho}{\partial t} + \frac{\partial(\rho u)}{\partial x} + \frac{\partial(\rho v)}{\partial y} = 0 \tag{22-1c}$$

$$\rho \frac{Dh}{Dt} - \frac{Dp}{Dt} = \frac{\partial}{\partial y}\left(\frac{\mu}{Pr}\frac{\partial h}{\partial y}\right) + \mu\left(\frac{\partial u}{\partial y}\right)^2 + \rho Q \tag{22-1d}$$

where

$$\frac{D}{Dt} = \frac{\partial}{\partial t} + u\frac{\partial}{\partial x} + v\frac{\partial}{\partial y}$$

In deriving these equations it has been assumed that the enthalpy depends on temperature only. In addition it is in general assumed (cf. Eq. 18-2) that

$$h = e + \frac{p}{\rho} = c_p T, \qquad c_p = \text{const} \tag{22-2a}$$

$$p = \mathcal{R}\rho T \tag{22-2b}$$

$$Pr = \frac{\mu c_p}{k} = \text{const} \tag{22-2c}$$

$$\mu = \text{fn}(T) \tag{22-2d}$$

$$e = c_v T, \qquad c_v = \text{const} \tag{22-2e}$$

and hence

$$\mathcal{R} = c_p - c_v = \text{const} \tag{22-2f}$$

$$\gamma = \frac{c_p}{c_v} = \text{const} \tag{22-2g}$$

(The symbol "fn" means "some function of.")

Special forms for the functional dependence of μ on T will be discussed later; reference may also be made to the discussion of state variables and transport coefficients in Art. 3.

By the methods used in Art. 3 alternative forms of the momentum and energy equations may be derived, such as

$$\frac{\partial(\rho u)}{\partial t} + \frac{\partial(\rho u^2)}{\partial x} + \frac{\partial(\rho uv)}{\partial y} + \frac{\partial p}{\partial x} = \frac{\partial}{\partial y}\left(\mu \frac{\partial u}{\partial y}\right) + \rho f_1 \tag{22-1a$'$}$$

$$\rho \frac{DH}{Dt} - \frac{\partial p}{\partial t} = \frac{\partial}{\partial y}\left[\frac{\mu}{Pr}\frac{\partial H}{\partial y} - \frac{\mu(1 - Pr)}{Pr}\frac{\partial(u^2/2)}{\partial y}\right] + \rho u f_1 + \rho Q \tag{22-1d$'$}$$

where $H = h + (u^2/2)$ is the total enthalpy (cf. Eq. 3-8b).

For flow past a solid surface, a typical set of boundary conditions are given by Eq. 22-3 below. The subscript $_w$ denotes values at the wall, that is, the solid surface. The subscript $_e$ denotes values of the external flow, that is, values of the nonviscous solution for $y = 0$:

$$\begin{aligned}
&(a)\quad u(x, 0) = 0\\
&(b)\quad v(x, 0) = 0\\
&(c)\quad \left(\frac{\partial h}{\partial y}\right)_w = 0 \text{ or } (c')\ h(x, 0) = h_w(x)\\
&(d)\quad u(x, \infty) = u_e(x, 0)\\
&(e)\quad p(x, \infty) = p_e(x, 0)\\
&(f)\quad h(x, \infty) = h_e(x, 0)
\end{aligned} \tag{22-3}$$

In general $v(x, \infty)$ may not be prescribed.

The parabolic nature of the equations (Eq. 22-1) requires that initial conditions also be prescribed at some line $x_0 = \text{const}$,

$$\begin{aligned} &(a)\ u(x_0, y) = \text{fn}(y) \\ &(b)\ h(x_0, y) = \text{fn}(y) \end{aligned} \tag{22-4}$$

These normally take the form

$$\begin{aligned} &(a)\ u(0, y) = u_e(0, 0) \\ &(b)\ h(0, y) = h_e(0, 0) \end{aligned} \tag{22-4'}$$

The second momentum equation (Eq. 22-1b) together with the boundary condition (Eq. 22-3e) implies that $p(x, y) = p_e(x)$. Hence, using well-known relations for the external flow one finds that

$$\frac{1}{\rho_e}\frac{dp}{dx} = -u_e\frac{du_e}{dx} = \frac{dh_e}{dx} \tag{22-5}$$

As mentioned earlier, two of the most important quantities to be determined from boundary layer theory are the skin friction, τ_w, and the heat transfer per unit area and unit time from the wall to the fluid, q_w. These are defined by

$$\tau_w = \left(\mu\frac{\partial u}{\partial y}\right)_{y=0} \tag{22-6a}$$

$$q_w = -\left(k\frac{\partial T}{\partial y}\right)_{y=0} = -\left(\frac{\mu}{Pr}\frac{\partial h}{\partial y}\right)_{y=0} \tag{22-6b}$$

Eq. 22-1 may be written in nondimensional form by referring u, v, h, etc. either to $u_e(x)$ and $h_e(x)$, i.e. the velocity and enthalpy in the external flow just outside the boundary layer, or to constant values, as for example, the free stream values U, h_∞, etc. Examples of the first method are given in Art. 23. Using the second method one obtains Eq. 22-7 below if the assumptions (Eq. 22-2) are used. Using the notation of Eq. 18-1a and 18-1b (cf. also Eq. 18-3b) and noting that $(\gamma - 1)M^2 = U^2/h_\infty$ one finds

$$\tilde{\rho}\left(\frac{\partial u^*}{\partial t^*} + u^*\frac{\partial u^*}{\partial x^*} + v^*\frac{\partial u^*}{\partial y^*}\right) + \frac{\partial p^*}{\partial x^*} = \frac{1}{Re}\frac{\partial}{\partial y^*}\left(\tilde{\mu}\frac{\partial u^*}{\partial y^*}\right) + \frac{\tilde{\rho}f_1L}{U^2} \tag{22-7a}$$

$$\frac{\partial\tilde{\rho}}{\partial t^*} + \frac{\partial(\tilde{\rho}u^*)}{\partial x^*} + \frac{\partial(\tilde{\rho}v^*)}{\partial y^*} = 0 \tag{22-7b}$$

$$\begin{aligned} \tilde{\rho}\left(\frac{\partial\tilde{h}}{\partial t^*} + u^*\frac{\partial\tilde{h}}{\partial x^*} + v^*\frac{\partial\tilde{h}}{\partial y^*}\right) - (\gamma - 1)M^2\left(\frac{\partial p^*}{\partial t^*} + u^*\frac{\partial p^*}{\partial x^*}\right) \\ = \frac{1}{Pr\,Re}\frac{\partial}{\partial y^*}\left(\tilde{\mu}\frac{\partial\tilde{h}}{\partial y^*}\right) + (\gamma - 1)\frac{M^2}{Re}\tilde{\mu}\left(\frac{\partial u^*}{\partial y^*}\right)^2 + \tilde{\rho}\frac{Q}{h_\infty}\frac{L}{U} \end{aligned} \tag{22-7c}$$

$$\tilde{\rho}\tilde{h} = 1 + \gamma M^2 p^* \tag{22-7d}$$

The general discussion of the meaning of compressibility in Art. 20 applies of course to the boundary layer equations, and may be elaborated further for this special case. A boundary layer is said to be compressible if the variation of ρ and T is not negligible. In this case μ will in general also have a non-negligible variation, a fact of importance for numerical computations. Since the pressure is independent of y, the density ρ is inversely proportional to T or to h for a fixed value of x. The energy equation, Eq. 22-1d or 22-7c, shows when it is necessary to assume that h (and hence ρ) is variable. Its right-hand side contains three terms, each of which may be responsible for addition or subtraction of heat in certain regions of the flow. The third term describes the influence of external heat sources in the fluid; in most applications this term is zero. The second term may be regarded as describing the effect of intrinsic heat sources, due to dissipation; this term represents the main Mach number effect. The first term describes the spreading of heat produced by either of these heat sources or transferred to or from the fluid at the wall. For flow at a negligible Mach number and without external heat sources compressibility may still be important if $T_e - T_w$ is not negligible compared to T_e. Low speed heat transfer is therefore an important application of the theory of compressible boundary layers.

Roughly speaking, the nondimensional form (Eq. 22-7c) of the energy equation shows that compressibility effects in the boundary layer are not negligible if the thermal energy introduced into (or removed from) the fluid through heat sources, conduction at the wall, or dissipation is not negligible compared to the intrinsic energy of the fluid as measured, for example, by h_∞. The Mach number effects in a boundary layer are quite different from the effects in inviscid flow. A main effect in a boundary layer is that the heat added by dissipation is proportional to M^2. The concept of Mach wave is of limited value because of the predominance of viscous stresses over pressure. Formally, this is shown in particular by the equation $p_y = 0$. It follows that there is no qualitative difference between a subsonic and a supersonic boundary layer per se. The larger the Mach number the greater the dissipation, but Mach number unity plays no critical role. However, the Mach number enters a boundary layer problem also through the boundary conditions: For a given body the external values u_e, p_e, and h_e may be strongly dependent on Mach number. In addition, interaction between the boundary layer and the external flow may be qualitatively different from subsonic and supersonic flow.

Since the momentum and continuity equations differ from the corresponding equations for the incompressible case only through the fact that ρ and μ may vary, one may expect that the variation of the skin friction is qualitatively the same in the two cases if the pressure gradient is the same. Viscosity increases with increasing temperature for gases. On the other hand, density decreases with increasing temperature which suggests that the boundary layer may become thicker with an attendant decrease in the slope of the velocity profile. Thus the two effects of temperature may be expected to cancel partially for the skin friction $(\mu\partial u/\partial y)_w$. For the special case of μ proportional to T, it is shown later that the skin friction on a flat plate with zero pressure gradient is in fact independent of compressibility. It is also shown that with the additional assumption $Pr = 1$ the problem of compressible flow with pressure gradient along an insulated wall may be reduced to an equivalent incompressible flow with a distorted pressure gradient.

Aerodynamic heating at the wall is of course an entirely new phenomenon when boundary layer theory is generalized to compressible flow. However, the kind of result to be expected may be easily estimated for certain cases. Consider the case when $\mathbf{f}$ and Q are zero and the external flow is stationary (although the boundary layer itself may be nonstationary). If the Prandtl number is unity, it is easily seen that an integral of the energy equation (Eq. 22-1d′) is

$$H = h + \frac{u^2}{2} = \text{const} = h_\infty + \frac{U^2}{2} = h_w \tag{22-8}$$

This equation implies that $(\partial h/\partial y)_w = (-u\partial u/\partial y)_w = 0$. This particular solution is thus valid for the case of zero heat transfer at the wall. Consequently if $Pr = 1$, stagnation due to viscous effects at an insulated wall raises the temperature by the same amount as adiabatic stagnation in the outer flow. Eq. 22-8 is the Busemann energy integral [*124*]. In Art. 27 other energy integrals are discussed.

ALTERNATIVE FORMS OF THE BOUNDARY LAYER EQUATIONS

For many purposes it is convenient to work with alternative differential equations obtained by making a change of variables in the original boundary layer equations (Eq. 22-1), or to work with integral relations and average quantities in the manner of von Kármán's momentum method for incompressible fluids. Some of these new equations exhibit much more clearly than the original equations the role of certain parameters and of the viscosity law, as well as the relation to the incompressible case. They are also more suitable for numerical computations.

Some of the most important transformed equations are derived in the article below. Motivations for the transformations introduced are not given here. However, subsequent articles will demonstrate the usefulness of the transformed equations. It is assumed below that the external forces, *f*, and heat sources, *Q*, are zero and only stationary flow is considered.

B,23. Change of Variables in the Differential Equations.[31]

von Mises transformation. This transformation was originally proposed by von Mises [*125*] for incompressible fluids. It was extended by von Kármán and Tsien to the case of compressible fluids in [*126*]. The underlying idea is that the substantial derivative D/Dt is a derivative along a streamline. Hence if one replaces the y coordinate by a coordinate which is constant along streamlines, such as the stream function ψ, then D/Dt may be written as some factor times $(\partial/\partial x)_{\psi=\text{const}}$. In the derivation of the transformed equations it is convenient to use the notation x, y for the original coordinates and ξ, ψ for the new ones. Here $\xi \equiv x$; however, $\partial/\partial x$ denotes a derivative with y kept constant and $\partial/\partial\xi$ a derivative with ψ kept constant. After the derivation has been completed ξ may be replaced by x again.

The following table defines the transformation by enumerating the variables to be used in the transformed equations.

$$\text{Independent variables: } \xi = x, \psi \qquad (23\text{-}1a)$$

$$\text{Dependent variable: } \quad u = \frac{\rho_\infty}{\rho}\frac{\partial\psi}{\partial y} \qquad (23\text{-}1b)$$

The basic formulas for transforming the derivatives are then

$$\frac{\partial}{\partial x} + \frac{v}{u}\frac{\partial}{\partial y} = \frac{\partial}{\partial\xi} \qquad (23\text{-}2a)$$

$$\frac{\partial}{\partial y} = \frac{\rho}{\rho_\infty}u\frac{\partial}{\partial\psi} \qquad (23\text{-}2b)$$

Note that flow quantities which are independent of y, such as p, h_e, etc. are also independent of ψ. Hence

$$\frac{dp}{dx} = \frac{dp}{d\xi}, \qquad \frac{dh_e}{dx} = \frac{dh_e}{d\xi} \qquad (23\text{-}2c)$$

The boundary layer equations in *von Mises' form* are thus

$$\rho u\frac{\partial u}{\partial x} + \frac{dp}{dx} = \frac{\rho}{\rho_\infty^2}u\frac{\partial}{\partial\psi}\left(\mu\rho u\frac{\partial u}{\partial\psi}\right) \qquad (23\text{-}3a)$$

$$\frac{\partial h}{\partial x} - \frac{1}{\rho}\frac{dp}{dx} = \frac{1}{Pr}\frac{1}{\rho_\infty^2}\frac{\partial}{\partial\psi}\left(\mu\rho u\frac{\partial h}{\partial\psi}\right) + \frac{\mu\rho u}{\rho_\infty^2}\left(\frac{\partial u}{\partial\psi}\right)^2 \qquad (23\text{-}3b)$$

[31] In Art. 17 it was pointed out that different flow fields may result when the boundary layer approximation is made with respect to different coordinate systems. The change of variables studied in the present article is of a different nature. After the boundary layer equations have been derived in one special system of coordinates, the variables are changed in the equations. While this procedure may simplify the formal solution of the equations, it does not alter the flow field obtained from the equations.

Note that v does not occur in these equations and that the continuity equation is not used explicitly. Its validity is assumed implicitly through the use of the stream function ψ.

If the stream function is normalized so that $\psi = 0$ corresponds to the surface of the body, the boundary conditions on $u(x, \psi)$ etc. corresponding to Eq. 22-3 are

$$u(x, 0) = 0 \tag{23-4a}$$

$$u(x, \infty) = u_e(x) \tag{23-4b}$$

$$h(x, 0) = h_w(x) \tag{23-4c}$$

$$h(x, \infty) = h_e(x) \tag{23-4d}$$

Eq. 23-4c may be replaced by a condition on the heat flux at the wall q_w, suitably expressed in the new coordinates with the aid of Eq. 23-5b below. Initial conditions at $x = 0$ are unchanged and need not be discussed further.

Special attention has to be paid to the expressions for shear and heat transfer at the wall because of the singular behavior of ψ there. For a fixed value of x in the neighborhood of $y = 0$, $u = a_1 y + a_2 y^2 + \cdots$; ρu has a similar form. Hence, $\psi = b_1 y^2 + b_2 y^3 + \cdots$, $y = c_1 \psi^{\frac{1}{2}} + c_2 \psi + \cdots$, and $u = d_1 \psi^{\frac{1}{2}} + d_2 \psi + \cdots$. At the wall, therefore, $\partial u/\partial \psi$ is infinite, but $\partial(u^2)/\partial \psi$ is in general finite so that τ may be written

$$\tau = \mu \frac{\partial u}{\partial y} = \frac{\mu \rho u}{\rho_\infty} \frac{\partial u}{\partial \psi} = \frac{\mu \rho}{\rho_\infty} \frac{\partial (u^2/2)}{\partial \psi} \tag{23-5a}$$

Similarly the heat flux q is

$$q = -k \frac{\partial T}{\partial y} = -\frac{\mu}{Pr} \frac{\partial h}{\partial y} = -\frac{1}{Pr} \frac{\mu \rho u}{\rho_\infty} \frac{\partial h}{\partial \psi} \tag{23-5b}$$

Evaluating q at the wall ($\psi = 0$) one obtains q_w, i.e. the rate of heat transfer from the wall to the fluid.

Crocco's transformation. This transformation was introduced and used by Crocco in [*127*]. The table for this transformation is:

$$\text{Independent variables: } \xi = x, u \tag{23-6a}$$

$$\text{Dependent variable: } \quad \tau = \mu \frac{\partial u}{\partial y} = \mu \frac{1}{y_u} \tag{23-6b}$$

Hence

$$u_x = -\frac{y_\xi}{y_u} \tag{23-7a}$$

$$\frac{\partial}{\partial x} = \frac{\partial}{\partial \xi} - \frac{y_\xi}{y_u} \frac{\partial}{\partial u} \tag{23-7b}$$

$$\frac{\partial}{\partial y} = u_y \frac{\partial}{\partial u} = \frac{\tau}{\mu} \frac{\partial}{\partial u} = \frac{1}{y_u} \frac{\partial}{\partial u} \tag{23-7c}$$

With this change of variables the momentum equation (Eq. 22-1a) becomes

$$\rho u(-y_\xi) + \rho v + y_u p_\xi = \tau_u \tag{a}$$

and the continuity equation

$$y_u \frac{\partial(\rho u)}{\partial \xi} - y_\xi \frac{\partial(\rho u)}{\partial u} + \frac{\partial(\rho v)}{\partial u} = 0 \tag{b}$$

The term ρv in (a) may be eliminated by taking the u derivative and subtracting (b). Since $y_u = \mu/\tau$ and $y_{\xi u} = y_{u\xi}$ one obtains Eq. 23-8a below.

The energy equation (Eq. 22-1d) is changed into

$$\mu\rho u h_\xi - \rho u y_\xi \tau h_u + \rho v \tau h_u - \mu u p_\xi = \frac{\tau}{Pr}(\tau h_u)_u + \tau^2 \tag{c}$$

Here ρv may be eliminated by subtracting (a) multiplied by τh_u:

$$\mu\rho u h_\xi + \tau\tau_u h_u - \tau h_u y_u p_\xi - \mu u p_\xi = \frac{\tau\tau_u h_u + \tau^2 h_{uu}}{Pr} + \tau^2 \tag{d}$$

From (d) one obtains Eq. 23-8b below.

Thus *Crocco's equations* are

$$-\frac{\partial}{\partial x}\left(\frac{\mu\rho u}{\tau}\right) + \frac{\partial}{\partial u}\left(\frac{\mu p_x}{\tau}\right) = \tau_{uu} \tag{23-8a}$$

$$(1 - Pr)\tau\tau_u h_u + \tau^2(h_{uu} + Pr) - Pr\mu\rho u h_x + Pr\mu(h_u + u)p_x = 0 \tag{23-8b}$$

As in von Mises' transformation, v has been eliminated with the aid of the continuity equation. If τ, h, and hence μ have been found as functions of x and u, one may compute y as a function of x and u by

$$y = \int_0^u y_u du = \int_0^u \frac{\mu}{\tau} du \tag{23-9}$$

Inverting this relation gives u as a function of x and y; v may then be found from the continuity equation.

The boundary conditions on h are found from the fact that $y = 0$ corresponds to $u = 0$ and $y = \infty$ to $u = u_e(x)$. The condition $u = u_e(x)$ for $y = \infty$ is replaced by $\tau = 0$ at $u = u_e$. From the momentum equation in its original form it follows that for $y = 0$:

$$p_x = \tau_y = \tau_u u_y$$

Hence the boundary conditions are

$$\tau\tau_u = \mu p_x \quad \text{for} \quad u = 0 \tag{23-10a}$$

$$\tau = 0 \quad \text{for} \quad u = u_e \tag{23-10b}$$

$$h(x, 0) = h_w(x) \tag{23-10c}$$

$$h(x, u_e) = h_e \tag{23-10d}$$

The heat flux at any point is

$$q = -\frac{\mu h_y}{Pr} = -\frac{\tau h_u}{Pr} \tag{23-11}$$

The rate of heat transfer from the wall to the fluid, q_w, is then q evaluated at $u = 0$.

The restricted form of Howarth's transformation. A third method of transforming the boundary layer equations was used by Howarth in [*128*]. The complete transformation involves a distortion of both the x and y coordinates. We first introduce only a restricted form of this transformation, in which only the y coordinate is modified. The complete transformation will be given later in Stewartson's modified form. In the case of zero pressure gradient the transformation to be introduced is equivalent to the full Howarth transformation and to the Howarth-Stewartson transformation.

The variables of Howarth's restricted transformation are

$$\text{Independent variables: } \xi = x,\ \eta = \int_0^y \frac{\rho}{\rho_\infty}\, dy \tag{23-12a}$$

$$\text{Dependent variables: } \quad u,\ \bar{v} = -\int_0^\eta u_\xi\, d\eta,\ h \tag{23-12b}$$

Hence

$$\frac{\partial}{\partial x} = \frac{\partial}{\partial \xi} - \frac{y_\xi}{y_\eta}\frac{\partial}{\partial \eta} = \frac{\partial}{\partial \xi} - \frac{\rho y_\xi}{\rho_\infty}\frac{\partial}{\partial \eta} \tag{23-13a}$$

$$\frac{\partial}{\partial y} = \frac{\rho}{\rho_\infty}\frac{\partial}{\partial \eta} \tag{23-13b}$$

The momentum equation may therefore be written

$$\rho u u_\xi + \frac{\rho^2}{\rho_\infty}(v - y_\xi u)u_\eta + p_\xi = \frac{\rho}{\rho_\infty^2}(\mu\rho u_\eta)_\eta \tag{e}$$

and the continuity equation

$$(\rho v)_\eta = -\rho_\infty u_\xi - \frac{\rho_\infty}{\rho} u \rho_\xi + y_\xi(\rho u)_\eta \tag{f}$$

Since $(y_\xi)_\eta = (y_\eta)_\xi = \left(\frac{\rho_\infty}{\rho}\right)_\xi = -\frac{\rho_\infty \rho_\xi}{\rho^2}$ and $-u_\xi = \bar{v}_\eta$, (f) may be written

$$(\rho v)_\eta = \rho_\infty \bar{v}_\eta + (\rho u)(y_\xi)_\eta + (\rho u)_\eta y_\xi \tag{g}$$

By integrating (g) a value for ρv is obtained, from which it follows that

$$\frac{\rho}{\rho_\infty}(v - y_\xi u) = \bar{v} \tag{h}$$

Inserting this value into (e) we obtain Eq. 23-14a below.

The energy equation becomes

$$\rho u h_\xi + \frac{\rho^2}{\rho_\infty}(v - u y_\xi)h_\eta - u p_\xi = \frac{1}{Pr}\frac{\rho}{\rho_\infty^2}(\mu\rho h_\eta)_\eta + \frac{\mu\rho^2}{\rho_\infty^2}u_\eta^2 \tag{i}$$

Using (h), one then obtains Eq. 23-14c below. The modified continuity equation (Eq. 23-14b below) follows directly from the definition of $\bar{v}$.

Thus *Howarth's equations* (in the restricted form) are

$$\rho_\infty(u u_x + \bar{v} u_\eta) + \frac{\rho_\infty}{\rho} p_x = \frac{1}{\rho_\infty}(\mu\rho u_\eta)_\eta \tag{23-14a}$$

$$u_x + \bar{v}_\eta = 0 \tag{23-14b}$$

$$\rho_\infty(u h_x + \bar{v} h_\eta) - \frac{\rho_\infty}{\rho} u p_x = \frac{1}{\rho_\infty}\left(\frac{\mu\rho}{Pr} h_\eta\right)_\eta + \frac{\mu\rho}{\rho_\infty} u_\eta^2 \tag{23-14c}$$

The boundary conditions for the system 23-14 are identical with those for the original system if y and v are replaced by η and $\bar{v}$, respectively. Expressions for τ_w and q_w in Howarth's coordinates are given by Eq. 23-20.

The momentum equation may be written

$$u u_x + \bar{v} u_\eta + \frac{1}{\rho} p_x = \frac{\nu_\infty p}{p_\infty}(C u_\eta)_\eta \tag{23-14a$'$}$$

where $C = \mu T_\infty / \mu_\infty T = \text{fn}(h)$.

Eq. 23-14c may be replaced by an equation for the total enthalpy $H = h + (u^2/2)$ (cf. Eq. 22-1d′)

$$u H_x + \bar{v} H_\eta = \frac{\nu_\infty p}{p_\infty}\frac{\partial}{\partial\eta}\left[\frac{C}{Pr}\frac{\partial H}{\partial\eta} - \frac{C(1 - Pr)}{Pr}\frac{\partial\left(\dfrac{u^2}{2}\right)}{\partial\eta}\right] \tag{23-14c$'$}$$

The Howarth-Stewartson transformation. We now introduce the full Howarth transformation of the independent variables [*128*]. However, we shall not use the modified stream function introduced by Howarth in [*128*]. Following Stewartson [*129*], we instead combine the use of Howarth independent variables with the use of the original stream function ψ. This choice of variables leads to the modification of Howarth's equations introduced by Stewartson in [*129*] for the purpose of correlating compressible boundary layers with $M \neq 0$ with incompressible boundary layers. For the convenience of derivation, the transformation is made in two steps. We first introduce

$$\text{Independent variables: } \zeta = \frac{a_e}{a_\infty}\eta = \frac{a_e}{a_\infty}\int_0^y \frac{\rho}{\rho_\infty}\,dy,\ X = x \tag{23-15a}$$

$$\text{Dependent variables: } \psi,\ H \tag{23-15b}$$

Later on (Eq. 23-17) a distorted x coordinate will be introduced. Eq. 23-15 implies (note that $da_e/dx = da_e/dX$)

$$\frac{\partial}{\partial x} = \frac{\partial}{\partial X} + \frac{\eta}{a_\infty}\frac{da_e}{dX}\frac{\partial}{\partial \zeta} \tag{23-16a}$$

$$\frac{\partial}{\partial \eta} = \frac{a_e}{a_\infty}\frac{\partial}{\partial \zeta} \tag{23-16b}$$

and

$$u = \frac{a_e}{a_\infty}\psi_\zeta; \qquad \bar{v} = -\left(\psi_X + \frac{\eta}{a_\infty}\frac{da_e}{dX}\psi_\zeta\right) \tag{23-16c}$$

$$u_x = \frac{a_e}{a_\infty}\left(\psi_{\zeta X} + \frac{\eta}{a_\infty}\frac{da_e}{dX}\psi_{\zeta\zeta}\right) + \frac{1}{a_\infty}\frac{da_e}{dX}\psi_\zeta \tag{23-16d}$$

The momentum equation (23-14a′) then becomes

$$\frac{a_e^2}{a_\infty^2}[\psi_\zeta\psi_{\zeta X} - \psi_X\psi_{\zeta\zeta}] + \frac{a_e}{a_\infty^2}\left(\frac{da_e}{dX}\right)\psi_\zeta^2 + \frac{1}{\rho}p_X = \nu_\infty\frac{a_e^3}{a_\infty^3}\frac{p}{p_\infty}(C\psi_{\zeta\zeta})_\zeta \tag{a}$$

Now, (cf. Eq. 23-5)

$$\frac{1}{\rho}\frac{dp}{dX} = \frac{\rho_e}{\rho}\frac{dh_e}{dX} = \frac{h}{h_e}\frac{dh_e}{dX} \tag{b}$$

and

$$\frac{a_e}{a_\infty^2}\left(\frac{da_e}{dX}\right)\psi_\zeta^2 = \frac{1}{a_e}\left(\frac{da_e}{dX}\right)u^2 = \frac{1}{h_e}\left(\frac{dh_e}{dX}\right)\frac{u^2}{2} \tag{c}$$

Hence,

$$\frac{p_\infty}{p}\frac{a_\infty}{a_e}\left[(\psi_\zeta\psi_{\zeta X} - \psi_X\psi_{\zeta\zeta}) + \frac{a_\infty^2}{a_e^2}\frac{1}{h_e}\left(\frac{dh_e}{dX}\right)\left(h + \frac{u^2}{2}\right)\right] = \nu_\infty(C\psi_{\zeta\zeta})_\zeta \tag{d}$$

Furthermore,

$$a_e^2 = (\gamma - 1)h_e; \qquad \frac{u_e^2}{h_e} = (\gamma - 1)M_e^2 \tag{e}$$

and hence

$$\begin{aligned} 2(\gamma - 1)M_e\frac{dM_e}{dX} &= \frac{2u_e}{h_e}\frac{du_e}{dX} - \frac{u_e^2}{h_e^2}\frac{dh_e}{dX} \\ &= -\frac{2}{h_e^2}\frac{dh_e}{dX}\left(\frac{u_e^2}{2} + h_e\right) \\ &= -\frac{2}{h_e^2}\left(\frac{dh_e}{dX}\right)H_\infty \end{aligned} \tag{f}$$

Hence

$$\frac{a_\infty^2}{a_e^2 h_e}\frac{dh_e}{dX} = \frac{a_\infty^2}{(\gamma - 1)}\frac{1}{h_e^2}\frac{dh_e}{dX} = -\frac{a_\infty^2}{H_\infty}M_e\frac{dM_e}{dX} \tag{g}$$

and the last term inside the bracket of (d) is

$$-\frac{H}{H_\infty}(a_\infty M_e)\frac{d(a_\infty M_e)}{dX} \tag{h}$$

Finally we introduce a distorted X variable such that

$$\frac{p_\infty}{p}\frac{a_\infty}{a_e}\frac{\partial}{\partial X} = \frac{\partial}{\partial \xi} \tag{23-17}$$

Since, in the external flow

$$\frac{p}{p_\infty} = \left(\frac{a_e}{a_\infty}\right)^{\frac{2\gamma}{\gamma-1}} \tag{i}$$

this definition is equivalent to

$$\xi = \int_0^X \left(\frac{a_e}{a_\infty}\right)^{\frac{3\gamma-1}{\gamma-1}} dX + \text{const} \tag{23-17'}$$

Introducing (h) and Eq. 23-17 into (d) one finally obtains Eq. 23-18a below.

Introducing the variable ζ as defined by Eq. 23-15a into the energy equation (Eq. 23-14c′) and using Eq. 23-16, one obtains

$$\frac{p_\infty}{p}\frac{a_e}{a_\infty}(\psi_\zeta H_X - \psi_X H_\zeta) = \frac{\nu_\infty}{Pr}\frac{a_e^2}{a_\infty^2}\frac{\partial}{\partial \zeta}\left[CH_\zeta - (1 - Pr)C\left(\frac{a_e}{a_\infty}\right)^2 \frac{1}{2}(\psi_\zeta^2)_\zeta\right] \tag{j}$$

Finally, introducing ξ as defined by Eq. 23-17 one obtains Eq. 23-18b below.

The *Howarth-Stewartson equations* are thus

$$\psi_\zeta\psi_{\zeta\xi} - \psi_\xi\psi_{\zeta\zeta} - \frac{H}{H_\infty}(a_\infty M_e)\frac{d(a_\infty M_e)}{d\xi} = \nu_\infty(C\psi_{\zeta\zeta})_\zeta \tag{23-18a}$$

$$\psi_\zeta H_\xi - \psi_\xi H_\zeta = \frac{\nu_\infty}{Pr}\frac{\partial}{\partial \zeta}\left[C\left\{H_\zeta - (1 - Pr)\left(\frac{a_e}{a_\infty}\right)^2\frac{1}{2}(\psi_\zeta^2)_\zeta\right\}\right] \tag{23-18b}$$

Note that ψ_ζ and $-\psi_\xi$ are distorted velocity components which satisfy the continuity equation for incompressible fluids. The following notation is used

$$\bar{U} = \psi_\zeta = \frac{a_\infty}{a_e}u \tag{23-18c}$$

$$\bar{V} = -\psi_\xi = \frac{p_\infty}{p_e}\frac{a_\infty}{a_e}\left(\frac{\rho}{\rho_\infty}v + \frac{\partial \xi}{\partial x}\frac{a_\infty}{a_e}u\right) \tag{23-18d}$$

Then

$$\bar{U}_\xi + \bar{V}_\zeta = 0 \tag{23-18e}$$

The boundary conditions are

$$\psi(\xi, 0) = \psi_\zeta(\xi, 0) = 0 \tag{23-19a}$$

$$\frac{a_e}{a_\infty}\psi_\zeta(\xi, \infty) = u_e(\xi) \tag{23-19b}$$

or

$$\psi_\zeta(\xi, \infty) = a_\infty M_e(\xi) = \bar{U}_e \tag{23-19b'}$$

The expressions for the skin friction τ_w and the rate of heat transfer from the wall to the fluid q_w are

$$\tau_w = \mu \frac{\partial u}{\partial y} = \mu \frac{\rho}{\rho_\infty} \frac{\partial u}{\partial \eta} = \mu_\infty C \left(\frac{a_e}{a_\infty}\right)^2 \psi_{\zeta\zeta} \tag{23-20a}$$

$$q_w = -k \frac{\partial T}{\partial y} = -\frac{\mu}{Pr} \frac{\partial h}{\partial y} = -\frac{\mu}{Pr} \frac{\rho}{\rho_\infty} \frac{\partial h}{\partial \eta} = -\frac{\mu_\infty C}{Pr} \frac{a_e}{a_\infty} h_\zeta \tag{23-20b}$$

where all quantities are evaluated at the wall ($y = \eta = \zeta = 0$).

The Illingworth transformation. These equations are a modification of von Mises' equations. They were derived in [*130*] for the purpose of correlating compressible and incompressible boundary layers in certain cases. In this respect they serve the same purpose as Stewartson's modification of Howarth's equations.

The starting point for Illingworth's transformation are the equations in von Mises' form (Eq. 23-3). The dependent variables occurring there are then made nondimensional by referring their values to the corresponding values of the external flow with the same x coordinates; ψ will be made nondimensional by dividing it by the free stream velocity U and a characteristic length L, and x is replaced by a nondimensional distorted variable s^*. The following notation is used:

$$\bar{u}(x, \psi) = \frac{u(x, \psi)}{u_e(x)} \tag{23-21a}$$

$$\bar{\rho}(x, \psi) = \frac{\rho(x, \psi)}{\rho_e(x)} \tag{23-21b}$$

$$\bar{h}(x, \psi) = \frac{h(x, \psi)}{h_e(x)} \tag{23-21c}$$

$$\bar{\mu}(x, \psi) = \frac{\mu(x, \psi)}{\mu_e(x)} \tag{23-21d}$$

$$\psi^* = \frac{\psi}{UL} \tag{23-21e}$$

$$s^* = \int_0^x \frac{\rho_e \mu_e u_e}{\rho^0 \mu^0 UL} \, dx \tag{23-21f}$$

Here the superscript ^{0} denotes isentropic stagnation values.

Introducing these definitions into Eq. 23-3a and 23-3b and using the relations

$$\frac{dp_e}{dx} = -\rho_e u_e \frac{du_e}{dx} \qquad \rho_e \frac{dh_e}{dx} = \frac{dp_e}{dx} \qquad \bar{\rho}\bar{h} = \frac{p(x, \psi)}{p_e(x)} = 1$$

$$(\gamma - 1)M_e^2 = \frac{u_e^2}{h_e}$$

one obtains the *Illingworth equations*,

$$\bar{u}\frac{\partial \bar{u}}{\partial s^*} + (\bar{u}^2 - \bar{h})\frac{d(\ln u_e)}{ds^*} = \frac{\bar{u}}{Re}\frac{\partial}{\partial \psi^*}\left(\bar{\rho}\bar{\mu}\bar{u}\frac{\partial \bar{u}}{\partial \psi^*}\right) \tag{23-22a}$$

$$\frac{\partial \bar{h}}{\partial s^*} = \frac{1}{PrRe}\frac{\partial}{\partial \psi^*}\left(\bar{\rho}\bar{\mu}\bar{u}\frac{\partial \bar{h}}{\partial \psi^*}\right) + (\gamma - 1)\frac{M_e^2}{Re}\bar{\rho}\bar{\mu}\bar{u}\left(\frac{\partial \bar{u}}{\partial \psi^*}\right)^2 \tag{23-22b}$$

The corresponding boundary conditions are

$$\bar{u}(s^*, 0) = 0 \tag{23-23a}$$

$$\bar{u}(s^*, \infty) = 1 \tag{23-23b}$$

$$\bar{h}(s^*, 0) = \frac{h_w}{h_e} = \bar{h}_w \tag{23-23c}$$

$$\bar{h}(s^*, \infty) = 1 \tag{23-23d}$$

Other transformations. There are of course many other transformations of the boundary layer equations which may be used to advantage for various purposes. Reference is made here only to the transformation introduced by Dorodnitsyn in [*131*].

B,24. Integral Relations for Two-Dimensional Boundary Layers. In Chap. 1, various integral conservation laws were discussed. These laws are, of course, considerably simplified if the assumptions of boundary layer theory are used. With the aid of these assumptions integral laws are derived below for momentum, energy, and also other quantities which are polynomials in the velocity. These equations are generalizations of von Kármán's integral law for incompressible boundary layers (Eq. 13-24). The domain of integration is a thin rectangular strip parallel to the y direction, more precisely the region $x_0 \leqq x \leqq x_0 + dx$, $0 \leqq y < \infty$. The integral relations are derived from the differential equations (Eq. 22-1). Only stationary flow is considered and f and Q are assumed to be zero. In Eq. 22-1a $\mu\partial u/\partial y$ is written as τ. The results to be derived below are then also valid for the turbulent boundary layer if τ is interpreted as the shear in turbulent flow.

We first note that the following relations follow from the continuity equation ($k = 0, 1, 2, \ldots$ etc.)

$$\rho(k+1)u^k \frac{Du}{Dt} = \frac{\partial}{\partial x}(\rho u u^{k+1}) + \frac{\partial}{\partial y}(\rho v u^{k+1}) \tag{a}$$

$$(\rho v u^{k+1})_{y=\infty} = -u_e^{k+1} \int_0^\infty \frac{\partial(\rho u)}{\partial x}\, dy$$
$$= \int_0^\infty \left[(k+1)\rho u u_e^k \frac{du_e}{dx} - \frac{\partial(\rho u u_e^{k+1})}{\partial x} \right] dy \tag{b}$$

Since also $\dfrac{dp}{dx} = -\rho_e u_e \dfrac{du_e}{dx}$, we obtain

$$(k+1)\int_0^\infty u^k \left(\rho \frac{Du}{Dt} + \frac{dp}{dx}\right) dy$$
$$= \int_0^\infty \left[\frac{\partial}{\partial x}(\rho u u^{k+1}) + (k+1)\rho u u_e^k \frac{du_e}{dx} - \frac{\partial}{\partial x}(\rho u u_e^{k+1}) - (k+1)\rho_e u_e u^k \frac{du_e}{dx} \right] dy \tag{c}$$

If one now multiplies the momentum equation (Eq. 22-1a) by u^k and integrates between $y = 0$ and $y = \infty$, one obtains from (c)

$$\frac{1}{k+1}\frac{d}{dx}\int_0^\infty \rho u(u^{k+1} - u_e^{k+1})dy + \frac{du_e}{dx}\int_0^\infty (\rho u u_e^k - \rho_e u_e u^k)dy = \int_0^\infty u^k \frac{\partial \tau}{\partial y}\, dy \tag{24-1}$$

It should be noted that the integrals used in the above equations are invariant, that is, independent of the choice of coordinates. This may be proved by the methods used in the discussion of the invariants of the incompressible boundary layer (cf. the discussion of invariants in Art. 17).

Eq. 24-1 is a general integral law of boundary layer theory based on the momentum and continuity equations. As pointed out earlier it is also valid for turbulent flow. In the special case $k = 0$ one obtains von Kármán's integral relation (cf. Eq. 13-24)

$$\frac{d}{dx}(\rho_e u_e^2 \delta^{**}) + \rho_e u_e \delta^* \frac{du_e}{dx} = \tau_w \tag{24-2}$$

where

$$\delta^* = \text{displacement thickness} = \frac{1}{\rho_e u_e}\int_0^\infty (\rho_e u_e - \rho u)dy \tag{24-3a}$$

and

$$\delta^{**} = \text{momentum thickness} = \frac{1}{\rho_e u_e^2}\int_0^\infty \rho u(u_e - u)dy \tag{24-3b}$$

and τ_w is the shear at the wall.

Eq. 24-3a and 24-3b define two length parameters associated with a boundary layer. In a similar way one may also suggest other length parameters formed with the aid of integrals involving higher powers of u. The differential equations describing the x variation of these parameters are then obtained from Eq. 24-1 by giving k the values 1, 2, etc.

An analogous procedure can obviously be applied to the various forms of the energy equation. As an example, if Eq. 22-1d′ with $f = 0$, $Q = 0$, $\partial/\partial t = 0$ is integrated between $y = 0$ and $y = \infty$, one obtains

$$\frac{d}{dx}\int_0^\infty \rho u(H_e - H)dy = \left(\frac{\mu}{Pr}\frac{\partial h}{\partial y}\right)_w = \left(k\frac{\partial T}{\partial y}\right)_w = -q_w \tag{24-4}$$

This relation is valid for laminar boundary layers only.

Von Kármán's integral relation (Eq. 24-1 with $k = 0$ or Eq. 24-2) relates the mean value of the flow of momentum across a line x = const to the stress exerted by the fluid on the wall τ_w. In a similar way Eq. 24-4 relates the mean flow of $(H_e - H)$ to the heat q_w transferred from the fluid to the wall per unit time and unit area. For an insulated wall $(q_w = 0)$ the weighted integral of $(H_e - H)$ must vanish. Thus, for example, if the surface temperature of an insulated wall is less than the isentropic stagnation temperature, there must exist a region within the boundary layer where the stagnation temperature is above the isentropic value. This example shows how the integral laws may be used to obtain qualitative information about the boundary layer. Applications of the integral laws will be given in Art. 32.

Integral laws based on the transformed equations. To illustrate the effect of the coordinate transformations described in Art. 23 various forms of Eq. 24-2 are given for the special case of constant external flow ($u_e = U$ = const, $\rho_e = \rho_\infty$, etc.).

von Mises' transformation.

$$\rho_\infty \frac{d}{dx}\int_0^\infty (U - u)d\psi = \tau_w \tag{24-5a}$$

Crocco's transformation.

$$\frac{d}{dx}\int_0^U \frac{\rho\mu u}{\tau}(U - u)du = \tau_w \tag{24-5b}$$

Howarth's transformation.

$$\rho_\infty \frac{d}{dx}\int_0^\infty u(U - u)d\eta = \tau_w \tag{24-5c}$$

The use of the Howarth-Stewartson transformation (Eq. 23-18) is now considered in more detail. In analogy with Eq. 24-3a and 24-3b one may define certain thickness parameters based on the distorted velocity components $\bar{U}$, $\bar{V}$ (Eq. 23-18c and 23-18d) and the distorted y variable ζ (Eq. 23-15a). This gives

$$\bar{\delta}^* = \int_0^\infty \left(1 - \frac{\bar{U}}{\bar{U}_e}\right) d\zeta \tag{24-6a}$$

$$\bar{\delta}^{**} = \int_0^\infty \frac{\bar{U}}{\bar{U}_e}\left(1 - \frac{\bar{U}}{\bar{U}_e}\right) d\zeta \tag{24-6b}$$

The corresponding integrals for the total enthalpy are

$$\bar{\vartheta}^* = \int_0^\infty \left(1 - \frac{H}{H_e}\right) d\zeta \tag{24-6c}$$

$$\bar{\vartheta}^{**} = \int_0^\infty \frac{\bar{U}}{\bar{U}_e}\left(1 - \frac{H}{H_e}\right) d\zeta \tag{24-6d}$$

For incompressible fluids $\bar{\delta}^* = \delta^*$ and for $M = 0$ one finds $\bar{\delta}^{**} = \delta^{**}$; these equalities are, however, not true in general.

Integrating Eq. 23-18a and 23-18b from $\zeta = 0$ to $\zeta = \infty$ one finds

$$\frac{d(\bar{U}_e^2\bar{\delta}^{**})}{d\xi} + \bar{U}_e\frac{d\bar{U}_e}{d\xi}(\bar{\delta}^* - \bar{\vartheta}^*) = \nu_\infty[C\bar{U}_\zeta]_{\zeta=0} \tag{24-7a}$$

$$\frac{d(\bar{U}_eH_e\bar{\vartheta}^{**})}{d\xi} = \frac{\nu_\infty}{Pr}[CH_\zeta]_{\zeta=0} \tag{24-7b}$$

Eq. 24-3a is the analogue of Eq. 24-2. While the latter equation is apparently simpler, the former equation is actually sometimes more convenient (cf. Art. 32). Eq. 24-7b is the analogue of Eq. 24-4.

THE FLAT PLATE WITH ZERO PRESSURE GRADIENT

B,25. Introductory Remarks. There is an extensive literature on the theory of boundary layers with zero pressure gradient. The relative simplicity of the flow allows the boundary layer equations to be solved for a variety of boundary conditions and assumptions regarding the transport properties of the fluid. In particular, it is possible to study the dependence of skin friction, heat transfer, and recovery temperature on Mach number, viscosity law, Prandtl number, and wall temperature. It is also possible to check approximate methods of calculation against more exact methods.

The discussion in this article is based on the differential equations, especially on the transformed systems introduced in Art. 23. The integral relations introduced in Art. 24 may also be used for very rapid approximate calculations of skin friction and heat transfer. However, a discussion of integral methods is postponed to Art. 32.

We assume that the external forces and heat sources are zero. Furthermore, the flow is assumed to be stationary.

By assumption all external parameters are constant:

$$u_e = U, \qquad h_e = h_\infty, \qquad p_e = p_\infty \tag{25-1}$$

The pressure is therefore constant everywhere in the boundary layer and Eq. 22-1 reduce to

$$\rho(uu_x + vu_y) = (\mu u_y)_y \tag{25-2a}$$

$$(\rho u)_x + (\rho v)_y = 0 \tag{25-2b}$$

$$\rho(uh_x + vh_y) = \left(\frac{\mu}{Pr} h_y\right)_y + \mu(u_y)^2 \tag{25-2c}$$

The boundary conditions are

$$u = v = 0,\ h = h_w(x) \qquad \text{for} \qquad y = 0 \tag{25-3a}$$

$$u = U,\ h = h_\infty \qquad \text{for} \qquad y = \infty \text{ or } x = 0 \tag{25-3b}$$

Note that in general h_w is a function of x. The condition $h(x, 0) = h_w(x)$ may sometimes be replaced by a condition on the heat transfer at the wall.

In studying the boundary layer on a flat plate it is convenient to use the transformed equations of Art. 23, rather than Eq. 25-2. When the pressure gradient is zero, the Illingworth equations and the Howarth-Stewartson equations do not differ essentially from the von Mises equations and the restricted form of Howarth's equations, respectively. Hence we only consider Eq. 23-3, 23-8, and 23-14. After putting $dp/dx = 0$ in these equations and canceling the common factor ρu in Eq. 23-3a one sees that all of the transformed equations (but not the original Eq. 25-2) have the important property that *μ and ρ enter only in the combination $\mu\rho$.* The velocity field is primarily determined by the momentum and continuity equations. However, if $\mu\rho$ depends on h the velocity distribution also depends on the temperature distribution. The velocity field enters the energy equation by way of the transport and dissipation terms. The energy equation is nonlinear in h whenever $\mu\rho$ depends on h. Thus, in general, the boundary layer equations form a complicated nonlinear system. There are, however, certain special assumptions which allow a considerable simplification of these equations to be achieved. One such assumption is that the wall temperature h_w is constant, in which case the number of independent variables may be reduced by similarity considerations. On the other hand, special assumptions may be made regarding the transport coefficients, especially the assumption that the Prandtl number Pr is equal to unity or that the viscosity μ is proportional to temperature. These special cases are discussed below in some detail.

B,26. The Boundary Condition h_w = const. Similarity Solutions.

The discussion in Art. 13 of the boundary layer equations for incompressible flow showed that information could often be derived about the form of the solution by application of similarity considerations. To a limited extent, the same arguments apply to compressible flow, since Eq. 25-2 have the same similarity as in the incompressible case. They are invariant under the mapping

$$x \to c^2x,\ y \to cy,\ v \to c^{-1}v,\ u \to u,\ h \to h,\ \rho \to \rho,\ \mu \to \mu \qquad (26\text{-}1)$$

where c is any positive constant. The boundary conditions on u and v are also invariant under the mapping (Eq. 26-1). However, the boundary conditions on h are invariant only if h_w is constant. It is seen later that this requirement is equivalent to the requirement $q_w = \text{const}/\sqrt{x}$, and thus includes zero heat transfer as a special case. In Art. 27 conditions are mentioned under which the energy equation may be uncoupled from the momentum and continuity equations. The velocity profile then has separate similarity even for variable surface temperature. However, in the present discussion it is assumed that

$$h_w = \text{const} \qquad (26\text{-}2)$$

From the invariance of Eq. 25-2 and 25-3 under the mapping (Eq. 26-1) it follows that Eq. 25-2 may be reduced to ordinary differential equations with $y/\sqrt{x}$ as the independent variable. In addition, the following functional relations follow directly from the similarity (the symbol "fn" stands for "some function of," different in each case),

$$u,\ \rho,\ h = \text{fn}\left(\frac{y}{\sqrt{x}}\right) \qquad (26\text{-}3a)$$

$$v,\ \tau,\ q = \frac{1}{\sqrt{x}}\,\text{fn}\left(\frac{y}{\sqrt{x}}\right) \qquad (26\text{-}3b)$$

$$\psi = \sqrt{x}\,\text{fn}\left(\frac{y}{\sqrt{x}}\right) \qquad (26\text{-}3c)$$

Assuming that u is a monotone function of $y/\sqrt{x}$, the relation $u =$ fn $(y/\sqrt{x})$ may be inverted and u used as independent variable instead of $y/\sqrt{x}$ to obtain the functional relations

$$\rho,\ h = \text{fn}\ (u) \qquad (26\text{-}3d)$$

$$v,\ \tau,\ q = \frac{1}{\sqrt{x}}\,\text{fn}\ (u) \qquad (26\text{-}3e)$$

These similarity laws go beyond dimensional analysis, although they will later be combined with dimensional analysis. Eq. 26-3 imply directly two important results: (*1*) *There exists an energy integral: that is, a relation between h and u which is independent of x and y;* (*2*) *The skin friction and heat transfer at the wall vary as* $1/\sqrt{x}$ *and are therefore proportional to each other* (*Reynolds' analogy*):

$$\tau_w = \frac{c_1}{\sqrt{x}}, \qquad q_w = \frac{c_2}{\sqrt{x}}, \qquad \therefore\ q_w = \frac{c_2}{c_1}\tau_w \tag{26-4}$$

where c_1 and c_2 obviously depend on the external parameters of the problem but are independent of x.

For zero heat transfer at the wall, h_w is constant; hence the relations (Eq. 26-4) are true for this case also. To see this, first let h_w be zero. Then q_w and hence c_2, are negative. As h_w increases to large values, q_w, and hence c_2, must eventually become positive. The case $q_w = 0$, i.e., $c_2 = 0$, must therefore be included as a special case for some intermediate value of h_w.

The transformed equations of Art. 23 have similarity properties equivalent to those of Eq. 25-2. The form of Eq. 23-14 in particular resembles that of Eq. 25-2. By a direct similarity argument one then finds that u is a function of $\eta/\sqrt{x}$. However, $\eta = \int_0^y \text{fn}\,(y/\sqrt{x})dy = \sqrt{x}\,\text{fn}\,(y/\sqrt{x})$ so that the statements $u = \text{fn}\,(\eta/\sqrt{x})$ and $u = \text{fn}\,(y/\sqrt{x})$ are equivalent. If one applies the similarity argument to Eq. 23-3 instead, one sees that $\psi/\sqrt{x}$ may be used as the sole independent variable. But by Eq. 26-3c $\psi/\sqrt{x}$ depends on $y/\sqrt{x}$ only, so that again the functional relations (Eq. 26-3a and 26-3b) follow. Finally, Eq. 23-8 (with $p_x = 0$) are invariant under the mapping

$$x \to c^2x, \qquad \tau \to \frac{\tau}{c} \tag{26-1'}$$

with all other variables unchanged. It follows that h, $\rho = \text{fn}\,(u)$, and $\tau = (1/\sqrt{x})\,\text{fn}\,(u)$, with other relations equivalent to Eq. 26-3. It is interesting to note that for Crocco's equations the similarity argument leads to a separation of variables.

In using Eq. 26-3 to reduce the partial differential equations of the laminar boundary layer to ordinary differential equations, it is convenient to work with the transformed equations. This reduction is now carried out after introducing nondimensional variables.

von Mises' equations. Rather than $\psi/\sqrt{x}$, the nondimensional variable

$$\zeta = \frac{\psi}{\sqrt{\nu_\infty Ux}} \tag{26-5}$$

is used. The previously defined nondimensional variables $u^* = u/U$,

$\tilde{\rho} = \rho/\rho_\infty$, $\tilde{\mu} = \mu/\mu_\infty$, $\tilde{h} = h/h_\infty$ are also appropriate. Since in general

$$\frac{\partial}{\partial x} f\left(\frac{\psi}{\sqrt{\nu_\infty U x}}\right) = -\frac{\zeta}{2x}\frac{df}{\partial \zeta}$$

Eq. 23-3 reduce to

$$-\frac{\zeta}{2}\frac{du^*}{d\zeta} = \frac{d}{d\zeta}\left(\tilde{\mu}\tilde{\rho}u^*\frac{du^*}{d\zeta}\right) \tag{26-6a}$$

$$-\frac{\zeta}{2}\frac{d\tilde{h}}{d\zeta} = \frac{1}{Pr}\frac{d}{d\zeta}\left(\tilde{\mu}\tilde{\rho}u^*\frac{d\tilde{h}}{d\zeta}\right) + (\gamma - 1)M^2\tilde{\mu}\tilde{\rho}u^*\left(\frac{du^*}{d\zeta}\right)^2 \tag{26-6b}$$

Crocco's equations. The factor $\mu\rho u$ in the first term of Eq. 23-8a depends on u only and may hence be taken outside the sign of differentiation. In Eq. 23-8b h_x is zero and the common factor τ may be canceled. Putting

$$\tau = \frac{g(u)}{\sqrt{2x}} \text{ or } \frac{\tau}{\rho_\infty U^2} = \tilde{g}\sqrt{\frac{\nu_\infty}{2Ux}} \text{ with } \tilde{g} = \frac{g}{\sqrt{\mu_\infty\rho_\infty U^3}} \tag{26-7}$$

in Eq. 23-8a and 23-8b one obtains

$$g\frac{d^2g}{du^2} + \mu\rho u = 0 \tag{26-8a}$$

$$\left(\frac{d^2h}{du^2} + Pr\right)g + (1 - Pr)\frac{dh}{du}\frac{dg}{du} = 0 \tag{26-8b}$$

or in nondimensional form,

$$\tilde{g}\frac{d^2\tilde{g}}{du^{*2}} + \tilde{\mu}\tilde{\rho}u^* = 0 \tag{26-8a'}$$

$$\left[\frac{d^2\tilde{h}}{du^{*2}} + (\gamma - 1)M^2Pr\right]\tilde{g} + (1 - Pr)\frac{d\tilde{h}}{du^*}\frac{d\tilde{g}}{du^*} = 0 \tag{26-8b'}$$

Howarth's equations. The continuity equation (Eq. 23-14b) implies the existence of a stream function $\psi(x, \eta)$, such that

$$\psi_\eta = u, \qquad \psi_x = -\bar{v} \tag{26-9a}$$

With $p_x = 0$, the momentum equation (Eq. 23-14a) becomes

$$\psi_{x\eta}\psi_\eta - \psi_{\eta\eta}\psi_x = \nu_\infty(\tilde{\rho}\tilde{\mu}\psi_{\eta\eta})_\eta \tag{26-9b}$$

Using similarity, Eq. 26-9b may be transformed into an ordinary differential equation by the change of variables (cf. Eq. 14-18 and 14-19)

$$\psi = \sqrt{2\nu_\infty U x}\, f(\theta) \tag{26-10a}$$

where

$$\theta = \eta\sqrt{\frac{U}{2\nu_\infty x}} \tag{26-10b}$$

with the result given in Eq. 26-11a below. By the same change of variables, the energy equation

$$u\tilde{h}_x + \bar{v}\tilde{h}_\eta = \frac{\nu_\infty}{Pr}(\tilde{\mu}\tilde{\rho}\tilde{h}_\eta)_\eta + (\gamma - 1)M^2\nu_\infty\tilde{\mu}\tilde{\rho}(u_\eta^*)^2 \qquad \text{(26-9c)}$$

may be written as Eq. 26-11b. The Howarth equations are thus

$$f(\theta)f''(\theta) + (\tilde{\mu}\tilde{\rho}f''(\theta))' = 0 \qquad \text{(26-11a)}$$

$$f(\theta)\tilde{h}'(\theta) + \frac{1}{Pr}(\tilde{\mu}\tilde{\rho}\tilde{h}'(\theta))' + (\gamma - 1)M^2\tilde{\mu}\tilde{\rho}(f''(\theta))^2 = 0 \qquad \text{(26-11b)}$$

Under the assumption of constant wall temperature the boundary layer equations for the flat plate may thus be reduced to a system of two nonlinear ordinary differential equations. Convenient forms of this system are given by Eq. 26-6, 26-8, or 26-11, respectively. However, even these simplified equations are not easily solved. In Art. 27 and 28 we discuss certain assumptions regarding the transport coefficients of the equations. In many cases these assumptions lead to comparatively simple methods for determining skin friction or heat transfer at the wall.

B,27. Prandtl Number Equal to Unity. Energy Integrals. The treatment of the shock layer in Art. 19 was found to be considerably simplified when the assumption $Pr'' = 1$ was made. The same assumption also simplified the treatment of the heat pulse in the discussion of linearized waves in Art. 21. These problems dealt principally with longitudinal waves, and the viscous stresses were those arising from compression. The boundary layer, on the other hand, is a transversal wave in which viscous stresses are associated with shear only. In this case the assumption $Pr = 1$ leads to a simplification of the equations. This assumption is about as realistic for air as the assumption $Pr'' = 1$ (cf. Art. 3 and 19) and has consequences which are sufficiently interesting to justify a close study. It is therefore assumed in the discussion below that $Pr = 1$.

Energy integrals for constant wall temperature. That the assumption of Prandtl number unity simplifies the mathematical problem may be seen directly from the energy equation in Crocco's form (Eq. 23-8b) or in the form of an equation for the total enthalpy (Eq. 22-1d′). It was proved in Art. 22 that certain nonsteady boundary layer flows with $Pr = 1$ and zero heat transfer have a Busemann energy integral; the total enthalpy $H = h + (u^2/2)$ is constant throughout the boundary layer. Crocco [*132*] was able to relax the restriction on heat transfer provided that the flow is steady, the pressure gradient is zero, and the wall temperature is constant. If the momentum equation (Eq. 25-2a) is multiplied by $u + AU$, where A is an arbitrary const, and added to the energy equation (Eq.

25-2c), there is obtained an equation which obviously has the integral $h + (u^2/2) + AUu = \text{const}$. Since $u = 0$ for $y = 0$, the constant must be h_w; hence this energy integral exists whenever the wall temperature is independent of x and t. The significance of the factor A may be seen by evaluating the energy integral at $y = \infty$. It follows that

$$A = \frac{h_w - h_\infty - \dfrac{U^2}{2}}{U^2} = \frac{h_w - H_\infty}{U^2}$$

where $H_\infty = h_\infty + (U^2/2)$ is the adiabatic stagnation enthalpy of the external flow. It also follows that the heat transfer at the wall q_w (cf. Eq. 26-4) is

$$q_w = -\left(k\frac{\partial T}{\partial y}\right)_w = -\frac{1}{Pr}\left(\frac{dh}{du}\right)_w\left(\mu\frac{\partial u}{\partial y}\right)_w = \frac{AU}{Pr}\tau_w = AU\tau_w$$

Note also that any constant value of h_w may be obtained by the proper choice of the parameter A. These results may be summarized: If $Pr = 1$ and if h_w, p, and U are independent of x and t, then

$$h + \frac{u^2}{2} + AUu = \text{const} = h_w = h_\infty + \frac{U^2}{2} + AU^2 \tag{27-1a}$$

and

$$q_w = AU\tau_w \tag{27-1b}$$

where

$$A = \frac{h_w - h_\infty + \dfrac{U^2}{2}}{U^2} = \frac{h_w - H_\infty}{U^2} \tag{27-1c}$$

The nondimensional form of the energy integral is

$$\tilde{h} + \frac{(\gamma - 1)}{2} M^2 u^{*2} + (\tilde{h}_w - \tilde{H}_\infty)u^* = \tilde{h}_w \tag{27-1'}$$

where

$$\tilde{H}_\infty = 1 + \frac{\gamma - 1}{2} M^2$$

It was shown in Art. 26 that $h_w = \text{const}$ implies similarity, that q_w is then proportional to τ_w, and that an energy integral exists. It has now been shown that *if one in addition assumes that $Pr = 1$, then the constant of proportionality and the form of the energy integral may be written explicitly.* There being no heat transfer at the wall if $h_w = H_\infty$, the temperature at the wall when $q_w = 0$ is the same as the adiabatic stagnation temperature T^0. In other words, for zero heat transfer the recovery temperature at the wall T_r, obtained by bringing the fluid to rest through the action of viscous forces, is identical with the stagnation temperature T^0, obtained in the free stream by bringing the fluid to rest through the action of pressure forces. Heat is transferred from the wall into the fluid if $h_w > H_\infty$ and in the opposite direction if $h_w < H_\infty$.

Eq. 27-1 may of course be derived from any of the transformed equations of Art. 23. Crocco's equation (Eq. 26-8b) is especially suitable. For $Pr = 1$ this equation simplifies to $d^2h/du^2 = -1$, from which one immediately obtains Eq. 27-1 by simply integrating twice.

For flow at zero Mach number with heat transfer (cf. Art. 20 and 22), the dissipation term $\mu(u_y)^2$ is absent from the energy equation (Eq. 25-2c). The procedure originally used in deriving Eq. 27-1 should then be modified. The momentum equation should be multiplied only by the constant AU and not by u. The corresponding energy integral is

$$h + AUu = h_w = h_\infty + AU^2 \tag{27-2}$$

or in nondimensional form,

$$\tilde{h} + (h_w - 1)u^* = \tilde{h}_w \tag{27-2'}$$

In this case the recovery enthalpy is $h_r = h_\infty$. If $h_w = h_\infty$, then h is constant throughout the fluid, i.e. the flow is incompressible.

The discussion so far may be restated as follows: By an energy integral is meant a relation between h and u which is valid throughout the boundary layer independent of x and y. For zero pressure gradient such an energy integral exists when $h_w =$ const. Its form is in general not known explicitly. However, if in addition $Pr = 1$, then the energy integral has the explicit form given in Eq. 27-1a. Even when the pressure gradient is different from zero, the relation $h + (u^2/2) = h_w = h_\infty + (U^2/2)$ is still valid for steady flow provided the heat transfer at the wall is zero and $Pr = 1$. For zero Mach number, Eq. 27-1a reduces to Eq. 27-2; if furthermore $h_w = h_\infty$, then the flow is incompressible in the sense that the temperature and density are constant everywhere. The existence of an energy integral implies Reynolds' analogy; heat transfer at the wall is proportional to skin friction. For $Pr = 1$, the constant of proportionality is explicitly given by $(h_w - H_\infty)/U^2$: thus for zero heat transfer $h_w = H_\infty$.

Solution of momentum and continuity equations. It has been shown that for $Pr = 1$ and $h_w =$ const the enthalpy distribution h and the heat transfer at the wall q_w may be expressed very simply in terms of the velocity distribution u and the skin friction τ_w. The problem is therefore reduced to that of determining u and τ_w. To fix the ideas, consider Howarth's form of the boundary layer equations (Eq. 26-11). (Similar reasoning may of course be applied to the corresponding equations in von Mises' or Crocco's form.) The variables $\tilde{\mu}$ and $\tilde{\rho}$ occurring in the combined momentum and continuity equation (Eq. 26-11a) may be expressed as functions of u^* with the aid of the energy integral, the equation of state, and any one of several formulas for the dependence of viscosity on temperature. Since $u^* = f'(\theta)$, Eq. 26-11a reduces to an equation with f as the only dependent variable and θ as the independent variable (for notation, cf. Eq. 26-10). This equation is a generalization to compressible flow of the Blasius equation. Its form depends on the viscosity-temperature relation assumed. It was studied by von Kármán and Tsien [*126*] for the special law $\mu/\mu_\infty = (T/T_\infty)^{0.76}$. If one instead assumes that μ is proportional to T and hence to $1/\rho$, Eq. 26-11a reduces to the Blasius equation Eq. 14-19a, whether $Pr = 1$ or not. In this case the problem simplifies considerably, even for arbitrary Prandtl number, as it will now be shown.

B,28. Solutions with Viscosity Coefficient Proportional to Temperature.

The Chapman Viscosity Relation. Above it was shown how the assumption $Pr = 1$ leads to certain integrals of the energy equation. This assumption is now dropped. Instead we make a special assumption regarding the dependence of the viscosity on the temperature. With this assumption the momentum and continuity equations may be solved, in a certain sense, independently of the energy equation. Furthermore, the energy equation admits simple solutions, even for general values of Pr and for variable h_w.

According to Art. 3 experimental data for the viscosity of dry air at ordinary temperatures and pressures are accurately represented by Sutherland's formula, Eq. 3-22. The use of this formula or of approximations, such as $\mu/\mu_\infty = (T/T_\infty)^\omega$, necessitates extensive numerical calculations (cf. [*126*]) except for $\omega = 1$. When $\omega = 1$, i.e. when μ is proportional to T, the boundary layer equations simplify considerably. On the other hand this viscosity law is not very realistic. Chapman (see [*133*] and references given there) therefore proposed the use of an average viscosity law (cf. Eq. 3-23) of the form

$$\frac{\mu T_\infty}{\mu_\infty T} = C = \text{const} \tag{28-1}$$

As may be seen from Fig. B,3, C can be chosen so that the curve representing Eq. 28-1 intersects the Sutherland curve at any point. In general, the two curves are not tangential at the point of intersection, and Eq. 28-1 is only locally correct. However, with a suitably chosen value of the const C, the error in computing τ_w and q_w may be very small. It is obviously important to have a correct viscosity coefficient near the wall where the shear is large. In the case of constant wall temperature we therefore define C by

$$C = \frac{\mu_w T_\infty}{\mu_\infty T_w} = \frac{\mu_w h_\infty}{\mu_\infty h_w} \tag{28-2}$$

As mentioned above, the boundary layer computations become considerably simplified if it is assumed that μ is strictly proportional to T. The case corresponding to $C = 1$ in Eq. 28-1 is discussed first. Then it is shown that, with a slight modification, the results may be applied to the case $C \neq 1$.

Solution of Momentum and Continuity Equations.

$C = 1$. Since for zero pressure gradient ρ is inversely proportional to T, Eq. 28-1 with $C = 1$ implies that $\mu\rho = \text{const} = \mu_\infty\rho_\infty$, i.e.

$$\tilde{\mu}\tilde{\rho} = 1 \tag{28-3}$$

It was pointed out earlier that in the transformed equations μ and ρ occur only in the combination $\mu\rho$. Hence these equations are considerably simplified if Eq. 28-3 is assumed to hold. We choose Howarth's form for purposes of illustration. If $dp/dx = 0$ and Eq. 28-3 holds, the momentum and continuity equations (Eq. 23-14a and 15-14b) are independent of the energy equation. Furthermore, the effect of compressibility does not appear explicitly in these equations. They are in fact formally identical with the corresponding equations for incompressible flow. The variables y and v in one problem correspond to η and $\bar{v}$ in the other. It is also known a priori that $y = 0$ and $y = \infty$ correspond to $\eta = 0$ and $\eta = \infty$, respectively. From this and the definition of $\bar{v}$ it follows that the boundary conditions are formally identical in the two cases. Hence the relations expressing u and $\bar{v}$ as functions of x and η in the compressible case are identical with the relations expressing u and v as functions of x and y in the incompressible case. Furthermore, Eq. 23-20a shows that $\tau_w = \mu_\infty(\partial u/\partial\eta)_{\eta=0}$ in the compressible case; the skin friction is therefore the same function of x in both cases. In deriving these results the energy equation (Eq. 23-14c) or the boundary conditions on h have not been used. The momentum and continuity equations are thus independent of the energy equation in the sense that the latter is not needed for determining u and $\bar{v}$ (as functions of x and η) or for determining τ_w. In particular, the similarity laws for u, v, τ, and ψ as expressed by Eq. 26-3 are valid even for variable h_w, provided that y is replaced by η. However, the relation between y and η (cf. Eq. 23-12a) and hence the relation between u and y cannot be evaluated until the temperature field has been determined.

The results just obtained may be summarized as follows: *If $\mu/\mu_\infty = T/T_\infty$ and $dp/dx = 0$, the skin friction is independent of Mach number, Prandtl number, and of the thermal boundary conditions. Its value is thus given by Eq. 14-28a with ρ and ν evaluated at free stream conditions.* The velocity profile is determined by the Blasius function, the argument of which includes a distorted coordinate η whose relation to the physical coordinates x and y depend on the temperature field.

To obtain this result, any of the transformations discussed in Art. 23 might have been used. As pointed out above, for $C = 1$ the momentum and continuity equations have the similarity expressed by Eq. 26-1 independent of the value of h_w. Hence Eq. 26-6a, 26-8a′, and 27-11a are valid. If $\tilde{\mu}\tilde{\rho} = 1$, the last of these equations is formally identical with the Blasius equation for incompressible flow; the other two are then equivalent to it.

The problem of determining the temperature field is dealt with later. First we consider the effect of assuming $C \neq 1$.

$C \neq 1$. We now assume the more general Chapman relation Eq. 28-1 where C is a constant, not necessarily unity. An effective coefficient of viscosity at infinity may then be defined by

$$\mu'_\infty = C\mu_\infty, \qquad \nu'_\infty = C\nu_\infty \tag{28-4}$$

It is then easily seen that the momentum and continuity equations may be solved as above, except that μ_∞ has to be replaced by μ'_∞. In particular the skin friction is now

$$\tau_w = C\mu_\infty \left(\frac{\partial u}{\partial \eta}\right)_{\eta=0} = \sqrt{\frac{C\mu_\infty\rho_\infty U^3}{2x}}\, B''(0) \tag{28-5a}$$

where $B''(0) = 0.46960$. For future reference we note that the expression for heat transfer at the wall is now

$$q_w = -C\frac{\mu_\infty}{Pr}\left(\frac{\partial h}{\partial \eta}\right)_{\eta=0} \tag{28-5b}$$

Assume now that h_w is constant and that C is determined from Eq. 28-2 and a viscosity-temperature relation. Eq. 28-5a then shows that the local friction coefficient $c_f = 2\tau_w/\rho_\infty U^2$ varies as $\sqrt{C}$ at a given value of the local Reynolds number Ux/ν_∞; it thus depends on h_w and h_∞ only. If the value of h_w is a prescribed boundary condition, c_f is then independent of M and Pr. At an insulated wall, however, h_w is the recovery enthalpy h_r which depends on M and Pr (cf. Eq. 28-14 below). In this sense the skin friction coefficient depends on M and Pr for an insulated wall.

The actual value of $\sqrt{C}$ may be determined from h_w and h_∞ if one assumes a suitable viscosity-temperature relation. For air at moderate temperatures one may use the Sutherland formula (Eq. 3-22). The experimental curve should be translated until the reference point (μ_0, T_0) corresponds to free stream conditions; the Chapman line should then be drawn through the point corresponding to wall conditions. It can be concluded that (1) skin friction at an insulated wall decreases slightly with increasing Mach number, this effect of compressibility being more marked in free flight than in a wind tunnel, (2) skin friction decreases slightly when heat is transferred from the wall to the fluid, and (3) the effect of changes in Prandtl number on friction is small near $Pr = 1$.

Solutions of the Energy Equations. The energy equation is now discussed under the assumption that the average viscosity law (Eq. 28-1) may be used instead of a more accurate law. The value of the Prandtl number is assumed to be constant but otherwise arbitrary. The solution for the case of constant wall temperature is given first. This includes the determination of the recovery temperature for zero heat transfer at the wall. Methods applicable to the case of variable wall temperature will then be introduced. These methods will be discussed further in Art. 30.

Linearity of the energy equation. Under the assumptions (Eq. 28-1) and $dp/dx = 0$ the energy equation in Howarth's form may be written:

$$uh_x + \bar{v}h_\eta = \frac{\nu'_\infty}{Pr} h_{\eta\eta} + \nu'_\infty u_\eta^2 \tag{28-6}$$

where ν'_∞ is defined by Eq. 28-4. The velocity components u and $\bar{v}$ may be determined as functions of x and η by the method discussed above. They involve derivatives of the Blasius function and may be regarded as known functions. Eq. 28-6 is hence a linear parabolic equation in h where the lowest order derivatives have variable coefficients. The equation is nonhomogeneous because of the term $\nu'_\infty u_\eta^2$. Mathematically speaking, this term is a known forcing function; its physical meaning as heat addition through dissipation has been pointed out previously. For $M = 0$ this term is absent. The general solution of Eq. 28-6 may be expressed as the sum of a particular solution and the general solution of the corresponding homogeneous equation. In other words this equation is solved for all values of M and all boundary conditions if one has the general solution for $M = 0$ and one solution for $M \neq 0$ for some special boundary conditions.

Eq. 28-6 may thus be regarded as a heat equation with constant coefficient of heat conduction subject to transport of heat by a known variable flow field and to addition of heat from a known external source distribution. The problem for the compressible fluid may be viewed as a problem for a fictitious incompressible fluid with constant ρ and μ but with variable temperature and with heat added by dissipation. Such a fluid has been studied by Pohlhausen [*134*] (cf. also the discussion in [*24*, p. 623ff.]). Whether fluids exist for which these conditions are approximately true is hardly relevant here, since Pohlhausen's results may still be applied to a compressible fluid (variable ρ and μ) if the relation (Eq. 28-1) is assumed.

h_w = *const.* For the present the discussion of Eq. 28-6 is restricted to flows with similarity; that is, flows for which h_w = const. Then Eq. 28-6 and the corresponding equations obtained by von Mises' and Crocco's transformations reduce to ordinary differential equations which have already been presented as Eq. 26-6b, 26-8b′, and 26-11b. With the assumption (Eq. 28-1) these equations are

$$\frac{1}{Pr}\frac{d}{d\zeta}\left(u^*\frac{d\bar{h}}{d\zeta}\right) + \frac{\zeta}{2}\frac{d\bar{h}}{d\zeta} + (\gamma - 1)M^2u^*\left(\frac{du^*}{d\zeta}\right)^2 = 0 \tag{28-7a}$$

$$\left[\frac{d^2\bar{h}}{du^{*2}} + (\gamma - 1)M^2Pr\right]\bar{g} + (1 - Pr)\frac{d\bar{h}}{du^*}\frac{d\bar{g}}{du^*} = 0 \tag{28-7b}$$

$$\frac{1}{Pr}\frac{d^2\bar{h}}{d\theta^2} + f\frac{d\bar{h}}{d\theta} + (\gamma - 1)M^2\left(\frac{d^2f}{d\theta^2}\right)^2 = 0 \tag{28-7c}$$

Whenever μ_∞ or ν_∞ enter into the definition of any variable used in the above equations, they should be replaced by μ'_∞ and ν'_∞, respectively (cf. Eq. 28-4). In particular,

$$\theta = \eta\sqrt{\frac{U}{2\nu'_\infty x}} \tag{28-7d}$$

Eq. 28-7c is, except for constants, identical with the equation studied by Pohlhausen in [*134*]; Eq. 28-7a and 28-7b are equivalent to this equation.

To obtain the solution of Eq. 28-7c we first consider the corresponding homogeneous equation

$$\bar{h}'' + PrB\bar{h}' = 0 \tag{28-8}$$

where a prime denotes a derivative with respect to θ. Observing that the same equation is satisfied by $(h - h_\infty)/(h_w - h_\infty)$ and using the method of the integrating factor, one obtains

$$\frac{h - h_\infty}{h_w - h_\infty} = \frac{\int_\theta^\infty e^{-Pr\int_0^\theta B(\zeta)d\zeta}d\theta}{\int_0^\infty e^{-Pr\int_0^\theta B(\zeta)d\zeta}d\theta} \tag{28-9}$$

Since $B''' + BB'' = 0$ and hence $B''(\theta) = B''(0)e^{-\int_0^\theta B(\zeta)d\zeta}$, one may write Eq. 28-9 as

$$\frac{h - h_\infty}{h_w - h_\infty} = \frac{\bar{h} - 1}{\bar{h}_w - 1} = \frac{\int_\theta^\infty [B''(\zeta)]^{Pr}d\zeta}{\int_0^\infty [B''(\zeta)]^{Pr}d\zeta} \tag{28-9'}$$

The heat transfer from the wall to the fluid is then (cf. Eq. 28-5b and 28-7d)

$$q_w = -\sqrt{\frac{\rho_\infty\mu'_\infty U}{2x}}\frac{h'(0)}{Pr} \tag{28-10a}$$

where

$$h'(0) = -\frac{(h_w - h_\infty)[B''(0)]^{Pr}}{\int_0^\infty [B''(\zeta)]^{Pr}d\zeta} \tag{28-10b}$$

Eq. 28-10a and 28-10b may be combined in the following form:

$$\frac{Prq_w}{\mu'_\infty(h_w - h_\infty)} = \sqrt{\frac{U}{\nu'_\infty x}}\, a(Pr) = \frac{\tau_w}{\mu'_\infty U}\frac{\sqrt{2}\, a(Pr)}{B''(0)} \tag{28-11a}$$

where

$$a(Pr) = \frac{[B''(0)]^{Pr}}{\sqrt{2}\int_0^\infty [B''(\zeta)]^{Pr} d\zeta},\ B''(0) \cong 0.46960 \tag{28-11b}$$

The values of the function $a(Pr)$ will be discussed later (Eq. 28-18b).

Since the nonhomogeneous term in Eq. 28-7c represents heat added by dissipation, the homogeneous equation (Eq. 28-8) is appropriate for heat transfer at zero Mach number. If $\tilde{h}_w = 1$ no heat transfer takes place and $\tilde{h} = 1$. The flow is then incompressible.

The right-hand side of Eq. 28-9′ is a function of θ and implicitly a function of u^* since $B'(\theta) = u^*$. The formula (Eq. 28-9′) is thus a special form of the energy integral which according to Eq. 26-3d must exist when $h_w =$ const. For $Pr = 1$ it reduces to the simple energy integral $\tilde{h} = \tilde{h}_w + (1 - h_w)u^*$ previously derived directly (Eq. 27-2′) without the assumption of a linear viscosity law.

We now turn to the nonhomogeneous equation (Eq. 28-7c). A particular integral satisfying $\tilde{h}(\infty) = 1$ may be found by using an integrating factor. If this solution is added to the solution of the homogeneous equation as given by Eq. 28-9′, one obtains the general solution of Eq. 28-7c subject only to the restriction that $\tilde{h}(\infty) = 1$

$$(\tilde{h} - 1) = A\int_\theta^\infty [B''(\xi)]^{Pr} d\xi + (\gamma - 1)M^2 Pr \int_\theta^\infty \left\{[B''(\xi)]^{Pr}\int_0^\xi [B''(\zeta)]^{2-Pr} d\zeta\right\} d\xi \tag{28-12}$$

This is an energy integral valid under the assumptions of a linear viscosity law and constant wall temperature. For $Pr = 1$ it reduces to the energy integral previously derived directly (Eq. 27-1′) without the assumption $\mu \sim T$.

The constant A in Eq. 28-12 is readily related to the heat transfer and temperature at the wall. Since

$$\tilde{h}'(0) = -A[B''(0)]^{Pr} \tag{28-13}$$

it follows that zero heat transfer at the wall corresponds to $A = 0$. The wall temperature is in this case the recovery temperature T_r, which hence is given by the formula

$$\tilde{T}_r - 1 = \tilde{h}_r - 1 = \frac{(\gamma - 1)M^2}{2} b(Pr) \tag{28-14a}$$

where

$$b(Pr) = 2Pr\int_0^\infty \left\{[B''(\xi)]^{Pr}\int_0^\xi [B''(\zeta)]^{2-Pr} d\zeta\right\} d\xi \tag{28-14b}$$

Comparing Eq. 28-14a and 28-12 at $\theta = 0$, the constant A is found to be

$$A = \frac{\bar{h}_w - \bar{h}_r}{\int_0^\infty [B''(\xi)]^{Pr} d\xi} \tag{28-15}$$

Eq. 28-10a, 28-13, and 28-15 imply the following formula for the heat transfer from the wall to the fluid

$$\frac{Prq_w}{\mu'_\infty(h_w - h_r)} = \sqrt{\frac{U}{\nu'_\infty x}}\, a(Pr) = \frac{\tau_w}{\mu'_\infty U} \frac{\sqrt{2}\, a(Pr)}{B''(0)} \tag{28-16}$$

where $a(Pr)$ is defined by Eq. 28-11b as before. For $M = 0$ the recovery enthalpy h_r is equal to h_∞. Thus it is seen that Eq. 28-16 is a simple generalization of Eq. 28-11a obtained by replacing h_∞ by h_r. Note also that q_w is proportional to τ_w as required by Eq. 26-4. The constant of proportionality is given explicitly in the present case.

The heat transfer q_w may be expressed as $-[(\mu/Pr)\partial h/\partial y]_w$. It is sometimes convenient to form a dimensionally similar expression from the parameters of the problem, for example $q_0 = -(\mu_\infty/Pr)(h_2 - h_1)/L$ and to give the law of heat transfer for the nondimensional quantity q_w/q_0, called the Nusselt number. Eq. 28-16 suggests that logical choices for h_2 and h_1 are h_w and h_r, respectively. An obvious choice for the length L is the boundary layer thickness δ, which is proportional to $\sqrt{\nu_\infty x/U}$. The constant of proportionality is, however, arbitrary. We therefore make the somewhat less natural choice $L = x$. Eq. 28-16 may then be written

$$(Nu_x)(Re_x)^{-\frac{1}{2}} = a(Pr) \tag{28-17a}$$

where

$$Nu_x = \frac{Prq_w x}{\mu'_\infty(h_w - h_r)} \tag{28-17b}$$

and

$$Re_x = \frac{Ux}{\nu'_\infty} \tag{28-17c}$$

Here Nu_x is the Nusselt number based on the distance x from the leading edge of the plate, and Re_x is a local Reynolds number also based on x. In both definitions the effective viscosity at infinity $\mu'_\infty = C\mu_\infty$ has been used.

Eq. 28-14 and 28-16 give the recovery temperature and the heat transfer at the wall in terms of the functions $b(Pr)$ and $a(Pr)$. These functions are known in principle since they involve integrals of the Blasius function. In particular, the value of $a(1)$ according to Eq. 28-11b and 14-28a is

$$a(1) = \frac{B''(0)}{\sqrt{2}} \cong 0.332 \tag{28-18a}$$

Pohlhausen ([*134*]; see also [*26*, p. 265ff.]) has evaluated $a(Pr)$ for several values of Pr. For Pr near unity an approximate formula is

$$a(Pr) \cong a(1)Pr^{1/3} \cong 0.332Pr^{1/3}, \qquad Pr = 1 \tag{28-18b}$$

The function $b(Pr)$ has been computed by Pohlhausen [*134*] and by Eckert and Drewitz [*135*]. The results of these computations are also given in [*24*, p. 630] and [*26*, p. 266]. The exact value for $Pr = 1$ is obviously $b(1) = 1$. An approximate formula for Pr near unity is

$$b(Pr) \cong Pr^{1/2}, \qquad Pr \cong 1 \tag{28-19}$$

Variable wall temperature. We shall now very briefly study the case when h_w is not constant. As remarked earlier, the general solution of Eq. 28-6 may be expressed as the sum of a particular solution, taken from Eq. 28-12, and the general solution of the homogeneous equation

$$uh_x + \bar{v}h_\eta = \frac{\nu'_\infty}{Pr} h_{\eta\eta} \tag{28-20}$$

Physically this equation represents heat transfer at zero Mach number. A convenient particular solution h_P is the solution corresponding to zero heat transfer at the wall (Eq. 28-12 with $A = 0$). Defining h_d by

$$h_d = h - h_P \tag{28-21a}$$

one sees that h_d satisfies Eq. 28-20 with boundary conditions

$$h_d(\infty) = 0 \tag{28-21b}$$

$$h_d(0) = h_{dw} = h_w - h_r \tag{28-21c}$$

Furthermore,

$$\left(\frac{\partial h}{\partial \eta}\right)_w = \left(\frac{\partial h_d}{\partial \eta}\right)_w \tag{28-21d}$$

The heat transfer at the wall thus depends indirectly on M since the boundary condition on h_d at the wall depends on h_r and hence on M.

An approximate method of solving Eq. 28-20 was proposed by Fage and Falkner in 1931 [*136*]. The approximation involved consists in replacing the actual velocity profile, i.e. the Blasius profile $u = UB'(\theta)$ by a linear profile u_1 that has the correct slope in η at the wall. The corresponding approximation to $\bar{v}$ is obtained from the continuity equation which in the present case is Eq. 23-14b. Thus

$$u_1 = \alpha_1 U\theta \tag{28-22a}$$

$$v_1 = \alpha_1 U \sqrt{\frac{\nu'_\infty}{2Ux}}\,\frac{\theta^2}{2} \tag{28-22b}$$

where

$$\theta = \eta \sqrt{\frac{U}{2\nu'_\infty x}} \tag{28-22c}$$

and

$$\alpha_1 = B''(0) \tag{28-22d}$$

A comparison with Weyl's iterative method of solving the Blasius equation (Art. 13) shows that the values used for the transport velocity are the same as those used by Weyl in obtaining his second approximation to the Blasius function.

The dependence of surface heat transfer on Prandtl number may be deduced immediately. Inserting the approximate velocity components from Eq. 28-22a and 28-22b into Eq. 28-20, one obtains

$$\alpha_1 U\theta h_x + \alpha_1 U \sqrt{\frac{\nu'_\infty}{2Ux}} \frac{\theta^2}{2} h_\eta = \frac{\nu'_\infty}{Pr} h_{\eta\eta} \tag{28-22e}$$

One may eliminate the Prandtl number from this equation by introducing the variable $\eta^* = Pr^{\frac{1}{3}}\eta$. Thus

$$\left(\frac{\partial h}{\partial \eta}\right)_w = Pr^{\frac{1}{3}} \left(\frac{\partial h}{\partial \eta^*}\right)_w \tag{28-23a}$$

and hence (cf. Eq. 28-5b)

$$q_w = -\frac{C\mu_\infty}{Pr^{\frac{2}{3}}} \left(\frac{\partial h}{\partial \eta^*}\right)_w \tag{28-23b}$$

When $M > 0$, Eq. 28-22e should be applied to h_d (Eq. 28-21a) rather than to h. The value of h_d at the wall depends on the recovery enthalpy h_r (Eq. 28-21c) and hence on the Prandtl number. Thus, although the Prandtl number may be eliminated from the equation for h_d, it enters in a boundary condition. As a result $(\partial h/\partial \eta^*)_w$, which is equal to $(\partial h_d/\partial \eta^*)_w$, in general depends on the Prandtl number. On the other hand, when the Mach number is negligible, the recovery enthalpy is equal to h_∞ and Eq. 28-22 is valid for h. Eq. 28-23b then shows that the heat transfer at the wall varies as $Pr^{-\frac{2}{3}}$.

Lighthill [*137*] has extended the method of Fage-Falkner to the case of a variable external pressure (in this case, however, the parameter $\bar{h}_w - 1$ must be assumed to be small) and found the general solution of the resulting differential equation. Any solution of Eq. 28-22 is then a special case of Lighthill's general solution, which is discussed in Art. 30. Here only one example of a solution of Eq. 28-22 is given.

We consider the special case of h_w = const and $M = 0$. The quantity

$$\mathcal{H} = \frac{h - h_w}{h_\infty - h_w} \tag{28-24}$$

then obeys Eq. 28-22, which because of the similarities of the problem may be written

$$\frac{d^2\mathcal{H}}{d\theta^{*2}} + \frac{\theta^{*2}}{2} \frac{d\mathcal{H}}{d\theta^*} = 0 \tag{28-25a}$$

where

$$\theta = \eta^* \alpha_1^{\frac{1}{3}} \sqrt{\frac{U}{2\nu'_\infty x}} \tag{28-25b}$$

The boundary conditions are

$$\mathcal{H}(0) = 0 \tag{28-26a}$$

$$\mathcal{H}(\infty) = 1 \tag{28-26b}$$

The function f_2', where f_2 is Weyl's second approximation to the Blasius function, obeys formally identical conditions (cf. Eq. 14-26). Hence

$$\left(\frac{d\mathcal{H}}{d\theta^*}\right)_{\theta^*=0} = [6^{\frac{1}{3}}\Gamma(\tfrac{4}{3})]^{-1} \tag{28-27a}$$

and

$$\frac{Prq_w}{\mu'_\infty(h_w - h_\infty)} = \frac{(\alpha_1 Pr)^{\frac{1}{3}}}{6^{\frac{1}{3}}\Gamma(\frac{4}{3})} \sqrt{\frac{U}{2\nu'_\infty x}} \tag{28-27b}$$

This equation may be compared with the more accurate Eq. 28-11a. We first note that Eq. 28-11a also shows that q_w varies as $Pr^{-\frac{2}{3}}$ provided that the approximate formula (Eq. 28-18b) is used for $a(Pr)$. If one furthermore uses Weyl's second approximation (Eq. 14-27) to the value of α_1, then Eq. 28-27b becomes identical with Eq. 28-11b.

In deriving Eq. 28-6 it was assumed that $\mu\rho$ = const. Formally, the Fage-Falkner method then gives an approximate solution for any constant value of the Prandtl number. Actually, as it will be seen in Art. 30, the Fage-Falkner method is theoretically an asymptotic method valid for large Prandtl numbers. The method gives, however, surprisingly good numerical accuracy even for, say, $Pr = 1$. This fact is illustrated by the particular example given above.

BOUNDARY LAYERS WITH PRESSURE GRADIENTS

B,29. Introduction. The preceding article dealt with the boundary layer on a flat plate; the dependence of skin friction, heat transfer, etc. on the Mach number, Prandtl number, wall temperature, and viscosity-temperature relation was studied. The same problems are now discussed for a boundary layer with a nonvanishing pressure gradient. The emphasis is again placed on a qualitative evaluation of the various effects, exact solutions for idealized cases, and on approximate but very rapid methods.

One result valid for arbitrary pressure gradient and Mach number was obtained in Art. 22: For Prandtl number unity the total enthalpy $H = h + (u^2/2)$ is constant in a boundary layer at an insulated wall. This implies in particular that for $Pr = 1$ the recovery temperature is the same as the isentropic stagnation temperature. On the other hand, the more general energy integral $h + (u^2/2) + cUu = \text{const}$, discussed in Art. 27, is no longer valid; the derivation of this integral was based on the assumption of a zero pressure gradient. The energy integral $H = \text{const}$ assumes a special significance in view of a correlation theorem due to Illingworth [*130*] and Stewartson [*129*]. These authors showed that under the assumptions that lead to this energy integral, i.e. $Pr = 1$ and $q_w = 0$, and the additional assumption $\mu \sim T$, the problem of solving the momentum equation with a given pressure gradient for $M > 0$ may be reduced to an equivalent problem for incompressible flow. A slightly more general form of this correlation theorem will be derived in Art. 31 and its consequences discussed in Art. 32.

If the wall is not insulated or if $Pr \neq 1$ the reduction to a problem in incompressible flow is not possible for a general pressure gradient. Other simplifying assumptions are therefore considered. Art. 30 deals with compressible boundary layers at $M = 0$. It is then assumed that the external velocity is small compared to the speed of sound so that dissipation effects are negligible and the external enthalpy h_e is practically constant; compressibility effects arise only from a difference between the wall temperature and the free stream temperature. Naturally the boundary layer equations are somewhat simplified for this case. However, the momentum equation cannot be uncoupled from the energy equation by any of the coordinate transformations of Art. 23. As an example, in the momentum equation in Howarth's form, the term dp/dx is multiplied by the factor h/h_∞; one may then say that the actual pressure gradient is replaced by an apparent temperature-dependent pressure gradient. If the wall temperature is constant and the external velocity field varies as a power of the distance from the leading edge, the usual similarity principles may be applied. The momentum and energy equations may then be reduced to ordinary differential equations. These are, however, still coupled, i.e. the velocity and the temperature fields interact.

The only case in which the velocity field is independent of the temperature field is the case of small heat transfer, i.e. when $|h_w - h_\infty|$ is small compared to h_∞. Then, to the first approximation, the velocity boundary layer may be considered to be incompressible, and may hence be computed separately. The energy equation then becomes a linear equation which states that the temperature is transported with the velocity field of the incompressible boundary layer, while at the same time diffusing with a constant coefficient of heat conductivity. This equation may be further simplified by the Fage-Falkner method; the general solution of the resulting equation has been found by Lighthill. His solution assumes $M = 0$, $\mu \sim T$ and that the skin friction is known. The pressure gradient and the distribution of wall temperature may be arbitrary. Formally, the solution is given for any value of the Prandtl number; actually the Fage-Falkner approximation is not valid when this parameter is too small.

The corresponding approximation for very small Prandtl number is also considered here and the general solution of the resulting equation is given.

The calculations for $M = 0$ may also be used for high speed flow with the aid of an extension of the Illingworth-Stewartson correlation theorem. In the proof of this theorem one may omit the assumption that the wall is insulated. It follows then that for $Pr = 1$ and $\mu \sim T$ a boundary layer with an arbitrary pressure gradient and at an arbitrary Mach number may be correlated with a boundary layer at zero Mach number. This is shown in Art. 31.

Even for an incompressible fluid, accurate solutions for boundary layers with pressure gradients require extensive numerical calculations. These difficulties are of course considerably increased for compressible boundary layers since the momentum equation is coupled to the energy equation (except in some very special cases). There is therefore a great need for rapid approximate methods. Several such methods have been developed. Only the most successful of these methods is given here, namely the Kármán-Pohlhausen momentum method (Art. 32). The classical form of this method together with some recent improvements are given for incompressible fluids. Extension of the method to other cases is then indicated.

A summarizing discussion of compressibility effects in boundary layers is also given in Art. 32. Various qualitative ideas are discussed there, which, it is hoped, will shed some light on the more analytical reasoning in earlier articles.

B,30. Compressible Boundary Layers at $M = 0$.

LIMITING FORM OF THE EQUATIONS FOR $M = 0$. We first discuss how the Howarth-Stewartson equations (Eq. 23-18) may be simplified when $M = 0$. It is then seen that the resulting equations could also have been obtained directly from the restricted form of Howarth's equations (Eq. 23-14). Some of the ideas used below may be clarified by a comparison with the nondimensional form of the original boundary layer equations (Eq. 22-7) and the discussion in Art. 20.

For $M = 0$ the effect of dissipation may be neglected. Hence the last term of Eq. 23-18b may be omitted. Furthermore,

$$\frac{H}{h_\infty} = \frac{h}{h_\infty} + \frac{u^2}{2h_\infty} = \tilde{h} + \frac{\gamma - 1}{2} M^2 u^{*2} \tag{a}$$

Hence for $M = 0$, H may be replaced by h in Eq. 23-18a and 23-18b. This is actually also a consequence of neglecting dissipation since the effect of dissipation is considered implicitly when H is used instead of h.

Note also that $h_e = h(x, \infty) = h_\infty = \text{const}$. This corresponds to the statement $H_e = \text{const}$ for $M > 0$.

As M tends to zero, a_e and a_∞ tend to infinity in such a manner that their ratio tends to unity, i.e.

$$\frac{a_e}{a_\infty} \rightarrow 1 \tag{b}$$

Consequently

$$M_e a_\infty \rightarrow u_e \tag{c}$$

and (Eq. 23-15a)

$$\zeta \rightarrow \eta \tag{d}$$

and (Eq. 23-17′)

$$\xi \rightarrow x \tag{e}$$

For $M = 0$ the role of pressure is purely dynamic, and not thermodynamic. This is the reason why Eq. 23-17′ rather than Eq. 23-17 is most convenient in deriving (e). Furthermore, the equation of state reduces to (cf. Eq. 20-1d and 22-7d)

$$\tilde{\rho}\tilde{h} = 1 \tag{f}$$

A comparison of Eq. 23-18a and 23-18b with the relations (a) to (f) shows that the boundary layer equations for $M = 0$ are

$$\psi_\eta \psi_{\eta x} - \psi_x \psi_{\eta\eta} - \frac{h}{h_\infty} u_e (u_e)_x = \nu_\infty (C\psi_{\eta\eta})_\eta \tag{30-1a}$$

$$\psi_\eta h_x - \psi_x h_\eta = \frac{\nu_\infty}{Pr} (C h_\eta)_\eta \tag{30-1b}$$

$$\rho h = \rho_\infty h_\infty \tag{30-1c}$$

where

$$\eta = \int_0^y \frac{\rho}{\rho_\infty} dy \tag{30-1d}$$

$$C = \frac{\mu h_\infty}{\mu_\infty h} \tag{30-1e}$$

$$\psi_\eta = u, \qquad \psi_x = -\bar{v} \tag{30-1f}$$

Note that here C is not assumed to be constant.

The boundary and initial conditions on the enthalpy are

$$h(x, 0) = h_w \tag{30-2a}$$

$$h(x, \infty) = h_e = h_\infty = \text{const} \tag{30-2b}$$

$$h(0, \eta) = h_\infty \tag{30-2c}$$

The corresponding conditions on u and $\bar{v}$ are the usual ones.

The above equations may obviously be derived directly from the restricted form of Howarth's equations (Eq. 23-12 and 23-14). Note that Eq. 30-1c implies that the factor h/h_∞ in Eq. 30-1a may be replaced by ρ_∞/ρ.

Even if it is assumed that $C = \text{const}$, the momentum equation (Eq. 30-1a) is coupled to the energy equation (Eq. 30-1b) through the factor h/h_∞ in the term which expresses the effect of pressure gradient. Thus the simple results obtained in Art. 28 no longer apply when the pressure gradient does not vanish. Eq. 30-1a and 30-1b form a system of coupled nonlinear partial differential equations. Hence one may expect to encounter great difficulties in obtaining quantitative solutions, unless further simplifications are made.

Generalizations of the Falkner-Skan Solutions. As in the case of incompressible flow, the simplest special solutions are the similarity solutions. In analogy with Eq. 14-6 ff. we therefore consider the case in which

$$u_e = Ax^l \tag{30-3a}$$

and

$$h_w = \text{const} \tag{30-3b}$$

A combination of the similarity arguments used in Art. 14 and 26 shows that Eq. 30-1a and 30-1b may be reduced to the following system of ordinary differential equations:

$$f''' + ff'' + \beta[\bar{h} - (f')^2] = 0 \tag{30-4a}$$

$$\bar{h}'' + Prf\bar{h}' = 0 \tag{30-4b}$$

where f is defined by

$$\psi = \sqrt{\frac{2A\nu'_\infty}{l+1}}\, x^{\frac{l+1}{2}} f(\theta) \tag{30-4c}$$

and

$$\theta = \sqrt{\frac{(l+1)A}{2\nu'_\infty}}\, x^{\frac{l-1}{2}} \eta = \sqrt{\frac{(l+1)u_e}{2\nu'_\infty x}}\, \eta \tag{30-4d}$$

and

$$\beta = \frac{2l}{l+1} \tag{30-4e}$$

The average viscosity law (Eq. 28-2) has been assumed above and ν'_∞ is defined by Eq. 28-4.

The momentum equation (Eq. 30-4a) is coupled to the energy equation (Eq. 30-4b) through the term $\bar{h}$. For incompressible fluids, $\bar{h} = 1$, so that Eq. 30-4a reduces to the ordinary Falkner-Skan equation (Eq. 14-6).

The existence of similarity solutions for compressible flow (including the case $M > 0$) was first pointed out by Stewartson [*129*]. The various similarity cases have been studied and classified by T. Y. Li and Nagamatsu [*138*], Cohen [*139*], Cohen and Reshotko [*140*], and Levy [*141*]. The equations have been solved with the aid of computing machines for

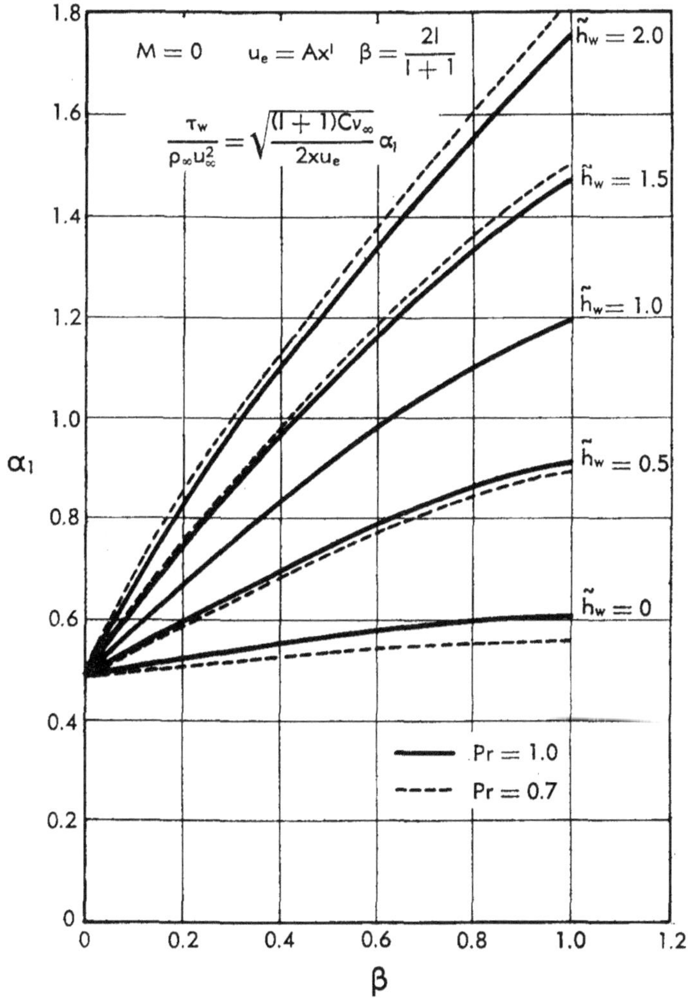

Fig. B,30a. Variation of skin friction with β, Pr, and wall temperature. From [*141*].

several cases (see in particular [*140*]). It has also been shown by Lal [*142*] that momentum methods (cf. Art. 32) give quite accurate results very rapidly. Some results of these calculations are shown in Fig. B,30a and B,30b. In using these graphs, the dependence of τ_w and q_w on C should be noted. A discussion of the results is given in Art. 32.

CALCULATION OF HEAT TRANSFER FROM SKIN FRICTION. WEAK HEAT TRANSFER.

Equations for weak transfer. The effect of heating or cooling the wall may be measured by $\Delta\tilde{h}$ as defined by (cf. Art. 20)

$$\Delta\tilde{h} = \frac{h_w - h_\infty}{h_\infty} = \tilde{h}_w - 1 \tag{30-5}$$

If $\Delta\tilde{h} = 0$, then the flow is incompressible, no heat transfer takes place

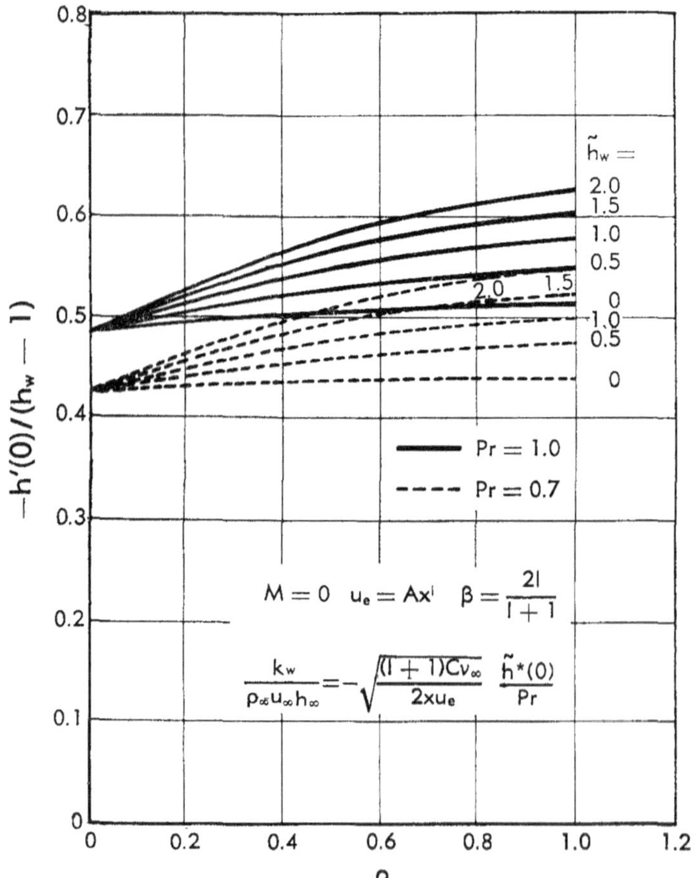

Fig. B,30b. Variation of heat transfer with β, *Pr*, and wall temperature. From [*141*].

and $h = \text{const} = h_w = h_\infty = h_r$. When $\Delta\tilde{h}$ is small one may then find perturbations of the solution for incompressible flow by assuming expansions of the form,

$$\frac{\psi}{UL} = \psi_0 + (\Delta\tilde{h})\psi_1 + (\Delta\tilde{h})^2\psi_2 + \cdots \tag{30-6a}$$

$$\tilde{h} = 1 + (\Delta\tilde{h})h_1 + (\Delta\tilde{h})^2h_2 + \cdots \tag{30-6b}$$

If h_w is not constant $\Delta\tilde{h}$ should of course be replaced by some suitable constant measuring the order of magnitude of $\Delta\tilde{h}$. Inserting these expansions into Eq. 30-1a and 30-1b one sees that ψ_0 obeys Eq. 30-1a with h/h_∞ and C replaced by unity, and hence is the solution for incompressible flow. To this approximation η may also be replaced by y. The function h_1 obeys Eq. 30-1b with $C = 1$ and ψ replaced by ψ_0. By linearity h and $\tilde{h}$ obey the same equation for $\Delta\tilde{h}$ small. Hence we find: *For $M = 0$ and $|h_\infty - h_w|$ small compared to h_∞ one may assume that the enthalpy is transported with the velocity field of an incompressible fluid.* The momentum and energy equations are then uncoupled. If we assume that the solution for ψ_0 has been found, then the equation for h is a linear equation with known coefficients. This equation is a generalization of Eq. 28-20. In the latter equation it was assumed that u and $\bar{v}$ represent a Blasius velocity field; in the present case we consider more general velocity distributions. An approximate method for solving the general equation is now given.

The Fage-Falkner-Lighthill method. The Fage-Falkner method of solving the heat equation for the flat plate was discussed at the end of Art. 28. Lighthill [*137*] has extended this method to the case of an arbitrary pressure gradient and obtained the general solution of the resulting differential equation.

As in Art. 28 we approximate to u by a linear velocity profile with correct slope at the wall.

$$u \rightarrow u_1 = \frac{\tau_w}{\mu'_\infty}\eta \tag{30-7a}$$

The continuity equation then gives the corresponding approximation to $\bar{v}$

$$\bar{v} \rightarrow v_1 = -\frac{(\tau_w)_x\eta^2}{2\mu'_\infty} \tag{30-7b}$$

With this substitution Eq. 30-1b becomes

$$\frac{\tau_w}{\mu'_\infty}\eta h_x - \frac{(\tau_w)_x\eta^2}{2\mu'_\infty}h_\eta = \frac{\nu'_\infty}{Pr}h_{\eta\eta} \tag{30-8}$$

This approximate equation is actually valid even if $\Delta\tilde{h}$ is not small. It may be used as long as the skin friction has been determined by one method or another; the function $(\tau_w)_x$ may, for example, be taken from experiments. With this generality in mind we have retained η and μ'_∞ in Eq. 30-8 instead of y and μ_∞. In the method of solution to be discussed it is assumed that C, and hence μ'_∞ and ν'_∞, are constants. Furthermore, for reasons to be discussed later the value of Pr cannot be too small. The Mach number is of course always assumed to be zero. In solving Eq. 30-8 one may first consider the case when h_w is a simple step function of the form,

$$\begin{aligned} h_w &= h_\infty = \text{const, for} && 0 \leqq x \leqq x_0 \\ h_w &= h_\infty + \Delta h = \text{const, for} && x_0 < x \end{aligned} \tag{30-9a}$$

Since there is no upstream influence according to the boundary layer approximation $h = h_\infty$ for $x \leqq x_0$ and we may use as initial condition:

$$h(x_0, \eta) = h_\infty \tag{30-9b}$$

If Eq. 30-8 has been solved with the boundary condition (Eq. 30-9) then the solution for an arbitrary distribution of wall temperature, but with the same skin friction, is obtained by a simple superposition. This follows from the linearity of Eq. 30-8.

The method of solution is based on the fact that by a transformation Eq. 30-8 and 30-9 may be brought into a standard form which is independent of the choice of the function $\tau_w(x)$ and the number x_0 and, incidentally, also the value of Pr. Thus the general solution follows from any one special solution. The following transformation may be used:

$$x_1 = \int_{x_0}^{x} \sqrt{\tau_w}\, dx \tag{30-10a}$$

$$\eta_1 = Pr^{\frac{1}{3}} \sqrt{\tau_w}\, \eta \tag{30-10b}$$

If these variables are used, Eq. 30-8 and 30-9 become

$$\frac{\eta_1}{\mu'_\infty} \frac{\partial h}{\partial x_1} = \nu'_\infty \frac{\partial^2 h}{\partial \eta_1^2} \tag{30-11a}$$

$$h(x_1, 0) = h_\infty + \Delta h \quad \text{for} \quad x_1 > 0 \tag{30-11b}$$

$$h(0, \eta_1) = h_\infty \tag{30-11c}$$

The existence of this standardized form means that the solution of any one heat transfer problem (of the type considered here) gives the solution to any other problem. The general solution of Eq. 30-8 and 30-9 may then be obtained from the previously derived solution for the flat plate (cf. the discussion of Eq. 28-27b). Alternatively, Eq. 30-11 may of course be solved directly by a similarity principle as follows.

Putting

$$\theta = \eta_1 (x_1 \mu'_\infty \nu'_\infty)^{-\frac{1}{3}} \tag{30-12}$$

one finds that Eq. 30-11a becomes

$$h'' + \frac{\theta^2}{3} h' = 0 \tag{30-13}$$

which has the solution (cf. Eq. 14-26 and 14-27)

$$h - h_\infty = -h'(0) \int_\theta^\infty e^{-\frac{\theta^3}{9}} d\theta \tag{30-14a}$$

where

$$-h'(0) = \frac{\Delta h}{9^{\frac{1}{3}}\Gamma(\frac{4}{3})} \tag{30-14b}$$

From Eq. 30-10, 30-12, and 30-14b one then finds the formula for the heat transfer from the wall to the fluid,

$$\frac{q_w Pr}{\mu'_\infty \Delta h} = \left(\frac{Pr}{9\mu'_\infty \nu'_\infty}\right)^{\frac{1}{3}} \frac{\sqrt{\tau_w}}{\Gamma(\frac{4}{3})} \left(\int_{x_0}^{x} \sqrt{\tau_w}\, dx\right)^{-\frac{1}{3}} \tag{30-15}$$

This formula, based on Eq. 30-8 and the boundary conditions (Eq. 30-9), was derived by Lighthill in [*137*] by a method different from the one employed here. The validity of Eq. 30-15 is discussed later. First we investigate the significance of the Fage-Falkner approximation more closely.

Asymptotic expansions for large Pr. In the Fage-Falkner method a velocity profile is used which is correct only near the wall. The method is then expected to be correct if most of the heat transfer takes place in this region. This means that the thermal boundary layer is much thinner than the velocity boundary layer, in other words that the Prandtl number is large. This idea may be checked by a formal expansion procedure. Let $h(x, \eta; Pr)$ be a solution of Eq. 30-1b. We assume that for Pr large h has an asymptotic expansion of the form,

$$h(x, \eta; Pr) = h^{(1)}(x, \eta^*) + Pr^{-\frac{1}{3}} h^{(2)}(x, \eta^*) + \cdots \tag{30-16a}$$

where

$$\eta^* = Pr^{\frac{1}{3}}\eta \tag{30-16b}$$

Inserting this expansion into Eq. 30-1b one finds indeed that $h^{(1)}$ obeys Eq. 30-8. The equation for the second approximation $h^{(2)}$ is also easily found; it involves the curvature of the velocity profile at the wall.

Thus, theoretically, the Fage-Falkner approximation is valid only for very large Pr. It gives, however, quite accurate results for surprisingly small values of Pr. Before discussing this we consider the corresponding approximation for Pr near zero.

Approximations for small Pr. The assumption of small Pr means that the coefficient of heat conductivity k is large compared to μc_p. On the other hand, it is assumed that the Péclet number $Pe = PrRe = \rho c_p UL/k$ is large so that it still is permissible to treat heat transfer with the aid of boundary layer theory.

For small Pr the thermal boundary layer is much thicker than the velocity boundary layer. It is then to be expected that only the values of the velocity in the outer part of the velocity layer will be of importance. As an approximation it should then be possible to replace u by u_e. Formally one may proceed as follows: We assume that h has an expansion, valid for Pr near zero, of the form,

$$h(x, \eta; Pr) = h^{(1)}(x, \bar{\eta}) + Pr^{\frac{1}{2}}h^{(2)}(x, \bar{\eta}) + \cdots \tag{30-17a}$$

where

$$\bar{\eta} = Pr^{\frac{1}{2}}\eta \tag{30-17b}$$

Inserting this expansion into Eq. 30-1b one finds that, to the first approximation, h obeys the equation,

$$u_e h_x - (u_e)_x \bar{\eta} h_{\bar{\eta}} = \nu'_\infty h_{\bar{\eta}\bar{\eta}} \tag{30-18}$$

Using the fact that for η large $\psi \cong u_e(\eta - \delta^*)$ one finds easily an equation for the second approximation $h^{(2)}$. This equation involves the displacement thickness δ^*.

The general solution of Eq. 30-18 is easily found by the same method that was used for Eq. 30-8. Let the boundary conditions be those given by Eq. 30-9a. Introducing the variables

$$x_2 = \int_{x_0}^{x} u_e dx \tag{30-19a}$$

$$\eta_2 = u_e \bar{\eta} = u_e Pr^{\frac{1}{2}}\eta \tag{30-19b}$$

one finds

$$h_{x_2} = \nu'_\infty h_{\eta_2\eta_2} \tag{30-20a}$$

$$h(x_2, 0) = h_\infty + \Delta h \quad \text{for} \quad x_2 > 0 \tag{30-20b}$$

Note that since $\psi \sim u_e\eta$ at the edge of the boundary layer, the transformation used is essentially a von Mises transformation, followed by a distortion of the x coordinate.

Eq. 30-20 corresponds to Eq. 30-11a and is independent of Pr and of the point x_0.

The solution of Eq. 30-20 is a well-known problem in the theory of heat conduction; it is actually also a similarity solution:

$$h - h_\infty = (\Delta h) \operatorname{erfc} \frac{\eta_2}{\sqrt{4\nu'_\infty x_2}} \tag{30-21}$$

Hence, the analogue of Eq. 30-15 is

$$\frac{Prq_w}{\mu'_\infty(h_w - h_\infty)} = \frac{u_e Pr^{\frac{1}{2}}}{\sqrt{\pi\nu'_\infty x_2}} \tag{30-22}$$

where x_2 is given by Eq. 30-19a.

Falkner-Skan case. Comparison of exact and approximate solutions. In order to evaluate the errors made in the various approximations proposed above, we discuss some concrete examples where an exact solution of Eq. 30-1b may be found, using the complete velocity profile. The simplest solutions are the similarity solutions. Hence we consider Eq. 30-1b in the special case given by Eq. 30-4b. By assumption the function f is then a solution of the Falkner-Skan equation for incompressible flow (Eq. 14-6). This function depends on the variable θ defined by Eq. 30-4d (cf. Eq. 14-11c) and the parameter β defined by Eq. 14-9a and 14-9c or Fig. B,14a. We sometimes exhibit these explicitly by writing $f(\theta; \beta)$.

The method of the integrating factor gives the solution

$$\mathcal{H} = \frac{h - h_\infty}{h_w - h_\infty} = \frac{G(\theta; Pr, \beta)}{G(0; Pr, \beta)} \tag{30-23a}$$

where

$$G(\theta; Pr, \beta) = \int_\theta^\infty e^{-Pr\int_0^\xi f(\zeta;\beta)d\zeta} d\xi \tag{30-23b}$$

The integral $G(0; Pr, \beta)$ is equal to $-1/\mathcal{H}'(0)$; hence it determines the heat transfer at the wall. The final formula for the heat transfer from the wall to the fluid may be written in the same form as Eq. 28-11.

$$\frac{Prq_w}{\mu'_\infty(h_w - h_\infty)} = \sqrt{\frac{(l+1)u_e}{\nu'_\infty x}}\, a(Pr; \beta) = \frac{\tau_w}{\mu'_\infty u_e} \frac{\sqrt{2}\, a(Pr; \beta)}{\alpha_1(\beta)} \tag{30-24a}$$

where $a(Pr; \beta)$ is defined by

$$a(Pr; \beta) = \frac{1}{\sqrt{2}\, G(0; Pr, \beta)} = -\frac{\mathcal{H}'(0)}{\sqrt{2}} \tag{30-24b}$$

As before, u_e is proportional to x^l, $l = \beta/(2 - \beta)$, and $\alpha_1(\beta) = f''(0;\beta)$. The relation between α_1 and the skin friction τ_w is given by Eq. 14-13a with ρ and ν replaced by ρ_∞ and ν'_∞, respectively. Since the functions $f(\theta; \beta)$ are tabulated (see end of Art. 14) the integral $G(0; Pr, \beta)$ may be evaluated numerically. Certain results may, however, be obtained analytically. For $Pr = 1$, $\beta = 0$ one finds $\sqrt{2}\, a = B''(0) = \alpha_1(0)$ (Eq. 28-18a). Using the fact that $f(\theta; \frac{1}{2})$ satisfies an equation similar to the Blasius equation (Eq. 14-29a) one may also evaluate $G(0; Pr, \frac{1}{2})$ and its derivative with respect to Pr for $Pr = 1$. This yields

$$\sqrt{2}\, a(1; \tfrac{1}{2}) = \frac{1}{2\alpha_1(\frac{1}{2})} \tag{30-25a}$$

and

$$\sqrt{2}\left[\frac{da(Pr; \frac{1}{2})}{dPr}\right]_{Pr=1} = \frac{2\alpha_1 - \alpha_2}{6[\alpha_1(\frac{1}{2})]^2} \tag{30-25b}$$

(α_1 and α_2 are defined by Eq. 14-12a, 14-12b and the numerical values given in Table B,14 and Fig. B,14b and B,14c).

For Pr near zero one finds

$$\sqrt{2}\, a(Pr; \beta) \cong \frac{Pr^{\frac{1}{2}}}{\sqrt{\pi/2} + \alpha_2(\beta)Pr^{\frac{1}{2}}}, \quad Pr \text{ small} \tag{30-26}$$

This expression may be obtained by finding the limit as Pr tends to zero of $Pr^{\frac{1}{2}}G(0; Pr, \beta)$, and of its derivative with respect to $Pr^{\frac{1}{2}}$. This method thus corresponds to using the first two terms of the expansion (Eq. 30-17). Similarly, one finds

$$\sqrt{2}\, a(Pr; \beta) = \frac{6^{\frac{2}{3}}[\alpha_1(\beta)]^{\frac{2}{3}}Pr^{\frac{2}{3}}}{6^{\frac{1}{3}}\Gamma(\frac{4}{3})[\alpha_1(\beta)]^{\frac{1}{3}}Pr^{\frac{1}{3}} + \beta\Gamma(\frac{5}{3})},\ Pr \text{ large} \qquad (30\text{-}27)$$

This formula corresponds to the use of the first two terms in the expansion (Eq. 30-16). If the second term in the denominator is neglected, the resulting formula is a special case of Lighthill's formula (Eq. 30-15) since

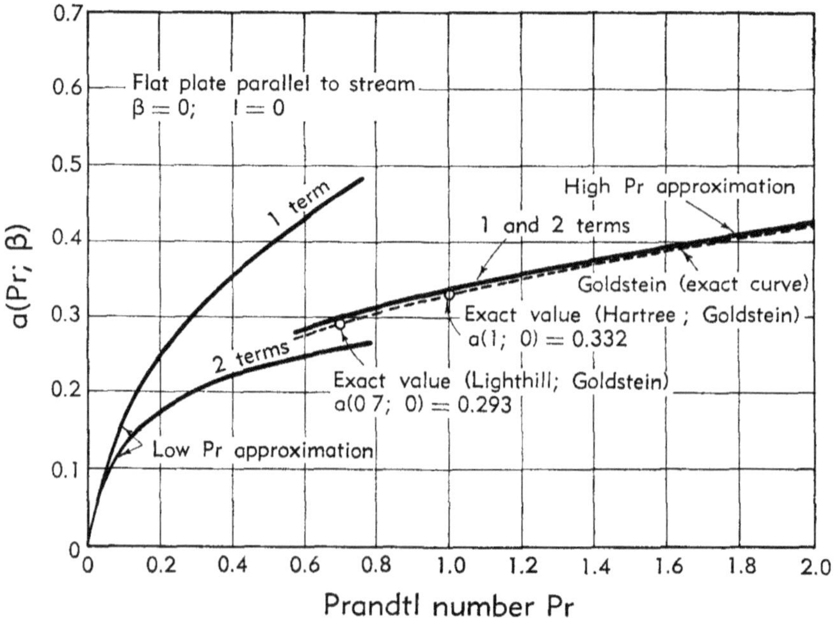

Fig. B,30c. Heat transfer coefficient vs. Prandtl number.

it corresponds to using only the first term of the expansion (Eq. 30-16).

Fig. B,30c, B,30d, and B,30e show the variation of $a(Pr; \beta)$ with Pr for $\beta = 0$, $\frac{1}{2}$ and 1, respectively. The approximate graphs are taken from the formulas just derived, with values of α_1 and α_2 based on [*66*]. The exact values for $Pr = 0.7$ are taken from Lighthill's previously cited paper [*137*]. Exact values of a for other values of Pr have been taken from Goldstein's book [*24*] for $\beta = 0$, and from Sebulkin's computations [*143*] for $\beta = \frac{1}{2}$. The exact value of a for $Pr = 1$, $\beta = \frac{1}{2}$ given by Eq. 30-25a, as well as the slope given by Eq. 30-25b, agree with Sebulkin's values.

Let us first consider the first approximation for very large Prandtl numbers (Eq. 30-27 with the second term in the denominator omitted).

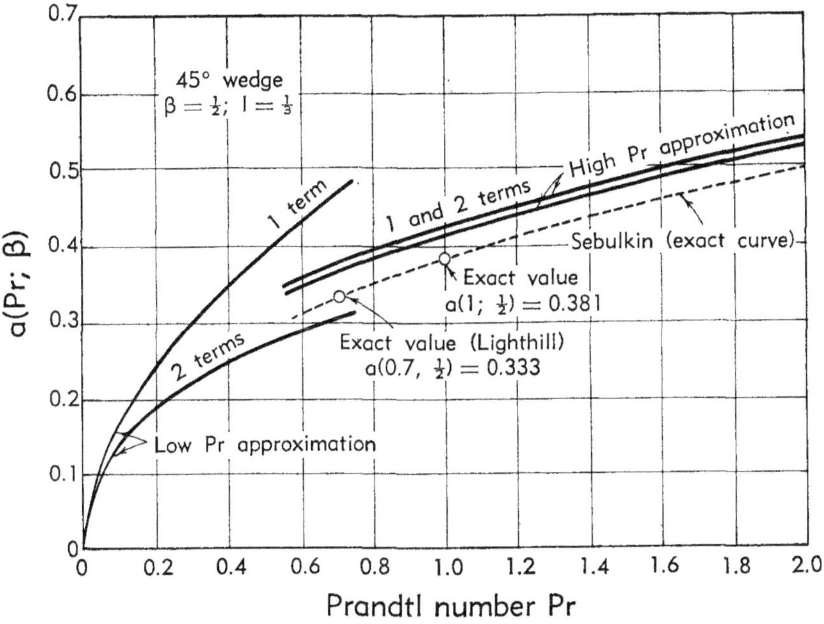

Fig. B,30d. Heat transfer coefficient vs. Prandtl number.

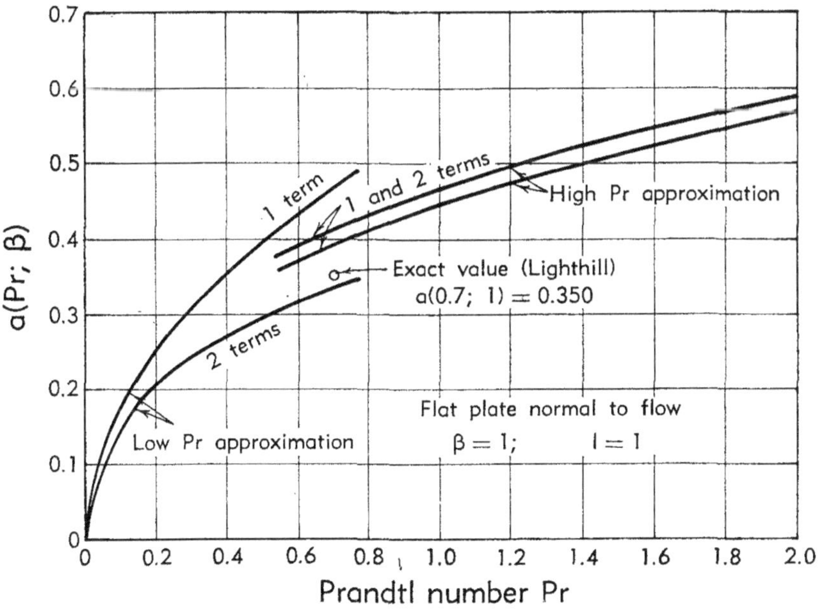

Fig. B,30e. Heat transfer coefficient vs. Prandtl number.

As mentioned earlier, this approximation is a special case of Lighthill's approximate formula for a general distribution of skin friction (Eq. 30-15). The graphs show that this approximation is quite good even near $Pr = 1$. However, it overestimates the heat transfer and the error increases with increasing value of β. This qualitative feature of the approximation may be easily understood. The heat transfer must increase with increasing velocity since the heat diffusing from the boundary is then washed away faster. The basis for the high Prandtl number approximation is the assumption of a velocity profile which at the boundary agrees with the exact profile, but which then continues to grow linearly and hence has larger velocities than the actual velocity profile. For $\beta = 0$ the actual profile is practically linear for some distance from the wall. This is the reason why the first approximation works best in this case. For $\beta > 0$, that is, when a negative pressure gradient exists, the exact velocity profile has a positive curvature, proportional to β, at the wall. Thus, except when β is very small, the assumption of a linear profile noticeably overestimates the velocity near the wall, and hence overestimates the heat transfer. This is the reason why the first approximation becomes increasingly inaccurate for higher values of β. The exact relation between the curvature at the wall and the pressure gradient is

$$\left(\frac{dp}{dx}\right) = \left[\frac{1}{\rho}\frac{\partial}{\partial y}\left(\mu\frac{\partial u}{\partial y}\right)\right]_w$$

For a Falkner-Skan profile this relation appears in the disguised form (cf. Eq. 14-6a) $\beta = -f'''(0)$. It is then seen that the function of the second approximation for high Pr (Eq. 30-27) is to correct for the curvature of the velocity profile at the wall. The second approximation is identical with the first for $\beta = 0$, but becomes increasingly more important for increasing values of β.

In the first approximation for low Prandtl numbers the transport velocity in the x direction is assumed to be u_e. This is also an overestimate. Furthermore, the error is greatest near the wall, that is, exactly in the region where most of the heat transfer takes place. Accordingly, the overestimate of the heat transfer is quite serious except for very low values of Pr. Fig. B,30c shows that for $Pr = 0.7$ and $\beta = 0$ the approximation derived for high Prandtl numbers is better than the one derived for low Prandtl numbers. While this may seem surprising at first sight, the reason is, however, the one stated above: The velocity profile near the wall is of crucial importance for the heat transfer. The second approximation for low Prandtl numbers (Eq. 30-26) makes use of the displacement thickness δ^* which for Falkner-Skan flow is represented by the constant α_2 (cf. Eq. 14-13b). Now, δ^* is a measure of the velocity deficiency in the boundary layer. Hence the low values of the velocity near the wall are accounted for, not in detail, but at least by an over-all value. As seen from the graphs, the second approximation gives reasonable agreement with exact values for $Pr < 0.7$.

Fage-Falkner approximation for arbitrary heat transfer. We now briefly consider the case in which $|\tilde{h}_w - 1|$ is not necessarily small. If the Prandtl number is very large the thermal boundary layer is well inside the velocity boundary layer. Hence in the larger part of the velocity layer $h = h_\infty$, i.e. the flow is incompressible there. As an approximation one may then give $\tilde{h}$ the value unity in the momentum equation, which then takes the same form as for incompressible flow. The skin friction to be used in Eq. 30-8 may then be computed. The same result is obtained formally if an expansion of the form (30-16) is inserted into the equations for a compressible boundary layer. In principle Lighthill's formula (Eq. 30-15) may then be used even when $|\tilde{h}_w - 1|$ is not small if Pr is sufficiently large. The value of τ_w occurring in this formula should be that corresponding to incompressible flow. However, the accuracy of the method for a fixed value of Pr obviously decreases with increasing $|\tilde{h}_w - 1|$.

In principle, asymptotic theories for Pr very large or very small may easily be developed. The analytical difficulties in obtaining reasonably accurate values for τ_w and q_w are, however, considerably increased when $|\tilde{h}_w - 1|$ cannot be considered small.

B,31. Correlation between Boundary Layers at $M > 0$ and $M = 0$ for $Pr = 1$, $\mu \sim T$. In Art. 26 it was shown that if it is assumed that $\mu \sim T$ then the study of a compressible boundary layer with zero pressure gradient may be reduced to the study of a correlated incompressible boundary layer. The proof follows from any of the transformed equations given in Art. 23. It involves essentially a distortion of the y coordinate as shown e.g. by Eq. 23-12a. Illingworth [*130*] and Stewartson [*129*] showed independently that if one in addition assumes that $Pr = 1$ and considers only the case of an insulated wall (so that the energy integral given by Eq. 22-8 is valid) then a boundary layer at any Mach number and with an arbitrary pressure gradient may be correlated with an incompressible boundary layer with, in general, a different pressure gradient. For this correlation a distortion of the x coordinate is also needed (as originally suggested by Howarth). Stewartson's proof is based on Eq. 23-18 whereas Illingworth uses Eq. 23-22.

The methods used by either of these authors actually show a more general correlation theorem: If $\mu \sim T$ and $Pr = 1$ then *any boundary layer, with arbitrary Mach number and pressure gradient, may be correlated with a boundary layer at zero Mach number.* If the wall of the first boundary layer is not insulated, then the flow in the correlated boundary layer is compressible, i.e. it has variable temperature. The correlation theorem is now proved from the Howarth-Stewartson equations. An equivalent proof could have been carried through on the basis of the Illingworth equations.

Consider a given flow at $M > 0$. Assume that the enthalpy at the wall, h_w, is prescribed (note that $h_w = H_w$). The boundary conditions then determine the constant H_∞ and also $\bar{U}_e \equiv a_\infty M_e$ and h_w as functions of ξ where ξ is given by Eq. 23-17′. Consider now an associated flow at $M = 0$ in the x, y plane, determined by the following boundary conditions (subscript $_a$ denotes quantities in the associated flow)

$$u_{ae}(x) = \bar{U}_e(\xi), \qquad h_{aw}(x) = h_w(\xi), \qquad h_{a\infty} = H_\infty \tag{31-1a}$$

Furthermore assume

$$C_a\nu_{a\infty} = C\nu_\infty \tag{31-1b}$$

and

$$\rho_{a\infty} = \rho_\infty \tag{31-1c}$$

The equations (Eq. 23-18a and 23-18b) as well as the boundary conditions for the original flow are then formally identical with the equations (Eq. 30-1a and 30-1b) and the boundary conditions for the associated flow, provided that the variables are correlated as follows:

$$\xi \to x,\ \zeta \to \eta,\ H \to h_a,\ \psi \to \psi_a,\ \bar{U} \to u_a \tag{31-1d}$$

Here ζ and η are defined by Eq. 23-15a and Eq. 30-1d respectively. It then follows that

$$\psi_a(x, y) = \psi(\xi, \zeta),\ h_a(x, y) = H(\xi, \zeta) \tag{31-1e}$$

In particular (cf. Eq. 23-20)

$$\left(\frac{a_e}{a_\infty}\right)^2 \tau_{aw}(x) = \tau_w(\xi),\ \left(\frac{a_e}{a_\infty}\right) q_{aw}(x) = q_w(\xi) \tag{31-1f}$$

Thus, if the solution for the associated flow has been found, the solution for the original flow is automatically given by Eq. 31-1e.

Note that the proof of the correlation theorem is valid only if one assumes an average viscosity law such that $C = \text{const}$ (Eq. 28-1). If C depends on h in Eq. 30-1b, the correlation would require it to depend on H in the same manner in Eq. 23-18b. The assumption (31-1b) is not strictly necessary; one could also have achieved formal identity by absorbing $C\nu_\infty$ in the spatial coordinates (cf. the discussion of Eq. 13-12). More generally, assumptions (Eq. 31-1b and 31-1c) may be omitted if Eq. 31-1e and 31-1f are expressed in a suitable nondimensional form. However, since the associated flow is a fictitious flow, one may as well assume a viscosity-temperature relation which makes the result formally simple.

In the special case $H_w = H_\infty$ (zero heat transfer at the wall) in the associated flow. Thus flow past an insulated wall may be correlated with an incompressible flow.

B,32. The Kármán-Pohlhausen Method and Related Methods.

It follows from the discussion in Art. 13 and in the last article that there is a great need for a rapid approximate method for treating boundary layers with an arbitrary pressure gradient. Such a method was developed by von Kármán and Pohlhausen in 1921 [*144,145*] for incompressible fluids. This method, together with some recent improvements, are now described. (The reader is also referred to Schlichting's presentation of the method [*26*].) It follows from the Illingworth-Stewartson correlation theorem that under certain conditions the results may be applied also to compressible fluids. Furthermore, the method is capable of generalization to more general cases of compressible flow.

The Classical Kármán-Pohlhausen Method for Incompressible Fluids. The essence of the Kármán-Pohlhausen method is to approximate to the velocity profile at any station by a profile described by a polynomial which is faired into a constant velocity profile. By imposing a suitable number of boundary conditions and matching conditions, these profiles may then be shown to depend on a single parameter $\lambda(x)$, which then may be determined from the integral momentum equation. We follow the original work of Pohlhausen in assuming a polynomial of fourth degree (polynomials up to eleventh degree have been used occasionally).

For each station x one assumes the existence of a length $\delta(x)$ such that u is a polynomial in y for $0 \leqq y \leqq \delta(x)$, and $u = u_e$, independent of y, for $y \geqq \delta(x)$. Thus $\delta(x)$ may be thought of as a boundary layer thickness; its actual value is determined in the course of the computations. The polynomial part of the profile may then be written:

$$\frac{u}{u_e} = a_1\eta + a_2\eta^2 + a_3\eta^3 + a_4\eta^4 \tag{32-1}$$

where $\eta = y/\delta(x)$, and a_i in general depends on x.

At the solid ($y = 0$) we assume that $u = 0$, and that the polynomial above may be used for computing the first and second derivatives of u. The first condition has already been used in eliminating the constant term in the polynomial in Eq. 32-1. The second condition states that in the expressions

$$\tau_w = \mu\left(\frac{\partial u}{\partial y}\right)_w, \quad \frac{dp}{dx} = \mu\left(\frac{\partial^2 u}{\partial y^2}\right)_w \tag{32-2}$$

one may compute the derivatives from Eq. 32-1.

At $y = \delta(x)$ we assume that the polynomial profile is faired into the constant profile in such a way that the value of u and its first two derivatives match, that is,

$$u = u_e, \frac{\partial u}{\partial y} = 0, \frac{\partial^2 u}{\partial y^2} = 0 \quad \text{for} \quad y = \delta(x) \tag{32-3}$$

Note that computations for third and higher derivatives will be incorrect. From the boundary layer equations one obtains easily that $(\partial^3 u/\partial y^3)_w = 0$. This is in general violated by Eq. 32-1. Similarly, the third derivative is discontinuous at $y = \delta(x)$. However, the proposed approximate formula for the velocity profile is used in the integral laws rather than in the differential equations. Thus, certain average properties of the profile, rather than detailed differential properties, are of importance.

By using the three matching conditions (Eq. 32-3), one finds that the four coefficients $a_i(x)$ may be expressed as functions of a single parameter which for convenience is taken to be $-2a_2$ and denoted by λ. Hence,

$$a_1 = \frac{\lambda + 12}{6}, \qquad a_2 = -\frac{\lambda}{2}, \qquad a_3 = \frac{\lambda - 4}{2}, \qquad a_4 = \frac{6 - \lambda}{6} \tag{32-4}$$

where λ is a function of x.

Furthermore, $\delta(x)$ may be expressed in terms of λ and the known pressure gradient dp/dx with the aid of Eq. 32-2:

$$\frac{1}{\rho}\frac{dp}{dx} = -u_e \frac{du_e}{dx} = -\frac{\nu u_e}{\delta^2}\lambda \tag{32-5}$$

Finally, the skin friction is

$$\tau_w = \mu u_e \frac{a_1}{\delta(x)} = \frac{\mu u_e}{\delta(x)} \frac{\lambda + 12}{6} \tag{32-6}$$

Now, von Kármán's integral momentum law (Eq. 13-24) may be written:

$$\frac{d\delta^{**}}{dx} + \frac{1}{u_e}\frac{du_e}{dx}(\delta^* + 2\delta^{**}) = \frac{\tau_w}{\rho u_e^2} \tag{32-7}$$

where the displacement thickness δ^* and the momentum thickness δ^{**} are defined by Eq. 13-21 and 13-23. It follows from the above discussion that all quantities in Eq. 32-7 may be expressed in terms of $\lambda(x)$ and the known external velocity distribution $u_e(x)$. Eq. 32-7 is then a differential equation for the unknown function $\lambda(x)$.

The solution for λ is sometimes multivalued. In this case that solution should be used which is in the physically meaningful range, namely

$$-12 \leqq \lambda \leqq +12 \tag{32-8}$$

The lower bound is necessary since $\lambda < -12$ would give a negative skin friction according to Eq. 32-6. The value $\lambda = -12$ thus represents separation of the flow. On the other hand $\lambda > 12$ is excluded since then there result values of $u/u_e > 1$ in the boundary layer, which is physically inadmissible.[32]

In a flow with similarity, u/u_e depends on one variable only, this variable being a combination of x and y. Eq. 32-7 must then reduce to an algebraic equation. As pointed out in Art. 14 the integral momentum equation for Falkner-Skan flow is the algebraic equation (14-15). In this case the polynomial approximation (Eq. 32-1, 32-4, 32-5) may be written as (for notation cf. Art. 14)

$$\frac{u}{u_e} = f'(\theta) = \frac{\lambda + 12}{6}\left(\frac{\theta}{\delta}\right) - \frac{\lambda}{2}\left(\frac{\theta}{\delta}\right)^2 + \frac{\lambda - 4}{2}\left(\frac{\theta}{\delta}\right)^3 - \frac{\lambda - 6}{6}\left(\frac{\theta}{\delta}\right)^4 \quad \text{(32-9a)}$$

where

$$\delta^2 = \frac{\lambda}{\beta} \quad \text{(32-9b)}$$

and the skin friction is determined from

$$\alpha_1(\beta) = f''(0; \beta) = \frac{\lambda + 12}{6\delta} \quad \text{(32-9c)}$$

The integral momentum equation (Eq. 32-7 or Eq. 14-15), then becomes

$$\lambda^3 + P\lambda^2 + Q\lambda + R = 0 \quad \text{(32-10a)}$$

where

$$P = 9.6 + 151.2\left(\frac{\beta}{2 + 2\beta}\right) \quad \text{(32-10b)}$$

$$Q = -2419.2\left(\frac{\beta}{2 + 2\beta}\right) - 1065.6 \quad \text{(32-10c)}$$

$$R = 36{,}288\left(\frac{\beta}{2 + 2\beta}\right) \quad \text{(32-10d)}$$

To make the solution of this equation unique, the condition (Eq. 32-8) has to be imposed on λ. As an example consider the case $\beta = 0$. The three roots of Eq. 32-10a in this case are

$$\lambda = 0, \ +28.2, \text{ and } -37.8$$

of which $\lambda = 0$ is the correct solution. Eq. 32-9b is, in this case, indeterminate and does not yield a value of δ. However, by using Eq. 32-9c and 14-15 (with $\beta = 0$) one finds

$$\alpha_1(0) = 0.485 \quad \text{(32-11a)}$$

[32] There are, however, cases where strong temperature effects may lead to velocities in the boundary layer which are $>u_e$. This may be seen by the asymptotic methods indicated at the end of Art. 30. They are not discussed further here.

Similarly, for $\beta = 1$, the correct solution for λ as obtained from Eq. 32-10a is $\lambda = 7.052$. By using Eq. 32-9b and 32-9c the value of α_1 is obtained as being

$$\alpha_1(1) = 1.196 \tag{32-11b}$$

A comparison with the exact values in Table B,14 shows that the error in α_1 is $+3.3$ per cent for $\beta = 0$ and -3.0 per cent for $\beta = 1$.

In the general case one initial condition, say the value of λ at $x = 0$, is needed for Eq. 32-7. If the wedge angle at the nose is $\beta\pi$ we require $\lambda(0)$ to have the value given by the similarity solution for the same β. This is equivalent to requiring $(d\lambda/dx)_{x=0}$ to be finite. For a blunt-nosed body the value of $\lambda(0)$ is then the value given by the similarity solution above for $\beta = 1$, namely $\lambda = 7.052$.

Special Methods for Solving the Integral Momentum Equation for Incompressible Fluids.

The method of Holstein, Bohlen, and Walz. If Eq. 32-7 is written out explicitly with λ as dependent variable, the resulting equation is quite complicated. Various special methods of solving this equation have therefore been proposed. The method of Holstein and Bohlen [*146*] starts from the integral momentum equation rearranged in the form,

$$\frac{u_e}{2\nu}\frac{d(\delta^{**})^2}{dx} = -\left(\frac{\delta^*}{\delta^{**}} + 2\right)\frac{u_e'(\delta^{**})^2}{\nu} + \frac{\tau_w\delta^{**}}{\mu u_e} \tag{32-7'}$$

This equation is still exact. If one in addition uses the Kármán-Pohlhausen approximation to the velocity profile, Eq. 32-7′ may be written

$$\frac{u_e}{\nu}\frac{d(\delta^{**})^2}{dx} = F(K) \tag{32-12a}$$

where

$$K = \frac{(\delta^{**})^2 u_e'}{\nu} \tag{32-12b}$$

The function $F(K)$ is a universal function of K, i.e. independent of the form of the pressure gradient. The universality follows from the fact that δ^*/δ^{**}, $\tau_w\delta^*/\mu_\infty u_e$, and K may be expressed as functions of the single parameter λ; hence the first two quantities may be expressed in terms of K and their values inserted in Eq. 32-7′ to give Eq. 32-12. While the function $F(K)$ has a complicated analytical form, it may be approximated to by simple functions. Table B,32a (taken from [*26*]) and Fig. B,32a, B,32b, and B,32c reproduce the computations of Holstein and Bohlen. It turns out that $F(K)$ is monotonically decreasing and only slightly concave. Hence it can often be approximated by a straight line of suitable slope. Walz [*148*], for example, has shown that the function $0.47 - 6K$ approximates $F(K)$ very well in the region of falling pressure

Table B,32a. Method of Holstein and Bohlen.

λ	K	$F(K)$	$\frac{\delta^*}{\delta^{**}}$	$\frac{\delta^{**}\tau_w}{\mu u_e}$
12	0.0948	−0.0948	2.250	0.356
11	0.0941	−0.0912	2.253	0.355
10	0.0919	−0.0800	2.260	0.351
9	0.0882	−0.0608	2.273	0.347
8	0.0831	−0.0335	2.289	0.340
7.8	0.0819	−0.0271	2.293	0.338
7.6	0.0807	−0.0203	2.297	0.337
7.4	0.0794	−0.0132	2.301	0.335
7.2	0.0781	−0.0051	2.305	0.333
7.052*	0.0770	0	2.308	0.332
7	0.0767	0.0021	2.309	0.331
6.8	0.0752	0.0102	2.314	0.330
6.6	0.0737	0.0186	2.318	0.328
6.4	0.0721	0.0274	2.323	0.326
6.2	0.0706	0.0363	2.328	0.324
6	0.0689	0.0459	2.333	0.321
5	0.0599	0.0979	2.361	0.310
4	0.0497	0.1579	2.392	0.297
3	0.0385	0.2300	2.427	0.283
2	0.0264	0.3004	2.466	0.268
1	0.0135	0.3820	2.508	0.252
0†	0	0.4698	2.554	0.235
− 1	−0.0140	0.5633	2.604	0.217
− 2	−0.0284	0.6609	2.647	0.199
− 3	−0.0429	0.7640	2.716	0.179
− 4	−0.0575	0.8698	2.779	0.160
− 5	−0.0720	0.9780	2.847	0.140
− 6	−0.0862	1.0877	2.921	0.120
− 7	−0.0999	1.1981	2.999	0.100
− 8	−0.1130	1.3080	3.085	0.079
− 9	−0.1254	1.4167	3.176	0.059
−10	−0.1369	1.5229	3.276	0.039
−11	−0.1474	1.6257	3.383	0.019
−12‡	−0.1567	1.7241	3.500	0

$$K = \frac{(\delta^{**})^2 u_e'}{\nu}$$

$$F(K) = -2K\left(\frac{\delta^*}{\delta^{**}} + 2\right) + \frac{\tau_w \delta^{**}}{\mu u_e}$$

* Stagnation point for blunt-nosed body.
† Point of maximum velocity in external flow.
‡ Point of separation.

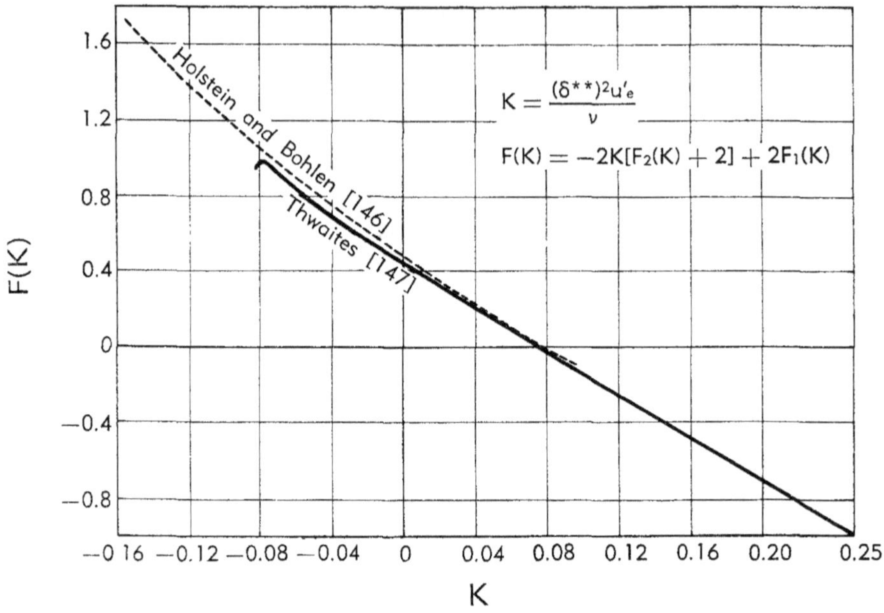

Fig. B,32a. Variation of $F(K)$ with K.

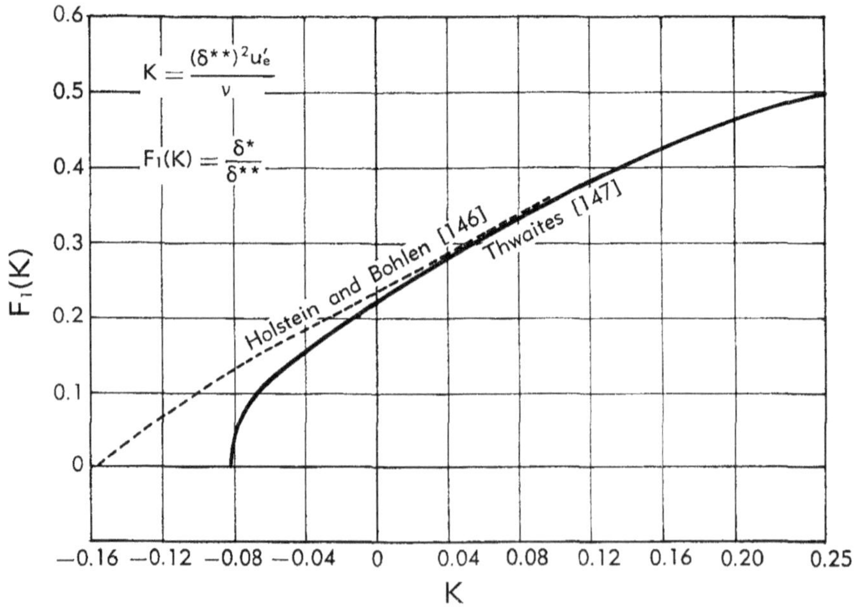

Fig. B,32b. Variation of $F_1(K)$ with K.

between the stagnation point and the point of maximum velocity. In this particular case, Eq. 32-12 integrates immediately. In more general terms, one may develop $F(K)$ in the neighborhood of $K = K_0$ and thus obtain an approximation valid for the neighborhood of K_0. Thus, an approximation to Eq. 32-12 is

$$\frac{d(\delta^{**})^2}{dx} \cong \frac{\nu}{u_e}[F(K_0) + F'(K_0)(K - K_0)] \tag{32-13}$$

In this equation one may regard $(\delta^{**})^2$ as the dependent variable and x as the independent variable. The arbitrary function $u_e(x)$ and its derivative $u_e'(x)$ also occurs in the equation. However, $[u_e(x)]^{-F'(K_0)}$

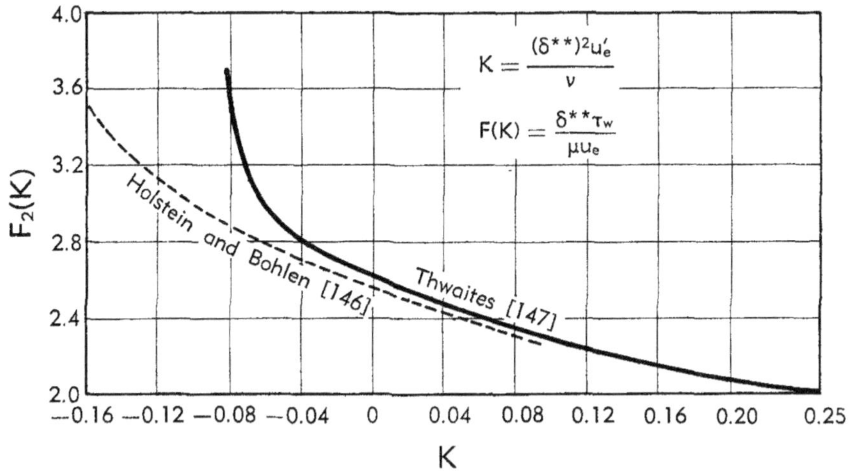

Fig. B,32c. Variation of $F_2(K)$ with K.

is an integrating factor for any choice of $u_e(x)$. Hence one finds the general solution,

$$(\delta^{**})^2 = \left[\frac{u_e(x)}{u_e(x_0)}\right]^{F'(K_0)} \left\{[\delta_0^{**}]^2 + \nu[F(K_0) - F'(K_0)K_0][u_e(x_0)]^{F'(K_0)} \int_{x_0}^{x} [u_e(\xi)]^{-F'(K_0)-1} d\xi\right\} \tag{32-14}$$

where K_0 and δ_0^{**} are the values at $x = x_0$. For Walz's approximation,

$$F(K) = 0.47 - 6K, \quad x_0 = 0 \tag{32-15}$$

Then, if $u_e(0) = 0$, Eq. 32-14 reduces to

$$(\delta^{**})^2 = \frac{0.47\nu}{u_e^6}\int_0^x u_e^5 dx \tag{32-16}$$

Hence when u_e is given, one may obtain δ^{**} from Eq. 32-14 or 32-16 (or from a more exact solution of Eq. 32-12). From $\delta^{**}(x)$ or $K(x)$ one then finds $\lambda(x)$ and hence the velocity profile at any x. The point where $\lambda = -12$ is the separation point. It should be borne in mind, however, that the method is not very accurate for the case of an adverse pressure gradient.

The method of Thwaites. The method proposed by Thwaites [*147*] resembles in many ways the method discussed just above. However, it does not assume a priori that the velocity profile may be approximately represented by a polynomial as in Eq. 31-1; it is instead based on an analysis of existing exact solutions of the boundary layer equations. An improvement in accuracy results from this method, especially for regions with a positive pressure gradient.

Generalizing the Kármán-Pohlhausen approach, Thwaites assumes the existence of an approximate universal function $G(x_1, x_2)$ such that any boundary layer profile may be written approximately as

$$\frac{u}{u_e} = G\left[\frac{y}{\delta(x)}, \lambda(x)\right] \tag{32-17}$$

The functions $\delta(x)$ and $\lambda(x)$ depend on the pressure gradient, i.e. on $u_e(x)$; the function G, on the other hand, is assumed to be universal, i.e. it is the same for all boundary layers. In the Kármán-Pohlhausen method the function $G(x_1, x_2)$ was assumed to be a polynomial of the fourth order in x_1 and linear in x_2 (cf. Eq. 31-1 and 32-4), and the significance of $\delta(x)$ and $\lambda(x)$ was also determined. In the method of Thwaites the form of G and the meaning of $\delta(x)$ and $\lambda(x)$ are not specified a priori. It is, however, assumed that the approximate form for u may be used to compute u, u_y, and u_{yy} at the wall ($y = 0$) and also the average quantities δ^* and δ^{**}. With these assumptions Eq. 32-17 leads to relations of the form (the prime denotes the derivative of G with respect to its first argument)

$$\frac{\tau_w \delta}{\mu u_e} = G'[0, \lambda(x)] \tag{32-18a}$$

$$\frac{u_e' \delta^2}{\nu} = -G''[0, \lambda(x)] \tag{32-18b}$$

$$\frac{\delta^*}{\delta} = G_1[\lambda(x)] \tag{32-18c}$$

$$\frac{\delta^{**}}{\delta} = G_2[\lambda(x)] \tag{32-18d}$$

Here G_1 and G_2 are universal functions obtained as certain integrals of G; G' and G'' are of course also universal.

The functions $\lambda(x)$ and $\delta(x)$ may now be eliminated in favor of functions with direct physical significance. From the above equations one may in principle find relations of the form,

$$\frac{\tau_w \delta^{**}}{\mu u_e} = F_1(K) \tag{32-19a}$$

$$\frac{\delta^*}{\delta^{**}} = F_2(K) \tag{32-19b}$$

where F_1 and F_2 are universal functions, and $K = (\delta^{**})^2 u_e'/\nu$ as in Eq. 32-12. If these relations are inserted into the exact integrated momentum equation in the form (Eq. 32-7′) one finds

$$\frac{u_e}{\nu}\frac{d(\delta^{**})^2}{dx} = -2[F_2(K) + 2]K + 2F_1(K) = F(K) \tag{32-20}$$

where $F(K)$ is also a universal function of K.

Thus the following has been proved: *If there exists an approximate universal function G which determines the velocity profile according to Eq. 32-17, and if this approximation may be used to compute u, u_y, and u_{yy} at $y = 0$ as well as δ^* and δ^{**}, then Eq. 32-19a, 32-19b, and 32-20 must be valid for some universal functions F_1, F_2, and F.*

The problem then remains to determine F_1 and F_2; F is a simple combination of these functions. In the Holstein-Bohlen method it is assumed that G is given by a Kármán-Pohlhausen polynomial. F_1 and F_2 may then be computed and F becomes the function used in Eq. 32-12 and tabulated in Table B,32. Thwaites, on the other hand, determines suitable functions F_1 and F_2 from various existing solutions of the boundary layer equations which have been computed by accurate numerical methods. Each such solution, or family of solutions, gives a curve $\tau_w\delta^{**}/\mu u_e$ vs. K. An average and approximate curve of $F_1(K)$ is then found from a curve that approximates all the accurate curves as closely as possible. In a similar fashion $F_2(K)$ is found from accurate computations of δ^*/δ^{**}. For favorable pressure gradients ($u_e' \geqq 0$, i.e. $K \geqq 0$) the accurate curves lie rather close together and also show good agreement with the Holstein-Bohlen values as given in Table B,32a. In the case of an adverse pressure gradient ($u_e' < 0$, $K < 0$) there is considerably more divergence between the accurate curves. Average approximating curves for F_1 and F_2 may, however, be found which show considerably better agreement with the accurate values than curves based on Table B,14. Values of F_1 and F_2, as determined by Thwaites by the method just described, are given in Table B,32b and compared with the values of Table B,32a in Fig. B,32a and B,32b. It is seen that the Kármán-Pohlhausen method gives quite inaccurate results for a strong adverse

Table B,32b. Method of Thwaites.

K	$F(K)$	$F_1(K)$	$F_2(K)$
0.25	−1.000	0.500	2.00
0.20	−0.702	0.463	2.07
0.14	−0.362	0.404	2.18
0.12	−0.251	0.382	2.23
0.10	−0.138	0.359	2.28
0.080	−0.028	0.333	2.34
0.075*	0	0.326	2.35
0.064	0.064	0.313	2.39
0.048	0.156	0.291	2.44
0.032	0.249	0.268	2.49
0.016	0.342	0.244	2.55
0†	0.440	0.220	2.61
−0.008	0.490	0.208	2.64
−0.016	0.539	0.195	2.67
−0.024	0.590	0.182	2.71
−0.032	0.640	0.168	2.75
−0.040	0.691	0.153	2.81
−0.048	0.744	0.138	2.87
−0.052	0.770	0.130	2.90
−0.056	0.797	0.122	2.94
−0.060	0.825	0.113	2.99
−0.064	0.854	0.104	3.05
−0.068	0.886	0.094	3.13
−0.070	0.902	0.089	3.17
−0.072	0.919	0.083	3.23
−0.074	0.936	0.076	3.30
−0.076	0.952	0.067	3.38
−0.078	0.963	0.055	3.47
−0.079	0.970	0.049	3.52
−0.080	0.971	0.039	3.58
−0.0804	0.969	0.035	3.59
−0.0808	0.967	0.030	3.61
−0.0812	0.962	0.024	3.63
−0.0816	0.956	0.016	3.66
−0.0818	0.953	0.011	3.69
−0.082‡	0.938	0	3.70

$$K = \frac{(\delta^{**})^2 u_e'}{\nu}, \qquad F_1(K) = \frac{\tau_w \delta^{**}}{\mu u_e}, \qquad F_2(K) = \frac{\delta^*}{\delta^{**}}$$

$$F(K) = -2K[F_2(K) + 2] + 2F_1(K)$$

* Stagnation point for blunt-nosed body.
† Point of maximum velocity in external flow.
‡ Point of separation.

pressure gradient. Finally $F(K)$, as obtained from the accurate numerical calculations, is plotted vs. K. In this case the points show very little scatter over the entire range of K; furthermore the result checks quite closely with the values obtained by Holstein and Bohlen (cf. Tables B,32a and B,32b and Fig. B,32c). Hence a linearizing approximation, similar to that of Walz (Eq. 32-15) may be used. Thwaites proposes

$$F(K) = 0.45 - 6K \tag{32-21}$$

Eq. 32-20 may then be integrated to yield Walz's solution (Eq. 32-16) with the factor 0.47 replaced by 0.45.

Thus in determining δ^{**} as a function of x the method of Thwaites agrees closely with that of Holstein, Bohlen, and Walz. To determine τ_w from δ^{**} the function F_1 is needed; here the two methods give different results for adverse pressure gradients, the method of Thwaites being presumably more accurate.

With the method of Thwaites one may thus determine $\tau_w(x)$ from $u_e(x)$ without a priori assumptions about function G. If one wishes to find an approximate velocity profile, the question about the form of G can no longer be dodged. Thwaites proposes to express u/u_e as an implicit function of y/δ by assuming y/δ to be a polynomial in u/u_e. Near the separation point, half-integer powers of u/u_e are needed. For these questions the reader is referred to [*147*].

Approximate, explicit expressions for pressure gradient effects. It is evident from Tables B,32a and B,32b that δ^*/δ^{**} is a slowly varying function of K, except near the separation point. Hence, for the purpose of fast, approximate computations it is permissible to replace δ^*/δ^{**} by a constant value, say A, corresponding to some value of K. In particular, let the external velocity field be of the form,

$$u_e = U[1 + \epsilon f(x)] \tag{a}$$

where $U = \text{const}$ and $\epsilon \ll 1$. Then Eq. 32-7 and 32-16 lead immediately to a simple approximate expression for τ_w:

$$\frac{\tau_w}{\rho U^2} = \sqrt{\frac{0.47\nu}{Ux}}\,[1 - (6 - A)\epsilon f'(x)x] \tag{32-22a}$$

or

$$\frac{\tau_w - \tau_{w_0}}{\tau_{w_0}} = -(6 - A)\epsilon x f'(x) \tag{32-22b}$$

where τ_{w_0} is the skin friction on a flat plate with velocity U. For A one may take the value corresponding to uniform external flow, namely $A = 2.55$. For the relative difference between the true skin friction drag D of a thin section with chord C and the skin friction drag D_0

computed for flat plate flow, it follows that

$$\frac{D - D_0}{D_0} = -(6 - A) \frac{\epsilon \int_0^C xf'(x)\tau_{w_0} dx}{\int_0^C \tau_{w_0} dx}$$

or

$$\frac{D - D_0}{D_0} = -(6 - A) \frac{\epsilon \int_0^C \sqrt{x}\, f'(x) dx}{2\sqrt{C}} \tag{32-23}$$

Eq. 32-23 is useful for many crude computations, for example, to establish the accuracy with which the external velocity has to be made uniform to allow reliable skin friction measurements corresponding to ideal flat plate flow.

Extension of Momentum Methods to Compressible Fluids. The various momentum methods described above may be readily extended to the case of compressible fluids. It is thereby desirable to make use of the coordinate transformations introduced in Art. 23. To see this, consider the case $M = 0$, $h_w = h_\infty$. Following the Kármán-Pohlhausen method one may use a polynomial approximation for u and an analogous polynomial for h. However, since $h \sim 1/\rho$ the integrand in the definition of δ^* (Eq. 24-3a) contains a ratio of two polynomials. Hence δ^* is not easily evaluated. This difficulty may be avoided by the use of distorted coordinates as we now show for a general Mach number. The Howarth-Stewartson transformation is used. It is assumed that $\bar{U}$ (defined by Eq. 23-18c) and H are the fourth order polynomials in ζ (defined by Eq. 23-15a). Thus

$$\frac{\bar{U}}{\bar{U}_e} = \sum_{i=1}^{4} a_i \left(\frac{\zeta}{\delta_V}\right)^i \tag{32-24a}$$

$$\frac{H}{H_e} = \sum_{i=0}^{4} b_i \left(\frac{\zeta}{\delta_H}\right)^i \tag{32-24b}$$

Here δ_V is the thickness of the velocity boundary layer and δ_H is the thickness of the total enthalpy layer. Both these thicknesses are assumed to be functions of ξ (Eq. 23-17′); they are in general not equal.

Eq. 32-24a and 32-24b generalize Eq. 31-1. By imposing the same type of matching conditions as in the incompressible case, one finds that the a_i, b_i, δ_V, and δ_H are functions of two independent parameters, say λ_V and λ_H; the skin friction and the heat transfer at the wall may also be related to these parameters (cf. Eq. 32-4, 32-5, and 32-6).

If now the expressions (Eq. 32-24a and 32-24b) are inserted into Eq. 24-6a, 24-6b, 24-6c, and 24-6d, one finds that the integrands are simple polynomials. The thickness parameters $\bar{\delta}^*$, $\bar{\delta}^{**}$, etc. may thus be determined as functions of a_i, b_i, δ_V, and δ_H and hence ultimately as functions of the two parameters, λ_V and λ_H. Inserting these functions into the integral relations (Eq. 24-7a and 24-7b), one obtains two coupled nonlinear ordinary differential equations with λ_V and λ_H as dependent variables and ξ as independent variable. Details of the procedure just described are given by Lal in [*142*]. The methods of solution of the differential equations resemble those used for the incompressible flow. In the Falkner-Skan case (Eq. 30-4a and 30-4b) they reduce to two simultaneous algebraic equations. The graphs in Fig. B,30a and B,30b are based on solutions of these equations and are taken from [*142*]. The method of Holstein, Bohlen, and Walz may be generalized to compressible fluids (see [*142*]).

The method of Thwaites, which is not based on a polynomial approximation, has been extended to compressible fluids by Cohen and Reshotko [*149*].

B,33. Summarizing Discussion of Compressibility Effects in Two-Dimensional Stationary Boundary Layers. A survey of some of the most important compressibility effects in boundary layers is now given. Various qualitative ideas are presented and illustrated by the quantitative results derived in preceding articles.

Compressibility effects in a boundary layer occur when the Mach number is greater than zero and also at zero Mach number if the wall temperature is different from the free stream temperature. Inside the boundary layer, heat addition due to dissipation and to conduction from the wall causes changes in temperature in the boundary layer and hence also changes in density, viscosity, and heat conductivity. Dissipation is negligible at zero Mach number. Another Mach number effect is the thermodynamic role of pressure: for $M > 0$ the full equation of state (cf. Eq. 22-7d) must be used and the substantial derivative of the pressure must be retained in the energy equation for enthalpy (cf. Eq. 22-7c). Finally, the external flow past a given object depends on the Mach number. Thus the external boundary conditions of the boundary layer depend on the Mach number; this is an indirect Mach number effect.

In the discussion below, we assume in general that the dependence of viscosity on temperature may be accounted for by using an average viscosity law of the form $\bar{\mu} = C\bar{h}$, $C = \text{const}$ (Eq. 28-1). It must be remembered, however, that the special value $C = \bar{\mu}_w/\bar{h}_w$ (Eq. 28-2) overemphasizes the influence of the wall temperature.

Compressibility Effects at Zero Mach Number. In this case compressibility effects are due only to the dependence of h, ρ, μ, and k on the wall temperature.

Zero pressure gradient. To fix the ideas, let us consider the case $h_w > h_\infty$. Heat conduction from the wall then tends to increase the viscosity in the boundary layer above its free stream value and hence, to increase skin friction. On the other hand, the decrease in density tends to thicken the boundary layer and hence to decrease the value of $\partial u/\partial y$ at the wall, i.e. to decrease the skin friction. The balance between these two effects can clearly be seen in Howarth's form of the momentum equation (Eq. 23-14a′; note that p/p_∞ in the right-hand side should be replaced by unity for $M = 0$). Assuming a constant C one finds (Eq. 28-5a)

$$\frac{\tau_w}{\tau_{w_{inc}}} = \sqrt{C} \tag{33-1}$$

where $\tau_{w_{inc}}$ is the skin friction in the corresponding incompressible flow. For $C = 1$, i.e. $\tilde{\mu} = \tilde{h}$, the effects mentioned above balance each other and the skin friction is independent of the wall temperature. If the viscosity increases more slowly than T, the second effect dominates the first. As an example, assume that $\tilde{\mu} = \tilde{h}^{\frac{3}{4}}$. If then the average viscosity law (Eq. 28-1) is used and C is evaluated from Eq. 28-2, one finds $C = \tilde{h}_w^{-\frac{1}{4}}$. In this case there is then a very slow decrease in τ_w with increase in h_w.

According to Eq. 33-1, or Eq. 28-5a, the value of τ_w is independent of Prandtl number, i.e. of the temperature distribution inside the boundary layer. This is due to the fact that in using the average viscosity law all temperature effects on skin friction are accounted for by the known wall temperature. Calculations using a more accurate viscosity law show that the skin friction is very little affected by the value of Prandtl number.

On the other hand, the actual velocity profile, i.e. u as a function of y, depends on the temperature distribution inside the fluid, even at constant C. The velocity as a function of η depends on C only; however, the relation between η and y depends on the density distribution and hence on the temperature distribution inside the boundary layer (Eq. 23-12a).

The heat transfer from the wall to the fluid is given by Eq. 28-11a for the case of constant wall temperature. This formula was derived using an average viscosity law. It shows that q_w is proportional to $h_w - h_\infty$ and to $\sqrt{C}$. Thus if C is equal to unity, q_w is linear in $h_w - h_\infty$ even for large temperature differences. If, however, a more realistic value of C is used the factor $\sqrt{C}$ gives a nonlinear dependence of q_w on h_w and h_∞ (cf. the above discussion of τ_w). As shown by Eq. 28-11a and Fig. B,30c the asymptotic formula $q_w \sim Pr^{-\frac{2}{3}}$ is approximately valid for, say, $Pr \geqq 0.7$. The asymptotic formula for small Prandtl numbers, $q_w \sim Pr^{-\frac{1}{2}}$ takes effect only for very small values, say, $Pr \leqq 0.2$.

Nonzero pressure gradients. Two different effects of wall temperature on skin friction were discussed above for the case $dp/dx = 0$. The net result of these two effects is that τ_w varies as $\sqrt{C}$. When $dp/dx \neq 0$

these two effects remain and there is an additional third effect. To fix the ideas assume that dp/dx is negative, i.e. the pressure gradient gives a positive acceleration, and that $h_w > h_\infty$. Since the fluid now is lighter than in the incompressible case the acceleration caused by the forcing function $-dp/dx$ is greater in the compressible case. The region of comparatively high velocities then moves closer to the wall and the skin friction increases. Fig. B,30a shows that even for $C = 1$ the skin friction is an increasing function of wall temperature for $dp/dx < 0$. The variation with Prandtl number shown by Fig. B,30a is also due to the same effect. The larger the heat conductivity, the further the region of low density (for $h_w > h_\infty$) extends from the wall, hence the larger is the acceleration due to pressure. For this reason τ_w is a decreasing function of Pr for $h_w > h_\infty$. Similarly, it follows that τ_w increases with Pr for $h_w < h_\infty$.

As mentioned above, the velocity u as a function of η is independent of h_w for $dp/dx = 0$ (except for the variation in C). However, this is no longer true when $dp/dx \neq 0$; Eq. 23-14a shows that a density dependence of acceleration due to pressure remains even after Howarth's η coordinate has been introduced. Thus, since there are essential changes in the velocity profile due to wall temperature, the transport of heat varies with the wall temperature and q_w is no longer linear in $h_w - h_\infty$.

Only when the parameter $\tilde{h}_w - 1$ is very small may one assume that the transporting velocity is independent of h_w, and is that of the corresponding incompressible flow. The comments about heat transfer made above for the case of $dp/dx = 0$ then still apply to the case $dp/dx \neq 0$. However, a comparison of Fig. B,30c, B,30d, and B,30e shows that the region of validity of the asymptotic formulas for Prandtl number dependence vary with the strength of the pressure gradient, i.e. with β. As β increases, any given value of Pr becomes effectively small and the approximation for low Prandtl number should be used. As an example, for $Pr = 0.7$ and $\beta = 0$, the asymptotic formula for large Prandtl numbers gives a reasonably accurate value of q_w. On the other hand, for $Pr = 0.7$ and $\beta = 1$, the best accuracy is obtained by using the formula for small Prandtl numbers.

When $\tilde{h}_w - 1$ is not small, the influence of h_w on the velocity profile must be taken into account. As a result, q_w is no longer linear in $h_w - h_\infty$, even for $C = 1$. As mentioned above, if $h_w > h_\infty$ a negative pressure gradient has a tendency to speed up the fluid near the wall. The heat transferred from the wall is then washed away faster and the heat transfer is increased. (Other effects of density changes are accounted for by the factor C, cf. Eq. 30-1b.) The resulting nonlinear effects are illustrated by Fig. B,30b.

Thus the decrease in density due to the wall temperature increases the velocity near the wall. This velocity increase leads to an increase of both q_w and τ_w. If a linear velocity profile is used near the wall the velocity profile and τ_w determine each other uniquely. In this case it must then be possible to express the nonlinear increase of q_w in terms of τ_w. This is done in Lighthill's formula for the heat transfer (Eq. 30-15). As mentioned earlier this formula may be applied even for $h_w - 1$ large, provided that τ_w is known and that Pr is sufficiently large.

Mach number effects. The effect of Mach number variations on boundary layer flow is seen most directly in the energy equation (14-7c) and the equation of state (22-7d). The first of these equations shows that for $M > 0$ the increase in temperature due to viscous dissipation must be taken into account. Furthermore, both equations show the thermodynamic role of pressure; pressure changes are coupled to changes in temperature and density. Since the Mach number affects the temperature and density distribution, it has an indirect effect on the velocity field, as seen from the momentum and continuity equations. A change in the velocity field of course in its turn affects the transport of temperature and hence the temperature distribution.

For flow past a given object the external flow is Mach number dependent. Hence changes in the Mach number also influence the boundary layer by way of the boundary conditions at the outer edge of the boundary layer.

There are essentially only two cases in which important analytical conclusions have been derived from the boundary layer equations at $M > 0$. The first case is that of a boundary layer on a flat plate with zero pressure gradient. This problem has been solved under quite general conditions. The second case is that of a boundary layer with the Prandtl number equal to unity. In this case the recovery temperature is known to be equal to the isentropic stagnation temperature. If furthermore $\mu \sim T$ the problem may be reduced to an equivalent problem for a boundary layer at $M = 0$ (Illingworth-Stewartson correlation theorem).

The flat plate boundary layer was discussed in Art. 25, 26, 27, and 28. It was seen there that the recovery temperature T_r (or the recovery enthalpy, h_r) is a significant quantity for flow at $M > 0$. The value of T_r was found under the assumption $\mu \sim T$: The difference $h_r - h_\infty$ is proportional to M^2 and its dependence on the Prandtl number is given by Eq. 28-14 and 28-19. For $Pr = 1$, h_r is simply equal to the isentropic stagnation enthalpy H_∞; for Pr near unity, $h_r - h_\infty$ varies as $Pr^{\frac{1}{2}}$. The significance of the recovery temperature is evident in the formula for heat transfer. A comparison of Eq. 28-11 and Eq. 28-16 shows that the formula for heat transfer at constant wall temperature may be easily generalized to an arbitrary Mach number by replacing h_∞ by h_r. This result is based on the assumption of a constant Prandtl number and an average temperature-viscosity law (Eq. 28-1). It is also seen that under the same assumptions, the estimated influence of the Mach number on the skin friction is slight. For the insulated plate, the wall temperature, i.e. the recovery temperature, is dependent on the Mach number; the wall temperature in its turn influences the skin friction (cf. the discussion after Eq. 28-5a and Eq. 25-5b).

The situation is considerably more complicated when the pressure gradient is not zero. However, some simplification of the problem is obtained when $Pr = 1$. In this case the energy equation for the total enthalpy H is formally identical with the corresponding equation for the enthalpy h in the case $M = 0$. The external boundary conditions in the two cases are $H = H_\infty = \text{const}$ and $h = h_\infty = \text{const}$ respectively. Thus the solution $h = \text{const}$ (incompressible flow) in the latter case corresponds to the solution $H = \text{const}$ in the former case. When H is constant the heat transfer at the wall is zero; in other words the recovery enthalpy at $Pr = 1$ is equal to the isentropic stagnation enthalpy.

The special solution of the energy equation just mentioned shows the advantage of using H instead of h. When using H one implicitly takes two Mach number effects into account, namely the effect of dissipation and that of pressure variation on the changes in enthalpy. A combination of this idea with some suitable coordinate distortions leads to the Howarth-Stewartson equations (Eq. 23-19). These equations, as well as the Illingworth equations (Eq. 23-23) are especially suited for the study of flow with pressure gradients at $M > 0$. Under the additional assumption $\mu \sim T$ these equations lead in a natural way to the correlation theorem derived in Art. 31. (See Eq. 31-1 and the discussion thereof.) The correlation theorem leads to especially simple results when the heat transfer at the body is zero. The associated boundary layer at $M = 0$ is then incompressible. Approximate solutions of the boundary layer equations may then be obtained for any given pressure gradient, say, with the aid of the momentum methods for incompressible flow discussed in Art. 32. The various Mach number effects may be seen from the analytical formulation of the correlation theorem. Certain effects tend to lower the skin friction while others increase it; hence it is difficult to make a general statement about the effect of the Mach number on skin friction for an arbitrary pressure gradient.

The correlation theorem is of importance even when the heat transfer at the body is not zero. Any solution for a compressible boundary layer at $M = 0$ may be used to generate a family of solutions at $M > 0$. The body geometry (or external flow) for the flow at $M > 0$ depends on M but not on the temperature conditions.

THREE-DIMENSIONAL BOUNDARY LAYERS

In Art. 20 equations valid for three-dimensional nonstationary boundary layers were derived. On the basis of these equations, three-dimensional boundary layer theory may be developed. In view of the detailed discussion of this subject which follows in Sec. C, only two topics will be considered in the following articles: In Art. 34, the application of plane-flow theory to axisymmetric boundary layer problems is described, and in Art. 35, the subject of displacement effects, introduced in Art. 13, will be generalized.

B,34. Bodies of Revolution. Before making any restrictions as to symmetry, one may observe that much of the general discussion of the two-dimensional boundary layer carries over to the three-dimensional case. An example of this is furnished by the energy equation, in which it will be assumed that the external forces and heat sources are zero, and, unless otherwise specified, that the flow is steady.

We consider the stagnation enthalpy defined by

$$H = h + \frac{u_1^2 + u_2^2}{2} \tag{34-1}$$

To begin with, stationarity of the flow will not be assumed. If the two momentum equations (Eq. 20-9b and 20-9c) are multiplied by u_1 and u_2 respectively and added, one obtains an equation for $(u_1^2 + u_2^2)/2$ which is simplified by the fact that the curvature terms vanish. If then this equation is added to the enthalpy equation (Eq. 20-9e) one obtains the generalization of Eq. 22-1d′:

$$\rho \frac{DH}{Dt} - \frac{\partial p}{\partial t} = \frac{1}{e_3}\frac{\partial}{\partial x_3}\left\{\frac{\mu}{e_3}\frac{\partial}{\partial x_3}\left[H + h\left(\frac{1 - Pr}{Pr}\right)\right]\right\} \tag{34-2}$$

where

$$\frac{D}{Dt} = \frac{\partial}{\partial t} + \sum_{i=1}^{3} \frac{u_i}{e_i}\frac{\partial}{\partial x_i}$$

If the external flow is independent of time (the boundary layer, however, may vary with time), and if $Pr = 1$, then a solution of Eq. 34-2 is the energy integral

$$H = h + \frac{u_1^2 + u_2^2}{2} = \text{const} \tag{34-3}$$

This equation implies $\partial h/\partial x_3 = 0$ at $x_3 = 0$. Hence, for steady external flow, Prandtl number unity and zero heat transfer at the wall, the stagnation enthalpy is constant throughout the boundary layer (cf. Eq. 22-8).

Consider now a body of revolution with the z axis as the axis of symmetry, the vertex at the origin and defined by an equation $r = r_0(z)$, $z \geqq 0$ (notation as in Eq. 5-8). The coordinates x_1, x_2, and x_3 are found such that the boundary layer equations assume a comparatively simple form. In order to utilize possible symmetries of the problem, we require x_2 to be a function of θ only, and x_1 and x_3 to be functions of r and z. If the body is at zero angle of attack, $\partial/\partial x_2$ is equal to zero. However, u_2 may still be different from zero if the body is spinning about its axis. The coordinates are chosen in such a way that, whenever $\partial/\partial x_2 = 0$ and $u_2 = 0$, the equations become formally identical with the two-dimensional boundary layer equations. This may be achieved for the first momentum equation and the energy equation (Eq. 20-9b and 33-3) if $e_1 = e_3^2$ and a new normal velocity component $\tilde{u}_3 = e_3 u_3$ is introduced. The continuity equation has then the two-dimensional form if $e_2 e_3 = 1$. For x_2 we choose the most natural value, namely θ. Then $e_2 = r_0(z)$; the relations just stated show that $e_3 = 1/r_0$ and $e_1 = 1/r_0^2$. It follows from the discussion in Art. 20 that it is sufficient to determine x_1 and x_2 on the surface of the solid and x_3 in an infinitesimal neighborhood of the solid. Thus x_1 is sufficiently determined by e_1 and an initial condition. If the arc length between x_1 and $x_1 + dx_1$ of the intersection of the surface with the plane $x_2 = 0$ is ds, then $ds = \sqrt{1 + (dr_0/dz)^2}\, dz = e_1 dx_1$. This, together with the requirement $x_1 = 0$ at the vertex, determines x_1 on the surface of the solid. x_2 was already defined everywhere as θ. x_3 is defined to be zero on the surface of the solid. At a point whose infinitesimal normal distance to the surface is dn, the value of x_3 is $dx_3 = (1/e_3)dn$. The definitions introduced above may be summarized in the following table:

$$e_1 = \frac{1}{r_0^2(z)}, \quad \frac{dx_1}{dz} = r_0^2 \sqrt{1 + \left(\frac{dr_0}{dz}\right)^2}, \qquad x_1 = 0 \quad \text{at} \quad z = 0 \tag{34-4a}$$

$$e_2 = r_0(z), \qquad x_2 = \theta \tag{34-4b}$$

$$e_3 = \frac{1}{r_0(z)}, \qquad x_3 = 0 \quad \text{when} \quad r = r_0(z) \tag{34-4c}$$

$$\tilde{u}_3 = e_3 u_3 \tag{34-4d}$$

With this metric, Eq. 20-9a, 20-9b, and 20-9c and Eq. 34-3 reduce to

$$\frac{\partial(\rho u_1)}{\partial x_1} + \frac{1}{r_0^3}\frac{\partial(\rho u_2)}{\partial x_2} + \frac{\partial(\rho \tilde{u}_3)}{\partial x_3} = 0 \tag{34-5a}$$

$$\rho\left[\frac{\tilde{D}u_1}{Dt} - u_2^2\frac{r_0'}{r_0}\right] + \frac{\partial p}{\partial x_1} = \frac{\partial}{\partial x_3}\left(\mu\frac{\partial u_1}{\partial x_3}\right) \tag{34-5b}$$

$$\rho\left[\frac{\tilde{D}u_2}{Dt} - u_1 u_2\frac{r_0'}{r_0}\right] + \frac{1}{r_0^3}\frac{\partial p}{\partial x_2} = \frac{\partial}{\partial x_3}\left(\mu\frac{\partial u_2}{\partial x_3}\right) \tag{34-5c}$$

$$\rho\frac{\tilde{D}H}{Dt} = \frac{\partial}{\partial x_3}\left[\mu\frac{\partial}{\partial x_3}\left(H + h\frac{1 - Pr}{Pr}\right)\right] \tag{34-5d}$$

where

$$\frac{\tilde{D}}{Dt} = u_1 \frac{\partial}{\partial x_1} + \frac{u_2}{r_0^2} \frac{\partial}{\partial x_2} + \tilde{u}_3 \frac{\partial}{\partial x_3}$$

and

$$r_0' = \frac{dr_0}{dx_1} = \frac{dr_0}{dz} \frac{1}{r_0^2 \sqrt{1 + \left(\frac{dr_0}{dz}\right)^2}}$$

If the body is at zero angle of attack, then $\partial/\partial x_2 = 0$ and the above equations reduce to

$$\frac{\partial(\rho u_1)}{\partial x_1} + \frac{\partial(\rho \tilde{u}_3)}{\partial x_3} = 0 \tag{34-6a}$$

$$\rho\left[u_1 \frac{\partial u_1}{\partial x_1} + \tilde{u}_3 \frac{\partial u_1}{\partial x_3} - u_2^2 \frac{r_0'}{r_0}\right] + \frac{\partial p}{\partial x_1} = \frac{\partial}{\partial x_3}\left(\mu \frac{\partial u_1}{\partial x_3}\right) \tag{34-6b}$$

$$\rho\left[u_1 \frac{\partial u_2}{\partial x_1} + \tilde{u}_3 \frac{\partial u_2}{\partial x_3} + u_1 u_2 \frac{r_0'}{r_0}\right] = \frac{\partial}{\partial x_3}\left(\mu \frac{\partial u_2}{\partial x_3}\right) \tag{34-6c}$$

$$\rho\left[u_1 \frac{\partial H}{\partial x_1} + \tilde{u}_3 \frac{\partial H}{\partial x_3}\right] = \frac{\partial}{\partial x_3}\left[\mu \frac{\partial}{\partial x_3}\left(H + h\frac{1 - Pr}{Pr}\right)\right] \tag{34-6d}$$

If, in addition, the body is not spinning, then $u_2 = 0$. Eq. 34-6a, 34-6b, and 34-6d are then formally identical with the two-dimensional boundary layer equations in accordance with the construction of the metric. This reduction to the two-dimensional form was first achieved by Mangler [*150*]. The above derivation follows the method of Hayes [*111*].

The form of the above equations suggests various approximate methods. If, for example, the term $u_2^2 r_0'/r_0$ is negligible in Eq. 34-6b, then Eq. 34-6a, 34-6b, and 34-6d may be solved separately as a two-dimensional problem. Eq. 34-6c then becomes a linear equation for u_2.

Computations for spinning bodies of revolution have been carried out by Illingworth [*151*] and by Howarth [*152*] and, with the aid of momentum methods, by Schlichting [*153*].

As an application we consider a circular cone in supersonic flow with zero angle of attack and zero rotation. Let the cone be defined by $r_0 = \lambda z$, and let s be the distance from the vertex measured along a conical ray. Then one obtains from Eq. 34-4a

$$x_1 = \lambda^2 \sqrt{1 + \lambda^2}\, \frac{z^3}{3} \tag{a}$$

and

$$s = \sqrt{1 + \lambda^2}\, z \tag{b}$$

In nonviscous flow the values of the x_1 component of the velocity and the state variables are constant on the surface of the cone. These are denoted by U_e, ρ_e, T_e, etc. Let μ_e be the value of the viscosity coefficient determined from T_e. Since $\partial p_e/\partial x_1 = 0$ and $u_2 = 0$, Eq. 34-6a, 34-6b, and 34-6d and the corresponding boundary conditions are formally identical with the corresponding equations and boundary conditions for two-dimensional flow past a flat plate with zero pressure gradient and with U_e, ρ_e, μ_e, etc. as values at infinity. Here x_1, x_3, u_1, $\bar{u}_3$ correspond formally to x, y, u and v as defined for the flat plate (cf. Art. 26). Only the case where a similarity solution exists is considered. Then

$$\tau_{\text{flat plate}} = \mu \left(\frac{\partial u}{\partial y}\right)_{y=0} = \frac{A}{\sqrt{x}} \tag{c}$$

where A is a function of U_e, ρ_e, μ_e. From the formal identity of the two cases it follows that

$$\mu \left(\frac{\partial u_1}{\partial x_3}\right)_{x_3=0} = \frac{A}{\sqrt{x_1}} \tag{d}$$

and, hence, using (a), (b), and Eq. 34-1c

$$\tau_{\text{cone}} = \frac{\mu}{e_3}\left(\frac{\partial u_1}{\partial x_3}\right) = \frac{\lambda z A}{\sqrt{x_1}} = \frac{\sqrt{3}\,A}{\sqrt{5}} \tag{e}$$

Hence,

$$\tau_{\text{cone}} = \sqrt{3}\,\tau_{\text{flat plate}} \tag{34-7}$$

where the skin friction is measured at the same distance from the vertex in the two cases and where the free stream conditions for the flat plate are identical with the nonviscous flow conditions on the surface of the cone.

The circular cone may also be treated as a special case in the theory of conical boundary layers [*111; 154*].

B,35. Displacement Effects. In Art. 13 (Eq. 13-15ff.) it was shown how one may determine a correction to boundary layer theory from the displacement thickness. These ideas were further elaborated in Art. 17. The discussion there was restricted to two-dimensional incompressible stationary flow; the ideas used there, however, may equally well be applied to general three-dimensional boundary layers.

The following notation is used: ρ, $\mathbf{u}$, u_i, $\mathbf{m} = \rho\mathbf{u}$, etc. denote the values of the flow quantities corresponding to the solution of the full Navier-Stokes equations. The subscript $_e$ denotes the corresponding values in nonviscous flow and is *not* restricted to the wall. The following perturbation quantities are introduced:

$$\rho' = \rho - \rho_e, \qquad \mathbf{m}' = \mathbf{m} - \mathbf{m}_e, \qquad u_i' = u_i - u_{i_e} \text{ etc.} \tag{35-1}$$

It is of course not assumed that these perturbations are small. The boundary layer limit with respect to a system of coordinates, $\zeta = (x_1, x_2, x_3)$ is denoted by subscript ζ (cf. Eq. 13-4 and 13-7). The boundary layer approximations to the full solutions and to the perturbations described by Eq. 35-1 are then related by the following equations:

$$\rho'_\zeta = \rho_\zeta - \rho_{ew} \tag{35-2a}$$

$$u'_{i\zeta} = u_{i\zeta} - u_{iew}, \qquad i = 1, 2 \tag{35-2b}$$

$$u'_{3\zeta} = u_{3\zeta} - \left(\frac{\partial u_{3e}}{\partial x_3}\right)_w x_3 \tag{35-2c}$$

$$m'_{i\zeta} = \rho u_{i\zeta} - \rho_{ew} u_{iew}, \qquad i = 1, 2 \tag{35-2d}$$

Since the continuity equation is linear in density and mass flow, it is also valid for the perturbation flow,

$$e_1e_2e_3\frac{\partial \rho'}{\partial t} + \frac{\partial(e_2e_3m'_1)}{\partial x_1} + \frac{\partial(e_1e_3m'_2)}{\partial x_2} + \frac{\partial(e_1e_2m'_3)}{\partial x_3} = 0 \tag{35-3}$$

If one takes the boundary layer limit of this relation and integrates from $x_3 = 0$ to $x_3 = \infty$, one obtains

$$(e_1e_2)_w(m'_{3\zeta})_\infty = -\int_0^\infty \left[(e_1e_2e_3)_w \frac{\partial \rho'_\zeta}{\partial t} + \frac{\partial}{\partial x_1}(e_{2_w}e_{3_w}m'_{1\zeta}) + \frac{\partial}{\partial x_2}(e_{1_w}e_{3_w}m'_{2\zeta})\right] dx_3 \tag{35-4}$$

or, using Eq. 35-2,

$$(e_1e_2)_w(m'_{3\zeta})_\infty = (e_1e_2e_3)_w \frac{\partial}{\partial t}\int_0^\infty (\rho_{ew} - \rho_\zeta)dx_3 + \frac{\partial}{\partial x_1}(e_{2_w}I_1) + \frac{\partial}{\partial x_2}(e_{1_w}I_2) \tag{35-4'}$$

where

$$I_1 = \int_0^\infty (\rho_{ew}u_{1ew} - \rho_\zeta u_{1\zeta})e_{3_w}dx_3 \tag{35-5a}$$

and

$$I_2 = \int_0^\infty (\rho_{ew}u_{2ew} - \rho_\zeta u_{2\zeta})e_{3_w}dx_3 \tag{35-5b}$$

The integrals I_1 and I_2 are obviously generalizations of the integral defining the two-dimensional displacement thickness (Eq. 24-3a). In spite of the cumbersome notation, Eq. 35-4 (or 35-4′) has a simple intuitive interpretation. Consider a rectangular column normal to the boundary extending between $x_3 = 0$ and $x_3 = \delta$, with base at $(x_1, x_2, 0)$ and base area $e_1dx_1\ e_2dx_2$. Here δ is a vaguely defined boundary layer thickness such that at $x_3 = \delta$ the perturbation boundary layer has almost reached the values which formally are assumed only at $x_3 = \infty$. Within the boundary layer approximation, the mass flow out of the top of this column is the mass flow given by the nonviscous solution plus $m'_{3\zeta}e_{1w}dx_1e_{2w}dx_2$. Since the perturbation flow obeys the continuity equation, this mass flow must be balanced by a rate of mass decrease in the column and by a net influx through the sides of the column. This balance is expressed by Eq. 35-4′. Thus I_1 and I_2 are the components of a two-dimensional mass flow vector whose two-dimensional divergence is the density

$$\frac{1}{(e_1e_2)_w}\left[\frac{\partial}{\partial x_1}(e_{2w}I_1) + \frac{\partial}{\partial x_2}(e_{2w}I_2)\right]$$

In the stationary case this density is equal to $(m'_{3\zeta})_\infty$ which is the flow density through the top surface of the column. Since the integrands in I_1 and I_2 depend on x_3 and ν only in the combination $x_3/\sqrt{\nu}$, it follows that the value of the integrals is of the form $\sqrt{\nu}\,fn(x_1, x_2)$. Knowing the value of the perturbation outflow at the edge of the boundary layer one may find a correction of order $\sqrt{\nu}$ to the flow outside the boundary layer. This is the flow due to displacement thickness. Formally one may define (cf. Eq. 13-19)

$$\mathbf{m}'_e = \lim_{\nu\to 0}\frac{\mathbf{m} - \mathbf{m}_e}{\sqrt{\nu}} \qquad \text{for} \qquad x_3 \neq 0 \tag{35-6}$$

and use similar definitions for other flow quantities. $\sqrt{\nu}\,m'_e$ is then flow due to displacement thickness. It satisfies a linearized equation for nonviscous flow and also the following boundary condition at the wall:

$$\sqrt{\nu}\,(m'_{3e})_w = (m'_{3\zeta})_\infty \tag{35-7}$$

This equation states that the normal flow at the wall of the first external perturbation is equal to the normal flow at infinity of the boundary layer perturbation. The boundary layer thus gives rise to an apparent source distribution at the wall. Eq. 35-7 may be proved formally in the same way as Eq. 13-20 or 17-20 (cf. Art. 17). More intuitively it means that, if one subtracts out $\mathbf{m}_e$ and divides by $\sqrt{\nu}$, then the normal mass flow at $x_3 = \delta$ is approximately $(m'_{3\zeta})_\infty/\nu$. In the limit $\nu \to 0$ one may let δ tend to zero so that this value is assumed arbitrarily near the wall.

From the integrals I_1 and I_2, one may define two quantities of dimension length which are analogous to the concept of displacement thickness in two dimensions. One may also define an apparent body shape which gives rise to the flow due to displacement thickness in nonviscous flow (cf. [*103*]). However, the discussion above shows that there is no urgent need for introducing these concepts. The mass flow vector (I_1, I_2) as introduced above has a direct physical meaning, and its components may be shown to enter naturally into the momentum integral equations. Furthermore, the divergence of this vector also has a direct physical meaning as perturbation outflow through the edge of the boundary layer, and as boundary condition at the wall for the first external perturbation. The latter is the flow due to displacement thickness. This concept, rather than that of displacement thickness, is the basic concept.

B,36. Cited References.

1. Kirkwood, J. G. *J. Chem. Phys. 14,* 180 (1946); *15,* 72 (1947); *17,* 998 (1949).
2. Grad, H. *Commun. on Pure and Appl. Math. 2,* 331 (1949).
3. Truesdell, C. *J. Rat. Mech. and Anal. 1,* 125 (1952).
4. Brillouin, L. *Les Tenseurs en Mécanique et en Elasticité,* 1st Amer. ed. Dover, 1946.
5. Michal, A. D. Matrix and tensor calculus. *Guggenheim Aeronaut. Lab., Calif. Inst. Technol. Aeronaut. Series,* 1st ed. Wiley, 1947.
6. Sommerfeld, A. *Mechanics of Deformable Bodies.* Academic Press, 1950.
7. Durand, W. F., ed. *Aerodynamic Theory,* Vol. III, G. Springer, Berlin, 1934–1936.
8. Lamb, H. *Hydrodynamics,* 6th ed. Cambridge Univ. Press, 1932.
9. Liepmann, H. W., and Roshko, A. *Elements of Gasdynamics.* Wiley, 1957.
10. Filon, L. N. G. *Proc. Roy. Soc. London A113,* 7 (1926).
11. Manual of the ICAO standard atmosphere. *NACA Tech. Note 3182,* 1954.
12. Epstein, P. S. *Thermodynamics.* Wiley, 1937.
13. Taylor, W. J., and Johnston, H. L. *J. Chem. Phys. 14,* 219 (1946).
14. Vasilesco, V. *Annales de Physique 20,* 137, 292 (1945).
15. Chapman, S., and Cowling, T. G. *The Mathematical Theory of Non-Uniform Gases.* Cambridge Univ. Press, 1939.
16. Partington, J. R. *An Advanced Treatise on Physical Chemistry.* Longmans, Green, 1951.
17. Keyes, F. G. The heat conductivity, viscosity, specific heat, and Prandtl numbers for thirteen cases. *Project Squid Tech. Rept. 37,* 1952.
18. Hoge, H. G. *Trans. Am. Soc. Mech. Engrs. 72,* 779 (1950).
19. *The NBS-NACA Tables of Thermal Properties of Gases.* Natl. Bur. Standards, 1949.
20. Moore, L. L. *J. Aeronaut. Sci. 19,* 505 (1952).
21. Chapman, D. R. *NACA Rept. 1259,* 1956. Supersedes *NACA Tech. Note 3226,* 1954.
22. Goldstein, S. *Proc. Roy. Soc. London A123,* 216 (1929).
23. Imai, I. *Proc. Roy. Soc. London A208,* 487 (1951).
24. Goldstein, S. *Modern Developments in Fluid Dynamics,* 1st ed. Oxford Univ. Press, 1938.
25. Oseen, C. W. *Neuere Methoden und Ergebnisse in der Hydrodynamik.* Akad. Verlagsgesellschaft, Leipzig, 1927.
26. Schlichting, H. *Grenzschicht-Theorie.* G. Braun, Karlsruhe, 1950. English transl. *Boundary Layer Theory.* McGraw-Hill, 1955.
27. Courant, R., and Hilbert, D. *Methoden der Mathematischen Physik,* 2nd ed. Interscience, 1937.
28. Stokes, G. G. *Cambridge Trans. 9,* 1850.
29. Carslaw, H. S., and Jaeger, J. C. *Conduction of Heat in Solids,* 1st ed. Oxford Univ. Press, 1947.
30. Erdelyi, A., Magnus, W., Oberhettinger, F., and Tricomi, F. *Higher Transcendental Functions, Vol. 1, 2, and 3.* McGraw-Hill, 1953.
31. Rayleigh, J. W. S. *Phil. Mag. 22,* 381 (1911).
32. Lagerstrom, P. A., and Cole, J. D. *J. Rat. Mech. and Anal. 4,* 817 (1955).
33. Batchelor, G. K. *Quart. J. Mech. and Appl. Math. 7,* 179 (1954).

34. Howarth, L. *Proc. Cambridge Phil. Soc. 46*, 127 (1950).
35. Sowerby, L. *Phil. Mag. 42*, 176 (1951).
36. Sowerby, L., and Cooke, J. L. *Quart. J. Mech. and Appl. Math. 6*, 50 (1953).
37. Poincaré, H. *Théorie des Tourbillons.* G. Carré, Paris, 1893.
38. Truesdell, C. *The Kinematics of Vorticity. Science Series 19.* Indiana Univ. Publs., 1954.
39. Ackeret, J. *J. Appl. Math. and Phys. 3*, 259 (1952).
40. Berker, R. *Sur quelques cas d'Intégration des Equations du Mouvement d'un Fluide Visqueux Incompressible.* Lille, Paris, 1936.
41. Oseen, C. W. *Opuscula Mathematica A. Wiman dedicata.* Lund, 1930.
42. Hiemenz, K. *Dingler's Polytech. J. 326*, 321 (1911).
43. Homann, F. The effect of high viscosity on the flow around a cylinder and around a sphere. *NACA Tech. Mem. 1334*, 1952.
44. Hamel, G. *Jahresber. Dt. Mathematiker Vereinigung 25*, 34 (1916).
45. Rosenblatt, A. Solutions exactes des équations du mouvement des liquides visqueux. *Mémorial des Sciences Mathématiques 72*, Paris, 1935.
46. von Mises, R., and Friedrichs, K. O. *Fluid Dynamics.* Brown Univ. Press, 1941.
47. Rosenhead, L. *Proc. Roy. Soc. London A175*, 436 (1940).
48. Lin, C. C. On the motion of vortices in two dimensions. *Univ. Toronto Studies, Appl. Math. Series 5*, 1943.
49. Birkhoff, G. *Hydrodynamics—A Study in Logic, Fact and Similitude*, 1st ed. Princeton Univ. Press, 1950.
50. Krakowski, M., and Charnes, A. *Carnegie Inst. Technol., Dept. of Math. Tech. Rept. 37*, 1953.
51. Stokes, G. G. On the effect of internal friction of fluids on the motion of a pendulum. *Trans. Cambridge Phil. Soc. ix (8), Papers iii*, 1851.
52. Proudman, I., and Pearson, J. R. A. Expansions at small Reynolds numbers for the flow past a sphere and a circular cylinder. *J. Fluid Mech. 2*, 237–262 (1957).
53. Lagerstrom, P. A., Cole, J. D., and Trilling, L. Problems in the theory of viscous compressible fluids. *Office of Nav. Research, Calif. Inst. Technol. Guggenheim Aeronaut. Lab. Rept.*, 1949.
54. Magnus, W., and Oberhettinger, F. *Formulas and Theorems for the Special Functions of Mathematical Physics.* Chelsea, New York, 1949.
55. Tomotika, S., and Aoi, T. *Quart. J. Mech. and Appl. Math. 3*, 140 (1950).
56. Weyl, H. *Annals of Math. 43*, 381 (1942).
57. Burgers, J. M. *Koninkl. Ned. Akad. Wetenschap. Proc. 23*, 1082 (1922).
58. Levinson, N. *Annals of Math. 51*, 428 (1950).
59. Wasow, W. *Annals of Math. 49*, 852 (1948).
60. Latta, G. E. *Proc. Eighth Intern. Congress Appl. Mech., Istanbul*, 479, 480 (1953).
61. Prandtl, L. *Verhandl. d. III. Intern. Math. Kongr., Heidelberg*, 1904. Reprinted in *Vier Abhdlg. zur Hydro-u. Aerodynamik 1*, Göttingen, 1927.
62. Goldstein, S. *Proc. Cambridge Phil. Soc. 35*, 338 (1939).
63. Mangler, W. *Z. angew. Math. u. Mech. 23*, 243 (1943).
64. Falkner, V. M., and Skan, S. W. *Phil. Mag. 12*, 865 (1931).
65. Hartree, D. R. *Proc. Cambridge Phil. Soc. 33, Part II*, 223 (1937).
66. Smith, A. M. O. *Inst. Aeronaut. Sci., S.M.F. Fund Paper FF-10*, 1954.
67. Blasius, H. *Z. Math. u. Phys. 56*, 1 (1908).
68. Howarth, L. *Proc. Roy. Soc. London A164*, 547 (1938).
69. Weyl, H. *Proc. Natl. Acad. Sci. 27*, 578 (1941); *28*, 100 (1942).
70. Toepfer, C. *Z. Math. Phys. 60*, 397 (1912).
71. Goldstein, S. *Quart. J. Mech. and Appl. Math. 1*, 43 (1948).
72. Stewartson, K. *Quart. J. Mech. and Appl. Math. 4*, 182 (1951).
73. Lighthill, M. J. *Proc. Roy. Soc. London A224*, 1 (1954).
74. Stewartson, K. *Quart. J. Mech. and Appl. Math. 13*, 113 (1955).
75. Glauert, M. B., and Lighthill, M. J. *Proc. Roy. Soc. London A230*, 188 (1955).
76. Seban, R. A., and Bond, R. *J. Aeronaut. Sci. 18*, 671 (1951).
77. Kelly, H. R. *J. Aeronaut. Sci. 21*, 634 (1954).

78. Roshko, A. On the development of turbulent wakes from vortex streets. *NACA Tech. Note 2913*, 1953; *NACA Rept. 1191*, 1954.
79. Roshko, A. On the drag and shedding frequency of bluff bodies. *NACA Tech. Note 3169*, 1954.
80. Roshko, A. *J. Aeronaut. Sci. 22*, 124 (1955).
81. Delany, N. K., and Sorensen, N. E. *NACA Tech. Note 3038*, 1953.
82. Finn, R. K. *J. Appl. Phys. 24*, 771 (1935).
83. Thom, A. The pressure on the front generator of a cylinder. *Brit. Aeronaut. Research Council Repts. and Mem. 1389*, 1931.
84. Fage, A., and Warsap, J. H. The effects of turbulence and surface roughness on the drag of a circular cylinder. *Brit. Aeronaut. Research Council Repts. and Mem. 1283*, 1930.
85. Schubauer, G. B. Air flow in a separating laminar boundary layer. *NACA Rept. 527*, 1935.
86. Black, J. *Proc. Roy. Aeronaut. Soc. 56*, 279 (1952).
87. Schubauer, G. B., and Skramstad, H. K. *J. Aeronaut. Sci. 14*, 69 (1947).
88. Tomotika, S., and Aoi, T. *Quart. J. Mech. and Appl. Math. 6*, 290 (1953).
89. Janssen, E. An analog solution of the Navier-Stokes equation for the case of flow past a flat plate at low Reynolds numbers. *Heat Transfer and Fluid Mech. Inst.*, Stanford Univ. Press, 1956.
90. Janour, Z. *NACA Tech. Mem. 1316*, 1951.
91. Schaaf, S. A., and Sherman, F. S. *J. Aeronaut. Sci. 22*, 85–90 (1954).
92. Dryden, H. L. Air flow in the boundary layer near a plate. *NACA Rept. 562*, 1936.
93. Liepmann, H. W. Investigations on laminar boundary-layer stability and transition on curved boundaries. *NACA Wartime Rept. W107*, 1943.
94. Liepmann, H. W., and Fila, G. H. *NACA Rept. 890*, 1947.
95. Dhawan, S. *NACA Tech. Note 2567*, 1952.
96. Kaplun, S. *Z. angew. Math. u. Mech. 5*, 111–135 (1954).
97. Illingworth, C. R. *Proc. Cambridge Phil. Soc. 46*, 469 (1950).
98. Courant, R., and Friedrichs, K. O. *Supersonic Flow and Shock Waves.* Interscience, 1948.
99. von Mises, R. *J. Aeronaut. Sci. 17*, 551 (1950).
100. Weyl, H. *Commun. on Pure and Appl. Math. 2*, 103 (1949).
101. Gilbarg, D. *J. Math. and Phys. 73*, 256 (1951).
102. Gilbarg, D., and Paolucci, D. *J. Rat. Mech. and Anal. 2*, 617 (1953).
103. Moore, F. K. Displacement effect of a three-dimensional boundary layer. *NACA Tech. Note 2722*, 1952.
104. Kaplun, S., and Lagerstrom, P. A. Asymptotic expansion of Navier-Stokes solution for small Reynolds numbers. *J. Math. and Mech. 6*, 585–594 (1957).
105. Kaplun, S. Low Reynolds number flow past a circular cylinder. *J. Math. and Mech. 6*, 595 (1957).
106. Sherman, F. S. *NACA Tech. Note 3298*, 1955.
107. Stalder, J. R., and Jukoff, D. *NACA Rept. 944*, 1949. Supersedes *NACA Tech. Note 1682*, 1948.
108. Stalder, J. R., Goodwin, G., and Creager, M. O. *NACA Rept. 1032*, 1951. Supersedes *NACA Tech. Note 2244*, 1950.
109. Stalder, J. R., Goodwin, G., and Creager, M. O. *NACA Rept. 1093*, 1952. Supersedes *NACA Tech. Note 2438*, 1951.
110. Mott-Smith, H. M. *Phys. Rev., Ser. 2, 82*, 885–892 (1951).
111. Hayes, W. D. The three-dimensional boundary layer. *Nav. Ord. Rept. 1313*, 1951.
112. Howarth, L. *Phil. Mag. 42*, 239, 1433 (1951).
113. Ludford, G. S. S. *Quart. Appl. Math. 10*, 1 (1952).
114. Burgers, J. A mathematical model illustrating the theory of turbulence. *Advances in Applied Mechanics.* Academic Press, 1948.
115. Wu, T. Y. *On Problems of Heat Conduction in Compressible Fluids. Ph.D. Thesis,* Calif. Inst. Technol., 1952.

116. Wu, T. Y. *J. Math. and Phys. 35*, (*1*), 13-27 (1956).
117. Van Dyke, M. D. *J. Appl. Math. and Phys. 3*, 343 (1952).
118. Cole, J. D., and Wu, T. Y. *J. Appl. Mech. 19*, 209 (1952).
119. Wu, T. Y., and Cole, J. D. *Heat Transfer and Fluid Mech. Inst.*, 139 (1952).
120. Truesdell, C. *J. Rat. Math. 2*, 644 (1953).
121. Howarth, L. *Quart. J. Mech. and Appl. Math. 4*, 157 (1951).
122. Cole, J. D. *Quart. Appl. Math. 9*, 225 (1951).
123. Hopf, E. *Commun. on Pure and Appl. Math. 3*, 201 (1950).
124. Busemann, A. *Handbuch der Exp. Physik, 4, Part 1*, 366. Akad. Verlag. M. B. H., Leipzig, 1931.
125. von Mises, R. *Z. angew. Math. u. Mech. 7*, 425 (1927).
126. von Kármán, Th., and Tsien, H. S. *J. Aeronaut. Sci. 5*, 227 (1938).
127. Crocco, L. Lo strato limite laminare nei gas. *Monografie Sci. Aeronaut. 3*, 1946.
128. Howarth, L. *Proc. Roy. Soc. London A194*, 16 (1948).
129. Stewartson, K. *Proc. Roy. Soc. London A200*, 84 (1949).
130. Illingworth, C. R. *Proc. Roy. Soc. London A199*, 533 (1949).
131. Dorodnitsyn, A. A. *Prikl. Mat. Mekh. 6*, 1942.
132. Crocco, L. *Aerotecnica 12*, 181 (1932).
133. Chapman, D. R., and Rubesin, M. W. *J. Aeronaut. Sci. 16*, 547 (1949).
134. Pohlhausen, E. *Z. angew. Math. u. Mech. 1*, 115 (1921).
135. Eckert, E., and Drewitz, O. *Luftfahrtforschung 19*, 189 (1942).
136. Fage, A., and Falkner, V. M. Relation between heat transfer and surface friction for laminar flow. *Brit. Aeronaut. Research Council Repts. and Mem. 1408*, 1931.
137. Lighthill, M. J. *Proc. Roy. Soc. London A202*, 359 (1950).
138. Li, T. Y., and Nagamatsu, H. T. *J. Aeronaut. Sci. 20*, 653 (1953).
139. Cohen, C. B. *J. Aeronaut. Sci. 21*, 281 (1954).
140. Cohen, C. B., and Reshotko, E. *NACA Tech. Note 3325*, 1955.
141. Levy, S. *J. Aeronaut. Sci. 21*, 459 (1954).
142. Lal, S. *J. Aeronaut. Soc. India, 8*, No. 1, 1 (1956), and *8* No. 2, 19 (1956).
143. Sebulkin, M. *J. Aeronaut. Sci. 19*, 570 (1952).
144. von Kármán, Th. *Z. angew. Math. u. Mech. 1*, 233 (1921).
145. Pohlhausen, K. *Z. angew. Math. u. Mech. 1*, 235 (1921).
146. Holstein, H., and Bohlen, T. *Lilienthalgesellschaft-Bericht 5*, 10 (1940).
147. Thwaites, B. *Aeronaut. Quart. 1*, 245 (1949).
148. Walz, A. *Ing.-Arch. 16*, 243 (1948).
149. Cohen, C. B., and Reshotko, E. *NACA Tech. Note 3326*, 1955.
150. Mangler, W. Boundary layers on bodies of revolution in symmetrical flow. *Gt. Brit. Ministry of Air Production Volkenrode Repts. and Transls. 55*, 1946.
151. Illingworth, C. R. *Phil. Mag. 45*, 1 (1954).
152. Howarth, L. *Phil. Mag. 43*, 1308 (1951).
153. Schlichting, H. *Ing.-Arch. 21*, 227 (1953).
154. Moore, F. K. Three-dimensional compressible laminar boundary-layer flow. *NACA Tech. Note 2279*, 1951.

SECTION C

THREE-DIMENSIONAL LAMINAR BOUNDARY LAYERS

A. MAGER

CHAPTER 1. GENERAL CONSIDERATIONS

C,1. Introduction. The boundary layer concept as introduced by Prandtl and the resulting approximations involved in the boundary layer theory for two-dimensional flow have been thoroughly discussed in B,13. These basic ideas are readily translatable to three-dimensional flow, but in this case the surface geometry and the curving of the external flow play essential parts, so that the mathematical and physical theory of the three-dimensional boundary layer is considerably more complicated.

By two-dimensional boundary layer we shall mean a boundary layer which is formed over a plane surface, infinite in lateral extent, when the projections of the streamlines of the outer flow on this surface are straight lines perpendicular to the leading edge. All other cases of boundary layer flow fall properly into the three-dimensional classification. Thus, for example, when the outer flow streamlines have straight-line projections but the surface is laterally curved, as in meridional flow over a cone, the boundary layer, though axially symmetric and solvable in any of the meridional planes, is nevertheless three-dimensional in character. This three-dimensional behavior is readily recognized, when it is realized that while flowing up the cone surface, the boundary layer is required to stretch so as to blanket a progressively greater circumference. This results in a net thinning effect relative to the corresponding purely two-dimensional flow. The surface geometry, in this case, causes a stretching in width of individual stream tubes, and this special three-dimensional effect is known as streamline divergence or surface dilation.

Another characteristically three-dimensional effect arises when the outer flow streamlines are curved as, for example, in a circular flow over a plane surface. In such a case the pressure gradient of the outer flow, at each radial position, is balanced by the centrifugal force corresponding to the local curvature of the outer flow streamlines. Furthermore, according to the boundary layer concept, at any given surface position, the outer flow pressure prevails also inside the boundary layer. However, the flow in the boundary layer is necessarily decelerated by viscous effects, so that the pressure cannot be balanced by the centrifugal force unless the streamline curvature inside the layer is correspondingly greater than that in the outer flow. This means that the surface projections of boundary layer streamlines are different from those of the outer flow, or in other words, the flow in the boundary layer differs from the outer flow not only in the magnitude, but also in the direction of the velocity. This change of direction is most easily represented by introduction of a transverse flow component, in a direction normal to the streamline of the outer flow. This new component is usually known as secondary flow.

There are certain other aspects of three-dimensional boundary layers which must be noted here. In two-dimensional flow, since the boundary layer is restricted to move in the direction of the outer flow, any sufficiently strong opposing pressure gradient ultimately results in separation of the flow from the surface and triggers important readjustments of the outer flow. In three-dimensional boundary layers, on the other hand, the flow retains the freedom of choosing the least difficult path, and strong unfavorable pressure gradients do not necessarily lead to detachment from the surface, but demonstrate themselves by drastic changes in flow direction. Consequently, not only is the boundary layer separation modified in three-dimensional flow, but also it carries different implications (regarding over-all effects) from those which apply in two-dimensional cases.

Also, in two-dimensional flow it is very convenient to describe the boundary layer behavior in terms of certain averages, characteristic of the momentum and mass flow deficiencies. These are known as the momentum and displacement thicknesses. In three-dimensional flow, however, because of the additional velocity component, one has two displacement thicknesses and four momentum thicknesses, so that simple representation of momentum and mass flow deficiencies at a given point is impossible, and these concepts forfeit some of their usefulness.

As can be seen from the above discussion, the understanding of two-dimensional boundary layers is of limited help, even in heuristic attempts to describe and explain three-dimensional boundary layer motion. Of course, the analysis of three-dimensional boundary layers is much more complex than that of two-dimensional layers, because of the additional component motion and also because of surface-geometry effects. For these reasons, while the theory of two-dimensional boundary layers has been relatively well developed and established for some time now, the corresponding three-dimensional theory is comparatively new and not nearly as complete. Still, the demands of modern flight, depending rather critically on boundary layer behavior on such aircraft and missile components as yawed wings, nose cones, inlet spikes, and on the internal parts of turbomachinery, have recently stimulated a considerable number of theoretical investigations into the three-dimensional boundary layer flow. As a result of these stimuli, the theory of three-dimensional boundary layers has been advancing rather rapidly over the last few years, as some of the recent reviews of this subject [*1*,*2*] indicate.

In this section, Art. 2 to 7 are devoted to some very general considerations applicable to almost any three-dimensional boundary layer, regardless of the specific surface shape or outer flow condition. Art. 8 to 13 illustrate various three-dimensional effects, as well as the methods of solution which have been developed to deal with them. It will be noted that the problems have been arranged, without disrupting certain natural divisions, in the order of increasing complexity. Thus, paralleling this order, the various methods of solution become more and more approximate. Finally, Art. 14 to 18 contain discussions of certain special problems, touching on areas where much research still needs to be done before an understanding of the present subject will be complete.

C,2. Boundary Layer Equations. The three laws governing the motion of a compressible fluid in any system express the conservation of mass

$$\frac{\partial \rho}{\partial t} + (\rho u_j)_{,j} = 0 \tag{2-1}$$

the conservation of momentum,

$$\rho\left[\frac{\partial u_i}{\partial t} + u_j u_{i,j} + \Lambda_{,i} + 2(\omega_j u_k - \omega_k u_j) - \tfrac{1}{2}\omega^2 R^2_{,i}\right] = -p_{,i} + \tau_{ij,j} \tag{2-2}$$

and the conservation of energy,

$$\rho\left(\frac{\partial I}{\partial t} + u_j I_{,j} - \frac{\partial \Lambda}{\partial t}\right) = \frac{\partial p}{\partial t} + (u_i\tau_{ij})_{,j} + (kT_{,j})_{,j} \tag{2-3}$$

where the shear and total enthalpy are defined as

$$\tau_{ij} \equiv (\mu' - \tfrac{2}{3}\mu)\delta_{ij}u_{k,k} + \mu(u_{i,j} + u_{j,i}) \tag{2-4}$$

$$I = c_pT + \tfrac{1}{2}u_iu_i + \Lambda - \tfrac{1}{2}(\omega R)^2 \tag{2-5}$$

with Λ being a body force potential and R denoting the perpendicular distance from the axis of rotation. These equations are written employing the usual summation convention and are applicable to nonsteady flows, acted on by external body force fields in a system which may or may not be rotating with constant angular velocity about some axis in space.

For the study of three-dimensional boundary layers it is convenient to put the appropriate equations into a form which *ab initio* admits the possibility that the surface and the streamlines of the external flow, may be curved in some arbitrary manner. Using an orthogonal system, which is not less general but simplifies matters considerably, we write the expression for a general element of length in the curvilinear coordinates ξ, η, and ζ (see Fig. C,2) as

$$(ds)^2 = h_1^2(d\xi)^2 + h_2^2(d\eta)^2 + h_3^2(d\zeta)^2 \tag{2-6}$$

In this system, the coordinates ξ and η are understood to lie and be defined in the surface over which the boundary layer is flowing, while ζ extends into the boundary layer. The length functions h_1, h_2, and h_3 are, in general, dependent on all three coordinates ξ, η, and ζ. However, as pointed

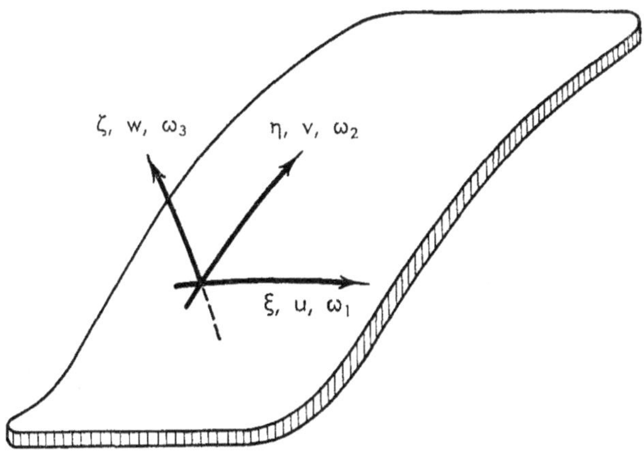

Fig. C,2. Orthogonal coordinates with corresponding velocities and angular rotations.

out in [3], when the surface is regular and not excessively curved in comparison to the boundary layer thickness, they may be considered to be functions of ξ and η alone. We shall see later that in certain problems this last simplification is not appropriate and the necessary changes in the form of the equations will then be pointed out.

One should note that if the coordinates ξ, η, and ζ are known functions of some Cartesian system x, y, z, then the length factors h_1, h_2, and h_3 are easily determinable by

$$\left(\frac{1}{h_1}\right)^2 = \left(\frac{\partial \xi}{\partial x}\right)^2 + \left(\frac{\partial \xi}{\partial y}\right)^2 + \left(\frac{\partial \xi}{\partial z}\right)^2$$
$$\left(\frac{1}{h_2}\right)^2 = \left(\frac{\partial \eta}{\partial x}\right)^2 + \left(\frac{\partial \eta}{\partial y}\right)^2 + \left(\frac{\partial \eta}{\partial z}\right)^2$$
$$\left(\frac{1}{h_3}\right)^2 = \left(\frac{\partial \zeta}{\partial x}\right)^2 + \left(\frac{\partial \zeta}{\partial y}\right)^2 + \left(\frac{\partial \zeta}{\partial z}\right)^2$$

Conversely, if x, y, z are known as functions of ξ, η, and ζ, then,

$$h_1^2 = \left(\frac{\partial x}{\partial \xi}\right)^2 + \left(\frac{\partial y}{\partial \xi}\right)^2 + \left(\frac{\partial z}{\partial \xi}\right)^2$$

$$h_2^2 = \left(\frac{\partial x}{\partial \eta}\right)^2 + \left(\frac{\partial y}{\partial \eta}\right)^2 + \left(\frac{\partial z}{\partial \eta}\right)^2$$

$$h_3^2 = \left(\frac{\partial x}{\partial \zeta}\right)^2 + \left(\frac{\partial y}{\partial \zeta}\right)^2 + \left(\frac{\partial z}{\partial \zeta}\right)^2$$

Applying the standard transformations to the conservation laws one obtains in this ξ, η, ζ system with corresponding velocities u, v, w the transformed form of the conservation equations. Then, by eliminating the terms of $O(\delta)$, or smaller, these are simplified to the boundary layer equations,

$$\frac{\partial \rho}{\partial t} + \frac{1}{h_1 h_2 h_3}\left[\frac{\partial}{\partial \xi}(h_2 h_3 \rho u) + \frac{\partial}{\partial \eta}(h_1 h_3 \rho v) + \frac{\partial}{\partial \zeta}(h_1 h_2 \rho w)\right] = 0 \tag{2-7}$$

$$\frac{\partial u}{\partial t} + \frac{u}{h_1}\frac{\partial u}{\partial \xi} + \frac{v}{h_2}\frac{\partial u}{\partial \eta} + \frac{w}{h_3}\frac{\partial u}{\partial \zeta} + \frac{uv}{h_1 h_2}\frac{\partial h_1}{\partial \eta} - \frac{v^2}{h_1 h_2}\frac{\partial h_2}{\partial \xi} + \frac{1}{h_1}\frac{\partial \Lambda}{\partial \xi} - 2\omega_3 v - \frac{1}{2}\frac{\omega^2}{h_1}\frac{\partial R^2}{\partial \xi} = -\frac{1}{\rho h_1}\frac{\partial p}{\partial \xi} + \frac{1}{\rho h_3}\frac{\partial}{\partial \zeta}\left(\frac{\mu}{h_3}\frac{\partial u}{\partial \zeta}\right) \tag{2-8}$$

$$\frac{\partial v}{\partial t} + \frac{u}{h_1}\frac{\partial v}{\partial \xi} + \frac{v}{h_2}\frac{\partial v}{\partial \eta} + \frac{w}{h_3}\frac{\partial v}{\partial \zeta} + \frac{uv}{h_1 h_2}\frac{\partial h_2}{\partial \xi} - \frac{u^2}{h_1 h_2}\frac{\partial h_1}{\partial \eta} + \frac{1}{h_2}\frac{\partial \Lambda}{\partial \eta} + 2\omega_3 u + \frac{1}{2}\frac{\omega^2}{h_2}\frac{\partial R^2}{\partial \eta} = -\frac{1}{\rho h_2}\frac{\partial p}{\partial \eta} + \frac{1}{\rho h_3}\frac{\partial}{\partial \zeta}\left(\frac{\mu}{h_3}\frac{\partial v}{\partial \zeta}\right) \tag{2-9}$$

$$\frac{1}{h_3}\frac{\partial \Lambda}{\partial \zeta} + 2(\omega_1 v - \omega_2 u) - \frac{1}{2}\frac{\omega^2}{h_3}\frac{\partial R^2}{\partial \zeta} = -\frac{1}{\rho h_3}\frac{\partial p}{\partial \zeta} \tag{2-10}$$

$$\frac{\partial H}{\partial t} + \frac{u}{h_1}\frac{\partial H}{\partial \xi} + \frac{v}{h_2}\frac{\partial H}{\partial \eta} + \frac{w}{h_3}\frac{\partial H}{\partial \zeta} = \frac{1}{\rho}\frac{\partial p}{\partial t} + \frac{1}{\rho h_3}\frac{\partial}{\partial \zeta}\left[\frac{\mu}{Pr h_3}\frac{\partial H}{\partial \zeta} - \frac{1}{2}\frac{\mu(1 - Pr)}{Pr h_3}\frac{\partial}{\partial \zeta}(u^2 + v^2)\right] \tag{2-11}$$

It should be noted that these equations contain Coriolis acceleration terms depending on the angular velocity ω_i, so as to permit their use when coordinates are fixed in a surface rotating about some axis in space, as, for instance, in the blading of turbomachinery. Time variations of angular velocity and of the body force potential have been neglected. Also, the kinetic energy of the motion in the direction normal to the wall was neglected in the definition of the modified total enthalpy H:

$$I = H + \Lambda - \tfrac{1}{2}(\omega R)^2; \; H \equiv c_p T + \tfrac{1}{2}(u^2 + v^2) \tag{2-12}$$

It is important to point out that the Coriolis accelerations or the normal derivative of the body force potential cause the pressure derivative in the direction normal to the surface to be of $O(1)$. Although this is larger than is usual for a boundary layer, the total change of the pressure in this direction is still small in such cases because of the small thickness of the layer, and may usually be neglected. As a consequence of this standard boundary layer approximation, the pressure gradient components may immediately be written as

$$\frac{\partial U}{\partial t} + \frac{U}{h_1}\frac{\partial U}{\partial \xi} + \frac{V}{h_2}\frac{\partial U}{\partial \eta} + \frac{UV}{h_1 h_2}\frac{\partial h_1}{\partial \eta} - \frac{V^2}{h_1 h_2}\frac{\partial h_2}{\partial \xi} + \frac{1}{h_1}\frac{\partial \Lambda_e}{\partial \xi} - 2\omega_3 V - \frac{1}{2}\frac{\omega^2}{h_1}\frac{\partial R^2}{\partial \xi} = -\frac{1}{\rho_e h_1}\frac{\partial p}{\partial \xi} \tag{2-13}$$

$$\frac{\partial V}{\partial t} + \frac{U}{h_1}\frac{\partial V}{\partial \xi} + \frac{V}{h_2}\frac{\partial V}{\partial \eta} + \frac{UV}{h_1 h_2}\frac{\partial h_2}{\partial \xi} - \frac{U^2}{h_1 h_2}\frac{\partial h_1}{\partial \eta} + \frac{1}{h_2}\frac{\partial \Lambda_e}{\partial \eta} + 2\omega_3 U - \frac{1}{2}\frac{\omega^2}{h_2}\frac{\partial R^2}{\partial \eta} = -\frac{1}{\rho_e h_2}\frac{\partial p}{\partial \eta} \tag{2-14}$$

where the capital letters and subscript $_e$ signify the values in the non-viscous external flow, and ω_3 is taken constant through the boundary layer.

Eq. 2-7, 2-8, 2-9, 2-10, and 2-11 should be solved subject to the conventional boundary conditions. This means that, in the absence of suction, one has at the surface ($\zeta = 0$) the no-slip condition:

$$u = v = w = 0 \qquad \text{and} \qquad T \text{ or } k\left(\frac{\partial T}{\partial \zeta}\right) \text{ is prescribed} \tag{2-15}$$

Similarly at the outer edge of the boundary layer ($\zeta \to \infty$) the velocity components and the temperature take on the external flow values,

$$u \to U(\xi, \eta), \qquad v \to V(\xi, \eta), \qquad T \to T_e(\xi, \eta) \tag{2-16}$$

C,3. Choice of Coordinates. The equations listed in Art. 2 are written employing a very general class of coordinates. The important issue which will now be discussed concerns the various alternatives that one has when choosing a particular coordinate system most suitable to the specific problem at hand. First, however, an additional simplification obviously applicable in boundary layer studies will be made. This simplification, suggested by Lin [*4*, pp. 103, 121], Howarth [*5*], and others, restricts the general orthogonal system defined in Eq. 2-6 by setting $h_3(\xi, \eta) = 1$, so that ζ represents an actual distance measured along a straight normal from the surface. As a result of this simplifying restriction, only the choice of the two remaining surface coordinates ξ and η needs to be made.

These two coordinates obviously should be so chosen that both the differential equations and the boundary conditions are in their simplest or most convenient forms. In this connection, the possibility of simplifying the differential equations will be considered first. Turning back, therefore, to the equations in Art. 2 one notes that their form is complicated not only because of the additional velocity component, but also because of the presence of the length functions h_1 and h_2 and their derivatives. It should now be noted that, while the selection of a particular orthogonal coordinate system can be made in a purely arbitrary manner, the length functions for the selected system are not arbitrary, and, in fact, insure that the surface geometry is complied with. This intimate connection with surface geometry becomes apparent when it is realized that the geodesic curvatures of the coordinate lines are given by

$$\kappa_1 = \frac{1}{h_1 h_2} \frac{\partial h_1}{\partial \eta}; \text{ for } \eta = \text{const} \tag{3-1}$$

$$\kappa_2 = \frac{1}{h_1 h_2} \frac{\partial h_2}{\partial \xi}; \text{ for } \xi = \text{const} \tag{3-2}$$

and, moreover, at any point on the surface the Gaussian curvature K (also known as the total curvature) may be expressed as

$$K = -\frac{1}{h_1 h_2}\left[\frac{\partial}{\partial \xi}(h_2 \kappa_2) + \frac{\partial}{\partial \eta}(h_1 \kappa_1)\right] \tag{3-3}$$

Regardless of the choice of the coordinate system, at a given point this total curvature must always be the same.

Keeping in mind all that has been said above, one now inquires under what conditions the length functions can be made to disappear from the three-dimensional boundary layer equations, and also for what coordinate systems this will occur. One case, that of Cartesian coordinates on a plane, is obvious and sufficiently elementary not to require any further discussion. The question nevertheless remains: For what other surfaces is this possible? To answer this, it will be recalled that a large class of surfaces has the property of being developable. Such surfaces are easily recognized, for they can be rolled out, without being stretched, into a plane after suitable cuts have been made. In differential geometry, this is equivalent to saying that their total curvature K is zero everywhere. Suppose now that one has a developable surface and rolls it out into a plane. In this position a Cartesian coordinate system $\xi = x$, $\eta = y$ may be defined and then the surface may be rolled back into its original shape. Since all these operations occur without stretching, the defined coordinates remain orthogonal and, moreover, the length functions remain unchanged (e.g. $h_1 = h_2 = 1$), thus disappearing from three-dimensional boundary layer equations. Consequently, for any developable surface it is possible to write the equations in a Cartesian form. It should be mentioned here that this last result has also been demonstrated by Sedney [6], who noted that the subtensor character of the boundary layer equations implies an invariance to certain types of transformations.

In practice, of course, it is desirable to know, without going through the mechanical process of unrolling, which set of coordinates in the developable surface is the one that may be given the Cartesian form. To deal with this question, we shall first digress a bit and recall from differential geometry that a geodesic is a surface curve for which the geodesic curvature is zero. Geodesics may easily be visualized by considering the line of contact of a piece of thread stretched over the surface, with constant tension, between two of its points. Now, an important property of developable surfaces is that the geodesics on such a surface become straight lines when the surface is rolled out into a plane. Therefore, it becomes fairly obvious that the desired coordinate system must consist of orthogonal geodesics.

It should be stressed at this point that, while the Cartesian form of the equations results in great simplifications, the complications which may thus be introduced in the boundary conditions may make its choice inconvenient. A good example of this is provided by the supersonic flow over a cone. As is well known, the outer flow in this case is conical (VI,H,14); that is, all the quantities (e.g., pressure and velocities) remain constant along the rays originating at the vertex of the cone. From the boundary-layer standpoint, this results in extremely simple boundary conditions when the cone generators serve for one family of the surface coordinates. On the other hand, the cone is also a developable surface, so that the boundary layer equations may be written in the Cartesian form, provided the chosen coordinates are orthogonal geodesics. Unfortunately, however, in general, such coordinates cannot be made to lie along the cone generators, so that when the equations are written in the Cartesian form the advantage of the very simple boundary conditions corresponding to the conical nature of the flow is lost.

The system of coordinates which are orthogonal geodesics can be formulated in general only in surfaces of zero total curvature. Consequently for surfaces which are not developable both length functions cannot be made to vanish from the boundary layer equations. However, following the well-known theorem of Gauss [7, p. 174]: "Given any curve C upon a surface and the geodesics orthogonal to C; when equal lengths are measured from C along these geodesics, the locus of their end points is an orthogonal trajectory of the geodesics," one can create in any surface the curves known as geodesic parallels to curve C. Thus if an orthogonal coordinate system is chosen so as to consist of the geodesics ξ, and the geodesic parallels η, then, from the definition of the geodesic parallel, the length function for the coordinates which are the geodesics becomes unity and consequently disappears from the three-dimensional boundary layer equations; that is,

$$(ds)^2 = (d\xi)^2 + h_2^2(d\eta)^2 + (d\zeta)^2 \tag{3-4}$$

This advantageous choice of coordinates has apparently been noted first by Moore [*8*] who, in some of his later work [*2*], suggests that three-dimensional boundary layer equations be written in general in this system, that is, employing only one of the length functions.

A good example of where such a system is obviously convenient is provided by the aforementioned flow, over the cone for which the geodesic coordinates are now taken to be the cone generators. The geodesic parallels are the circles swept by the meridional angle, and the corresponding length function is the local radius of the body. Similarly convenient is the application to bodies of revolution and to surfaces having conical symmetry. However, it should also be noted that the application of this system in a general surface that is void of any special properties, while possible, does not result in particularly advantageous boundary conditions, which in turn might sometimes make the solution of the equations rather difficult. Consequently, while the system suggested by Moore is probably the best from the standpoint of simplifying the differential equations, one should also consider the advantages which may be gained by choosing the coordinates so as to simplify the boundary conditions.

One such system, known as the intrinsic coordinate system, has been suggested by Hayes [*3*] and also by Timman [*9*]. The essence of the intrinsic system lies in choosing the surface projections of the outer flow streamlines as one family of the coordinate lines. The other coordinates thus become the orthogonal trajectories of the outer flow streamlines, which, for irrotational flow, are the outer flow potential lines. Because of this choice, the boundary conditions are considerably simplified since the outer flow velocity component in the direction normal to the streamlines vanishes in the whole flow field. This allows one, by the use of Eq. 2-13, 2-14, and 3-1, to express the pressure gradient components for steady flow as

$$-\frac{1}{\rho_e h_1}\frac{\partial p}{\partial \xi} = \frac{1}{h_1}\frac{\partial}{\partial \xi}\left(\frac{U^2}{2} + \Lambda_e\right) - \frac{1}{2}\frac{\omega^2}{h_1}\frac{\partial R^2}{\partial \xi} \tag{3-5}$$

$$-\frac{1}{\rho_e h_2}\frac{\partial p}{\partial \eta} = -U^2\kappa_1 + 2\omega_3 U + \frac{1}{h_2}\frac{\partial \Lambda_e}{\partial \eta} - \frac{1}{2}\frac{\omega^2}{h_2}\frac{\partial R^2}{\partial \eta} \tag{3-6}$$

where the ξ coordinate is taken along the outer flow streamlines. One should note that although the velocity V has thus been eliminated, the corresponding component v nevertheless persists in the boundary layer and is, in fact, known as the secondary flow. However, when $\Lambda_e = \Lambda_e(\xi)$, $\omega = 0$, and $\kappa_1 = 0$, Eq. 2-9 and the boundary conditions (Eq. 2-15) indicate that $v = 0$ is a solution. Consequently, in the absence of lateral body force and for nonrotating coordinates, if the outer flow streamlines are geodesics of the surface, then the secondary flow vanishes. Of course, if the deviations from these requirements are small, one may expect that the secondary flow v will also be small when compared to the principal flow u. This indicates the possibility of a "small secondary flow" approximation in which products of small quantities may be neglected, thus simplifying certain three-dimensional boundary layer problems.

Aside from these advantages, the use of an intrinsic coordinate system permits a relatively simple determination of the length functions h_1, and h_2, if the outer flow is irrotational and its potential Φ is known. When the coordinate ξ is measured along streamlines of the outer flow, so that the distances can be expressed in terms of the potential, then from the definition of a potential (e.g. $\Phi_x^2 + \Phi_y^2 + \Phi_z^2 \equiv U^2$) and from the expression for $1/h_1^2$ given in Art. 2, it is immediately apparent that since $\xi = \Phi$, then $1/h_1 = U$. Furthermore, the coordinate η must now be measured along potential lines, so that distances along it can be expressed in terms of mass flow, that is, in terms of the surface stream function of the external flow (i.e., $\eta = \Psi$). Under these conditions one has for distance along a potential line $ds = h_2 d\Psi$, but $d\Psi = \rho_e U L ds$, and therefore $h_2 = 1/(\rho_e U L)$. For such a system, then, Eq. 2-6 becomes

$$(ds)^2 = \left(\frac{d\Phi}{U}\right)^2 + \left(\frac{d\Psi}{\rho_e U L}\right)^2 \tag{3-7}$$

The function $L(\Phi)$ used in these expressions describes the characteristic width of an elemental stream tube, and consequently must be obtained from the continuity equation. In particular, as $\zeta = \delta$ where $V = W = 0$, one obtains

$$\frac{1}{h_1 h_2}\frac{\partial}{\partial \Phi}(h_2 \rho_e U) + \frac{\partial}{\partial \zeta}(\rho_e W) = \rho_e U^2 L \frac{\partial}{\partial \Phi}\left(\frac{1}{L}\right) + \frac{\partial}{\partial \zeta}(\rho_e W) = 0$$

which leads to the equation determining $L(\Phi)$,

$$\frac{d \ln L}{d\Phi} = \frac{1}{\rho_e U^2}\left[\frac{\partial(\rho_e W)}{\partial \zeta}\right]_{\zeta=\delta} \tag{3-8}$$

Similar results were obtained from somewhat different considerations by Timman [*9*] (in that reference $T \equiv U^2$, and their ρ is our $(\rho_e L)^2$).

In certain special cases there exists a possibility of combining the advantages of the geodesic parallel and the intrinsic coordinate systems. An example of such an application is given in [*10*] where the geodesic parallels are also the streamlines of the outer flow. In a general case, a system of this type cannot be formulated because only one of the outer flow streamlines can be made to coincide with the geodesic parallel; however, sometimes for the purpose of an approximate solution in the immediate neighborhood of a chosen streamline, such an attack on the problem [*11*] can also result in substantial simplifications.

The discussion in this article is intended to emphasize that, before tackling any three-dimensional boundary layer problem, the investigator should give serious considerations to the particular surface geometry and the special choice of coordinate system most suitable in his case. Although the "best choice" applicable to all problems does not exist, often considerable simplification of the differential equations and the boundary conditions can be obtained by using one or more of the schemes outlined above. Also, in many cases, when the surface under consideration is developable (or nearly so), the results obtained over a surface of seemingly different geometry can be directly applied if the boundary conditions (e.g. outer flow) and the location of the leading edge correspond.

C,4. Two-Component Vector Potential. In steady, compressible, two-dimensional boundary layer problems, the number of dependent variables appearing in the differential equations may be conveniently reduced by the introduction of the well-known Stokes stream function (B,2). This stream function is defined so as to satisfy identically the conservation of mass; thus one of the equations is eliminated from further consideration. The useful concept of the Stokes stream function, aside from mathematical simplification, also has an important physical meaning, since the lines on which this function is a constant are streamlines of the flow, and thus are a measure of the mass flow passing through any given control plane.

The conventionally defined Stokes stream function can also be used in certain special three-dimensional boundary layer problems, that is, specifically, when the product $\rho h_1 h_3 v$ is *not* a function of the η coordinate so that the term involving it in Eq. 2-7 vanishes. In general, however, the two-dimensional definition does not allow for the additional velocity component present in three-dimensional boundary layer problems and thus it may not be used. Actually, the Stokes stream function can be shown (B,2) to be a special case of two, more general, scalar functions, say S and N, which form a vector potential (S grad N) defined so that its curl equals the mass flow vector. This use of functions S and N again eliminates one of the dependent variables and thus, in principle, leads to a simplification of the equations. In practice, however, such definition requires that each mass flow component be expressed in terms of the derivatives of both of the functions S and N. For this reason the above procedure is not as convenient for boundary layer investigations as the one suggested by Moore [*8*], which is based on the use of a two-component vector potential. The two scalar functions (say ψ and φ) are now defined in terms of the mass flow components by writing

$$\rho h_2 h_3 u \equiv \rho_0 \psi_\zeta; \qquad \rho h_1 h_3 v \equiv \rho_0 \varphi_\zeta; \qquad \rho h_1 h_2 w = -\rho_0(\psi_\xi + \varphi_\eta) \quad (4\text{-}1)$$

so that the steady flow continuity (Eq. 2-7) is again identically satisfied, and one of the dependent variables is eliminated from the equations. It should be noted that Eq. 4-1 do not represent the only way in which two functions can be defined so as to satisfy the steady state form of Eq. 2-7 identically. However, as pointed out by Moore, this particular arrangement reduces for two-dimensional flow to the conventional definition of the Stokes stream function, and it also provides a certain symmetry of ψ and ξ against φ and η.

The existence of the functions ψ and φ follows from a well-known theorem of vector analysis which states that any continuously differentiable vector having a vanishing divergence may be expressed as the curl of a vector potential. As stated by the equation for the conservation of mass (Eq. 2-7), the divergence of the mass flow vector vanishes, so that it follows that a vector potential exists such that its curl equals the mass flow vector. This definition of the vector potential in terms of its curl has been shown in [*8*] and B,2 to be sufficient for the specification of the two functions ψ and φ, and this specification in terms of the vector potential components has been shown in turn to be equivalent to Eq. 4-1.

It should be noted that in general the functions S and N, whose constant values are true stream surfaces, are not the same as ψ and φ. This is clearly apparent when incompressible flow with $h_1 = h_2 = h_3 = 1$ is considered. One then has:

$$u = \psi_z = S_y N_z - N_y S_z \quad (4\text{-}2)$$

$$v = \varphi_z = S_z N_x - S_x N_z \quad (4\text{-}3)$$

$$w = -(\psi_x + \varphi_y) = S_x N_y - S_y N_x \quad (4\text{-}4)$$

and

$$\psi = \int (S_y N_z - N_y S_z) dz + f_1(x, y) \quad (4\text{-}5)$$

$$\varphi = \int (S_z N_x - S_x N_z) dz + f_2(x, y) \quad (4\text{-}6)$$

where, from Eq. 4-4, f_1 and f_2 must satisfy:

$$f_{1x} + f_{2y} = 0 \quad (4\text{-}7)$$

and both can be taken identically zero if the integrations are carried from the wall. ψ and φ are thus seen to be quite different from S and N and the intersections of ψ-const and φ-const do not represent streamlines. This lack of direct physical meaning is, however, greatly outweighed by the comparative simplicity which the so-defined scalar functions bring into the description of the velocity components, as is clearly apparent from Eq. 4-2, 4-3, and 4-4.

C,5. Integral Equations. In boundary layer investigations it is often convenient to obtain the first integral of the momentum and energy equations (see B,23). For two-dimensional flow, equations written in this form lend themselves to the approximate but powerful Kármán-Pohlhausen technique (B,31) which permits a relatively easy treatment of complicated boundary layer problems. Furthermore, when placed in this form, the equations reveal the existence of certain thicknesses representing the momentum and mass flow deficiencies which provide an easily visualized characterization of boundary layer effects.

In three-dimensional flow one may expect similar advantages from the integral forms of Eq. 2-8, 2-9, and 2-11. These integral forms have been obtained for various coordinate systems in [*12,3,9,11*]. Although there are a number of ways by which such integral forms may be derived, it is perhaps most direct to integrate the equations with respect to ζ from the body surface to some fixed distance d, and then eliminate by use of the continuity equation (Eq. 2-7) the velocity w normal to the surface. The pressure-gradient terms are, of course, known and given by Eq. 2-13 and 2-14. If one takes for simplicity's sake $h_3 = 1$ and $\omega_3 = \Lambda = 0$, this procedure applied to Eq. 2-8 and 2-9 results in the momentum-integral equations:

$$\begin{aligned}\frac{1}{h_1}\frac{\partial\theta_{11}}{\partial\xi} + \frac{1}{h_2}\frac{\partial\theta_{12}}{\partial\eta} + \theta_{11}F_1 + \theta_{12}F_2 - \kappa_2(\theta_{11}+\theta_{22}) + \frac{\delta_1}{h_1Q}\frac{\partial U}{\partial\xi}\\ + \frac{\delta_2}{Q}\left(\frac{1}{h_1}\frac{\partial V}{\partial\xi} + \operatorname{curl}_3 Q\right) + \frac{1}{\rho_eQ^2}\left\{\frac{\partial}{\partial t}\left[\int_0^d (U-u)\rho d\zeta\right]\right.\\ \left. + \frac{\partial U}{\partial t}\int_0^d(\rho_e-\rho)d\zeta\right\} - \frac{\tau_{1w}}{\rho_eQ^2} = 0 \quad (5\text{-}1)\end{aligned}$$

$$\begin{aligned}\frac{1}{h_1}\frac{\partial\theta_{21}}{\partial\xi} + \frac{1}{h_2}\frac{\partial\theta_{22}}{\partial\eta} + \theta_{21}F_1 + \theta_{22}F_2 - \kappa_1(\theta_{11}+\theta_{22}) + \frac{\delta_2}{h_2Q}\frac{\partial V}{\partial\eta}\\ + \frac{\delta_1}{Q}\left(\frac{1}{h_2}\frac{\partial U}{\partial\eta} + \operatorname{curl}_3 Q\right) + \frac{1}{\rho_eQ^2}\left\{\frac{\partial}{\partial t}\left[\int_0^d (V-v)\rho d\zeta\right]\right.\\ \left. + \frac{\partial V}{\partial t}\int_0^d(\rho_e-\rho)d\zeta\right\} - \frac{\tau_{2w}}{\rho_eQ^2} = 0 \quad (5\text{-}2)\end{aligned}$$

In these equations, τ_{1w} and τ_{2w} represent the surface shear components in the ξ and η directions respectively. Also, the following definitions have been used for the displacement thicknesses,

$$\delta_1 \equiv \int_0^d\left(\frac{U}{Q} - \frac{\rho u}{\rho_eQ}\right)d\zeta; \qquad \delta_2 \equiv \int_0^d\left(\frac{V}{Q} - \frac{\rho v}{\rho_eQ}\right)d\zeta \quad (5\text{-}3)$$

and the momentum thicknesses,

$$\left.\begin{aligned}\theta_{11} &\equiv \int_0^d \left(\frac{U}{Q} - \frac{u}{Q}\right)\frac{\rho u}{\rho_e Q}\,d\zeta & \theta_{21} &\equiv \int_0^d \left(\frac{V}{Q} - \frac{v}{Q}\right)\frac{\rho u}{\rho_e Q}\,d\zeta \\ \theta_{12} &\equiv \int_0^d \left(\frac{U}{Q} - \frac{u}{Q}\right)\frac{\rho v}{\rho_e Q}\,d\zeta & \theta_{22} &\equiv \int_0^d \left(\frac{V}{Q} - \frac{v}{Q}\right)\frac{\rho v}{\rho_e Q}\,d\zeta\end{aligned}\right\} \tag{5-4}$$

Furthermore the outer flow functions F_1, F_2, and $\text{curl}_3\, Q$ are given by

$$F_1 \equiv \frac{1}{h_1}\frac{\partial \ln (\rho_e Q^2 h_2^2)}{\partial \xi} = \frac{1}{h_1}\left(\frac{1}{\rho_e}\frac{\partial \rho_e}{\partial \xi} + \frac{2}{Q}\frac{\partial Q}{\partial \xi}\right) + 2\kappa_2 \tag{5-5}$$

$$F_2 \equiv \frac{1}{h_2}\frac{\partial \ln (\rho_e Q^2 h_1^2)}{\partial \eta} = \frac{1}{h_2}\left(\frac{1}{\rho_e}\frac{\partial \rho_e}{\partial \eta} + \frac{2}{Q}\frac{\partial Q}{\partial \eta}\right) + 2\kappa_1 \tag{5-6}$$

$$\text{curl}_3\, Q = \frac{1}{h_1 h_2}\left[\frac{\partial}{\partial \xi}(h_2 V) - \frac{\partial}{\partial \eta}(h_1 U)\right] \tag{5-7}$$

As can be seen from Eq. 5-3 and 5-4, the three-dimensional flow reveals the vectorlike nature of the displacement thickness and the tensorlike character of the momentum thickness. Because of this, the easy way in which the two-dimensional boundary layer can be visualized in terms of these quantities is lost when a third dimension is added. Nevertheless the simplification achieved by the integration of the momentum equations is sufficient to provide a basis for approximate methods of three-dimensional boundary layer calculation. It should be noted that Eq. 5-3 and 5-4 are not the only ways in which the various momentum and displacement thicknesses can be defined. One may also, for instance, normalize the various thicknesses in terms of the velocity U or V or both. However, the particular form used here leads to an inherent symmetry of Eq. 5-1 and 5-2, and it is also convenient because it is easily adaptable to cases where U or V vanishes; that is, to an intrinsic coordinate system.

It should be noted that for steady and irrotational outer flow, Eq. 5-1 and 5-2 reduce to

$$\frac{1}{h_1}\frac{\partial \theta_{11}}{\partial \xi} + \frac{1}{h_2}\frac{\partial \theta_{12}}{\partial \eta} + \theta_{11}F_1 + \theta_{12}F_2 - \kappa_2(\theta_{11} + \theta_{22}) + \frac{\delta_1}{h_1 Q}\frac{\partial U}{\partial \xi} + \frac{\delta_2}{h_1 Q}\frac{\partial V}{\partial \xi} - \frac{\tau_{1w}}{\rho_e Q^2} = 0 \tag{5-8}$$

$$\frac{1}{h_1}\frac{\partial \theta_{21}}{\partial \xi} + \frac{1}{h_2}\frac{\partial \theta_{22}}{\partial \eta} + \theta_{21}F_1 + \theta_{22}F_2 - \kappa_1(\theta_{11} + \theta_{22}) + \frac{\delta_1}{h_2 Q}\frac{\partial U}{\partial \eta} + \frac{\delta_2}{h_2 Q}\frac{\partial V}{\partial \eta} - \frac{\tau_{2w}}{\rho_e Q^2} = 0 \tag{5-9}$$

which are quite obviously related to the two-dimensional Kármán momentum-integral relation in the absence of cross flow components and curvatures of coordinates.

Application of the foregoing procedure to Eq. 2-11 yields after some manipulation the energy-integral equation:

$$\frac{1}{h_1}\frac{\partial \Theta_1}{\partial \xi} + \frac{1}{h_2}\frac{\partial \Theta_2}{\partial \eta} + \Theta_1 F_3 + \Theta_2 F_4 + \frac{\delta_1}{h_1}\frac{\partial \ln H_e}{\partial \xi} + \frac{\delta_2}{h_2}\frac{\partial \ln H_e}{\partial \eta}$$
$$+ \frac{1}{\rho_e Q H_e}\left\{\frac{\partial}{\partial t}\left[\int_0^d (H_e - H)\rho d\zeta\right] + \frac{\partial H_e}{\partial t}\int_0^d (\rho_e - \rho)d\zeta\right\} - C_H = 0 \tag{5-10}$$

In this equation C_H represents a coefficient characterizing the heat transfer from the fluid to the surface,

$$C_H \equiv \frac{\left[k\dfrac{\partial T}{\partial \zeta}\right]_{\zeta=0}}{\rho_e Q H_e} \tag{5-11}$$

The total enthalpy deficiency thicknesses are defined as

$$\Theta_1 \equiv \int_0^d \left(1 - \frac{H}{H_e}\right)\frac{\rho u}{\rho_e Q}d\zeta; \qquad \Theta_2 \equiv \int_0^d \left(1 - \frac{H}{H_e}\right)\frac{\rho v}{\rho_e Q}d\zeta \tag{5-12}$$

and the outer flow functions F_3 and F_4 are given by

$$F_3 = \frac{1}{h_1}\frac{\partial \ln (\rho_e Q H_e h_2)}{\partial \xi} \tag{5-13}$$

$$F_4 = \frac{1}{h_2}\frac{\partial \ln (\rho_e Q H_e h_1)}{\partial \eta} \tag{5-14}$$

For steady and isoenergetic outer flow, (Eq. 5-10) simplifies to

$$\frac{1}{h_1}\frac{\partial \Theta_1}{\partial \xi} + \frac{1}{h_2}\frac{\partial \Theta_2}{\partial \eta} + \Theta_1 F_3 + \Theta_2 F_4 - C_H = 0 \tag{5-15}$$

C,6. Correlated Flows.

TRANSFORMATION TO INCOMPRESSIBLE FORM. For two-dimensional steady flow the computation of compressible boundary layer development is greatly facilitated by introduction of a transformation suggested by Illingworth [*13*] and Stewartson [*14*] which establishes the correlated incompressible flow (B,22). The key idea of this transformation lies in the stretching of the actual physical system into an imaginary one, in which the properties of the fluid are evaluated at certain reference conditions, but the mass flow is unchanged. The correlation of the two-dimensional compressible and incompressible flow is accomplished, in the strict sense, only when the total enthalpy is constant in the boundary layer; i.e. when there is no heat transfer between the wall and the fluid, and when the Prandtl number is unity. However, even in cases when these restrictions do not hold exactly, the use of the transformation eliminates density variation from most of the terms and thus simplifies considerably the solution of the problem.

The essential ideas of the two-dimensional correlation may also be applied to three-dimensional flow. If the Prandtl number is unity, as Hayes [*3*] and Moore [*8*] first observed, one notes that the solution of Eq. 2-11 for steady flow, which satisfies the zero heat transfer requirement, is again given by the constant total enthalpy:

$$H = C_pT + \tfrac{1}{2}(u^2 + v^2) = H_e \tag{6-1}$$

We now seek to define a new coordinate system in which the boundary layer equations are reduced to incompressible forms. This process can be carried out in a relatively easy manner by taking advantage of the previously introduced length functions. In what follows, we consider steady flow and set $h_3 = 1$, $\omega = \Lambda = 0$, although these last restrictions, introduced for the sake of simplicity, are not essential. The transformation is defined by

$$\left.\begin{aligned} &\bar{\xi} = \xi; \qquad \bar{\eta} = \eta; \qquad \bar{\zeta} = A(\xi, \eta)\int_0^{\zeta}\frac{\rho}{\rho_0}\,d\zeta \\ &\bar{h}_1 = \frac{h_1}{L_1}; \qquad \bar{h}_2 = \frac{h_2}{L_2} \\ &\bar{\psi} = \psi; \qquad \bar{\varphi} = \varphi \end{aligned}\right\} \tag{6-2}$$

where the barred quantities pertain to the transformed system and A, L_1, L_2 are functions of ξ and η only and are, as yet, undetermined. It should be noted at this point that, unless $L_1 = L_2 = 1$, the transformed system must necessarily correspond to a distorted, imaginary surface whose geodesic curvature and total curvature are defined by the new functions $\bar{h}_1$, and $\bar{h}_2$. If now the components of the vector potential are defined in the incompressible manner:

$$\bar{\psi}_{\bar{\zeta}} = \bar{h}_2\bar{u}; \qquad \bar{\varphi}_{\bar{\zeta}} = \bar{h}_1\bar{v}; \; -(\bar{\psi}_{\bar{\xi}} + \bar{\varphi}_{\bar{\eta}}) = \bar{h}_1\bar{h}_2\bar{w} \tag{6-3}$$

then Eq. 6-2 and 4-1 lead to the relations between the velocities in both systems,

$$u = \frac{A}{L_2}\,\bar{u}; \qquad v = \frac{A}{L_1}\,\bar{v}; \qquad \rho w = \frac{\rho_0}{L_1L_2}\left[\bar{w} - \frac{\bar{u}}{\bar{h}_1}\frac{\partial\bar{\zeta}}{\partial\xi} - \frac{\bar{v}}{\bar{h}_2}\frac{\partial\bar{\zeta}}{\partial\eta}\right] \tag{6-4}$$

which reduce the continuity relation (Eq. 2-7) to its incompressible form,

$$\frac{\partial}{\partial\bar{\xi}}\,(\bar{h}_2\bar{u}) + \frac{\partial}{\partial\bar{\eta}}\,(\bar{h}_1\bar{v}) + \frac{\partial}{\partial\bar{\zeta}}\,(\bar{h}_1\bar{h}_2\bar{w}) = 0 \tag{6-5}$$

Furthermore, using Eq. 6-2, 6-4, 2-13, and 2-14, the momentum relations (Eq. 2-8 and 2-9) become, after some manipulation,

$$\frac{\bar{u}}{\bar{h}_1}\frac{\partial \bar{u}}{\partial \bar{\xi}} + \frac{\bar{v}}{\bar{h}_2}\frac{\partial \bar{u}}{\partial \bar{\eta}} + \bar{w}\frac{\partial \bar{u}}{\partial \bar{\zeta}} + \frac{\bar{v}\bar{u}}{\bar{h}_2}\left\{\frac{\partial \ln [(L_1/L_2)A\bar{h}_1]}{\partial \bar{\eta}}\right\} - \frac{\bar{v}^2}{\bar{h}_1}\left(\frac{L_2}{L_1}\right)^2\left[\frac{\partial \ln (A\bar{h}_2)}{\partial \bar{\xi}}\right]$$
$$= F_1\left\{\frac{\bar{U}}{\bar{h}_1}\frac{\partial \bar{U}}{\partial \bar{\xi}} + \frac{\bar{V}}{\bar{h}_2}\frac{\partial \bar{U}}{\partial \bar{\eta}} + \frac{\bar{V}\bar{U}}{\bar{h}_2}\left\{\frac{\partial \ln [(L_1/L_2)A\bar{h}_1]}{\partial \bar{\eta}}\right\} - \frac{\bar{V}^2}{\bar{h}_1}\left(\frac{L_2}{L_1}\right)^2\left[\frac{\partial \ln (A\bar{h}_2)}{\partial \bar{\xi}}\right]\right\} + (L_2L_1AB)\nu_0\frac{\partial^2 \bar{u}}{\partial \bar{\zeta}^2} \quad (6\text{-}6)$$

$$\frac{\bar{u}}{\bar{h}_1}\frac{\partial \bar{v}}{\partial \bar{\xi}} + \frac{\bar{v}}{\bar{h}_2}\frac{\partial \bar{v}}{\partial \bar{\eta}} + \bar{w}\frac{\partial \bar{v}}{\partial \bar{\zeta}} + \frac{\bar{v}\bar{u}}{\bar{h}_1}\left\{\frac{\partial \ln [(L_2/L_1)A\bar{h}_2]}{\partial \bar{\xi}}\right\} - \frac{\bar{u}^2}{\bar{h}_2}\left(\frac{L_1}{L_2}\right)^2\left[\frac{\partial \ln (A\bar{h}_1)}{\partial \bar{\eta}}\right]$$
$$= F_2\left\{\frac{\bar{U}}{\bar{h}_1}\frac{\partial \bar{V}}{\partial \bar{\xi}} + \frac{\bar{V}}{\bar{h}_2}\frac{\partial \bar{V}}{\partial \bar{\eta}} + \frac{\bar{V}\bar{U}}{\bar{h}_1}\left\{\frac{\partial \ln [(L_2/L_1)A\bar{h}_2]}{\partial \bar{\xi}}\right\} - \frac{\bar{U}^2}{\bar{h}_2}\left(\frac{L_1}{L_2}\right)^2\left[\frac{\partial \ln (A\bar{h}_1)}{\partial \bar{\eta}}\right]\right\} + (L_2L_1AB)\nu_0\frac{\partial^2 \bar{v}}{\partial \bar{\zeta}^2} \quad (6\text{-}7)$$

where

$$B \equiv \frac{\rho\mu}{\rho_0\mu_0} \cong \frac{\rho_e\mu_e}{\rho_0\mu_0} \quad (6\text{-}8)$$

$$F_1 \equiv \frac{\left[\frac{1}{\rho} + \frac{\partial \ln (A/L_2)}{\partial p}(u^2 + v^2)\right]}{\left[\frac{1}{\rho_e} + \frac{\partial \ln (A/L_2)}{\partial p}(U^2 + V^2)\right]} \quad (6\text{-}9)$$

and

$$F_2 \equiv \frac{\left[\frac{1}{\rho} + \frac{\partial \ln (A/L_1)}{\partial p}(u^2 + v^2)\right]}{\left[\frac{1}{\rho_e} + \frac{\partial \ln (A/L_1)}{\partial p}(U^2 + V^2)\right]} \quad (6\text{-}10)$$

One now has the freedom of choosing the functions L_1, L_2, and A so as to bring Eq. 6-6 into convenient, preferably incompressible form. A number of different alternatives, each having its own advantages, is possible and we shall discuss these separately.

Case 1. $L_1 = L_2 = 1$

This case, which retains the original surface shape, has been discussed in [3]. Eq. 6-6 and 6-7 can be put into a nearly incompressible form if A is defined by

$$\frac{dA}{A} = \frac{\mathcal{R}}{2c_p}\frac{dp}{p} = \frac{\gamma - 1}{2\gamma}\frac{dp}{p} \quad (6\text{-}11)$$

so that, because of the constant total enthalpy equation (Eq. 6-1), one obtains $F_1 = F_2 = 1$. The solution of Eq. 6-11, with the requirement that

$A = 1$ at the reference condition, gives

$$A = \left(\frac{p}{p_0}\right)^{\frac{\gamma-1}{2\gamma}} = \frac{a_e}{a_0} = \left(\frac{T_e}{T_0}\right)^{\frac{1}{2}} \tag{6-12}$$

and thus

$$B = \frac{\mu_e}{\mu_0} A^{\frac{2}{\gamma-1}} \tag{6-13}$$

With these relations, Eq. 6-6 and 6-7 can be written as

$$\frac{\bar{u}}{h_1}\frac{\partial \bar{u}}{\partial \xi} + \frac{\bar{v}}{h_2}\frac{\partial \bar{u}}{\partial \eta} + \bar{w}\frac{\partial \bar{u}}{\partial \bar{\zeta}} + \frac{\bar{v}\bar{u}}{h_2}\left[\frac{\partial \ln (Ah_1)}{\partial \eta}\right] - \frac{\bar{v}^2}{h_1}\left[\frac{\partial \ln (Ah_2)}{\partial \xi}\right] = \frac{\bar{U}}{h_1}\frac{\partial \bar{U}}{\partial \xi}$$
$$+ \frac{\bar{V}}{h_2}\frac{\partial \bar{U}}{\partial \eta} + \frac{\bar{V}\bar{U}}{h_2}\left[\frac{\partial \ln (Ah_1)}{\partial \eta}\right] - \frac{\bar{V}^2}{h_1}\left[\frac{\partial \ln (Ah_2)}{\partial \xi}\right] + \left(\frac{\mu_e}{\mu_0} A^{\frac{\gamma+1}{\gamma-1}}\right)\nu_0 \frac{\partial^2 \bar{u}}{\partial \bar{\zeta}^2} \tag{6-14}$$

$$\frac{\bar{u}}{h_1}\frac{\partial \bar{v}}{\partial \xi} + \frac{\bar{v}}{h_2}\frac{\partial \bar{v}}{\partial \eta} + \bar{w}\frac{\partial \bar{v}}{\partial \bar{\zeta}} + \frac{\bar{v}\bar{u}}{h_1}\left[\frac{\partial \ln (Ah_2)}{\partial \xi}\right] - \frac{\bar{u}^2}{h_2}\left[\frac{\partial \ln (Ah_1)}{\partial \eta}\right] = \frac{\bar{U}}{h_1}\frac{\partial \bar{V}}{\partial \xi}$$
$$+ \frac{\bar{V}}{h_2}\frac{\partial \bar{V}}{\partial \eta} + \frac{\bar{V}\bar{U}}{h_1}\left[\frac{\partial \ln (Ah_2)}{\partial \xi}\right] - \frac{\bar{U}^2}{h_2}\left[\frac{\partial \ln (Ah_1)}{\partial \eta}\right] + \left(\frac{\mu_e}{\mu_0} A^{\frac{\gamma-1}{\gamma+1}}\right)\nu_0 \frac{\partial^2 \bar{v}}{\partial \bar{\zeta}^2} \tag{6-15}$$

It may be seen from Eq. 6-14 and 6-15 that this form differs from the incompressible one because the viscous terms are now multiplied by a function of ξ and η, and also because the curvature terms are changed. The correlation is thus incomplete. On the other hand, the equations retain the original surface coordinates so that the change from the transformed to the physical system can be accomplished with ease.

Case 2. $L_2 = 1, L_1 = 1/(AB)$

In this case only one of the coordinates is distorted. Consequently we may expect that this type of modification would be most suitable when the undistorted coordinate is unimportant, that is, when the variations with respect to this coordinate vanish from the equations. For such a three-dimensional flow, Eq. 6-6 and 6-7 can be put into the incompressible form by choosing A so that the function F_1, multiplying the nonvanishing outer flow terms, becomes unity. Since $L_2 = 1$, this results in values of A and B the same as in Case 1, that is, given by Eq. 6-12 and 6-13 respectively. Furthermore, since all the variations with respect to the coordinate η are assumed to vanish, it is convenient to define a new coordinate ξ^*, such that

$$\xi^* \equiv \int_0^\xi \bar{h}_1 d\xi = \int_0^\xi ABh_1 d\xi \tag{6-16}$$

and write the equations using the original velocity $v = A^2B\bar{v}$. For Cartesian coordinates ($h_2 = h_1 = 1$), these changes result in a transformation identical to that of Stewartson [*14*]. In this form, also, the transformation has been used in [*15*] to treat the compressible boundary layer flow over a yawed infinite cylinder. For curvilinear coordinates, however, Eq. 6-6 and 6-7 become

$$\bar{u}\,\frac{\partial \bar{u}}{\partial \xi^*} + \bar{w}\,\frac{\partial \bar{u}}{\partial \bar{\zeta}} = \bar{U}\,\frac{\partial \bar{U}}{\partial \xi^*} - \frac{(V^2 - v^2)}{A^2}\left[\frac{\partial \ln (Ah_2)}{\partial \xi^*}\right] + \nu_0\,\frac{\partial^2 \bar{u}}{\partial \bar{\zeta}^2} \tag{6-17}$$

$$\bar{u}\,\frac{\partial v}{\partial \xi^*} + \bar{w}\,\frac{\partial v}{\partial \bar{\zeta}} + v\bar{u}\left[\frac{\partial \ln h_2}{\partial \xi^*}\right] = \nu_0\,\frac{\partial^2 v}{\partial \bar{\zeta}^2} \tag{6-18}$$

and have, except for the changed curvature term, a truly incompressible form. Even so, it should be noted that this change in the curvature term may cause considerable complication because, in Cartesian coordinates, for instance, it creates a certain dependence of velocity $\bar{u}$ on v, while in the truly incompressible case ($A = B = 1$) such dependence does not exist. Also, it should be remarked that when $V = 0$ a solution which satisfies Eq. 6-18 and the boundary conditions (Eq. 2-15 and 2-16) is $v = 0$. A practical example of such a case is purely meridional flow over a body of revolution; also, under these conditions, Eq. 6-17 becomes completely correlated without any troublesome curvature terms.

Case 3. $L_1 = L_2 = (AB)^{-\frac{1}{2}}$

This is the most general case, where, for the purpose of correlation, the shape of the physical surface is completely distorted. To put Eq. 6-6 and 6-7 into an incompressible form, one again needs to define A so that both F_1 and F_2 become unity. Proceeding as in Case 1, this leads to

$$d \ln (A^{\frac{3}{2}}B^{\frac{1}{2}}) = \frac{\gamma - 1}{2\gamma}\, d \ln p \tag{6-19}$$

which, when integrated, gives

$$A = \left(\frac{p}{p_0}\right)^{\frac{\gamma-1}{3\gamma}} (B)^{-\frac{1}{3}} = \left(\frac{T_e}{T_0 B}\right)^{\frac{1}{3}} = \left[\frac{\mu_0}{\mu_e}\left(\frac{T_0}{T_e}\right)^{\frac{2-\gamma}{\gamma-1}}\right]^{\frac{1}{3}} \tag{6-20}$$

$$\bar{h}_1 = \left[\left(\frac{\mu_e}{\mu_0}\right)^2\left(\frac{T_e}{T_0}\right)^{\frac{\gamma+1}{\gamma-1}}\right]^{\frac{1}{3}} h_1; \qquad \bar{h}_2 = \left[\left(\frac{\mu_e}{\mu_0}\right)^2\left(\frac{T_e}{T_0}\right)^{\frac{\gamma+1}{\gamma-1}}\right]^{\frac{1}{3}} h_2 \tag{6-21}$$

Using these relations, Eq. 6-6 and 6-7 can be written in an incompressible form, with curvature terms again changed:

$$\frac{\bar{u}}{\bar{h}_1}\frac{\partial \bar{u}}{\partial \xi} + \frac{\bar{v}}{\bar{h}_2}\frac{\partial \bar{u}}{\partial \eta} + \bar{w}\,\frac{\partial \bar{u}}{\partial \bar{\zeta}} + \frac{\bar{v}\bar{u}}{\bar{h}_2}\left[\frac{\partial \ln (A\bar{h}_1)}{\partial \eta}\right] - \frac{\bar{v}^2}{\bar{h}_1}\left[\frac{\partial \ln (A\bar{h}_2)}{\partial \xi}\right] = \frac{\bar{U}}{\bar{h}_1}\frac{\partial \bar{U}}{\partial \xi}$$
$$+ \frac{\bar{V}}{\bar{h}_2}\frac{\partial \bar{U}}{\partial \eta} + \frac{\bar{V}\bar{U}}{\bar{h}_2}\left[\frac{\partial \ln (A\bar{h}_1)}{\partial \eta}\right] - \frac{\bar{V}^2}{\bar{h}_1}\left[\frac{\partial \ln (A\bar{h}_2)}{\partial \xi}\right] + \nu_0\,\frac{\partial^2 \bar{u}}{\partial \bar{\zeta}^2} \tag{6-22}$$

$$\frac{\bar{u}}{\bar{h}_1}\frac{\partial \bar{v}}{\partial \xi} + \frac{\bar{v}}{\bar{h}_2}\frac{\partial \bar{v}}{\partial \eta} + \bar{w}\,\frac{\partial \bar{v}}{\partial \bar{\zeta}} + \frac{\bar{v}\bar{u}}{\bar{h}_1}\left[\frac{\partial \ln (A\bar{h}_2)}{\partial \xi}\right] - \frac{\bar{u}^2}{\bar{h}_2}\left[\frac{\partial \ln (Ah_1)}{\partial \eta}\right] = \frac{\bar{U}}{\bar{h}_1}\frac{\partial \bar{V}}{\partial \xi}$$
$$+ \frac{\bar{V}}{\bar{h}_2}\frac{\partial \bar{V}}{\partial \eta} + \frac{\bar{V}\bar{U}}{\bar{h}_1}\left[\frac{\partial \ln (A\bar{h}_2)}{\partial \xi}\right] - \frac{\bar{U}^2}{\bar{h}_2}\left[\frac{\partial \ln (Ah_1)}{\partial \eta}\right] + \nu_0\,\frac{\partial^2 \bar{v}}{\partial \bar{\zeta}^2} \tag{6-23}$$

Although the equations are thus correlated in terms of the modified length functions and the changed curvature terms, one should note that they do not permit easy construction of compressible flow solutions from incompressible ones, as may be done in two-dimensional flow. This is especially apparent when the physical surface is developable and the original coordinates Cartesian. In such case Eq. 6-22 and 6-23 still retain not only all the curvature terms, but also both length functions $\bar{h}_1$, and $\bar{h}_2$. Fortunately, however, Eq. 6-22 and 6-23 can be further simplified by changing them into a nearly Cartesian form, so that their solution becomes considerably easier.

Transformation to Nearly Cartesian Form. As explained in Art. 3, the three-dimensional boundary layer equations may be written in the Cartesian form only when the flow takes place over surfaces which are developable. However, similarly to the incompressible flow correlation, it is possible to distort the actual physical system into an imaginary one in which all, or most, of the explicit surface geometry effects are absent. Of course, in order to insure that the flows in the real and the distorted systems correspond, the mass flow remains invariant through the transformation.

In what follows we shall consider the correlated equations (Eq. 6-22 and 6-23), which are already written in distorted system, but which, for $A = 1$ are also the conventional incompressible equations for flow over a real surface. We shall show how these equations, through a further transformation, can be brought into the desired nearly Cartesian form. In practice, of course, these two transformations can be conveniently combined. Here they are separated so as to indicate more clearly the mechanics and the possibilities of each. The new transformation is exactly analogous to that given in Eq. 6-2,

$$\left.\begin{aligned} X = \xi; \qquad Y = \bar{\eta}; \qquad Z = G(\xi, \bar{\eta})\bar{\zeta} \\ h_1^* = \frac{\bar{h}_1}{L_1}; \qquad h_2^* = \frac{\bar{h}_2}{L_2} \\ \psi^* = \bar{\psi}; \qquad \varphi^* = \bar{\varphi} \end{aligned}\right\} \tag{6-24}$$

where the starred quantities pertain now to the transformed system and G, L_1, and L_2 are again functions of ξ, $\bar{\eta}$ only and are as yet undetermined. The components of the vector potential are defined exactly as in Eq. 6-3:

$$\psi_Z^* = h_2^* u^*; \qquad \varphi_Z^* = h_1^* v^*; \qquad -(\psi_X^* + \varphi_Y^*) = h_1^* h_2^* w^* \tag{6-25}$$

leading to the analogous relation between velocities in the two systems as given by Eq. 6-4,

$$\bar{u} = \frac{G}{L_2} u^*; \qquad \bar{v} = \frac{G}{L_1} v^*; \qquad \bar{w} = \frac{1}{L_1 L_2}\left(w^* - \frac{u^*}{h_1^*}\frac{\partial Z}{\partial \xi} - \frac{v^*}{h_2^*}\frac{\partial Z}{\partial \bar{\eta}}\right) \tag{6-26}$$

This, of course, also implies that the continuity equation becomes

$$\frac{\partial}{\partial X}(h_2^* u^*) + \frac{\partial}{\partial Y}(h_1^* v^*) + \frac{\partial}{\partial Z}(h_1^* h_2^* w^*) = 0 \tag{6-27}$$

but, now, Eq. 6-22 and 6-23 are put into a form which is slightly different from that of Eq. 6-6 and 6-7:

$$\begin{aligned}\frac{u^*}{h_1^*}\frac{\partial u^*}{\partial X} + \frac{v^*}{h_2^*}\frac{\partial u^*}{\partial Y} + w^*\frac{\partial u^*}{\partial Z} + \frac{u^{*2}}{h_1^*}\frac{\partial \ln (G/L_2)}{\partial X} + \frac{v^* u^*}{h_2^*}\frac{\partial \ln [GAh_1^*(L_1/L_2)]}{\partial Y} \\ - \frac{v^{*2}}{h_1^*}\left(\frac{L_2}{L_1}\right)^2 \frac{\partial \ln (AL_2 h_2^*)}{\partial X} = \frac{U^*}{h_1^*}\frac{\partial U^*}{\partial X} + \frac{V^*}{h_2^*}\frac{\partial U^*}{\partial Y} + \frac{U^{*2}}{h_1^*}\frac{\partial \ln (G/L_2)}{\partial X} \\ + \frac{V^* U^*}{h_2^*}\frac{\partial \ln [GAh_1^*(L_1/L_2)]}{\partial Y} - \frac{V^*}{h_1^*}\left(\frac{L_2}{L_1}\right)^2 \frac{\partial \ln (AL_2 h_2^*)}{\partial X} \\ + (L_1 L_2 G)\nu_0 \frac{\partial^2 u^*}{\partial Z^2}\end{aligned} \tag{6-28}$$

$$\begin{aligned}\frac{u^*}{h_1^*}\frac{\partial v^*}{\partial X} + \frac{v^*}{h_2^*}\frac{\partial v^*}{\partial Y} + w^*\frac{\partial v^*}{\partial Z} + \frac{v^{*2}}{h_2^*}\frac{\partial \ln (G/L_1)}{\partial Y} + \frac{u^* v^*}{h_1^*}\frac{\partial \ln [GAh_2^*(L_2/L_1)]}{\partial X} \\ - \frac{u^{*2}}{h_2^*}\left(\frac{L_1}{L_2}\right)^2 \frac{\partial \ln (AL_1 h_1^*)}{\partial Y} = \frac{U^*}{h_1^*}\frac{\partial V^*}{\partial X} + \frac{V^*}{h_2^*}\frac{\partial V^*}{\partial Y} + \frac{V^{*2}}{h_2^*}\frac{\partial \ln (G/L_1)}{\partial Y} \\ + \frac{U^* V^*}{h_1^*}\frac{\partial \ln [GAh_2^*(L_2/L_1)]}{\partial X} - \frac{U^{*2}}{h_2^*}\left(\frac{L_1}{L_2}\right)^2 \frac{\partial \ln (AL_1 h_1^*)}{\partial Y} \\ + (H_1 H_2 G)\nu_0 \frac{\partial^2 v^*}{\partial Z^2}\end{aligned} \tag{6-29}$$

The functions L_1, L_2, and G may now be so chosen that the length functions h_1^* and h_2^* become unity and the shear terms retain their conventional form. This means that $L_1 = \bar{h}_1$, $L_2 = \bar{h}_2$, $G = 1/(\bar{h}_1\bar{h}_2)$ and the transformation (Eq. 6-24) is similar to that first suggested by Struminskii [16]. Eq. 6-28 and 6-29 become

$$\begin{aligned}u^*\frac{\partial u^*}{\partial X} + v^*\frac{\partial u^*}{\partial Y} + w^*\frac{\partial u^*}{\partial Z} - u^{*2}\frac{\partial \ln (\bar{h}_1\bar{h}_2^2)}{\partial X} + v^* u^*\frac{\partial \ln (A/\bar{h}_2^2)}{\partial Y} \\ - v^{*2}\left(\frac{\bar{h}_2}{\bar{h}_1}\right)^2 \frac{\partial \ln (A\bar{h}_2)}{\partial X} = U^*\frac{\partial U^*}{\partial X} + V^*\frac{\partial U^*}{\partial Y} - U^{*2}\frac{\partial \ln (\bar{h}_1\bar{h}_2^2)}{\partial X} \\ + V^* U^*\frac{\partial \ln (A/\bar{h}_2^2)}{\partial Y} - V^{*2}\left(\frac{\bar{h}_2}{\bar{h}_1}\right)^2 \frac{\partial \ln (A\bar{h}_2)}{\partial X} + \nu_0\frac{\partial^2 u^*}{\partial Z^2}\end{aligned} \tag{6-30}$$

$$\begin{aligned}u^*\frac{\partial v^*}{\partial X} + v^*\frac{\partial v^*}{\partial Y} + w^*\frac{\partial v^*}{\partial Z} - v^{*2}\frac{\partial \ln (\bar{h}_2\bar{h}_1^2)}{\partial Y} + u^* v^*\frac{\partial \ln (A/\bar{h}_1^2)}{\partial X} \\ - u^{*2}\left(\frac{\bar{h}_1}{\bar{h}_2}\right)^2 \frac{\partial \ln (A\bar{h}_1)}{\partial Y} = U^*\frac{\partial V^*}{\partial X} + V^*\frac{\partial V^*}{\partial Y} - V^{*2}\frac{\partial \ln (\bar{h}_2\bar{h}_1^2)}{\partial Y} \\ + U^* V^*\frac{\partial \ln (A/\bar{h}_1^2)}{\partial X} - U^{*2}\left(\frac{\bar{h}_1}{\bar{h}_2}\right)^2 \frac{\partial \ln (A\bar{h}_1)}{\partial Y} + \nu_0\frac{\partial^2 v^*}{\partial Z^2}\end{aligned} \tag{6-31}$$

and although they are not fully reduced to the Cartesian form, still the simplifications thus obtained are considerable.

It should be noted that in certain special cases a complete reduction to Cartesian form is possible. For example, if only one length function is present, say, $\bar{h}_1 = 1$, if the velocity $\bar{v}$ is identically zero and none of the quantities varies with $\bar{\eta}$, then one may apply the transformation suggested by Mangler [*17*]. The object of this transformation is to eliminate the explicit dependence on the length factor $\bar{h}_2$. This means that since $h_2^* = 1$ is sought, one must have $L_2 = \bar{h}_2$ and also $G = L_2$. But G must equal $(L_1L_2)^{-1}$ if the viscous term is to remain unaffected, so that $L_1 = (L_2)^{-2}$ and $h_1^* = \bar{h}_2^2$. With these choices it is convenient to define

$$\bar{X} = \int_0^X h_1^* dX = \int_0^\xi \bar{h}_2^2 d\xi \tag{6-32}$$

and furthermore, the velocities obtained from Eq. 6-26 are,

$$\bar{u} = u^*; \qquad \bar{w} = \bar{h}_2\left(w^* - Z\frac{\partial \bar{h}_2}{\partial \xi}\bar{u}\right) \tag{6-33}$$

while Eq. 6-27 and 6-28 reduce to Cartesian form:

$$\frac{\partial u^*}{\partial \bar{X}} + \frac{\partial w^*}{\partial Z} = 0 \tag{6-34}$$

$$u^*\frac{\partial u^*}{\partial \bar{X}} + w^*\frac{\partial u^*}{\partial Z} = U^*\frac{\partial U^*}{\partial \bar{X}} + \nu_0\frac{\partial^2 u^*}{\partial Z^2} \tag{6-35}$$

It should be mentioned that many other choices of the functions L_1, L_2, and G are possible, these depending on the particular case under consideration. The method presented here provides a convenient insight into the nature of the transformation which is required to give a certain desired result.

C,7. Similarity Requirements. When confronted with a set of difficult and complicated partial differential equations, one may obtain useful information by solving the equations in an exact manner, subject to certain very special and restrictive boundary conditions, and then finding the actual physical situations in which these boundary conditions apply. In boundary layer theory, the so-called similarity solutions result from such an exploratory approach. For two-dimensional flows, investigations of such solutions have been discussed in B,25 and B,29. These investigations showed that in general, for two dimensional flows, similarity solutions apply when the outer flow velocity varies as a power of the distance along the surface or as an exponential of that distance. For three-dimensional, incompressible flows the conditions under which similarity solutions can be obtained have been studied by Hansen and Herzig for Cartesian [*18*] and polar [*19*] coordinates, and by Geis [*20*] and Hansen [*21*] for general curvilinear coordinates.

These investigations have led to the development of a technique somewhat different from that discussed in B,25, which permits the systematic study of the conditions governing the existence of similarity solutions. However, as will be shown below, the problem of finding a similarity solution is quite analogous to the previously discussed problem of obtaining a correlated flow. In fact, study of the above references reveals that the similarity solutions are obtained through a transformation identical to that given by Eq. 6-24. This realization leads to an approach to the similarity which is somewhat different from that of [*21*], although some phases of the technique introduced in [*21*] must also be used. Furthermore, aside from unifying the various types of transformations used in boundary layer studies, the present approach also permits one to extend the study of similarity requirements to compressible flows.

The conventional statement of similarity is that the transformed velocities u^* and v^* in Eq. 6-26 are to be functions of a single similarity variable, which for our purposes can be chosen to be the coordinate Z. It is convenient, at this point, to set

$$u^* = F'(Z); \qquad v^* = S'(Z) \tag{7-1}$$

and furthermore, in order to simplify the boundary conditions, one may require that the outer flow velocities be given by

$$\bar{U} = \frac{G}{L_2}; \qquad \bar{V} = \frac{G}{L_1} \tag{7-2}$$

With this relation between the stretching functions L_1 and L_2 and the outer flow velocities $\bar{U}$ and $\bar{V}$, the transformed outer velocities become, according to Eq. 6-26, identically equal to unity, and their derivatives disappear from the equations of motion, Eq. 6-28 and 6-29. Should an intrinsic coordinate system be used where one of the coordinates is chosen along $\bar{U}$, then the quantity G/L_1 will have the meaning of a scaling function by which $S'(Z)$ must be multiplied to give the actual physical velocity $\bar{v}$. Thus one can still retain for G/L_1 the symbol $\bar{V}$; however, it must be remembered that $\bar{V}$, in this special case, does not represent an outer flow velocity. The boundary conditions at the surface require that $F'(0) = S'(0) = w^*(0) = 0$. Similarly, at the outer edge of the boundary layer, one has $F'(\infty) = S'(\infty) = 1$, except when the intrinsic coordinates are used, in which case $S'(\infty) = 0$. The problem at hand is to find the form of L_1, L_2, h_1^*, h_2^*, G, and A (or, equivalently, $\bar{U}$, $\bar{V}$, $\bar{h}_1$, $\bar{h}_2$, G, and A) which permits such restrictions on velocities $\bar{u}$, $\bar{v}$, and $\bar{w}$.

To solve this problem the statement of similarity is substituted into Eq. 6-27, 6-28, and 6-29. First, integrating Eq. 6-27 with respect to Z one obtains

$$w^*(Z) = \frac{1}{h_1^*}\frac{\partial \ln h_2^*}{\partial X}F + \frac{1}{h_2^*}\frac{\partial \ln h_1^*}{\partial Y}S \tag{7-3}$$

where, in conformity with the surface boundary conditions, it is understood that $F(0) = S(0) = 0$. Next, substituting this result into Eq. 6-28 and 6-29 one obtains after some slight rearrangement:

$$\begin{aligned}(L_1L_2G\nu_0)F''' &- \left(\frac{1}{h_1^*}\frac{\partial \ln h_2^*}{\partial X}\right)FF'' - \left(\frac{1}{h_2^*}\frac{\partial \ln h_1^*}{\partial Y}\right)SF'' \\ &+ \left[\frac{1}{h_1^*}\frac{\partial \ln (G/L_2)}{\partial X}\right](1 - F'^2) + \left\{\frac{1}{h_2^*}\frac{\partial \ln [GAh_1^*(L_1/L_2)]}{\partial Y}\right\}(1 - S'F') \\ &- \left[\frac{1}{h_1^*}\left(\frac{L_2}{L_1}\right)^2\frac{\partial \ln (AL_2h_2^*)}{\partial X}\right](1 - S'^2) = 0\end{aligned} \tag{7-4}$$

$$\begin{aligned}(L_1L_2G\nu_0)S''' &- \left(\frac{1}{h_1^*}\frac{\partial \ln h_2^*}{\partial X}\right)FS'' - \left(\frac{1}{h_2^*}\frac{\partial \ln h_1^*}{\partial Y}\right)SS'' \\ &+ \left[\frac{1}{h_2^*}\frac{\partial \ln (G/L_1)}{\partial Y}\right](1 - S'^2) + \left\{\frac{1}{h_1^*}\frac{\partial \ln [GAh_2^*(L_2/L_1)]}{\partial X}\right\}(1 - F'S') \\ &- \left[\frac{1}{h_2^*}\left(\frac{L_1}{L_2}\right)^2\frac{\partial \ln (AL_1h_1^*)}{\partial Y}\right](1 - F'^2) = 0\end{aligned} \tag{7-5}$$

One now observes that, if the proposed similarity is to be truly valid, then Eq. 7-4 and 7-5 must be total differential equations with Z as the independent variable. Consequently, the coefficients of all the terms must be mutually proportional so that they can be divided out of the equations. This means that the following nine terms when multiplied by arbitrary constants must be equal to each other:

(1) $L_1L_2G\nu_0 = \dfrac{\nu_0G^3}{\bar{U}\bar{V}}$

(2) $\dfrac{1}{h_1^*}\dfrac{\partial \ln h_2^*}{\partial X} = \dfrac{G}{\bar{V}\bar{h}_1}\dfrac{\partial \ln (\bar{U}\bar{h}_2/G)}{\partial X}$

(3) $\dfrac{1}{h_2^*}\dfrac{\partial \ln h_1^*}{\partial Y} = \dfrac{G}{\bar{U}\bar{h}_2}\dfrac{\partial \ln (\bar{V}\bar{h}_1/G)}{\partial Y}$

(4) $\dfrac{1}{h_1^*}\dfrac{\partial \ln (G/L_2)}{\partial X} = \dfrac{G}{\bar{V}\bar{h}_1}\dfrac{\partial \ln \bar{U}}{\partial X}$

(5) $\dfrac{1}{h_2^*}\dfrac{\partial \ln (G/L_1)}{\partial Y} = \dfrac{G}{\bar{U}\bar{h}_2}\dfrac{\partial \ln \bar{V}}{\partial Y}$

$$(6)\quad \frac{1}{h_1^*}\frac{\partial \ln [GAh_2^*(L_2/L_1)]}{\partial X} = \frac{G}{\bar{V}\bar{h}_1}\frac{\partial \ln (\bar{V}A\bar{h}_2)}{\partial X}$$

$$(7)\quad \frac{1}{h_2^*}\frac{\partial \ln [GAh_1^*(L_1/L_2)]}{\partial Y} = \frac{G}{\bar{U}\bar{h}_2}\frac{\partial \ln (\bar{U}A\bar{h}_1)}{\partial Y}$$

$$(8)\quad \frac{1}{h_1^*}\left(\frac{L_2}{L_1}\right)^2\frac{\partial \ln (AL_2h_2^*)}{\partial X} = \frac{G}{\bar{V}\bar{h}_1}\left(\frac{\bar{V}}{\bar{U}}\right)^2\frac{\partial \ln (A\bar{h}_2)}{\partial X}$$

$$(9)\quad \frac{1}{h_2^*}\left(\frac{L_1}{L_2}\right)^2\frac{\partial \ln (AL_1h_1^*)}{\partial Y} = \frac{G}{\bar{U}\bar{h}_2}\left(\frac{\bar{U}}{\bar{V}}\right)^2\frac{\partial \ln (A\bar{h}_1)}{\partial Y}$$

These proportionality requirements amount to a set of eight differential equations and are exactly equivalent to the conditions established in [*21*], although the nine terms given there are not the same as those given above. However, it can be shown that the differences are immaterial, since they can be traced to a common factor $G/(\bar{U}\bar{V})$ and also to the fact that some of the above terms contain parts which individually must also be proportional.

The eight differential equations determine the forms which the functions $\bar{U}$, $\bar{V}$, $\bar{h}_1$, $\bar{h}_2$, A, and G must have when a similarity solution is applicable. Although in principle one could proceed to the successive general solutions of the eight equations, in practice it is easier to solve the equations under certain simplifying assumptions. Two such rather obvious assumptions are that the velocities $\bar{U}$ and $\bar{V}$ are proportional to each other and that the flow is incompressible; that is, $A = 1$. These are the restrictions imposed in [*21*] and the results of this study, expressed in our nomenclature (with all the bars omitted), are summarized in Table C,7.

It should be noted that if the surface is developable, additional conditions, arising from the fact that for such a surface the Gaussian curvature K vanishes, must be applied to the length functions h_1 and h_2. The question of whether the surface is developable or not is important, since it has been shown in [*22*] that the two resulting total differential equations may be separable (e.g. one may be solved independently of the other) only for developable surfaces and then only if the coordinates are Cartesian ($h_1 = h_2 = 1$). Also, if the outer flow is irrotational, the velocity components must satisfy

$$\frac{\partial(h_2V)}{\partial\xi} = \frac{\partial(h_1U)}{\partial\eta} \tag{7-6}$$

and this condition should be used in addition to the proportionality requirements outlined above. Irrotational outer flows were investigated in [*20*], as well as cases for which U is not proportional to V.

Hansen [*22*] has also investigated the general nature of the non-developable surfaces for which the flow obeying similarity requirements

Table C,7. Similarity requirements after [21].

U	V	$\nu_0 G^2$	h_1	h_2	K	Irrotational flow
$C_1\xi^n\eta^m$	C_2U	$C_3\xi^{n-1}\eta^{m-1}$	η	ξ	0	$V \neq 0$: $(m+1) = C_2(n+1)$ $V = 0$: $m = -1$
$C_1\xi^n e^{m\eta}$	C_2U	$C_3\xi^{n-1}e^{m\eta}$	1	ξ	"	$V \neq 0$: $m = C_2(n+1)$ $V = 0$: $C_1m = 0$
$C_1e^{n\xi}$	$C_2e^{m\xi}$	$C_3e^{\eta\xi}$	1	1	"	$C_2m = 0$
$C_1e^{n\eta}$	$C_2e^{m\eta}$	$C_3e^{m\eta}$	1	1	"	$C_1n = 0$
$C_1e^{n\xi}e^{m\eta}$	$C_2e^{n\xi}e^{m\eta}$	$C_3e^{n\xi}e^{m\eta}$	1	1	"	$C_1m = C_2n$
$C_1\xi^n$	$C_2\xi^m$	$C_3\xi^{n-1}$	1	1	"	$C_2m = 0$
$C_1\eta^n$	$C_2\eta^n$	$C_3\eta^{m-1}$	1	1	"	$C_1n = 0$
$C_1(a\xi + b\eta)^n$	C_2U	$C_3(a\xi + b\eta)^{n-1}$	1	1	"	$bC_1 = aC_2$
$C_1e^{n\xi}\eta^{m-1}$	$C_2e^{n\xi}\eta^m$	$C_3e^{n\xi}\eta^{m-1}$	1	1	"	$C_1(m-1) = C_2n = 0$
$C_1\xi^n e^{m\eta}$	$C_2\xi^{n-1}e^{m\eta}$	$C_3\xi^{n-1}e^{m\eta}$	1	1	"	$C_1m = C_2(n-1) = 0$
$C_1\xi^n\eta^{m-1}$	$C_2\xi^{n-1}\eta^m$	$C_3\xi^{n-1}\eta^{m-1}$	1	1	"	$C_1(m-1) = C_2(n-1) = 0$
C_1	C_2U	$C_3(a\xi + b\eta + c)^{-(n+1)}$	$(a\xi + b\eta + c)^n$	$(a\xi + b\eta + c)^n$	$\neq 0$	$C_2a = b (V \neq 0)$
C_1	C_2U	$C_3e^{-(a\xi+b\eta+c)}$	$e^{(a\xi+b\eta+c)}$	$e^{(a\xi+b\eta+c)}$	"	$V \neq 0$: $C_2a = b$
$C_1e^{n\xi}$	C_2U	$C_3Ue^{-(a\xi+b\eta+c)}$	$e^{(a\xi+b\eta+c)}$	$e^{(a\xi+b\eta+c)}$	"	$b = (n + a)$
$C_1e^{n(a\xi+b\eta+c)}$	C_2U	$C_3Ue^{-(a\xi+b\eta+c)}$	$e^{(a\xi+b\eta+c)}$	$e^{(a\xi+b\eta+c)}$	"	$b = C_2a$
$C_1e^{n(a\xi+b\eta+c)}$	C_2U	$C_3U(a\xi + b\eta + c)^{-(n+1)}$	$(a\xi + b\eta + c)^n$	$(a\xi + b\eta + c)^n$	"	$b = C_2a$
$C_1e^{n\xi}e^{m\eta}$	C_2U	$C_3Ue^{-(a\xi+b\eta+c)}$	$e^{(a\xi+b\eta+c)}$	$e^{(a\xi+b\eta+c)}$	"	$m + b = C_2(n + a)$
$C_1\xi^m$	C_2U	$C_3\xi^{m-1}$	1	ξ^n	"	$m = -n$
$C_1e^{n\xi}$	C_2U	$C_3e^{n\xi}$	1	$e^{m\xi}$	"	$m = -n$
C_1	C_2	C_3	1	$e^{n\xi}$	"	$V = 0$
C_1	C_2	$C_3\xi^{-1}$	1	ξ^n	"	$V = 0$

may be obtained. Specifically, he showed that in general this can be done only on surfaces generated by revolving a curve $z(x)$ about the x axis, where $z(x)$ either has the form

$$z = \int \left[\left(\frac{1}{am}\right)^2 \left(\frac{x}{a}\right)^{\frac{2(1-m)}{m}} - 1\right]^{\frac{1}{2}} dx \tag{7-7}$$

with a and m constants, or, when $z(x)$ is a tractrix. Of course, all other surfaces which can be obtained from these two by deformation, without stretching, lend themselves to similarity solutions as well.

It should be pointed out that when the flow is compressible, the length functions $\bar{h}_1$ and $\bar{h}_2$ and the compressibility transformation function A depend on the local outer flow temperature, which in turn depends on the local Mach number. This causes additional restrictions on similarity, which, in fact, may be impossible to satisfy. Finally, it also should be mentioned that the form of the function G determines the location where the boundary layer has zero thickness. This can be seen from Eq. 6-24: If the outer edge of the boundary layer corresponds to some finite value of Z, then the actual boundary layer thickness must vanish whenever G happens to be infinite. As can be seen from Table C,7, for many of the flows there listed, the location of the zero boundary layer thickness is not a definite line, but occurs either at the origin of the coordinates or at infinity. This fact may sometimes cause difficulties in finding a real counterpart of the postulated similarity flow.

CHAPTER 2. ILLUSTRATIVE SOLUTIONS

C,8. Surfaces of Revolution. The Divergence Effect. In the present age of high speed flight, the study of purely meridional flow over axially symmetric bodies has taken on considerable practical significance. Furthermore, this type of flow offers the simplest example of a three-dimensional effect, so that it is appropriate to consider it first. As mentioned in Art. 3, when the outer flow streamlines coincide with the surface geodesics, as in this case, no secondary flow occurs. The only three-dimensional effect is one of flow divergence; that is, the surface geometry causes a stretching (or contraction) in width of the individual stream tubes which results in a thinning (or thickening) of the boundary layer. Following the discussion of Art. 3, the appropriate coordinate system to be chosen in this case is quite obvious and consists of geodesics (η = const), laid out in the meridional direction, and geodesic parallels (ξ = const) in the circumferential direction (see Fig. C,8). The local radius of the body $r(\xi)$ gives the one length function h_2 which, according to Eq. 3-4, occurs in the definition of the elemental length ds. Since $v = 0$, the appropriate equations for steady flow in the absence of body forces or rotation, with Prandtl number of unity, are

$$\frac{\partial(\rho r u)}{\partial \xi} + \frac{\partial(\rho r w)}{\partial \zeta} = 0 \tag{8-1}$$

$$u\frac{\partial u}{\partial \xi} + w\frac{\partial u}{\partial \zeta} = -\frac{1}{\rho}\frac{\partial p}{\partial \xi} + \frac{1}{\rho}\frac{\partial}{\partial \zeta}\left(\mu\frac{\partial u}{\partial \zeta}\right) \tag{8-2}$$

$$u\frac{\partial H}{\partial \xi} + w\frac{\partial H}{\partial \zeta} = \frac{1}{\rho}\frac{\partial}{\partial \zeta}\left(\mu\frac{\partial H}{\partial \zeta}\right) \tag{8-3}$$

It should be noted that, in line with previous remarks, the length function r, which is indicative of the surface geometry, appears only in the

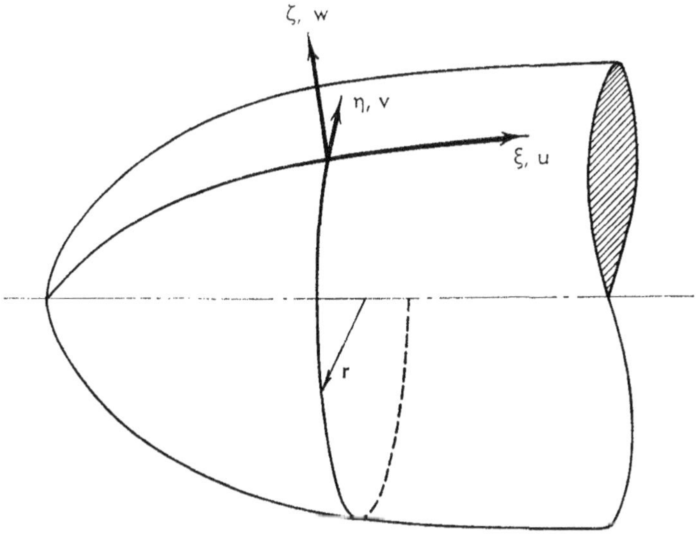

Fig. C,8. Coordinates and velocities for an axially symmetric body.

continuity equation, thus indicating the divergence effect expected in this case.

One could now, following the discussion of Art. 6, combine all the transformations to obtain the correlated incompressible plane flow. In fact, we shall combine the various transformations when discussing the flow with heat transfer. At present, however, in the absence of heat transfer, and in order to indicate more clearly the divergence effect, we shall apply the transformation giving the correlated incompressible flow only as suggested in [*14*]. Using, therefore, Eq. 6-1, 6-2, 6-12, 6-13, and 6-16, one obtains

$$\left.\begin{aligned} \xi^* &= \int_0^\xi \frac{\rho_e\mu_e a_e}{\rho_0\mu_0 a_0}\,d\xi; \qquad \bar{\zeta} = \frac{a_e}{a_0}\int_0^\zeta \frac{\rho}{\rho_0}\,d\zeta \\ \frac{\bar{u}}{a_0} &= \frac{u}{a_e}; \qquad \rho w = \rho_0\left(\bar{w} - \bar{u}\frac{\partial \bar{\zeta}}{\partial \xi^*}\right) \end{aligned}\right\} \tag{8-4}$$

which, in turn, transform Eq. 8-1, 8-2, and 8-3 to their incompressible form:

$$(r\bar{u})_{\xi^*} + (r\bar{w})_{\bar{\zeta}} = 0 \tag{8-5}$$

$$\bar{u}\bar{u}_{\xi^*} + \bar{w}\bar{u}_{\bar{\zeta}} = \bar{U}\bar{U}_{\xi^*} + \nu_0\bar{u}_{\bar{\zeta}\bar{\zeta}} \tag{8-6}$$

$$H = H_e = \text{const} \tag{8-7}$$

The examination of Table C,7 reveals now that similarity solutions are possible if: (1) $\bar{U} \sim (\xi^*)^m$ and $r \sim (\xi^*)^n$, or (2) if both $\bar{U}$ and r are exponential functions of ξ^*. Let us assume that case 1 is applicable. This permits us to define the similarity variable Z and the nondimensional stream function F:

$$\bar{\psi} \equiv r\sqrt{\bar{U}\nu_0\xi^*}\,F(Z); \qquad Z \equiv \bar{\zeta}\sqrt{\bar{U}/\nu_0\xi^*} \tag{8-8}$$

so that Eq. 8-6 becomes

$$F''' + \tfrac{1}{2}FF''(2n + m + 1) + m[1 - (F')^2] = 0 \tag{8-9}$$

Since the velocities are now given by

$$\bar{u} = \bar{U}F'; \qquad \bar{w} = \frac{1}{2}\sqrt{\frac{\bar{U}\nu_0}{\xi^*}}\,(m - 1)[(2n + m + 1)F - \eta F'] \tag{8-10}$$

therefore the boundary conditions which function F must satisfy are

$$F(0) = F'(0) = 0; \qquad F'(\infty) = 1 \tag{8-11}$$

Using these boundary conditions, Eq. 8-9 could be solved numerically for various values of n and m. Such a solution has in fact been obtained in [23] for the special case of the flow near the forward stagnation point where $n = m = 1$. In general, however, it is more convenient at this point to employ a transformation which is equivalent to those discussed in Art. 6 and which rather clearly points to the effect of n (and thus r) on the similarity solutions. If one sets

$$F(Z) = \frac{1}{\sqrt{2n + 1}}f(\bar{Z}); \qquad \bar{Z} = Z\sqrt{2n + 1} \tag{8-12}$$

then Eq. 8-9 becomes

$$f''' + \tfrac{1}{2}ff''(\bar{m} + 1) + \bar{m}[1 - (f')^2] = 0 \tag{8-13}$$

with

$$\bar{m} \equiv \frac{m}{2n + 1}; \qquad f(0) = f'(0) = 0; \qquad f'(\infty) = 1 \tag{8-14}$$

Eq. 8-13 was first given for plane flow by Falkner and Skan (B,29) and was solved in detail by Hartree [24]. Consequently, it is possible to obtain the similarity solutions for bodies of revolution from those for a plane flow. Eq. 8-12 and its derivatives, which establish the connection between these two flows, make the effects of the body radius r quite plain. The value of $\bar{u}/\bar{U}$ $(= F')$ at any Z is given by f', now taken, however, at $Z\sqrt{2n+1}$ and computed with the modified pressure-gradient parameter $\bar{m}$. Of these two changes, the change from Z to $\bar{Z}$ is by far more important in determining the velocity profile. Consequently, when $n > 0$, that is, when the body tends to increase in diameter as ξ^* increases, these modifications will in general tend to "thin-out" the boundary layer, thus accounting for its spreading over a progressively larger circumference. On the other hand, when $-\frac{1}{2} < n < 0$, that is, when the body circumference gets progressively smaller, the opposite tends to occur.

The above development and previous discussion have indicated the desirability of achieving completely correlated flows. There are, however, instances where incomplete correlation is useful in suggesting certain physical approximations. A good example of this is provided by the flow over a body of revolution, with heat transfer, but with a body temperature much lower than that outside the boundary layer. A situation fitting these conditions concerns the blunt-nosed body in hypersonic flow, and is of tremendous importance to the design of modern missiles.

Flow of this type has been treated by Lees [*25*], who combined some of the transformations discussed previously with that given by Levy [*26*]. In particular, Lees defines the similarity variable Z:

$$Z \equiv (2X)^{-\frac{1}{2}}\rho_e U \int_0^\zeta r \frac{\rho}{\rho_e}\, d\zeta \tag{8-15}$$

and modifies the meridional coordinate,

$$X \equiv \int_0^\xi \rho_e\mu_e U r^2 d\xi \tag{8-16}$$

but leaves the velocities and the enthalpy unchanged. This means that the continuity equation has to be satisfied by a conventionally defined stream function ψ,

$$\rho u r = \psi_\zeta; \qquad \rho w r = -\psi_\xi \tag{8-17}$$

This stream function must then be expressed in terms of the new variables X and Z. Taking

$$\psi \equiv (2X)^{\frac{1}{2}}F(Z); \qquad H \equiv H_e G(Z) \tag{8-18}$$

so that $u = UF'$, one obtains from Eq. 8-2 and 8-3,

$$F''' + FF'' + \frac{2X}{U}\frac{dU}{dX}\left[\frac{\rho_e}{\rho} - (F')^2\right] = 0 \tag{8-19}$$

and

$$G'' + FG' = 0 \tag{8-20}$$

where it has been assumed that Prandtl number is unity and $\rho\mu = \rho_e\mu_e$ as before. On first glance, it would appear that no great advantage has been obtained by use of the transformations (Eq. 8-15 and 8-16). To solve Eq. 8-19 and 8-20 one still has to have U and H varying in a very special manner. Actually, however, as noted by Lees, this is not necessary. The examination of the term in the brackets of Eq. 8-19 at the boundaries; e.g. at $Z = 0$ and for $Z \rightarrow \infty$ indicates that it vanishes when the wall is very cold. This is fairly obvious at the edge of the boundary layer and becomes also obvious at $Z = 0$ when it is realized that, for a cold wall, the density at the wall tends to become very large. Lees now argues that aside from vanishing at the boundaries, the term in the brackets is actually very small everywhere in the boundary layer. This argument is based on the fact that should the term be zero, then an obvious solution for G would be F', modified by an appropriate constant, and consequently the actual value of the term in the brackets can be estimated and shown to be rather small. Consequently Lees argues, the effect of the pressure gradient on the velocity profile as well as on the enthalpy distribution is in such cases negligible and both are given to sufficiently good approximation by the well known Blasius function. Actually, as discussed by Moore (H,8), this argument is not as sound as would appear at first glance, and the success of this neglect may not be justified on purely theoretical grounds in all cases. Still, it is clear, that because the transformation Eq. 8-15 and 8-16 do not correlate the compressible and the incompressible flows completely, they tend to put the equation in such a form that approximate solutions become readily available. One should further note that the divergence effect previously mentioned demonstrates itself now through the presence of r in the stretched coordinates X and Z. The form in which this occurs is identical with that suggested in [*17*] and was also given in Eq. 6-24 and 6-32.

C,9. Effects of Rotation. The Secondary Flow, Rotating Disk, and Related Problems. Perhaps the first study involving secondary flows was carried out by von Kármán [*27*] in connection with the rotating disk problem. Although this study actually gives a complete solution of the Navier-Stokes equations of motion, the flow is very similar to a boundary layer, and in fact, the solution confirms the main assumptions of the boundary layer theory. Also the disk solution illustrates the characteristics of secondary motion in three-dimensional boundary layers. The problem can be stated as follows: Consider the fixed cylindrical coordinate system r, θ, z illustrated in Fig. C,9a, and imagine that a surface, infinite in extent, and coplanar with the plane $z = 0$, is rotating with an angular velocity Ω_0 about the axis $r = 0$. Furthermore, imagine that above this surface there is an incompressible fluid which would be at rest except for the motion caused by the viscous drag of the rotating surface. The problem is to find this motion.

Taking the velocity components u, v, and w in the r, θ, and z directions

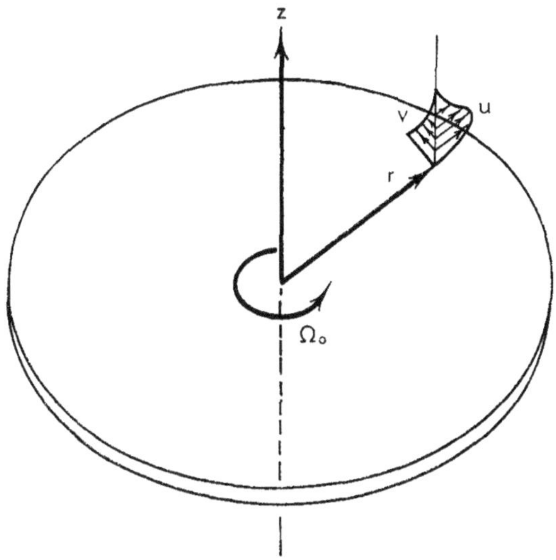

Fig. C,9a. Rotating disk problem.

respectively, one has for the equations of motion

$$\frac{1}{r}(ru)_r + w_z = 0 \tag{9-1}$$

$$uu_r + wu_z - \frac{v^2}{r} = -\frac{1}{\rho}p_r + \nu_0\left(u_{rr} + \frac{1}{r}u_r + u_{zz} - \frac{u}{r^2}\right) \tag{9-2}$$

$$uv_r + wv_z - \frac{uv}{r} = \nu_0\left(v_{rr} + \frac{1}{r}v_r + v_{zz} - \frac{v}{r^2}\right) \tag{9-3}$$

$$uw_r + ww_z = -\frac{1}{\rho}p_z + \nu_0\left(w_{rr} + \frac{1}{r}w_r + w_{zz}\right) \tag{9-4}$$

It is important to note at this point that, because a fixed nonrotating coordinate system has been chosen, the Coriolis terms do not enter the equations of motion. For this reason, too, the boundary conditions are

$$\left.\begin{aligned} &\text{at } z = 0\text{: } u = w = 0 \quad \text{and} \quad v = \Omega_0 r \\ &\text{at } z \to \infty\text{: } u = v = 0 \end{aligned}\right\} \tag{9-5}$$

To solve these equations, von Kármán set

$$u = r\Omega_0 F(Z);\ v = r\Omega_0 G(Z);\ w = (\nu_0\Omega_0)^{\frac{1}{2}}H(Z);\ p = \rho\nu_0\Omega_0 P(Z) \tag{9-6}$$

with the similarity variable Z defined by

$$Z = \left(\frac{\Omega_0}{\nu_0}\right)^{\frac{1}{2}} z \tag{9-7}$$

and obtained from Eq. 9-1, 9-2, 9-3, and 9-4 a set of simultaneous total differential equations:

$$2F + H' = 0 \tag{9-8}$$

$$F^2 - G^2 + F'H = F'' \tag{9-9}$$

$$2FG + G'H = G'' \tag{9-10}$$

$$HH' = P' + H'' \tag{9-11}$$

with the corresponding boundary conditions,

$$\left.\begin{aligned} F(0) &= H(0) = 0; G(0) = 1 \\ F(\infty) &= G(\infty) = 0 \end{aligned}\right\} \tag{9-12}$$

Eq. 9-8, 9-9, and 9-10 permit us now to solve for the functions F, G, and H, while Eq. 9-11 gives the appropriate variation of the pressure. Such a solution was carried out first, in an approximate manner, by von Kármán and later somewhat more accurately in [*28*] and is illustrated in Fig. C,9b.

These solutions show that F and G tend to zero exponentially and are nearly zero at $Z \cong 5.0$. Thus, if ν_0/Ω_0 is small, the velocities u and v

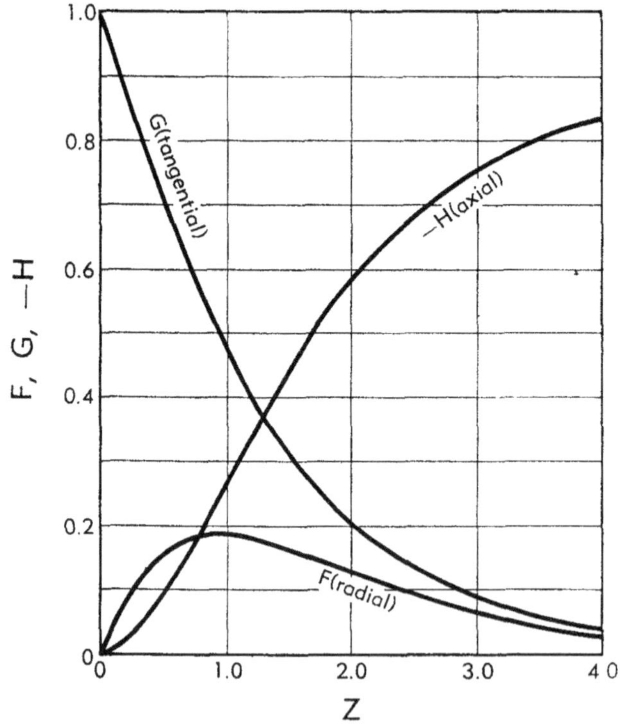

Fig. C,9b. Velocity profiles for rotating disk.

are important only in a layer immediately adjacent to the disk, of thickness $5(\nu_0/\Omega_0)^{\frac{1}{2}}$. Furthermore, the pressure change across this layer is also small—of the order of $\nu_0\rho\Omega_0$. Consequently, the main assumptions of the boundary layer theory appear to be well substantiated by this classical example.

As to the physical description of the motion, the above solutions indicate that the fluid is being continuously transported in spiral paths along the disk's surface from the axis of rotation to the outer edges. The disk thus acts very much like a centrifugal pump. Of course, in order to maintain this motion, the boundary layer must be fed by an "inflow" and, in fact, it may be seen from Fig. C,9b that $-w(\infty) \cong 0.9(\nu_0\Omega_0)^{\frac{1}{2}}$.

This basic disk problem has been considerably extended by various other investigators. For example, Bödewadt [*29*] considered the "reverse" situation where the surface is stationary but the fluid is in an axially symmetric rotary flow. Again a similarity solution can be formulated, but now the results indicate that the flow near the surface moves inward, that is, toward the axis of rotation. Bödewadt's solution was criticized by Stewartson [*30*] who argued on the basis of a linearized analysis that, for this case, no solution exists with finite w velocity far away from the surface. Actually, however, as pointed out in [*2*], regardless of whether such a solution can be found, the realism of the case considered by Bödewadt is open to serious question, since the boundary layer must start at infinity under very singular conditions involving infinite values of u and v. Furthermore, since the boundary layer moves toward the axis, one may expect an "eruption" near the axis, and this most certainly cannot be described by a similarity solution, which, after all, is a key point in Bödewadt's approach to the problem.

Batchelor [*31*], Squire [*32*], and Stewartson [*30*] have also considered a number of variations of the disk problem. They concern themselves with both the disk and the fluid rotating in the same and also in opposite directions. Furthermore, they also consider two coaxial disks spaced a fixed distance apart. This work, while concerned with an interesting class of three-dimensional boundary layer problems, has been rather thoroughly discussed by Moore [*2*] and consequently will not be reviewed again here. It will suffice to say that these investigations tend to show what has already been mentioned, that since the pressure variation in the direction normal to the surface is small, the balance between the local centrifugal pressure gradient and that imposed externally to the boundary layer is achieved through substantial changes in local streamline curvature. Consequently, depending on the relative magnitude of the angular velocities of the disk ($= \Omega_0$) and the external fluid ($= \Omega_1$), one obtains flow toward the axis of rotation or toward the edges of the disk. This tendency to centrifuge the fluid or to concentrate it at the center results, respectively, in normal inflow, or outflow at the outer edge of the boundary layer so that the law of mass conservation is satisfied. The situation is somewhat more complicated in the case of opposite rotations where at a certain height above the disk the radial velocity must undergo a reversal in direction, or in the case of two parallel disks, where under certain conditions the outflow out of one layer balances the inflow into the other.

The dependence of inflow on the relative magnitude of Ω_0 and Ω_1 has been rather nicely shown by Stewartson [30] (we consider here the case where Ω_0 and Ω_1 are in the same direction) by expanding in powers of $\beta = (\Omega_0 - \Omega_1)/\Omega_1$, and then assuming that β is small. In this way, he was able to obtain

$$w(\infty) = -(\nu_0\Omega_1)^{\frac{1}{2}}(\beta - \tfrac{3}{10}\beta^2 + \tfrac{7}{80}\beta^3 - \cdot\cdot\cdot) \tag{9-13}$$

so that for $\beta > 0$ there is an "inflow" and for $\beta < 0$ there is an "outflow" from the boundary layer.

It has been pointed out by Proudman [33], that this technique of expansion in terms of a small parameter is especially valuable in those problems where the fluid flow may be considered only a small perturbation upon a rigid body rotation. Thus, for example, should the two angular velocities be related by $\Omega_1 = \Omega_0(1 + \epsilon)$ where $\epsilon \ll 1$, then taking

$$ru = -\epsilon a\Omega_0\psi_z; \qquad rv = a\Omega_0 r^2 + \epsilon a\Omega_0\varphi; \qquad rw = \epsilon a\Omega_0\psi_r \tag{9-14}$$

with a being a characteristic length, one has, because of the axial symmetry, the continuity equation identically satisfied. Furthermore, for $\epsilon = 0$, the Navier-Stokes equations are also satisfied if $\psi = \varphi = 0$ and the motion is a rigid rotation. Consequently, for $\epsilon \neq 0$, retaining only first order terms, one obtains

$$2\left(\frac{a^2\Omega_0}{\nu_0}\right)\frac{\partial\varphi}{\partial z} = \left\{\frac{\partial^2}{\partial r^2} - \frac{1}{r}\frac{\partial}{\partial r} + \frac{\partial^2}{\partial z^2}\right\}^2\psi \tag{9-15}$$

$$-2\left(\frac{a^2\Omega_0}{\nu_0}\right)\frac{\partial\psi}{\partial z} = \left\{\frac{\partial^2}{\partial r^2} - \frac{1}{r}\frac{\partial}{\partial r} + \frac{\partial^2}{\partial z^2}\right\}\varphi \tag{9-16}$$

and the equations are linear, regardless of the Reynolds number $(a^2\Omega_0/\nu_0)$.

Using this method, Proudman [33] was able to analyze the flow between two concentric spheres rotating in the same sense and with nearly the same angular velocity (see Fig. C,9c). He found that for large Reynolds numbers, the cylindrical surface touching the inner sphere and parallel to the axis of rotation divides the fluid into two distinctly different regimes. Everywhere outside this surface the fluid rotates as a rigid body with the angular velocity of the outer sphere. Inside this surface, on the other hand, the angular velocity of the fluid is a function of the distance from the axis of rotation, and is determined by balancing the fluid which enters the boundary layer of the faster moving sphere against the fluid which leaves the boundary layer of the slower moving sphere. The cylindrical shear surface serves for the return of the fluid so that the motion can maintain itself. We thus see that in spite of the considerably different boundary conditions, the motion in the central core is qualitatively similar to that which has been described in connection with disk problems. In fact, in order to determine the nature of the cylindrical shear surface, Stewartson [*34*] foregoes the Proudman spheres in favor of a cleverly modified double-disk configuration. In this way he is able to establish that the change in the angular velocity of the main body of

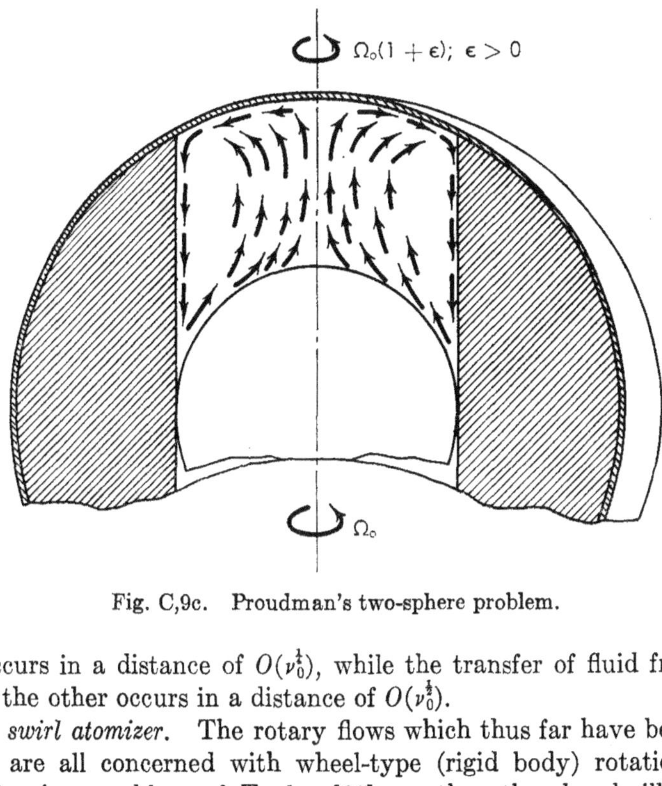

Fig. C,9c. Proudman's two-sphere problem.

fluid occurs in a distance of $O(\nu_0^{\frac{1}{4}})$, while the transfer of fluid from one disk to the other occurs in a distance of $O(\nu_0^{\frac{1}{3}})$.

The swirl atomizer. The rotary flows which thus far have been considered are all concerned with wheel-type (rigid body) rotation. The swirl atomizer problem of Taylor [*35*], on the other hand, illustrates the three-dimensional boundary layer behavior which will occur when the outer flow is in a vortex-type motion (see Fig. C,9d). The swirl atomizer is a conical, convergent, nonrotating discharge nozzle, such as may be used in fuel injectors. The fluid which enters the large end of this nozzle is given a strong swirling motion which tends to promote the jet dispersion which is so necessary for efficient combustion processes. In his formulation of the problem, Taylor represents the nozzle as a frustrum of a cone, and assumes that the inviscid fluid behavior outside the boundary layer may be represented by a classical plane vortex coaxial with the cone. The axial motion which would normally be superposed is assumed small enough to be neglected.

Now, in a vortex flow, the swirl velocity is inversely proportional to the distance from the axis, that is $V = \Gamma/(2\pi\xi \sin\alpha)$, where Γ is the circulation. Consequently the pressure acting on the boundary layer is easily obtainable from Bernoulli's equation as $p = p_\infty - \rho\Gamma^2/(8\pi^2\xi^2 \sin^2\alpha)$ and it is seen to diminish toward the vertex of the atomizer. One may therefore expect that in the boundary layer, where the velocity must decrease due to the viscous action, the fluid will be driven by the pressure gradient toward the vertex, thus causing an effective outflow through the discharge orifice. Since the flow outside the boundary layer has no axial motion, this discharge is a result of secondary flow in the boundary layer

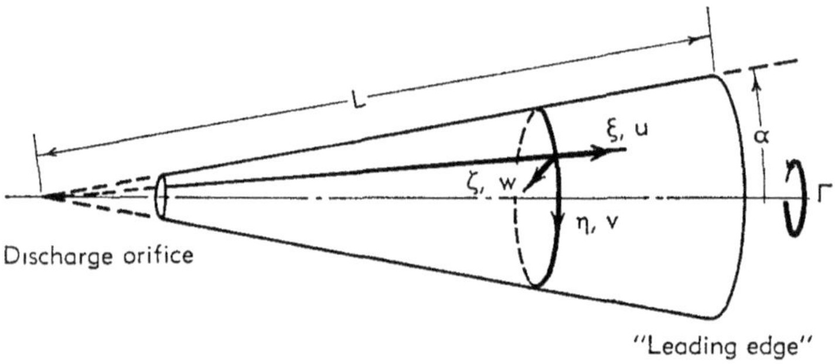

Fig. C,9d. Taylor's swirl atomizer problem.

and represents a typical three-dimensional effect caused by rotation. The problem therefore is to analyze the boundary layer flow.

Considering steady incompressible flow with axial symmetry in the illustrated nonrotating coordinate system ξ, η, ζ, one has $h_1 = h_3 = 1$ and $h_2 = \xi \sin\alpha$. The continuity equation may thus be obtained from Eq. 2-7 as

$$(u\xi \sin\alpha)_\xi + (w\xi \sin\alpha)_\zeta = 0 \tag{9-17}$$

while the momentum relations, Eq. 2-8 and 2-9, give

$$uu_\xi + wu_\zeta - \frac{v^2}{\xi} = -\frac{\Gamma^2}{4\pi^2\xi^3 \sin^2\alpha} + \nu_0 u_{\zeta\zeta} \tag{9-18}$$

$$uv_\xi + wv_\zeta + \frac{uv}{\xi} = \nu_0 v_{\zeta\zeta} \tag{9-19}$$

To solve these equations, Taylor used an adaptation of the Kármán-Pohlhausen procedure (B,31). In this procedure, the momentum-integral equations are solved approximately by choosing convenient expressions for the velocities u and v. Instead of retracing Taylor's steps we obtain the momentum-integral equations directly from Eq. 5-1 and 5-2 by observing that $U = 0$, $Q = V = \Gamma/(2\pi\xi \sin \alpha)$, $\partial/\partial\eta = 0$ and $\rho_e = \rho =$ const. This gives $F_1 = F_2 = \text{curl}_3\, Q$ and $\kappa_1 = 0$ and $\kappa_2 = -(1/Q)(\partial Q/\partial\xi) = (1/\xi)$. Thus the appropriate equations are

$$\frac{\partial\theta_{11}}{\partial\xi} - \frac{1}{\xi}(\theta_{11} + \theta_{22} - \delta_2) = \frac{\tau_{1w}}{\rho V^2} \tag{9-20}$$

$$\frac{\partial\theta_{21}}{\partial\xi} = \frac{\tau_{2w}}{\rho V^2} \tag{9-21}$$

with

$$\delta_2 = \int_0^d \left(1 - \frac{v}{V}\right) d\zeta; \qquad \theta_{11} = -\int_0^d \left(\frac{u}{V}\right)^2 d\zeta;$$

$$\theta_{22} = \int_0^d \left(1 - \frac{v}{V}\right)\frac{v}{V}\, d\zeta; \qquad \theta_{21} = \int_0^d \left(1 - \frac{v}{V}\right)\frac{u}{V}\, d\zeta \tag{9-22}$$

These equations differ only in unimportant details from those of Taylor who used spherical polar coordinates and who did not express his integrals in terms of the various momentum and displacement thicknesses. If now the two velocities u and v are expressed in terms of two characteristic parameters, then Eq. 9-20 and 9-21 can be solved for the determination of these parameters and a complete solution of the problem. Since the method hinges on an approximate description of the velocities, the validity of such an approximation is an important issue in the solution. To insure that the approximate velocity profiles closely represent those that actually occur, a number of conditions derived from the differential equations and boundary conditions are imposed at the surface as well as at the junction with the external flow. These conditions insure that, at least at the boundaries, the approximate velocity profiles have the appropriate behavior. Consequently the greater the number of conditions the better the approximation within certain limits.

The accuracy of the approximation is also dependent to some extent on the characteristic parameters selected by the investigator, and the boundary conditions do not offer any clue as to what these should be. Now, in two-dimensional boundary layer problems, it is conventional to use some measure of boundary layer thickness as one such parameter. This means that the momentum thickness, or the displacement thickness, or even the boundary layer thickness itself (defined as the value of ζ where $u = U$) can be and have been used. In three-dimensional problems, however, the additional velocity component requires that a second parameter be specified. If it is assumed that a single boundary layer thickness defines a certain viscous region, that is, that both velocity profiles are equally thick, then it is logical to seek the second parameter at the other extremity of the boundary layer; for example, many investigators have chosen the ratio of the wall frictional forces. On the other hand, if it is assumed that boundary layer thickness is that value of ζ at which $u = U$, perhaps one should not expect that the velocity v will reach V at the same value of ζ. Consequently, a second thickness suggests itself as a parameter. The choice between the two methods outlined above cannot be decided as a general matter, but depends on the particular problem.

Taylor assumed that there is only one boundary layer thickness, and chose the other parameter as describing the ratio of the frictional forces at the wall. The approximation thus constructed for the velocity profiles was criticized by Cooke [36], who pointed out that Taylor's choice prevented him from satisfying the appropriate values of second derivatives of velocities at the wall (this is apparent from Eq. 9-18 and 9-19 when $u = v = w = 0$). Cooke argued for two separate thicknesses, suggested by the work of Wild (Art. 10), as being a better set of characteristic parameters. While Cooke's argument concerning the introduction of an additional thickness is not entirely convincing, there is little doubt that his expressions for the velocity profiles are preferable, since they do satisfy more boundary conditions. For this reason, we now revert to Cooke's modified treatment of the swirl atomizer problem.

Now, the boundary conditions which the velocities u and v must satisfy are

$$\text{at } \zeta = 0\text{:} \quad u = v = 0; \qquad u_{\zeta\zeta} = \frac{\Gamma^2}{(4\pi^2\nu_0\xi^3 \sin^2 \alpha)}; \qquad v_{\zeta\zeta} = 0 \tag{9-23}$$

$$\text{at } \zeta = \delta\text{:} \qquad u = u_\zeta = 0 \tag{9-24}$$

$$\text{at } \zeta = \Delta\text{:} \qquad v = \frac{\Gamma}{(2\pi\xi \sin \alpha)}; \qquad v_\zeta = 0 \tag{9-25}$$

where δ and Δ are the two boundary layer thicknesses, that is, the values of ζ for which the u and v velocities, respectively, reach the external flow values. To satisfy these conditions, Cooke proposed the following third order polynomials:

$$f = -\frac{\zeta}{4\delta}\left(1 - \frac{\zeta}{\delta}\right)^2; \qquad g = \frac{\zeta}{2\Delta}\left(3 - \frac{\zeta^2}{\Delta^2}\right) \tag{9-26}$$

and set

$$\frac{u}{V} = \left(\frac{V\delta}{\nu_0}\right)\left(\frac{\delta}{\xi}\right)f; \qquad \frac{v}{V} = g \tag{9-27}$$

These expressions may be substituted into the definitions for the momentum and displacement thicknesses (Eq. 9-22) and integrated across the boundary layer. Note, however, that since Eq. 9-26 for f and g hold only for $\zeta \leqq \delta$ and $\zeta \leqq \Delta$ respectively, it is understood that u is identically zero for all $\zeta \geqq \delta$ and similarly v is identically unity for all $\zeta \geqq \Delta$. Consequently, if the integrals are to be taken across the boundary layer, special care must be exercised to carry this integration properly, for mixed thickness such as θ_{21} will have different values depending on whether

δ is larger or smaller than Δ. This is conveniently carried out by setting

$$\Delta = k\delta \quad \text{and} \quad \zeta = z\delta \tag{9-28}$$

so that, as $\zeta \to \Delta$, $z \to k$ and, as $\zeta \to \delta$, $z \to 1$. To illustrate the procedure, assuming that $\Delta < \delta$ (this is the only situation applicable here), one has

$$\left.\begin{aligned} \theta_{21} &= \int_0^d \left(1 - \frac{v}{V}\right)\frac{u}{V}\,d\zeta = \int_0^\delta \frac{u}{V}\,d\zeta - \int_0^\Delta \frac{v}{V}\frac{u}{V}\,d\zeta - \int_\Delta^\delta \frac{u}{V}\,d\zeta \\ &= \left(\frac{V\delta}{\nu_0}\right)\left(\frac{\delta}{\xi}\right)\delta\left[\int_0^1 f(z)dz - \int_0^k g\left(\frac{z}{k}\right)f(z)dz - \int_0^1 f(z)dz\right] \end{aligned}\right\} \tag{9-29}$$

so that θ_{21} is a function of both parameters δ and k. Evaluation of the integrals (Eq. 9-22) and substitution into Eq. 9-20 and 9-21 results in two total differential equations which in turn may be substantially simplified if one takes, according to Cooke's suggestion,

$$I \equiv \left(\frac{V\delta}{\nu_0}\right)^2\left(\frac{\delta}{\xi}\right)^2 \quad \text{and} \quad \xi = xL \tag{9-30}$$

where L is the value of ξ at the "leading edge." With these Cooke obtains

$$5I'x = 48(35 - 72k) \tag{9-31}$$

$$120IBk'x = 10080 - A(4I + 3I'x) \tag{9-32}$$

where the primes indicate differentiation with respect to x, and the quantities A, B are

$$A = k^3(9k^2 - 35k + 42)$$

$$B = k^2(12k^2 - 35k + 28)$$

Cooke solves these with the initial condition $I = 0$ (but $k \neq 1$) at the "leading edge," that is, at $x = 1$. The solution obtained is illustrated in Fig. C,9e.

It is interesting to note, from this figure, that the boundary layer becomes thinner as it proceeds toward the discharge orifice of the atomizer. Apparently the strong, favorable pressure gradient is able to provide more than enough acceleration to overcome the retarding effect of friction at the wall. Furthermore, one should note from Fig. C,9e that the thickness Δ of the swirl boundary layer is never very much smaller than the thickness δ of the flow toward the vertex. The greatest difference between the two occurs right at the "leading edge" where $k = 0.879$. For small values of x the two become nearly equal.

It was noted by Cooke that the behavior of I is fairly well represented by

$$I = 2722(1 - x^{0.1})x^{0.125} \tag{9-33}$$

and this, in turn, as pointed out in [2], means that as $x \to 1$, $\delta \sim (1 - x)^{\frac{1}{4}}$. Consequently, the radial boundary layer has a $\frac{1}{4}$ order singularity at the "leading edge," just like the singularity occurring in the free convection flow near a heated vertical plate located in a fluid at rest (D,3). This analogy could be expected if it is recognized that the centrifugal pressure gradient of the swirl atomizer is like the hydrostatic pressure gradient of the heated vertical plate, and that the flow in the direction of these

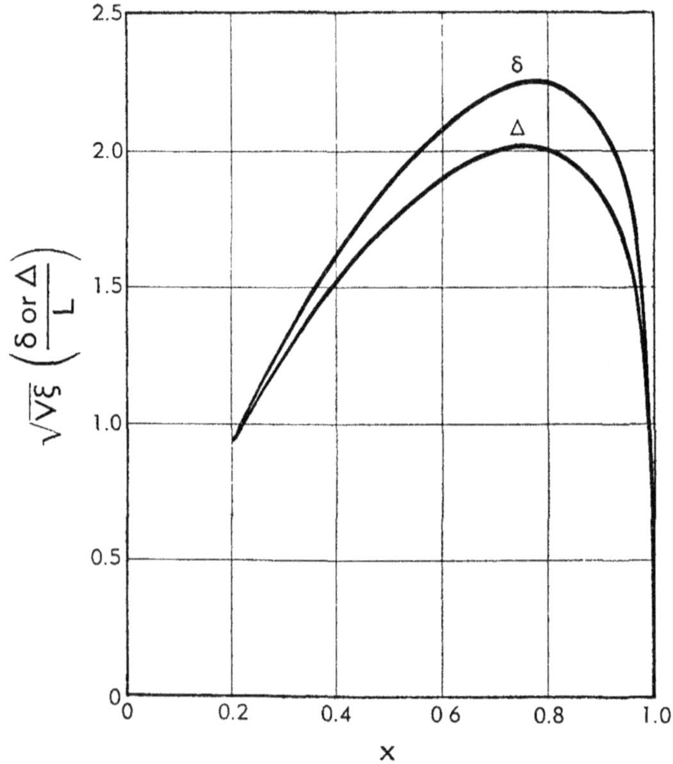

Fig, C,9e. Boundary layer thickness in the swirl atomizer.

gradients, for both cases, occurs inside the boundary layer only. In discussing the swirl atomizer problem, Moore [2] also points out that near the vertex ($x \to 0$) there are two possible solutions of Cooke's equations (Eq. 9-31 and 9-32),

$$\delta \sim x; \qquad \Delta \sim x \tag{9-34}$$

and

$$\delta \sim x^{-0.0061}; \qquad \Delta \sim x^{-0.0307} \tag{9-35}$$

The first of these suggests that the boundary layer vanishes at the origin, while the second indicates that, although there is a singularity at $x = 0$

giving an "eruption" of the flow, this singularity is very weak and indeed somewhat questionable in view of the approximate nature of the momentum-integral method. Which one of these solutions actually applies is, of course, a somewhat academic question for the swirl atomizer (but may be of real interest in such a related case as the tornado) and cannot readily be ascertained from Cooke's results.

In this connection, it should be pointed out that the solution of the swirl-atomizer problem may also be attempted in terms of strict similarity. As a matter of fact, realizing that the cone is a developable surface, one obtains the total differential equations directly from the work of [19], by substituting the appropriate constants. Thus one has for the velocities and the similarity variable,

$$u = VF'(Z); \qquad v = VG'(Z); \qquad w = \sqrt{\frac{\nu_0 V}{\xi}}\,(ZF' - F);$$

$$Z = \frac{\zeta}{\xi}\sqrt{\frac{V\xi}{\nu_0}} \tag{9-36}$$

while the equations become

$$F''' + FF'' + (F')^2 + (G')^2 - 1 = 0 \tag{9-37}$$

$$G''' + FG'' = 0 \tag{9-38}$$

with the boundary conditions,

$$F(0) = F'(0) = G'(0) = F'(\infty) = 0; \qquad G'(\infty) = 1 \tag{9-39}$$

To test the correctness of a similarity approach, we now try to devise an asymptotic "outer" solution, valid for large Z,

$$F = A + F_0; \qquad G' = 1 + G_0' \tag{9-40}$$

where A is a constant $= F(\infty)$, F_0/A and $G_0' \ll 1$, and where F_0 and $G_0' \to 0$ as $Z \to \infty$. Substituting Eq. 9-40 into Eq. 9-37 and 9-38, one obtains after elimination of second order terms:

$$F_0''' + AF_0'' + 2G_0' = 0 \tag{9-41}$$

$$G_0''' + AG_0'' = 0 \tag{9-42}$$

and these give as solutions valid for large Z:

$$F_0 = \frac{2B}{A^4}\left[AZ + \left(2 + \frac{CA^2}{2B}\right)\right]e^{-AZ};$$

$$F_0' = -\frac{2B}{A^3}\left[AZ + \left(1 + \frac{CA^2}{2B}\right)\right]e^{-AZ} \tag{9-43}$$

$$G_0' = -\frac{B}{A}\,e^{-AZ} \tag{9-44}$$

where B and C are constants.

We note from these solutions that if they are to vanish as $Z \to \infty$ then A must be positive. But $A = F(\infty)$, and consequently it is apparent from the expression for the w velocity (Eq. 9-36) that the vanishing of the asymptotic solution requires that the flow at the outer edge of the boundary layer be spiraling toward the surface. The boundary layer is thus drawing steadily on the external flow and if, as a result of the pressure gradient, there is a secondary flow toward the origin, one may expect some sort of "eruption" near $x = 0$. But from the definition of Z (Eq. 9-36), one also notes that the boundary layer thickness must vary linearly with x and consequently vanishes at $x = 0$. In addition, if the secondary-flow velocity distribution F' has a form nearly like that given by Cooke's f (Eq. 9-26), that is, it is negative over the whole boundary layer, then F which starts with zero value at $Z = 0$ can never reach a positive value of $F(\infty) = A$. This rather contradictory behavior has apparently led Moore [2] to conclude that no valid solutions of Eq. 9-37 and 9-38 can be found. Actually, however, one should also note that all of these contradictions disappear if the function F' changes sign somewhere in the boundary layer. In that case, not only can F reach th appropriate positive value A, but also, a mechanism is provided by which some of the secondary flow could be flowing away from the origin so that the absence of an "eruption" at $x = 0$ is conceivable. In this connection, it is interesting to note that the computations of Browning [37, pp. 157, 158] who has treated Bödewadt's solid-body-rotation case, show definite reversals of the u velocity near the outer fringes of the boundary layer.

An important question which now arises is whether Taylor and Cooke, having also assumed similarity profiles (Eq. 9-27) can obtain valid solutions, especially when their boundary layer thickness is forced to behave in a manner different than required by the strict similarity considerations, that is, it is given the zero value at $x = 1$. In answer to this question one can argue that since the momentum-integral method is an approximate way of treating the problem, what these investigators have tacitly assumed is that the flow is not quite similar but that the deviations from the similarity are sufficiently slight to be taken care of by adjustment of the boundary layer thickness alone. For this reason, too, the assumptions of Cooke, who uses two different thicknesses and thus has two separate similarity parameters, afford greater flexibility in such an adjustment and are apt to represent the physical situation more closely. In connection with this greater flexibility it will be noted that the more accurate Cooke solution tends to a similarity behavior, that is, to a nearly linear variation of δ and Δ with x as $x \to 0$ where the effect of the imposed boundary condition ($\delta = 0$ at $x = 1$) is felt less and less. Taylor's solution, on the other hand, gives values of δ which are very nearly constant even for small values of x.

Rotating body of revolution amid axial flow. The boundary layer over a body of revolution rotating about its axis, with its axis parallel to the direction of flow ahead of the body, is important in connection with the flight of spin-stabilized missiles, for example. This type of boundary layer flow has been studied by Illingworth [*38*], Chu and Tifford [*39*], and by Schlichting [*40*]. Invoking axial symmetry with a nonrotating coordinate system as illustrated in Fig. C,8, the steady boundary layer equations for Prandtl number of unity may be written as

$$\frac{\partial(\rho r u)}{\partial \xi} + \frac{\partial(\rho r v)}{\partial \zeta} = 0 \tag{9-45}$$

$$\rho\left[u\frac{\partial u}{\partial \xi} + w\frac{\partial u}{\partial \zeta} - \frac{v^2}{r}\frac{dr}{d\xi}\right] = -\frac{\partial p}{\partial \xi} + \frac{\partial}{\partial \zeta}\left(\mu\frac{\partial u}{\partial \zeta}\right) \tag{9-46}$$

$$\rho\left[u\frac{\partial(rv)}{\partial \xi} + w\frac{\partial(rv)}{\partial \zeta}\right] = \frac{\partial}{\partial \zeta}\left[\mu\frac{\partial(rv)}{\partial \zeta}\right] \tag{9-47}$$

$$\rho\left[u\frac{\partial H}{\partial \xi} + w\frac{\partial H}{\partial \zeta}\right] = \frac{\partial}{\partial \zeta}\left(\mu\frac{\partial H}{\partial \zeta}\right) \tag{9-48}$$

with boundary conditions at the surface,

$$u = w = 0; \qquad v = \Omega r \quad \text{and} \quad H(0) \text{ or } k\left(\frac{\partial T}{\partial \zeta}\right)_0 \text{ prescribed} \tag{9-49}$$

and at the outer edge of the boundary layer,

$$u = U, \qquad v = 0 \quad \text{and} \quad H = H_e \tag{9-50}$$

where Ω is the angular velocity of spin.

As pointed out by Chu and Tifford [*39*] and clearly apparent from Eq. 9-47 and 9-48, one particular solution of the energy equation is given by

$$H = K_1 + K_2 r\Omega v \tag{9-51}$$

where K_1 and K_2 are constants. One of these constants may be readily evaluated from the conditions at the outer edge of the boundary layer, which, because $v = 0$ there, gives $K_1 = H_e$. Furthermore, by considering the derivative of Eq. 9-51 with respect to ζ at $\zeta = 0$, one obtains

$$\left(C_p\frac{\partial T}{\partial \zeta}\right)_{\zeta=0} = \left(r\Omega(K_2 - 1)\frac{\partial v}{\partial \zeta}\right)_{\zeta=0} \tag{9-52}$$

thus showing that the constant K_2 plays a role in determining the heat transfer rate. In particular, defining the heat transfer coefficient, C_h,

$$C_h[H(0) - H_e] \equiv \left(k\frac{\partial T}{\partial \zeta}\right)_{\zeta=0} \tag{9-53}$$

the spin-retarding friction coefficient $C_{f,\eta}$,

$$\frac{1}{2}\rho_e(r\Omega)^2 C_{f,\eta} \equiv \left(\mu\frac{\partial v}{\partial \zeta}\right)_{\zeta=0} \tag{9-54}$$

and the Stanton number, St,

$$\rho_e(r\Omega)St \equiv C_h \tag{9-55}$$

one finds by simple manipulation that for Prandtl number of unity Eq. 9-52 leads to a modified "Reynolds analogy" based on the circumferential flow:

$$St = \left(\frac{K_2 - 1}{K_2}\right)\frac{C_{f,\eta}}{2} \tag{9-56}$$

With the physical meaning of the constants K_1 and K_2 thus clarified, the surface enthalpy is given by:

$$\left(c_pT\right)_{\zeta=0} = H_{e,r} + (K_2 - 1)(r\Omega)^2 \tag{9-57}$$

where $H_{e,r}$ is the free stream enthalpy modified by the addition of the kinetic energy of rotation:

$$H_{e,r} = H_e + \tfrac{1}{2}(r\Omega)^2 \tag{9-58}$$

Since H_e is a constant it is clear from Eq. 9-57 that the solution (Eq. 9-51) is applicable only when the surface enthalpy varies parabolically with r. Note that for adiabatic flow, when $K_2 = 1$, the surface enthalpy is given exactly by $H_{e,r}$, as one may have expected from the basic principles. Observe here that the same result follows directly from Eq. 2-3 (taken relative to the rotating body) whose solution in that case is I = constant. Since the value of the total enthalpy in the free stream is specified and must be invariant regardless of the rotation of the coordinate system, therefore one has immediately $I = H_e$. On the other hand, at the wall, according to Eq. 2-5, the surface enthalpy is $I + \frac{1}{2}(\Omega r)^2$ which again equals $H_{e,r}$.

The correlation between the compressible and incompressible form of Eq. 9-45, 9-46, and 9-47 was first accomplished by Illingworth [*38*] and later also by Chu and Tifford [*39*]. This correlation is carried out much in the same manner as discussed in Art. 6, except that now, since the energy integral is given by Eq. 9-51, rather than by Eq. 6-1, the equations derived in Art. 6 are strictly applicable only in a very special case when $K_2 = 0$ (Chu and Tifford do not make this restriction, but then their flow is not correlated in the same sense either) and the surface temperature is

$$\left(c_pT\right)_{\zeta=0} = H_e - \tfrac{1}{2}(r\Omega)^2 \tag{9-59}$$

Of course, as noted by Illingworth, for very slow rotation when $r\Omega \ll U$, one may neglect the square of the spin and then the surface temperature becomes the total temperature, as it should for Prandtl number of unity. Furthermore, Eq. 6-5, 6-16, 6-17, and 6-18 give for such a case

$$\frac{\partial(r\bar{u})}{\partial\xi^*} + \frac{\partial(r\bar{w})}{\partial\bar{\zeta}} = 0 \tag{9-60}$$

$$\bar{u}\frac{\partial\bar{u}}{\partial\xi^*} + \bar{w}\frac{\partial\bar{u}}{\partial\bar{\zeta}} = \bar{U}\frac{\partial\bar{U}}{\partial\xi^*} + \nu_0\frac{\partial^2\bar{u}}{\partial\bar{\zeta}^2} \tag{9-61}$$

$$\bar{u}\frac{\partial v}{\partial\xi^*} + \bar{w}\frac{\partial v}{\partial\bar{\zeta}} + \frac{v\bar{u}}{r}\frac{\partial r}{\partial\xi^*} = \nu_0\frac{\partial^2 v}{\partial\bar{\zeta}^2} \tag{9-62}$$

It will be noted that these transformed equations not only are in a completely correlated incompressible form, but also are separated, that is, the velocity v does not enter into the determination of $\bar{u}$ or $\bar{w}$. This independence of the flow in the ξ direction from that in the η direction occurs also in other cases of three-dimensional boundary layer flow and will be discussed more fully in the next article.

If the spin is large, the simplifications suggested by Illingworth cannot be made. In such cases, however, as shown in [*39*], the energy integral (Eq. 9-51) may be used to advantage in obtaining a correlation with incompressible flow. In fact it is only necessary to recall that the function F_1 in Art. 6 is really the total enthalpy ratio H/H_e. Since in the present case this enthalpy ratio is greater than unity by $K_2r\Omega v/H_e$, all terms which were multiplied by F_1 must be appropriately increased. One thus obtains from Eq. 6-5, 6-17, and 6-18 directly with a_e and a_0 being the external and reference speeds of sound respectively:

$$\frac{\partial(r\bar{u})}{\partial\xi^*} + \frac{\partial(r\bar{w})}{\partial\bar{\zeta}} = 0 \tag{9-63}$$

$$\bar{u}\frac{\partial\bar{u}}{\partial\xi^*} + \bar{w}\frac{\partial\bar{u}}{\partial\bar{\zeta}} - \left(\frac{a_0}{a_e}\right)^2 v^2\left[\frac{\partial\ln(a_e r/a_0)}{\partial\xi^*}\right] = \bar{U}\frac{\partial\bar{U}}{\partial\xi^*}\left[1 + \frac{K_2r\Omega v}{H_e}\right] + \nu_0\frac{\partial^2\bar{u}}{\partial\bar{\zeta}^2} \tag{9-64}$$

$$\bar{u}\frac{\partial v}{\partial\xi^*} + \bar{w}\frac{\partial v}{\partial\bar{\zeta}} + v\bar{u}\frac{\partial\ln r}{\partial\xi^*} = \nu_0\frac{\partial^2 v}{\partial\bar{\zeta}^2} \tag{9-65}$$

It should be noted that, while the correlation with an incompressible flow is thus possible for many surface temperature distributions, independence between the flows in the ξ and η directions is now lost. For this reason, Illingworth [*38*], when treating the problem of the circular cone in supersonic flow with large spin velocity, obtains a solution for u as a Taylor series in Ω^2r/U and a solution for v as a series in odd powers of Ω. We shall consider such expansions in series elsewhere in this section. Still, it is appropriate to point out that for a more general axially symmetric body, the series expansion method is very tedious, so that one would like to have a somewhat more rapid and more convenient procedure. Unfortunately, however, although such a procedure has been devised by Schlichting [40], it applies only to incompressible flow. Although Eq. 9-63, 9-64, and 9-65 contain no varying fluid properties or density, still they are not identical with the true incompressible flow equations for the corresponding case. For this reason, Schlichting's results may not be interpreted directly in terms of their compressible equivalents by using the inverse of the transformation of Art. 6. One would need a separate, approximate solution of Eq. 9-63, 9-64, and 9-65.

The Schlichting procedure [40] is very similar to that used by Cooke [36] in solving the swirl atomizer problem. Like Cooke, he also defines two separate thicknesses and uses polynomial representations for the approximate u and v velocity distributions. When introduced into the momentum-integral equations, these give two simultaneous first order, total differential equations, which Schlichting solves numerically. In this way he is able to analyze the flow over a number of axially symmetric bodies, and, in particular, he presents the two friction coefficients, as well as the effect of spin on separation. Typical of these are the results for the sphere, for which centrifugal effects shift the separation point upstream by some 3 degrees, when $\Omega r/U_0$ is of order one (U_0 represents velocity in the undisturbed field ahead of the sphere). The effect of Ωr on the spin moment is also very small, causing only a 2 per cent reduction when $\Omega r/U_0$ is increased from zero to unity.

C,10. Solutions of Separable Equations. Yawed Infinite Cylinders.

We shall now discuss a certain special class of three-dimensional boundary layer problems in which the differential equations are separable, meaning that two of the velocities can be determined quite independently of the third. The central problem in this class is that of a yawed infinite cylinder. The incompressible flow over such a cylinder was solved independently by Prandtl [41], Struminskii [42], Jones [43], and Sears [44], who all noted the above-mentioned independence and its consequences. Later, some additional but basic work on this problem was also carried out by Cooke [45], who considered the flow over a yawed infinite wedge, and by Hansen [22] and Nickel [46], who inquired what conditions are necessary to attain such separability in general. It should also be mentioned in these preliminary remarks that the yawed infinite cylinder is very closely related to a rotating body of revolution in axial flow, particularly if the spin is small, so that the centrifugal effects may be neglected. This correspondence has apparently been first noted by Tifford [47] and is a natural extension of the well-known approximate representation of a turbomachinery stage by an infinite cascade of blades.

Consider a cylindrical wing, with an infinite aspect ratio, which is yawed with respect to the direction of the undisturbed flow far ahead of the wing. A wing of this type is a developable surface, and consequently we choose a Cartesian coordinate system with the coordinate x lying normal to the leading edge and the coordinate y lying along the wing generators just as illustrated in Fig. C,10a. Since the wing is cylindrical and infinite, none of the flow quantities may vary along y. This

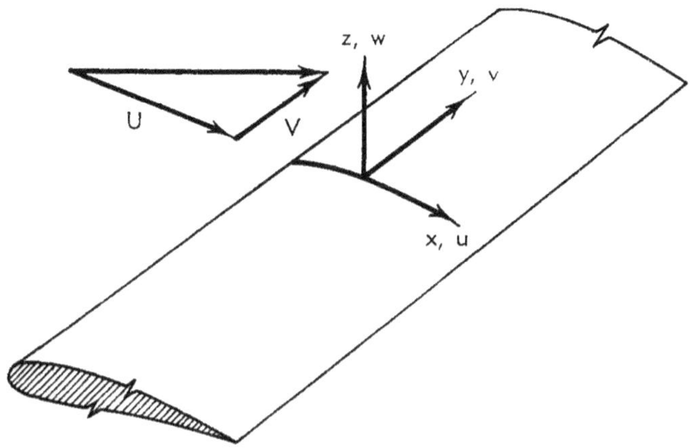

Fig. C,10a. Coordinates and velocities for yawed, infinite wing problem.

fact introduces substantial simplifications in the differential equations, which for compressible flow and Prandtl number of unity become

$$\frac{\partial(\rho u)}{\partial x} + \frac{\partial(\rho w)}{\partial z} = 0 \tag{10-1}$$

$$u\frac{\partial u}{\partial x} + w\frac{\partial u}{\partial z} = U\frac{\partial U}{\partial x} + \frac{1}{\rho}\frac{\partial}{\partial z}\left(\mu\frac{\partial u}{\partial z}\right) \tag{10-2}$$

$$u\frac{\partial v}{\partial x} + w\frac{\partial v}{\partial z} = \frac{1}{\rho}\frac{\partial}{\partial z}\left(\mu\frac{\partial v}{\partial z}\right) \tag{10-3}$$

$$u\frac{\partial H}{\partial x} + w\frac{\partial H}{\partial z} = \frac{1}{\rho}\frac{\partial}{\partial z}\left(\mu\frac{\partial H}{\partial z}\right) \tag{10-4}$$

with boundary conditions,

$$u(0) = v(0) = w(0) = 0; \qquad H(0) \quad \text{or} \quad \left(k\frac{\partial H}{\partial z}\right)_{z=0} \text{specified} \tag{10-5}$$

and

$$u(\infty) = U; \qquad v(\infty) = V; \qquad H(\infty) = H_e \tag{10-6}$$

Inspection immediately reveals that for incompressible adiabatic flow Eq. 10-1 and 10-2 are identical to those for a two-dimensional flow and can be solved for the u and w velocities quite independently of the spanwise flow. Thus, for example, if the airfoil is wedgelike, as considered by Cooke [45], i.e. when $U = Ax^n$, then the solution may be obtained by use of a conventional Stokes stream function defined by

$$u \equiv \psi_z; \qquad w \equiv -\psi_x \tag{10-7}$$

and nondimensionalized by

$$\psi = \sqrt{\frac{2\nu_0 Ux}{n+1}}\, f(Z); \qquad Z = z\sqrt{\frac{(n+1)U}{2\nu_0 x}} \tag{10-8}$$

which converts Eq. 10-2 to

$$f''' - ff'' + \frac{2n}{n+1}(1 - f'^2) = 0 \tag{10-9}$$

with the boundary conditions $f(0) = f'(0) = 0$ and $f'(\infty) = 1$. The solutions of Eq. 10-9 are, of course, the well-known Falkner-Skan profiles B,29 and need not be computed again. Setting now $v = Vg(Z)$, one obtains from Eq. 10-3

$$g'' + fg' = 0 \tag{10-10}$$

with the boundary conditions $g(0) = 0$ and $g(\infty) = 1$. Since f is a known function of Z, Eq. 10-10 is obviously separable and the corresponding value of g is easily obtained by a numerical integration. Such computations have been carried out in [45]. It may be worthwhile to remark that for $n = 0$, that is, for a flat plate, the solution of Eq. 10-9 and 10-10 is clearly given by the well-known Blasius function, a fact which was first noted in [43] and [44]. For a flat plate, however, as pointed out by Moore [8], neither the yaw, nor even a curved leading edge, have any effect on the boundary layer. In other words, both the u and v velocity profiles are given by the same function, so that if an intrinsic coordinate system were chosen there would be no secondary flow. For values of $n \neq 0$, Rott and Crabtree [48] have shown, graphically, that Cooke's solutions [45], when considered as a function of distance z made nondimensional by the characteristic boundary layer momentum thickness, all have the same shape as for $n = 0$. Consequently, as far as spanwise flow is concerned, the effect of the chordwise pressure gradient is primarily restricted to changes of boundary layer thickness but not of the shape of the v velocity profile.

For a more realistic airfoil shape which, unlike the wedge, does not conform to strict similarity requirements, the procedure suggested by Sears [44] may be used. In this method one expresses the outer flow as a polynomial:

$$U = U_0(X - X^3 + \cdots); \qquad V = V_0; \qquad X = \frac{x}{c} \tag{10-11}$$

where U_0 and V_0 are constants and c is the chord, so that one has at his disposal a known two-dimensional solution [49] of the form:

$$\left.\begin{aligned} u &= U_0[Xf_1'(Z) - X^3f_3'(Z) + \cdots] \\ w &= U_0(Re)^{-\frac{1}{2}}\cdot[-f_1(Z) + 3X^2f_3(Z) - \cdots] \\ \text{with} \qquad Z &= z\sqrt{\frac{U_0}{\nu_0 c}} \end{aligned}\right\} \tag{10-12}$$

Setting now

$$v = V_0 \sum_{n=0}^{\infty} X^n G_n(Z) \tag{10-13}$$

and substituting this together with the known solution (Eq. 10-12) into the incompressible form of Eq. 10-3, Sears obtains, after equating the coefficients of equal powers of X, the following differential equations for the functions G_n:

$$G_n'' + f_1G_n' - nf_1'G_n = 3f_3G_{(n-2)}' - (n-2)f_3'G_{(n-2)} - 5f_5G_{(n-4)} + \cdots \tag{10-14}$$

where all the functions with negative subscript are understood to be identically zero, so that the equations become nonhomogeneous for all $n \geqq 2$. The appropriate boundary conditions are

$$G_n(0) = 0; \qquad G_0(\infty) = 1; \qquad G_{n\neq 0}(\infty) = 0 \tag{10-15}$$

It is easy to see from Eq. 10-14 and Eq. 10-15 that all G functions corresponding to odd values of n have a trivial solution equal identically to zero. However, since the f functions are known only in tabular form, the G functions corresponding to even values of n must be computed numerically. The thus-obtained solution of Sears is shown in Fig. C,10b for various distances from the leading edge. As may be seen from this figure, although the chordwise velocity separates at X = 0.69, the spanwise flow v remains attached there. This means that the direction of the limiting "streamline": $dy/dx = \lim_{z\to 0}(v/u)$ at X = 0.69 is purely spanwise and differs substantially from the potential flow streamline $dy/dx = V_0/U$, as shown in Fig. C,10c. Of course, since $v(0) = u(0) = 0$ everywhere, l'Hospital's rule provides that the direction of the "limiting streamline" is the same as the direction in which surface shear acts. We thus have a phenomenon, possible in three-dimensional flow but never encountered in two-dimensional boundary layers, where the fluid near the surface, being unable to negotiate the unfavorable pressure gradient, may turn and flow along an isobar, rather than separate and leave the wall immediately. Although, as will be seen in subsequent discussions in Art. 18, this behavior also leads eventually to a separation; never-

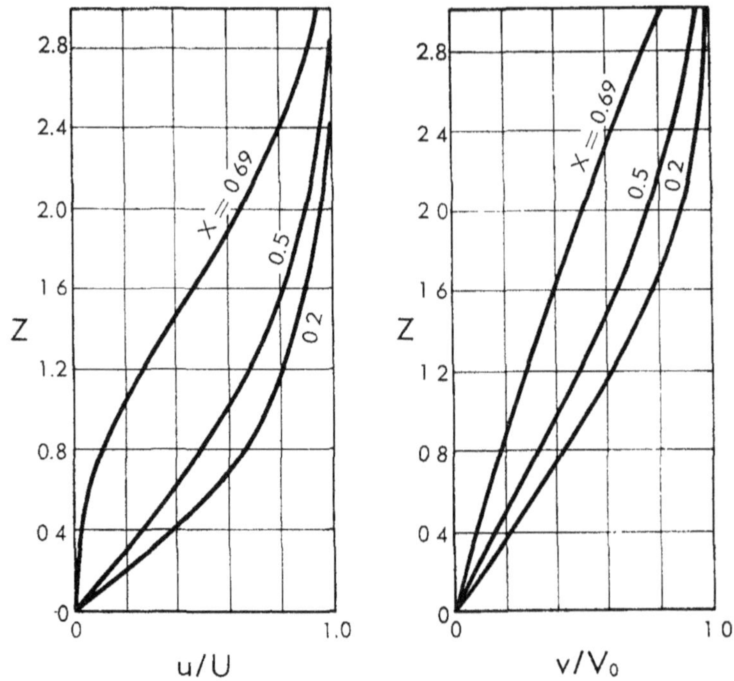

Fig. C,10b. Sears' solution for velocity distributions on a yawed infinite wing.

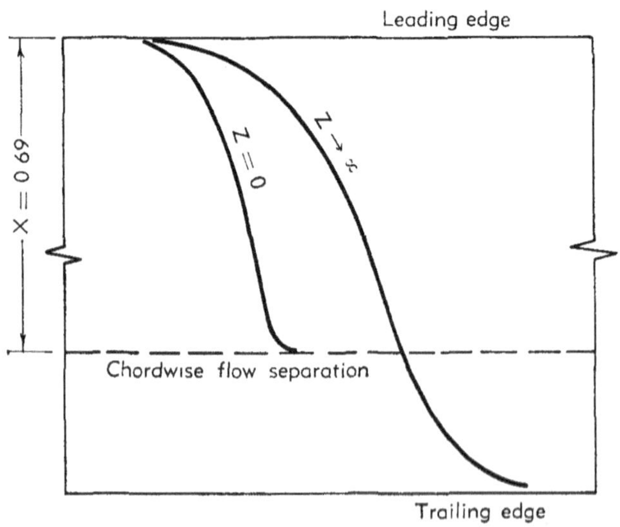

Fig. C,10c. Limiting and external streamlines on a yawed wing.

theless, the differences at the onset of a two- and three-dimensional separation are here clearly apparent.

While the method proposed in [*44*] could be applied to most any U velocity distribution (by including a sufficient number of terms in the expansion), the labor involved in such computations is obviously too tedious for common engineering practice. For this reason it is worthwhile to call attention to the momentum-integral methods proposed by Wild [*50*] and by Rott and Crabtree [*48*]. The essence of Wild's method is to use two different boundary layer thicknesses in the chordwise and spanwise directions. Wild's original idea of characterizing the crossflow by a different thickness has also been used in [*36*] and [*40*] in treatments of the swirl atomizer and rotating body of revolution (Art. 9) respectively. Wild chooses fourth order polynomials for the velocity profiles and thus, in addition to all the conditions previously discussed in connection with the swirl atomizer problem, he is able to specify also the vanishing of the second derivative of the velocity at the outer edge of the boundary layer. His expressions are

$$\frac{u}{U} = F\left(\frac{z}{\delta}\right) + \frac{\delta^2}{\nu_0}\frac{dU}{dx}G\left(\frac{z}{\delta}\right); \qquad \frac{v}{V} = F\left(\frac{z}{\Delta}\right) \tag{10-16}$$

$$F = \frac{z}{\delta}\left[2 - 2\left(\frac{z}{\delta}\right)^2 + \left(\frac{z}{\delta}\right)^3\right]; \qquad G = \frac{1}{6}\frac{z}{\delta}\left(1 - \frac{z}{\delta}\right)^2 \tag{10-17}$$

Since the independence principle holds, δ is presumed known and the problem is to determine the second thickness Δ which characterizes the scale of the v velocity profile. Now, from Eq. 5-2, when the flow is incompressible and all the derivatives with respect to y vanish, one has

$$\frac{d\theta_{21}}{dx} + 2\frac{\theta_{21}}{Q}\frac{dQ}{dx} = \frac{d}{dx}(Q^2\theta_{21}) = \frac{\tau_{2w}}{\rho_0 Q^2} \tag{10-18}$$

It is convenient here to redefine the momentum thickness θ_{21},

$$\theta_{21} = \int_0^d \left(\frac{V}{Q} - \frac{v}{Q}\right)\frac{u}{Q}\,dz = \frac{UV}{Q^2}\int_0^\Delta \left(1 - \frac{v}{V}\right)\frac{u}{U}\,dz \equiv \frac{UV}{Q^2}\theta_{yx} \tag{10-19}$$

and obtain upon substitution into Eq. 10-18,

$$\frac{d}{dx}(U\theta_{yx}) = \frac{\tau_{2w}}{\rho_0 V} = \frac{\nu_0}{V}\left(\frac{\partial v}{\partial z}\right)_{z=0} = \frac{2\nu_0}{\Delta} \tag{10-20}$$

Since the connection between θ_{yx} and Δ may easily be obtained upon substitution of the assumed velocity distribution (Eq. 10-16), Eq. 10-20 may now be numerically integrated to obtain the characteristic thickness and thus the spanwise flow. Of course, as in Art. 9, one has to be careful to set u/U equal to unity for $\delta \leqq z \leqq \Delta$ if $\delta < \Delta$. It should also be noted, as Wild pointed out, that in this method not only is u independent of v, but moreover the shape of the v profile is independent of u. In other words, the velocity v is always described by the same function F and all that changes is the scale Δ. Although such shape independence cannot be exactly correct, it is approximately borne out by the similarity solutions of [*45*]. For more general flows, as evidenced by the results of Sears, the idea of a universal v profile seems to be invalid near chordwise flow separation.

Wild's procedure has been improved upon by Rott and Crabtree [*48*], who used the momentum thicknesses for the characteristic lengths. They noted that one may write

$$s \equiv \frac{\theta_{yx}}{\theta_x} = \int_0^\infty \left[1 - g\left(\frac{z}{q\theta_x}\right)\right] f\left(\frac{z}{\theta_x}, m\right) d\left(\frac{z}{\theta_x}\right) \tag{10-21}$$

with

$$\theta_x = \int_0^\infty \left(1 - \frac{u}{U}\right)\frac{u}{U}\,dz; \qquad \theta_y = \int_0^\infty \left(1 - \frac{v}{V}\right)\frac{v}{V}\,dz \tag{10-22}$$

$$\frac{u}{U} = f\left(\frac{z}{\theta_x}, m\right); \qquad \frac{v}{V} = g\left(\frac{z}{\theta_y}\right) = f_{m=0}\left(\frac{z}{\theta_y}\right) \tag{10-23}$$

and

$$m = -\frac{\theta_x^2}{\nu_0}\frac{dU}{dx}; \qquad q = \frac{\theta_y}{\theta_x} \tag{10-24}$$

The ratio s is thus a universal function of m and q, which can easily be evaluated once the function f is specified. (A similar observation was also made in [*40*].) This universal relationship, which may be plotted once and for all, not only aids greatly in a detailed step-by-step solution, but also points the way toward further simplification. Examination of accurate solutions (which may be superimposed on the graph of $s = s(q, m)$) suggested to Rott and Crabtree that the dependence of s upon m is much smaller than the dependence upon q. This was particularly true in regions of favorable pressure gradient. Consequently, for such regions, the relationship $s = q^2$ was advocated and with it the momentum equation (Eq. 10-20) may be integrated to give

$$s = \frac{1}{U\theta_x}\left[0.337\nu_0 \int_0^x \left(\frac{U}{\theta_x}\right)^{\frac{1}{2}} dx\right]^{\frac{2}{3}} \tag{10-25}$$

Once s is known, the value of $q(x)$ may be obtained from $q = (s)^{\frac{1}{2}}$ or from the plot of the universal relation $s = s(q, m)$. If the latter determination is used, then a successive integration of Eq. 10-20 with $q(x)$ known gives

$$s = \frac{0.225\nu_0}{U\theta_x}\int_0^x \frac{dx}{q\theta_x} \tag{10-26}$$

and leads to an iteration procedure which may also be used in regions of adverse pressure gradient.

All the solutions discussed so far consider adiabatic, incompressible flow. Of course, yawed wings are of primary interest at high flight speeds where such assumptions often cannot be made. In the compressible case, ρ, T, and μ may not be assumed constant, and, in addition to the momentum equations, the energy equation (Eq. 10-4) must also be solved. However, as first noted in [*8*], the energy equation is identical in form to the equation for v (Eq. 10-3). This means that, just as in Art. 9, one possible solution of Eq. 10-4 is

$$H = K_1 + K_2 \frac{v}{V} \tag{10-27}$$

Substituting the boundary conditions, one now finds that $K_1 = H(0)$ and $K_2 = H_e - H(0)$. It follows that, while Eq. 10-27 is capable of representing heat transfer, it can do so only under isothermal surface conditions. Of course, if the flow is adiabatic ($K_2 = 0$), then the enthalpy remains constant through the whole boundary layer.

Using the above solution (Eq. 10-27) and the compressible flow transformation discussed in Art. 6, one can again obtain a correlated incompressible flow. Such equations were first given by Crabtree [*15*] when $K_2 = 0$ and by Moore [*2*] when there is heat transfer and $K_2 \neq 0$. Here again, we simply set the function F_1 of Art. 6 equal to H/H_e and multiply by it the appropriate terms in Eq. 6-17 and 6-18 with $x^* = \xi^*$, $\bar{z} = \bar{\zeta}$, and $h_2 = 1$.

This gives

$$\frac{\partial \bar{u}}{\partial x^*} + \frac{\partial \bar{w}}{\partial \bar{z}} = 0 \tag{10-28}$$

$$\begin{aligned} \bar{u}\frac{\partial \bar{u}}{\partial x^*} + \bar{w}\frac{\partial \bar{u}}{\partial \bar{z}} = \bar{U}\frac{\partial \bar{U}}{\partial x^*}\left[\frac{v}{V} + \frac{H(0)}{H_e}\left(1 - \frac{v}{V}\right)\right] \\ - \left(\frac{V}{A}\right)^2 \frac{\partial \ln A}{\partial x^*}\left(1 - \frac{v}{V}\right)\left(\frac{H(0)}{H_e} + \frac{v}{V}\right) + \nu_0 \frac{\partial^2 \bar{u}}{\partial \bar{z}^2} \end{aligned} \tag{10-29}$$

$$\bar{u}\frac{\partial v}{\partial x^*} + \bar{w}\frac{\partial v}{\partial \bar{z}} = \nu_0 \frac{\partial^2 v}{\partial \bar{z}^2} \tag{10-30}$$

However, from the definition of A, using the transformation of Art. 6, one can show that

$$-\frac{V^2}{A^2}\frac{\partial \ln A}{\partial x^*} = \bar{U}\frac{\partial \bar{U}}{\partial x^*}\frac{(\gamma - 1)V^2}{2\left(a_e^2 + \frac{\gamma - 1}{2}U^2\right)} \equiv B\bar{U}\frac{\partial \bar{U}}{\partial x^*} \tag{10-31}$$

so that after some manipulation one obtains from Eq. 10-30 the equation given in [*2*]:

$$\bar{u}\frac{\partial \bar{u}}{\partial x^*} + \bar{w}\frac{\partial \bar{u}}{\partial \bar{z}} = \bar{U}\frac{\partial \bar{U}}{\partial x^*}\left\{1 + \beta\left[1 - \left(\frac{v}{V}\right)^2 - 2\frac{K_2}{V^2}\left(1 - \frac{v}{V}\right)\right]\right\} + \nu_0\frac{\partial^2 \bar{u}}{\partial \bar{z}^2} \tag{10-29a}$$

It is quite apparent that Eq. 10-29a and 10-30 are not independent except in the very special case of the flat plate $\bar{U} \neq \bar{U}(x)$. One should note, though, that the parameter β is a constant and, furthermore, that it will be usually quite small even for moderate Mach number. This fact led Crabtree [*15*] to propose a solution by expansion in powers of β. Even a zero order approximation ($\beta = 0$) will then represent a considerable improvement over the incompressible flow, since the effects of compressibility are partly accounted for by the transformation. It should also be mentioned that, as Moore [*2*] has already noted, Eq. 10-28, 10-29a, and 10-30 may be solved exactly, even if β is not small, provided the external flow satisfies the strict similarity requirements. A procedure similar to that used in [*45*] can easily be extended to such a case, although the resulting system will consist of two simultaneous differential equations which are not as easily solvable as when the flow is incompressible.

Separable equations satisfying similarity requirements. Since, as has been demonstrated, the ability to separate the three-dimensional boundary layer equations so obviously aids in their solution, it is important to know under what general conditions such separation is possible. For steady, incompressible flow, this question has been answered by Nickel [*46*]. In addition, Hansen [*22*] considered the further practical restrictions, usually arising in the process of the solution, by demanding that the equations be not only separable, but that they conform to strict similarity requirements as well. The general equations for such flows, before any similarity restrictions have been made, were given in Art. 7 (Eq. 7-4 and 7-5). Hansen found that these equations will satisfy similarity requirements, that is, will become total differential equations and will also be separable if and only if:

1. The flow surface is developable and the coordinates are Cartesian.
2. The outer flow velocities and the similarity variable have one of the two following forms:

(a) $$U = Ae^{nx}; \qquad V = Be^{mx}; \qquad Z = z\sqrt{\frac{cU}{\nu_0}}$$

(b) $$U = Ax^n; \qquad V = Bx^m; \qquad Z = z\sqrt{\frac{cU}{\nu_0 x}}$$

Setting $u = UF'(Z)$ and $v = VS'(Z)$, Eq. 7-4 and 7-5 subject to conditions (a) give the separated equations

$$cF''' + n[\tfrac{1}{2}FF'' - (F')^2 + 1] = 0 \tag{10-32}$$

$$cS''' + \frac{n}{2}FS'' - m(F'S' - 1) = 0 \tag{10-33}$$

Similarly, if conditions (b) are used, the corresponding total differential equations become

$$cF''' + \frac{(n+1)}{2}FF'' - n[(F')^2 - 1] = 0 \tag{10-34}$$

$$cS''' + \frac{(n+1)}{2}FS'' - m[F'S' - 1] = 0 \tag{10-35}$$

with boundary conditions (valid also for set (a)),

$$F(0) = F'(0) = S(0) = S'(0) = 0; \qquad F'(\infty) = S'(\infty) = 1 \tag{10-36}$$

It should be noted that for both cases, since the outer flow is a function of one coordinate only, the external streamlines are all "translates." In other words, the whole outer flow streamline system may be obtained by translating any one streamline along the y coordinate, that is, along the leading edge of the surface. Of course, the similarity conditions indicate that the streamlines inside the boundary layer are also "translates." Solutions of Eq. 10-34 and 10-35 with $n = 0$ and various values of m were obtained and discussed in [*51,52,53,54*]. Such solutions taken from [*54*] are shown in Fig. C,10d.

In addition, in [*55*, pp. 21-33, and *22*], following the procedure suggested in [*3*], approximate solutions for such a system were found. These approximate solutions are obtained by dividing the boundary layer flow into two separate inner and outer regions. In the outer region, where S''' and S'' must necessarily be small, the relative importance of various terms depends on the ratio $m/(n + 1)$ (if desired, the constant c may be set equal to $n + 1$). Consequently if $m/(n + 1)$ is large, the approximate solution of Eq. 10-35, in this outer region, is given by $S' = 1/F'$. This outer region solution must be matched with that obtained in the inner layer, where, for small values of Z, a valid approximation for function F is easily obtainable by a few terms of a Taylor series. One may thus assume, as was done in [*55*, pp. 21–33], that, in the inner region, S is also representable by a polynomial whose coefficients are evaluated by the matching conditions and the values at the surface. Alternately one may solve Eq. 10-35 with the approximate representation of F, as was done in [*22*]. This latter procedure leads to a Bessel equation and appears to give results agreeing somewhat better with the exact solutions for $n = 0$ ([*22*]). In any case, as shown in [*55*, pp. 21–33], the agreement with the exact computations for $n = 0$ depends principally on the value of m. For large values of m the approximate solution is very good indeed. The physical reasons behind this fact are easily understood when it is realized that $U = A$, $V = Bx^m$ represent outer flow streamlines of the form: $y = \{B/[A(m + 1)]\}x^{m+1} + \text{const}$, so that as m increases, larger and larger portions of these streamlines lie in the x direction. Consequently, the boundary layer has an opportunity to thicken before any significant turning occurs, and thus the assumed division into inner and outer layers is more valid. Furthermore, as m becomes larger, the stronger

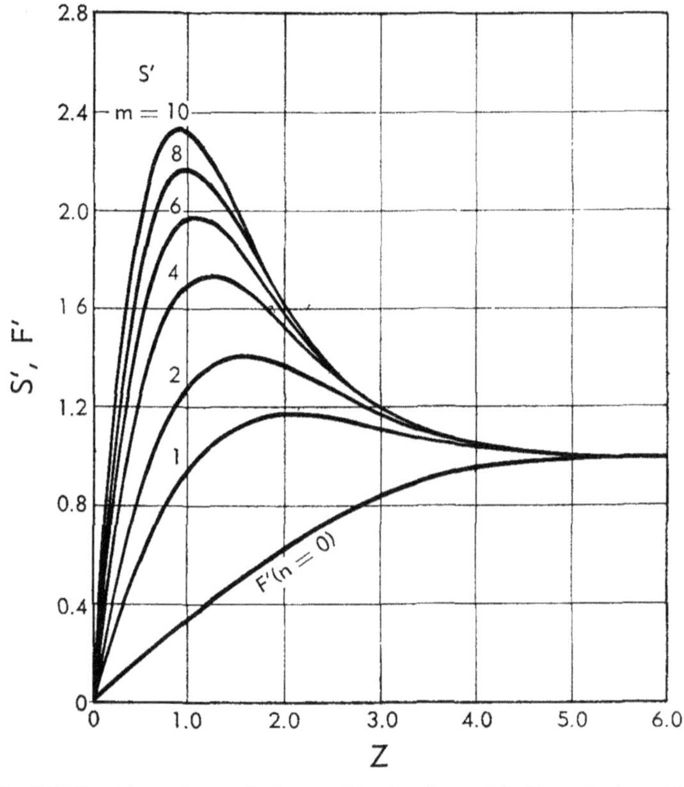

Fig. C,10d. Cross flow velocity profiles for flow with $U = A$; $V = Bx^m$.

lateral pressure gradient exercises more influence than the no-slip condition at the wall, so that the viscous inner region becomes thinner, which also makes "inner and outer" assumptions more valid. It should be remarked that the outer region solution $S' = 1/F'$ may also be obtained directly from the equations for inviscid but rotational motion. Such flows, usually designated by internal aerodynamicists as secondary, are of great importance in the design of curving ducts and turbomachinery blading. Of practical interest, therefore, is the fact that the secondary inviscid flow solution may, under certain circumstances, be comfortably extended into the boundary layer where the effect of the viscous retarding action is restricted to a very thin inner layer immediately adjacent to the surface.

Finally it should also be pointed out that Eq. 10-35 is linear in S. This, of course, permits solutions to be combined and thus the analysis may easily be extended to flows with cross flow velocity given by $V = \sum_m B_m x^m$, as demonstrated by the author in [*53*] and also by Hansen and Herzig in [*54*]. Furthermore, as noted in [*55*, pp. 21–33], the solutions thus obtained may also be used to give the temperature distribution in two-dimensional boundary layers with distributed heat sources [*56*], which is of great interest in connection with the after-shutdown cooling of nuclear reactors.

C,11. Solutions by Parametric Expansion. As shown in Art. 10, a three-dimensional boundary layer which is independent of one of the surface coordinates can be relatively easily analyzed, since the two components of cross flow are unaffected by the three-dimensional nature of the boundary layer and are thus given by a two-dimensional solution. Such convenient simplifications, unfortunately, occur only in a very few three-dimensional problems. In general, both surface coordinates play a role in the boundary layer development, and consequently the independence principle does not apply. It is possible, however, in many cases to formulate the problem so that the three-dimensional effects present but a small departure from some known, preferably two-dimensional, flow. This suggests that the solutions of many three-dimensional boundary layer problems may be obtained in the form of expansions about known solutions. In this article we shall discuss examples of this type.

Stagnation point flow. We start by considering the incompressible boundary layer flow near the nose of a blunted, but not necessarily axially symmetric, body. It was Howarth [*57*] who first observed that, since only the immediate vicinity of the stagnation point is considered, the problem may be formulated in the Cartesian coordinates fixed in a plane osculating the stagnation point. This is illustrated in Fig. C,11a.

If the outer flow is irrotational, its components, upon suitable rotation of coordinates, are then given by $U = Ax$, $V = By$, where the constants A and B are related to the shape of the body near the stagnation point. For an outer flow of this type, a similarity solution,

$$u = Uf'(Z); \qquad v = Vg'(Z); \qquad Z = z\sqrt{\frac{U}{\nu_0 x}} \tag{11-1}$$

is possible and the appropriate differential equations become

$$f''' + ff'' - (f')^2 + 1 = -cgf'' \tag{11-2}$$

$$g''' + fg'' = c[(g')^2 - gg'' - 1] \tag{11-3}$$

where $c = B/A$. The shapes of the outer flow streamline patterns suggest (see Fig. C,11b) that, when $c = 0$, the equations should correspond to two-dimensional flow and, when $c = 1$, the equations should describe the flow over a body of revolution. Actually, for $c = 1$, as may be seen from Eq. 11-2, 11-3, and 8-9 (for $n = m = 1$), the functions f and g are both given by the Homann solution ([*23*] and Art. 8) for an axially

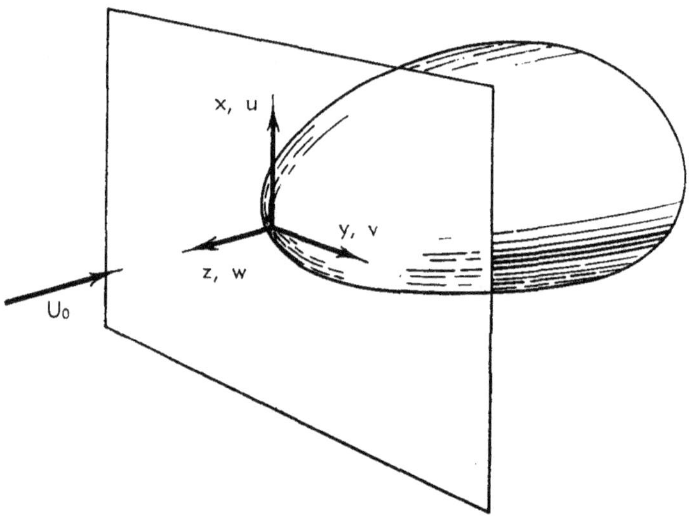

Fig. C,11a. Howarth's formulation of stagnation flow.

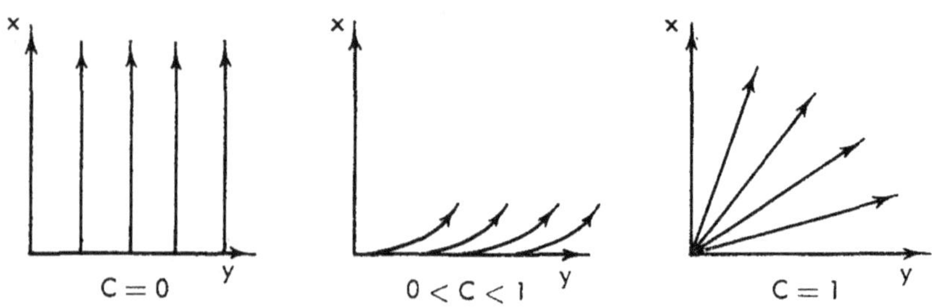

Fig. C,11b. External streamlines in Howarth's formulation of stagnation flow.

symmetric stagnation point. Consequently, for $c = 1$, there is no secondary flow in the boundary layer and three-dimensionality demonstrates itself in divergence effects only.

Of special interest, however, are the solutions for small values of c, since these values result in the sharpest turning of the external flow streamlines. For small values of c, one may expand the functions f and g:

$$f = f_0 + cf_1 + c^2f_2 + \cdots \tag{11-4}$$

$$g = g_0 + cg_1 + c^2g_2 + \cdots \tag{11-5}$$

and upon substitution into Eq. 11-2 and 11-3 obtain the differential equations for each order of c. The zeroeth order equations are

$$f_0''' + f_0 f_0'' + 1 - (f_0')^2 = 0 \tag{11-6}$$

$$g_0''' + f_0 g_0'' = 0 \tag{11-7}$$

and the first order equations are

$$f_1''' + f_0 f_1'' + f_1 f_0'' - 2f_0' f_1' = -g_0 f_0'' \tag{11-8}$$

$$g_1''' + f_0 g_1'' + f_1 g_0'' = (g_0')^2 - g_0 g_0'' - 1 \tag{11-9}$$

The boundary conditions, of course, conform to the original definition of velocities, and thus may be written as

$$f_0(0) = f_1(0) = f_0'(0) = f_1'(0) = g_0(0) = g_1(0) = g_0'(0) = g_1'(0) = 0 \tag{11-10}$$

$$f_0'(\infty) = g_0'(\infty) = 1 \quad \text{and} \quad f_1'(\infty) = g_1'(\infty) = 0 \tag{11-11}$$

In this way the problem is not only considerably simplified, but also the equations for functions f and g are effectively separated. In this connection it should be pointed out that for $c = 0$, the cross flow V vanishes and, consequently, only Eq. 11-2 need be considered. That equation, when taken alone, gives the two-dimensional stagnation point flow. Considering the vanishing of c as a continuous limiting process however, that is, considering Eq. 11-6 and 11-7, one notes that these do not result in two-dimensional stagnation-point flow. As may easily be seen by comparison with Eq. 10-34 and 10-35 (with $n = 1$ and $m = 0$), the functions f_0 and g_0 represent the flow for a yawed stagnation line. This is the reason why Howarth [57] (who obtained solutions to second order) found the surface flow deflection (from the direction of the outer flow) greatest in the limit $c \to 0$. In other words, the very sharp turning of the external flow streamlines corresponding to small values of c gives rise to strong secondary flows in the boundary layer.

The solution of [57] applies in the immediate vicinity of the stagnation point, that is, for very small values of x and y. Somewhat farther away from the stagnation point, the formulation in the osculating plane is, of course, invalid, since there the effect of surface curvatures is changing and must be felt in the differential equations themselves. One can, however, still formulate an expansion procedure:

$$h_1 = 1 + D_1\xi + D_2\eta; \qquad h_2 = 1 + E_1\xi + E_2\eta \tag{11-12}$$

$$U = A_{10}\xi + A_{02}\eta + A_{11}\xi^2 + A_{12}\xi\eta + A_{22}\eta^2 \tag{11-13}$$

$$V = B_{10}\xi + B_{02}\eta + B_{11}\xi^2 + B_{12}\xi\eta + B_{22}\eta^2 \tag{11-14}$$

whereby the first-order equations would then be identical (upon suitable rotation) to those used in [57], and consequently their solution is known.

However, the second-order flow will depend then on a solution of four simultaneous differential equations. Because of this, Squire [*58*], who has solved this problem, decided instead to use a nonorthogonal coordinate system. In such coordinates, the rotation of coordinates suggested by Howarth may again be used to advantage, although the appropriate equations carry certain additional terms resulting from nonorthogonality. The process of solution, for the second order flow, is, however, made considerably easier, since some of the equations are separated, and, at most, a simultaneous solution of only two differential equations is required.

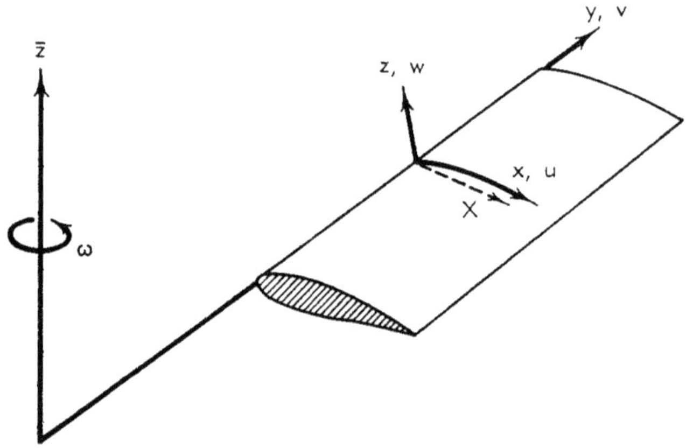

Fig. C,11c. Coordinates and velocities for rotating blade problem.

Rotating blade and related problems. Fogarty [*59*] considered the incompressible flow over a rotating blade (cylinder) as shown schematically in Fig. C,11c. Again, since the surface is developable, Cartesian coordinates may be used. If these coordinates are fixed in the rotating blade, the appropriate equations for incompressible flow become (see Art. 2)

$$\frac{\partial u}{\partial x} + \frac{\partial v}{\partial y} + \frac{\partial w}{\partial z} = 0 \tag{11-15}$$

$$u\frac{\partial u}{\partial x} + v\frac{\partial u}{\partial y} + w\frac{\partial u}{\partial z} - 2\omega_3 v - \frac{1}{2}\omega^2\frac{\partial R^2}{\partial x} = -\frac{1}{\rho_0}\frac{\partial p}{\partial x} + \nu_0\frac{\partial^2 u}{\partial z^2} \tag{11-16}$$

$$u\frac{\partial v}{\partial x} + v\frac{\partial v}{\partial y} + w\frac{\partial v}{\partial z} + 2\omega_3 u - \frac{1}{2}\omega^2\frac{\partial R^2}{\partial y} = -\frac{1}{\rho_0}\frac{\partial p}{\partial y} + \nu_0\frac{\partial^2 v}{\partial z^2} \tag{11-17}$$

with boundary conditions,

$$u(0) = v(0) = w(0) \quad \text{and} \quad u(\infty) = U,\ v(\infty) = V \tag{11-18}$$

In these equations, R is the distance from the axis of rotation measured in the plane of the rotation, and ω_3 is the component of the angular velocity vector in the direction normal to the surface. Before attempting the solution of the boundary layer problem, one must know the corresponding flow outside the boundary layer. Fortunately, in this case, such an outer flow has been found by Sears [60]:

$$U = \omega y \frac{\partial \varphi}{\partial X}; \qquad V = \omega(\varphi - 2X) \tag{11-19}$$

where $\varphi(X, \bar{z})$ is the velocity potential for plane flow of the same airfoil (cylinder) section and $\bar{z}$ is taken in the direction of the axis of rotation while $X = (R^2 - y^2)^{\frac{1}{2}}$. It is rather remarkable that, as this exact solution indicates, only the U component depends on the spanwise coordinate y and furthermore that this dependence is linear.

Fogarty [59] now assumes that the airfoil is sufficiently "flat" so that $x \cong X$ and $\omega_3 \cong \omega$. Furthermore, he notes from Eq. 11-19 that V/U is of order x/y which, for large values of y, will be quite small. This order of magnitude presumably applies also to the velocities inside the boundary layer, so that one is justified in expanding them in powers of x/y. This has been formalized by Tan [61] and we follow here his development, introducing only minor changes in the mode of description of the outer flow. Thus one sets

$$u = u_0 + u_1 + u_2 + \cdots u_n \tag{11-20}$$

$$v = v_1 + v_2 + \cdots v_n \tag{11-21}$$

$$w = w_0 + w_1 + w_2 + \cdots w_n \tag{11-22}$$

where the subscripts pertain to the order of x/y. Substituting these into Eq. 11-15, 11-16, and 11-17 and expressing the pressure gradients in terms of the outer flow components, one obtains

$$\frac{\partial u_n}{\partial x} + \frac{\partial v_{(n-1)}}{\partial y} + \frac{\partial w_n}{\partial z} = 0 \tag{11-23}$$

$$\sum_{S=0}^{n}\left[u_S \frac{\partial u_{(n-S)}}{\partial x} + w_S \frac{\partial u_{(n-S)}}{\partial z}\right] + \sum_{S=1}^{n-1} v_S \frac{\partial u_{(n-1-S)}}{\partial y} + 2\omega[V_{(n-1)} - v_{(n-1)}]$$
$$= \sum_{S=0}^{n} U_S \frac{\partial U_{(n-S)}}{\partial x} + \sum_{S=1}^{n-1} V_S \frac{\partial U_{(n-1-S)}}{\partial y} + \nu_0 \frac{\partial^2 u_n}{\partial z^2} \tag{11-24}$$

$$\sum_{S=0}^{n}\left[u_S \frac{\partial v_{(n-S)}}{\partial x} + w_S \frac{\partial v_{(n-S)}}{\partial z}\right] + \sum_{S=1}^{n-1} v_S \frac{\partial v_{(n-1-S)}}{\partial y} - 2\omega[U_{n-1} - u_{(n-1)}]$$
$$= \sum_{S=0}^{n} U_S \frac{\partial V_{(n-S)}}{\partial x} + \sum_{S=1}^{n-1} V_S \frac{\partial V_{(n-1-S)}}{\partial y} + \nu_0 \frac{\partial^2 v_n}{\partial z^2} \tag{11-25}$$

together with the boundary conditions,

$$u_n(0) = v_n(0) = w_n(0) = 0; \qquad u_n(\infty) = U_n; \qquad v_n(\infty) = V_n \quad \text{(11-26)}$$

Thus, for example, if the "airfoil" is a flat plate, $\varphi = x$ and $U = -\omega x$. For $n = 0$ Eq. 11-24 and 11-25 thus describe the standard, two-dimensional Blasius flat-plate problem so that

$$Z = \frac{z}{2}\sqrt{\frac{U}{\nu_0 x}}; \qquad u_0 = \frac{U}{2}F'(Z); \qquad w_0 = \frac{1}{2}\sqrt{\frac{U\nu_0}{x}}\,(ZF' - F) \quad \text{(11-27)}$$

For $n = 1$, Eq. 11-25 and the boundary conditions indicate that $u_1 = 0$

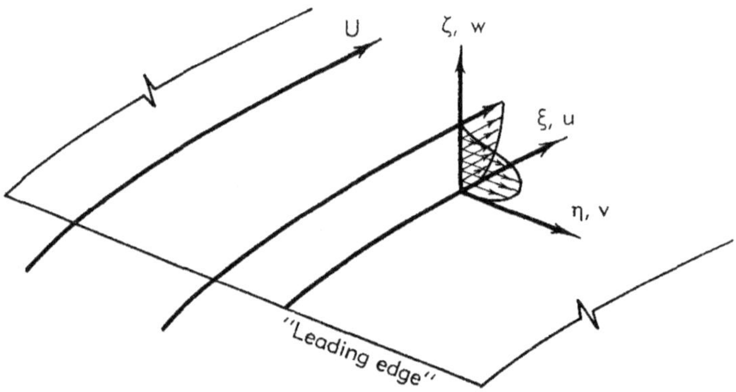

Fig. C,11d. Coordinates and velocities for circular flow over a flat plate.

is a solution. Eq. 11-26, however, now gives the first order equation for the spanwise velocity $v_1 = -0.5VG'(Z)$,

$$G''' + FG'' - 2F'G' = 8(F' - 1) \qquad \text{(11-28)}$$

with the boundary conditions,

$$G(0) = G'(0) = 0; \qquad G'(\infty) = -2 \qquad \text{(11-29)}$$

This last relation (Eq. 11-28) has been solved in [*59*] and it gives the expected outward secondary flow.

The reader will realize, in view of the discussion of von Kármán's [*27*] and Bödewadt's [*29*] solutions in Art. 9, that the secondary flow obtained above is not the same as would exist if the plate were stationary but the outer flow were rotating. In fact, Mager and Hansen [*10*] solved this second problem, also using an ordered expansion. They employed a coordinate system consisting of geodesics and geodesic-parallels, choosing one of the outer flow streamlines as the reference line from which the geodesics are measured and starting this reference line at the leading edge (see Fig. C,11d). It may be shown that for such coordinates $h_1 = 1 + c\eta$ and $h_2 = h_3 = 1$ where $c(\xi)$ is the local curvature of the reference line.

Of course, in this particular problem, where the outer flow is in rotary motion, the curvature c is a constant and, furthermore, since all the outer flow streamlines are circular, the postulated coordinates are also intrinsic. Note now that the turning of the outer flow is given by $c\xi$ and, for small values of this quantity, one again may select the terms in the differential equations according to the order of $(c\xi)$. Thus the zero-order terms give the Blasius equation for the flow in the outer flow direction, while the first order equation for the cross flow is given by

$$G''' + \tfrac{1}{2}FG'' - F'G' + (F')^2 - 1 = 0 \tag{11-30}$$

with

$$u = UF'(Z); \qquad v = c\xi UG'(Z); \qquad Z = \zeta\{U/[\nu_0\xi(1 + c\eta)]\}^{\frac{1}{2}} \tag{11-31}$$

and

$$G(0) = G'(0) = G'(\infty) = F(0) = F'(0) = 0; \qquad F'(\infty) = 1 \tag{11-32}$$

It is interesting to note that Loos [*50*] has found this equation to be involved in his solution for the secondary flow with simple, parabolic outer flow streamlines: U = const, $V = a + bx$. As Loos pointed out, this occurs because, for a flow with $a = 0$ and close to the leading edge, the circular streamlines coincide, to the first order, with the parabolic ones.

In Fig. C,11e the numerically computed solution of Eq. 11-30 for the inward secondary flow is compared to the results of Fogarty (after necessary adjustment of the similarity variables) and one sees that the secondary flow which results when the outer flow rotates is different in direction, and also much larger in magnitude, than that which results when the surface rotates. In many turbomachinery problems, both the surface (blade) and the outer flow are in rotation. Moore [*2*] found that for such a mixed problem one may simply superimpose Fogarty's and Mager and Hansen's results after, of course, the necessary substitution of the relative angular velocity of plate and fluid in place of ω has been made. Moore also found that the deflection at the surface vanishes when the plate and the fluid are rotating in the opposite sense, with the fluid rotating 0.92 times as fast as the plate. The corresponding secondary velocity profile is also shown in Fig. C,11e. Clearly, for outer flow rotation greater than the above value, the secondary flow in the boundary layer must undergo a reversal of flow direction.

In addition to working out the solution for a rotating flat plate, Fogarty [*59*] has also treated a rotating cylinder with outer flow $U = \omega y[x/L - (x/L)^2]$. In this case, it turns out that the outward secondary flow is very weak. As Moore [*2*] pointed out, this is due to the fact that the outer flow has a "secondary flow" of its own (see equation for V:

Eq. 11-19) which is directed toward the axis. The boundary layer, having thus to conform to this outer flow as well as to the Coriolis acceleration, undergoes, at any surface position, flow direction reversals, so that only a small net secondary flow results.

The rotating blade problem has also been treated by Smith [*62*], who found solutions for rotating infinite wedges (e.g. with $U = \omega y x^m$),

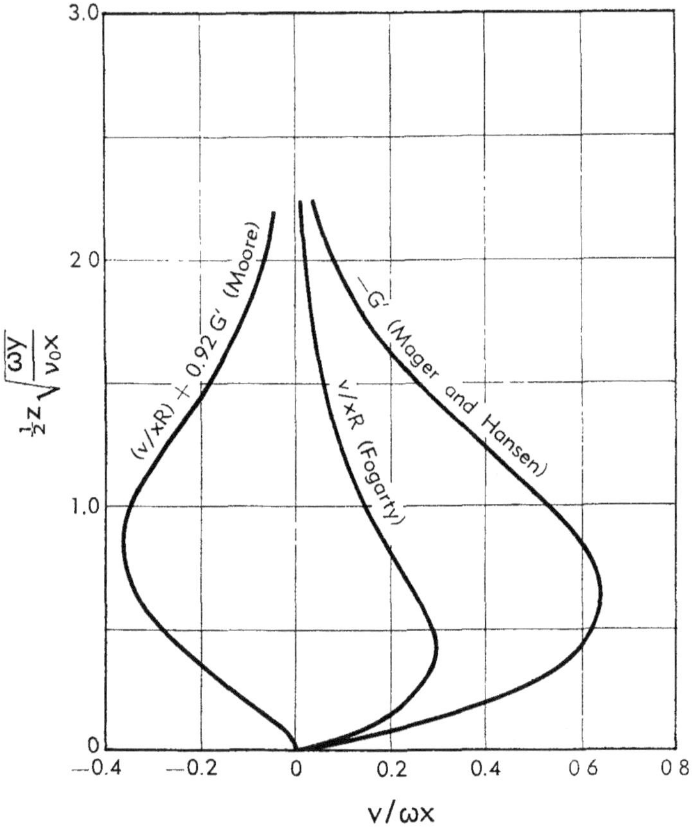

Fig. C,11e. Secondary flow velocities due to rotation and in the presence of a leading edge.

and by Graham [*63*] who used Smith's results as a basis for a momentum-integral method, thus paralleling the work of [*48*] in the yawed wing case. In all of these investigations the leading edge was assumed to extend radially from the axis of rotation, although Graham indicated that some small eccentricities are admissible in her method. The effect of such eccentricity was also specifically investigated by Liu [*64*] who found that, to the first order, the location of the leading edge has no effect on the boundary layer development over the flat plate.

The extension of all these results to compressible flow encounters the same sort of difficulty as in the yawed wing case. Mager and Hansen [*10*], however, using Howarth's [*65*] (also B,22) transformation and assuming adiabatic flow with a Prandtl number of unity, were able to obtain a solution valid in the immediate neighborhood of the reference streamline. This transformation is quite similar to Case 1 discussed in Art. 6, except that here, not only the surface coordinates, but also the velocities, are required to remain invariant. Thus, since the coordinate normal to the surface is distorted so as to account for the expansion (or contraction) of the boundary layer due to internal heating, the mass flow must be appropriately adjusted. Specifically, designating by a bar the quantities in the transformed system, one has

$$\bar{\xi} = \xi; \qquad \bar{\eta} = \eta; \qquad \bar{\zeta} = k\left(\frac{p}{p_0}\right)^{-\frac{1}{2}} \int_0^{\zeta} \rho\, d\zeta \tag{11-33}$$

and

$$\bar{\psi} = k\left(\frac{p}{p_0}\right)^{-\frac{1}{2}} \psi; \qquad \bar{\varphi} = k\left(\frac{p}{p_0}\right)^{-\frac{1}{2}} \varphi \tag{11-34}$$

where k is a constant and ψ and φ are the components of the vector potential (Art. 4). Using the above transformation, Eq. 11-30 with M_e evaluated on the reference line (and therefore constant) becomes

$$\bar{G}''' + \frac{1}{2}\bar{F}\bar{G}'' - \bar{F}'\bar{G}' + \left(1 + \frac{\gamma - 1}{2} M_e^2\right)[(\bar{F}')^2 - 1] = 0 \tag{11-35}$$

for which the solution may immediately be obtained from the incompressible case,

$$\bar{G}' = \left(1 + \frac{\gamma - 1}{2} M_e^2\right) G'(\bar{Z}) \tag{11-36}$$

This solution indicates that, aside from the usual thickening caused by aerodynamic heating, compressibility also causes a substantial increase of the secondary flow.

Cone at small angle of attack. From the analytical standpoint, the inviscid supersonic flow over general conical bodies (VI, H) has, as first pointed out by Busemann, certain very advantageous properties: In such a flow, all physical quantities (e.g. pressure, density, velocities, etc.) are constant along rays originating at the apex of the cone. One may suspect, therefore, that in the boundary layer this fact also leads to substantial simplifications. Indeed, Moore [*8*] and Hayes [*3*], independently, pointed out that as a consequence of the constancy of the pressure along the rays, the boundary layer flowing over such a general conical body, which incidentally need not have a circular cross section, exhibits a conventional, Blasius-type, parabolic similarity. It is important to note that this similarity occurs only along rays—it may, or may not, hold in the circumferential direction. In this case, therefore, the previously discussed (Art. 7) strict similarity requirements do not apply.

Using the coordinates consisting of geodesics and geodesic parallels (see Fig. C,11f) which, as discussed in Art. 3, are especially applicable in such a case, one has for a circular cone $h_2 = r(\xi) = \xi \sin \theta$, and the applicable equations for Prandtl number of unity become

$$\frac{\partial(\rho r u)}{\partial \xi} + \frac{\partial(\rho r v)}{\partial \eta} + \frac{\partial(\rho r w)}{\partial \zeta} = 0 \tag{11-37}$$

$$u \frac{\partial u}{\partial \xi} + \frac{v}{r} \frac{\partial u}{\partial \eta} + w \frac{\partial u}{\partial \zeta} - v^2 \frac{d \ln r}{d\xi} = \frac{1}{\rho} \frac{\partial}{\partial \zeta}\left(\mu \frac{\partial u}{\partial \zeta}\right) \tag{11-38}$$

$$u \frac{\partial v}{\partial \xi} + \frac{v}{r} \frac{\partial v}{\partial \eta} + w \frac{\partial v}{\partial \zeta} + uv \frac{d \ln r}{d\xi} = -\frac{1}{\rho r} \frac{\partial p}{\partial \eta} + \frac{1}{\rho} \frac{\partial}{\partial \zeta}\left(\mu \frac{\partial v}{\partial \zeta}\right) \tag{11-39}$$

$$u \frac{\partial H}{\partial \xi} + \frac{v}{r} \frac{\partial H}{\partial \eta} + w \frac{\partial H}{\partial \zeta} = \frac{1}{\rho} \frac{\partial}{\partial \zeta}\left(\mu \frac{\partial H}{\partial \zeta}\right) \tag{11-40}$$

Clearly, for adiabatic flow, the obvious solution of Eq. 11-40 is $H = \text{const.}$ Also, for slender cones, e.g. in the limit when $r/\xi \to \theta \to 0$,

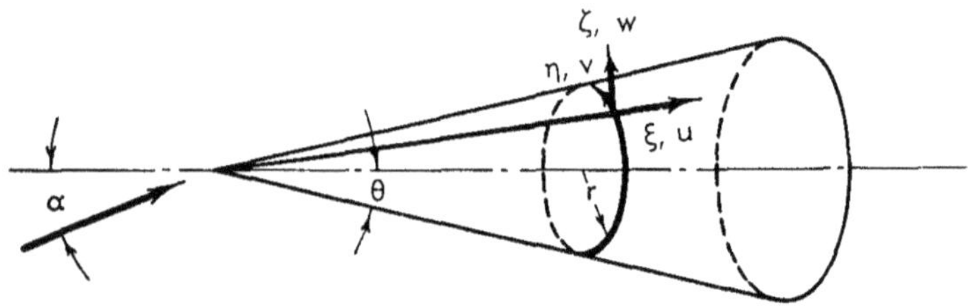

Fig. C,11f. Coordinates and velocities for a cone at angle of attack.

it will be noted that these equations become equivalent to those for a yawed cylinder (Art. 10). Making use of the similarity properties and employing simultaneously Howarth's [*65*] and Mangler's [*17*] transformations (Eq. 11-33, 11-34, 6-32, and 6-33, respectively), Moore [*66,67*] defines the similarity variable as

$$Z = \left(3 \frac{p_0}{p} \frac{C\mu_0}{\rho_0 U_0 \xi}\right)^{\frac{1}{2}} \int_0^{Re/C} \frac{\rho}{\rho_0} d(Re/C) \tag{11-41}$$

where Re is the reference Reynolds number $\zeta \rho_0 U_0/\mu_0$ and C is a constant giving the dependence of viscosity on temperature. Furthermore, he relates the normalized components of the vector potential to the velocities by

$$\frac{u}{U_0} = \frac{\partial F}{\partial Z}; \qquad \frac{v}{U_0} = \frac{\partial G}{\partial Z} \tag{11-42}$$

so that Eq. 11-38 and 11-39 become

$$L\frac{\partial^2 F}{\partial Z^2} - \frac{2}{3\sin\theta}\frac{\partial G}{\partial Z}\frac{\partial^2 F}{\partial Z\partial\eta} + \frac{2}{3}\left(\frac{\partial G}{\partial Z}\right)^2 + 2\frac{\partial^3 F}{\partial Z^3} = 0 \tag{11-43}$$

$$L\frac{\partial^2 G}{\partial Z^2} - \frac{2}{3\sin\theta}\frac{\partial G}{\partial Z}\frac{\partial^2 G}{\partial Z\partial\eta} - \frac{2}{3}\frac{\partial G}{\partial Z}\frac{\partial F}{\partial Z} + 2\frac{\partial^3 G}{\partial Z^3} = \frac{2}{3U_0^2\sin\theta\rho}\frac{dp}{d\eta} \tag{11-44}$$

where

$$L \equiv F + \frac{1}{3\sin\theta}\frac{d\ln p}{d\eta}G + \frac{2}{3\sin\theta}\frac{\partial G}{\partial\eta} \tag{11-45}$$

and the density ρ has been included in Eq. 11-44 only for the sake of clarity, since it can be eliminated by the use of the equation of state and the solution H = const. The boundary conditions are

$$F(0,\eta) = \left.\frac{\partial F}{\partial Z}\right|_{0,\eta} = G(0,\eta) = \left.\frac{\partial G}{\partial Z}\right|_{0,\eta} = 0 \tag{11-46}$$

$$\left.\frac{\partial F}{\partial Z}\right|_{\infty,\eta} = \frac{U(\eta)}{U_0}; \qquad \left.\frac{\partial G}{\partial Z}\right|_{\infty,\eta} = \frac{V(\eta)}{U_0} \tag{11-47}$$

with the outer flow velocities $U(\eta)$ and $V(\eta)$ known from the inviscid solution and obtainable, say, from M.I.T. Tables [*68*].

Moore [*66*] now notes that for small angle of attack, that is, for $\alpha \ll \theta$, the outer flow velocities may be written as

$$U = 1 - \alpha A_1\cos\eta + \cdots; \qquad V = \alpha A_2\sin\eta + \cdots \tag{11-48}$$

where the quantities A_1 and A_2 are dependent only on θ and Mach number and are obtainable from [*68*]. This approximate representation, as noted by Ferri [*69*], is valid if the outer flow is isentropic, that is, when neither θ nor the Mach number are too large. Still, Eq. 11-48 indicates that under these circumstances the outer circumferential flow is small when compared to the meridional flow. Furthermore when $\alpha = 0$, that is, for axially symmetric flow, obviously $G = \partial/\partial\eta = 0$ and Eq. 11-43 and 11-44 reduce to the Blasius flat plate equation, so that F in such a case is a function of Z alone. Consequently for small angle of attack, Moore [*66*] proposes an expansion in powers of α:

$$F = f_0(Z) - (\alpha A_1\cos\eta)f_1(Z) + \cdots; \qquad G = (\alpha A_2\sin\eta)g_1(Z) + \cdots \tag{11-49}$$

which, upon proper separation of orders of magnitude, gives the Blasius equation for f_0 and a system of linear equations for f_1 and g_1

$$2f_1''' + f_0''f_1 + f_0f_1'' - \frac{2A_2}{3A_1\sin\theta}f_0''g_1 = 0 \tag{11-50}$$

$$2g_1''' + f_0g_1'' - \frac{2}{3}f_0'g_1' + 2\left\{1 + \frac{\gamma-1}{2}M_0^2[1-(f_0')^2]\right\} = 0 \tag{11-51}$$

with boundary conditions,

$$f_1(0) = f_1'(0) = g_1(0) = g_1'(0) = 0; \qquad f_1'(\infty) = g_1'(\infty) = 1 \quad (11\text{-}52)$$

The numerical solution of these functions [66] indicated that for small angle of attack the boundary layer on the lee side of the cone, although appreciably thickened, still remains attached to the surface. This occurs in spite of the fact that the secondary flow tends to transport

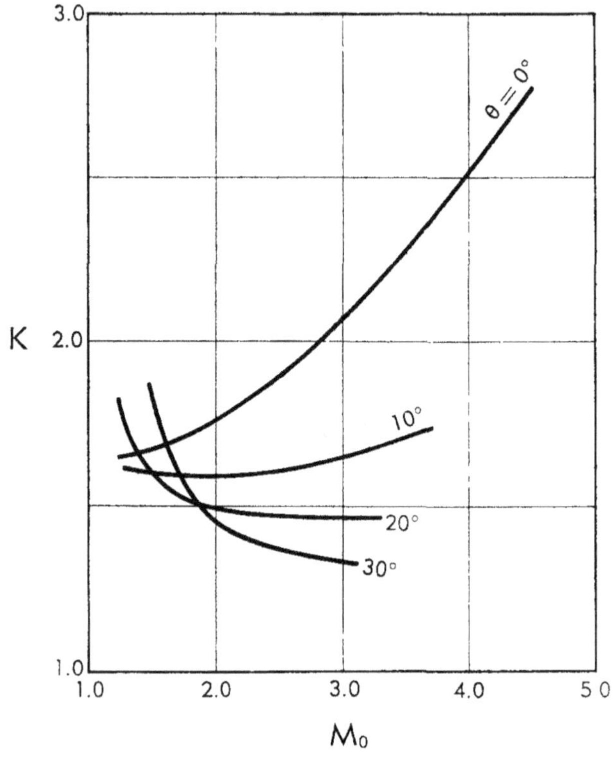

Fig. C,11g. Moore's solution for K (Eq. 11-53).

the fluid to this side, and one could therefore expect some sort of eruption there. Apparently, at small angles of attack, the boundary layer is simply able to augment the meridional flow on the top side of the cone, without any undesirable separation effects. The degree of asymmetry, however, which is produced by even a very small angle of attack is rather large. This is particularly apparent from Moore's [67] expression for the displacement surface[1] appropriate for three-dimensional flow,

$$\delta^* = \delta^*\Big|_{\alpha=0}\left[1 - \frac{\alpha}{\theta} K \cos\eta + \cdots\right] \qquad (11\text{-}53)$$

where K, a function of θ and M_0, is shown in Fig. C,11g.

[1] Definition of this surface will be discussed in Art. 13.

Eq. 11-53 indicates, then, not only the expected thickening of the boundary layer on the lee side of the cone ($\eta = \pi$), but also that the change in δ^* is directly proportional to the expansion parameter; that is, the ratio α/θ.

Flow with small turning. The expansion procedure discussed above may be generalized to study any type of three-dimensional boundary layer flow, as long as the three-dimensional effects are small when compared to the principal, two-dimensional flow. Such a generalization has been carried out for a developable surface in [*53*], where the boundary conditions causing the three-dimensional flow (e.g. outer flow velocities, or surface rotation) were assumed to be only slightly different from those for the Blasius-type, flat plate flow. This, of course, implies boundary layer solutions with only small pressure gradients and only a very slight turning of the outer flow streamlines. Denoting, therefore, the boundary layer flow velocities u^*, v^*, and w^* as linearly composed of the principal, two-dimensional flat plate flow and a perturbation flow, one has

$$\left.\begin{aligned} u^* &= u^0(x, y) + u(x, y, z) \\ v^* &= v(x, y, z) \\ w^* &= w^0(x, y) + w(x, y, z) \end{aligned}\right\} \tag{11-54}$$

so that u^0 and w^0 are given by the Blasius solution:

$$u^0 = U^0F', \qquad w^0 = \left(\frac{1}{2}\right)\sqrt{\frac{\nu_0 U^0}{x}}\,(ZF' - F), \qquad Z = z\sqrt{\frac{U^0}{(\nu_0 x)}}$$

while the first order incompressible flow equations for u, v, and w become

$$u^0\frac{\partial u}{\partial x} + u\frac{\partial u^0}{\partial x} + w^0\frac{\partial u}{\partial z} + w\frac{\partial u^0}{\partial z} = U^0\frac{\partial U}{\partial x} + \nu_0\frac{\partial^2 u}{\partial z^2} \tag{11-55}$$

$$u^0\frac{\partial v}{\partial x} + w^0\frac{\partial v}{\partial z} + 2\omega_3(U^0 - u^0) = U^0\frac{\partial V}{\partial x} + \nu_0\frac{\partial^2 v}{\partial z^2} \tag{11-56}$$

$$\frac{\partial u}{\partial x} + \frac{\partial v}{\partial y} + \frac{\partial w}{\partial z} = 0 \tag{11-57}$$

If, now, the external flow perturbation velocity components and ω_3 are expressible as powers of the x coordinate,

$$U = \sum_m A_m(y)x^m; \qquad V = \sum_n B_n(y)x^n; \qquad \omega_3 = \sum_r E_r(y)x^r \tag{11-58}$$

and furthermore, if the perturbation velocities in the boundary layer are written as

$$u = \sum_m A_m x^m f'_m + \sum_n \frac{\partial B_n}{\partial y} x^{n+1} f'_n + 2 \sum_r \frac{\partial E_r}{\partial y} x^{r+2} f'_r \tag{11-59}$$

$$v = \sum_n B_n x^n g'_n + 2 \sum_r E_r x^{r+1} g'_r \tag{11-60}$$

$$\begin{aligned} w = \frac{1}{2}\sqrt{\frac{\nu_0}{U^0 x}} \Bigg\{ & \sum_m A_m x^m [(1+2m) f_m - f'_m Z] \\ & + \sum_n \frac{\partial B}{\partial y} x^{n+1} [(2n+3) f_n + 2 g_n - f'_n Z] \\ & + 2 \sum_r \frac{\partial E_r}{\partial y} x^{r+2} [(2r+5) f_r + 2 g_r - f'_r Z] \Bigg\} \end{aligned} \tag{11-61}$$

where the f_m, f_n, f_r, g_n, and g_r are functions of the similarity variable Z only, then, in order to make Eq. 11-55, 11-56, and 11-57 uniformly valid for all values of x and y, one must satisfy the following set of differential equations:

$$f'''_m + \frac{F}{2} f''_m - mF' f'_m + \left(m + \frac{1}{2}\right) F'' f_m + m = 0 \tag{11-62}$$

$$f'''_n + \frac{F}{2} f''_n - (n+1) F' f'_n + \left(n + \frac{3}{2}\right) F'' f_n + F'' g_n = 0 \tag{11-63}$$

$$f'''_r + \frac{F}{2} f''_r - (r+2) F' f'_r + \left(r + \frac{5}{2}\right) F'' f_r + F'' g_r = 0 \tag{11-64}$$

$$g'''_n + \frac{F}{2} g''_n - nF' g'_n + n = 0 \tag{11-65}$$

$$g'''_r + \frac{F}{2} g''_r - (r+1) F' g'_r - (1 - F') = 0 \tag{11-66}$$

with the boundary conditions,

$$\begin{aligned} f_m(0) = f'_m(0) = f_n(0) = f'_n(0) = f_r(0) = f'_r(0) = g_n(0) = g'_n(0) \\ = g_r(0) = g'_r(0) = f'_n(\infty) = f'_r(\infty) = g'_r(\infty) = 0 \end{aligned} \tag{11-67}$$

$$f'_m(\infty) = g'_n(\infty) = 1 \tag{11-68}$$

As may be seen from Eq. 11-65 and 11-66, the functions g are completely independent of the functions f, which means that, to a first approximation, the cross flow v is completely independent of the principal flow perturbation u, and is affected only by the gradients of the external flow V and the component of the angular velocity normal to the surface, ω_3. Furthermore, when the cross velocity v is not a function of the y coordinate, the equations for the principal flow perturbation and the cross flow perturbation are completely uncoupled. This result is similar to those discussed in Art. 10 but applies (to this order of approximation) even when the u velocity varies in the y direction. In fact, as may be seen from Eq. 11-59, the only time the cross flow v does affect the principal flow (through functions f'_n and f'_r) occurs when that cross flow is itself a function of the crosswise coordinate y.

Solutions of Eq. 11-62 to 11-66 are easily obtainable by numerical methods and are given in [*53*]. In this connection it should be noted that Eq. 11-65 is identical to Eq. 10-35 (with n in the last equation set to zero and c set to unity). This occurs because, when V is not a function of y, the terms neglected in the small turning approximation would vanish, even if such approximation were not made. In that special case, therefore, the small perturbation solution is exact.

For compressible adiabatic flow, the Howarth transformation (Eq. 11-33 and 11-34) may again be used. In that case and in the absence of surface rotation, it is shown in [*53*] that the compressible flow solutions are given by

$$\bar{f}_m = f_m(\bar{Z}) + \frac{\gamma - 1}{2} M^2[f_m(\bar{Z}) - F(\bar{Z})] - \frac{M^2}{2}\left[m(2\gamma - 1) + \frac{\gamma - 1}{2}\right]\bar{f}_{m,c} \tag{11-69}$$

$$\bar{f}_n = f_n(\bar{Z}) + \frac{\gamma - 1}{2} M^2[f_n(\bar{Z}) - \bar{f}_{m,c}] \text{ (with } m = n + 1) \tag{11-70}$$

$$\bar{g}_n = g_n(\bar{Z}) + \frac{\gamma - 1}{2} M^2[g_n(\bar{Z}) - F(\bar{Z})] \tag{11-71}$$

where the bar signifies the transformed, compressible flow, the Mach number M is that of the principal flow, and the auxiliary function $\bar{f}_{m,c}$ satisfies

$$\bar{f}'''_{m,c} + \frac{F}{2}\bar{f}''_{m,c} - mF'\bar{f}'_{m,c} + \left(\frac{1}{2} + m\right)F''\bar{f}_{m,c} + FF'' = 0 \tag{11-72}$$

with the boundary conditions

$$\bar{f}_{m,c}(0) = \bar{f}'_{m,c}(0) = \bar{f}'_{m,c}(\infty) = 0 \tag{11-73}$$

It is apparent from Eq. 11-71 and Fig. C,10d that the effects of compressibility result not only in the characteristic stretching of the z coordinate, but also accentuate the cross flow.

The small perturbation procedure has also been used by Wilkinson

[*70*] to investigate the case where

$$U^* = U^0 x^m \quad \text{and} \quad V^* = \alpha U^0 y \quad \text{with} \quad \alpha \ll 1.$$

In other words, rather than perturbing the Blasius equation, Wilkinson instead chooses to consider the Falkner-Skan relation (B,29) as describing the principal flow. Furthermore, as already indicated in the discussion above, since V^* is a function of the y coordinate, one may expect an interaction between the cross flow and the principal flow. Wilkinson, in fact, finds that the skin friction is increased as a result of the cross flow and although the solution tends to break down for the critical value of m (at which the two-dimensional separation would be reached), the indications are that separation is delayed. This in fact has been demonstrated by Moore [*71*] who, by perturbing slightly not just the velocity profiles, but also the value of the pressure parameter m, was able to repair the breakdown of Wilkinson's solution. This permitted him to calculate the actual shift of the separation point towards a more negative value of m. As Moore points out, this is the direct consequence of flow divergence which tends to thin out the boundary layer. It should be noted that similar results were also found in [*53*] where, for a flow in a simulated S-shaped duct, the cross flow was shown to delay the separation in regions where it increases with y, but to accelerate it in regions where V^* decreases with y.

The general small perturbation procedure has also been extended to spinning and yawing bodies of revolution [*72*], for which the surface shape is given as a power function of the meridional coordinate. It must, however, be noted here that this procedure will, in general, be invalid very near the stagnation point, where the perturbation velocities due, say, to the angle of attack, will necessarily be of the same order of magnitude as the principal, axially symmetric flow.

For circular cylinders and cones, the small perturbation method has also been extended to include second order terms by Martin [*73*] and by Sedney [*74*], respectively, and even higher order terms by Kelly and Thacker [*75*]. All of these investigators were principally interested in a so-called "Magnus effect," which causes a side force and a yawing moment on spinning projectiles, thus influencing their dispersion. This Magnus effect arises because of an interaction between the flow changes due to an angle of attack and those due to body spin, which demonstrates itself by causing certain asymmetries in the boundary layer displacement thickness distribution. Since such an interaction does not occur in the first order theory, consideration of higher order terms is absolutely essential to a study of the Magnus effect. In fact, Martin's [*73*] results indicate that the Magnus force is a linear function of the product of the angle of attack and spin rate. This result has recently been confirmed experimentally by Thorman [76] who also found that at greater spin rates the inclusion of higher order terms [75] gave results which, although not quite as accurate, nevertheless were still qualitatively correct.

In addition to checking the theories describing the Magnus effect, Thorman's [76] measurements also shed some light on the validity of

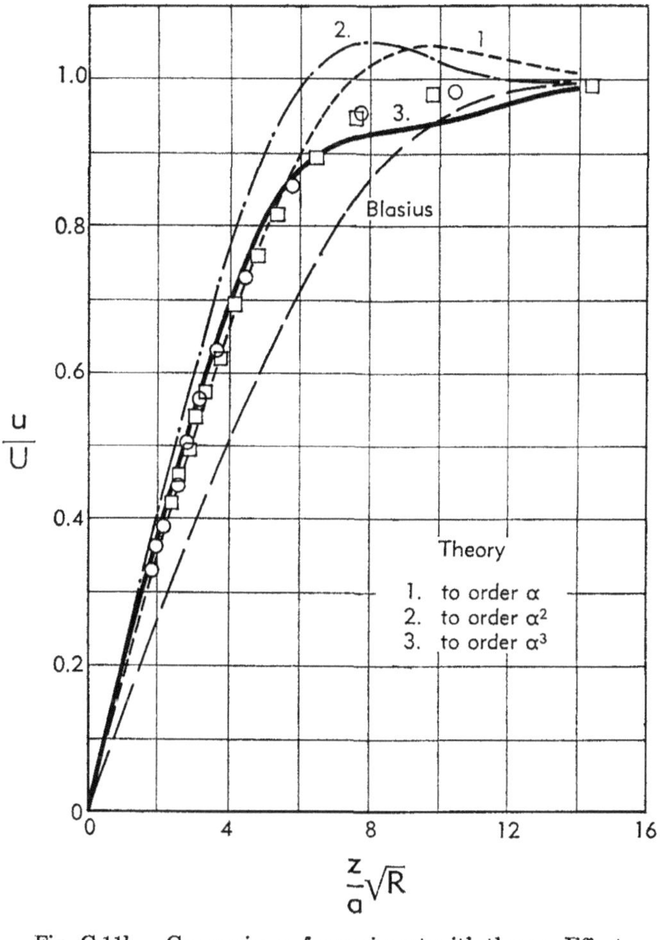

Fig. C,11h. Comparison of experiment with theory. Effect of angle of attack on meridional flow (from [76]).

the expansion procedure in general. For example, one set of his measurements concerns the meridional velocity profiles on an axially symmetric body at 3° angle of attack. These measurements, taken on the windward side in the plane of symmetry, together with approximations of various order, are shown in Fig. C,11h.

As may be seen from this figure, while the third order approximation is best, the first order solution describes the velocity profile somewhat more accurately than the second order solution. Also, for a spinning body at zero angle of attack, the crosswise velocity profile is described quite well by the first order theory. This is shown in Fig. C,11i.

It should be noted that in this case, since the measurements were taken on a cylindrical portion of the body far from the nose, the first order solution is in fact the exact solution and the good agreement with

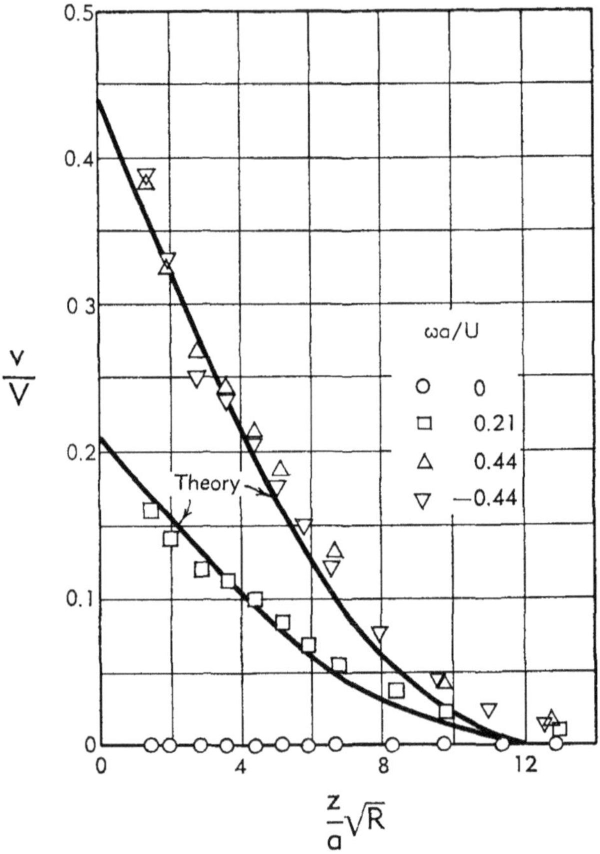

Fig. C,11i. Comparison of experiment with theory. Effect of body spin on crosswise flow (from [76]).

experiment is thus not too surprising. Still, the meridional velocity for the same case, as may be seen from Fig. C,11j, did not show as good an agreement with the theory, the experiment indicating a velocity decrease with increasing spin. As Thorman [76] noted, while this disagreement may have been due in part to some uncertainties in the location of the virtual origin for the boundary layer, it is unlikely that different origins would apply to the transverse and meridional flows. For this reason one must conclude that here the theories appear to fall short of describing the experimental data properly. It is conceivable that this disagreement may be due to the pressure gradient normal to the surface (see Eq. 2-10) which, although neglected by all the theories, may, for a thick spinning boundary layer, be appreciable.

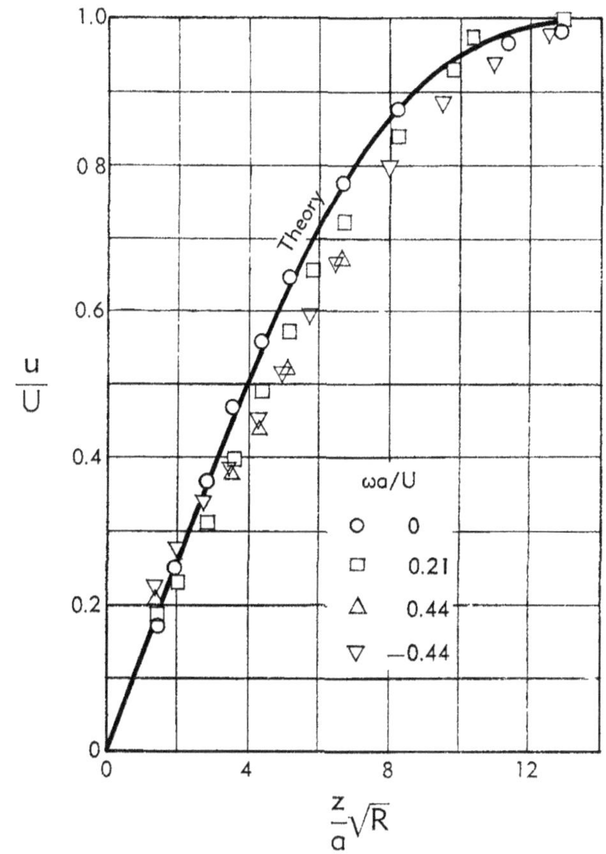

Fig. C,11j. Comparison of experiment with theory. Effect of body spin on meridional flow (from [76]).

C,12. Restricted Solutions. In many important cases it happens that boundary layer flow, while generally three-dimensional in character, is, in certain restricted regions of the surface, very nearly two-dimensional. We recall from the discussion in Art. 3 that, in the absence of body and Coriolis forces and when the outer flow streamlines are coincident with the geodesics of the surface, the secondary flow in the boundary layer vanishes. As will be seen below, this means that in such cases the boundary layer is very nearly two-dimensional along the streamlines of the outer flow. In general, of course, such a situation does not prevail everywhere on the surface; however, sometimes it holds in the most critical regions, precisely where the solution is of greatest interest. It was first pointed out by Moore [77] and later by Squire [58] that such local, nearly two-dimensional boundary layer flow may be used to advantage in the study of cases which otherwise would be difficult, if not impossible, to analyze.

We start by considering the steady, incompressible boundary layer flow equations in intrinsic coordinates, free of any body or Coriolis forces, for which $\kappa_1 = 0$; that is, when the external flow streamline (e.g. ξ coordinate) is a geodesic. Making the appropriate simplifications in Eq. 2-7, 2-8, 2-9, 3-5, and 3-6, one obtains

$$\frac{1}{h_1 h_2}\left[\frac{\partial(h_2 u)}{\partial \xi} + h_1 \frac{\partial v}{\partial \eta} + h_1 h_2 \frac{\partial w}{\partial \zeta}\right] = 0 \tag{12-1}$$

$$\frac{u}{h_1}\frac{\partial u}{\partial \xi} + \frac{v}{h_2}\frac{\partial u}{\partial \eta} + w\frac{\partial u}{\partial \zeta} - \frac{v^2}{h_1 h_2}\frac{\partial h_2}{\partial \xi} = \frac{U}{h_1}\frac{\partial U}{\partial \xi} + \nu_0 \frac{\partial^2 u}{\partial \zeta^2} \tag{12-2}$$

$$\frac{u}{h_1}\frac{\partial v}{\partial \xi} + \frac{v}{h_2}\frac{\partial v}{\partial \eta} + w\frac{\partial v}{\partial \zeta} + \frac{uv}{h_1 h_2}\frac{\partial h_2}{\partial \xi} = \nu_0 \frac{\partial^2 v}{\partial \zeta^2} \tag{12-3}$$

As pointed out in Art. 3, since $v(0) = v(\infty) = 0$, the homogeneity of Eq. 12-3 indicates that $v = 0$ is a solution. It should be noted, however, that inasmuch as one now considers the case where not all of the external flow streamlines are geodesics, $\partial v/\partial \eta$ may not be set equal to zero. It is, in fact, precisely because of this term that the remaining Eq. 12-1 and 12-2 are not two-dimensional, even along the ξ coordinate. Furthermore, one now notes that, since Eq. 12-3 vanishes, the two remaining equations are not sufficient to determine the flow because they contain the troublesome term $\partial v/\partial \eta$. As a way out of these difficulties, Squire [58] suggested that the velocities u, v, and w be expanded in power series in η with the coefficients being functions of ξ and ζ. The geodesic curvature κ_1 may also be locally represented in such a way, so that if the geodesic is coincident with the $\eta = 0$ coordinate, one takes

$$\left.\begin{aligned} u &= u_0(\xi, \zeta) + \eta u_1(\xi, \zeta) + \cdots \\ v &= \qquad\qquad\;\; \eta v_1(\xi, \zeta) + \cdots \\ w &= w_0(\xi, \zeta) + \eta w_1(\xi, \zeta) + \cdots \\ \kappa_1 &= \qquad\qquad\;\; \eta \kappa_{11}(\xi, \zeta) + \cdots \end{aligned}\right\} \tag{12-4}$$

and obtains from Eq. 2-7 and 2-8 the zero-order equations:

$$\frac{1}{h_1 h_2}\left[\frac{\partial(h_2 u_0)}{\partial \xi} + h_1 v_1 + h_1 h_2 \frac{\partial w_0}{\partial \zeta}\right] = 0 \tag{12-5}$$

$$\frac{u_0}{h_1}\frac{\partial u_0}{\partial \xi} + w_0 \frac{\partial u_0}{\partial \zeta} = \frac{U_0}{h_1}\frac{\partial U_0}{\partial \xi} + \nu_0 \frac{\partial^2 u_0}{\partial \zeta^2} \tag{12-6}$$

and from Eq. 2-9 the equation for v_1

$$\frac{u_0}{h_1}\frac{\partial v_1}{\partial \xi} + \frac{v_1^2}{h_2} + w_0\frac{\partial v_1}{\partial \zeta} + \frac{u_0 v_1}{h_1 h_2}\frac{\partial h_2}{\partial \xi} - u_0^2\kappa_{11} = -U_0^2\kappa_{11} + \nu_0\frac{\partial^2 v_1}{\partial \zeta^2} \quad (12\text{-}7)$$

Eq. 12-5, 12-6, and 12-7 may now be solved for the velocities u_0, w_0, and v_1 in terms of the two independent variables ξ, ζ by using the standard two-dimensional boundary layer methods. In fact Squire [*58*], to illustrate the procedure, carried out such a solution along the line of symmetry of an ellipsoid. It should be noted, however, that the expansion process used, while giving meaningful results independent of the rest of the boundary layer, is in a certain sense not uniform. This is true because the equation determining the v velocity, in the neighborhood of the $\eta = 0$ coordinate, is always by one order higher than the equations determining the u and w components. This fact suggests another method, which actually antedated the work of Squire and was first proposed by Moore [*77*].

Moore was interested in obtaining a solution of flow over a cone at supersonic speeds and at a large angle of attack. The appropriate relations (Eq. 11-43, 11-44, and 11-45) for such a flow were discussed before in Art. 11 and a linearized solution applicable to small angles of attack was outlined there. For a large angle of attack, however, such a linearization is invalid and a general solution of the equations is very difficult, if not impossible, to obtain. Fortunately, at large angle of attack, as Moore noted, the most interesting flow region is in the plane of symmetry, and by restricting consideration to that plane only, one is able to simplify the problem considerably. At this point, it should be noted that in the plane of symmetry (e.g. at $\eta = 0, \pi$) the surface projections of the external flow streamlines coincide with the geodesics and one has exactly the condition discussed previously; namely $v = \partial G/\partial Z = 0$ and $dp/d\eta = 0$. Moore recognized these two simplifications, and, in addition, argued that since $u = \partial F/\partial Z$ is even about the plane of symmetry (and may be expected to be regular), $\partial^2 F/(\partial Z \partial \eta)$ vanishes there. The net result of all these simplifications is, as mentioned above, that every term in Eq. 11-44 vanishes, and one is unable to determine F and G through the one remaining Eq. 11-43, which unfortunately contains the derivatives of both functions. To solve this dilemma, Moore proposed to differentiate the cross flow equation (Eq. 11-44) with respect to η first, and only then to neglect the terms which vanish at $\eta = 0, \pi$. Defining a new dependent variable $S(Z)$,

$$\left.\frac{\partial V}{\partial \eta}\right|_{0,\pi} S(Z) \equiv U_0(0, \pi) \left.\frac{\partial G}{\partial \eta}\right|_{0,\pi} \quad (12\text{-}8)$$

and a convenient parameter k,

$$U_0(0, \pi)k \equiv \frac{2}{3 \sin \theta} \left.\frac{\partial V}{\partial \eta}\right|_{0,\pi} \quad (12\text{-}9)$$

this procedure yields the following results:

$$(F + kS)F'' + 2F''' = 0 \tag{12-10}$$

$$(F + kS)S'' + 2S''' - k(S')^2 - \frac{2}{3}S'F' + \left(k + \frac{2}{3}\right)\left\{1 + \frac{\gamma - 1}{2}M^2[1 - (F')^2]\right\} = 0 \tag{12-11}$$

$$F(0) = F'(0) = S(0) = S'(0) = 0; F'(\infty) = S'(\infty) = 1 \tag{12-12}$$

where the primes indicate total differentiation with respect to the similarity variable Z, and the Mach number M is taken at the outer edge of the boundary layer. It should be noted that the parameter k depends essentially on the angle of attack; the greater the angle, the greater the magnitude of $\partial V/\partial\eta$, and hence k. Furthermore, at the bottom of the cone k is positive, while at the top it is negative.

Eq. 12-10 and 12-11 must, of course, be solved separately at $\eta = 0$ and at $\eta = \pi$. Such numerical solutions at the bottom of the cone ($\eta = 0$), carried out in [77], show the expected thinning of the boundary layer as the angle of attack is increased. At the top of the cone, however, one encounters difficulties in trying to obtain numerical solutions for a large angle of attack. These difficulties led Moore [77] to consider the behavior of the asymptotic solutions for large values of Z. In other words, he assumed that $F = Z + f(Z)$, $S = Z + s(Z)$ with $f'(\infty) = s'(\infty) = 0$, and substituted these expressions into Eq. 12-10 and 12-11. As a result of this simplification and with a certain change of the dependent variable, Eq. 12-11 may be shown to reduce to Weber's equation [78] and thus gives two complementary solutions for s'. For $k > -\frac{1}{3}$ (e.g. at the bottom of the cone or at the top of the cone but for very small α) one of these asymptotic solutions must be rejected, since it does not vanish for large values of Z as required by the boundary condition. Moore was thus able to conclude that the solutions of the complete equations (Eq. 12-10 and 12-11) for $k > -\frac{1}{3}$ may be unique (a necessary but not sufficient condition). For $k < -\frac{1}{3}$, however, both of the solutions must be retained and in addition an undetermined constant appears. This seems to indicate, then, that for $k < -\frac{1}{3}$ (e.g. at the top of the cone and at large α) the solution of Eq. 12-10 and 12-11 is not unique. This lack of uniqueness appears to reflect the fact that for large angles of attack the flow on the top of the cone is strongly affected by the circumferential flow in the rest of the boundary layer, and one cannot expect a solution restricted to the top region (with no account taken of the development history of the boundary layer elsewhere) to hold. The value of $k = -\frac{1}{3}$ thus determines the limits of the applicability of the restricted solution. Of course, if one had a solution of the complete equations (Eq. 11-43 and

11-44), no such question of uniqueness would ever appear. Furthermore, since the limiting value $k = -\frac{1}{3}$ was obtained from an asymptotic solution at large Z, one may expect that, due to the appreciable transfer of fluid by secondary motion, the minimum limiting value of k for an actual solution of Eq. 12-10 and 12-11 should be greater than $-\frac{1}{3}$. In fact, Moore argues that a better limit is given by $k > (-\frac{1}{3})(S'/F')_{max}$.

So far, nothing has been said about the behavior of the solution for f', which may be obtained from the asymptotic form of Eq. 12-10. This solution is given by $f'' = \exp[-(1 + k)/4]Z^2$, which indicates that $f''(\infty) \to 0$ only if $k > -1$. For $k \leqq -1$ no solution exists which satisfies the proper boundary conditions, and therefore Moore concluded that the usual boundary layer assumptions fail to apply as the boundary layer erupts in separation. This limiting value of the angle of attack for the existence of a boundary layer at the top of the cone was first noted (and similarly explained) by Hayes [*3*], who arrived at his result from consideration of boundary layer divergence in that region. One may note that the existence of such a limiting value of angle of attack is fairly obvious on purely intuitive grounds: For a large angle of attack, the flow over the top of the cone must be similar to that on the wake side of a yawed cylinder, and the failure of the boundary layer equations as a result of separation may be expected to occur there first.

C,13. Momentum-Integral Methods. The momentum-integral methods which have such wide use in two-dimensional boundary layer calculations may also be extended to three-dimensional flows. Examples of such extensions have, in fact, already been discussed in Art. 9 and 10. In the present article, we shall concern ourselves not so much with the many details which are involved in the application of such methods to various three-dimensional flows, but, rather, with a discussion of the more general features of such analyses.

As is well known, any momentum-integral method implies that one is able to write, in an approximate fashion, the expressions for the velocities u and v. In general, this may be conveniently done by introducing a characteristic parameter for each velocity profile, such as (but not necessarily) a boundary layer thickness, and describing both velocities in terms of these parameters so that they satisfy the greatest number of boundary conditions. Such a description has been used, for example, by Wild [*50*] and also by Cooke [*36*] in their procedures for solution of three-dimensional momentum-integral equations. Some other investigators [*41,9*] found it more convenient to use both parameters when describing the cross flow, but employ only one in the description of the primary flow. Such different uses of the parameters will introduce only nonessential changes in the arguments which are to follow.

Now, if $P_1(\xi, \eta)$ and $P_2(\xi, \eta)$ are parameters corresponding to the velocities u and v respectively, then each one of the momentum and displacement thicknesses may be expressed as some function of P_1 and/or P_2 so that, for adiabatic incompressible flow, the two relations, Eq. 5-8 and 5-9, are sufficient to determine these parameters completely. A direct substitution into these equations results in

$$\left(\frac{1}{h_1}\frac{\partial\theta_{11}}{\partial P_1}\right)\frac{\partial P_1}{\partial\xi} + \left(\frac{1}{h_2}\frac{\partial\theta_{12}}{\partial P_1}\right)\frac{\partial P_1}{\partial\eta} + \left(\frac{1}{h_2}\frac{\partial\theta_{12}}{\partial P_2}\right)\frac{\partial P_2}{\partial\eta} + D_1 = 0 \quad (13\text{-}1)$$

$$\left(\frac{1}{h_1}\frac{\partial\theta_{21}}{\partial P_1}\right)\frac{\partial P_1}{\partial\xi} + \left(\frac{1}{h_1}\frac{\partial\theta_{21}}{\partial P_2}\right)\frac{\partial P_2}{\partial\xi} + \left(\frac{1}{h_2}\frac{\partial\theta_{22}}{\partial P_2}\right)\frac{\partial P_2}{\partial\eta} + D_2 = 0 \quad (13\text{-}2)$$

where D_1 and D_2 may be functions of P_1, P_2, h_1, h_2, U, V and the variables ξ and η, but contain no derivatives of either of the parameters P_1 and P_2. The form of the above equations has prompted Timman [*9*] to suggest that the solutions of the momentum-integral equations be carried out by the method of characteristics. This suggestion, however, is impractical, because for small cross flow, Eq. 13-1 and 13-2 may be shown [*79*] to be parabolic so that only one characteristic exists. (Incidentally, a parabolic nature of the turbulent boundary layer momentum-integral equations has also been indicated by evaluation of experimental data in [*11*].) Timman himself apparently recognized the futility of applying the method of characteristics, since he does not employ it in his later work [*80*]. In a perfectly general case, the solution of Eq. 13-1 and 13-2 must thus be obtained by rather involved numerical methods which usually require the use of modern electronic computers. Such a procedure goes against the very essence of the momentum-integral method, which, after all, is at best approximate and should not result in complicated computations.

In many cases, however, one can again make use of the small cross flow approximation [*11,70,79*]. One thus considers each velocity as described by its own characteristic parameter and writes

$$u = u_0(\zeta, P_{10}) + u_1(\zeta, P_{11}) + \cdots; \qquad v = v_1(\zeta, P_{21}) + \cdots \quad (13\text{-}3)$$

Here, of course, consistent with the very idea of small cross flow, P_{10} is a function of ξ only, but P_{11} and P_{21} may be functions of both ξ and η. All the characteristic integral thicknesses may now be similarly expanded. For example, the momentum thicknesses thus become

$$\left.\begin{aligned}\theta_{11} &= \theta_{11}^0(P_{10}) + \theta_{11}^1(P_{10}, P_{11}) + \cdots\\ \theta_{12} &= \theta_{12}^1(P_{10}, P_{21}) + \cdots\\ \theta_{21} &= \theta_{21}^1(P_{10}, P_{21}) + \cdots\\ \theta_{22} &= O(v^2)\end{aligned}\right\} \quad (13\text{-}4)$$

A substitution of such expanded quantities into Eq. 5-8 and 5-9 gives the zero order total differential equation:

$$\left(\frac{1}{h_1}\frac{\partial\theta_{11}^0}{\partial P_{10}}\right)\frac{dP_{10}}{d\xi} + D_{10} = 0 \tag{13-5}$$

where now D_{10} contains terms of zero order only and thus may not be a function of η. The integration of Eq. 13-5 thus yields $P_{10}(\xi)$, so that this parameter may henceforth be considered known. In addition, the first order equations become

$$\left(\frac{1}{h_1}\frac{\partial\theta_{11}^1}{\partial P_{11}}\right)\frac{\partial P_{11}}{\partial\xi} + \left(\frac{1}{h_2}\frac{\partial\theta_{12}^1}{\partial P_{21}}\right)\frac{\partial P_{21}}{\partial\eta} + D_{11} = 0 \tag{13-6}$$

$$\left(\frac{1}{h_1}\frac{\partial\theta_{21}^1}{\partial P_{21}}\right)\frac{\partial P_{21}}{\partial\xi} + D_{21} = 0 \tag{13-7}$$

where the derivatives of P_{10} were incorporated into the functions D_{11} and D_{21}. As may be seen from Eq. 13-7, the parameter P_{21} can be evaluated by integration along the ξ coordinate. By carrying this out for a number of neighboring coordinates, one is then in a position to determine numerically the derivative $\partial P_{21}/\partial\eta$ which enters into Eq. 13-6.

Once the values of this derivative along ξ are known, the integration of Eq. 13-6 for P_{11} may be accomplished. As we see then, while P_{21} may be found relatively simply, the determination of P_{11} is, at best, quite complex and one would like, if possible, to avoid it. This may be accomplished, sometimes, by a proper choice of coordinates, such that not only v but also its variation with respect to the crosswise coordinate is vanishingly small. In those cases, of course, the cross flow does not interact with the primary flow, so that θ_{11}^1 may be considered negligible and need not be computed. This proper choice of coordinates is often apparent even before any numerical solution is attempted. Thus, for example, if the outer flow streamlines have nearly the shape of surface geodesics, one may expect the intrinsic coordinates (Art. 3) to be most suitable for the purpose. In fact, Zaat [*81*] has carried out computations in such an intrinsic system (and neglecting θ_{11}^1) for a yawed ellipsoid. His results agree rather well with more accurate computations [*80*] of the same case, although v is not completely independent of the crosswise coordinate. In this connection, Zaat [*81*] also found that the integration of Eq. 13-5 along the streamline (e.g. in the system given by Eq. 3-7) could be considerably simplified by the use of an approximate linear relation, analogous to that given by Thwaites (B,31) for two-dimensional flow. In this way he obtained

$$P_{10} = P_{10,i}\left(\frac{L}{L_i}\right)^2\left(\frac{U_i}{U}\right)^4 + 5.08\left(\frac{L}{U^2}\right)^2\int_{\Phi_i}^{\Phi}\left(\frac{U}{L}\right)^2 d\Phi \tag{13-8}$$

where P_{10} represents the boundary layer thickness of the principal flow, the function L has the meaning explained in Art. 3, and subscript $_i$ signifies the initial values at Φ_i.

Once the various parameters are known, using Eq. 13-3 and 13-4, all the integral thicknesses as well as the velocities u and v may be computed. In addition, it is often desirable to determine the amount by which the boundary layer modifies the shape of the body by deflecting the external flow. Now, in two-dimensional problems, this information is given directly by the displacement thickness; however, in three-dimensional flow, one has two separate displacement thicknesses (Eq. 5-3) and it is not clear how these are related to the amount of deflection we are seeking. In this connection, Moore [*82*] found the equation governing a displacement surface which defines the desired effective modification of the body shape. If δ^* is the value of ζ locating the displacement surface, then the defining equation may be written in terms of the total external velocity Q as

$$\frac{\partial}{\partial \xi}\left[h_2\rho_e Q\left(\frac{U}{Q}\,\delta^* - \delta_1\right)\right] + \frac{\partial}{\partial \eta}\left[h_1\rho_e Q\left(\frac{V}{Q}\,\delta^* - \delta_2\right)\right] = 0 \quad (13\text{-}9)$$

so that once the displacement thicknesses δ_1 and δ_2 are known, δ^* may also be found.

CHAPTER 3. SPECIAL PROBLEMS

C,14. Sudden Lateral Perturbation. In all the examples considered so far, the boundary layer flows in the principal and crosswise directions have been affected by the actions of viscosity and pressure gradient to similar degrees. As a result of this, the thicknesses of the principal and the crosswise flows tended to remain roughly the same. There exists, however, a class of three-dimensional boundary layer problems of great interest in turbomachinery design, where such "parallel" development of both flows does not exist. These problems occur when a boundary layer remains two-dimensional for a considerable distance, and then is suddenly perturbed laterally. Such a perturbation may be caused either by a sudden lateral movement of the surface over which the flow takes place, or by an abrupt curvature of the outer flow streamlines. In turbomachinery this happens, for example, at the junction of the rotor and stator hubs, and also when a casing boundary layer enters a rotor stage. The resulting cross flow boundary layer will then grow, starting not from the origin of the principal flow but from the point where the abrupt action took place. Problems of this type have been investigated in [*55*, pp. 21-33] and a thorough and illuminating discussion of this work is also given in [*2*].

Sudden lateral movement of the surface. Imagine two coplanar flat plates (see Fig. C,14a) of which one is stationary but the other moves with the lateral velocity V. The sliding contact between the two plates occurs at $x = L$. A normal, Blasius-type, two-dimensional boundary layer develops over the stationary plate, starting from the leading edge ($x = 0$). Since the moving plate drags fluid with it, crossflow must develop downstream of the sliding junction, i.e. for all $x > L$. But also, since none of the quantities varies with the crosswise coordinate y, the appropriate boundary layer equations describing the flow over the moving plate are separable (Art. 10). This means that, even for $x > L$,

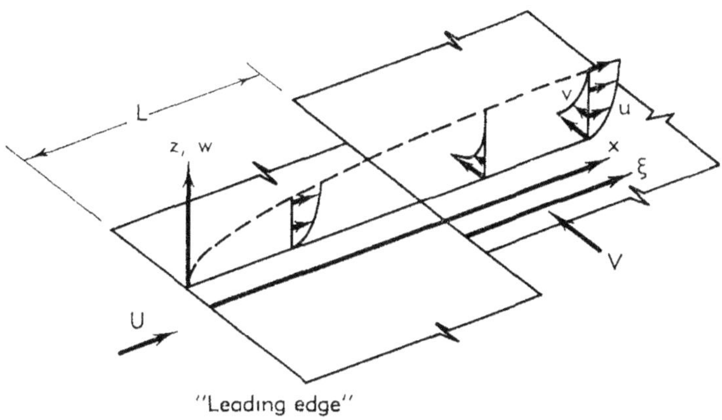

Fig. C,14a. Coordinates and velocities for sudden lateral movement of the surface problem.

the principal flow is completely independent of the lateral movement of the surface and is still given by the Blasius solution:

$$u = UF'(Z), \qquad w = 0.5\sqrt{\frac{\nu_0 U}{x}}\,(ZF' - F), \qquad Z \equiv z\sqrt{\frac{U}{(\nu_0 x)}}$$

The only velocity which now remains to be determined is v, which is governed by the linear relation, Eq. 10-3, with the new boundary conditions $v(0) = V$ and $v(\infty) = 0$. Except for this minor change of the boundary conditions it would seem, therefore, that the present problem is quite similar to those discussed in Art. 10. Actually, however, this is not the case, for one must insure that the cross flow start not at the leading edge but at the sliding contact $x = L$. In practice, this means that the Blasius similarity variable Z may not be used to describe the cross flow and a new similarity variable based on the distance from the onset of the cross flow $\xi = x - L$ must be introduced. In this connection, it is convenient to define $X \equiv [1 - (L/x)]^{\frac{1}{2}} = (\xi/x)^{\frac{1}{2}}$ so that as ξ grows from zero to infinity, X varies between zero and unity. For $X = 1$, the boundary layer is either very far away from the sliding contact, or the cross flow starts immediately at the original leading edge.

One could now seek a solution of Eq. 10-3 by substituting the Blasius expressions for the u and w velocities and solving the resulting equations numerically, assuming, of course, that v is a function of Z as well as of X. However, since the cross flow starts developing only when the primary layer is already relatively thick (thus forming sort of a "sublayer"), one may instead introduce the leading terms in the Taylor series for F

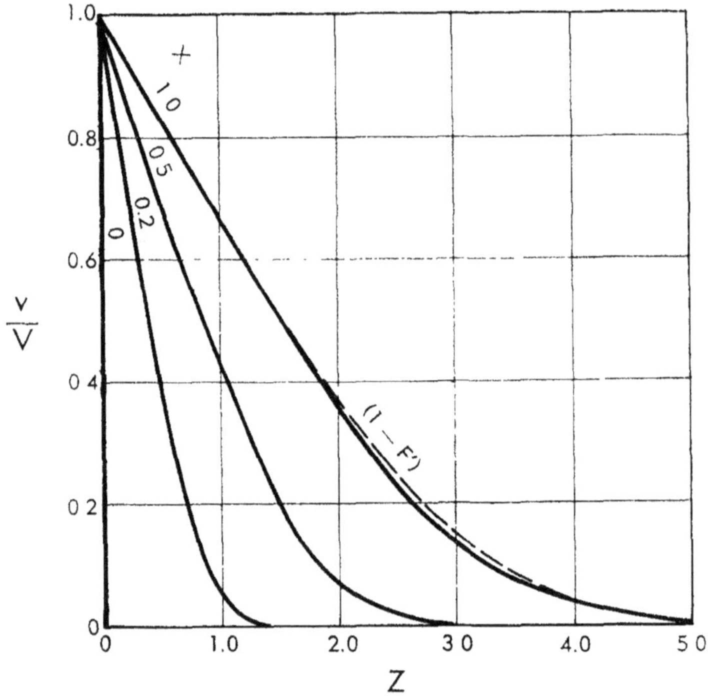

Fig. C,14b. Cross flow due to sudden lateral movement of the surface.

and F' near $z = 0$. Such a procedure was followed in [*55*, pp. 21–33] and it leads after the determination of a proper similarity variable to the Whittaker's total differential equation. The solution of this equation with our boundary conditions (which, incidentally, differ from those used in [*55*, pp. 21–33) becomes

$$\frac{v}{V} = 1 - E_3 \left\{ \left[0.0553 \left(\frac{1}{X^2} - \frac{1}{2} \right) \right]^{\frac{1}{2}} Z \right\} \tag{14-1}$$

where E_3 is a function related to the incomplete gamma function as defined in [*83*, p. 23]. This solution is illustrated in Fig. C,14b for various values of X.

Because of the expansion used in approximating the u and w velocities, one may expect that the greatest inaccuracies will arise when the cross flow is thick, that is, when $X \to 1$. On the other hand, for $X = 1$, as may be inferred from the discussion in Art. 10, the correct solution for the present boundary conditions is $v/V = 1 - F'$. As may be seen from Fig. C,14b the approximate solution seems to come rather close to that result. One should point out that since $E_3(1.5) \cong 1.0$, the cross flow boundary layer thickness is given by

$$\frac{\delta}{x} \cong \frac{1.5}{\left[0.0553\left(\frac{1}{X^2} - \frac{1}{2}\right)\right]^{\frac{1}{2}}} Re^{-\frac{1}{2}} \tag{14-2}$$

and consequently $\delta/x \to 0$ as $(\xi/x)^{\frac{1}{2}} \to 0$, while in the conventional laminar boundary layer problems, $\delta \to 0$ as $x^{\frac{1}{2}} \to 0$. Furthermore, for $X = 1$, Eq. 14-2 gives $\delta/x \cong 5Re^{-\frac{1}{2}}$, which is in rather good agreement with the Blasius value.

It is also important to note that Goldstein [*49*], who investigated the termination of the two-dimensional flat plate boundary layer in a wake, has also found a sort of "sublayer" starting at the trailing edge. Here the sublayer represents, of course, the gradual spreading of the effects arising from the abrupt replacement of the no-slip by a zero-shear condition. It is therefore of interest that in Goldstein's sublayer the velocity profiles are functions of $z\xi^{\frac{1}{2}}$ (with ξ measured from the trailing edge) similarly to the results obtained in [*55*, pp. 21–33] for the abruptly formed cross flow.

Abrupt curvature of outer flow streamlines. When the abrupt change originates in the external flow, the procedure is similar although not identical. For example, imagine again that, up to $x = L$, a two-dimensional flow corresponding to the constant external flow velocity U exists. At $x = L$, however, the external flow encounters a set of turning vanes of negligible thickness, so that while U still remains constant, a cross flow component $V = A\xi^m$ is abruptly created. As discussed in Art. 10, the equations are separable, but now, since there is a crosswise pressure gradient, two separate solutions must be carried out. First, the "nonviscous" solution, which applies in the outer part of the boundary layer, owing to the presence of a crosswise pressure gradient, is $(v/V)_{\text{outer}} = U/u$. This solution obviously does not satisfy the boundary conditions at the surface, and thus must be joined at some suitable point to an inner viscous solution applicable only near the surface. The viscous solution, therefore, may be obtained by considering only a few terms of a power series in Z, substituting these into Eq. 10-33, and applying boundary and joining conditions. It must be noted, however, that while the outer solution represents a redistribution of vorticity which is essentially independent of X, the inner solution, arising because of a viscous shearing action, must be a function of both X and Z. Such an inner solution was found in [*55*, pp. 21–33] and may be written here as

$$\left(\frac{v}{V}\right)_{\text{inner}} = B\left(\frac{Z}{X}\right) - \frac{m}{2}\left(\frac{Z}{X}\right)^2 + D\left(\frac{Z}{X}\right)^3 \tag{14-3}$$

where the coefficients B and D are given by

$$B = \left(\frac{X}{Z^*}\right)\left[\frac{1}{0.332Z^*} + \frac{m}{2}\left(\frac{Z^*}{X}\right)^2 - D\left(\frac{Z^*}{X}\right)^3\right] \tag{14-4}$$

$$D = \left(\frac{X}{Z^*}\right)^3\left[\frac{1}{0.996} + (Z^*)^2\left(\frac{m}{6X} - \frac{1}{72}\right) - \frac{0.332}{864}(Z^*)^5\right] \tag{14-5}$$

and Z^* is the value of Z at which the inner and outer solutions are joined. This junction point may be conveniently approximated by

$$Z^* \cong \left[\frac{24}{0.332\left(1 + \dfrac{3m}{2X^2}\right)}\right]^{\frac{1}{3}} \tag{14-6}$$

It is therefore apparent from Eq. 14-6 that, as X tends to zero, the extent of the viscous inner solution diminishes. This occurs, of course, because the no-slip condition has less length (ξ) in which to act, while the suddenly applied crosswise pressure gradient causes an immediate redistribution of vorticity. For small values of X, therefore, the outer nonviscous solution is applicable practically throughout the boundary layer. This in turn means that for $X \to 0$ one thus obtains very large cross flows, especially near the surface, since u tends to zero there. It should be noted, however, that Z^* tends to zero as $(\xi/x)^{\frac{1}{3}}$ or, in other words, the viscous action is qualitatively the same as when the abrupt perturbation occurred due to the sudden lateral movement of the surface.

C,15. Flow with Heat Transfer to a Very Cold Surface. As discussed in Art. 9 and 10, for flows involving surface rotation, or for yawed wings, the determination of the convective heat transfer rate is fairly straightforward when Prandtl number is unity. Thus, for example, the heat transfer of a yawed cylinder was analyzed in [*84*] by obtaining solutions of Eq. 10-27, 10-29a, and 10-30. Also, for flows which are somewhat more complex, the methods and simplifications discussed in Art. 11 and 12 are equally useful in the computations of heat transfer. For instance, Reshotko [*85*], by employing a restricted solution, was able to determine the heat transfer at the most windward streamline (which, incidentally, is where heat transfer rate should be highest) for a cone at a large angle of attack. However, rather than go into the details of these investigations, it is perhaps more worthwhile to call attention to a special, but important, case of three-dimensional boundary layer flow;

namely that over a very cold surface. Flow of this type has been first treated by Reshotko [*86*] in his analysis of the heat transfer to a general stagnation point. Later, Vaglio-Laurin [*87*] extended this to even more general solutions for flow and heat transfer to such cold surfaces.

Consider the general three-dimensional boundary layer equations under the transformations (Eq. 6-2 and 6-4), but with $L_1 = L_2 = A = 1$. Under these conditions the equations are not correlated with the incompressible flow, and $F_1 = F_2 = \rho_e/\rho$ (Eq. 6-9 and 6-10). If now intrinsic coordinates are chosen, so that $V = 0$, then Eq. 6-7 (omitting all bars except over ζ and w) may be written as

$$\frac{u}{h_1}\frac{\partial v}{\partial \xi} + \frac{v}{h_2}\frac{\partial v}{\partial \eta} + \bar{w}\frac{\partial v}{\partial \bar{\zeta}} + vu\kappa_2 - B\nu_0\frac{\partial^2 v}{\partial \bar{\zeta}^2} = \kappa_1 U^2\left[\left(\frac{u}{U}\right)^2 - \frac{\rho_e}{\rho}\right] \tag{15-1}$$

where κ_1 and κ_2 are the geodesic curvatures (Art. 3) and $B \equiv \rho\mu/\rho_0\mu_0 \approx \rho_e\mu_e/\rho_0\mu_0$. One now notes that the term on the right-hand side of Eq. 15-1 is negligibly small. This, of course, is particularly true at high Mach numbers when the external flow streamlines tend to become geodesics but the argument has actually been used (Art. 8) even at relatively low Mach numbers by suggesting that near a very cold surface $\rho_e/\rho \cong (u/U)^2$. Therefore setting the right-hand side equal to zero, one immediately sees, upon considering the boundary conditions $v(0) = v(\infty) = 0$, that $v = 0$ *is a solution and consequently the secondary flow vanishes.* In other words, in line with these approximations, the three-dimensional boundary layer flow over a cold surface, in spite of nongeodesic streamlines, should reflect only the divergence effects so that its solution is much easier. Actually, as discussed in greater detail by Moore (H,8), there is a bit of good luck connected with the relative success of this method—particularly near a stagnation point—and a lot of caution should be used when applying it to three-dimensional boundary layer flow. Of course, should the neglect of the nonhomogeneous term not be completely justifiable, then one could consider $v = 0$ as only the zero-order approximation and obtain the higher order terms by an expansion procedure similar to those discussed in Art. 11.

With $v \equiv 0$ the remaining equations for a Prandtl number of unity are

$$\frac{\partial(h_2 u)}{\partial \xi} + \frac{\partial(h_1 h_2 \bar{w})}{\partial \bar{\zeta}} = 0 \tag{15-2}$$

$$\frac{u}{h_1}\frac{\partial u}{\partial \xi} + \bar{w}\frac{\partial \bar{u}}{\partial \bar{\zeta}} = \frac{\rho_e}{\rho}\frac{U}{h_1}\frac{\partial U}{\partial \xi} + B\nu_0\frac{\partial^2 u}{\partial \bar{\zeta}^2} \tag{15-3}$$

$$\frac{u}{h_1}\frac{\partial H}{\partial \xi} + \bar{w}\frac{\partial H}{\partial \bar{\zeta}} = B\nu_0\frac{\partial^2 H}{\partial \bar{\zeta}^2} \tag{15-4}$$

It should now be noted that, inasmuch as all the variations with respect to η have disappeared from the equations, the integrations may be performed along each individual streamline, quite independently of all the others. This, in turn, means that the length-stretching factor h_1 can be arbitrarily chosen. Accordingly, one may choose $h_1 = 1$ so that ξ represents the actual arc length along each streamline.

Additional simplifications may now be obtained by using the transformation 6-24: $x = \xi$, $z = G\zeta$, $L_1h_1^* = h_1 = 1$ and $L_2h_2^* = h_2$. We require now that $u^* = u/U$ and $h_2^* = 1$ which in turn gives $h_1^* = Uh_2^2B, G = Uh_2$, and $U^* = 1$. Furthermore, defining for convenience a new coordinate X,

$$X \equiv \int_0^x h_1^* dx = \int_0^\xi BUh_2^2 d\xi \tag{15-5}$$

one obtains directly from Eq. 6-27 and 6-28 the transformed form of Eq. 15-2 and 15-3,

$$\frac{\partial u^*}{\partial X} + \frac{\partial w^*}{\partial z} = 0 \tag{15-6}$$

$$u^* \frac{\partial u^*}{\partial X} + w^* \frac{\partial u^*}{\partial z} - \nu_0 \frac{\partial^2 u^*}{\partial z^2} = \frac{\partial \ln U}{\partial X}\left(\frac{\rho_e}{\rho} - u^{*2}\right) \tag{15-7}$$

and after some slight manipulation also the transformation of Eq. 15-4,

$$u^* \frac{\partial H}{\partial X} + w^* \frac{\partial H}{\partial z} - \nu_0 \frac{\partial^2 H}{\partial z^2} = 0 \tag{15-8}$$

It is immediately apparent that in this zero-order approximation one should, for the sake of consistency, consider again the right-hand side of Eq. 15-7 as vanishingly small. If it is taken as zero, then the solution of Eq. 15-8 which satisfies the proper boundary conditions is given by

$$H = H(0) + [H_e - H(0)]u^* \tag{15-9}$$

It remains, therefore, to solve for u^* and w^*. Observe, however, that these velocities satisfy the flat plate equations. Consequently, the solution may immediately be written as

$$\frac{u}{U} = u^* = f'(Z) \quad \text{with} \quad Z = z/\sqrt{2\nu_0 X} \tag{15-10}$$

where the normalized stream function $f\ (= \psi/\sqrt{2\nu_0 X})$ satisfies

$$f''' + ff'' = 0 \quad \text{with} \quad f(0) = f'(0) = 0, f'(\infty) = 1 \tag{15-11}$$

Once the velocity and enthalpy distributions are known, the heat transfer to the surface may easily be found. Specifically, using the heat transfer coefficient C_h defined in Eq. 9-53, it may easily be shown (with Prandtl number of unity) that

$$C_h = f''(0) \frac{\rho_0 U h_2 B}{\sqrt{\frac{2X}{\nu_0}}} \tag{15-12}$$

The problem of three-dimensional boundary layer flow near a very cold surface has thus been reduced to the determination of the external flow streamline pattern. Once this pattern is known, the stretching factor $h_2(\xi)$ can be found (by actual measurement of arc length along normals to the streamlines, or otherwise) and the correlation with the flat plate boundary layer may easily be made.

C,16. Boundary Regions. In accordance with the Prandtl boundary layer concept, all the boundary layers discussed so far have been assumed to occur next to regularly curved surfaces, where the radii of curvature were large in comparison to the boundary layer thickness. There are, however, many obviously important cases of boundary layer flow where such assumptions may not be made. Consider, for example, the boundary layer over a very long and slender circular cylinder—a form which is so common in modern missiles. Certainly, far enough downstream, a region will be reached where the boundary layer thickness is comparable to or greater than the radius of the cylinder, so that our previous assumptions cease to be valid there. It is, therefore, important to know what, if any, changes in the nature of the flow result and how such regions should be treated. A similar problem arises when the surface curvature does not vary in a regular manner. A common example of such a situation is provided by the boundary layer flow in the corner region between two orthogonally intersecting flat plates, so that along the intersection line the surface curvature is infinite. Clearly the development of the boundary layer on one of the plates may not be assumed independent of that on the other; if it were, velocity profiles would be discontinuous, contrary to the laws of viscous diffusion. In the region near the intersection, therefore, the assumptions of the conventional boundary layer theory do not hold, and some modifications are necessary. It has been suggested by Kemp [*88*] that flow in the regions where large or irregularly varying surface curvatures are important be called boundary region flow as distinguished from the conventional boundary layer flow.

Boundary regions due to large surface curvature. It is clear from the foregoing remarks that one must now reevaluate the original simplifications of the boundary layer theory, and reexamine which terms may, and which may not, be neglected in writing the conservation laws. For this purpose, consider a developable cylindrical surface shown in Fig. C,16a.

We choose the coordinate x along the generators, the coordinate η as a reference line coincident with the leading edge and coordinate z normal to the surface. It is fairly easy to see that the length-stretching factors are now given by $h_1 = 1$, $h_2 = 1 + cz$, $h_3 = 1$ where c, the lateral curvature of the surface, is a function of η only. Observe that, since h_2 is now a function of z as well as of η, the equations may not be written in a Cartesian form even though the surface is developable. However, by

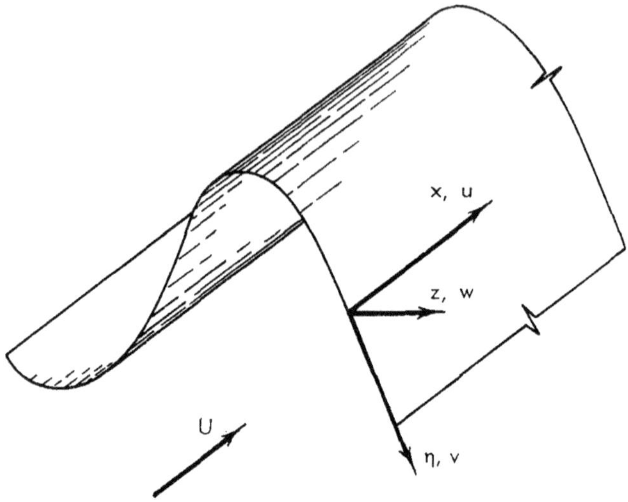

Fig. C,16a. Coordinates and velocities for flow over cylindrical surface with large curvature.

use of the standard transformation formulas (B,5) the equations of continuity and momentum for incompressible flow can easily be written as

$$\frac{\partial[(1+cz)u]}{\partial x}+\frac{\partial v}{\partial \eta}+\frac{\partial[(1+cz)w]}{\partial z}=0 \tag{16-1}$$

$$\begin{aligned} u\frac{\partial u}{\partial x}+\frac{v}{1+cz}\frac{\partial u}{\partial \eta}+w\frac{\partial u}{\partial z} &= -\frac{1}{\rho_0}\frac{\partial p}{\partial x}+\nu_0\left[\frac{\partial^2 u}{\partial x^2}+\frac{\partial^2 u}{\partial z^2}\right.\\ &\quad \left.+\frac{1}{(1+cz)^2}\frac{\partial^2 u}{\partial \eta^2}+\frac{c}{1+cz}\frac{\partial u}{\partial z}-\frac{z}{(1+cz)^3}\frac{\partial u}{\partial \eta}\frac{dc}{d\eta}\right] \end{aligned} \tag{16-2}$$

$$\begin{aligned} u\frac{\partial v}{\partial x}+\frac{v}{1+cz}\frac{\partial v}{\partial \eta}+w\frac{\partial v}{\partial z}+\frac{cvw}{1+cz} &= -\frac{1}{\rho_0}\frac{1}{(1+cz)}\frac{\partial p}{\partial \eta}\\ &\quad +\nu_0\left[\frac{\partial^2 v}{\partial x^2}+\frac{\partial^2 v}{\partial z^2}+\frac{1}{(1+cz)^2}\frac{\partial^2 v}{\partial \eta^2}-\frac{z}{(1+cz)^3}\frac{\partial v}{\partial \eta}\frac{dc}{d\eta}+\frac{2c}{(1+cz)^2}\frac{\partial w}{\partial \eta}\right.\\ &\quad \left.+\frac{w}{(1+cz)^2}\left(1-\frac{cz}{1+cz}\right)\frac{dc}{d\eta}+\frac{c}{1+cz}\frac{\partial v}{\partial z}-\frac{c^2 v}{(1+cz)^2}\right] \end{aligned} \tag{16-3}$$

$$\begin{aligned} u\frac{\partial w}{\partial x}+\frac{v}{1+cz}\frac{\partial w}{\partial \eta}+w\frac{\partial w}{\partial z}-\frac{cv^2}{1+cz} &= -\frac{1}{\rho_0}\frac{\partial p}{\partial z}+\nu_0\left[\frac{\partial^2 w}{\partial x^2}+\frac{\partial^2 w}{\partial z^2}\right.\\ &\quad +\frac{1}{(1+cz)^2}\frac{\partial^2 w}{\partial \eta^2}-\frac{2c}{(1+cz)^2}\frac{\partial v}{\partial \eta}-\frac{c^2 w}{(1+cz)^2}+\frac{c}{1+cz}\frac{\partial w}{\partial z}\\ &\quad \left.+\left(\frac{cz}{1+cz}-1\right)\frac{v}{(1+cz)^2}\frac{dc}{d\eta}-\frac{z}{1+cz}\frac{\partial w}{\partial \eta}\frac{dc}{d\eta}\right] \end{aligned} \tag{16-4}$$

Applying certain of the boundary layer approximations, that is, considering w and z to be of $O(\delta)$ but c to be $O(1/\delta)$ and neglecting all terms of order δ or smaller, Eq. 16-2, 16-3, and 16-4 may be simplified to

$$u\frac{\partial u}{\partial x}+\frac{v}{1+cz}\frac{\partial u}{\partial \eta}+w\frac{\partial u}{\partial z}=-\frac{1}{\rho_0}\frac{\partial p}{\partial x}+\nu_0\left(\frac{\partial^2 u}{\partial z^2}+\frac{c}{1+cz}\frac{\partial u}{\partial z}\right) \tag{16-5}$$

$$u\frac{\partial v}{\partial x}+\frac{v}{1+cz}\frac{\partial v}{\partial \eta}+w\frac{\partial v}{\partial z}+\frac{cvw}{1+cz}=-\frac{1}{\rho_0}\frac{1}{(1+cz)}\frac{\partial p}{\partial \eta}$$
$$+\nu_0\left(\frac{\partial^2 v}{\partial z^2}+\frac{c}{1+cz}\frac{\partial v}{\partial z}-\frac{c^2 v}{(1+cz)^2}\right) \tag{16-6}$$

$$-\frac{cv^2}{1+cz}=-\frac{1}{\rho_0}\frac{\partial p}{\partial z} \tag{16-7}$$

while the equation of continuity remains unchanged. Observe that these new equations contain additional viscous terms which vanish when the lateral curvature becomes small.

Consider now a uniform external flow, parallel to the generators so that $\partial p/\partial x = \partial p/\partial \eta = V = 0$ and $U = \text{const.}$ Since all the velocities must still vanish at $z = 0$, it becomes immediately obvious that $v = 0$ is a solution of Eq. 16-6. To deal with the remaining equations, it is profitable to apply the transformation,

$$\bar{x}\equiv x;\qquad \bar{z}\equiv\int_0^z(1+cz)dz;\qquad \bar{u}\equiv u;\qquad \bar{w}\equiv(1+cz)w \tag{16-8}$$

which leaves the mass flow unchanged but results in a considerably simplified form:

$$\frac{\partial \bar{u}}{\partial \bar{x}}+\frac{\partial \bar{w}}{\partial \bar{z}}=0 \tag{16-9}$$

$$\bar{u}\frac{\partial \bar{u}}{\partial \bar{x}}+\bar{w}\frac{\partial \bar{u}}{\partial \bar{z}}=\nu_0\frac{\partial}{\partial \bar{z}}\left[(1+2c\bar{z})\frac{\partial \bar{u}}{\partial \bar{z}}\right] \tag{16-10}$$

In this form the equations are similar to those obtained by Mark [*89*] in his studies of boundary layers on slender axially symmetric bodies. Furthermore, by analogy to conventional boundary layer solutions, it is convenient to define two new variables and a stream function:

$$X\equiv 4c\sqrt{\frac{\nu_0\bar{x}}{\bar{U}}};\qquad Z\equiv\frac{\bar{z}}{2}\sqrt{\frac{\bar{U}}{\nu_0\bar{x}}};\qquad \bar{\psi}\equiv\sqrt{\nu_0\bar{x}\bar{U}}\,F(X,Z) \tag{16-11}$$

so that $2\bar{u} = U(\partial F/\partial Z)$ and Eq. 16-9 and 16-10 become

$$\frac{\partial}{\partial Z}\left[(1+XZ)\frac{\partial^2 F}{\partial Z^2}\right]+F\frac{\partial^2 F}{\partial Z^2}+X\left(\frac{\partial F}{\partial X}\frac{\partial^2 F}{\partial Z^2}-\frac{\partial F}{\partial Z}\frac{\partial^2 F}{\partial Z\partial X}\right)=0 \tag{16-12}$$

with the boundary conditions,

$$\text{for}\qquad Z=0: F=\frac{\partial F}{\partial Z}=\frac{\partial F}{\partial X}=0;\qquad \text{for } Z\to\infty: \frac{\partial F}{\partial Z}=2 \tag{16-13}$$

Eq. 16-12 and the boundary conditions (Eq. 16-13) were obtained first by Seban and Bond [*90*], who investigated the boundary layer development on a very slender circular cylinder. It should be pointed out, however, that this correspondence is not surprising, when it is realized that, for $v \equiv 0$, Eq. 16-1 and 16-5 with their associated boundary conditions become those for a slender circular cylinder, after one sets

$$r \equiv \frac{1}{c} + z \tag{16-14}$$

This fact means that the boundary region along the surface generators of a cylinder having a regular, but otherwise arbitrary, cross-sectional shape develops exactly like that on an equivalent circular cylinder whose radius is $a = 1/c$. As a direct consequence, one may now carry over the results of Seban and Bond, who obtained the solution for small values of X by expanding in a Taylor series,

$$F(X, Z) = F_0(Z) + XF_1(Z) + X^2F_2(Z) + \cdots \tag{16-15}$$

and by solving the resulting set of total differential equations numerically. (This numerical solution was later corrected slightly in [*91*].) One observes that the extreme lateral curvature destroys the boundary layer similarity and that the velocity profile in a boundary region continuously changes its shape as the flow proceeds along the generators away from the leading edge. Thus, in comparison with the flat plate case, differences in boundary layer quantities occur at the same streamwise position, owing to the lateral curvature of the surface. The ratio of skin friction coefficients [*91*] is

$$\frac{c_f}{c_{f_{\text{Blasius}}}} = 1 + 0.525X - 0.120X^2 + \cdots \tag{16-16a}$$

and it is immediately obvious that, while a convex surface ($c > 0$) tends to increase the skin friction above the Blasius flat-plate value, the opposite happens when the surface is concave ($c < 0$). A similar reversal of trend occurs for the displacement thickness [*91*],

$$\frac{\delta_1}{\delta_{1_{\text{Blasius}}}} = 1 - 0.204X + 0.146X^2 - \cdots \tag{16-16b}$$

which a concave surface increases, but a convex surface decreases.

An approximate solution of Eq. 16-1 and 16-5 for $v = \partial p/\partial x = 0$ was also found in a somewhat different way in [*79*]. That solution, while tending to overestimate the effect of the curvature on the skin friction (by 43 per cent to $O(X)$), and to underestimate the effect on the displacement thickness (by some 9 per cent to $O(X)$), nevertheless shows rather clearly the changes that occur in the velocity distribution. Using the present variables, it may be written as

$$\frac{u}{U} \cong \frac{F_0' + XZ}{2 + XZ} \tag{16-17}$$

so that the normalized slope at the wall becomes $0.5(F_0'' + X)_{z=0}$. Consequently we see that the concave curvature tends to enhance a separation profile, while the convex curvature does just the opposite. These effects are, of course, in line with the variations indicated by Eq. 16-16a and 16-16b.

Eq. 16-14, expressing the correspondence between a laterally, but arbitrarily, curved surface and a very slender circular cylinder cannot be used with a concave curvature ($c < 0$) very far from the leading edge (e.g. for large x, where the boundary region is thick). This is so because, for $cz < 1$, r becomes negative, and also the system illustrated in Fig. C,16a leads to double values. This indicates that actually, for $cz < 1$, there will occur additional interactions in the flow which were not accounted for by the simplified equations. For convex curvature, however, no such difficulties occur. One is therefore able to use Eq. 16-14 to interpret the results of Stewartson [*92*] who studied the axial flow over a slender circular cylinder far downstream from the leading edge. Specifically, by obtaining an asymptotic solution, Stewartson found that, for large x, the skin friction may be represented by a double power series in descending powers of $\ln (X^2/8)$ and $(X^2/8)$. The important leading term in this series is proportional to $[\ln (X^2/8)]^{-1}$, so that again the effect of the convex curvature shows up as an increase of the skin friction over that of a flat plate at the same Reynolds number. Furthermore, the convex curvature also causes a decrease of boundary layer thickness, although this decrease is rather slight, thus indicating that the main changes in the velocity distribution occur very close to the surface.

If the effect of the sharp lateral curvature is to increase the skin friction for flows without pressure gradients, one may expect a similar tendency to persist even when U is not a constant but varies with x. This means that for an unfavorable pressure gradient the sharp lateral curvature should delay the onset of separation. This, in fact, has been shown to be true by Stewartson [*92*], who also investigated the case where $U = x^m$ and found that the skin friction will ultimately be positive if $m > -0.5$. This is clearly a delaying effect since in the absence of curvature the separation profile is obtained when $m = -0.0904$ (B,29). Observe also that this delay of separation is similar to the one computed on a flat surface by Moore [*71*], who suggested as its cause the boundary layer thinning effect of the divergent outer flow streamlines.

Boundary regions in corners. If the surface curvature does not vary in a regular manner, additional difficulties appear. An excellent example

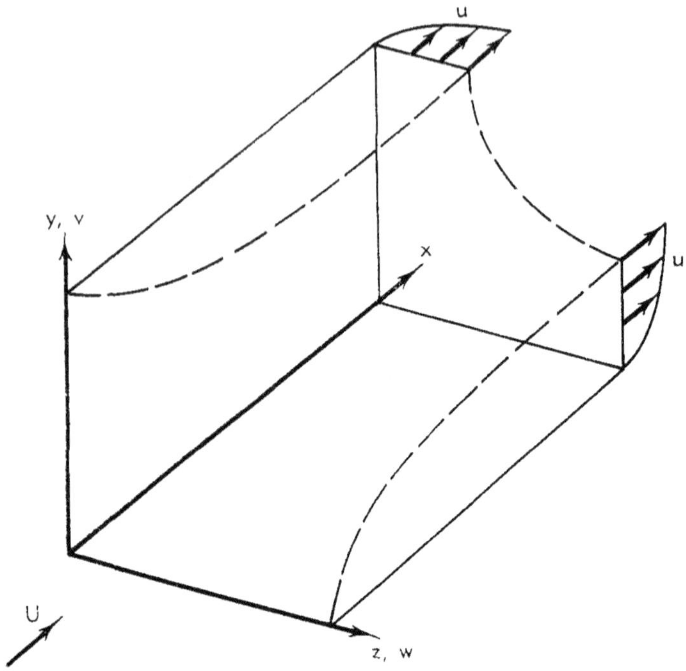

Fig. C,16b. Carrier's corner flow problem.

of this is furnished by the flow in the corner between two orthogonally intersecting flat plates (see Fig. C,16b), which was first treated by Carrier [93]. If Cartesian coordinates are chosen, and an order of magnitude argument is made, then the appropriate equations (when U = const) become

$$\frac{\partial u}{\partial x} + \frac{\partial v}{\partial y} + \frac{\partial w}{\partial z} = 0 \tag{16-18}$$

$$u \frac{\partial u}{\partial x} + v \frac{\partial u}{\partial y} + w \frac{\partial u}{\partial z} = \nu_0 \left(\frac{\partial^2 u}{\partial y^2} + \frac{\partial^2 u}{\partial z^2} \right) \tag{16-19}$$

Additional terms thus appear on the right-hand side of Eq. 16-19, just as when the curvature is varying in a regular manner. Observe, however, that two equations (Eq. 16-18 and 16-19) are insufficient for the determination of the three unknown velocities u, v, and w. Consequently, in spite of the order of magnitude argument, one must now consider the momentum equations in the y and z directions with the pressure p as the fourth unknown:

$$u \frac{\partial v}{\partial x} + v \frac{\partial v}{\partial y} + w \frac{\partial v}{\partial z} = -\frac{1}{\rho_0} \frac{\partial p}{\partial y} + \nu_0 \left(\frac{\partial^2 v}{\partial y^2} + \frac{\partial^2 v}{\partial z^2} \right) \tag{16-20}$$

$$u \frac{\partial w}{\partial x} + v \frac{\partial w}{\partial y} + w \frac{\partial w}{\partial z} = -\frac{1}{\rho_0} \frac{\partial p}{\partial z} + \nu_0 \left(\frac{\partial^2 w}{\partial y^2} + \frac{\partial^2 w}{\partial z^2} \right) \tag{16-21}$$

By cross-differentiation, this pressure may be eliminated from the last two equations giving instead the relation describing the diffusion of the x component of the vorticity Ω ($= \partial v/\partial z - \partial w/\partial y$),

$$u \frac{\partial \Omega}{\partial x} + v \frac{\partial \Omega}{\partial y} + w \frac{\partial \Omega}{\partial z} = \nu_0 \left(\frac{\partial^2 \Omega}{\partial y^2} + \frac{\partial^2 \Omega}{\partial z^2} \right) \tag{16-22}$$

The complete system for the velocities u, v, and w is thus represented by Eq. 16-18, 16-19, and 16-22.

It is possible now, following the procedure of [*93,2*], to define the similarity variables

$$Y \equiv y \sqrt{\frac{U}{\nu_0 x}}; \qquad Z \equiv z \sqrt{\frac{U}{\nu_0 x}} \tag{16-23a}$$

and the nondimensional velocities,

$$u \equiv UF(Y, Z); \quad v \equiv \sqrt{\frac{U\nu_0}{x}}\, G(Y, Z); \quad w \equiv \sqrt{\frac{U\nu_0}{x}}\, J(Y, Z) \tag{16-23b}$$

which, when substituted into our system, give

$$\frac{1}{2}\left(Y \frac{\partial F}{\partial Y} + Z \frac{\partial F}{\partial Z} \right) = \frac{\partial G}{\partial Y} + \frac{\partial J}{\partial Z} \tag{16-24}$$

$$\left[\left(G - \frac{Y}{2} F \right) \frac{\partial}{\partial Y} + \left(J - \frac{Z}{2} F \right) \frac{\partial}{\partial Z} - \left(\frac{\partial^2}{\partial Y^2} + \frac{\partial^2}{\partial Z^2} \right) \right] F = 0 \tag{16-25}$$

$$\left[\left(G - F - \frac{Y}{2} \right) F \frac{\partial}{\partial Y} + \left(J - \frac{Z}{2} F \right) \frac{\partial}{\partial Z} - \left(\frac{\partial^2}{\partial Y^2} + \frac{\partial^2}{\partial Z^2} \right) \right] \left(\frac{\partial G}{\partial Z} - \frac{\partial J}{\partial Y} \right) = 0 \tag{16-26}$$

with the boundary conditions $F = G = J = 0$ when either Y or Z equal 0 and $F \to 1$ when both Y and Z tend to infinity.

Until the present time, there appears to be no complete published solution of the system, Eq. 16-24, 16-25, and 16-26. Carrier [*93*] has constructed a solution to Eq. 16-24 and 16-25 by noting that the continuity equation (Eq. 16-24) will be satisfied by a single function g, if one defines

$$F \equiv \frac{\partial^2 g}{\partial Z \partial Y}; \quad G \equiv \frac{1}{2}\left(Y \frac{\partial^2 g}{\partial Y \partial Z} - \frac{\partial g}{\partial Z} \right); \quad J \equiv \frac{1}{2}\left(Z \frac{\partial^2 g}{\partial Y \partial Z} - \frac{\partial g}{\partial Y} \right) \tag{16-27}$$

These expressions may then be used in Eq. 16-25 and g can thus be determined. Specifically, Carrier set $g = F_0(Y)F_0(Z) + F_1(Y, Z)$ where F_0 is the usual Blasius function and determined the correction function F_1 by numerical integration. This correction turns out to be relatively small, so that g represents essentially hyperbolas in the $F_0(Y) - F_0(Z)$ plane and the boundary layer for large Y or Z grows very much like that on a regular flat plate.

Observe, however, that using a single function g for the definition of all three velocities u, v, and w does not satisfy Eq. 16-26. For this reason Carrier's [93] solution cannot be strictly correct. According to the remarks made in [1,2], Carrier and Dowdell have subsequently modified the approach of [93] and found a solution which satisfies the complete system. The differences between these new results and those of [93] are said to be small—this work, however, is not published as yet, so that the exact nature of these differences could not be ascertained.

Approximate methods. The boundary region flow may also be analyzed by employing certain approximations which have proved to be so useful in the study of the boundary layers. One of these is based on the so-called Rayleigh problem (B,7). One may draw the analogy of time-dependent heat conduction in an infinite solid, having some internal cylindrical boundary of arbitrary shape and being initially at zero temperature. At zero time, the temperature of the internal boundary is suddenly raised to some constant, nonzero value. If now the time is interpreted as x/U and the thermal diffusivity as ν_0, then the temperature becomes analogous to the departure of the velocity u from U in the fluid boundary region. By this analogy, the problem is linearized, since the velocity U acts in the convective process rather than the local velocity u. Examples of such an attack on the boundary region problem may be found in [94, 95,96,97]. Unfortunately, however, this approach seriously overestimates the skin friction and underestimates the boundary region thickness, except far downstream, where the conditions in the large part of the boundary region are very nearly like those outside.

Another more useful way to treat boundary region problems is by the use of the momentum-integral method. This method was pioneered for such problems by Loitsianskii and Bolshakov [98] and more recently was generalized to arbitrarily shaped cylindrical bodies by Varley [99]. One assumes a Cartesian coordinate system fixed in a cylindrical body, but not necessarily on the surface, as illustrated for a particular case in Fig. C,16c. For the external flow with constant velocity U, Eq. 16-18 and 16-19 are to be solved under the boundary conditions $u = v = w = 0$ on S and $u \to U$ as $y, z \to \infty$ or $x \to 0$. If now Eq. 16-19 is integrated, at any x, over the whole area A outside the boundary S, then Eq. 16-18 may be used to eliminate the velocities v and w. This process requires, of course, that certain arbitrary functions be set identically equal to zero, and thus suffers from the same defect as the previously discussed treatment of Carrier. Still, one obtains the correct momentum-integral equation which, incidentally, may also be derived by direct physical argument:

$$\rho_0 U^2 \frac{d\bar{\theta}}{dx} = F \tag{16-28}$$

In this equation the momentum deficiency area is defined by

$$\bar{\theta} \equiv \int_A \left(1 - \frac{u}{U}\right) \frac{u}{U}\, dydz \tag{16-29}$$

and the frictional force F per unit length of the cylinder is given by

$$F \equiv \mu_0 \oint_S \frac{\partial u}{\partial n}\, dS \tag{16-30}$$

where n is the outward normal to the perimeter S.

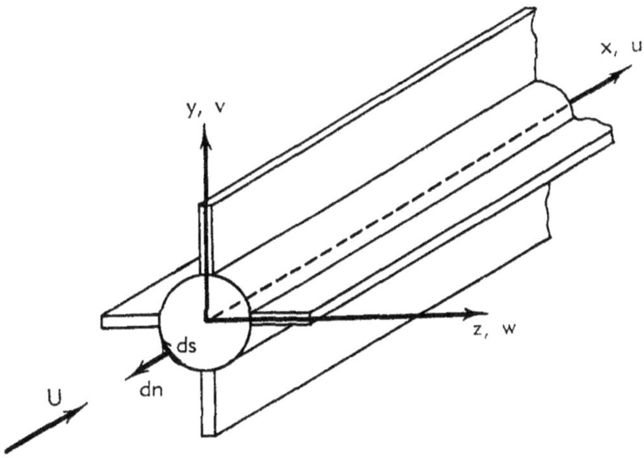

Fig. C,16c. Coordinates and velocities for application of the momentum-integral method to boundary region flow.

To solve Eq. 16-28, it is now necessary to approximate the velocity distribution in the boundary layer. Varley suggests

$$\left.\begin{aligned} \frac{u}{U} &= \frac{\phi(y, z)}{\delta(x)} \quad \text{for} \quad 0 < \phi \leqq \delta \\ \frac{u}{U} &= 1 \quad \text{for} \quad \phi \geqq \delta \end{aligned}\right\} \tag{16-31}$$

(a similar form has also been used in [*98*]) with ϕ satisfying

$$\frac{\partial^2 \phi}{\partial y^2} + \frac{\partial^2 \phi}{\partial z^2} = 0 \tag{16-32}$$

and having a zero value on S. Note that this form of the velocity distribution leads to a discontinuous change of the slope at the edge of the boundary region (e.g. at $\phi = \delta$). However, Glauert and Lighthill [*100*], who investigated the slender circular cylinder by the momentum integral method, have shown that this discontinuity does not, in itself, lead to serious errors in such an approximate treatment.

The particular solution of Eq. 16-32 which gives $\phi = 0$ on S is the electrostatic potential of a cylinder at zero potential and unit charge per unit length. This means that

$$F = \frac{\mu_0 U}{\delta} \oint_S \frac{\partial \phi}{\partial n} dS = 2\pi\mu_0 \frac{U}{\delta} \tag{16-33}$$

Furthermore, if the boundary S is circular, of radius a, then the solution is given by $\phi = \ln (r/a)$, or equivalently, by $\zeta = ae^{\xi}$, where $\zeta = re^{i\theta}$ and $\xi = \phi + i\theta$. For any other boundary shape, then, one needs only to find the conformal mapping:

$$z + iy \equiv W = f(\zeta) \quad \text{with} \quad f(\zeta) \sim \zeta \text{ for large } \zeta \tag{16-34}$$

which maps the exterior of the boundary S in the W plane on to the exterior of the equivalent circle of radius a in the ζ plane. Once this mapping is determined, ϕ (the real part of ξ) will be known as a function of y and z. Note also that since

$$dydz = \left|\frac{dW}{d\xi}\right|^2 d\theta d\phi \tag{16-35}$$

the momentum-deficiency area may now be written as

$$\bar{\theta} = \int_0^{\delta} \frac{(\delta - \phi)\phi}{\delta^2} d\phi \int_0^{2\pi} \left|\frac{dW}{d\xi}\right|^2 d\theta \tag{16-36}$$

and this again may be evaluated when the appropriate mapping is known. By use of Eq. 16-33 and 16-36, Eq. 16-28 may now be written in the form

$$\frac{d}{dx}\left[\int_0^{\delta} \frac{(\delta - \phi)\phi}{\delta^2} d\phi \int_0^{2\pi} \left|\frac{dW}{d\xi}\right|^2 d\theta\right] = \frac{2\pi\nu_0}{U\delta} \tag{16-37}$$

and it is the solution of this last equation that yields the desired boundary region thickness δ as a function of $(\nu_0 x/Ua^2)$.

As examples of the application of his method, Varley computed closed-form solutions for a flat plate of finite width and for the elliptic cylinder. Furthermore, he also considered several polygonal shapes and an n pronged cross. By expanding the closed-form solutions for small values of X (where $X \equiv \nu_0 x/Ul^2$ and $2\pi l$ is the perimeter), Varley was able to compare the first coefficient with the value given by Blasius, since, as may be expected from [*90*], these two should agree. The results of such an expansion for the flat plate of finite width may be given as

$$\frac{F}{F_{\text{Blasius}}} = 0.967 + 4.22X + 56.6X^2 + \cdots \tag{16-38}$$

showing that even at $X = 0$ (where the accuracy might be expected to be poor) the frictional force is given with an error of 3 per cent, which is considerably better than the error of approximately 70 per cent which results from Rayleigh's method. Observe also that the finite width of the flat plate causes an increase of the skin frictional force, as does sharp convex curvature of a cylinder.

In regard to the n-pronged cross, Varley found that only for reentrant angles greater than 90° did his method yield, near the leading edge edge ($X \to 0$), the expected square-root growth of the boundary region with distance. This then indicates that for reentrant angles smaller than 90° this method is probably not applicable.

C,17. Boundary Region with Cross Flow. There are many physical examples of flow where the situation in the boundary regions is further complicated by the presence of a crosswise velocity V in the external flow. A case in point is the flow in the corners between the blading and the hub of turbomachinery. Even if nonrotating stages are considered, such as stators or guide vanes, the external nonviscous flow has usually an appreciable secondary motion (from the blades onto the hub and vice versa) so that for the corner flow, the V component may not be neglected. In what follows we shall indicate, in line with the method of [*53*], how such further complications may be handled when the surface is regular and when the external crosswise velocity V is small when compared to the principal velocity U.

One imagines a cylindrical surface and chooses coordinates as represented schematically in Fig. C,16a. For simplicity's sake it is also convenient to take $U = U^0$ and $V = Ax^n$, where both U^0 and A are constants. Since $V/U^0 \ll 1$, the small perturbation procedure discussed in Art. 11 is again applicable. Accordingly one sets (bar signifying a total value),

$$\bar{u} = u^0 + u; \qquad \bar{v} = v; \qquad \bar{w} = w^0 + w; \qquad \bar{p} = p^0 + p \tag{17-1}$$

and obtains the appropriate equations by substitution into Eq. 16-1, 16-5, 16-6, and 16-7. This procedure yields the zero-order equations:

$$\frac{\partial[(1 + cz)u^0]}{\partial x} + \frac{\partial[(1 + cz)w^0]}{\partial z} = 0 \tag{17-2}$$

$$u^0 \frac{\partial u^0}{\partial x} + w^0 \frac{\partial u^0}{\partial z} = \frac{\nu_0}{(1 + cz)} \frac{\partial}{\partial z}\left[(1 + cz) \frac{\partial u^0}{\partial z}\right] \tag{17-3}$$

and the first order equations,

$$\frac{\partial[(1 + cz)u]}{\partial x} + \frac{\partial v}{\partial \eta} + \frac{\partial[(1 + cz)w]}{\partial z} = 0 \tag{17-4}$$

$$u^0 \frac{\partial u}{\partial x} + u \frac{\partial u^0}{\partial x} + \frac{v}{(1 + cz)} \frac{\partial u^0}{\partial \eta} + w^0 \frac{\partial u}{\partial z} + w \frac{\partial u^0}{\partial z} = \frac{\nu_0}{(1 + cz)} \frac{\partial}{\partial z}\left[(1 + cz) \frac{\partial u}{\partial z}\right] \tag{17-5}$$

$$u^0 \frac{\partial v}{\partial x} + w^0 \frac{\partial v}{\partial z} + \frac{cw^0 v}{(1 + cz)} = -\frac{1}{\rho_0} \frac{1}{(1 + cz)} \frac{\partial p}{\partial \eta} + \nu_0 \frac{\partial}{\partial z} \left\{ \frac{1}{(1 + cz)} \frac{\partial}{\partial z} [(1 + cz)v] \right\} \quad (17\text{-}6)$$

$$\frac{\partial p}{\partial z} = O(v^2) \quad (17\text{-}7)$$

The zero-order equations are, of course, identical to those considered in Art. 16, and consequently, their solution is known. One needs, therefore, to solve only the first order equations, and here the process is essentially a combination of the procedures used in Art. 11 and 16. Specifically, after defining the components of the vector potential,

$$(1 + cz)u \equiv \frac{\partial \psi}{\partial z}; \qquad v \equiv \frac{\partial \varphi}{\partial z}; \qquad (1 + cz)w \equiv -\left(\frac{\partial \psi}{\partial x} + \frac{\partial \varphi}{\partial \eta}\right) \quad (17\text{-}8)$$

and the two normalizing variables,

$$Z \equiv \frac{z}{2} \sqrt{\frac{U^0}{\nu_0 x}}; \qquad X \equiv 4c \sqrt{\frac{\nu_0 x}{U^0}} \quad (17\text{-}9)$$

Having in mind the solution for the zero-order flow, one writes

$$\psi \equiv x \frac{\partial X}{\partial \eta} \frac{V}{U^0} \sqrt{\nu_0 x U^0} \sum_{i=0} X^i f_{n,i}(Z) \quad (17\text{-}10)$$

$$\varphi \equiv \frac{V}{U^0} \sqrt{\nu_0 x U^0} \sum_{i=0} X^i g_{n,i}(Z) \quad (17\text{-}11)$$

and substitutes this, together with Eq. 16-15 (or if a more approximate answer is satisfactory, Eq. 16-17) into Eq. 17-5, 17-6, and 17-8. As a result of such substitutions, for each power of X and each value of n, two total differential equations are obtained which define, together with the appropriate boundary conditions, the functions $f_{n,i}$ and $g_{n,i}$. It is immediately obvious that, since F_0 (Eq. 16-15) is the Blasius function, $g_{n,0}$ is identical with the function given by Eq. 11-65 (note, however, a slight change in the definition of Z) and thus may be presumed known. All other functions must, however, be computed numerically. Such computations were carried out in [53] (here again the definitions of X and Z differ slightly) for $n = 0$, 1 and $i = 0$, 1. Of course, once $f_{n,i}$ and $g_{n,i}$ are known, any surface whose curvature is a regular function of η, may be analyzed. Thus, for example, the previously mentioned flow in a filleted blade corner was treated in [53]. The results of these computations indicate that crosswise velocities can cause considerable asymmetries in boundary region flow, even if the curvature c is a symmetrical function of η. Furthermore, these asymmetries were found to be more pronounced for $n = 1$ than for $n = 0$.

C,18. Separation. In two-dimensional flows the customary definition of boundary layer separation requires that the derivative of the velocity in the direction normal to the surface vanish at the surface. This definition is quite satisfactory for such cases and it implies the start of flow reversal. In three-dimensional boundary layers, however, where in general the velocity must be described by two components, the two-dimensional definition is obviously inadequate since the vanishing of the derivative of one of the components can mean simply a local change of the flow direction at the surface. It is necessary, therefore, to establish a criterion for three-dimensional boundary layer separation anew, in some different, more precise manner. A number of investigators [*3,77,101, 102,103*] have recognized and discussed this problem and nearly all of them propose ideas which revolve about the same qualitative concept, although, of course, they do differ in details.

The roots of this concept lie in the recognition, first apparently made by Prandtl [*41*] that, at the surface, the three-dimensional boundary layer flow tends to a definite limiting direction. This suggests that the velocity components be expanded as a power series in the coordinate normal to the wall. At the surface then, as is obvious from the continuity equation and the no-slip condition, only the velocities parallel to the surface have first order terms. These terms allow one to describe the flow closest to the wall and determine the set of so-called limiting streamlines. Thus, for example, with a Cartesian system and a developable surface, one obtains

$$\frac{dy}{dx} = \left(\frac{\frac{\partial v}{\partial z}}{\frac{\partial u}{\partial z}}\right)_{z=0} \tag{18-1}$$

The various ideas concerning the separation of the three-dimensional flow are based on the examination of the pattern of these limiting streamlines. In particular, a definition is sought of a line which divides all the limiting streamlines into those coming from separated, and those coming from unseparated, portions of the flow.

To obtain such a definition, Eichelbrenner and Oudart [*102*] argued that the separation implies an existence of a surface which may be visualized as impermeable to the flow inside the boundary layer. Since the streamlines cannot end in the interior of the fluid, such an impermeable separation surface must be an envelope of the steady flow streamlines. Furthermore, if the fluid on the downstream side of the separation surface is not at rest, then the separation surface must necessarily also be an envelope of the streamlines of this second region, which usually contains reversed motion. The intersection of the so-defined separation surface with the body is the desired separation line, marking the beginning of the flow reversal, just as in two-dimensional flow. Using these concepts, the separation line may be defined as an envelope of the limiting streamlines and the separation criterion may thus be written

$$\left(\frac{\partial q}{\partial z}\right)_{z=0} = 0 \tag{18-2}$$

where q signifies the component of the velocity in the direction perpendicular to the separation line.

It is obvious that, in two-dimensional flow, where there is only one velocity component and where the flow may not change its direction, Eq. 18-2 becomes just the customary definition. On the other hand, in three-dimensional flow, Eq. 18-2 implies that the limiting streamlines become tangent to the separation line and the reversed flow region forces the streamlines everywhere on the upstream side of the separation surface to pass around it. The separation concept advanced by Eichelbrenner and Oudart [*102*] does not indicate then that the limiting streamlines actually leave the surface. This causes certain conceptual difficulties, since it is not clear by what means the limiting streamlines, while bypassing the reversed flow region, can avoid a process of "piling up." Such a pile-up would eventually have to advance upstream into the flow and prevent at least some of the streamlines from coming in actual contact with the separation surface.

A slightly different variation of these ideas, but clarifying the above objections, was advanced by Maskell [*103*]. Maskell proposed that the separation surface be considered as actually composed of the separating limiting streamlines, coming from the forward (upstream) and reversed (downstream) flow regions. Consequently, he argued that the separation occurs at a point only when two such distinct limiting streamlines converge and meet at that point. Conversely, of course, when such streamlines diverge from a point, then this point is a point of attachment. Having once met, the two streamlines combine and leave the surface as a single separation streamline (see Fig. C,18a). In this way the line of separation, that is, the intersection of the separation surface with the body, is both the envelope and the cusp locus of the surface limiting streamlines. In addition, for an ordinary separation, it is also the envelope of the separation streamlines. In other words, the two limiting streamlines, after combining, form a separation streamline which at first is still tangent to the body but then leaves the body and climbs in the separation surface a short time thereafter (see Fig. C,18b).

In addition to the ordinary separation described above, Maskell also postulates a singular separation, where the separation streamline is at once inclined to the body. This is the situation which, for example, occurs in two-dimensional flow (see Fig. C,18c) and is, of course, also the reverse of what happens at the stagnation point of a blunt body.

The separation surface containing all the separation streamlines can, in Maskell's picture, bend back and intersect the body again. At such subsequent intersection the separation streamlines undergo the reversal of the separation process, that is, they become attached. These are the

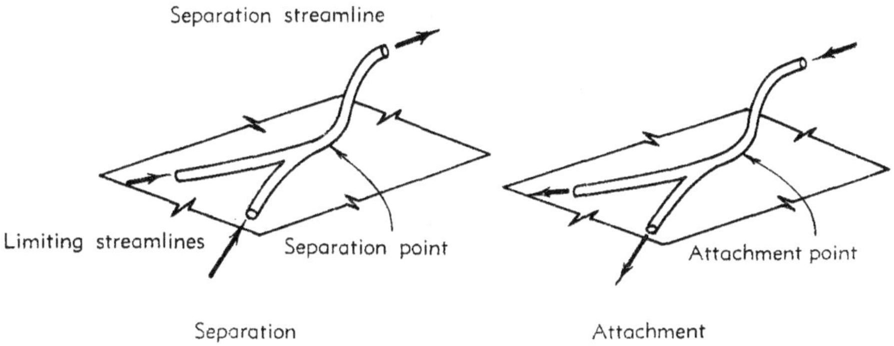

Fig. C,18a. Maskell's concept of ordinary separation and attachment.

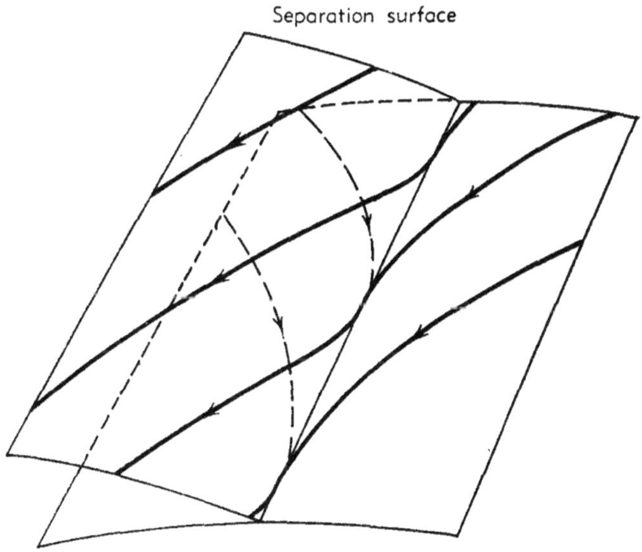

Fig. C,18b. Maskell's concept of ordinary separation surface.

cases when a reversed flow "bubble" is formed on the body. It may also happen, of course, that the separation surface never comes again in contact with the body and then it becomes the boundary of the main flow wake, or in other words, a vortex layer. As Maskell visualizes it, both of these phenomena can and do occur simultaneously, the separation surface often forming both the bubble and the free vortex layer.

It is clear from the above discussion that Maskell's model is considerably more satisfactory than that given by Eichelbrenner and Oudart, since now the limiting streamlines actually leave the surface and do not "pile up" in bypassing the reversed flow region. Still it should be pointed out that Maskell has not devised an actual mathematical criterion for the separation line. Since the separation is thought to occur only when two limiting streamlines meet, Eq. 18-2 becomes a necessary, but not quite sufficient, condition for the occurrence of separation. This is clearly so, since it fits the Eichelbrenner and Oudart model just as well

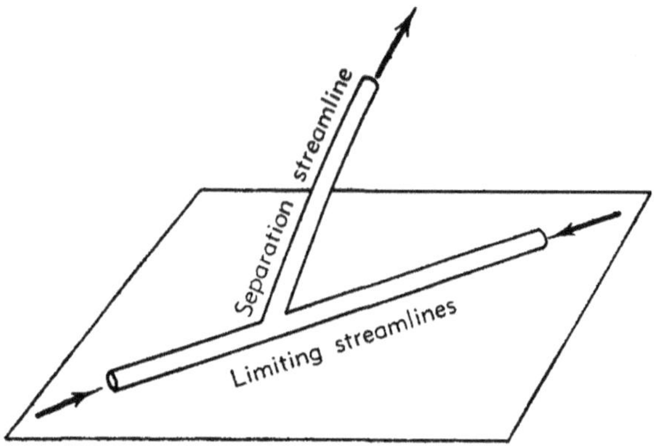

Fig. C,18c. Maskell's concept of singular separation.

as it fits Maskell's model. Nevertheless, in the absence of a better criterion, one probably can postulate the existence of the separation line when Eq. 18-2 is satisfied. This means that the three-dimensional boundary layer theory, which provides information about the pattern of the limiting streamlines (see for example Fig. C,10c) in the still thin, attached, forward flow region can be used to predict the location of the separation line.

C,19. Cited References.

1. Sears, W. R. Boundary layers in three-dimensional flow. *Appl. Mech. Rev. 7*, 281–285 (1954).
2. Moore, F. K. Three-dimensional boundary layer theory. *Advances in Applied Mechanics 4*. Academic Press, 1956.
3. Hayes, W. D. The three-dimensional boundary layer. *Nav. Ord. Rept. 1313*, 1951.
4. Michal, A. D. *Matrix and Tensor Calculus*. Wiley, 1947.
5. Howarth, L. The boundary layer in three-dimensional flow, Part I. *Phil. Mag. 42*, 239–243 (1951).
6. Sedney, R. Some aspects of three-dimensional boundary layer flows. *Quart. Appl. Math. 15*, 113–122 (1957).
7. Eisenhart, L. P. *An Introduction to Differential Geometry*. Princeton Univ. Press, 1947.

8. Moore, F. K. Three-dimensional compressible laminar boundary layer flow. *NACA Tech. Note 2279*, 1951.
9. Timman, R. A calculation method for three-dimensional laminar boundary layers. *Nat. Luchtvaartlab. Rept. F66*, 1951.
10. Mager, A., and Hansen, A. G. Laminar boundary layer over flat plate in a flow having circular streamlines. *NACA Tech. Note 2658*, 1952.
11. Mager, A. Generalization of boundary-layer momentum integral equations to three-dimensional flows including those of rotating system. *NACA Rept. 1067*, 1952.
12. Tetervin, N. Boundary-layer momentum equations for three-dimensional flow. *NACA Tech. Note 1479*, 1947.
13. Illingworth, C. R. Steady flow in the laminar boundary layer of a gas. *Proc. Roy. Soc. London A199*, 533 (1949).
14. Stewartson, K. Correlated incompressible and compressible boundary layers. *Proc. Roy. Soc. London A200*, 84 (1949).
15. Crabtree, L. F. The compressible laminar boundary layer on a yawed infinite wing. *Aeronaut. Quart. 5*, 85–100 (1954).
16. Struminskii, V. V. Equations of three-dimensional boundary layer in a compressible gas for arbitrary surfaces. *Doklady Akad. Nauk S.S.S.R. 114*, 271–274 (1957). (In Russian.)
17. Mangler, W. Boundary layers on bodies of revolution in symmetrical flow. *Ministry Aircraft Production (Völkenrode) Repts. and Transl. 55*, 1946.
18. Hansen, A. G., and Herzig, H. Z. On possible similarity solutions for three-dimensional incompressible laminar boundary layers. I: Similarity with respect to stationary rectangular coordinates. *NACA Tech. Note 3768*, 1956.
19. Herzig, H. Z., and Hansen, A. G. On possible similarity solutions for three-dimensional incompressible laminar boundary layers. II: Similarity with respect to stationary polar coordinates. *NACA Tech. Note 3832*, 1956.
20. Geis, T. Ähnliche Dreidimensionale Grenzschichten. *J. Rat. Mech. and Anal. 5*, (1956).
21. Hansen, A. G. Possible similarity solution of the laminar incompressible boundary layer equations. *Am. Soc. Mech. Engrs. Paper 57A79*, 1957.
22. Hansen, A. G. Similarity solutions of the laminar, incompressible, three-dimensional boundary-layer equations. *Ph. D. Thesis*. Case Inst. Technol., 1958.
23. Homann, F. Der Einfluss grosser Zähigkeit bei der Strömung um den Zylinder und um die Kugel. *Z. angew. Math. u. Mech. 16*, 153–164 (1936).
24. Hartree, D. R. On an equation occurring in Falkner and Skan's approximate treatment of the equations of the boundary layer. *Cambridge Phil. Soc. Proc. 33*, 223–239 (1937).
25. Lees, L. Laminar heat transfer over blunt-nosed bodies at hypersonic flight speeds. *Jet Propul. 26*, 259–269 (1956).
26. Levy, S. Effect of large temperature changes (including viscous heating) upon laminar boundary layers with variable free-stream velocity. *J. Aeronaut. Sci. 21*, 459–474 (1954).
27. von Kármán, Th. Über laminare und turbulente Reibung. *Z. angew. Math. u. Mech. 1*, 244–247 (1921).
28. Cochran, W. G. The flow due to a rotating disk. *Proc. Cambridge Phil. Soc. 30*, 365–375 (1934).
29. Bödewadt, U. T. Die Drehströmung über festem Grunde. *Z. angew. Math. u. Mech. 20*, 241–253 (1940).
30. Stewartson, K. On the flow between two rotating coaxial disks. *Proc. Cambridge Phil. Soc. 49*, 333–341 (1953).
31. Batchelor, G. K. Note on a class of solutions of the Navier-Stokes equations representing steady rotationally symmetric flow. *Quart. J. Mech. and Appl. Math. 4*, 29–41 (1951).
32. Squire, H. B. *Brit. Aeronaut. Research Council Rept. 16021*, 1953.
33. Proudman, I. The almost-rigid rotation of viscous fluid between concentric spheres. *J. Fluid Mech. 1*, 505–516 (1956).

34. Stewartson, K. On almost rigid rotations. *J. Fluid Mech. 3*, 17–26 (1957).
35. Taylor, G. I. The boundary layer in the converging nozzle of a swirl atomizer. *Quart. J. Mech. and Appl. Math. 3*, 129–139 (1950).
36. Cooke, J. C. On Pohlhausen's method with application to a swirl problem of Taylor. *J. Aeronaut. Sci. 19*, 486–490 (1952).
37. Schlichting, H. *Boundary Layer Theory.* McGraw-Hill, 1955.
38. Illingworth, C. R. The laminar boundary layer of a rotating body of revolution. *Phil. Mag. 44*, 389–403 (1953).
39. Chu, S. T., and Tifford, A. N. The compressible laminar boundary layer on a rotating body of revolution. *J. Aeronaut. Sci. 21*, 345–346 (1954).
40. Schlichting, H. Die laminare Strömung um einen axial angeströmten rotierenden Drehkörper. *Ing.-Arch. 21*, 227–244 (1953).
41. Prandtl, L. On boundary layers in three-dimensional flow. *Ministry Aircraft Production (Völkenrode) Repts. and Transl. 64*, 1946.
42. Struminskii, V. V. Sideslip in a viscous compressible gas. *Doklady Akad. Nauk S.S.S.R. 54*, 1946. Transl. *NACA Tech. Mem. 1276*, 1951.
43. Jones, R. T. Effects of sweepback on boundary layer and separation. *NACA Tech. Note 1402*, 1947. *NACA Rept. 884*, 1949.
44. Sears, W. R. The boundary layer of yawed cylinders. *J. Aeronaut. Sci. 15*, 49–52 (1948).
45. Cooke, J. C. The boundary layer of a class of infinite yawed cylinders. *Proc. Cambridge Phil. Soc. 46*, 645–648 (1950).
46. Nickel, K. Das Unabhängigkeitsprinzip bei dreidimensionalen Grenzschichten. *Arch. der Math. 9*, 313–320 (1958).
47. Tifford, A. N. *J. Aeronaut. Sci. 15*, 280–281 (1948).
48. Rott, N., and Crabtree, L. F. Simplified laminar boundary layer calculations for bodies of revolution and for yawed wings. *J. Aeronaut. Sci. 19*, 553–565 (1952).
49. Goldstein, S., ed. *Modern Developments in Fluid Dynamics.* Oxford Univ. Press, 1939.
50. Wild, J. M. The boundary layer of yawed infinite wings. *J. Aeronaut. Sci. 16*, 41–45 (1949).
51. Loos, H. G. A simple laminar boundary layer with secondary flow. *J. Aeronaut. Sci. 22*, 35–40 (1955).
52. Sowerby, L. Secondary flow in a boundary layer. *Roy. Aircraft Establishment Rept. Aero. 2512*, 1954.
53. Mager, A. Three-dimensional laminar boundary layer with small cross-flow. *J. Aeronaut. Sci. 21*, 835–845 (1954).
54. Hansen, A. G., and Herzig, H. Z. Cross flows in laminar incompressible boundary layers. *NACA Tech. Note 3651*, 1956.
55. Mager, A. Thick laminar boundary layer under sudden lateral perturbation. *50 Jahre Grenzschichtforschung*, Vieweg, Braunschweig, 1955.
56. Chambré, P. L. The laminar boundary layer with distributed heat sources or sinks. *Appl. Sci. Research A*, 393–401 (1957).
57. Howarth, L. The boundary layer in three-dimensional flow. II: The flow near a stagnation point. *Phil. Mag. 7*, 1433–1440 (1951).
58. Squire, L. C. The three-dimensional boundary layer equations and some power series solutions. *Brit. Aeronaut. Research Council Tech. Rept. R. M. 3006*, 1957.
59. Fogarty, L. E. The laminar boundary layer on a rotating blade. *J. Aeronaut. Sci. 18*, 247–252 (1951).
60. Sears, W. R. Potential flow around a rotating cylindrical blade. *J. Aeronaut. Sci. 17*, 183 (1950).
61. Tan, H. S. On laminar boundary layer over a rotating blade. *J. Aeronaut. Sci. 20*, 780–781 (1953).
62. Smith, W. E. Some examples of laminar boundary layer flow on rotating blades. *M. Aeronaut. Eng. Thesis.* Cornell Univ., 1952.
63. Graham, M. E. Calculation of the laminar boundary layer on rotating blades. *Ph.D. Thesis.* Cornell Univ., 1954.

64. Liu, S. W. The laminar boundary layer flow on rotating cylinders. I: Flat plate. *Air Force Office Sci. Research Tech. Note 57-298*, 1957.
65. Howarth, L. Concerning the effect of compressibility on laminar boundary layers and their separation. *Proc. Roy. Soc. London A194*, 16–42 (1948).
66. Moore, F. K. Laminar boundary layer on a circular cone in supersonic flow at small angle of attack. *NACA Tech. Note 2521*, 1951.
67. Moore, F. K. Three-dimensional laminar boundary layer flow. *J. Aeronaut. Sci. 20*, 525–534 (1953).
68. Kopal, Z. Tables of supersonic flow around yawing cones. *Mass. Inst. Technol. Tech. Rept. 3*, 1947.
69. Ferri, A. Supersonic flow around circular cones at angles of attack. *NACA Tech. Note 2236*, 1950.
70. Wilkinson, J. Some examples of three-dimensional effects in boundary layer flow. *Aeronaut. Quart. 5*, 73–85 (1954).
71. Moore, F. K. The unsteady laminar boundary layer of a wedge, and a related three-dimensional problem. *1957 Heat Transfer and Fluid Mechanics Institute*, 99–117 (1957).
72. Mager, A. Incompressible, non-meridional boundary layer flow on bodies of revolution. *Nav. Ord. Rept. 3366*, 1954.
73. Martin, J. C. On Magnus effects caused by the boundary layer displacement thickness on bodies of revolution at small angles of attack. *J. Aeronaut. Sci. 24*, 421–429 (1957).
74. Sedney, R. Laminar boundary layer on a spinning cone at small angles of attack in a supersonic flow. *J. Aeronaut. Sci. 24*, 430–436 (1957).
75. Kelly, H. R., and Thacker, G. R. The effect of high spin on the Magnus force on a cylinder at small angles of attack. *Nav. Ord. Rept. 5036*, 1956.
76. Thorman, C. H. Boundary layer measurements on an axisymmetric body with spin and yaw. *Ph.D. Thesis*. Calif. Inst. Technol., 1958.
77. Moore, F. K. Laminar boundary layer on cone in supersonic flow at large angle of attack. *NACA Tech. Note 2844*, 1952.
78. Whittaker, E. T., and Watson, G. N. *Modern Analysis*. Macmillan, 1943.
79. Mager, A. Laminar boundary layer problems associated with flow through turbomachines. *Ph.D. Thesis*. Calif. Inst. Technol., 1953.
80. Zaat J. A., Van Spiegel, E., and Timman, R. The three-dimensional boundary layer flow about a yawed ellipsoid at zero incidence. *Nat. Luchtvaartlab. Rept. F165*, 1955.
81. Zaat, J. A. A simplified method for the calculation of three-dimensional laminar boundary layers. *Nat. Luchtvaartlab. Rept. F184*, 1956.
82. Moore, F. K. Displacement effect of a three-dimensional boundary layer. *NACA Tech. Note 2722*, 1952.
83. Jahnke, E., and Emde, F. *Tables of Functions*. Dover, 1945.
84. Reshotko, E., and Beckwith, I. E. Compressible laminar boundary layer over a yawed infinite cylinder with heat transfer and arbitrary Prandtl number. *NACA Tech. Note 3986*, 1957.
85. Reshotko, E. Laminar boundary layer with heat transfer on a cone at angle of attack in a superonic stream. *NACA Tech. Note 4152*, 1957.
86. Reshotko, E. Heat transfer to a general three-dimensional stagnation point. *Jet Propul. 28*, 58–60 (1958).
87. Vaglio-Laurin, R. Laminar heat transfer on blunt nosed bodies in three dimensional hypersonic flow. *Wright Air Develop. Center Tech. Note 58-147*, 1958.
88. Kemp, N. H. The laminar three-dimensional boundary layer and a study of the flow past a side edge. *M. Aeronaut. Eng. Thesis*. Cornell Univ., 1951.
89. Mark, R. M. Laminar boundary layers on slender bodies of revolution in axial flow. *Ph.D. Thesis*. Calif. Inst. Technol., 1954.
90. Seban, R. A., and Bond, R. Skin friction and heat transfer characteristics of a laminar boundary layer on a cylinder in axial incompressible flow. *J. Aeronaut. Sci. 18*, 671–675 (1951).

91. Kelly, H. R. A note on the laminar boundary layer on a circular cylinder in axial incompressible flow. *J. Aeronaut. Sci. 21*, 634 (1954).
92. Stewartson, K. The asymptotic boundary layer on a circular cylinder in axial incompressible flow. *Quart. Appl. Math. 13*, 113–122 (1955).
93. Carrier, G. F. The boundary layer in a corner. *Quart. Appl. Math. 4*, 367–370 (1946).
94. Hashimoto, H. Note on Raleigh's problem for a plate of finite breadth. *Proc. First Japan. Natl. Congress Appl. Mech.*, 447–452 (1951).
95. Howarth, L. Raleigh's problem for a semi-infinite plate. *Proc. Cambridge Phil. Soc. 46*, 127–140 (1950).
96. Sowerby, L., and Cooke, J. C. The flow of fluid along corners and edges. *Quart. J. Mech. and Appl. Math. 6*, 50–70 (1953).
97. Batchelor, G. K. *Quart. J. Mech. and Appl. Math. 7*, 179 (1954).
98. Loitsianskii, L. G., and Bolshakov, V. P. On motion of fluid in boundary layer near line of intersection of two planes. *Central Aero-Hydrodynam. Inst. Trans. Rept. 279*, 3–18 (1936). Transl. *NACA Tech. Mem. 1308*, 1951.
99. Varley, E. An approximate boundary layer theory for semi-infinite cylinders of arbitrary cross-section. *J. Fluid Mech. 3*, 601–614 (1958).
100. Glauert, M. B., and Lighthill, M. J. The axisymmetric boundary layer on a long thin cylinder. *Proc. Roy. Soc. London A230*, 188–203 (1955).
101. Eichelbrenner, E. A., and Oudart, A. Observations sur un critère de décollment laminaire dans la couche-limite tridimensionelle. *Recherche aéronaut. Paris 40*, 1954.
102. Eichelbrenner, E. A., and Oudart, A. Le décollment laminaire en trois dimensions. *Recherche aéronaut. Paris 47*, 1955.
103. Maskell, E. C. Flow separation in three dimensions. *Roy. Aircraft Establishment Rept. Aero 2565*, 1955.

SECTION D

THEORY OF TIME-DEPENDENT LAMINAR FLOWS

NICHOLAS ROTT

CHAPTER 1. TIME-DEPENDENT SOLUTIONS OF THE NAVIER-STOKES EQUATIONS

D,1. Introduction. Many important aerodynamic problems involve a degree of unsteadiness that has important effects on the boundary layer. The heat transfer to the stagnation region of a blunt object undergoing oscillations in pitch during flight furnishes an example of such an unsteady boundary layer problem. In addition to problems arising from flight dynamics, the viscous aspects of acoustic and shock wave phenomena often require the study of time-dependent laminar flows.

In the present treatment, we shall consider that unsteadiness in the boundary layer is caused by some well-defined time dependence of the boundary conditions governing the viscous flow. Thus, we specifically exclude questions of hydrodynamic stability (to be discussed in Sec. G) or turbulence.

First we shall extend the classical theory of time-dependent solutions of the Navier-Stokes equations introduced in B,6. Here, the emphasis will be on unsteady motions of plane surfaces bounding a viscous fluid, including those for which Couette flow and stagnation-point flow are special, steady cases. The typical phenomena of unsteady viscous flow will be illustrated by these examples with particularly simple flow geometry.

Next, the time-dependent boundary layer will be treated. Many problems in this category have the Blasius flat-plate boundary layer as the special steady case; here, we imagine that temporal changes occur in the outer flow or in the motion of the plate. Some aspects of unsteady boundary layers with pressure gradient will also be described.

Before proceeding, it may perhaps be helpful to anticipate certain general features of the theory, in a qualitative way. When the imposed unsteadiness has a very gradual character, the diffusion process in the viscous part of the flow is able to keep pace, so to speak, with the changing boundary conditions. In such a case, the flow is called quasi-steady, and it behaves at each instant very nearly as the appropriate steady flow. At the other extreme, if changes are very rapid compared with the rate of adjustment possible by diffusion, the unsteadiness is concentrated in a layer which is very thin compared with the viscous layer as a whole. These two extremes of fast and slow (relative to the capacity of the flow to adjust) are qualitatively so different that only in certain particular cases is it possible to construct solutions valid for all rates of change. In other cases, a distinction must be made between "quite fast" and "quite slow."

D,2. Solutions of Stokes and Rayleigh and Their Generalizations. Time-dependent solutions of the Navier-Stokes equations were first investigated by Stokes in 1851 [*1*] (see also B,7); his solution is of fundamental importance for our understanding of time-dependent viscous phenomena. Presently, Stokes' solution will be found as a member of a more general family of flows, which will be discussed in this article.

Let the incompressible flow of a viscous fluid (with density ρ and viscosity μ, $\nu = \mu/\rho$) have Cartesian components u, v, and w in an x, y, z system. Consider a field in which the velocity is everywhere parallel to the x axis, and its magnitude depends on y, z, and the time t only:

$$u = u(y, z, t), \qquad v = 0, \qquad w = 0 \tag{2-1}$$

Continuity is satisfied, and the Navier-Stokes equation for the u component, given in Sec. B, is simplified to the following expression:

$$\frac{\partial u}{\partial t} = -\frac{1}{\rho}\frac{\partial p}{\partial x} + \nu\left(\frac{\partial^2 u}{\partial y^2} + \frac{\partial^2 u}{\partial z^2}\right) \tag{2-2}$$

The equations of motion in y and z direction evidently give $\partial p/\partial y = \partial p/\partial z = 0$, so that p can be a function of x and t only. As Eq. 2-2 has to yield solutions independent of x, the pressure gradient $\partial p/\partial x$ can be a function of t only, so that it is possible to put, say,

$$-\frac{1}{\rho}\frac{\partial p}{\partial x} = \frac{d}{dt}[U^*(t)] \tag{2-3}$$

If a new velocity u' is introduced, making use of the quantity U^* defined above, by putting

$$u' = u(y, z, t) - U^*(t) \tag{2-4}$$

Eq. 2-2 is changed into a homogeneous equation for u',

$$\frac{\partial u'}{\partial t} = \nu\left(\frac{\partial^2 u'}{\partial y^2} + \frac{\partial^2 u'}{\partial z^2}\right) \tag{2-5}$$

This simple transformation shows the equivalence of a spatially uniform pressure gradient to a spatially uniform, but time-dependent, velocity affecting the whole field (including boundaries) which may be thought of as caused by "pistons at infinity." The equivalence is naturally only true in incompressible flow.

The case under consideration is one of the few where the exact Navier-Stokes equations lead to a linear problem. Furthermore, u (or u') is exactly analogous to the temperature in the time-dependent problem of heat conduction in two spatial dimensions; the full wealth of the many known solutions to the latter problem can be used here.

D,3. The Fundamental Solutions in a Plane. The first member of this flow family, which will be discussed, is Stokes' case. Let a half-infinite field of fluid be limited by a rigid infinite plane at $y \equiv 0$, which performs some translatory motion in the x direction. Only the case of zero pressure gradient is physically meaningful. The velocity u is independent of z, and fulfills the equation

$$\frac{\partial u}{\partial t} = \nu \frac{\partial^2 u}{\partial y^2} \tag{3-1}$$

Boundary conditions are given at $y = 0$ by the prescribed motion of the plane, and the fluid is at rest for $y = \infty$. Stokes' case is obtained for the periodic motion of the plane; the boundary conditions are

$$u(0, t) = u_0 e^{i\omega t}, \qquad u(\infty, t) = 0 \tag{3-2}$$

and the solution is

$$u = u_0 e^{i\omega t} e^{-\sqrt{i\omega/\nu}y} \tag{3-3}$$

or in real form ($\sqrt{2i} = 1 + i$)

$$u = u_0 \cos\left(\omega t - \sqrt{\frac{\omega}{2\nu}}\, y\right) e^{-\sqrt{\frac{\omega}{2\nu}}y} \tag{3-4}$$

The solution represents a strongly damped wave (Fig. D,3) whose "penetration depth" into the fluid is proportional to $(\nu/\omega)^{\frac{1}{2}}$, i.e., inversely proportional to the square root of the frequency. The shear at the wall is

$$\tau_0 = \mu \left(\frac{\partial u}{\partial y}\right)_0 = u_0 \sqrt{\mu\rho\omega}\, \sin\left(\omega t - \frac{\pi}{4}\right) \tag{3-5}$$

showing a 45° phase difference between velocity and shear.

In Stokes' solution, the dependence of the penetration depth on the kinematic viscosity is the same as in boundary layer theory, namely, proportional to $\nu^{\frac{1}{2}}$. The boundary layer character of the flow is even better illustrated in a case first discussed by Rayleigh [*2*]: the wall initially at

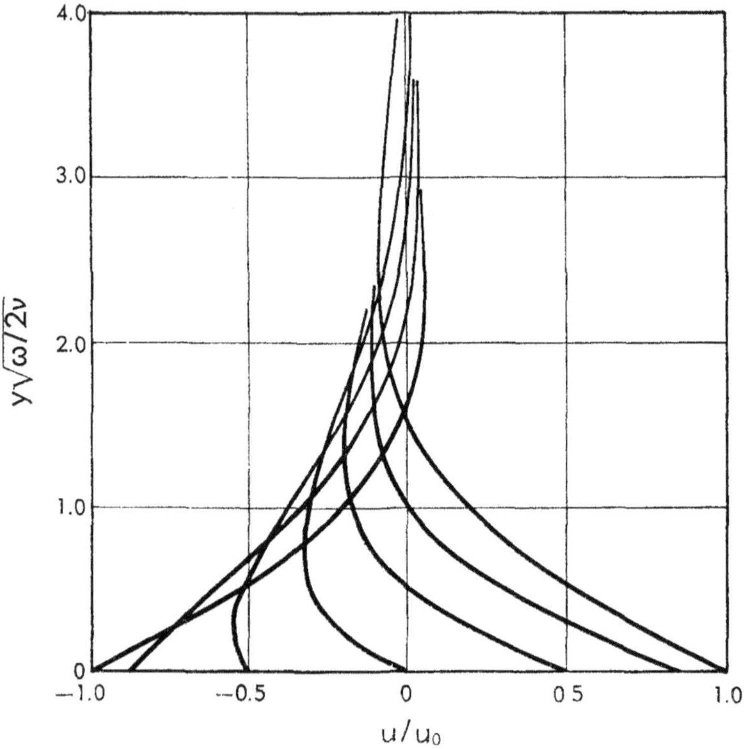

Fig. D,3. The velocity distribution for Stokes' solution (oscillating wall) at different phases of the motion. Profiles are shown for a half period. The phase difference is $\pi/6$.

rest acquires instantaneously, at $t = 0$, a constant velocity U. The boundary conditions are

$$u(0, t) = U1(t), \qquad u(\infty, t) = 0 \tag{3-6}$$

The solution is well known from the theory of heat conduction:

$$\begin{aligned} u &= U \frac{2}{\sqrt{\pi}} \int_{y/2\sqrt{\nu t}}^{\infty} e^{-\eta^2} d\eta \\ &\equiv U\left[1 - \operatorname{erf}\left(\frac{y}{2\sqrt{\nu t}}\right)\right] \end{aligned} \tag{3-7}$$

The shear at the wall varies as $t^{-\frac{1}{2}}$,

$$\tau_0 = U\sqrt{\frac{\rho\mu}{\pi t}} \tag{3-8}$$

The "penetration depth" is now proportional to $(\nu t)^{\frac{1}{2}}$. If t is replaced by U_0x (= velocity × distance from the leading edge) the same law is found as for the growth of Blasius' boundary layer on a flat plate, and the shear equation (Eq. 3-8) differs from Blasius' result only by a numerical factor. It may be noted, however, that the usage of the word "boundary layer" in Prandtl's sense is generally restricted to flows where the convective (nonlinear) terms do play a role, and where the dependence on the kinematic viscosity ν agrees with the law quoted above only to a first approximation in the limit of large Reynolds numbers.

Next we consider the case when the flow is limited by a second plane at $y = h$. If the plane at $y = 0$ performs a periodic motion while the plane at $y = h$ is at rest, i.e.,

$$u(0, t) = u_0 e^{i\omega t}, \qquad u(h, t) = 0 \tag{3-9}$$

the solution of Eq. 3-1 is easily obtained by the classical method of the separation of the variables. For the shear on the wall $y = 0$, the following expression is obtained:

$$\left.\begin{aligned} \tau_0 &= -\frac{\mu u_0 e^{i\omega t}}{h} \varphi\left(\frac{i\omega h^2}{\nu}\right) \\ \varphi(s) &= \sqrt{s} \coth \sqrt{s} \end{aligned}\right\} \tag{3-10}$$

This result is readily discussed by considering the expansions

$$\sqrt{s} \coth \sqrt{s} = 1 + \frac{1}{3} s + \frac{1}{45} s^2 + \cdots \tag{3-11}$$

for small arguments, while for large arguments the form

$$\sqrt{s} \coth \sqrt{s} = \sqrt{s}\,(1 + 2e^{-2\sqrt{s}} + 2e^{-4\sqrt{s}} + \cdots) \tag{3-12}$$

is used. Introducing Eq. 3-11 into Eq. 3-10 leads to an expansion appropriate for low frequencies, which starts with a "quasi-steady" term. The leading term obtained by the use of Eq. 3-12 is Stokes' value, Eq. 3-5. The criterion that determines the choice between the two expansions is obtained by comparing Stokes' "penetration depth" with the channel width h.

The case when, for the same geometry, the wall motion at $y \equiv 0$ is given by a step function, i.e.,

$$u(0, t) = U1(t), \qquad u(h, t) = 0 \tag{3-13}$$

will be considered as a typical example for obtaining the "indicial" type of motion from the periodic solution, i.e., the "Rayleigh-type" solution from the "Stokes-type" results. A standard method is to find for any given geometry the periodic solution first, as this problem is soluble (if an appropriate coordinate system is used) by separation of the variables. Next, the response to a step function is found by the well-known Fourier integral representation of this function, viz.,

$$1(t) = \frac{1}{2\pi} \lim_{\beta=0} \int_{-\infty}^{+\infty} \frac{e^{i\omega t}}{\beta + i\omega}\, d\omega \tag{3-14}$$

By the principle of superposition the whole flow field can be determined, using this formula. The present application will be restricted to the shear at $y = 0$. If the shear for the boundary condition, Eq. 3-13, is called $\bar{\tau}_0$, its integral representation is, by use of Eq. 3-10 and 3-14,

$$\bar{\tau}_0 = -\frac{\mu U}{2\pi h} \lim_{\beta=0} \int_{-\infty}^{+\infty} \frac{e^{i\omega t}}{\beta + i\omega} \varphi\left(\frac{i\omega h^2}{\nu}\right) d\omega \tag{3-15}$$

Put

$$\frac{i\omega h^2}{\nu} + \frac{\beta h^2}{\nu} = s \tag{3-16a}$$

Also, we introduce the dimensionless time:

$$\theta = \frac{\nu t}{h^2} \tag{3-16b}$$

and a new constant

$$\frac{\beta h^2}{\nu} = c \tag{3-16c}$$

so that Eq. 3-15 becomes, taking the limit $\beta = c = 0$ everywhere except for the integration path,

$$\bar{\tau}_0 = -\frac{\mu U}{2\pi i h} \int_{-i\infty+c}^{+i\infty+c} \frac{\varphi(s)}{s} e^{\theta s} ds \tag{3-17}$$

This is the so-called Bromwich integral for the inverse Laplace transform of $\varphi(s)/s$, and the standard textbooks and tables will have the answers listed in many important cases. If φ is given by Eq. 3-10, the result is a theta function; here it will be given in series form. By use of the expansion (Eq. 3-12) in Eq. 3-17, it is found that

$$\bar{\tau}_0 = -\frac{\mu U}{h\sqrt{\pi\theta}}\left(1 + 2\sum_{n=1}^{\infty} e^{-n/\theta}\right) \tag{3-18}$$

The first term agrees with the Rayleigh solution (Eq. 3-8). An alternate way to obtain Eq. 3-18 is given, e.g. by Schlichting [*8*]: an infinite series of Rayleigh's solution (Eq. 3-7) is superimposed with the origin shifted for each term by $2h$ (in both positive and negative y direction), so that by the principle of images the sum satisfies the boundary condition at $y = h$.

The solution (Eq. 3-18) is particularly suitable for obtaining results for small times θ; this is connected with the fact that the expansion (Eq. 3-12) is valid for large values of s. Now one could try the expansion (Eq. 3-11) for small s in connection with the formula (Eq. 3-17). The first term indeed corresponds to the correct steady (Couette-flow) value for $\theta = \infty$, but the regular terms of the expansion of $\varphi(s)/s$ do not contribute to any information about the inverse Laplace transform of the function. A successful asymptotic expansion of $\bar{\tau}_0$ for large values of θ can be obtained, however, by representing $\varphi(s)$ by a sum of singular terms obtained from the well-known infinite sum of partial fractions for the cotangent function, which yields

$$\varphi(s) = \sqrt{s} \coth \sqrt{s} = 1 + 2 \sum_{n=1}^{\infty} \frac{s}{s + (\pi n)^2} \tag{3-19}$$

which, introduced in Eq. 3-17, gives

$$\bar{\tau}_0 = -\frac{\mu U}{h}\left(1 + 2 \sum_{n=1}^{\infty} e^{-\pi^2 n^2 \theta}\right) \tag{3-20}$$

This expression converges very fast for large values of θ.

The shear at $y = h$ is found to be

$$\bar{\tau}_h = -\frac{\mu U}{h}\left(1 + 2 \sum_{n=1}^{\infty} (-1)^n e^{-\pi^2 n^2 \theta}\right) \tag{3-21}$$

The result for any motion of both walls is obtained, starting from the known response to the step function, by the use of Duhamel integrals.

A modification of the Stokes solution due to a uniform motion of the plate combined with uniform suction was investigated by Stuart [*4*]. The steady part is the so-called asymptotic suction boundary layer profile, the unsteady part a modified Stokes solution. Watson [*5*] considered the corresponding case of impulsive change and the more general motions of the plate.

D,4. Channel Flow in a Plane. The case of the flow caused by a time-dependent pressure gradient can in principle be reduced to the case of moving walls, by the use of the definition (Eq. 2-3) and the relation (Eq. 2-4). It is, however, more convenient to attack the "inhomogeneous" problem

$$\frac{\partial u}{\partial t} = -\frac{1}{\rho}\frac{dp}{dx} + \nu \frac{\partial^2 u}{\partial y^2} \tag{4-1}$$

directly. The only case of interest is the flow between two walls; let these planes be assumed at rest at $y = +h$ and $y = -h$. By symmetry, it is sufficient to consider the problem between $y = 0$ and $y = h$, and to require vanishing shear at $y = 0$. The problem for a periodic pressure gradient will be formulated as follows:

$$-\frac{1}{\rho}\frac{dp}{dx} = Ke^{i\omega t}; \qquad y = 0, \quad \frac{\partial u}{\partial y} = 0; \qquad y = h, \quad u = 0 \tag{4-2}$$

The solution is

$$u = \frac{Ke^{i\omega t}}{i\omega}\left(1 - \frac{\cosh\sqrt{\frac{i\omega}{\nu}}\,y}{\cosh\sqrt{\frac{i\omega}{\nu}}\,h}\right) \tag{4-3}$$

In the limit $\omega = 0$, this reduces to

$$u = \frac{K}{2\nu}(h^2 - y^2) \tag{4-4}$$

i.e., to the Poiseuille flow. For $\omega \to \infty$, it is appropriate to put

$$\cosh\sqrt{\frac{i\omega}{\nu}}\,h \cong \frac{1}{2}\,e^{\sqrt{i\omega/\nu}\,h}$$

so that Eq. 4-3 becomes

$$u = \frac{K}{i\omega}\{e^{i\omega t} - \exp[\sqrt{i\omega/\nu}\,(y - h) + i\omega t] - \exp[-\sqrt{i\omega/\nu}\,(y + h) + i\omega t]\} \tag{4-5}$$

Here, the first term is the "inviscid" response to the fluctuating pressure; the second and the third terms can be interpreted as "boundary layers" at $y = h$ and at $y = -h$, respectively, of the Stokes type, Eq. 3-3. Consistently with the approximations already made, which imply that the Stokes-layer thickness is small compared to the half width h, it is sufficient to consider the second term at the wall $y = h$; the "symmetrical" solution is given by the third term alone at $y = -h$. For the discussion of the results, it is interesting to consider the "mean square" of the velocity u. The result is

$$\overline{u^2} = \frac{K^2}{2\omega^2}(1 - 2e^{-\eta}\cos\eta + e^{-2\eta}) \tag{4-6}$$

where $\eta = (h - y)\sqrt{\omega/2\nu}$ for the part near the wall $y = h$. It is seen (Fig. D,4) that the mean square velocity "overshoots" the very nearly uniform distribution at the center ($\eta \to \infty$) by about 13 per cent; the maximum occurs at $\eta \cong 2.28$.

This effect was first discovered experimentally in the analogous case of round channels by Richardson and Tyler [6]; hence the name "annular effect." The theory was given by Sexl [7] in the axially symmetric case, which will be mentioned later (p. 406). As is to be expected from the boundary layer character of the solution (Eq. 4-5), there will be no difference between the solutions near the wall in a "round" and a "flat" (plane) channel in the limit $\omega \to \infty$.

The problem of the response to a step function of the pressure gradient is soluble by the formalism employed in the previous examples. A different method will be given now, which, however, can be considered as a certain physical interpretation of the steps taken for the evaluation of the "inverse Laplace transform" discussed previously.

Let the "homogeneous" case with vanishing pressure gradient be reconsidered, and let some initial disturbance $u(y, 0)$ be assumed at the time $t = 0$ in the channel $y = \pm h$. Disturbances symmetrical in y will be treated only; thus, the boundary conditions $u(h) = 0$, $(\partial u/\partial y)_0 = 0$ apply. The method consists of finding "eigensolutions" of Eq. 3-1 by

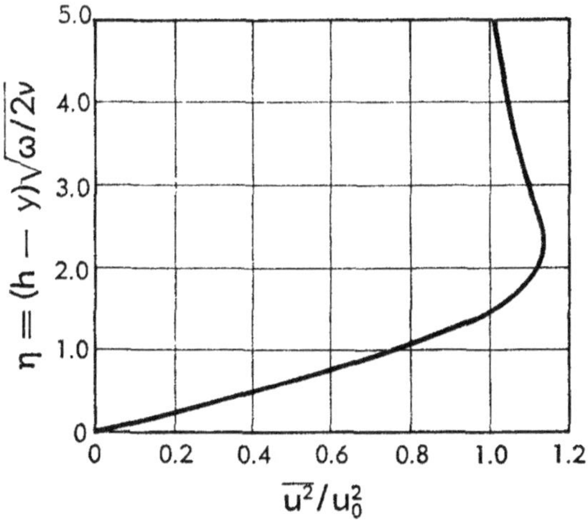

Fig. D,4. **The mean square of the velocity distribution near a wall with oscillating free stream (annular effect).**

separation of the variables, and to consider an infinite sum of these solutions with arbitrary coefficients (a_n, say). The outcome of these elementary operations is the following series for the "transient":

$$u_t = \sum_{n=1}^{\infty} a_n \exp\left[-\pi^2(n - \tfrac{1}{2})^2 \frac{\nu t}{h^2}\right] \cos\left[\pi(n - \tfrac{1}{2}) \frac{y}{h}\right] \tag{4-7}$$

Each term fulfills Eq. 3-1 and the boundary conditions stated above. For a given initial disturbance $u_t(y, 0)$, the coefficients a_n can be determined by Fourier's theorem.

The transient in the case of a step function in the pressure gradient will be found as follows: let $-(1/\rho)(dp/dx)$ jump to the value K at $t = 0$. The steady state reached at $t = \infty$ is the Poiseuille flow:

$$u_s = \frac{K}{2\nu}(h^2 - y^2) \tag{4-8}$$

The transient is obtained by assuming an initial disturbance which exactly cancels the Poiseuille flow at $t = 0$:

$$u_t(y, 0) = -u_s = -\frac{K}{2\nu}(h^2 - y^2) \tag{4-9}$$

and solving for the a_n in Eq. 4-7 for $t = 0$, by Fourier's theorem, using Eq. 4-9. The result is

$$a_n = \frac{Kh^2}{2\nu}\frac{4(-1)^n}{\pi^3(n - \frac{1}{2})^3} \tag{4-10}$$

The total response to the step function is the sum of u_s and u_t, with a_n given above.

The connection between the present analysis and the method of evaluating the inverse Laplace transform is found when it is realized that the homogeneous solutions, i.e., the terms in the series (Eq. 4-7), are the singular cases of the inhomogeneous solution (Eq. 4-3). Thus, the series (Eq. 4-7) corresponds to the evaluation of the Bromwich integral by use of partial fractions.

D,5. Nonplanar Problems. Proceeding now to cases where u is a function of both y and z, many interesting solutions have been given in the more recent literature.

Consider, first, cases of axial symmetry, where the "moving wall" is an infinite circular cylinder. A solution depending only on $r = \sqrt{x^2 + y^2}$ fulfills the equation

$$\frac{\partial u}{\partial t} = -\frac{1}{\rho}\frac{dp}{dx} + \nu\left(\frac{\partial^2 u}{\partial r^2} + \frac{1}{r}\frac{\partial u}{\partial r}\right) \tag{5-1}$$

For the "outer" solution between the cylinder radius a and $r = \infty$, only the homogeneous case with no pressure gradient is of interest. The periodic solution is expressed by the appropriate Hankel function (vanishing at infinity) of the argument $r\sqrt{i\omega/\nu}$, which replaces the exponential in the corresponding plane solution (Eq. 3-3). The Rayleigh problem of indicial motion for the circular cylinder was treated by Batchelor [*8*], whose discussion is based on an earlier solution of the corresponding heat conduction problem by Carslaw and Jaeger [*9*]. The expansions for small and large times, describing the behavior of the shear, will be quoted. If $t \ll a^2/\nu$,

$$\tau_0 = \mu U\left(\frac{1}{\sqrt{\pi\nu t}} + \frac{1}{2a} - \sqrt{\frac{\nu t}{\pi}}\frac{1}{4a^2} + \frac{\nu t}{a^3}\cdots\right) \tag{5-2}$$

The first term of this series coincides with the formula for planar flow (Eq. 3-8); this is physically clear from the boundary layer character of the flow. It was shown by Batchelor that the next term is valid on the outside of a cylinder with arbitrary smooth cross section (i.e., no corners) if a is replaced by the local radius of curvature R. Hasimoto [10] later proved that even the third term is valid for all smooth cylinders with R replacing a, and that the fourth term is generalized by replacing a^{-3} by $R^{-3} + \partial^2(R^{-1})/\partial S^2$, where S is the arc length. Cooke [11] gave a simplified discussion of Hasimoto's method and determined one more term.

In the case of $t \gg a^2/\nu$, Batchelor gives for the circular cylinder the expression

$$\tau_0 = \frac{\mu U/a}{\ln(\sqrt{4\nu t}/a) - \gamma} \tag{5-3}$$

where $\gamma = 0.577 \ldots$ is Euler's constant. The flow region at large times is large compared to a and changes slowly except at the outer edge of this region. Again, Hasimoto [12] was able to show that a formula of the form (Eq. 5-3) is valid for any cross section; his method permits an expansion in $a/\sqrt{\nu t}$, which joins smoothly the expansion for small times and is valid for arbitrary cross sections which may include corners.

The Rayleigh problem for the corner has received particular attention by many authors, partly because of the connection with the difficult steady boundary layer (or rather "boundary region") problem in the corner (C,16). The semi-infinite plate (corner with zero opening angle) was treated by Howarth [13]; the arbitrary angle was solved by Sowerby [14], Hasimoto [15,16], and Sowerby and Cooke [17]. The known solutions for the infinite corners enabled Batchelor [8] and Hasimoto [12] to compute a correction to the drag of cylinders with corners to the order of the second term in the expansion (Eq. 5-2). These corrections are based on the fact that if the shear is integrated on one side of a plate, forming a corner at $r = 0$, one finds

$$\lim_{t=0} \int_0^\infty \left(\frac{\tau_0}{\mu U} - \frac{1}{\sqrt{\pi\nu t}}\right) dr = K(\alpha) \tag{5-4}$$

i.e., the integral exists and gives a constant depending on the corner angle α. Howarth found $K = \frac{1}{2}$ for the semi-infinite plate; Hasimoto and Sowerby calculated tables of $K(\alpha)$. For "inside" corners, K is negative; for instance, $K = -2/\pi$ for the inside of the right-angled corner.

It is seen that the Rayleigh problem for a cylinder with arbitrary cross section was most successfully treated by methods of direct expansions for small and large times. The classical method for simple shapes consists of finding the periodic (Stokes) solution first by separation of the variables, and then applying the Laplace transform methods in order to obtain the Rayleigh case. Besides the classical case of the circle, this method was applied to rectangular cross-sections by Davies [18], in a different context. For the special case of an infinitely thin rectangle, i.e., for a strip of finite width, Levine [19] obtained an expression for the drag which is usable for all times.

As to the "inner" problem, only the channel flow inside a circular pipe will be considered. A periodic pressure gradient causes a flow which has already been mentioned; the solution, given by Sexl, which corresponds to Eq. 4-3 is

$$u = \frac{Ke^{i\omega t}}{i\omega}\left[1 - \frac{J_0(\sqrt{-i\omega/\nu}\, r)}{J_0(\sqrt{-i\omega/\nu}\, a)}\right] \tag{5-5}$$

The analogue of the "transient" velocity distribution (Eq. 4-7) is obtained by replacing the trigonometric function by a Bessel function of order zero:

$$u_t = \sum_{n=1}^{\infty} a_n \exp\left(-\frac{k_n^2 \nu t}{a^2}\right) J_0\left(\frac{k_n r}{a}\right)$$

$$J_0(k_n) = 0 \tag{5-6}$$

The steady state is the well-known Poiseuille flow:

$$u_s = \frac{K}{\nu}(a^2 - r^2) \tag{5-7}$$

If the pressure gradient $-(1/\rho)(dp/dx)$ jumps from 0 to the constant value K at $t = 0$, the solution is obtained by superposing Eq. 5-7 and the transient, Eq. 5-6, with

$$a_n = -\frac{Ka^2}{\nu}\frac{4J_2(k_n)}{[k_n^2 J_1(k_n)]^2} \tag{5-8}$$

D,6. Vortex-Type Solutions. A further group of exact solutions of the time-dependent Navier-Stokes equations can be obtained by considering plane flows whose streamlines are concentric circles (see also B,7). Let v be the circumferential velocity, a function of the radius r and of the time t. For incompressible flow, v fulfills the equation (see p. 71):

$$\frac{\partial v}{\partial t} = \nu\left(\frac{\partial^2 v}{\partial r^2} + \frac{1}{r}\frac{\partial v}{\partial r} - \frac{v}{r^2}\right) \tag{6-1}$$

Circumferential variation of the pressure is physically impossible under the assumptions made, and the radial pressure change is obtained from

$$\frac{\partial p}{\partial r} = \rho\frac{v^2}{r} \tag{6-2}$$

If the variable $\Gamma = 2\pi r v$ is introduced, Eq. 6-1 becomes

$$\frac{\partial \Gamma}{\partial t} = \nu\left(\frac{\partial^2 \Gamma}{\partial r^2} - \frac{1}{r}\frac{\partial \Gamma}{\partial r}\right) \tag{6-3}$$

while the vorticity $\omega = \partial(rv)/r\partial r$ fulfills

$$\frac{\partial \omega}{\partial t} = \nu \left(\frac{\partial^2 \omega}{\partial r^2} + \frac{1}{r} \frac{\partial \omega}{\partial r} \right) \tag{6-4}$$

The regular motion in- and outside a rotating cylinder or between two concentric cylinders is similar in character to flows that have been treated before. Here, only those cases will be discussed which are initially ($t = 0$) singular and correspond to a vortex-type motion.

The fundamental solution of Eq. 6-1 of this type was apparently first discussed by Oseen [*20*] and has the form

$$v = \frac{\Gamma_\infty}{2\pi r} (1 - e^{-\frac{r^2}{4\nu t}}) \tag{6-5}$$

For $r^2 \gg 4\nu t$, the distribution of a potential vortex with the circulation Γ_∞ is found. For small values of r, v exhibits solid rotation:

$$v \cong \frac{\Gamma_\infty r}{8\pi \nu t}$$

The solution (Eq. 6-5) provides a smooth connection between the steady outside flow and the growing core; the two domains "meet" at a radius $r^* \cong \sqrt{4\nu t}$, and the time-history of r^* describes the decay law of the vortex.

The vorticity connected with the solution (Eq. 6-5) viz.,

$$\omega = \frac{\Gamma_\infty}{4\pi \nu t} e^{-\frac{r^2}{4\nu t}} \tag{6-6}$$

is well known from the theory of heat conduction as the solution corresponding to a "heat pulse" at $t = 0$, $r = 0$.

A series of further solutions can be obtained by differentiating Eq. 6-5 with respect to time, and reinterpreting the result as a new velocity distribution (an interpretation which is inadmissible for spatial derivatives). It was noted by de Neufville [*21*] that the corresponding set of functions in the vorticity ω forms an orthogonal set; they are expressible by Laguerre polynomials of the argument $r^2/4\nu t$. It is possible to treat the decay for a general initial vortex distribution with the aid of this series.

Next, let a new flow field with a radial velocity u and an axial velocity w be superposed to the present flow such that

$$u = -ar, \qquad w = 2az \tag{6-7}$$

The field (Eq. 6-7) represents an axially symmetric stagnation point flow with the "inflow gradient" a (for $a > 0$, u is directed inward). It is easy to verify that the superposition of the field (Eq. 6-1) still permits solution of the Navier-Stokes equation where v is a function of r and t only; the equation for v is modified to

$$\frac{\partial v}{\partial t} - ar\left(\frac{v}{r} + \frac{\partial v}{\partial r}\right) = \nu\left(\frac{\partial^2 v}{\partial r^2} + \frac{1}{r}\frac{\partial v}{\partial r} - \frac{v}{r^2}\right) \tag{6-8}$$

The steady solution of Eq. 6-8 was discussed by Burgers [*22*]:

$$v = \frac{\Gamma_\infty}{2\pi r}\left(1 - e^{-\frac{ar^2}{2\nu}}\right) \tag{6-9}$$

This has the form of Oseen's solution, only made steady (or "frozen"), evidently by the convection of ω directed toward the core, which opposes the diffusion effect. A generalization to an unsteady solution was given by Rott [*23*]:

$$v = \frac{\Gamma_\infty}{2\pi r}\left(1 - e^{-\frac{ar^2}{2\nu}F(t)}\right)$$

$$F(t) = (1 + \beta e^{-2at})^{-1} \tag{6-10}$$

where β is an arbitrary constant. The solution represents the history of a field in which the circulation is restricted to the outside of a certain radius, which shrinks in accordance with the convection-diffusion law of vorticity. Again, the time derivatives of Eq. 6-10 can be used as new solutions for v, and rather general initial distributions of vorticity can be treated.

Flow problems in which a steady line sink was superimposed to a time-dependent vortex flow were investigated by Preston [*24*].

D,7. Stagnation Point Flow. A case of unsteady stagnation point flow will be considered, which (although an exact solution of the Navier-Stokes equations) is primarily of interest from the point of view of boundary layer theory. In general, unsteady boundary layer problems require the solution of a nonlinear problem. An important method of solution considers a steady "fundamental" nonlinear solution, with a linearized time-dependent part superimposed. For stagnation point flow, a case exists where the "split" in a steady and unsteady part is possible such that the unsteady part exactly fulfills a linear equation.

The case under consideration consists of a plane stagnation point flow bounded by a plate $y = 0$, where the plate performs motion in its own plane (the x direction), leaving the flow outside the boundary layer steady. The solution is obtained (for periodic motion) by assuming for the velocity component parallel to the plate the expression

$$u = axf'(\eta) + be^{i\omega t}g(\eta) \tag{7-1}$$

where η is the boundary layer variable: $\eta = y\sqrt{a/\nu}$; a is the velocity gradient of the inviscid flow, and f is the steady (Hiemenz) solution, discussed in B,8. With the velocity component v, the pressure p, and the Hiemenz solution $f(\eta)$ for steady flow left unchanged, an exact solution of the Navier-Stokes equation is found, with the function g for the unsteady profile fulfilling the linear equation

$$g'' + fg' - f'g - iKg = 0 \tag{7-2}$$

where $K = \omega/a$ is a "reduced" frequency, and the boundary conditions are: $g(0) = 1$, $g(\infty) = 0$.

Solutions were given by Rott [25] and by Glauert [26]. The procedure, which will also be used later (for the linearized case), involves two expansions with respect to the reduced frequency K, one for small and one for large values of K.

If $K \ll 1$, put

$$g = g_0 + iKg_1 + (iK)^2g_2 + \cdots \tag{7-3}$$

which leads to the set of equations

$$g_0'' + fg_0' - f'g_0 = 0 \tag{7-4}$$

$$g_1'' + fg_1' - f'g_1 = g_0 \tag{7-5}$$

etc.

The solution of Eq. 7-4 is found to be

$$g_0 = \frac{1}{f''(0)} f''(\eta) = 0.8113f''(\eta) \tag{7-6}$$

i.e., g_0 is proportional to the shear of the Hiemenz flow. The proof is obtained by checking the boundary conditions and by comparing Eq. 7-4 with the Hiemenz equation (p. 80) differentiated once with respect to η. The function g_0 represents the complete solution in the special case of a steady wall motion ($K = 0$) with the velocity b.

The functions g_1, g_2, etc., must be determined by numerical calculation. For the time-dependent part of the wall shear,

$$\tau_w = -\rho\sqrt{a\nu}\, be^{i\omega t}g'(0) \tag{7-7}$$

Glauert gives the numbers

$$-g'(0) = 0.8113 + 0.4932iK + 0.0947K^2 + \cdots \tag{7-8}$$

For $K \gg 1$, Rott uses a WBK-method, while Glauert's solution involves an expansion which is more generally applicable and will be given here. First, the independent variable will be changed from the Hiemenz variable η to the "Stokes" variable

$$Y = \sqrt{\frac{i\omega}{\nu}}\, y = \sqrt{\frac{i\omega}{a}}\, \eta \equiv \alpha^{-1}\eta \tag{7-9}$$

Denoting differentiations with respect to Y with an asterisk, (Eq. 7-2) now reads

$$g^{**} - g = \alpha(f^*g - fg^*) \tag{7-10}$$

The Hiemenz function f, originally given as a function of η, has to be expressed in terms of Y; this will be done by considering the expansion for $f(\eta)$ near $\eta = 0$. It is found that

$$\begin{aligned} f &= \tfrac{1}{2}A\eta^2 - \tfrac{1}{6}\eta^3 + \tfrac{1}{120}A^2\eta^5 - \cdots \\ &= \tfrac{1}{2}A\alpha^2Y^2 - \tfrac{1}{6}\alpha^3Y^3 + \tfrac{1}{120}A^2\alpha^5Y^5 \end{aligned} \tag{7-11}$$

where $A = f''(0) = 1.2326$. A solution of Eq. 7-10 is sought in the form

$$g = \sum_{n=0}^{\infty} \alpha^n \varphi_n(Y) \tag{7-12}$$

where, for the purpose of identifying proper powers of α, the expansion (Eq. 7-11) for f is inserted in Eq. 7-10.

The leading term of Eq. 7-12 fulfills

$$\varphi_0^{**} - \varphi_0 = 0 \tag{7-13}$$

solved by

$$\varphi_0 = e^{-Y} \tag{7-14}$$

This is Stokes' solution. It is characteristic for all high frequency expansions that a Stokes solution (completely independent of the "outside" flow) appears as the leading term.

For the subsequent terms, Glauert finds $\varphi_1 = \varphi_2 = 0$, and φ_3 fulfills

$$\varphi_3^{**} - \varphi_3 = Ae^{-Y}(Y + \tfrac{1}{2}Y) \tag{7-15}$$

which is solved (with $\varphi_3(0) = \varphi_3(\infty) = 0$) by

$$\varphi_3 = -Ae^{-Y}(\tfrac{3}{8}Y + \tfrac{3}{8}Y^2 + \tfrac{1}{12}Y^3) \tag{7-16}$$

All solutions φ_n have the form $e^{-Y}\sum a_K Y^K$. It is seen that the method actually involves an expansion both in α and in Y.

The wall shear is given by Eq. 7-7 with

$$-g'(0) = \sqrt{iK} + \frac{0.4622}{iK} + \cdots \tag{7-17}$$

where the first term is Stokes' value.

The indicial flow problem is soluble in principle by the methods discussed earlier (p. 399), except that difficulties may arise due to the fact that only expansions with respect to the frequency parameter are known. If the expansion for $K \gg 1$ is used, a term-by-term Laplace transform gives the correct expansion of the indicial problem, valid for small times. On the other hand, the expansion $K \ll 1$ around the "regular" point $K = 0$ does not contain information that is directly usable for the description of the asymptotic behavior of the indicial flow. What would be needed is the knowledge of the singularities in the complex K plane, for a method which was used to obtain the solution (Eq. 3-20). The singular solutions can be found from the solutions to the homogeneous problem, so that the method used in D,4 would be best suited for obtaining the asymptotic behavior. This would involve the determination of all eigenfunctions for real negative values of $p = iK$, for which Eq. 7-2 has solutions with the homogeneous boundary conditions $g(0) = g(\infty) = 0$.

In view of the complications of these methods, a practical approach was proposed by Rott and Rosenzweig [*27*], which essentially involves a curve-fitting to the plot of $-g'(0)$ against $iK = p$. The expression

$$-g'(0) = \frac{p + 1.3164}{\sqrt{p + 2.6328}} + \frac{0.4622p}{(p + 3.860)^2} + \cdots \tag{7-18}$$

has its first term so chosen that the first terms in the series (Eq. 7-8 and 7-17) are exactly reproduced; the second term of the expression (Eq. 7-18) corrects for the exact value of the second term in each series. The principle of this "fitting" is obvious; it is easily extended to further terms and adapted to different series. The determination of the transform of Eq. 7-18, which gives the shear for the transient, is an elementary problem. Once the shear is known, the profiles can be constructed by simple approximate methods; further details are given by Rosenzweig [*28*].

It was proposed by Rott and Rosenzweig that these procedures also be adapted to the general linearized solution, which will be discussed presently. Low and high frequency expansions will be found for the periodic case; this is all that is needed for the construction of expressions corresponding to Eq. 7-18.

The indicial problem was also considered by Watson [*29*], who determined a series for the whole profile valid for small t, based on the transform of the high frequency expansion. As pointed out before, the transform method does not cause any fundamental difficulties in this case, but in general an impractically large number of terms is needed for a good description of the results for large t.

Further cases of linear time-dependent problems are found in connection with three-dimensional stagnation point flow. A case originally discussed by Wuest [*30*] considers a plane steady stagnation flow as before, but with the plate moving in the third (z) dimension.

CHAPTER 2. THE TIME-DEPENDENT LAMINAR BOUNDARY LAYER

D,8. The Fundamental Equations of the Time-Dependent Laminar Boundary Layer. The problem of the laminar time-dependent boundary layer in two-dimensional compressible flow involves the solution of the following equations, which were derived previously (Sec. B, Eq. 20-12), using boundary layer coordinates x, y:

$$\frac{\partial \rho}{\partial t} + \frac{\partial(\rho u)}{\partial x} + \frac{\partial(\rho v)}{\partial y} = 0 \tag{8-1}$$

$$\rho\left(\frac{\partial u}{\partial t} + u\frac{\partial u}{\partial x} + v\frac{\partial u}{\partial y}\right) + \frac{\partial p}{\partial x} = \frac{\partial}{\partial y}\left(\mu\frac{\partial u}{\partial y}\right) \tag{8-2}$$

$$\frac{\partial p}{\partial y} = 0 \tag{8-3}$$

$$\rho c_p\left(\frac{\partial T}{\partial t} + u\frac{\partial T}{\partial x} + v\frac{\partial T}{\partial y}\right) - \left(\frac{\partial p}{\partial t} + u\frac{\partial p}{\partial x}\right) = \frac{1}{Pr}\frac{\partial}{\partial y}\left(\mu\frac{\partial T}{\partial y}\right) + \mu\left(\frac{\partial u}{\partial y}\right)^2 \tag{8-4}$$

The energy equation is already specialized for a perfect gas with a constant Prandtl number Pr; the equation of state

$$p = \mathcal{R}T\rho \tag{8-5}$$

reduces in view of Eq. 8-3 to $T\rho$ = const across the boundary layer.

Great simplifications are found in the case of incompressible flow, where density and viscosity are taken as constants. The assumption ρ = const and μ = const inside the time-dependent boundary layer is actually permissible in a large class of problems where the boundary conditions derive from phenomena connected with the compressibility of the medium. For instance, the flow near an obstacle which is small compared to the wavelength of a progressing sound wave, or the boundary layer caused by a standing wave in a tube, will be treated with the assumption of constant density across the boundary layer. The simplification ρ = const is certainly not admissible in a gas with a high heat transfer rate across the boundary layer. There exists, however, an important class of problems for which even the high heat transfer case is reducible to the solution of the incompressible equations. It was noted by Moore [*31*] that the introduction of the Howarth variable $\int_0^y \rho dy$ (see B,23) will reduce Eq. 8-1 and 8-2 to the incompressible case, even for time-dependent flow. The transformation is successful if the product $\rho\mu$ is constant (or $\mu \sim T$) and if the pressure gradient is zero. The first restriction is not considered to be serious in view of the interpretation offered by Chapman and Rubesin (see B,28) namely, that $\rho\mu$ = const is a "fit" to the actual law, which can be optimized by the proper choice of the constant. As to the second restriction, the combined effect of pressure gradient and heat transfer has no elegant and simple treatment even in the steady case.

In view of its simplicity and importance, the ρ = const and μ = const approximation will be treated first. In this case, the velocity boundary layer problem can be solved independently of the energy equation. Generalizations by the Howarth-Moore transformation and other compressibility effects are discussed in D,14.

Even the simple "incompressible" equations represent a set which requires rather involved methods of solution. Following the pattern of steady incompressible boundary layer theory, one might first try to find "similar" solutions for which the problem can be reduced to an ordinary differential equation for the profile shape. In an investigation by Schuh [*32*], solutions were found, most of which were already known by other methods; a new case was found and discussed, in which the free stream velocity varied proportional to x/t.

In the general case, some method of successive approximation is needed. Historically, the first investigations of time-dependent boundary layer were concerned with problems involving the start from rest; these cases are discussed in D,11. Here, the problem of changes affecting an already established steady boundary layer will be treated first. This more recent development begins with a paper by Moore [*31*], who treated the boundary layer on a flat plate with a velocity given as an arbitrary function of time. Moore made a series expansion with the terms $x\dot{u}/u^2$, $x^2\ddot{u}/u^3$ . . . , ($\dot{u} \equiv du/dt$), and ascending powers of each of these terms for higher approximations. A generalization to the case of the wedge flow was also given by Moore [*33*]. A necessary condition for the applicability of his method is that the terms of his series form a decreasing sequence. This will be fulfilled only for nearly quasi-steady flow; that is, changes occur gradually enough that the boundary layer is nearly, but not quite, a succession of steady states. Moore's expansion is in terms of departure from quasi-steadiness, and, as such, is not concerned with amplitude of unsteady fluctuations. Changes of large amplitude are permitted, if sufficiently gradual, whereas abrupt changes, even of small amplitude, are not permitted.

The following articles will treat cases with special well-defined time-dependences, and discuss the particular phenomena connected with them. The actual expansions and the functions to be determined will sometimes be the same as those given by Moore. One general result of Moore may be quoted, namely, the shear on the surface of a flat plate:

$$\frac{\tau_w}{\frac{1}{2}\rho_\infty U^2} = 0.664\sqrt{\frac{\nu}{ux}}\left[1 + 2.555\frac{x\dot{u}}{u^2} - 1.414\frac{x^2\ddot{u}}{u^3} + \cdots + 0.816\left(\frac{x\dot{u}}{u^2}\right)^2 + \cdots\right] \tag{8-6}$$

D,9. Linearized Periodic Solutions of the Boundary Layer Problem. For the linearized treatment of small disturbances of an arbitrary steady boundary layer, the fundamental case of periodic time-dependence was treated by Lighthill [*34*] and in the special case of the flat plate by Gibellato [*35*]. Lighthill considers the incompressible laminar boundary layer where the free stream velocity U has periodic fluctuations with an amplitude that is proportional to the steady distribution:

$$U(x, t) = U_0(x)(1 + \epsilon e^{i\omega t}) \tag{9-1}$$

Put

$$u = u_0 + u_1\epsilon e^{i\omega t}, \qquad v = v_0 + v_1\epsilon e^{i\omega t} \tag{9-2}$$

in the boundary layer equations for ρ = const, and using

$$-\frac{1}{\rho}\frac{\partial p}{\partial x} = \frac{\partial U}{\partial t} + U\frac{\partial U}{\partial x} \tag{9-3}$$

the equations for u_1 and v_1 are obtained by keeping terms of order ϵ:

$$\frac{\partial u_1}{\partial x} + \frac{\partial v_1}{\partial y} = 0$$

$$i\omega u_1 + u_0\frac{\partial u_1}{\partial x} + \frac{\partial u_0}{\partial x}u_1 + v_0\frac{\partial u_1}{\partial y} + \frac{\partial u_0}{\partial y}v_1 - \nu\frac{\partial^2 u_1}{\partial y^2} = i\omega U_0 + 2U_0U_0' \tag{9-4}$$

with the boundary conditions $u_1(0) = v_1(0) = 0$, $u_1(\infty) = U_0$. Now the "low frequency expansion"

$$u_1 = \sum_{n=0}^{\infty}(i\omega)^n u_{1n}, \qquad v_1 = \sum_{n=0}^{\infty}(i\omega)^n v_{1n} \tag{9-5}$$

is made. For the "quasi-steady" terms u_{10} and v_{10} which fulfill Eq. 9-4 with $\omega = 0$, explicit solutions in terms of the steady basic flow u_0, v_0 were given by Lighthill:

$$\begin{aligned} u_{10} &= u_0 + \frac{1}{2}y\frac{\partial u_0}{\partial y} \\ v_{10} &= \frac{1}{2}\left(v_0 + y\frac{\partial v_0}{\partial y}\right) \end{aligned} \tag{9-6}$$

These expressions are obtained by differentiating u_0 and v_0 with respect to the free stream velocity $U_0(x)$, a procedure which is well understandable by the interpretation that u_{10} and v_{10} are due to a "quasi-steady" change of $U_0(x)$ to $U_0(x)(1 + \epsilon)$.

Lighthill bases the determination of u_{11} on the Kármán-Pohlhausen method and proposes, after many further simplifications (justified by rather lengthy error-estimates), the expression

$$u_{11} = \frac{U_0\delta^2}{\nu}(1 - \eta)^2[A\eta + (2A - \tfrac{1}{2})\eta^2] \tag{9-7}$$

where δ is the Pohlhausen boundary layer thickness, $\eta \equiv y/\delta$, and with the Pohlhausen parameter $\Lambda = U'\delta^2/\nu$,

$$A = \frac{(36 - \Lambda)}{240} \tag{9-8}$$

The shape of the profile (Eq. 5-7) was chosen by Lighthill such that $\nu\partial^2 u_{11}/\partial y^2 = -U_0$ at $y = 0$ for all values of Λ. The further significant properties of Eq. 9-7 are the following:

1. If both the velocities u_{1n} and the frequency ω in Eq. 9-5 are to be made dimensionless, Eq. 9-7 indicates that the proper frequency parameter K is

$$K = \frac{\omega\delta^2}{\nu} \tag{9-9}$$

2. The profile has a shape parameter depending only on the local steady profile shape, and its scale is given by the local steady boundary layer thickness.

As the order of magnitude of δ in steady flow is given by

$$\delta^2 \sim \frac{\nu x}{U_0}$$

an alternate possibility for Eq. 9-9 is

$$K' = \frac{\omega x}{U_0} \tag{9-10}$$

For similar steady (Falkner-Skan) profiles, the two definitions agree for all x, except for a constant factor. It was shown by Rott and Rosenzweig [27] that, if the fundamental steady solution belongs to the Falkner-Skan family, the expansion (Eq. 9-5) leads to similar solutions u_{1n}, v_{1n} if the frequency parameter (Eq. 9-10) is used instead of ω. This means a simultaneous expansion in ω and x, with the exception of the stagnation point flow, where $U_0/x = \text{const}$ and Eq. 9-10 agrees with the definition of K used in E,3.

In the light of this discussion, it becomes clear that Lighthill's expression for u_{11} can be thought of as a result of replacing the actual steady "background" boundary layer development by a similar solution fitted at each station to the actual local steady flow.

For high frequencies, the leading term will be obtained by retaining in Eq. 9-4 the terms with the factor ω together with the derivative of the highest order; the resulting equation,

$$i\omega u_1 - \nu \frac{\partial^2 u_1}{\partial y^2} = i\omega U_0 \tag{9-11}$$

has the solution $u_1^{(0)}$ (say)

$$u_1^{(0)} = U_0(x)(1 - e^{-y\sqrt{i\omega/\nu}}) \tag{9-12}$$

This is Stokes' solution, with a locally varying amplitude.

Further terms were given by Rott and Rosenzweig [*24*], using the series

$$u_1 = \sum_{n=0}^{\infty} (i\omega)^{-\frac{n}{2}} u_1^{(n)}(x, Y)$$

$$v_1 = \sum_{n=0}^{\infty} (i\omega)^{-\frac{n+1}{2}} v_1^{(n)}(x, Y) \tag{9-13}$$

where

$$Y = y\sqrt{\frac{i\omega}{\nu}} \tag{9-14}$$

The method of Glauert is used, which was previously (p. 409) presented in detail: u_0 and v_0 are expanded in function of y around $y = 0$; the terms of equal power in $(i\omega)^{-\frac{1}{2}}$ are gathered after the substitution (Eq. 9-14) is made in the u_0 and v_0 series. It is found that $u_1^{(0)}$ is given by Eq. 9-12, that $u_1^{(1)}$ is zero and that u_1 up to and including $u_1^{(2)}$ is obtained by solving the equation

$$i\omega u_1 - \nu \frac{\partial^2 u_1}{\partial y^2} = i\omega U_0 + 2U_0U_0' \tag{9-15}$$

Of the two solutions to the homogeneous equation, e^{+Y} has to be rejected as it diverges for $y \to \infty$; thus, the solution of Eq. 9-15 which fulfills $u_1(0) = 0$ is

$$u_1 = [U_0 + (i\omega)^{-1}2U_0U_0'](1 - e^{-Y}) \tag{9-16}$$

It is seen, however, that the boundary condition $u_1(\infty) = U_0$ is not fulfilled. (For $U_0' = 0$, the difficulty is only delayed to the next term.) There is no doubt that the term e^{+Y} has to be rejected for physically meaningful solutions; this condition supersedes and in general disagrees with the "old" boundary condition $u_1(\infty) = U_0$. This difficulty was first encountered by Illingworth [*36*]. It is actually not surprising that a method involving an expansion at $y = 0$ will in general not permit a term-by-term fulfillment of the boundary conditions at $y = \infty$. (Only in very special cases—like the one treated by Glauert and Rott—does this difficulty not arise.) The further development of the method shows that the appearance of terms which are proportional to Y, Y^2, etc. cannot be avoided. The general term $u_1^{(n)}$ will contain a part that is multiplied by e^{-Y}, and an additional polynomial in Y. The first part is restricted to a thickness of the order $\sqrt{\nu/\omega}$, which for large ω is only a small fraction of the boundary layer thickness (of the order $\sqrt{\nu x/U_0}$). The second (polynomial) part has to be interpreted as a high frequency influence on the outer layer, for which an expansion at $Y = 0$ is obtained. It is remarkable that nowhere is a "degree of freedom" found to make use of the boundary condition at infinity. It is therefore a nontrivial assertion that the expansion represents a flow which fulfills the free stream boundary condition.

The proof of this statement was given by Lam and Rott [*37*], who gave a new kind of expansion in which each term was a sum of an "internal" solution multiplied by e^{-Y} and an external solution given in function of the original steady boundary layer variable; the boundary condition at infinity was fulfilled by each term.

Returning to the calculation of $u_1^{(3)}$, an "interaction" term has to be calculated which involves all the terms of Eq. 9-4 that are not contained in Eq. 9-15. It suffices to approximate U_0 and V_0 by just one term in the expansion near the wall; u_1 and v_1 can be replaced by $u_1^{(0)}$ and $v_1^{(0)}$. With these values, the "interaction" term is a new inhomogeneous part of Eq. 9-15. The result can be summed up in the following formula for the wall shear $\tau_1 = \mu(\partial u_1/\partial y)_w$, which uses the terms of the high frequency expansion up to $u_1^{(3)}$:

$$\tau_1 = (i\omega\rho\mu)^{\frac{1}{2}}[U_0 + (i\omega)^{-1}2U_0U_0'] - (i\omega)^{-1}[\tfrac{1}{4}U_0'\tau_0 + \tfrac{5}{8}U_0\tau_0'] \quad (9\text{-}17)$$

The high frequency expansion as outlined above presupposes that the time-dependent part of the boundary layer is essentially thinner than the steady basic part. In general, this condition will be fulfilled only far enough "downstream" along the body, where the steady boundary layer is thick enough. In particular, if the basic flow is a similar solution starting with zero thickness (e.g., the Blasius solution for a flat plate), a region "upstream" will always exist where the high frequency expansion is not valid.

Under these circumstances, it is amazing that the proposed high frequency expansion does not contain "free" constants which provide a possibility of matching with the time-dependent boundary layer upstream, which, in view of the fact that the fundamental expansion parameter is given by Eq. 9-10, is identified as the low frequency expansion. Lam and Rott [37] have shown that the high frequency solution admits exponentially small "asymptotic eigenfunctions" which formally could account for the difference between the two expansions in their region of overlapping validity. Lam and Rott further have shown that the low frequency expansion has an infinite radius of convergence, and that the time-dependent linearized boundary layer problem has a unique solution. Uniqueness together with the validity of the assumptions underlying the high frequency expansion in the downstream region provide an almost complete proof of the matching of the two expansions. A complete formal proof of the fact that the high frequency expansion is valid independently of upstream conditions, with the "asymptotic eigenfunctions" providing the only "weak" link between the two regions, is still missing.

A method to obtain the response to an impulsive change of the free stream velocity was described previously (p. 399). It was based on the knowledge of expansions for the shear for low and high frequencies, which have been discussed for the general linearized case in the present chapter.

D,10. Nonlinear Effects in Periodic Problems (Acoustic Streaming). Almost the entire literature of nonlinear effects in periodic boundary layers is restricted to the case when the mean velocity of the initial inviscid motion is zero, a case which is realized, e.g., for a body oscillating in a fluid at rest. With the inviscid velocity along the body given in the form

$$U_1(x, t) = \hat{U}(x) \cos \omega t \tag{10-1}$$

the linearized boundary layer equation

$$\frac{\partial u_1}{\partial t} - \nu \frac{\partial^2 u_1}{\partial y^2} = \frac{\partial U_1}{\partial t} \tag{10-2}$$

is completely solved by the "local" Stokes distribution, which will be given in the real form:

$$u_1 = \hat{U} [\cos \omega t - e^{-\eta} \cos (\omega t - \eta)] \tag{10-3}$$

where

$$\eta = y \sqrt{\frac{\omega}{2\nu}} \tag{10-4}$$

The second order term (quadratic in the velocity amplitude) fulfills the equation

$$\frac{\partial u_2}{\partial t} - \nu \frac{\partial^2 u_2}{\partial y^2} = U_1 \frac{\partial U_1}{\partial x} - u_1 \frac{\partial u_1}{\partial x} - v_1 \frac{\partial u_1}{\partial y} \tag{10-5}$$

where the right-hand side is calculated from Eq. 10-3, with v_1 determined from continuity. The calculation of this quadratic term must be based naturally on the real expression for u_1, and the result can be expressed by terms having the time factor cos $2\omega t$ and sin $2\omega t$ and also terms that are time-independent. For the solution of the inhomogeneous linear equation (Eq. 10-5) for u_2, put

$$u_2 = \omega^{-1}\hat{U}\hat{U}'[\varphi_{A_1}(\eta) \cos 2\omega t + \varphi_{A_2}(\eta) \sin 2\omega t + \varphi_B(\eta)] \qquad (10\text{-}6)$$

The boundary conditions to be fulfilled by the periodic part are that $\varphi_{A_1}(0) = \varphi_{A_2}(0) = 0$, $\varphi_{A_1}(\infty) = \varphi_{A_2}(\infty) = 0$, and such a solution can be found, making use also of the time-dependent homogeneous solution of Eq. 10-5. On the other hand, for the most interesting steady part, once the no-slip condition $\varphi_B(0) = 0$ is fulfilled, one can demand only that $\varphi_B(\infty)$ remain finite. The result obtained for the steady part is, after a lengthy but straightforward calculation,

$$\varphi_B = -\tfrac{3}{4} + \tfrac{1}{4}e^{-2\eta} + 2e^{-\eta} \sin \eta + \tfrac{1}{2}e^{-\eta} \cos \eta - \tfrac{1}{2}\eta e^{-\eta}(\cos \eta - \sin \eta) \qquad (10\text{-}7)$$

Thus, at great distances from the wall, a stationary influence on the free stream is found, which has the magnitude

$$u_2(\infty) = -\frac{3}{4\omega}\hat{U}\frac{d\hat{U}}{dx} \qquad (10\text{-}8)$$

Before discussing this remarkable result, the conditions for the validity of the procedure described so far will be investigated. First, the validity of the expansion in ascending powers of $\hat{U}$ must be established. This will be permissible if in the free stream

$$\left|\frac{\partial U_1}{\partial t}\right| \gg \left|U_1\frac{\partial U_1}{\partial x}\right|$$

With $|U_1| \sim \omega A$, where A is the amplitude of the oscillating motion of the body, and $|\partial U_1/\partial x| \sim \omega A/L$, where L is a characteristic body dimension, this condition is fulfilled if

$$\frac{A}{L} \ll 1 \qquad (10\text{-}9)$$

Next, the validity of the boundary layer approximation has to be considered. This method is applicable only if the boundary layer thickness of the fundamental Stokes solution is small compared to the body dimension, i.e., if the square of this ratio fulfills

$$\frac{\nu}{\omega L^2} \ll 1 \qquad (10\text{-}10)$$

Under these circumstances, the result (Eq. 10-8) states that a steady flow —the "acoustic streaming"—is "induced" in the inviscid flow by the oscillation of a body. The magnitude of this flow is independent of viscosity provided, naturally, that the condition (Eq. 10-10) is fulfilled. The acoustic streaming is zero at the stagnation point and at the extrema of the inviscid basic velocity; it is directed toward the stagnation point.

The analysis of this effect, which is presented here, is due to Schlichting [*38*]. Previously, Rayleigh [*39*] gave a (not quite complete) solution in the particular case of the standing sound wave in a tube, which was based on the same principle. The flow pattern calculated by Schlichting for a circular cylinder is shown in Plate D,10, together with an experimental verification. The flow past an oscillating cylinder in water is made visible by a time exposure picture taken with a camera that is moving with the cylinder. Streamlines are made visible by metal dust on the water surface. Similar flow patterns past cylinders and spheres were observed by Andrade, who used smoke for flow visualization.

In the case of a standing sound wave in a tube, Schlichting's result immediately gives the "secondary" flow effect, which is seen to be directed towards the node. This explains why dust particles are transported toward the node (as was observed by Kundt). If, however, the "Kundt tube" is filled with a big number of small particles (e.g., cork dust) a surprising regular pattern of discrete "ridges" is formed at and around the anti-node, which does not move away. A qualitative discussion of this effect can be found in papers by Andrade [*40*] and Schuster and Motz [*41*].

The third- and fourth order effects of the periodic boundary layer problem were calculated by Andres and Ingard [*42*]. The third order effects are periodic with frequencies ω and 3ω, while the fourth order introduces an additional steady term. The ratio of this correction to the solution (Eq. 10-8) is of the order A^2/L^2.

If the problem is to be solved for the case of "low Reynolds number" flow, i.e., when the condition (Eq. 10-10) is reversed ($\nu/\omega L^2 \gg 1$), the boundary layer solution is naturally not a suitable starting point. Andres and Ingard [*43*] obtained a solution for the circular cylinder by using the Stokes-Oseen approximation to the Navier-Stokes equations. They found an acoustic streaming which is directed opposite to the one given by Schlichting. Qualitatively it can be understood how the two solutions can "join" by observing the streamline patterns given by Schlichting (Plate D,10). It is seen that the innermost region of the boundary layer flow is directed opposite to the "free stream" acoustic streaming. The extension of such an inner region over the whole flow field gives the behavior of the low Reynolds number solution.

Experimentally, early observations by Carrière [*44*] indicated acoustic streaming away from the stagnation points in the strongly viscous domain. Andres and Ingard [*43*] established experimental limits between the two types of motion.

Further investigations on acoustic streaming were carried out for the

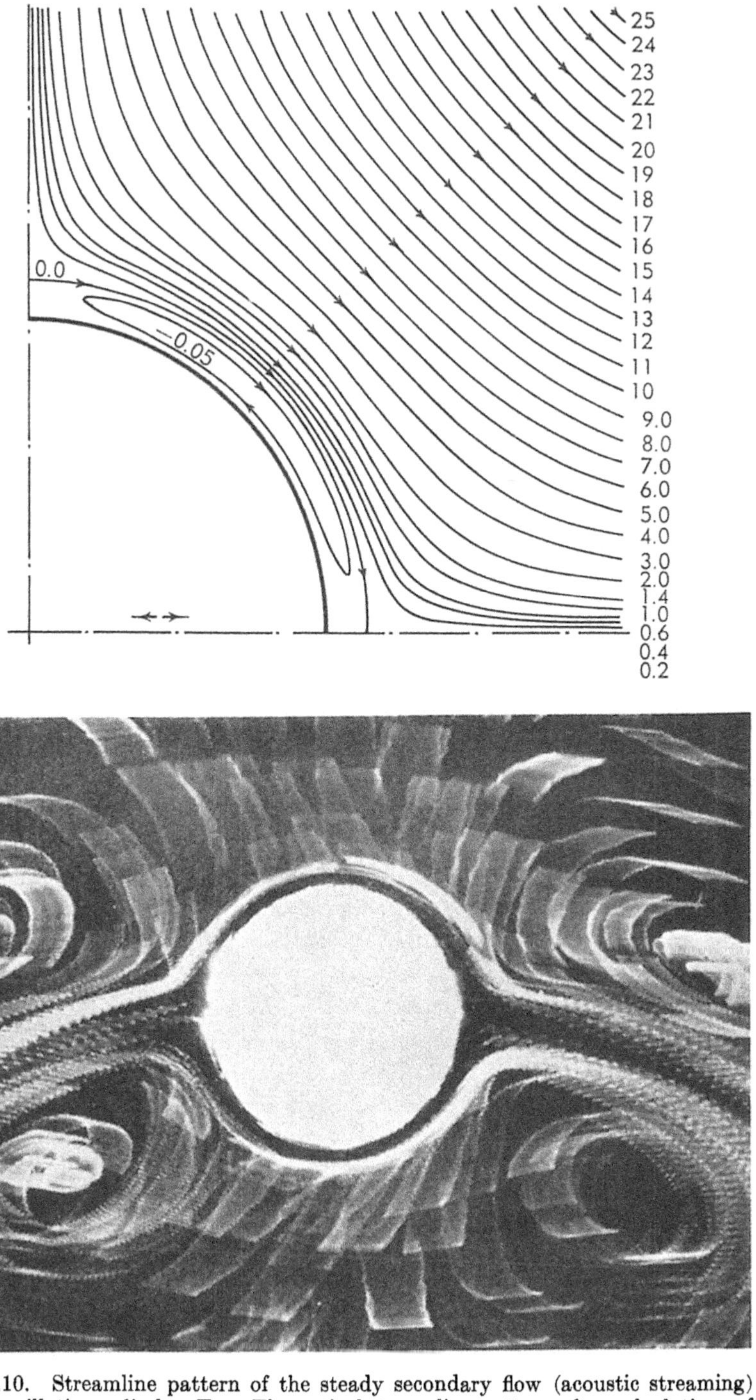

Plate D,10. Streamline pattern of the steady secondary flow (acoustic streaming) near an oscillating cylinder. Top: Theoretical streamline pattern after calculations of Schlichting. Numbers are values of the stream function. Bottom: Visualization obtained by a time exposure taken with a camera moving with the cylinder. Streamlines are made visible by metal dust on the water surface (from [38]).

special case of the cylinder and the sphere, taking advantage of the symmetries of the problem and using essentially a high Reynolds number expansion, in a series of papers by Holtsmark, et al. [*45*]; Raney, et al. [*46*]; and Lane [*47*]. The solution near the boundary was investigated by Nyborg [*48,49*].

In the case of periodic disturbances superimposed on a steady flow, Gibson [*50*] solved the problem of acoustic streaming (second approximation in the amplitude) for high frequencies. His results also include a solution to the linear high frequency expansion problem (discussed in Art. 9), which is valid across the whole basic steady boundary layer.

D,11. Boundary Layer Development Starting Impulsively from Rest. The investigation of the boundary layer development past bodies that are put in motion impulsively was started by Blasius in 1908. The approach proposed by Blasius was followed by many authors in subsequent years; a survey of these results is given by Schlichting [*3*].

The flow pattern immediately following the start of an impulsive motion in an incompressible viscous fluid is necessarily the same as would occur in an inviscid fluid, except for the presence of (infinitely thin) vortex sheets along all solid boundaries. This follows from the fact that, in the case of a suddenly "created" velocity, the dominant terms in the Navier-Stokes equation are

$$\frac{\partial u}{\partial t} = -\frac{1}{\rho}\operatorname{grad} p \qquad (11\text{-}1)$$

so that curl $u = 0$, i.e., a potential flow pattern will be found except near solid boundaries where the no-slip condition causes the initial vortex sheet. An interesting generalization of this result was pointed out by Sears (private communication), who considered an impulsive change superposed on an already established viscous flow pattern. He showed that the initial motion following any impulsive change of the boundary conditions consists of the superposition of the velocity pattern existing just before the change and the inviscid flow velocity pattern due to the impulsive boundary values (together with the corresponding infinitely thin wall vortex sheets). The statement is evident if one realizes that Eq. 11-1 dominates the pattern at the moment of change and that the terms of this equation are all linear.

Returning to the problem of the start from rest, let the inviscid velocity along the body be given:

$$U(x, t) = \bar{U}(x)1(t) \qquad (11\text{-}2)$$

Initially, when the diffusion of the infinitely thin vortex sheet is considered, the quadratic (convective) terms will be small, and as $\partial U/\partial t = 0$

for $t > 0$ in the special case given by Eq. 11-2, the velocity distribution u_1 (say) fulfills the equation

$$\frac{\partial u_1}{\partial t} - \nu \frac{\partial^2 u_1}{\partial y^2} = 0 \tag{11-3}$$

with $u_1(0, t) = 0$, $u_1(\infty, t) = \bar{U}$. The solution is

$$u_1 = \bar{U}(x) \frac{2}{\sqrt{\pi}} \int_0^\eta e^{-\eta^2} d\eta \tag{11-4}$$

where

$$\eta = \frac{y}{2\sqrt{\nu t}} \tag{11-5}$$

Eq. 11-4 represents a Rayleigh solution, with a variable free stream value adjusted to the local conditions.

The second approximation (for larger times) can be calculated iteratively. Putting

$$\begin{aligned} u &= u_1 + u_2 + \cdots \\ &= \bar{U}\zeta_1(\eta) + t\bar{U}\frac{d\bar{U}}{dx}\zeta_2(\eta) + \cdots \end{aligned} \tag{11-6}$$

the equation for u_2 is the same as the one considered in the previous paragraph (see Eq. 10-5). The function $\zeta_2(\eta)$ was determined by Blasius [*51*] in closed form. A further term of the expansion was calculated by Goldstein and Rosenhead [*52*]. Here, only the wall derivatives of ζ_1 and ζ_2 (needed for the shear calculation) will be quoted, as given by Blasius:

$$\zeta_1'(0) = \frac{2}{\sqrt{\pi}}, \qquad \zeta_2'(0) = \frac{2}{\sqrt{\pi}}\left(1 + \frac{4}{3\pi}\right) \tag{11-7}$$

It is now possible to discuss the time t_s which will elapse until the wall shear becomes zero at a given point, i.e., until "separation." Taking only two terms of the series, it follows from Eq. 11-6 and 11-7 that t_s is given by the equation

$$1 + \left(1 + \frac{4}{3\pi}\right)\frac{d\bar{U}}{dx} t_s = 0 \tag{11-8}$$

Using Goldstein and Rosenhead's result [*52*], t_s is determined by the equation

$$t_s^{-1} = -0.7122\bar{U}' + (0.727\bar{U}'^2 + 0.0597\bar{U}\bar{U}'')^{\frac{1}{2}} \tag{11-9}$$

It is seen that, according to the first approximation, separation only occurs if $\bar{U}'$ is negative; the second approximation gives a finite time at a point where $\bar{U}' = 0$. For a circular cylinder, $-\bar{U}^{-1}$ has its greatest value at the rear stagnation point ($\bar{U}' = -2U_\infty/R$) and the minimum time for separation occurs there; it is found to be $t_s = 0.351R/U_\infty$, using the first approximation; the value is about 9 per cent less, according to the second approximation. The maximum negative value of $\bar{U}'$ does not necessarily occur at the rear stagnation point. As pointed out by Tollmien [53] and Görtler [54], for an elliptic cylinder with its major axis across the stream, the extremum of $-\bar{U}'$ is not the rear stagnation point if the ratio of the axes exceeds $2/\sqrt{3}$. If this ratio tends to infinity, i.e., for a flat plate across the stream, $t_s = 0$ at the edge.

The applicability of these methods to bluff bodies evidently will not hold beyond a short time, and the calculation of too many further approximations is of little interest. For streamlined bodies, where this objection does not hold, the question of the "joining" of the expansion (Eq. 11-6) with the final steady state appears to be a difficult problem, which has not been attacked so far. In the limiting case of a flat plate, when $\bar{U}$ = const, the expansion (Eq. 11-6) gives zero for all terms except the first, a paradoxical result which will be discussed presently.

The case when the velocity of the body grows proportional to time was treated by Blasius [51], and Görtler [55] considered growth with t^n up to $n = 4$. The leading term in these expansions can be obtained from the Rayleigh solution by the Duhamel integral method. Analogous problems for bodies of revolution were solved by Boltze [56]. Tollmien [53] treated the case of a circular cylinder impulsively starting to move and to rotate. Indicial problems connected with the Kármán-Cochran problem (rotating infinite plane) were treated by Thiriot [57,58]. This problem was also solved by Nigam [59]; Probstein [60] treated the corresponding transient heat transfer phenomena. Nigam and Rangasami [61] investigated the growth of the boundary layer on a rotating sphere. The growth on arbitrary spinning bodies was treated by Illingworth [62]. Wundt [63] considered boundary layer growth on a cylinder with yawing motion.

We now consider the impulsive start of a flat plate; the "Blasius method" of expansion apparently leads to a blind alley. To make further progress, Stewartson [64] solved two strongly simplified problems. First, the problem was treated by the Rayleigh method, as a solution of the linear Oseen-type equation which can be used as a substitute to the boundary layer equation only for qualitative discussions. The equation is

$$\frac{\partial u}{\partial t} + U\frac{\partial u}{\partial x} = \nu\frac{\partial^2 u}{\partial y^2} \tag{11-10}$$

and the boundary conditions for impulsive start to a velocity U are, with a coordinate origin fixed at the leading edge of the plate,

$$\begin{aligned} u = v = 0 &\text{ for } y = 0, \quad x > 0,\ t > 0 \\ u = U \quad &\text{ for } y = \infty,\ x \geqq 0,\ t \geqq 0 \\ u = U \quad &\text{ for } y > 0, \quad x \geqq 0,\ t = 0 \\ &\text{ and } x = 0,\ t \geqq 0 \end{aligned} \tag{11-11}$$

Stewartson finds that the solution is everywhere a Rayleigh profile, but with a scale depending on the domain under consideration:

$$u = U \frac{2}{\sqrt{\pi}} \int_0^\eta e^{-\eta^2}\, dy \tag{11-12a}$$

where

$$\begin{aligned} \eta &= \frac{y}{2\sqrt{\nu x/U}} \quad \text{for} \quad x \leqq Ut \\ \eta &= \frac{y}{2\sqrt{\nu t}} \quad \text{for} \quad x \geqq Ut \end{aligned} \tag{11-12b}$$

Both solutions fulfill "individually" Eq. 11-10, and only the question of the proper joining at the point $x = Ut$ has to be discussed. Stewartson shows that no analytic joining at the point $x = Ut$ is possible.

Next, Stewartson used the momentum integral method for the approximate treatment of the problem; the same approach for this and several other related problems was also made by Schuh [*65*]. According to Stewartson, the boundary layer thickness grows proportional to $\sqrt{\nu x/U}$ for $x \leqq 0.377Ut$, while for larger x the growth is with $\sqrt{\nu t}$. Schuh finds, using a different profile shape, that the limit is $x/Ut = 0.385$. Again, two domains without analytic joining are found, but the "transition" is shifted to a point nearer to the leading edge. It will be seen that the real behavior exhibits a gradual transition in the region $0 < x < Ut$, while for $x > Ut$ the Rayleigh solution is exact.

Returning to the full time-dependent boundary layer problem, Stewartson investigates the use of new coordinates. On dimensional grounds, u can be expressed as a function of only two variables, e.g.,

$$\zeta = \frac{y}{\sqrt{\nu t}}, \qquad \theta = \frac{Ut}{x} \tag{11-13}$$

so that if a stream function ψ is defined by

$$u = U \frac{\partial \psi}{\partial \zeta} \tag{11-14}$$

it is found to fulfill the equation

$$\frac{\partial^3 \psi}{\partial \zeta^3} + \frac{1}{2}\zeta \frac{\partial^2 \psi}{\partial \zeta^2} = \left(\theta - \theta^2 \frac{\partial \psi}{\partial \zeta}\right) \frac{\partial^2 \psi}{\partial \zeta \partial \theta} + \theta^2 \frac{\partial^2 \psi}{\partial \zeta^2} \frac{\partial \psi}{\partial \theta} \tag{11-15}$$

Alternately, the pair of variables

$$\eta = y\sqrt{\frac{U}{\nu x}}, \qquad \xi = \frac{x}{Ut} \tag{11-16}$$

can be used; a stream function defined by

$$u = U \frac{\partial \phi}{\partial \eta} \tag{11-17}$$

fulfills

$$\frac{\partial^3\phi}{\partial\eta^3} + \frac{1}{2}\phi\frac{\partial^2\phi}{\partial\eta^2} = \left(\xi\frac{\partial\phi}{\partial\eta} - \xi^2\right)\frac{\partial^2\phi}{\partial\eta\partial\xi} - \xi^2\frac{\partial^2\phi}{\partial\eta^2}\frac{\partial\phi}{\partial\xi} \tag{11-18}$$

For small times, the use of Eq. 11-15 is appropriate and a power series in θ with coefficients depending on ζ is the obvious choice for a solution. The leading term, independent of θ, is easily found to be the error function; however, all subsequent terms vanish—this is the "blind alley" discussed before. Stewartson showed that a nonanalytic joining, or an essential singularity, will be found at $\theta = 1$, $\eta \to \infty$, and he also argued that no such singularity will occur in the region $0 \leqq \theta < 1$. Thus, in this domain ($x > Ut$), the one-term Rayleigh solution is the full solution, with a nonanalytic joining at $x = Ut$ to the remaining field.

For small x (large t), Eq. 11-18 will be used, and a power series in ξ with coefficients depending on η is assumed as a solution. The leading term, which is independent of ξ, is recognized as the Blasius solution for the flat plate. Again, all further terms of the series are identically zero. Stewartson showed, however, that an essential singularity can be found at $\xi = 0$, and that an additional solution is possible which behaves as $\exp(-\lambda/\xi^3)$ at $\xi = 0$, where λ is a positive constant. This term is exponentially small at $\xi = 0$, and cannot be expressed by a power series in ξ—a fundamental property of essential singularities. Thus, in contrast to the former case, the "one-term" series in ξ is nowhere exact, but for small ξ the error is exponentially small.

Stewartson considered his results to some extent as tentative only, but a uniqueness proof by Lam and Crocco [*66*] supported them completely. This subject will be discussed in the next paragraph in connection with a more general problem.

If the flat plate starts moving at $t = 0$ with some arbitrary time dependence $U(t)$, in particular proportional to t^n, the general solution of Moore [*31*] and a similar approach by Cheng and Elliott [*67*] gives the equivalent of a series expansion for small ξ. In general, this series does not "end," except for U changing impulsively at $t = 0$; in this case, the previous "one-term series" is recovered. These expansions naturally do not include possible singular solutions of the type discussed above, but this does not exclude a "practical" range of convergence where these solutions can be used.

D,12. Boundary Layer Development behind a Shock Progressing along a Flat Plate. A boundary layer problem with practical importance for shock tube flow is obtained by considering a shock wave moving into a gas at rest in a half-space limited by an infinite plane parallel to the direction of the shock propagation. The gas behind the shock is brought into motion, and a boundary layer develops, beginning at the shock.

This boundary layer problem is only "trivially" unsteady, as a change from the wall-fixed (laboratory) system to shock-fixed coordinate reduces it to a steady case. From this point of view, the gas and the wall move together ahead of the shock with the same velocity (U_1). After the shock, the gas velocity is reduced to U_2, whereas the wall continues to move with the velocity U_1, so that a boundary layer develops (see Fig. D,12a).

The pressure is constant behind the shock, and the case under consideration differs from the classical compressible "flat plate" problem only by the boundary condition at the wall. If the viscosity is assumed to be proportional to the absolute temperature, the compressible boundary

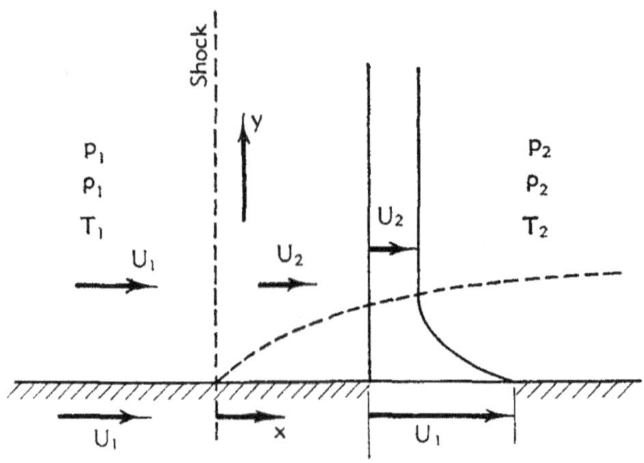

Fig. D,12a. Shock-fixed coordinate system for the treatment of the boundary layer development behind a shock normal to an infinite plate.

layer problem can be reduced to the incompressible case (see p. 225), and the only remaining task is to solve the Blasius equation for the flat plate with modified boundary conditions accounting for the wall motion.

This problem was first formulated and solved by Hollyer [*68*]; earlier, Donaldson and Sullivan [*69*] used momentum methods to obtain the effect of this boundary layer on the shock attenuation in the shock tube. Further treatments of this case, and the associated heat transfer problems, were made by Mirels [*70*], Rott and Hartunian [*71*], Bromberg [*72*], and Bershader and Allport [*73*]. Tables of the resulting profile function (using the Howarth variable) were calculated by Hollyer and Mirels. In the particular case of a weak shock, when U_2 differs only slightly from U_1 the linearization of the problem becomes permissible and a Rayleigh profile is obtained in the limit $U_2 \to U_1$.

The question of the heat transfer to the wall, with consideration of the heat conduction in the solid, can be solved easily if boundary layer assumptions are made for the conduction in the wall: the flux along the wall is neglected compared to the flux in the perpendicular direction.

For laminar boundary layer in the gas, the wall surface temperature jumps to a constant value T_w behind the shock, which is, however, only very little above the initial common temperature T_1 of the gas and the wall, even if the gas temperature T_2 after the shock is very high. An approximate formula given by Rott and Hartunian is based on the assumption of a Rayleigh profile, which is exact for weak shocks but turns out to be very well usable even for strong shocks:

$$\frac{T_w - T_1}{T_2 - T_1} = \left[1 + Pr^{0.364}\left(\frac{U_1 - U_2}{U_1 + U_2}\right)\right]\left[1 - \frac{U_1 - U_2}{U_1}(\sqrt{2} - 1)\right]^{\frac{1}{2}} \times \left[\frac{(k\rho c_p)_{\text{gas}}}{(k\rho c)_{\text{wall}}}\right]^{\frac{1}{2}} \tag{12-1}$$

where Pr is the Prandtl number and k is the heat conductivity. It is the

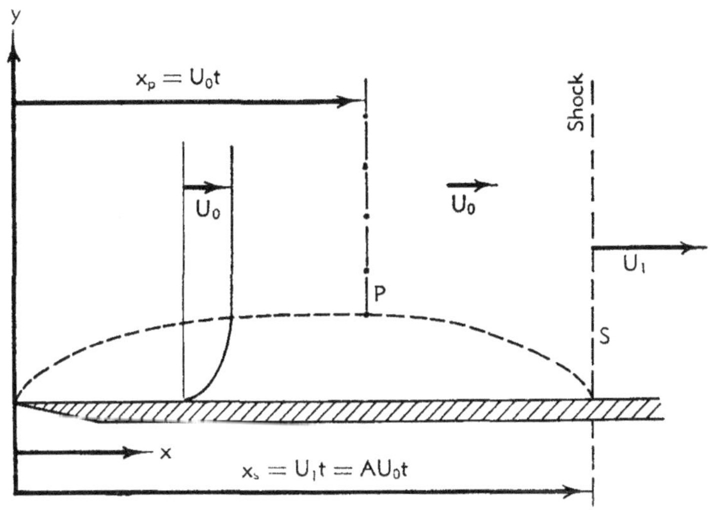

Fig. D,12b. Boundary layer development behind a shock moving past a half-infinite flat plate.

denominator in the last term, the high density (or high heat capacity per unit volume, ρc) in the solid that makes $T_w - T_1$ small, even if the wall is made of a good insulator.

A far more complicated problem arises when a shock strikes the leading edge of a flat plate. The boundary layer development that is to be expected is sketched in Fig. D,12b, after the shock has traveled a distance $x = U_1T$ along the plate. Behind the shock, the boundary layer has to be just the same as in the simple case of an infinite plane treated before, but there is also a boundary layer growth beginning at the leading edge $x = 0$, with free stream velocity $U_0 = U_1 - U_2$, which has to "join" the other part. (The line $x = U_0t$ shows the position of the free stream particles which were situated on the axis $x = 0$ at $t = 0$, when the shock struck the leading edge.) The limiting case $U_0 \ll U_1$ reduces to Stewartson's problem; the general case was treated by Lam and Crocco [66] (see also Crocco and Lam [74]). The problem admits an elegant formulation by use of the Crocco transformation of the boundary layer equations. The dependent variable will be the shear τ, or the dimensionless quantity

$$f = \sqrt{\frac{x}{\rho\mu U_0}}\,\frac{\tau}{U_0} \tag{12-2}$$

As in Stewartson's case, dimensional arguments show that f can be assumed to depend only on the two variables

$$\eta = \frac{u}{U_0}, \qquad \xi = \frac{x}{U_0 t} \tag{12-3}$$

It is found that f fulfills for $\rho\mu = \text{const}$ the equation

$$ff_{\eta\eta} + \tfrac{1}{2}\eta = \xi(\eta - \xi)\frac{\partial \ln f}{\partial \xi} \tag{12-4}$$

For $\xi = 0$, at the leading edge, Eq. 12-4 reduces to the Crocco form of the Blasius equation. It can be shown that the "shock tube" solution found previously also fulfills Eq. 12-4: Put $U_1/U_0 = A$ and

$$f = \sqrt{\frac{\xi(A - 1)}{A - \xi}}\,g(\eta) \tag{12-5}$$

then an easy calculation shows that g fulfills the equation

$$gg_{\eta\eta} + \frac{A - \eta}{A - 1} = 0 \tag{12-6}$$

This is the Blasius equation modified so that when η varies from 0 to 1, the variable $(A - \eta)/(A - 1)$ varies according to the proper boundary conditions of the shock tube problem in shock-fixed coordinates; for $\eta = 0$, $A/(A - 1) = U_1/U_2$. Note that the solution Eq. 12-5 fulfills Eq. 12-4 without any restrictions on ξ, and the question arises: What will limit the region of the validity of this solution? The limit is found to be Stewartson's essential singularity at the point $\xi = \eta = 1$, so that the solution (Eq. 12-5) will be restricted to $1 \leqq \xi \leqq A$, or, in the physical picture, to the region between P and S (Fig. D,12b). Within this region, Lam and Crocco proved the uniqueness of the solution, so that an exact independence of the shock-tube-type flow from leading edge conditions is found for $1 \leqq \xi \leqq A$.

For the remaining part $0 \leqq \xi \leqq 1$, the problem is the following: for $\xi \to 0$, f becomes the Blasius solution; for $\xi = 1$, $f = g$ as defined by Eq. 12-5. Other boundary conditions are: $f_\eta = 0$ for $\eta = 0$ (no slip), and $f = 0$ for $\eta = 1$ (free stream). The unusual situation arises that boundary conditions are specified on all four sides of the square $0 \leqq \eta \leqq 1$, $0 \leqq \xi \leqq 1$, and in particular that the "downstream" condition $f = g$ has to be met by the solution that "develops" from the leading edge. The question of the existence of a function f fulfilling $f = g$ for $\xi = 1$ has to be considered.

The character of Eq. 12-4 is parabolic; if linearized, an equation of the diffusion type is found with ξ playing the role of the time. The term f_ξ is multiplied by $\eta - \xi$, which changes sign for $\eta = \xi$, thereby changing also the sign of the "diffusivity." It is known, however, that for "negative diffusivity" stable solutions of parabolic equations can be obtained only if "time's arrow" is also reversed. This is a first argument for a possible "upstream influence," starting from the conditions at $\xi = 1$. A second argument is that the boundary condition at $\xi = 0$ is given for a value of ξ where the "diffusivity" is zero, as is seen in Eq. 12-4. It is known from the theory of parabolic differential equations that "initial" conditions given at a "time" of zero diffusivity are too weak to determine the subsequent developments. Thus, given f on the line $\xi = 0$ only, the solution would not be unique.

Although Lam and Crocco did not give a complete proof of existence (except for an almost complete proof in the linearized case), the above arguments leave very little doubt about the assertion that the problem for $0 \leqq \xi \leqq 1$ is well set.

To find the solution in this domain, Lam and Crocco use an iterative procedure, beginning with inserting an f_ξ value in Eq. 12-4 which in parts of the domain is assumed to have the Blasius value (= 0), and in the other parts is given by Eq. 12-5. The first iterative steps indicate good convergence, particularly for the wall shear (which varies smoothly between $0 \leqq \xi \leqq 1$, without showing the "kink" obtained by the momentum methods in the preceding paragraph). Every step leads to a Blasius-type equation, which is solved numerically. After a big number of steps, however, a divergence of the solution begins to develop. This is understandable as the numerical methods are not suitable for producing the fine details of the Stewartson singularities at $\xi = 0$ and at $\xi = \eta = 1$. (The line $\xi = \eta$ is regular.) A practical remedy can be found by readjusting the shear in every step to the exact momentum equation. Further progress would be possible only if the singularities were analytically treated. Results of practical significance would change only very little by such an improvement.

Finally, an interesting problem of shock wave-boundary layer interaction will be mentioned, which is of practical importance for shock tube applications. Suppose a plane shock is traveling down a straight tube, causing most of the gas behind the shock to move with constant velocity in the same direction. At the base of a boundary layer on the tube walls, the gas remains at rest, however. Now, if the tube has a closed end, the shock will reflect and proceed back upstream, canceling the previous downstream motion so that the bulk of the gas is brought to rest in a highly compressed state. The question arises, what happens to the gas in the boundary layer when the main "outer flow" is brought to rest? Mark [*75,76*] solved a simplified version of this problem, neglecting viscosity. The velocity distribution encountered by the reflected shock is known; in the system relative to the reflected shock, the boundary layer will show up as a defect in the oncoming velocity distribution, and the defect in Mach number distribution can also be given. Now it is assumed that each stream tube of the boundary layer (treated as steady and inviscid) will reach the pressure behind the shock; the corresponding velocity distribution after the shock can be determined. However, it is clear that if the velocity defect (or, more significantly, the Mach number defect) in the velocity field encountered by the reflected shock is too big, then not even the stagnation pressure in the stream tube will be high enough to meet the free stream pressure behind the reflected shock. If this happens, backflow and separation with important shock wave interactions are to be expected. Mark calculated that, for incident shock Mach numbers higher than $M_s = 1.33$ for $\gamma = 1.4$ ($M_s = 1.57$ for $\gamma = 1.67$), there will be a stream tube with a stagnation pressure that is too low to cross the shock. Experimentally, big interaction regions were found for M_s larger than about 1.5 in air; in these cases a big "bubble" was found at the foot of the reflected shock, which developed a λ-shaped bifurcation very similar to the interaction found in the steady case for laminar boundary layers (see [*77*]).

Carrying the investigation into regions of strong shocks, Mark found that the ability of the boundary layer to overcome the shock pressure jump improves again. The physical reason for this phenomenon is that the boundary layer is strongly cooled, and stream tubes with velocity defect can compensate for their lack of momentum with high density. An alternate way to express these facts is to state that the Mach number of the flow relative to the reflected shock determines its maximum attainable pressure, and cooling evidently increases this Mach number. The importance of this cooling effect is big for strong shocks, when the temperature behind the shock rises sharply while the wall temperature remains practically unchanged, as was pointed out before. For the layer nearest the wall, Mark found that it could overcome the pressure rise of the reflected shock if the incident shock Mach number is higher than $M_s = 6.45$, for a perfect gas with constant specific heat ratio $\gamma = 1.4$; the limit is $M_s = 2.8$ for $\gamma = 1.67$. A decrease in γ increases this limiting Mach number strongly, and real gas effects connected with strong shocks have to be considered.

A later investigation by Byron and Rott [*78*] showed that it would be too optimistic to assume that no strong boundary layer interaction would be found beyond the limits determined by Mark. The reason is that the minimum Mach number relative to the reflected shock is found somewhere inside the boundary layer, and not at its edge. Calculations for such cases are complicated by a strong Prandtl number effect; it is not admissible to put $Pr = 1$ for simplicity. For a low Prandtl number, when the thermal boundary layer is thicker than the velocity boundary layer, the cooling effect spreads further into the stream and increases the minimum Mach number. Byron and Rott found that, while for $\gamma = \frac{5}{3}$ Mark's limit changes only little, the increase of the limiting Mach number with decreasing γ is much sharper than predicted by Mark. If γ is less than about 1.54, the strong interaction region does not even disappear for infinite M_s, according to calculations of Byron and Rott. Thus, only noble gases—at most slightly contaminated—will have "clean" reflected shocks for strong incident shock Mach numbers. Such cases were actually observed by Mark in argon with only a small content of air. The strong interaction of the reflected shock with the boundary layer is of practical importance for shock tube investigation of high temperature gases, as the interaction causes a change with time of the reflected shock strength and of the state of the gas behind it.

D,13. On the Separation Problem in Unsteady Flow. Little is known on questions of separation in unsteady flow, even if attention is restricted to the problem of laminar separation. The only general statement which can be made is that, in the limit of infinite Reynolds number, separation must occur at sharp edges. Even this result must be qualified: the limit in steady flow, i.e., the effect of high but finite Reynolds number (finite boundary layer thickness) on the Kutta-Joukowsky condition is well understood. No such full quantitative investigations exist in unsteady flow, but it was often conjectured that high frequencies would affect the viscous modification of the Kutta-Joukowsky condition in periodically varying flow. Maybe even more important is the effect of transients. These phenomena have great practical importance for questions like stall of airplanes in fast maneuvers, wing flutter at high reduced frequencies, and rotating stall. The only attack on these problems with the tools of time-dependent boundary layer theory was made by Moore [*79*]. An account of his work, together with a general survey of the state of the art has been given by Sears [*80*].

In unsteady flow, it is not known whether the vanishing of the wall shear has the same physical importance as in the steady case. It seems that the significance of the so-called laminar separation point in steady flow is that it marks the beginning of convection which is direction upstream, so that the problem of further boundary layer developments may become indeterminate without specifying downstream conditions.

This criterion may be generalized to quasi-steady flow, but the limit is not well understood. For high frequencies, it is clear that vanishing wall shear may occur at a point with no special significance for laminar separation in the sense defined above.

Even in the simple case of steady flow with boundary layer development along a moving wall, the meaning of laminar separation has to be reconsidered. Moore [81] and, independently, Sears and Rott (unpublished) came to the conclusion that a sort of stagnation point, $u = v = 0$, within the boundary layer (viewed in the steady system, the walls moving) takes the significance of the laminar separation point. With boundary layer approximations, the conditions for such a point are also very nearly $u = \partial u/\partial y = 0$.

D,14. Heat Transfer and Compressibility Effects. Heat transfer coefficients were calculated by Lighthill [34], based on the linear approximation for periodic disturbances, which was discussed in Art. 9. Lighthill treated the problem using the incompressible approximation and neglecting dissipation, but the fundamental steady boundary layer was allowed to have an arbitrary pressure gradient.

In the particular case of the flat plate, it has already been pointed out in Art. 8 that the Howarth-Moore transformation reduces the case $\rho\mu = \text{const}$ to the $\rho = \text{const}$, $\mu = \text{const}$ problem for the velocity boundary layer. The reduction occurs only if the pressure gradient is zero. Care should be taken in verifying whether this assumption is appropriate for any given physical situation. For example, if the flat plate is moving in still air with a velocity that is some given function of time, the assumption of zero pressure gradient is correct (in boundary layer approximation). Velocity and shear calculated for incompressible flow is directly applicable for the compressible case; in particular, Moore [31] and Ostrach [82] also calculated, for this case, the recovery temperature T_r and the heat transfer rate q_w from a plate at constant temperature T_w. For a value of the Prandtl number equal to 0.72, their results are

$$\frac{T_r}{T_\infty} = 1 + 0.848\,\frac{\gamma - 1}{2}\,M_\infty^2\left[1 - 2.852\,\frac{x\dot{u}}{u^2} + 3.266\,\frac{x^2\ddot{u}}{u^3} + \cdots + 1.784\left(\frac{x\dot{u}}{u^2}\right)^2 + \cdots\right] \tag{14-1}$$

$$q_w = 0.411\rho_\infty c_p(T_w - T_\infty)\sqrt{\frac{\nu u}{x}}\left[1 - 0.424\Phi - 0.069(1 + 0.275\Phi)\,\frac{x\dot{u}}{u^2} - 0.423(1 - 0.550\Phi)\,\frac{x^2\ddot{u}}{u^3}\cdots + 0.240(1 + 3.92\Phi)\left(\frac{x\dot{u}}{u^2}\right)^2\cdots\right] \tag{14-2}$$

where $\Phi \equiv (\gamma - 1)M_\infty^2 T_\infty/(T_w - T_\infty)$.

An example where the Howarth-Moore transformation does not simply reduce the velocity boundary layer to the incompressible problem was given by Illingworth [*36*], who treated the case of a flat plate at rest in a uniform stream on which a sound wave progressing parallel to the plate is superposed. Now the pressure gradient is not zero, and shear values will not only exhibit a direct effect of this pressure gradient, but an effect of the mean heat transfer rate also is found in the shear formula; this shows that the Moore-Howarth transformation (which is still used by Illingworth to simplify the problem) does not achieve the reduction to the ρ = const case. Illingworth calculated the shear and the heat transfer due to the oscillating free stream in linearized approximation for low and high frequencies of the harmonic wave.

There exist time-dependent heat transfer phenomena which are connected with strange and violent effects; a classical example is the Rijke tube, described by Lord Rayleigh in his *Theory of Sound.* Modern examples—strongly complicated by chemical problems—occur in combustion chambers and rockets. Qualitative explanations are based on the time lag found between the fluctuations of the pressure and of the heat transfer. It is by no means assured, however, that the existing linearized approximations to the theory are adequate for the description of these phenomena.

Returning to the theory of simple cases, the Rayleigh problem, i.e., the impulsive motion of an infinite plane, will be discussed for compressible flow. An exact solution of the Navier-Stokes equation was found in the incompressible case. Now, the same solution is easily extended to the boundary layer approximation for compressible flow. This was done by Illingworth [*83*], who noticed that in this case a boundary layer solution can be obtained for an arbitrary viscosity-temperature law. If, however, the special case $\rho\mu$ = const is assumed, the Rayleigh profile, i.e., the error function, Eq. 3-7, is found again, but now in function of the Howarth variable.

If the Prandtl number is 1, it is easy to show that for this particular flow the Busemann-Crocco relation between enthalpy and velocity, which is well known for steady flow, still holds:

$$c_pT = c_pT_\infty + Cu - \tfrac{1}{2}u^2 \tag{14-3}$$

where T_∞ is found far from the plate. The constant C equals the velocity U of the plate in the case of the heat-insulated wall, whose temperature jumps instantaneously to the recovery temperature when the plate is put into motion. Other values of C can account for different wall temperatures, which remain constant after the impulsive start; such a boundary condition is actually appropriate for a homogeneous conducting wall which fills the "solid" half of the infinite space.

If the Prandtl number is not 1, but $\mu \sim T$, the enthalpy distribution can be calculated. An analysis applicable to the heat-insulated case was given by Emmons [84] who found an explicit expression for the recovery temperature:

$$c_pT_r = c_pT_\infty + r\frac{U^2}{2}$$

$$r = \frac{4}{\pi}\sqrt{\frac{Pr}{2 - Pr}}\sin^{-1}\sqrt{1 - \frac{Pr}{2}} \cong Pr^{\left(1-\frac{2}{\pi}\right)} = Pr^{0.364} \qquad (14\text{-}4)$$

The last line of this formula represents an approximation to Emmons' exact result for Pr values near one.

For other wall temperatures, Illingworth gives the distribution of T; the additional profile is proportional to $u(\eta Pr^{\frac{1}{2}})$, where η is the Howarth variable.

These results are of practical importance for heat transfer behind a shock to the walls of a shock tube. It was pointed out in Art. 13 that the Rayleigh profile is the exact limit for the velocity distribution behind weak shocks; Rott and Hartunian [71] used this shape to approximate results for all shock strength. It is also of interest to note that for $Pr = 1$ the Busemann-Crocco relation holds in the shock-fixed system, even for the unsteady Lam-Crocco problem discussed in Art. 13.

Finally, it will be recalled that the Rayleigh solution is exact in the case of incompressible flow; it may be interesting, therefore, to consider the compressible case also as a solution of the Navier-Stokes equations rather than as a boundary layer problem. One method of attack was initiated by Lagerstrom, Cole, and Trilling [85], generalized and carried to an explicit solution by Howarth [86]. Their work is based on the linearization of the Navier-Stokes equations for small values of the Mach number M. A different approach was given by Van Dyke [87], who started with a boundary layer solution, as discussed before. The next step is the investigation of the influence of the boundary layer on the free stream. Due to compressibility effects, a displacement thickness proportional to M^2 is found which grows with the square root of the time. This causes a v velocity, which is associated with an acoustic wave—and thereby with a pressure change—propagating perpendicular to the plate. The modified free stream conditions call for a recalculation of the boundary solution; this again modifies the displacement effect, etc. For the pressure on the plate surface, Van Dyke finds as a first correction, in agreement with Howarth:

$$\frac{p_w - p_\infty}{p_\infty} = \frac{\gamma(\gamma - 1)}{\sqrt{2\pi}}\frac{M^2}{a_\infty}\sqrt{\frac{\nu_\infty}{t}} \qquad (14\text{-}5)$$

where a_∞ equals the (undisturbed) sound speed. The next correction equals $(\gamma + 1)/4\gamma$ times the square of the value (Eq. 14-5). Van Dyke noticed that further terms cannot be calculated unless the acoustic approximation is abandoned for the treatment of the outer flow field.

The limiting case of very high Mach numbers, where the displacement effect causes a strong shock, was solved by Stewartson [*88*].

D,15. Cited References.

1. Stokes, G. G. On the effect of the internal friction of fluids on the motion of a pendulum. *Trans. Cambridge Phil. Soc. 9*, 20 (1851). Also *Math. Phys. Papers 3* (Cambridge, 1850).
2. Lord Rayleigh (J. W. Strutt). On the motion of solid bodies through a viscous fluid. *Phil. Mag. 21*, (*6*), 697 (1911).
3. Schlichting, H. *Grenzschicht-Theorie*. Braun, 1951. Also *Boundary Layer Theory*, McGraw-Hill, 1955.
4. Stuart, J. T. A solution of the Navier-Stokes and energy equations illustrating the response of skin friction and temperature of an infinite plate thermometer to fluctuations in the stream velocity. *Proc. Roy. Soc. London A231*, 116 (1955).
5. Watson, J. A solution of the Navier-Stokes equations illustrating the response of a laminar boundary layer to a given change in the external stream velocity. *Quart. J. Mech. and Appl. Math. 11*, 302–325, 1958.
6. Richardson, E. G., and Tyler, E. The transverse velocity gradients near the mouths of pipes in which an alternating or continuous flow of air is established. *Proc. Roy. Soc. London 42*, 1 (1929).
7. Sexl, T. Über den von E. G. Richardson entdeckten Annulareffekt. *Z. Phys. 61*, 349 (1930).
8. Batchelor, G. K. The skin friction of infinite cylinders moving parallel to their length. *Quart. J. Mech. Appl. and Math. 7*, 179 (1954).
9. Carslaw, H. S., and Jaeger, J. G. *Conduction of Heat in Solids*. Oxford Univ. Press, 1947.
10. Hasimoto, H. Rayleigh's problem for a cylinder of arbitrary shape II. *J. Phys. Soc. Japan 10*, 397 (1955).
11. Cooke, J. C. On Rayleigh's problem for a general cylinder. *J. Phys. Soc. Japan 11*, 1181 (1956).
12. Hasimoto, H. Rayleigh's problem for a cylinder of arbitrary shape I. *J. Phys. Soc. Japan 9*, 611 (1954).
13. Howarth, L. Rayleigh's problem for a semi-infinite plate. *Proc. Cambridge Phil. Soc. 46*, 127 (1950).
14. Sowerby, L. The unsteady flow of a viscous incompressible fluid inside an infinite channel. *Phil. Mag. 42*, 176 (1951).
15. Hasimoto, H. Note on Rayleigh's problem for a bent flat plate. *J. Phys. Soc. Japan 6*, 400 (1951).
16. Hasimoto, H. Rayleigh's problem for a plate of finite breadth. *Proc. First Japan. Natl. Congress Appl. Mech.*, 447 (1951).
17. Sowerby, L., and Cooke, J. C. The flow of fluids along corners and edges. *Quart. J. Mech. and Appl. Math. 6*, 50-70 (1953).
18. Davies, D. R. Three-dimensional turbulence and evaporation in the lower atmosphere. *Quart. J. Mech. and Appl. Math. 3*, 51 (1950).
19. Levine, H. Skin friction on a strip moving parallel to its length. *J. Fluid Mech. 3*, 145 (1957).
20. Oseen, C. W. Über Wirbelbewegung in einer reibenden Flüssigkeit. *Ark. f. Mat. Astron. Fys. 7*, 14 (1912).
21. de Neufville, A. The dying vortex. *Proc. Fifth Midwest. Conference Fluid Mech.*, Univ. Michigan, 365 (1957).
22. Burgers, J. M. A mathematical model illustrating the theory of turbulence. *Advances in Applied Mechanics 1*, Academic Press, 1948.
23. Rott, N. On the viscous core of a line vortex. *Z. Angew. Math. u. Phys. 9b*, 543 (1958).
24. Preston, J. H. Non-steady flows under asymptotic suction conditions. *Quart. J. Mech. and Appl. Math. 3*, 435 (1950).

25. Rott, N. Unsteady viscous flow in the vicinity of a stagnation point. *Quart. Appl. Math. 13*, 444 (1956).
26. Glauert, M. B. The laminar boundary layer on oscillating plates and cylinders. *J. Fluid Mech. 1*, 97 (1956).
27. Rott, N., and Rosenzweig, M. L. On the response of the laminar boundary layer to small fluctuations of the free-stream velocity. *J. Aeronaut. Sci. 27*, (1960).
28. Rosenzweig, M. L. *The Response of the Laminar Boundary Layer to Impulsive Motions.* Thesis, Cornell University, 1959. Also *USAF OSR TN-59-92.*
29. Watson, J. The two-dimensional laminar flow near the stagnation point of a cylinder which has an arbitrary transverse motion. *Quart. J. Mech. and Appl. Math. 12*, 175 (1959).
30. Wuest, W. Grenzschichten an zylindrischen Körpern mit nichtstationärer Querbewegung. *Z. Angew Math. u. Mech. 32*, 172 (1952).
31. Moore, F. K. Unsteady laminar boundary layer flow. *NACA Tech. Note 2471*, 1951.
32. Schuh, H. Über die ähnlichen Lösungen der instationären laminaren Grenzschichtgleichung in inkompressibler Strömung. *50 Jahre Grenzschichtforschung.* Vieweg, 1955.
33. Moore, F. K. Aerodynamic effects of boundary layer unsteadiness. *Proc. Sixth Anglo-Am. Aero. Conf.*, 1957.
34. Lighthill, M. J. The response of laminar skin friction and heat transfer to fluctuations in the stream velocity. *Proc. Roy. Soc. London A224*, 1 (1954).
35. Gibellato, S. Strato limite attorno ad una lastra piana investita da un fluido incompressible dotato di una velocità che è samma di una parte costante e di una parte alternata. *Atti. accad. sci. Torino Classe-sci., fis., mat., e nat. 89*, 180 (1955) and *91*, 152 (1957).
36. Illingworth, C. R. The effects of a sound wave on the compressible boundary layer on a flat plate. *J. Fluid Mach. 3*, 471 (1958).
37. Lam, S. H., and Rott, N. Theory of time-dependent boundary layers. *Cornell Univ. GSAE Rept.*, 1960. See also Proc. *Tenth Int. Congress of Appl. Mech.*, 239, Stresa 1960, Elsevier Publ. Co., 1962.
38. Schlichting, H. Berechnung ebener periodischer Grenzschichtströmungen. *Phys. Z. 33*, 327 (1932).
39. Lord Rayleigh. On the circulation of air observed in Kundt's tubes, and on some allied acoustical problems. *Phil. Trans. A175*, 1 (1883).
40. Andrade, E. N. da C. On the circulation caused by the vibrations of air in a tube. *Proc. Roy. Soc. London A134*, 445 (1931).
41. Schuster, K., and Motz, W. *Akust. Z. 5*, 349 (1940).
42. Andres, J. M., and Ingard, U. Acoustic streaming at high Reynolds numbers. *J. Acoust. Soc. Amer. 25*, 928 (1953).
43. Andres, J. M., and Ingard, U. Acoustic streaming at low Reynolds numbers. *J. Acoust. Soc. Amer. 25*, 932 (1953).
44. Carrière, Z. Analyse ultra-microscopique des vibrations aeriénnes. *J. Phys. Radium 10*, (*1*), 198 (1929).
45. Holtsmark, J., Johnson, I., Sikkeland, T., and Skuvlem, S. Boundary layer flow near a cylindrical obstacle in an oscillating incompressible fluid. *J. Acoust. Soc. Amer. 26*, 26 (1954).
46. Raney, W. P., Corelli, J. C., and Westervelt, P. J. Acoustical streaming in the vicinity of a cylinder. *J. Acoust. Soc. Amer. 26*, 1006 (1954).
47. Lane, C. A. Acoustical streaming in the vicinity of a sphere. *J. Acoust. Soc. Amer. 27*, 1082 (1955).
48. Nyborg, L. W. Acoustic streaming equations: laws of rotational motion for fluid elements. *J. Acoust. Soc. Amer. 25*, 398 (1953).
49. Nyborg, L. W. Acoustic streaming near a boundary. *J. Acoust. Soc. Amer. 30*, 329 (1958).
50. Gibson, W. *Unsteady Laminar Boundary Layers. Thesis*, Mass. Inst. Technol., 1957.
51. Blasius, H. Grenzschichten in Flüssigkeiten mit kleiner Reibung. *Z. Math. Phys. 56*, 1 (1908).

52. Goldstein, S., and Rosenhead, L. Boundary layer growth. *Proc. Cambridge Phil. Soc. 32*, 392 (1936).
53. Tollmien, W. Grenzschicht-Theorie. *Handb. Exp.-Phys. 4*, (*1*), 274 (1931).
54. Görtler, H. Grenzschichtentstehung an Zylindern bei Anfahrt aus der Ruhe. *Arch. Math. 1*, 138 (1948).
55. Görtler, H. Verdrängungswirkung der laminaren Grenzschicht und Druckwiderstand. *Ing.-Arch. 14*, 286 (1944).
56. Boltze, E. *Grenzschichten an Rotationskörpern in Flüssigkeiten mit kleiner Reibung. Thesis*, Göttingen, 1908.
57. Thiriot, K. H. Über die laminare Anlaufströmung einer Flüssigkeit über einem rotierenden Boden bei plötzlicher Änderung des Drehungszustandes. *Z. Angew. Math. u. Mech. 20*, 1 (1940).
58. Thiriot, K. H. Grenzschichtströmung kurz nach dem plötzlichen Anlauf bzw. Abstoppen eines rotierenden Bodens. *Z. Angew. Math. u. Mech. 30*, 390 (1950).
59. Nigam, S. D. Rotation of an infinite plane lamina: boundary layer growth: motion started impulsively from rest. *Quart. Appl. Math. 9*, 89 (1951).
60. Probstein, R. F. On a solution of the energy equation for a rotating plate started from rest. *Quart. Appl. Math. 11*, 240 (1953).
61. Nigam, S. D., and Rangasami, K. S. I. Growth of boundary layer on a rotating sphere. *Z. Angew. Math. u. Phys. 4*, 221 (1953).
62. Illingworth, C. R. Boundary-layer growth on a spinning body. *Phil. Mag. 45*, 1 (1954).
63. Wundt, H. Wachstum der laminaren Grenzschicht an schräg angeströmten Zylindern bei Anfahrt aus der Ruhe. *Ing.-Arch. 23*, 212 (1955).
64. Stewartson, K. On the impulsive motion of a flat plate in a viscous fluid. *Quart. J. Mech. and Appl. Math. 4*, 182 (1951).
65. Schuh, H. Calculation of unsteady boundary layers in two-dimensional flow. *Z. Flugwiss. 1*, 121 (1953).
66. Lam, S. H., and Crocco, L. Shock-induced unsteady laminar compressible boundary layers on a semi-infinite flat plate. *Princeton Univ. Rept. 428*, 1958.
67. Cheng, S. I., and Elliott, D. The unsteady laminar boundary layer on a flat plate. *Princeton Univ. Rept. 318*, 1955. Also Cheng, S. I. Some aspects of unsteady laminar boundary layer flows. *Quart. Appl. Math. 14*, 337 (1957).
68. Hollyer, R. N. A study of attenuation in the shock tube. *Univ. Michigan Rept.*, 1953.
69. Donaldson, C. duP., and Sullivan, R. D. The effect of wall friction on the strength of shock waves in tubes and hydraulic jumps in channels. *NACA Tech. Note 1942*, 1949.
70. Mirels, H. Attenuation in a shock tube due to unsteady boundary-layer action. *NACA Tech. Note 3278*, 1956.
71. Rott, N., and Hartunian, R. On the heat transfer to the walls of a shock tube. *Cornell Univ. GSAE Rept.*, 1955.
72. Bromberg, R. Use of the shock tube wall boundary layer in heat transfer studies. *J. Am. Rocket Soc. 26*, 737 (1956).
73. Bershader, D., and Allport, J. On the laminar boundary layer induced by a travelling shock wave. *Princeton Univ. Rept.*, 1956.
74. Lam, S. H., and Crocco, L. Note on the shock-induced unsteady laminar boundary layer on a semi-infinite flat plate. *J. Aero Space Sci. 26*, 54 (1959).
75. Mark, H. The interaction of a reflected shock wave with the boundary layer in a shock tube. *Cornell Univ. GSAE Rept.*, 1957. Also *NACA Tech. Mem. 1418.*
76. Mark, H. The interaction of a reflected shock wave with the boundary layer in a shock tube. *J. Aeronaut. Sci. 24*, 307 (1956).
77. Ackeret, J., Feldmann, F., and Rott, N. Untersuchungen an Verdichtungsstössen und Grenzschichten in schnell bewegten Gasen. *Mitt. Inst. für Aerod. ETH Zürich*, 1946.
78. Byron, S., and Rott, N. On the interaction of the reflected shock wave with the laminar boundary layer on the shock tube walls. *Proc. Heat Transfer and Fluid Mech. Inst.*, Stanford Univ. Press, 1961.

79. Moore, F. K. Lift hysteresis at stall as an unsteady boundary-layer phenomenon. *NACA Tech. Note 3571*, 1955.
80. Sears, W. R. Some recent developments in airfoil theory. *J. Aeronaut. Sci. 23*, 490 (1956).
81. Moore, F. K. On the separation of the unsteady laminar boundary layer. *Boundary Layer Research* (H. Görtler, Ed.) Symp. Freiburg, Springer, 1958.
82. Ostrach, S. Compressible laminar boundary layer and heat transfer for unsteady motions of a flat plate. *NACA Tech. Note 3569*, 1955.
83. Illingworth, C. R. Unsteady laminar flow of a gas near an infinite flat plate. *Proc. Cambridge Phil. Soc. 46*, 603 (1950).
84. Emmons, H. W. Note on aerodynamic heating. *Quart. Appl. Math. 8*, 402 (1951).
85. Lagerstrom, P. A., Cole, J. D., and Trilling, L. Problems in the theory of viscous compressible fluids. *Calif. Inst. Technol. GALCIT Rept.*, 1949.
86. Howarth, L. Some aspects of Rayleigh's problem for a compressible fluid. *Quart. J. Mech. and. Appl. Math. 4*, 157 (1951).
87. Van Dyke, M. D. Impulsive motion of an infinite plate in a viscous compressible fluid. *Z. Angew. Math. u. Phys. 3*, 343 (1952).
88. Stewartson, K. On the motion of a flat plate at high speed in a viscous compressible fluid. *Proc. Cambridge Phil. Soc. 51*, 202 (1955).

SECTION E

HYPERSONIC BOUNDARY LAYER THEORY

F. K. MOORE

E,1. Introduction. Boundary layers in the high speed flow of a gas must be considered compressible, to the degree that the square of the Mach number differs from zero. For compressible flow, the thermodynamics of the gas and the fluid motion are connected, and must be treated concurrently, rather than in the separate way appropriate for low speed "incompressible" flow. So long as the gas may be assumed to obey the ideal equation of state and to have constant specific heats, the theoretical complications owing to compressibility can often be effectively removed by the introduction of coordinate transformations such as those due to Dorodnitzyn, Stewartson, and others (B,23).

Since diffusion rather than wave processes govern the boundary layer, boundary layer theory, unlike inviscid flow theory, does not undergo profound qualitative change when the flow speed is changed from low to moderately high Mach number; the present volume abounds in analyses first applied to incompressible boundary layer problems, and subsequently extended to account for compressibility effects. This section is intended to deal with problems of the laminar boundary layer for which effects associated with high Mach number do not succumb to the classical transformations. These problems appear chiefly in the context of hypersonic aerodynamics.

Throughout this article, it will be assumed not only that the mechanics and thermodynamics of the continuum apply, but also that the molecular mean free path can be neglected in comparison with boundary layer thickness. Thus, a basic regime of flow associated with hypersonic flight will not be discussed; namely, the "superaerodynamics," or rarefied gas flow, encountered at very high altitude. This subject is discussed, however, by Schaaf in another volume of this Series (III,H) and in [*1*], and by Hayes and Probstein [*2*].

The basic problem of hypersonic flight is that of high temperature. For illustrative purposes, we may consider the classical compressible solution for the boundary layer of a semi-infinite flat plate aligned parallel to a high Mach number free stream. The Kármán-Tsien solution of the energy equation (assuming a perfect gas with Prandtl number equal to 1) states that an insulated plate surface would reach a temperature greater than that of the free stream by a factor $1 + \frac{1}{2}(\gamma - 1)M_\infty^2$. At a hypersonic Mach number of 15, this ideal gas, if originally at a temperature of 200°K, would reach 9200°K at the plate. Even if the plate were cooled at a rate sufficient to maintain the plate at the same low temperature as the outer stream, the deceleration in the outer part of the boundary layer would produce a maximum temperature of $1 + \frac{1}{8}(\gamma - 1)M_\infty^2$ times the stream value, or 2450°K, in the middle of the boundary layer. The very high temperatures expected in the hypersonic boundary layer have two consequences which will set the direction of this section: First, we realize that boundary layer theory must take some account of the complicated physics and chemistry of a gas, such as air, elevated to very high temperature; and second, from the aeronautical engineering standpoint, anxiety concerning surface heat transfer rates provides the chief motivation for boundary layer theory.

The complexity of the gas, air, at high temperature is illustrated in Fig. E,1. In broadest terms, the gas is no longer calorically perfect at high temperature (and the foregoing example is quantitatively incorrect) because the molecules of highly excited air can absorb energy in many more degrees of freedom than can the same gas at low temperature. At room temperature, only the translational and rotational modes of the predominating diatomic molecules are excited. As temperature increases to about 1000°K with static pressure held at $\frac{1}{10}$ atm, the vibrational modes of the diatomic molecules become appreciably excited, so that specific heats increase. The difference between specific heats, being constant, becomes proportionately smaller, so that the ratio of specific heats decreases below 1.4. Further elevation of temperature above about 2000°K results in dissociation of oxygen molecules, and a corresponding absorption of energy. This mode depends on density as well as temperature—at sufficiently low density, oxygen dissociation could become important at a temperature lower than that for which vibrational excitation is complete. The presence of oxygen atoms favors the formation of nitric oxide, which leaves the picture again at higher temperatures, by dissociation. Nitrogen finally dissociates. Finally, the electronic excitation of the atoms and molecules causes a "dissociation" of electrons from their atoms (and molecules) to provide a plasma of positive and negative ions.

Naturally, the field of high temperature gas physics is beyond the scope of this section. However, it is evidently necessary for boundary layer theory to deal with an adequately realistic model of high temperature air if hypersonic problems are to be solved. In particular, it is clear that high temperature air must be regarded as a multicomponent substance in which specific heats are not constant, and in which heats of chemical reaction are important. Further, in such a mixture, processes of diffusion are obviously complicated by the gradients of composition that must be expected in a boundary layer subject to large temperature differences. It is regrettable that, while the diffusion process is absolutely fundamental to the boundary layer, quantitative determination of the appropriate high temperature diffusion coefficients is a most difficult problem of high temperature gas physics, for which results are not complete, either experimentally or theoretically.

The heat transfer to a surface exposed to a boundary layer involving high temperature air can, of course, be enormously large. In addition, the

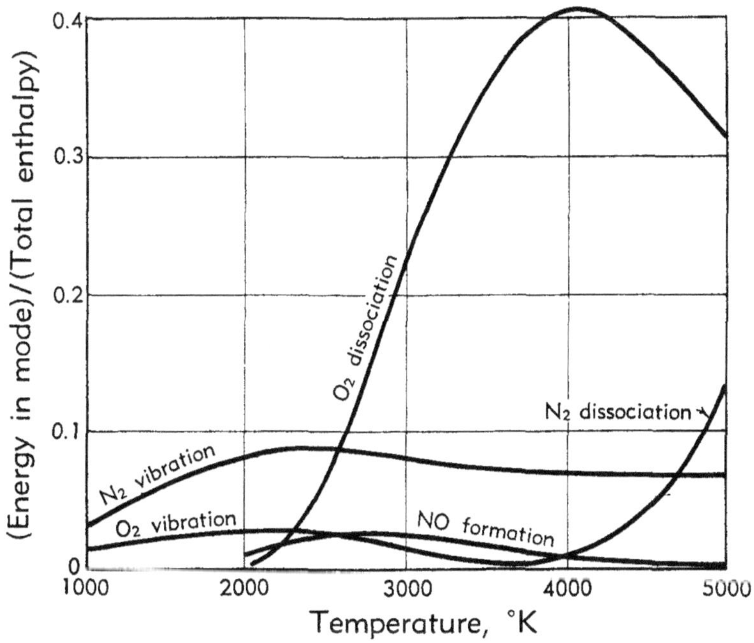

Fig. E,1. Distribution of energy among degrees of freedom of heated air, at a pressure of $\frac{1}{10}$ atmosphere.

mechanisms of heat transfer can be remarkably complicated; for example, one must consider whether or not oxygen atoms can diffuse to the surface and, taking into account the effectiveness of the surface as a chemical catalyst, whether recombination of the atoms and the consequent release of chemical energy may then take place there.

So far, the problem that has been outlined refers to the determination of heat transfer from a boundary layer of air which is, at least in part, at elevated temperature, and it has been noted that the air must be regarded as a multicomponent mixture. A generalization of this problem arises from the natural desire to provide thermal protection of such a surface. It may often be useful to accomplish this protection by the introduction of a coolant at the surface, either by blowing a fluid through a porous boundary or by ablation of the surface itself. Thus, owing to the activities of the engineer, the hypersonic boundary layer may have gaseous components in addition to those pointed out in Fig. E,1.

The high temperatures encountered in the hypersonic boundary layer have other aerodynamic consequences apart from the necessity to consider real-gas physics and chemistry. These concern the interaction between the boundary layer and the outer inviscid flow. Consider the boundary layer developing from the leading edge of a semi-infinite flat plate at zero angle of attack to a hypersonic outer flow. The sudden rise in temperature of fluid particles passing over the leading edge and into the boundary layer occurs roughly at constant pressure. Thus, the density suddenly drops, and one observes something like a constant pressure explosion. This effect increases in strength with Mach number, and is responsible for a Mach number dependence of the displacement thickness, which is a measure of effectiveness of the boundary layer growth in deflecting the outer stream. Large deflections of the outer stream result in wave disturbances that cause a pressure rise on the plate itself. In turn, the development of the boundary layer is affected and surface heat transfer and skin friction are increased. Another type of interaction of boundary layer and outer flow, to which special interest attaches in the hypersonic regime, is the "vorticity interaction." For reasons of heat transfer control, the forward portion of a hypersonic configuration is, to some degree, blunt. Therefore, a strong shock, highly curved near the nose, must be expected. In the inviscid flow in the "shock layer," a wide variation of entropy and vorticity occurs from streamline to streamline, depending on the curvature of the shock where each streamline passes through. Vorticity and entropy gradients appearing in the inviscid flow just outside the boundary layer can have a substantial effect on surface heat transfer rate.

Hypersonic boundary layer theory is a subject of strenuous research activity at the present time and may reasonably be defined to have a scope embracing magneto-aerodynamics, multiphase boundary layers associated with ablative heat protection, and other topics. In the present section, no attempt will be made to achieve this scope. Rather, our field of view will be limited to one within which the boundary layer theory may be regarded as reasonably complete.

With this limitation in mind, we shall recall the accounts of classical boundary layer theory provided, for example, by Goldstein [*3*] and Schlichting [*4*], and, in a sense, go over similar ground, introducing the extensions necessary to account for hypersonic effects. Typically, the hypersonic considerations of this section will fall into two classes: First, real-gas complications due purely to high temperature—in this connection, we shall emphasize a binary reactive mixture of atoms and molecules with suitably varying vibrational specific heat—and second, effects produced, even in an ideal gas, by high Mach number (these somewhat more purely "aerodynamic" effects include the pressure and vorticity interactions already mentioned).

E,2. Equations of Motion for a Multicomponent Gas. Looking toward the development of boundary layer equations suitable for the analysis of hypersonic laminar flows, we begin by writing the equations of motion for a gas consisting of n components. In a particular sample of this gas, the mass of each component is taken to be a fraction c_i of the mass of the sample, so that

$$\sum_{i=1}^{n} c_i = 1 \tag{2-1}$$

The gas will be assumed to be perfect; that is, sufficiently dilute that the van der Waals corrections to the usual state equation are unnecessary. For a multicomponent perfect gas, static pressure p, mass density ρ, and temperature T are related through the equation of state

$$p = \left(\sum_i \frac{c_i}{\mathfrak{M}_i}\right) \rho R T \tag{2-2}$$

where R is the universal gas constant, and $\mathfrak{M}_i$ is the molecular weight of the ith component.

The conservation of mass for the mixture is expressed by the usual equation

$$\frac{\partial \rho}{\partial t} + \nabla \cdot (\rho \mathbf{u}) = 0 \tag{2-3}$$

where $\mathbf{u}$ is the average velocity. Mass may be exchanged between the various forms c_i, owing to dissociation, for example. Therefore, when n components are present, we shall be concerned with n dependent variables (the c_i). Accordingly, an additional set of $n - 1$ equations is necessary for a full statement of mass conservation. These equations may be introduced by expressing conservation of mass for each constituent separately:

$$\frac{\partial}{\partial t}(\rho c_i) + \nabla \cdot [\rho(\mathbf{u} + \mathbf{u}_i')c_i] = W_i \tag{2-4}$$

where W_i is the mass rate of formation of component i per unit volume. Of course, this rate of formation for each constituent must be balanced by an equal decrease in mass of the other constituents, so that

$$\sum_i W_i = 0 \tag{2-5}$$

The quantity $\mathbf{u}_i'$ is the "diffusion velocity" vector of the ith component and represents the fact that the mass identified with any single component may be transported not only by the average velocity of the gas sample, but also by diffusion. Now, if we make a precise definition of $\mathbf{u}$ as being the mass-averaged velocity vector, it follows that

$$\sum_i \mathbf{u}_i' c_i = 0 \tag{2-6}$$

We note that summing Eq. 2-4 over the n components and applying Eq. 2-1, 2-5, and 2-6 yields Eq. 2-3. Thus, Eq. 2-4 provides the $n - 1$ new equations needed to account for the variables c_i. We have, however, acquired two new sets of variables, the $\mathbf{u}_i'$ and W_i. The $\mathbf{u}_i'$ must be specified in terms of the transport properties of the mixture. This rather formidable task will be deferred to another article, as will the specification of the W_i, which must be found from the applicable laws of chemical kinetics. In boundary layer problems, derivatives in a direction (y) normal to the surface predominate. Then, making the further assumption that the components[1] of $\mathbf{u}_i'$ are all of the same magnitude, so that, for example, $\partial v_i'/\partial y \gg \partial u_i'/\partial x$, we may subtract Eq. 2-3 from Eq. 2-4, finding a simpler equation to replace Eq. 2-4:

$$\rho\left(\frac{\partial c_i}{\partial t} + \mathbf{u} \cdot \nabla c_i\right) = W_i - \frac{\partial}{\partial y}(\rho c_i v_i') \tag{2-7}$$

The conservation of mass-averaged momentum is expressed by the usual set of equations (B,3). For plane flow, we may then immediately write the boundary layer momentum equation in its customary form, omitting the body-force term:

$$\rho\left(\frac{\partial u}{\partial t} + u\frac{\partial u}{\partial x} + v\frac{\partial u}{\partial y}\right) = -\frac{\partial p}{\partial x} + \frac{\partial}{\partial y}\left(\mu\frac{\partial u}{\partial y}\right) \tag{2-8}$$

where μ denotes coefficient of viscosity.

Similarly, the conservation of energy for the mixture may be written as follows, omitting external energy addition, and immediately adopting the boundary layer assumptions in regard to gradients of all properties:

$$\rho\left(\frac{\partial \bar{e}}{\partial t} + \mathbf{u} \cdot \nabla \bar{e}\right) + p\frac{\partial u}{\partial y} = \mu\left(\frac{\partial u}{\partial y}\right)^2 - \frac{\partial q}{\partial y} \tag{2-9}$$

or

$$\rho\left(\frac{\partial \bar{h}}{\partial t} + \mathbf{u} \cdot \nabla \bar{h}\right) - \frac{\partial p}{\partial t} - \mathbf{u} \cdot \nabla p = \mu\left(\frac{\partial u}{\partial y}\right)^2 - \frac{\partial q}{\partial y} \tag{2-10}$$

where $\bar{e}$ is the internal energy of the mixture per unit mass, including energy associated with internal degrees of freedom and a "potential" energy associated with chemical change, and $\bar{h}$ is the corresponding

[1] Velocity components will hereafter be denoted by u, v, and w, in the x, y, z directions, respectively.

enthalpy. That is,

$$\bar{e} \equiv \sum_i c_i(e_i + h_i^{(0)}); \qquad \bar{h} \equiv \sum_i c_i(h_i + h_i^{(0)}) \tag{2-11}$$

where $h_i^{(0)}$ is the heat of reaction at the absolute zero of temperature absorbed in the formation of a unit mass of species i. The last term of Eq. 2-9 is the heat flux due to diffusion processes proceeding, for the boundary layer case, in a direction normal to the surface.

It is, of course, not obvious that Eq. 2-8, 2-9, and 2-10 apply for a reactive multicomponent mixture. These equations may be derived [*5*, Chap. 7] from kinetic theory, making use of the Enskog equation of change, which is based on the Boltzmann equation for molecular velocity distribution.

Two diffusion effects now require specification, namely, the diffusion velocity v_i' and the heat flux q. From the kinetic theory of multicomponent gases [*5*, Chap. 7]

$$v_i' = \left(\sum_i \frac{c_i}{\mathfrak{M}_i}\right)^2 \frac{\mathfrak{M}_i}{c_i} \sum_j \mathfrak{M}_j D_{ij} \left\{ \frac{\partial}{\partial y}\left(\frac{c_j}{\sum_i \frac{\mathfrak{M}_j}{\mathfrak{M}_i} c_i}\right) + \left[\frac{\sum_i c_i\left(1 - \frac{\mathfrak{M}_j}{\mathfrak{M}_i}\right)}{\sum_i \frac{\mathfrak{M}_j}{\mathfrak{M}_i} c_i}\right] \frac{\partial \ln p}{\partial y} \right\} + \frac{D_i^T}{\rho c_i} \frac{\partial \ln T}{\partial y} \tag{2-12}$$

where the D_{ij} are the coefficients of diffusion and D_i^T is the coefficient of thermal diffusion. Inspection of the curly bracket in Eq. 2-12 shows that the diffusion owing to the concentration gradient, and that owing to the pressure gradient, are grouped together in a single term. Actually, it may be shown that the curly bracket really represents a gradient of hydrostatic imbalance, and it is this imbalance that causes diffusion, rather than the concentration or pressure gradients separately. For example, if one considers a mass of gas in equilibrium undergoing a uniform constant acceleration, a detailed balance of forces with the acceleration shows that the curly bracket of Eq. 2-12 vanishes. In a boundary layer, however, $\partial(\ln p)/\partial y$ is very small. Therefore, the gas will be out of "hydrostatic balance" whenever a significant gradient of concentration exists, and we may write

$$v_i' = \left(\sum_i \frac{c_i}{\mathfrak{M}_i}\right)^2 \frac{\mathfrak{M}_i}{c_i} \sum_j D_{ij} \frac{\partial}{\partial y}\left(\frac{c_j}{\sum_i \frac{c_i}{\mathfrak{M}_i}}\right) - \frac{D_i^T}{\rho c_i} \frac{\partial \ln T}{\partial y} \tag{2-13}$$

Again from kinetic theory, we may write q as follows, neglecting radiative heat transfer:

$$q = -kT_y + \rho \sum_i c_i v_i'(h_i + h_i^{(0)}) + \frac{RT}{\sum_i \frac{c_i}{\mathfrak{M}_i}} \sum_{i,j} \frac{c_j D_i^T}{\mathfrak{M}_i \mathfrak{M}_j \mathfrak{D}_{ij}} (v_i' - v_j') \qquad (2\text{-}14)$$

where k is the coefficient of heat conduction, D_i^T is the coefficient of thermal diffusion, and the $\mathfrak{D}_{ij}$ are the binary diffusion coefficients which, for a two-component gas, are equal to the D_{ij}. The first term of Eq. 2-14 represents the familiar effect of heat conduction. The second term represents the enthalpy (including heat of reaction) transported by the diffusion velocities. The third term, representing the so-called Dufour effect [*5*, Chap. 8], represents an energy flux due to the interaction of normal and thermal diffusion.

E,3. Various Properties of a Binary Gas Mixture. So far, we have displayed the basic conservation equations for a mixture of gases, and outlined the framework within which mixing processes may be analyzed. Before proceeding to the study of particular flow problems, we must specify the thermodynamics and chemistry of the gas mixture, and determine suitable values for the various transport coefficients. These questions are intolerably complex in the general case so far considered. Therefore, before proceeding further, we recall that a wide class of hypersonic real-gas problems concerns air with some degree of dissociation, or air into which another gas is injected at a boundary without chemical interaction. Further, under assumptions which will be discussed subsequently, we may regard air as a binary gas, composed of diatomic "air molecules" and, in case of dissociation, "air atoms." Therefore, we shall hereafter restrict consideration to binary mixtures.

Diffusion velocities and heat transfer. First, we rewrite Eq. 2-13 for a binary mixture, using Eq. 2-1:

$$c_1 v_1' = \left(-D_{12} + \frac{\mathfrak{M}_1}{\mathfrak{M}_2} D_{11}\right) \frac{\partial c_1}{\partial y} - \frac{D_1^T}{\rho} \frac{\partial \ln T}{\partial y} \qquad (3\text{-}1)$$

$$c_2 v_2' = \left(-D_{21} + \frac{\mathfrak{M}_2}{\mathfrak{M}_1} D_{22}\right) \frac{\partial c_2}{\partial y} - \frac{D_2^T}{\rho} \frac{\partial \ln T}{\partial y} \qquad (3\text{-}2)$$

By Eq. 2-6, we know that the sum of Eq. 3-1 and 3-2 must vanish. Thus we immediately have that $D_2^T = -D_1^T$. In the general theory of multicomponent diffusion, the definition of convenience is made that $D_{ii} = 0$ [*5*, Chap. 7]. Then, noting that $\partial c_1/\partial y + \partial c_2/\partial y = 0$, we see that $D_{12} = D_{21}$. These equations show how the specification $D_{ii} = 0$ (which should not, of course, be taken to mean that the coefficient of self-diffusion for a single-component gas is zero) symmetrizes the matrix D_{ij}. Thus we may write

$$v_1' = -\mathfrak{D}_{12}\frac{\partial \ln c_1}{\partial y} - \frac{D_1^T}{\rho c_1}\frac{\partial \ln T}{\partial y} \tag{3-3}$$

$$v_2' = -\mathfrak{D}_{12}\frac{\partial \ln c_2}{\partial y} + \frac{D_1^T}{\rho c_2}\frac{\partial \ln T}{\partial y} \tag{3-4}$$

Specializing Eq. 2-14 for a binary mixture, we find

$$-q = kT_y - \rho(\bar{h}_1 v_1' + \bar{h}_2 v_2') - \frac{2RT}{c_1\mathfrak{M}_2 + c_2\mathfrak{M}_1}\frac{D_1^T}{\mathfrak{D}_{12}}(v_2' - v_1') \tag{3-5}$$

The relative simplicity of Eq. 3-5 enables us to show that the last term may generally be neglected for a reacting mixture. Redefining the thermal diffusion coefficient [5, Chap. 8],

$$k_T \equiv \frac{\mathfrak{M}_1\mathfrak{M}_2}{\rho(c_1\mathfrak{M}_2 + c_2\mathfrak{M}_1)^2}\frac{D_1^T}{\mathfrak{D}_{12}} \tag{3-6}$$

and applying Eq. 2-1, 2-6, and 2-11 yields the equations:

$$\bar{h}_1 v_1' + \bar{h}_2 v_2' = (h_1 + h_1^{(0)} - h_2 - h_2^{(0)})c_1 v_1' \tag{3-7}$$

$$v_2' - v_1' = -\frac{v_1'}{c_2} \tag{3-8}$$

which, when substituted into Eq. 3-5, yield the conclusion that the third term of Eq. 3-5 may be neglected relative to the second, if

$$\frac{h_1 + h_1^{(0)} - h_2 - h_2^{(0)}}{2RT\left(\dfrac{c_1}{\mathfrak{M}_1} + \dfrac{c_2}{\mathfrak{M}_2}\right)} \gg \frac{k_T}{c_1 c_2} \tag{3-9}$$

We shall see that for a mixture of diatomic molecules and atoms at high temperature, $k_T/(c_1c_2) \sim 0.2$. Now, if the mixture is reacting, energies of reaction $h_1^{(0)} - h_2^{(0)}$ are usually of the order of 100 times $RT/\mathfrak{M}$, and Eq. 3-9 shows that the Dufour effect may certainly be neglected. However, if the mixture is nonreacting, so that $h_1^{(0)}$, $h_2^{(0)}$ do not enter into the calculation (and may formally be set equal to zero), then the left side of Eq. 3-9 is of order 1, and it is perhaps doubtful that the Dufour effect can be neglected in comparison with the preferential diffusion of enthalpy. Of course, in nonreacting mixtures, conduction (the first term of Eq. 3-5) tends to dominate heat transfer, so that the Dufour effect remains of slight importance. Hereafter we shall neglect this term entirely. However, at this point we do not neglect thermal diffusion in Eq. 3-3 and 3-4, because we cannot compare $\partial \ln c_1/\partial y$ and $\partial \ln T/\partial y$ except in the context of a specific problem. Substituting Eq. 3-3 and 3-4 into Eq. 3-5, and neglecting the Dufour effect, we find

$$-q = kT_y + \mathfrak{D}_{12}\left[\rho\left(\bar{h}_1\frac{\partial \ln c_1}{\partial y} + \bar{h}_2\frac{\partial \ln c_2}{\partial y}\right) + \frac{D_1^T}{\mathfrak{D}_{12}}\left(\frac{\bar{h}_1}{c_1} - \frac{\bar{h}_2}{c_2}\right)\frac{\partial \ln T}{\partial y}\right] \tag{3-10}$$

Transport coefficients. The proper specification of the transport coefficients for use in multicomponent boundary layer theory suffers from two difficulties. First, at the present time, the theoretical calculations of the necessary coefficients, while extensive, do not cover the enormous variety of collision processes which can, in principle, be involved; further, for high temperatures, the existing calculations have not been subjected to the necessary experimental verification owing to the difficulty of making meaningful experiments at temperatures of interest. Second, we know that, even if our information were complete, the transport properties must be complicated functions of temperature and concentration, for example. Thus we must look for extreme simplifications which permit boundary layer analysis to proceed by customary methods of analysis.

Limiting our attention to the binary mixture, we must specify four transport coefficients: viscosity μ, binary diffusion $\mathfrak{D}_{12}$, thermal conductivity k, and thermal diffusion D_1^T or k_T. We may write the basic formula [*5*, Chap. 8]

$$\mathfrak{D}_{12} = 0.002628 \frac{\sqrt{\dfrac{\mathfrak{M}_1 + \mathfrak{M}_2}{2\mathfrak{M}_1\mathfrak{M}_2} T^3}}{p\sigma_{12}^2\Omega_{12}^{(1,1)*}} \tag{3-11}$$

where $\mathfrak{D}_{12}$ is given in cm²/sec, pressure p in atmospheres, and temperature T in °K. The "collision diameter" σ_{12} is given in Å. The "collision integral" $\Omega_{12}^{(1,1)*}$ is a function of the dimensionless temperature $\mathfrak{k}T/\epsilon_{12}$, where ϵ_{12} is a measure of the forces exerted during an interaction and $\mathfrak{k}$ is the Boltzmann constant. The collision integral depends on the interaction potential assumed—examples are the rigid-sphere model (for which $\Omega_{12} = 1$) the Sutherland potential, and the more sophisticated and, often, more realistic Lennard-Jones "6-12" (so-called because of the power laws assumed for the repulsive and attractive parts of the potential). For this last potential, $\Omega_{12}^{(1,1)*}$ is tabulated [*5*], and to apply Eq. 3-11 we need only to specify the "interaction parameters" σ_{12} and $\epsilon_{12}/\mathfrak{k}$.

We may compute μ for a binary mixture from the semiempirical formula [*5*, p. 533]

$$\mu = \frac{c_1\mu_1}{c_1 + \dfrac{1.385}{1 + \dfrac{\mathfrak{M}_2 c_1}{\mathfrak{M}_1 c_2}}\left(\dfrac{\mu_1}{\rho\mathfrak{D}_{12}}\right)} + \frac{c_2\mu_2}{c_2 + \dfrac{1.385}{1 + \dfrac{\mathfrak{M}_1 c_2}{\mathfrak{M}_2 c_1}}\left(\dfrac{\mu_2}{\rho\mathfrak{D}_{12}}\right)} \tag{3-12}$$

where μ_1 and μ_2 are the viscosities of the two components taken separately:

$$\mu_i = 0.00002669 \frac{\sqrt{\mathfrak{M}_i T}}{\sigma_i^2\Omega^{(2,2)*}} \tag{3-13}$$

$\Omega^{(2,2)*}$ is a function of $\mathfrak{k}T/\epsilon_i$, tabulated [*5*] for the Lennard Jones "6-12" potential.

As regards thermal conductivity k, we are not able to present a well-recommended formula for a binary mixture. For a pure gas, the Eucken formula states the proportionality of k and μ in the form [5, p. 534]

$$k_i = \mu_i \frac{15}{4} \frac{R}{\mathfrak{M}_i}\left(\frac{4}{15} \mathfrak{M}_i \frac{c_{v_i}}{R} + \frac{3}{5}\right) \tag{3-14}$$

For boundary layer theory, we shall be interested in k chiefly as it appears in the Prandtl number

$$Pr \equiv \frac{c_p \mu}{k} \tag{3-15}$$

Suppose we consider that our binary mixture is composed of diatomic molecules $\mathfrak{M}_2$ and the corresponding monatomic molecules $\mathfrak{M}_1$. Then, for the undissociated gas ($c_2 = 1$), Eq. 3-13 and 3-14 provide that $Pr = 0.783$ when vibration is fully excited, and 0.737 when only rotation is excited. For the fully dissociated gas, $Pr = 0.667$. The indications are [6,7] that Prandtl number is rather a smooth function of composition. Therefore, since the total variation of Pr between the extremes cited in the foregoing example amounts only to about 15 per cent, we may expect in most cases to be able to choose a constant value for Prandtl number with good accuracy.

The coefficient of thermal diffusion D_1^T is generally quite small; however, we cannot safely neglect this effect without knowing the relative magnitudes of concentration and thermal gradients. An approximate formula for binary isotropic mixtures [5, p. 542] should be generally satisfactory for our purposes: The formula, applicable if $(\mathfrak{M}_1 - \mathfrak{M}_2)/(\mathfrak{M}_1 + \mathfrak{M}_2) < 0.15$, may be written

$$k_T = c_1 c_2 \frac{\mathfrak{M}_1 - \mathfrak{M}_2}{\mathfrak{M}_1 + \mathfrak{M}_2} K_T \tag{3-16}$$

where K_T may be computed from the tables of [5]. Beyond $\mathfrak{k}T/\epsilon_1 \cong 10$, the quantity K_T is nearly constant at a value of about 0.55. For a partly dissociated diatomic gas, $(\mathfrak{M}_1 - \mathfrak{M}_2)/(\mathfrak{M}_1 + \mathfrak{M}_2) = \frac{1}{3}$, which, though not small enough to insure good accuracy for Eq. 3-16, may still be small enough that Eq. 3-16 may be used for simple calculations. The maximum value that c_1c_2 can have is $\frac{1}{4}$, for $c_1 = \frac{1}{2}$. Thus, for the dissociated gas under discussion, $k_T \leqq 0.047$, approximately.

The approximation of lumped constituents. Before proceeding further in the matter of transport properties, we must be more specific concerning the identification of air as a binary mixture of atoms and molecules. As indicated in Art. 1, high temperature air produced by hypersonic flight consists of a mixture of diatomic oxygen and nitrogen, together with the corresponding monatomic molecules existing by reason of dissociation. Also, nitric oxide is important under some circumstances. A strict theory for this mixture would have to take into account at least four constituents and two chemical-kinetic processes affecting relative concentrations. Development of such a theory would be a formidable task indeed, and, in fact, is not really necessary. It is customary [8,9] to take advantage of the fact that, in terms of interaction phenomena, the gases O_2 and N_2 are quite similar, as are the gases O and N. To a good approximation, we may define an air molecule A_2 having properties given by a suitable average of those of O_2 and N_2, and similarly an air atom A representing a composite of O and N. The molecule A_2 could also include NO when appropriate. Accordingly, the binary system A-A_2 may be adopted for the high temperature boundary layer of air, so long as it is uncontaminated by, for example, combustible vapors released at an ablating surface.

A number of considerations affect the validity of the foregoing model for air. First, we may ask under what circumstances a gas having two constituents may be represented as a single pure gas for purposes of fluid flow studies. Certainly, classical aerodynamics makes this assumption regarding air, and we may well suspect that in many circumstances it is not particularly important that O_2 and N_2 have nearly the same molecular weights. The "lumped" equations should be valid even for, say, the mixture N_2-He. Turning to the equations for a binary mixture, we see that a lumped gas may be assumed if the concentrations (c_i) remain constant. From Eq. 2-7 it is clear that, in general, c_i remain constant if $W_i = 0$ and $v_i' = 0$. The first requirement is that there be no chemical interchange between the constituents either in the gas or at a surface (just as true for N_2-He as for N_2-O_2, at ordinary temperatures), and the second requirement, by Eq. 3-3, is that thermal diffusion may be neglected.[2] That is,

$$\frac{\partial}{\partial y}\left(D_1^T \frac{\partial \ln T}{\partial y}\right) \cong 0 \tag{3-17}$$

At ordinary temperatures, both D_1^T and $\partial \ln T/\partial y$ are small, and, even at elevated temperatures, D_1^T is very small if $\mathfrak{M}_1 - \mathfrak{M}_2$ is small (Eq. 3-16). For the O_2-N_2 mixture, $k_T \leq 0.007$. Thus, we see that all components of a nonreacting gas mixture may be lumped if thermal diffusion is negligible for all pairs of components, even if they are quite different gases. For such a case, ease of specification of the lumped transport properties is a second, and subsidiary, consideration. If all the two gases have nearly the same molecular weights and other properties of importance, then the lumped transport coefficients may be determined by simple averages. If the gases are widely different, proper values of the necessary coefficients might be difficult to determine, of course.

The requirements for lumping nonreacting pairs in a reacting mixture

[2] In general, pressure diffusion would also have to be negligible. We have already neglected this effect on the basis of the boundary layer assumptions.

are more strict. If, in the case of dissociated air, for example, we wish to lump the nonreactive pairs N_2-O_2 and N-O (leaving out of account NO formation) to form a binary mixture A_2-A, Eq. 2-11 tells us that the specific enthalpy, both internal and chemical, of the lumped constituents must be nearly the same. This implies that the corresponding molecular weights, specific heats and heats of dissociation must also be nearly the same. By Eq. 3-16, the previous requirement of negligible thermal diffusion between lumped constituents is thus automatically met. We may still, however, consider themal diffusion between A_2-A if the level of dissociation is sufficiently high. In the case of dissociating air, it is fortunate that the gases O_2 and N_2 are similar in their properties, and the binary grouping is well justified.

Coefficients for slight dissociation. Proceeding to the formulation of transport properties for the A_2-A mixture, we must first specify the interaction parameters. For use with the Lennard-Jones "6-12" potential, Lees [*8*] proposed the values

$$\sigma_{12} = 3.3\text{Å}; \qquad \sigma_2 = 3.69\text{Å}; \qquad \frac{\epsilon_{12}}{\mathfrak{k}} = \frac{\epsilon_2}{\mathfrak{k}} = 84°\text{K} \tag{3-18}$$

where we denote the A molecule by subscript 1, and A_2 by subscript 2. A purely repulsive power-law potential of the form Cr^{-n} was chosen by Bauer and Zlotnick [*80*] to describe collision processes in high temperature air. They found that inverse powers of either $n = 6$ or 8, together with the corresponding values of C (either 25.6 or $4.02 \cdot 10^{-11}$ for A_2-A_2 collisions, and 19.7 or $3.08 \cdot 10^{-10}$ for A-A), gave good agreement with available experimental data.

Of course, the A-A interaction is complicated by the valence attraction exerted during collisions, and it is doubtful that any simple potential function can accurately represent the forces involved. This difficulty need not concern us, however, provided we consider cases with rather low levels of dissociation: If, instead of using Eq. 3-12, we turn to the complete kinetic-theory formula [*5*, p. 531], expand in powers of c_1, and make a reasonable estimate of a certain collision quantity (A_{12}^*) which does not depend much on the choice of interaction potential, we find the following simple formula applies when μ_{12} differs only slightly from μ_2:

$$\mu = \mu_2\left\{1 + \left[0.15 + 3.0\left(1 - \frac{\mu_2}{\mu_{12}}\right)\right] c_1 + \mathcal{O}\,(c_1^2)\right\} \tag{3-19}$$

Here, the mass-ratio has been taken to be 2. Physically, the first order contribution shown in Eq. 3-19 is a diffusion term arising from A_2-A collisions occurring at a rate in linear proportion to atom concentration. A-A collisions, however, occur less frequently, at a rate proportional to c_1^2. Thus, μ_1 does not appear in Eq. 3-19. Moreover, the A_2-A "viscosity" μ_{12} is usually less than 10 per cent larger than μ_2, and therefore the coefficient of c_1 in Eq. 3-19 is quite small—typically between 0.2 and 0.5. Thus, in the many cases when c_1 is small, we may take $\mu = \mu_2$ for the A_2-A mixture, using only Eq. 3-13 to calculate μ_2. For boundary layer calculations we shall be chiefly interested in the ratio

$$J \equiv \frac{\rho\mu}{(\rho\mu)_e} \cong (1 - c_1 + c_{1_e}) \frac{(\sqrt{T^*}\,\Omega^{(2,2)*})_e}{\sqrt{T^*}\,\Omega^{(2,2)*}} + \mathcal{O}(c_1^2) \qquad (3\text{-}20)$$

In this equation, concentration dependence enters through application of the state equation (Eq. 2-2). Subscript $_e$ denotes conditions at the outer edge of the boundary layer.

In [9], Fay and Riddell cite calculations of transport coefficients by Penner and Litvak. We may select from those calculations the quantity J_w evaluated[3] by means of a Sutherland potential, taking $T_w = 300°K$ and $c_{1_w} = 0$:

T_e	3510	5580	6730	7410
c_{1_e}	0.054	0.234	0.375	0.536
J_w [9]	2.5	3.8	4.8	5.8

and compare the results obtained from Eq. 3-20:

J_w (3-20)	2.4	3.6	5.0	6.7

We see that up to a dissociation level of about 0.4, the two calculations agree within 10 per cent.

For boundary layer calculations, a great analytical simplification is accomplished by setting

$$J = \text{const} \qquad (3\text{-}21)$$

For temperature changes of a few hundred degrees, Eq. 3-21 (due to Chapman and Rubesin [10]) is satisfactory. For changes of the order of 1000°K, especially if changes in composition occur, the numerical examples given above indicate that Eq. 3-21 must not be used. However, it will appear later that there are circumstances in which Eq. 3-21, and even the special case $J = 1$, may be used with good accuracy.

We have noted that Prandtl number may, with good justification, be taken as constant. A representative value would be

$$Pr = \tfrac{3}{4} \qquad (3\text{-}22)$$

Another important quantity which will appear in the boundary layer equations is the Lewis number $Le \equiv Pr\rho\mathfrak{D}_{12}/\mu$. Using Eq. 3-11, 3-13, 3-18, and 3-19 and the fact that $\Omega_{12}^{(1,1)*} \cong 0.9\Omega_2^{(2,2)*}$ for all temperatures of interest, we find, for a slightly dissociated mixture A_2-A,

$$Le = 1.53(1 - c_1) \qquad (3\text{-}23)$$

[3] Subscript w signifies evaluation at the surface.

While Le tends to decrease at high temperature (cf. [7]), the change is not as great as that experienced by J, and the common assumption that Le = const may be employed with some confidence.

Energy expressions. To complete the picture of an A_2-A mixture, we must specify the thermodynamics and chemistry of the system. Returning to the definitions of internal energy and enthalpy (Eq. 2-11) and referring to standard works [5, Chap. 2] or [11, Chap. 9], we find, for internal energy per unit mass of the mixture,

$$\bar{e} = c_1\left(\frac{3}{2}\frac{R}{\mathfrak{M}_1}T + h_1^{(0)}\right) + c_2\left(\frac{3}{2}\frac{R}{\mathfrak{M}_2}T + \frac{R}{\mathfrak{M}_2}T + \frac{\frac{R}{\mathfrak{M}_2}T_v}{e^{T_v/T} - 1}\right) \tag{3-24}$$

where the five terms represent, respectively, translational energy of A atoms, energy of dissociation of a unit mass of A_2 molecules, translational, rotational (fully excited), and vibrational energies of A_2 molecules. Electronic contributions to specific heats are neglected, being generally small for present applications. The assumption of rotational equilibrium is well justified except in cases involving hydrogen: hydrogen-bearing molecules tend to have a high characteristic temperature for rotation. The characteristic temperature for vibration (T_v) is 2260°K for O_2 and 3380°K for N_2. The latter value might be most appropriate for air. When $T \gg T_v$, the vibrational energy term becomes RT, and when $T \ll T_v$, it vanishes. The quantity $h_1^{(0)}$ is the usual dissociation energy D divided by molecular weight: For oxygen, $h_1^{(0)} = \frac{117}{32} = 3.66$ kcal/g.

We may rewrite Eq. 3-24 in the form

$$\bar{e} = 3\frac{RT}{\mathfrak{M}_2}\left[c_1 + c_2\left(\frac{5}{6} + \frac{1}{3}\frac{T_v/T}{e^{T_v/T} - 1}\right)\right] + c_1 h_1^{(0)} \tag{3-25}$$

and observe that the quantity in parentheses has a rather narrow range of variation, between $\frac{5}{6}$ at low temperature and $\frac{7}{6}$ at high temperature. Lighthill [12] has proposed that, for the sake of mathematical simplicity, this quantity be approximated by its mean value, unity, as though molecules always carried vibrational energy of one half the fully excited value. This approximation (which also will simplify the mass action law for the mixture) is desirable for analytical work, but has little point when high speed computing programs are used. The specific enthalpy of the mixture is obtained by adding $[(c_1/\mathfrak{M}_1 + (c_2/\mathfrak{M}_2)]RT$ to Eq. 3-25:

$$\bar{h} = 5\frac{RT}{\mathfrak{M}_2}\left[c_1 + \frac{4}{5}c_2\left(\frac{7}{8} + \frac{1}{4}\frac{T_v/T}{e^{T_v/T} - 1}\right)\right] + c_1 h_1^{(0)} \tag{3-26}$$

Again, under Lighthill's approximation, the parentheses would be set equal to unity.

Mass action law. Turning to the chemical kinetics of an A_2-A mixture, we first note that when the rates of the reaction

$$A_2 \rightleftharpoons 2A \tag{3-27}$$

are sufficiently fast, then chemical equilibrium prevails, governed by the mass-action law [*11*, Chap. X], which relates relative concentration to variables of state. We may write this law in the form [*12*]

$$\frac{c_{1_{eq}}^2}{1 - c_{1_{eq}}} = \frac{\rho_d}{\rho} \exp\left[\frac{-2h_1^{(0)}}{R/\mathfrak{M}_1} T\right] \tag{3-28}$$

where subscript $_{eq}$ is used to denote the equilibrium state, and where ρ_d is a "characteristic density," related to the more usual "equilibrium constant" $K_p(T)$, by

$$K_p = 2\frac{R}{\mathfrak{M}_1}\rho_d T \exp\left[\frac{-2h_1^{(0)}}{R/\mathfrak{M}_1} T\right] \tag{3-29}$$

Lighthill [*12*] has pointed out that ρ_d is nearly constant in the range $1000°K < T < 7000°K$, having values of about 150 for O_2 and 130 for N_2. Now, apart from constants, the temperature dependence of ρ_d (obtained from Van't Hoff's equation) is approximately

$$\rho_d \sim \sqrt{T}\,(1 - e^{-T_v/T}) \tag{3-30}$$

By differentiating Eq. 3-30 we see that $\rho_d(T)$ has a relative maximum at the temperature for which $(T_v/T)/(e^{T_v/T} - 1) = \frac{1}{2}$. This temperature has already been cited as that for which the parentheses in Eq. 3-25 and 3-26 vanish. The existence of this relative maximum helps the accuracy of the simplifying assumption [*12*] that ρ_d is constant.

Kinetics for slight dissociation. In the event that reaction rates are not sufficiently fast to insure chemical equilibrium, one requires an expression for the production term W_1 found in the continuity equation (Eq. 2-7). W_1 is the mass rate of formation of A per unit volume by dissociation of A_2. It is generally accepted that recombination of atoms of O or N (hence A) occurs by three-body collisions involving two atoms of A and some third particle, which might be either an A or A_2, which then helps to absorb the energy released by the chemical reaction $2A \rightarrow A_2$. For each type of third body, the occurrence of successful collisions is proportional to the number density of molecules of each of the three particles and to a somewhat uncertain function of temperature, which recent experimental studies [*13,14*] suggest should be taken to be $T^{-3/2}$, when the third body is an A_2. Taking account of the reverse reaction (dissociation), one finds that the net mass rate of formation of A per unit volume is

$$W_1 = \rho\kappa(p, T)(c_{1_{eq}}^2 - c_1^2); \qquad \kappa = Kp^2T^{-7/2} \tag{3-31}$$

where $c_{1_{eq}}$, the mass fraction that would exist at the local pressure and temperature if chemical equilibrium prevailed, may be determined from the mass action law (Eq. 3-28). Eq. 3-31 is essentially in the form given by Fay and Riddell [9], except that we have assumed both c_1 and $c_{1_{eq}}$ to be small compared with unity. The quantity κ is, in effect, the reciprocal of a relaxation time, which shortens as pressure increases or temperature decreases. The constant K may be estimated from the results of experimental studies of the relaxation zones behind strong shock waves [13,14].

Now, the expression for W_1 given in Eq. 3-31 refers to the kinetics of colliding gaseous particles. The chemical rate at a solid surface is quite a different matter. In a manner that depends on the specific properties of the surface, atoms A colliding with the surface are adsorbed, recombine at the surface, and then are desorbed as molecules A_2. The reverse reaction may also take place. This process of surface catalysis is quite different from the gas-phase reaction. For example, a collision resulting in recombination need not involve three bodies—in fact, it is necessary only that a single atom collide with the surface, there to react with other atoms that previously collided and are held at the surface. In chemical terms, the surface reaction of recombination is found experimentally to be of "first order"—that is, proceeding at a rate proportional to the first power of concentration.

Rather than modify W_1 to account for surface catalysis, we may interpret the surface effect as, in effect, a boundary condition on the diffusional velocity v_1' at the surface [15]. For a noncatalytic surface, one would employ the obvious kinematic boundary condition that $v_{1_w}' = 0$. If, on the other hand, the surface were catalytic, $-v_{1_w}'$ should represent the rate at which the surface is able to adsorb atoms. To formulate this boundary condition, we note that kinetic theory gives the mass rate of collision of atoms with the surface as $\rho c_1 \sqrt{RT/\pi\mathfrak{M}_2}$. Only a certain fraction Γ of this mass attaches to the surface and recombines. The "catalytic efficiency" Γ depends on the characteristics of the surface. Further, molecules arrive at the surface at a mass rate $\rho c_2 \sqrt{RT/2\pi\mathfrak{M}_2}$ and dissociate in the proportion Γ_2. Thus the boundary condition we need is

$$-(\rho c_1 v_1')_w = \Gamma(\rho c_1)_w \sqrt{\frac{RT_w}{\pi\mathfrak{M}_2}} - \Gamma_2(\rho c_2)_w \sqrt{\frac{RT_w}{2\pi\mathfrak{M}_2}} \tag{3-32}$$

The nature of the coefficient Γ_2 is not altogether clear; however, Eq. 3-32 can be simplified by noting that, at chemical equilibrium, the net flux of atoms must vanish, and $c_{1_{eq}}$ must be given by the mass action law (Eq. 3-28). Then, assuming that c_1 is small compared with unity, we may write

$$-(c_1 v_1')_w = \Gamma \sqrt{\frac{RT_w}{\pi\mathfrak{M}_2}} (c_1 - c_{1_{eq}})_w \tag{3-33}$$

without any further assumption about Γ_2 except that it be a regular function of composition. Depending on the surface, Γ ranges between 10^{-5} and 1. It should be noted that Eq. 3-33 must be retained even when Γ is near the lower end of this range, since the velocity $\sqrt{RT_w/\pi\mathfrak{M}_2}$ is of the order of sound speed, which is very large compared with what we expect for diffusion velocity v_1' in the boundary layer. Finally, we should note that, in making application of Eq. 3-33 to unsteady problems, we would assume that the process of establishing layers or centers of adsorbed atoms proceeds much more rapidly than do changes in the flow situation. This assumption is difficult to verify.

E,4. The Boundary Layer Equations. We are now in a position to write the compressible laminar boundary layer equations for a reactive mixture of atoms and molecules. First, consideration will be restricted to steady plane or axially symmetric flows. We immediately apply the scale transformations that relate axially symmetric and plane flows (Mangler), and that partially account for compressibility (Levy, Dorodnitzin), in the form suggested by Lees [*8*], defining

$$\eta \equiv \frac{(\rho\mu)_e}{\sqrt{2\xi}} \int_0^y r^k \frac{\rho}{\rho_e}\, dy \tag{4-1}$$

$$\xi \equiv \int_0^x (\rho\mu u)_e r^{2k} dx \tag{4-2}$$

where u is the velocity component parallel to the surface, y is the distance measured normal to the surface, and x is measured along the surface. The length $r(x)$ defines body shape, being the distance measured from the axis of symmetry to the surface. The index $k = 0$ or 1 for plane or axially symmetric flow, respectively, and the subscript $_e$ refers to conditions at the outer edge of the boundary layer. The equation of overall continuity (Eq. 2-3) is identically satisfied by the definition of a stream function

$$\rho u r^k \equiv \psi_y; \qquad \rho v r^k \equiv -\psi_x \tag{4-3}$$

where v is the velocity component normal to the surface. Then, if we set $\psi \equiv \sqrt{2\xi}\, f(\xi, \eta)$,

$$\frac{u}{u_e} = f_\eta(\xi, \eta) \tag{4-4}$$

Next, we define a total enthalpy ratio

$$\frac{H}{H_e} \equiv g(\xi, \eta) \tag{4-5}$$

where

$$H \equiv \bar{h} + \tfrac{1}{2}u^2 \tag{4-6}$$

Substitution of Eq. 4-1 through 4-4 and 3-20 into Eq. 2-8 yields the momentum equation in the form

$$(Jf_{\eta\eta})_\eta + ff_{\eta\eta} + \frac{2\xi u_{e_\xi}}{u_e}\left(\frac{\rho_e}{\rho} - f_\eta^2\right) = 2\xi(f_\eta f_{\xi\eta} - f_\xi f_{\eta\eta}) \tag{4-7}$$

Making the appropriate substitutions into Eq. 2-7 for c_1 (and dropping the subscript for convenience) yields the continuity equation for atoms:

$$\left[J\frac{Le}{Pr}\left(c_\eta - \frac{K_T}{6}c\frac{T_\eta}{T}\right)\right]_\eta + fc_\eta + W_1\frac{2\xi}{\rho r^{2k}(\rho\mu u^2)_e} = 2\xi(f_\eta c_\xi - f_\xi c_\eta) \tag{4-8}$$

Here, the thermal diffusion term has been simplified by assuming $c \ll 1$. Similarly, the energy equation (Eq. 2-10) becomes, after Eq. 3-5, 4-5, and 4-6, and other appropriate relations are used,

$$\begin{aligned}\left(\frac{J}{Pr}g_\eta\right)_\eta + fg_\eta + \frac{u_e^2}{H_e}\left[J\left(1 - \frac{1}{Pr}\right)f_\eta f_{\eta\eta}\right]_\eta + \Bigg\{\frac{J}{Pr}\Bigg[(Le - 1)c_\eta \\ - \frac{LeK_T}{6}c\frac{T_\eta}{T}\Bigg]\left(\frac{h_1 - h_2 + h_1^{(0)}}{H_e}\right)\Bigg\}_\eta = 2\xi(f_\eta g_\xi - f_\xi g_\eta)\end{aligned} \tag{4-9}$$

while for heat flux at the surface, Eq. 3-10 yields

$$-\frac{\sqrt{2\xi}\,Prq_w}{H_e u_e(\rho\mu)_w r^k} = \left\{g_\eta + \left[(Le - 1)c_\eta - \frac{LeK_T}{6}c\frac{T_\eta}{T}\right]\left(\frac{h_1 - h_2 + h_1^{(0)}}{H_e}\right)\right\}_w \tag{4-10}$$

Using Eq. 3-3 and 3-33 we may express the catalytic wall condition in the form

$$\left(c_\eta - \frac{K_T}{6}c\frac{T_\eta}{T}\right)_w = \left(\frac{Pr}{Le}\right)_w \frac{\sqrt{2\xi}}{\mu_w u_e r^k}\Gamma\sqrt{\frac{RT_w}{\pi\mathfrak{M}_2}}(c - c_{eq})_w \tag{4-11}$$

The remaining boundary conditions are the customary ones, and will be mentioned in relation to the specific problems to be discussed. To Eq. 4-7 through 4-11 we must add a specification of the relation between static enthalpy and temperature. Eq. 3-26, with or without the Lighthill simplifications [*12*], provides the necessary relation and, together with the state equation (Eq. 2-2), yields a density-enthalpy relationship for inclusion in Eq. 4-7.

The foregoing steady flow equations are especially useful when boundary layer similarity can be invoked. We shall also require a corresponding set of equations applicable to the study of one-dimensional diffusion waves of the boundary layer type. In such cases, y and t are the only independent variables, and the transformations

$$\zeta \equiv \frac{\rho_\infty}{\mu_\infty}\int_0^y \frac{\rho}{\rho_\infty}\,dy \tag{4-12}$$

$$\tau \equiv \frac{\rho_\infty}{\mu_\infty}t \tag{4-13}$$

may be applied, where subscript ∞ denotes a constant reference condition. From the continuity equation (Eq. 2-3), we notice that

$$-\frac{\rho}{\rho_\infty}v = \frac{\partial\zeta}{\partial\tau} \tag{4-14}$$

and the boundary layer equations, corresponding to Eq. 4-7 through 4-11 become, respectively,

$$u_\tau = (Ju_\zeta)_\zeta \tag{4-15}$$

where, now, $J \equiv \rho\mu/(\rho_\infty\mu_\infty)$,

$$c_\tau = \frac{\mu_\infty}{\rho\rho_\infty}W_1 + \left[\frac{JLe}{Pr}\left(c_\xi - \frac{K_T}{6}c\frac{T_\zeta}{T}\right)\right]_\zeta \tag{4-16}$$

$$H_\tau = \frac{1}{\rho}p_\tau + \left(\frac{J}{Pr}\left\{H_\zeta + (Pr-1)uu_\zeta + \left[(Le-1)c_\zeta - \frac{LeK_T}{6}c\frac{T_\zeta}{T}\right](h_1 - h_2 + h_1^{(0)}\right\}\right)_\zeta \tag{4-17}$$

$$-\frac{\mu_\infty Pr}{(\rho\mu)_w}q_w = \left\{H_\zeta - uu_\zeta + \left[(Le-1)c_\zeta - \frac{LeK_T}{6}c\frac{T_\zeta}{T}\right](h_1 - h_2 + h_1^{(0)})\right\}_w \tag{4-18}$$

$$\left(c_\zeta - \frac{K_T}{6}c\frac{T_\zeta}{T}\right)_w = \left(\frac{Pr}{Le}\right)_w\frac{\mu_\infty}{\mu_w}\Gamma\sqrt{\frac{RT_w}{\pi\mathfrak{M}_2}}(c - c_{eq})_w \tag{4-19}$$

It should be recalled that in using the catalytic condition stated in Eq. 4-19 for a nonsteady problem, one assumes that the activation of the surface as a catalyst proceeds much more rapidly than other changes taking place in the flow.

E,5. Couette Flow of a Dissociated Gas.

General considerations. We turn now to a consideration of various classical problems of viscous flow theory, in each case bringing out the particular consequences of hypersonic flow conditions.

The simplest viscous flow is Couette flow, or the motion set up in a fluid contained between two parallel infinite plane walls, a distance l apart, which are in relative motion. Such a flow constitutes a primitive model for the boundary layer. With this thought in mind, we shall regard the lower wall (subscript $_w$) to be fixed, and the upper one (subscript δ) perhaps in motion with a very high velocity. The upper plate will be considered a reservoir of heat at a specified high stagnation temperature, while the lower plate will be a heat reservoir at some specified moderate temperature that would be suitable for survival of the material surface. Steady flow will be assumed, with gradients of flow properties occurring only in the y direction normal to the plate surface. The foregoing problem is one which is sufficiently simple mathematically that the complications resulting from real-gas effects such as chemical activity and component diffusion may be studied quite thoroughly, in the interest of qualitative understanding.

The appropriate differential equations are Eq. 4-15 through 4-19, with all time derivatives set equal to zero. We identify the reference state (∞) with conditions at the upper plate (subscript δ). We shall also use the symbol δ to designate the value of ζ corresponding to $y = l$; from Eq. 4-12, $\delta \equiv (\rho/\mu)_\delta \int_0^l (\rho/\rho_\delta)dy$. Following Clarke [16], who has made a thorough study of Couette flow in a dissociating gas, we may at once integrate the momentum equation to give

$$Ju'(\zeta) = J_w u'(0) \tag{5-1}$$

where we take $J \equiv \rho\mu/(\rho\mu)_\delta$. If the function inside the curly bracket of Eq. 4-18 is identified as $q(\zeta)$, integration of the energy Eq. 4-17 yields

$$J(q + Pr\, uu') = J_w q_w \tag{5-2}$$

Here we assume constant Pr but allow J to be variable, for the time being. Clarke [16] observed that the variable J may be eliminated between Eq. 5-1 and 5-2. This done, Eq. 5-2 may be integrated again to give the basic relation between energy and velocity profiles:

$$\frac{\bar{h} - \bar{h}_w + h_1^{(0)}\left[(Le - 1)(c - c_w) - \frac{LeK_T}{6}\int_0^\zeta cd(\ln T)\right] + \frac{1}{2}Pr\, u^2}{\bar{h}_\delta - \bar{h}_w + h_1^{(0)}\left[(Le - 1)(c_\delta - c_w) - \frac{LeK_T}{6}\int_0^\delta cd(\ln T)\right] + \frac{1}{2}Pr\, u_\delta^2} = \frac{u}{u_\delta} \tag{5-3}$$

as well as an explicit formula for heat transfer to the lower plate:

$$-\frac{Pr}{\rho_\delta J_w}\frac{u_\delta}{u'(0)}q_w = \bar{h}_\delta - \bar{h}_w + h_1^{(0)}\left[(Le - 1)(c_\delta - c_w) - \frac{LeK_T}{6}\int_0^\delta cd(\ln T)\right] + \frac{1}{2}Pr\, u_\delta^2 \tag{5-4}$$

In the foregoing two equations, the customary assumption is made that the energy conveyed by preferential diffusion of atoms consists chiefly of dissociational energy; that is, $h_1^{(0)} \gg h_1 - h_2$. Finally, we may note that for "frozen" flow ($W_1 = 0$), Eq. 4-16 may be integrated and combined with Eq. 4-19 to yield

$$c = c_\delta - \frac{K_T}{6} \int_\zeta^\delta c\, d\,(\ln T) - \frac{u_\delta - u}{u'(0)} \Gamma \mathfrak{U}(c - c_{eq})_w \qquad (5\text{-}5)$$

where we have written the right-hand side of Eq. 4-19 as $\Gamma\mathfrak{U}(c - c_{eq})_w$, noting that $\mathfrak{U}$ has the dimensions of velocity.

Let us now, using Eq. 5-1 through 5-5 and various other relations discussed in previous paragraphs, explore certain formulations of the Couette flow problem that provide analogies to real-gas boundary layer flows.

Upper plate moving at hypersonic speed. First, we consider the upper surface to be moving at hypersonic speed, holding both upper and lower surfaces at moderate temperatures (taken to be equal, for simplicity). This problem is somewhat analogous to that of a hypersonic boundary layer over a semi-infinite flat plate. We assume $Le = 1$, neglect thermal diffusion, and choose $T_w = T_\delta$ to be sufficiently low that no vibrational excitation exists at the plates. Thus, at the plates, but not in the gas between, the ratio of specific heats has its ideal-gas value of $\gamma = 1.4$. Then, directly from Eq. 5-3, we find the static enthalpy profile to be

$$\frac{h}{h_w} = 1 + \frac{\gamma - 1}{2} Pr\, M_\delta^2 \left(\frac{u}{u_\delta} - \frac{u^2}{u_\delta^2}\right) \qquad (5\text{-}6)$$

Taking $M_\delta = 15$ and $Pr = \frac{3}{4}$, we find that viscous dissipation of energy causes $\bar{h}$ to reach a maximum value of $9.44\bar{h}_w$ at the point where $u = \frac{1}{2}u_\delta$.

Next, we use Eq. 3-26 for static enthalpy of the mixture to determine the temperature T_m at the maximum enthalpy level. We choose $T_\delta = T_w = 500°K$, $T_v = 3380°K$ (the value for N_2), and, first, consider frozen flow, for which $c = 0$ in this problem. If we were to neglect vibrational excitation, we should find, corresponding to $\bar{h} = 9.44\bar{h}_w$, that $T_m = 4720°K$. In contrast, if the term in Eq. 3-26 that refers to vibration is included, a lower temperature, $T_m = 4000°K$, results. The temperature at maximum "aerodynamic heating" is lowered still further if equilibrium dissociation is considered to occur in the gas between the plates. In using Eq. 3-28 to specify equilibrium dissociation, let us consider that the temperature is in a range in which only oxygen is available for dissociation. Nitrogen dissociation would be assigned to a higher temperature range in view of its higher energy of dissociation. On this basis, if a mixture of 4 parts nitrogen and 1 part oxygen is assumed, the maximum value that c can achieve is $\frac{1}{5}$, and Eq. 3-28 takes the form

$$\left[\frac{(5c)^2}{(1 - 5c)(1 + c)}\right]_{eq} = 2050 \frac{T}{p_{atm}} e^{-59{,}000/T} \qquad (5\text{-}7)$$

where the equation of state (Eq. 2-2) has been used, the value of ρ_d = 150 gm/cm^3 suggested for oxygen by Lighthill [*12*] is adopted, and the usual value of 117 kcal/mole is used for the dissociation energy of oxygen. Fig. E,5a shows curves of dissociation level calculated from Eq. 5-7. Choosing $p = \frac{1}{20}$ atm and incorporating Eq. 5-7 into Eq. 3-26, we find that the temperature corresponding to $\bar{h} = 9.44\bar{h}_w$ is lowered considerably, to a level of 2940°K, owing to absorption of energy in dissociation, even though c_{eq} at that temperature is only 8 per cent. If we follow Lighthill [*12*] and set the parenthesis in Eq. 3-26 equal to zero, we get, instead, T_m = 3090°K and c_{eq} = 11 per cent.

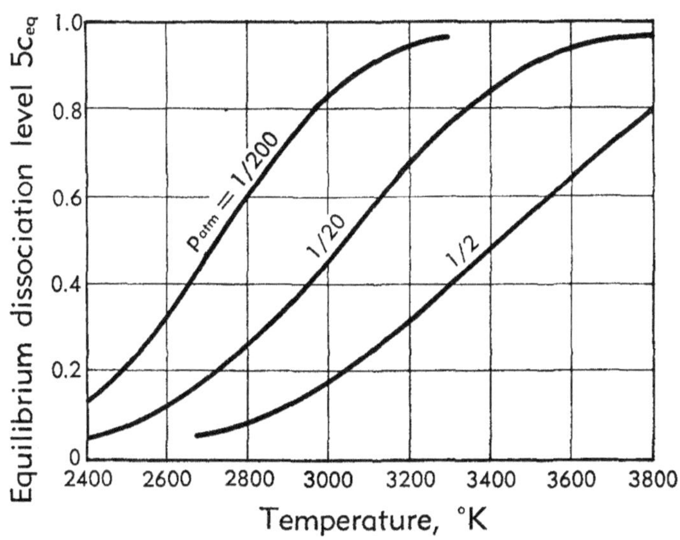

Fig. E,5a. Equilibrium dissociation level for a 4:1 mixture of oxygen and nitrogen in which only oxygen dissociates.

In the foregoing problem, the right-hand side of Eq. 5-4 is simply $\frac{1}{2}Pr\, u_\delta^2$, so that surface heat transfer is unaffected by real-gas considerations, except insofar as variation of J between the plates affects the value of $u'(0)$.

Upper plate stationary. Effect of variable J. We now turn to a problem analogous to that of the boundary layer at the stagnation point of a blunt body in hypersonic flow; as before, we take the stationary lower surface to be rather cool (T_w = 500°K), but now the upper plate will be hot (T_δ = 4000°K), at rest ($u_\delta = 0$) and in chemical equilibrium ($c_\delta = c_{eq_\delta}$). We note that setting $u_\delta = 0$ does not relieve us of the necessity to solve Eq. 5-1, which has been used to eliminate J in Eq. 5-3, 5-4, and 5-5. Of course, if $J = 1$, the result is trivial: $u/u_\delta = \zeta/\delta$ and $u_\delta/u'(0) = \delta$. Let us first complete the solution for an ideal gas, for which $\bar{h} \sim T$, in order to see the effect of variable J. We use the rigid sphere potential, for which $\Omega^{(2,2)*} = 1$, to simplify Eq. 3-20, so that

$$J = \sqrt{\frac{\bar{h}_\delta}{\bar{h}}} \tag{5-8}$$

Combining and solving Eq. 5-1, 5-8, and Eq. 5-3, for which the left-hand side is now simply $(\bar{h} - \bar{h}_w)/(\bar{h}_\delta - \bar{h}_w)$, we find the solution

$$\frac{\bar{h}}{\bar{h}_\delta} = \frac{\bar{h}_w}{\bar{h}_\delta} + 2\sqrt{\frac{\bar{h}_w}{\bar{h}_\delta}}\left(1 - \sqrt{\frac{\bar{h}_w}{\bar{h}_\delta}}\right)\frac{\zeta}{\delta} + \left(1 - \sqrt{\frac{\bar{h}_w}{\bar{h}_\delta}}\right)^2 \frac{\zeta^2}{\delta^2} \tag{5-9}$$

and, after using Eq. 4-12 to find the relation between δ and the distance l between the plates,

$$l = \frac{\mu_\delta}{\rho_\delta}\left(1 + \sqrt{\frac{\bar{h}_w}{\bar{h}_\delta}} + \frac{\bar{h}_w}{\bar{h}_\delta}\right)\frac{\delta}{3} \tag{5-10}$$

we obtain

$$-q_w = \frac{\mu_\delta}{Pr}\frac{2}{3}\frac{1 - (\bar{h}_w/\bar{h}_\delta)^{\frac{3}{2}}}{1 - \bar{h}_w/\bar{h}_\delta}\frac{\bar{h}_\delta - \bar{h}_w}{l} \tag{5-11}$$

Now, if $J = 1$ had been assumed,

$$\frac{\bar{h}}{\bar{h}_\delta} = \frac{\bar{h}_w}{\bar{h}_\delta} + \left(1 - \frac{\bar{h}_w}{\bar{h}_\delta}\right)\frac{\zeta}{\delta} \tag{5-12}$$

$$-q = \frac{\mu_\delta}{Pr}\frac{1}{2}\left(1 + \frac{\bar{h}_w}{\bar{h}_\delta}\right)\frac{\bar{h}_\delta - \bar{h}_w}{l} \tag{5-13}$$

Comparing Eq. 5-9 and 5-12 we see that the increase of J as the lower plate is approached imparts a parabolic bulge to the enthalpy profile. Comparing expressions for heat transfer, we first must note that writing μ_δ in Eq. 5-13 represents, in effect, an arbitrary choice; having assumed $J = 1$, we might just as well have written $(\rho\mu)_w/\rho_\delta$ there. Actually, μ_δ is a good choice, as we may see by taking the ratio of Eq. 5-13 to 5-11:

$$\frac{(q_w)_{J=1}}{(q_w)_{J\neq 1}} = 0.75\,\frac{1 - (\bar{h}_w/\bar{h}_\delta)^2}{1 - (\bar{h}_w/\bar{h}_\delta)^{\frac{3}{2}}} \tag{5-14}$$

The right-hand side of Eq. 5-14 approaches unity as $\bar{h}_w \to \bar{h}_\delta$; however, our interest is in those cases when $\bar{h}_w \ll \bar{h}_\delta$. We see that, at worst (when $\bar{h}_w/\bar{h}_\delta \to 0$), the heat transfer is in error by 25 per cent when $J = 1$ is assumed, even though $J_w = \infty$. When $\bar{h}_w/\bar{h}_\delta = \frac{1}{2}$, the error is about 15 per cent. Even in the most extreme case, the greater portion of the gas layer has a J not too different from unity, and, as one might suppose, this thick layer controls the rate at which heat is delivered to the surface. A similar conclusion is reached in a subsequent discussion of stagnation-point heat transfer.

Effect of variable J for moving upper plate. A similar analysis may be applied to the problem of the cold, fast-moving upper plate. Combining the static enthalpy profile given in Eq. 5-6 with the assumed variation of J, Eq. 5-8, and with the momentum Eq. 5-1, yields

$$u' = u'(0)\left[1 + \frac{\gamma - 1}{2} Pr\, M_\delta^2 \left(\frac{u}{u_\delta} - \frac{u^2}{u_\delta^2}\right)\right]^{\frac{1}{2}} \tag{5-15}$$

From Eq. 4-18, the surface heat transfer is

$$-\frac{2}{\rho_\delta u_\delta} q_w = J_w u'(0) \tag{5-16}$$

and we note that $J_w = 1$ for this problem. Now, for a perfect gas, inversion of Eq. 4-12 gives $\rho_\delta l/\mu_\delta = \int_0^\delta (\bar{h}/\bar{h}_\delta)d\zeta$ which, when combined with Eq. 5-6 and 5-15, yields

$$u'(0)\frac{\rho_\delta l}{(u\mu)_\delta} = \int_0^1 \left[1 + \frac{\gamma - 1}{2} Pr\, M_\delta^2 \left(\frac{u}{u_\delta} - \frac{u^2}{u_\delta^2}\right)\right]^{\frac{1}{2}} d\left(\frac{u}{u_\delta}\right) \tag{5-17}$$

Thus a simple integration provides the quantity $u'(0)$ needed to evaluate heat transfer with variable J. The result, expressed in terms of the parameter

$$\theta \equiv \frac{\gamma - 1}{2} Pr\, M_\delta^2 \tag{5-18}$$

is

$$u'(0)\frac{\rho_\delta l}{(u\mu)_\delta} = \frac{1}{2} + \frac{4 + \theta}{4\sqrt{\theta}} \sin^{-1}\sqrt{\frac{\theta}{4 + \theta}} \tag{5-19}$$

If we had assumed $J = 1$, then the simpler result

$$u'(0)\frac{\rho_\delta l}{(u\mu)_\delta} = 1 + \frac{\theta}{6} \tag{5-20}$$

would apply. The ratio of Eq. 5-20 and 5-19 gives, in view of Eq. 5-16,

$$\frac{(q_w)_{J=1}}{(q_w)_{J\neq 1}} = \frac{1 + \theta/6}{\frac{1}{2} - \frac{4 + \theta}{4\sqrt{\theta}} \sin^{-1}\sqrt{\frac{\theta}{4 + \theta}}} \tag{5-21}$$

In the limit $\theta \to 0$, viscous heating vanishes, and the ratio described in Eq. 5-21 approaches unity. In the hypersonic limit $\theta \to \infty$, the value $4\sqrt{\theta}/3\pi$ is approached. To interpret this result, we note that Eq. 5-6 shows that the maximum value of $\bar{h}/\bar{h}_\delta$ is $\theta/4$, when θ is large, and, therefore, the minimum J is $J_m = 2/\sqrt{\theta}$. Accordingly, the hypersonic limit of Eq. 5-21 is

$$\frac{(q_w)_{J=1}}{(q_w)_{J\neq 1}} = \frac{0.85}{J_m} \tag{5-22}$$

Thus, the assumption $J = 1$ is inadmissible for this problem. Referring to Eq. 5-16, however, we see that if we follow Chapman and Rubesin [*10*] by selecting an appropriate constant value of J (here, we choose $J = J_m$), then $(q_w)_{J=J_m} = J_m(q_w)_{J=1}$, and, by Eq. 5-22, an error of only 15 per cent is incurred. An alternative route to the same result is to carry out the analysis with $J = 1$, but then in the final solution replace the product $(\rho\mu)_\delta$ by $(\rho\mu)_m$.

Summarizing the results of the foregoing two Couette problems in regard to J variation, we may conclude that reasonable accuracy is obtained by using $J = 1$ in the boundary layer analysis, if, in the solution for heat transfer, the product $\rho\mu$ is evaluated in the hottest part of the boundary layer. Henceforth, we shall adopt this procedure when convenient, and will find further justification by comparison with calculations of flat-plate and stagnation-point boundary layers done with variable $\rho\mu$.

Real-gas solution for stationary upper plate. In the remainder of this article we shall explore the stationary heat conduction problem, paying attention to real-gas effects. We assume $J = 1$, and assume vibrational energy to be one half its fully excited value [*12*]. We neglect T whenever it is grouped with dissociation level, and since $c \leqq 0.2$, we retain c only when multiplied by dissociation energy. Also, we take $K_T = \frac{1}{2}$. Thus Eq. 5-3 becomes

$$\frac{T - T_w - 14{,}800Le\left[c - c_w - \frac{1}{12}\int_0^\zeta c\,d(\ln T)\right]}{T_\delta - T_w + 14{,}800Le\left[c_\delta - c_w - \frac{1}{12}\int_0^\delta c\,d(\ln T)\right]} = \frac{\zeta}{\delta} \qquad (5\text{-}23)$$

and, from Eq. 5-4, introducing a Nusselt number,[4]

$$-q_w = \frac{\mu_\delta}{Pr}\frac{H_\delta - H_{w(\text{therm})}}{l}Nu \qquad (5\text{-}24)$$

one finds, for the present case of $T_\delta - T_w = 3500°\text{K}$,

$$Nu = \frac{1 + 4.23Le\left[c_\delta - c_w - \frac{1}{12}\int_0^\delta c\,d(\ln T)\right]}{1 + 4.23c_\delta}\int_0^1 \frac{T}{T_\delta}\,d\left(\frac{\zeta}{\delta}\right) \qquad (5\text{-}25)$$

From the foregoing general solution we may derive various effects of dissociation.

Equilibrium dissociation. First, if we assume that c is in equilibrium throughout the gas layer, c is obtained from Fig. E,5a (choosing $p = \frac{1}{20}$ atm), and we solve Eq. 5-23, for the moment neglecting thermal diffusion, with the result shown in Fig. E,5b. The temperature profile has a pronounced S-shape, the flatness of the central part of the curve resulting from the fact that, when the dissociation level is changing, a

[4] The driving potential $H_\delta - H_w$ is defined to exclude dissociational energy at the surface, so that, in case of frozen flow, $H_\delta - H_w$ is still known a priori.

large amount of energy can be transferred with only a small temperature gradient, by diffusion of atoms. This effect is somewhat accentuated by an increase in Lewis number (i.e., larger diffusion rate). Heat transfer rate is obtained from Eq. 5-25:

$$(Nu)_{eq} = 0.65(Le = 1) \\ = 0.71(Le = 1.5) \tag{5-26}$$

So far, we have automatically satisfied Eq. 4-16 and 4-19 by assuming that the reaction is so fast that $c = c_{eq}$, and by taking the lower plate temperature sufficiently low that c_{eq} vanishes at the plate and also for

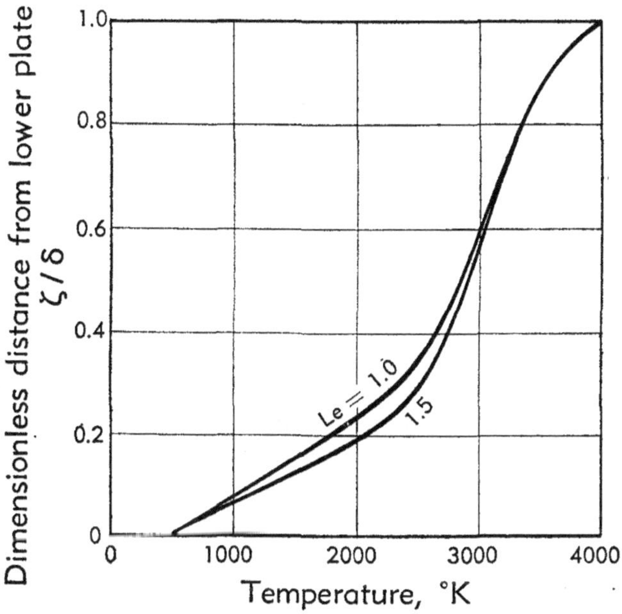

Fig. E,5b. Temperature profile between stationary parallel plates, the gas being in equilibrium at $p = \frac{1}{10}$ atmosphere.

some distance above the plate, so that $c_w = c'_w = 0$. The results of the foregoing example are profoundly altered when these assumptions are not met.

Chemically "frozen" dissociation. Consider the situation when the reaction rate is very slow, so that composition is frozen so far as gas-phase chemistry is concerned, though composition may vary by diffusion processes. We note that the width of the channel is involved in the definition of "slowness": Reactions cannot as readily go to equilibrium in a narrow channel as a wide one. In this extreme, noting that $(u_\delta - u)/u'(0) = \delta - \zeta$ when $J = 1$, we see from Eq. 5-5 that c has a linear profile, thermal diffusion being neglected. Under the assumptions of the present problem, we find

$$\frac{c}{c_\delta} = 1 - \frac{\Gamma Re\left(1 - \frac{\zeta}{\delta}\right)}{\int_0^1 \frac{T}{T_\delta}\, d\left(\frac{\zeta}{\delta}\right) + \Gamma Re}; \qquad Re \equiv \frac{\rho_\delta \mathcal{U} l}{\mu_\delta} \tag{5-27}$$

Clearly, even when T_w is very small, c_w cannot vanish unless the "effective catalycity" ΓRe is very large. On the other hand, if ΓRe is very small, c is constant at the value c_δ. It is important that, although the efficiency factor Γ is usually small and in no case can exceed 1, it is grouped with a Reynolds number that can be very large; we recall that the velocity $\mathcal{U}$ is of the order of sound speed in the gas. Therefore, it often makes sense to speak of a surface as "fully catalytic" in the sense $\Gamma Re \to \infty$, even though Γ may be rather small.

Suppose that the wall is noncatalytic. Then, $c = c_\delta = 0.2$, and, by Eq. 5-23, the temperature profile is linear. From Eq. 5-25,

$$(Nu)_{\mathrm{fr,noncat}} = 0.30 \tag{5-28}$$

We see by comparison with Eq. 5-26 that the absence of heat release by recombination in a frozen gas over a noncatalytic surface causes a great reduction in heat transfer, which now occurs purely by conduction.

If, on the other hand, the surface is fully catalytic, dissociation energy will be released at the surface, even if the gas above is frozen. In this case, Eq. 5-27 says that $c/c_\delta = \zeta/\delta$, and by use of Eq. 5-23 we again find a linear temperature profile. Even though, in this case, atoms are diffusing and carrying potential energy of recombination toward the surface, the frozen character of the flow provides that none of this energy is released on the way; therefore, the energy in internal degrees of freedom is unaffected. At the catalytic surface, where the energy is released, we find from Eq. 5-25

$$\begin{aligned}(Nu)_{\mathrm{fr,cat}} &= 0.56(Le = 1)\\ &= 0.69(Le = 1.5)\end{aligned} \tag{5-29}$$

which is almost double the rate experienced by the noncatalytic wall, and nearly the same as that provided by a gas in full equilibrium (Eq. 5-26). Again, at higher Le, atoms diffuse more freely to the surface causing a higher rate of heat release there.

The nonequilibrium sublayer. For the wall temperature chosen here (500°K), the equilibrium result is the same whether the wall is catalytic or not, because both c_w and c_w' vanish, and Eq. 4-19 is satisfied identically, whatever the value of Γ. However, at low pressures and high wall temperatures, this is no longer true. Let us, by way of illustration, revise our previous example by raising the wall temperature to 3000°K. At a pressure of 0.05 atm, c_{eq_w} is then equal to 0.09. Still assuming an equilibrium boundary layer, we may find by direct differentiation of Eq. 5-23, when Eq. 5-7 is included, that $c'_{\mathrm{eq_w}} = (0.13)/\delta$. This result has nothing to do with the catalytic condition (Eq. 4-19), and, in fact, is inconsistent with Eq. 4-19 unless the wall is fully catalytic. If the wall is completely noncatalytic, then Eq. 4-19 sets the boundary condition $c'_{\mathrm{w}} = 0$. The inconsistency of this boundary condition with the value $(0.13)/\delta$ tells us that, at least near the wall, the gas cannot be in chemical equilibrium. In fact, Hirschfelder [*15*] has pointed out that the necessary adjustment takes place in a thin "sublayer" near the wall, governed by Eq. 4-16. Near the wall we may write

$$c''\left(\frac{\zeta}{\delta}\right) = K_{\mathrm{r}}\left[c - \left(0.09 + 0.13\,\frac{c}{\delta}\right)\right] \tag{5-30}$$

where

$$K_{\mathrm{r}} \equiv 0.18\,\frac{\mu_\delta \kappa \delta^2}{\rho_\delta} \tag{5-31}$$

assuming that the adjustment will not cause c to depart substantially from its equilibrium value. Noting that $\delta \sim \rho l/\mu$, we observe that K_{r} is a ratio of the time required for diffusion to act between the plates and the chemical "relaxation time" $1/\kappa$. The parameter K_{r} has, in effect, already been assumed to be large, because we have postulated chemical equilibrium except for the influence of Eq. 4-19. The appropriate exact solution of Eq. 5-30 is

$$c = 0.09 + 0.13\left(\frac{\zeta}{\delta} + \frac{e^{-\sqrt{K_{\mathrm{r}}}\frac{\zeta}{\delta}}}{\sqrt{K_{\mathrm{r}}}}\right) \tag{5-32}$$

which is valid near the wall, and satisfies the requirement that $c'_{\mathrm{w}} = 0$. Since K_{r} is large, the adjustment of the gradient c' takes place in a very small interval, $0 < \zeta/\delta \lesssim 1/\sqrt{K_{\mathrm{r}}}$, and the surface value

$$c_{\mathrm{w}} = 0.09 + \frac{0.13}{\sqrt{K_{\mathrm{r}}}} \tag{5-33}$$

differs only slightly from the catalytic result, $c_{\mathrm{eq_w}} = 0.09$. A consequence of the adjustment of c'_{w} is that the temperature gradient must increase steeply within the same thin layer in order to keep a balance of energy flux. The net heat transfer rate remains nearly unaffected, however.

We have discussed only the extremes of chemical equilibrium and frozenness, and fully catalytic and noncatalytic walls. It is obviously possible to carry out solutions of the appropriate equations for all intermediate situations. For example, Broadwell [*17*] has analyzed Couette flow with finite reaction rate by linearizing the right-hand side of Eq. 3-31. Also, Hirschfelder [*15*] has studied the near-equilibrium case, paying particular attention to the catalytic wall effect. Liepmann and Bleviss [*18*] have studied the equilibrium Couette problem and have included the effect of ionization, which they conclude may be neglected for $M < 25$.

Thermal diffusion. Finally, we return to a previous assumption which we are now in position to evaluate. We have neglected thermal diffusion, and we may inquire whether, in the previous series of examples, this neglect was proper. Thermal diffusion occurs in Eq. 5-23 and 5-25 grouped with concentration diffusion in the following way (taking $K_T = \frac{1}{2}$ and $c \ll 1$):

$$c - c_w - \tfrac{1}{12} \int_0^{\zeta} c\, d(\ln T) \tag{5-34}$$

and we inquire whether the second term in Eq. 5-34 may be neglected relative to the difference $c - c_w$. Referring to the problem of stationary plates with large temperature difference, we first consider the equilibrium case, $c = c_{eq}(T)$. The integral in Eq. 5-34 is then $\int_{T_w}^{T_\delta} c_{eq}(T)dT/T$, and we observe from Fig. E,5a that $c_{eq}(T)$ differs substantially from zero only at rather high temperature (say, 3000°K for $p = \frac{1}{20}$ atm). Typically then, Eq. 5-34 takes on the extreme value

$$c_\delta \left[1 - \frac{1}{12} \ln \left(\frac{T_\delta}{3000} \right) \right] \tag{5-35}$$

and the neglect of thermal diffusion would result in an error of only about 5 per cent when $T_\delta \cong 5000$°K.

Eq. 5-34 suggests that thermal diffusion would have its greatest effect when the flow is frozen, the lower plate is noncatalytic ($c_w = c_\delta$), and the upper plate temperature is very high, so that the constant value of c, as well as the temperature gradient appearing in the integral, are both as large as possible. In this case, however, one must take into account the fact that c_w itself depends on thermal diffusion: One may show from Eq. 5-5 that

$$c_w = \frac{c_\delta - \frac{1}{12} \int_0^{\delta} c\, d(\ln T)}{1 + \delta \Gamma \mathfrak{U}} \tag{5-36}$$

From Eq. 5-36, taking both the temperature and concentration profiles to be linear, one finds

$$c_\delta - c_w - \tfrac{1}{12} \int_0^{\delta} c\, d(\ln T) \cong c_\delta \left[\frac{\delta \Gamma \mathfrak{U}}{1 + \delta \Gamma \mathfrak{U}} - \frac{1}{12} \frac{\delta \Gamma \mathfrak{U}}{(1 + \delta \Gamma \mathfrak{U})^2} \ln \left(\frac{T_\delta}{T_w} \right) \right] \tag{5-37}$$

Referring to Eq. 5-25, we see that thermal diffusion has its greatest effect on heat transfer rate when the second term in the square bracket of Eq. 5-37 is largest; that is, when $\delta \Gamma \mathfrak{U} = 1$. For this intermediate degree of catalycity, the right side of Eq. 5-37 is

$$\frac{1}{2} c_\delta \left[1 - \frac{1}{24} \ln\left(\frac{T_\delta}{T_w}\right)\right] \tag{5-38}$$

Thus, if $T_\delta = 5000°K$ and $T_w = 500°K$, the error incurred by neglecting thermal diffusion would be about 10 per cent.

In general, it seems proper to neglect thermal diffusion, even in cases of frozen flow, as long as errors of about 10 per cent can be tolerated. This neglection is usually made in hypersonic theory, though Scala [*19*] has included thermal diffusion in certain calculations of stagnation-point heat transfer.

E,6. Rayleigh Flow. Classical Rayleigh flow of an incompressible fluid [*3*, pp. 137, 138] is concerned with the boundary layer established

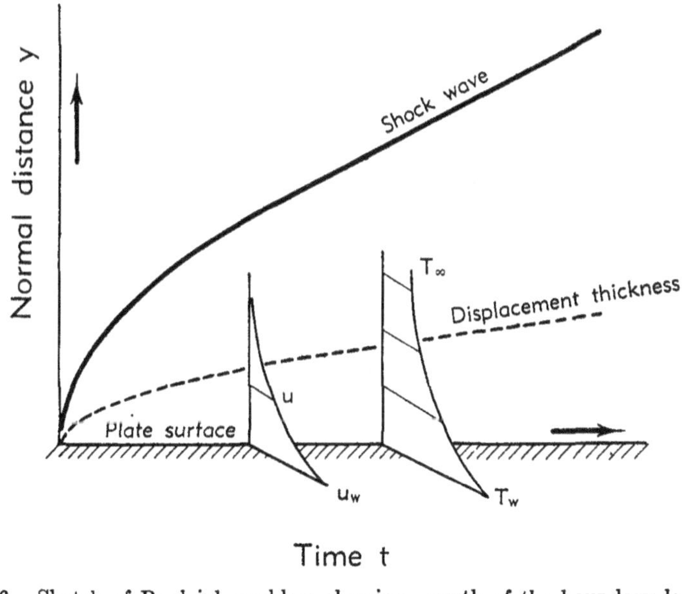

Fig. E,6. Sketch of Rayleigh problem showing growth of the boundary layer and, when applicable, the shock wave. Illustrative boundary layer profiles are shown.

when a doubly infinite plane surface is suddenly set in motion at a constant velocity in a direction parallel to the plane. The fluid is initially at rest. The boundary layer produced in this way increases in thickness as the square root of time, as illustrated in Fig. E,6. Of course, an equivalent problem is that of fluid and plate both moving at a constant initial velocity, and the plate being suddenly stopped. Apart from its academic importance, Rayleigh flow is qualitatively equivalent to the boundary layer of steady flow over a semi-infinite flat plate, and, being linear, provides a simple analogy to steady problems without pressure gradient.

Hypersonic Rayleigh flow may profitably be studied with this analogy in mind. Of course, in the hypersonic case, we should consider not only cases in which the suddenly imposed velocity is very large, but also the imposition of sudden changes of thermal boundary conditions at the plate surface.

Basic solutions. We now write the differential equations, specifying that all derivatives with respect to x (parallel to the plate) are zero. In the classical case, a similarity law connects the remaining variables t and y (normal to the plate). For a multicomponent mixture this is not possible in general, as we shall see. By way of trial, however, we shall assume that all quantities are functions of

$$z \equiv \frac{\zeta}{\sqrt{\tau}} \tag{6-1}$$

Then, taking $J = 1$ and neglecting thermal diffusion, the momentum and energy equations (Eq. 4-15 and 4-17) become

$$u'' + \tfrac{1}{2}zu' = 0 \tag{6-2}$$

$$H'' + (Pr - 1)(uu')' + (Le - 1)h_1^{(0)}c'' + \tfrac{1}{2}PrzH' = 0 \tag{6-3}$$

The continuity equation for atoms (Eq. 4-16) can be consistent with similarity only in the two extremes of equilibrium ($c = c_{eq}$) or frozen flow; for the latter case

$$c'' + \frac{1}{2}\frac{Pr}{Le}c' = 0 \tag{6-4}$$

Similarly, the catalytic wall condition (Eq. 4-19) can be consistent with similarity only in the fully catalytic ($c_w - c_{eq_w}$) or noncatalytic ($c_w' = 0$) extremes.

The solution of Eq. 6-2 for which $u(0) = u_w$, $u(\infty) = 0$ is

$$\frac{u}{u_w} = \frac{1}{\sqrt{\pi}}\int_z^\infty e^{-\omega^2/4}\,d\omega \equiv \operatorname{erfc}(2z) \tag{6-5}$$

The assumptions $Pr = Le = 1$ clearly simplify Eq. 6-3, providing the solution

$$\frac{H - H_\infty}{H_w - H_\infty} = \operatorname{erfc}(2z) \tag{6-6}$$

If we apply the definition of Nusselt number given in Eq. 5-24, replacing l by the quantity $\sqrt{\mu_\infty t/\rho_\infty}$ which, by Eq. 4-12, 4-13, and 6-1, is proportional to boundary layer thickness in the present problem,

$$Nu = (0.56)\left[1 + \frac{u_w^2 - c_w h_1^{(0)}}{H_\infty - H_{w_{\text{therm}}}}\right] \tag{6-7}$$

Eq. 6-5 through 6-7 are formally the same as those for an ideal compressible gas, and again we see the simplifying effect of the assumption of $Le = 1$. In the special case of frozen flow, this assumption is not needed; keeping $Pr = 1$, we may use Eq. 6-4 to show that

$$c = c_\infty + (c_w - c)_\infty \operatorname{erfc}\left(2 \frac{Z}{Le}\right) \tag{6-8}$$

and, in combination with Eq. 6-3, that

$$\frac{H - H_\infty - h_1^{(0)}(c - c_\infty)}{H_w - H_\infty - h_1^{(0)}(c_w - c_\infty)} = \operatorname{erfc}(2z) \tag{6-9}$$

It then may be shown that the Nusselt number in the frozen case is given by Eq. 6-7 for any Lewis number.

Sudden temperature change at stationary surface. So far, the results given for Rayleigh flow are general. Now, let us suppose that the plate remains at rest ($u_w = 0$), and that the gas, initially very hot, is quenched by a sudden drop in surface temperature. In principle, this situation can be produced experimentally by causing a strong shock to reflect from the cold end-plate of a shock-tube. As in the preceding discussion of Couette flow, we take $T_w = 500°K$ and $T_\infty = 4000°K$, and assume that $c \ll 1$.

If we assume equilibrium, so that $c_\infty \cong 0.02$ and $c_w = 0$, then Eq. 6-7 yields simply that $Nu = 0.56$, which may be compared with the corresponding Couette flow result, 0.65, in Eq. 5-26. Assuming frozen flow in an ideal dissociating gas [*12*], Eq. 6-7 becomes

$$(Nu)_{fr} = (0.56)(1 - 2.29c_w) \tag{6-10}$$

from which we find for the catalytic ($c_w = 0$) and noncatalytic ($c_w \cong 0.2$) cases

$$(Nu)_{fr} = 0.56 \text{ (cat) and } 0.31 \text{ (noncat)} \tag{6-11}$$

which are almost exactly the results obtained for Couette flow, Eq. 5-28 and 5-29 except that, in the present case, there is no dependence on Le. In this connection, it should be noted that Rayleigh flow, unlike Couette flow, has a boundary layer thickness controlled by diffusion processes. In fact, comparing Eq. 6-8 with Eq. 6-5 and 6-9, we see that, when $Le > 1$, the concentration profile is thicker than the velocity and enthalpy profiles. This thickening compensates for the greater atom diffusion coefficient, so that the number of atoms actually reaching the wall is unaffected by changes in Le.

Plate suddenly set in motion. Weak displacement effect. Next, we consider that the gas is originally cool, and a velocity is suddenly imparted to the plane surface. If the Mach number based on this velocity is high, then, just as in the Couette flow problem, there will be a region of hot, low density gas produced by viscous heating within the boundary layer. In the hypersonic limit $u_w^2/H_w = 2$, Eq. 6-7 gives $Nu = -0.56$. In the Rayleigh problem, the heated layer grows with time—one may liken the effect to a constant pressure explosion. The power of this explosion is, of course, diminished by cooling the plate, and also by the participation of vibrational excitation or dissociation which absorb part of the energy of the explosion. Even apart from real-gas effects, one may expect that the "explosion" described would generate a pressure wave in the gas outside the viscous layer, propagating in a direction normal to the surface. Further, the inertia of the gas moving outward under the action of this wave would be expected to cause a pressure rise on the surface.

It is convenient to describe the outward motion due to the boundary layer in terms of a displacement thickness. For this unsteady problem, the appropriate definition of displacement thickness is [20]

$$\delta^* \equiv \int_0^\infty \left(1 - \frac{\rho}{\rho_\infty}\right) dy = \sqrt{\frac{\mu_\infty t}{\rho_\infty}} \int_0^\infty \left(\frac{T}{T_\infty} - 1\right) dz \tag{6-12}$$

neglecting any dissociation. Assuming a completely ideal gas, with $\gamma = 1.4$, and defining a Mach number $M_\infty \equiv u_w/a_\infty$, we find from Eq. 6-5 and 6-6

$$\frac{T}{T_\infty} - 1 = \left(\frac{1}{5} M_\infty^2 + \frac{T_w}{T_\infty} - 1\right) \frac{u}{u_w} - \frac{1}{5} M_\infty^2 \left(\frac{u}{u_w}\right)^2 \tag{6-13}$$

or, in terms of the "recovery temperature" T_r that would be attained at an insulated surface,

$$\frac{T}{T_\infty} - 1 = \left(\frac{2}{5} M_\infty^2 + \frac{T_w - T_r}{T_\infty}\right) \frac{u}{u_w} - \frac{1}{5} M_\infty^2 \left(\frac{u}{u_w}\right)^2 \tag{6-14}$$

Inserting Eq. 6-5 into 6-14, and carrying out the integration indicated in Eq. 6-12 yields

$$\delta^* = \frac{2}{\sqrt{\pi}} \sqrt{\frac{\mu_\infty t}{\rho_\infty}} \left(\frac{T_w - T_r}{T_\infty} + \frac{\sqrt{2}}{5} M_\infty^2\right) \tag{6-15}$$

The outward velocity of this "displacement plane" is

$$v \equiv \frac{d\delta^*}{dt} = \frac{1}{\sqrt{\pi}} \sqrt{\frac{\mu_\infty}{\rho_\infty t}} \left(\frac{T_w - T_r}{T_\infty} + \frac{\sqrt{2}}{5} M_\infty^2\right) \tag{6-16}$$

Now, by simple acoustic theory, this value of v at the surface corresponds to a sound wave for which $v \sim (t - y/a_\infty)^{\frac{1}{2}}$, and for which the pressure rise is $\rho_\infty a_\infty v$. Using this information, we calculate an acoustic pressure rise at the plate surface of

$$\frac{p - p_\infty}{p_\infty} = \frac{\gamma(\gamma - 1)}{\sqrt{2\pi}} \left(1 - 0.71 \frac{T_r - T_w}{T_r - T_\infty}\right) \frac{M_\infty^3}{\sqrt{Re_t}}; \; Re_t \equiv \frac{\rho_\infty u_w^2 t}{\mu_\infty} \tag{6-17}$$

Apart from the term in $T_r - T_w$, this is the result obtained by Howarth [*21*] and by Van Dyke [*22*], who point out that the validity of Eq. 6-17 is restricted to large times (i.e., large Re_t).

A parameter equivalent to $M_\infty^3/\sqrt{Re_t}$ will appear subsequently in connection with the interaction between the boundary layer and leading edge shock for steady hypersonic flow over a flat plate. We see here a clear example of the importance in hypersonic flow of such interactions. If, say, $Re_t = 10^6$, the parameter $M_\infty^3 = \sqrt{Re_t}$ rises from the negligible value of 0.01 at $M_\infty = 2$ to unity at $M_\infty = 10$. It is important to note that the pressure rise is diminished if the wall is cooled ($T_w < T_r$). However, an overpressure remains even if $T_w/T_r \to 0$. In that case, the parenthesis of Eq. 6-17 equals 0.293 for large M_∞, as compared with unity in the case of an insulated surface.

Strong displacement effect. The result given in Eq. 6-17 is the leading term of an expansion in powers of $M_\infty^3/\sqrt{Re_t}$, and applies when this parameter is rather small. Higher order terms in the expansion have been obtained by Van Dyke [*22*]. As we have noted, the above parameter is not always small. At a Mach number of 22 and a Reynolds number of 10^6, the parameter has the value 10. Closer to the leading edge, the parameter can rise to very large values indeed. Therefore, it is important to obtain a boundary layer solution, valid when $M_\infty^3/\sqrt{Re_t}$ is large, to complement the information represented in Eq. 6-17. We shall provide here a brief account of such a development, essentially following the analysis of Stewartson [*23*]. However, we will not give a complete account of this so-called "strong interaction" problem in the context of Rayleigh flow. The rather arduous business of evaluating and justifying the approximations made is perhaps best saved for the problem of steady flow over a semi-infinite flat plate, which is a problem of more direct technical interest. Rather, so far as Rayleigh flow is concerned, we shall be chiefly concerned with the way in which the strong interaction similarity appears.

For short times, that is, shortly after the abrupt beginning of the plate motion, the shock wave produced by the thermal expansion of the boundary layer gas must be regarded as quite strong and certainly not acoustic. Curiously enough, however, there is no need to abandon the boundary layer assumption, even though Eq. 6-17 is invalid. Roughly, we are entitled to suspect that this is the case when we note that, even though the "interaction parameter" $M_\infty^3/\sqrt{Re_t}$ is very large, the streamline inclinations within the boundary layer, indicated by Eq. 6-16 to depend on $M_\infty^2/\sqrt{Re_t}$, still may be small.

In the paragraphs to follow we shall make the thin boundary layer assumption, deferring an assessment of its validity until a later section on steady flow over a flat plate. We shall assume that $Pr = 1$ and that viscosity is proportional to temperature, so that J is constant across the boundary layer. However, we must allow pressure to vary with time. Thus,

$$\rho\mu = (\rho\mu)_\infty \frac{p}{p_\infty} \quad \text{or} \quad J = \frac{p}{p_\infty} \tag{6-18}$$

where, as previously, subscript ∞ refers to conditions in the undisturbed gas.[5]

The boundary layer equations (Eq. 4-15 and 4-17) for momentum and energy conservation then become, for an ideal gas,

$$u_\tau = \frac{p}{p_\infty} u_{\zeta\zeta} \tag{6-19}$$

$$H_\tau - \frac{1}{\rho} p_\tau = \frac{p}{p_\infty} H_{\zeta\zeta}; \qquad H \cong c_p T + \tfrac{1}{2}u^2 \tag{6-20}$$

The appearance of pressure as a variable in Eq. 6-19 and 6-20 provides the essential difference between the present problem and the "weak interaction" problem discussed previously. We now attempt to form a similarity solution on the supposition that pressure varies as some power of τ:

$$\frac{p}{p_\infty} \equiv C\tau^n; \qquad \hat{\zeta} \equiv \tau^m C^{-\frac{1}{2}} \zeta \tag{6-21}$$

The exponents m and n will be determined subsequently. By inclusion of Eq. 6-21 into Eq. 6-19 and 6-20, we find the following differential equations

$$u'' - m\hat{\zeta}u' = 0 \tag{6-22}$$

$$H'' - m\hat{\zeta}H' + n\frac{\gamma - 1}{\gamma}(H - \tfrac{1}{2}u^2) = 0 \tag{6-23}$$

provided

$$2m + n + 1 = 0 \tag{6-24}$$

The boundary conditions are

$$\left.\begin{aligned} u(0) &= u_w;\; u(\infty) = H(\infty) = 0 \\ H(0) &= c_p T_w + \tfrac{1}{2}u_w^2 \end{aligned}\right\} \tag{6-25}$$

where, as is usual in boundary layer theory, we assume that we may represent the outer edge of the boundary layer as the limit $\zeta \to \infty$.

Furthermore, we make the crucial assumption that, to sufficient accuracy, H vanishes at the outer edge of the boundary layer. Of course, it can only vanish in the sense that the temperature at the outer edge of the boundary layer is much lower than temperatures occurring within

[5] Here we must take specific account of the pressure dependence of J, rather than setting $J = 1$. Nevertheless, a reference value different from $(\rho\mu)_\infty$ should be used in the final result, for best accuracy.

the boundary layer owing to viscous heating. A detailed justification of this assumption, which Eq. 6-23 shows is obviously necessary if a simple similarity rule is to hold, will be provided in the subsequent discussion of steady, flat plate flow.

The boundary layer described by Eq. 6-22 through 6-25 will have a displacement thickness

$$\delta^* = \int_0^\infty \left(1 - \frac{\rho}{\rho_1}\right) dy = \frac{\mu_\infty}{\rho_1} C^{\frac{1}{2}} \tau^{-m} \int_0^\infty \left(\frac{T}{T_1} - 1\right) d\hat{\zeta} \qquad (6\text{-}26)$$

where subscript 1 denotes conditions at the outer edge of the boundary layer. It is interesting that specifying $H_\infty = 0$ (or, that $T/T_1 \gg 1$ over most of the boundary layer) gives the "boundary layer thickness" a definite value. That is, inverting Eq. 4-12, and comparing with Eq. 6-26,

$$(y)_{\zeta=\infty} = \frac{\mu_\infty}{\rho_1} \int_0^\infty \frac{\rho_1}{\rho} d\zeta = \frac{\mu_\infty}{\rho_1} \int_0^\infty \frac{T}{T_1} d\zeta \cong \delta^*$$

In effect, under the hypersonic condition $T/T_1 \gg 1$, viscous heating is the chief determinant of boundary layer thickness, overshadowing the growth due to diffusion at the cool outer edge.

The growth of δ^* with time acts as a piston to drive a shock wave which, in the present instance, must be considered strong. Although the inviscid flow field between the boundary layer and the shock may be dealt with on a similarity basis in rather exact fashion, as shown by Stewartson [*23*], for illustrative purposes we shall use the "tangent-wedge approximation," also cited for this problem by Stewartson, in which it is assumed that the pressure rise on the piston face is approximately determined by the instantaneous velocity of the piston face, even though that velocity actually varies with time. In the limit of a very strong shock, we may write this pressure rise in the following form:

$$\frac{p}{p_\infty} \cong \frac{\gamma(\gamma + 1)}{2a_\infty^2} \left(\frac{d\delta^*}{dt}\right)^2 \qquad (6\text{-}27)$$

The solution we seek is essentially one of compatibility between the boundary layer solution, Eq. 6-22 to 6-25, and the inviscid flow model, Eq. 6-27.

We may evaluate the integral appearing in Eq. 6-26 by integrating Eq. 6-23 between $\hat{\zeta} = 0$ and ∞. Substituting the resulting relation into Eq. 6-26 yields

$$\frac{d\delta^*}{dt} = (\gamma - 1) \frac{m + n}{m + n \dfrac{\gamma - 1}{\gamma}} \frac{\tau^{-(m+n+1)}}{\sqrt{c}} \left[-\frac{H'(0)}{a_\infty^2} + \frac{m}{2} \int_0^\infty \left(\frac{u}{a_\infty}\right)^2 d\zeta \right] \qquad (6\text{-}28)$$

We see from Eq. 6-27 and 6-28 that the pressure should vary as $\tau^{-2(m+n+1)}$. However, in Eq. 6-21 we postulated that p varies as τ^n. Further, from Eq. 6-24, we already have one relation between m and n. From these three considerations we find that consistency requires

$$n = -\tfrac{1}{2}; \qquad m = -\tfrac{1}{4} \tag{6-29}$$

Thus the pressure varies with τ in the same way as found in Eq. 6-17. However, the boundary layer similarity is different. The "strong interaction" boundary layer is not parabolic, but quartic.

To evaluate the square bracket in Eq. 6-28 we first note that, having a definite value for m, we may write the solution of Eq. 6-22 as

$$\frac{u}{u_w} = \operatorname{erfc}\left(\frac{\hat{\zeta}}{2\sqrt{2}}\right) \tag{6-30}$$

Carrying out the integration indicated in Eq. 6-28, inserting the proper values of m and n, and substituting Eq. 6-28 and 6-21 into the shock relation (Eq. 6-27), we find finally that

$$\frac{p}{p_\infty} = 0.279\left[1 - 2\sqrt{\pi}\,\frac{c_pT'(0)}{u_w^2}\right]\frac{M_\infty^3}{\sqrt{Re_t}} \tag{6-31}$$

Apart from the heat transfer term in the square brackets, this is the result obtained by Stewartson [*23*] for the tangent-wedge approximation. In his more exact solution of the inviscid flow field, Eq. 6-31 applies except that the numerical constant is 0.304 rather than 0.279. Further, Stewartson has obtained the next term in the expansion. The numerical evaluation has been done with $\gamma = 1.4$.

In order to put Eq. 6-31 into a form more closely corresponding to Eq. 6-17, and to evaluate $T'(0)$, it is necessary to solve Eq. 6-23. First, we may rewrite Eq. 6-23 in terms of the variable T, making use of Eq. 6-30:

$$T'' + \tfrac{1}{4}\hat{\zeta}T' - \frac{\gamma-1}{2\gamma}\,T + \frac{\gamma-1}{2\pi}\,T_\infty M_\infty^2 e^{-\hat{\zeta}^2/4} = 0 \tag{6-32}$$

Stewartson has solved this equation exactly. For illustrative purposes, we may obtain nearly his result by the following approximation: We note that, in general, $(\gamma - 1)/2$ is a rather small quantity, and while it turns out to be quite inaccurate to neglect the third term of Eq. 6-32 on this account, one may successfully approximate this term by simply arranging that T go from its wall value to zero at $\hat{\zeta} = \infty$ in some suitable way. We choose the variation to be according to $e^{-\hat{\zeta}^2/4}$ in order that, when combined with the fourth term, a simple modification of the coefficient ensues.

$$T'' + \tfrac{1}{4}\hat{\zeta}T' \cong \frac{\gamma-1}{2\gamma}\left[T_w - \frac{\gamma M_\infty^2}{\pi}\,T_\infty\right]e^{-\hat{\zeta}^2/4} \tag{6-33}$$

One may quickly write down the solution of Eq. 6-33:

$$\frac{T}{T_\infty} \cong \left[\left(1 - \pi\frac{\gamma - 1}{2\gamma}\right)\frac{T_w}{T_\infty} + \frac{\gamma - 1}{2}M_\infty^2\right]\frac{u}{u_w} + \pi\frac{\gamma - 1}{2\gamma}\left(\frac{T_w}{T_\infty} - \frac{\gamma M_\infty^2}{\pi}\right)\left(\frac{u}{u_w}\right)^2 \tag{6-34}$$

If the conductive heat transfer to the surface is zero, the wall (recovery) temperature is, for $\gamma = 1.4$,

$$\frac{T_r}{T_\infty} = \frac{\gamma - 1}{2 + \pi\dfrac{\gamma - 1}{\gamma}}M_\infty^2 = 0.138M_\infty^2 \tag{6-35}$$

The exact value of this quantity, computed by Stewartson, has the numerical coefficient 0.132. From Eq. 6-34 and 6-35, we may rewrite Eq. 6-31 in a form corresponding to Eq. 6-17:

$$\frac{p}{p_\infty} = 0.279\left[1 - 0.362\left(1 - \frac{T_w}{T_r}\right)\right]\frac{M_\infty^3}{\sqrt{Re_t}} \tag{6-36}$$

Eq. 6-17 and 6-36 are of exactly the same form, provided that the recovery temperature is much greater than the temperature at the outer edge of the boundary layer, and the undisturbed pressure is much less than plate pressure; both these conditions are met in hypersonic flow. We see that maintaining the plate at a cold temperature below the recovery temperature (a practical necessity in hypersonic flow) has a smaller thinning effect on the boundary layer in the case of strong interaction than it does in the case of weak interaction.

Of course, there are uncertainties connected with the analysis described above. In particular, one must be sure of the validity of the boundary layer assumption for the strong interaction problem, and one must also evaluate the effect of the assumption that $H_\infty = 0$. Stewartson [*23*] has discussed these points, remarking that the question of the proper boundary condition on H at the outer edge of the boundary layer concerns the failure of the analysis at the very earliest times after initiation of the plate motion. Immediately after the plate begins to move, the rate of increase in boundary layer thickness is, in effect, too fast to permit a shock wave to form; that is, the velocity of the diffusion boundary is of the order $\sqrt{\mu_\infty/\rho_\infty t}$ which, for very small times, is larger than the sound speed a_∞ which controls the rate of formation of a shock wave. Thus, the present study does not apply when $\rho_\infty a_\infty^2 t/\mu_\infty$ is very small. During this early time, the thermal expansion of the heated air in the boundary layer is not able to keep pace with the growth of the boundary layer, and one has what amounts to a constant volume explosion corresponding, as Stewartson points out, to the "incompressible" solution of

Howarth applied for very small times. Also, during this period, rarefied gas effects may be anticipated.

E,7. Weak Nonequilibrium Waves. We return now to the study of real-gas effects in viscous flows, and recall that in our discussion of the Couette and Rayleigh problems we have considered that dissociation was either frozen or in equilibrium. Except for the brief discussion (p. 466) of the Hirschfelder sublayer, we have not displayed any results that apply for conditions intermediate between the frozen and the equilibrium state. The difficulty, of course, lies in finding sufficiently general analytical solutions of the equation for atom production (Eq. 4-16). Further, it does not seem possible to make any good approximations to the troublesome term involving W_1 that would insure tractability and yet preserve the general features of concentration profile in a large temperature gradient.

For one-dimensional unsteady cases, combining Eq. 3-31 and 4-16 yields

$$c_\tau = \frac{\mu_\infty}{\rho_\infty} \kappa (c_{eq}^2 - c^2) + c_{\zeta\zeta} \tag{7-1}$$

If $J = Pr = Le = 1$ and thermal diffusion is neglected. As we have seen (Eq. 5-31), the parameter governing the importance of the production term is

$$K_r = \frac{c_{eq}\mu_\infty \kappa \delta^2}{\rho_\infty} \tag{7-2}$$

where δ is the thickness of the viscous region measured in the ζ coordinate. If $K_r \gg 1$ (diffusion time $\ll$ relaxation time), then frozen flow occurs, while if $K_r \ll 1$ (diffusion time $\gg$ relaxation time), equilibrium prevails. Now, in many cases, including the Rayleigh problem, δ increases from zero indefinitely as time progresses. In such cases, there must therefore always be a period of frozenness and a subsequent period of equilibrium, these periods being joined by a nonequilibrium process during which K_r changes from small to large. In the present article, we shall analyze such equilibration processes in a class of weak waves for which Eq. 7-1 may be linearized.

As previously mentioned, Broadwell [*17*] has treated Couette flow and nonlinear boundary layer problems under a linearization of Eq. 7-1. In the present discussion, which follows [*24*], we shall deal with the consistently linearized problem of weak, one-dimensional, unsteady, viscous disturbances of a gas originally having an equilibrium level of dissociation at some appropriately high temperature. We shall imagine that the disturbance is introduced by some thermal action at a surface. In following this course, one may recall that linearized viscous flows are important for qualitative understanding; together with sound waves, they constitute the basic types of weak disturbance in a gas. A discussion of incompressible viscous waves from this standpoint has been provided by Lagerstrom elsewhere in this volume (B,6,2). Also, of course, weak viscous waves do occur in practical problems, particularly in connection with turbulence.

Perturbation equations. As previously, we shall assume that the potential energy $h_1^{(0)}$ is much greater than the energy in the internal degrees of freedom of the molecule, and that the dissociation level c is quite low. We shall write all physical quantities as perturbations from the initial state of the gas. For example, the perturbation in total enthalpy may be written as follows, using the symbol ~ to denote small quantities:

$$\tilde{H} = h_1^{(0)}\tilde{c} + c_p\tilde{T} \tag{7-3}$$

where, since small perturbations are involved, we may write c_p as a constant identified with the mean state of the gas. If small velocity disturbances are assumed to be of the same order as thermal changes, they make only second order contributions to energy, and therefore are not coupled with the thermochemical disturbance. We shall linearize the production term of Eq. 7-1 by writing

$$\tilde{c}_{eq} = \left(\frac{\partial c_{eq}}{\partial T}\right)\tilde{T} \equiv \tilde{\theta} \tag{7-4}$$

$$c - c_{eq} = \tilde{c} - \tilde{\theta} \tag{7-5}$$

where $\tilde{\theta}$ is a symbol introduced, for convenience, to represent a dimensionless temperature disturbance, and $\partial c_{eq}/\partial T$ is the local slope of the appropriate curve in Fig. E,5a. Then the basic perturbation forms of Eq. 7-1 and 4-17 are

$$\tilde{c}_\tau + 2\left(\frac{c_{eq}\mu\kappa}{\rho}\right)_\infty(\tilde{c} - \tilde{\theta}) - \tilde{c}_{\zeta\zeta} = 0 \tag{7-6}$$

$$\tilde{H}_\tau - \tilde{H}_{\zeta\zeta} = 0 \tag{7-7}$$

There are three boundary conditions to which the foregoing equations are subject: First, we produce the disturbance by some change of temperature at the surface $\zeta = 0$:

$$\tilde{\theta}(\tau, 0) = \tilde{\theta}_w(\tau) \tag{7-8}$$

Second, we require all disturbances to die out at large distances from the plate:

$$\tilde{H}(\tau, \infty) = \tilde{c}(\tau, \infty) = \tilde{\theta}(\tau, \infty) = 0 \tag{7-9}$$

and, third, we impose the catalycity condition (Eq. 4-19) at the surface:

$$\tilde{c}_\zeta(\tau, 0) = \Gamma\mathfrak{U}(\tilde{c} - \tilde{\theta})_w \tag{7-10}$$

Now, we introduce a set of transformations which will simplify the foregoing boundary-value problem:

$$\Phi \equiv \tilde{c} - \tilde{\theta}; \qquad \Psi \equiv \frac{\tilde{H}}{h_1^{(0)}} = \tilde{c} + \nu\tilde{\theta} \tag{7-11}$$

where

$$\nu \equiv \frac{c_p}{h_1^{(0)}(\partial c_{eq}/\partial T)_\infty} \tag{7-12}$$

and

$$\tilde{\tau} \equiv \sigma\tau; \qquad \tilde{\zeta} \equiv \sqrt{\sigma}\,\zeta; \qquad G \equiv \frac{(1+\nu)\Gamma\mathfrak{U}}{\sqrt{\sigma}} \tag{7-13}$$

where

$$\sigma \equiv \frac{1+\nu}{\nu} 2\left(\frac{c_{eq}\mu\kappa}{\rho}\right)_\infty \tag{7-14}$$

Under these transformations, Eq. 7-6 through 7-10 become

$$\Phi + \Phi_{\tilde{\tau}} - \Phi_{\tilde{\zeta}\tilde{\zeta}} = 0 \tag{7-15}$$

$$\Psi_{\tilde{\tau}} - \Psi_{\tilde{\zeta}\tilde{\zeta}} = 0 \tag{7-16}$$

$$(\Psi - \Phi)_w = (1+\nu)\tilde{\theta}_w(\tilde{\tau}) \tag{7-17}$$

$$(\Psi_{\tilde{\zeta}} + \nu\Phi_{\tilde{\zeta}})_w = G\Phi_w \tag{7-18}$$

$$\Phi_\infty = \Psi_\infty = 0 \tag{7-19}$$

The ideal-gas version of the problem set out above would be a simple question of heat conduction governed by the classical heat equation (Eq. 7-16) with a specified value of Ψ_w. The nonequilibrium complications involve a new dependent variable Φ, which represents the excess of atoms at any instant, and the parameters G and ν. G is proportional to the catalytic efficiency of the surface, while ν compares the energy absorbed, during a small temperature rise, by internal degrees of freedom ($cp\Delta T$) with that absorbed chemically ($h_1^{(0)}\Delta c_{eq}$) in an equilibrium process. Now, if the rest state of the gas is on the steep slope of one of the curves shown in Fig. E,5a, ν can be quite small; when $p_\infty = \frac{1}{4}$ atm, ν can be as small as about $\frac{1}{3}$. At lower pressure, the equilibrium curve is steeper and ν can be smaller; e.g., about $\frac{1}{8}$ when $p_\infty = \frac{1}{200}$ atm. Accordingly, we look for analytical simplifications based on an assumption that $\nu \ll 1$.

Moderately catalytic surface. While the foregoing argument may suffice to justify neglecting ν relative to unity, as in Eq. 7-17, Eq. 7-18 presents difficulty; we see from Eq. 7-11 that when $\tilde{c}$ is small (frozen flow), Ψ and $\nu\Phi$ are of the same order. For a reasonably catalytic wall ($G \gg \nu$), we can avoid this difficulty by expressing the problem in terms of $\tilde{c}$, Φ rather than Ψ, Φ. In this formulation, it is correct to set $\nu = 0$ wherever it appears:

$$\tilde{c}_{\tilde{\tau}} - \tilde{c}_{\xi\xi} = 0 \tag{7-20}$$

$$\left(\tilde{c} - \frac{1}{G}\tilde{c}_{\xi}\right)_{w} = \tilde{\theta}_{w} \tag{7-21}$$

$$\Phi_{w} = (\tilde{c} - \tilde{\theta})_{w} \tag{7-22}$$

These equations replace Eq. 7-16, 7-17, and 7-18. We realize the additional advantage that Eq. 7-21 may be solved first for $\tilde{c}$ without reference to Φ. Then, using Eq. 7-15 and 7-22, Φ may be found. The enthalpy fluctuation can then be reconstructed as $\Psi = \tilde{c} - \nu\Phi$, in which Φ should be evaluated

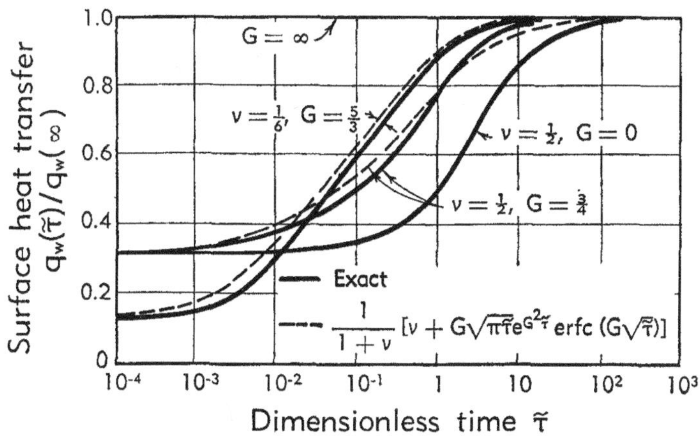

Fig. E,7a. Heat transfer at catalytic surface, for a small step-rise of surface temperature [24].

in the "frozen" limit, because it is only in that condition that $\nu\Phi$ can be appreciable in comparison with $\tilde{c}$.

Applying this theory to a "Rayleigh" problem, $\tilde{\theta}_w = 1(t)$, we find [24], for example, that the wall value of concentration equilibrates according to the formula

$$\tilde{c}_w = 1 - e^{G^2\tilde{\tau}} \operatorname{erfc}\,(G\sqrt{\tilde{\tau}}) \tag{7-23}$$

and the heat transfer at the surface is proportional to

$$-(\Psi_{\xi})_w = \frac{\nu}{\sqrt{\pi\tilde{\tau}}} + Ge^{G^2\tilde{\tau}} \operatorname{erfc}\,(G\sqrt{\tilde{\tau}}) \tag{7-24}$$

Clearly, large values of catalycity (G) tend to promote equilibration and increase heat transfer rate. Fig. E,7a shows the equilibration of heat transfer for various values of ν and G, compared with exact solutions [24] of Eq. 7-15 through 7-20. Agreement is excellent even for $\nu = \frac{1}{2}$, provided the quantity $1 + \nu$ is employed to make Eq. 7-24 correct in the limit $G \to \infty$.

Noncatalytic surface. When the wall is noncatalytic ($G \ll \nu$), we return to Eq. 7-15 to 7-19, setting $G = 0$. In seeking simple solutions in this case, it appears necessary to consider early and late times separately. For early time, when $\tilde{t}$ is small, Eq. 7-11 suggests that $\Psi \ll \Phi$. Making this assumption, the simple boundary condition $\Phi_w = -1(\tilde{\tau})$

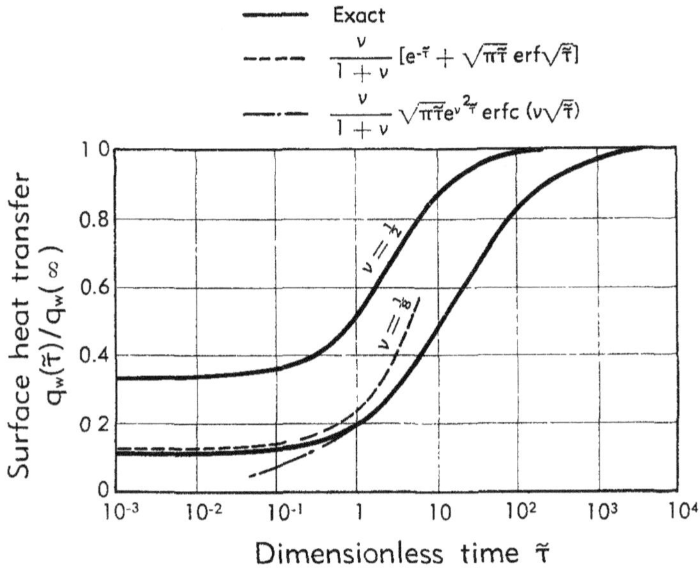

Fig. E,7b. Heat transfer at a noncatalytic surface, for a small step-rise of surface temperature [24].

applies to Eq. 7-15, and the solution yields the heat transfer rate

$$-(\Psi_\zeta)_w = \nu(\Phi_\zeta)_w = \nu\left(\frac{e^{-\tilde{\tau}}}{\sqrt{\pi\tilde{\tau}}} + 1 - \mathrm{erfc}\,\sqrt{\tilde{\tau}}\right) \tag{7-25}$$

In Fig. E,7b, Eq. 7-25 is compared with the exact solution, and is quite satisfactory for early time.

For late time, we have to deal with a Hirschfelder sublayer [15], wherein Φ may be expected to differ from zero only in the close vicinity of the surface. In such a situation, we may assume that normal derivatives dominate Eq. 7-15, so that $\Phi - \Phi_{\zeta\zeta} \cong 0$. Accordingly, Φ must vary like $e^{-\zeta}$ in the sublayer. This information permits the combination of Eq. 7-17 and 7-18 to yield a single boundary condition on Ψ:

$$\left(\Psi - \frac{1}{\nu}\Psi_\zeta\right)_w = \tilde{\theta}_w(\tilde{\tau}) \tag{7-26}$$

The boundary-value problem for Ψ presented by Eq. 7-16 and 7-26 is precisely that found for $\bar{c}$ in the catalytic case, if G is replaced by ν. By reference to Eq. 7-24 we therefore find

$$-(\psi_{\xi})_w = \nu e^{\nu^2\tau} \operatorname{erfc} (\nu \sqrt{\tau}) \tag{7-27}$$

Fig. E,7b shows good agreement between this formula and the heat transfer rate calculated exactly for this problem.

Fig. E,7a and E,7b show how greatly the speed of equilibration is affected by catalycity and the value of ν. For small values of ν (low pressures), a moderate value of G causes equilibration of heat transfer to occur in $\frac{1}{100}$ the time taken when $G = 0$.

E,8. Local Similarity in Steady Flow. So far, this section has been concerned with certain classical problems of fluid mechanics, especially Couette and Rayleigh flows. Analysis of these simple cases under hypersonic conditions provides information useful in the consideration of nonlinear hypersonic boundary layers. We proceed now to the study of plane or axially symmetric steady boundary layers governed by Eq. 4-7 through 4-11. In the present article, an approximate method of calculation which is suitable for a wide class of hypersonic boundary layers will be discussed. Then, in articles to follow, the specific problems of flat-plate and stagnation-point flows will be described.

Complete similarity. Eq. 4-7 through 4-11 are written in a form particularly suited to the analysis of "similar" flows, in which physical quantities are functions of a single independent variable, combining normal and streamwise coordinates. For such flows, the right-hand sides of Eq. 4-7 through 4-9 vanish and analysis is simplified. The conditions for similarity have been discussed by a number of writers [*8,25*] and the matter is reviewed quite thoroughly by Hayes and Probstein [*2*]. If f, g, and c/c_e are to be functions of η alone, Eq. 4-7 tells us that $2\xi(u_{e_\xi}/u_e)[(\rho_e/\rho) - f_\eta^2]$ must be a function of η alone. Also, J must be a function of η; this is satisfied if one has already decided to make the approximation $J = 1$. Similar restrictions apply to Lewis and Prandtl number. Eq. 4-8 is consistent with similarity only if the production term involving W_1 is constant. This is possible in general only for stagnation-point flow, for which $u_e^2 \sim \xi$. For other problems, it is necessary that dissociation be either frozen ($W_1 \cong 0$) or in equilibrium ($W_1 \cong \infty$). Eq. 4-8 also requires either that c_e be constant or some power of ξ.

Eq. 4-9 yields no new requirements if $Pr = Le = 1$. However, if $Pr \neq 1$, then u_e^2/H_e must be constant. This is true if the flow at the outer edge of the boundary layer (not just in the free stream) is hypersonic; then $u_e^2/H_e = 2$. If $Le \neq 1$, but we assume $h_1^{(0)}$ (a constant) $\gg h_1 - h_2$, then if H_e is constant, c_e must be constant. Eq. 4-11, like Eq. 4-8, is consistent with similarity only for stagnation-point flow, unless Γ is effectively zero or infinity.

The most crucial of the foregoing similarity requirements is that $2\xi(u_{e_\xi}/u_e)[(\rho_e/\rho) - f_\eta^2]$ be a function of η alone. In incompressible flow, $\rho_e/\rho = 1$, and u_e must be some power of ξ for similarity. In this way, the classical Falkner-Skan [*3*] (and B,8.2) family of problems is defined. For a compressible real gas, the equation of state (Eq. 2-2) and the enthalpy definitions (Eq. 3-26, 4-5, and 6-4) permit a fairly simple expression of ρ_e/ρ in terms of c and g, if the ideal gas assumption of Lighthill [*12*] is used. For the special case of a perfect gas, $\bar{h} = c_pT$, and

$$2\xi \frac{u_{e_\xi}}{u_e}\left(\frac{\rho_e}{\rho} - f_\eta^2\right) = \beta(\xi)(g - f_\eta^2) \tag{8-1}$$

where the quantity

$$\beta \equiv \left(\frac{H}{\bar{h}}\right)_e 2\xi \frac{u_{e_\xi}}{u_e} \tag{8-2}$$

must be constant for similarity. In this case, Eq. 8-2 may be solved to yield

$$\frac{u_e}{a_e} \sim \xi^{\beta/2} \tag{8-3}$$

provided H_e is, as usual, constant. Further, when the flow at the outer edge of the boundary layer is hypersonic ($H_e \cong \frac{1}{2}u_e^2$) and isentropic, it may be shown that Eq. 8-2 implies the following power law for pressure:

$$p \sim \left(\int_0^x r^{2k} dx\right)^{-\frac{1}{1+(\gamma-1)/\beta\gamma}} \tag{8-4}$$

In summary, for similar boundary layers in an ideal gas, setting $J = Pr = 1$, one has the two ordinary differential equations derived from Eq. 4-7 and 4-9:

$$f''' + ff'' + \beta(g - f'^2) = 0 \tag{8-5}$$

$$g'' + fg' = 0 \tag{8-6}$$

These equations, which are correct only if the outer flow happens to satisfy Eq. 8-3 or, when appropriate, Eq. 8-4, are subject to the boundary conditions

$$f(0) = f'(0) = 0; \qquad f'(\infty) = 1 \tag{8-7}$$

$$g(0) = H(0)/H_e; \qquad g(\infty) = 1 \tag{8-8}$$

Eq. 8-5 through 8-8 have been solved by Cohen and Reshotko [*26*] for various values of β ranging from $\beta = 0$ (the flat plate with zero-pressure gradient) through $\beta = \frac{1}{2}$ (axially symmetric stagnation point) and $\beta = 1$ (plane stagnation point) up to $\beta = 2$. Levy [*27*] and Li and Nagamatsu

[28] have also provided solutions of this system. Cohen and Reshotko's results for surface friction and heat transfer are shown as solid lines in Fig. E,8a and E,8b.

Apart from the specific cases of the flat plate and the stagnation point, the Falkner-Skan family of similar solutions has had only indirect importance in incompressible boundary layer theory, owing to the peculiar requirements of the incompressible condition $u_e \sim x^m$. There is no technically important surface configuration for which a constant fractional value of m applies over a significant range of x. For incompressible

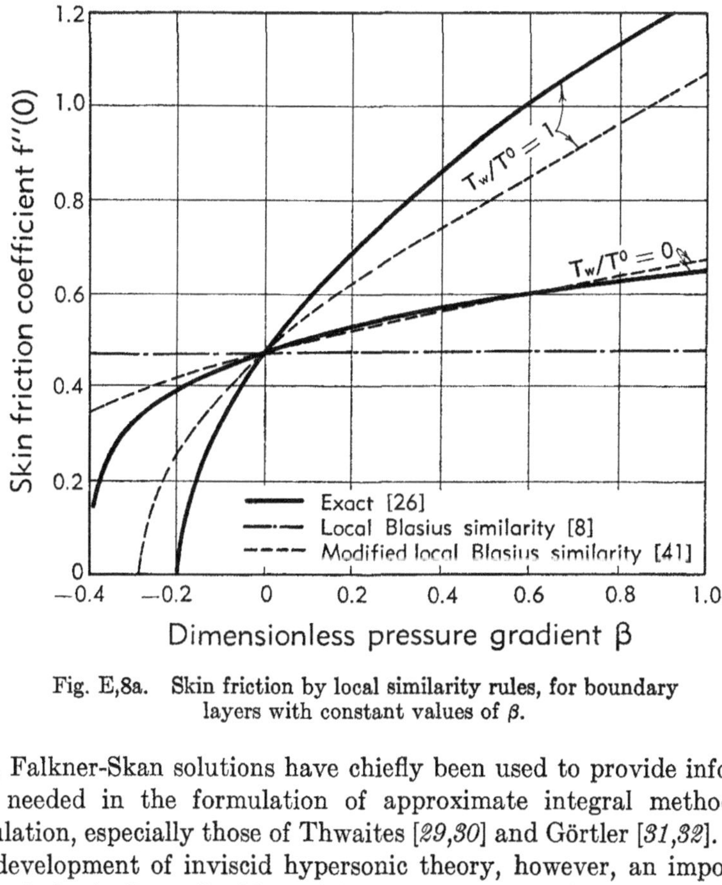

Fig. E,8a. **Skin friction by local similarity rules, for boundary layers with constant values of β.**

flow, Falkner-Skan solutions have chiefly been used to provide information needed in the formulation of approximate integral methods of calculation, especially those of Thwaites [29,30] and Görtler [31,32]. With the development of inviscid hypersonic theory, however, an important class of physically realizable problems appeared, to which the similarity theory, including fractional values of β, applies directly. Of course, the stagnation point is a member of this class; however, we refer more particularly to the studies of blunt power-law shapes intermediate between the wedge or cone and (for axially symmetric flow) the parabola [33]. According to "hypersonic small disturbance theory," such shapes produce inviscid flows with pressure variations of the type given in Eq. 8-4. Hayes and Probstein [2, Chap. 2] give a complete review of this body of theory. We see, then, that the theory of similar boundary layers corresponds in application to the inviscid small disturbance theory of power-law bodies. This correspondence extends to the "strong interaction" problem of a flat plate, wherein boundary layer displacement produces an equivalent body growing according to a $\frac{3}{4}$-power law. In addition to the direct applications of similarity solutions, these solutions also provide useful information for the construction of approximate methods for hypersonic boundary layer calculation [2]; examples are the integral method of Hayes [34] and the series expansion method of Ho and Probstein [35].

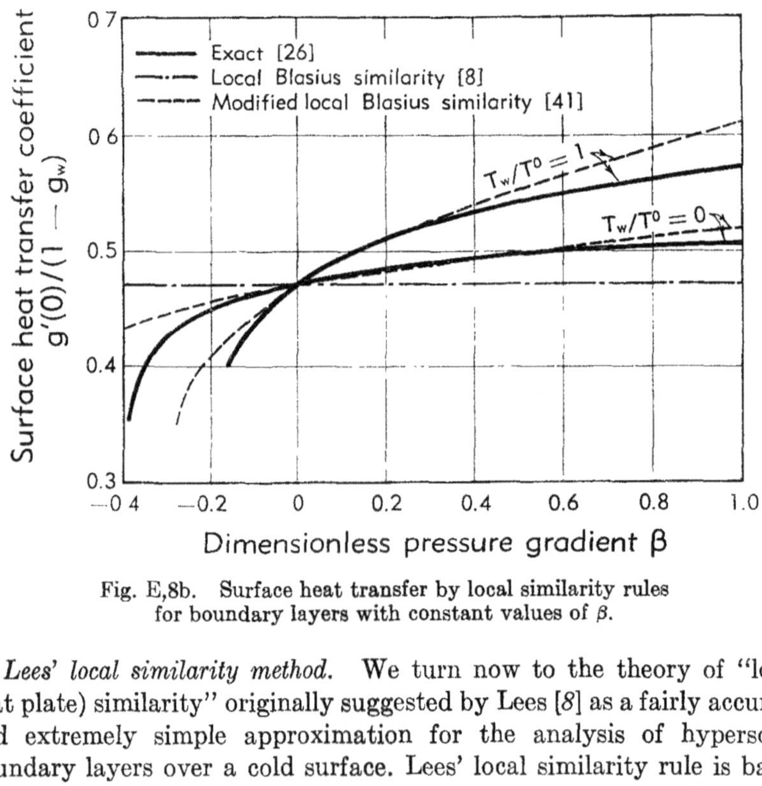

Fig. E,8b. Surface heat transfer by local similarity rules for boundary layers with constant values of β.

Lees' local similarity method. We turn now to the theory of "local (flat plate) similarity" originally suggested by Lees [8] as a fairly accurate and extremely simple approximation for the analysis of hypersonic boundary layers over a cold surface. Lees' local similarity rule is based on the idea that, in the typical hypersonic boundary layer, the wall is necessarily much cooler than gas in the interior of the boundary layer. Thus, at the surface, a high gas density prevails, and this dense gas is expected to be rather insensitive to the action of pressure gradient. The pressure gradient term of Eq. 4-7, rewritten in Eq. 8-1, is clearly diminished when the wall is assumed cold; in that case, the enthalpy profile function g has the same boundary values as f_η^2 (see Eq. 8-7 and 8-8), tending to make the difference $g - f_\eta^2$ small. Further, in certain cases, β may be quite small, even though the pressure gradient is substantial. Eq. 8-4 shows this effect clearly; when $(\gamma - 1)/\beta\gamma$ is unity, β itself, for $\gamma = 1.4$, is only 0.286. We shall return to this point subsequently.

The flat plate similarity rule, then, consists in neglecting the effect of pressure gradient in Eq. 4-7, or 8-5, so that boundary layer profiles are given by the Blasius function at each location on the surface. Thus, from Eq. 8-5 and 8-6, skin friction and surface heat transfer are given by the classical values:

$$f''(0) = \frac{g'(0)}{(1 - g_w)} = 0.470 \tag{8-9}$$

If we compare this flat plate similarity rule with the more general similarity calculations of Cohen and Reshotko [*26*], illustrated in Fig. E,8a and E,8b, we see that Eq. 8-9 is rather accurate for heat transfer to a cold wall over a range of β that includes stagnation-point values. Accuracy is not so satisfactory for skin friction, however. Extensive use has been made of Lees' local (flat plate) similarity rule for boundary layers on blunt bodies by Lees [*8,36*]; Kemp, Rose, and Detra [*37*], who included real-gas considerations; Cheng, et al. [*38*]; and, for three-dimensional flows, by Vaglio-Laurin [*39*], and Cheng [*40*]. In problems of the three-dimensional boundary layer, local Blasius similarity, which expresses the insensitivity of the boundary layer to streamwise pressure gradient, also implies insensitivity to pressure gradients transverse to streamlines, and therefore suggests that secondary flows (C,1) may be neglected.

Before proceeding further, we should observe that local Blasius similarity does not imply that the body is effectively a flat plate upon which the boundary layer grows parabolically. Rather, the given body is related to an equivalent flat plate by a coordinate transformation involving the local pressure. This transformation, in effect, accounts for variations of local Reynolds number per foot caused by changes in pressure, and may be derived from Eq. 4-1 and 4-2. For convenience, we shall specialize to plane flows ($k = 0$). Now, for flat plate flow, the similarity variable corresponding to η would be

$$\eta = \sqrt{\frac{(\rho\mu)_e}{2\mu_e x^*}} \int_0^y \frac{\rho}{\rho_e}\, dy \tag{8-10}$$

where x^* is measured along the flat plate. Comparing Eq. 4-1 and 8-10, and maintaining the assumption $J = 1$ so that $(\rho\mu)_e \sim p_e$, we find that distance along the equivalent flat plate is related to distance along the actual body by

$$x^* = \frac{1}{(pu)_e} \int_0^x (pu)_e dx \tag{8-11}$$

In a certain limiting sense which we shall now discuss, local flat plate similarity is a bona fide part of hypersonic flow theory. To develop this idea, following [*41*], we first write β in terms of physical variables, assuming that the boundary layer is sufficiently thin that its outer edge is effectively a streamline. On this basis, Eq. 8-2 yields

$$\beta = -2 \frac{1 + \frac{\gamma - 1}{2} M_e^2}{\gamma M_e^2 - 1} \left[\frac{(pu)' \int_0^x pu dx}{(pu)^2} \right]_e \tag{8-12}$$

The local flat plate similarity rule requires that the foregoing quantity be negligible. Let us assume that the quantity in square brackets is of unit order. Clearly then, β cannot be vanishingly small unless M_e is very large, and even then, only if

$$\frac{1}{M_e^2} + \frac{\gamma - 1}{2} \ll 1 \tag{8-13}$$

that is, both $1/M_e^2$ and $(\gamma - 1)/2$ must be very small. Now, $1/M_e^2 \ll 1$ for hypersonic flows over configurations that are not too blunt. $(\gamma - 1)/2$ is never zero, of course. For a diatomic gas, roughly, $0.1 < (\gamma - 1)/2 < 0.2$, the lower value corresponding to high temperature with many degrees of freedom excited. Nevertheless, the approximation $(\gamma - 1)/2 \ll 1$ plays an important role in the inviscid theory of hypersonic shock layers on blunt bodies (Freeman [*42*]), pointed bodies (Cheng [*43*]), and bodies that are slender in the sense of hypersonic small disturbance theory (Cole [*44*]).

Thus, under the assumption (Eq. 8-13), local flat plate similarity is the viscous counterpart of an extensive body of inviscid theory. Cheng's study [*38*] of the combined effects of bluntness and viscosity on a flat plate appeals specifically to this correspondence. On blunt bodies [*42*], local flat plate similarity, as theory, is excluded because M_e is not large in such problems. As we have seen, application of the rule to blunt bodies is nevertheless quite successful in a practical sense. At the other extreme, when the body is so slender that flow deflections are small [*44*], $[(\gamma - 1)/2]M_e^2 \ll 1$, and, by Eq. 8-13, the value of β is dominated by $(\gamma - 1)/2$.

A modified local similarity method. In inviscid theory, it is understood that the accuracy of the assumption $(\gamma - 1)/2 \ll 1$ requires verification. Various extensions of shock layer theory have been made to include higher order terms in this quantity [*42,43*], and the assumption has been found to be quite accurate. It is possible to add such a correction to the flat plate similarity rule, providing a test of the accuracy of Eq. 8-13, and, incidentally, improving the agreement of the rule with certain known solutions.

We return to the momentum equation (Eq. 4-7), with the pressure gradient term expressed by Eq. 8-1 for a perfect gas, and look for a new scale transformation $\xi^* \equiv D\xi$, which does a better job than ξ (Eq. 4-2) when β is small. In terms of the new variable, Eq. 4-7 becomes, for $J = 1$,

$$f''' + (D + \xi D' + ED)ff'' + D[\beta(g - f'^2) - Eff''] = 0 \quad (8\text{-}14)$$

in which a quantity $EDff''$ has been added and subtracted within the equation. Now, we may determine E so that the average across the boundary layer of the term in square brackets vanishes. Within that bracket we take f to be the Blasius function and g to be the Crocco integral of Eq. 8-6 and 8-8:

$$g = f' + g_w(1 - f') \quad (8\text{-}15)$$

permitting easy evaluation of E in term of β and the properties of the Blasius function. Next, we arrange that f satisfy the Blasius equation (on the average) by requiring

$$D + \xi D' + ED = 1 \quad (8\text{-}16)$$

Using Eq. 8-2, we may solve Eq. 8-16 to yield the corrected transformation

$$\xi^* = M_e^{-2\left(1+2.60\frac{T_w}{T_0}\right)} \int_0^\xi M_e^{2\left(1+2.60\frac{T_w}{T_0}\right)} d\xi \quad (8\text{-}17)$$

or, in place of Eq. 8-11,

$$x^*_{\text{mod}} = \frac{1}{(\rho u)_e M_e^{2\left(1+2.60\frac{T_w}{T_0}\right)}} \int_0^x (\rho u)_e M_e^{2\left(1+2.60\frac{T_w}{T_0}\right)} dx \quad (8\text{-}18)$$

as the scale transformation under which flat plate similarity applies locally.

Comparing Eq. 8-11 and 8-18, we see that the two forms do not approach each other in the limit of a cold wall $T_w/T_0 \to 0$, but, rather, when M_e tends to remain constant. By inviscid theory, this happens when the body is not too blunt and $(\gamma - 1)/2 \to 0$; and here again we see that local flat plate similarity is theoretically justified under the hypersonic assumptions of Eq. 8-13, rather than the assumption of a cold wall.

Eq. 8-18 applies only to velocity and skin friction. A different, and weaker, correction applies for enthalpy similarity. One finds [*41*] that the corrected scale function ξ^{**} is given by

$$\sqrt{\frac{\xi^{**}}{\xi}} = 1 - \frac{2}{5}\left(\sqrt{\frac{\xi}{\xi^*}} - 1\right) + \frac{1}{9}\,\xi^{-\frac{1}{3}} \int_0^\xi \left(\sqrt{\frac{\xi}{\xi^*}} - 1\right) \xi^{-\frac{2}{3}} d\xi \quad (8\text{-}19)$$

when ξ and ξ^* are nearly the same. Eq. 8-19 applies in the determination of surface heat transfer, in the same way that Eq. 8-17 applies for skin friction.

In view of Eq. 4-1, the relations between the modified [*41*] and original [8] flat plate similarity rules for the coefficients of skin friction $c_f \equiv 2\mu_w(u_y)_w/(\rho u^2)_e$ and surface heat transfer $c_h \equiv \mu_w(g_y)_w/[(\rho u)_e(1 - g_w)]$ are, respectively,

$$c_{f_{\text{mod}}} = (c_f)_{\text{orig}} \sqrt{\frac{\xi}{\xi^*}} = 0.470 \sqrt{\frac{2\mu_e}{(\rho u)_e x^*}} \sqrt{\frac{\xi}{\xi^*}} \tag{8-20}$$

$$c_{h_{\text{mod}}} = (c_h)_{\text{orig}} \sqrt{\frac{\xi}{\xi^{**}}} = 0.470 \sqrt{\frac{\mu_e}{2(\rho u)_e x^*}} \sqrt{\frac{\xi}{\xi^{**}}} \tag{8-21}$$

where x^* is the original quantity of Eq. 8-11. Fig. E,8a and E,8b show how the application of Eq. 8-20 and 8-21 improves the agreement of the flat plate similarity rule with the calculations of Cohen and Reshotko [26] for the similar "wedge flows" $(u/a)_e \sim \xi^{\beta/2}$. The improvement is especially evident for nonzero values of T_w/T_0, and is significant even for stagnation-point flow (to which the flat plate similarity theory does not really apply) and, to a degree, for adverse pressure gradient. Similar comparisons and conclusions may be drawn for other boundary layer quantities, such as displacement thickness [41], which, in the special case of small deflections in the outer flow $([(\gamma - 1)/2]M_e^2 \gg 1)$, may be written under modified local flat plate similarity as follows:

$$\delta^* = 0.470 \frac{\gamma - 1}{2} M_e^2 \sqrt{\frac{2\mu_e x^*}{(\rho u)_e}} \left[\left(1 + 2.60 \frac{T_w}{T_0}\right) \sqrt{\frac{\xi^{**}}{\xi}} - 3.60 \left(\sqrt{\frac{\xi^{**}}{\xi}} - \sqrt{\frac{\xi^*}{\xi}}\right)\right] \tag{8-22}$$

We shall have occasion to refer to the ideas of local flat plate similarity in subsequent articles, having in mind the extreme simplicity of the method and its surprising accuracy for hypersonic flow problems. Simplicity, especially, is a great virtue for an approximate method of hypersonic boundary layer calculation; in such cases, the appearance of chemical and physical complications, which cannot be described too confidently, often makes boundary layer theory exceedingly complicated and, at the same time, quite uncertain. In such cases, economy of theoretical effort at the expense of mathematical accuracy is well justified.

E,9. Steady Flow over a Semi-Infinite Flat Plate. So far, we have discussed a number of simple viscous flows in which chemical changes and other effects of heating by strong shock waves may be rather simply analyzed. Making appropriate use of the qualitative information furnished by these idealized problems and of the theory of local flat plate similarity, we now proceed to consider the two basic problems of the steady hypersonic boundary layer; the semi-infinite flat plate and, in the next section, the stagnation point. In each of these problems, we shall first consider ideal-gas aerodynamic interactions between the outer flow and the boundary layer, and then, separately, the boundary layer of a binary gas mixture with no interactions.

The hypersonic displacement effect. We begin our consideration of the hypersonic flat plate boundary layer with the problem of interaction between the boundary layer and outer flow. We have seen (Art. 6) that, in Rayleigh flow, the displacement effect of the boundary layer produced on a plane surface suddenly set in motion at a hypersonic speed causes a shock wave to form. This shock wave is, at first, very strong, and subsequently weakens to become a sound wave. In both the "strong" and

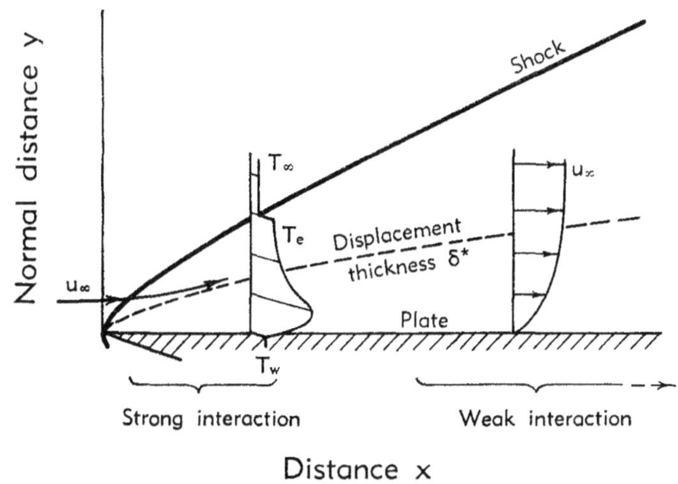

Fig. E,9a. Sketch of hypersonic flow over sharp-edged flat plates. Illustrative profiles of temperature and velocity are shown. Heavy line indicates typical streamline entering flow field through strong part of shock.

"weak" limits, surface pressure, skin friction, and heat transfer are affected by the presence of the shock. This description of the Rayleigh problem applies also to the steady flow problem illustrated in Fig. E,9a. Stewartson [*23,45*], in particular, has made effective use of the physical correspondence between these two problems.

Before proceeding with the details of strong and weak interactions, a few general observations concerning hypersonic displacement effects may be made. From Eq. 8-22 we may observe that displacement thickness can be very large indeed, owing to the high temperature and correspondingly low density prevailing through much of the boundary layer. That is,

$$\delta^* \sim (\gamma - 1) \frac{M_e^2}{\sqrt{Re_{e,x}}} x \tag{9-1}$$

where $Re_{e,x}$ is the Reynolds number based on conditions at the outer edge of the boundary layer and distance aft of the leading edge. It may also be shown that a consequence of the extreme viscous heating is that the outer edge of the boundary layer very nearly coincides with the displacement thickness, provided that the Mach number at the edge of the boundary layer is large. Making use of Eq. 4-1, the state equation, and the constancy of pressure across the boundary layer, we may write

$$y \cong \frac{\sqrt{2\xi}}{(\rho u)_e} \frac{T^0}{T_e} \int_0^\eta \frac{T}{T^0} \, d\eta \tag{9-2}$$

The general formula for displacement thickness, under the same transformation, is

$$\delta^* \equiv \int_0^\infty \left[1 - \frac{\rho u}{(\rho u)_e}\right] dy = \frac{\sqrt{2\xi}}{(\rho u)_e} \frac{T^0}{T_e} \int_0^\infty \left[\frac{T}{T^0} - \frac{u}{u_e}\frac{T_e}{T^0}\right] d\eta \tag{9-3}$$

Now, within the boundary layer, the temperature is of the same order as stagnation temperature T^0, whereas the temperature at the outer edge of the boundary layer is much smaller than the stagnation temperature, provided $M_e^2 \gg 1$. Thus, in Eq. 9-3, the contribution of the second term to the integral may be neglected in comparison to that of the first. Effectively, the temperature ratio T/T^0 approaches zero at the outer edge of the boundary layer, and we see that Eq. 9-2 and 9-3 are identical in the hypersonic limit $T_e/T^0 \to 0$, when $\eta \to \infty$ in Eq. 9-2. The result, according to which the "outer edge" of the boundary layer occurs at a definite height above the surface, is, of course, peculiar to hypersonic flow. Even in the hypersonic case, there must be an indefinitely thick part of the boundary layer associated with the fact that the quantity T/T^0 does not precisely approach zero at the outer edge of the boundary layer. However, this "indefinite part" is very much thinner than the "definite part" in which the temperature is very high and the density low.

Coincidence of the outer edge of the boundary layer and the displacement thickness means that, while the boundary layer entrains fluid from the outer flow in the usual manner, the growth of the boundary layer from this cause is negligible in comparison with the explosive growth due to viscous heating. Thus, the hypersonic boundary layer tends to be something rather like a cavity extending downstream from the leading edge, very hot, and containing very little mass flow in comparison with that which would be carried by a stream tube of the same size situated in the outer flow.

One must, of course, consider the possibility that with such a very thick boundary layer the Prandtl boundary layer theory may not be applicable. However, Shen [*46*] has shown that the assumptions of the Prandtl theory are met so long as $(\delta^*/x)^2$ remains sufficiently small. From Eq. 9-1, therefore, it is necessary that, when the Mach number is very high, the Reynolds number must be sufficiently high that

$$\left[(\gamma - 1)\frac{M_e^2}{\sqrt{Re_{e,x}}}\right]^2 \ll 1 \tag{9-4}$$

Weak interaction. Even though the assumptions of boundary layer theory may still apply in the hypersonic case, it will certainly be true in general that the large displacement effect of the boundary layer will induce important pressure changes on the flat plate. If we go sufficiently far downstream that this induced effect of displacement thickness is rather weak, we may use the Ackeret formula for the pressure increment associated with small flow deflections produced, in this case, by boundary layer displacement:

$$\frac{p}{p_\infty} = 1 + \gamma M_\infty \frac{d\delta^*}{dx} \tag{9-5}$$

into which we substitute, as a first approximation, Eq. 8-22 evaluated with all quantities at the outer edge of the boundary layer held constant at their free stream values:

$$\frac{p}{p_\infty} = 1 + 0.664\,\frac{\gamma(\gamma - 1)}{4}\left(1 + 2.60\,\frac{T_w}{T^0}\right)\chi \tag{9-6}$$

where the "interaction parameter" χ is defined as

$$\chi \equiv \frac{M_\infty^3}{\sqrt{Re_{\infty,x}}} \tag{9-7}$$

From Eq. 9-6, it is evident that the displacement effect on pressure is almost four times as great for an insulated plate surface as it is for a surface at a temperature low compared with stagnation temperature. The latter case, of course, is the more practical one for hypersonic flow.

We proceed now to calculate the changes in surface heat transfer and skin friction resulting from the self-induced pressure gradient described by Eq. 9-6. This problem was first treated by Lees and Probstein [*47*], and is discussed quite thoroughly (and more accurately) by Hayes and Probstein [*2*]. First, we should take note of the fact that what we propose to do is not generally proper in boundary layer theory, but has a special justification for hypersonic flow. It is indeed always possible to calculate the pressure gradient induced by displacement; however, for low speed flow, Kuo [*48*] and others show that calculations of induced changes in skin friction and heat transfer cannot, in general, be carried out within the confines of Prandtl boundary layer theory. However, in the hypersonic weak interaction problem, the boundary layer equations hold when $(M_\infty^2/\sqrt{Re_{\infty,x}})^2 = \chi^2/M_\infty^2 \ll 1$, while induced changes along the plate are of the order χ. Thus, if M_∞ is sufficiently large, induced effects depending on χ may be correctly treated with boundary layer theory.

We adopt Eq. 8-1 for use in Eq. 4-7 and, in Eq. 8-12, set $M_e = \infty$ and $u_e = \text{const}$, according to hypersonic small deflection theory [2, Chap. 2]. thus,

$$\beta \cong -\frac{\gamma - 1}{\gamma} \frac{p' \int_0^x p dx}{p^2} \tag{9-8}$$

Then, using Eq. 4-2 and 9-6, we find the approximate result

$$\beta \cong 0.664 \frac{(\gamma - 1)^2}{\gamma} \left(1 + 2.60 \frac{T_w}{T^0}\right) \frac{u_\infty M_\infty^3}{\sqrt{\xi}} \equiv \beta_1 \xi^{-\frac{1}{2}} \tag{9-9}$$

where β_1 is introduced for convenience. In view of the form of Eq. 9-9, we expand f and g in powers of $\xi^{-\frac{1}{2}}$:

$$f = f_0(\eta) + \beta_1 \xi^{-\frac{1}{2}} f_1(\eta) + \cdots \tag{9-10}$$

$$g = g_0(\eta) + \beta_1 \xi^{-\frac{1}{2}} g_1(\eta) + \cdots \tag{9-11}$$

The boundary conditions are that

$$f_n(0) = f_n'(0) = 0 \quad (\text{all } n) \tag{9-12}$$

$$f_0'(\infty) = g_0(\infty) = 1; \qquad g_0(0) = g_w \tag{9-13}$$

$$f_n'(\infty) = g_n(\infty) = g_n(0) = 0 \quad (n > 0) \tag{9-14}$$

Substituting Eq. 9-10 and 9-11 into Eq. 4-7 and 4-9 and assuming $J = Pr = 1$, we find that f_0 satisfies the Blasius equation

$$f_0''' - f_0 f_0'' = 0 \tag{9-15}$$

and

$$g_0 = g_w + (1 - g_w) f_0' \tag{9-16}$$

while the first order solution is governed by

$$f_1''' + f_0 f_1'' + f_0' f_1' + g_0 - f_0'^2 = 0 \tag{9-17}$$

$$g_1'' + f_0 g_1' + f_0' g_1 = 0 \tag{9-18}$$

The homogeneous character of the problem set by Eq. 9-14 and 9-18 implies the solution

$$g_1(\eta) = 0 \tag{9-19}$$

and, as pointed out by Maslen [49], Eq. 9-17 has the exact solution

$$f_1' = f_0' - f_0'' \int_0^\eta [g_w \eta + K + (1 - g_w) f_0] \frac{d\eta}{f_0''} \tag{9-20}$$

K is an arbitrary constant of integration of the type that would normally be fixed by the boundary condition at infinity. However, for large values of η,

$$f_1' \overset{a}{=} -\frac{1.22 g_w + K}{\eta - 1.22} \tag{9-21}$$

which satisfies the boundary condition $f_1'(\infty) = 0$ whatever the value of K. Of course, this situation is not unfamiliar in incompressible boundary layer theory; in particular, the Falkner-Skan problem for adverse pressure gradient exhibits such an indeterminacy. Various grounds may be cited for choosing K to make the numerator of Eq. 9-21 vanish, so that $f_1' \to 0$ exponentially, rather than as a simple pole. This question takes on a particular importance in hypersonic flow, which will be discussed in the next paragraph. If $K \equiv -1.22g_w$, Eq. 9-20 becomes

$$f_1' = f_0' - f_0'' \int_0^\eta [g_w(\eta - 1.22) + (1 - g_w)f_0] \frac{d\eta}{f_0''} \tag{9-22}$$

which may be used in the formula for velocity profile yielded by Eq. 9-9 and 9-10:

$$\frac{u}{u_\infty} = f_0' + 0.664 \frac{(\gamma - 1)^2}{8} \left(1 + 2.60 \frac{T_w}{T_0}\right) f_1' \chi \tag{9-23}$$

Eq. 9-19 and 9-23 essentially represent the solutions cited by Hayes and Probstein [2] for the weak interaction problem.

Let us return to the question of the ambiguous solution represented by the constant K. It is clear that the solution is one which asymptotically (Eq. 9-21) satisfies the inertia terms of the differential equations separately, that is, the second and third terms of Eq. 9-17, and thus includes an inviscid, constant pressure, vortical disturbance which should be considered a part of the given outer flow. If the outer flow is known not to include such a rotational disturbance, then we are justified in setting $K = -1.22g_w$. However, in hypersonic flow problems, one may not take the outer flow to be irrotational; in the flat plate problem, the boundary layer begins its development under the influence of a strong and highly curved shock wave which may be expected to produce large gradients of entropy and vorticity normal to the plate surface. These gradients are then convected downstream along the outer edge of the boundary layer, and may affect the boundary layer in an important way. It may easily be shown from Eq. 4-7 that a pattern of vorticity that is constant along streamlines parallel to the displacement thickness (i.e., a function of $y - \delta^*$) is always a possible solution of the boundary layer equations, and therefore, in principle, it is always possible to incorporate into a boundary layer solution the effects of vorticity existing at the outer edge of the boundary layer.

It is not always possible, however, to accomplish this matching of boundary layer solution with given outer vorticity patterns within a special similarity scheme for the boundary layer. Vortical outer flow is impossible under the strict similarity of the Blasius problem (Eq. 9-15). That is, it is not possible within the similarity assumption to have a part of the velocity distribution dependent on $y - \delta^*$ alone. To first order, Eq.

9-10 shows that such a solution is possible, though it must have the particular form $f_1' \sim 1/(\eta - 1.22)$, so that the quantity $\xi^{-\frac{1}{2}}$ is cancelled to give a function of $y - \delta^*$. Such a solution, however, is not necessarily consistent with the vorticity pattern as it is produced by the shock wave. In fact, it appears that the proper distribution of vorticity would vary like $(y - \delta^*)^{-\frac{3}{2}}$ and be proportional to χ^3. Accordingly, in the weak interaction problem, we should not expect a vorticity correction to be necessary up to second order in χ, and such a correction would be developed from terms like $\xi^{-\frac{3}{4}}$ in Eq. 9-10.

Finally, we may apply Eq. 9-19 and 9-23 to the calculation of weak interaction effects on the coefficients (based on free stream conditions) of skin friction and surface heat transfer:[6]

$$C_{f_\infty} = \frac{0.664}{\sqrt{Re_{\infty,x}}}\left[1 + 0.664\,\frac{(\gamma - 1)^2}{8}\left(1 + 2.60\,\frac{T_w}{T_0}\right)^2 \chi + \cdots\right] \tag{9-24}$$

$$C_{h_\infty} = 0 + \cdots \tag{9-25}$$

Thus, to first order in χ, surface heat transfer rate is unaffected by the weak interaction, while the skin friction increases in proportion to χ, about ten times more rapidly for an insulated surface as for the more practical case of a cold surface. The second order solution developed by Lees and Probstein [*47*] and Hayes and Probstein [*2*], will not be discussed here.

The problem of weak interaction is one for which we might expect that application of the modified local flat plate similarity scheme presented in the preceding section (Art. 7) would apply rather well. To apply Eq. 8-20 and 8-21, we first note that the coefficients based on free stream conditions (Eq. 9-24 and 8-25) are different from those based on outer edge conditions; if $u_e \cong u_\infty$ and $\rho\mu \sim p$,

$$\sqrt{Re_{\infty,x}}\, C_{f_\infty} = \sqrt{\frac{p}{p_\infty}}\,(\sqrt{Re_{e,x}}\, C_f) \tag{9-26}$$

Taking account of the fact that the perturbation dp/p_∞ is proportional to $x^{-\frac{1}{2}}$ (or $\xi^{-\frac{1}{2}}$) in the case of weak interaction (Eq. 9-6), we quickly find from Eq. 8-11 that

$$x^* = x\left(1 + \frac{dp}{p_\infty}\right) \tag{9-27}$$

Since the outer velocity is constant and the flow along the outer edge of the boundary layer is isentropic, $d(M_e^2)/M_e^2 = -[(\gamma - 1)/\gamma]dp/p_\infty$,

[6] As previously mentioned, we may expect to improve the accuracy of the result by selecting an appropriate reference condition for $\rho\mu$, modifying the values of $Re_{\infty,x}$ and χ.

and Eq. 8-17 yields

$$\xi^* = \xi\left[1 - \frac{\gamma - 1}{\gamma}\left(1 + 2.60\frac{T_w}{T_0}\right)\frac{dp}{p_\infty}\right] \tag{9-28}$$

From Eq. 8-19,

$$\xi^{**} = \xi \tag{9-29}$$

Combining Eq. 9-6 and 9-26 through 9-28 with Eq. 8-20 and 8-21, we obtain Eq. 9-24 and 9-25 precisely. Thus, the modified flat plate similarity rule is not only simple, but accurate, for weak pressure gradient. The modification correctly provides the term of Eq. 9-24 involving χ, which would not appear if the original rule were used.

Strong interaction. The foregoing analysis has been concerned with the flow so far downstream from the leading edge of the plate that the quantity χ is small, and interaction between bow shock and boundary layer is "weak." However, it is not necessary for the validity of the boundary layer assumption that χ be small. In fact, χ might be very large nearer the leading edge of the flat plate, and, provided the Mach number is sufficiently high, the criterion $(\delta^*/x)^2 \ll 1$, or Eq. 9-4, would still be satisfied. We may anticipate, from the previous discussion of the strong-shock Rayleigh problem (Art. 6), that we shall find the overpressure on the plate to be of the form $p/p_\infty \sim \chi$ (cf. Eq. 6-36). Now, we know that

$$Re_{e,x} = Re_{\infty,x}\left(\frac{\rho_e}{\rho_\infty}\right)^2\left(\frac{p_\infty}{p}\right) \tag{9-30}$$

when $J = 1$. Also, $M_e^2 \cong M_\infty^2(T_\infty/T_e)$, approximately. Substituting the foregoing information into Eq. 9-4, we find the criterion for satisfaction of the Prandtl assumptions to be

$$\left(\frac{\delta^*}{x}\right)^2 \sim \frac{(\gamma - 1)M_\infty}{\sqrt{Re_{\infty,x}}} = \frac{(\gamma - 1)\chi}{M_\infty^2} \ll 1 \tag{9-31}$$

Thus, we shall be concerned with flows in which χ is large compared with unity, but small compared with M_∞^2. This combination of requirements implies that the Reynolds number must be rather large. One may also infer from Eq. 9-31 that δ^* increases like $x^{\frac{3}{4}}$ in the strong-interaction zone. In the case when χ is large, Li and Nagamatsu [*28*] first showed that boundary layer similarity would hold, though under a law different from that which applies in the case of no interaction. Lees [*50*], Stewartson [*45*], and, more recently, Oguchi [*51*] have corrected and extended the theory of strong interaction.

In order to reach most expeditiously a first order strong-interaction solution, we shall essentially follow Lees' approximate analysis [*50*]. Inasmuch as the interaction is strong, we must employ a relation between pressure and displacement thickness appropriate for a strong shock. A useful approximation in this case is the tangent-wedge formula which we may write as follows

$$\frac{p}{p_\infty} = \frac{\gamma(\gamma+1)}{2} M_\infty^2 (\delta_x^*)^2 \tag{9-32}$$

Expecting that a similarity law will hold to first order for this problem, we anticipate that the applicable boundary layer equations will take the form given in Eq. 8-5 through 8-8, assuming $Pr = J = 1$. For such a similar boundary layer, making the hypersonic small deflection assumptions yields

$$\delta^* = \frac{\gamma-1}{2} M_\infty^2 \sqrt{\frac{2}{Re_{\infty,x}} \frac{p_\infty}{p}} \sqrt{x \int_0^\infty \frac{p}{p_\infty} dx} \int_0^\infty (g - f'^2) d\eta \tag{9-33}$$

which is, in a sense, equivalent to Eq. 8-22, except that, here, we do not attempt to represent g and f by means of the Blasius function. Eq. 9-32 and 9-33 are consistent under the supposition that $p \sim x^{-\frac{1}{2}}$, which permits elimination of δ^* between the two equations:

$$\frac{p}{p_\infty} = \frac{3}{8}(\gamma-1)\sqrt{2\gamma(\gamma+1)}\, \chi \int_0^\infty (g - f'^2) d\eta \tag{9-34}$$

Further, inasmuch as $p \sim x^{-\frac{1}{2}}$, Eq. 8-4 directly gives

$$\beta = \frac{(\gamma-1)}{\gamma} \tag{9-35}$$

for plane flow. The expressions for skin friction and surface heat transfer coefficients referred to free stream conditions

$$\left.\begin{matrix} C_{f_\infty} \\ 2C_{h_\infty} \end{matrix}\right\} = \sqrt{\frac{2}{Re_{\infty,x}} \frac{p}{p_\infty}} \sqrt{\frac{x}{\int_0^x \frac{p}{p_\infty} dx}} \left\{\begin{matrix} f''(0) \\ g'(0)/(1-g_w) \end{matrix}\right. \tag{9-36}$$

are analogous to Eq. 8-20 and 8-21 after Eq. 9-26 is applied. Now, the boundary-value problem posed by Eq. 8-5 through 8-8 and Eq. 9-35 is precisely that solved in much detail by Cohen and Reshotko [*26*]. For the case of a cold wall ($g_w = 0$), which is technically the most important, we find by interpolation of their results that the integral appearing in Eq. 9-34 has the value 0.354 for air ($\beta = 0.286$), and 0.329 for helium ($\beta = 0.4$). For air and a cold wall, therefore, Eq. 9-34 gives

$$\frac{p}{p_\infty} = 0.138\chi \tag{9-37}$$

and Eq. 9-34 yields

$$\left.\begin{matrix} C_{f_\infty} \\ 2C_{h_\infty} \end{matrix}\right\} = \left\{\begin{matrix} 0.20 \\ 0.18 \end{matrix}\right\} \sqrt{\frac{\chi}{Re_{\infty,x}}} \tag{9-38}$$

The results given in Eq. 9-37 and 9-38 may be regarded as the leading terms of asymptotic expansions in reciprocal powers of χ. For an insulated plate surface, Hayes and Probstein [*2*] suggest

$$\frac{p}{p_\infty} = 0.514\chi + 0.759 + \cdots \tag{9-39}$$

which shows the stronger interaction to be expected when the wall is not cold, and includes two terms of the asymptotic expansion.

In Lees' study [*50*], the viscosity formula

$$\rho\mu = (\rho\mu)_\infty C \frac{p}{p_\infty} \tag{9-40}$$

is used rather than Eq. 6-18, where C is to be chosen in such a way as to minimize the error entailed in assuming a linear viscosity-temperature dependence. This parameter may be accounted for [*10*] by replacing $Re_{\infty,x}$ by $Re_{\infty,x}/C$ throughout the present analysis. This change amounts to selecting an improved reference condition for the quantity $\rho\mu$, in accordance with previous discussion.

Concerning the application of local flat plate similarity to this problem, we find that local similarity in the sense originally used by Lees [*8,50*] overestimates the overpressure due to viscous interaction by about 30 per cent for the cold wall. On the other hand, the application of the modified local flat plate similarity (Eq. 8-22) of the previous section underestimates the pressure interaction by about 15 per cent. Thus, it is clear that even the modified local flat plate similarity rule is not altogether satisfactory for pressure gradients as strong as those encountered in the strong interaction problem.

It is obvious that a theory attempting to deal with the rather close vicinity of the leading edge of a flat plate must be carefully examined for error. One class of errors has to do with the application of Prandtl boundary layer theory. We have already noted that these are of the order $(\gamma - 1)M_\infty/\sqrt{Re_{\infty,x}}$, and we may add that errors due to slip flow may be expected to the same order [*50*]. The other errors that should be discussed are connected with the assumptions made concerning the flow outside the boundary layer. It is fortunate in this problem that similar solutions for hypersonic flow over power-law bodies exist [*45,33*], applicable for plane flow in a range of powers which includes that which describes the displacement surface during strong interaction, namely $\delta^* \sim x^{\frac{3}{4}}$. Means are therefore at hand to explore the details of the process whereby inner and outer flows may be correctly matched. This has been done by Stewartson [*45*], Oguchi [*51*], and Lees [*52*].

It develops from these studies that use of the tangent-wedge approximation for pressure (Eq. 9-32) results in an underestimate of pressure by about 10 per cent, for the insulated case.

Near the leading edge of the flat plate, the displacement interaction must produce a very strong, highly curved shock, and we should expect that streamlines coming from the strong-shock region proceeding downstream along the outer edge of the boundary layer would carry sharp gradients of vorticity and entropy. These effects have been neglected in the simple analysis described in the foregoing paragraphs. Further, by writing Eq. 9-33, one assumes that the velocity at the outer edge of the boundary layer is approximately constant and that $M_e^2 \gg 1$. This last assumption is equivalent to the neglect of "shock wave heating"; since $M_e^2 \sim T^0/T_e$ for hypersonic flow, one assumes that, even though the shock is strong, the temperature behind it (T_e) remains small compared with stagnation temperature (T^0). Now, if we trace an outer streamline back through the inviscid layer to the place where it crosses the shock (the dotted line in Fig. E,9a) we see that the "shock heating" is likely to be controlled by a part of the shock which is much stronger than that directly above the point of interest. Lees [*52*] has shown that the shock-heating error and the other errors mentioned above are physically related, and, in fact, are all of the order $(M_\infty/\sqrt{Re_{\infty,x}})^{1-2/(3\gamma)}$.

For an insulated plate, Oguchi [*51*] has corrected the strong-interaction theory to include terms of the order $(M_\infty/\sqrt{Re_{\infty,x}})^{1-2(3\gamma)}$:

$$\frac{p}{p_\infty} = 0.555\left[1 + 0.35\left(\frac{M_\infty}{\sqrt{Re_{\infty,x}}}\right)^{0.52}\right]\chi + \cdots \tag{9-41}$$

for air, in which $1 - 2/(3\gamma) = 0.52$. Apparently, this exponent was misprinted as 0.82 in [*51*]. The difference between the leading coefficients in Eq. 9-39 and 9-41 represents the error involved in using the tangent-wedge approximation (Eq. 9-32) for the flow between shock and boundary layer. For the strong-interaction case, validity of the Prandtl boundary layer theory requires that $(\gamma - 1)M_\infty/\sqrt{Re_{\infty,x}} \ll 1$. Thus, the order of the entropy correction is about half that of the correction that would be necessary to take account of errors involved in applying the Prandtl boundary layer theory, and should therefore be retained as part of the first order, strong-interaction solution.

Fig. E,9b shows the application of Eq. 9-6 and 9-37 to the problem of overpressure due to displacement on a sharp-edged flat plate with a cold surface, at a free stream Mach number of 20. The pressure falls off as $Re_{\infty,x}$ (distance from the leading edge) increases, changing over from the strong to the weak-interaction laws in a certain intermediate range of χ. Values of $(\delta^*/x)^2$ were computed directly from Eq. 8-22, using Eq. 9-30 and 9-37. For this case, $(\delta^*/x)^2$ reaches a value of $\frac{1}{10}$ at $Re_{\infty,x} \cong 10^3$; upstream of this location, the boundary layer theory would not apply, and slip effects may appear. The shock-heating factor, which represents the order of entropy effects, ranges between 10 and 100 per cent in the range of applicability of the strong-interaction theory, showing the importance that must be ascribed to corrections of this type.

Of course, leading edge bluntness also produces overpressure [*53*] and related effects on heat transfer. Combined effects of bluntness and displacement have been discussed by Lees [*52*], and, more recently, by Cheng, et al. [*38*], who used a combination of detached shock-layer theory

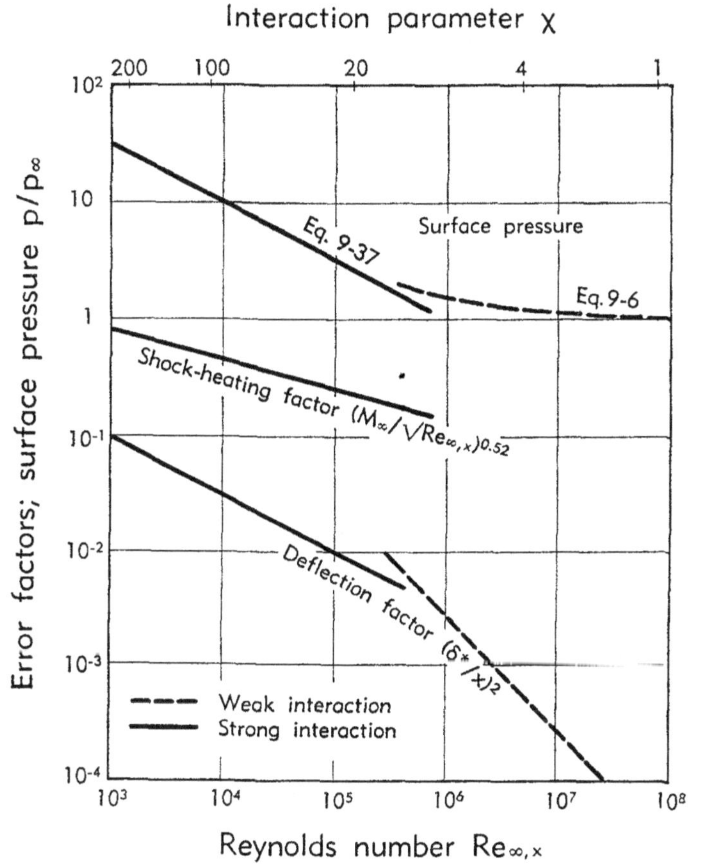

Fig. E,9b. Surface pressure caused by interaction of shock and boundary layer on semi-infinite flat plate, for $M_\infty = 20$. Also shown are pertinent error factors.

and the local flat plate similarity rule. Also, in [*38*], interaction theory for surface heat transfer has been compared with shock tunnel experiments, with rather good agreement. Yasuhara [*54*] has given a solution for strong interaction for a $\frac{3}{4}$-power body of revolution (see also [*2*]). In this particular case, as in the flat plate problem, boundary layer and outer flow [*33*] are both compatible with a single similarity scheme. When bluntness is involved, the "shock-heating" correction is larger than when the plate is sharp [*52*]. Generally, the more extensive strong-shock region associated with bluntness greatly enhances the importance of vorticity and entropy interactions, as emphasized by Ferri [*55*].

A considerable body of literature is concerned with the effects on the flat plate boundary layer of an outer shear flow, as might be produced by a curved shock. These effects are important (Ferri and Libby [*56*]) and, as we have seen, contribute to strong-interaction theory [*53*]. This class of problem is, of course, difficult, and many writers have attacked the apparently simple, idealized problem of incompressible flow over a sharp flat plate in the presence of a uniform outer shear flow, imagined to be produced by some agency upstream of the plate. This problem has not proved to be simple, however, and the literature on the subject contains much controversy. Recent and careful studies of this problem have been provided by Li [*57*], Murray [*58*], and Ting [*59*]. We shall not discuss this problem here; instead, we shall, in the next section, consider vorticity interaction at an axially symmetric stagnation point. This latter problem contains the features typical of vorticity interaction, and is more specifically related to hypersonic flow.

Multicomponent boundary layer of a flat plate. We continue our consideration of the flat plate boundary layer, turning now to problems of a multicomponent gas mixture. As we have seen in the more idealized Couette (Art. 5) and Rayleigh-type (Art. 6 and 7) problems, dissociation reactions and preferential diffusion of components produce many ramifications in boundary layer theory. Accordingly, we shall neglect all displacement effects and consider the pressure to be constant outside and within the boundary layer. Of course, we may anticipate that a theory based on these assumptions may often find application to cases involving pressure gradients through the theory of local flat plate similarity (Art. 8). Following the work of Baron [*60*] and Lees [*36*], we shall consider that gas may be added by blowing from the surface, as might be done in practice for thermal protection of the surface.

For present purposes, we return to Eq. 4-7 through 4-11, assuming $u_e = \text{const}$ and $K_T = 0$. It is clear from Eq. 4-8 that full similarity (e.g., $c = c(\eta)$) can be assumed only in the two extreme cases of frozen flow ($W_1 = 0$) or equilibrium flow ($c = c_{eq}(\rho, T)$). A similar situation exists in regard to the wall condition (Eq. 4-11) where, again, the form of the equation is compatible with similarity only in two limiting cases: the noncatalytic wall ($\Gamma = 0$) and the fully catalytic wall ($\Gamma = \infty$). We shall assume that J, Le, and Pr may be regarded as constants. The pertinent equations for similar flow on a flat plate under the foregoing restrictions are

$$Jf''' + ff'' = 0 \tag{9-42}$$

$$\frac{JLe}{Pr} c'' + fc' = 0 \text{ (froz), or } c = c_{eq} \text{ (eq)} \tag{9-43}$$

$$\frac{J}{Pr} g'' + fg' + \frac{h_1^{(0)}}{H_e} \frac{J}{Pr} (Le - 1)c'' + \frac{u_e^2}{H_e} J \left(1 - \frac{1}{Pr}\right) (f'f'')' = 0 \quad (9\text{-}44)$$

$$-\frac{q_w}{H_e} = \frac{(\rho\mu)_w u_e}{Pr\sqrt{2\xi}} \left[g' + \frac{h_1^{(0)}}{H_e} (Le - 1)c' \right]_w \quad (9\text{-}45)$$

$$c_w' = 0 \text{ (noncat), or } c_w = c_{eq_w} \text{ (fully cat)} \quad (9\text{-}46)$$

wherein we neglect $h_1 - h_2$ relative to $h_1^{(0)}$, as previously. For frozen flow, following Lees [36], we may replace Eq. 9-44 by a simpler equation for the partial enthalpy $g^* \equiv g - (h_1^{(0)}/H_e)c$ obtained by combining Eq. 9-43 and 9-44:

$$\frac{J}{Pr} g^{*\prime\prime} + fg^{*\prime} + \frac{u_e^2}{H_e} \left(1 - \frac{1}{Pr}\right) (f'f'')' = 0 \quad (9\text{-}47)$$

In the case of mass addition at the surface, we encounter the familiar restriction that the similarity law assumed determines the streamwise distribution of mass flow at the surface: the rate of efflux must vary inversely as $\sqrt{\xi}$; that is,

$$f(0) = -\frac{\rho_w \sqrt{2\xi}}{(\rho\mu u)_e} v_w \quad (9\text{-}48)$$

Of course, Eq. 8-48 deals with the total flow rate across the flat plate surface. We must, in addition, specify what gas it is that is being introduced. Throughout this section, a binary mixture will be assumed, and, in the event of mass addition at the surface, the added gas will be that designated by subscript 1. Thus, toward the surface, we require that the diffusional velocity of component 2 exactly balance the mass-averaged velocity v_w:

$$\left(v - \mathfrak{D}_{12} \frac{c_{2y}}{c_2}\right)_w = 0 = (1 - c) \frac{Pr}{JLe} f(0) - c_w' \quad (9\text{-}49)$$

If $J = 1$, Eq. 9-42 is simply the Blasius equation, and we take it to be subject to a blowing boundary condition $f(0) = \text{const}$, together with the usual conditions $f'(0) = 0$, $f'(\infty) = 1$. This boundary-value problem has been solved by Emmons and Leigh [61], who give a tabulation of the function f and its derivatives for various values of $f(0)$. Assuming also that $Le = Pr = 1$, and having in mind either a purely frozen or a purely equilibrium gas mixture in the boundary layer, Eq. 9-43 and 9-44 have the simple solutions

$$c = c_e + (c_w - c_e)(1 - f'), \text{ or } c = c_{eq} \quad (9\text{-}50)$$

$$g = 1 + (g_w - 1)(1 - f') \quad (9\text{-}51)$$

We first consider the hypersonic boundary layer of a diatomic gas, with no mass addition at the surface and with an outer temperature sufficiently low that $c(\infty) = 0$. In addition, the wall temperature is sufficiently low that $c_{eq}(0) = 0$. Let us contrast the solutions then for the equilibrium and the frozen boundary layers over a fully catalytic surface. Since atom concentration vanishes at the wall and in the stream, Eq. 9-50 shows that c is identically zero throughout the boundary layer in the frozen case. In the equilibrium case, c would be determined by the mass action law. From Eq. 9-51, the solution for g will be the same, regardless of whether the flow is frozen or equilibrium. From Eq. 9-45, we see that the heat transfer is also the same for the two cases if $Le = 1$. Of course, in the interior of the boundary layer, there are differences between the frozen and equilibrium solutions. In the equilibrium case, there is an exchange between dissociation energy and internal energy which, while it does not affect the surface heat transfer rate, does affect boundary layer quantities such as displacement thickness. The insensitivity of surface heat transfer to reaction rate, so long as the surface is fully catalytic, has previously been noted in the cases of Couette (Art. 5) and Rayleigh (Art. 6) flows.

Let us now imagine that a gas species 1 is introduced at the surface. This species will not be considered to react with the primary gas in the boundary layer, however, and we may consider the mixture to be chemically frozen, with $c(\infty) = 0$. Combining Eq. 9-49 and 9-50, we find the solution for c to be

$$c = \frac{1}{1 + \dfrac{f''(0)}{-f(0)}} [1 - f'(\eta)] \tag{9-52}$$

Thus, the surface concentration of the added species increases with $-f(0)$, as shown in Fig. E,9c. The value $c_w = 1$ occurs at the "blow-off" limit $f''(0) = 0$, which corresponds to $f(0) \cong -0.88$. Also shown is a calculation of Baron [*60*] showing the effect of variable J when the two gas components are quite dissimilar. Now, the surface heat transfer for this problem is

$$-\frac{q_w}{H_e} = \frac{(\rho\mu)_w u_e}{\sqrt{2\xi}} (1 - g_w) f''(0) \tag{9-53}$$

as blowing increases, $f''(0)$ decreases, and mass addition thus tends to reduce heat transfer as well as skin friction [*62*]. If species 1 is a very light gas, and the blowing rate is sufficient to produce a substantial surface concentration, then heat transfer is materially reduced by the fact that $\rho\mu$ is very small at the surface. This result is indicated by Eq. 9-53; however, this may not be taken to be a result of the foregoing analysis, in which the assumption $J = 1$ has been made. Lightness of the added gas also tends to give a large value of g_w, for a given surface temperature, and thus a reduced enthalpy potential $g_e - g_w$. This effect further tends to reduce surface heat transfer rate.

The system represented by Eq. 9-40 and 9-45 has been studied extensively, with variable properties taken into account; that is, with Pr and J under the differentiation sign, as in Eq. 4-7 and 4-9. Work by Crocco [*63*], Van Driest [*64*], Romig and Dore [*65*], and others[7] shows that variability of J is most important, while constant values of Pr and specific heats are quite acceptable. Now, in our discussion of Couette flow (Art. 5) between two cool plates in relative motion, we showed that heat transfer is rather accurately given by an analysis with $J = 1$, provided $\rho\mu$ is evaluated at

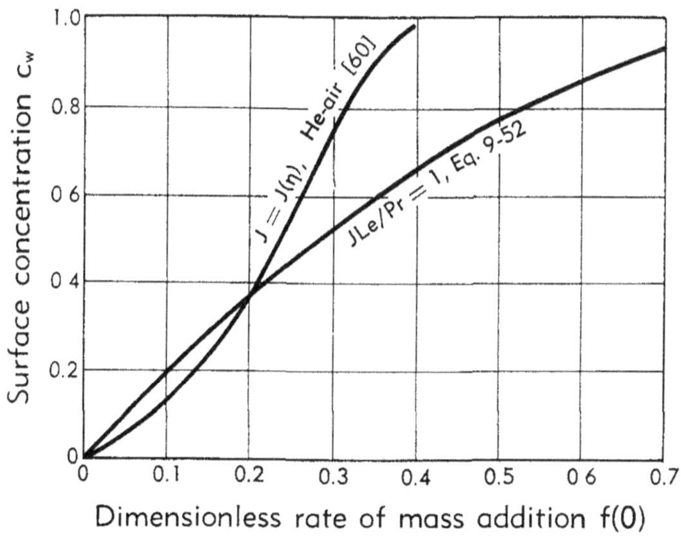

Fig. E,9c. Surface concentration of species introduced into boundary layer by efflux from surface of flat plate.

nearly the maximum enthalpy in the boundary layer. This result corresponds almost exactly with that obtained by application by Eckert's recommendation [*66*] that $\rho\mu$ be evaluated at the reference enthalpy

$$\bar{h}_m \equiv 0.5(\bar{h}_e - \bar{h}_w) + 0.22(\bar{h}_r - \bar{h}_e) \tag{9-54}$$

Further, Romig and Dore [*65*] found that their heat transfer results for an equilibrium real-gas boundary layer were reproduced quite accurately by the "incompressible" solution with $\rho\mu$ evaluated at $\bar{h}_m$. The foregoing reference method cannot be said to apply to a mixture of highly dissimilar gases, and we do know from Baron's work [*60*] that large errors are incurred when $J = 1$ is assumed for such a mixture.

By rather simple calculations, we may indicate the roles played by Prandtl number and Lewis number in the constant pressure boundary layer, usually found by numerical computation. We keep the specification $J = 1$ and consider first the frozen boundary layer. If we neglect the

[7] See the review of Hayes and Probstein [*2*].

third term of Eq. 9-47; i.e., assume $u_e^2/H_e \ll 1$, the solution of Eq. 9-47 for $f(0) = 0$ is approximately

$$g^* = (g_w^* - g_e^*)[1 - f'(Pr^{1/3}\eta)]g_e^* \tag{9-55}$$

and similarly for c,

$$c = (c_w - c_e)\left[1 - f'\left(\frac{Pr^{1/3}}{Le^{1/3}}\eta\right)\right] + c_e \tag{9-56}$$

These solutions are not exact. They represent a well-known approximation in compressible boundary layer theory which we may justify in the following way: If we introduce into Eq. 9-47 the transformation $\hat{\eta} \equiv Pr^{1/3}\eta$, we may write the differential equation in the form

$$g^{*\prime\prime}(\hat{\eta}) + f(\hat{\eta})g^{*\prime}(\hat{\eta}) = [f(\hat{\eta}) - Pr^{2/3}f(Pr^{-1/3}\hat{\eta})]g^{*\prime}(\hat{\eta}) \tag{9-57}$$

In effect, we may say that solution Eq. 9-53 is obtained by neglecting the right-hand side of Eq. 9-57. Let us show that the average across the boundary layer of the right-hand error term is small, if we assume that the Prandtl number is near one. For this purpose, we may assume that $g^{*\prime}$ is nearly proportional to f'', and by expanding in powers of $1 - Pr$, we find an expression for the right side of Eq. 9-57 in the form

$$-\tfrac{2}{3}(1 - Pr)\left(f - \frac{\hat{\eta}}{2}f'\right)f'' + \text{order } [(1 - Pr)^2] \tag{9-58}$$

which, when integrated across the boundary layer, yields an average value of $-0.045\ (1 - Pr)$. For Pr about 0.7, this average is very small indeed as compared with the average values of the terms on the left of Eq. 9-57, which are of unit order. Eq. 9-56 may be justified in a similar manner.

Now, let us estimate the effect of the third term of Eq. 9-47 in the hypersonic limit $u_e^2/H_e = 2$. For $1 - Pr \ll 1$, we may write

$$g^{*\prime\prime} + fg^{*\prime} = 2(1 - Pr)(f'f'')' \tag{9-59}$$

where, using the previous result, $\hat{\eta}$ is the independent variable. Eq. 9-59 has a simple exact solution, by which one finds

$$g^{*\prime}(0) = [g_e^* - \tfrac{1}{2}(1 - Pr) - g_w^*]f''(0) \tag{9-60}$$

Using Eq. 9-60 and 9-56 in Eq. 9-45, and making the identification

$$1 - \tfrac{1}{2}(1 - Pr) \cong \sqrt{Pr} \equiv g_r \tag{9-61}$$

one finds, for zero mass addition [*8*,*36*],

$$-\frac{q_w}{H_e} = \frac{\rho\mu}{\sqrt{2\xi}}u_e f''(0)Pr^{-2/3}(g_r - g_w)\left[1 + (Le^{2/3} - 1)\frac{h_1^{(0)}}{H_e}\frac{c_e - c_w}{g_r - g_w}\right] \tag{9-62}$$

Here, the usual "recovery factor" $\sqrt{Pr}$ defining g_r is found only for small values of $1 - Pr$, whereas, of course, numerical calculations give this result over an extended range of Pr. Eq. 9-62 applies also in the low speed case $u_e^2/H_e \ll 1$, except that g_r should be replaced by 1. This result may be verified by comparing the derivative of Eq. 9-55 with Eq. 9-62. In Eq. 9-62, and in subsequent expressions for constant pressure heat transfer, we have in mind that $\rho\mu$, appearing both explicitly and in the variable ξ, is to be evaluated at the appropriate reference condition. The quantity $c_e - c_w$ may be determined by combining the catalytic condition (Eq. 9-44) and the concentration equation (Eq. 9-56). Eq. 9-61, when compared with Eq. 9-53, shows the effects of Prandtl and Lewis number variations for frozen flow.

For an equilibrium boundary layer over a cold wall, so that $c'(0) = 0$, a similar procedure, beginning in this case with Eq. 9-44, gives for zero mass addition

$$-\frac{q_w}{H_e} = \frac{\rho\mu}{\sqrt{2\xi}} u_e f''(0) Pr^{-\frac{2}{3}}(g_r - g_w) \tag{9-63}$$

which is the same as the frozen result (Eq. 9-62) with the term bearing $Le^{\frac{2}{3}} - 1$ omitted. Actually, it is possible to include, approximately, the Lewis number effect in Eq. 9-63. We shall defer making this change until reaching the next section on stagnation-point flow, where a standard of comparison is available from the work of Fay and Riddell [*9*].

Returning to the frozen boundary layer, now with mass addition at the rate $-f(0) = 0.5$, we may repeat the previous procedure, defining $\hat{\eta} \equiv Pr^n\eta$ and deriving a value of n for which g^* depends approximately on $f'(\hat{\eta})$. We find, in this case, that $n \cong 0$, which leads to Lees' result for "moderate" blowing [*36*]:

$$-\frac{q_w}{H_e} = \frac{\rho\mu}{\sqrt{2\xi}} u_e f''(0) Pr^{-1}(1 - g_w) \left[1 + (Le - 1)\frac{h_1^{(0)}}{H_e}\frac{c_e - c_w}{1 - g_w}\right] \tag{9-64}$$

In [*36*], a more detailed discussion of heat transfer from boundary layers with mass addition or chemical reaction is provided, with emphasis on problems for which Blasius similarity applies. In Baron's study of nonreacting mixtures with mass addition [*60*], it is pointed out that, when mass addition occurs by an evaporative process, one must apply a boundary condition specifying a connection between surface mass transfer and heat transfer, of the form

$$(g^{*\prime} - \Lambda f)_w = 0 \tag{9-65}$$

where Λ is a coefficient having to do with heat of evaporation, for example. Eq. 9-65 would replace the previous specification of $f(0)$ alone. Use of Eq. 9-65 is simple enough so far as boundary layer theory is concerned, but making the proper choice of Λ for a given problem is a difficult problem of physical chemistry.

Finally, we may reemphasize that the problem of steady, nonequilibrium flat plate flow has not been solved except under the stringent assumption of a linearized rate law made by Broadwell [*17*].

E,10. Steady Flow at a Stagnation Point. One of the classical problems of laminar boundary layer theory concerns the flow against a vertical plane, producing a two-dimensional or axially symmetric stagnation point (B,63) depending upon the symmetry assumed. The boundary layer solution concerns the viscous flow away from the stagnation point along the vertical surface, and in this problem the boundary layer has nonvanishing thickness at its starting point. Theoretically, one of the most important features of the boundary layer in the vicinity of the stagnation point is that it may be analyzed by separation of variables. That is, flow quantities near the stagnation point are regular functions of distance x away from the starting point, and, hence, have Taylor series about $x = 0$. The boundary layer solution then is simply a matter of finding those functions of normal distance, y, which constitute the leading terms of a Taylor series in x. This Taylor-series solution does not depend on the Prandtl boundary layer assumptions; exact solutions of the Navier-Stokes equations for this problem may be found which, in the limit of small viscosity, are consistent with the Prandtl assumptions.

Stagnation-point flow is of special practical interest for hypersonic flight, because of the necessity to provide some degree of leading edge or nose bluntness in order to control maximum surface heat transfer rate. As in the case of the flat plate, we shall begin by discussing a modification of classical stagnation-point theory that depends simply on high Mach number in the free stream, even if the boundary layer is composed of an ideal gas. Then, we shall consider the stagnation-point boundary layer in which the gas may be dissociated.

Vorticity interaction at a stagnation point. In the problem of the hypersonic flow over a flat plate (Art. 9), we found that even if the gas in the boundary layer may be considered ideal, certain interactions with the outer flow must be considered when the free stream Mach number is very high. In the flat plate problem, a pressure interaction predominated, with entropy and vorticity interactions playing a smaller role. The chief interaction of interest for hypersonic stagnation-point flow is the response of the boundary layer to vorticity produced by a curved bow shock. According to Fig. E,10a, we imagine a blunt body having a local radius of curvature L at the stagnation point[8] with a very strong shock

[8] Subsequently, it will appear that L may be interpreted by including both body radius and displacement thickness of the boundary layer. We exclude cases of flat-faced bodies, about which little boundary layer information exists.

wave lying close to the body surface, "standing off" by a distance approximately equal to ϵL, where $\epsilon \equiv (\gamma - 1)/(\gamma + 1)$, which will be regarded as a rather small quantity. The gas density behind the shock is nearly constant at the strong-shock limit, and the flow between shock and boundary layer may be approximated by incompressible flow about a sphere [*2*,*12*].

Now, when fluid elements cross the shock wave, they acquire vorticity, owing to the local curvature of the shock. The circulation about each fluid particle is then conserved as it travels the subsequent subsonic

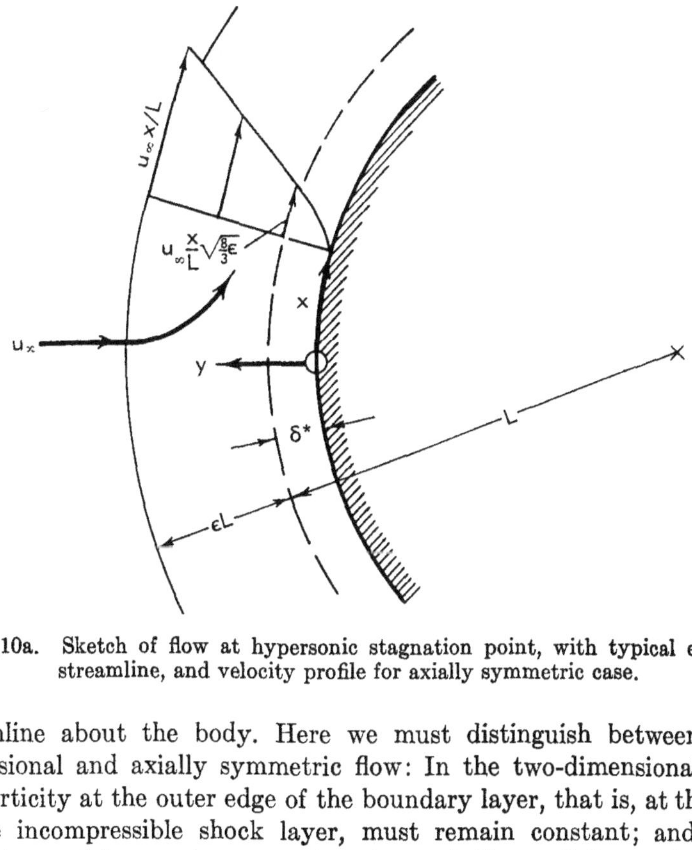

Fig. E,10a. Sketch of flow at hypersonic stagnation point, with typical entering streamline, and velocity profile for axially symmetric case.

streamline about the body. Here we must distinguish between two-dimensional and axially symmetric flow: In the two-dimensional case, the vorticity at the outer edge of the boundary layer, that is, at the base of the incompressible shock layer, must remain constant; and since vorticity must be zero by symmetry at $x = 0$, the vorticity at the outer edge of the boundary layer must remain zero everywhere. In the axially symmetric case, however, we may imagine a toroidal fluid element passing through the shock, acquiring a certain circulation and then spreading out incompressibly over the body. We notice that, by continuity, the cross-sectional area of the torus must diminish as its diameter increases. Since its circulation must remain constant, the vorticity in the torus, which is given by the ratio of circulation to cross-sectional area of the vortex filament, must increase. Thus we may expect a strong vorticity interaction for the axially symmetric stagnation point and a very weak vorticity interaction for the plane case.

For the axially symmetric problem, the velocity profile in the shock layer may be represented fairly well as decreasing linearly from a value equal to the tangential component of the outer flow just behind the shock $u_\infty x/L$ to a value smaller in proportion to $\sqrt{\epsilon}$,

$$u_e \cong u_\infty \frac{x}{L}\sqrt{\frac{8}{3}\epsilon} \tag{10-1}$$

just at the base of the shock layer [2,12].

Following Hayes and Probstein [2], we compare this shock-produced vorticity with that arising by viscous action in the boundary layer. Using Eq. 10-1 and Eq. 4-1 and 4-2 evaluated at an axially symmetric stagnation point, we may invert Eq. 4-1 to give

$$\frac{y}{L} = \left(\frac{8}{3}\epsilon\right)^{-\frac{1}{4}} \sqrt{\frac{u_e}{2\rho_e u_\infty L}} \int_0^\eta \frac{\rho_e}{\rho}\, d\eta \tag{10-2}$$

A boundary layer thickness may be defined by setting the integral appearing in Eq. 10-2 equal to unity. Then, identifying y as δ, we estimate the boundary layer vorticity as u_e/δ, and form the vorticity ratio

$$\Omega \equiv \frac{u_\infty x/\epsilon L^2}{u_e/\delta} \cong \frac{1}{\epsilon\left(\frac{8}{3}\epsilon\right)^{\frac{3}{4}}} \sqrt{\frac{\mu_e}{2\rho_e u_\infty L}} \tag{10-3}$$

This is the quantity that may be expected to govern the importance of interaction of external vorticity with the boundary layer. Now the quantities described in Eq. 10-3 may be expressed in various ways. For example, if we assume that the surface is sufficiently cool that the wall temperature is of the same order as the stream temperature, and hence that $\mu_w \cong \mu_\infty$, and if we express Reynolds number in terms of free stream quantities, assuming a linear variation of viscosity with temperature, we may employ the state equation and the Newtonian estimate $p_w/p_\infty \cong M_\infty^2$ to yield

$$\Omega \sim \frac{\epsilon^{-\frac{3}{4}} M_\infty}{\sqrt{Re_{\infty,L}}} \tag{10-4}$$

Now, the presence of a negative power of the small quantity ϵ in Eq. 10-4 suggests that vorticity interaction must be considered, even in cases for which the ratio $M_\infty/\sqrt{Re_{\infty,L}}$ is small. Rott and Lenard [67] have pointed out that the vorticity interaction effect is the same order in reciprocal Reynolds number as various other corrections for slip effects and certain effects of boundary layer displacement. For small ϵ, however, it may be proper to expect vorticity interaction to predominate among the class of low Reynolds number corrections. On this basis, we justify study of the vorticity effect separately.

We now proceed to indicate the boundary layer solution for vorticity interaction. Eq. 8-5 applies for the present purpose, with $\beta = \frac{1}{2}$ for an axially symmetric stagnation point:

$$f''' + ff'' + \tfrac{1}{2}(1 - f'^2) = 0 \tag{10-5}$$

where, for simplicity, attention is restricted to the incompressible case [*67,68*] $g = 1$. The boundary conditions given in Eq. 8-7 apply, except that the condition $f'(\infty) = 1$ must be replaced by a new boundary condition at the outer edge of the boundary layer which connects the velocity profile with a given level of outer vorticity. Before specifying such a boundary condition, it is prudent to inquire what boundary conditions are permitted by the differential equation (Eq. 10-5). For example, it is clear that the usual condition $f'(\infty) = 1$ constitutes a solution of Eq. 10-5. Therefore, it is generally proper to prescribe this behavior as a boundary condition at infinity. Next, we inquire whether a function f describing a simple shear flow is a possible asymptotic solution of Eq. 10-5. We may quickly verify that the polynomial

$$f = A + B\eta + \tfrac{1}{2}\Omega\eta^2 \tag{10-6}$$

is an exact solution of Eq. 10-5, provided that

$$B^2 = 1 + 2A\Omega \tag{10-7}$$

We see then that it is proper to require, as a boundary condition on Eq. 10-5, that the shear approach a finite value:

$$f''(\infty) = \Omega \tag{10-8}$$

If the shear at the outer edge of the boundary layer is $u_\infty x/\epsilon L^2$, as previously described, we find by use of Eq. 4-1, 4-2, and 10-1 that the Ω defined in Eq. 10-6, 10-7, and 10-8 is exactly that given in Eq. 10-3.

Boundary condition (Eq. 10-8) may be put in a slightly different form by combining Eq. 10-6 and 10-7:

$$f'^2 = 1 + 2\Omega f \tag{10-9}$$

This boundary condition, proposed by Hayes [*34*], is attractive in that, when Ω is set equal to zero, the usual boundary condition is recovered, whereas Eq. 10-8 would state that $f''(\infty) = 0$ when $\Omega = 0$. Eq. 10-8 is a more basic boundary condition, however, because it does not embody information obtained from the differential equation, except for a consideration which shows that this boundary condition is a possible one to apply to Eq. 10-5. It may be noted that, in the case $\Omega = 0$, the differential equation shows that the only solution satisfying $f''(\infty) = 0$ is the one for which $f'(\infty) =$ const., reducible to 1 by a transformation.

Velocity profiles determined by solution of Eq. 10-5, subject to Eq. 10-8, would have the character sketched in Fig. E,10b. Kemp [*68*] has solved the foregoing problem for the incompressible stagnation point,

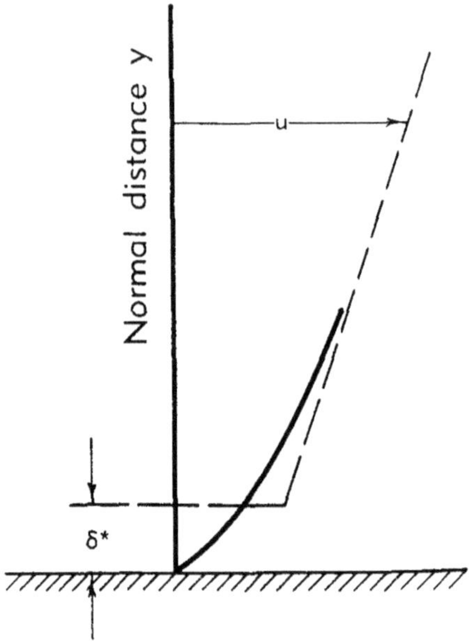

Fig. E,10b. Sketch of velocity profile with vorticity interaction. Solid line refers to the boundary layer, dashed line to the inviscid shear flow.

finding results indicated in Fig. E,10c. He suggests as formulas for the effect of vorticity on skin friction and heat transfer

$$f''(0) = 0.93[1 + 0.78\Omega] \tag{10-10}$$

$$g'(0) = 0.44[1 + 0.36\Omega](Pr = 0.71) \tag{10-11}$$

suitable in the range[9] $0 \leqq \Omega \leqq 0.6$. Thus, we see that vorticity interaction increases skin friction and heat transfer, the effect for skin friction being about twice that for heat transfer. For the compressible case in which the surface temperature is low, Hayes and Probstein [*2*] report considerably smaller interaction effects, replacing the foregoing numbers 0.78 and 0.36 by 0.49 and 0.19, respectively.

Before proceeding further, we should discuss two questions posed by the foregoing analysis. First, a definite velocity (Eq. 10-1) has been

[9] Kemp's analysis is derived from the Navier-Stokes equations, without making the boundary layer assumption. Therefore, he is not concerned with the question, which we shall discuss shortly, of whether finite values of Ω can be contemplated under boundary layer theory.

used to normalize the boundary layer profile, whereas the profile, as sketched in Fig. E,10b, has no definite "outer velocity." We may ask, then, what meaning may properly be attached to the quantity u_e. The meaning is implied in the differential equation (Eq. 10-5): u_e is the velocity at which inertia and pressure forces balance, so that $1 - f'^2 = 0$. In

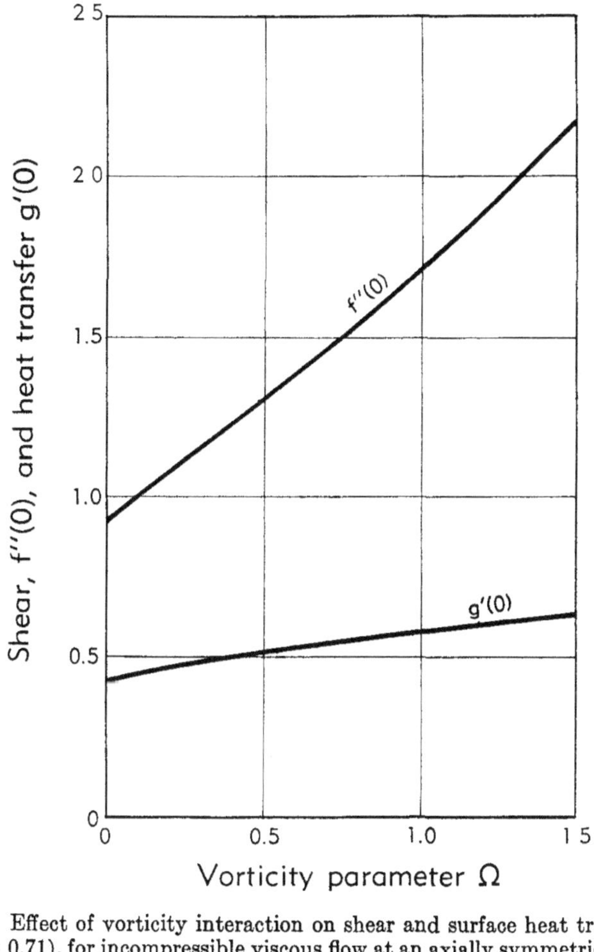

Fig. E,10c. Effect of vorticity interaction on shear and surface heat transfer coefficients ($Pr = 0.71$), for incompressible viscous flow at an axially symmetric stagnation-point after Kemp [*68*].

other words, u_e is simply a quantity which defines the external streamwise pressure gradient, presumed to be known. This point will be discussed further in subsequent paragraphs.

A second question, related to the first, concerns the displacement effect. In applying the boundary condition (Eq. 10-8 or Eq. 10-9), we have no control of the velocity level at which the constant shear solution takes over. We can show, however, that the shift of velocity level exactly compensates for boundary layer displacement, so that the displaced outer flow still has at its base the velocity given by Eq. 10-1: In view of Eq. 10-6, the equation defining displacement for this problem,

$$\delta^*(B + \tfrac{1}{2}\Omega\delta^*) = \int_0^\infty (B + \Omega\eta - f')d\eta \tag{10-12}$$

yields the statement

$$A + B\delta^* + \tfrac{1}{2}\Omega\delta^{*2} = 0 \tag{10-13}$$

that is, at $\eta = \delta^*$, the asymptotic stream function vanishes, and therefore, by Eq. 10-9, $u_e^* = u_e$ in the displaced outer flow, and L should be measured to the displacement surface, as in Fig. E,10a. Of course, displacement effects modify the skin friction and heat transfer results indicated in Eq. 10-10 and 10-11. These effects are of higher order in ϵ than is Ω, and may be neglected if ϵ is sufficiently small.

General discussion of vorticity interaction. We now proceed to a more general discussion of certain aspects of vorticity interaction according to boundary layer theory, making use of insight obtained from the foregoing problem. First, we may observe that the asymptotic solution (Eq. 10-6), which justifies imposition of boundary conditions (Eq. 10-8 or 10-9), exactly satisfies an inviscid boundary layer equation, so to speak; that is, the viscous term (f''') vanishes for this solution. Thus, in the stagnation-point problem, the matching of boundary layer flow with the external flow is accomplished by asking that the boundary layer approach a form in which the normal gradient of shear stress vanishes, and which therefore may be regarded as inviscid. A commonplace example of such an "inviscid" boundary layer is Couette flow between two parallel planes in relative motion. In such a case, viscosity is not neglected by physical assumption, and does, in fact, act on the surface; however, viscosity does not affect the mechanics of the velocity profile.

If, as in the stagnation-point example, the outer flow that the boundary layer approaches is itself a solution of the boundary layer equations, we are under no necessity to consider that the interaction parameter Ω is small. That is, so long as the vortical flow has constant shear, its shear may even be large compared with the average shear in the boundary layer, and still the boundary layer equations may apply throughout, provided, of course, that the normal velocity remains small.

For purposes of further discussion, let us consider the plane incompressible boundary layer equation with the viscous term omitted:

$$uu_x + vu_y = -\frac{1}{\rho}p_x \equiv u_e u_{e_x} \tag{10-14}$$

Here, u_e is not a limiting velocity, but simply a convenient way of expressing pressure. This equation is the one that must be obeyed in the outer region of the boundary layer. In terms of a stream function for which

$u = \psi_y$, $v = -\psi_x$, Eq. 10-14 may readily be integrated to yield

$$u^2 = u_e^2 + F(\psi) \tag{10-15}$$

where F is an arbitrary function. Eq. 10-15 gives the form which must be taken asymptotically by the boundary layer solution, including for u^2 a part depending on pressure and a part representing a vortical flow with constant pressure. In the event that the shear u_y is constant with respect to y at the outer edge of the boundary layer F may be determined, with the result

$$u^2 = u_e^2 + 2(u_y)_\infty\psi \tag{10-16}$$

Eq. 10-9 is, of course, a special case of Eq. 10-16. The boundary condition that $(u_y)_\infty$ be constant would perhaps be of most common interest for hypersonic flows. Classical boundary layer theory does, however, provide a number of examples which may be included within the framework of the more general Eq. 10-15.

Consider the class of Falkner-Skan flows (B,8.2) for which $u_e \sim x^m$, and for which, under the transformation

$$\eta \equiv \sqrt{\frac{m+1}{2\mu/\rho}}\, x^{\frac{1}{2}(m-1)}y \tag{10-17}$$

the stream function

$$\psi \equiv \sqrt{\frac{2\mu/\rho}{m+1}}\, x^{\frac{1}{2}(m+1)}f(\eta) \tag{10-18}$$

satisfies the differential equation

$$f''' + ff'' + \frac{2m}{m+1}(1 - f'^2) = 0 \tag{10-19}$$

In general, the asymptotic solution of Eq. 10-19 has the largest term

$$f \overset{a}{=} \eta^{(1+m)/(1-m)} \tag{10-20}$$

from which we may quickly verify that, apart from a constant,

$$u^2 \overset{a}{=} x^{2m} + (\text{const})\psi^{4m/(1+m)} \tag{10-21}$$

Thus, in these cases, $F(\psi)$ takes on a particular form, involving m, fixed by the similarity rule adopted in Eq. 10-17. Accordingly, we have an illustration of a point made by Hayes [34] that, in a given problem, the assumption of complete similarity may prevent application of a desired boundary condition such as Eq. 10-16.

In the special case $m = \frac{1}{3}$, which corresponds to the axially symmetric stagnation point, Eq. 10-21 happens to be consistent with constant shear as described in Eq. 10-16. For the problem of the flat plate ($m = 0$), Eq. 10-21 says that no vortical solution is possible within the similarity assumption. For adverse pressure gradient, for which m may be slightly negative, the special power-law vorticity profile is one that diminishes to zero as one proceeds out away from the plate. This peculiar circumstance has been described in the literature [*69*] as one of indeterminate solution when the boundary condition $f'(\infty) = 1$ is imposed. Various arguments have been advanced for rejecting this extra solution. From the foregoing discussion, it would appear that the important point is that the power-law behavior for large values of η represents an outer flow with vorticity, and should be rejected if such a vorticity distribution is not considered to be a proper part of the description of outer-flow velocity.

Finally, we return to the observation that u_e is a quantity which describes pressure. It is important to add that we may not always know the external pressure a priori; in fact, in the case of shear flow over a flat plate, a pressure gradient arises due to the interaction, and is proportional to Ω [*57,58,59*]. This result is due physically to boundary layer displacement. Mathematically, the cause is in the looseness of the basic shear boundary condition on $(u_y)_\infty$. We have seen that this is the most general boundary condition permitted for the boundary layer equation; as regards velocity level far from the plate, we must be satisfied with what the boundary layer solution gives us, if pressure has been taken as known. Conversely, if we wish to impose a condition on velocity level as well as shear, we must determine during the analysis the particular pressure gradient that will cause the boundary layer solution to meet the added condition far from the plate. In effect, this is the procedure carried out by Li [*57*], Murray [*58*] and Ting [*59*] in the flat plate case. In the case of the axially symmetric stagnation point, this complication is absent, as we have seen; for that particular problem the simple shear condition yields a correct far solution, properly displaced by boundary layer growth.

Stagnation-point flow with dissociation. Introduction. We return now to the stagnation-point problem, and in the remainder of this section, consider the boundary layer to contain a binary mixture of gas molecules and atoms. Earlier, it was mentioned that boundary layer profiles of the various physical quantities at a stagnation point may be obtained as leading terms of a Taylor series in distance away from the stagnation point along the surface. This feature of stagnation-point flow permits inclusion of a wide range of physical phenomena in boundary layer theory. In particular, the stagnation-point boundary layer is the only realistic steady boundary layer for which available theory considers nonequilibrium chemistry in an adequate way. We refer especially to the work of Fay and Riddell [*9*], who solved, in effect, Eq. 4-7, 4-8, and 4-9 for the axially symmetric stagnation point. Of course, these equations apply to a binary mixture of atoms and molecules, which is a reasonable approximation for dissociated air (Art. 3). For the special case of chemical equilibrium, Fay and Riddell also solved more general equations based on more exact specification of the components of high temperature air.

At the stagnation point, full similarity prevails (because of the existence of a Taylor series in x), and the terms on the right of Eq. 4-7, 4-8, and 4-9 may be discarded. In the momentum equation (Eq. 4-7), the quantity $2\xi u_{e_\xi}/u_e = \frac{1}{2}$ for the axially symmetric case, and in the energy equation (Eq. 4-9), the second term may be discarded because $u_e^2/H_e = 0$ at a stagnation point. In [9], and in this discussion, thermal diffusion is neglected ($K_T = 0$). The rate of production of atoms, the last term on the left of Eq. 4-8, is written by Fay and Riddell [9] in the form (cf. Eq. 3-31):

$$\text{Rate} = -C_1\left(\frac{T}{T_e}\right)^{-\frac{7}{2}}\frac{(c^2 - c_{eq}^2)}{(1 - c)} \tag{10-22}$$

where C_1, called the "recombination-rate parameter," is

$$C_1 \equiv K p_e^2 T_e^{-\frac{7}{2}} \mathfrak{R}^{-2}\left(\frac{\partial u_e}{\partial x}\right)^{-1} \tag{10-23}$$

and may be identified [9] as a ratio of time required for diffusion across the stagnation-point boundary layer to the normal lifetime of an atom.

Equilibrium solutions. For equilibrium flow, Fay and Riddell [9] compare numerical solutions obtained from Eq. 4-7 and 4-9 and from the more complicated system involving more complete equilibrium air chemistry. By correlating their machine calculations,[10] they found the following formula for surface heat flux at an axially symmetric stagnation point.

$$-\frac{q_w}{H_e} = 0.76 Pr^{-06}(\rho_w\mu_w)^{0.1}(\rho_e\mu_e)^{0.4}(1 - g_w)\left[1 + (Le^{0.52} - 1)\frac{c_e h_1^{(0)}}{H_e}\right]\sqrt{\frac{\partial u_e}{\partial x}} \tag{10-24}$$

The foregoing result may nearly be achieved by use of the ideas of local Blasius similarity (Art. 8), even though, as we have seen, Blasius similarity has no theoretical justification for stagnation-point flow. By use of Eq. 4-2 and the fact that $u_e = x\partial u_e/\partial x$ at the stagnation point, the flat plate formula (Eq. 9-62), applied under the modified local Blasius similarity rule for $\beta = \frac{1}{2}$ and a cold wall, yields

$$-\frac{q_w}{H_e} = 0.70(\rho_e\mu_e)^{\frac{1}{2}} Pr^{-\frac{2}{3}}(1 - g_w)\sqrt{\frac{\partial u_e}{\partial x}} \tag{10-25}$$

This result, embodying assumptions that $J = Le = 1$, was given by Lees [36].[11] If the surface is not very cold, the numerical coefficient is higher, in

[10] Made for $Pr = 0.71$; $1.0 \leq Le \leq 2.0$; $5800 \leq u_\infty \leq 22{,}800$ ft/sec; altitudes up to 120,000 ft; $300 \leq T_w \leq 3000$°K.

[11] Lees [36] gives the coefficient 0.70 as "about" the Blasius value. Actually, the Blasius value is 0.66. If the correction given in Fig. E,8b at $\beta = \frac{1}{2}$ for a cold wall is applied to the value 0.66, the result 0.70 is obtained.

the proportion indicated by Fig. E,8b. Eq. 10-25 compares quite favorably with the more exact Eq. 10-24.

We shall subsequently improve Eq. 10-25 by estimating the effect of Lewis number as slightly different from one. First, however, the assumption $J = 1$ should be discussed. In Eq. 10-25, the dependence on viscosity is close to that found in Eq. 10-24. However, since Eq. 10-25 and 9-60 are based on the assumption that $J = 1$, that is, that the product $\rho\mu$ is constant throughout the boundary layer, it would appear to be a matter of indifference where the product $\rho\mu$ was evaluated. However, it is certainly necessary to make a choice, as is evident from the wide variation of $\rho\mu$ across the stagnation-point boundary layer. In Fig. E,10d, Fay and Riddell's calculation [9] of this quantity, based on the Sutherland law, is displayed. In effect, Eq. 10-25 embodies the assumption that the outer-flow viscosity dominates the boundary layer. This assumption may be justified by the following argument: We notice from Fig. E,10d that $\rho\mu$ tends to remain near its stream value down to the near proximity of the surface, and then rises rather abruptly as the surface is closely approached. Now suppose we imagine an extreme limit of this situation, that is, a boundary layer composed of two fluid layers: a very thin layer with a high value of $\rho\mu$ and a very much thicker layer with a lower constant value of $\rho\mu$. If the inner layer is imagined to become indefinitely thin, we note that the flux of heat to the surface would be nearly equal to the flux coming into this layer from the thicker layer above. Hence, the heat transferred to the wall would be equal to the heat flux out of the thicker layer, and therefore would be controlled by the outer value of $\rho\mu$, quite independently of the value of $\rho\mu$ near the surface. Fig. E,10d shows that a tendency exists for the stagnation-point boundary layer to have a character approaching the extreme situation described above, and thus a certain rough justification is provided for the use of $(\rho\mu)_e$ n Eq. 10-25. The more exact equation (Eq. 10-24) verifies the approximate correctness of this point of view, as does the example provided by Couette flow (Eq. 5-14), in which the maximum error entailed in the assumption $J = 1$ is about 25 per cent.

We now seek to include the effect of $Le \neq 1$ in the Blasius-rule result (Eq. 10-25). For this purpose, we write the energy equation (Eq. 4-9) in an appropriate form

$$g'' + Prfg' + (Le - 1)\frac{h_1^{(0)}}{H_e}c'' = 0 \tag{10-26}$$

As we have seen in the case of the flat plate, it is a simple matter to indicate the effect of Prandtl number by a transformation of the independent variable; and in the case of a frozen boundary layer manipulations involving the homogeneous form of Eq. 4-8 provide, by the same means, an assessment of the Lewis number effect. For the equilibrium boundary

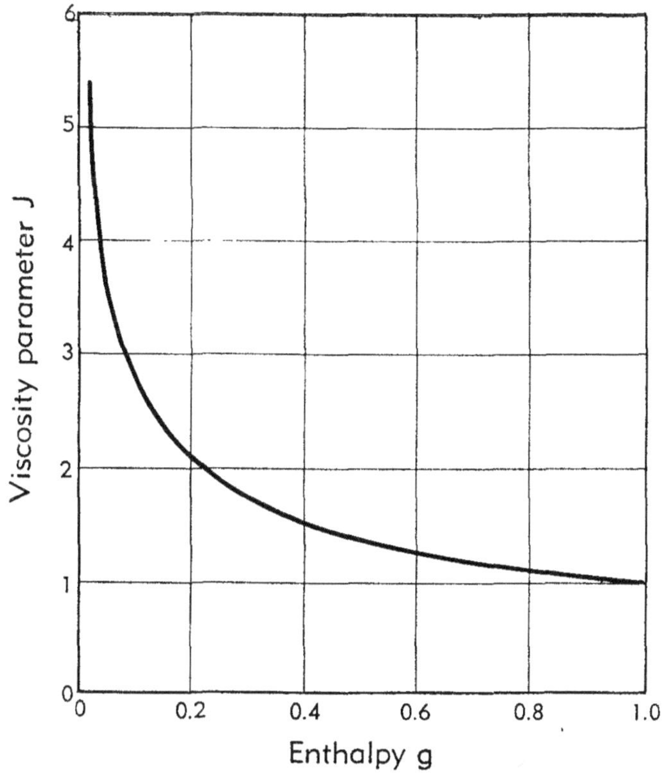

Fig. E,10d. Variation of $J \equiv \rho\mu/(\rho\mu)_e$ through stagnation-point boundary layer, after Fay and Riddell [9], for $u_\infty = 19{,}700$ ft/sec and $T_w = 300°$K.

layer, however, c is determined by the mass-action law for the mixture, which is difficult to incorporate into a simple boundary layer theory. For purposes of estimation, let us suppose that the equilibrium boundary layer is one for which the dissociation level is approximately proportional to the total enthalpy within the boundary layer:

$$c \cong c_e g \tag{10-27}$$

Eq. 10-26 would then become

$$\frac{1}{Pr}\left[1 + (Le - 1)\frac{c_e h_1^{(0)}}{H_e}\right] g'' + fg' = 0 \tag{10-28}$$

Then, following the previous procedure of introducing a transformation based on the $\frac{1}{3}$ power of the coefficient of the leading term, and assuming that Le and Pr are close to one, we write

$$g = 1 + (g_w - 1)\left\{1 - f'\left(\left[1 - \tfrac{1}{3}(Le - 1)\frac{c_e h_1^{(0)}}{H_e}\right] Pr^{\frac{1}{3}}\eta\right)\right\} \tag{10-29}$$

whence, in place of Eq. 10-25, we find

$$-\frac{q_w}{H_e} = 0.70(\rho_e\mu_e)^{\frac{1}{2}}Pr^{-\frac{2}{3}}\left[1 + (Le^{\frac{2}{3}} - 1)\frac{c_e h_1^{(0)}}{H_e}\right](1 - g_w)\sqrt{\frac{\partial u_e}{\partial x}} \tag{10-30}$$

where we have used the fact that, when Lewis number is near one, $\frac{2}{3}(Le - 1) \cong Le^{\frac{2}{3}} - 1$. Comparison of Eq. 10-30 and 10-24 shows that application of the Blasius similarity rule reproduces the main features of the more exact result for the equilibrium heat transfer at an axially symmetric stagnation point.

Frozen solutions. Let us turn now to the case of frozen flow. In this problem, a condition on wall catalycity must be imposed. For present purposes, Eq. 4-11 may be written in the form

$$c'(0) = Sc(0) \tag{10-31}$$

where it is assumed that the wall is sufficiently cool so that the equilibrium atom concentration at the wall is zero, and where S is a catalytic coefficient which may be identified by inspection of Eq. 4-11. For a fully catalytic wall, $c(0) = 0$, Fay and Riddell [9] find

$$-\frac{q_w}{H_e} = 0.76Pr^{-0.6}(\rho_e\mu_e)^{0.4}(\rho_w\mu_w)^{0.1}\left[1 + (Le^{-0.63} - 1)\frac{c_e h_1^{(0)}}{H_e}\right](1 - g_w)\sqrt{\frac{\partial u_e}{\partial x}} \tag{10-32}$$

Here, we see that, as usual, the frozen boundary layer over a fully catalytic wall delivers just about the same heat transfer as does the equilibrium boundary layer: The only difference between Eq. 10-24 and 10-32 is in the slightly greater exponent according to which Lewis number enters. For a noncatalytic wall, the result suggested by Fay and Riddell's calculations [9] is

$$-\frac{q_w}{H_e} = 0.76Pr^{-0.6}(\rho_e\mu_e)^{0.4}(\rho_w\mu_w)^{0.1}\left(1 - \frac{c_e h_1^{(0)}}{H_e}\right)\sqrt{\frac{\partial u_e}{\partial x}} \tag{10-33}$$

In this case, no mechanism exists either within the boundary layer or on the wall for atom concentration to change, and hence the atom concentration remains constant throughout the boundary layer. As a result, the heat transfer rate is decreased owing to the amount of heat of dissociation retained in the gas.

The application of local Blasius similarity to the frozen stagnation-point boundary layer may be accomplished by use of Eq. 9-55 and 9-56. Combining Eq. 9-56 and 10-31 yields atom concentration at the surface:

$$\frac{c_w}{c_e} = \left(1 + S\frac{Le^{\frac{1}{3}}}{Pr^{\frac{1}{3}}f''(0)}\right)^{-1} \tag{10-34}$$

The heat transfer rate obtained from Eq. 9-62 and 10-34 is

$$-\frac{q_w}{H_e} = 0.70(\rho_e\mu_e)^{\frac{1}{2}}Pr^{-\frac{2}{3}}\left[1 - g_w + (Le^{\frac{2}{3}} - 1)\frac{c_e h_1^{(0)}}{H_e}\frac{Le^{\frac{2}{3}}S}{Le^{\frac{2}{3}}S + 0.47Pr^{\frac{2}{3}}}\right]\sqrt{\frac{\partial u_e}{\partial x}} \tag{10-35}$$

In the limit $S \to \infty$, Eq. 10-35 gives a result for the fully catalytic wall that is identical with Eq. 10-31, previously obtained for the equilibrium boundary layer, provided the assumption is made that the total enthalpy in the stream is much greater than that at the wall ($g_w \ll 1$). Under this same assumption, Eq. 10-35 agrees well with the more exact equation (Eq. 10-32). In the noncatalytic case, the square bracket in Eq. 10-35 agrees with the corresponding factor in Eq. 10-33, becoming

$$[1 - g_w] = [g_e^* - g_w^*] \cong g_e^* = \left[1 - \frac{c_e h_1^{(0)}}{H_e}\right] \tag{10-36}$$

provided the external temperature is much greater than the surface temperature.

Nonequilibrium solution. We have previously noted that the stagnation-point boundary layer is one for which the rules of similarity permit consideration of flows in a condition intermediate between frozen and equilibrium. Fay and Riddell [9] have carried out calculations assuming finite values of the recombination rate parameter (Eq. 10-23), with results indicated in Fig. E,10e. We notice that, for a catalytic wall, heat transfer rate is practically constant over a wide range of recombination rate parameters. Comparing Eq. 10-24 for equilibrium flow with Eq. 10-32 for frozen flow over a fully catalytic wall shows that the slightly higher exponent applied to Lewis number in Eq. 10-32 would mean, for $Le > 1$, a slightly greater heat transfer rate in the frozen case, as Fig. E,10e shows. For a noncatalytic wall, as the recombination rate parameter decreases, one would expect to shift from heat transfer at the rate given by Eq. 10-24 to the considerably smaller value given in Eq. 10-33. Fig. E,10e indicates that the transition between these two expressions occurs at quite small values of the recombination rate parameter.

Fay and Riddell [9] also show the conductive part of heat transfer rate for a catalytic wall. For large values of recombination rate parameter, as Fig. E,10e suggests, heat transfer is entirely by conduction, because the wall temperature is sufficiently low that the normal gradient of equilibrium atom concentration is zero. The curve indicated for the noncatalytic wall is also purely conductive, because at the wall the normal gradient of concentration vanishes as a consequence of the chemical inertness of the wall. Comparing these two processes, one concludes that equilibrium, so to speak, is favored by a chemically inert wall. This is explained by Fay and Riddell as due to a piling-up of atoms near the wall in the noncatalytic case, causing locally large rates of recombination.

It is interesting that Fig. E,7b and E,10e are qualitatively similar in describing equilibrium in two quite different situations. There is also a quantitative analogy to be noted: Let us compare the cooling effects of a noncatalytic surface in the two cases, both in the frozen limit. For

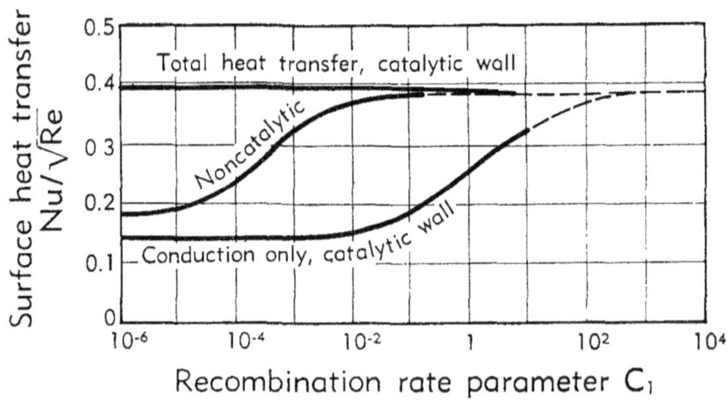

Fig. E,10e. Nonequilibrium heat transfer at axially symmetric stagnation point, after Fay and Riddell [9], according to the definition $q_w/H_e = (Nu/\sqrt{Re})\sqrt{\rho_w\mu_w\partial u_e/\partial x}$ $(1 - g_w)/Pr$ with $Le = 1.4$, $Pr = 0.71$, $T_w = 300°K$, $g_w = 0.0123$.

frozen heat transfer at the stagnation point, the ratio of noncatalytic to catalytic values is obtained from Eq. 10-33 and 10-32 as

$$\frac{(q_w)_{\text{noncat}}}{(q_w)_{\text{cat}}} \cong \frac{H_e^*}{H_e} \tag{10-37}$$

when it is assumed that $Le = 1$, $g_w \ll 1$. From Eq. 7-25, 7-27, and 7-12, the equivalent ratio in the Rayleigh-type problem is

$$\frac{(q_w)_{\text{noncat}}}{(q_w)_{\text{cat}}} \cong \frac{\nu}{1+\nu} = \left(\frac{c_p\partial T}{c_p\partial T + h_1^{(0)}\partial c}\right)_\infty \tag{10-38}$$

Eq. 10-38 may be interpreted to give the differential ratio $\partial H_e^*/\partial H_e$. This differential ratio apparently plays a similar role in the linearized heat conduction problem, as does the ratio of the quantities themselves in the nonlinear boundary layer of a stagnation point.

Other stagnation-point problems. Additional information concerning the stagnation-point boundary layer is contained in a large body of literature consisting largely of numerical solutions of equations we have already described. Since this article is concerned with theory, we shall not attempt a thorough review of this work. However, it should be remarked that Goulard [*70*] and Scala [*19*] have made calculations of the Fay-Riddell type, assuming finite values of surface catalytic efficiency, with the chief purpose of setting requirements for heat protection by use of noncatalytic coatings. The very extensive calculations of Scala [*19*] take into account thermal diffusion, which has been discussed here only for Couette flow (Art. 5).

Reshotko and Cohen [*71*] have solved the compressible stagnation-point equations (in effect, Eq. 4-7 and 4-9) with $J = Le = 1$, $\beta = \frac{1}{2}$, 1, $K_T = 0$, $u_e^2/H_e = 0$, and with $Pr = 0.7$. They include the effect of mass addition at the surface. Through the local Blasius similarity rule, a correspondence might be inferred between this work and the constant pressure solutions of Baron [*60*] discussed earlier (Art. 9).

Finally, brief mention should be made of the extension of boundary layer theory to cases of multiphase stagnation-point flows. This work stems from practical interest in the possibilities of thermal protection by absorption of heat during an ablation process. The physical chemistry of ablation can be extremely complex, involving a number of alternative processes. Lees [*72*] has provided a general discussion of ablation with an outline of applicable theory.

The theory of ablation so far developed by Sutton [*73*], Scala and Sutton [*74*], Lees [*75*], Roberts [*76*], Bethe and Adams [*77*], and others, includes consideration of surface melting to form a liquid boundary layer beneath the gaseous boundary layer. If the liquid is highly viscous, the gas layer is mechanically unaffected by the liquid motion. At the gas-liquid interface, vaporization or sublimation may occur, possibly followed by chemical reaction in the gas layer [*78*; 79, pp. 539–545]. Heats of fusion, vaporization, and chemical reaction, radiation from the hot interface, and the insulating effect of mass addition (Art. 9) all may contribute to lowering the heat transfer rate.

E,11. Concluding Remarks. In the present discussion of hypersonic laminar boundary layer theory, certain viewpoints and limitations have been adopted which, in closing, should perhaps be reviewed. First, no serious attempt has been made to give a balanced review of the state of knowledge of hypersonic viscous flows. Instead, we have discussed, in more than usual detail, the linearized "analogy" problems carried from classical theory into correspondence with hypersonic cases. In choosing this course, we have slighted many useful calculations of more direct application to questions of hypersonic flight.

From the theoretical standpoint, it should be emphasized that the analogy problems of Couette and Rayleigh flows (Art. 5 and 6) are just as useful and suggestive in the hypersonic regime as in classical incompressible flow. In fact, they should be consulted even more carefully in the hypersonic case, because of the profusion of physical and chemical effects that require qualitative understanding. For example, Couette-type solutions give simple formulas for the effect of variable $\rho\mu$ (Eq. 5-14 and 5-22), and from these, conclusions about viscosity evaluation in nonlinear boundary layer theory may be drawn with substantial accuracy (Art. 9 and 10). Also, the kinetic process in a nonequilibrium Rayleigh-type boundary layer (Art. 7), as displayed in Fig. E,7a and E,7b, corresponds surprisingly well to the stagnation-point solutions [*9*] shown in Fig. E,10e. Qualitatively, the roles of catalycity are the same, and an analogy exists between the influences of the ratio of thermal to chemical energy increments, in the Rayleigh case, and the ratio of thermal to chemical energy at the edge of the boundary layer, in the stagnation-point case.

We have not mentioned, so far, the matter of experimental verification for the various theoretical ideas discussed in this article. Actually, many of the cited references refer to pertinent experimental evidence, and it is not believed that any of the theoretical material presented here is flatly contradicted by reliable experiment. Of course, the limits of applicability of certain assumptions, such as those underlying strong interaction theory (Art. 9), are under experimental study at the present time. In certain instances, experimental evidence has served to emphasize the importance of theoretical ideas, for example in the matter of entropy and vorticity interactions [*55*].

Practical obstacles have made difficult the performance of hypersonic boundary layer experiments sufficiently reliable and precise to constitute definitive checks of theory. These obstacles are being overcome and, as progress continues in this direction, increasing interest will attach to the subject of viscous hypersonic similitude, which has already been studied by Hayes and Probstein [*51*] and by Cheng, et al. [*38*], and which has an important bearing on the planning of experimental research.

E,12. Cited References.

1. Schaaf, S. A. Recent progress in rarefied gasdynamics. *J. Am. Rocket Soc. 30*, 443–447 (1960).
2. Hayes, W. D., and Probstein, R. F. *Hypersonic Flow Theory*. Academic Press, 1959. Chap. 13, 9, unless otherwise noted.
3. Goldstein, S. *Modern Developments in Fluid Mechanics*, Vol. I. Clarendon Press, Oxford, 1938.
4. Schlichting, H. *Boundary-Layer Theory* (transl. by J. Kestin). McGraw-Hill, 1955.
5. Hirschfelder, J. O., Curtiss, C. F., and Bird, R. B. *Molecular Theory of Gases and Liquids*. Wiley, 1954.
6. Hansen, C. F. Note on the Prandtl number for dissociated air. *Readers' Forum, J. Aeroneut. Sci. 20*, 789–90 (1953).
7. Scala, S. M., and Baulknight, C. W. Transport and thermodynamic properties in a hypersonic laminar boundary layer. Part I—Properties of the pure species. *J. Am. Rocket Soc. 29*, 39–45 (1959).
8. Lees, L. Laminar heat transfer over blunt-nosed bodies at hypersonic flight speeds. *Jet Propul. 26*, 259–269 (1956).
9. Fay, J. A., and Riddell, F. R. Theory of stagnation point heat transfer in dissociated air. *J. Aeronaut. Sci. 25*, 73–85 (1958).

10. Chapman, D. R., and Rubesin, M. W. Temperature and velocity profiles in the compressible laminar boundary layer with arbitrary distribution of surface temperature. *J. Aeronaut. Sci. 16*, 547–565 (1949).
11. Slater, J. C. *Introduction to Chemical Physics.* McGraw-Hill, 1939.
12. Lighthill, M. J. Dynamics of a dissociating gas. Part I. Equilibrium flow. *J. Fluid Mech. 2*, 1–32 (1957).
13. Byron, S. R. Measurement of the rate of dissociation of oxygen. *J. Chem. Phys. 30*, 1380–1392 (1959).
14. Mathews, D. L. Interferometric measurement in the shock tube of the dissociation rate of oxygen. *Phys. Fluids 2*, 170–178 (1959).
15. Hirschfelder, J. O. Heat transfer in chemically reacting mixtures. I. *J. Chem. Phys. 26*, 274–281 (1957).
16. Clarke, J. F. Energy transfer through a dissociated diatomic gas in Couette flow. *J. Fluid Mech. 4, Pt. 5*, 441–465 (1958).
17. Broadwell, J. E. A simple model of the nonequilibrium dissociation of a gas in Couette and boundary-layer flows. *J. Fluid Mech. 4*, 113–139 (1958).
18. Liepmann, H. W., and Bleviss, Z. O. The effects of dissociation and ionization on compressible Couette flow. *Douglas Rept. SM-19831*, 1956.
19. Scala, S. M. Hypersonic stagnation point heat transfer to surfaces having finite catalytic efficiency. *Proc. Third U.S. Nat'l Congress of Appl. Mech., Am. Soc. Mech. Engrs.*, 799–806 (1958).
20. Moore, F. K., and Ostrach, S. Displacement thickness of the unsteady boundary layer. *Readers' Forum, J. Aeronaut. Sci. 24*, 77–78 (1957).
21. Howarth, L. Some aspects of Rayleigh's problem for a compressible fluid. *Quart. J. Mech. and Appl. Math. 4, Part 2*, 157–169 (1951).
22. Van Dyke, M. D. Impulsive motion of an infinite plate in a viscous compressible fluid. *Z. Angew. Math. u. Phys. 3*, 343–353 (1952).
23. Stewartson, K. On the motion of a flat plate at high speed in a viscous compressible fluid. I. Impulsive motion. *Proc. Cambridge Phil. Soc. 51, Part 1*, 202–219 (1955).
24. Moore, F. K., and Curtis, J. T. Weak, nonequilibrium, diffusive waves in a dissociated gas. *Proc. 1960 Heat Transfer and Fluid Mech. Inst. 164–175*, Stanford Univ. Press, 1960.
25. Cohen, C. B., and Reshotko, E. The compressible laminar boundary layer with heat transfer and arbitrary pressure gradient. *NACA Tech. Note 3326*, 1955.
26. Cohen, C. B., and Reshotko, E. Similar solutions for the compressible laminar boundary layer with heat transfer and pressure gradient. *NACA Rept. 1293*, 1956. Also *NACA Tech. Note 3325*, 1955.
27. Levy, S. Effect of large temperature changes (including viscous heating) upon laminar boundary layers with variable free-stream velocity. *J. Aeronaut. Sci. 21*, 459–474 (1954).
28. Li, T-Y, and Nagamatsu, H. T. Shock-wave effects on the laminar skin friction of an insulated flat plate at hypersonic speeds. *J. Aeronaut. Sci. 20*, 345–355 (1953).
29. Thwaites, B. Approximate calculation of the laminar boundary layer. *Aeronaut. Quart. 1*, 245–280 (1949).
30. Rott, N., and Crabtree, L. F. Simplified laminar boundary-layer calculations for bodies of revolution and for yawed wings. *J. Aeronaut. Sci. 19*, 553–565 (1952).
31. Görtler, H. A new series for the calculation of steady laminary boundary layer flows. *J. Math. and Mech. 6*, 1–66 (1957).
32. Sparrow, E. M. Application of Görtler's series method to the boundary layer energy equation. *Readers' Forum, J. Aeronaut. Sci. 25*, 71–72 (1958).
33. Lees, L., and Kubota, T. Inviscid hypersonic flow over blunt-nosed slender bodies. *J. Aeronaut. Sci. 24*, 195–202 (1957).
34. Hayes, W. D. On laminar boundary layers with heat transfer. *Jet Propul. 26*, 270–274 (1956).
35. Ho, T-H, and Probstein, R. F. A series solution to the laminar heat transfer problem at hypersonic speeds. *Wright Air Develop. Center Tech. Note 58-263, AD-202333*, 1958.

36. Lees, L. Convective heat transfer with mass addition and chemical reactions. *Combustion and Propulsion, Third AGARD Colloquium.* Pergamon Press, 1958.
37. Kemp, N. H., Rose, P. H., and Detra, R. W. Laminar heat transfer around blunt bodies in dissociated air. *J. Aeronaut. Sci. 26*, 421–430 (1959).
38. Cheng, H. K., Hall, J. G., Golian, T. C., and Hertzberg, A. Boundary-layer displacement and leading-edge bluntness effects in high temperature hypersonic flow. *J. Aeronaut. Sci. 28*, 353–381 (1961).
39. Vaglio-Laurin, R. Laminar heat transfer on blunt-nosed bodies in three-dimensional hypersonic flow. *J. Am. Rocket Soc. 29*, 123–129 (1959).
40. Cheng, H. K. The shock-layer concept and three-dimensional hypersonic boundary layers. *Cornell Aeronaut. Lab. Rept. AF-1285-A-3*, 1960.
41. Moore, F. K. On local flat-plate similarity in the hypersonic boundary layer. *J. Aeronaut. Sci. 28*, 753–762 (1961).
42. Freeman, N. C. On the theory of hypersonic flow past plane and axially symmetric bluff bodies. *J. Fluid Mech. 1, Pt. 4*, 366–387 (1956).
43. Cheng, H. K. Hypersonic flows past a yawed circular cone and other pointed bodies. *J. Fluid Mech. 12*, 169–191 (1962).
44. Cole, J. D. Newtonian flow theory for slender bodies. *J. Aeronaut. Sci. 24*, 448–455 (1957).
45. Stewartson, K. On the motion of a flat plate at high speed in a viscous compressible fluid. II. Steady motion. *J. Aeronaut. Sci. 22*, 303–309 (1955).
46. Shen, S. F. An estimate of viscosity effect on the hypersonic flow over an insulated wedge. *J. Math. and Phys. 31*, 192–205 (1952).
47. Lees, L., and Probstein, R. F. Hypersonic viscous flow over a flat plate. *Princeton Univ. Dept. Aeronaut. Eng. Rept. 195*, 1952.
48. Kuo, Y. H. On the flow of an incompressible viscous fluid past a flat plate at moderate Reynolds numbers. *J. Math. and Phys. 32*, 83–101 (1953).
49. Maslen, S. H. Second approximation to laminar compressible boundary layer on flat plate in slip flow. *NACA Tech. Note 2818*, 1952.
50. Lees, L. On the boundary-layer equations in hypersonic flow and their approximate solutions. *Readers' Forum, J. Aeronaut. Sci. 20*, 143–145 (1953).
51. Oguchi, H. First-order approach to a strong interaction problem in hypersonic flow over an insulated flat plate. *Univ. Tokyo Aeronaut. Research Inst. Rept. 330, 24*, 1 (1958).
52. Lees, L. Influence of the leading-edge shock wave on the laminar boundary layer at hypersonic speeds. *J. Aeronaut. Sci. 23*, 594–600 (1956).
53. Bogdonoff, S. M., and Hammitt, A. G. Fluid dynamic effects at speeds from $M = 11$ to 15. *J. Aeronaut. Sci. 23*, 108–116 (1956).
54. Yasuhara, M. On the hypersonic viscous flow past slender bodies of revolution. *J. Phys. Soc. Japan 11*, 878–886 (1956).
55. Ferri, A. Some heat transfer problems in hypersonic flow. *Wright Air Develop. Center Tech. Note 59-308, PIBAL Rept. 473*, 1959.
56. Ferri, A., and Libby, P. A. Note on an interaction between the boundary layer and the inviscid flow. *Readers' Forum, J. Aeronaut. Sci. 21*, 130 (1954).
57. Li, T. Y. Effects of free-stream vorticity on the behavior of a viscous boundary layer. *Readers' Forum, J. Aeronaut. Sci. 23*, 1128–29 (1956).
58. Murray, J. D. The boundary layer on a flat plate in a stream with uniform shear. *J. Fluid Mech. 11*, 309–316 (1961).
59. Ting, L. Boundary layer over a flat plate in presence of shear flow. *Physics of Fluids 3*, 78–81 (1960).
60. Baron, J. R. The heterogeneous, laminar, boundary layer. *AFOSR TN 57-392, AD 132-474, Mass. Inst. Technol. Naval Supersonic Lab. Tech. Rept. 236*, 1957.
61. Emmons, H. W., and Leigh, D. Tabulation of the Blasius function with blowing and suction. *Aeronaut. Research Council Fluid Motion Sub-Committee Rept. 15966*, 1953.
62. Low, G. M. The compressible laminar boundary layer with fluid injection. *NACA Tech. Note 3404*, 1955.
63. Crocco, L. Lo Strato Limite Laminare nei Gas. *Monografie Sci. Aeronaut. 3*,

Rome, 1946. Transl. as *North Amer. Aviation Aerophys. Lab. Rept. APL/NAA/CF-1038*, 1948.

64. Van Driest, E. R. Investigation of the laminar boundary layer in compressible fluids using the Crocco method. *NACA Tech. Note 2597*, 1952.
65. Romig, M. F., and Dore, F. J. Solutions of the compressible laminar boundary layer including the case of a dissociated free stream. *Convair Rept. ZA-7-012*, 1954.
66. Eckert, E. R. G. Engineering relations for friction and heat transfer to surfaces in high velocity flow. *Readers' Forum, J. Aeronaut. Sci. 22*, 585–587 (1955).
67. Rott, N., and Lenard, M. Vorticity effect on the stagnation-point flow of a viscous incompressible fluid. *Readers' Forum, J. Aero/Space Sci. 26*, 542–3 (1959).
68. Kemp, N. H. Vorticity interaction at an axisymmetric stagnation point in a viscous incompressible fluid. *J. Aero/Space Sci. 26*, 543–44 (1959).
69. Hartree, D. R. On an equation occurring in Falkner and Skan's approximate treatment of the equations of the boundary layer. *Proc. Cambridge Phil. Soc. 33*, 223–239 (1937).
70. Goulard, R. On catalytic recombination rates in hypersonic stagnation heat transfer. *Jet Propul. 28*, 737–745 (1958).
71. Reshotko, E., and Cohen, C. B. Heat transfer at the forward stagnation point of blunt bodies. *NACA Tech. Note 3513*, 1955.
72. Lees, L. Ablation in hypersonic flows. *Proc. 7th Anglo-Am. Aeronaut. Conf.* 344–362 (1959).
73. Sutton, G. W. The hydrodynamics and heat conduction of a melting surface. *J. Aeronaut. Sci. 25*, 29–32 (1958).
74. Scala, S. M., and Sutton, G. W. The two-phased hypersonic laminar boundary layer—A study of surface melting. *Proc. Heat. Transfer and Fluid Mech. Inst.*, 231–240 (1958).
75. Lees, L. Similarity parameters for surface melting of a blunt nosed body in a high velocity gas stream. *J. Am. Rocket. Soc. 29*, 345–354 (1959).
76. Roberts, L. On the melting of a semi-infinite body of ice placed in a hot stream of air. *J. Fluid Mech. 4*, 505–528 (1958).
77. Bethe, H. A., and Adams, M. C. A theory for the ablation of glassy materials. *J. Aero/Space Sci. 26*, 321–328 (1959).
78. Cohen, C. B., Bromberg, R., and Lipkis, R. P. Boundary layers with chemical reactions due to mass additions. *Jet Propul. 28*, 659–668 (1958).
79. Sutton, G. W. Combustions of a gas injected into a hypersonic boundary layer. *Seventh Symposium on Combustion.* Butterworths, London, 1958.
80. Bauer, E., and Zlotnick, M. Transport coefficients of air to 8000°K. *J. Am. Rocket Soc. 29*, 721–728 (1959).

SECTION F

LAMINAR FLOWS WITH BODY FORCES

SIMON OSTRACH

CHAPTER 1. GENERAL CONSIDERATIONS

F,1. Introduction. The forces exerted on an element of fluid are of two general types. Forces of internal stress such as pressure and viscous forces result from the interaction of contiguous fluid elements. Field or body forces arise from action at a distance, that is, from vector fields (e.g. gravity and magnetic fields) and are independent of the interactions of adjacent fluid elements; the magnitudes of body forces depend on a property of the medium (e.g. mass density or electrical conductivity) on which they act.

Although the forces due to the interactions of fluid particles are distributed throughout the medium, the net sum of these forces on all the particles is equal to the force in the bounding surface alone. This follows from the fact that the sum of the internal action and reaction forces is zero, even though local interior effects of these stresses are not zero. These forces due to particle interactions are, therefore, called surface forces. In contradistinction to the above, the net sum of the body forces at any interior point in the fluid is not zero since the reactions to these forces are exerted on bodies exterior to the system under consideration; these forces can therefore be regarded as being effective throughout the interior of the fluid body.

Most aerodynamicists are already familiar with the manner in which surface (pressure and viscous) forces cause, and alter, fluid motions. Less familiar, however, are body forces such as the gravitational, centrifugal, Coriolis, and electromagnetic forces which can also generate or alter fluid motions.

Fluid motions due entirely to the action of a gravity field are usually called natural flows, in contrast to forced flows brought about by surface forces. If natural flow is not constrained to a finite region by boundaries, we refer to it as free convection. In such cases, pressures are generally almost the same as those due to the body force in a fluid at rest. Although this distinction between natural and free flows is not universal, it will be retained herein. Hydromagnetics or such terms as magnetohydrodynamics, magnetoaerodynamics, and the like are used to denote fluid flow problems in which effects of electric and magnetic forces are important.

Until the past decade or so, flows with body forces had not been studied as intensively as most other branches of fluid mechanics. With a few notable exceptions which will be discussed subsequently, the principal investigations were those of meteorologists, geophysicists, and astrophysicists, who were interested in the special case of convection in a gravity field in which density variations result from pressure or temperature differences. The apparent lack of association between body force flows and aeronautics, the field to which so much of the fluid dynamics research was directed, and the inherent complexities of the phenomenon due to the inseparable interaction of its dynamics and thermodynamics, are possibly the reasons for its slow development.

However, with the advent of nuclear power, jet propulsion, and high speed flight, the three prime factors governing thermal natural convection, namely, the body force, the fluid volumetric expansion, and the temperature difference, all could exceed their previously considered bounds. When it was further realized that body forces other than the gravitational force can influence forced flows as well as generate natural flows, and that appreciable density variations can be due to such other causes as the nonhomogeneous composition of a medium (as, for example, the concentration of admixtures dissolved in a fluid), it became apparent that in many present-day practical problems in the fields of aeronautics, nuclear power, electronics, geophysics, and chemical engineering the body force effects can be important. Accordingly, more interest has been directed toward problems of this sort and numerous research papers have been published. The most widely studied body force flow problems reported are still those dealing with natural convection due to the gravitational force and temperature variations in the fluid. It is not difficult to relate this work to the diffusion (or concentration) type of convection. More recently there have appeared papers dealing with flows generated by centrifugal, Coriolis, and electromagnetic forces. It is, therefore, the purpose of this section to present a unified account of the effects of body forces on fluid flows, in order to show the physical characteristics of the phenomenon and to indicate that conscious utilization of body forces renders possible a rational control of flow and transfer problems. In so doing, it should be remarked that, because of space limitations, not all the important work in the field can be given here. Instead, representative work on most of the major aspects of body force phenomena will be discussed.

In order not to be lost in too great generality, the influences of electromagnetic forces will be deferred to the final chapter of this section. We shall, therefore, first be concerned with natural and free flows. Here too, since most of the published work deals with problems wherein the density variations result from temperature changes, we shall deal primarily with problems in thermal convection. Wherever possible, analogies with problems in which density variations are due to other causes will be made. Within the framework of the present volume, we shall here be concerned with laminar flows. However, remarks about turbulence will be interjected whenever appropriate.

F,2. Fundamental Equations and Parameters. The basic equations expressing the conservation of mass, momentum, and energy for a compressible, viscous, and heat-conducting fluid subject to a body force are derived in Sec. B of this volume. These equations, together with an equation of state, describe the velocity and temperature distributions of flows with body forces and can be written in Cartesian tensor notation, respectively, as

$$\frac{\partial \rho}{\partial t} + \frac{\partial}{\partial X_j}(\rho U_j) = 0 \tag{2-1}$$

$$\rho \frac{DU_i}{Dt} = \rho f_i + \frac{\partial}{\partial X_j}\left[\mu\left(\frac{\partial U_i}{\partial X_j} + \frac{\partial U_j}{\partial X_i}\right)\right] - \frac{2}{3}\frac{\partial}{\partial X_i}\left(\mu \frac{\partial U_j}{\partial X_j}\right) - \frac{\partial P}{\partial X_i} \tag{2-2}$$

$$\rho c_p \frac{DT}{Dt} = U_j \frac{\partial P}{\partial X_j} + \frac{\partial}{\partial X_j}\left(k \frac{\partial T}{\partial X_j}\right) + \mu\left[\frac{\partial U_i}{\partial X_j}\left(\frac{\partial U_i}{\partial X_j} + \frac{\partial U_j}{\partial X_i}\right) - \frac{2}{3}\left(\frac{\partial U_j}{\partial X_j}\right)^2\right] \tag{2-3}$$

$$\rho = \rho(P, T) \tag{2-4}$$

where the material or substantive derivative D/Dt is defined by

$$\frac{D}{Dt} \equiv \frac{\partial}{\partial t} + U_j \frac{\partial}{\partial X_j} \tag{2-5}$$

where U_i denotes the velocity components in the corresponding X_i coordinate direction, ρ is the density, t denotes time, f_i is the ith component of the body force per unit mass, μ is the absolute viscosity coefficient, P is the pressure, c_p is the specific heat at constant pressure, T is temperature, and k is the thermal conductivity coefficient. The terms in the momentum equations (Eq. 2-2) from left to right represent inertia, body, viscous, and pressure forces. In the energy equation, the first term represents the heat convection, the second is related to the work of compression, and the last two terms, respectively, denote the heat conduction and viscous dissipation.

The first step toward the understanding of a phenomenon is to find the dimensionless parameters pertaining thereto. Such a procedure is simplified in the present case if we consider that the body force effects are due to temperature differences, that the flow is steady, and the viscosity and thermal conductivity coefficients are constants. In addition we will assume that the state equation is independent of pressure, so that we can write

$$\rho = \rho_r[1 - \beta(T - T_r)] \tag{2-6}$$

where $\beta = \rho[\partial(1/\rho)/\partial T]_P$ is the volumetric expansion coefficient of the fluid, and the subscript $_r$ denotes the unheated no-flow state. The justification and limitations of this last assumption will be further discussed in subsequent articles. If we now introduce the above stated assumptions and Eq. 2-6 into Eq. 2-3 and 2-4, and also let

$$u_i = \frac{U_i}{\bar{U}}, \ \vartheta = \frac{\theta}{\theta_w}, \ x_i = \frac{X_i}{d} \text{ and } p = \frac{P}{\rho \bar{U}^2} \tag{2-7}$$

the momentum and energy equations for steady flows become, respectively,

$$u_j \frac{\partial u_i}{\partial x_j} = \frac{Gr}{Re^2}\vartheta + \frac{1}{Re}\left[\frac{\partial}{\partial x_j}\left(\frac{\partial u_i}{\partial x_j} + \frac{\partial u_j}{\partial x_i}\right) - \frac{2}{3}\frac{\partial}{\partial x_i}\left(\frac{\partial u_j}{\partial x_j}\right)\right] - \frac{\partial p}{\partial x_i} \tag{2-8}$$

$$u_j \frac{\partial \vartheta}{\partial x_j} = \frac{1}{PrRe}\frac{\partial^2 \vartheta}{\partial x_j \partial x_j} + Eu_j\frac{\partial p}{\partial x_j} + \frac{E}{Re}\left[\frac{\partial u_i}{\partial x_j}\left(\frac{\partial u_i}{\partial x_j} + \frac{\partial u_j}{\partial x_i}\right) - \frac{2}{3}\left(\frac{\partial u_j}{\partial x_j}\right)^2\right] \tag{2-9}$$

where the capital and lower case letters denote dimensional and dimensionless quantities, respectively; $\theta = T - T_r$ is a temperature difference; d is a characteristic dimension; and the subscript $_w$ denotes wall conditions. The reference velocity $\bar{U}$ denotes a unique velocity that characterizes the flow. For combined forced and natural convection flows, $\bar{U}$ is usually taken to be the free stream velocity U_∞.

From Eq. 2-8 and 2-9, we can thus see that solutions of forced convection problems including body forces depend on four parameters: the Reynolds number Re which represents the ratio of inertia to viscous forces; the Prandtl number Pr; the Eckert number E; and the Froude number, which is Re^2/Gr where Gr is the Grashof number which denotes the ratio of body or bouyancy forces to viscous forces. The Froude number thus compares inertia and body forces. Specifically these parameters are given by

$$Re = \frac{\bar{U}d}{\nu}; \qquad Gr = \frac{\beta f_i \theta_w d^3}{\nu^2}; \qquad Pr = \frac{c_p \mu}{k},$$

$$\text{and} \qquad E = \frac{\bar{U}^2}{c_p \theta_w} = \frac{2\theta_a}{\theta_w} \tag{2-10}$$

where ν is kinematic viscosity coefficient and θ_a denotes the adiabatic temperature difference.

For the case of natural or free convection flows (i.e., where the flow is generated solely by the action of a body force), however, there exists no unique prescribed characteristic velocity like the free stream velocity, so that the Reynolds and Eckert numbers lose their significance. In this case, a new characteristic velocity must be defined which, as was first explicitly pointed out by Ostrach [1], involves the three prime physical quantities of this phenomenon, namely, the fluid volumetric expansion coefficient, the body force, and a representative temperature difference. Unfortunately, as will presently be seen, the magnitude of the velocities in natural convection flows cannot be determined from the boundary conditions, so that the characterization of the reference velocity must come from the differential equations. The explicit form of this velocity will depend on the specific problem under consideration.

Therefore, in order to obtain nondimensional equations for natural convection which are analogous with those for forced convection, we let

$$\bar{U} = \sqrt{\beta f_i \theta_w d} \tag{2-11}$$

The equations for natural convection thus are

$$u_j \frac{\partial u_i}{\partial x_j} = \vartheta + \frac{1}{\sqrt{Gr}} \left[\frac{\partial}{\partial x_j} \left(\frac{\partial u_i}{\partial x_j} + \frac{\partial u_j}{\partial x_i} \right) - \frac{2}{3} \frac{\partial}{\partial x_i} \left(\frac{\partial u_j}{\partial x_j} \right) \right] - \frac{\partial p}{\partial x_i} \tag{2-12}$$

$$u_j \frac{\partial \vartheta}{\partial x_j} = \frac{1}{Pr \sqrt{Gr}} \frac{\partial^2 \vartheta}{\partial x_j \partial x_j} + K u_j \frac{\partial p}{\partial x_j} + \frac{K}{\sqrt{Gr}} \left[\frac{\partial u_i}{\partial x_j} \left(\frac{\partial u_i}{\partial x_j} + \frac{\partial u_j}{\partial x_i} \right) - \frac{2}{3} \left(\frac{\partial u_j}{\partial x_j} \right)^2 \right] \tag{2-13}$$

It is thus evident that, for flows generated entirely by body forces, three parameters are of significance: the Prandtl and Grashof numbers, already defined; and the new parameter K, which was first presented in [1] and more fully discussed in [2].

Comparison of Eq. 2-8 and 2-9 with Eq. 2-12 and 2-13 shows that the Grashof number is analogous to the Reynolds number and the new dimensionless group $K = \beta f_i d / c_p$ is analogous to the Eckert number. In fact, Gr is simply the square of the Reynolds number based on $\bar{U}$ as given by Eq. 2-11 and K is the Eckert number based on $\bar{U}$ as given by Eq. 2-11. The parameter K is important in the theory of natural convection, because the effects of the last two terms in the energy equation (the first related to the compression work and the second to the frictional heating) are determined by it and not by the Eckert number. Previous to [1], these effects were neglected on the basis of arguments showing that E was always small; in order to do so, a guess had to be made as to the magnitude of the reference velocity relative to the physically imposed temperature difference. When it is realized that in modern applications the body force and volumetric expansion coefficient, which are in the numerator of K, can easily be many times the usual values (for example, in a centrifugal field associated with rotating machinery, accelerations can be as much as 10^5 times that due to gravity), it is clear that the compression work and the frictional heating can influence natural convection flow and heat transfer. We can perhaps anticipate that the frictional heating provides heat sources in the fluid and results in higher fluid velocities and temperatures.

Having determined the parameters on which thermal flows with body forces depend and their physical significance, let us now discuss the mathematical character of the equations. The essential feature of thermal flows with body forces is that the velocity and temperature distributions are interdependent. Indeed, for thermal flows generated solely by body forces (natural flows), the fluid velocity is due entirely to the action of buoyancy (Archemedian) forces arising from the density variations, and this same distribution of density or, equivalently, of temperature, is modified as soon as the fluid moves. Thus distributions of velocity and density cannot be considered separately. They are both part of the problem and must be considered together. This fact, together with the nonlinearity of the basic governing equations, make natural convection problems very difficult. Since the remarks of this paragraph can be generalized, we can state that some sort of coupling must exist in problems in which body forces are important. This basic characteristic is what makes mathematical analysis of such flows so complex.

For theoretical analysis of this type of problem, it is, therefore, necessary to limit oneself in one or more of several ways. First of all, restrictions can sometimes be made on how the fluid properties vary. Secondly, specific configurations can be chosen so that the equations become simplified, and sometimes even linearized. Finally, the parameters of the problems can be considered to approach limiting values in order to obtain mathematical simplification. One or more of these assumptions is made in all the problems discussed herein. The justification and limitations of these assumptions will be explicitly pointed out in each case.

Before turning to the specific problems, it should be noted that the above discussion can easily be generalized to problems in which the density variations are due either solely or additionally to the concentration of admixtures. For if, instead of Eq. 2-6, we write the state equation more accurately as

$$\rho = \rho_r + \frac{\partial\rho}{\partial\theta}\theta + \frac{1}{2}\frac{\partial^2\rho}{\partial\theta^2}\theta^2 + \cdots + \frac{\partial\rho}{\partial C}C + \frac{1}{2}\frac{\partial^2\rho}{\partial C^2}C^2 + \cdots + \frac{\partial^2\rho}{\partial\theta\partial C}\theta C + \cdots + \frac{\partial C}{\partial P}P + \cdots \tag{2-14}$$

where C is the concentration of an admixture, and if we once again assume that the pressure variations and the second order effects are negligible, we can write the buoyancy term as $\rho_r(\beta\theta + \beta_1 C)$ where $\beta_1 = (1/\rho)(\partial\rho/\partial C)$. We would then have to add the diffusion (Fick) equation:

$$\frac{DC}{Dt} = \mathfrak{D}\frac{\partial^2 C}{\partial x_j \partial x_j}$$

to our list of basic equations where $\mathfrak{D}$ is the diffusion coefficient. The diffusion equation is easily recognized to be identical in form to the incompressible energy equation. Because of the symmetry between thermal and diffusive natural convection and the fact that most experimental work has treated thermal convection, we will here give consideration only to the thermal problems. The phenomena of thermodiffusive (combined thermal and diffusive) convection require independent investigation.

CHAPTER 2. FREE CONVECTION. THE EXTERNAL PROBLEM

Because problems in free or natural convection are complicated by the nonlinearity and high order of the governing partial differential equations, one must be rather discerning about the problems he chooses to analyze. If we examine other nonlinear problems on which considerable work has been done and which have a mathematical character similar to that discussed herein, we are quite naturally led to the techniques of boundary layer theory. Therefore, we shall first consider flows generated external to surfaces by body forces (free convection), and shall adapt the solution methods developed for boundary layers to our present needs. Essentially two approaches are available to us. In the first, asymptotic forms of the partial differential equations are obtained and then reduced to nonlinear ordinary equations by "similarity" transformations. The second method is an integral method, in which the equations and boundary conditions are satisfied at the surfaces and only the integrated equations are satisfied away from the surfaces.

ISOTHERMAL SURFACES

Let us in the following articles consider the free convection about an isothermal vertical surface (or, more generally, a surface parallel to the direction of the body force). The flow about such a surface will be produced in the following manner: When the surface is at the same temperature as the surrounding fluid, the body forces acting on the fluid are in equilibrium with the hydrostatic pressure and no flow exists in the steady state. If a temperature gradient normal to the body force direction is imposed by heating (or cooling) the surface, there will exist a defect (or excess) of body force because of the decreased (or increased) density, with the fluid closer to the surface having the greater defect (or excess) than that away from the surface. This unbalance of forces causes the fluid to be accelerated with the particles nearer the surface moving more rapidly than those farther away. Viscosity, of course, modifies this picture.

F,3. Vertical Flat Plate. The analysis and experiments by Schmidt and Beckmann [*3*] of the free convection flow of air subject to the gravitational force about an isothermal, vertical flat plate (the simplest configuration of the type to be discussed in this article) constitute one of the earliest comprehensive studies of this subject. Eckert and Soehngen [*4*], as well as others, verified and extended the experimental work, and Schuh [*5*] extended the numerical calculations by computing velocity and temperature distributions for several Prandtl numbers different from that of air. The present writer [*6*] reformulated the theoretical problem in a more general and formal manner, starting directly from the basic equations for a compressible, viscous, and heat-conducting fluid. The boundary layer equations were developed as asymptotic forms (for large Grashof numbers) of the basic equations in a rigorous manner. Although the final equations are identical with those in [*3*], a more detailed study of the significance of the various important parameters and assumptions was made. In addition, calculations were made to cover a more complete range of parameters. Accordingly, the method of [*6*] will be outlined below.

Particular consideration is given to the steady two-dimensional free convection about a semi-infinite flat plate. The X_1 axis of the coordinate system is taken along the plate and the X_2 axis normal to it. The body force is taken to be constant and parallel to the plate. In order to define the problem clearly, a choice must still be made of the position of the origin of the coordinate system. Before making a definite decision, note that for constant plate temperatures there are four permutations of the body force direction (say either upward or downward) and the plate thermal condition (either heated or cooled), which will lead to free convection flows.

Once the position of the edge of the plate, which is also to be the origin of the coordinate system, is decided, there are two combinations of the body force direction and plate thermal condition that will yield flows that proceed away from the edge. It is this type of flow that is amenable to the type of analysis to be made here. This point will be discussed more fully subsequently. If, for example, the edge of the (semi-infinite) plate is taken at bottom of the plate (i.e., the plate extends to $+\infty$ in the X_1 direction), the two combinations leading to flows in the proper direction (in the positive X_1 direction) are: the body force acting downward with a heated plate and the body force acting upward with a cooled plate. The equations developed for one of the cases reduce directly to those for the other. The remaining two permutations, namely, the body force acting downward with a cooled plate and the body force acting upward with a heated plate, would yield flows that proceed toward the edge of the plate if this edge were taken at the bottom. This type of flow would violate the physical condition that the flow starts at the plate edge. Therefore, for these cases the origin must be fixed at the top of the plate, and again the equations reduce to the previous form. Thus, we need only to be careful to analyze one consistent combination of body force direction and surface thermal condition, and then the results will be applicable to all the others. Here, we then take the origin to be at the bottom of a heated plate with the body force acting in the negative X_1 direction (see Fig. F,3a).

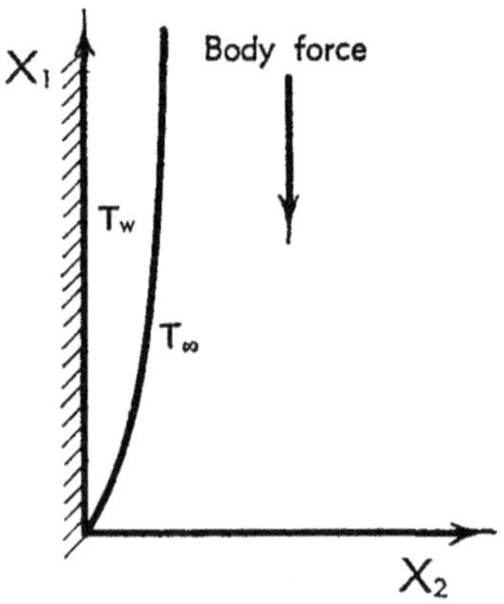

Fig. F,3a. Schematic sketch of free convection boundary layer.

We shall now assume that the viscosity and thermal conductivity coefficients are functions of the temperature alone and obey the following laws:

$$\left.\begin{aligned} \mu &= \mu_\infty \left(\frac{T}{T_\infty}\right)^m \\ k &= k_\infty \left(\frac{T}{T_\infty}\right)^n \end{aligned}\right\} \tag{3-1}$$

where the m and n are arbitrary exponents and the subscript ∞ denotes evaluation at undisturbed conditions.

The substitution of Eq. 3-1 into the basic dimensional equations, Eq. 2-1 to 2-3, yields

$$\frac{\partial}{\partial X_j}(\rho U_j) = 0 \tag{3-2}$$

$$\begin{aligned} \rho U_j \frac{\partial U_i}{\partial X_j} = -\rho(-f_i) + \frac{\mu_\infty}{T_\infty^m} \frac{\partial}{\partial X_j}\left[T^m\left(\frac{\partial U_i}{\partial X_j} + \frac{\partial U_j}{\partial X_i}\right)\right] \\ - \frac{2}{3}\frac{\mu_\infty}{T_\infty^m}\frac{\partial}{\partial X_i}\left(T^m \frac{\partial U_j}{\partial X_j}\right) - \frac{\partial P}{\partial X_i} \end{aligned} \tag{3-3}$$

$$\begin{aligned} \rho c_p U_j \frac{\partial T}{\partial X_j} = U_j \frac{\partial P}{\partial X_j} + \frac{k_\infty}{T_\infty^n}\frac{\partial}{\partial X_j}\left(T^n \frac{\partial T}{\partial X_j}\right) \\ + \frac{\mu_\infty}{T_\infty^m} T^m \left[\frac{\partial U_i}{\partial X_j}\left(\frac{\partial U_i}{\partial X_j} + \frac{\partial U_j}{\partial X_i}\right) - \frac{2}{3}\left(\frac{\partial U_j}{\partial X_j}\right)^2\right] \end{aligned} \tag{3-4}$$

The equations express, respectively, the conservation of mass, momentum, and energy. Note that the only nonzero component of the body force is the X_1 component. To complete the system of equations, we must add an equation of state

$$\rho = \rho(P, T) \tag{3-5}$$

For later use, this equation is written as

$$d\rho = \rho(\bar{K}dP - \beta dT) \tag{3-5a}$$

where $\bar{K} = -\rho[\partial(1/\rho)/\partial P]_T$ is the isothermal compressibility coefficient. In addition to a general state equation as given in Eq. 3-5 and 3-5a, it will be convenient at times to refer to the equation of state for an ideal gas which is

$$P = \rho \mathcal{R} T \tag{3-5b}$$

where $\mathcal{R}$ is the gas constant.

The boundary conditions associated with the given problem require that:

a. The fluid must adhere to the plate (the no-slip condition of viscous flows), and the plate must be a streamline, or mathematically:

$$U_1(X_1, 0) = U_2(X_1, 0) = 0 \tag{3-6}$$

b. The temperature of the fluid at the plate must be equal to the plate temperature, that is,

$$T(X_1, 0) = T_w \tag{3-7}$$

c. The fluid at large distances from the plate must remain undisturbed, or

$$U_1(X_1, \infty) = 0 \tag{3-8}$$

d. The temperature at large distances from the plate must be equal to the undisturbed fluid temperature, or,

$$T(X_1, \infty) = T_\infty \tag{3-8a}$$

Note that the velocity boundary conditions are homogeneous so that no velocity magnitude can be inferred from them, as was previously mentioned.

As a first step in the simplification of the boundary-value problem defined by Eq. 3-2 to 3-8, we define a small quantity ϵ as

$$\epsilon = \beta(T_w - T_\infty) \tag{3-9}$$

which is a measure of the magnitude of temperature variation in the flow field. The coefficient of volumetric expansion β is generally of an order of magnitude between 10^{-2} and 10^{-4}, and for gases $\beta = 1/T$. The coefficient β will be assumed constant herein. (Thus, for gases, $\epsilon = (T_w - T_\infty)/T_\infty$; that is, ϵ is the relative temperature difference.) Because, in the steady state, flow ensues only when there is a temperature variation in the fluid, the free convection velocity should depend directly on ϵ, and variations in pressure and density (from the static, $\epsilon \equiv 0$, case) due to temperature differences should also depend on ϵ. Thus

$$U_i = \epsilon \left(\frac{\rho_\infty f_X l^2}{\mu_\infty} \right) u_i \tag{3-10}$$

$$P = P_s + P_\infty \epsilon \sigma \tag{3-11}$$

$$\rho = \rho_s + \rho_\infty \epsilon \varphi \tag{3-12}$$

$$T = T_\infty (1 + \epsilon \theta) \tag{3-13}$$

where $-f_X$ denotes the X_1 component of the body force per unit mass; u_i, σ, φ and θ denote dimensionless functions (which, in general, can be functions of ϵ); l is some characteristic length (for example, the distance from the edge of the plate to the point of interest); P_s and ρ_s are the pressure and density, respectively, for the static case ($U_i \equiv 0$ or $\epsilon \equiv 0$); and P_∞ and ρ_∞ denote constant values of the pressure and density (that is, values if no force field were present) defined by the state equation (in particular, for a gas $P_\infty = \rho_\infty \mathcal{R} T_\infty$). Note that although, for convenience, the velocity is not made dimensionless by the characteristic velocity as given by Eq. 2-11, the characteristic velocity used in Eq. 3-10 does contain β, f_X, and $(T_w - T_\infty)$, as is essential according to the discussion in Art. 2.

In order to determine the static quantities, let us first consider the particular case of a gas flow. The problem is then considered with the temperature uniform throughout the entire flow field at the value T_∞ (therefore, there will be no flow and $U_i \equiv 0$). For this situation, Eq. 3-5b and 3-3 become

$$P_s = \rho_s \mathcal{R} T_\infty \tag{3-14}$$

and

$$\left. \begin{aligned} \frac{\partial P_s}{\partial X_1} + \rho_s f_X &= 0 \\ \frac{\partial P_s}{\partial X_2} &= 0 \end{aligned} \right\} \tag{3-15}$$

Note that Eq. 3-15 expresses the physical fact previously stated: that the body force and hydrostatic pressure are in equilibrium for the static case. Substitution of Eq. 3-14 into Eq. 3-15 leads to

$$P_s = P_\infty \exp\left(- \frac{f_X}{\mathcal{R} T_\infty} X_1 \right) \tag{3-16}$$

and Eq. 3-16, together with Eq. 3-14 and the equation defining P_∞ and ρ_∞, yields

$$\rho_s = \rho_\infty \exp\left(- \frac{f_X}{\mathcal{R} T_\infty} X_1 \right) = \rho_\infty \exp\left(- \frac{f_X \rho_\infty}{P_\infty} X_1 \right) \tag{3-17}$$

If the exponential in Eq. 3-17 is expressed in terms of its series expansion, it becomes

$$\rho_s = \rho_\infty \left(1 - \frac{\rho_\infty f_X}{P_\infty} X_1 + \cdots \right) \tag{3-18}$$

A computation of the second term in the parentheses of Eq. 3-18, for the case of air under normal conditions with $f_X = g$ (the acceleration of gravity) and where X_1 is of the order of magnitude l, shows that $\rho_\infty gl/P_\infty \sim 10^{-5}l/\text{ft}$. For the type of problem under consideration, l will always be of unit order of magnitude so that, even if the body force represents a centrifugal force many times that of gravity, the inequality $\rho_\infty f_X X_1/P_\infty \ll 1$ can be satisfied. Thus, in the subsequent development, it will be assumed that $\rho_s \cong \rho_\infty$. This assumption, which was justified by the computation for the case of a gas, is expected to be reasonable for other fluids as well. The physical interpretation of this assumption is that, under static conditions ($\epsilon \equiv 0$), the density (or pressure) is not affected by the body force field.

If we now let $x_i = X_i/l$ and apply Eq. 3-10 to 3-13 into Eq. 3-2 to 3-4 and Eq. 3-5a, noting Eq. 3-15 and that $\rho_s \cong \rho_\infty$, we obtain the following nondimensional equations to the first order in ϵ:

$$\frac{\partial u_1}{\partial x_1} + \frac{\partial u_2}{\partial x_2} = 0 \tag{3-19}$$

$$Gr\left(u_1 \frac{\partial u_1}{\partial x_1} + u_2 \frac{\partial u_1}{\partial x_2}\right) = \Delta u_1 - NGr\frac{\partial \sigma}{\partial x_1} - \varphi \tag{3-20}$$

$$Gr\left(u_1 \frac{\partial u_2}{\partial x_1} + u_2 \frac{\partial u_2}{\partial x_2}\right) = \Delta u_2 - NGr\frac{\partial \sigma}{\partial x_2} \tag{3-21}$$

$$PrGr\left(u_1 \frac{\partial \theta}{\partial x_1} + u_2 \frac{\partial \theta}{\partial x_2}\right) = \frac{\gamma}{\gamma - 1} PrGr\left(u_1 \frac{\partial \sigma}{\partial x_1} + u_2 \frac{\partial \sigma}{\partial x_2}\right) + \Delta\theta \tag{3-22}$$

$$d\varphi = \bar{K}P_\infty d\sigma - \beta T_\infty d\theta \tag{3-23}$$

where Δ is the Laplacian operator, N is a number defined by $P_\infty/\rho_\infty f_X l = NGr$, and γ is the ratio of specific heats. The property values in the Prandtl and Grashof numbers are evaluated at the undisturbed condition (subscript ∞). The coefficients $\bar{K}$ and β in Eq. 3-23 can be taken to be constants (see [7, pp. 26, 27]).

The boundary conditions, Eq. 3-6 to 3-9, in dimensionless form are

$$u_1(x_1, 0) = u_2(x_1, 0) = 0 \tag{3-24}$$

$$\theta(x_1, 0) = \frac{1}{\beta T_\infty} \tag{3-25}$$

$$u_1(x_1, \infty) = 0 \tag{3-26}$$

$$\theta(x_1, \infty) = 0 \tag{3-27}$$

Note that for gases $\beta T_\infty = 1$.

The prime assumption made in the analysis thus far is that the higher order terms in ϵ are negligible, which implies that ϵ is small, and consequently, that either the temperature difference or β is moderately small. It is a consequence of this assumption alone that the basic equations were simplified to Eq. 3-19 to 3-26, wherein the viscous dissipation term in the energy equation is neglected and the only coupling of the momentum and energy equations occurs by means of the body force term in Eq. 3-20. As a result of this assumption, the variations of viscosity and heat conductivity coefficients with temperature are also negligible.

It is now convenient to revert to the more familiar notation where $x = x_1$, $y = x_2$, $u = u_1$, and $v = u_2$. Eq. 3-19 implies the existence of a stream function ψ such that

$$u = \frac{\partial \psi}{\partial y} = \psi_y \quad \text{and} \quad v = -\frac{\partial \psi}{\partial x} = -\psi_x \tag{3-28}$$

Applying Eq. 3-28 to Eq. 3-20 to 3-22 yields, respectively:

$$\frac{1}{Gr}(\Delta\psi_y - \varphi) = \psi_y\psi_{xy} - \psi_x\psi_{yy} + N\sigma_x \tag{3-29}$$

$$-\frac{1}{Gr}(\Delta\psi_x) = -\psi_y\psi_{xx} + \psi_x\psi_{xy} + N\sigma_y \tag{3-30}$$

$$\frac{1}{PrGr}(\Delta\theta) = \psi_y\theta_x - \psi_x\theta_y - \frac{\gamma}{\gamma - 1}(\psi_y\sigma_x - \psi_x\sigma_y) \tag{3-31}$$

The boundary conditions, Eq. 3-24 to 3-27, become

$$\psi_y(x, 0) = \psi_x(x, 0) = 0 \tag{3-32}$$

$$\theta(x, 0) = \frac{1}{\beta T_\infty} \tag{3-33}$$

$$\psi_y(x, \infty) = \theta(x, \infty) = 0 \tag{3-34}$$

Eq. 3-29 to 3-31 and Eq. 3-23 are the system of equations for the four unknown functions ψ, θ, φ, and σ of the problem. The system is nonlinear, and therefore further simplification of the equations is desirable. Just as in the case of forced convection flows where the Reynolds number determines the type of flow or solution, the Grashof number is the prime factor for free convection flows. Recall that the Grashof number represents the ratio of body forces to viscous forces. For the case of small Grashof number, it can be seen from Eq. 3-29 to 3-31 that a perturbation in the small parameter Gr will yield a system of linear equations. For Grashof numbers of unit order of magnitude, no further important simplification is possible and the solutions would have to be obtained numerically. For the other limiting case, that of large Grashof numbers (which is the case of interest herein), it would, at first thought, appear that some simplification could be obtained by performing a perturbation in the small parameter $1/Gr$. However, this would imply that the terms containing the highest order derivatives (the left side of Eq. 3-29 to 3-31) could, among others, be neglected. The omission of the highest order derivatives from consideration, however, would lead to solutions that would not satisfy all the boundary conditions. Problems of this type are called singular perturbation problems (see [8], [9] for further discussion). This approach also would imply that the body force term φ in Eq. 3-29, which is essentially causing the flow, could also be neglected.

Equations in which a small parameter multiplies the highest order terms are said to be of the boundary layer type, because, in order for solutions which satisfy all the boundary conditions to be obtained, the highest order terms must be considered near the boundary. This fact implies the existence of a thin region, called the boundary layer, wherein the functions vary rapidly from the value at the boundary to that in the flow outside this layer. The conclusion to be drawn from this discussion is that for large Grashof numbers the flow is of the boundary layer type. Schmidt and Beckmann [3] also made the boundary layer assumptions in their theoretical development, and these assumptions were justified on the basis of their experimental observations. The Grashof numbers for their experiments were of the order of 8×10^6.

Since we know that the derivatives across the boundary layer must be large enough to make the highest order terms of magnitude equal to the other significant terms in the equations, we scale the normal coordinate and dependent variables so that all terms are of same order in Gr. To this end, let $\bar{y} = Gr^r y$, $\psi = Gr^s \bar{\psi}$, $\sigma = Gr^t \bar{\sigma}$, $\varphi = \bar{\varphi}$, and $\theta = \bar{\theta}$. Applying this transformation to Eq. 3-29 to 3-31 and 3-23, and retaining only the dominant part (that is, that multiplied by Gr to the highest power) of each individual term, leads to the somewhat simplified set

$$Gr^{3r+s-1}\bar{\psi}_{\bar{y}\bar{y}\bar{y}} - Gr^{-1}\bar{\varphi} = Gr^{2(r+s)}(\bar{\psi}_{\bar{y}}\bar{\psi}_{x\bar{y}} - \bar{\psi}_x\bar{\psi}_{\bar{y}\bar{y}}) + NGr^t\bar{\sigma}_x \tag{3-35}$$

$$-Gr^{2r+s-1}\bar{\psi}_{x\bar{y}\bar{y}} = Gr^{r+2s}(-\bar{\psi}_{\bar{y}}\bar{\psi}_{xx} + \bar{\psi}_x\bar{\psi}_{x\bar{y}}) + NGr^{r+t}\bar{\sigma}_{\bar{y}} \tag{3-36}$$

$$\frac{Gr^{2r-1}}{Pr}\bar{\theta}_{\bar{y}\bar{y}} = Gr^{r+s}(\bar{\psi}_{\bar{y}}\bar{\theta}_x - \bar{\psi}_x\bar{\theta}_{\bar{y}}) - \frac{\gamma}{\gamma - 1}Gr^{r+s+t}(\bar{\psi}_{\bar{y}}\bar{\sigma}_x - \bar{\psi}_x\bar{\sigma}_{\bar{y}}) \tag{3-37}$$

$$d\bar{\varphi} = \bar{K}P_\infty Gr^t d\bar{\sigma} - \beta T_\infty d\bar{\theta} \tag{3-38}$$

By a proper choice of the exponents r, s, and t we can make the important terms in the differential equations of equal order. Thus if $r = \frac{1}{4}$, $s = -\frac{3}{4}$, and $t = -1$, Eq. 3-35 to 3-38 become

$$\bar{\psi}_{\bar{y}\bar{y}\bar{y}} - \bar{\varphi} = \bar{\psi}_{\bar{y}}\bar{\psi}_{x\bar{y}} - \bar{\psi}_x\bar{\psi}_{\bar{y}\bar{y}} + N\bar{\sigma}_x \tag{3-39}$$

$$N\bar{\sigma}_{\bar{y}} = \mathcal{O}(Gr^{-\frac{1}{4}}) \cong 0 \tag{3-40}$$

$$\bar{\theta}_{\bar{y}\bar{y}} = Pr(\bar{\psi}_{\bar{y}}\bar{\theta}_x - \bar{\psi}_x\bar{\theta}_{\bar{y}}) \tag{3-41}$$

$$d\bar{\varphi} + \beta T_\infty d\bar{\theta} = 0 \tag{3-42}$$

More generally, if N is very much different from unit order of magnitude, a value of t can always be chosen such that Eq. 3-39 to 3-42 are obtained. For any negative t less than -1, the last term of Eq. 3-39 will also disappear.

There are now several important points to be discussed concerning the transformation just made and the resulting simplified equations. First, we note that the transformation is merely a formal expression of the boundary layer assumptions first made by Prandtl; hence the solutions will be asymptotic for large Gr. Second, the second equation of motion reduces to a statement that the pressure is constant across the boundary layer. Third, the terms in the energy and state equations containing the pressure are here found to be negligible. Thus the a priori assumptions made by others purely on a physical basis are verified in a formal manner. Finally, note that integration of the general state equation (independent of pressure), as given by Eq. 3-42, leads to

$$\bar{\varphi} + \beta T_\infty \bar{\theta} = 0 \tag{3-43}$$

where the integration constant has been taken to be zero without loss of generality.

The boundary conditions now can be written as

$$\bar{\psi}_{\bar{y}}(x, 0) = \bar{\psi}_x(x, 0) = 0 \tag{3-44}$$

$$\bar{\theta}(x, 0) = \frac{1}{\beta T_\infty} \tag{3-45}$$

$$\bar{\psi}_{\bar{y}}(x, \infty) = \bar{\theta}(x, \infty) = 0 \tag{3-46}$$

Since we are here considering a flat plate, $\bar{\sigma}_x$ in Eq. 3-39 is zero and, if we eliminate φ in Eq. 3-39 by use of Eq. 3-43, there results

$$\bar{\psi}_{\bar{y}\bar{y}\bar{y}} + \beta T_\infty \bar{\theta} = \bar{\psi}_{\bar{y}} \bar{\psi}_{x\bar{y}} - \bar{\psi}_x \bar{\psi}_{\bar{y}\bar{y}} \tag{3-47}$$

$$\bar{\theta}_{\bar{y}\bar{y}} = Pr(\bar{\psi}_{\bar{y}} \bar{\theta}_x - \bar{\psi}_x \bar{\theta}_{\bar{y}}) \tag{3-48}$$

Thus the problem has been reduced to the solution of two simultaneous partial differential equations subject to the boundary conditions given by Eq. 3-44 to 3-46.

Final simplification of the equations is made by application of a type of similarity transformation familiar in boundary layer theory. Thus, let

$$\eta = \frac{\bar{y}}{(4x)^{\frac{1}{4}}} \tag{3-49}$$

$$\bar{\psi} = (4x)^{\frac{3}{4}} F(\eta) \tag{3-50}$$

$$\bar{\theta} = \frac{H(\eta)}{\beta T_\infty} \tag{3-51}$$

Then Eq. 3-47 and 3-48 reduce to the following ordinary differential equations:

$$F''' + 3FF'' - 2F'^2 + H = 0 \tag{3-52}$$

$$H'' + 3PrFH' = 0 \tag{3-53}$$

where the primes denote differentiations with respect to η. The reciprocal one-fourth power similarity as given in Eq. 3-49 is characteristic of free convection flows, just as the reciprocal square-root type is characteristic of forced convection flows.

The boundary conditions become

$$F'(0) = F(0) = 0 \tag{3-54}$$

$$H(0) = 1 \tag{3-55}$$

$$F'(\infty) = H(\infty) = 0 \tag{3-56}$$

The use of a transformation like Eq. 3-49 essentially specifies an additional boundary condition, namely, that the conditions satisfied at $y = \infty$ (or $\eta = \infty$) should also be satisfied at $x = 0$. It is for this reason that the flows previously discussed which would flow toward the edge (downward) are not amenable to this type of analysis, for such flows would violate this additional condition, which essentially states that the boundary layer development starts at the edge of the plate.

The solutions of the simplified equations, Eq. 3-52 and 3-53 satisfying the boundary conditions, Eq. 3-54 to 3-56, were obtained by use of an IBM Card Programmed Electronic Calculator and are presented in tabular form in [*6*]. A detailed account of the forward-integration procedure is presented in Appendix B of [*6*]. The functions F and H, as well as their derivatives, are given in tabular form in [*6*] for Prandtl numbers of 0.01, 0.72, 0.733, 1, 2, 10, 100, and 1000. Even though the Prandtl number for air is taken as 0.72 herein, the solutions for $Pr = 0.733$ were also computed and are presented as a check with the Schmidt and Beckmann calculations [*3*], wherein $Pr = 0.733$ was used. The other particular values of Prandtl numbers were chosen to correspond to those for liquid metals, gases, liquids (such as water and oil), and very viscous liquids (such as glycerin or oils at very low temperatures).

We can relate the physical quantities to the dimensionless functions F and H, which are now known by means of the various transformations made in the analysis. Thus

$$\frac{U}{2\sqrt{\beta(T_w - T_\infty)f_X X}} = \frac{UX/\nu_\infty}{2\sqrt{Gr_X}} - F'(\eta) \tag{3-57}$$

and

$$\frac{T - T_\infty}{T_w - T_\infty} = H(\eta) \tag{3-58}$$

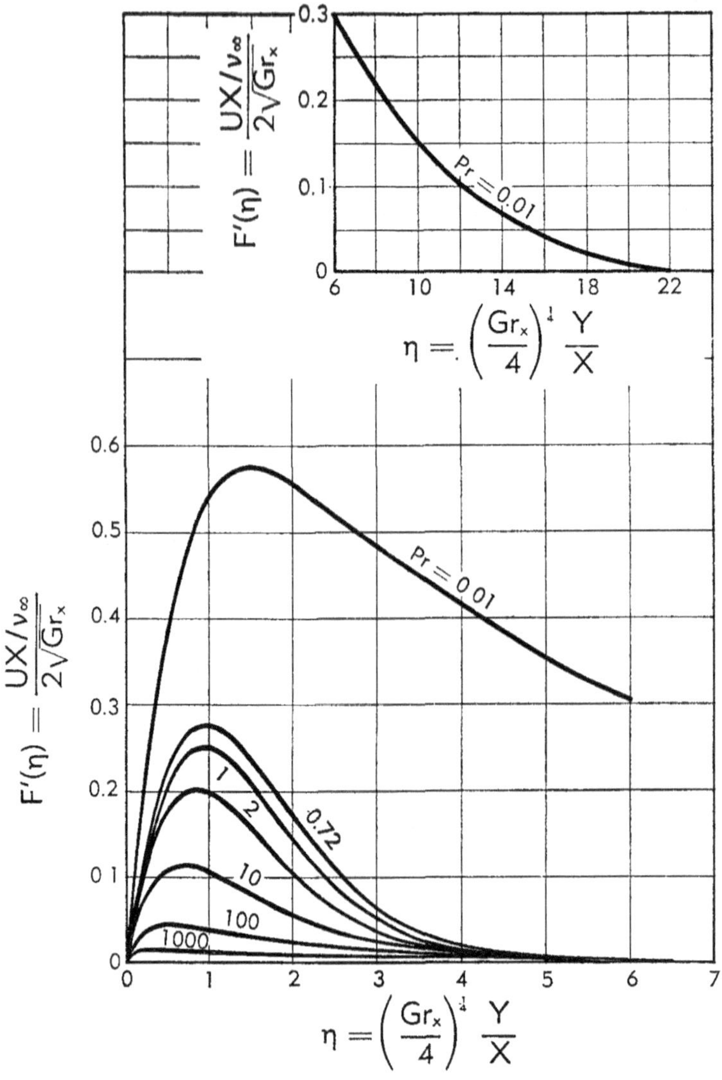

Fig. F,3b. Dimensionless velocity distributions for various Prandtl numbers (Ostrach [6]).

where

$$\eta = \left[\frac{\beta f x (T_w - T_\infty)}{4\nu_\infty^2 X}\right]^{\frac{1}{4}} Y = \left(\frac{Gr_X}{4}\right)^{\frac{1}{4}} \frac{Y}{X} \tag{3-59}$$

and Gr_X is the local Grashof number (based on X). The dimensionless velocity and temperature distributions as given by Eq. 3-57 and 3-58 are presented in Fig. F,3b and F,3c for various values of Prandtl number. The computations made here for $Pr = 0.733$ agree with those of [3] up to the third significant figure. For $Pr = 10$, 100, and 1000, the results

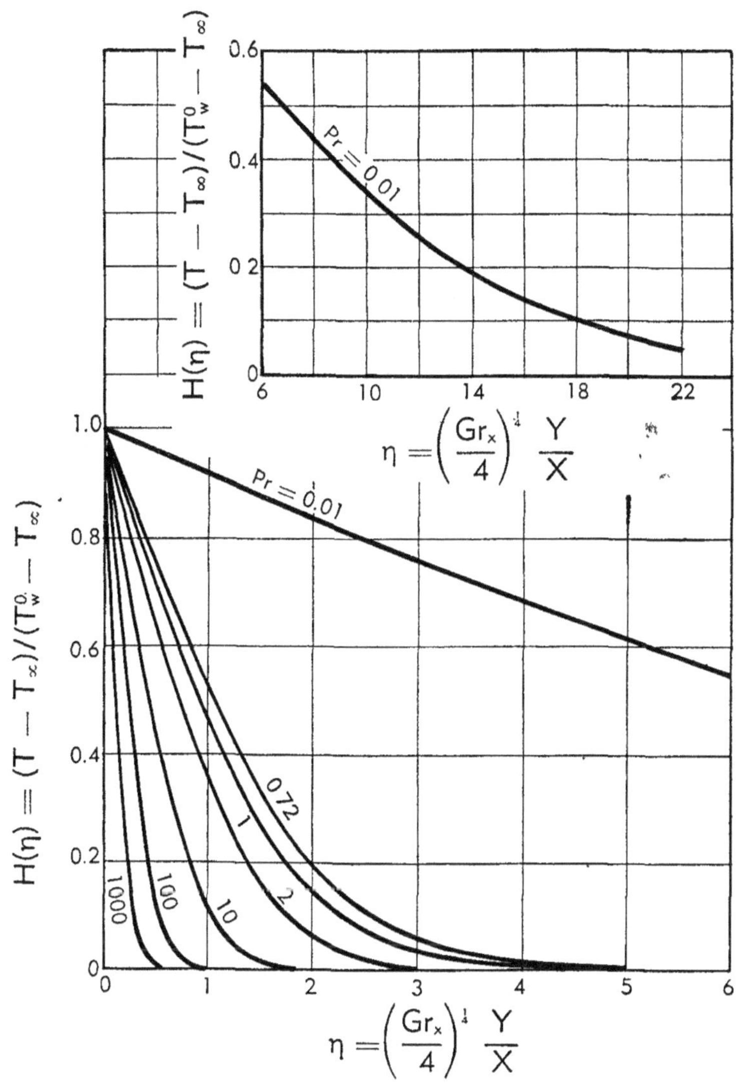

Fig. F,3c. Dimensionless temperature distributions for various Prandtl numbers (Ostrach [*6*]).

agree in general with those of [*5*]. Since only graphical results are presented in [*5*], the precision of the agreement cannot be stated.

The maximum values of the dimensionless velocity distributions can be seen to occur at larger values of the argument η as the Prandtl number decreases and the velocities decrease with increasing *Pr*. The dynamic and thermal boundary layer thicknesses can be estimated from the abscissas of Fig. F,3b and F,3c; for $Pr \gg 1$ the velocity boundary layer is much thicker than the thermal boundary layer.

Careful experiments with free convection flows (generated by the gravitational force) about vertical flat plates were made by Schmidt and Beckmann [*3*], in which velocity measurements at various points along the plate were made by means of a quartz-filament anemometer and the temperature measurements were obtained by means of manganese-constantan thermocouples. Eckert and Soehngen [*4*] performed similar experiments in which the measurements were made by means of a Mach-Zehnder interferometer. Both sets of experimental data are in good agreement, but since the results presented in [*3*] appear in more detail, these data are used for comparison with the theory. Before proceeding with the comparison, it should be pointed out that an excellent description of the instrumentation for free and natural convection experiments together with discussions of techniques, procedures, and results are given by Kraus [*10*].

The experiments of [*3*] were performed on two different (in that the edges were smoothed either symmetrically or not) 12- by 25-centimeter plates and on one 50- by 50-centimeter plate. The results for the two smaller plates were almost identical, and the flow was entirely laminar except near the outer edge of the boundary layer where the slight turbulence of the room air disturbed the measurements somewhat. (This effect was also observed by Eckert and Soehngen.) Large periodic oscillations of the flow near the downstream edge of the larger plate were observed, as well as the slight turbulence near the outer edge of the boundary layer. Hence, the data from the larger plate should not be expected to yield completely satisfactory agreement with the laminar theory.

Since the physical quantities can be expressed in terms of a single variable as in Eq. 3-57 and 3-58, the data taken at various points along the plates should be expected all to lie on a single line if the data are correlated according to Eq. 3-57 and 3-58. Thus for the smaller plates, where $(T_w - T_\infty) = 95.22°R$ and $T_\infty = 518.68°R$, Eq. 3-57 and 3-58 become

$$\frac{U}{4.862\sqrt{X}} = F'(\eta) \tag{3-60}$$

$$\frac{T - 518.68}{95.22} = H(\eta) \tag{3-61}$$

$$\eta = 88.26\frac{Y}{X^{\frac{1}{4}}} \tag{3-62}$$

The velocity and temperature distributions are so plotted in Fig. F,3d and F,3e, as are the curves computed theoretically for $Pr = 0.72$. The agreement is in general very good for small values of η and somewhat less satisfactory, although still rather good, for larger values of η. The scatter in the range of the larger η values is attributed to the room turbulence. It should also be noted that the points farthest away from the theoretical are those measured near the plate leading edge. These points should not, of course, be expected to agree too well with the theory since the boundary layer assumptions made in the theoretical development imply that the distance along the plate is large as compared with the

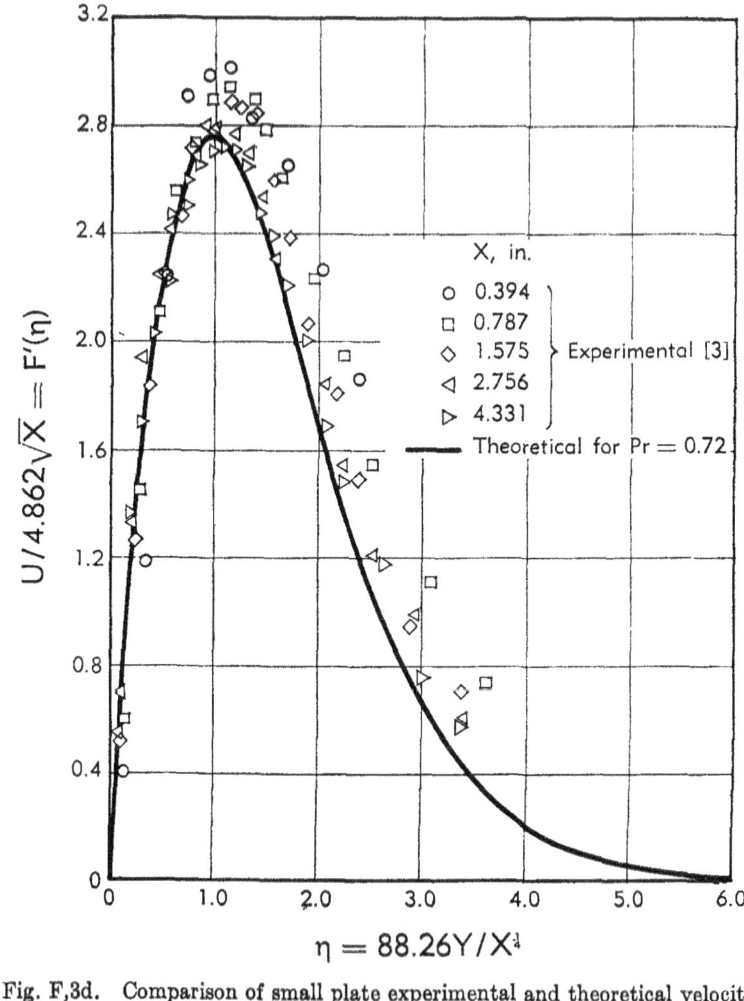

Fig. F,3d. Comparison of small plate experimental and theoretical velocity distributions for Prandtl number of 0.72 (Ostrach [6]).

boundary layer thickness. This assumption is invalid near the leading edge. Schmidt and Beckmann obtained closer agreement between theory and experiments for the temperature data, and poorer agreement for the velocity data, by basing the kinematic viscosity coefficient in Eq. 3-59 on the plate temperature rather than on the undisturbed air temperature

For the larger plate, $(T_w - T_\infty) = 83.7°R$ and $T_\infty = 527.14°R$, so that Eq. 3-57 to 3-59 become

$$\frac{U}{4.522\sqrt{X}} = F'(\eta) \tag{3-63}$$

$$\frac{T - 527.14}{83.7} = H(\eta) \tag{3-64}$$

$$\eta = 83.93\frac{Y}{X^{\frac{1}{4}}} \tag{3-65}$$

The velocity distribution for this experiment is plotted on Fig. F,3f

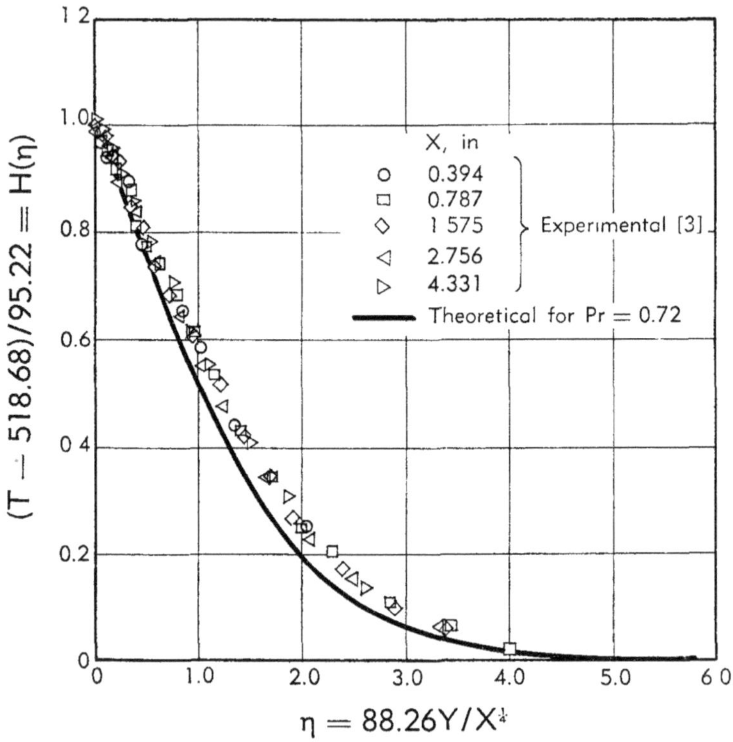

Fig. F,3e. Comparison of small plate experimental and theoretical temperature distributions for Prandtl number of 0.72 (Ostrach [*6*]).

and again the theoretical curve for $Pr = 0.72$ is included. The temperature distribution is not shown because it looks very similar to that for the small plate (Fig. F,3e). In Fig. F,3f the agreement for large η can be seen to be rather poor, particularly for the data for both small and large values of X. The poor agreement for small X values is again due to the theory limitation near the plate and, for large X values, because the flow was becoming turbulent there.

In addition to the velocity and temperature distributions, it is often

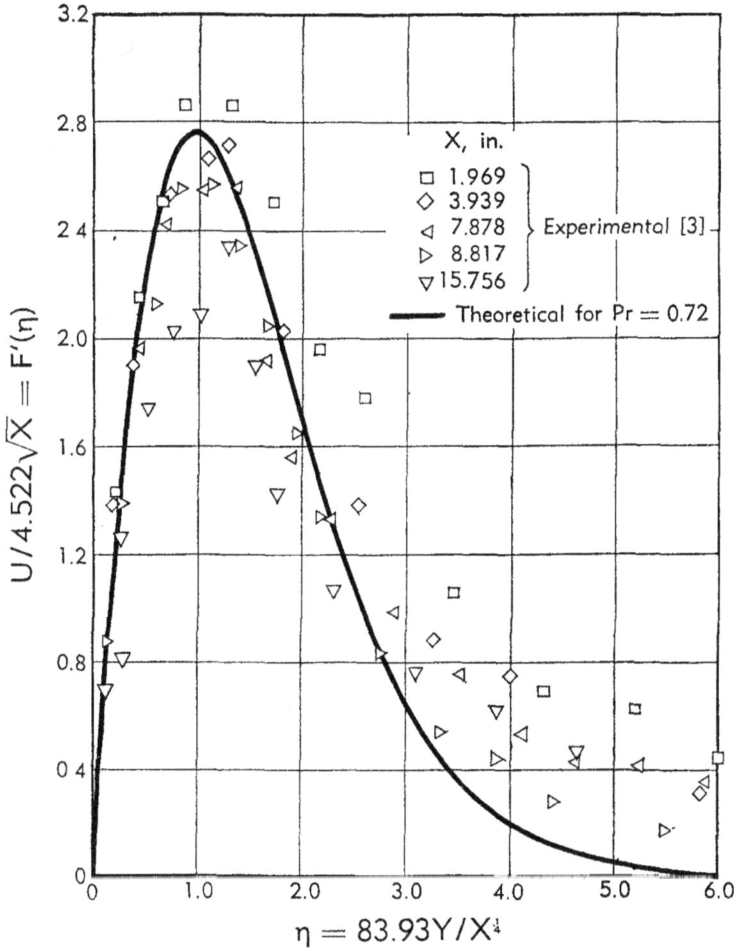

Fig. F,3f. Comparison of large plate experimental and theoretical velocity distributions for Prandtl number of 0.72 (Ostrach [6]).

desirable to compute other physically important quantities associated with free convection flow such as shear stress, drag, heat transfer rate, and heat transfer coefficient. For this purpose, we derive two parameters, a flow parameter and a heat transfer parameter, respectively, to be

$$\frac{\tau}{(4Gr_X^3)^{\frac{1}{4}}(\nu_\infty \mu_w / X^2)} = F''(0)$$

and

$$\frac{Nu}{(Gr_X/4)^{\frac{1}{4}}} = -H'(0)$$

where $H'(0)$ is a function of the Prandtl number, $\tau = \mu_w(\partial U/\partial Y)_{Y=0}$ is the shear stress, and $Nu = hX/k$ is the local Nusselt number with h

denoting the heat transfer coefficient. The flow parameter is presented as a function of the Prandtl number in Fig. F,3g, and all other flow quantities can easily be determined from it. The local heat transfer parameter is also plotted in Fig. F,3g as a function of Prandtl number.

On the basis of a simplified theory (that is, by use of integrated momentum and energy equations and assumed velocity and temperature distributions), Eckert [*11*, p. 162] obtained the approximate relation

$$\frac{Nu}{(Gr_X/4)^{\frac{1}{4}}} = \frac{0.718(Pr)^{\frac{1}{2}}}{(0.952 + Pr)^{\frac{1}{4}}}$$

The curve representing this equation is also presented in Fig. F,3g; it

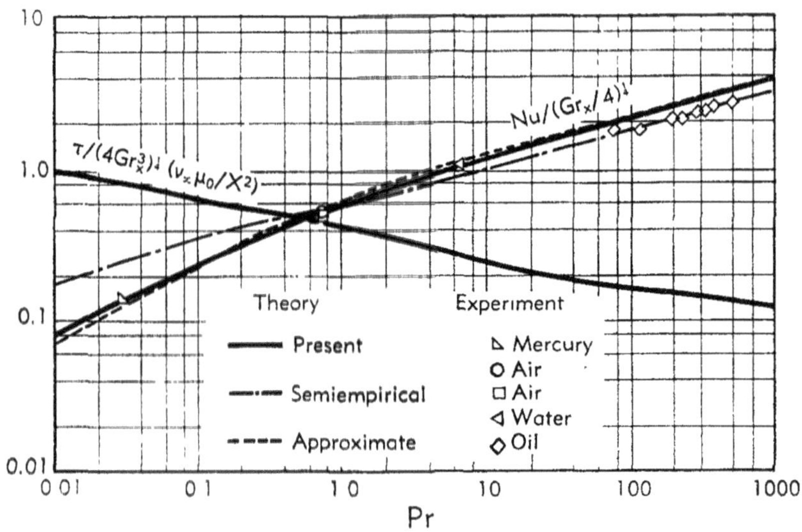

Fig. F,3g. Dimensionless flow and heat-transfer parameters as function of Prandtl number (Ostrach [*6*]).

approximates closely over the entire Prandtl number range (to within about 10 per cent) the curve determined by the more exact considerations. A semiempirical equation, as given in McAdams [*12*, p. 242], relating the average (over the length X) Nusselt number to the Prandtl and Grashof numbers, is

$$\overline{Nu} = 0.548[(Pr)Gr]^{\frac{1}{4}}$$

The constant 0.548 pertains specifically to air: for oil, it should be 0.555 (see [*13*]) and for mercury, approximately 0.33 (see [*14*]).

To obtain local values from the average one, it can be shown from the definitions that it is merely necessary to multiply the average values by 0.75. Thus, in terms of the local quantities, the semiempirical relation becomes for air

$$Nu = 0.411[(Pr)Gr_x]^{\frac{1}{4}}$$

or

$$\frac{Nu}{(Gr_x/4)^{\frac{1}{4}}} = 0.581(Pr)^{\frac{1}{4}}$$

The curve given by this equation is also presented in Fig. F,3g, and the agreement with the theoretical curve is very good for Prandtl numbers near unity, not so good for large Prandtl numbers, and quite poor for small Prandtl numbers. The use of the proper constant at the extreme values (as given above) would, of course, improve the agreement there.

The relatively poor agreement of the approximate Nusselt number formulas, which contain the Prandtl number explicitly with the exact expressions at the extremes of the Prandtl number range, caused Le Fèvre [*15*] to solve the asymptotic forms of Eq. 3-52 and 3-53 for vanishing and infinite values of the Prandtl number. On the basis of these solutions, he developed the following formula for the mean Nusselt number which agrees to within one unit in the third decimal place with the exact results of Ostrach [*6*].

$$Nu = \left(\frac{GrPr^2}{2.43478 + 4.884\sqrt{Pr} + 4.95283\ Pr}\right)^{\frac{1}{4}}$$

Recently, Sparrow and Gregg [*16,17*] obtained accurate numerical solutions for the free convection about a vertical flat plate for Prandtl numbers of 0.003, 0.008, 0.01, 0.02, and 0.03. Their results agreed to within one per cent with those of Ostrach [*6*] for $Pr = 0.01$ and with those of Sugawara and Michiyoshi [*18*] for $Pr = 0.03$. If the local Nusselt number for the Prandtl number range studied in [*16*] is represented as

$$Nu = 0.565(Gr_xPr^2)^{\frac{1}{4}}$$

it will agree with the numerical results of [*16*] to within three per cent.

The local values of the heat transfer parameter obtained experimentally for mercury ($Pr = 0.03$), air ($Pr = 0.72$), water ($Pr = 7$), and oil ($Pr = 75.5$, 115, 190, 224, 275, 318, 368, and 442) are here reduced by the factor 0.75 from the average values reported. The value for mercury is an average taken of four readings from a curve, since this experiment was the only one not reported in tabular form. From Fig. F,3g it can be seen that all of the experimental values, except those for the oil experiments, are in close agreement with the theoretically computed values. The data for oil, though not so good, show reasonable agreement (maximum error of approximately 20 per cent) with the theoretical curve and good agreement, as is to be expected, with the semiempirical curve. The difference between the theoretical and experimental results for oil is possibly because the viscosity changes in oil are large, even for small temperature differences, and can also be due to end effects in the experiments. We thus see that adaptation of boundary layer theory to the rather simple configuration just discussed has yielded analytical results that are in good agreement with experimental data.

It should here be remarked that the approximate method of Kármán and Pohlhausen for solving boundary layer equations in which the boundary conditions and the equations themselves are satisfied at the boundaries, and integrals of the basic equations are satisfied in the interior of the flow, was also applied to the present problem (see [*19*, pp. 671–673; *11*, pp. 158–163]). The heat transfer results of those calculations are in reasonable agreement with the exact solutions just presented.

F,4. Inclined Plate. If the plate is not parallel to the body force direction but is inclined an angle φ from it, the differential equations for the vertical plate are modified only in that the cosine of φ appears as a multiplicative factor in the body force term. Rich [*20*] showed that the Nusselt number for free convection about an inclined plate is

$$Nu = f(Pr, Gr \cos \varphi) \tag{4-1}$$

whereas we have just seen that for the vertical plate

$$Nu = f(Pr, Gr) \tag{4-2}$$

By means of a Mach-Zehnder interferometer, Rich [*20*] determined the heat transfer and temperature distributions at various positions along inclined plates. The Nusselt numbers for inclined plates predicted by modification of vertical plate results according to Eq. 3-66 and 3-67 were found to agree with experimental values to within 10 per cent.

F,5. Vertical Cylinder. To find the effect of surface curvature transverse to the flow direction, Sparrow and Gregg [*21*] studied the free convection for the outer surface of a vertical cylinder. The configuration investigated is shown in Fig. F,5a, and the surface temperature of the cylinder is assumed to be uniform. (Note that the case of a cooled wall with the flow proceeding down from the cylinder top would be completely analogous to this case.) We may anticipate that, when the boundary layer is so thin that the radial coordinate of any point in the boundary layer is not appreciably different from the radius of the cylinder, the results for the cylinder should be close to those for a flat plate. As the boundary layer thickness increases relative to the cylinder radius, we can expect the cylinder results to depart further and further from those of the flat plate.

The analysis for the free convection flow follows directly from the similar one for forced convection given in [*22*]. The equations expressing conservation of mass, momentum, and energy in cylindrical coordinates are, respectively,

$$\frac{\partial(RU)}{\partial X} + \frac{\partial(RV)}{\partial R} = 0 \tag{5-1}$$

$$U\frac{\partial U}{\partial X} + V\frac{\partial U}{\partial R} = \beta g(T - T_\infty) + \frac{\nu}{R}\frac{\partial}{\partial R}\left(R\frac{\partial U}{\partial R}\right) \tag{5-2}$$

$$U\frac{\partial T}{\partial X} + V\frac{\partial T}{\partial R} = \frac{\alpha}{R}\frac{\partial}{\partial R}\left(R\frac{\partial T}{\partial R}\right) \tag{5-3}$$

where X is the axial, R the radial coordinate, and α the thermal diffusivity. Subject to the same assumptions made for the flat plate, the density is considered to vary only in the formulation of the buoyancy term, and

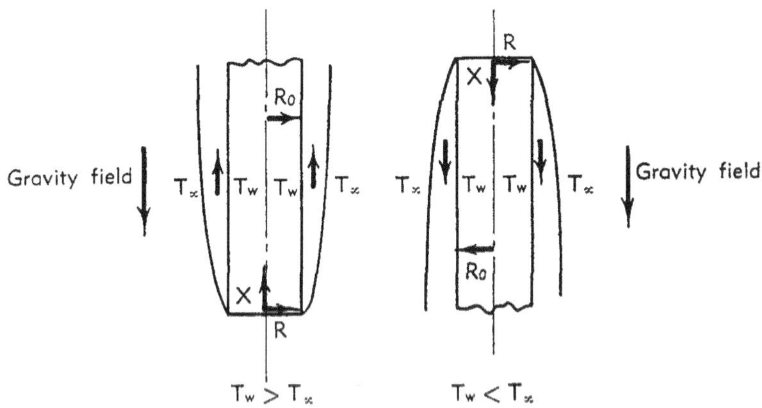

Fig. F,5a. Schematic sketch of vertical cylinder configurations (Sparrow and Gregg [21]).

all other fluid properties are taken to be constant. Viscous dissipation and compression work have also been neglected.

Eq. 5-1 is identically satisfied by introduction of a stream function ψ where

$$U = \frac{1}{R}\frac{\partial\psi}{\partial R} \qquad V = -\frac{1}{R}\frac{\partial\psi}{\partial X} \tag{5-4}$$

Introducing Eq. 5-4, together with the transformations,

$$\eta = \frac{Gr^{\frac{1}{4}}}{2^{\frac{3}{2}}R_0^{\frac{3}{4}}}\frac{(R^2 - R_0^2)}{X^{\frac{1}{4}}} \tag{5-5}$$

$$\xi = \frac{2^{\frac{3}{2}}}{(R_0 Gr)^{\frac{1}{4}}}X^{\frac{1}{4}} \tag{5-6}$$

$$f(\xi, \eta) = \frac{2^{\frac{3}{2}}}{\nu(R_0 Gr)^{\frac{3}{4}}X^{\frac{1}{4}}}\psi \tag{5-7}$$

$$\theta(\xi, \eta) = \frac{T - T_\infty}{T_w - T_\infty} \tag{5-8}$$

into Eq. 5-2 and 5-3 yields

$$\xi\left(\frac{\partial f}{\partial \eta}\frac{\partial^2 f}{\partial \xi \partial \eta} - \frac{\partial f}{\partial \xi}\frac{\partial^2 f}{\partial \eta^2}\right) - f\frac{\partial^2 f}{\partial \eta^2} = \xi^2 \frac{\partial}{\partial \eta}\left[(1 + \xi\eta)\frac{\partial^2 f}{\partial \eta^2}\right] + \xi^4 \theta \quad (5\text{-}9)$$

$$\xi\left(\frac{\partial \theta}{\partial \xi}\frac{\partial f}{\partial \eta} - \frac{\partial \theta}{\partial \eta}\frac{\partial f}{\partial \xi}\right) - f\frac{\partial \theta}{\partial \eta} = \frac{\xi^2}{Pr}\frac{\partial}{\partial \eta}\left[(1 + \xi\eta)\frac{\partial \theta}{\partial \eta}\right] \quad (5\text{-}10)$$

where R_0 is the cylinder radius and Gr is the Grashof number based on this radius (i.e., $\beta g(T_w - T_\infty)R_0^3/\nu^2$). Note that Eq. 5-9 and 5-10 are coupled through the buoyancy term, thus showing that the free convection flow is generated by density differences caused by temperature differences.

To find f and θ, we express these functions in a power series of ξ as

$$f(\xi, \eta) = \xi^2[f_0(\eta) + \xi f_1(\eta) + \xi^2 f_2(\eta) + \cdots] \quad (5\text{-}11)$$

$$\theta(\xi, \eta) = \theta_0(\eta) + \xi\theta_1(\eta) + \xi^2\theta_2(\eta) + \cdots \quad (5\text{-}12)$$

Eq. 5-11 and 5-12 are then substituted into Eq. 5-9 and 5-10 and the terms are grouped according to the power of ξ which multiplies them. Since the resulting equations must hold for all values of ξ, the factor of each power of ξ is equated to zero yielding the following set of equations for the functions f_n and θ_n

$$\left.\begin{aligned} f_0''' + 3f_0f_0'' - 2(f_0')^2 + \theta_0 &= 0 \\ \theta_0'' + 3Prf_0\theta_0' &= 0 \end{aligned}\right\} \quad (5\text{-}13)$$

$$\left.\begin{aligned} f_1''' + 3f_0f_1'' - 5f_0'f_1' + 4f_0''f_1 + f_0'' + \eta f_0''' + \theta_1 &= 0 \\ \theta_1'' - Pr(f_0'\theta_1 - 3f_0\theta_1' - 4f_1\theta_0') + \theta_0' + \eta\theta_0'' &= 0 \end{aligned}\right\} \quad (5\text{-}14)$$

$$\left.\begin{aligned} f_2''' + 3f_0^2f_2'' - 6f_0'f_2' + 5f_0''f_2 + \eta f_1''' + f_1'' - 3(f_1')^2 & \\ + 4f_1f_1'' + \theta_2 &= 0 \\ \theta_2'' - Pr(2f_0'\theta_2 - 3f_0\theta_2' + f_1'\theta_1 - 5f_2\theta_0' - 4f_1\theta_1') + \theta_1' + \eta\theta_1'' &= 0 \end{aligned}\right\} \quad (5\text{-}15)$$

The boundary conditions at $R = R_0$ are that $U = 0$, $V = 0$, and at $R = \infty$, $U = 0$; in terms of the function f these become

$$\left.\begin{aligned} &\text{at } \eta = 0 \quad \begin{cases} f = 0 \\ \dfrac{\partial f}{\partial \eta} = 0 \\ \dfrac{\partial f}{\partial \xi} = 0 \end{cases} \\ &\text{at } \eta = \infty \quad \left\{ \dfrac{\partial f}{\partial \eta} = 0 \right. \end{aligned}\right\} \quad (5\text{-}16)$$

From the definition of θ, Eq. 5-8, it follows that

$$\text{at } \eta = 0,\ \theta = 1 \quad \text{at} \quad \eta = \infty,\ \theta = 0 \quad (5\text{-}17)$$

These boundary conditions can be written in terms of the various functions given in Eq. 5-11 and 5-12 as

$$\left.\begin{aligned} &\text{at } \eta = 0 \quad \begin{cases} f_0 = f_1 = f_2 \cdots = 0 \\ f_0' = f_1' = f_2' \cdots = 0 \\ \theta_0 = 1,\ \theta_1 = \theta_2 \cdots = 0 \end{cases} \\ &\text{at } \eta = \infty \quad \begin{cases} f_0' = f_1' = f_2' \cdots = 0 \\ \theta_0 = \theta_1 = \theta_2 \cdots = 0 \end{cases} \end{aligned}\right\} \tag{5-18}$$

The differential equations for f_0 and θ_0, Eq. 5-13 together with their boundary conditions, are identical to those for a flat plate. Thus, as we anticipated, the succeeding f_n and θ_n give the departures from the flat plate solutions due to the cylindrical geometry. Numerical solutions of Eq. 5-13 to 5-15 subject to the conditions given by Eq. 5-18 were obtained on an IBM Card Programmed Calculator for Prandtl numbers of 0.72 and 1. The calculation procedure is outlined in an Appendix of [*21*], and more details can be found in the Appendix of [*6*].

The local Nusselt number for free convection about a vertical cylinder can be written in terms of the θ_n functions as

$$Nu_{\text{cyl}} = \frac{hX}{k} = -\frac{1}{\sqrt{2}} Gr_X^{\frac{1}{4}} \theta_0'(0) \left[1 + \xi \frac{\theta_1'(0)}{\theta_0'(0)} + \xi^2 \frac{\theta_2'(0)}{\theta_0'(0)} + \cdots \right] \tag{5-19}$$

where h is the local heat transfer coefficient and Gr_X denotes the Grashof number based on X. We have, however, previously seen that the factor of the bracketed term in Eq. 5-19 is the local Nusselt number for the flat plate; therefore, we can write

$$\frac{Nu_{\text{cyl}}}{Nu_{\text{fp}}} = 1 + \xi \frac{\theta_1'(0)}{\theta_0'(0)} + \xi^2 \frac{\theta_2(0)}{\theta_0'(0)} + \cdots \tag{5-19a}$$

Eq. 5-19a compares the local Nusselt number for a cylinder at a location X with that for a flat plate at the same X, under the condition that $(T_w - T_\infty)$ is the same for the flat plate and the cylinder. The $\theta_n'(0)$ depend only on the Prandtl number. Therefore, for a fixed value of the Prandtl number, the ratio given by Eq. 5-19a is a function only of ξ where explicitly

$$\xi = \frac{2^{\frac{3}{4}}}{Gr_X^{\frac{1}{4}}} \left(\frac{X}{R_0} \right) \tag{5-20}$$

After the various $\theta_n'(0)$ determined from the calculations are substituted into Eq. 5-19a, the ratio of local Nusselt numbers is plotted as a function of ξ in Fig. F,5b for Prandtl numbers of 0.72 and 1. Once the ratio is obtained from the figure, $Nu_{x,\text{cyl}}$ can be determined as soon as $Nu_{x,\text{fp}}$ is given. From [*6*] it is, for example, found that

$$\left.\begin{aligned} Nu_{\text{fp}} &= 0.3568 Gr_X^{\frac{1}{4}} \quad \text{for} \quad Pr = 0.72 \\ Nu_{\text{fp}} &= 0.4010 Gr_X^{\frac{1}{4}} \quad \text{for} \quad Pr = 1 \end{aligned}\right\} \tag{5-21}$$

It can immediately be seen from Fig. F,5b that the cylinder Nusselt number deviates more and more from that for the flat plate as ξ increases

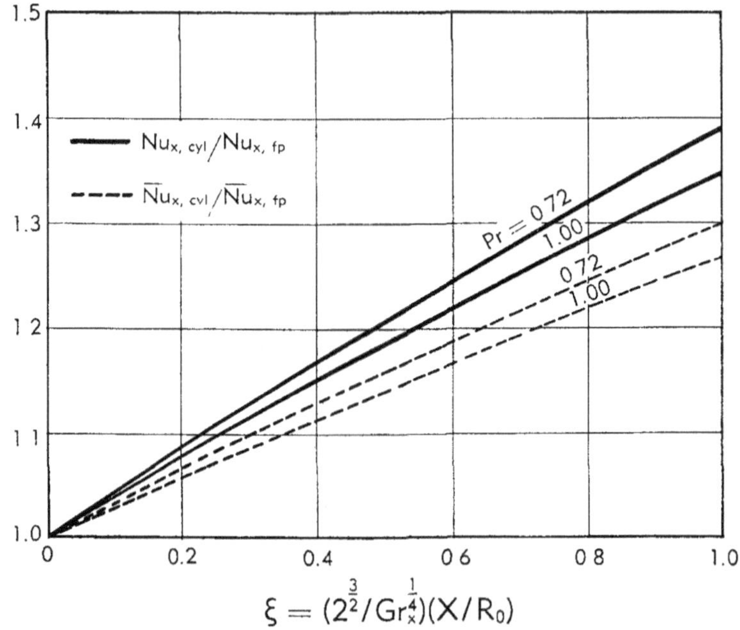

Fig. F,5b. Comparison of local and average Nusselt number for vertical cylinder with that for flat plate (Sparrow and Gregg [21]).

for a fixed Prandtl number. Also, at a fixed value of ξ, the deviations increase as the Prandtl number decreases. These trends become more plausible as soon as the physical interpretation of the parameter ξ becomes evident. For the flat plate, the thermal boundary layer thickness Δ is given by

$$\frac{\Delta}{X} = \frac{\lambda(Pr)}{Gr_X^{\frac{1}{4}}}$$

where $\lambda(Pr)$ is a number which increases with decreasing Prandtl number. It then follows that

$$\frac{\Delta}{R_0} = \frac{\Delta}{X}\frac{X}{R_0} = \frac{\lambda(Pr)}{Gr_X^{\frac{1}{4}}}\frac{X}{R_0} = \xi\frac{\lambda(Pr)}{2^{\frac{3}{2}}}$$

so that it is clear that ξ is proportional to Δ/R_0. For a fixed Prandtl number, when Δ is small relative to R_0 (small ξ), the cylinder results are close to those for a flat plate; as Δ increases relative to R_0 (increasing ξ), the cylinder results deviate increasingly from those for a flat plate. At a fixed ξ, Δ/R_0 increases with decreasing Prandtl number so that the cylinder results again deviate more from the flat plate results.

The average heat transfer coefficient $\bar{h}$ for a length of cylinder between $X = 0$ and X is

$$\bar{h} = \frac{1}{X}\int_0^X h dX$$

which upon integration of Eq. 5-19 can be written in terms of an average Nusselt number $\overline{Nu} \equiv \bar{h}X/k$ as

$$\frac{\overline{Nu}_{\text{cyl}}}{\overline{Nu}_{\text{fp}}} = 1 + \frac{3}{4}\,\xi\,\frac{\theta_1'(0)}{\theta_0'(0)} + \frac{3}{5}\,\xi^2\,\frac{\theta_2'(0)}{\theta_0'(0)} + \cdots \tag{5-22}$$

Eq. 5-22 thus compares average Nusselt numbers for the cylinder and flat plate. The results of calculations from Eq. 5-22 are presented in Fig. F,5b. Once again from the solutions of [*6*], it is found that

$$\left.\begin{aligned} \overline{Nu}_{X,\text{fp}} &= 0.4757 Gr_X^{1/4} \quad \text{for} \quad Pr = 0.72 \\ \overline{Nu}_{X,\text{fp}} &= 0.5347 Gr_X^{1/4} \quad \text{for} \quad Pr = 1 \end{aligned}\right\} \tag{5-23}$$

so that the average cylinder Nusselt numbers can be determined by inserting the results of Eq. 5-23 into Eq. 5-22.

The general trends for the average Nusselt number are the same as those for the local ones. It can be noted, however, that for a given ξ and Pr the ratio of average Nusselt numbers is less than the ratio of local Nusselt numbers or, in other words, the cylindrical geometry has a smaller effect on the average Nusselt number than on the local Nusselt number. This result is also reasonable because, in the average Nusselt number, the effects upstream of the station X, where the boundary layer is thinner, are integrated into the results.

For practical purposes, it is desirable to establish a criterion for distinguishing among those conditions under which the flat plate results are sufficiently accurate for calculating the heat transfer from a cylinder, and those conditions under which the Nusselt number results derived for the cylinder must be used. The analysis first described provides the basis for such a criterion. If it is decided that the flat plate results are sufficiently accurate if they yield errors of five per cent or less in calculating heat transfer parameters for the cylinder, then the flat plate results may be used under the following conditions:

For local Nusselt numbers

$$\frac{2^{3/2}}{Gr_X^{1/4}}\left(\frac{X}{R_0}\right) \leqq 0.11 \quad \text{for} \quad Pr = 0.72$$

$$\frac{2^{3/2}}{Gr_X^{1/4}}\left(\frac{X}{R_0}\right) \leqq 0.13 \quad \text{for} \quad Pr = 1$$

For average Nusselt numbers

$$\frac{2^{3/2}}{Gr_X^{1/4}}\left(\frac{X}{R_0}\right) \leqq 0.15 \quad \text{for} \quad Pr = 0.72$$

$$\frac{2^{3/2}}{Gr_X^{1/4}}\left(\frac{X}{R_0}\right) \leqq 0.17 \quad \text{for} \quad Pr = 1$$

The use of flat plate Nusselt numbers for computing the heat transfer from the cylinder will always give conservative results.

Carne [*23*] performed free convection experiments for vertical cylinders of various diameters and lengths, but these experiments did not yield sufficient data in the laminar flow region to permit a check of the analytical results presented here. Hence there remains a need for a set of systematic experiments on laminar free convection about vertical cylinders.

The analysis presented above is an important step in showing how the cylinder results start to deviate from those for the flat plate. However, it still gives no information on what happens when the heated layer becomes equal to, or greater than, the cylinder radius because of its essential assumption that ξ (ratio of boundary layer thickness to cylinder radius) is small. The other end of the problem (i.e., where the transverse curvature effects are large) is treated experimentally and theoretically by Hama and Recesso [*24*]. The theoretical analysis follows closely the approximate method of Glauert and Lighthill [*25*], which is applicable along the entire cylinder, and was developed for forced flow boundary layers. This method is essentially a Kármán-Pohlhausen integral procedure. It is pointed out in [*24*] that for axially symmetric problems the assumed profiles must be more carefully chosen; the appropriate ones used in [*24*] and [*25*] are logarithmic. To make this type of solution compatible with the previous analysis near the leading edge of the cylinder [*21,22*], corrections then have to be made to the results of [*24,25*].

In [*24*], temperature profiles were measured at several sections along an electrically heated brass cylinder which was 10 feet long and one-fourth inch in diameter. The experimentally determined, local heat transfer coefficients agreed reasonably well with the theory. It is interesting to note that, whereas it is difficult to obtain axially symmetric boundary layers in forced flow [*26*], in free convection the problem is quite simple. Furthermore, since the thickness of a free convection boundary layer (varies like $X^{-\frac{1}{4}}$) is greater than that in forced flow (varies like $X^{-\frac{1}{2}}$), the effect of transverse curvature is more significant.

F,6. Horizontal Cylinder. Free convection about horizontal cylinders is of particular importance because, next to the flat plate, it represents the simplest two-dimensional configuration. This configuration is important over a wide range of conditions, from the flow over the finest electrically heated wires to that over large pipes heated by fluids flowing through them. It is then not surprising that perhaps the greatest amount of free convection experimentation has centered around this configuration. One need only refer to McAdams [*12*], wherein there appears a rather comprehensive summary of this work, to see the variety of cylinder sizes and heat transfer fluids used in the experiments including air, hydrogen, carbon dioxide, oxygen, water, aniline, carbon tetrachloride, glycerin, toluene, and olive oil. Altogether, in these experiments, a 7300-fold range of diameters, a 10,000-fold range of pressures, an 11,000-fold range in the Prandtl number, and a range of temperature difference from a few degrees to 2850°F are covered. An excellent correlation of this data is presented in [*12*].

The good agreement between theory and experiment for free convection about a flat plate and the fact that schlieren photographs of free convection of air about horizontal cylinders (see Fig. 23 and 25 of [*27*]) showed that, for Grashof numbers above about 10^4, the flow is of a boundary layer type prompted Hermann [*28*] to analyze the cylinder flow according to boundary layer theory. The theoretical development is as follows.

We once again assume that the relative temperature differences are small so that, except for the essential density variation in the body force, all fluid properties can be taken to be constants. For treating the two-dimensional flow about a cylinder, it will be convenient to write the basic equations in terms of a (s, n) coordinate system, where s is the arc length on the cylinder perimeter measured in a clockwise direction with the lower stagnation point as the origin and n is the normal distance from the cylinder surface. The velocity components parallel and normal to the surface are U and V. Under the restriction on the fluid properties stated above we can then write the continuity, Navier-Stokes, and energy equations

$$\frac{R}{R+n}\frac{\partial U}{\partial s}+\frac{\partial V}{\partial n}+\frac{V}{R+n}=0 \tag{6-1}$$

$$\begin{aligned}\frac{R}{R+n}U\frac{\partial U}{\partial s}+V\frac{\partial U}{\partial n}+\frac{UV}{R+n}&=-\frac{1}{\rho}\left(\frac{R}{R+n}\right)\frac{\partial P_d}{\partial s}+g\beta\theta\sin\frac{s}{R}\\&+\nu\left[\frac{\partial^2 U}{\partial n^2}+\frac{1}{R+n}\frac{\partial U}{\partial n}+\frac{R^2}{(R+n)^2}\frac{\partial^2 U}{\partial s^2}-\frac{U}{(R+n)^2}+\frac{2R}{(R+n)^2}\frac{\partial V}{\partial s}\right]\end{aligned} \tag{6-2}$$

$$\begin{aligned}\frac{R}{R+n}U\frac{\partial V}{\partial s}+V\frac{\partial V}{\partial n}-\frac{U^2}{R+n}&=-\frac{1}{\rho}\frac{\partial P_d}{\partial n}-g\beta\theta\cos\frac{s}{R}\\&+\nu\left[\frac{\partial^2 V}{\partial n^2}+\frac{1}{R+n}\frac{\partial V}{\partial n}+\frac{R^2}{(R+n)^2}\frac{\partial^2 V}{\partial s^2}-\frac{V}{(R+n)^2}-\frac{2R}{(R+n)^2}\frac{\partial U}{\partial s}\right]\end{aligned} \tag{6-3}$$

$$\frac{R}{R+n}U\frac{\partial\theta}{\partial s}+V\frac{\partial\theta}{\partial n}=\alpha\left[\frac{\partial^2\theta}{\partial n^2}+\frac{1}{R+n}\frac{\partial\theta}{\partial n}+\frac{R^2}{(R+n)^2}\frac{\partial^2\theta}{\partial s^2}\right] \tag{6-4}$$

where all the symbols are defined as in the previous articles, except that now $\theta = T - T_\infty$ is the difference between the local and undisturbed fluid temperatures, R is the cylinder radius, and $P_d = P - P_\infty$ is the dynamic pressure. Note that the dynamic pressure equals zero where there is no motion and no temperature difference with respect to the undisturbed fluid. As will be seen in a subsequent chapter, the present author independently found it convenient to introduce the dynamic pressure for internal free convection flows. The buoyancy terms (second on the right side of Eq. 6-2 and 6-3) are formed by combining the body force term with the hydrostatic pressure gradient.

The boundary conditions for the problem are

$$\left.\begin{aligned} U(s, 0) &= V(s, 0) = 0,\ \theta(s, 0) = \theta_w \\ U(s, \infty) &= V(s, \infty) = \theta(s, \infty) = P_d(s, \infty) = 0 \end{aligned}\right\} \tag{6-5}$$

The boundary layer approximations could now be made in a formal manner as was done for the flat plate, but for expediency we will here use the order of magnitude arguments made originally by Prandtl. Thus, we assume that the cylinder radius R, the arc length s, and U are all of unit order. This is clearly not valid near the lower stagnation point. The normal distance n is taken to be of the order of magnitude of the boundary layer thickness ϵ where $\epsilon \ll 1$. It immediately follows from the continuity equation that V is also of order ϵ. In order that the friction and inertia terms in Eq. 6-2 and the conduction and convection terms in Eq. 6-4 be of equal order (as they must in boundary layers), ν and α must be of order ϵ^2. Further, since the buoyancy is generating the flow, the buoyancy term in Eq. 6-2 must be of the same order as the inertia and viscous forces. Therefore, $g\beta\theta = \mathcal{O}(1)$. From the two momentum equations, Eq. 6-2 and 6-3, it can be seen that the pressure terms are at most of unit order. Since P_d is zero at the outer edge of the boundary layer, integration of the expression

$$\frac{1}{\rho}\frac{\partial P_d}{\partial n} \cong 1$$

from the cylinder surface to the outer edge of the boundary layer shows that P_d/ρ in the boundary layer is at most of order ϵ. Then in Eq. 6-2 $(1/\rho)\partial P_d/\partial s$ is also of order ϵ.

If we now apply these arguments to Eq. 6-1 to 6-4 and retain only terms of unit order of magnitude, we find that the boundary layer equations for free convection about an horizontal cylinder are

$$\left.\begin{aligned} \frac{\partial U}{\partial s} + \frac{\partial V}{\partial n} &= 0 \\ U\frac{\partial U}{\partial s} + V\frac{\partial U}{\partial n} &= \nu\frac{\partial^2 U}{\partial n^2} + g\beta\theta \sin\frac{s}{R} \\ U\frac{\partial \theta}{\partial s} + V\frac{\partial \theta}{\partial n} &= \alpha\frac{\partial^2 \theta}{\partial n^2} \end{aligned}\right\} \tag{6-6}$$

$$\frac{U^2}{R} = \frac{1}{\rho}\frac{\partial P_d}{\partial n} + g\beta\theta\cos\frac{s}{R} \tag{6-7}$$

We have thus achieved some mathematical simplification. Note that the momentum and energy equations are coupled through the buoyancy term and that U, V, and θ can be determined from Eq. 6-6. Thereafter, if desired, the dynamic pressure can be computed from Eq. 6-7.

If we had nondimensionalized Eq. 6-1 to 6-4 before estimating the sizes of the terms, we would have found that the Grashof number based on cylinder diameter would be of order of ϵ^{-4}. Therefore, if we want our results to be accurate to within 10 per cent, the maximum value of ϵ would be 0.1, and we would have to consider flows in which $Gr = 10^4$ as the lower limit. As is to be expected, the approximation improves as the Grashof number increases (i.e., as ϵ or the boundary layer thickness decreases). The upper limit on Gr is, of course, attained when the flow becomes turbulent. According to Part IV of [*28*], turbulent flow appears at the equator of the cylinder when $Gr = 3 \times 10^9$. The laminar boundary layer results for a horizontal cylinder should, therefore, be valid for Grashof numbers in the range between 10^4 and 10^9.

Let us now, in accord with boundary layer theory, attempt to reduce Eq. 6-6 to ordinary differential equations. To this end we first define the stream function ψ,

$$U = \frac{\partial \psi}{\partial n}, \qquad V = -\frac{\partial \psi}{\partial s} \tag{6-8}$$

and also nondimensional quantities as

$$x = \frac{s}{R}, \qquad y = \frac{n}{R} Gr_R^{\frac{1}{4}}, \qquad \chi = \frac{\psi}{\nu} Gr_R^{-\frac{1}{4}}, \qquad \vartheta = \frac{\theta}{\theta_w} \tag{6-9}$$

where the subscript R on the Grashof number means that the radius, and not the diameter, is the characteristic length. Introducing these expressions into Eq. 6-6 yields

$$\frac{\partial \chi}{\partial y}\frac{\partial^2 \chi}{\partial x \partial y} - \frac{\partial \chi}{\partial x}\frac{\partial^2 \chi}{\partial y^2} = \frac{\partial^3 \chi}{\partial y^3} + \vartheta \sin x \tag{6-10}$$

$$\frac{\partial \chi}{\partial y}\frac{\partial \vartheta}{\partial x} - \frac{\partial \chi}{\partial x}\frac{\partial \vartheta}{\partial y} = \frac{1}{Pr}\frac{\partial^2 \vartheta}{\partial y^2} \tag{6-11}$$

The transformation given by Eq. 6-9 is not an arbitrary one, but is definitely determined to eliminate the Grashof number from the resulting equations, Eq. 6-10 and 6-11. We can, therefore, from Eq. 6-9 determine how the boundary layer characteristics (e.g., velocity, temperature, boundary layer thickness, heat transfer coefficient) vary with Grashof number without finding the solution to Eq. 6-10 and 6-11. If we let

$$\bar{U} = \frac{\partial \chi}{\partial y} \quad \text{and} \quad \bar{V} = -\frac{\partial \chi}{\partial x} \tag{6-12}$$

it follows from Eq. 6-9 that

$$\left.\begin{aligned} \frac{n}{R} &= yGr_R^{-\frac{1}{4}}, & \frac{UR}{\nu} &= \bar{U}Gr_R^{\frac{1}{2}} \\ \frac{VR}{\nu} &= \bar{V}Gr_R^{\frac{1}{4}}, & \frac{R}{\theta_w}\frac{\partial\theta}{\partial n} &= \frac{\partial\vartheta}{\partial y}Gr_R^{\frac{1}{4}} \end{aligned}\right\} \tag{6-13}$$

Thus, we see that the nondimensional boundary layer thickness decreases as $Gr^{-\frac{1}{4}}$, the velocities increase as $Gr^{\frac{1}{2}}$, and the temperature gradient increases as $Gr^{\frac{1}{4}}$. From the last, there follows directly the well-known, one-fourth-power law of heat transfer which states that the Nusselt number is proportional to $Gr^{\frac{1}{4}}$ for free convection from isothermal surfaces. These results agree with those for a flat plate and are not restricted to that case and the present one. They are valid, in general, for all two dimensional free convection flows about isothermal surfaces which are of the boundary layer type, that is, where the thickness of the heated layer is small compared with the distance from the leading edge and with the radius of curvature of the wall, and the latter does not change discontinuously in the flow direction. Flow over sharp edges is, therefore, excluded. We have already seen that these general conclusions have been confirmed in the previous sections for the flat plate and vertical cylinder (in which not all the restrictions are satisfied).

The solution of Eq. 6-10 and 6-11, of course, gives information on the Prandtl number and space-coordinate dependence of the flow and, once determined, describes the temperature and velocity fields for all values of the Grashof number through Eq. 6-9. Let us, therefore, continue our development to find these solutions. Following the idea of similar profiles in the boundary layer, we next introduce the transformation

$$\eta = yg(x) \tag{6-14a}$$

$$\chi(x, y) = F(\eta)q(x) \tag{6-14b}$$

$$\vartheta(x, y) = H(\eta) \tag{6-14c}$$

into Eq. 6-10 and 6-11, so that

$$\left.\begin{aligned} (F')^2(q^2gg' + qq'g^2) - FF''qq'g^2 &= F'''qg^3 + H\sin x \\ \frac{1}{Pr}(gH'') + q'FH' &= 0 \end{aligned}\right\} \tag{6-15}$$

where primes on g and q denote differentiations with respect to x, and on F and H denote differentiations with respect to η.

The separation into functions of two independent variables x and η, so that Eq. 6-15 become ordinary differential equations, is possible only if the functions $g(x)$ and $q(x)$ can be uniquely determined from the four equations:

$$q'(x) = ag(x) \tag{6-16a}$$

$$q^2(x)g(x)g'(x) = b \sin x \tag{6-16b}$$

$$q(x)q'(x)g^2(x) = c \sin x \tag{6-16c}$$

$$q(x)g^3(x) = d \sin x \tag{6-16d}$$

where a, b, c, and d are, as yet, unknown constants. If such a determination is possible, the functions $F(\eta)$ and $H(\eta)$ can be determined from

$$\left.\begin{aligned}(b + c)(F')^2 - cFF'' &= dF''' + H \\ H'' + aPrFH' &= 0\end{aligned}\right\} \tag{6-17}$$

We shall now see if it is possible to find uniquely the two functions $g(x)$ and $q(x)$ from the four Eq. 6-16. From physical considerations we can determine the boundary conditions required for solving Eq. 6-16. Thus, from

$$\frac{\partial\vartheta}{\partial y} = \frac{dH}{d\eta}\, g(x)$$

it follows that

$$g(0) = g_0 \neq 0 \tag{6-18a}$$

Otherwise, the temperature along the normal from the cylinder at the lower stagnation point would be a constant so that the surface temperature would equal that in the free stream and, hence, there would be no flow at all (since we are treating the isothermal surface case here). From Eq. 6-16a and 6-18, we obtain

$$q'(0) = ag_0 \tag{6-18b}$$

The dimensionless tangential velocity can by means of Eq. 6-14 be written as

$$\bar{U} = \frac{\partial\chi}{\partial y} = g(x)q(x)F'(\eta)$$

Since the flow is to be symmetrical about the plane $x = 0$, the tangential velocity at the lower stagnation point ($x = 0$) must vanish so that

$$q(0) = 0 \tag{6-18c}$$

The system of Eq. 6-16, together with the boundary conditions, Eq. 6-18, thus contain five constants which are to be determined. By introducing

$$G(x) = \frac{g(x)}{g_0} \quad \text{and} \quad Q(x) = \frac{q(x)}{ag_0} \tag{6-19}$$

the system is transformed to

$$G = Q' \tag{6-20a}$$

$$Q^2GG' = \frac{b}{a^2g_0^4}\sin x \tag{6-20b}$$

$$QQ'G^2 = \frac{c}{a^2g_0^4}\sin x \tag{6-20c}$$

$$QG^3 = \frac{d}{ag_0^4}\sin x \tag{6-20d}$$

$$Q(0) = 0,\ Q'(0) = G(0) = 1 \tag{6-21}$$

The two functions G and Q could be uniquely determined from the four Eq. 6-20 if we could prove that the four equations are not independent of one another, but that two equations follow from the other two. This proof is partly possible only insofar as one equation (for a suitable choice of the constants) is a consequence of the remaining equations. Thus, elimination of G by means of Eq. 6-20a, gives

$$Q^2Q'Q'' = \frac{b}{a^2g_0^4}\sin x \tag{6-22a}$$

$$QQ'^3 = \frac{c}{a^2g_0^4}\sin x \tag{6-22b}$$

$$QQ'^3 = \frac{d}{ag_0^4}\sin x \tag{6-22c}$$

Eq. 6-22b thus follows from Eq. 6-22c if we set

$$d = \frac{c}{a} \tag{6-23}$$

For the determination of Q, we are thus left with the two equations

$$QQ'^3 = \alpha \sin x \tag{6-24a}$$

$$Q^2Q'Q'' = \beta \sin x \tag{6-24b}$$

where, here, for brevity $\alpha = c/a^2g_0^4$ and $\beta = b/a^2g_0^4$. If the introduction of Eq. 6-14 is to lead to the ordinary differential equations, Eq. 6-17, it must now be shown that the two equations, Eq. 6-24a and 6-24b, have two equal (or at least approximately equal) solutions.

This proof is attempted in [*28*] by solving each of the differential equations Eq. 6-24a and 6-24b by means of a powers series development about the equator $x = \pi/2$. The two solutions Q^I and Q^{II} (corresponding to Eq. 6-24a and 6-24b, respectively) were found to agree at the equator in the value of the function and its first two derivatives. The higher derivatives differ to a small degree. The functions $Q^I(x)$, $Q^{II}(x)$, and $G(x)$

are presented in Table F,6a, and the agreement in the range $x = 30°$ to 150° and even at 165° is seen to be reasonable. The negative value of Q^{II} for $x = 0$ has no significance because of the singular behavior of the equation there. Therefore, since Q^{I} behaves properly at the stagnation point, it is used in the subsequent computations.

Table F,6a. Solutions of the system of Eq. 6-20 and 6-21.

X	0°	30°	60°	90°	120°	150°	165°	180°
Q^{I}	0.0007	0.518	1.025	1.504	1.936	2.298	2.433	2.560
Q^{II}	−0.23	0.456	1.018	1.504	1.941	2.352	2.55	2.763
G	1.000	0.989	0.945	0.873	0.765	0.602	0.473	0

Thus, Q^{I} and Q^{II} are approximately, but not rigorously, equal. The system, Eq. 6-16, is, strictly speaking, overspecified and has only approximately a unique solution. The solution for the flow and heat transfer obtained using Eq. 6-14 is not rigorously, but approximately, valid. Because of the agreement of Q^{I} and Q^{II} at the equator, up to and including, second derivatives, the method is quite good except near the upper and lower stagnation points where the boundary layer assumptions are also not justified. The method used thus gives an approximate solution of the boundary layer equations within the limits imposed by the physical approximations of the boundary layer assumptions.

In the process of finding G and Q, values of α and β (as defined after Eq. 6-24b) were determined to be $\alpha = 1$ and $\beta = -\frac{1}{3}$. From the definitions and Eq. 6-23, the constants in Eq. 6-17 can be written as

$$b = -\frac{a^2 g_0^4}{3}, \quad c = a^2 g_0^4, \quad d = a g_0^4$$

so that the equation itself becomes

$$\left.\begin{aligned} F''' + aFF'' - \left(\frac{2}{3}\right) a(F')^2 + \left(\frac{1}{a g_0^4}\right) H &= 0 \\ H'' + aPrFH' &= 0 \end{aligned}\right\} \tag{6-25}$$

with the boundary conditions

$$\left.\begin{aligned} F(0) = F'(0) = 0, \; H(0) &= 1 \\ F'(\infty) = H(\infty) &= 0 \end{aligned}\right\} \tag{6-26}$$

The constants a and g_0 are as yet still free, so that if we let

$$a = 3 \quad \text{and} \quad g_0 = (3)^{-\frac{1}{4}} \tag{6-27}$$

Eq. 6-25 and the boundary conditions become identical to those for the vertical plate, Eq. 3-52 to 3-56. It is thus unnecessary to do any further integrating of the resulting equations, because the extensive calculations of Pohlhausen [3] and Ostrach [6] appear in tabular form in the original references indicated and are given herein in Fig. F,3b and F,3c.

Finally, the azimuthal functions are now uniquely determined from Eq. 6-19 and 6-27 to be

$$g(x) = 0.760G(x) \quad \text{and} \quad q(x) = 2.280Q(x) \tag{6-28}$$

With values of $g(x)$ and $q(x)$ determined by Eq. 6-28 and Table F,6a, the values of $F(\eta)$ and $H(\eta)$ from tables of flat plate solutions, and the transformations (Eq. 6-8, 6-9, and 6-14), there is obtained the complete solution of the boundary layer equations (Eq. 6-6) for the velocity (U and V) and temperature (θ) fields. Collecting all these results and also using Eq. 6-12 and 6-13, we obtain

$$U(s, n) = \frac{\nu}{R}\left(\frac{Gr}{8}\right)^{\frac{1}{2}} g(x)q(x)F'(\eta) \tag{6-29}$$

$$V(s, n) = -\frac{\nu}{R}\left(\frac{Gr}{8}\right)^{\frac{1}{4}}\left[q'(x)F(\eta) + \frac{q(x)g'(x)}{g(x)}\eta F'(\eta)\right] \tag{6-30}$$

$$\theta(s, n) = \theta_w H(\eta), \quad \frac{\partial\theta}{\partial n}(s, n) = \frac{\theta_w}{R}\left(\frac{Gr}{8}\right)^{\frac{1}{4}} g(x)H'(\eta) \tag{6-31}$$

with

$$x = \frac{s}{R} \quad \text{and} \quad \eta = \frac{n}{R}\left(\frac{Gr}{8}\right)^{\frac{1}{4}} g(x) \tag{6-32}$$

Note that $Gr = 8Gr_R$ and all Grashof number dependence agrees with our previous discussion.

The azimuthal variations of the physical quantities presented in Eq. 6-29, 6-31, and 6-32 are described by the functions $g(x)$ and $g(x)q(x)$ which are presented in Fig. F,6a. For the variation of the normal velocity component V given by Eq. 6-30, we can find $q'(x)$ from Eq. 6-16a and can obtain $g'(x)$ by graphical differentiation of $g(x)$. We shall discuss this variation qualitatively shortly.

The azimuthal variation of the tangential velocity at a fixed η, for example the maximum velocity position, is described by $g(x)q(x)$. This velocity increases from zero at the lower stagnation point (approximately linearly up to 60°) to a maximum at about 128° and, on approaching the upper stagnation point, where the flow on either side converges, rapidly decreases to zero. This physically correct behavior of the solution at the upper stagnation point was not fixed by a boundary condition but is obtained as a consequence of the theory.

From Eq. 6-32 it can be seen that the azimuthal variation of any characteristic distance as, for example, the maximum velocity position or boundary layer thickness, is given by $1/g(x)$. Thus, the boundary layer thickness at the lower stagnation point possesses a finite approximately constant value over a large range of x, but increases at first gradually, and then more and more rapidly, and attains a theoretically infinite value at the upper stagnation point corresponding to an upward current of warm air. The function $g(x)$, since it is proportional to the normal temperature gradient, Eq. 6-31, also represents the variation of the local heat transfer coefficient over the cylinder perimeter. At the lower stagnation point ($x = 0$), $g(x)$ is constant and, with constant slope, connects with values on the other side of the cylinder. At the upper stagnation point ($x = \pi$) this is no longer the case; $g(x)$ has a sharp peak which is associated with the fact that the boundary layer assumptions at this

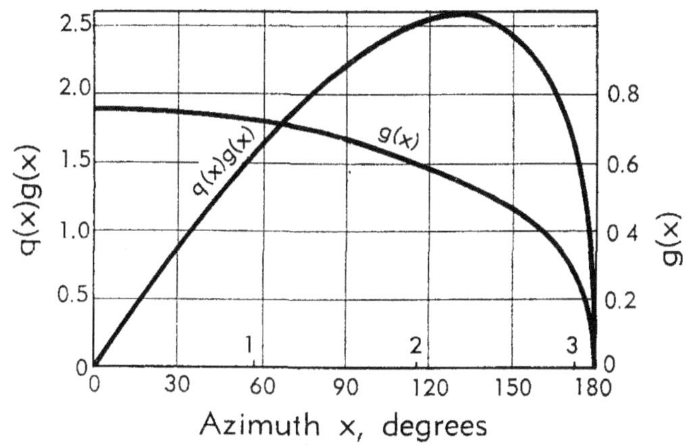

Fig. F,6a. Azimuth function $g(x)$ giving variation of local heat transfer coefficient or of a reciprocal distance from the wall which is characteristic for profile (for example, boundary layer thickness) along cylinder perimeter (Hermann [28]).

point sharply deviate from actual conditions (normal velocity large compared with tangential velocity).

An over-all picture of the velocity and temperature fields is given by the streamlines and isotherms. These are computed for various values of the constants from the following equations developed from Eq. 6-9 and 6-14:

$$\frac{\psi(x, n/R)}{\nu} = Gr_R^{\frac{1}{4}} q(x) F\left[Gr_R^{\frac{1}{4}} \frac{n}{R} g(x)\right] = \text{const}$$

$$\frac{\theta(x, n/R)}{\theta_w} = H\left[Gr_R^{\frac{1}{4}} \frac{n}{R} g(x)\right] = \text{const}$$

where the arguments are contained in the brackets. The streamline picture presented in Fig. F,6b was computed for azimuth intervals of 15° and, in addition, for 5°, 10°, 170°, and 175°. The isotherm field given in Fig. F,6c was calculated at 30° intervals and also at 165°, 170°, and 175°. The scale of the representations holds for $Gr = 10^4$; for $Gr = 10^6$ the distances from the cylinder surface should be reduced by a factor of

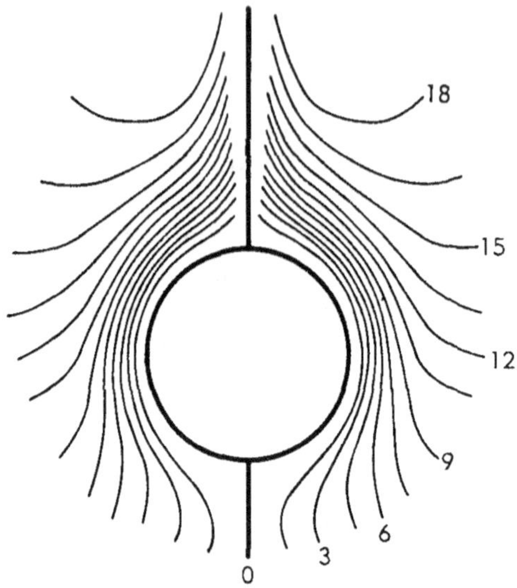

Fig. F,6b. Theoretical streamlines for horizontal cylinder. The numbers give the values for ψ/ν. Scale is for $Gr = 10^4$: For $Gr = 10^6$, the distances from the surface are to be shortened by 1/3.16; for $Gr = 10^8$, by 1/10 (Hermann [28]).

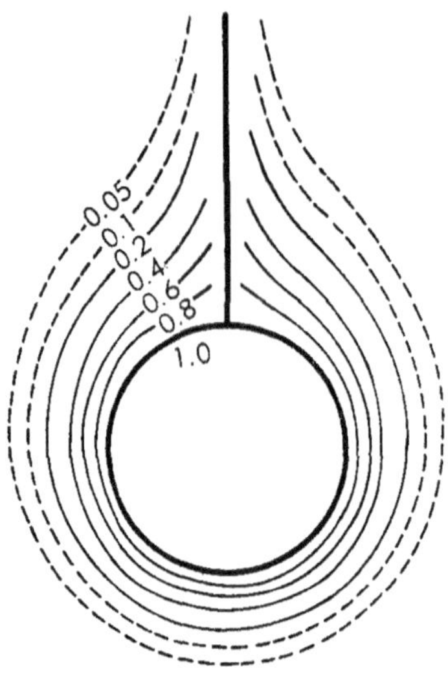

Fig. F,6c. Theoretical isotherms for horizontal cylinder. The numbers are values of θ/θ_w. The scale is for $Gr = 10^4$, to be shortened as in Fig. F,6b for other values of Gr (Hermann [28]).

3.16, and for $Gr = 10^8$ by 10. At the lower part and equatorial neighborhood of the cylinder up to an azimuth of 105°, the flow near the surface is directed toward the surface, while, above, it is directed away from the wall upwards. In this respect, the theoretical solution for the cylinder differs fundamentally from that for the flat plate, which yields only a flow toward the plate. The isotherms at the lower part form almost concentric circles. At the upper stagnation point, where there is an upward warm current, the isotherms theoretically go to infinity, whereas actually (corresponding to the gradual diffusion and dispersion of the warm upward flow) they close at a relatively large distance above the cylinder. The maximum tangential velocity occurs at the same position as the 0.53 isotherm.

It will be convenient for the subsequent discussion to write the Nusselt number explicitly for free convection from a horizontal cylinder. From Eq. 6-31, the local Nusselt number is determined to be

$$Nu(x) \equiv \frac{h(x)d}{k} = -(2)^{\frac{1}{4}} g(x) Gr^{\frac{1}{4}} H'(0) \tag{6-33a}$$

where $h(x)$ is the local heat transfer coefficient and for $Pr = 0.733$, $H'(0) = -0.508$. Thus, for air, the mean heat transfer coefficient averaged over the cylinder perimeter is

$$\overline{Nu} \equiv \frac{\bar{h}d}{k} = 0.372 Gr^{\frac{1}{4}} \tag{6-33b}$$

where

$$\bar{h} = \frac{1}{\pi} \int_0^\pi h(x) dx$$

If this result is generalized (see [26]), it becomes

$$\overline{Nu} = 0.4 (Pr\, Gr)^{\frac{1}{4}} \tag{6-33c}$$

The theoretically predicted velocity and temperature distributions are compared with those obtained experimentally by Jodlbauer [29]. The best agreement would, of course, be obtained for the largest laminar Grashof number (about 3.5×10^8). Here, however, we present representative results for $Gr = 3.76 \times 10^6$ in Fig. F,6d and F,6e, and will also discuss results for $Gr = 6.54 \times 10^5$ which are presented in detail by Hermann [28].

From Fig. F,6d, it can be seen that there is a regular deviation between measured and computed velocity profiles, in that the experimental values are higher than the theoretical ones. The deviation from the theory of the experimental maximum velocity amounts, on the average, to 22 per cent for the smaller and 17 per cent for the larger value of Gr. Part of the difference should, however, be attributed to the difficulties in experimentally measuring such small velocities. This is particularly true at large distances from the cylinder where agreement is poorest. The inconsistency in measurement accuracy is perhaps best illustrated by noting that the data points obtained at the 60° azimuth for $Gr =$

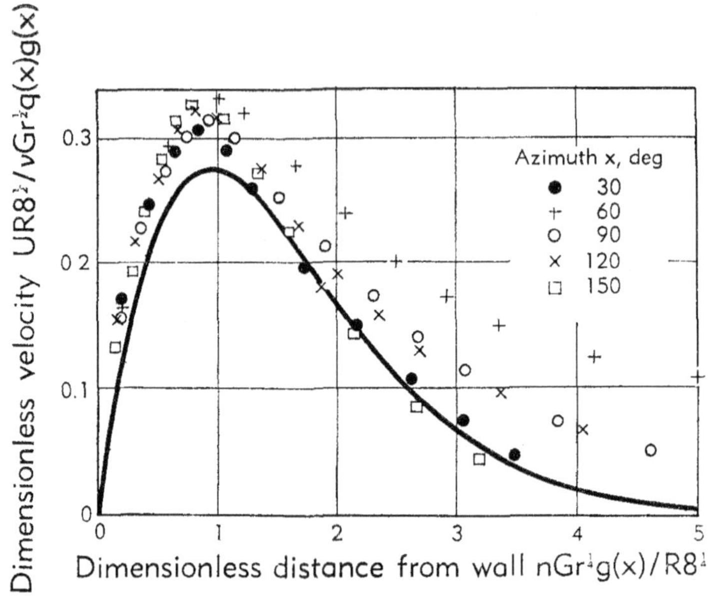

Fig. F,6d. Velocity profiles near cylinder referred to nondimensional coordinates. Continuous curve represents theoretical solution. Points according to measurements of Jodlbauer [29] for $2R = 9$ centimeters; $T_w = 99.2°C$; $T_\infty = 18.1°C$; $Gr = 3.76 \times 10^5$ (Hermann [28]).

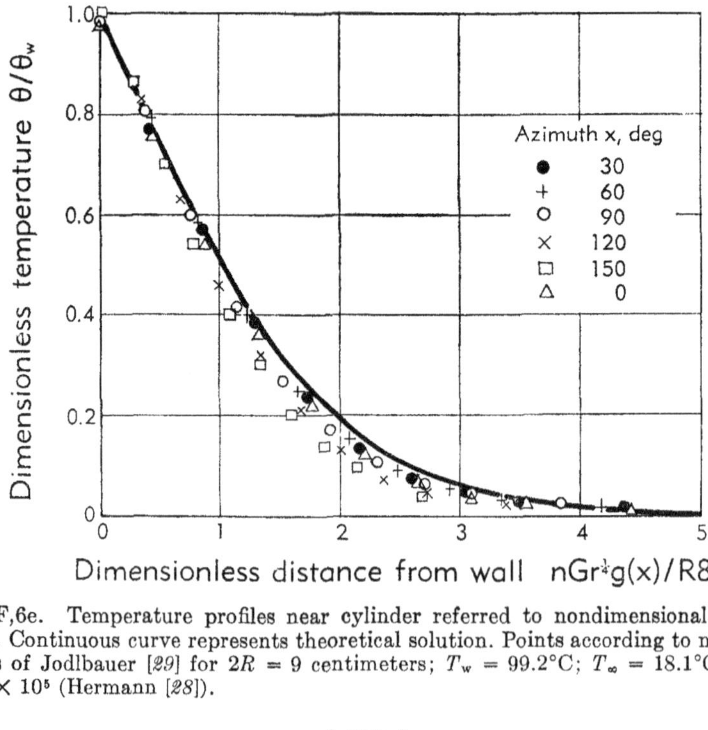

Fig. F,6e. Temperature profiles near cylinder referred to nondimensional coordinates. Continuous curve represents theoretical solution. Points according to measurements of Jodlbauer [29] for $2R = 9$ centimeters; $T_w = 99.2°C$; $T_\infty = 18.1°C$; $Gr = 3.76 \times 10^5$ (Hermann [28]).

3.76×10^6 lie considerably above the theoretical curve (Fig. F,6d), whereas for 6.54×10^5 they lie on the theoretical curve. On the other hand, the velocities measured at $x = 30°$ for 6.54×10^5 are considerably greater than the theoretical, whereas for $Gr = 3.76 \times 10^6$ they agree with the theory. The measured distance from the surface to the position of maximum velocity is everywhere in good agreement with the theory.

The measured temperature profiles agree very well with the theory even for azimuths of 165° (see Fig. F,6e). A small regular deviation can be noted for all azimuths and Grashof numbers for medium distances from the wall. The small differences near the wall are partly due to the fact that the surface was not quite isothermal in the tests. The Nusselt numbers determined from Jodlbauer's [*29*] field measurements are, on the average, about four per cent higher than those determined theoretically from Eq. 6-33b. Other experiments, in which only heat transfer measurements (and not detailed profiles) were made, gave Nusselt numbers somewhat higher than Jodlbauer's.

Now that theoretical boundary layer solutions for free convection about the vertical plate and the horizontal cylinder have been found, it is of interest to compare the flow and heat transfer about these two standard two-dimensional bodies. The results of the forced flow over a plate (Blasius solution) will be introduced as a reference wherever possible.

The boundary layer thickness of the plate in forced flow increases as the square root of the distance from the leading edge (according to the parabolic similarity found by Blasius); that of the vertical plate in free convection increases, however, as the fourth root (see, for example, Eq. 3-49). Both start with zero thickness, which for the forced convection case implies infinite velocity gradients at the leading edge and for the free convection case implies an infinite temperature gradient normal to the surface and, hence, an infinite local heat transfer coefficient at the leading edge. In the case of the horizontal cylinder, on the contrary, the boundary layer starts at the lower stagnation point with a finite thickness, and the velocity and temperature gradients and the local heat transfer coefficient are all finite there also (see Eq. 6-29 to 6-32 and Fig. F,6a). The boundary layer thickness increases as $1/g(x)$, Eq. 6-32, and theoretically becomes infinite at the upper stagnation point where the velocity, normal temperature gradient, and local heat transfer coefficient all vanish.

In spite of these fundamental differences it is possible to relate all three of these cases in a number of ways. In this way, it is possible to show on one hand a close hydrodynamic and thermal kinship which is due to the boundary layer character of the differential equations underlying the theoretical solutions. On the other hand, they give practical viewpoints with regard to the application of rectangular plates or horizontal cylinders for heat transfer. We shall not here go into the details of the specific relationships which are presented in [*28*], but shall rather summarize the conclusions that can be drawn from them.

Although the boundary layer growth for a vertical plate and a horizontal cylinder in free convection differ fundamentally, as has already been remarked, we shall compare the boundary layer thicknesses at the upper edge (away from the leading edge) of the vertical plate and in the upper stagnation region of the horizontal cylinder. The usual definition of boundary layer thickness is multiplied by the tangential velocity in both cases, in order that the singularity at the upper stagnation point of the cylinder may be eliminated. For equal Grashof number for the plate and the cylinder, that is, for the case where the plate height equals the cylinder diameter, the boundary layer thickness at the upper stagnation point of the cylinder is 22 per cent greater than that at the upper edge of the plate. This result gives no information about the hydrodynamic relation between the two, because the comparison refers to different characteristic lengths. It is useful only for the rapid comparison of plate and cylinder for the usual values of Gr. It is then known that, for equal Grashof numbers, the cylinder has the greater boundary layer thickness so that we might anticipate the flow over the cylinder to become turbulent before that of the plate. For this case, with all conditions on the plate and cylinder equal, the mean heat transfer coefficient of a cylinder is only 78 per cent of that of a plate whose height is equal to the cylinder diameter. The total heat transfer from a horizontal cylinder is 22 per cent greater than that from both sides of a vertical plate, when the two Grashof numbers are equal (i.e., when the plate height is equal to the cylinder diameter under otherwise equal conditions).

The two characteristic lengths are equal when the plate height equals $\pi/2$ times the cylinder diameter. In that case, the boundary layer thickness in the upper stagnation region of the cylinder is 13 per cent smaller than that of a plate whose height is equal to the semicircumference of the cylinder. This retardation of the boundary layer development as compared with the vertical plate is due to the fact that the boundary layer at the lower stagnation point of the cylinder already has a finite thickness, whereas for the plate the thickness must increase from zero. The total heat transfer of a cylinder is 13 per cent smaller than that of the two sides of a plate of height equal to the developed semicircumference of the cylinder (equal characteristic lengths) under otherwise equal conditions.

The same boundary layer thickness will be obtained at the upper stagnation point of the cylinder and upper edge of a vertical plate if the Grashof number of the cylinder (based on the diameter) is 0.44 times that of the plate (based on the plate height). This permits a rapid conversion of the flow data for the plate into those for the cylinder, and vice versa. For example, if it is known from experiments on a plate that the departure from the laminar $Gr^{\frac{1}{4}}$ heat transfer law occurs at a Grashof number (based on plate height) of 8.8×10^8, then it can be concluded without further tests that the laminar results break down at a Grashof number (based on the cylinder diameter) of 3.5×10^8. The mean heat transfer coefficient of a cylinder is, under otherwise equal conditions, equal to that of a plate whose height is 2.76 times that of the cylinder diameter. This factor of 2.76 is, however, not essentially significant, because of the small decrease of local heat transfer coefficient at large plate heights. Good correlation with experimental data for cylinders was obtained by replacing the plate height by two (rather than 2.76) times the cylinder diameter in the mean heat transfer coefficient expression for flat plates. The values so determined were only about 8.3 per cent above the experimental data.

The total heat transfer of a cylinder, under otherwise identical conditions, is the same as that from two sides of a vertical plate if the plate height is 1.31 times that of the cylinder diameter. Therefore, it can be stated that a horizontal cylinder and a vertical plate, which transfers heat on both sides, of height 1.31 times the cylinder diameter are equivalent, both thermally (with regard to total heat transfer) and hydrodynamically (with regard to the boundary layer development at the upper stagnation point and at the upper edge, respectively).

Before dealing with additional complexities of free convection flows, it should be remarked that most analysis and experiments of the phenomenon are concerned with the three basic configurations (flat plate and vertical and horizontal cylinders) which we have discussed up to now. For example, although the sphere is geometrically a simpler shape than the vertical cylinder, not much technical interest has been shown in free convection from spherical bodies. According to Elenbaas [*30*], the heat transfer rate from a sphere is about 25 per cent greater than that from a cylinder of the same diameter with the same driving temperature difference. However, data and analysis for the sphere and other bodies are relatively scarce.

F,7. Effect of Variable Fluid Properties. The presence of a buoyancy force is a requirement for the existence of a free convection flow. In the problems that we have considered up to now, the buoyancy was due to density differences which resulted from temperature gradients in the fluid. Any analytical treatment of free convection must, therefore, include density variations at least to describe the buoyancy force. In all the free convection analyses that we have discussed thus far, all fluid property (the thermal conductivity, viscosity, specific heat, and density) variations have been neglected except for the essential density differences just noted. This assumption was formally verified by Ostrach [*6*] as long as temperature differences are small relative to the absolute temperature.

It is of interest to investigate the importance of the fluid property variations for problems in which the temperature differences are large.

The problem of free convection with variable fluid properties about an isothermal vertical plate was analyzed independently by Sparrow [31,32] and Tanaev [33]. Since the former work is more extensive, we will here essentially follow that development. The configuration studied is that shown in Fig. F,3a.

The laws expressing the conservation of mass, momentum, and energy for the steady state, free convection boundary layer are, respectively,

$$\frac{\partial}{\partial X}(\rho U) + \frac{\partial}{\partial Y}(\rho V) = 0 \tag{7-1}$$

$$\rho\left(U\frac{\partial U}{\partial X} + V\frac{\partial U}{\partial Y}\right) = g(\rho_\infty - \rho) + \frac{\partial}{\partial Y}\left(\mu\frac{\partial U}{\partial Y}\right) \tag{7-2}$$

$$\rho c_p\left(U\frac{\partial T}{\partial X} + V\frac{\partial T}{\partial Y}\right) = \frac{\partial}{\partial Y}\left(k\frac{\partial T}{\partial Y}\right) \tag{7-3}$$

Viscous dissipation and compression work have been neglected. The buoyancy term $g(\rho_\infty - \rho)$ results from a combination of the pressure gradient $\partial P/\partial X$ with the body force ρg, taking cognizance of the fact that $\partial P/\partial Y = 0$ for boundary layers.

The boundary conditions appropriate to the problem are

$$\left.\begin{aligned} U(X, 0) = V(X, 0) = 0, \qquad T(X, 0) = T_w \\ U(X, \infty) = 0, \qquad T(X, \infty) = T_\infty \end{aligned}\right\} \tag{7-4}$$

where T_w and T_∞ are prescribed constants.

In the constant property treatment of the problem, the density difference is expressed as a temperature difference by means of an approximate state equation and ρ, μ, c_p, and k are taken to be constant. The present analysis, however, continues without any restrictions on the property variations. To satisfy Eq. 7-1, we introduce a stream function ψ defined as

$$\frac{\rho U}{\rho_w} = \frac{\partial \psi}{\partial Y}, \qquad \frac{\rho V}{\partial_w} = -\frac{\partial \psi}{\partial X} \tag{7-5}$$

where the density at the surface ρ_w is assumed constant. Let us now define a new independent variable η by combining the similarity transformation with a transformation introduced by Dorodnitzyn [34] and Howarth [35] to reduce compressible forced flows to incompressible ones

$$\eta = CX^{-\frac{1}{4}}\int_0^Y \frac{\rho}{\rho_w}\,dY, \quad C = \frac{g(\rho_\infty - \rho_w)/\rho_w}{4\nu_w^2} \tag{7-6}$$

We also introduce the new dependent variable

$$\left.\begin{aligned} F(\eta) &= \frac{1}{4\nu_{\mathrm{w}}C}\frac{\psi}{X^{\frac{3}{4}}} \\ \theta(\eta) &= \frac{T - T_\infty}{T_{\mathrm{w}} - T_\infty} \end{aligned}\right\} \tag{7-7}$$

where θ is a dimensionless temperature, and the velocity components are related to F by

$$U = 4\nu_{\mathrm{w}}C^2X^{\frac{1}{2}}F', \qquad V = \left(\frac{\rho_{\mathrm{w}}}{\rho}\right)\left(\frac{\nu_{\mathrm{w}}C}{X^{\frac{1}{4}}}\right)(\eta F' - 3F) \tag{7-8}$$

where the primes denote differentiation with respect to η. Substituting Eq. 7-5 to 7-7 into Eq. 7-1 to 7-4 transforms the original partial differential equations into the following ordinary differential equations:

$$\frac{d}{d\eta}\left(\frac{\rho\mu}{\rho_{\mathrm{w}}\mu_{\mathrm{w}}}F''\right) + 3FF'' - 2(F')^2 + \frac{(\rho_\infty/\rho) - 1}{(\rho_\infty/\rho_{\mathrm{w}}) - 1} = 0 \tag{7-9}$$

$$\frac{d}{d\eta}\left(\frac{\rho k}{\rho_{\mathrm{w}}k_{\mathrm{w}}}\theta'\right) + 3Pr_{\mathrm{w}}\left(\frac{c_p}{c_{p_{\mathrm{w}}}}\right)F\theta' = 0 \tag{7-10}$$

with the boundary conditions

$$\left.\begin{aligned} F(0) = F'(0) = 0, \qquad \theta(0) = 1 \\ F'(\infty) = \theta(\infty) = 0 \end{aligned}\right\} \tag{7-11}$$

where Pr_{w} is the Prandtl number evaluated at the wall ($Y = 0$) conditions.

We note that the resulting ordinary differential equations can be reduced directly to the well-known equations for the constant property problem by making the usual approximations. Since we wish here to depart from the previous restrictions, we must now consider specific fluids in order to simplify Eq. 7-9 and 7-10.

For the special case of fluids that obey the perfect gas law $P = \rho \mathcal{R} T$ (where $\mathcal{R}$ is the gas constant), there are certain simplifications which can be introduced into the analysis. Consider first the buoyancy force $g(\rho_\infty - \rho)$ which appears in Eq. 7-2. For a perfect gas, this density difference can be expressed in terms of a temperature difference without making any assumptions, simply by noting that both ρ_∞ and ρ are to be evaluated at the same static pressure (same X, since $P_Y = 0$). It then follows directly that

$$\rho_\infty - \rho = \frac{\rho}{T_\infty}(T - T_\infty) \tag{7-12}$$

Recall that in the constant property analysis we wrote

$$\rho_\infty - \rho = \beta\rho(T - T_\infty)$$

where β is the fluid volumetric coefficient. Thus, the buoyancy force used in the constant property analysis is precisely correct, provided that β is replaced by $1/T_\infty$. Within the limits of the constant property case (i.e., for small relative temperature differences), this replacement is permissible since, from the definition of β, we find that $\beta = 1/T$ for perfect gases.

The dimensionless buoyancy term is further simplified by

$$\frac{(\rho_\infty/\rho) - 1}{(\rho_\infty/\rho_w) - 1} = \frac{T - T_\infty}{T_w - T_\infty} = \theta \tag{7-13}$$

Eq. 7-9 then becomes

$$\frac{d}{d\eta}\left(\frac{\rho\mu}{\rho_w\mu_w} F''\right) + 3FF'' - 2(F')^2 + \theta = 0 \tag{7-9a}$$

We now note that, for a rather special kind of variable property gas, Eq. 7-9a and 7-10 reduce directly to the equations for the constant property problem. Explicitly, if the properties of a gas vary in such a way that $P = \rho\mathcal{R}T$, $\rho\mu = \text{const}$, $\rho k = \text{const}$, and $c_p = \text{const}$, Eq. 3-9a and 3-10 become

$$\left.\begin{aligned} F''' + 3FF'' - 2(F')^2 + \theta &= 0 \\ \theta'' + 3PrF\theta' &= 0 \end{aligned}\right\} \tag{7-14}$$

We have thus reduced the variable property problem identically to the constant property problem, and all previous solutions obtained for the constant property problem now apply for this special variable property fluid. As has already been mentioned, a similar reduction was made for forced convection flows in [*34,35*]. For fluids whose properties do not vary in the manner described to lead to the reduction to the constant property case, Eq. 7-9a and 7-10 must be solved without further simplification. Sparrow and Gregg [*32*] obtained numerical solutions of these equations for five perfect gases, each obeying different property laws, and for liquid mercury. On the basis of this data, there is then developed in [*32*] a reference temperature relation that permits good approximations of variable property results to be obtained quite simply from constant property results.

F,8. "Similar" Flows about Two-Dimensional and Axially Symmetric Bodies. It is now of interest to see how the methods of the preceding articles can be applied to other types of bodies. Schuh [*5*] considered the problem of similar free convection flows over two-dimensional bodies in a manner analogous to that of Falkner and Skan [*36*] for forced flows. Schuh discussed in detail only the flow over a two-dimensional body having everywhere finite curvature and a boundary layer of constant thickness. Later, Merk and Prins [*37*] developed the general relations describing axially symmetric bodies as well. They showed that the cone is one axially symmetric body having a similarity flow. In [*38*) the previous analyses are generalized and a considerable number of two-dimensional bodies and two more of the axially symmetric bodies are developed. These bodies are limited to having closed lower ends. They are open above, but usually only of finite extent (see Fig. F,8a). Isothermal and variable surface temperatures are considered. Since the last work includes the results of the previous investigators, it will be presented below.

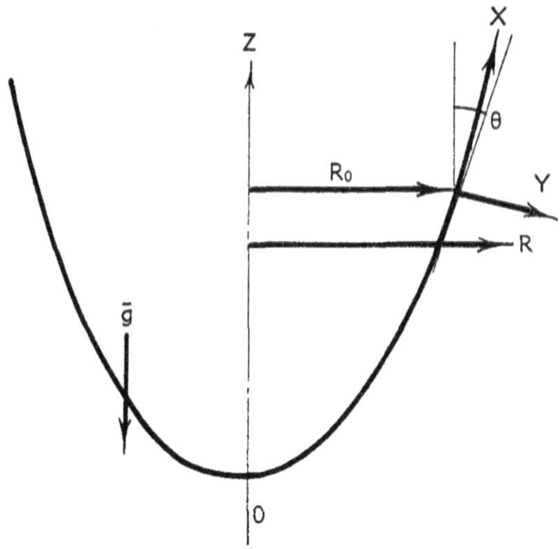

Fig. F,8a. Closed-end body configuration for "similar" flows.

From Goldstein [*19*, p. 120], the equations for a steady boundary layer in a curved two-dimensional surface are

$$UU_X + VU_Y = -\frac{1}{\rho}P_X + \nu U_{YY} + F_X \tag{8-1}$$

$$-\kappa U^2 = -\frac{1}{\rho}P_Y + F_Y \tag{8-2}$$

$$\rho c_p(UT_X + VT_Y) = kT_{YY} \tag{8-3}$$

$$U_X + V_Y = 0 \tag{8-4a}$$

where F_X and F_Y are the components of the body force parallel and normal to the body, respectively (all other subscripts denote differentiation), and κ is the radius of curvature. The pressure difference across the boundary layer will be of the order of the boundary layer thickness, so that the second equation may be neglected.

From [*19*, p. 130] we find that the same equations hold for axially symmetric bodies, except that the continuity equation (Eq. 8-4a) becomes

$$(RU)_X + (RV)_Y = 0 \tag{8-4b}$$

The forms of F_X and F_Y are obtained resolving the axial body force $\bar{g}$ in the X and Y directions using the relations of [*19*, p. 128]:

$$\frac{\partial R}{\partial X} = \sin \theta \qquad \frac{\partial R}{\partial Y} = \sin \theta$$

For the static case ($U \equiv 0$ and T = const), the above equations reduce to

$$\frac{\partial P_s}{\partial X} = \rho_s F_X; \qquad \frac{\partial P_s}{\partial Y} = \rho_s F_Y \tag{8-5}$$

where on the body

$$F_X = -g\sqrt{1 - (R_0')^2}; \qquad F_Y = gR_0'$$

and the prime denotes differentiation with respect to X. If we now subtract Eq. 8-5 from Eq. 8-1 and use the state equation (Eq. 2-6), we obtain

$$\rho(UU_X + VU_Y) = \mu U_{YY} + \rho\beta g(T - T_0)\sqrt{1 - (R_0')^2} \tag{8-6}$$

The basic equations can be nondimensionalized by letting $X = Lx$, $Y = Ly$, $R = Lr$, $U = U^*u$, $V = U^*v$, $T - T_s = \Delta T t$ where L, U^*, and ΔT are a characteristic dimension, velocity and temperature difference respectively. Eq. 8-6, 8-3, and 8-4 thus become

$$uu_x + vu_y = \frac{\nu}{U^*L} u_{yy} + \frac{\beta g \Delta T L}{U^{*2}} \sqrt{1 - (r_0')^2}\, t \tag{8-7}$$

$$ut_x + vt_y = \frac{1}{Pr}\frac{\nu}{U^*L} t_{yy} \tag{8-8}$$

$$u_x + v_y = 0 \text{ (two-dimensional)} \tag{8-9a}$$

$$(r_0u)_x + (r_0v)_y = 0 \text{ (axially symmetric)} \tag{8-9b}$$

At any station in the boundary layer, the fundamental length is the thermal boundary layer thickness, because it is over this length that the buoyancy force acts. Hence, to make the thermal boundary layer of unit order, we expand the y coordinate as

$$\bar{y} = \sqrt{Pr \frac{U^*L}{\nu}}\, y \tag{8-10}$$

Correspondingly,

$$\bar{v} = \sqrt{Pr \frac{U^*L}{\nu}}\, v \tag{8-11}$$

Then Eq. 8-7 and 8-9b become

$$uu_x + \bar{v}u_{\bar{y}} = Pru_{\bar{y}\bar{y}} + \frac{\beta g \Delta TL}{U^{*2}} \sqrt{1 - (r_0')^2}\, t \tag{8-7a}$$

$$ut_x + \bar{v}t_{\bar{y}} = t_{\bar{y}\bar{y}} \tag{8-8a}$$

$$u_x + \bar{v}_{\bar{y}} = 0 \tag{8-9c}$$

$$(r_0u)_x + (r_0\bar{v})_{\bar{y}} = 0 \tag{8-9d}$$

If we are to consider the entire range of Prandtl numbers (as, for example, from the order of 0.01 for liquid metals to thousands for very viscous fluids), we must carefully consider what the characteristic velocity

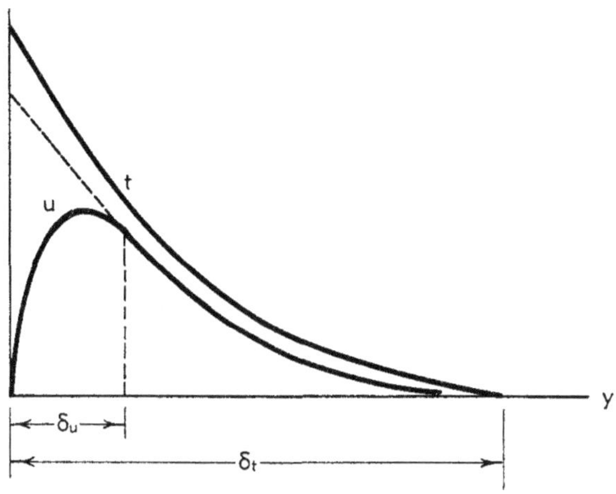

Fig. F,8b. Representative velocity and temperature profiles for small Prandtl numbers.

U^* should be. For small Prandtl numbers, the viscous boundary layer thickness is much less than the thermal boundary layer (see Fig. F,8b). Thus the main body of fluid can be considered to be inviscid, and the buoyancy and inertia terms must be of the same order of magnitude. From Eq. 8-7a it then follows that, for small Pr,

$$U^* = \sqrt{\beta g \Delta TL} \tag{8-12a}$$

Although the thermal boundary layer is small for large Prandtl numbers (see Fig. F,8c), all of the buoyancy force acts there. Hence, the viscous forces are stronger near the wall and the buoyancy force must be of the same order of magnitude as the viscous forces. From Eq. 8-7a, we thus find that

$$U^* = \sqrt{\frac{\rho g \Delta TL}{Pr}} \tag{8-12b}$$

The appropriate U^* as given either in Eq. 8-12a or 8-12b must be used in Eq. 8-10.

In the development that follows, the U^* will remain unspecified; it should also be kept in mind that the $\bar{y}$ can take on two different forms, as was just mentioned. For Prandtl numbers of unit order of magnitude, the use of Eq. 8-12a and Eq. 8-10 without Pr in it will lead to the familiar forms of the differential equations.

We shall now consider two-dimensional problems first and the axially symmetric cases later.

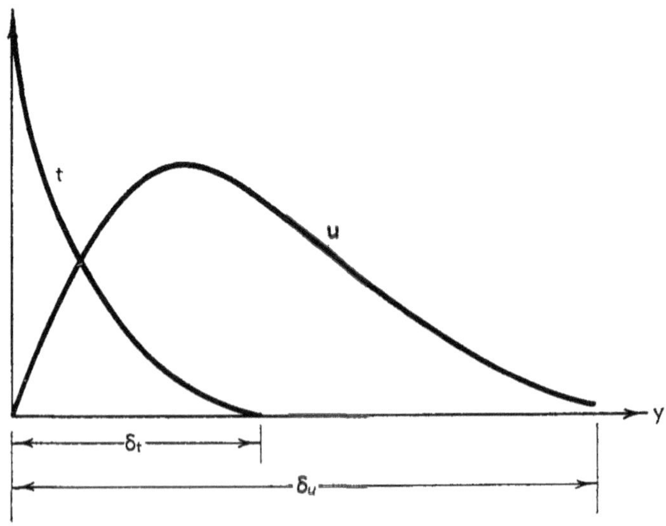

Fig. F,8c. Representative velocity and temperature profiles for large Prandtl numbers.

TWO-DIMENSIONAL CASES. After defining a stream function ψ from Eq. 8-9c, we substitute

$$\psi = f(x)F(\eta), \qquad \eta = \frac{\bar{y}}{g(x)}, \qquad t = h(x)H(\eta) \tag{8-13}$$

into Eq. 8-7a and 8-8a, which yields

$$\frac{f'}{f}(F'^2 - FF'') - \frac{g'}{g}F'^2 = \frac{Pr}{fg}F''' + \frac{\beta g\Delta TL}{U^{*2}}\sqrt{1-(r_0')^2}\,h\left(\frac{g}{f}\right)^2 H \tag{8-14}$$

$$\frac{h'}{h}F'H - \frac{f'}{f}FH' = \frac{1}{fg}H'' \tag{8-15}$$

where f is the scaling of x of velocity, g is the boundary layer thickness function, h is the temperature distribution along the surface, and the primes denote differentiation with respect to the arguments given in

Eq. 8-13. A like procedure is described by Schuh [5]. The system of Eq. 8-14 and 8-15 will be reduced to ordinary differential equations in $F(\eta)$ and $H(\eta)$ if

$$\frac{f'}{f} = C_1 \frac{g'}{g} = C_2 \frac{1}{fg} = C_3 \sqrt{1 - (r_0')^2}\, h \left(\frac{g}{f}\right)^2 = C_4 \frac{h'}{h} \tag{8-16}$$

where the C_i are constants. These then are four conditions for the four functions: f, g, h, and r_0. For simplicity, we shall restrict ourselves to isothermal surfaces so that $h = 1$ and $h' = 0$. The class of bodies can also be found if this restriction on $h(x)$ is not imposed. Thus, Eq. 8-16 reduces to

$$\frac{f'}{f} = C_1 \frac{g'}{g} = C_2 \frac{1}{fg} = C_3 \sqrt{1 - (r_0')^2} \left(\frac{g}{f}\right)^2 \tag{8-17}$$

Integration of this system yields

$$f = A_1 x^{1/M+1}, \; g = A_2 x^{M/M+1}, \text{ and } r_0' = \pm \sqrt{1 - Cx^{\frac{2(1-3M)}{M+1}}} \tag{8-18}$$

For each value of $M (M \leq \frac{1}{3})$ we get a different closed body for which there will be a "similar" flow. Thus, the exponent M is analogous to the exponent m in the class of potential flow velocities ($U = AX^m$) shown by Falkner and Skan [36] to yield "similar" forced flows.

Wedge flows. If $M = \frac{1}{3}$, $r_0' = \pm \sqrt{1 - C}$ so that $r_0 = C_5 \pm \sqrt{1 - C}x$, which is the equation for a wedge. The parameter C determines the slope for a wedge. In the special case $C = 1$, we obtain the vertical flat plate; $C = 0$ gives a horizontal plate. For this case ($M = \frac{1}{3}$), f behaves like $x^{\frac{3}{4}}$ and g like $x^{\frac{1}{4}}$. Thus all wedge flows are given by the flat plate flow with the Grashof number modified to take into account the proper component of the body force, as was discussed in the case of the inclined plate.

Parabolic-nose bodies. If we take $M = 0$ so that $(1 - 3M)/(M + 1) = 1$, $f = x$ and $g = 1$. From Eq. 8-13 it can be seen that, if $g = 1$, the boundary layer thickness will be constant over the entire surface. Note that in the stagnation region this result is consistent with that of Hermann [28] for a horizontal cylinder. To determine the body shape, we note that now Eq. 8-18 can be written as $r_0' = \pm (1/a) \sqrt{a^2 - x^2}$ where $a = C^{-2}$. Integration of this expression yields

$$\frac{r_0}{a} = \frac{1}{2}\left[\frac{x}{a}\sqrt{1 - \left(\frac{x}{a}\right)^2} + \sin^{-1}\left(\frac{x}{a}\right)\right] \tag{8-19}$$

By definition we also have $(z')^2 + (r_0')^2 = 1$, so that

$$\frac{z}{a} = \frac{1}{2}\left(\frac{x}{a}\right)^2 \tag{8-20}$$

Eq. 8-19 and 8-20 thus are the parametric equations for the body. The nondimensional radius of curvature at any point on this body is

$$r_k = a\sqrt{1 - \left(\frac{x}{a}\right)^2}$$

from which it follows that at the nose of the body ($x = 0$), the radius of curvature is equal to a. Eq. 8-19 and 8-20 thus represent a family of bodies with parabolic noses. The boundary layer equations for this class of bodies are thus

$$(F')^2 - FF'' = PrF''' + \frac{\beta g \Delta T L}{U^{*2}} H \tag{8-21}$$

$$-FH' = H'' \tag{8-22}$$

where the primes denote differentiation with respect to $\eta = y$. The L appearing in Eq. 8-21 is the dimensional nose radius of curvature, since the a that was the nondimensional nose radius of curvature can be set equal to unity.

Pointed-nose bodies. If we set $2(1 - 3M)/(M + 1) = 1$, then $M = \frac{1}{7}$ and $f = b_1x^{7/8}$, $g = b_2x^{1/8}$ where b_1 and b_2 are constants. Eq. 8-18 reduces to $r_0' = \pm\sqrt{1 - Cx}$, from which

$$r_0 = b_3 - \frac{3}{2C}(1 - Cx)^{3/2}$$

and for $r_0(0) = 0$, $b_3 = \frac{2}{3}C$, so that

$$r_0 = \frac{2}{3C}[1 - (1 - Cx)^{3/2}] \tag{8-23}$$

Again from its definition we find

$$z = \frac{2\sqrt{C}}{3} x^{3/2} \tag{8-24}$$

The radius of curvature is

$$r_k = 2\sqrt{\frac{x(1 - Cx)}{C}}$$

and vanishes at $x = 0$, so that Eq. 8-23 and 8-24 are the parametric equations for a class of pointed-nose bodies. From the above equations we find that the slope dz/dr_0 is

$$\frac{dz}{dr_0} = \sqrt{\frac{Cx}{1 - Cx}}$$

which also vanishes at the nose. A representative body of the class discussed here is presented in Fig. F,8d.

If for convenience we set $b_1 = 8/7b_2$ and $b_2 = (\frac{8}{7})^{\frac{1}{4}}C^{-\frac{1}{4}}$ the momentum and energy equations for this class of bodies are

$$\frac{6}{7}(F')^2 - FF'' = PrF''' + \frac{\beta g \Delta TL}{U^{*2}} H \tag{8-25}$$

$$-FH' = H'' \tag{8-26}$$

where $\eta = y/b_2(x)^{\frac{1}{4}}$, and the parameter C is free and can be fixed to specify one of the family of bodies given by Eq. 8-23 and 8-24.

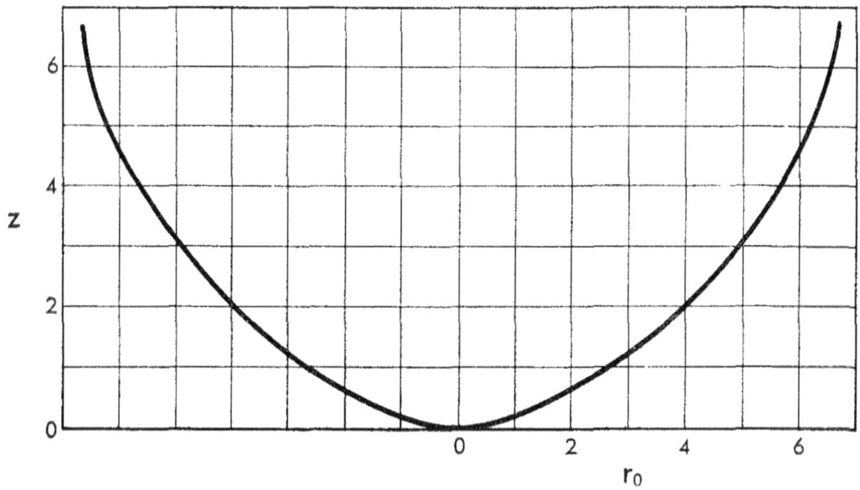

Fig. F,8d. A typical two-dimensional, pointed-nose body computed from Eq. 8-23 and 8-24 with $C = 1$.

Axially Symmetric Cases. For axially symmetric bodies, we define the stream function from Eq. 8-9b as

$$\psi_y = r_0 u \quad \text{and} \quad \psi_x = -r_0 v$$

where we have taken $r_0 \cong r$ in the boundary layer. Once again, we make the transformations as given by Eq. 8-13. The momentum and energy equations thus become, respectively,

$$\left(\frac{f'}{f} - \frac{r_0'}{r_0} - \frac{g'}{g}\right)(F')^2 - \frac{f'}{f}FF'' = Pr\frac{r_0}{fg}F''' + \frac{\beta f x \Delta TL}{U^{*2}}\sqrt{1-(r_0')^2}\left(\frac{r_0 g}{f}\right)^2 hH \tag{8-27}$$

$$\frac{h'}{h}F'H - \frac{f'}{f}FH' = \frac{r_0}{fg}H'' \tag{8-28}$$

To reduce these equations to ordinary differential form, the coefficient of each term must be proportional to that of the other terms (see Eq.

8-17). We again restrict ourselves to isothermal surfaces so that $h = 1$ and $h' = 0$. The proportionality of the first two coefficients of Eq. 8-27 immediately tells us that $f^m = A_1 g r_0$ where m is as yet undetermined.

Cones. If we now assume a solution of the form $f = B_1 x^M$, then from the last expression above, $x^{mM} = A_2 g r_0$. The proportionality between the second and third terms of Eq. 8-27 yields $f' = (A_3 r_0)/g$ or $x^{M-1} = (A_4 r_0)/g$. The product of the two expressions for x gives $x^{mM+M-1} = A_5 r_0^2$, so that

$$g = A_6 x^{\frac{mM-M+1}{2}}$$

From the proportionality between the coefficients of the two terms on the right of Eq. 8-27, there is obtained

$$A_7 = \frac{r_0 g^3}{f} \sqrt{1 - (r_0')^2}$$

or using previous results,

$$A_8 = x^{2mM-2M+1} \sqrt{1 - Cx^{mM+M-3}}$$

Since both exponents of x must vanish, we find that $M = \frac{7}{4}$, $m = \frac{5}{7}$. Therefore, $f = A_9 x^{\frac{7}{4}}$, $g = A_{10} x^{\frac{1}{4}}$, and $r_0 = A_{11} x$, where all the A's are constants. The last equation defines a family of cones. The momentum and energy equations for these bodies thus are

$$\frac{1}{2}(F')^2 - \frac{7}{4}FF'' = PrF''' + \frac{\beta f_X \Delta T L}{U^{*2}} H \tag{8-29}$$

$$-\tfrac{7}{4}FH' = H'' \tag{8-30}$$

Kettle shapes. If we start with the proportionalities in the form:

$$f^m f' = B_1 r_0^2, \qquad f' = \frac{B_2 r_0}{g}, \quad \text{and} \quad ff'' = B_3 r_0^2 g^2 \sqrt{1 - (r_0')^2}$$

we obtain, by combining the first two of these, $f^{-m} f' = B_4 g^{-2}$. Combining this result with the first and third proportionalities gives

$$f^{1-2m} f' = B_5 \sqrt{1 - (r_0')^2}$$

If we set $m = \frac{1}{3}$ in this expression and the first proportionality, we obtain an equation for r_0 alone. $r_0^2 = a^2 \sqrt{1 - (r_0')^2}$ or $r_0' = \sqrt{1 - (r_0/a)^4}$ which can be written as

$$d(x/a) = \frac{d(r_0/a)}{\sqrt{1 - (r_0/a)^4}}$$

To integrate this expression we let $r_0/a \equiv \cos\varphi$, so that

$$\int_0^x \frac{dx}{a} = -\frac{1}{\sqrt{2}} \int_{\pi/2}^{\varphi} \frac{d\varphi}{\sqrt{1 - \frac{1}{2}\sin^2\varphi}}$$

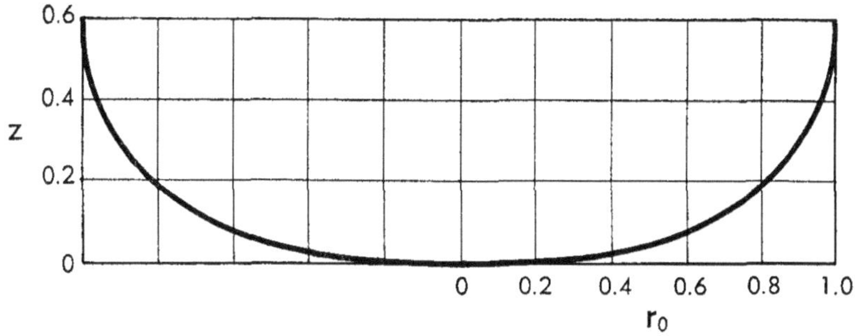

Fig. F,8e. A typical axially symmetric kettle-shape body computed from Eq. 8-32 and $r_0/a = \cos\phi$ with $a = 1$.

Thus

$$x = \frac{1}{\sqrt{2}}[K(45°) - F(45°, \varphi)] \tag{8-31}$$

where K and F are elliptic functions of the first kind (see [*39*])

Now, from the definition, $(z')^2 + (r')^2 = 1$, we obtain

$$\frac{dz}{a} = \frac{(r_0/a)^2 d(r_0/a)}{\sqrt{1 - (r_0/a)^4}}$$

so that

$$\begin{aligned}\frac{z}{a} &= -\frac{1}{2}\int_{\pi/2}^{\varphi} \frac{\cos^2\varphi\, d\varphi}{\sqrt{1 - \frac{1}{2}\sin^2\varphi}} \\ &= \frac{1}{\sqrt{2}}\left\{2\left[E\left(45°, \frac{\pi}{2}\right) - E(45°, \varphi)\right] - \left[F\left(45°, \frac{\pi}{2}\right) - F(45°, \varphi)\right]\right\}\end{aligned} \tag{8-32}$$

where $E(\alpha, \varphi)$ is the elliptic function of the second kind. Eq. 8-31 and 8-32 thus are the parametric equations for the bodies and

$$f = \left(\frac{4}{3}B_6 a^2\right)^{\frac{3}{4}} z^{\frac{3}{4}}, \qquad g = \frac{B_7 f^{\frac{1}{3}}}{r_0} = \frac{(\frac{4}{3}B_8 a^2)^{\frac{1}{4}} z^{\frac{1}{4}}}{\cos\varphi}$$

A representative kettle-shape body is shown in Fig. F,8e. The governing differential equations for this class of bodies are (setting $B_6 = B_8 = 1/a^2$):

$$\frac{2}{3}(F')^2 - FF'' = PrF''' + \frac{\beta f_X \Delta TL}{U^{*2}} H \tag{8-33}$$

$$-FH' = H'' \tag{8-34}$$

It is important to note that by setting $F(\eta) = (3)^{\frac{3}{4}}(Pr)^{\frac{1}{4}}\chi(\zeta)$ and $\eta = (3)^{\frac{1}{4}}(Pr)^{\frac{1}{4}}\zeta$, Eq. 8-33 and 8-34 are identically the equations for free convection about a flat plate [*6*] (see Eq. 3-52 and 3-53), and the solutions for various Prandtl numbers are already known.

In the above, we have left the number a unspecified. We can take $a = 1$, so that the maximum value of the body radius is unity. That is, we use the maximum value of the radius of the body as the characteristic length. Thus, we now have

$$f = \left(\frac{4z}{3}\right)^{\frac{3}{4}}, \qquad g = \frac{\left(\frac{4z}{3}\right)^{\frac{1}{4}}}{\cos\varphi}; \qquad r_0 = \cos\varphi$$

The function g is a measure of the boundary layer thickness, and we note that it is indeterminate at the nose of the body. Using L'Hospital's rule, we find that the boundary layer thickness at the nose is infinite so that our boundary layer equations are invalid there. This appears to be reasonable because the body is so flat near the nose. In this region, the configuration takes on the characteristics of a horizontal flat plate heated from above; in that case conduction predominates.

Parabolic-nose bodies. We have already obtained a class of bodies with infinite curvature at the nose (cone) and zero curvature at the nose (kettle). We would now like to find a class of bodies with a finite but nonvanishing curvature at the nose. To obtain such a class, we first recall that the parameter in the proportionality relations takes on the value $\frac{5}{7}$ for the cone and $\frac{1}{3}$ for the kettle. We thus choose to try $m = \frac{1}{2}$ which is in between the previous values. Setting the constants equal to unity and combining the proportionality conditions, we find that

$$f = \frac{r_0^4}{1 - (r_0')^2} \quad \text{and} \quad f' = \sqrt{1 - (r_0')^2}$$

These can be combined and written in terms of z as

$$\frac{dz}{dr_0} = \frac{d}{dr_0}\left\{\frac{r_0^4\left[1 + \left(\frac{dz}{dr_0}\right)^2\right]}{\left(\frac{dz}{dr_0}\right)^2}\right\}$$

whence

$$z = \frac{r_0^4\left[1 + \left(\frac{dz}{dr_0}\right)^2\right]}{\left(\frac{dz}{dr_0}\right)^2}$$

or finally

$$\left(\frac{dz}{dr_0}\right)^2 = \frac{r_0^4}{z - r_0^4}$$

From which, the body shapes are given by

$$z = D\left[\frac{1}{a}r_0^2 + \frac{1}{5}r_0^4 + \frac{12}{7 \cdot 25}ar_0^6 + \frac{16 \cdot 13}{63 \cdot 5^3}a^2r_0^8 + \frac{64 \cdot 163}{33 \cdot 49 \cdot 5^4}a^3r_0^{10} + \cdots\right]$$

where D is a constant and $a = (4)^{\frac{1}{4}}$. In the vicinity of the nose, the body shape is obviously parabolic.

The differential equations for these bodies reduce to

$$(F')^2 - 2FF'' = PrF''' + \frac{\beta f_X \Delta TL}{U^{*2}} H$$

$$-2FH' = H''$$

NONISOTHERMAL SURFACES

In numerous technical applications of the free convection phenomenon, the surface thermal conditions are not uniform; often the temperature variation along the surface is prescribed, or else the variation of heat flux along the surface is specified. These two types of conditions are related because, if for a given flow the temperature variation is prescribed, the heat flux variation is fixed, and vice versa. Relatively speaking, considerably less study has been made of free convection about nonisothermal surfaces than isothermal surfaces. The most noteworthy of the more general analyses are those of Sparrow and Gregg [*40,41,42*] and Millsaps and Pohlhausen [*43*]. We shall give here a more or less unified account of this work.

F,9. Vertical Flat Plate. We start with the configuration shown in Fig. F,3a, except for the thermal boundary conditions, and again make all the same assumptions as were made for the case of the isothermal vertical plate. The dimensional forms of the governing boundary layer equations (corresponding to the dimensionless equations (Eq. 3-47 and 3-48)) are

$$\psi_Y \psi_{XY} - \psi_X \psi_{YY} = \beta g(T - T_\infty) + \nu \psi_{YYY} \tag{9-1}$$

$$\psi_Y T_X - \psi_X T_Y = \alpha T_{XY} \tag{9-2}$$

where the subscripts denote partial differentiation.

The boundary conditions for the present problem are

$$U(X, 0) \equiv \psi_Y(X, 0) = 0; \qquad V(X, 0) \equiv -\psi_X(X, 0) = 0 \tag{9-3}$$

$$T(X, 0) = T_w(X) \quad \text{or} \quad T_Y(X, 0) = \frac{-q(X)}{k} \tag{9-4}$$

$$U(X, \infty) = 0, \qquad T(X, \infty) = T_\infty \tag{9-5}$$

where q is the heat flux rate. Note that the difference between the present problem and that for an isothermal plate appears in the boundary condition given by Eq. 9-4.

In order to solve the boundary-value problem defined above, we resort to the usual similarity transformation of boundary layer theory. It is, however, not possible to apply this transformation for any arbitrary wall-temperature variation. Therefore, we restrict our consideration to two families of wall-temperature variations which permit reduction of the partial differential equations (Eq. 9-1 and 9-2) to ordinary differential equations by means of the similarity transformation. These families are

$$T_w - T_\infty = NX^n \tag{9-6a}$$

$$T_w - T_\infty = Me^{mX} \tag{9-6b}$$

where M, N, m, and n are constants.

Reduction for the power law case. For a wall-temperature variation of the type given by Eq. 9-6a, we define the similarity variable η as

$$\eta = \left(\frac{\beta g N}{4\nu^2}\right)^{\frac{1}{4}} YX^{(n-1)/4} \tag{9-7}$$

We introduce this variable, together with the new dependent variables F and θ given by

$$F(\eta) = \frac{\psi}{(4^3\nu^2\beta gNX^{n+3})^{\frac{1}{4}}} \tag{9-8}$$

$$\theta(\eta) = \frac{T - T_\infty}{T_w - T_\infty} \tag{9-9}$$

into Eq. 4-1 to 4-5 to obtain

$$F''' + (n + 3)FF'' - 2(n + 1)(F')^2 + \theta = 0 \tag{9-10}$$

$$\theta'' + Pr[(n + 3)F\theta' - 4nF'\theta] = 0 \tag{9-11}$$

$$F(0) = F'(0) = 0 \tag{9-12}$$

$$\theta(0) = 1 \tag{9-13a}$$

$$F'(\infty) = \theta(\infty) = 0 \tag{9-14}$$

where the primes denote differentiation with respect to η.

If instead of specifying the wall-temperature variation, as was done above, we had chosen to prescribe the wall heat-flux variation according to the second of Eq. 9-4, we would replace the denominator of Eq. 9-9 by $-(\text{const})q(X)/k$. The relation between the two types of boundary conditions given in Eq. 9-4 is graphically presented in [*40*]. For similarity, the $q(X)$ would also have to be given by one of Eq. 9-6 so that the only difference in the boundary-value problem would be that the condition

$$\theta'(0) = 1 \tag{9-13b}$$

would replace Eq. 9-13a. For uniform heat flux at the wall ($q(X) = \text{const}$), X^n would also have to multiply the constant in the denominator of Eq. 9-9. It can then be determined (as will be done subsequently) that, for uniform wall heat flux, $n = \frac{1}{5}$ (which agrees with [*40*]). Explicitly, for the case of uniform wall heat flux, it is found in [*40*] that if

$$\eta = C_1 \frac{Y}{X^{\frac{1}{5}}}; \qquad F(\eta) = \frac{\psi}{C_2 X^{\frac{4}{5}}}; \qquad \theta(\eta) = \frac{C_1(T_\infty - T)}{qX^{\frac{1}{5}}/k}$$

where

$$C_1 = \left(\frac{\beta g q}{5k\nu^2}\right)^{\frac{1}{5}} \quad \text{and} \quad C_2 = \left(\frac{5^4 \beta g q \nu^3}{k}\right)^{\frac{1}{5}}$$

the equations are

$$F''' + 4FF'' - 3(F')^2 - \theta = 0$$

$$\theta'' + Pr(4F\theta' - F'\theta) = 0$$

and

$$U = C_1 C_2 X^{\frac{3}{5}} F'(\eta); \; V = \frac{C_2}{5X^{\frac{1}{5}}}[\eta F'(\eta) - 4F(\eta)]$$

$$T_w - T_\infty = -\frac{5^{\frac{1}{5}} qX}{k}\left(\frac{\beta q g X^4}{\nu^2 k}\right)^{-\frac{1}{5}} \theta(0)$$

Coming back to the case where the surface temperature is specified, we find that the dimensionless stream function is related to the velocity components by

$$U = (4\beta g N X^{n+1})^{\frac{1}{2}} F'(\eta) \tag{9-15}$$

$$V = -\left(\frac{\beta g N \nu^2 X^{n-1}}{4}\right)^{\frac{1}{4}} [(n+3)F + (n-1)\eta F'] \tag{9-16}$$

It is worth noting that, because of the nature of the transformation given by Eq. 9-7, boundary conditions cannot be independently specified at $X = 0$, once conditions at $Y = 0$ and $Y = \infty$ have been specified. For example, if $n > 1$, the conditions at $X = 0$ are identical with those at $Y = 0$, that is, the velocity is zero and the temperature is equal to the plate temperature along $X = 0$.

Reduction for exponential case. For a wall-temperature variation given by Eq. 9-6b we must consider separately the cases of positive and negative exponents m. For positive m, we transform Eq. 9-1 and 9-2 by introducing

$$\xi = \left(\frac{\beta g m M e^{mX}}{4\nu^2}\right)^{\frac{1}{4}} Y \tag{9-17}$$

$$G(\xi) = \frac{\psi}{(4^3 M \nu^2 \beta g m^5 e^{mX})^{\frac{1}{4}}}; \qquad \varphi(\xi) = \frac{T - T_\infty}{T_w - T_\infty} \tag{9-18}$$

There is thus obtained

$$G''' + GG'' - 2(G')^2 + \varphi = 0 \tag{9-19}$$

$$\varphi'' + Pr[G\varphi' - 4G'\varphi] = 0 \tag{9-20}$$

where the primes here denote differentiation with respect to ξ. The boundary conditions of the functions G and φ are identical with those for F and θ. These equations can be treated in a manner similar to that for the power-law variation if the wall heat flux rather than the wall-temperature variation is specified. The case of uniform heat flux can only be obtained if $m = 0$ so that, for exponential variations, no wall-temperature variations are possible for this case.

As in the previous section, the conditions at $X = 0$ cannot be specified once conditions at $Y = 0$ and $Y = \infty$ have been given. In this connection, we note from the relations

$$U = \left(\frac{4\beta g M e^{mX}}{m}\right)^{\frac{1}{2}} G'(\xi); \qquad V = -\left(\frac{\beta g m \nu^2 M e^{mX}}{4}\right)^{\frac{1}{4}} (\xi G' + G) \quad (9\text{-}21)$$

and Eq. 9-17 that U is not zero and not uniform (since G' varies with ξ) along $X = 0$. This fact that U is nonzero at $X = 0$ may lead one to interpret solutions of Eq. 9-19 and 9-20 as belonging to problems of combined forced and free convection flows. The temperature also is not uniform along $X = 0$.

For negative values of m, the transformation given by Eq. 9-17 fails because the coefficient becomes imaginary. A real transformation was found in [*31*] to reduce the equations for negative m. However, study of the resulting equations showed that their solutions had rather unusual characteristics. In particular, negative velocities and/or temperatures less than ambient are indicated. These findings are attributed to the unusual nature of the conditions at $X = 0$ and have not been pursued further. In this regard, it should be mentioned that a negative m implies that, at least near the plate, heavier fluid is being piled on top of lighter fluid. This situation leads to thermal instabilities which will be discussed in a later chapter.

Solutions. The solution of both boundary-value problems depends on the Prandtl number. It is, however, somewhat surprising that, although the parameter n which specifies the shape of the variation of the thermal conditions appears explicitly in Eq. 9-10 and 9-11 for the power-law case, the corresponding parameter for the exponential case does not appear in Eq. 9-19 and 9-20.

Numerical solutions of the two problems for $Pr = 0.7$ and 1.0 were obtained on an IBM Card Programmed Calculator using the technique presented in detail in Appendix B of [*6*]. The calculations were made for $n = 3, 2, 1, 0.5, 0.2, 0, -0.2, -0.5$, and -0.8 for the power-law variation case.

Profiles. Dimensionless velocity and temperature distributions for the power-law surface temperature (given by Eq. 9-6a) case are presented in Fig. F,9a and F,9b, respectively, for $Pr = 0.7$ and several representative values of n. The velocity profiles vary regularly with the parameter n. However, the temperature profiles for negative values of n differ notably in shape from the uniform wall temperature case ($n = 0$).

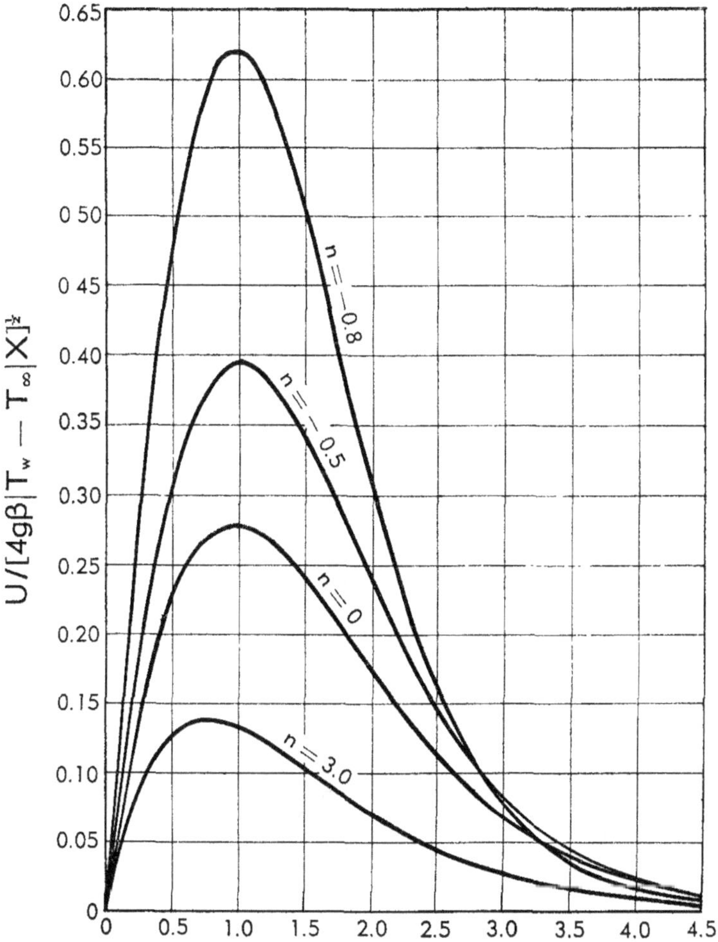

$$\frac{Y}{X}\left(\frac{Gr_x}{4}\right)^{\frac{1}{4}} = \frac{Y}{X}\left[\frac{g\beta|T_w - T_\infty|X^3}{4\nu^2}\right]^{\frac{1}{4}}$$

Fig. F,9a. Dimensionless velocity distributions for several values of $n(T_w - T_\infty = NX^n)$. $Pr = 0.7$ (Sparrow and Gregg [*42*]).

An inflection point can be seen for the $n = -0.5$ curve, and the $n = -0.8$ curve has a peak so that $T > T_w$. Similar behavior was found by Schuh [*5*] for forced convection over a flat plate with a power-law surface temperature variation. For positive n, the temperature distributions are similar in shape to that for $n = 0$.

The profiles for $Pr = 1$ depend in the same way on n as was discussed above, and the velocity curves for this case lie slightly below those for $Pr = 0.7$. To show the effect of Prandtl number on the temperature distribution, characteristic profiles are presented in Fig. F,9c. The rela-

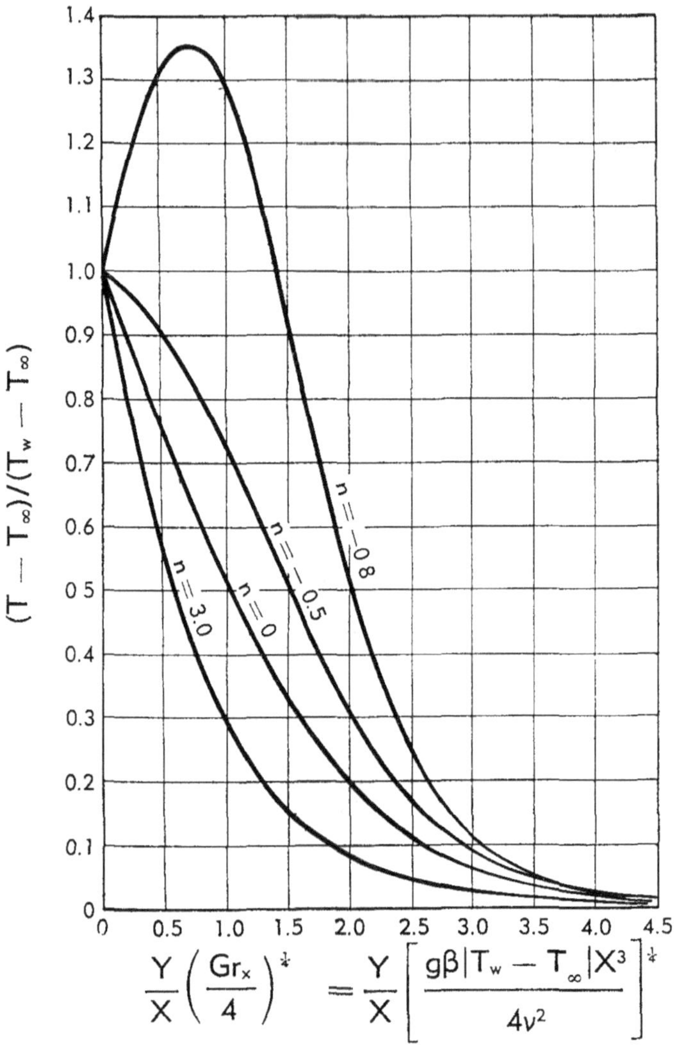

Fig. F,9b. Dimensionless temperature distributions for several values of $n(T_w - T_\infty = NX^n)$. $Pr = 0.7$ (Sparrow and Gregg [42]).

tive orientation of the curve for $n = 0.5$ is typical for all others studied, except for temperature profiles for $n = -0.8$. Therefore, the temperature curves for $n = -0.8$ are also presented.

For $n < -0.6$, investigation of the mathematical model shows that there is an infinite source of energy at the leading edge. The fluid is thus able to transfer an infinite amount of heat to a finite length of the plate. Since such sources are not physically realistic, there is an uncertainty as to the region where the results for $n < -0.6$ may be applied (i.e., how far from the leading edge).

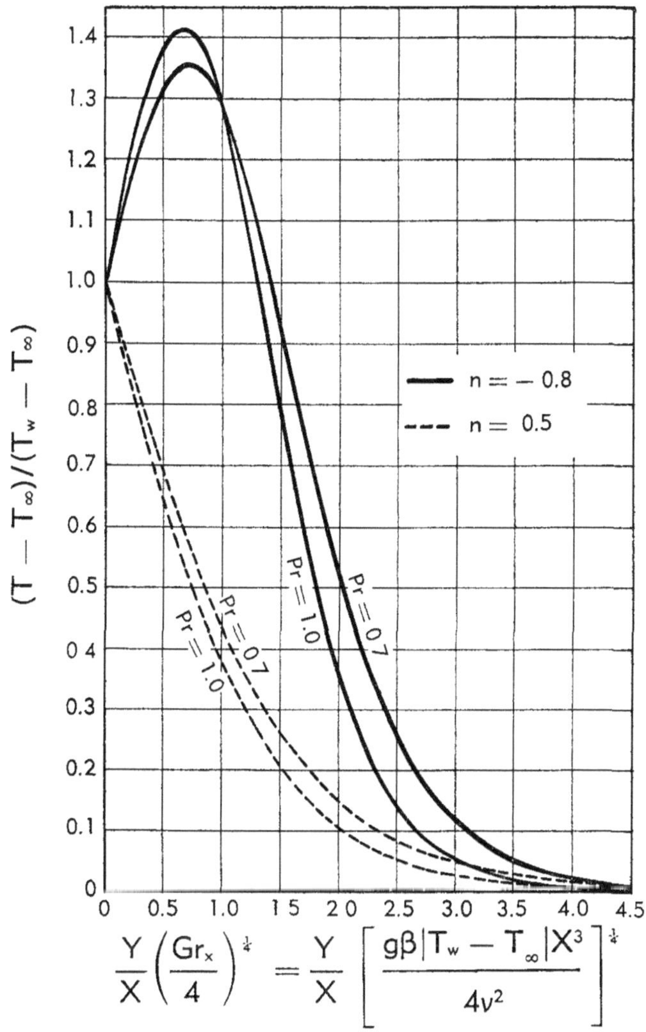

Fig. F,9c. Typical curves showing relationship between dimensionless temperature distributions for $Pr = 0.7$ and $Pr = 1.0$. The relative orientation of the curves for $n = 0.5$ is typical for the other cases in the range $-0.5 \leqq n \leqq 3.0$ $T_w - T_\infty = NX^n$ (Sparrow and Gregg [42]).

The dimensionless velocity and temperature distributions for exponential surface temperature variations (i.e., given by Eq. 9-6b) for different Prandtl numbers are similar in appearance to those for power-law surface temperature variations (Eq. 9-6a). The temperature profiles, however, exhibit no behavior such as at $n = -0.8$. When $X = 0$, it is found (as already noted) that nonuniform profiles exist along this line. The question immediately arises as to how strongly the heat transfer results are bound up with the nonuniform conditions imposed at the leading edge. Experiments or additional analyses that permit other conditions to be specified at $X = 0$ are needed to answer this question. The utility of the results for the exponential wall-temperature variation cannot, therefore, be stated with certainty herein.

Calculations presented in [40] for the case of uniform heat flux at the plate were made for a number of different Prandtl numbers. The dimensionless profiles for this case are presented in Fig. F,9d and F,9e. Recall that the difference between the surface and ambient temperatures is proportional to $X^{\frac{1}{5}}$, so that the rate of temperature increases with

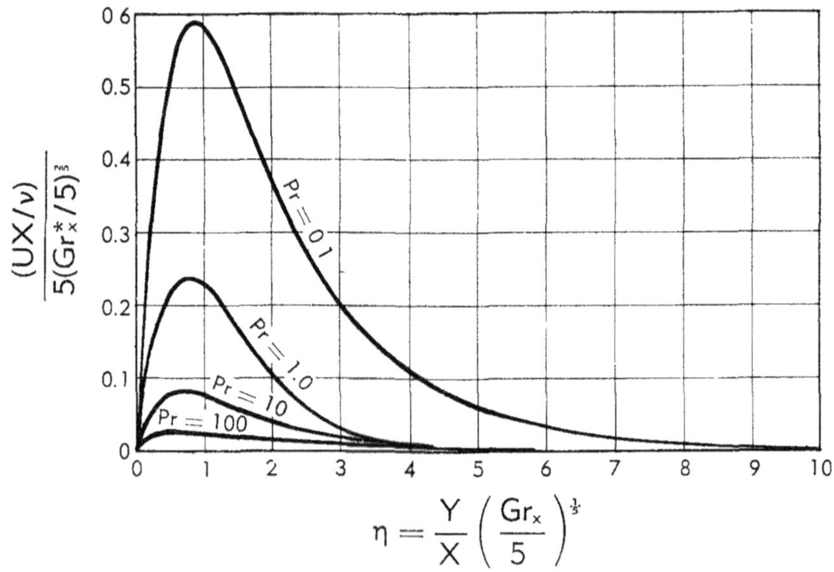

Fig. F,9d. Dimensionless velocity distributions for several Prandtl numbers (Sparrow and Gregg [40]).

distance is greatest near the plate leading edge. The dimensionless surface temperature variation $(T_w - T_\infty)/Gr^*)^{\frac{1}{5}}/(qX/k)$ is plotted as a function of Prandtl number in Fig. F,9f where $Gr^* = \beta g q X^4/\nu^2 k$ is a modified Grashof number. Note that the curve between $Pr = 0.1$ and 0.01 is an extrapolation.

Heat transfer. The local heat transfer from the surface to the fluid is, for the power-law surface temperature case, given by

$$q = -k\left(\frac{\partial T}{\partial Y}\right)_{Y=0} = -k\left(\frac{\beta g N^5}{4\nu^2}\right)^{\frac{1}{4}} X^{(5n-1)/4}\left(\frac{d\theta}{d\eta}\right)_{\eta=0} \qquad (9\text{-}22)$$

The derivative $(d\theta/d\eta)_{\eta=0} = \theta'(0)$ is a function of both Prandtl number and n and is found from the solution of Eq. 9-10 and 9-14.

The dependence of the local heat flux on X is clear from Eq. 9-22.

Corresponding to a variation of $T_w - T_\infty$ proportional to X^n, there is a heat flux variation proportional to $X^{(5n-1)/4}$. The inverse is also true, i.e., if a heat flux variation proportional to X^r were prescribed, then the corresponding variation of $T_w - T_\infty$ would be proportional to $X^{(4r+1)/5}$.

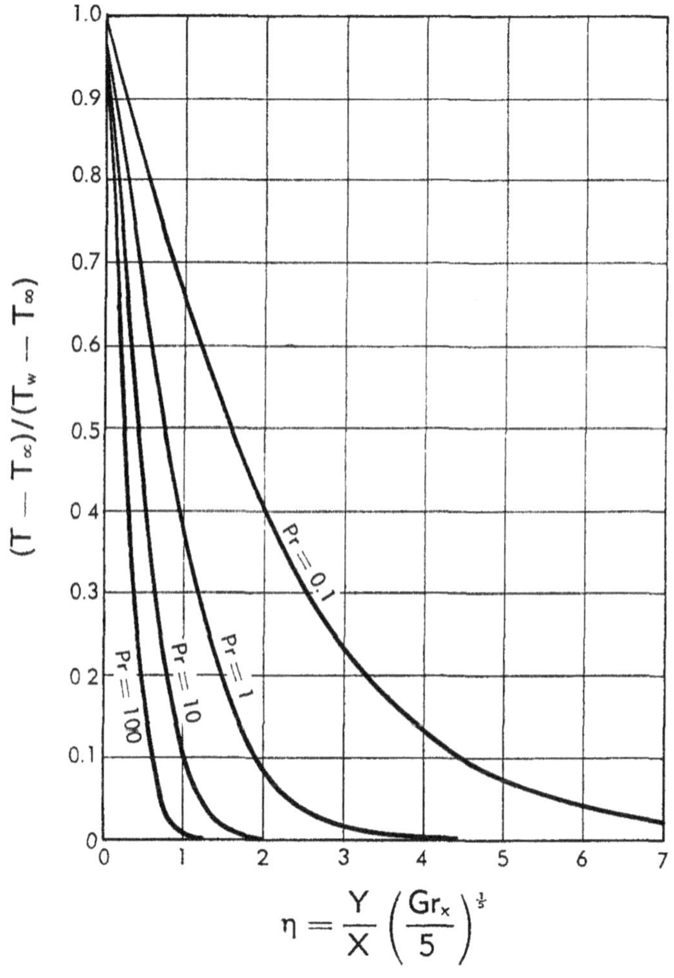

Fig. F,9e. Dimensionless temperature distributions for several Prandtl numbers with uniform surface heat flux (Sparrow and Gregg [*40*]).

It is thus evident that, for the important case of uniform heat flux ($r = 0$), $T_w - T_\infty$ varies as $X^{\frac{1}{5}}$, as has already been pointed out.

The local Nusselt number thus is

$$\frac{Nu_x}{Gr_X^{\frac{1}{5}}} = -\frac{\theta'(0)}{\sqrt{2}} \tag{9-23}$$

where the local Grashof number is $(\beta g|T_w - T_\infty|X^3)/\nu^2$. The use of the absolute magnitude of the temperature difference eliminates separate consideration of the two cases $T_w > T_\infty$ and $T_w < T_\infty$. A plot of the Nusselt-Grashof relation given by Eq. 9-23 is shown in Fig. F,9g over

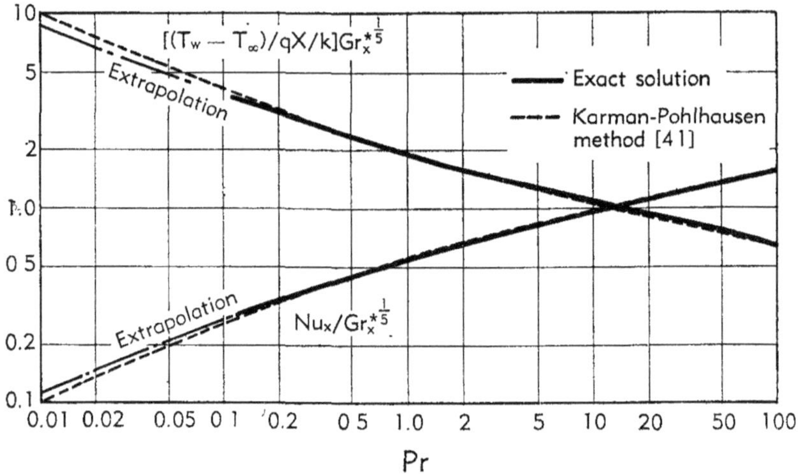

Fig. F,9f. Variation of $\left(\frac{T_w - T_\infty}{\frac{qX}{k}}\right) Gr_X^{*\frac{1}{5}}$ and $\frac{Nu_X}{Gr_X^{*\frac{1}{5}}}$ with Prandtl number. ($T_w - T_\infty$) is the local temperature difference between the wall and ambient. Nu_X is the local Nusselt number. Gr_X^* is the local value of the modified Grashof number (Sparrow and Gregg [40]).

the range $-0.8 \leq n \leq 3.0$ for $Pr = 0.7$ and 1.0. For values of n less than -0.6, it is seen that $Nu_X/(Gr_X)^{\frac{1}{4}}$ is negative. Physically, this corresponds

Table F,9a. Comparison of isothermal and nonisothermal local heat transfer. $\frac{q(\text{variable } T_w)}{q(\text{constant } T_w)}$

n	$Pr = 0.7$	$Pr = 1.0$
3.0	1.94	1.92
2.0	1.75	1.74
1.0	1.49	1.49
0.5	1.30	1.30
0.2	1.14	1.14
0	1.00	1.00
−0.2	.81	.81
−0.5	.31	.32
−0.8	−1.51	−1.68

to a heat transfer from the fluid to the wall even though $T_w > T_\infty$. The reason for this result has already been discussed (infinite energy at the leading edge) in connection with the temperature distribution.

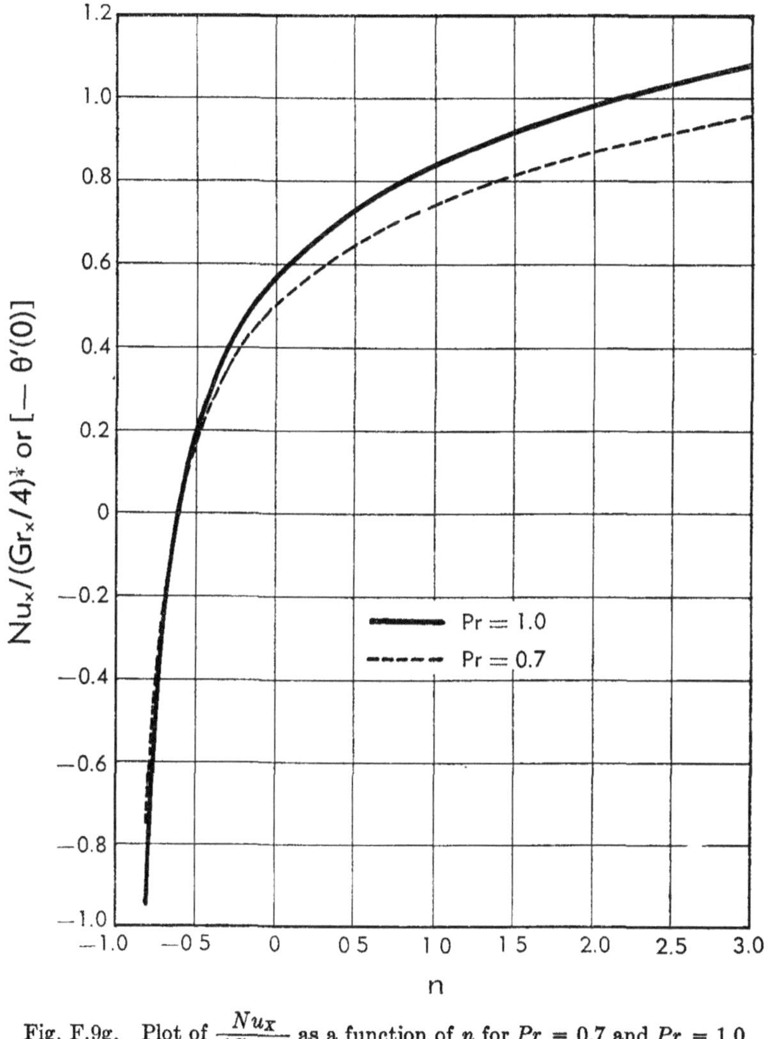

Fig. F,9g. Plot of $\dfrac{Nu_x}{\left(\dfrac{Gr_x}{4}\right)^{\frac{1}{4}}}$ as a function of n for $Pr = 0.7$ and $Pr = 1.0$. $T_w - T_\infty = NX^n$ (Sparrow and Gregg [40]).

It is of practical interest to inquire as to how well the heat flux for the nonisothermal wall could be predicted by local application of the isothermal wall results. Such a procedure might conceivably be resorted to in the absence of other information about the variable wall-temperature problem. A comparison of local heat transfer for the isothermal and non-isothermal walls is given in Table F,9a.

Except for very gradual variations of the wall temperature, i.e., n is very close to zero, the procedure of locally applying isothermal wall results to a nonisothermal surface appears unsatisfactory. A similar conclusion applies to forced convection over a plate with a power-law, surface-temperature variation [44]. The over-all heat transfer for a plate of length L and width b is computed from Eq. 9-23 in [41].

For the exponential surface-temperature variation case, the local heat transfer is

$$q = -k\left(\frac{\beta g m}{4\nu^2}\right)^{\frac{1}{4}} (Me^{mX})^{\frac{5}{4}}\varphi'(0) \tag{9-24}$$

On the basis of Eq. 9-24, the problem, for this case, can be rephrased in terms of a prescribed heat flux proportional to e^{SX} and a resulting wall-temperature variation proportional to $e^{4SX/5}$. Thus, if we specify that the heat flux is uniform ($S = 0$), the wall temperature will also be constant, as has been pointed out.

For this case (i.e., a variation given by Eq. 9-6b) the dimensionless local heat flux is

$$\frac{Nu_X}{(Gr_X)^{\frac{1}{4}}} = -(mX)^{\frac{1}{4}}\frac{\varphi'(0)}{\sqrt{2}}, \quad m > 0 \tag{9-25}$$

The calculated values of $\varphi'(0)$ for $Pr = 0.7$ and 1.0 are -0.735 and -0.823, respectively.

For the case of uniform wall heat flux, the local Nusselt number is

$$\frac{Nu_X}{(Gr_X^*)^{\frac{1}{5}}} = -\frac{1}{5^{\frac{1}{5}}\theta(0)} \tag{9-26}$$

This equation is plotted as a function of Prandtl number in Fig. F,9f. In order to obtain a measure of the influence of upstream conditions on local heat transfer parameters, a comparison is made in the Appendix of [40] of the local Nusselt numbers for the cases of uniform surface heat flux and uniform surface temperature. The comparison is made for two separate conditions, viz. that local values of $(T_w - T_\infty)$ are equal and that local values of q are equal. In these two cases, either the temperature difference upstream of the comparison station is smaller than that at the station or the heat flux is greater upstream, respectively. Yet the local Nusselt numbers are only moderately affected by these differences in upstream conditions; the Nusselt numbers for the uniform heat transfer case are between 10 to 15 per cent higher than those for the uniform temperature plate.

Since there is no one temperature difference that is characteristic of the problem, the choice of a temperature difference in defining an average heat transfer coefficient (or average Nusselt number) is purely arbitrary. In the literature of experimental work on free convection about surfaces with uniform heat flux, average Nusselt numbers have been defined using the average plate minus ambient temperatures (Dotson [*45*]) and also by using the temperature difference halfway along the plate (Saunders [*46*]). We will here define average Nusselt numbers on these bases and then compare them with those for an isothermal plate derived by Ostrach [*6*].

If the average temperature difference is

$$\overline{T_w - T_\infty} = \frac{1}{L}\int_0^L (T_w - T_\infty)dX = \frac{5}{6}(T_w - T_\infty)_L$$

where L is the plate length, the average Nusselt number is

$$\overline{Nu} \equiv \frac{\bar{h}L}{k} = -(Gr_L)^{\frac{1}{4}}\left[\frac{6}{5^{\frac{6}{5}}\theta(0)}\right]^{\frac{5}{4}} \tag{9-27}$$

where

$$Gr_L = \frac{\beta g(\overline{T_w - T_\infty})L^3}{\nu^2}$$

The comparison of this expression with that of Ostrach [*6*] is given in Table F,9b.

Table F,9b. Comparison of average Nusselt numbers for uniform wall heat flux and constant wall temperature.

Pr	$\overline{Nu}/(Gr_L)^{\frac{1}{4}}$ (Eq. 9-27)	$\overline{Nu}/(Gr_L)^{\frac{1}{4}}$ (Ostrach)	$\overline{Nu}/(Gr_L)^{\frac{1}{4}}$ (Eq. 9-28)
0.1	0.237	0.219	0.224
1.0	.573	.535	.543
10.0	1.17	1.10	1.11
100.0	2.18	2.07	2.07

It is seen from the first and second columns of the table that $\overline{Nu}/(Gr_L)^{\frac{1}{4}}$, evaluated at the average value of $(T_w - T_\infty)$ for a plate having uniform surface heat flux, differs at most by 8 per cent from the value for a plate having uniform surface temperature. The deviation is often within the limits of experimental accuracy. Thus, we see why experimentally determined Nusselt numbers based on the average value of $(T_w - T_\infty)$ for a uniform heat flux plate have correlated with results for a uniform wall-temperature plate.

If we base the average heat transfer coefficient on the temperature difference between the plate and ambient at $L/2$

$$(T_w - T_\infty)_{L/2} = \left(\frac{1}{2}\right)^{\frac{1}{5}}(T_w - T_\infty)_L$$

we can write

$$\frac{\overline{Nu}}{(Gr_L)^{\frac{1}{4}}} = \left[-\left(\frac{2}{5}\right)^{\frac{1}{5}}\frac{1}{\theta(0)}\right]^{\frac{5}{4}} \tag{9-28}$$

The comparison of this expression with that for the isothermal plate can be made from the last two columns of Table F,9b. These values agree more closely with those for the uniform wall-temperature case. Thus, the agreement between experimental results for uniform surface heat flux with those for uniform wall temperature is made plausible.

In concluding our discussion of exact solutions of the boundary layer equations for free convection about nonisothermal surfaces, we should note an important difference between nonisothermal surface condition problems in forced and free convection. For forced convection about a flat plate, the velocity and temperature problems are independent, either when the properties are constant or when $\rho\mu$ and ρk are constant. Under either of these conditions, the energy equation is linear provided that the Prandtl number is constant. Thus, the superposition principle can be applied to solve the energy equation. In free convection, however, the velocity and temperature problems are always interrelated, no matter whether the properties are variable or constant.

It has not been proved that Eq. 9-6 are the only wall-temperature variations that permit application of the similarity transformation. However, experience with the similar problem of forced convection with variable free stream velocity suggests that no other temperature variations will permit a similarity transformation. For wall-temperature variations other than those given by Eq. 9-6, different and less exact methods of solution must be used. Such a treatment is presented in [*41*] and will be discussed subsequently.

Experimental verification. Experiments are reported by Dotson [*45*] for free convection heat transfer to air from a vertical aluminum sheet $2\frac{2}{3}$ ft high, 10 ft wide, and 0.005 in. thick. An electric current passing through the sheet yielded an essentially uniform power dissipation in the material, so that comparison can be made with the theoretical results for uniform wall heat flux. Temperatures were recorded by 63 thermocouples soldered to the back of the sheet. The influence of the lateral edges on the data was reduced by taking the voltage measurements at positions away from these edges.

Although essentially uniform power dissipation took place within the aluminum, the heat transferred to the air by free convection was made nonuniform by radiation and by conduction through the insulation at the rear surface of the sheet. When these radiation and conduction losses were computed, it was found that their influence in promoting a nonuniform heat transfer rate to the fluid was felt principally in the vicinity of the leading edge.

The quantitative effect of the longitudinal variation of the heat transfer to the fluid has been studied [*41*]. The results of that analysis were applied to the longitudinal heat flux variations encountered by Dotson's experiments. It was found that the local heat transfer coefficient would differ at most by 2 per cent from those on a plate with uniform heat flux $\bar{q}$, provided that $\bar{q}$ is chosen at some appropriate value intermediate to the maximum and minimum heat transfer rates on the plate with longitudinal heat flux variation.

Local Nusselt numbers computed from Dotson's data evaluating fluid properties at ambient conditions are compared in Fig. F,9h with theoretical values obtained from Eq. 9-26. The agreement is good over the entire range $10^5 \leqq Gr_X^* \leqq 10^{11}$. The greatest deviation between theoretical and experimental values are for the lower heat transfer rates.

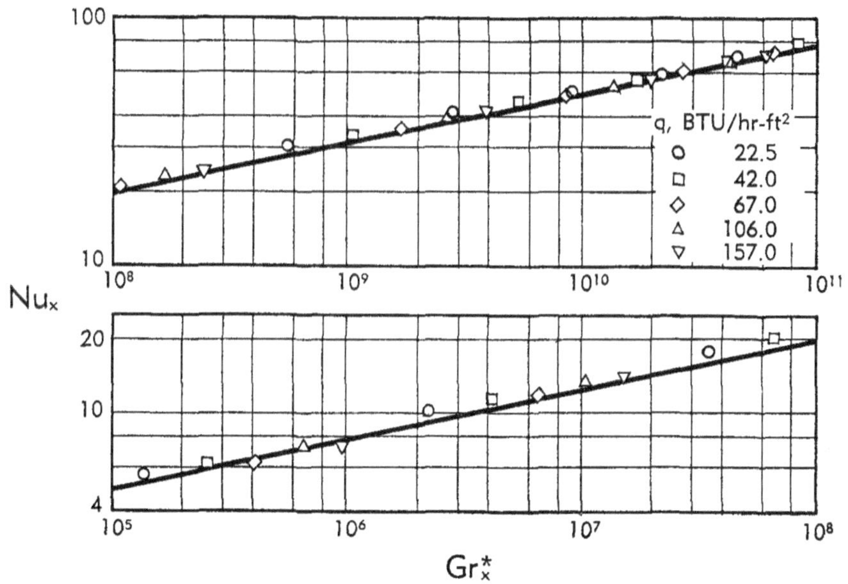

Fig. F,9h. Comparison of the predicted Nusselt numbers with those calculated from the data of Dotson [45] (Sparrow and Gregg [38]).

Dotson indicates that the experimental data become less accurate as the heat transfer rate decreases.

It is to be noted that, although the value of the conventional Grashof number Gr at which transition from laminar to turbulent flow occurs is roughly 10^9 for air, it would seem that considerably higher values of Gr^* (modified Grashof number in that qX/k replaces $(T_w - T_\infty)$) are associated with transition. This is borne out by the good agreement of the experimental data with the laminar theory up to $Gr_X^* = 10^{11}$, and also by Dotson's observations of the location at which thermocouple readings fluctuated with time.

Insufficient data are given by Saunders [46] to permit comparison with theory.

Kármán-Pohlhausen method. The same problems as discussed above, with somewhat more general boundary conditions, are treated in a more approximate manner by the Kármán-Pohlhausen method in [41]. The thermal boundary conditions are restricted to vary monotonically from the leading edge value as either

$$\frac{q_w}{q_0} = 1 \pm \epsilon \left(\frac{X}{X_l}\right)^r \tag{9-29}$$

or

$$\frac{\theta_w}{\theta_0} = 1 \pm \epsilon \left(\frac{X}{X_L}\right)^r \tag{9-30}$$

where the zero subscript denotes the leading edge ($X = 0$) value and the L denotes the value at the position at the end of a plate of length L. The parameter ϵ represents the maximum (percentage) deviation of the function in the region $X = 0$ to $X = X_L$.

In the Kármán-Pohlhausen method, integrals of the differential equations

$$\frac{d}{dX}\int_0^\delta U^2 dY = \beta g \int_0^\delta (T - T_\infty)dY - \nu\left(\frac{\partial U}{\partial Y}\right)_{Y=0} \tag{9-31}$$

$$\frac{d}{dX}\int_0^\delta U(T - T_\infty)dY = -\alpha\left(\frac{\partial T}{\partial Y}\right)_{Y=0} \tag{9-32}$$

are satisfied in the interior (away from the boundaries) of the region. δ is the boundary layer thickness. The velocity distribution is assumed to be

$$U = \omega \frac{Y}{\delta}\left(1 - \frac{Y}{\delta}\right)^2 \tag{9-33}$$

where $\omega(X)$ is proportional to the maximum velocity at any X station and the temperature distribution appropriate to Eq. 9-29 is taken as

$$\theta = \frac{q\delta}{2k}\left(1 - \frac{Y}{\delta}\right)^2 \tag{9-34}$$

and for Eq. 9-30

$$\theta = \theta_w\left(1 - \frac{Y}{\delta}\right)^2 \tag{9-35}$$

These profiles satisfy the boundary conditions

$$U(X, 0) = 0; \text{ either } \frac{\partial T}{\partial Y}(X, 0) = -\frac{q(X)}{k} \text{ or } T(X, 0) = T_w(X) \tag{9-36}$$

$$T(X, \delta) = T_\infty; \qquad U(X, \delta) = 0 \tag{9-37}$$

and the smooth-fit conditions

$$\frac{\partial U}{\partial Y}(X, \delta) = \frac{\partial T}{\partial Y}(X, \delta) = 0 \tag{9-38}$$

Substitution of the polynomial profiles, Eq. 9-33 and either 9-34 or 9-35, into the integrated equations, Eq. 9-31 and 9-32, yields two first order, ordinary differential equations for the two unknown functions $\omega(X)$ and $\delta(X)$. These equations are nondimensionalized, and the dependent variables are then expanded in Maclaurin series in terms of the parameter ϵ (defined by Eq. 9-29 or 9-30) with the first term corresponding to the case of uniform thermal boundary conditions. The succeeding terms give the influence of the nonuniformity of the thermal conditions. The first five terms of the series have been calculated in [*41*].

The accuracy of the Kármán-Pohlhausen method must always be confirmed in new problems. We have already seen that good agreement with theoretical heat transfer results over a large Prandtl number range was obtained (Fig. F,3g) for the case of free convection about an isothermal plate. Since, for nonuniform thermal conditions, the theory and experiments for the special case of uniform wall heat flux are readily available, we shall compare the Kármán-Pohlhausen results with those. First of all, we note on Fig. F,9f that the surface temperature variation with Prandtl number computed from numerical solutions and the Kármán-Pohlhausen method agrees very well with some deviation at the lower Prandtl numbers. The local Nusselt number for the case of uniform wall heat flux as derived by the Kármán-Pohlhausen method is

$$\frac{Nu_X}{(Gr_X^*)^{\frac{1}{5}}} = 2\left[\frac{Pr^2}{360(0.8 + Pr)}\right]^{\frac{1}{5}} \tag{9-39}$$

This expression is compared to that found from the exact solution, Eq. 9-26, in Fig. F,9f, and again the agreement is good everywhere except at the lower Prandtl numbers. Since the experimental values for air also agree with the numerical results (see Fig. F,9h), we see that the heat transfer can be predicted with reasonable accuracy by the approximate method. The fact that such good agreement is obtained on the basis of such simple assumed quadratic profiles as given by Eq. 9-33 to 9-35 is perhaps somewhat surprising in view of the rather sophisticated profiles assumed in the application of the Kármán-Pohlhausen method for forced convection problems. However, in forced convection it is necessary to approximate not only the heat transfer results but also the flow characteristics (e.g., skin friction). The above procedure does not give good agreement with flow parameters for free convection. Braun and Heighway [*47*] have modified the Kármán-Pohlhausen method, not only to give better agreement with flow characteristics for free convection but also to improve the accuracy of the heat transfer results for very low and high Prandtl numbers.

F,10. Vertical Cylinder. We have already seen in Art. 5 that, for the case of free convection about an isothermal vertical cylinder, no similarity solution could be found and the problem was analyzed by a direct treatment of the partial differential equations by a power series expansion [21] or by the Kármán-Pohlhausen method [48]. If the cylinder surface temperature is permitted to vary linearly with the axial coordinate, however, a similarity solution is possible and has been developed by Millsaps and Pohlhausen [43].

The basic equations for this case of a linear surface temperature variation are identical to those for the isothermal surface, Eq. 5-1 to 5-3. If we substitute into those equations the transformations

$$\left.\begin{aligned} X &= xd,\ R = rd,\ U = \frac{\nu}{rd}\psi_r = \frac{\nu x}{rd}f'(r) \\ V &= -\frac{\nu}{rd}\psi_x = -\frac{\nu}{rd}f(r),\ T - T_\infty = \frac{\nu^2 x}{g\beta r^4 d^3}h(r) \end{aligned}\right\} \tag{10-1}$$

we obtain the ordinary differential equations

$$r^3f''' + r^2ff'' - r^2f'' - r^2(f')^2 - rff' + rf' + h = 0 \tag{10-2}$$

$$r^2h'' + (Prf - 7)rh' + (16 - 4Prf - Prrf')h = 0 \tag{10-3}$$

where d is the cylinder diameter, ψ is the dimensionless stream function appropriate to Eq. 5-1, and the primes denote differentiation with respect to r. Eq. 10-2 and 10-3 can be further simplified by letting

$$r = e^t \tag{10-4}$$

so that the final equations are

$$F''' + (F - 4)F'' + (4 - 2F - F')F' + H = 0 \tag{10-5}$$

$$H'' + (PrF - 8)H' + (16 - 4PrF - PrF')H = 0 \tag{10-6}$$

where the primes now denote differentiation with respect to t and $H(t) = h(r)$ and $F(t) = f(r)$.

The physical boundary conditions when transformed to the dimensionless variables are

$$F(0) = F'(0) = 0,\ H(0) = H_w = \text{const} \tag{10-7}$$

and

$$F'(\infty) = H(\infty) = 0 \tag{10-8}$$

The condition $H(0) = \text{const}$ specifies that the temperature distribution on the cylinder surface varies linearly with the axial distance x. Note that $h(1)$ or $H(0)$ is a modified Grashof number.

To obtain estimates of the heat transfer simply and to provide initial values for numerical solutions of the boundary-value problem defined by Eq. 10-5 to 10-6, the Kármán-Pohlhausen integral method was first applied to the basic equations. The procedure and results of this approximation appears in [43].

Exact solutions of the boundary-value problem were obtained on a large scale digital computer for Prandtl numbers of 0.733, 1, 10, and 100. Fig. F,10a and F,10b are representative graphs of the exact solutions for the velocity and temperature profiles for parametric values of the Prandtl and modified Grashof numbers. (The latter can also be interpreted as the axial temperature gradient along the cylinder surface.)

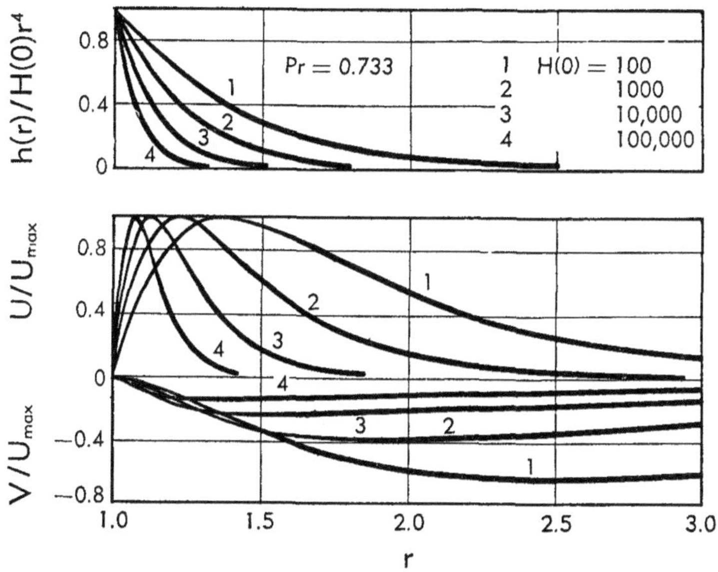

Fig. F,10a. Exact solutions for the dynamic and thermal profiles for parametric values of the modified Grashof number and for Prandtl number 0.733 (Millsaps and Pohlhausen [*41*]).

The Nusselt number based on the cylinder radius in terms of the exact solution is

$$Nu = \left| \frac{H'(0)}{H(0)} - 4 \right| \tag{10-9}$$

This function is presented in Fig. F,10c as a function of the Prandtl and modified Grashof numbers. The initial slopes of the velocity profiles (which are proportional to the surface shear) are also included in this figure. Approximate values of these quantities obtained by the Kármán-Pohlhausen method appears in this figure as the solid lines for comparison purposes. The exact values are represented by the points. The effect of the surface temperature variation on the heat transfer can be seen by comparing the isothermal results given in Fig. F,5b with the present results shown on Fig. F,10c.

It should be noted that the results of [*43*] presented here were concerned only with the case in which the surface temperature increases

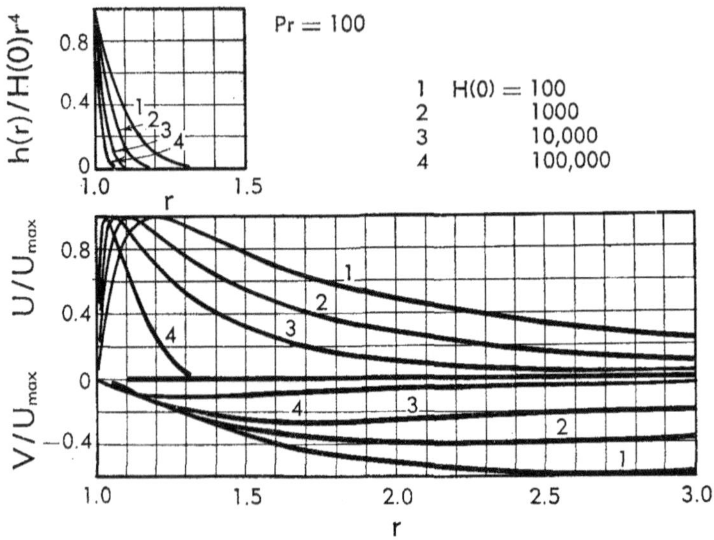

Fig. F,10b. Exact solutions for the dynamic and thermal profiles for parametric values of the modified Grashof number and for Prandtl number 100.0 (Millsaps and Pohlhausen [41]).

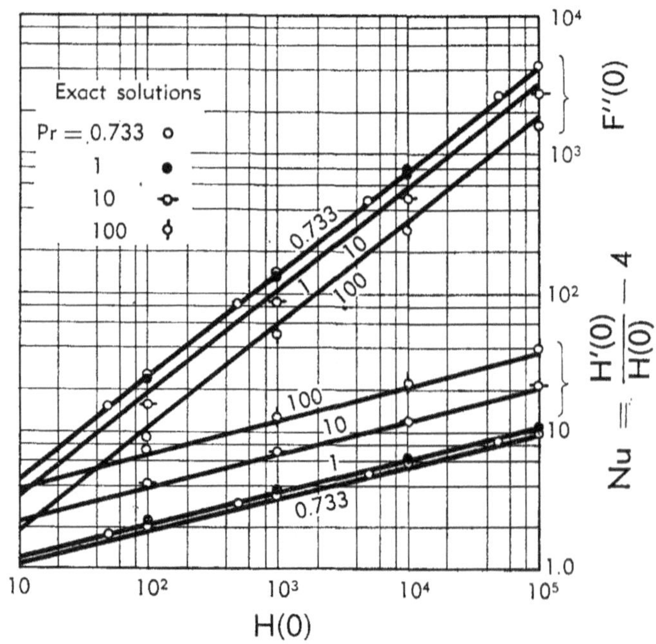

Fig. F,10c. Nusselt number and the initial slope of the dynamic profiles for parametric values of the Prandtl number. (The full lines represent the parabolic approximations; the points represent the values of the exact solutions; Millsaps and Pohlhausen [41].)

along the axis. The interesting situation where the temperature decreases axially corresponding to heating from below (and possibly thermal instability) is yet to be analyzed. Treatment of the flow around the two-dimensional and axially symmetric closed end bodies with variable wall temperatures is given in [*38*].

COMBINED FREE AND FORCED FLOWS

Recently there has been considerable interest in the effect of body forces on forced convection phenomena. Primarily the concern is with the action of electromagnetic forces so that a great deal of activity in the field of magnetohydrodynamics has been stimulated. This problem will be treated in a subsequent chapter, but it is well here to review the work along these lines which has been done on related problems exclusive of those in the domain of magnetohydrodynamics.

Illingworth [*49*] was concerned with the effect of gravity on boundary layer flows and found specific conditions for neglecting gravitational influences. Ostrach [*2*] developed the general parameters relating to combined flow for his treatment of superposed, fully developed natural and forced flows and flows due to more than one body force, to be discussed in a later chapter. The essential features of these works will be outlined below.

F,11. Basic Equations. The fundamental parameter which determines whether the body force effects are important in a forced flow problem was shown in [*2*] to be the Froude number (see Eq. 2-8). The Froude number (ratio of square of Reynolds number to Grashof number) compares the inertia to the buoyancy or body forces. Thus, qualitatively speaking, the influence of a body force on a flow field is significant if the Froude number is of unit order of magnitude or less, i.e.,

$$\frac{Re^2}{Gr} = \frac{U_\infty^2}{\beta f_X \theta_w d} = \mathcal{O}(1) \tag{11-1}$$

By considering the three-dimensional boundary layer produced by the steady flow of a perfect gas past a cylindrical obstacle whose generators are perpendicular to the incident stream, Illingworth [*49*] investigated the specific situations where Eq. 11-1 would hold and then determined the effects of gravity on the given flow field. To formulate the general equations, we let the Z axis coincide with the foremost generator of the cylinder, X be the distance measured round a normal section of the cylinder, in the plane Z = const, from the foremost point, and Y be the distance from the cylinder in the plane Z = const (see Fig. F,11). The continuity equation is thus

$$\frac{\partial}{\partial X}(\rho U) + \frac{\partial}{\partial Y}(\rho V) + \frac{\partial}{\partial Z}(\rho W) = 0 \tag{11-2}$$

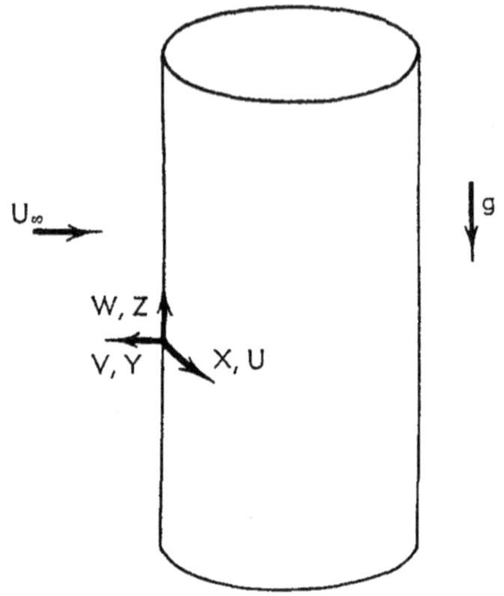

Fig. F,11. Cylindrical configuration for study of combined free and forced flows.

The momentum equations with the boundary layer approximations are

$$\rho DU = -\rho g l - \frac{\partial P}{\partial X} + \frac{\partial}{\partial Y}\left(\mu \frac{\partial U}{\partial Y}\right) \tag{11-3}$$

$$\frac{\partial P}{\partial Y} = \mathcal{O}(1) \tag{11-4}$$

$$\rho DW = -\rho g n - \frac{\partial P}{\partial Z} + \frac{\partial}{\partial Y}\left(\mu \frac{\partial W}{\partial Y}\right) \tag{11-5}$$

where gravity is taken as the only body force acting, so that g is the body force per unit mass, and where

$$D \equiv U \frac{\partial}{\partial X} + V \frac{\partial}{\partial Y} + W \frac{\partial}{\partial Z}$$

and l, m, n are the direction cosines of the upward vertical relative to the orthogonal triad formed by the principal tangent, the normal, and the generator at the point $(X, 0, Z)$. The energy equation is

$$\rho Dh - DP = \frac{1}{Pr}\frac{\partial}{\partial Y}\left(\mu \frac{\partial h}{\partial Y}\right) + \mu\left[\left(\frac{\partial U}{\partial Y}\right)^2 + \left(\frac{\partial W}{\partial Y}\right)^2\right] \tag{11-6}$$

and the equation of state is

$$P = \rho(h - e) = \rho \mathcal{R} T \tag{11-7}$$

where

$$h = \int_0^T c_p(\vartheta)d\vartheta \quad \text{and} \quad e = \int_0^T c_v(\vartheta)d\vartheta$$

are the enthalpy and internal energy of the gas, respectively. The specific heats, c_p and c_v, are functions of the temperature T alone, and $c_p - c_v$ is the gas constant $\mathcal{R}$. In Eq. 11-6 the Prandtl number Pr has been assumed to be independent of the state of the gas. The variation of c_p and c_v with temperature is much less important than that of μ, which is also a function of T alone. In some cases, therefore, it will be assumed that c_p and c_v are constant.

At the outer edge of the boundary layer, where quantities will be denoted by the subscript ∞, the velocity at the section X has the single component $U_\infty(X)$ in the direction of the tangent to the cylinder at $(X, 0, Z)$. The quantities P_∞, ρ_∞, etc., are functions of X and Z only. The main stream is described by the equations

$$-\rho_\infty U_\infty \frac{dU_\infty}{dX} = \rho_\infty gl + \frac{\partial P_\infty}{\partial X} \tag{11-8}$$

$$0 = \rho_\infty gn + \frac{\partial P_\infty}{\partial Z} \tag{11-9}$$

corresponding to Eq. 11-3 and 11-4 and

$$\rho_\infty \frac{\partial h_\infty}{\partial X} = \frac{\partial P_\infty}{\partial X} \tag{11-10}$$

corresponding to Eq. 11-6, together with the state equation,

$$P_\infty = \rho_\infty(h_\infty - e_\infty) = \rho_\infty \mathcal{R} T_\infty \tag{11-11}$$

From Eq. 11-4 we deduce in the usual way that we may take $P = P_\infty$, and it follows, by adding Eq. 11-8 to Eq. 11-3 and Eq. 11-9 to Eq. 11-5, that the momentum equations can be written as

$$\rho DU - \rho U_\infty \frac{dU_\infty}{dX} = (\rho_\infty - \rho)gl + \frac{\partial}{\partial Y}\left(\mu \frac{\partial U}{\partial Y}\right) \tag{11-12}$$

$$\rho DW = (\rho_\infty - \rho)gn + \frac{\partial}{\partial Y}\left(\mu \frac{\partial W}{\partial Y}\right) \tag{11-13}$$

The energy equation (Eq. 11-6) with the help of Eq. 11-10 may be written as

$$\rho Dh - U\rho_\infty \frac{\partial h_\infty}{\partial x} - W\frac{\partial P_\infty}{\partial Z} = \frac{1}{Pr}\frac{\partial}{\partial Y}\left(\mu \frac{\partial h}{\partial Y}\right) + \mu\left[\left(\frac{\partial U}{\partial Y}\right)^2 + \left(\frac{\partial W}{\partial Y}\right)^2\right] \tag{11-14}$$

Eq. 11-2 and Eq. 11-8 to 11-14 are general, but in order to solve them we shall henceforth assume all quantities to be independent of Z, i.e., the flow is unaffected by the ends of the cylinder. This assumption, however, also implies certain conditions on the effect of gravity on the main flow. These implications will have to be considered before solutions are found.

Gravity causes a variation of the state of the main stream with altitude, and in particular, unless $n = 0$, a variation of P_∞, ρ_∞, etc., with Z. In the special case where $n = 0$, the generators of the cylinder are horizontal. Unless $n = 0$, gravity also causes a transverse flow in the boundary layer with velocity W parallel to the generators (or axis); the pressure across a section of the boundary layer is approximately constant and equal to P_∞, and the gradient of P_∞ prevents motion parallel to the generators in a fluid of uniform density ρ_∞ (Eq. 11-9); but the density varies rapidly across a section of the boundary layer. Inside the layer the pressure gradient does not balance the weight of the fluid. In addition to the transverse drift W, unless $l = 0$, the buoyancy force sets up a convection current that contributes to the boundary layer flow parallel to the main stream; when $l = 0$ the generators of the cylinder are vertical. Thus unless the cylinder is horizontal ($n = 0$) we will have to make certain that the Z variations of P_∞, ρ_∞, and h_∞ due to gravity, and of T due to the transverse velocity, are relatively unimportant before we attempt to find solutions independent of Z.

Primary flow. If gravity is neglected altogether, Eq. 11-2 and the modified equations (Eq. 11-8 to 11-14) describe a flow in the boundary layer induced solely by, and nearly parallel to, the given mainstream. This flow will be called the primary flow. The difference due to gravity, between the actual flow and the primary flow, will be termed secondary flow. The equations being nonlinear, the secondary flow is not obtained correctly by neglecting the main stream and taking account of the gravity terms alone.

The primary flow (gravity neglected) is independent of Z as required, and satisfies the equations

$$\frac{\partial}{\partial X}(\rho U) + \frac{\partial}{\partial Y}(\rho V) = 0 \tag{11-15}$$

$$\rho\left(U\frac{\partial U}{\partial X} + V\frac{\partial U}{\partial Y}\right) - \rho_\infty U_\infty \frac{dU_\infty}{dX} = \frac{\partial}{\partial Y}\left(\mu \frac{\partial U}{\partial Y}\right) \tag{11-16}$$

$$\rho\left(U\frac{\partial h}{\partial X} + V\frac{\partial h}{\partial Y}\right) - U\rho_\infty \frac{dh_\infty}{dX} = \frac{1}{Pr}\frac{\partial}{\partial Y}\left(\mu \frac{\partial h}{\partial Y}\right) + \mu\left(\frac{\partial U}{\partial Y}\right)^2 \tag{11-17}$$

with

$$\frac{dh_\infty}{dX} + U_\infty \frac{dU_\infty}{dX} = 0 \tag{11-18}$$

It is now necessary to distinguish between flows that can be closely approximated by the primary flow alone and flows in which the secondary flow must be taken into account. For simplicity, the only cases considered in detail will be those in which the cylinder is either horizontal or vertical;

an inclined cylinder will exhibit a combination of the features of these two extreme cases.

F,12. Horizontal Cylinder. When the cylinder is horizontal (n = 0), the quantities P_∞, ρ_∞, etc. are independent of Z and $W = 0$, even when gravity is taken into account. For this case the exact boundary layer equations for the total flow are Eq. 11-15 and 11-17, together with

$$\rho\left(U\frac{\partial U}{\partial X} + V\frac{\partial U}{\partial Y}\right) - \rho_\infty U_\infty \frac{dU_\infty}{dX} = (\rho_\infty - \rho)gl + \frac{\partial}{\partial Y}\left(\mu\frac{\partial U}{\partial Y}\right) \tag{12-1}$$

and

$$\frac{dh_\infty}{dX} + gl + U_\infty\frac{dU_\infty}{dX} = 0 \tag{12-2}$$

in place of Eq. 11-16 and 11-18. We see that here the gravity terms in Eq. 12-1 and 12-2 can be neglected if

$$U_\infty^2 \gg g\left|\int_0^X l dX\right| \tag{12-3}$$

When this inequality holds, the secondary flow is very much smaller than the primary flow; the expression (Eq. 12-3) essentially follows also from Eq. 11-1. The condition (Eq. 12-3) will hold when X is not too large, except perhaps near the foremost generator of a blunt-nosed cylinder, where U_∞ vanishes.

Before pursuing this last situation more closely, let us for a moment consider a somewhat simple case, namely, that of a flat plate in a stream in its own plane. The above notation and equations apply when the Z axis is along the leading edge; the plate is to be considered as a horizontal cylinder when the Z axis is horizontal, since the "generators" are to be taken parallel to the leading edge. When the whole plane of the plate is horizontal, $l = n = 0$, and there is no secondary flow at all; Eq. 11-15 to 11-18 with U_∞ = const exactly describe the flow. If the plate is inclined at an angle α above the horizontal $(-\pi/2 \leqq \alpha \leqq \pi/2)$, then Eq. 12-3 becomes

$$U_\infty^2 \ll gX \sin|\alpha|$$

which is satisfied for a distance of, say, $U_\infty^2/(10g \sin|\alpha|)$ from the leading edge. A similar remark applies to any sharp-nosed cylinder for which $U_\infty(0) \neq 0$; the secondary flow will be negligible for at least some distance along the boundary.

It is important to note that the discussions given above dealing with the inequality (Eq. 12-3) are limited to the case where the body force is just the gravitational force. Similar inequalities can be derived for other body forces (see Eq. 11-1) which can be many times larger than the gravitational force. In such cases, it is entirely possible that

Eq. 12-3 could never be satisfied and the effects of these body forces must always be taken into account.

Let us now return to the case of a horizontal blunt-nosed cylinder in a stream inclined at an angle α above the horizontal. Then $U_\infty(0) = 0$ and the application of the inequality (Eq. 12-3) near the stagnation line must be examined more closely. If the distance X is measured from the stagnation line along the upper side of the cylinder, we can write for small X

$$l \cong \cos(\alpha - \kappa X)$$

where κ is the curvature of a normal cross section at the front stagnation point, and with $U_\infty \cong CX$ the inequality (Eq. 12-3) becomes

$$(CX)^2 \gg \frac{g}{\kappa} |\sin\alpha - \sin(\alpha - \kappa X)|$$

or

$$C^2X \gg |g\cos\alpha + \tfrac{1}{2}\kappa X \sin\alpha|$$

If $\alpha \neq \pm(\pi/2)$, this requires

$$C^2X \gg g\cos|\alpha|$$

which cannot be satisfied when X is small. If, however, $\alpha = \pm(\pi/2)$ so that the cylinder is in a vertical stream (i.e. parallel to the direction of the gravitational force), the inequality (Eq. 12-3) is satisfied if

$$C^2 \gg \frac{g\kappa}{2} \tag{12-4}$$

Near the front stagnation point of a section of a horizontal blunt-nosed cylinder, therefore, the primary flow equations are invalid except when the stream is vertical and the inequality (Eq. 12-4) is satisfied. Except in this case, the exact Eq. 11-15, 11-17, 12-1, and 12-2 are required. Illingworth [*49*] shows that if Eq. 12-4 is not satisfied, the equations for the flow near the foremost generator of a horizontal blunt-nosed cylinder in a vertical stream, including the gravitational effects, can be treated in a manner identical with that for the primary flow, except that a computation is required for each body shape. No solutions have been obtained as yet for blunt horizontal cylinders in a nonvertical stream.

For a smooth cylinder, C will depend on the curvature κ, and the condition (Eq. 12-4) needs further consideration. Evidence is provided by Ferri's experiments [*50*] on a circular cylinder of diameter d at Mach numbers $M_\infty \cong 2$, for which $C/a_\infty \cong 3/d$ where a_∞ is the velocity of sound in the incident undisturbed stream. If we accept that a similar order of magnitude is to be expected for any blunt-nosed cylinder at comparable Mach numbers, with d replaced by twice the radius of curvature at the nose, then $C \cong 1.5a_\infty\kappa$ and the inequality (Eq. 12-4) becomes

$$4.5\kappa \gg \frac{g}{a_\infty^2}$$

This condition is satisfied unless the radius of curvature at the nose is very large (or, in a generalized sense, if the body force is very large).

F,13. Vertical Cylinder and Plate. For a vertical cylinder $l = 0$. To find solutions independent of Z we must start by neglecting the Z variation of the state variables P_∞, ρ_∞, and h_∞ in the main stream. This is a good approximation (see [*6*]) if conditions are studied on a local rather than on a meteorological scale, as they usually are in boundary layer theory. There will be a cross-drift W, but if we assume that it is small relative to the main stream velocity U_∞, the term $W(\partial W/\partial Z)$ in the momentum equation and $W(\partial T/\partial Z)$ in the energy equation will be negligible, relative to the other inertia and convection terms, respectively. Also, we assume that the terms $W(\partial P_\infty/\partial Z)$ and $\mu(\partial W/\partial Y)^2$ in the energy equation can also be neglected. Illingworth [*49*] shows a posteriori that the omission of these terms is permissible, not only for a flat plate and for a sharp-nose cylinder but also for a blunt-nose cylinder.

The gravitational effects become important (i.e., those terms cannot be neglected) only if the values of X are very large, of order $M_\infty a_\infty^2/g$, for sharp-nosed bodies, and if $g/Ca_{\infty 0}$ is not small for blunt-nosed bodies where $a_{\infty 0}$ is the free stream stagnation velocity of sound. Thus none of the dependent variables will depend on Z. The continuity Eq. 11-2 reduces to Eq. 11-15 and, since $l = 0$, the general Eq. 11-12 reduces to Eq. 11-16, and the general Eq. 11-8 becomes Eq. 11-18. The energy Eq. 11-14 reduces to Eq. 11-17 so that the flow in the X direction is very nearly the primary flow obtained from Eq. 11-15 to 11-18. The remaining general Eq. 11-13 now becomes

$$\rho\left(U\frac{\partial W}{\partial X} + V\frac{\partial W}{\partial Y}\right) = (\rho_\infty - \rho)g + \frac{\partial}{\partial Y}\left(\mu\frac{\partial W}{\partial Y}\right) \tag{13-1}$$

Eq. 11-15 to 11-18 are independent of Eq. 13-1; once their solution is obtained, the drift velocity W can be determined from Eq. 13-1.

The primary Eq. 11-15 to 11-18 are solved by Illingworth [*49*] for several configurations, including ones that we will use to find the drift due to gravity. To obtain the solutions, the equations are first transformed by the Mises transformation in which the independent variables are changed from (X, Y) to (X, ψ), where ψ is the stream function

$$\rho U = \frac{\partial\psi}{\partial Y}, \qquad \rho V = -\frac{\partial\psi}{\partial X} \tag{13-2}$$

which satisfies Eq. 11-15. Nondimensional functions (denoted by the lower-case letters or asterisks) are then introduced where $U = U_\infty u$, $\rho = \rho_\infty \rho^*$, etc., together with $\varphi = \mu\rho^* u$. A similarity variable $\eta = \psi/(2s)^{\frac{1}{2}}$ where

$$s = \int_0^X \varphi_\infty(\tau)d\tau$$

is then applied to yield ordinary differential equations. Finally, the independent variable is changed from η to ζ where

$$\zeta = \int_0^\eta \frac{d\vartheta}{u(\vartheta)}$$

so that $u = d\eta/d\zeta$.

The Mises transformation changes Eq. 13-1 into

$$U \frac{\partial W}{\partial X} = g(T^* - 1) + U \frac{\partial}{\partial \psi}\left(\mu\rho U \frac{\partial W}{\partial \psi}\right) \tag{13-3}$$

Let us now determine the drift velocities for several cases. We shall, therefore, consider first the case of a flat plate with its leading edge vertical in a horizontal stream. In this case $U_\infty = \text{const}$, $\varphi_\infty = \text{const}$, $U = U_\infty \eta'(\zeta)$, $h = h_\infty h^*(\zeta)$ where $\eta = \psi/(2\varphi_\infty X)^{\frac{1}{2}}$ and η and h^* are solutions found by Illingworth. If c_p is constant, so that $h^* = T^*$, and if $\chi(\zeta) = (T^* - 1)/[h^*(0) - 1]$, then

$$W = \frac{2g[h^*(0) - 1]X}{U_\infty} w(\zeta) \tag{13-4}$$

where $w(\zeta)$ is the solution of

$$\frac{d}{d\zeta}(\theta w') + \eta w' - 2\eta' w + \chi = 0 \tag{13-5}$$

where $\theta = \mu^*\rho^*$ (so that $\varphi = \theta u$). Eq. 13-5 is subject to the boundary conditions

$$w(0) = w(\infty) = 0$$

Illingworth computed w for the easiest case, when $Pr = 1$ and $\theta = 1$, the plate being insulated. If we write

$$\eta(\zeta) = \frac{1}{\sqrt{2}} f_0\left(\frac{\zeta}{\sqrt{2}}\right), \quad \eta'(\zeta) = \frac{1}{2} f_0'\left(\frac{\zeta}{\sqrt{2}}\right)$$

the function η is quite readily obtained since f_0 satisfies

$$f_0''' + f_0 f_0'' = 0, \; f_0(0) = f_0'(0) = 0, \; f_0'(\infty) = 2$$

and f_0 and its derivatives are tabulated (see, Howarth [*51*], for example). Also,

$$h^* + \frac{\gamma - 1}{2} M_\infty^2 u^2 = 1 + \frac{\gamma - 1}{2} M_\infty^2 \text{ so that } \chi = 1 - (\eta')^2$$

where γ is the ratio of specific heats evaluated at free stream conditions and M_∞ is the free stream Mach number. The values of w are given in Table F,13a, along with the values of u.

Table F,13a. Velocity components in the boundary layer of a heat-insulated vertical flat plate including the effects of gravity, when $\mu\alpha T$, and $Pr = 1$.

$\xi = \frac{\zeta}{\sqrt{2}}$	$u = \frac{1}{2}f'(\xi)$	w	$\xi = \frac{\zeta}{\sqrt{2}}$	$u = \frac{1}{2}f'(\xi)$	w
0	0	0	1.2	0.729	0.216
0.1	0.061	0.099	1.3	0.772	0.187
0.2	0.133	0.178	1.4	0.812	0.159
0.3	0.199	0.238	1.5	0.846	0.133
0.4	0.265	0.280	1.6	0.876	0.109
0.5	0.330	0.307	1.7	0.902	0.087
0.6	0.394	0.319	1.8	0.923	0.069
0.7	0.456	0.320	1.9	0.941	0.054
0.8	0.517	0.310	2.0	0.956	0.041
0.9	0.575	0.293	2.5	0.992	0.008
1.0	0.630	0.270	3.0	0 999	0.000
1.1	0.681	0.244			

It is important to note from Eq. 13-4 that the maximum drift velocity across any section X is proportional to the distance from the leading edge. The relative drift is given by

$$\frac{W}{U} = \frac{2g[h^*(0) - 1]X}{U_\infty^2} \frac{w}{u}$$

where w and u are given in Table F,13a.

For the case of a vertical blunt-nose cylinder, the values of η and h_* were computed by Illingworth [*49*] for $U = CX$, $Pr = 0.725$ and $h^*(0) = 0.8$ which corresponds to the flow of air at the speed of sound against a circular cylinder that is maintained at the temperature of the undisturbed part of the incident stream. Sutherland's law was used for the variation of viscosity with enthalpy. Using these values of η and h^*, together with the assumption that $h^* = T^*$ and the same notation for $\chi(\zeta)$ as above, the solution of Eq. 13-3 is

$$W = \frac{2g[h^*(0) - 1]}{C} w(\zeta) \tag{13-6}$$

where $w(\zeta)$ satisfies

$$\frac{d}{d\zeta}(\theta w') + \eta w' + \frac{1}{2}\chi = 0 \tag{13-7}$$

and $w(0) = w(\infty) = 0$. Note that here the maximum drift velocity across a section is independent of the distance from the foremost generator. This solution together with those for η and h^* are tabulated in Table F,13b.

Table F,13b. Velocity components in the boundary layer of a vertical cylinder maintained at a constant temperature including the effects of gravity, when the viscosity-enthalpy variation is given by the Sutherland formula and $Pr = 0.725$.

ζ	h^*	u	w
0	0.800	0	0
0.2	0.819	0.202	0.06
0.4	0.838	0.374	0.11
0.6	0.858	0.519	0.14
0.8	0.876	0.636	0.16
1.0	0.894	0.731	0.16
1.2	0.912	0.805	0.15
1.4	0.928	0.861	0.14
1.6	0.942	0.903	0.12
1.8	0.954	0.934	0.10
2.0	0.965	0.956	0.08
2.5	0.985	0.986	0.04
3.0	0.995	0.996	0.01
3.5	0.999	1.000	0.00
4.0	1.000	1.000	0.00

The relative drift for this case is

$$\frac{W}{U} = \frac{2g[h^*(0) - 1]}{C^2X} \frac{w}{u}$$

Note that the primary flow solutions presented above were obtained by neglecting viscous dissipation. From these considerations in which the Froude number was large (inertia forces large relative to gravitational force), we see that, even though the gravitational effects do not greatly influence forced flows, the relative drift due to gravity can be of some consequence under certain conditions. As is clear from the discussion above, not even all of the large Froude number cases have as yet been analyzed. For other body forces, which can be orders of magnitude larger than the gravitational force, the Froude number decreases. Therefore, studies for unit order and smaller Froude numbers must be made to determine the influence of body forces on forced flows in such circumstances. That this influence could be significant in practical cases is evident.

CHAPTER 3. THERMAL INSTABILITY

F,14. Introduction. After studying all the different cases of free convection in the previous chapters, one might, at this point, wonder whether consideration of a seemingly simple configuration, namely, that of free convection about a horizontal flat plate, has been inadvertently omitted. The omission has, however, been quite deliberate because this case is one of the most difficult to think about and to analyze. The primary reason for the difficulty is that a new question, that of stability of free convection flows, appears when the density (or temperature) gradient is parallel to the body force direction. Although this feature is common to all flows generated by body forces under conditions analogous to those explicitly presented below, we shall here focus our attention on the case of gravitational convection from a heated horizontal flat plate, in order to describe this new problem. With this configuration in mind, it is clear that the fluid at rest above and below a horizontal plate is in a state of static equilibrium. As the plate is heated, a problem of stability arises, requiring a determination of conditions under which the equilibrium will be upset by a small disturbance before the flow itself can be determined. This stability problem is distinct from the familiar problem of hydrodynamic (laminar-turbulent) stability of homogeneous fluids in the absence of body forces. Thus, the seemingly simple configuration under consideration herein leads us to a wider class of problems in which other factors (specifically here, body forces) influence the stability of fluid motions. It should here be remarked, however, that the problem of hydrodynamic stability (i.e., the determination of when laminar flow breaks down) also exists for flows generated or influenced by body forces. This problem is treated in a later chapter.

To gain some physical insight into the nature of the instability with which we are presently concerned, let us for the moment consider what happens if we have two layers of fluid of different densities in a gravitational field with the heavier above the lighter (i.e., the density decreases in the direction of the body force). These fluid layers are in a state of unstable equilibrium, and any disturbance of this configuration will establish a transient motion in which the heavier fluid moves to the bottom. The limit of this motion is a rest state with the heavier fluid below (i.e., for stability, the density must increase in the body force direction).

Returning now to our heated horizontal flat plate, we see that in the fluid below the plate the density gradient is a stable one, so that the fluid essentially remains at rest and is stratified with the lighter fluid above the heavier. Above the plate (i.e., if there is heating from below), on the other hand, the density decreases in the body force direction so that it appears that we have a situation analogous to that of the two layers, with the heavier above. However, it has been found that, as the plate is heated, the fluid above the plate remains stationary until a certain critical temperature gradient (depending on the fluid and specific configuration) is reached. The equilibrium then becomes unstable for infinitely small disturbances and the fluid starts to move. The heat from below, moreover, is a source of energy which serves to maintain a motion once started, since the descending heavier fluid is heated and becomes lighter. In this case, the final state is not one of stable equilibrium but is an essentially steady motion. As has already been stated, this type of stability problem also occurs for flows subject to body forces other than gravitational.

Until recently, this type of instability has been of interest primarily to astrophysicists, geophysicists, and meteorologists. Since most of their studies were made for the heating of horizontal fluid layers from below in a gravitational field, the phenomenon is usually described as a thermal instability. Although there are many interesting aspects of the thermal instability problem, it would ordinarily be beyond the scope of a volume such as the present one to discuss this phenomenon. However, Ostrach [*52*] pointed out that thermal instabilities could be important in many present-day problems in the fields of aeronautics, propulsion, atomic power, electronics, and chemical engineering. In [*53*], he relates the thermal instabilities obtained in modern fluid flow configurations to those of an astrophysical and meteorological nature. We have already mentioned flows heated from below in our discussion of the variable surface temperature plate; since in our consideration of internal (natural convection) flows generated by body forces we shall again encounter configurations in which the fluid is heated from below in problems of current interest in aeronautics, we will here present the highlights of the work on thermal instabilities from the meteorological and astrophysical viewpoint. With this as a background, perhaps some of the unusual results encountered in the next chapter will be more understandable.

The research on fluid layers heated from below takes on additional significance because of the analogy between such problems and others in applied mechanics. For example, Low [*54*] first pointed out the similarity between the flow bounded by two conducting, parallel, horizontal rigid surfaces, the lower of which is heated, and the flow between two coaxial cylinders rotating at different speeds [*55*]. Comparison of the mathematical developments for the two problems reveals that the two sets of boundary conditions are identical, but the differential equations differ somewhat. Even though the analogy is not exact, a general qualitative correspondence exists between the two problems, which is discussed in [*56,57*], as well as elsewhere. Certain aspects of the boundary layer flow over a slightly wavy wall [*58*], and of the flow in horizontal fluid layers heated from below, also appear to be similar.

The reader is referred to the surveys of Ostrach [*53*] and Batchelor [*59*] for a general account of the meteorologist's and astrophysicist's work. Rather complete reference lists of the original papers are given in the surveys. A detailed account of the mathematical analysis of the thermal instability problem is presented by Lin [*60*, pp. 103–115]; he also includes an account of the more general considerations of Chandrasekhar (see

[*61*] for example) in which the effects of Coriolis and electromagnetic forces on the instability are determined.

F,15. Cellular Flows. Bénard [*62*] performed the first laboratory-controlled experiments on thermally unstable liquid layers. He worked with very thin liquid layers (on the order of several millimeters) lying on a metallic plate which was heated and maintained at a uniform temperature. The upper surface of the liquid layer was free and at a lower temperature than at the plate surface, since it was in contact with the ambient air. The detailed development of the flow was verified by many other experimenters; it is described in numerous sources. Basically, the flow occurs in two phases as follows.

The fluid remains at rest until the vertical temperature difference becomes sufficiently large. A random motion of the fluid then results. Shortly thereafter, the first phase of relatively short duration appears, in which the fluid forms cells of semiregular shape. In this phase, which lasts from a few seconds to several minutes for more viscous fluids, the cellular cross sections are nearly regular polygons of, in general, four to seven sides. During the second phase, the cells become equal and regular and align themselves. The limit of the second phase is thus a permanent regime of cells with vertical boundaries and hexagonal cross sections (see Fig. F,15), in which the liquid rises in the core of the cell, moves outward at the top, descends at the outer periphery (i.e., in the vertical boundary between adjacent cells) and moves inward at the bottom. These cells are called Bénard cells. Plate F,15 is a reproduction of a photograph of the cellular pattern in a layer of spermaceti taken during Bénard's experiments.

In the mathematical formulation of so complicated a physical phenomenon, simplifying assumptions must be made. All existing theoretical analyses are based on the assumption that density variations are negligible except insofar as they modify the action of the body force. Justification of this assumption can be found in many papers (see [*1*], for example). In addition, for the problem at hand, several different configurations can be specified. The top and bottom surfaces (see Fig. F,15) of the fluid are taken to be plane and parallel, and these surfaces may further be rigid or free and conducting or insulated. The vertical boundary or cell wall can either be a rigid boundary or a surface of symmetry between two adjacent cells. In all the studies of horizontal layers the vertical cell boundaries are taken to be insulated, and with but one recent exception to be noted below it is also assumed that the fluid temperature decreases linearly. Batchelor [*59*, p. 354] raises the question of the validity of the assumption employed in some of the theoretical work that the fluid temperature at large distances from the heated surface is uniform. If this were invalid, then a quasi-steady state could not exist in the case of a semi-infinite fluid layer above a heated boundary. However, the results of those studies are in essential agreement with those for fluid layers bounded above and below by rigid surfaces, so that the assumption may well be realistic.

Lord Rayleigh [*64*] was the first to investigate the problem theoretically; his work has since been extended and verified by many authors,

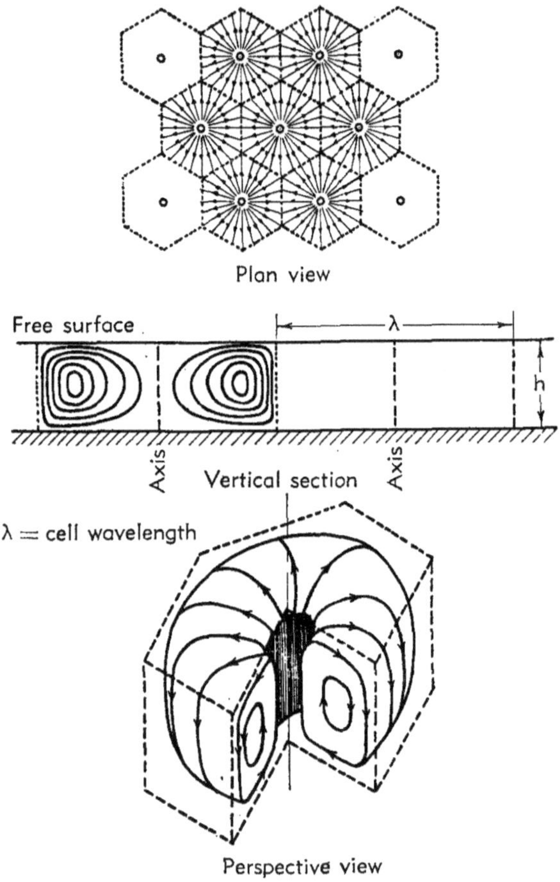

Fig. F,15. Schematic sketch of Bénard cell structure.

e.g., [*56,57*]. Each analysis is made in two parts: (1) the investigation of the geometry of the convection, and (2) the determination of a criterion for the onset of free convection in a fluid layer of limited depth. It was assumed that the equations for neutral stability are obtained by neglecting time-dependent and second order velocity terms and, hence, the analysis deals with only the first phase of the flow development. In this way, assuming further that the cells were rectangular in cross section, the form of small disturbances which are neither amplified nor damped was determined, as was also the relation of the fundamental parameters as a function of cell size.

Some of the limitations of the analyses due to the many assumptions will be discussed shortly, but the analyses did yield the fundamental result that a top-heavy fluid layer was statically stable under the joint influence of viscosity and conduction until the vertical temperature gradient became sufficiently large. In particular, it was found that the sole parameter determining stability was the product of the Prandtl and Grashof numbers. This product, called the Rayleigh number, is independent of cell shape and is given by $\beta g h^3 \Delta T/\kappa\nu$ where h is the layer depth, ΔT is the temperature difference between the bounding surface, κ is the thermal diffusivity, and the other symbols have the same meaning as in previous chapters. For a fluid layer bounded by two rigid conducting parallel surfaces, for example, the critical Rayleigh number (for flow to start) is approximately 1700.

Although, after the work of Bénard, numerous careful experiments have not only verified the qualitative picture presented above but also substantiated quantitatively the predicted critical Rayleigh number, numerous refinements and extensions of the earlier work have appeared in the literature. We shall now discuss the more important of these. In 1940 Pellew and Southwell [*65*] reformulated the theory. In a rigorous manner, they showed that any motion with a temporal periodicity will decay or subside so that the basic equations used in the early theories were reconfirmed. They further point out that, for vertical boundaries that are thermal insulators, the two horizontal boundaries must be conductors for maintained motion. However, the most novel aspect of the Pellew and Southwell analysis is that the shape of the cell is left indefinite in terms of a cell wavelength. In the previous analyses, the cell was assumed to be rectangular in cross section. Pellew and Southwell thus reduced the problem to (a) relating the Rayleigh number to the wavelength, and (b) evaluating the wavelength for a specified cell shape for the purpose of determining the "preferred" cell shape. However, it was found that the wavelength was essentially dependent on the ratio of the depth of the fluid layer to a characteristic length that defines the cell size. Therefore, for a fluid layer of indefinite horizontal extent, any symmetrical cell pattern can be formed in which the cells can be of any shape. Consequently, all wavelengths are possible for such configurations, and the cell pattern that forms will be the one associated with the smallest possible Rayleigh number. The apparent preference for a hexagonal pattern must then possibly be explained on the basis of a nonlinear theory.

Contours of the vertical velocity component computed from an exact solution of the linearized equations for a hexagonal cell were shown to have all the basic qualities of the flow observed experimentally. Finally, it was demonstrated that cells bounded by two rigid vertical boundaries could not be analyzed in the same manner as for layers of indefinite horizontal extent. In this regard, an interesting recent note by Soberman [*66*] should be mentioned. Soberman showed that the cell shape at the onset of convection is definitely influenced by lateral boundaries by placing first a cylindrical, then a rectangular, boundary around a fluid heated from below. With the former configuration, the initial cellular structure resembles a series of concentric circles, and with the latter the cells are rectangular. After a long time, however, the influence of the lateral boundaries is attenuated and the cells are nearly hexagonal. It is thus reasonable to expect that the temperature difference required to initiate convection would also be influenced by the spacing and the shape of the lateral boundaries, so that research to determine the magnitude of this result is necessary.

Although the linear theory yields a definitive stability criterion and provides a qualitative picture of the detail and form of a convective cell, it deals only with the first phase of the motion. That is, it demonstrates that the fluid remains at rest until the critical Rayleigh number is reached. The fluid then becomes unstable to small disturbances and a semiregular cellular pattern forms. Above the critical Rayleigh number, an increasing range of wavelengths will be amplified. One of the wavelengths establishes itself by being selectively amplified until nonlinear effects intervene in a stabilizing manner and prevent further growth. The steady motion so attained then corresponds to the second or final phase of the flow development. It therefore becomes clear that the actual heat transfer and fluid flow can only be determined from a nonlinear theory.

Pillow [*67*] accordingly formulates the nonlinear problem for two-dimensional flow between two horizontal plates. For Rayleigh number magnitudes of the order sufficient for steady flows, the basic equations are in a form analogous to the classical boundary layer equations and hence can be studied by a singular perturbation technique, that is, the effects of viscosity and heat conduction need be considered only near the cell boundaries. Solutions for the cell interior (i.e., away from the boundaries) indicate that both the vorticity and temperature are constant there. The latter has been verified experimentally by Schmidt and Saunders [*68*].

A satisfactory solution for the flow near the cell boundaries has not as yet been found.

A generalization of the small perturbation analyses was made by Morton [*69*] who considered the effect of a quasi-steady temperature distribution which is nonlinear in a viscous conducting fluid layer with free surfaces, i.e., no shear on the upper and lower boundaries. He finds a method for calculating stationary values (with respect to variations of the estimated velocity distribution from the true velocity) of the growth rate of the perturbations and finds that overstability (oscillatory instability, or, oscillations with increasing amplitude) does not arise. Further, an approximate analysis show that departures from a linear temperature distribution may be expected to increase the critical Rayleigh number.

Goldstein [*70*] studies the case of a fluid bounded above and below by two parallel rigid planes (i.e., no slip conditions imposed), with the lower one heated unsteadily. In addition to the gravitational force, a time-dependent body force (e.g., due to deceleration) is considered to be acting. This configuration simulates an internal cooling system that could be applied to a reentry vehicle. Goldstein finds that the critical Rayleigh numbers based on the over-all density (rather than temperature) differential can be much higher than for the static case, are dependent only on the shape of the density profile, and are independent of the Prandtl number and the rate of change of either the temperature distribution or of the body force field. The initial motion corresponds to approximately the same cell shape as for the static case. Growth of the velocity perturbation is initially proportional to the product of the Prandtl number and the rate of increase of the body-force field, whereas the temperature perturbation growth rate is independent of these parameters. For a steady force field the velocity growth rate is initially zero, but later increases at a rate dependent on the Prandtl number. The increase of the critical Rayleigh number with unsteady heating found by Goldstein [*70*] is much greater than that reported by Morton [*69*]. This difference results from the fact that, in Goldstein's configuration, the temperature gradient is near the lower surface, right where the fluid motion is inhibited by viscosity; in Morton's case, a free lower boundary, the motion can start much more easily.

All the research described up to this point, except Pillow's [*67*], was aimed at determining the criterion for the onset of a flow and the form of this flow. Schmidt and Milverton [*71*] showed that there was a sudden increase in heat transfer when the motion started. Schmidt and Saunders [*68*] undertook an investigation to see if there was any change in the motion of air or water between two parallel and horizontal plates for temperature differences in excess of those required to start a flow. They found that the length of the horizontal side of a cell was twice the layer depth when the flow was cellular. As the Rayleigh number was increased beyond the critical, the temperature gradients at the plane, midway between the bounding surfaces, decreased to negligible values, as has already been mentioned, indicating that all the heat was being convected. On further heating, the layer becomes completely turbulent at Rayleigh numbers of approximately 45,000. The turbulence developed suddenly in the tests with water, but in air the transition was much more gradual, the first signs of turbulence appearing at a Rayleigh number of 5000.

The increase in the rate of heat transfer at the change from equilibrium to cellular motion was greater than that at the change from cellular to turbulent flow.

The relative influence of the thermal instability on heat transfer can be seen by comparing the following simplified equations given by McAdams [*12*]: For an upward-facing horizontal plate,

$$h = 0.38(\Delta T)^{\frac{1}{4}}$$

For a downward-facing horizontal plate,

$$h = 0.2(\Delta T)^{\frac{1}{4}}$$

For a vertical plate more than 1 ft high,

$$h = 0.27(\Delta T)^{\frac{1}{4}}$$

where h is the heat transfer coefficient in Btu/(hr)(ft^2)(°F) and ΔT is the temperature difference between the plate and the ambient air.

Because of space limitations we shall not discuss in detail Chandrasekhar's important work (see [*61*]) dealing with the effects of Coriolis forces and magnetic fields on thermal instabilities occurring in astrophysical problems. However, some of his significant findings are that the magnetic field is in general stabilizing, i.e., it delays the cellular motion; overstability (oscillatory instability) can occur as a result of either Coriolis or magnetic forces; and with both rotation and a magnetic field the critical wavelength can change abruptly.

F,16. Columnar Flows. Chandra [*72*], in trying to corroborate earlier work, varied the depth of the fluid layer with which he worked and, in so doing, shed a completely new light on the problem. A comparison of Chandra's results with those analytically determined by Jeffreys [*56*] is shown in Fig. F,16. Note that $\Delta T/\kappa\nu T$ is plotted against the layer depth, where ΔT denotes the critical temperature difference between the two plates, κ denotes the thermal diffusivity, ν is the kinematic viscosity, and T is the mean temperature. From the definition of the Rayleigh number it is clear that the critical temperature difference varies inversely as the cube of the layer depth (from Jeffreys' criterion, then, $\Delta T/\kappa\nu T = 1709/gh^3$).

The experimental results show reasonable agreement with the theoretical in layers of depth exceeding 10 mm. However, for more shallow layers, Chandra observed initial motions whose character differed from that for the deeper layers. For example, in layers 6 mm or less in depth, the flow patterns bore no resemblence to the previously described polygonal prismatic cells. Instead, after a brief interval, vertical columns of rising air appeared, and this ascending motion seemed to continue indefinitely. The centers of convection drifted about so that, after a

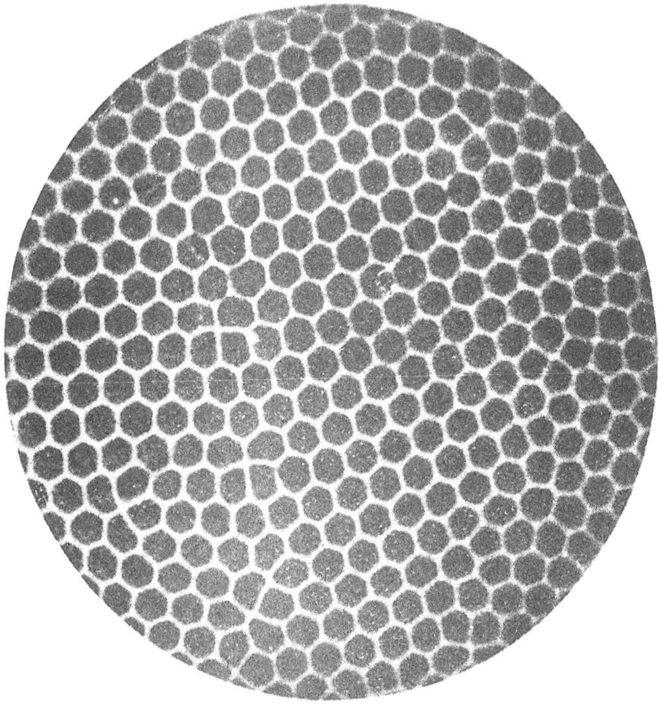

Plate F,15. Bénard cells in layer of spermaceti (from [63]).

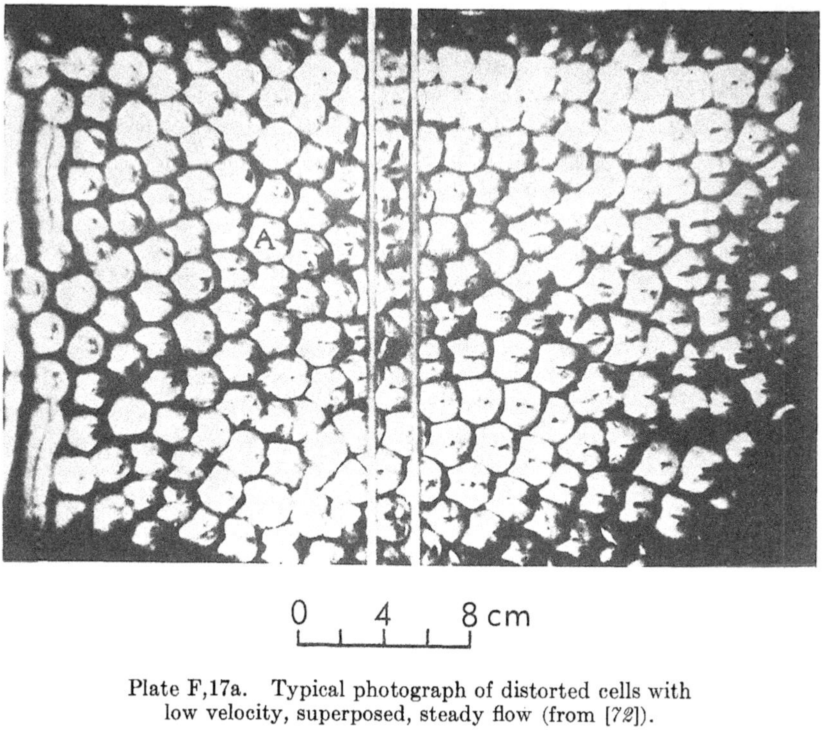

Plate F,17a. Typical photograph of distorted cells with low velocity, superposed, steady flow (from [72]).

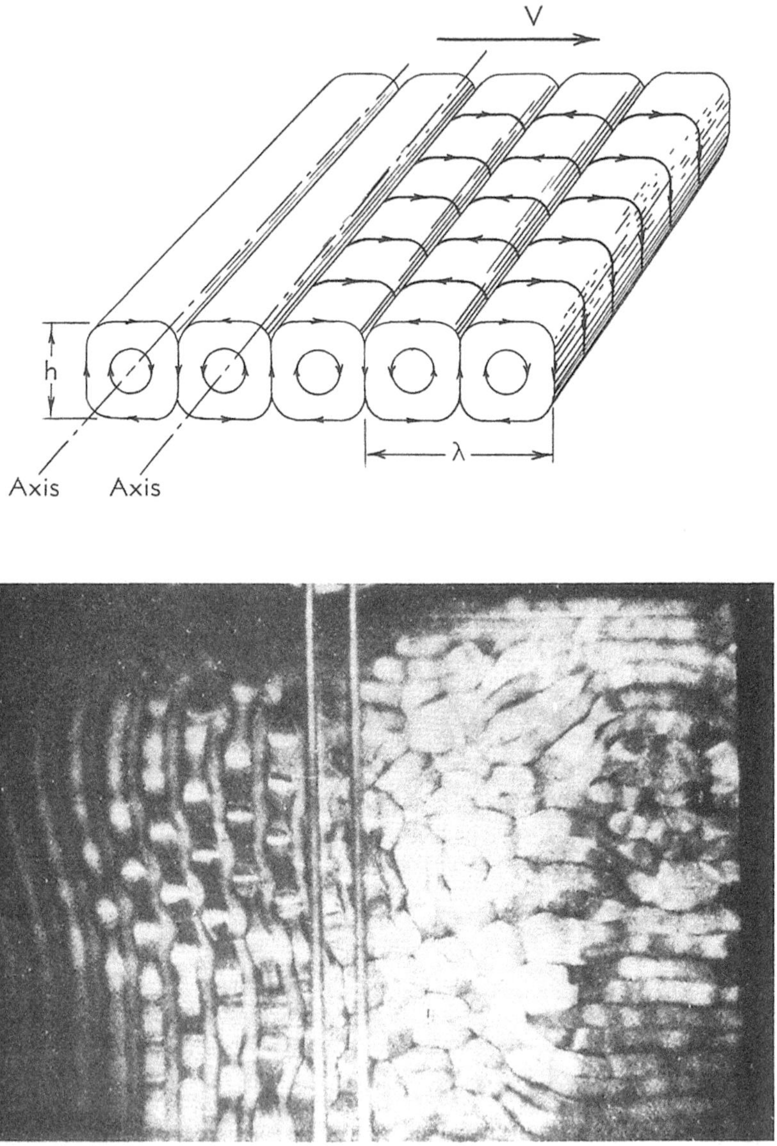

Plate F,17b. Typical schematic sketch and photograph of transverse rolls with superposed steady flow of intermediate velocity. h = fluid layer depth, λ = cell wavelength, V = superposed velocity. (From [*72*,*63*].)

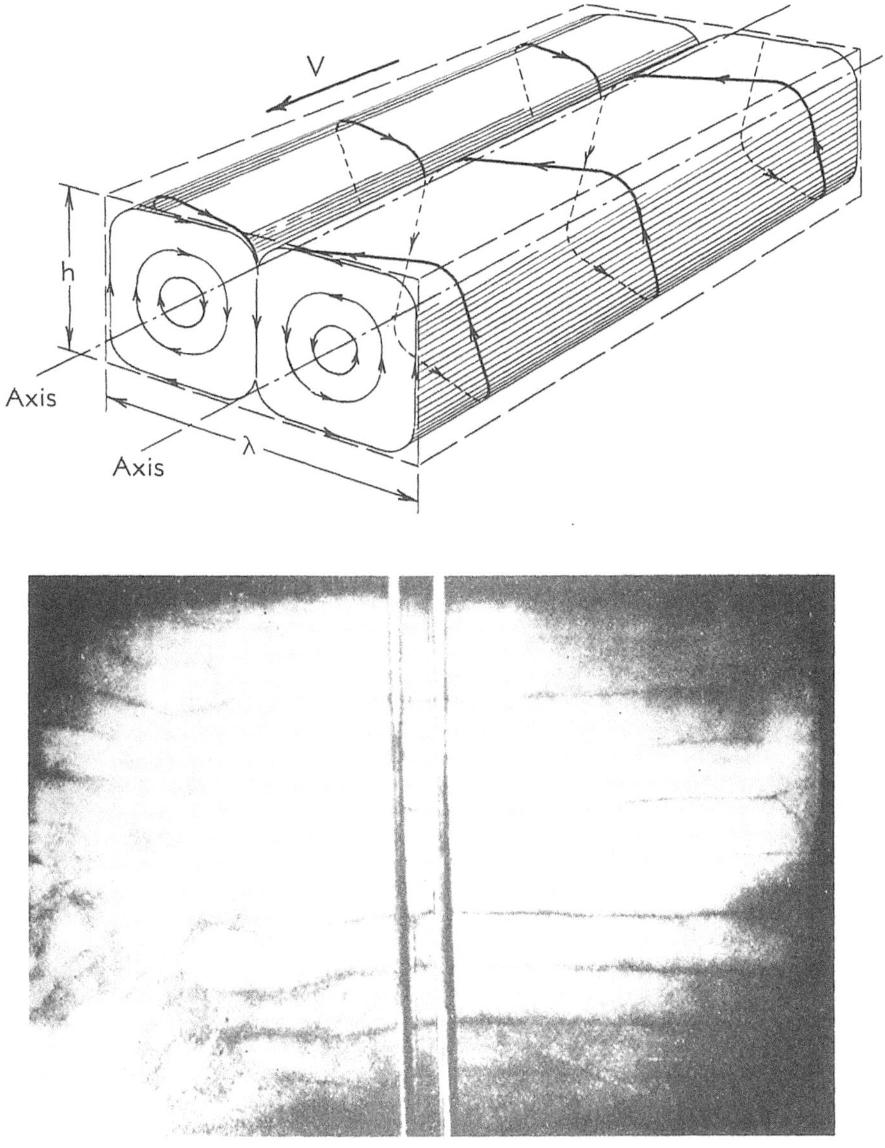

Plate F,17c. Typical schematic sketch and photograph of longitudinal rolls with high velocity, superposed, steady flow. h = fluid layer depth, λ = cell wavelength, V = superposed velocity. (From [*72,63*].)

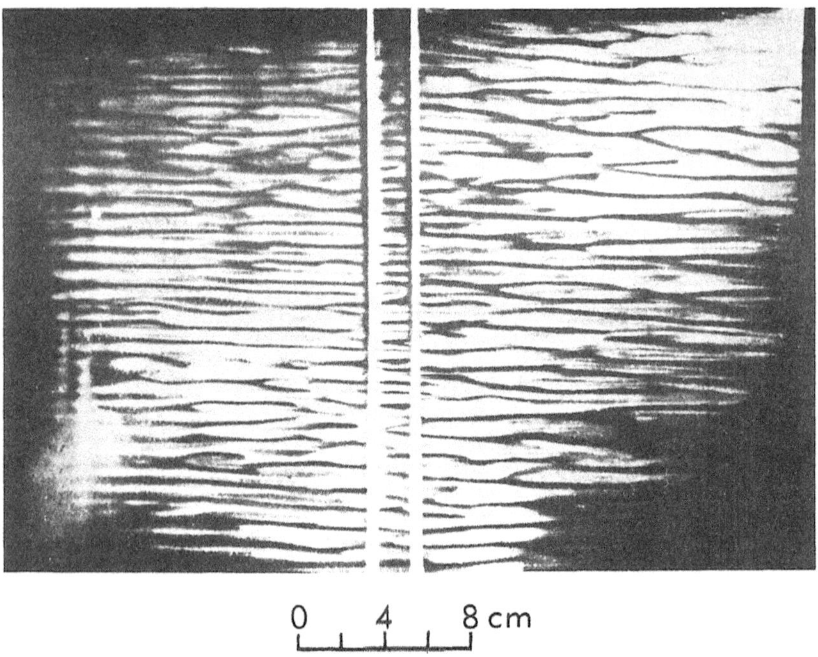

Plate F,17d. Typical photograph of spindlelike structure obtained by superposing steady flow on columnar-type flow (from [72]).

time, it was difficult to recognize any definite pattern. This type of motion is described as being "columnar," in contradistinction to the cellular. It was not possible to obtain cellular convection in the shallow layers no matter how large the temperature difference was made by heating the lower plate.

With fluid layers between 7 and 10 mm, columnar convection first appears for relatively small temperature differences. For larger temperature differences, the motion essentially becomes cellular. The minimum temperature differences for the transition from columnar to cellular are

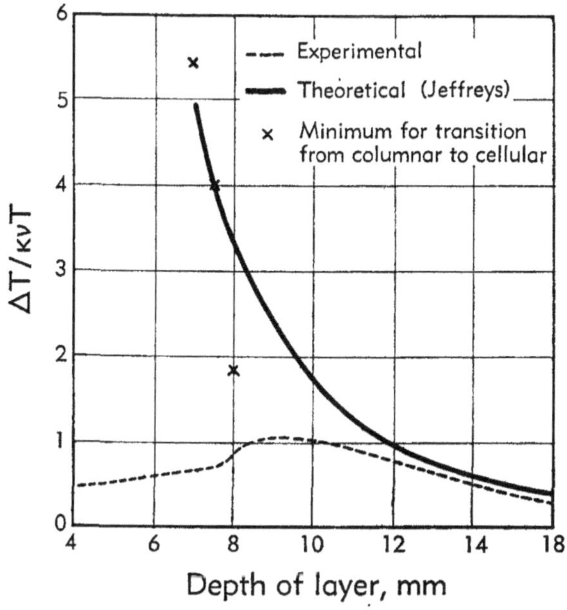

Fig. F,16. Comparison of Chandra's experimental and Jeffreys' theoretical critical temperature differences (from [72]). ΔT = critical temperature difference, κ = thermal diffusivity, ν = kinematic viscosity, T = mean temperature.

indicated by crosses in Fig. F,16. There is excellent agreement between these observations and Jeffreys' theory, particularly for layers of 7 and 7.5 mm.

Detailed descriptions and photographs of the flows can be found in [72]. Thus Chandra confirmed the theories for relatively deep layers and showed that for relatively shallow layers a distinct new type of instability is encountered which was not predicted by the earlier research. Chandra also showed that the columnar-type flow passes into the cellular type for layers of medium depth. In any event, these experiments show that significant natural convection flows can be obtained in thin fluid layers. In addition, observations of the cellular flow with air showed that the motion was the reverse of that found by Bénard for liquids; that is, the fluid descends in the central core and rises in the outer peripheries. It can thus be concluded that the flow direction in the cells depends on the particular configuration and fluid. The theories give no indication of the flow direction.

From the definition of the critical Rayleigh number, the theory implies that the critical temperature (or density) difference required for sustained convection increases indefinitely as the layer depth approaches zero. However, Chandra's experiments showed a more complicated state of affairs for shallow layers for which there was no theoretical treatment. Sutton [*73*], therefore, to resolve the differences between the theory and Chandra's experiments, critically reexamined the three fundamental assumptions made in the theories. These assumptions are that (1) density variations are negligible except for their influence in the buoyancy, (2) the initial motion is slow, and (3) a linear temperature profile extends throughout the entire fluid layer at the start of the motion. Sutton accepts the first two as reasonable, but on the basis of Chandra's and other experiments argues that the validity of the third assumption is doubtful. Sutton suggests that a small boundary layer with a large temperature gradient forms as the heating of the lower surface begins. As the heating continues, the boundary layer becomes unstable and yields the columnar flow observed by Chandra. Further heating then leads to the establishment of the condition postulated by the theories, and hence the cellular flow and agreement for large temperature difference.

Sutton does point out that the theoretical formulation would correspond to an experimental one in which the heat is added so slowly as to preclude the possibility of the formulation of a boundary layer. He dismisses the possibility that the experimental procedures were of this kind, and reexamines the problem according to his postulate that the columnar motion occurs because of an instability of a shallow boundary layer whose depth is unrelated to the distance between the upper and lower boundaries. On this basis, Sutton derives a criterion for the columnar flow which involves only the ratio of the absolute temperatures of the upper and lower rigid boundaries and agrees with Chandra's observations. Another expression is derived and experimentally verified for the critical temperature difference at which transition takes place from the columnar to the cellular flow. Finally, it is shown, in agreement with Chandra, that the cellular type will occur for fluid layers exceeding a certain depth, but that for more shallow layers the columnar type is generated initially, passing to the cellular for increased temperature differences.

F,17. Unstable Layers with Superposed Steady Flow. In order to explain various cloud formations in the presence of winds, many of the previous experiments were also run with a steady superposed motion in the horizontal direction (see [*72,74,75*]). The superposed motion was established either by moving one of the boundaries in its plane or by fans and pumps. Since it is conceivable that analogous situations might be encountered in present-day applications in fields other than meteorology, we will give a brief summary of that research.

The results are divided into two categories: viz., the effects of steady flow on (1) cellular convection patterns, and (2) on columnar flows. For a configuration in which cellular flows could occur, the steady flow did not alter the stability condition; that is, if the temperature difference was lower than the critical, no flow pattern (other than the steady) was discernible. When the critical temperature difference was exceeded, polygonal cells were, of course, obtained with no superposed flow. However, when the steady flow rate was maintained below a certain low limiting value, say u_1, the original polygonal pattern was distorted as shown in Plate F,17a. When the steady flow exceeded u_1 but was below another limit, say u_2, the fluid aligned itself into a series of vortices or rolls with axes transverse to the steady flow direction (see Plate F,17b). For steady velocities greater than u_2, the flow pattern was a series of longitudinal rolls; that is, vortices with axes parallel to the steady flow direction, Plate F,17c.

For a fluid layer of a given depth, u_1 and u_2 both increase with the temperature difference between the top and bottom boundaries. It is not possible to state the values of u_1 and u_2 for given configurations a priori.

When a columnar flow was subjected to a steady flow, the fluid appeared to be filled with irregular short spindlelike structures as shown in Plate F,17d. As the steady flow velocity was increased, the lengths of the spindlelike structures increased and the structures appeared to be more regular, but long continuous rolls or transverse rolls were never encountered.

To the best of the author's knowledge, no more details (such as relative heat transfer rates and the like) are available for the superposed flows described, and no theoretical treatment has been attempted.

CHAPTER 4. NATURAL CONVECTION. THE INTERNAL PROBLEM

F,18. Introduction. Most of the early work on internal flows with body forces was of an experimental or semiempirical nature. An extensive study of the latter type was made by Elenbaas [*76*] for the natural convection heat transfer between two parallel plates heated to the same temperature. He also described experiments made with air in that reference. In [*77*] Elenbaas compared semiempirical Nusselt numbers for air flow in vertical tubes of different cross section, and in [*78*] (primarily a summary paper) functional equations for the Nusselt number were obtained essentially by dimensional analysis. These equations were then rewritten by correlation with experimental or approximately computed results to yield semiempirical formulas for the Nusselt numbers. None of this work, however, gave detailed information on the velocity and temperature distributions; in some cases the semiempirical formulas deviated from the experimental data.

The successful application of natural convection flows in hollow passages in turbine rotor blades for cooling, as demonstrated by Schmidt [79], and the use of this type of heat transfer in some of the many schemes for extracting heat energy from an atomic pile, not only increased the interest in internal flows with body forces but also pointed out a number of problems for which new answers were needed. For example, none of the early works could be applied to fluids containing internal heat sources as would be encountered in the nuclear power applications. Therefore, starting about 1945, increased attention was given to the internal natural convection problems.

We have already seen how difficult the analysis of this phenomenon was, due to the interaction of the dynamics and thermodynamics in the external problems. The influence of boundary surfaces further complicates matters in the internal problems. To analyze internal flows, we invoke one of the simplifications noted in the first chapter, viz., we choose configurations that are simple enough to lead to tractable mathematics and, at the same time, retain the essential physical elements of the natural convection phenomenon. The simplest of such configurations is the fully developed flow between two parallel planes. Consideration of such flows is justified on two accounts. Firstly, the basic assumptions underlying such flows are actually typical for a wide range of experimentally produced thermal convection phenomena. Secondly, there are many flows that deviate somewhat from the fully developed flows. These small but not negligible deviations can be taken into account as nonlinear corrections to the fully developed solutions. A chimney is perhaps the most obvious example of nearly fully developed natural convection.

F,19. Fully Developed Flow in Vertical Channels. Consideration is given to the fully developed laminar flow of a viscous fluid between two parallel planes oriented parallel to the direction of the generating body force (see Fig. F,19a). For simplicity we will consider the body force to be that due to gravity, so that the channel formed by the two plates is vertical. By "fully developed" we mean that the velocity components are independent of the axial (vertical) distance, or, in other words, the solutions obtained are valid away from the channel ends and, hence, apply to channels with large length-gap ratios. The temperatures of the two surfaces are, in general, taken to be functions of the vertical coordinate and can differ by a constant. It is further assumed that the physical properties of the fluids are constant, except that the essential influence of density variations on the flow is accounted for by the introduction of the fluid volumetric expansion coefficient β in the body force term. We have already seen, in connection with the flat plate problem, the justification of this assumption for small temperature differences. This assumption is made in almost all the work in this field, but it is discussed in only a few

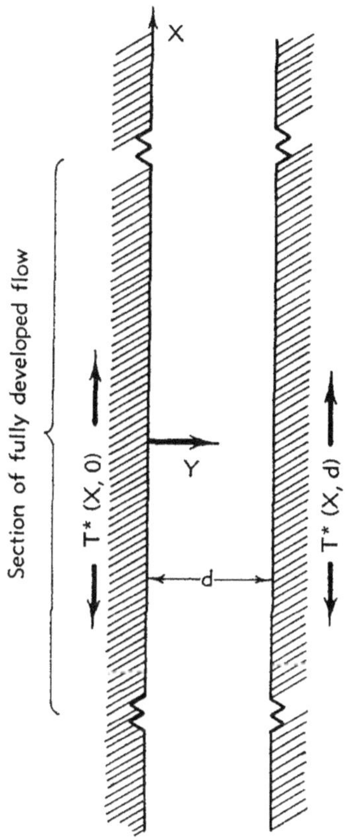

Fig. F,19a. Schematic sketch of vertical channel configuration.

papers. Recently, Maslen [80] carefully reexamined the problem, including the effects of compressibility as well as fluid property variations. The limitations of the above assumptions which he found will be discussed in detail in a subsequent article of this section.

Under the conditions stated, it follows from the (incompressible) continuity equation that the only nonzero velocity component is the axial one, U, and that it must be a function only of the transverse coordinate Y. From the Y momentum equation the pressure is found to be a function of the axial coordinate alone. Thus, rewriting the body force term as a buoyancy term, as is indicated in Appendix B of [1], the equations governing the fully developed flow are

$$\frac{d^2U}{dY^2} + \frac{\rho\beta f_X}{\mu}(T^* - T_r^*) = \frac{1}{\mu}\left(\frac{dP}{dX} + \rho_r f_X\right) \tag{19-1}$$

$$\frac{\partial^2 T^*}{\partial X^2} + \frac{\partial^2 T^*}{\partial Y^2} = \frac{\rho c_v}{k} U \frac{\partial T^*}{\partial X} - \frac{\mu}{k}\left(\frac{dU}{dY}\right)^2 - \frac{Q}{k} \tag{19-2}$$

where f_X is the negative of the X component of the body force per unit mass; U and T^* are the dimensional velocity and temperature, respectively; and $_r$ denotes a reference condition which, for the channel with isothermal walls, is the hydrostatic condition, and for the linearly varying channel wall cases is the condition at the wall at $Y = 0$. The heat due to heat sources in the fluid is denoted by Q.

The viscous fluid boundary condition that the fluid must adhere to the channel walls is written as

$$U(0) = U(d) = 0 \tag{19-3}$$

where d is the distance between the two bounding planes. Two distinct sets of thermal boundary conditions will be specified. The first,

$$\theta(0) = T_{w_0}^* - T_s^*,\ \theta(d) = T_{w_1}^* - T_s^* \tag{19-4}$$

corresponding to a channel with isothermal (but, in general, different) surfaces, and the second

$$T^*(X, 0) = AX + T(0),\ T^*(X, d) = AX + T(d)$$

or

$$\theta(0) = 0,\ \theta(d) = \theta_{w_1} \tag{19-5}$$

where $\theta = T^* - T_r^*$, the subscripts $_{w_0}$ and $_{w_1}$ refer to conditions at the surface at $Y = 0$ and $Y = d$, respectively, and A is the axial temperature gradient $\partial T^*/\partial X$. Note that $A = 0$ for the isothermal surface case so that we can say that $\theta = T - T_r$. In order to satisfy Eq. 19-1 and the boundary conditions given by Eq. 19-5, the temperature must be of the form

$$T^*(X, Y) = AX + T(Y) \tag{19-6}$$

The substitution of Eq. 19-6 into Eq. 19-1 and 19-2 yields

$$\frac{d^2U}{dY^2} + \frac{\rho\beta f_X}{\mu}\theta = \frac{1}{\mu}\left(\frac{dP}{dX} + \rho_r f_X\right) \tag{19-7}$$

$$\frac{d^2\theta}{dY^2} - \frac{\rho c_v A}{k} U + \frac{\mu}{k}\left(\frac{dU}{dY}\right)^2 + \frac{Q}{k} = 0 \tag{19-8}$$

Since the left side of Eq. 19-7 is a function of Y alone, and the right side a function of X alone, it is clear that each side must be equal to a constant. Thus Eq. 19-7 can be written as

$$\frac{d^2U}{dY^2} + \frac{\rho\beta fx}{\mu}\theta = \tilde{C} \tag{19-9a}$$

or

$$\frac{1}{\mu}\left(\frac{dP}{dX} - \rho_r fx\right) = \tilde{C} \tag{19-9b}$$

The function θ is thus independent of the axial coordinate X.

If we now let

$$u = U\sqrt{\frac{KPr}{c_p\theta_r}}, \qquad \vartheta = \frac{K\theta}{\theta_r}, \qquad y = \frac{Y}{d}$$

the momentum and energy Eq. 19-9a and 19-8, respectively, become

$$u'' + \vartheta = CK \tag{19-10}$$

$$\vartheta'' - Ra_A u + (u')^2 + \alpha K = 0 \tag{19-11}$$

where the primes denote differentiations with respect to y and

$$\left.\begin{aligned} C &= \tilde{C}\sqrt{\frac{Prd^4}{c_p\theta_r K}}, & K &= PrGr\left(\frac{\beta fxd}{c_p}\right) \\ Gr &= \frac{\beta fx\theta_{w_0}d^3}{\nu^2}, & Ra_A &= PrGr_A\alpha = \frac{Qd^2}{k\theta_{w_0}} \end{aligned}\right\} \tag{19-12}$$

The Gr_A denotes a modified Grashof number defined with Ad as the characteristic temperature difference, rather than θ_{w_0}. In the varying wall temperature cases, Ad replaces θ_{w_0} in K and α also. From Eq. 19-10 we see that C essentially specifies the temperature level of the problem.

The terms from left to right in Eq. 19-10 denote viscous, buoyancy, and axial pressure forces, and in Eq. 19-11 represent conduction, convection, frictional heating, and heat due to internal heat sources. It is clear from Eq. 19-11 that, for the channel with isothermal surfaces, the convection term vanishes identically. Thus, if the frictional heating and heat due to sources are disregarded, the temperature distribution will be just that due to conduction. The parameter C merely specifies the temperature level or axial pressure gradient.

The elimination of ϑ between Eq. 19-10 and 19-11 yields a fourth order nonlinear differential equation:

$$u^{iv} + Ra_A u - (u')^2 - \alpha K = 0 \tag{19-13}$$

which applies both to channels with isothermal walls and to channels with linear temperature variations along the walls. It should be added here that, instead of specifying the surface temperatures, we could as easily have specified the wall heat flux. Specific problems are defined by the modified Rayleigh number, Ra_A, through the axial temperature gradient A. Thus, for the isothermal wall case $A = 0$, $Ra_A = 0$; if the wall temperature increases upward, A and Ra_A are both positive. For A and Ra_A, both negative, the temperature decreases upwards so that the fluid is essentially being heated from below. We should, therefore, not be too surprised if some of the unusual aspects associated with thermal instabilities should appear in the latter problems. Few other examples are known in which a single parameter has as much significance as Ra in the class of problems treated here. For not only is the fluid thermal stability determined by Ra, but also the hydrodynamic stability of the fluid motion is a function of Ra, as will be seen in greater detail in a later chapter.

In terms of nondimensional functions the boundary conditions are

$$u(0) = u(1) = 0 \tag{19-14}$$

in all cases. Making use of Eq. 19-10, for channels with constant (but not necessarily equal) surface temperatures,

$$u''(0) = (C - 1)K \tag{19-15a}$$

$$u''(1) = m(C - 1)K \tag{19-16a}$$

where m which is essentially a measure of the difference between the two wall temperatures is

$$m = \frac{C - \dfrac{\theta_{w_1}}{\theta_{w_0}}}{C - 1} \tag{19-17a}$$

The thermal boundary conditions for channels with linearly varying wall temperatures are

$$u''(0) = CK \tag{19-15b}$$

$$u''(1) = mCK \tag{19-16b}$$

where

$$m = 1 - \frac{\theta_{w_1}}{CAd} \tag{19-17b}$$

More explicit developments of the above equations can be found in [*1*] and [*52*]. We have now obtained boundary-value problems defined by Eq. 19-13 to 19-17 which are readily amenable to rather simple methods of solution. Nevertheless, the results will enable us to obtain considerable information on confined flows with body forces. For example, we will obtain detailed velocity and temperature distributions from which we can compute various flow and heat transfer coefficients. In addition, we will be able to determine the effects of heating fluids below on vertical fluid columns and of frictional heating on flows with body forces.

An analytical iteration procedure was used to solve the problems because of the nonlinearity of the governing differential equation (Eq. 19-13). The zeroth order solution (i.e., neglecting the frictional heating term $(u')^2$ in Eq. 19-13) is given by

$$u_0 = K \sum_{i=1}^{4} \frac{a_i}{i!} y^i \tag{19-18}$$

for the constant wall-temperature case (see [*1*]), where

$$a_1 = \tfrac{1}{24}(4m + \alpha + 8),\ a_2 = -1,\ a_3 = -\left(m + \frac{\alpha}{2} - 1\right),\ a_4 = \alpha$$

For the linear wall-temperature cases, the solutions depend on the sign of the modified Rayleigh number, Ra_A, and are given by

$$u_0 = \pm \left\{ \frac{\alpha K}{Ra_A}(1 - v_{00}) + \frac{CK}{2\sqrt{Ra_A}}[(m-1)v_{01} + (m+1)v_{02}] \right\} \tag{19-19}$$

For positive Rayleigh numbers (see [*52*]), the plus sign is used and

$$\left.\begin{aligned} v_{00} &= \frac{1}{\cosh^2 \bar{R} + \cos^2 \bar{R} - 1} (\cosh \bar{R} \cos \bar{R} \cosh \bar{R}\eta \cos \bar{R}\eta \\ &\qquad + \sinh \bar{R} \sin \bar{R} \sinh \bar{R}\eta \sin \bar{R}\eta) \\ v_{01} &= \frac{1}{\cos^2 \bar{R} - \cosh^2 \bar{R}} (\cosh \bar{R} \sin \bar{R} \sinh \bar{R}\eta \cos \bar{R}\eta \\ &\qquad - \sinh \bar{R} \cos \bar{R} \cosh \bar{R}\eta \sin \bar{R}\eta) \\ v_{02} &= \frac{1}{\cosh^2 \bar{R} + \cos^2 R - 1} (\cosh \bar{R} \cos \bar{R} \sinh \bar{R}\eta \sin \bar{R}\eta \\ &\qquad - \sinh \bar{R} \sin \bar{R} \cosh \bar{R}\eta \cos \bar{R}\eta) \end{aligned}\right\} \tag{19-20}$$

with

$$\bar{R} = \frac{(Ra_A)^{\frac{1}{4}}}{2\sqrt{2}},\ \eta = 2y - 1$$

For negative Rayleigh numbers (see [*81*]), the minus sign in Eq. 19-19 is used, and

$$\left.\begin{aligned} v_{00} &= \frac{1}{2}\left(\frac{\cosh \sqrt{2}\,\bar{R}\eta}{\cosh \sqrt{2}\,\bar{R}} + \frac{\cos \sqrt{2}\,\bar{R}\eta}{\cos \sqrt{2}\,\bar{R}}\right) \\ v_{01} &= \frac{1}{2}\left(\frac{\sinh \sqrt{2}\,\bar{R}\eta}{\sinh \sqrt{2}\,\bar{R}} - \frac{\sin \sqrt{2}\,\bar{R}\eta}{\sin \sqrt{2}\,\bar{R}}\right) \\ v_{02} &= \frac{1}{2}\left(\frac{\cosh \sqrt{2}\,\bar{R}\eta}{\cosh \sqrt{2}\,\bar{R}} - \frac{\cos \sqrt{2}\,\bar{R}\eta}{\cos \sqrt{2}\,\bar{R}}\right) \end{aligned}\right\} \tag{19-21}$$

In the above equations, Eq. 19-19 to 19-21, Ra_A is always positive. The corresponding zeroeth order temperature distributions can be obtained from Eq. 19-10 by using Eq. 19-18 and 19-19. Higher order approximations for u and ϑ are presented in [*1,52*] wherein it is also shown that the zeroeth order solutions are good approximations to exact solutions obtained on high speed computing machines, as long as $\beta f_X d/c_p$ (the new frictional heating parameter discussed in the first chapter) is much less than unity.

The solutions presented above are all given in terms of the parameter C. Therefore, to specify a given problem completely, the axial pressure gradient must be known or the mass flow in the channel must be specified. The parameter C may be related to the nondimensional mass flow in the channel M which is defined as

$$M = \int_0^1 u_0 dY$$

Thus we find that, for the linear variation of wall temperature,

$$C = \frac{\theta_{w1}}{2Ad} + \frac{1}{\sqrt{2}\, Ra_A^{\frac{1}{4}}} \left(\frac{\cosh 2\bar{R} + \cos 2\bar{R}}{\sin 2\bar{R} - \sinh 2\bar{R}}\right) \left[\frac{M Ra_A}{K_A} - \alpha + \frac{\sqrt{2}\, \alpha}{Ra^{\frac{1}{4}}} \left(\frac{\sin 2\bar{R} + \sinh 2\bar{R}}{\cosh 2\bar{R} + \cos 2\bar{R}}\right)\right] \tag{19-22}$$

The mass flow is an independent parameter of the problem, and hence could be due in part to a forced flow. To simulate the case of a completely enclosed channel, we need only set $M = 0$ in Eq. 19-22 (i.e., the net mass flow is zero). This special case is discussed in detail in [*52,81*].

For the constant wall-temperature channel, the zeroeth order Nusselt number (i.e., neglecting frictional heating) for the wall at $Y = 0$ is given by

$$Nu_0 = \frac{1}{m-1}\left[m + \frac{\alpha}{2} - 1 + \frac{K}{48(2520)} \times (51\alpha^2 + 444m\alpha + 564\alpha + 1008m^2 + 2352m + 1680)\right]$$

and when the wall temperatures are equal, $m = 1$

$$Nu_0 = \left[\frac{\alpha}{2} + \frac{K}{48(2520)}(51\alpha^2 + 1008\alpha + 5040)\right]$$

The corresponding relation for a linearly increasing wall temperature (positive Rayleigh number) is

$$Nu_0 = \frac{1}{2\sqrt{2}\,(Ra)^{\frac{1}{4}}} \left\{ \frac{1}{\cosh^2 \bar{R} + \cos^2 \bar{R} - 1} \left[\alpha(\sinh 2\bar{R} + \sin 2\bar{R}) + \frac{C(m+1)Ra^{\frac{1}{4}}}{2} (\sinh 2\bar{R} - \sin 2\bar{R}) \right] + \frac{C(m-1)Ra^{\frac{1}{4}}}{2(\cos^2 \bar{R} - \cosh^2 \bar{R})} (\sinh 2\bar{R} + \sin 2\bar{R}) \right\}$$

The first objective of this series of problems, namely, to find detailed information on the flow and heat transfer of confirmed fluids subject to a body force, is fulfilled by the solutions and relations given above. Representative profiles will be presented in conjunction with the next topic.

Heating from below. An interesting aspect of natural convection becomes evident if it is noted from Eq. 19-20 and 19-21 that the solutions for negative Rayleigh numbers are quite different from those for positive Rayleigh numbers. In particular, the negative Rayleigh number solutions become meaningless for $Ra_A = (n\pi)^4$ where n denotes the integers; for $n = 1$ and 2 the critical values of Ra_A are 97.41 and 1558.55, respectively. Further, it can be seen on Fig. F,19b that the flow for negative Ra_A is of a different character for Rayleigh numbers on opposite sides of the critical values. To explain these seemingly unusual results, we note that the modified Rayleigh number can change sign if either the axial temperature gradient A or the volumetric expansion coefficient changes sign or if the body force direction changes. If, for example, we assume that the sign change of Ra_A is due to a change in A, the physical interpretation of the results becomes clearer. A negative axial temperature gradient implies that the fluid is heated from below. This situation, as we have already seen, leads to an instability due to the piling of heavy fluid on lighter fluid. Analogous situations, of course, follow directly for changes in body force direction or in the sign of the volumetric expansion coefficient.

From our discussion in Chapter 3 of fluids heated from below, recall that the fluid motion does indeed depend on the critical values of the Rayleigh number. The fluid remains at rest until a critical value is attained. It then has a columnar mode of motion which changes to a cellular mode at another critical Rayleigh number. This behavior as well as the changes in flow type with superposed velocities are not unlike the changes with Rayleigh number shown for negative Ra_A on Fig. F,19b. Hence, stability characteristics similar to those found for horizontal fluid layers are displayed in the present problem.

In [*82*] the problem discussed above is also treated for the cases where one and both walls are thermally insulated, and heat due to friction and sources are neglected. The results for one insulated wall are qualitatively similar to those for the channel with linearly varying wall temperatures, but the critical values of the Rayleigh number are somewhat different. When both walls are taken to be insulated, the problem becomes homogeneous, and the eigensolutions indicate that no flow ensues unless the Rayleigh number satisfies a given eigenrelation. This is qualitatively just what was predicted from the stability analyses for horizontal fluid layers heated from below and with insulated ends, and was actually found in the associated experiments. The effects of heat sources and frictional heating on this configuration are discussed in [*81*].

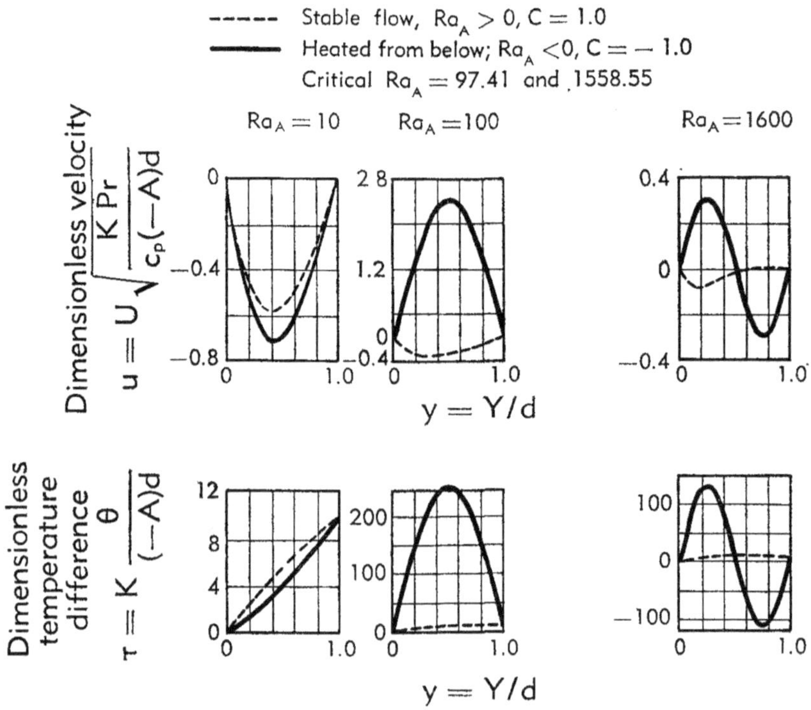

Fig. F,19b. Representative velocity and temperature distributions for stable flows ($Ra_A > 0$) and flows heated from below ($Ra_A < 0$). Walls at different temperatures ($m = 0$) and $K = 10$. Critical $Ra_A = 97.41$ and 1558.55 (Ostrach [*82*]).

Ostrach [*82*] shows that the mathematical formulation of the fully developed channel flow problem, neglecting frictional heating and with a negative axial temperature gradient, yields an equation that is identical with that describing the vibrations of uniform beams or of rotating shafts (see [*83*, pp. 286–290]). The two insulated surfaces case has its counterpart in the vibration problems if the ends of the shaft are fixed with no shear and the critical values of the Rayleigh number correspond to the critical or whirling speeds of the shaft. If the shaft has initial deflections or bending moments at the ends, the rotating shaft becomes dynamically unstable and the solutions become infinite at the critical speeds. This situation is analogous to that found in the present problem with linear side wall temperatures.

A representative set of velocity and temperature profiles computed for a positive Rayleigh number ($A > 0$), and all other conditions identical to those for the negative Rayleigh number case ($A < 0$) presented in Fig. F,19b, are superposed on that figure. These profiles are for the case of no heat due to heat sources of the fluid ($\alpha = 0$). The heat sources increase the velocities and temperatures; their quantitative effects are presented in [*1*,*52*,*81*]. Comparison of the profiles presented in Fig. F,19b for the two cases shows that the flow and heat transfer associated with the heating from below is greater, and the flow and heat transfer at the walls can be in opposite directions in the two problems.

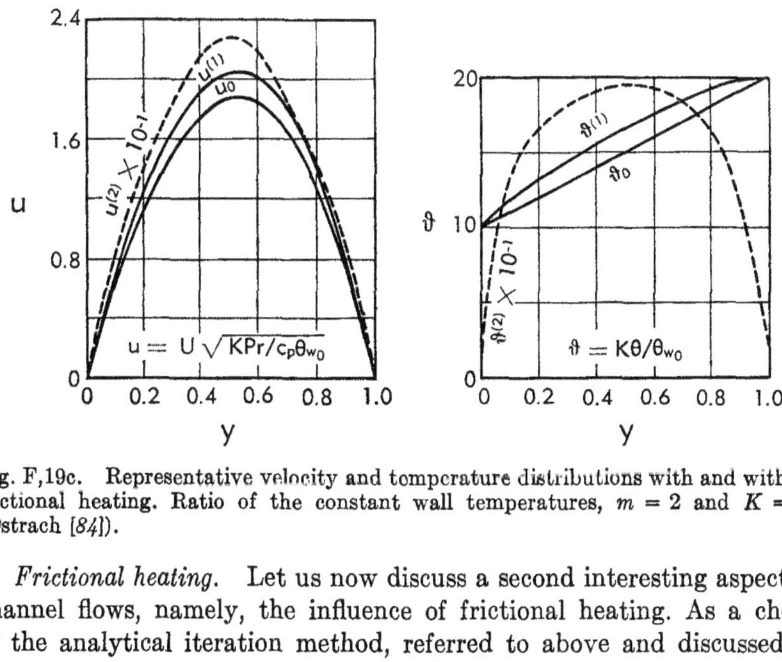

Fig. F,19c. Representative velocity and temperature distributions with and without frictional heating. Ratio of the constant wall temperatures, $m = 2$ and $K = 10$ (Ostrach [*84*]).

Frictional heating. Let us now discuss a second interesting aspect of channel flows, namely, the influence of frictional heating. As a check of the analytical iteration method, referred to above and discussed in [*1*,*52*], the same nonlinear problems (specified by Eq. 19-13 and the appropriate boundary conditions) were solved by numerical integration on a high speed computing machine. It was then discovered that each problem had either a pair of solutions (denoted herein by the superscripts $^{(1)}$ and $^{(2)}$ or no solution. Examples of the pairs of solutions are given in [*1*,*52*,*81*]. The solutions obtained analytically displayed no such behavior although they did approximate one of each pair of solutions in limited ranges of the parameters (see Fig. F,19c). A somewhat analogous situation was noted by Hartree [*85*] in his study of forced boundary layer flows with adverse pressure gradients. The second solutions to Hartree's problem are presented and discussed in [*86*,*87*].

Examination of the representative profiles in Fig. F,19c shows that the frictional heating acts like heat sources in the fluid and, hence, increases velocities and alters the temperature distributions so that the heat transfer at the wall is changed. The analytical solutions denoted by the subscript zero are computed by neglecting frictional heating. Other examples of profiles computed, including frictional heating and discussions of the conditions under which the analytical solutions are good approximations to the exact solutions, are presented in [*1,52,81*].

Ostrach and Albers [*88*] present more details on the pairs of solutions and their behavior for systematic variations of the physical parameters in order to interpret the two unusual results obtained by including frictional heating, viz., that (1) beyond certain parametric limits, no fully developed solutions exist, and (2) distinct pairs of solutions are found for a given set of conditions. Careful study of the machine solutions shows that the probability of obtaining any solutions decreases with increased heat input to the channel, as, for example, from frictional heating. It thus appears that the limit beyond which no fully developed solutions exist depends on the heat input to the channel. The reason for this result is not as yet evident. However, there may be an analogy to the familiar one-dimensional forced channel flow in which thermal choking occurs.

A representative pair of profiles is presented in Fig. F,19c. We have already pointed out that the flow denoted by the superscript 1 agrees reasonably well with that computed by the analytic iteration method. In general, the analytic solution agrees with the one of the pair that has the lower velocity extremum (maximum or minimum), because the iteration method started with the no-frictional-heating case. Because less frictional heating is associated with the less intense solution, this flow is the one to be expected in practice. The other solution, however, denoted by the superscript 2, represents a flow with velocities and temperature differences of more than an order of magnitude greater than the first. The second solutions are connected intimately with the regenerative action of frictional heating in natural convection, wherein the frictional heating acts as a heat source in the fluid and thereby leads to larger flow velocities.

As an illustration of the role that the dissipation plays in these problems, consider the simple case wherein the walls are kept at two different constant temperatures and no heat sources are in the fluid. For this case, which is treated in [*1*], the less intense solution (little frictional heating) yields essentially the conduction (linear) temperature distribution (see the ϑ_0 curve on Fig. F,19c). That is, no matter what kind of temperature profile exists in the fluid as it enters the channel, the profile becomes the conduction distribution by the time the flow has become fully developed, because heat is extracted at the walls in the entrance region. For the more intense solution, the entering profile becomes parabolic (see the $\vartheta^{(2)}$ curve on Fig. F,19c) and differs markedly from the conduction profile.

It is thus evident that the frictional heating acts as a heat source in the fluid. Another significant fact that can be seen from F,19c and the other examples of the pairs of solutions presented elsewhere (see [*1,52*], for example) is that the mass flow and energy associated with the more intense solution are greater than for the less intense solution. Therefore, is seems that, if the intense flow is to be obtained in practice, the flow would have to be started with the mass flow or energy at the higher level. This might be possible, for example, by heating the fluid considerably as it enters the channel.

The next question to arise in connection with the second solution concerns its stability; that is, since the frictional heat is dependent on the large velocities, would a small decrease in velocity, for example, decrease the frictional heating, and would this process keep repeating until the second solution attenuates to the first? The stability of the symmetric case ($m = 1$) has been analyzed, and the results, which will be presented in the chapter on hydrodynamic stability, show that this flow is stable.

The combined effects of heating from below and frictional heating can be seen on Fig. F,19d. Again we see that the approximate solution (neglecting frictional heating) approximates one of the pair of solutions including frictional heating (because $K/Gr \ll 1$) except near the critical Rayleigh number. Further, the approximate solution corresponds to a different member of each pair of the exact solutions on either side of the critical Rayleigh number, specifically, to the one with the smaller velocity extremum. The question then arises: Since the exact solutions do not display a discontinuity (with Ra_A), is the flow actually of a different type on either side of the critical Ra_A as indicated by the linearized solutions (see also Fig. F,19b)? Although a more general analysis of the present configuration would be necessary (perhaps relaxing the condition of fully developed flow) to answer this question conclusively, it seems reasonable that the actual flow is in fact of a different type, as indicated. The reason for this statement is that the linearized equations are approximate only in that the frictional heating is neglected. In starting a flow, the frictional heating is not important until velocities of appreciable magnitude are encountered. Therefore, since the larger velocity flows result from the regenerative action of frictional heating, it is felt that for a given set of conditions the flow first established at any Rayleigh number would be that with the smaller velocities.

The appreciable quantitative and qualitative differences between the approximate and exact solutions near the critical Ra_A shown in Fig. F,19d indicate that near that value the effects of frictional heating are very important. These effects may even play an important role in the transition from one type of flow to the other.

Although many interesting and unusual results, including the pairs of solutions, were obtained for the three somewhat different problems discussed above, all the analyses are restricted to fully developed flows of fluids that are only "quasi-incompressible." To find the limitations of the assumption of quasi-incompressibility, and to see if any of the above-discussed results are altered when the fluid is real, Maslen [80] studied the same configurations as discussed above, but he considered the fluid to be compressible and to have appropriate property variations. The essential conclusions that can be drawn from Maslen's analysis are as follows.

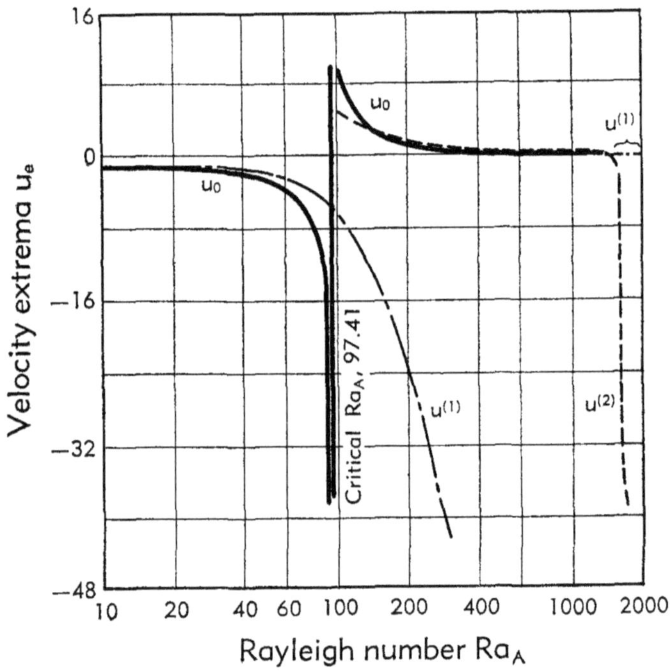

Fig. F,19d. Velocity extrema with and without frictional heating as functions of the Rayleigh number for flows heated from below. Identical wall temperatures ($m = 1$), $K = 10$, and $C = -1$ (Ostrach [81]).

The solutions of [1] for constant channel wall temperatures are exact for liquids with small temperature variations and, for this case, two solutions do, in fact, exist even if the appropriate property variations are taken into account. Maslen develops an analytic approximation method that yields both solutions in good agreement with the machine solutions. Although Maslen considers variable viscosity, he finds that, for cases of practical interest, wherein the fluid velocities must be small enough for the flow to be laminar, the temperature variations are small and the solutions of [1] are accurate representations of actual conditions.

When consideration is given to configurations in which there are streamwise temperature and pressure variations, but in which the velocity is virtually independent of the distance along the channel (as is the case in [*52,81,82*]), the solutions are approximate. For a gas, the mere presence of a pressure gradient requires a temperature gradient, while for a liquid the presence or absence of a streamwise temperature variation is governed by the wall-temperature conditions. In these cases one of two situations occurs. If the temperature variation across the channel is of the order of the temperature level, then in order that "channel flow" be maintained, the convection terms in the energy equation must be negligible. This implies that the mean flow Mach number is small and also that the frictional heating is negligible. For such circumstances the equations are separated and can readily be integrated for any case of interest. Only one solution exists and the solutions of the above-cited references are inapplicable under these conditions. On the other hand, if the temperature variations are small, more complicated effects occur. This situation of very small temperature changes admits large flow velocities. Then both frictional heating and the thermal convection effects can be significant. In such cases (when frictional heating matters), two solutions occur. These flows qualify as quasi-incompressible in that the only place where compressibility has an effect is in the buoyancy term in the streamwise momentum equation, as was assumed in [*52,81,82*]. Thus, all those cases are valid for small relative temperature variations.

It can therefore be stated from the work of Ostrach and Maslen that two-flow and heat transfer states are predicted for a fixed set of parametric values within certain limits for fully developed, combined, forced- and natural convection flows of certain real fluids. The only restriction of the analyses that has not been evaluated is that of assuming the flow to be fully developed. Definite experiments will have to be performed to evaluate this limitation, in order to establish definitely the physical existence of the second state of flow and heat transfer indicated by the analyses.

Han [*89*] and Lu [*90*] have generalized Ostrach's work by considering the fully developed flow in channels that are completely bounded on the sides. Han treats rectangular channels explicitly and obtains solutions for the profiles and Nusselt number in terms of double infinite series. Lu formulates the problem for channels of arbitrary cross section and presents solutions explicitly for circular annuli, rectangular pipes, and pipes of ring-sector cross sections. Solutions are given in terms of Bessel functions in double infinite series. Neither of these authors makes any calculations from these solutions. The effects of frictional heating are also ignored in [*89,90*]. Both of these papers point out an interesting analogy between some of the convection problems and elasticity problems in the theory of thin plates.

Some interesting experiments on two- and three-dimensional channels of the type discussed above are reported in [*91*]. However, the thermal boundary condition imposed was that of uniform heat input at the walls, and the flow was turbulent, so that no comparison can be made with the theoretical results.

F,20. Fully Developed Flow in Vertical Tubes. A rather extensive treatment of natural convection in cylindrical channels appears in a book by Ostroumov [*92*]. Although Ostroumov's original formulation is for channels of arbitrary cross section, most of the detailed analysis is for channels of circular section. Some of the new aspects of natural convection flows which the present writer treated in his consideration of the flow in a channel formed by parallel plates were also independently recognized by Ostroumov. For example, Ostroumov noted the different behavior of the solutions for positive and negative axial temperature gradients and presented the two types of solutions in detail but did not interpret them further; he also showed that the solutions presented were valid for combined forced and free flows, and that the only difference was in the change in the temperature levels of the flow. Ostroumov, however, never considered the effects of frictional heating or internal heat sources. Since a great deal of Ostroumov's motivation came from geophysical problems, in which the channels were drilled into solid blocks of various materials, he considers more general thermal boundary conditions than appear in other work on this subject.

In general, however, although the mathematics become a bit more elegant due to the different configuration (e.g., the biharmonic equation governs the flow instead of Eq. 19-13 and the solutions are given in terms of Bessel and Neumann functions), the results are essentially the same as those presented for the channel flows in the preceding article. Therefore, Ostroumov's work will not be discussed in detail herein. Hallman [*93*] extended Ostroumov's considerations of fully developed flow in circular pipes to include the effects of internal heat sources. Experimental verification of the predicted flow and heat transfer with no heat sources in a vertical tube with uniform wall heat flux is given in [*94*].

F,21. Closed-End Tubes. The flow in partially or fully enclosed regions can be treated as special cases of the solutions of the predicting articles by specifying that there is no net mass flow in the channel. This is explicitly done in [*52,81*]. However, the effects of rigid surfaces at the channel ends are not evident in this way. Therefore, to pursue further the effects of confining walls on natural convection flows, Lighthill [*95*] analyzed the flow in a closed-end cylindrical tube (see Fig. F,21a) with the walls at constant temperature and the body force acting toward the closed end. An orifice that supplies cool fluid is assumed at the open end. His motivation for this problem was the previously mentioned turbine cooling problem of Schmidt.

ity is virtually independent of the distance along the channel (as is the case in [*52*,*81*,*82*]), the solutions are approximate. For a gas, the mere presence of a pressure gradient requires a temperature gradient, while for a liquid the presence or absence of a streamwise temperature variation is governed by the wall-temperature conditions. In these cases one of two situations occurs. If the temperature variation across the channel is of the order of the temperature level, then in order that "channel flow" be maintained, the convection terms in the energy equation must be negligible. This implies that the mean flow Mach number is small and also that the frictional heating is negligible. For such circumstances the equations are separated and can readily be integrated for any case of interest. Only one solution exists and the solutions of the above-cited references are inapplicable under these conditions. On the other hand, if the temperature variations are small, more complicated effects occur. This situation of very small temperature changes admits large flow velocities. Then both frictional heating and the thermal convection effects can be significant. In such cases (when frictional heating matters), two solutions occur. These flows qualify as quasi-incompressible in that the only place where compressibility has an effect is in the buoyancy term in the streamwise momentum equation, as was assumed in [*52*,*81*,*82*]. Thus, all those cases are valid for small relative temperature variations.

It can therefore be stated from the work of Ostrach and Maslen that two-flow and heat transfer states are predicted for a fixed set of parametric values within certain limits for fully developed, combined, forced- and natural convection flows of certain real fluids. The only restriction of the analyses that has not been evaluated is that of assuming the flow to be fully developed. Definite experiments will have to be performed to evaluate this limitation, in order to establish definitely the physical existence of the second state of flow and heat transfer indicated by the analyses.

Han [*89*] and Lu [*90*] have generalized Ostrach's work by considering the fully developed flow in channels that are completely bounded on the sides. Han treats rectangular channels explicitly and obtains solutions for the profiles and Nusselt number in terms of double infinite series. Lu formulates the problem for channels of arbitrary cross section and presents solutions explicitly for circular annuli, rectangular pipes, and pipes of ring-sector cross sections. Solutions are given in terms of Bessel functions in double infinite series. Neither of these authors makes any calculations from these solutions. The effects of frictional heating are also ignored in [*89*,*90*]. Both of these papers point out an interesting analogy between some of the convection problems and elasticity problems in the theory of thin plates.

Some interesting experiments on two- and three-dimensional channels of the type discussed above are reported in [*91*]. However, the thermal boundary condition imposed was that of uniform heat input at the walls, and the flow was turbulent, so that no comparison can be made with the theoretical results.

F,20. Fully Developed Flow in Vertical Tubes. A rather extensive treatment of natural convection in cylindrical channels appears in a book by Ostroumov [*92*]. Although Ostroumov's original formulation is for channels of arbitrary cross section, most of the detailed analysis is for channels of circular section. Some of the new aspects of natural convection flows which the present writer treated in his consideration of the flow in a channel formed by parallel plates were also independently recognized by Ostroumov. For example, Ostroumov noted the different behavior of the solutions for positive and negative axial temperature gradients and presented the two types of solutions in detail but did not interpret them further; he also showed that the solutions presented were valid for combined forced and free flows, and that the only difference was in the change in the temperature levels of the flow. Ostroumov, however, never considered the effects of frictional heating or internal heat sources. Since a great deal of Ostroumov's motivation came from geophysical problems, in which the channels were drilled into solid blocks of various materials, he considers more general thermal boundary conditions than appear in other work on this subject.

In general, however, although the mathematics become a bit more elegant due to the different configuration (e.g., the biharmonic equation governs the flow instead of Eq. 19-13 and the solutions are given in terms of Bessel and Neumann functions), the results are essentially the same as those presented for the channel flows in the preceding article. Therefore, Ostroumov's work will not be discussed in detail herein. Hallman [*93*] extended Ostroumov's considerations of fully developed flow in circular pipes to include the effects of internal heat sources. Experimental verification of the predicted flow and heat transfer with no heat sources in a vertical tube with uniform wall heat flux is given in [*94*].

F,21. Closed-End Tubes. The flow in partially or fully enclosed regions can be treated as special cases of the solutions of the predicting articles by specifying that there is no net mass flow in the channel. This is explicitly done in [*52,81*]. However, the effects of rigid surfaces at the channel ends are not evident in this way. Therefore, to pursue further the effects of confining walls on natural convection flows, Lighthill [*95*] analyzed the flow in a closed-end cylindrical tube (see Fig. F,21a) with the walls at constant temperature and the body force acting toward the closed end. An orifice that supplies cool fluid is assumed at the open end. His motivation for this problem was the previously mentioned turbine cooling problem of Schmidt.

The type of flow in the tube depends primarily on the length-radius ratio for given Prandtl and Rayleigh numbers. For very small values of the l/R ratio, the flow is just like the free convection about a vertical plate, i.e., the effect of the confining walls is negligible if the boundary layer thickness is much smaller than the tube radius. For somewhat larger l/R values, however, this effect is no longer negligible. The three flow regimes to be expected if the walls influence each other are shown in Fig. F,21b. The first, for small values of l/R, is just the free convection boundary layer flow modified to account for a downward flow outside

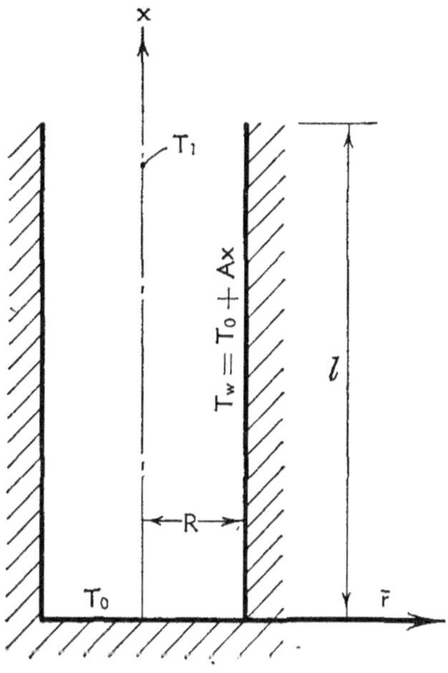

Fig. F,21a. Schematic sketch of closed-end tube configuration (Ostrach and Thornton [96]).

the boundary layer, equal to the upward flow. This regime breaks down at a value of l/R for which there is no longer a maximum volume flow of cool fluid at the orifice cross section. When l/R exceeds this critical value, the boundary layer mixes with the central flow and, when steady state is attained, the profiles fill the whole tube. This type of flow may be difficult to perceive intuitively, so let us consider the extreme case $l/R \gg 1$. In this case the tendency of the upward layer to thicken with axial distance disappears. Then the velocity and temperature distributions are similar at each cross section; only their scale increases as the orifice is approached. Therefore, in the intermediate l/R range, the velocity and temperature profiles fill the tube completely, but vary along the tube. Since the closed-end tube is a more complicated configuration than was considered in the previous articles, Lighthill used an integral method to obtain solutions that behave as was just described.

For the similarity regime (large l/R) he found that the flow fills the entire tube only for one value of l/R. If l/R is greater than this critical value, the motion stops near the closed end so that the effective l/R becomes equal to the critical. This result leads to one of the interesting and, at the same time, distressing aspects of natural convection in an enclosed region, because the stagnation of part of the fluid renders it ineffective as a coolant.

To study further the stagnation of fluids in closed-end tubes, Ostrach and Thornton [*96*] considered a configuration identical with Lighthill's except that now the temperature is taken to vary along the tube wall. This generalization of the wall temperature condition may be more realistic, since, for example, the temperature in a turbine blade varies

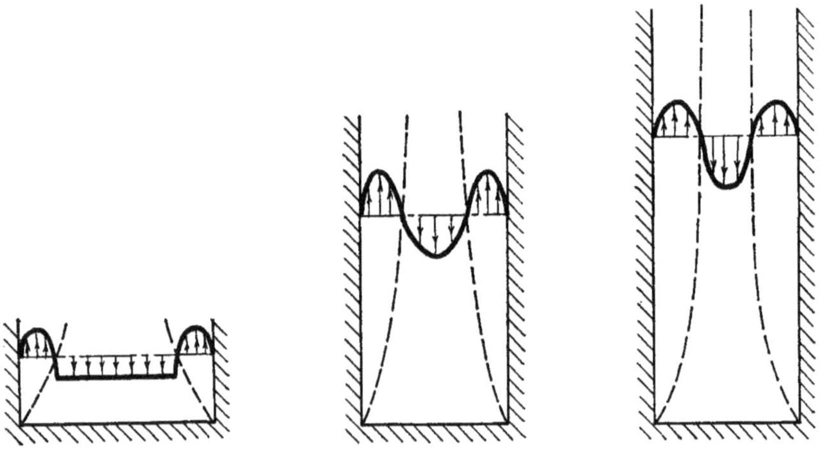

Fig. F,21b. The three flow regimes in a closed-end tube (Ostrach [*84*]).

in the spanwise direction. Examination of the mathematics of the problem will indicate the details of Lighthill's analysis as well as the results of [*96*]. The basic equations expressing consideration of mass, momentum, and energy are just the same as for free convection boundary layer flows, except that the wall rather than the hydrostatic condition is used as a reference. The integral method employed is to satisfy integrated forms of the differential equations in the tube interior and the equations themselves at the tube axis and walls. The nondimensional integrated equations are, respectively,

$$\int_0^1 rudr = 0 \tag{21-1}$$

$$\frac{1}{Pr}\frac{\partial}{\partial x}\int_0^1 ru^2dr = -\int_0^1 r(\vartheta - \vartheta_w)dr + \frac{1}{2}\left(\frac{\partial u}{\partial r} - \frac{\partial^2 u}{\partial r^2}\right)_{r=1} \tag{21-2}$$

$$\frac{\partial}{\partial x}\int_0^1 ru\vartheta dr = \left(\frac{\partial \vartheta}{\partial r}\right)_{r=1} \tag{21-3}$$

and at the wall and axis

$$\left(\frac{\partial^2\vartheta}{\partial r^2} + \frac{1}{r}\frac{\partial\vartheta}{\partial r}\right)_{r=1} = 0 \tag{21-4}$$

$$\left(u\frac{\partial\vartheta}{\partial x}\right)_{r=0} = \left(\frac{\partial^2\vartheta}{\partial r^2} + \frac{1}{r}\frac{\partial\vartheta}{\partial r}\right)_{r=0} \tag{21-5}$$

$$\frac{1}{Pr}\left(u\frac{\partial u}{\partial x}\right)_{r=0} = -(\vartheta - \vartheta_w)_{r=0} + \left(\frac{\partial^2 u}{\partial r^2} + \frac{1}{r}\frac{\partial u}{\partial r}\right)_1^0 \tag{21-6}$$

where $u = R^2U/\kappa l$, $\vartheta = (\beta f_X R^4/\nu\kappa l)(T_0 - T)$, $x = X/l$, $r = \bar{r}/R$, and $\vartheta_w \equiv \vartheta(x, 1)$. The zero subscript denotes conditions at the closed end. The boundary conditions are

$$u(x, 1) = v(x, 1) = u(0, r) = v(0, r) = 0 \tag{21-7}$$

$$\vartheta(0, r) = 0; \vartheta(x, 1) = -Ra_A x \tag{21-8}$$

$$\vartheta(1, 0) = \left(\frac{R}{l}Ra\right) \tag{21-9}$$

Eq. 21-9 defines the temperature at the orifice. The effects of frictional heating are disregarded in this problem.

Because we are particularly interested in a configuration wherein the flow can stagnate, we will restrict ourselves to only the large, but finite, l/R cases. Examination of the equations shows that u and ϑ must both vary linearly with x, and on the basis of symmetry and the boundary conditions at $x = 0$, u and ϑ must be of the form $xf(r^2)$ where $f(r^2)$ denotes third degree polynomials in r^2. For $Ra_A = 0$, these equations reduce to an eigenvalue problem (homogeneous equations and boundary conditions), which is the problem solved by Lighthill. The eigensolution obtained is valid only for a single value of $(R/l)Ra$. Therefore, the condition that the axial temperature should rise from its value T_1 at the orifice to T_0 at the closed end determines the one l/R, previously discussed, for this flow. In the variable wall temperature case [96], $Ra_A \neq 0$, there is an additional parameter in the problem so that an eigenrelation between the two parameters ($(R/l)Ra$ and Ra_A) is obtained. This relation and its derivation is explicitly presented in [96]. Accordingly, similar flows are possible for a range of parametric values rather than for discrete values, as in Lighthill's case.

The representative velocity and temperature distributions are shown in Fig. F,21c. For each $(R/l)Ra$, it can be seen that there is an Ra_A for which a "similar" flow exists in the entire tube. Lighthill's result appears as the special case with $Ra_A = 0$. From Fig. F,21c it can be seen that velocities greater than those obtained for constant wall temperatures are indicated with positive values of Ra_A and the associated smaller values of $(R/l)Ra$. The total Nusselt number over the tube length is given as a function of $(R/l)Ra$ in Fig. F,21d for the three flow regimes treated by

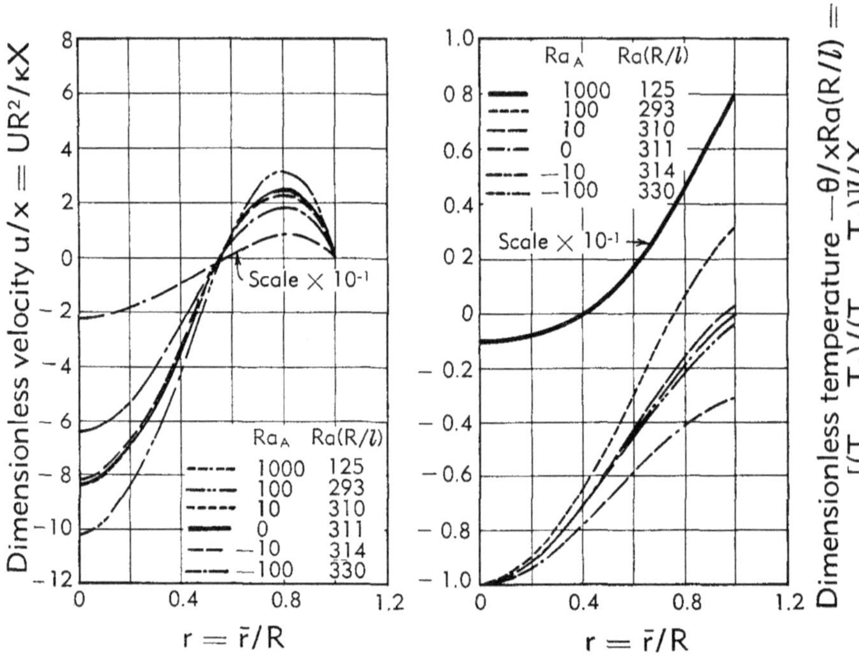

Fig. F,21c. Representative velocity and temperature distributions in a closed-end tube with linearly varying wall temperature (Ostrach and Thornton [*96*]).

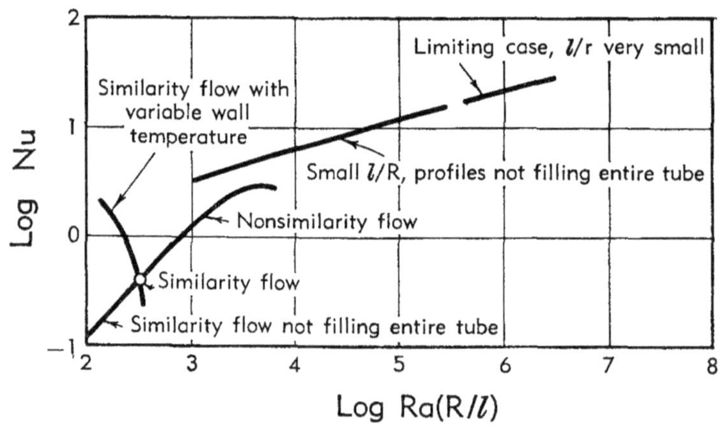

Fig. F,21d. Nusselt number as function of the product of the Rayleigh number and the radius-length ratio for closed-end tube flows (Ostrach [*84*]).

Lighthill. The result of [*96*] for variable wall temperatures and similar flows is superposed on this figure. It can thus be seen that higher heat transfer rates will be obtained for large (l/R) if the wall temperature is not constant.

Experimental verification of the three laminar flow regimes is presented in [*97*], and Lighthill's heat transfer results presented in Fig. F,21d are also essentially confirmed therein. Lighthill [*95*] also considered the case of turbulent flow and indicated that essentially the same types of flow regimes would occur as described for the laminar case. However, Martin found that some discrepancies existed between theory and experiment for the turbulent boundary layer regime. Hartnett and Welsh [*98*] performed experiments similar to Martin's, but with the wall boundary condition of uniform heat flux rather than uniform temperature. They found that as long as $Pr > 0.1$ the average heat transfer performance in a vertical tube closed at the bottom is equivalent for the cases of uniform wall temperature and uniform wall heat flux.

Hammitt [*99*] has extended Lighthill's boundary layer regime to include the effects of internal fluid heat sources, arbitrary axial wall-temperature distribution, and having both tube ends closed.

F,22. Flows in a Rotating Container. The body force in the analyses of the previous articles was always taken to be a constant. However, in many practical applications of the natural convection phenomenon, the body force need not be a constant. In Schmidt's turbine [*79*] the blades were cooled by the circulation of water, driven by centrifugal force, through them. One might wonder whether the same idea could also be employed in another difficult problem, namely, that of cooling a high speed vehicle entering the earth's atmosphere. In such a case, two (nonconstant) body forces are available, the force of retardation by the atmosphere and the centrifugal force generated by the rotation of the vehicle about its axis.

Despite the practical demonstration of the basic idea in [*79*], relatively little information on natural convection exists for internal flows and for body forces that vary temporally or spatially. The problem in a reentering vehicle is further complicated by the simultaneous action of the two kinds of body forces in different directions.

To gain some insight into problems of this sort Ostrach and Braun [*100*] analyzed a somewhat idealized configuration which, nevertheless, contained some of the complications associated with actual problems. In this way, the primary factors influencing the phenomenon were determined, and the relative importance of some of the interacting effects could be evaluated. A preliminary account of this work was first presented in [*84*], as well as a summary of all the natural convection work discussed thus far. Considerable work on problems similar to that discussed here has been done by meteorologists and astrophysicists. However, their configurations differ markedly from those of interest to the fluid dynamicist and aerodynamicist.

Specific consideration is given in [*84,100*] to the flow and heat transfer of a fluid subject to an axial body force (simulating the retardation force in a reentry vehicle) inside a rotating right circular cylinder of small height (see Fig. F,22a), which is heated at its lower surface. In this way, the rotational effects can be evaluated, and internal conditions in the vicinity of the external stagnation region of blunt reentry-type bodies are simulated. Considerably more discussion on the relationship of the configuration of Fig. F,22a with the reentry vehicle can be found in [*100*]. In order to avoid the complications introduced by unsteady effects, we shall assume that the axial retardation force is constant and concentrate on the effects of rotation of the container on the internal flow. In particular, two cases will be examined: (1) Rotation predominates, i.e., the internal motion is essentially due to the rotation of the container,

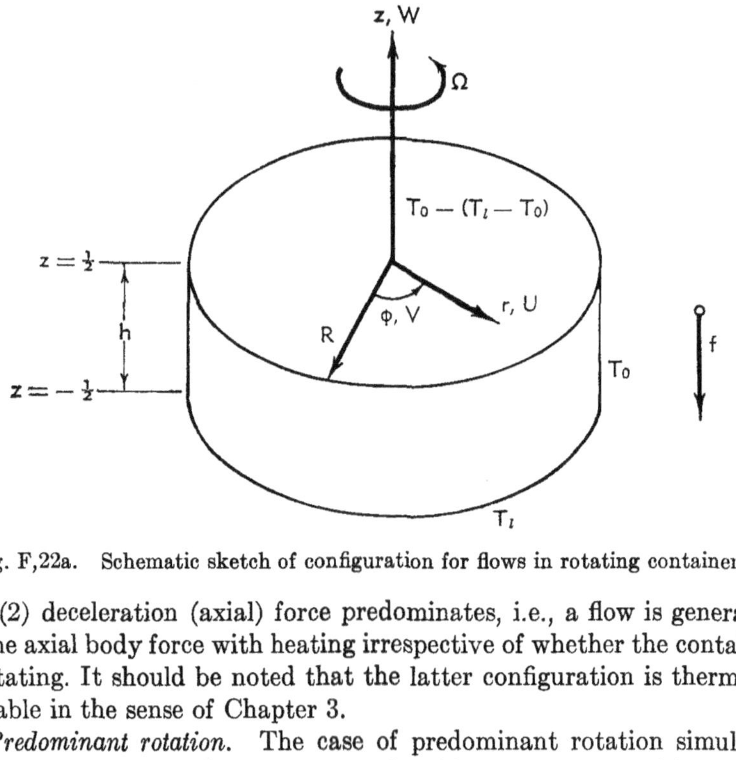

Fig. F,22a. Schematic sketch of configuration for flows in rotating containers.

and (2) deceleration (axial) force predominates, i.e., a flow is generated by the axial body force with heating irrespective of whether the container is rotating. It should be noted that the latter configuration is thermally unstable in the sense of Chapter 3.

Predominant rotation. The case of predominant rotation simulates conditions encountered by an internal fluid in a reentry vehicle, both during the time shortly after reentry when the deceleration force is still small and in the stagnation region before a flow has been established by the instability. Recall that the flow in an unstable configuration starts only after the critical Rayleigh number is exceeded. One might wonder whether, during these periods, rotating the vehicle would start a flow by centrifugal force as in the Schmidt turbine.

The state equation

$$\rho = \rho(T, P)$$

for a fluid may be approximated near a reference condition (0) by

$$\rho = \rho_0[1 - \beta(T - T_0) + \bar{k}(P - P_0)]$$

where β is the volumetric expansion coefficient and $\bar{k}$ is the isothermal compressibility coefficient. It can be shown from the conservation of mass in the closed vessel considered here (Fig. F,22a) that, if the heating (temperature difference) on the bottom and the cooling on the top are identical functions of the radius, and if there are no azimuthal or axial gradients of pressure, the pressure difference (from the reference state) vanishes identically. It is assumed, then, that the fluid in the container is heated and cooled symmetrically as described and that, even though some pressure gradient may exist, it has little effect on the density. (If a net heating of the contained fluid is desired, a small bulging of the container, properly distributed, will compensate the pressure increment.) Then, as for unconfined flows,

$$\rho = \rho_0[1 - \beta(T - T_0)]$$

Axially symmetric flow is assumed in the cylinder of Fig. F,22a. The characteristic velocity of the fluid must be proportional to ΩR (where Ω is the angular velocity and R is the radius), and it should also depend on a representative temperature difference between the heated and unheated states. Let T_0 be the temperature of the unheated state and T_l the temperature at the center of the lower (heated) plate. By the condition of symmetric heating and cooling, the temperature at the center of the upper plate is $T_0 - (T_l - T_0)$. If the temperature and density of the fluid are written as

$$T = T_0 + 2(T_l - T_0)\vartheta$$

and

$$\rho = \rho_0[1 - 2\beta(T_l - T_0)\vartheta]$$

it is apparent that the perturbation parameter for the system is $\epsilon = 2\beta(T_l - T_0)$.

It is now convenient to transform to dimensionless variables in a rotating system as follows:

$$\left.\begin{aligned} \bar{r} = Rr,\ \bar{\varphi} = \varphi + \Omega t,\ \bar{z} = hz \\ U = \epsilon\Omega Ru,\ V = \Omega R(r + \epsilon v),\ W = \epsilon\Omega R\left(\frac{h}{R}\right)w \\ P = \rho_0\Omega^2R^2\,(p_0 + \epsilon p) \end{aligned}\right\} \tag{22-1}$$

where the capital and barred letters denote dimensional quantities and t denotes time. The ϵ dependence in Eq. 22-1 was determined in [*100*] to be the only one consistent with the dynamics of the problem. The governing equations are just those for compressible, viscous, heat-conducting fluids in cylindrical coordinates subject to a body force. Fluid properties μ, k, etc., are taken to be constant. Under the transformation indicated in Eq. 22-1 these become, to zero order in ϵ,

$$\frac{\partial p_0}{\partial r} = r; \qquad \frac{\partial p_0}{\partial z} = -\frac{fh}{(\Omega R)^2}$$

where f is the negative of the z component of the body force per unit mass. These equations imply a solid-body rotation of the fluid when there is no heating ($\epsilon = 0$). To the first order in ϵ, i.e., including the perturbations from solid-body rotation due to heating, the basic equations are

$$\frac{\partial}{\partial r}(ru) + \frac{\partial}{\partial z}(rw) = 0 \tag{22-2}$$

$$\begin{aligned}\frac{Gr_\Omega}{Re_\Omega^2}\left(u\frac{\partial u}{\partial r} + w\frac{\partial u}{\partial z} - \frac{v^2}{r} + 2v\vartheta\right) + r\vartheta - 2v = -\frac{\partial p}{\partial r}\\ + \frac{1}{Re_\Omega}\left[\left(\frac{h}{R}\right)^2\left(\frac{\partial^2 u}{\partial r^2} + \frac{1}{r}\frac{\partial u}{\partial r} - \frac{u}{r^2}\right) + \frac{\partial^2 u}{\partial z^2}\right]\\ + \frac{1}{3}\frac{Gr_\Omega}{Re_\Omega^3}\left(\frac{h}{R}\right)^2\frac{\partial}{\partial r}\left(u\frac{\partial\vartheta}{\partial r} + w\frac{\partial\vartheta}{\partial z}\right)\end{aligned} \tag{22-3}$$

$$\begin{aligned}\frac{Gr_\Omega}{Re_\Omega^2}\left(u\frac{\partial v}{\partial r} + w\frac{\partial v}{\partial z} + \frac{uv}{r} - 2u\vartheta\right) + 2u\\ = \frac{1}{Re_\Omega}\left[\left(\frac{h}{R}\right)^2\left(\frac{\partial^2 v}{\partial r^2} + \frac{1}{r}\frac{\partial v}{\partial r} - \frac{v}{r^2}\right) + \frac{\partial^2 v}{\partial z^2}\right]\end{aligned} \tag{22-4}$$

$$\begin{aligned}\frac{Gr_\Omega}{Re_\Omega^2}\left(u\frac{\partial w}{\partial r} + w\frac{\partial w}{\partial z}\right) = -\left(\frac{R}{h}\right)^2\frac{\partial p}{\partial z} + \frac{f}{\Omega^2 h}\vartheta\\ + \frac{1}{Re_\Omega}\left[\left(\frac{h}{R}\right)^2\left(\frac{\partial^2 w}{\partial r^2} + \frac{1}{r}\frac{\partial w}{\partial r}\right) + \frac{\partial^2 w}{\partial z^2}\right]\\ + \frac{1}{3}\frac{Gr}{Re_\Omega^3}\left(\frac{h}{R}\right)\frac{\partial}{\partial z}\left(u\frac{\partial\vartheta}{\partial r} + w\frac{\partial\vartheta}{\partial z}\right)\end{aligned} \tag{22-5}$$

$$\begin{aligned}\frac{1}{\gamma}\frac{Gr_\Omega}{Re_\Omega^2}\left\{u\frac{\partial\vartheta}{\partial r} + w\frac{\partial\vartheta}{\partial z} - \frac{\gamma K_\Omega}{Re_\Omega}\left[\left(\frac{\partial u}{\partial z}\right)^2 + \mathcal{O}\left(\frac{h^2}{R^2}\right)\right]\right\}\\ = \frac{1}{PrRe_\Omega}\left[\left(\frac{h}{R}\right)^2\left(\frac{\partial^2\vartheta}{\partial r^2} + \frac{1}{r}\frac{\partial\vartheta}{\partial r}\right) + \frac{\partial^2\vartheta}{\partial z^2}\right] - \frac{Gr_\Omega}{Re_\Omega^2}K_\Omega p_0\left(u\frac{\partial\vartheta}{\partial r} + w\frac{\partial\vartheta}{\partial z}\right)\end{aligned} \tag{22-6}$$

where $Gr_\Omega = 2\beta\Omega^2 h^4(T_i - T_0)/\nu^2$, $Re_\Omega = \Omega h^2/\nu$, and $K_\Omega = \beta(\Omega R)^2/c_p$ are the Grashof and Reynolds numbers and the frictional heating parameter, respectively, based on the angular velocity.

The ratio Re_Ω^2/Gr_Ω is a Froude number and represents the ratio of inertia to buoyancy forces. In this particular case, the Froude number is large; in fact, from the definitions of the Reynolds and Grashof numbers based on Ω, the reciprocal of the Froude number is just ϵ. Therefore, the dominant inertia terms are the buoyancy and Coriolis terms, i.e., the last two terms on the left side, respectively, of Eq. 22-3 and the last on the left in Eq. 22-4. In the energy equation, the convection and frictional heating terms are negligible compared to the conduction term, so that the heat transfer is not affected by the motion and is merely due to conduction. All this implies, of course, that large velocities cannot be obtained by applying a temperature gradient transverse to the centrifugal body force in such a rotating configuration. To understand this seemingly unusual result, let us solve the equations that result by omitting the negligible terms and assuming $h/R \ll 1$. The parameter K_Ω will also be assumed small.

The continuity equation, Eq. 22-2, is unchanged, and Eq. 22-3 to 22-6 become

$$\frac{1}{Re_\Omega}\frac{\partial^2 u}{\partial z^2} = \frac{\partial p}{\partial r} + r\vartheta - 2v \tag{22-7}$$

$$\frac{1}{Re_\Omega}\frac{\partial^2 v}{\partial z^2} = 2u \tag{22-8}$$

$$\frac{\partial p}{\partial z} = 0 \tag{22-9}$$

$$\frac{\partial^2 \vartheta}{\partial z^2} = 0 \tag{22-10}$$

The boundary conditions on the upper and lower surfaces ($z = \pm\frac{1}{2}$) are

$$u = v = w = 0$$

At the upper surface, also,

$$\vartheta = -1$$

The thermal boundary conditions on the lower surface will be discussed subsequently.

The velocity and temperature distributions can be written as (see [*101*], one of the meteorological papers)

$$u = \frac{2}{\alpha}J_1(\alpha r)\bar{U}(z) \qquad w = J_0(\alpha r)\bar{W}(z)$$

$$v = \frac{2}{\alpha}J_1(\alpha r)\bar{V}(z) \qquad \vartheta = \frac{2}{\alpha r}J_1(\alpha r)\bar{T}(z)$$

where $J_0(\alpha r)$ and $J_1(\alpha r)$ are zero and first order Bessel functions, respectively, and α is the first root of $J_1(\alpha) = 0$. Qualitatively, the assumed radial variations in u, v, and w lead to the loop flow shown in Fig. F,22b. The viscous boundary conditions at the radial boundary (end) of the cylinder are satisfied by u and v, but for the w component the viscous

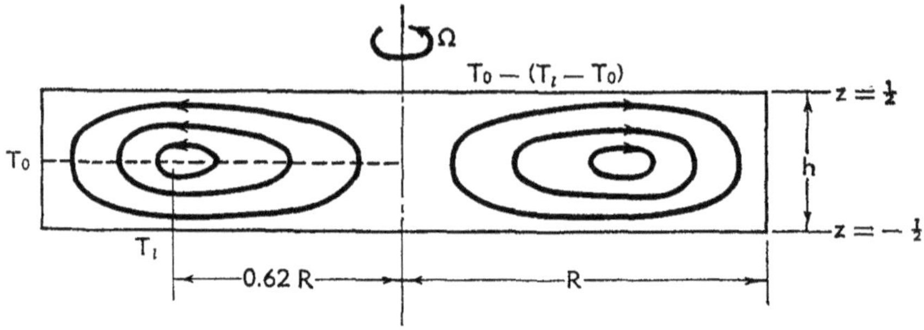

Fig. F,22b. Coolant flow loop in cylindrical shield (Ostrach and Braun [*100*]).

condition is replaced by a slip condition. This is consistent with the omission of terms of order $(h/R)^2$ in the differential equation.

The system of Eq. 22-2 and 22-7 to 22-10 can be combined to yield an ordinary differential equation for $\bar{V}$

$$(Re_\Omega)^2\bar{V}^{v} + 4\bar{V}' = -2$$

where the superscripts denote differentiations with respect to z. The boundary conditions become

$$\bar{V}(\tfrac{1}{2}) = \bar{V}(-\tfrac{1}{2}) = \bar{V}'(\tfrac{1}{2}) = \bar{V}'(-\tfrac{1}{2}) = \bar{V}''(\tfrac{1}{2}) = \bar{V}''(-\tfrac{1}{2}) = 0$$

Solution of this boundary-value problem yields

$$u = \frac{2}{\alpha} J_1(\alpha r)(C_1 \sinh \sqrt{Re_\Omega}\, z \cos \sqrt{Re_\Omega}\, z - C_2 \cosh \sqrt{Re_\Omega}\, z \sin \sqrt{Re_\Omega}\, z) \tag{22-11}$$

$$v = \frac{2}{\alpha} J_1(\alpha r)\left(C_1 \cosh \sqrt{Re_\Omega}\, z \sin \sqrt{Re_\Omega}\, z + C_2 \sinh \sqrt{Re_\Omega}\, z \cos \sqrt{Re_\Omega}\, z - \frac{z}{2}\right) \tag{22-12}$$

$$\begin{aligned} w = \frac{J_0(\alpha r)}{\sqrt{Re_\Omega}} \Bigg[& C_1\left(\sinh \frac{\sqrt{Re_\Omega}}{2} \sin \frac{\sqrt{Re_\Omega}}{2} + \cosh \frac{\sqrt{Re_\Omega}}{2} \cos \frac{\sqrt{Re_\Omega}}{2}\right) \\ & + C_2\left(\cosh \frac{\sqrt{Re_\Omega}}{2} \cos \frac{\sqrt{Re_\Omega}}{2} - \sinh \frac{\sqrt{Re_\Omega}}{2} \sin \frac{\sqrt{Re_\Omega}}{2}\right) \\ & - C_1(\sinh \sqrt{Re_\Omega}\, z \sin \sqrt{Re_\Omega}\, z + \cosh \sqrt{Re_\Omega}\, z \cos \sqrt{Re_\Omega}\, z) \\ & - C_2(\cosh \sqrt{Re_\Omega}\, z \cos \sqrt{Re_\Omega}\, z - \sinh \sqrt{Re_\Omega}\, z \sin \sqrt{Re_\Omega}\, z)\Bigg] \end{aligned} \tag{22-13}$$

$$\vartheta = \frac{2}{\alpha r} J_1(\alpha r)(-z) \tag{22-14}$$

$$p = 0$$

where

$$C_1 = \frac{\cosh \frac{\sqrt{Re_\Omega}}{2} \sin \frac{\sqrt{Re_\Omega}}{2}}{4\left(\sinh^2 \frac{\sqrt{Re_\Omega}}{2} + \sin^2 \frac{\sqrt{Re_\Omega}}{2}\right)}$$

$$C_2 = \frac{\sinh \frac{\sqrt{Re_\Omega}}{2} \cos \frac{\sqrt{Re_\Omega}}{2}}{4\left(\sinh^2 \frac{\sqrt{Re_\Omega}}{2} + \sin^2 \frac{\sqrt{Re_\Omega}}{2}\right)}$$

The dimensionless temperatures of the plates $\vartheta(r, \pm\frac{1}{2})$ are, from Eq. 22-14, equal to $\mp(1/\alpha r)J_1(\alpha r)$. Any other temperature distribution could be specified at the lower surface by expanding $\alpha r\vartheta(r, -\frac{1}{2})$ in a series in J_1.

The velocity distributions given by Eq. 22-11 to 22-13 are shown in Fig. F,22c. A strong Reynolds number effect is evident. For low Reynolds numbers, the viscous effects are present over the entire cross section of the cylinder, but at high Reynolds numbers a boundary layer appears in all three velocity component profiles. Accompanying the boundary layer is a diminution of the radial and axial components of velocity in the main body of the fluid while the azimuthal velocity distribution is nearly linear, lagging the rotation of the cylinder in the upper half and leading in the lower half.

Since the equations of motion have been solved, it is possible to explain why the velocities remain small. The driving force $r\vartheta$ in Eq. 22-7 is opposed by the pressure gradient, the Coriolis force $(-2v)$, and the viscous force. The pressure gradient balances that part of the driving force which is independent of z. At low Reynolds numbers the remainder of the driving force is opposed mainly by the viscous force, but at large Reynolds numbers the Coriolis force grows relatively larger and inhibits the flow. Consequently, the heat convection is negligible both for large and small Reynolds numbers, and the fluid essentially rotates as a rigid body.

A graphical illustration of the Coriolis force near the top of the cylinder is presented in Fig. F,22d. The Coriolis force has negative components in both the radial and angular directions. It opposes the azimuthal rotation of the fluid due to the viscous force and the outward radial flow set up by the buoyancy force.

It may be that flows of a cellular type could be obtained in this configuration if h/R is not negligibly small and the condition of axial symmetry is relaxed. However, it is doubtful even then that the heat transfer would be affected by the fluid motion. Thus, rotating a fluid container like that just considered does not significantly affect the heat transfer when the deceleration force is inoperative. In fact, the rotation

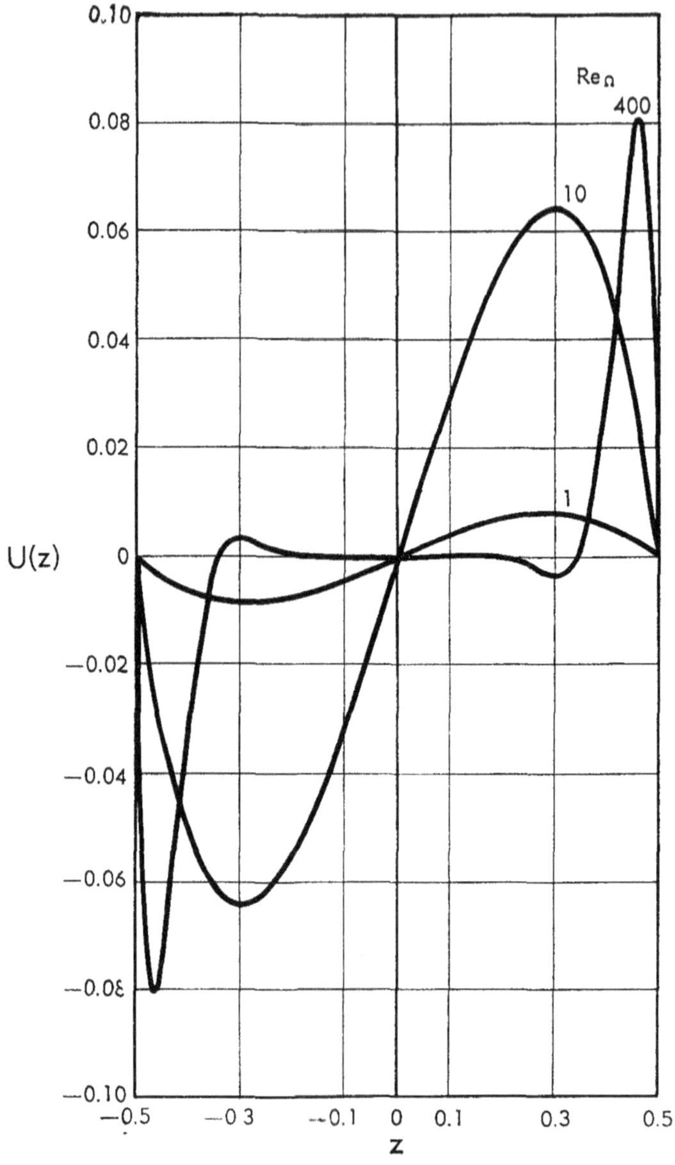

Fig. F,22c. Velocity profiles in rotating cylinder (Ostrach and Braun [*100*]). Radial components.

may even be detrimental, for Chandrasekhar [*102*] has shown that in a thermally unstable configuration the fluid motion is delayed by the action of the Coriolis forces.

Since the Coriolis force due to the circumferential velocity component opposes the radial buoyancy force and thus negates convection effects, and may also impede the unstable motion, it appears that significant convection could be obtained if the Coriolis forces were made negligible compared to the centrifugal force. The insertion of radial vanes appropriately spaced would accomplish this by reducing the azimuthal velocity component (and, hence, the corresponding Coriolis term). The buoyancy effects due to centrifugal force would, in this way, be unopposed (excepting viscous forces) and could lead to large heat convection as in the turbine blades investigated by Schmidt [*79*].

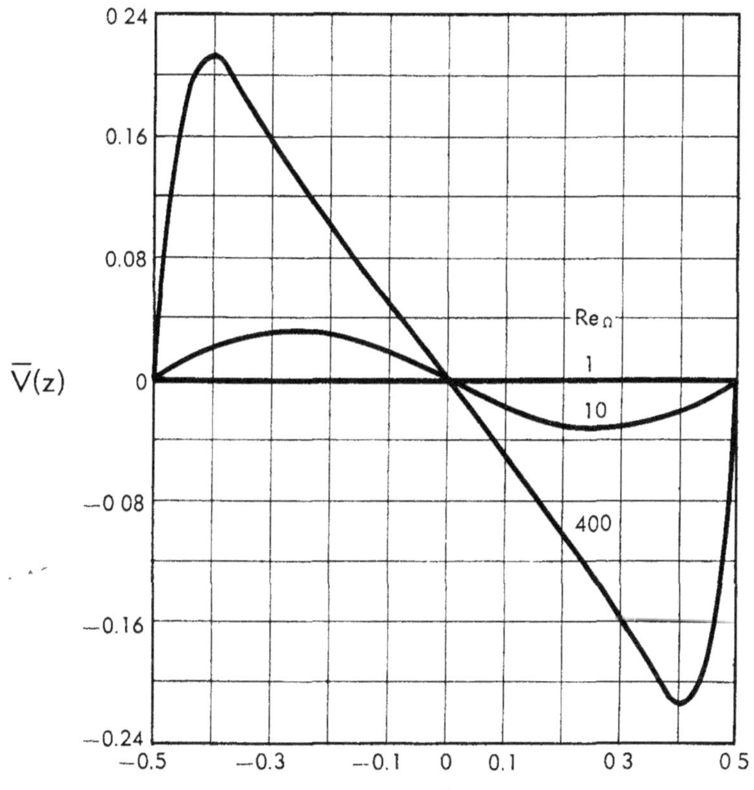

Fig. F,22c (continued). Azimuthal components.

In [*100*] the flow in a sector of the cylinder bounded by two radial vanes is analyzed to substantiate the above point and to determine the governing parameters. The resulting equations look like those for natural convection over a heated vertical plate where the inertia and heat convection terms are important. However, the driving force in this case is the centrifugal force which can be many times larger than the earth's gravitational force. Hence, the convective heat transfer in such a rotating configuration can be much larger than that due to gravitational effects, as was demonstrated by the Schmidt turbine.

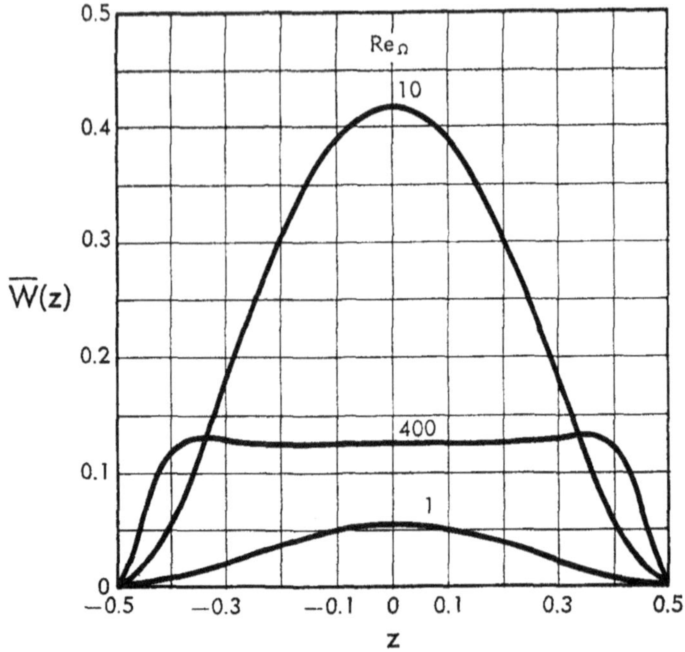

Fig. F,22c (concluded). Axial components.

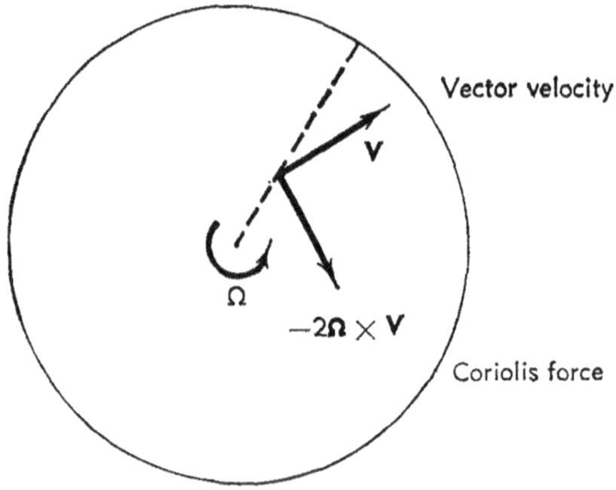

Fig. F,22d. Coriolis force in upper half of rotating cylinder (Ostrach and Braun [*100*]).

The equations show that the Coriolis and viscous forces in a rotating system will always have an adverse effect. All this means, however, is that convective heat transfer produced by centrifugal forces in a rotating sector can never be as large as that over a flat plate in a gravitational field of *corresponding magnitude*. It is shown in [*100*] that an optimum geometrical arrangement (for proper balance between Coriolis and viscous forces) would probably be one in which

$$\frac{h}{R} \leq \Delta\varphi \leqq \sqrt{\epsilon}$$

where $\Delta\varphi$ is the angle formed by the radial vanes.

Predominant deceleration force. The effects of rotation on a flow generated by an axial force (like the deceleration force in a reentry vehicle) must still be evaluated, because a reentry vehicle may be rotating about its axis for aerodynamic stability reasons. A deceleration force alone will establish flows in a flat cylindrical container (Fig. F,22a) after the critical Rayleigh number is surpassed. Also, if the retardation force is transverse to a temperature gradient, as it would be, for example, in a curved container like a spherical shell, it would immediately generate a conventional natural convection flow. Let us now, therefore, analyze the case where the flow is essentially due to the (constant) retardation force.

For inertia and buoyancy forces to be of an equal order of magnitude, the characteristic velocity must be $\sqrt{\epsilon fh}$. Accordingly, we make the following transformation from the inertial coordinates to dimensionless rotating coordinates:

$$U = \sqrt{\epsilon fh}\, u; \qquad V = \sqrt{\epsilon fh}\, v + \Omega R r; \qquad W = \frac{h}{R}\sqrt{\epsilon fh}\, w$$

$$\bar{r} = Rr; \qquad \bar{\varphi} = \varphi + \Omega t; \qquad \bar{z} = hz$$

$$P = \rho_0 fh(p_0 + \epsilon p); \qquad \rho = \rho_0(1 - \epsilon\vartheta); \qquad \vartheta = \frac{(T - T_0)}{2(T_l - T_0)}$$

In general, u, v, and w are functions of the angular velocity as well as the coordinates. Note that, if Ω is zero, the flow is due entirely to the action of the axial body force and the heating. On the other hand, for no heating the flow is again a solid-body rotation. This is clear from the equations since, for $\epsilon = 0$, the basic equations reduce to

$$\frac{\partial p_0}{\partial r} = \frac{(\Omega R)^2}{fh} r, \qquad \frac{\partial p}{\partial \varphi} = 0, \qquad \frac{\partial p_0}{\partial z} = -1$$

When $\epsilon \neq 0$, and again small compressibility and dissipation terms are omitted, the equations, to lowest order in ϵ, become

$$\frac{\partial}{\partial r}(ru) + \frac{\partial v}{\partial \varphi} + \frac{\partial}{\partial z}(rw) = 0 \tag{22-15}$$

$$u\frac{\partial u}{\partial r} + \frac{v}{r}\frac{\partial u}{\partial \varphi} + w\frac{\partial u}{\partial z} - \frac{v^2}{r} - 2\omega v = -\frac{\partial p}{\partial r} + \frac{1}{\sqrt{Gr}}\frac{R}{h}\left[\frac{\partial^2 u}{\partial z^2} + \left(\frac{h}{R}\right)^2\left(\frac{\partial^2 u}{\partial r^2} + \frac{1}{r}\frac{\partial u}{dr} + \frac{1}{r^2}\frac{\partial^2 u}{\partial \varphi^2} - \frac{u}{r^2} - \frac{2}{r^2}\frac{\partial v}{\partial \varphi}\right)\right] \tag{22-16}$$

$$u\frac{\partial v}{\partial r} + \frac{v}{r}\frac{\partial v}{\partial \varphi} + w\frac{\partial v}{\partial z} + 2\omega u + \frac{uv}{r} = -\frac{1}{r}\frac{\partial p}{\partial \varphi}$$

$$+ \frac{1}{\sqrt{Gr}}\frac{R}{h}\left[\frac{\partial^2 v}{\partial z^2} + \left(\frac{h}{R}\right)^2\left(\frac{\partial^2 v}{\partial r^2} + \frac{1}{r}\frac{\partial v}{\partial r} + \frac{1}{r^2}\frac{\partial^2 v}{\partial \varphi^2} + \frac{2}{r^2}\frac{\partial u}{\partial \varphi} - \frac{v}{r^2}\right)\right] \quad (22\text{-}17)$$

$$\left(\frac{h}{R}\right)^2\left(u\frac{\partial w}{\partial r} + \frac{v}{r}\frac{\partial w}{\partial \varphi} + w\frac{\partial w}{\partial z}\right) = -\frac{\partial p}{\partial z} + \vartheta$$

$$+ \left(\frac{h}{R}\right)^2\left(\frac{1}{\sqrt{Gr}}\frac{R}{h}\right)\left[\frac{\partial^2 w}{\partial z^2} + \left(\frac{h}{R}\right)^2\left(\frac{\partial^2 w}{\partial r^2} + \frac{1}{r}\frac{\partial w}{\partial r} + \frac{1}{r^2}\frac{\partial^2 w}{\partial \varphi^2}\right)\right] \quad (22\text{-}18)$$

$$u\frac{\partial \vartheta}{\partial r} + \frac{v}{r}\frac{\partial \vartheta}{\partial \varphi} + w\frac{\partial \vartheta}{\partial z}$$

$$= \frac{\gamma}{Pr}\frac{1}{\sqrt{Gr}}\frac{R}{h}\left[\frac{\partial^2 \vartheta}{\partial z^2} + \left(\frac{h}{R}\right)^2\left(\frac{\partial^2 \vartheta}{\partial r^2} + \frac{1}{r}\frac{\partial \vartheta}{\partial r} + \frac{1}{r^2}\frac{\partial \vartheta}{\partial \varphi^2}\right)\right] \quad (22\text{-}19)$$

In these equations

$$\omega \equiv \frac{\Omega R}{\sqrt{\epsilon f h}}, \qquad \sqrt{Gr} \equiv \frac{\sqrt{\epsilon f h}\, h}{\nu}$$

The parameter ω is proportional to the Froude number. In the case being considered here, ω assumes small values by definition because the numerator depends on the rotation and the denominator on the (predominant) deceleration force. (In the case of predominant rotation, the inertia and buoyancy effects both depended only on the rotation, with the consequence that the Froude number was independent of rotation.) Therefore, with $\omega \lesssim 1$, the inertia terms in Eq. 22-16 and 22-17 having ω as a coefficient are, at most, of the same order as the other like terms. The equations look like the conventional boundary layer equations, and the inertia and convection effects are important. The heat transfer will be greater than that due to conduction alone.

An examination of the foregoing equations shows that a steady cellular flow, as occurs by heating a horizontal surface from below [*64*] is affected by a small rotation to a different order of magnitude than an axially symmetric flow. Consider first the cellular flow. With $\Omega = 0$, the three velocity components of each cell are of the same order of magnitude. The first momentum equation (Eq. 22-16) shows that the rotation will affect this flow to the first order in ω. The preceding analysis of the flat cylinder with rotation only, and the analysis of the unstable static configuration [*64*], suggest that the effect will be adverse. Thus it is expected that, in cellular flow generated by an axial body force, rotation will reduce the heat transfer.

Second, consider a steady axially symmetric flow with zero azimuthal velocity component. When a small rotation is imposed, the azimuthal velocity is properly represented by

$$v = \omega v_1$$

Since the property of axial symmetry is retained ($\partial/\partial\varphi \equiv 0$), it is evident that the equations are affected (through the Coriolis and centrifugal force terms) only to order ω^2. Consequently, the rotation will have a negligible effect on this type of natural convection flow.

To summarize the results of [*100*], we can then say that the Froude number is the parameter that determines the relative effect of body rotation on internal natural convection flows. In flows with large Froude numbers (e.g., motions induced solely by rotation of a heated cylinder), the action of the Coriolis force prevents appreciable convection of heat. Although the Coriolis force inhibits convective heat transfer somewhat, even for Froude numbers of unit order of magnitude (rotating sector), considerable convection is obtained because the driving force can be greatly increased. Flows that have been set up by a strong external body force have a small Froude number and are not greatly affected by rotation of the container, especially if they have axial symmetry.

F,23. Completely Enclosed Flows. With the exception of the last article (Art. 22), none of the other configurations discussed in this chapter on natural convection flows were completely enclosed regions. The study of natural convection flows that are completely bounded by surfaces leads to a number of difficult problems, particularly if the Grashof number is large. Some of the more interesting aspects of these problems will be discussed below.

For large Grashof numbers, the flow will obviously be of a boundary layer type, but in a completely enclosed region the flow in the interior, i.e., away from the boundaries, is not readily determined from external boundary conditions, but depends on the boundary layer flow.

The "interior" and boundary layer flows are thus essentially coupled. Therefore, unless the problem is so idealized that solutions for all values of the governing parameter are easily obtained, the interior and the boundary layer solutions must be properly matched. The case of predominant rotation in Art. 22 represents a situation in which solutions for all values of the parameter (Re_Ω in that case) were readily obtained; for large Re_Ω the flow is of a boundary layer type with the radial and axial velocities reduced and the aximuthal velocity linear in the interior (see Fig. F,22c). The first attempt at treating a more general configuration appears to be the present writer's treatment of the natural convection in a cylindrical container (Doctoral Thesis, Brown University, 1950, and described by Carrier [*103*]). The interior flow in that problem was found to be a solid-body rotation at constant temperature. Pillow [*67*], as mentioned previously, considered the flow in a rectangular cell bounded by two rigid and two free surfaces in a similar manner. He found that in the interior flow the vorticity and temperature were constant, and he treated the boundary layer flow by an approximate method. A somewhat related problem is the ocean-current problem described by Carrier [*103*] wherein flow in a triangular region is considered. The flow in that problem is driven by wind forces which can be assumed known.

Until more is learned about how to handle problems of this sort, it would appear that a reasonable (and conservative) approximation of a flow generated by a body force and its associated heat transfer can be obtained by assuming that, in the interior of the region, the fluid has no motion and is at some constant temperature. In this way the internal problem is reduced to an external one. Thus, for example, the flow and heat transfer *inside* containers of the shapes given in Art. 8 can be immediately estimated.

F,24. Natural Convection at the Critical State. No discussion of the natural convection phenomenon would be complete without mention of an extremely interesting, but nevertheless, little explored area, namely, natural convection near the critical state of a liquid [*104*].

Normally, we define the state of a material as solid, liquid, or gas, and there is a clear conception about the vapor state that is passed in the transition between the liquid and gaseous state. If a liquid, say water, at a given pressure p and a specific volume v is heated at constant pressure, it will at a certain temperature (boiling temperature) begin to assume a state with the same temperature but increased specific volume. The temperature will remain constant, despite the addition of heat, until the water has been turned completely into steam. If still more heat is added to the steam, it becomes superheated, or speaking more generally, the vapor approaches the gaseous state more and more closely.

With increasing pressure, the specific volumes of water and steam at the respective boiling temperatures will come closer and closer together, until at a certain pressure, called the critical pressure, there will be a direct transition from water to steam with the intermediate range deleted completely. This point at which the specific volumes of the liquid and its vapor become equal, or where there is a direct transition from the liquid to the vapor state, is called the critical state. For example, the critical state of pure water is defined by the critical values $p_{cr} = 218.3$ atm, $T_{cr} = 374.2°C$, and $v_{cr} = 3.04 \times 10^{-3}$ m³/kg.

It is not quite consistent, in the case of water, to refer to the liquid in the critical state as "water," because its volume at that point has increased to roughly three times the normal volume, and in fact, no distinction can any longer be made between water and steam or, more generally, between the liquid and vapor states at or above the critical state.

From a T, v or p, v diagram (see [*104*]) we find that, at the critical state,

$$\left(\frac{\partial p}{\partial v}\right)_T = 0; \quad \left(\frac{\partial^2 p}{\partial v^2}\right)_T = 0 \quad \text{and} \quad \left(\frac{\partial T}{\partial v}\right)_p = 0; \quad \left(\frac{\partial^2 T}{\partial v^2}\right)_p = 0$$

The specific heat at constant pressure c_p and the volumetric expansion coefficient $\beta = (1/v)(\partial v/\partial T)_p$ both become infinite at the critical state. Furthermore, the viscosity becomes a minimum at this point.

In order to see how these peculiar properties of a fluid at or near the critical state affect the natural convection phenomenon, we recall that the flow and heat transfer are functions of the Rayleigh number,

$$Ra = \frac{c_p}{\mu k} (\beta f \rho^2 l^3 \theta)$$

Thus, near the critical state, the fluid acquires a large buoyancy due to the large increase in the volumetric expansion coefficient, whereas the inhibiting viscous force is minimized. At the same time, each unit of mass carries a comparatively large amount of heat because of the high specific heat. Thus a liquid at or near its critical state is ideally suited for heat transfer by natural convection.

To verify the above considerations experimentally, Schmidt, Eckert, and Grigull [*105*] introduced ammonia at its critical state into a closed oval-shaped ring of tubes. One section of the ring was then heated and another cooled to generate the natural convection flow. Measurements indicated that, near critical conditions, the heat transferred was approximately 100 times that which would be transferred by water at 60°C.

Schmidt [*79*] describes another experiment in which ammonia at the critical state circulates inside a hollow steel rod by natural convection. The apparent heat conductivity of this apparatus was measured; the results are shown in Fig. F,24. At critical conditions, this rod behaved like a metal having 20 times the heat conductivity of copper.

All of the above considerations are further enhanced if the body force is larger than the gravitational. Thus, application of this idea to rotating machinery is evident. Schmidt [*79*] in his water-cooled turbine arranged to have the critical conditions reached at the blade tip (the success of this cooling method has already been discussed).

A more recent investigation was conducted by Doughty and Drake [*106*] in which the free convection heat transfer from a horizontal right circular cylinder to Freon 12 was measured near the critical state. In the experiment, a horizontal platinum cylinder was supported at the midpoint of a vertical steel vessel. This test chanber was constructed in such a way as to permit charging with a test fluid. The charge, pressure, and temperature were then varied. The heat transfer coefficient was determined from an energy balance on the cylinder and a measurement of the temperature difference between the cylinder surface and the test fluid. The cylinder was also used as a resistance thermometer to determine both the fluid temperature and the cylinder surface temperature. Results indicate that the heat transfer coefficients increased by a ten-fold factor in the region of the critical state.

To conclude, it seems well established that very high heat transfer rates can be obtained by the free or natural convection of a fluid at a critical or slightly supercritical condition. However, there is much fundamental knowledge that must be obtained before more practical applications can be made of this phenomenon. Knowledge of the critical state itself is not very extensive; it also still remains to be determined over what range near the critical point the high heat transfer rates will be obtained. Although most organic and inorganic substances have critical pressures and temperatures within the strength limits covered by normal

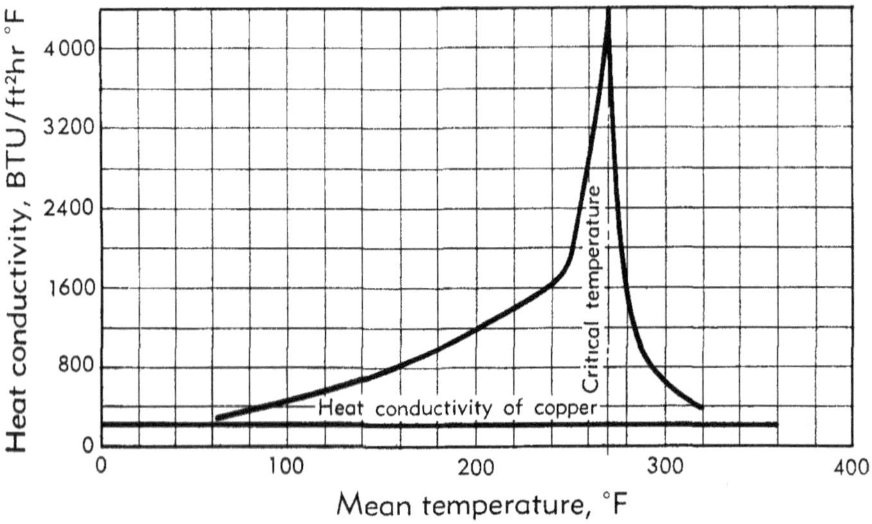

Fig. F,24. Apparent thermal conductivity of ammonia as measured by Schmidt [79].

or good steel, critical heat transfer rates have been determined for only a few.

CHAPTER 5. UNSTEADY FLOWS WITH BODY FORCES

Although transient flows driven by body forces are of considerable practical importance, relatively little attention has been given to problems of this type. The complications encountered in even the simple steady flow problems have served to discourage consideration of unsteady natural convection flows. What little work that has been done on transient flows will now be discussed in the pattern of the previous articles of this section, viz., the external problems will be discussed first and then the internal problems will be outlined. At the conclusion of this chapter, the body force effects on unsteady forced flows will be discussed.

F,25. Free Convection. The one-dimensional free convection about an infinite flat plate, whose temperature at time $t = 0$ is changed to a value different from that of the surrounding ambient fluid, was treated by Illingworth [*107*, Sec. 6]. Applying the Mises transformation and same notation as discussed in Art. 13, Illingworth derived the governing ordinary differential equation for flows with variable property values. Then, for the special case $c_p = \text{const}$ and $\theta = \rho^*\mu^* = 1$, where the asterisks denote quantities nondimensionalized with respect to ambient conditions, he presented the solution $u = U\{g \sin \alpha\, [h^*(0) - 1]t\}^{-1} = J(\eta)$ where, for $Pr \neq 1$

$$J = (1 + 2\eta^2)K(\eta) - K(\infty)\left[(1 + 2\eta^2) \operatorname{erf} \eta + \frac{2}{\sqrt{\pi}} \eta e^{-\eta^2}\right] + 1 - \operatorname{erf}\left(\frac{\eta}{\sqrt{Pr}}\right)$$

with

$$K = 2\sqrt{\frac{Pr}{\pi}}\left(\int_0^\eta \frac{e^{-Pr\xi^2}}{1 + 2\xi^2}\, d\xi - \frac{2}{1 - Pr}\int_0^\eta \frac{e^{-Pr\xi^2}}{(1 + 2\xi^2)^2}\, d\xi\right) - 1$$

and, for $Pr = 1$,

$$J = \frac{2}{\sqrt{\pi}} \eta e^{-\eta^2} - 2\eta^2(1 - \operatorname{erf} \eta)$$

The enthalpy distribution is

$$h^*(\eta) = h^*(0) + [1 - h^*(0)] \operatorname{erf} (\eta \sqrt{\sigma})$$

where h^* denotes the dimensionless enthalpy (temperature for constant c_p), t is time, α is the inclination of the plate to horizontal, g the acceleration of gravity, and $\eta = \psi/[2(\theta_\infty t)^{\frac{1}{2}}]$. ψ is the stream function and $\theta_\infty = \rho_\infty\mu_\infty$.

Sugawara and Michiyoshi [*108*] treated a similar step-function change in wall temperature for a semi-infinite plate by a method of successive approximations. For the first approximation, the heat transfer was considered to be due to conduction alone. The velocity and temperature distributions obtained by this premise were then substituted back into the differential equations to obtain a second approximation. The corrections resulting from the second approximation are given only for very short times, and it is not known if the second approximation will converge sufficiently well at large times to the steady state solution.

The method of successive approximations in [*108*] is similar to that used by Blasius (see [*19*], p. 181) for determining the boundary layer growth on a body started impulsively from rest. Schuh [*109*] investigated the same kind of problem as Blasius but he used a momentum-integral (Kármán-Pohlhausen) method and obtained solutions by the method of characteristics. Siegel [*110*] used Schuh's approach to study transient free convection on a vertical flat plate of infinite width and semi-infinite length. Two cases are considered for a plate initially at ambient temperature in quiescent fluid: (a) The plate is suddenly raised to a higher uniform temperature, and (b) the plate suddenly begins to generate a uniform heat flux at its surface. We shall here outline Siegel's analysis for case (a) and discuss his results for both cases.

The method proceeds as described in Art. 9, except that the time-dependent terms must now be retained. Integration of the momentum and energy equations thus yields equations identical in form with Eq. 9-31 and 9-32 with the addition of the term $\partial/\partial t \int_0^\delta U dY$ on the left side of Eq. 9-31 and $\partial/\partial t \int_0^\delta (T - T_\infty) dY$ on the left side of Eq. 9-32.

To solve these equations, approximate time-dependent velocity and temperature distributions like Eq. 9-33 and 9-35 are assumed to be valid throughout the entire transient flow development. When these profiles, $U/U_1 = (Y/\delta)[1 - (Y/\delta)]^2$ and $(T - T_\infty)/(T_w - T_\infty) = [1 - (Y/\delta)]^2$, are inserted into the integrated momentum and energy equations, the results are (in dimensionless form)

$$\frac{1}{12Pr}\frac{\partial}{\partial t^*}(u_1\delta^*) + \frac{1}{105Pr}\frac{\partial}{\partial x}(\delta^* u_1^2) = \frac{1}{3} Ra_L \delta^* - \frac{u_1}{\delta^*} \tag{25-1}$$

$$\frac{\partial \delta^*}{\partial t^*} + \frac{1}{10}\frac{\partial}{\partial x}(u_1\delta^*) = \frac{6}{\delta^*} \tag{25-2}$$

where the velocity is nondimensionalized by the ratio of the characteristic length L to the thermal diffusivity α, the dimensionless boundary layer thickness δ^* is defined as the ratio of the actual thickness to the characteristic length, time is nondimensionalized by α/L^2, x is the dimensionless coordinate along the plate, and Ra_L is the Rayleigh number based on L. Eq. 25-1 and 25-2 are to be solved simultaneously for u_1 and δ^* in terms of x and t^*. Siegel shows that Eq. 25-1 and 25-2 are hyperbolic, and hence a solution can be found by the method of characteristics as outlined in [*111*, Sec. 22]. The slopes of the characteristic lines are found to be proportional to the velocity u_1. Letting I_+ and I_- designate the two families of characteristics, we have

$$\mathrm{I}_+ \quad \frac{dx}{dt^*} = 0.155 u_1 \tag{25-3}$$

$$\mathrm{I}_- \quad \frac{dx}{dt^*} = 0.0739 u_1 \tag{25-4}$$

The characteristic equations that apply along the characteristic lines are designated by II_+ and II_-:

$$\text{II}_+ \quad \delta^* \frac{du_1}{dt^*} + 0.261 u_1 \frac{d\delta^*}{dt^*} + (4.43 + 12Pr)\frac{u_1}{\delta^*} - 4Ra_L Pr \delta^* = 0 \tag{25-5}$$

$$\text{II}_- \quad \delta^* \frac{du_1}{dt^*} - 0.547 u_1 \frac{d\delta^*}{dt^*} + (9.28 + 12Pr)\frac{u_1}{\delta^*} - 4Ra_L Pr \delta^* = 0 \tag{25-6}$$

Solutions of characteristic equations. When a vertical plate is suddenly raised to a uniform temperature, the heat transfer to the surrounding fluid is initially by pure conduction and, hence, is equivalent to the heat conducted into a semi-infinite solid when its surface temperature is suddenly increased. This situation arises because the fluid sufficiently far from the leading edge behaves as if the plate were infinite in length, so that the velocity distribution in this region is independent of x and the convective heat transfer is zero. The two-dimensional influence that causes boundary layer growth to vary with x gradually propagates from the leading edge and begins to alter the one-dimensional flow configuration at a different time for each position along the plate. Thus, for each x, it would be expected that, for sufficiently small times, the solution would be dependent on time only. On this basis a trial solution of the form

$$\delta^* = C_1(t^*)^m \tag{25-7}$$

$$u_1 = C_2(t^*)^n \tag{25-8}$$

is substituted into Eq. 25-5 and 25-6. For the resulting equations to be independent of t^*, it is found that $m = \frac{1}{2}$ and $n = 1$. The equations are then solved for C_1 and C_2, yielding

$$\delta^* = \sqrt{12}\,(t^*)^{\frac{1}{2}} \quad \text{or} \quad \delta = \sqrt{12}\,(\alpha t^*)^{\frac{1}{2}} \tag{25-9}$$

$$u_1 = \frac{4Ra_L Pr}{1.5 + Pr} t^* \quad \text{or} \quad U_1 = \frac{4g\beta\theta_w}{1.5 + Pr} t^* \tag{25-10}$$

where θ_w is the difference between the wall and ambient temperatures. These results can also be obtained directly from the original equations, Eq. 25-1 and 25-2, if the derivatives with respect to x are set equal to zero. The characteristic solution is necessary, however, to determine the domain of the t^*, x plane in which the solution applies.

At sufficiently large times, the characteristic equations should also yield a steady state solution in which u_1 and δ^* are functions of x alone; To obtain this result Eq. 25-5 and 25-6 are first rewritten in the form

$$\text{II}_+ \quad \delta^* \frac{du_1}{dx}\frac{dx}{dt^*} + 0.261 u_1 \frac{d\delta^*}{dx}\frac{dx}{dt^*} + (4.43 + 12Pr)\frac{u_1}{\delta^*} - 4Ra_L Pr \delta^* = 0 \tag{25-11}$$

$$\text{II}_- \quad \delta^* \frac{du_1}{dx}\frac{dx}{dt^*} - 0.547 u_1 \frac{d\delta^*}{dx}\frac{dx}{dt^*} + (9.28 + 12Pr)\frac{u_1}{\delta^*} - 4Ra_L Pr \delta^* = 0 \tag{25-12}$$

The dx/dt^* is eliminated from Eq. 25-11 and 25-12 by means of Eq. 25-3 and 25-4, respectively. A solution of the form

$$\delta^* = C_3 x^r \tag{25-13}$$

$$u_1 = C_4 x^s \tag{25-14}$$

is substituted into Eq. 25-11 and 25-12, from which it is found that $r = \frac{1}{4}$ and $s = \frac{1}{2}$ so that

$$\delta^* = 3.93(0.952 + Pr)^{\frac{1}{4}}(Ra_L Pr)^{-\frac{1}{4}} x^{\frac{1}{4}} \tag{25-15}$$

$$u_1 = 5.17(0.952 + Pr)^{-\frac{1}{2}}(Ra_L Pr)^{\frac{1}{2}} x^{\frac{1}{2}} \tag{25-16}$$

These are in agreement with the steady state solution for free convection on a vertical flat plate at uniform temperature as given in [*11*, p. 161].

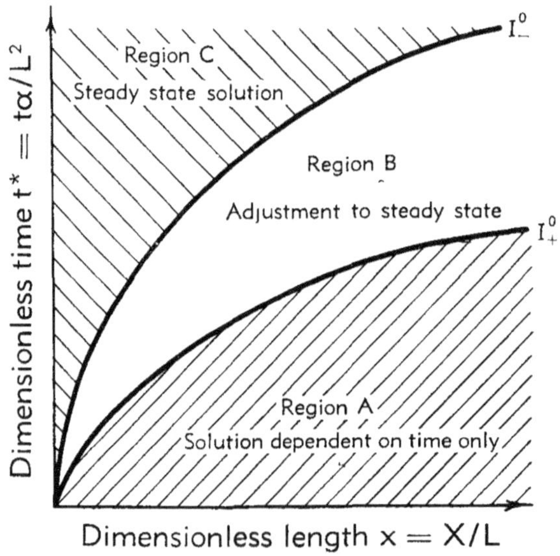

Fig. F,25. Regions on the t^*, x plane in which particular solutions apply (Siegel [*110*]).

The characteristics will now be used to determine the domain of the t^*, x plane in which these solutions are valid. Consider the two characteristic lines which pass through the origin of the t^*, x plane as shown in Fig. F,25. From Eq. 25-3 and 25-4, we note that the I_+ lines in the t^*, x plane have smaller slopes than the I_- lines. Hence, the two lines passing through the origin (designated by I_+^0 and I_-^0) divide the plane into three regions.

In region A, along and below the I_+^0 line, values of u_1 and δ^* can be obtained at the intersections of sets of characteristics which originate on the x axis. The solutions of the characteristic equations must satisfy the boundary conditions that u_1 and δ^* are zero for all x's along the x axis ($t^* = 0$). These requirements are satisfied by the solution given in Eq. 25-9 and 25-10 which is dependent only on time. Hence, in region A, the heat transfer is due solely to one-dimensional conduction, and the I_+^0 line gives the time, corresponding to each position along the plate, at which the one-dimensional process ends and the effect of the leading edge begins to influence the velocity and temperature distributions.

The equation of the I_+^0 line can be found by integrating Eq. 25-3 with u_1, obtained from Eq. 25-10, and the condition that $x = 0$ at $t^* = 0$. After rearrangement, this yields

$$t^* = 1.80(1.5 + Pr)^{\frac{1}{2}}(Ra_L Pr)^{-\frac{1}{2}}x^{\frac{1}{2}}$$

or

$$t = 1.80(1.5 + Pr)^{\frac{1}{2}}(\beta g \theta_w)^{-\frac{1}{2}}x^{\frac{1}{2}} \tag{25-17}$$

Eq. 25-17 then gives the time at which purely one-dimensional heat conduction is terminated for each position along the plate.

In region C along and above the I_-^0 line, values of u_1 and δ^* are found at intersections of characteristic lines originating along the t^* axis. The solutions of the characteristic equations must satisfy the boundary conditions that, for any t^*, at $x = 0$ (the plate leading edge), the values of u_1 and δ^* must vanish. These conditions are satisfied by the steady state solution, and thus the equation of the I_-^0 line gives the time at which steady state is reached as a function of position along the plate.

The equation of I_-^0 is found by integrating Eq. 25-4 with u_1, given by Eq. 25-16, subject to the condition $x = 0$ at $t^* = 0$. This yields

$$t^* = 5.24(0.952 + Pr)^{\frac{1}{2}}(Ra_L Pr)^{-\frac{1}{2}}x^{\frac{1}{2}}$$

or

$$t = 5.24(0.952 + Pr)^{\frac{1}{2}}(\beta g \theta_w)^{-\frac{1}{2}}X^{\frac{1}{2}} \tag{25-18}$$

which is the time required to reach the steady state.

As a numerical example, we shall calculate the time required to establish the flow pattern in air over the first foot of a plate which has suddenly been raised from 70° to 270°F. Then

$$t = 5.24(0.952 + 0.70)^{\frac{1}{2}}\left[\frac{530}{32.2(200)}\right]^{\frac{1}{2}}(1)^{\frac{1}{2}} = 1.93 \text{ sec}$$

To gain some insight into the sensitivity of the results with respect to the assumed profile shapes, Siegel [*110*] replaced the velocity distribution given just before Eq. 25-1 by $U/U_1 = (Y/\delta)[1 - (Y/\delta)]^4$, retained the temperature distribution, and repeated the above procedure. He then compared the results with the exact solution of Illingworth [*107*] for the transient period and with the exact solution of Ostrach [*6*] for the steady state. As previously mentioned, the entire analysis was also done for the case where the plate at ambient temperature suddenly begins to produce a uniform heat flux at its surface. For this case, Siegel [*110*] also derives the exact velocity distribution in the transient stage analogous to Illingworth [*107*] for the constant temperature case. The results for the two cases follow.

Results. Comparison of the exact profiles for the initial one-dimensional transient flow with those for the steady flow shows that they are similar in shape (see [*110*]). Thus, the assumption made in the integral method that the profiles remain similar in shape throughout the transient period appears to be reasonable. The approximate temperature profiles for the two cases where the plate is changed to another uniform temperature or to uniform heat flux agree quite well with the corresponding exact profiles. The approximate velocity profiles, however, show a larger deviation from the exact profiles. However, the two approximate velocity profiles for each case fall on either side of the corresponding exact profile, so that a better fit to the exact velocity profile would be expected to yield results between the values computed by Siegel. Since the two constant temperature computations yield times required to reach steady state that are in agreement within from 16 to 30 per cent for all Prandtl numbers, it is concluded that the results are not too sensitive to the velocity profile. (Siegel recommends the use of an average equation, which he presents in [*110*].)

It is found that an increase in $PrRa_L$ (where for the uniform heat flux case the Rayleigh number is based on q rather than a temperature difference) will cause a decrease in the value of t^* required to reach a steady state. Thus, less time is required to develop the convective flow if any of the factors β, g, q, or θ_w are increased. This agrees with the results of [*108*].

The analysis indicates a rather unusual result, namely, that during the transient flow development the boundary layer thickness for a time exceeds the steady state value. For $Pr = 1$, the transient boundary layer thickness is about 25 per cent larger than that for steady state. This implies that, at a given position along the plate, the heat conduction process is evidently sufficiently rapid to enable the boundary layer to grow beyond the steady state thickness, before the constraints introduced by the leading edge can propagate to that location and prevent the growth from continuing as if the plate were infinite in length. The overgrowth of the boundary layer causes a minimum in the heat transfer coefficient. In the paper, Siegel suggests that this minimum may not be a physical reality but may be due to the approximate method of his analysis. In one of the printed discussions [*110*] of the paper, this point is further explored by assuming profiles that are combinations of exponentials and error functions rather than polynomials. These profiles not only approximate the steady state solution well, but yield the exact transient solution. The minimum in the transient heat transfer coefficient is also found with these profiles, and agreement with Siegel's other results is generally good.

Another discussion in the printed comments on Siegel's paper refers to work of Ostroumov [*112*] for qualitative support of Siegel's surprising result. In that work, the thermal response of a wire immersed in either ethyl alcohol, water, or air to a suddenly applied electric current was studied. For the wire in alcohol, a plot of wire temperature, as a function of time, displayed a maximum in the latter part of the transient period that exceeded the steady state temperature. For water, the overshoot was less pronounced and, for air, no overshoot was observed. Ostroumov offers no explanation for these results, and it would thus appear that further experimental evidence is needed. In the author's closure to the discussion [*110*], Siegel offers further support of his result by referring to a thesis by H. Klei of the Massachusetts Institute of Technology, in which a study of the transient heat transfer from a vertical metallic foil heated electrically in a stepwise manner also showed a minimum in the heat transfer coefficient during the transient period.

The present writer feels that the entire matter of the unusual result found by Siegel in the transient-flow development is still an open question. He agrees with Siegel's original suggestion that it may be due to the method of analysis, because it is, in general, not certain that two asymptotic solutions (here for small and large time) can be joined in an intermediate region without considerable careful analysis. This is because a singularity sometimes occurs in the intermediate region due to the omission of terms, which are significant there, in obtaining the asymptotic solutions. The occurrence of a singularity (in the complex plane) in such a procedure is known as Stokes' phenomenon and is discussed by Watson [*113*, p. 201]. The Stokes' phenomenon occurs in hydrodynamic stability theory (see Lin [*114*]) due to the existence of a critical layer. Stewartson [*115*], in his analysis of the flow about a semi-infinite plate which is impulsively set into motion, also encountered a singularity when he tried to match the boundary-layer type flow near the leading edge with the Rayleigh type far downstream. To match the asymptotic solutions properly in the intermediate region the analytic form of solution there must be found so that the singularity can be avoided. Lin was able to do this successfully in his work. Even though the basic differential equations in Siegel's analysis are not satisfied at each point in the field but only on the average, because he employed an integral method of solution, it can be seen from Eq. 25-6 that a singularity can exist in his problem when the first two terms of that equation are equal. The conditions under which this happens should be investigated to see whether they do influence the matching of the asymptotic solutions, which are the only ones explicitly obtained by Siegel. It is, of course, possible that the qualitative experimental support of Siegel's unusual result indicates that it is a reality. However, this point merits further investigation.

F,26. Natural Convection. Izumi [*116*] presents one of the only two theoretical analyses of unsteady natural convection known to the present writer. Consideration is given to the fully developed flow that results in a tube when, at a given instant, heat is introduced at the tube walls.

The governing equations are

$$\frac{\partial u}{\partial t} = \nu \left(\frac{\partial^2 u}{\partial r^2} + \frac{1}{r} \frac{\partial u}{\partial r} \right) + \beta g \theta \tag{26-1}$$

$$\frac{\partial \theta}{\partial t} = \alpha \left(\frac{\partial^2 \theta}{\partial r^2} + \frac{1}{r} \frac{\partial \theta}{\partial r} \right) \tag{26-2}$$

where α is the thermal diffusivity and the other symbols have their usual meanings. The initial and boundary conditions are

$$\left. \begin{aligned} u(r, \mathrm{o}) &= \theta(r, 0) = 0 \\ u(0, t) &\text{ and } \theta(0, t) \text{ are finite} \\ u(R, t) = 0, \left(\frac{\partial \theta}{\partial r} \right) (R, t) &= \frac{-q_0}{k} = \text{const} \end{aligned} \right\} \tag{26-3}$$

Nondimensional solutions of this boundary-value problem are found by Laplace transforms to be

$$\frac{u}{\left[\dfrac{\beta g q_0 R^3}{k\nu(1 - Pr)} \right]} = 2 \sum_{n=0}^{\infty} \frac{1 - \left[\dfrac{J_0(\alpha_n \epsilon)}{J_0(\alpha_n)} \right]}{(\alpha_n)^4} Pr(1 - e^{-\alpha_n^2 F_0}) - \frac{1 - \epsilon^2}{2} F_0 Pr$$

$$+ \sum_{n=0}^{\infty} \frac{4 J_0(\beta_n \epsilon)}{\beta_n^4 J_1(\beta_n) Pr} e^{-\beta_n^2 F_0 Pr} [1 - e^{-\beta_n^2 F_0 Pr}(1 + \beta_n^2 F_0 Pr)]$$

$$+ \sum_{n=0}^{\infty} \frac{J_0(\beta_n \epsilon)}{2 \beta_n^2 J_1(\beta_n)} (1 - e^{-\beta_n^2 F_0 Pr})$$

$$- \sum_{n=0}^{\infty} \sum_{n=0}^{\infty} \left[\frac{4 J_0(\beta_n \epsilon) J_0(\alpha_n \epsilon)}{\beta_n \alpha_n^2 J_0(\alpha_n) J_1(\beta_n)} \right] \frac{Pr e^{-\beta_n^2 F_0 Pr}}{Pr \beta_n^2 + \alpha_n^2} (1 - e^{-\beta_n^2 F_0 Pr - \alpha_n^2 F_0}) \tag{26-4}$$

$$\frac{\theta}{q_0 R/k} = \frac{\epsilon^2}{2} + 2F_0 - \frac{1}{4} - 2 \sum_{n=0}^{\infty} \frac{J_0(\alpha_n \epsilon)}{\alpha_n^2 J_0(\alpha_n)} e^{-\alpha_n^2 F_0} \tag{26-5}$$

where $\epsilon = r/R$, $F = \alpha t/R^2$ is the Fourier number, the α_n's are the successive roots of $J_1(\alpha_n) = 0$, and the β_n's are the successive roots of $J_0(\beta_n) = 0$. The velocity and temperature distributions computed from Eq. 26-4 and 26-5 are presented in Fig. F,26a and F,26b, respectively, as functions of the Fourier number for various radii r/R.

If instead of applying the boundary conditions of uniform heat flux at the wall (the last of Eq. 26-3), it is specified that the wall temperature

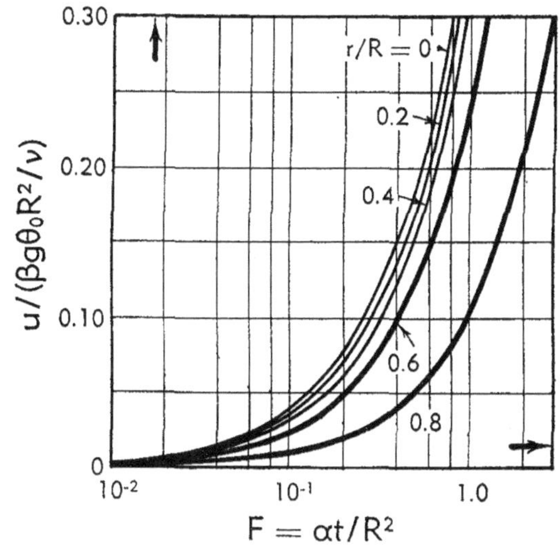

Fig. F,26a. Unsteady fully developed velocity distributions (Izumi [*116*]).

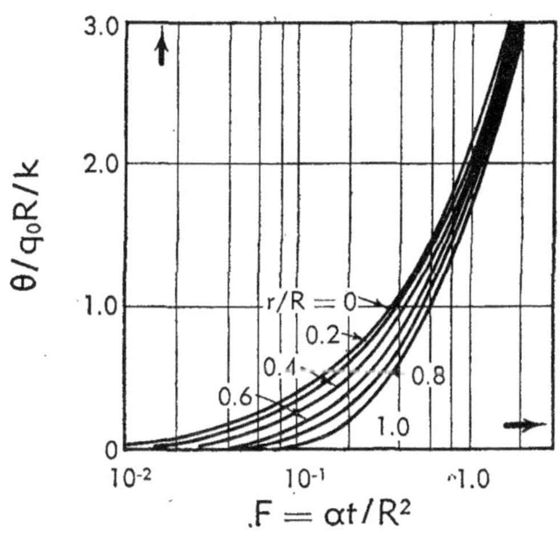

Fig. F,26b. Unsteady fully developed temperature distributions (Izumi [*116*]).

is uniform, i.e.

$$\theta(R, t) = \theta_0 = \text{const}$$

the velocity and temperature distributions are given by

$$\frac{u}{(\beta g \theta_0 R^2/\nu)} = \frac{2}{1 - Pr} \sum_{n=0}^{\infty} \frac{J_0(\beta_n \epsilon)}{\beta_n^3 J_1(\beta_n)} [(1 - e^{-\beta_n^2 Pr F_0}) - Pr(1 - e^{-\beta_n^2 F_0})] \tag{26-6}$$

$$\frac{\theta}{\theta_0} = 1 - 2 \sum_{n=0}^{\infty} \frac{J_0(\beta_n \epsilon)}{\beta_n J_0(\beta_n)} e^{-F_0 \beta_n^2} \tag{26-7}$$

From these equations, the mean values of the temperature and velocity are obtained as

$$\frac{u_m}{(\beta g\theta_0 R^2/\nu)} = \frac{2}{1 - Pr} \sum_{n=0}^{\infty} \frac{2}{\beta_n^4} [(1 - e^{-F_0 Pr \beta_n^2}) - Pr(1 - e^{-F_0\beta_n^2})] \quad (26\text{-}8)$$

and

$$\frac{\theta_m}{\theta_0} = 1 - 2 \sum_{n=0}^{\infty} \frac{2e^{-\beta_n^2 F_0}}{\beta_n^2} \quad (26\text{-}9)$$

The mean velocity and temperature for the uniform wall-temperature case given by Eq. 26-8 and 26-9 are presented in Fig. F,26c.

The other existing theoretical analysis of an unsteady natural convection problem is that of Ostroumov [*92*]. The problem considered deals

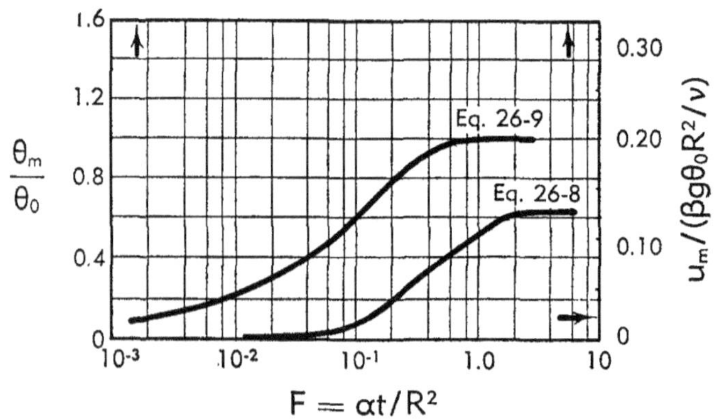

Fig. F,26c. Mean velocity and temperature distributions from Eq. 26-8 and 26-9 (Izumi [*116*]).

with a more complicated configuration than that of Izumi—that of a hot fluid poured into a closed duct or channel which is either drilled into a large piece of material or formed by pipe walls. Since the channel walls are relatively colder than the fluid, thermal convection in the fluid may be set up under certain conditions. This configuration simulates the casting of a molten metal in a mold. In that case, it is important to know the redistribution of the temperature due to convection because it could influence the process of solidification of the metal.

In order to simplify the analysis, the process is assumed to develop in three distinct stages. These stages are considered typical, although other processes could be possible.

The first stage of the process is that of filling the channel with a strongly turbulent fluid, which, on coming in contact with the channel walls, is cooled. Because of the strong turbulence of a nonthermal origin, effective thermal conductivity and diffusivity of the fluid will be much higher than those usually given on a molecular basis. Such a fluid is, therefore, equivalent to a strong heat-conducting solid body which is cooled at the boundaries. Such a thermal process has been well investigated. Particularly simple relations are obtained in the case where the form into which the fluid is poured is not a homogeneous body of infinite extent, but may be considered as a thin-walled vessel more or less thermally insulated from an infinite reservoir of constant temperature.

As a whole, the first stage is characterized by the fact that its violent mechanical nonthermal processes exclude the possibility of the occurrence of a more or less definite thermal convection motion. The duration of the first stage is determined by the decrease in the fluid of turbulent motion and of the thermal diffusivity of the fluid.

The second stage sets in when the turbulent motion is essentially ended, and the thermal convective motion is initiated and develops within the frame of the thermal pattern produced during the first stage. The transfer of heat by convection is not, however, large at this time, and no essential redistribution of temperature is produced. The duration of the second stage is determined by the magnitude of the kinematic viscosity of the fluid, that is, by the ratio of the viscous and inertia forces.

Finally, the forces of inertia in the convective motion practically dwindle to nothing. A more or less stable convective motion is established, which gradually dies down as the temperature differences produced diminish. As a result, the orderly convective transfer of heat distributes the temperature, the upper part becoming warmer than the lower. The distribution proceeds slowly as compared with the rate of cooling of the fluid in a cross section. This distribution constitutes the third, and last, stage of the process. The character of the third stage depends on the cross-sectional mean temperature distribution produced at the start of this stage. It is possible that, in the course of the third stage, the character of the process will slowly change as a result of the upward heat transfer. Moreover, the character of the third stage depends on the absolute rate of cooling, that is, on the intensity of the heat removed from the fluid by the channel walls. In this connection, two variants are investigated, one of which is rapid and the other slow.

The three-stage process just described yields linear equations for each step of the analysis. Although the details are presented in [*92*] for the cases where the channel is formed in a large mass of homogeneous material or by a thin-walled tube, the discussion is quite sketchy and seems to have suffered somewhat in translation. The reader is, therefore, referred to the original paper for a more precise treatment.

In addition to unsteady effects imposed essentially by the boundary conditions, it should be mentioned that, in a number of natural convection experiments, periodic unsteady flows were encountered under special conditions (see Siegel and Norris [*91*], Martin [*97*], and Sparrow and

Kaufman [*117*]). This latter unsteadiness is attributed to instabilities between different types of flows, but it has not, as yet, been studied extensively.

F,27. Effect of Body Forces on Unsteady Forced Flows. In a manner analogous to that of Art. 11 to 13 in Chap. 2 of this section, Illingworth [*107*] examines the influence of a body force on unsteady forced flows. He supposes that a flat plate, in contact with a perfect gas initially at rest, is set into motion in its own plane at time $t = 0$ in the direction perpendicular to its leading edge. At a later time, because of the action of viscosity, conductivity, and, to some extent, gravity, the gas will be moving inside a layer next to the plate, and will be at rest outside this layer. Only one side of the plate is considered and Cartesian coordinates, fixed with respect to the undisturbed gas, are used such that $Y = 0$ is the plane of the plate and the X axis is in the direction of motion of the plate, the Z axis being the initial position of the leading edge.

If it is assumed that the leading edge of the plate is horizontal, the motion of the gas is in the X, Y plane, and conditions are independent of Z; in other, more general, cases, because of gravity, all three components of velocity and variation with all three coordinates may have to be taken into account, though some may be reduced, approximately, to one of those considered here.

After a time t_s, the thickness of the layer will be $O[(\nu_\infty t_s)^{\frac{1}{2}}]$, where $\nu_\infty = \mu_\infty/\rho_\infty$ and V will be of the same order of magnitude. At stations in the layer with coordinate X_s, the distance from the leading edge at time t_s is $\int_0^{t_s} U_\infty(t)dt - X_s \equiv \xi_s$, $U_\infty(t)$ being the velocity of the plate. At stations at times t_s for which $\xi_s \gg (\nu_\infty t_s)^{\frac{1}{2}}$, $U_\infty(t)\xi_s \gg \nu_\infty$, $a_\infty \xi_s \gg U_\infty(t)(\nu_\infty t_s)^{\frac{1}{2}}$, $a_\infty^2 t_s \gg \nu_\infty$, and $a_\infty^2 \gg g(\nu_\infty t_s)^{\frac{1}{2}}$, the flow when gravity is the only body force is governed by the equations:

$$\frac{\partial \rho}{\partial t} + \frac{\partial}{\partial X}(\rho U) + \frac{\partial}{\partial Y}(\rho V) = 0 \tag{27-1}$$

$$\rho \frac{DU}{Dt} = -g\rho \sin \alpha - \frac{\partial P}{\partial X} + \frac{\partial}{\partial Y}\left(\mu \frac{\partial U}{\partial Y}\right) \tag{27-2}$$

$$\frac{\partial P}{\partial Y} = 0 \tag{27-3}$$

$$\rho \frac{Dh}{Dt} - \frac{DP}{Dt} = \frac{1}{Pr}\frac{\partial}{\partial Y}\left(\mu \frac{\partial h}{\partial Y}\right) + \mu \left(\frac{\partial U}{\partial Y}\right)^2 \tag{27-4}$$

where

$$\frac{D}{Dt} = \frac{\partial}{\partial t} + U\frac{\partial}{\partial X} + V\frac{\partial}{\partial Y}$$

and α is the inclination of the plate from the horizontal, a is the velocity of sound, and the enthalpy h is defined after Eq. 11-7. The conditions expressed by the six inequalities can be satisfied except at, and in, the immediate vicinity of the plate leading edge, and at, and immediately after, the moment the motion starts. The equation of state is

$$P = \rho(h - e) = \bar{\mathfrak{R}}\rho T \tag{27-5}$$

where the internal energy of the gas is also explicitly defined below Eq. 11-7. Eq. 27-3 leads to the fundamental boundary layer property that $P = P_\infty$ and the pressure P_∞ satisfies the equilibrium equations

$$\frac{dP_\infty}{dX} + g\rho_\infty \sin \alpha = 0 \tag{27-6}$$

$$P_\infty = \bar{\mathfrak{R}}\rho_\infty T_\infty \tag{27-7}$$

Employing Eq. 27-6, we can write the momentum equation, Eq. 27-2, as

$$\rho \frac{DU}{Dt} = g \sin \alpha(\rho_\infty - \rho) + \frac{\partial}{\partial Y}\left(\mu \frac{\partial U}{\partial Y}\right) \tag{27-8}$$

The variables of state P_∞, ρ_∞, h_∞, etc., are strictly functions of X except when the plate is horizontal, in which case they are constants. Their variation is neglected here, however, even when the plate is not horizontal. This means that attention is localized to a part of the plate that is not so large that the surrounding gas must be regarded as an atmosphere with a nonuniform structure. Moreover, this part of the plate is assumed to be sufficiently far from the leading edge for end-effects to be negligible. Under these circumstances, conditions depend only on t and Y, and the boundary layer equations simplify to

$$\frac{\partial \rho}{\partial t} + \frac{\partial}{\partial Y}(\rho V) = 0 \tag{27-9}$$

$$\rho\left(\frac{\partial U}{\partial t} + V\frac{\partial U}{\partial Y}\right) = g(\rho_\infty - \rho) \sin \alpha + \frac{\partial}{\partial Y}\left(\mu \frac{\partial U}{\partial Y}\right) \tag{27-10}$$

$$\rho\left(\frac{\partial h}{\partial t} + V\frac{\partial h}{\partial Y}\right) = \frac{1}{Pr}\frac{\partial}{\partial Y}\left(\mu \frac{\partial h}{\partial Y}\right) + \mu\left(\frac{\partial U}{\partial Y}\right)^2 \tag{27-11}$$

The solutions of these equations are asymptotic solutions of Eq. 27-1 to 27-4, valid for large negative values of X_s.

Eq. 27-9 to 27-11 are now subjected to von Mises's transformation, whereby Eq. 27-9 is replaced by

$$\frac{\partial \psi}{\partial Y} = \rho, \qquad \frac{\partial \psi}{\partial t} = -\rho V$$

involving a function $\psi(t, Y)$ analogous to the stream function in steady plane flow. The independent variables t and Y are replaced by t and ψ, so that $\partial/\partial t + V(\partial/\partial Y)$ and $\partial/\partial Y$ become $\partial/\partial t$ and $\rho(\partial/\partial\psi)$, respectively. At the same time, it is convenient to introduce nondimensional functions u, $\bar{\rho}$, τ, etc., defined by $U = U_\infty u$, $\rho = \rho_\infty\bar{\rho}$, and $T = T_\infty\tau$. Since $P_\infty = P$, it follows that $\bar{\rho}T = 1$ and the equations are transformed into

$$\frac{dU_\infty}{dt} u + U_\infty \frac{\partial u}{\partial t} = g(\tau - 1) \sin \alpha + \theta_\infty U_\infty \frac{\partial}{\partial\psi}\left(\theta \frac{\partial u}{\partial\psi}\right) \tag{27-12}$$

$$\frac{\partial h}{\partial t} = \frac{\theta_\infty}{Pr} \frac{\partial}{\partial\psi}\left(\theta \frac{\partial h}{\partial\psi}\right) + \theta_\infty b\theta\left(\frac{\partial u}{\partial\psi}\right)^2 \tag{27-13}$$

where $\theta = \rho\mu$ and $b = U_\infty^2/h_\infty$.

In these equations, u and h are the two dependent variables. Using a linear viscosity-temperature law implies that $\theta = 1$ which considerably simplifies Eq. 27-12 and 27-13.

There are two boundary conditions for each of u and h. Since $U = U_\infty$ at the plate, $u = 1$ when $\psi = 0$, $t > 0$. Since $U = 0$, $h = h_\infty$ at the edge of the boundary layer, $u \to 0$ and $h \to 1$ as $\psi \to \infty$. For the thermal condition at the plate, the temperature may be given, and then, at $\psi = 0$, $h = h_0/h_\infty$ is known; alternatively, the rate q_0 of heat transfer per unit area from the plate to the gas may be given, and then the enthalpy gradient at $\psi = 0$ must satisfy

$$q_0 = -\left(\frac{\mu c_p}{Pr} \frac{\partial T}{\partial Y}\right)_{Y=0} = -\frac{\theta_\infty h_\infty}{Pr}\left(\theta \frac{\partial h}{\partial\psi}\right)_{\psi=0} \tag{27-14}$$

The object is to find simple solutions of Eq. 27-12 and 27-13 subject to the boundary conditions, by reducing them to ordinary differential equations. The gravity term (which could, more generally, be any body force acting in the same direction) in Eq. 27-12 complicates the equations, and it is therefore helpful to see under what conditions it can be neglected.

Gravity, although now neglected in the undisturbed fluid, may still be important in the boundary layer because of the rapid variation in density. It produces a secondary flow, the primary flow being that which is produced in the X direction by the movement of the plate. If the leading edge of the plate were not horizontal, the secondary flow would also have a component in the Z direction; its importance could be estimated by similar methods.

When the plate is horizontal, $\alpha = 0$ and gravity is absent from the equations, so that there is only primary flow. If $\alpha \neq 0$, then we must compare the first terms on the two sides of Eq. 27-12. Since u and $(\tau - 1)$ are each of unit order of magnitude, it follows that gravity may be neglected in Eq. 27-12 if $dU_\infty/dt \gg g \sin \alpha$. If, for example, the plate starts from rest with a constant acceleration f, the condition for the primary flow to be a good approximation to the total flow in the X direction is $g \sin \alpha \ll f$. (This is one case where the total flow in the X direction can easily be found. It will be discussed subsequently.) If $U_\infty = kt^m$, where k is a constant and $m > 1$, gravity is not negligible in Eq. 27-12 until $t \gg [g \sin \alpha/(mk)]^{1/(m-1)}$. If $dU_\infty/dt = 0$, U_∞ is constant. The left-hand side of Eq. 27-12 is then of the form U_∞/t_s times a nondimensional quantity that is of the unit order of magnitude, where t_s is a representative time as before, and the gravity term is negligible if $gt_s \sin \alpha/U_\infty$ is very small.

The importance of the gravity term may, therefore, be expressed in terms of the magnitude of t_s. In order that Eq. 27-12 and 27-13 may be valid, the conditions $t_s \gg \nu_\infty/a_\infty^2$ and $t_s \ll a_\infty^4/(g^2\nu_\infty)$ have previously been imposed on t_s, and it has just been shown that gravity is negligible if $t_s \ll U_\infty(t_s)/(g \sin \alpha)$. The first two of these conditions are not important since, for air, for example, with $\nu_\infty = 1.4 \times 10^{-4}$ ft²/sec and $a_\infty =$ 1100 ft/sec, $\nu_\infty/a_\infty^2 = 1.2 \times 10^{-10}$ sec and $a_\infty^4/(g^2\nu_\infty) = 1.0 \times 10^{13}$ sec. The last one, however, may be relevant; for example, if $U_\infty =$ const $=$ 100 ft/sec, and the plate is moving in a vertical direction, the last condition states that secondary flow is negligible as long as $t_s \ll 3$ sec; for faster motion and more nearly horizontal plates, the condition is less restrictive. All the above arguments are, of course, modified if the body force is larger than that due to gravity.

The total flow (including gravitational effects) in the direction of motion of a uniformly accelerating isothermal plate has been determined by Illingworth [*107*]. If f is the constant acceleration, $U_\infty = $ ft, and dissipation is negligible for $t \ll a_\infty/f$, since then $b = U_\infty^2/h_\infty \ll 1$. Neglecting the dissipation, the energy equation (Eq. 27-13) can be reduced to the ordinary differential equation:

$$\frac{d}{d\eta}(\theta h') + 2Pr\eta h' = 0 \tag{27-15}$$

by letting

$$\eta = \frac{\psi}{2(\theta_\infty t)^{\frac{1}{2}}} \tag{27-16}$$

The primes denote differentiation with respect to η. For the simplest case, $\theta = 1$, the solution of Eq. 27-15 is

$$h(\eta) = h(0) + [1 - h(0)] \operatorname{erf}(\eta \sqrt{Pr}) \tag{27-17}$$

If we now assume that c_p is a constant so that $T = h$, the equation for u in the total flow is, from Eq. 27-12,

$$u'' + 2\eta u' - 4u + 4A[1 - \operatorname{erf}(\eta - \sqrt{Pr})] = 0 \tag{27-18}$$

where $A = g \sin \alpha[h(0) - 1]/f$. The solution of Eq. 27-18 is

$$u = (1 + 2\eta^2)(1 - \operatorname{erf} \eta) - \frac{2}{\sqrt{\pi}}\eta e^{-\eta^2} + AJ(\eta) \tag{27-19}$$

where $J(\eta)$ is explicitly given at the start of Art. 25.

If the acceleration of the plate is large enough, $A \ll 1$ and gravity is negligible, so that the total flow is indistinguishable from the primary flow. This may easily be seen, for example, for the case where $Pr = 1$ directly from the solution

$$u = (A - 1)J(\eta) + 1 - \text{erf}\ \eta$$

No profiles have yet been computed from these results.

CHAPTER 6. HYDRODYNAMIC STABILITY

In Sec. G of the present volume there appears an extensive treatment of the difficult problem of hydrodynamic stability of viscous fluids, i.e., the problem of finding the first departure from laminar flow of a fluid due to infinitesimal disturbances. When the flow is influenced by body forces, the hydrodynamic stability problem is further complicated by the interaction of the dynamics and thermodynamics of the phenomenon. A brief description of a few problems including body forces is given in G,21 and a qualitative discussion of thermal instabilities of fluids subject to body forces is presented in Chap. 3 of the present section. Because of the increased interest in flows with body forces, further studies of their stability have recently been made (most of which are not as yet published elsewhere) so that a more detailed discussion is warranted here. Furthermore, because transition in free convection occurs at low velocities, the onset of turbulence can be more easily observed experimentally than in forced flows. Since the physical aspects of transition in both types of flow are similar, information obtained in free convection studies may prove useful in increasing the understanding of the forced flow transition problem.

F,28. Introduction. One of the earliest attempts to find the stability of a flow subject to a body force is that of Prandtl [*118*] in which he considered the flow of a fluid whose density increased upward over a surface. Taylor [*119*], Goldstein [*120*] and Schlichting [*121*] each added to Prandtl's work somewhat as discussed in G,21. Schlichting's analysis is the most detailed and, although even it is not as general as possible, it bears a strong resemblance to the work that will be presented below. This relationship is noted in some detail in [*122*, Appendix 2]. The studies of Taylor [*55*] and Görtler [*123*], which deal with flows influenced by centrifugal forces, are also discussed in Sec. G.

The actual transition from laminar to turbulent flow of a free convection boundary layer about a vertical flat plate was studied experimentally by Eckert and Soehngen [*124*]. This work clearly showed the instability waves that precede transition. Wavelengths and velocities were determined from interference photographs. Birch [*125*] refined the experiments of Eckert and Soehngen by maintaining a uniform plate temperature and introducing controlled disturbances. Plapp [*122*] treats the stability of this flow theoretically; it is his analysis, together with unpublished work undertaken by the present writer on the stability of fully developed flows with body forces, that will be presented in what follows. Insofar as possible the theoretical and experimental studies will be compared.

F,29. Disturbance Differential Equations. The method of small disturbances employed to obtain the differential equations from which the flow instability can be determined is discussed in Sec. G and [*122*]. Plapp [*122*] carefully derives these equations, including the effects of body forces for surfaces inclined at an arbitrary angle with respect to the body force direction. In order not to be lost in too much generality, these equations will be written here only for surfaces parallel to the body force direction. These will be applicable to both free convection (vertical flat plate) and natural convection (fully developed channel) flows. We thus have

$$(\bar{u} - c)(\varphi'' - \alpha^2\varphi) - \bar{u}''\varphi + \frac{i\epsilon}{\alpha Fr}\tau' + \frac{i}{\alpha Re}(\varphi^{iv} - 2\alpha^2\varphi'' + \alpha^4\varphi) = 0 \quad (29\text{-}1)$$

$$(\bar{u} - c)\tau - \bar{\theta}'\varphi + \frac{i\gamma}{Pr\alpha Re}(\tau'' - \alpha^2\tau) + \frac{2i\gamma E\bar{u}'}{\alpha Re}(\varphi'' + \alpha^2\varphi) = 0 \quad (29\text{-}2)$$

Eq. 29-2 differs somewhat from Plapp's in that the c_v form of the energy equation was used, so that the ratio of specific heats appears in the last two terms in Eq. 29-2. The last term is the viscous dissipation term which is omitted completely by Plapp. In the above equations $\bar{u}$ is the dimensionless mean flow velocity, parallel to the surface, and is defined as the ratio of the local mean velocity component to the maximum dimensional velocity U_m in the flow; $\bar{\theta}$ is the nondimensional mean fluid temperature equal to the difference between the local mean temperature of the fluid and a reference temperature (the ambient temperature for free convection and the wall temperature for natural convection) divided by the absolute value of a characteristic temperature difference (e.g., wall to ambient for free convection). Both $\bar{u}$ and $\bar{\theta}$ are taken to be functions of only the coordinate y measured normal to the surface. The functions φ and τ, similarly nondimensionalized, are associated with a disturbance stream function ψ and a temperature perturbation θ, such that $\psi(x, y, t) = \varphi(y) \exp[i\alpha(x - ct)]$ and $\theta(x, y, t) = \tau(y) \exp[i\alpha(x - ct)]$, where x is the coordinate measured along the surface and nondimensionalized by a characteristic length (the boundary layer thickness δ for free convection and the channel width d for natural convection), y is the similarly nondimensionalized coordinate normal to the surface, t is time made dimensionless by dividing by the ratio of the characteristic length and the maximum velocity, c is the nondimensional phase velocity or velocity of propagation of the disturbance, and α is the wave number of the disturbance equal to $2\pi\delta/\Lambda$ or $2\pi d/\Lambda$, Λ being the disturbance wavelength. By definition $\epsilon = \beta\Delta T$, where β is the volumetric expansion coefficient and ΔT is the characteristic temperature difference. The Froude number, Fr, is defined as $U_{\mathrm{m}}^2/g\delta$ or U_{m}^2/gd, respectively, for free and natural convection, with g being the magnitude of the body force vector. Correspondingly, the Reynolds number Re is either $U_{\mathrm{m}}\delta/\nu$ or $U_{\mathrm{m}}d/\nu$ with ν the kinematic viscosity, E is the Eckert number, and $Pr = c_p\mu/k$ is the Prandtl number. (Recall that the Grashof number is some power of the Reynolds number based on the proper U_{m} for flows generated by body forces.) The primes denote differentiation with respect to y.

In accord with stability theory, only the linear disturbance terms appear in Eq. 29-1 and 29-2. As shown in [*122*, Appendix 2], these simplified equations hold strictly only for the case of parallel flow with all fluid properties, other than the density, taken to be constant. The effect of the density variation or body force appears only in the term $i\epsilon\tau'/\alpha Fr$ in Eq. 29-1 which couples that equation with Eq. 29-2. The justification for all the simplifications made in obtaining the above equations are given by Plapp [*122*].

For flows generated by body forces, the Froude and Reynolds numbers, along with ϵ, can be related by

$$\frac{\epsilon}{Fr} = \frac{\mathfrak{F}(Pr)}{Re} \tag{29-3}$$

where $\mathfrak{F}(Pr)$ is a dimensionless function of the velocity profile shape, and is presented in [*122*] for free convection about a plate. For channel flows, it will be shown below that $\mathfrak{F}$ is of unit order. If we apply Eq. 29-3 to Eq. 29-1, we can write

$$(\bar{u} - c)(\varphi'' - \alpha^2\varphi) - \bar{u}''\varphi + \frac{i}{\alpha Re}(\varphi^{iv} - 2\alpha^2\varphi'' + \alpha^4\varphi + \mathfrak{F}\tau') = 0 \tag{29-1a}$$

It is thus evident that the size of the buoyancy (coupling) term is proportional to $\mathfrak{F}$.

Since the general disturbance equations Eq. 29-1a and 29-2 contain the mean velocity and temperature profiles, it is necessary, at this point, to consider specific problems. Plapp's analysis [*122*] of free convection stability about a vertical flat plate will be considered first.

F,30. Free Convection Stability. In general, six boundary conditions must be specified to make the problem determinate. We thus require that all disturbances attenuate far from the plate, i.e., we consider that the disturbances associated with incipient turbulence originate in the layer in which there is flow. At the plate surface, the parallel and normal components of the disturbance velocity must vanish. From the definition of the disturbance stream function, it then follows that at the plate the boundary conditions are

$$\varphi'(0) = \varphi(0) = 0 \tag{30-1}$$

For an isothermal plate, the thermal boundary condition can be stated as

$$\tau(0) = 0 \tag{30-2}$$

and for an insulated plate the condition would be

$$\tau'(0) = 0 \tag{30-2a}$$

The general procedure used by Plapp to determine the stability of the flow, using Eq. 29-1a and 29-2 together with the boundary conditions, will now be outlined. The function τ is eliminated between the two governing equations to yield a single sixth order equation in φ. Because the equation is linear and homogeneous, its general solution φ is written as a linear combination of six linearly independent solutions φ_j as

$$\varphi = \sum_{j=1}^{6} C_j\varphi_j(Pr, c, \alpha, \alpha Re) \tag{30-3}$$

Applying the boundary conditions to Eq. 30-3, we obtain a set of six linear equations in the C_j's. For nontrivial solutions C_j of this set to exist, the determinant of the coefficients of the C_j's must vanish, or symbolically

$$D(Pr, c, \alpha, \alpha Re) = 0 \tag{30-4}$$

The expression $D(Pr, c, \alpha, \alpha Re)$ is the sum of a real and an imaginary part, each of which must separately be equal to zero. For a given fluid the Prandtl number is, of course, known. Also, the imaginary part of c is determined when a rate of amplification or damping of the disturbance is chosen, since the disturbances are proportional to $\exp [i\alpha(x - ct)]$. The only free parameters, then, which appear in Eq. 30-4, are R.P.(c), α, and αRe. The final step in solving the problem thus becomes the choosing of values of R.P.(c) and finding corresponding pairs of values of α and αRe that simultaneously satisfy the equations:

$$\text{R.P.}[D(Pr, c, \alpha, \alpha Re)] = 0 \tag{30-5a}$$

$$\text{I.P.}[D(Pr, c, \alpha, \alpha Re)] = 0 \tag{30-5b}$$

Enough pairs of values of α and αRe are found to determine a curve in the α, Re plane for the particular value of I.P.(c) chosen. Usually, only the case of neutral stability is investigated in which the disturbances are neither amplified nor damped; therefore, I.P.$(c) = 0$ is chosen. In this case, the curve in the α, *Re* plane divides pairs of values of α and *Re* into those associated with unstable or amplified oscillations. Such a curve is called an "indifference" or "neutral stability" curve.

Although the over-all procedure is quite straightforward, obtaining exact solutions for φ and τ and solving Eq. 30-5 would be practically impossible. Accordingly, as in the classical theory, approximations for solving the stability problem are employed, based on the assumption that the Reynolds number is large.

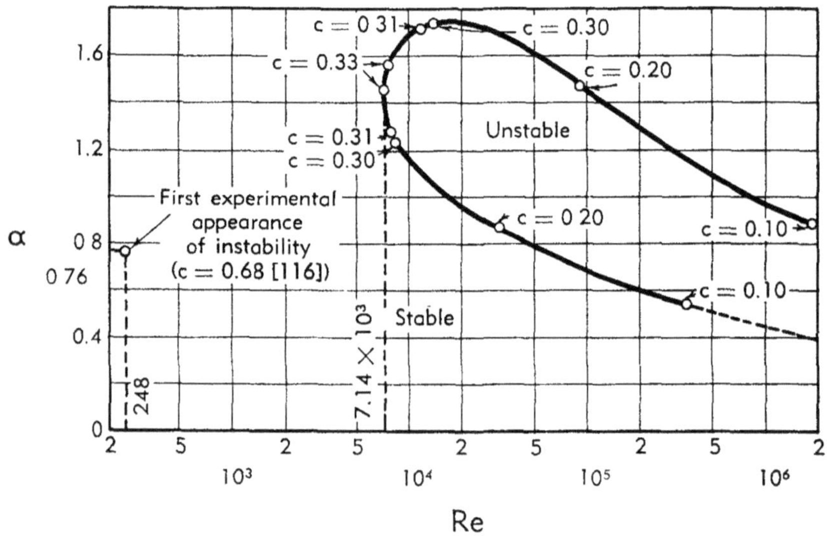

Fig. F,30a. Indifference curve for cubic polynomial velocity profile for $Pr = 0.72$ (Plapp [*122*]).

After deriving the approximate solutions and formulating the secular (determinant) equation, Plapp [*122*] finds that extensive high speed computing facilities would be necessary to complete the problem. Since such facilities were not available to him, he further simplified the problem by making various assumptions regarding the effects of viscosity, heat conduction, and the body force. The details are extensive and available in [*122*], so they will be omitted here. (In the development for natural convection, however, which will follow, comparisons will be made with Plapp's free-convection work.) In this way, five more distinct cases were treated; Plapp found that only two of these cases could be treated without the use of high speed computing machines. The two cases are the uncoupled, nonviscous case and the uncoupled, viscous case, where by uncoupled is meant that the body force is neglected so that only the momentum equation is considered. Thus, these cases differ from past work on incompressible flows without body forces only in that the free convection velocity profile is used in the calculations.

For the uncoupled, viscous case, Plapp [*122*] computed an "indifference" curve by approximating the mean velocity profile by a cubic polynomial, Fig. F,30a, and a significant portion of such a curve was determined for the exact profile, Fig. F,30b. On the basis of his mathematical

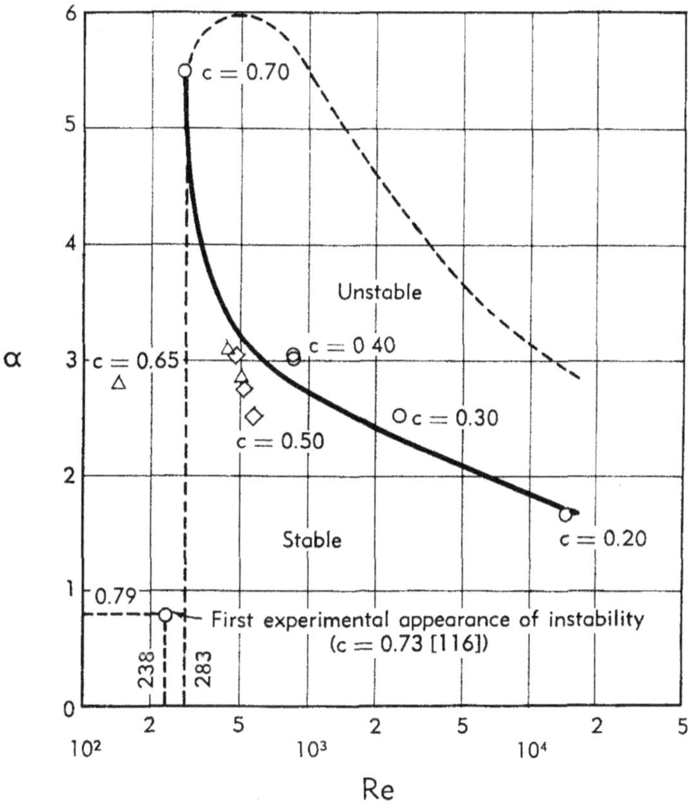

Fig. F,30b. Indifference curve for "exact" or tabulated velocity profile for Pr = 0.72 (Plapp [*122*]).

procedures, Plapp believes that the curve for the cubic polynomial profile is a valid approximation of the problem, but that the portion of the curve obtained for the exact profile may be considerably displaced from its true position. Both of these curves agree qualitatively with observations of the appearance of instability in free convection in that they predict that a minimum Reynolds number exists, below which the flow is stable for oscillations of all wave numbers. However, the curve for the cubic profile, Fig. F,30a, indicates that the instability waves that appear in the flow do so for values of the Reynolds number, the wave number, and the phase velocity of the disturbance approximately 29, 1.9, and 0.49, respectively, times the experimental values of these parameters.

The extreme discrepancy between the theoretical and experimental values is attributed to the poor approximation of the exact velocity profile by the cubic polynomial. The indifference curve for the exact profile, Fig. F,30b, gives values of the minimum critical Reynolds number, wave number, and disturbance phase velocity which are 1.19, 7.0, and 0.96 times the observed values. In this case, the agreement for Reynolds number and phase velocity was considered satisfactory, but the discrepancy in the wave number is far larger than what it should be if the analysis were valid and the observations were accurate.

F,31. Natural Convection Stability. The present writer and his coworker, Maslen, undertook the analysis of the fully developed flows discussed in Chap. 4 of the present section not only to determine the minimum critical Grashof (Reynolds) number for laminar stability, but also to see if the second solutions associated with frictional heating were stable. This work was done independently of that of Plapp, and it is instructive to compare the two procedures. Since the natural convection work has not been published as yet,* it will be presented here in some detail. In this way, the comparison with Plapp's study will also be clearer.

The first difference to be noted between the free convection study (of Plapp) and the natural convection study is the size of the body force term in Eq. 29-1a as determined by the velocity profile shape parameter $\mathfrak{F}$. Plapp [*122*] showed that $\mathfrak{F}$, which is a function of Prandtl number, has a minimum value of approximately 20. With the nondimensionalization appropriate for fully developed flows as indicated in Art. 19, $\mathfrak{F}$ is of unit order for that case. Thus, in the boundary layer problem (treated by Plapp) the body force terms are larger than the viscous terms, whereas in the channel problem they are of the same order.

To obtain the six solutions to the system of Eq. 29-1a and 29-2 for natural convection flows, we proceed as for forced flows, making appropriate modifications where necessary. Thus, as discussed in G,7, for large αRe, we first consider the "inviscid" parts of Eq. 29-1a and 29-2

$$(\bar{u} - c)(\varphi'' - \alpha^2\varphi) - \bar{u}''\varphi = 0 \tag{31-1}$$

$$(\bar{u} - c)\tau - \bar{\theta}'\varphi = 0 \tag{31-2}$$

Note that, in Plapp's case, the body force term would also appear in Eq. 31-1 because the body force terms are larger than the viscous terms.

* Note added in proof: Now published as: Ostrach, S. and Maslen, S. H. Stability of laminar viscous flows with a body force. *International Developments in Heat Transfer, Pt. V,* International Heat Trans. Conf., Boulder, Colo., Aug., 1961. (Publ. ASME, 1963)

In neither of the two cases for which he obtained indifference curves did Plapp include this term, although he did so in his discussion of the complete problem. What effect this would have on his computations must still be evaluated.

For small α, the solutions to Eq. 31-1 are

$$\varphi_1 = \bar{u} - c \quad \text{and} \quad \varphi_2 = (\bar{u} - c)\int \frac{dy}{(\bar{u} - c)^2} \tag{31-3}$$

Thus from Eq. 31-2 we find that

$$\tau_1 = \bar{\theta}' \quad \text{and} \quad \tau_2 = \bar{\theta}' \int \frac{dy}{(\bar{u} - c)^2} \tag{31-4}$$

Because Eq. 31-1 and 32-2 are of a lower order than Eq. 29-1a and 29-2, we must obtain additional solutions from the latter. Asymptotic solutions for large values of αRe are obtained by letting

$$\varphi = e^{-\int u dy} \qquad \tau = f e^{-\int u dy} \tag{31-5}$$

Substituting Eq. 31-5 into Eq. 29-1a and 29-2 (with $\mathfrak{F} = 1$), we obtain

$$(\bar{u} - c)(u^2 - u' - \alpha^2) - \bar{u}'' = -\frac{i}{\alpha Re}[u^4 - 6u^2u' + 4uu'' - u''' + 3u'^2 - 2\alpha^2(u^2 - u') + \alpha^4 + f' - fu] \tag{31-6}$$

$$(\bar{u} - c)f - \bar{\theta}' = -\frac{i\gamma}{\alpha Re}\left[\frac{1}{Pr}(f'' - 2uf' - u'f + u^2f - \alpha^2 f) + 2E\bar{u}'(u^2 - u' + \alpha^2)\right] \tag{31-7}$$

Now, generalizing the usual procedure, let

$$u = (\alpha Re)^n u_0 + (\alpha Re)^{n-\frac{1}{2}} u_1 + \cdots$$
$$f = (\alpha Re)^m f_0 + (\alpha Re)^{m-\frac{1}{2}} f_1 + \cdots$$

where from Eq. 31-6 we see that n must be positive. Retaining the largest terms in Eq. 31-6 and 31-7, we have

$$(\alpha Re)^{2n} u_0^2(\bar{u} - c) - \bar{u}'' = -i[(\alpha Re)^{4n-1}u_0^4 - (\alpha Re)^{m+n-1}u_0 f_0] \tag{31-8}$$

$$(\alpha Re)^m f_0(\bar{u} - c) - \bar{\theta}' = -i\gamma\left[\frac{(\alpha Re)^{2n+m-1}}{Pr} u_0^2 f_0 + 2E\bar{u}'(\alpha Re)^{2n-1}u_0^2\right] \tag{31-9}$$

To retain the viscous and heat conduction terms, we require that (a) $(4n - 1)$ must equal the largest of 0, $2n$, or $(m + n - 1)$ and (b) $(2n + m - 1)$ must equal the largest of m, 0, or $(2n - 1)$. We can immediately rule out $(4n - 1) = 0$, for then $2n > 0$ and the first term would dominate Eq. 31-8. From $(4n - 1) = 2n$, we find $n = \frac{1}{2}$ and requirement (a) is satisfied if $m \leqq \frac{3}{2}$. It then follows from requirement (b) that $m \geqq 0$, so that

$$n = \frac{1}{2}, \; 0 \leqq m \leqq \frac{3}{2} \tag{31-10}$$

are admissible. From the third possibility, viz., $(4n - 1) = (m + n - 1)$, we find that to satisfy the two requirements $n = \frac{1}{2}$ and $m = \frac{3}{2}$, which are included in Eq. 31-10. If, however, $0 < m < \frac{3}{2}$, the dominant terms in Eq. 31-8 and 31-9 are the first on each side of each equation. Under these conditions, there are two equations for u_0^2 with f_0 not appearing. Hence we are left with only two possibilities:

$$n = \frac{1}{2}, \; m = 0 \quad \text{or} \quad n = \frac{1}{2}, \; m = \frac{3}{2}$$

Our original expansion can thus be written as

$$\left.\begin{aligned} u &= \sqrt{\alpha Re}\, u_0 + u_1 + \cdots \\ f &= (\alpha Re)^m f_0 + (\alpha Re)^{m-\frac{1}{2}} f_1 + \cdots \end{aligned}\right\} \tag{31-11}$$

Substituting these equations into Eq. 31-6 and 31-7 and retaining the first two terms, we obtain

$$\begin{aligned} \left(u_0^2 + \frac{2u_0u_1 - u_0'}{\sqrt{\alpha Re}}\right)(\bar{u} - c) = -i\Bigg[& u_0^4 + \frac{4u_0^3u_1 - 6u_0^2u_0'}{\sqrt{\alpha Re}} \\ & - \begin{Bmatrix} 0 \;\text{ if }\; m = 0 \\ f_0u_0 + \dfrac{f_1u_0 + f_0u_1 - f_0'}{\sqrt{\alpha Re}} \;\text{ if }\; m = \dfrac{3}{2} \end{Bmatrix}\Bigg] \end{aligned} \tag{31-12}$$

$$\begin{aligned} \left(f_0 + \frac{f_1}{\sqrt{\alpha Re}}\right)(\bar{u} - c) - \begin{Bmatrix} \bar{\theta}' \;\text{ if }\; m = 0 \\ 0 \;\text{ if }\; m = \frac{3}{2} \end{Bmatrix} \\ = -\frac{i\gamma}{Pr}\left[u_0^2f_0 + \frac{(u_0^2f_1 + 2u_0u_1f_0 - 2u_0f_0' - u_0'f_0)}{\sqrt{\alpha Re}}\right] \\ - 2iE\bar{u}'\gamma \begin{Bmatrix} u_0^2 + \dfrac{(2u_0u_1 - u_0')}{\sqrt{\alpha Re}} \;\text{ if }\; m = 0 \\ 0 \;\text{ if }\; m = \frac{3}{2} \end{Bmatrix} \end{aligned} \tag{31-13}$$

Let us now consider the case $m = 0$. By equating terms of equal powers in αRe, we obtain from Eq. 31-12:

$$u_0^2 = i(\bar{u} - c) \tag{31-14}$$

$$(\bar{u} - c)(2u_0u_1 - u_0') = 2iu_0^2(3u_0' - 2u_0u_1) \tag{31-15}$$

from which we find that

$$u_1 = \frac{5u_0'}{2u_0} = \frac{5}{4}[\ln(\bar{u} - c)]' \tag{31-16}$$

If we apply the same procedure to Eq. 31-13, we obtain

$$f_0 = -\frac{1}{\left(\frac{\gamma}{Pr} - 1\right)}\left(\frac{\bar{\theta}'}{\bar{u} - c} + 2\gamma E\bar{u}'\right) \tag{31-17}$$

and

$$f_1(\bar{u} - c) = -\frac{i\gamma}{Pr}(u_0^2 f_1 + 2u_0 u_1 f_0 - 2u_0 f_0' - u_0' f_0) - 2i\gamma E\bar{u}'(2u_0 u_1 - u_0') \tag{31-18}$$

The function f_1 can be explicitly determined from Eq. 31-17 and 31-18.

Combining Eq. 31-5, 31-11, 31-14, 31-16, and 31-17, we find

$$\varphi_3 = (\bar{u} - c)^{-\frac{5}{4}} e^{-\int_0^y \sqrt{i\alpha Re(\bar{u} - c)}dy} \tag{31-19}$$

$$\varphi_4 = (\bar{u} - c)^{-\frac{5}{4}} e^{+\int_0^y \sqrt{i\alpha Re(\bar{u} - c)}dy} \tag{31-20}$$

$$\frac{\tau_3}{\varphi_3} = \frac{\tau_4}{\varphi_4} = \frac{1}{\left(1 - \frac{\gamma}{Pr}\right)}\left(\frac{\bar{\theta}'}{\bar{u} - c} + 2\gamma E\bar{u}'\right) \tag{31-21}$$

The remaining solutions are obtained by letting $m = \frac{3}{2}$ in Eq. 31-12 and 31-13. In the same manner, we now find that

$$u_0^2 = i\frac{Pr}{\gamma}(\bar{u} - c) \tag{31-22}$$

$$f_0 = u_0^3\left(1 - \frac{\gamma}{Pr}\right) \tag{31-23}$$

$$u_1 = \frac{f_0'}{f_0} + \frac{u_0}{2u_0} = \frac{7}{4}[\ln(\bar{u} - c)]' \tag{31-24}$$

Then

$$\varphi_5 = (\bar{u} - c)^{-\frac{7}{4}} e^{-\int_0^y \sqrt{\alpha Re i\frac{Pr}{\gamma}(\bar{u} - c)}dy} \tag{31-25}$$

$$\varphi_6 = (\bar{u} - c)^{-\frac{7}{4}} e^{+\int_0^y \sqrt{\alpha Re i\frac{Pr}{\gamma}(\bar{u} - c)}dy} \tag{31-26}$$

$$\frac{\tau_5}{\varphi_5} = \frac{-\tau_6}{\varphi_6} = -\left(1 - \frac{\gamma}{Pr}\right)\left[\frac{iPr\alpha Re}{\gamma}(\bar{u} - c)\right]^{\frac{3}{2}} \tag{31-27}$$

These solutions, which are all valid far from the critical points (where $\bar{u} = c$), agree with those given by Plapp [*122*]. Rather than proceed now with the development near the critical points, we shall here assume that the solutions behave properly there. Verification of the assumption will follow later.

Now that we have obtained the solutions, we must specify the boundary conditions. For simplicity we shall assume that the two walls of the channel are both maintained at the same temperature so that the basic profiles are symmetric. Then, from Eq. 29-1a we see that we have two cases to consider: (a) φ even, τ odd, and (b) φ odd, τ even. The boundary conditions are, then, at the wall ($y = 0$):

$$\sum_{j=1}^{6} a_j\varphi_j = 0, \sum_{j=1}^{6} a_j\varphi_j' = 0, \sum_{j=1}^{6} a_j\tau_j = 0 \tag{31-28}$$

and at the channel axis (say $y = 1$) either for case (a)

$$\sum_{j=1}^{6} a_j\varphi_j' = 0, \sum_{j=1}^{6} a_j\varphi_j''' = 0, \sum_{j=1}^{6} a_j\tau_j = 0 \tag{31-29a}$$

or for case (b)

$$\sum_{j=1}^{6} a_j\varphi_j = 0, \sum_{j=1}^{6} a_j\varphi_j'' = 0, \sum_{j=1}^{6} a_j\tau_j' = 0 \tag{31-29b}$$

For nontrivial solutions, the secular equation for case (a) is

$$\begin{vmatrix} \varphi_{10} & \varphi_{20} & \varphi_{30} & \varphi_{40} & \varphi_{50} & \varphi_{60} \\ \varphi_{10}' & \varphi_{20}' & \varphi_{30}' & \varphi_{40}' & \varphi_{50}' & \varphi_{60}' \\ \tau_{10} & \tau_{20} & \tau_{30} & \tau_{40} & \tau_{50} & \tau_{60} \\ \varphi_{11}' & \varphi_{21}' & \varphi_{31}' & \varphi_{41}' & \varphi_{51}' & \varphi_{61}' \\ \varphi_{11}''' & \varphi_{21}''' & \varphi_{31}''' & \varphi_{41}''' & \varphi_{51}''' & \varphi_{61}''' \\ \tau_{11} & \tau_{21} & \tau_{31} & \tau_{41} & \tau_{51} & \tau_{61} \end{vmatrix} = 0 \tag{31-30a}$$

where the first subscript refers to the particular solution and the second to the value of the argument at which the boundary condition is imposed. The equation for case (b) is

$$\begin{vmatrix} \varphi_{10} & \varphi_{20} & \varphi_{30} & \varphi_{40} & \varphi_{50} & \varphi_{60} \\ \varphi_{10}' & \varphi_{20}' & \varphi_{30}' & \varphi_{40}' & \varphi_{50}' & \varphi_{60}' \\ \tau_{10} & \tau_{20} & \tau_{30} & \tau_{40} & \tau_{50} & \tau_{60} \\ \varphi_{11} & \varphi_{21} & \varphi_{31} & \varphi_{41} & \varphi_{51} & \varphi_{61} \\ \varphi_{11}'' & \varphi_{21}'' & \varphi_{31}'' & \varphi_{41}'' & \varphi_{51}'' & \varphi_{61}'' \\ \tau_{11}' & \tau_{21}' & \tau_{31}' & \tau_{41}' & \tau_{51}' & \tau_{61}' \end{vmatrix} = 0 \tag{31-30b}$$

We now note from the solutions that $\varphi_{61}'/\varphi_{60}$, $\varphi_{61}'''/\varphi_{60}$, and τ_{61}/φ_{60} grow exponentially as $e^{+\sqrt{\alpha Re}}$, whereas $\varphi_{60}/\varphi_{60}$, $\varphi_{60}'/\varphi_{60}$, and τ_{60}/φ_{60} vary much slower with $\sqrt{\alpha Re}$. Hence, the top three terms in the right-hand column of Eq. 31-30a are negligible relative to the bottom three. The same argument can be applied to other columns in Eq. 31-30a and also to Eq. 31-30b to yield, respectively

$$\begin{vmatrix} \varphi_{10} & \varphi_{20} & \varphi_{30} & 0 & \varphi_{50} & 0 \\ \varphi_{10}' & \varphi_{20}' & \varphi_{30}' & 0 & \varphi_{50}' & 0 \\ \tau_{10} & \tau_{20} & \tau_{30} & 0 & \tau_{50} & 0 \\ \varphi_{11}' & \varphi_{21}' & 0 & \varphi_{41}' & 0 & \varphi_{61}' \\ \varphi_{11}''' & \varphi_{21}''' & 0 & \varphi_{41}''' & 0 & \varphi_{61}''' \\ \tau_{11} & \tau_{21} & 0 & \tau_{41} & 0 & \tau_{61} \end{vmatrix} = 0 \tag{31-31a}$$

and

$$\begin{vmatrix} \varphi_{10} & \varphi_{20} & \varphi_{30} & 0 & \varphi_{50} & 0 \\ \varphi'_{10} & \varphi'_{20} & \varphi'_{30} & 0 & \varphi'_{50} & 0 \\ \tau_{10} & \tau_{20} & \tau_{30} & 0 & \tau_{50} & 0 \\ \varphi_{11} & \varphi_{21} & 0 & \varphi_{41} & 0 & \varphi_{61} \\ \varphi''_{11} & \varphi''_{21} & 0 & \varphi''_{41} & 0 & \varphi''_{61} \\ \tau'_{11} & \tau'_{21} & 0 & \tau'_{41} & 0 & \tau'_{61} \end{vmatrix} = 0 \tag{31-31b}$$

We now note that $\varphi''_{61}/\varphi_{61}$ is proportional to αRe, whereas τ'_{61}/φ_{61} varies as $(\alpha Re)^2$. We can therefore neglect φ''_{61} and φ_{61} relative to τ'_{61}. Similarly, we can omit φ_{50}, φ'_{50}, φ'_{41}, φ'_{61}, τ_{41} and φ_{41} so that Eq. 31-31a and 31-31b become

$$\begin{vmatrix} \varphi_{10} & \varphi_{20} & \varphi_{30} & 0 \\ \varphi'_{10} & \varphi'_{20} & \varphi'_{30} & 0 \\ \tau_{10} & \tau_{20} & \tau_{30} & \tau_{50} \\ \varphi'_{11} & \varphi'_{21} & 0 & 0 \end{vmatrix} = 0 \tag{31-32a}$$

and

$$\begin{vmatrix} \varphi_{10} & \varphi_{20} & \varphi_{30} & 0 \\ \varphi'_{10} & \varphi'_{20} & \varphi'_{30} & 0 \\ \tau_{10} & \tau_{20} & \tau_{30} & \tau_{50} \\ \varphi_{11} & \varphi_{21} & 0 & 0 \end{vmatrix} = 0 \tag{31-32b}$$

We now define

$$\left.\begin{aligned} f_1 &= \begin{vmatrix} \varphi_{10} & \varphi_{20} \\ \varphi_{11} & \varphi_{21} \end{vmatrix} & f_2 &= \begin{vmatrix} \varphi_{10} & \varphi_{20} \\ \varphi'_{11} & \varphi'_{21} \end{vmatrix} \\ f_3 &= \begin{vmatrix} \varphi'_{10} & \varphi'_{20} \\ \varphi_{11} & \varphi_{21} \end{vmatrix} & f_4 &= \begin{vmatrix} \varphi'_{10} & \varphi'_{20} \\ \varphi'_{11} & \varphi'_{21} \end{vmatrix} \\ f_5 &= \begin{vmatrix} \tau_{10} & \tau_{20} \\ \varphi_{11} & \varphi_{21} \end{vmatrix} & f_6 &= \begin{vmatrix} \tau_{10} & \tau_{20} \\ \varphi'_{11} & \varphi'_{21} \end{vmatrix} \end{aligned}\right\} \tag{31-33}$$

Eq. 31-32a and 31-32b can then be written as

$$\frac{f_2}{f_4} = \frac{\varphi_{30}}{\varphi'_{30}} \tag{31-34a}$$

and

$$\frac{f_1}{f_3} = \frac{\varphi_{30}}{\varphi'_{30}} \tag{31-34b}$$

Eq. 31-34a and 31-34b are identical to those equations given by Lin [*114*] for an antisymmetrical and symmetrical disturbance, respectively, in a symmetrical incompressible viscous flow between solid walls. Plapp was not able to simplify his secular equation because he combined his momentum and energy disturbance equations. As a result, his analysis is in terms of a single dependent variable and each element of the determinant was quite complicated.

At this point, it may be noted that the remark about the behavior of the solution across the singular point is not a matter of concern, because the functions φ_1 through φ_4 are identical with those given by Lin. These solutions are already known to cross the singular point satisfactorily. The other solutions, φ_5 and φ_6, can be, at worst, such that

$$\varphi_5^+ = b_1\varphi_5^- + b_2\varphi_6^-$$
$$\varphi_6^+ = b_3\varphi_5^- + b_4\varphi_6^-$$

where the superscripts + and − denote the function on the two sides of the critical point.

When the thermal boundary conditions on the two walls are not the same, the basic flow profiles are asymmetric. Hence, the disturbance boundary conditions cannot be specified on the channel axis (as in Eq. 31-29) but must be given on both walls. For this case, the same procedure may be followed as in obtaining the secular equations (31-34), but now there results for asymmetric basic-flow profiles the secular equation

$$\frac{f_1}{f_3} - \frac{f_2}{f_3}\frac{\varphi_{42}}{\varphi_{42}'} = \frac{\varphi_{30}}{\varphi_{30}'} \tag{31-35}$$

This equation is also familiar since it is the same as the one given for Couette flow by Lin [*114*]. Aside from the fact that the solutions φ_i are different because of the basic profiles, one important difference must be noted between the problems treated herein and by Lin. The asymmetric profile considered by Lin is monotonic so that only one critical point (i.e., where $\bar{u} = c$) is obtained. In the present case of a channel with asymmetric velocities, the boundary conditions that must be used are that the velocities vanish at both walls. Thus two critical points will exist. In the derivation of Eq. 31-35 it was tacitly assumed that no difficulty results in crossing two critical points. Plapp, in his analysis of external flows, was confronted by the same problem and he showed that two critical points present no insurmountable problems, although they do complicate the numerical procedure. The present writer is at present working out the details of this problem.

We have thus shown that the body force does not affect the stability directly for natural convection, a result which Plapp [*122*] assumed for free convection in order to justify his calculations of the uncoupled cases. However, from his discussion of the complete coupled problem and the remarks presented herein concerning the size of the viscous and body force terms (through $\mathfrak{F}$) in the free convection problem, the question still remains as to whether Plapp's assumption is justified. In the case of free convection, the inclusion of the body force term in the inviscid equation should lead to solutions φ_1 and φ_2, which differ from those used here and by Plapp in his calculations. Whether this difference is, in fact, significant can be determined by carrying out the procedure presented.

To conclude this discussion, it is clear that to find the stability of fully developed, natural convection flows one has merely to use the appropriate profile in classical stability theory. The results are identical to that for forced Poiseuille flows. In this regard, the second solutions determined by Ostrach [*1*] also are stable below certain Grashof numbers.

CHAPTER 7. MAGNETOHYDRODYNAMICS

F,32. Introduction. We have, up to now, been concerned with the influence of body forces on fluid flows where these forces (gravitational, centrifugal, and Coriolis) were more or less beyond our control. Electromagnetic forces are also body forces which can, however, be controlled through changes in the current and magnetic field. Thus, if a fluid were a conductor of electricity, it would be possible that an electric body force produced in it could affect the flow in a significant way. The terms magnetohydrodynamics, hydromagnetics, magnetogasdynamics, and magneto-aerodynamics all refer to the branch of fluid dynamics that deals with the motion of electrically conducting fluids in the presence of electric and magnetic fields. Until recently, magnetohydrodynamics was of interest primarily to geophysicists and astrophysicists. A review of their work in this field is given by Elsasser [*126*].

Although it was pointed out by Hartmann [*127*] in 1937 that the principles of magnetohydrodynamics could be utilized in a practical way to pump liquid metals in pipes, it took the imminence of hypersonic and space flight to stimulate great interest in this field. The importance of magnetohydrodynamic effects in aeronautics follows directly from the fact that the temperatures encountered in high speed flight (near Mach numbers of 15) are sufficient to ionize air appreciably. It thus appears feasible to alter the motion of this conducting gas by a magnetic field rather than a solid surface. Modifying the drag of a reentry body in this way, for example, could reduce its aerodynamic heating. Some of the studies of the influence of magnetic fields on external aerodynamics have been made by Rossow [*128,129*], Neuringer and McIlroy [*130,131*], Meyer [*132*], Kemp [*133*], and Kemp and Petschek [*134*].

Considerable attention is currently being given to the development of nonchemical space propulsion systems, in which electric and magnetic fields are used to accelerate ions and plasmas. Numerous internal flows including magnetohydrodynamic effects, which have some bearing on the propulsion problems, have been studied by Resler and Sears [*135,136*], Sears [*137*], Shercliffe [*138,139*], as well as others. These internal magnetohydrodynamics flows are also related to new systems for generating electricity such as, for example, thermal fusion.

It is obviously beyond the scope of this section to discuss the vast amount of magnetohydrodynamic research directed along the numerous lines mentioned briefly above. However, because of the basic interaction of the fluid dynamics and electrodynamics in magnetohydrodynamics, problems in this field are, in a sense, generalizations of the work that has been presented in the preceding parts of this section. It thus appears appropriate to present here the fundamentals of magnetohydrodynamics and to discuss some representative problems that have already been treated. In the course of such a treatment, we shall attempt, whenever possible, to relate the magnetohydrodynamic work to corresponding studies in gravitational convection, with a view to obtaining insights and techniques for new problems. In this way, the present section on laminar flows with body forces will not only indicate the present state of knowledge of this branch of fluid mechanics, but also point the way to new and challenging problems and possibilities.

F,33. Fundamental Equations. The fundamental equations, Eq. 2-1 to 2-4, for a compressible, viscous, heat-conducting fluid subject to a body force, must be generalized to include the effects of electrical conductivity of the fluid and imposed electric and magnetic fields. The MKS system will be used throughout what follows.

The continuity equation (Eq. 2-1) remains unchanged and a state equation like Eq. 2-4 still is required. However, in the momentum equations (Eq. 2-2), the body force per unit volume ρf_i must be replaced by that due to the imposed electromagnetic fields (see [*140*], for example)

$$F_i = \rho_e E_i + (\mathbf{j} \times \mathbf{B})_i \tag{33-1}$$

where ρ_e is the excess charge density, E_i denotes the electric field components, $\mathbf{j}$ is the total (due to conduction and convection) electric current density, and $\mathbf{B}$ is the magnetic induction. The work done on the fluid by electromagnetic fields must be included in the energy equation; this can be shown [*141*] to add the so-called Joule heating term, j_c^2/σ, to Eq. 2-3, where j_c denotes the conduction current and σ the electrical conductivity, which is a property of the medium. We note immediately that, by introducing these terms in the usual hydrodynamic equations, we have increased the number of dependent variables by nine (viz., three components each of the electric field E_i, magnetic induction B_i, and electric current density j_i). In electrically neutral fluids, as will be considered herein, $\rho_e = 0$. We must, therefore, add nine more equations to our system. Six of these are Maxwell's equations for ponderable media

$$\nabla \times \mathbf{E} = -\frac{\partial \mathbf{B}}{\partial t} \tag{33-2}$$

$$\nabla \times \frac{\mathbf{B}}{\mu_e} = \mathbf{j} + \frac{\partial \epsilon \mathbf{E}}{\partial t} \tag{33-3a}$$

where t denotes time and μ_e and ϵ are, respectively, the permeability and dielectric constant, which depend on the medium under consideration. In problems of interest, to a good approximation μ_e and ϵ can be given their values in a vacuum (denoted by the subscript zero) so that Eq. 33-3a becomes

$$\nabla \times \mathbf{B} = \mu_{e0}\left(\mathbf{j} + \epsilon_0 \frac{\partial \mathbf{E}}{\partial t}\right) \tag{33-3b}$$

The remaining three equations come from (the empirical) Ohm's law

$$\mathbf{j} = \sigma(\mathbf{E} + \mathbf{U} \times \mathbf{B}) \tag{33-4}$$

where $\mathbf{U}$ is the fluid velocity vector.

We thus have 15 dependent variables (three components each of velocity, electric field strength, magnetic induction, and electric current density, together with the three state properties of the fluid). To determine these unknowns we have the six hydrodynamic equations (Eq. 2-1 to 2-4) and the nine electromagnetic equations (Eq. 33-2, 33-3a, and 33-4).

Before proceeding further, it will be instructive to compare the magnetohydrodynamic equations with the hydrodynamic equations including body forces. We note, first of all, from a combination of Eq. 33-1 and 33-4, that the body force in magnetohydrodynamics is proportional to the electrical conductivity, which is analogous to the dependence on density in the hydrodynamical case. However, there is now an additional dependence of this force on the fluid velocity. There is also another basic phenomenon in magnetohydrodynamics which can be of equal or, in some cases, greater importance than the electric body force. This is the Joule heating, which appears in the energy equation. This phenomenon is very familiar to us in electricity, particularly in metallic conductors. However, since the conductivities of fluids are considerably less than those of metals, the Joule heating j_0^2/σ may be even more significant in magnetohydrodynamics.

F,34. Basic Parameters. The previous articles of this section have clearly demonstrated how difficult it is to solve relatively simple hydrodynamic problems with body forces. The added complexities introduced by the electromagnetic terms make magnetohydrodynamic problems even more formidable. Let us, therefore, nondimensionalize the magnetohydrodynamic equations in order to see what simplifications can be made. In this way, we can study the relative magnitudes of the various terms, so that we will be able to retain the significant effects and discard others under certain conditions. This procedure will also lead to an appreciation of the practical importance of magnetohydrodynamic effects.

We shall limit our study here by imposing two restrictions: (a) We shall not consider high frequency phenomena, so that the time scale of our problem will be of the same order of magnitude as the ratio of the characteristic length to characteristic velocity, L/U_0; (b) we shall neglect relativistic effects, i.e., the velocity of fluid flow is negligible with respect to the velocity of light.

For simplicity only, we shall consider the density and other fluid properties constant. Let us now introduce the nondimensional quantities (denoted by the lower case letters and τ, θ, and φ):

$$\left.\begin{aligned} &\mathbf{U} = U_0\mathbf{u}, \quad X_i = Lx_i, \quad \mathbf{B} = B_0\mathbf{b}, \quad t = \frac{L}{U_0}\tau, \\ &P = \rho U_0^2 p, \quad T = \Delta T\theta, \quad \Phi = \frac{U_0^2}{L^2}\varphi \end{aligned}\right\} \tag{34-1}$$

into the fundamental magnetohydrodynamic equations. All of the symbols have their usual meaning with Φ denoting the viscous dissipation and the subscript zero denoting a characteristic or reference condition. The continuity, momentum, and energy equations can thus be written as

$$\nabla \cdot \mathbf{u} = 0 \tag{34-2a}$$

$$\frac{D\mathbf{u}}{D\tau} = -\nabla p + \frac{1}{Re}\nabla^2\mathbf{u} + R_m \frac{B_0^2/\mu_e}{\rho U_0^2}\left(\frac{\mathbf{E}}{U_0B_0} + \mathbf{u} \times \mathbf{b}\right) \times \mathbf{b} \tag{34-3a}$$

$$\frac{D\theta}{D\tau} = \frac{1}{PrRe}\nabla^2\theta + \frac{\mathcal{E}}{Re}\varphi + R_m\mathcal{E}\frac{B_0^2/\mu_e}{\rho U_0^2}\left(\frac{\mathbf{E}}{U_0B_0} + \mathbf{u} \times \mathbf{b}\right)^2 \tag{34-4a}$$

where $D/D\tau$ is the material derivative. The electromagnetic equations, Eq. 33-2 to 33-4, can be combined to yield a field equation

$$\nabla^2\mathbf{b} = R_m\left[\frac{\partial \mathbf{b}}{\partial \tau} - \nabla \times (\mathbf{u} \times \mathbf{b}) + \frac{U_0^2}{c^2}\frac{\partial^2 \mathbf{b}}{\partial \tau^2}\right] \tag{34-5a}$$

where c denotes the velocity of light. In Cartesian tensor notation these equations are

$$\frac{\partial u_i}{\partial x_i} = 0 \tag{34-2b}$$

$$\frac{Du_i}{D\tau} = \frac{1}{Re}\frac{\partial^2 u_i}{\partial x_\alpha \partial x_\alpha} - \frac{\partial p}{\partial x_i} + R_m\frac{B_0^2/\mu_e}{\rho U_0^2}(e_{i\alpha\beta}E_\alpha b_\beta + u_\beta b_\beta b_i - u_i b_\beta b_\beta) \tag{34-3b}$$

$$\frac{D\theta}{D\tau} = \frac{1}{PrRe}\frac{\partial^2\theta}{\partial x_\alpha \partial x_\alpha} + \frac{\mathcal{E}}{Re}\varphi + R_m\frac{B_0^2/\mu_e}{\rho c_p \Delta T}\left(\frac{E_i}{U_0B_0} + e_{i\alpha\beta}u_\alpha b_\beta\right)^2 \tag{34-4b}$$

$$\frac{\partial b_i}{\partial \tau} = \frac{1}{R_m}\left(\frac{\partial^2 b_i}{\partial x_\alpha \partial x_\alpha} - \frac{U_0^2}{c^2}\frac{\partial^2 b_i}{\partial \tau^2}\right) + b_\alpha\frac{\partial u_i}{\partial x_\alpha} - u_\alpha\frac{\partial b_i}{\partial x_\alpha} \tag{34-5b}$$

where the permutation symbol $e_{i\alpha\beta}$ is defined by

$$e_{i\alpha\beta} = \tfrac{1}{2}(i - \alpha)(\alpha - \beta)(\beta - i)$$

In addition to the familiar Reynolds Re, Prandtl Pr, and Eckert $\mathcal{E}$ (or Mach) numbers, which appear in the equations, we find three new dimensionless groups. These are the magnetic Reynolds number $R_m = \sigma\mu_e LU_0$, the so-called magnetic pressure number $(B_0^2/\mu_e)/\rho U_0^2$, sometimes denoted by R_H, and the parameter $(B_0^2/\mu_e)/(\rho c_p \Delta T)$. In many other similar discussions, the magnetic Reynolds number does not appear as a parameter. This is a result of writing the body force term in a form different from that given by the last term on the right of Eq. 34-5; the magnetic Reynolds number does not appear if Maxwell's equation (Eq. 33-3) instead of Ohm's law, Eq. 33-4, is combined with Eq. 33-1 to yield the explicit form of the body force term. The procedure followed herein, however, is considered to be more significant, because it introduces a property of the medium (electrical conductivity) into the problem and indicates clearly important simplifications for small R_m.

The magnetic Reynolds number is the interaction parameter for these flows; it appears before the interaction term in each equation. Pai [*142*, pp. 361, 362] interprets the magnetic Reynolds number to be either the ratio of fluid velocity and a characteristic (diffusion) velocity of the magnetic field, or the ratio of the characteristic lengths of the flow field and the magnetic field moving through a conductor. Resler and Sears [*135*] regard R_m to be the ratio of the induced to applied magnetic field strengths. In any event, in magnetohydrodynamic problems where the conductivity is high or the dimensions or fluid velocity is large so that R_m is very large, it is clear from Eq. 34-5 that the magnetic field is greatly influenced by the fluid motion. In fact, it can be shown that the magnetic field is convected with the fluid. Under these conditions, which are usually encountered in astrophysical problems, it is at times possible to neglect the applied magnetic field entirely, relative to the motion-induced field. Many cosmic problems are, therefore, treated with the assumption that the conductivity is infinite; considerable simplification of the equations is obtained thereby.

On the other hand, in hypersonic flows, the conductivity, as we shall presently see, is much smaller than unity. For unit order lengths, then, the magnetic Reynolds number never exceeds 0.1, even at satellite speeds. In this case the flow does not affect the magnetic field. If, however, the pressure number which represents the ratio of magnetic to kinetic energy per unit volume is large (as it is for strong magnetic fields), and the field is not parallel to the flow direction, we can see from Eq. 33-3 that the magnetic field can appreciably influence the flow. The parameter $(B_0^2/\mu_e)/(\rho c_p \Delta T)$ represents the ratio of magnetic to thermal energy per unit volume. Thus, for strong magnetic fields, this parameter, too, will be large so that the field can also affect the temperature distribution if $\mathbf{u}$ is not parallel to $\mathbf{B}$.

In order to obtain an appreciation of the magnitudes of these parameters in aeronautical problems, let us see what values of electrical conductivity and magnetic induction B are practically feasible. To this end, we shall follow the discussion of Resler and Sears [*135*] closely. Electrical conductivity in a gas is due to a drift of charged particles. Conductivity is, therefore, increased by increasing the number of charged particles and reduced by increasing the probability of their collisions with impeding particles, i.e., atoms, molecules, and charged particles.

If we let α denote the "fractional ionization" of the gas, then α is nearly zero for air at room temperature and $\alpha = 1$ represents a completely ionized gas in which each original particle has given up one electron. Although α can actually exceed unity, the state of complete ionization represents a relatively large value for most applications, attained only at very high temperatures. When α is small, so that charged particles for the most part collide with neutral molecules, we find that

$$\sigma \sim \alpha T^{\frac{1}{2}} \tag{34-6}$$

This situation is probably typical of aeronautical cases, but, in general, the form of relation (Eq. 34-6) is misleading, since α itself is strongly dependent on the temperature. Thus, even under these conditions, σ varies rapidly with temperature.

At the other extreme, when collisions between charged particles are dominant, the conductivity becomes independent of α, and instead of Eq. 34-6 we have

$$\sigma \sim T^{\frac{3}{2}}$$

For practical application, the strong dependence of conductivity on temperature may be minimized by introducing an easily ionizable substance, such as sodium or potassium, into the gas. This is called "seeding." In this way, even though the temperature may not be sufficient to ionize the gas itself appreciably, it may be sufficiently high to ionize the seeding substance completely. The conductivity thus becomes relatively insensitive to temperature.

Some data on the conductivity of air with, and without, seeding taken from [*143,144*] are presented in Fig. F,34a. The Mach numbers spotted along the curves are only approximate and are intended only for orientation with regard to the relationship of stagnation temperature to flight speed. It can be seen from Fig. F,34a that σ for air does not exceed about 0.1 mho/cm, even at the temperature corresponding to satellite speeds. On the other hand, with seeding, σ approaches unity even at Mach numbers contemplated for manned aircraft.

For the purpose of comparison, it might be helpful to list the conductivities of other similar substances, in the same units. For example,

Copper	60,000,000 mho/m
Mercury	1,080,000 mho/m
Salt water (saturated at 25°C)	25 mho/m
Pure water	0.00002 mho/m

To obtain an estimate of the magnetic Reynolds number to be expected in aeronautical problems, we can, on the basis of Fig. F,34a, take $\sigma = 1.0$ mho/cm. Noting further that $\mu_e = 1.257 \times 10^{-6}$ henry/m and $L = 1$ meter, we find that

$$R_m = \sigma\mu_e U_0 L = 1.257 \times 10^{-4}\ U_0(\text{m/sec})$$

so that, even at satellite speeds, R_m is only of unit order. Obviously, the

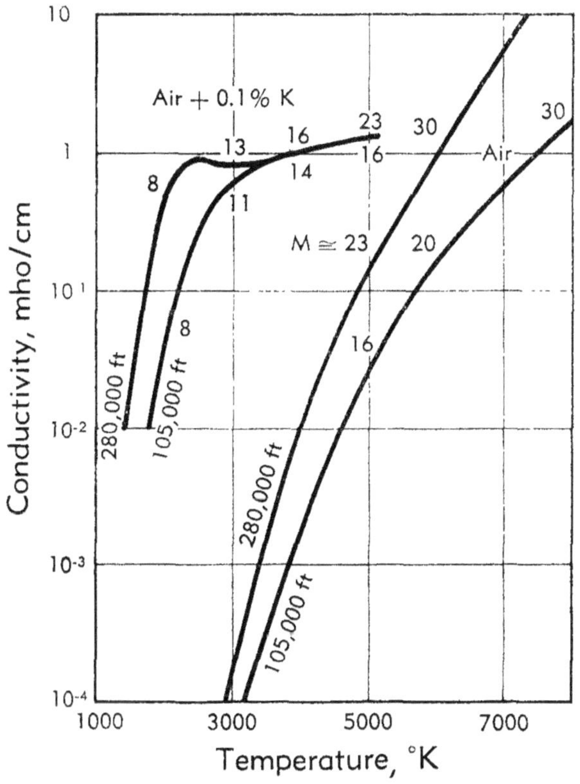

Fig. F,34a. Calculated conductivity of air and of air seeded with 0.1 per cent potassium (by weight) as functions of temperature for two altitudes. Corresponding Mach numbers M giving approximately these stagnation temperatures are indicated (Resler and Sears [*135*]).

assumption of infinite conductivity made for astrophysical problems is not appropriate for aeronautics. Larger magnetic Reynolds numbers will, of course, be associated with ion and plasma propulsion devices because of the higher conductivities.

There is, unfortunately, no firm basis from which to estimate what practical magnetic inductions are feasible. We can, however, cite some values that are presently known and use these as a guide for our estimates.

An induction of about 2 weber/m² saturates an iron core at room temperature. However, because of their weight, it is not certain that iron cores can be utilized in aeronautics; if they are used we can be certain that their saturation value will be maximum. This establishes only a very high upper limit even though in principle an air core magnet could be made stronger.

Another indication of feasible values may be suggested from the fact that for good modern materials the induction in a permanent magnet is approximately 0.15 to 0.2 weber/m².

It seems probable that, in aeronautical applications, magnetic fields will be produced by air core electromagnets whose currents are generated magnetoaerodynamically. The dimensions of such solenoids can be estimated from the vehicle size, but since the number of turns will be determined from such matters as weight and resistance, we still do not have an estimate of B. We shall thus arbitrarily use the value $B = 10^{-1}$ weber/m² in the calculations.

To find the effect of electromagnetic fields on the flow, we must estimate the magnitude of the product of the magnetic Reynolds and pressure numbers. This product represents the ratio of electromagnetic body forces to the dynamic forces. Resler and Sears [*135*] present values of this ratio as a function of altitude and Mach number; these results are presented in Fig. F,34b. (This figure includes a small correction to the original which appeared in [*135*].) From the figure it is clear that the electromagnetic effects will be large enough at high velocities and altitudes to affect the flow significantly. Similar arguments can be given to show that the temperature distribution can also be appreciably influenced by the electromagnetic forces.

By multiplying the ratio of electromagnetic body forces to dynamic forces (given by $R_m R_H$) by the ratio of dynamic to viscous forces (the Reynolds number), we obtain the ratio of body forces to viscous force (given by $R_m R_H Re$). It is thus evident that the body forces are much larger than the viscous forces except in boundary layers, wakes, and other such viscous regions where the Reynolds number is quite small. It follows immediately that an inviscid magnetohydrodynamic theory is generally useful, but it must be interpreted in light of viscous flow theory. In other words, boundary layers and similar viscous regions retain their significance in magnetohydrodynamic flows.

As a result of the recent great research activity in this new field, magnetohydrodynamic problems similar to all types of the usual fluid mechanics problems have already been treated. The aforementioned summary papers can give the reader an idea of the scope of the current work. As is appropriate, we shall here restrict our discussion to magnetohydrodynamic problems which are analogous to the body force problems previously discussed in this section.

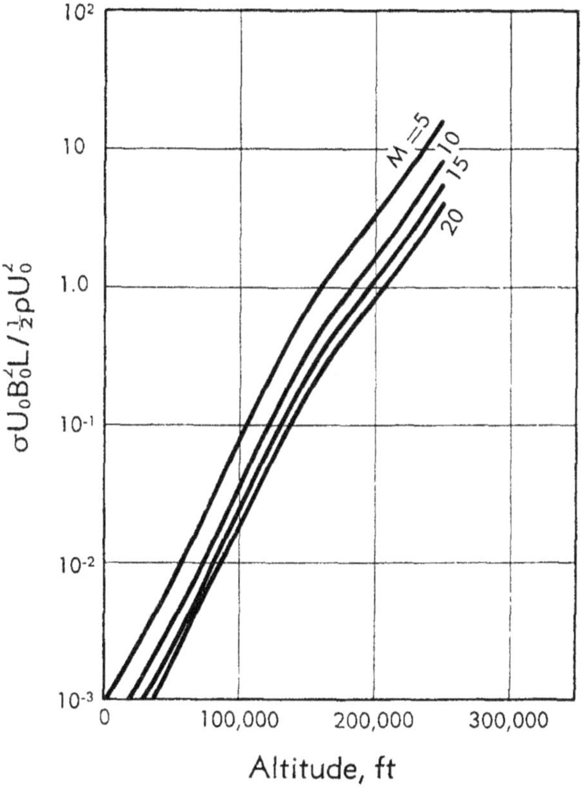

Fig. F,34b. Values of the ratio of typical electric body force to dynamic force as function of Mach number and altitude, assuming $\sigma = 1$ mho/cm, $H = 1000$ gauss, and $L = 1m$ (Resler and Sears [*135*]).

F,35. Boundary Conditions. Before proceeding to the specific problems a word must be said about boundary conditions. Clearly, the usual flow and thermal conditions must still be satisfied. However, since more dependent variables have been introduced in the problems, we need electromagnetic boundary conditions. For the additional conditions, we know that Maxwell's equations are valid only for those points in whose neighborhood the physical properties of the medium vary continuously. On the boundary of the flow field, however, the physical properties of the medium may have discontinuities. For example, at a solid boundary the electromagnetic properties change abruptly. Across such a surface of discontinuity of electromagnetic properties, the following four conditions obtained from integral forms of Maxwell's equation must be satisfied:

The normal component of magnetic induction is continuous, i.e.,

$$\mathbf{n} \cdot (\mathbf{B}_2 - \mathbf{B}_1) = 0 \tag{35-1}$$

where **n** is the unit normal to the surface of discontinuity. The jump condition of the magnetic induction at the boundary is

$$\mathbf{n} \times \left(\frac{\mathbf{B}_2}{\mu_{e2}} - \frac{\mathbf{B}_1}{\mu_{e1}}\right) = \mathbf{J}_s \tag{35-2}$$

where the surface current density $\mathbf{J}_s$ is zero for finite electrical conductivity. For infinite conductivity ($\sigma = \infty$), $\mathbf{J}_s$ may be different from zero.

The tangential component of the electric field is continuous, i.e.,

$$\mathbf{n} \times (\mathbf{E}_2 - \mathbf{E}_1) = 0 \tag{35-3}$$

and at the boundary also

$$\mathbf{n} \cdot (\epsilon_2 \mathbf{E}_2 - \epsilon_1 \mathbf{E}_1) = \rho_s \tag{35-4}$$

where ϵ denotes the dielectric constant and ρ_s is the surface free charge density.

F,36. External Problems. Since treatment of magnetohydrodynamic problems requires the simultaneous solution of the hydrodynamic and electromagnetic equations, simplifying assumptions obviously must be made for tractability. Rossow [*128*], in his study of the magnetohydrodynamic flow about a semi-infinite flat plate, applied the boundary layer assumptions to simplify the governing equations. However, to define the problem completely, he prescribed conditions at the outer edge of the boundary layer which led to considerable ambiguity as to the actual physical configuration being considered. In other external magnetohydrodynamic problems (particularly the stagnation flow treatments in [*129* to *133*]), the complete implications of the assumptions and limitations of the analyses have not been clearly delineated. As a result, seemingly contradictory results concerning the effect of magnetic fields on the nose heat transfer rates of missiles appear in the literature. Accordingly, Hess [*145*] set out to clarify all this work. The reader is referred to Hess' paper for the involved and detailed discussion of the stagnation flow work. We shall here concentrate on the flat plate problem which is somewhat analogous to the free convection analysis presented in Art. 3. The problem will be formulated from Hess's viewpoint and then Rossow's work will be discussed.

In view of the great simplification obtained in the nonmagnetic case by separating the flow into a viscous rotational boundary layer region and an inviscid potential flow outside the layer, Hess [*145*] investigated the conditions under which such a separation is possible in the magnetohydrodynamic case. For this purpose he started with the equation of motion for steady flow

$$\mathbf{U} \cdot \nabla \mathbf{U} = -\frac{1}{\rho} \nabla P + \frac{1}{\rho} (\mathbf{j} \times \mathbf{B}) + \nu \nabla^2 \mathbf{U} + \frac{1}{3} \nu \nabla \nabla \cdot \mathbf{U} \tag{36-1}$$

To this equation he applied the vector relations

$$\mathbf{U} \cdot \nabla \mathbf{U} = \nabla \frac{U^2}{2} - \mathbf{U} \times (\nabla \times \mathbf{U})$$

and

$$\nu \nabla^2 \mathbf{U} = \nu(\nabla \nabla \cdot \mathbf{U} - \nabla \times \nabla \times \mathbf{U})$$

so that Eq. 36-1 becomes

$$\frac{1}{\rho}(\mathbf{j} \times \mathbf{B}) = \nabla \frac{\mathbf{U}^2}{2} + \frac{1}{\rho} \nabla P - \mathbf{U} \times \nabla \times \mathbf{U} + \nu \nabla \times \nabla \times \mathbf{U} - \frac{4}{3} \nu (\nabla \nabla \cdot \mathbf{U})$$

For potential flow then, we must have

$$\frac{1}{\rho}(\mathbf{j} \times \mathbf{B}) = \nabla \frac{U^2}{2} + \frac{1}{\rho} \nabla P \tag{36-2}$$

From Eq. 36-2 it can be seen that, if the velocity distribution is to have a potential, $(\mathbf{j} \times \mathbf{B})/\rho$ must be expressible as a gradient or, equivalently:

$$\nabla \times \frac{\mathbf{j} \times \mathbf{B}}{\rho} = 0 \tag{36-3}$$

It is interesting to note that, in order to satisfy the condition of potential flow, the velocity term in Eq. 36-2 must not contain the density ρ, because potential flow refers to the potential of the velocity rather than its product with ρ. As a consequence, the condition for potential flow requires that the curl of $(\mathbf{j} \times \mathbf{B})/\rho$ be zero. Since the induced currents produce Joule heat, the density of the flow could vary if compressibility effects were considered. Thus the conditions for the existence of magnetogasdynamic potential flow will be even more restrictive than those for magnetohydrodynamic potential flow, where the density is assumed constant. We shall restrict ourselves in this article only to the case of constant density.

It can also be seen from Eq. 36-2 that, when the magnetic force does not have a potential, the resulting curl would have to be asborbed by the velocity field, which is capable of rotational behavior, whereas the scalar pressure field is not. The reverse is also true, viz., a rotational flow will produce a magnetic force that is not derivable from a potential.

By these arguments Hess has answered the criticism, sometimes heard, that "potential" implies a conservative force system, and since Joule heating is produced by the currents this requirement cannot be met in magnetohydrodynamics. Hess [*145*] has shown that, consistent with more recent usage, the existence of a potential depends on the vanishing of the curl of the particular quantity, be it a force or a velocity.

Hess [*145*] also noted the similarity between magnetohydrodynamic flows and free convection flows pointed out herein and he used this fact to obtain deeper insight into magnetohydrodynamic boundary layer flow problems. He showed that, if the magnetic force has a potential (see Eq. 36-2), then by analogy with the free and natural convection problems it is possible to separate the pressure into two parts, one a hydrodynamic and the other a magnetic part. Note that the term magnetic pressure does not refer to $B^2/2\mu_e$. Thus Eq. 36-2 can be written as

$$\frac{1}{\rho}(\mathbf{j} \times \mathbf{B}) = \nabla \frac{U^2}{2} + \frac{1}{\rho}\nabla(P_d + P_m) \tag{36-4}$$

Compare this with the split into hydrostatic and dynamic parts in free (Eq. 3-11) and natural convection [*1*, Appendix B]. From Eq. 36-4 it can be seen that the magnetic pressure gradient is balanced by the magnetic body force, i.e.,

$$\mathbf{j} \times \mathbf{B} = \nabla P_m \tag{36-5}$$

In contrast to the gravitational pressure, the magnetic pressure does not depend on the density. However, as previously remarked, the existence of a magnetic force potential would be extremely rare if compressibility effects were considered, because the Joule heat produces density variations. Thus, the simplicity of Eq. 36-5 alone is not enough to permit pressure splitting for compressible flow.

From Eq. 36-3, Hess [*145*] argues that the maintenance of potential flow outside the boundary layer, in most cases, is possible only when the deformation of the magnetic lines is small, but even then potential flow does not always exist. Potential flow should, however, be possible when the magnetic lines are normal to the flow and the curvature of the streamlines is small. All these conditions were specified by Rossow [*128*] for the different purpose of making the magnetic force dependent only on the longitudinal velocity.

Under these conditions, we can write

$$\mathbf{j} \times \mathbf{B} = \sigma[\mathbf{B}(\mathbf{B} \cdot \mathbf{U}) - \mathbf{U}(\mathbf{B} \cdot \mathbf{B})] \tag{36-6}$$

or

$$\frac{1}{\rho}(\mathbf{j} \times \mathbf{B}) = -\frac{\sigma B_0^2}{\rho}\mathbf{U} = -m_1\mathbf{U} \tag{36-7}$$

Here B_0 denotes the original imposed magnetic induction. With this expression for the body force, the magnetohydrodynamic boundary layer equation can be written in terms of the velocity components U and V as

$$U\frac{\partial U}{\partial X} + V\frac{\partial U}{\partial Y} + \frac{1}{\rho}\frac{\partial}{\partial X}(P_d + P_m) + m_1 U = \nu\frac{\partial^2 U}{\partial Y^2} \tag{36-8}$$

In the potential flow outside the boundary layer, the magnetic force is balanced by the magnetic pressure gradient, or

$$\frac{1}{\rho}\frac{\partial P_m}{\partial X} + m_1 U_P = 0 \tag{36-9}$$

As in conventional boundary layer flows, $\partial P_m/\partial X$ acts uniformly across the layer. An alternate form of Eq. 36-8 can be obtained by the use of Eq. 36-9 to be

$$U\frac{\partial U}{\partial X} + V\frac{\partial U}{\partial Y} + \frac{1}{\rho}\frac{\partial P_d}{\partial X} + m_1(U - U_P) = \nu\frac{\partial^2 U}{\partial Y^2} \qquad (36\text{-}10)$$

The analogy between the gravitational and magnetohydrodynamic problems can be seen by comparing Eq. 7-2, for example, with Eq. 36-10.

On the basis of Eq. 36-8 and 36-10, Hess *[145]* states that it is convenient to divide the great variety of boundary layer flows into two main types. For one, the distribution of the pressure (hydrodynamic plus magnetic) is assumed to be constant, whereas for the other the velocity distribution in the potential flow outside the boundary layer is taken to be constant and independent of the magnetic field. In this sense only, these are "flat-plate" flows. The value of such classification depends on whether these types, individually or in a combined form, lead to physically realizable boundary conditions. For example, for the boundary layer flow along a flat plate, it should be possible to maintain constant pressure independently of the field strength, if certain conditions are satisfied. If a magnetic field extends a finite distance above the plate and the resulting changes in flow direction are considered small (generally necessary for potential flow), it should be possible to transmit a constant pressure to the plate.

When, on the other hand, the velocity distribution of the potential flow is maintained independently of the magnetic field, the boundary layer flow requires a magnetic pressure gradient that balances the magnetic force. The maintenance of the velocity distribution may require the boundary condition of a pump, or may arise indirectly from more involved boundary conditions. In the literature *[128]*, this type of problem has been treated only for the case in which a given initial flow acceleration outside the boundary layer is maintained against a magnetic force. The case in which the velocity distribution of the potential flow is maintained is also of interest for certain aspects of stagnation flow. Details are given in [*145*, Appendix].

The advantage of the split pressure approach of Hess is evident when one considers problems treated by Rossow *[128]* in detail. For example, to obtain Rossow's equation for the magnetohydrodynamic flow over a semi-infinite flat plate with the magnetic field fixed relative to the plate (see Fig. F,36a), we must set

$$\frac{\partial}{\partial X}(P_d + P_m) = 0$$

in Eq. 36-8, i.e., constant pressure is maintained in the potential flow. It is immediately clear that both the hydrodynamic and magnetic pressure gradients can vary. (This is the first of the two main types of problems classified by Hess.) The resulting equation,

$$U\frac{\partial U}{\partial X} + V\frac{\partial U}{\partial Y} + m_1 U = \nu\frac{\partial^2 U}{\partial Y^2} \tag{36-11}$$

together with the boundary conditions

$$U(X, 0) = V(X, 0) = 0$$

$$\frac{\partial U}{\partial X}(X, \infty) = -m_1, \qquad U(X, \infty) = U_\infty$$

are the ones treated by Rossow [*128*] for the case described. To solve

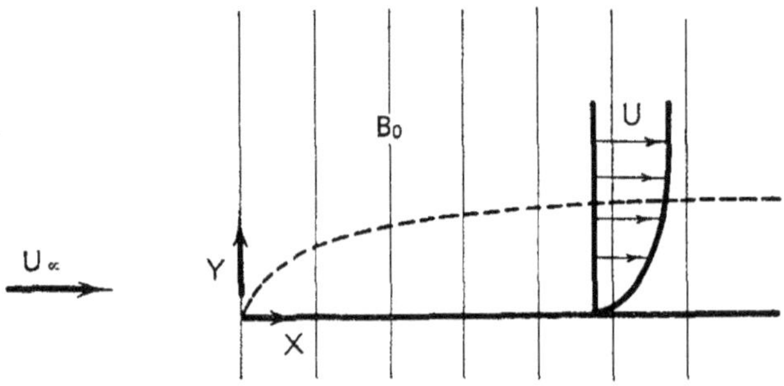

Fig. F,36a. Fluid flowing past a semi-infinite flat plate in the presence of a transverse magnetic field (Rossow [*128*]).

these equations, he introduces the following transformations in accord with boundary layer theory

$$\xi = X, \qquad \eta = Y\sqrt{\frac{U_\infty}{\nu X}}$$

$$\psi = \sqrt{U_\infty \nu X}\,[f_0 + \sqrt{mX}\,f_1 + mXf_2 + (mX)^{\frac{3}{2}}f_3 + \cdots]$$

where the f_i are functions of η alone and $m = m_1/U_\infty$. Equating equal powers of mX, he obtains a sequence of linear ordinary differential equations similar to those obtained in like conventional boundary layer problems. The same procedure is applied to the energy equation

$$U\frac{\partial T}{\partial X} + V\frac{\partial T}{\partial Y} - \frac{\sigma B_0^2}{\rho c_p}U^2 = \frac{\nu}{Pr}\frac{\partial^2 T}{\partial Y^2} + \frac{\mu}{\rho c_p}\left(\frac{\partial U}{\partial Y}\right)^2$$

for the cases of a uniformly cooled plate and an insulated plate.

The results of the numerical integrations are presented in Fig. F,36b. Examination of this figure shows that the skin friction and heat transfer will be reduced by the action of the magnetic field. These results could have been anticipated for the following reasons: Eq. 36-11 is similar to the well-known equation for boundary layer flow under an unfavorable

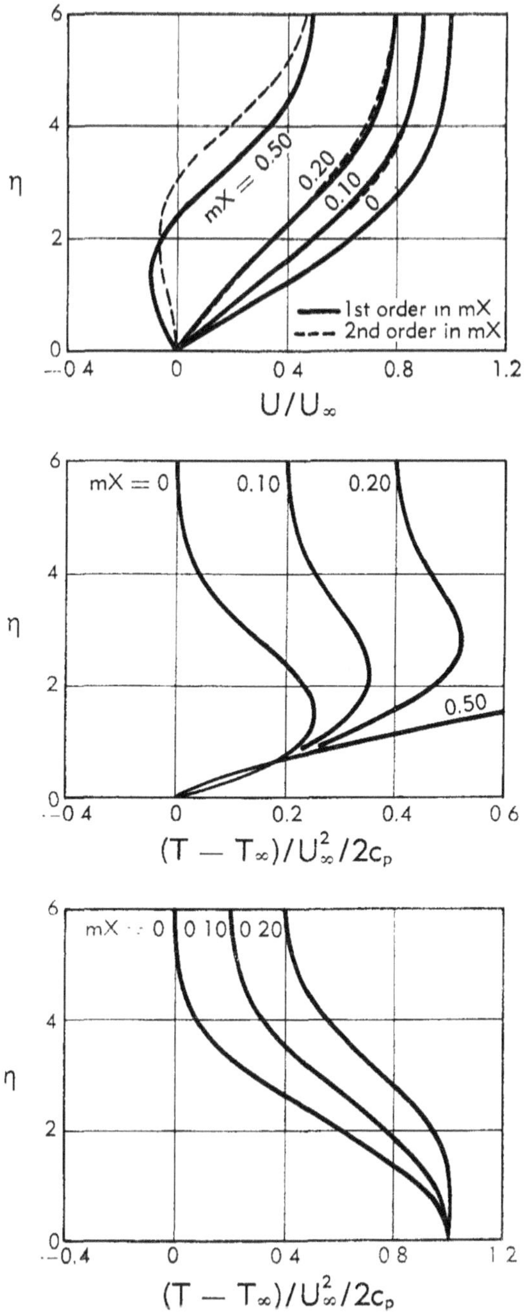

Fig. F,36b. Boundary layer profiles for fluid flowing over a semi-infinite flat plate with a transverse magnetic field fixed relative to the plate (Rossow [*128*]). Top: Velocity. Middle: Temperature distribution for a cool wall, $T_w = T_\infty$. Bottom: Temperature distribution for an insulated wall.

pressure gradient. The essential difference is that the pressure gradient acts uniformly across the boundary layer, whereas the magnetic force m_1U decreases with the reduction of the velocity toward the wall. Therefore, the reduction in skin friction will be less than that for the corresponding pressure gradient. Estimates of skin friction reduction due to the magnetic force could be developed on this basis.

Rossow [*128*] points out that the velocity profile with negative velocities (implying flow separation) resulted from lack of numerica accuracy. Hess [*145*] adds that, since the magnetic force vanishes at the wall and viscous effects are important there, too, an increasing amount of boundary layer material approaches zero velocity. Continuity requires that the flow be lifted from the surface, but reverse flow cannot occur.

The heat transfer results follow from the fact that the total energy of the fluid remains constant under the assumed conditions and for $Pr = 1$. Rossow [*128*], as well as others, have shown this. Thus, since deviations from Reynolds analogy (between skin friction and heat transfer) are well established for conditions of constant total energy [*87*], the effects of the magnetic field on heat transfer can be anticipated from the skin friction results. Like the reduction in skin friction, the reduction in heat transfer rates for the magnetic case can be expected to be less than for the corresponding pressure gradient case.

In this connection it is interesting to note that the body force in free convection acts like a driver, whereas in the problem just discussed the body force retards the flow.

Note the large effect of the magnetic field on the temperature profiles in Fig. F,36, middle, and F,36, bottom. From Fig. F,36, bottom, we can see that the recovery temperature is not altered by the magnetic field. Such a result is to be expected because the magnetic field and the fluid at the wall do not interact.

Another advantage of Hess' pressure split-up approach appears when we examine a second problem treated by Rossow [*128*], viz., the same one as shown in Fig. F,36a, but now the field is to be fixed relative to the fluid. For this case, Rossow uses Eq. 36-10 without the pressure gradient term, instead of Eq. 36-11. Hess [*145*], however, shows that Eq. 36-10 is applicable to a whole class of flows, namely, the second of his proposed two-main types where the velocity in the flow outside the boundary layer is taken to be constant and independent of the magnetic field. If we thus set $(\partial P_d/\partial X) = 0$, Eq. 36-10 becomes

$$U\frac{\partial U}{\partial X} + V\frac{\partial U}{\partial Y} + m_1(U - U_P) = \nu\frac{\partial^2 U}{\partial Y^2} \tag{36-12}$$

It is again of interest to compare this magnetohydrodynamic equation with the corresponding boundary layer equation with a pressure gradient.

Since $U_P > U$, the magnetic force term now corresponds to a favorable pressure gradient. This correspondence occurs because, when the velocity distribution is assumed to be constant, the magnetic pressure gradient which balances the magnetic force $-m_1U$ must "drag" the flow along. The effect of the magnetic force $m_1(U - U_P)$ on the shear is evidently smaller than the effect of a corresponding pressure gradient. As a matter of fact, the magnetic force is even zero in the potential flow but increases toward the wall. However, as the wall is approached, the retarding effects of the viscosity oppose the trend toward an increase in $\partial U/\partial Y$ and the shear. The increase in skin friction and heat transfer anticipated in this case from the analogy with the favorable pressure gradient boundary layer flow is shown quantitatively by Rossow [*128*]. In this case, the magnetic body force affects the flow like conventional body forces in free convection.

In one case of this second type of problem (where the velocity distribution outside the boundary layer is maintained independent of the magnetic field), Hess shows that a boundary layer flow occurs which approaches a fixed velocity profile (and shear) in an asymptotic manner. This happens for the flat plate boundary layer with a constant outside flow velocity, because usual diminishing shear (without a magnetic field) is counteracted by the magnetic force which acts like a favorable pressure gradient in increasing the shear.

All the work discussed above was based on the assumption that the electrical conductivity was either a constant or proportional to the velocity decrement in the boundary layer. Since the electrical conductivity is more accurately a function of temperature, Rossow [*146*] uses such an empirical relation in a reexamination of the two problems just discussed. As in the earlier paper [*128*], the skin friction and heat transfer are reduced when the magnetic field is fixed relative to the plate, and increased when it is fixed relative to the fluid; however, the magnetic effects are much larger in the variable conductivity case. The result follows from the specific empirical relation assumed, which yields high electrical conductivity in a portion of the boundary layer.

Rossow [*128*] and Hess [*145*] also discuss unsteady magnetohydrodynamic problems of the Rayleigh type.

F,37. Internal Problems. Of all the various types of internal magnetohydrodynamics problems already treated (see, for example [*135*] to [*139*] we shall here discuss essentially the work of Hartmann [*127*]. This study of fully developed magnetohydrodynamic flow in channels not only resembles the natural convection problems discussed in Chap. 4, but also brings out many of the physical aspects of magnetohydrodynamics while it remains relatively simple mathematically. Resler and Sears [*135*], Cowling [*147*] and Pai [*148*] also discuss Hartmann's paper.

The configuration to be studied is shown in Fig. F,37a. The flow proceeding from left to right is two-dimensional, steady, incompressible, and fully developed (i.e., independent of X). A uniform magnetic field B_0 is applied normal to the two parallel insulating walls. Under these conditions it follows that all physical velocities are functions of Y alone except, of course, the pressure gradient which can vary linearly with X.

The basic differential equations in dimensional form for this case will be written somewhat differently from those presented in Art. 34.

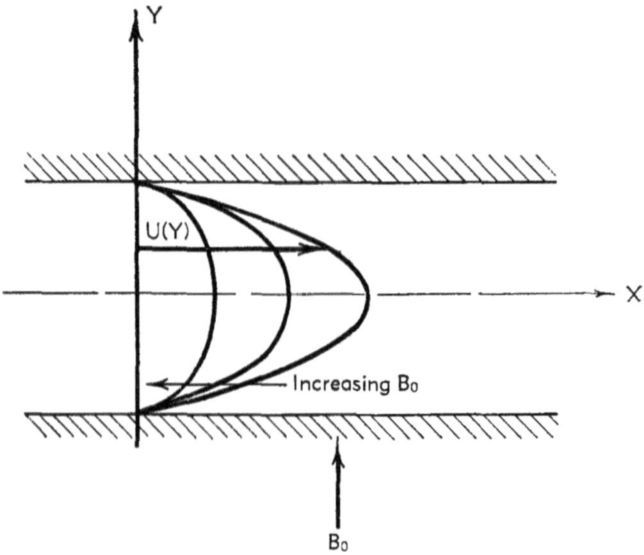

Fig. F,37a. Poiseuille flow of a conducting liquid. Magnetic field applied in plane of flow (Resler and Sears [*135*]).

Here the body force term, Eq. 33-1, will be explicitly formulated by using Maxwell's equation (Eq. 33-3) rather than Ohm's law (Eq. 33-4). Thus

$$F_i = -\frac{\partial}{\partial X_i}\left(\frac{1}{2\mu_e} B_j B_j\right) + \frac{B_j}{\mu_e}\frac{\partial B_i}{\partial X_j} \tag{37-1}$$

which has the advantage that one less dependent variable appears (the electric field is absent). Thus, corresponding to Eq. 34-3b, we have for fully developed steady flow

$$\rho\nu\frac{\partial^2 U}{\partial Y^2} - \frac{\partial P}{\partial X} + \frac{B_Y}{\mu_e}\frac{\partial B_X}{\partial Y} = 0 \tag{37-2}$$

$$-\frac{\partial P}{\partial Y} + \frac{1}{2\mu_e}\frac{\partial}{\partial Y}(B_Y^2 - B_X^2) = 0 \tag{37-3}$$

The field equations (Eq. 34-5b) are now

$$\frac{1}{\mu_e\sigma}\frac{\partial^2 B_X}{\partial Y^2} + B_Y\frac{\partial U}{\partial Y} = 0 \tag{37-4}$$

$$\frac{1}{\mu_e\sigma}\frac{\partial^2 B_Y}{\partial Y^2} = 0 \tag{37-5}$$

$$\frac{1}{\mu_e\sigma}\frac{\partial^2 B_Z}{\partial Y^2} = 0 \tag{37-6}$$

where the subscripts denote components. The continuity equation (Eq. 34-2b) is identically satisfied.

The boundary conditions are

$$\left.\begin{aligned} U(X, -L) &= U(X, L) = 0 \\ B(X, -L) &= B(X, L) = B_0 \end{aligned}\right\} \tag{37-7}$$

From Eq. 37-5 to 37-7 we find that $B_Y = B_0$ and $B_Z = 0$. Therefore, our final system of equations is

$$\rho\nu\frac{\partial^2 U}{\partial Y^2} - \frac{\partial P}{\partial X} + \frac{B_0}{\mu_e}\frac{\partial B_X}{\partial Y} = 0 \tag{37-8}$$

$$\frac{\partial}{\partial Y}\left(P + \frac{1}{2\mu_e}B_X^2\right) = 0 \tag{37-9}$$

$$\frac{1}{\mu_e\sigma}\frac{\partial^2 B_X}{\partial Y^2} + B_0\frac{\partial U}{\partial Y} = 0 \tag{37-10}$$

Let us first find the pressure. If we differentiate Eq. 37-8 with respect to X, we have

$$\frac{\partial^2 P}{\partial X^2} = 0$$

from which we know that

$$P = M(Y) - N(Y)X$$

Also, from Eq. 37-9, we have

$$P = -\frac{1}{2\mu_e}B_X^2 + f(X)$$

so that

$$P = P_0 - \frac{1}{2\mu_e}B_X^2(Y) - \frac{\partial P}{\partial X}X \tag{37-11}$$

where P_0 is the pressure at $X = Y = 0$. Thus, P is a linear function of X. It is thus clear that the pressure gradients across the channel in the present problem arise from the induced fields.

If we integrate the field equation (Eq. 37-10) once, we have

$$\frac{1}{\mu_e \sigma} \frac{\partial B_X}{\partial Y} = -B_0 U + \text{const}$$

which we can recognize to be the Z component of Maxwell's equation (Eq. 33-3a). Thus the constant must just be the induced electric field E_0, so that the field equation is

$$\frac{1}{\mu_e \sigma} \frac{\partial B_X}{\partial Y} = -B_0 U - E_0 \qquad (37\text{-}12)$$

We now combine Eq. 37-12 with Eq. 37-8 to obtain

$$\rho \nu \frac{\partial^2 U}{\partial Y^2} - \frac{\partial P}{\partial X} - B_0 \sigma (E_0 + U B_0) = 0 \qquad (37\text{-}13)$$

To nondimensionalize Eq. 37-13, let

$$U = U^* u(y), \; Y = Ly, \; E = U^* B_0 e, \; B_X = B_0 b(y) \qquad (37\text{-}14)$$

where U^* denotes a characteristic velocity as yet undefined. Substituting Eq. 37-14 into Eq. 37-13 yields

$$\frac{\rho \nu U^*}{L^2} \frac{\partial^2 u}{\partial y^2} - P_X - \sigma B_0^2 U^* [e + u(y)] = 0 \qquad (37\text{-}15)$$

Since we are interested in cases where the body forces balance the pressure gradient, we let

$$U^* = -\frac{P_X}{\sigma B_0^2} \qquad (37\text{-}16)$$

so that Eq. 37-15 becomes

$$\frac{d^2 u}{dy^2} + (Ha)^2 (1 - e - u) = 0 \qquad (37\text{-}17)$$

where the Hartmann number relating the electromagnetic body forces to the viscous forces is

$$Ha = B_0 L \sqrt{\frac{\sigma}{\rho \nu}} \qquad (37\text{-}18)$$

The solution of Eq. 37-17 subject to the boundary conditions is

$$u = (1 - e) \left[1 - \frac{\cosh (Ha) y}{\cosh (Ha)} \right] \qquad (37\text{-}19)$$

To eliminate e and also get a clear picture of the physics of the problem, we must now solve for the induced magnetic field B_X. To this end, we write Eq. 37-12 in nondimensional form as

$$\frac{db}{dy} = -R_m (e + u) \qquad (37\text{-}20)$$

where R_m is the magnetic Reynolds number. Integration of Eq. 37-20 with the use of Eq. 37-19 and the boundary conditions yields

$$b = -R_m \left[y - \frac{\sinh (Ha)y}{\sinh (Ha)} \right] \tag{37-21}$$

In this process we find that

$$e = 1 - (Ha) \coth (Ha)$$

so that

$$u = \frac{Ha}{\tanh (Ha)} \left[1 - \frac{\cosh (Ha)y}{\cosh (Ha)} \right] \tag{37-22}$$

Let us now examine the velocity, magnetic, and electric fields in more detail. As the Hartmann number approaches zero (weak field, therefore, small interaction), the velocity u, from Eq. 37-22, approaches $\frac{1}{2}(Ha)^2(1 - y)^2$. Dimensionally

$$U = U^*u = -\frac{P_X L^2}{2\rho\nu} (1 - y^2)$$

which is just the Poiseiulle velocity distribution obtained with no magnetic field (see Fig. F,37a). As the Hartmann number approaches infinity (large magnetic field), $u \to Ha$ and

$$U = -\frac{P_X L^2}{\rho\nu(Ha)} \to 0$$

Thus the effect of the magnetic field is to reduce the velocity as is shown in Fig. F,37a. In conventional, pure, natural convection flow, on the other hand, the body force is the driver. In combined forced- and natural flows, of course, the body force can, in some cases, retard the flow.

From Eq. 37-21 it can be seen that, for small values of the Hartmann number, the induced magnetic field vanishes. As $Ha \to \infty$, on the other hand, $b \to -R_m y$. The profiles of the magnetic induction are shown in Fig. F,37b, as is also a line of force.

For vanishing Hartmann numbers, e approaches zero and, for large Ha, e approaches $(-Ha)$. Thus

$$\frac{j}{\sigma U^* B_0} \to 0 \quad \text{as} \quad Ha \to 0$$

and

$$\frac{j}{\sigma U^* B_0} \to 1 \quad \text{as} \quad \text{Ha} \to \infty$$

One final point still needs to be made concerning this problem. From Fig. F,37a it can be seen that, for small Ha, the viscosity dominates over the induction drag and the velocity profiles are nearly parabolic. However, if Ha is very large, viscosity is unimportant except in a thin bound-

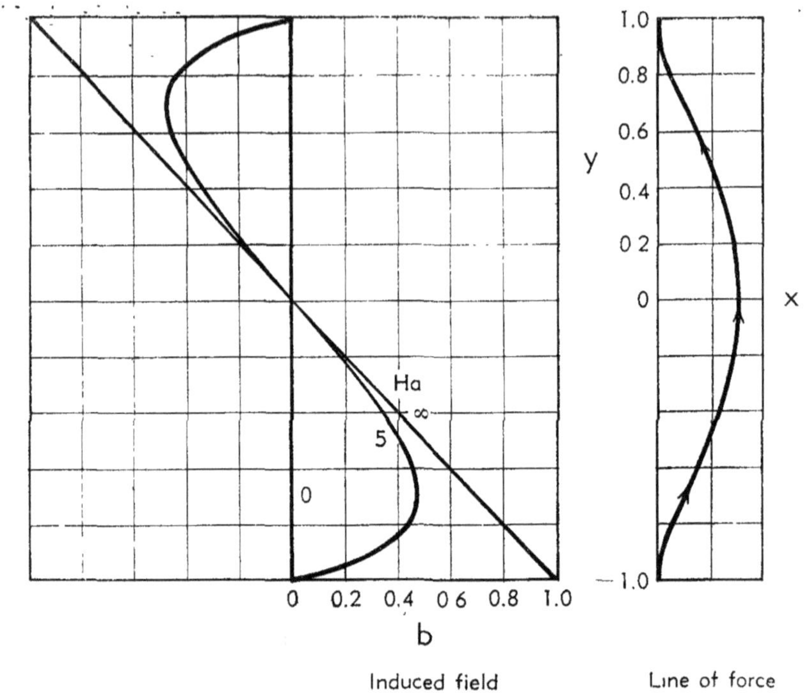

Fig. F,37b. Induced magnetic field for various Hartmann numbers in a fully developed flow and a line of force for $Ha = 5$, magnetic Reynolds number equals unity. Left: Induced field. Right: Line of force.

ary layer near the walls; away from the walls the velocity is essentially constant.

Experiments in a configuration similar to that just analyzed are described in [*147*].

F,38. Cited References

1. Ostrach, S. Laminar natural-convection flow and heat transfer of fluids with and without heat sources in channels with constant wall temperatures. *NACA Tech. Note 2863*, 1952.
2. Ostrach, S. New aspects of natural-convection heat transfer. *Trans. Am. Soc. Mech. Engrs. 75*, 1287–1290 (1953).
3. Schmidt, E., and Beckmann, W. Das Temperatur-und Geschwindigkeitsfeld vor einer Wärme abgebenden senkrechter Platte bei natürlicher Konvektion. *Tech. Mech. u. Thermodynamik, Bd. 1, (10)*, 341–349 (1930). Continued in *Bd. 1, (11)*, 391–406 (1930).
4. Eckert, E. R. G., and Soehngen, E. E. Studies on heat transfer in laminar free convection with Zehnder-Mach interferometer. *Tech. Rept. 5747, ATI 44580, Air Materiel Command* (Dayton, Ohio), Dec. 1948.
5. Schuh, H. Boundary layer of temperature. *Gt. Brit. Ministry of Air Production Volkenrode Repts. and Transls. 1007*, Apr. 1948.
6. Ostrach, S. An analysis of laminar free-convection flow and heat transfer about a flat plate parallel to the direction of the generating body force. *NACA Rept. 1111*, 1953. (Supersedes *NACA Tech. Note 2635.*)

7. Zemansky, M. W. *Heat and Thermodynamics*, 2nd ed. McGraw-Hill, New York, 1943.
8. Lagerstrom, P., Cole, J. D., and Trilling, L. Problems in the theory of viscous compressible fluids. *Calif. Inst. Technol. Guggenheim Aeronaut. Lab.*, 38–40, Mar. 1949.
9. Friedrichs, K. O., and Wasow, W. R. Singular perturbations of non-linear oscillations. *Duke Math. Jour. 13*, 367–381 (1946).
10. Kraus, W. *Messung des Temperatur-und Geschwindigkeitsfeldes bei freier Konvektion. Buch IV C 3*, Verlag G. Braun, Karlsruhe, 1955.
11. Eckert, E. R. G. *Introduction to the Transfer of Heat and Mass.* McGraw-Hill, 1950.
12. McAdams, W. H. *Heat Transmission.* McGraw-Hill, 1942.
13. Lorenz, H. H. Die Warmeübertragung von einer ebenen senkrechten Platte an Öl bei natürlicher Konvektion. *Z. tech. Phys. 15*, 362–366 (1934).
14. Saunders, O. A. Natural convection in liquids. *Proc. Roy. Soc. London A172*, 55–71 (1939).
15. Le Fèvre, E. J. Laminar free convection from a vertical plane surface. *Ninth Intern. Congress of Appl. Mech., Brussels, Belgium, Paper I-168*, 1956.
16. Sparrow, E. M., and Gregg, J. L. Low Prandtl number free convection. *Z. angew. Math. u. Phys. 9*, 383–387 (1958).
17. Sparrow, E. M., and Gregg, J. L. Details of exact low Prandtl number boundary-layer solutions for forced and free convection. *NASA Mem. 2-27-59E*, 1959.
18. Sugawara, S., and Michiyoshi, I. Effects of Prandtl number on the heat transfer by natural convection. The case where Prandtl number is comparatively small. *Proc. Third Japan National Congress for Appl. Mech.*, 315–318 (1953).
19. Goldstein, S., Ed. *Modern Developments in Fluid Dynamics.* Clarendon Press, Oxford, 1938.
20. Rich, B. R. An investigation of heat transfer from an inclined flat plate in free convection. *Am. Soc. Mech. Engrs. Paper 52-F-20*, Sept., 1952.
21. Sparrow, E. M., and Gregg, J. L. Laminar free convection heat transfer from the outer surface of a vertical circular cylinder. *Trans. Am. Soc. Mech. Engrs. 78*, 1823 (1956).
22. Seban, R. A., and Bond, R. Skin friction and heat transfer characteristics of a laminar boundary layer on a cylinder in axial incompressible flow. *J. Aeronaut. Sci. 18*, 671–675 (1951).
23. Carne, J. B. Heat loss by natural convection from vertical cylinders. *Phil. Mag. J. of Sci. 24*, 634–653 (1937).
24. Hama, F. R., and Recesso, J. V. Axisymmetric free-convection temperature field along a vertical thin cylinder. *Univ. Maryland Tech. Note BN116, AFOSR TN-58-25, ASTIA AD-148064*, Jan. 1958.
25. Glauert, M. B., and Lighthill, M. J. The axisymmetric boundary layer on a long thin cylinder. *Proc. Roy. Soc. London A230*, 188–203 (1955).
26. Richmond, R. L. Experimental investigation of thick axially symmetric boundary layers on cylinders at subsonic and hypersonic speeds. *Calif. Inst. Technol. Guggenheim Aeronaut. Lab., Hypersonic Research Project Mem. 39*, June 1957.
27. Schmidt, E. Schlierenaufnahmen des Temperaturfeldes in der Nähe warmeabgebender Körper. *Forsch. Ing.-Wes. 3*, 181 (1932).
28. Hermann, R. Wärmeübergang bei freier Strömung am wagrechten Zylinder in zweiatomigen Gasen. *VDI Forschungsheft 379*, 1–24 (1936). (Also translated as *NACA Tech. Mem. 1366*, 1954.)
29. Jodlbauer, K. Das Temperatur-und Geschwindigkeitsfeld um ein geheiztes Rohr bei freier Konvektion. *Forsch. Ing.-Wes. 4*, 157 (1933).
30. Elenbaas, W. The dissipation of heat by free convection from spheres and horizontal cylinders. *Physica 9*, (3), 285–296 (1942).
31. Sparrow, E. M. *Free Convection with Variable Properties and Variable Wall Temperature. Ph.D. Thesis*, Harvard Univ., 1956.
32. Sparrow, E. M., and Gregg, J. L. The variable fluid property problem in free convection. *Trans. Am. Soc. Mech. Engrs. 80*, 879 (1958).

33. Tanaev, A. A. Heat interchange under conditions of free laminar motion with a gas with variable viscosity at a vertical wall. *Zhurnal Tekhnicheskoi Fiziki 26*, 2714 (1956).
34. Dorodnitzyn, A. A. Boundary layer in a compressible gas. *Prikl. Mat. Mech. 6*, 449–85 (1942).
35. Howarth, L. Concerning the effect of compressibility on laminar boundary layers and their separation. *Proc. Roy. Soc. London A194*, 16–42 (1948).
36. Falkner, V. M., and Skan, S. W. Some approximate solutions of the boundary layer equations. *Phil. Mag. 12*, 865 (1931).
37. Merk, H. J., and Prins, J. A. Thermal convection in laminar boundary layers. *Appl. Sci. Res. A4*, 11–24, 195–206, 207–222 (1953–54).
38. Braun, W. H., Ostrach, S., and Heighway, J. E. Free-convection similarity flows about two-dimensional and axisymmetric bodies with closed lower ends *Inter. Jour. Heat Mass Trans. 2*, 121–135 (1961).
39. Jahnke, E., and Emde, F. *Tables of Functions with Formulae and Curves*. 4th ed. Dover Publications, New York, 1945.
40. Sparrow, E. M., and Gregg, J. L. Laminar free convection from a vertical plate with uniform surface heat flux. *Trans. Am. Soc. Mech. Engrs. 78*, 435 (1956).
41. Sparrow, E. M. Laminar free convection on a vertical plate with prescribed nonuniform wall heat flux or prescribed nonuniform wall temperature. *NACA Tech. Note 3508*, 1955.
42. Sparrow, E. M., and Gregg, J. L. Similar solutions for free convection from a nonisothermal vertical plate. *Trans. Am. Soc. Mech. Engrs. 80*, 379 (1958).
43. Millsaps, K., and Pohlhausen, K. The laminar free-convective heat transfer from the outer surface of a vertical circular cylinder. *J. Aeronaut. Sci. 25*, 357 (1958).
44. Levy, S. Heat transfer to constant-property laminar boundary layer flows with power-function free-stream velocity and wall-temperature variation. *J. Aeronaut. Sci. 19*, 341 (1952).
45. Dotson, J. P. *Heat Transfer from a Vertical Plate by Free Convection. MS Thesis*, Purdue Univ., 1954.
46. Saunders, O. A. Natural convection in liquids. *Proc. Roy. Soc. London A172*, 55–71 (1939).
47. Braun, W. H., and Heighway, J. E. An integral method for natural-convection flows at high and low Prandtl numbers. *NASA Tech. Note D-292*, 1960.
48. LeFèvre, E. J., and Ede, A. J. Laminar free convection from the outer surface of a vertical circular cylinder. *Ninth Intern. Congress of Appl. Mech., Brussels, Belgium, Paper I167*, 1956.
49. Illingworth, C. R. Steady flow in the laminar boundary layer of a gas. *Proc. Roy. Soc. A199* 533–557 (1949).
50. Ferri, A. Influenza del Numero di Reynolds ai Grandi Numeri di Mach. *Atti di Guidonia, No. 67, 68, 69*, 1942.
51. Howarth, L. On the solution of the laminar boundary layer equations. *Proc. Roy. Soc. London A164*, 547 (1938).
52. Ostrach, S. Combined natural- and forced-convection laminar flows and heat transfer of fluids with and without heat sources in channels with linearly varying wall temperatures. *NACA Tech. Note 3141*, 1954.
53. Ostrach, S. Convection phenomena in fluids heated from below. *Trans. Am. Soc. Mech. Engrs. 79*, 299–307 (1957).
54. Low, A. R. Instability of viscous fluid motion. *Nature 115*, 299–300 (1925).
55. Taylor, G. I. Stability of a viscous fluid between two rotating cylinders. *Phil. Trans. Roy. Soc. A223*, 289–343 (1923).
56. Jeffreys, H. Some cases of instability in fluid motion. *Proc. Roy. Soc. London A118*, 195–208 (1928).
57. Low, A. R. On the criterion for stability of a layer of viscous fluid heated from below. *Proc. Roy. Soc. London A125*, 180–195 (1929).
58. Görtler, H. Influence of a slight wall undulation on the distribution of the laminar boundary layer. *Kaiser Wilhelm Inst. for Aerodynamical Research,*

Göttingen, Germany, TPA3/TIB Transl. GDC 10/S556T, 1949. (Issued by Ministry of Supply.)

59. Batchelor, G. K. Heat convection and buoyancy effects in fluids. *Quart. J. Roy. Meteorol. Soc. 80*, 339–359 (1954).
60. Lin, C. C. *The Theory of Hydrodynamic Stability*. Cambridge Univ. Press, 1955.
61. Chandrasekhar, S. The instability of a layer of fluid heated from below and subject to the simultaneous action of a magnetic field and rotation. *Proc. Roy. Soc. London A225*, 1954.
62. Bénard, H. Tourbillons cellulaires dan une nappe liquide. *Rev. gén. sci. 11*, 1261–1271; 1309–1328 (1900).
63. Avsec, D. Tourbillons thermoconvectifs dans l'air application a la météorologie. *Publ. Sci. Tech. Minis. de l'Air 155*, 1939.
64. Lord Rayleigh (J. W. Strutt). On convection currents in a horizontal layer of fluid, when the higher temperature is on the under side. *Phil. Mag. 32*, 529–546 (1916).
65. Pellew, A., and Southwell, R. V. On maintained convective motion in a fluid heated from below. *Proc. Roy. Soc. London A176*, 312–343 (1940).
66. Soberman, R. K. Effect of lateral boundaries on natural convection. *J. Appl. Phys. 29*, 872–873 (1958).
67. Pillow, A. F. The free convection cell in two dimensions. *Aero. Research Lab. Rept. A79*, Melbourne, Australia, 1952.
68. Schmidt, R. J., and Saunders, O. A. On the motion of a fluid heated from below. *Proc. Roy. Soc. London A165*, 216–228 (1938).
69. Morton, B. R. On equilibrium of a stratified layer of fluid. *Quart. J. Mech. and Appl. Math. 10*, 433–447 (1957).
70. Goldstein, A. W. Stability of a fluid layer with unsteady heating and time dependent body force. *NASA Rept. 4*, 1959.
71. Schmidt, R. J., and Milverton, S. W. On the instability of a fluid when heated from below. *Proc. Roy. Soc. London A152*, 586–594 (1935).
72. Chandra, K. Instability of fluids heated from below. *Proc. Roy. Soc. London A164*, 231–242 (1938).
73. Sutton, O. G. On the stability of a fluid heated from below. *Proc. Roy. Soc. London A204*, 297–309 (1950–1951).
74. Terada, T. Some experiments on periodic columnar formation of vortices caused by convection. *Tokyo Imperial Univ., Aero. Research Inst. Rept. 3, (31)*, 1928.
75. Graham, A. Shear patterns in an unstable layer of air. *Phil. Trans. Roy. Soc. London A232*, 285–296 (1933).
76. Elenbaas, W. Heat dissipation of parallel plates by free convection. *Physica 9*, 1–28 (1942).
77. Elenbaas, W. The dissipation of heat by free convection—The inner surface of vertical tubes of different shapes of cross section. *Physica 9*, 865–874 (1942).
78. Elenbaas, W. Dissipation of heat by free convection. *Philips Research Repts. 3*, 338–360; 450–465 (1948).
79. Schmidt, E. H. W. Heat transmission by natural convection at high centrifugal acceleration in water-cooled gas turbine blades. *Proc. General Discussion on Heat Transfer*, 1951.
80. Maslen, S. H. On fully developed channel flows: Some solutions and limitations, and effects of compressibility, variable properties, and body forces. *NACA Tech. Note 4319*, 1958.
81. Ostrach, S. Unstable convection in vertical channels with heating from below including the effects of heat sources and frictional heating. *NACA Tech. Note 3458*, 1955.
82. Ostrach, S. On the flow, heat transfer, and stability of viscous fluids subject to body forces and heated from below in vertical channels. *50 Jahre Grenzschichtforschung*. (H. Görtler, Ed.) Vieweg, Braunschweig, 1955.
83. von Kármán, Th., and Biot, M. A. *Mathematical Methods in Engineering*, McGraw-Hill, 1940.

84. Ostrach, S. Internal viscous flows with body forces. *Boundary Layer Research*, Proc. International Union of Theoretical and Applied Mechanics, Freiburg, 1957. Springer-Verlag, Berlin, pp. 185–209 (1958).
85. Hartree, D. R. On an equation occurring in Falkner and Skan's approximate treatment of the equations of the boundary layer. *Proc. Cambridge Phil. Soc. 33*, 223–237 (1937).
86. Stewartson, K. Further solutions of the Falkner-Skan equation. *Proc. Cambridge Phil. Soc. 50*, 454–465 (1954).
87. Cohen, C. B., and Reshotko, E. Similar solutions for the compressible boundary layer with heat transfer and pressure gradient. *NACA Tech. Note 3325*, 1955.
88. Ostrach, S., and Albers, L. U. On pairs of solutions of a class of internal viscous flow problems with body forces. *NACA Tech. Note 4273*, 1958.
89. Han, L. S. Laminar heat transfer in rectangular tubes with combined free- and forced-convection. *J. Am. Soc. Nav. Eng. 67*, (*1*), 1955.
90. Lu, P. C. *A Theoretical Investigation of Combined Free- and Forced Convection Heat-Generating Laminar Flow inside Vertical Pipes with Prescribed Wall Temperatures. Master of Science Thesis*, Dept. Mech. Eng., Kansas State College of Agriculture and Applied Science, Manhattan, Kansas, 1958.
91. Siegel, R., and Norris, R. H. Tests of free convection in a partially enclosed space between two heated vertical plates. *Trans. Am. Soc. Mech. Engrs. 79*, 663–673 (1957).
92. Ostroumov, G. A. Svobodnaya convectzia v ousloviakh vnoutrennei zadachi. State Publishing House, Technio-Theoretical Literature, Moscow-Leningrad, 1952. Transl. as: Free convection under conditions of the internal problem. *NACA Tech. Mem. 1407*, 1958.
93. Hallman, T. M. Combined forced- and free-laminar heat transfer in vertical tubes with uniform internal heat generation. *Trans. Am. Soc. Mech. Engrs. 78*, 1831 (1956).
94. Hallman, T. M. *Combined Forced and Free Convection in a Vertical Tube. Ph.D. Thesis*, Purdue University, 1958.
95. Lighthill, M. J. Theoretical considerations on free convection in tubes. *Quart. J. Mech. and Appl. Math. 6*, 398–439 (1953).
96. Ostrach, S., and Thornton, P. R. On the stagnation of natural-convection flows in closed-end tubes. *Trans. Am. Soc. Mech. Engrs. 80*, 363–367 (1958).
97. Martin, B. W. Free convection in an open thermosyphon, with special reference to turbulent flow. *Proc. Roy. Soc. London A230*, 502–531 (1955).
98. Hartnett, J. P., and Welsh, W. E. Experimental studies of free convection heat transfer in a vertical tube with uniform wall heat flux. *Trans. Am. Soc. Mech. Engrs. 79*, 1551–1557 (1957).
99. Hammitt, F. G. Modified boundary layer type solution for free-convection flow in a vertical closed tube with arbitrarily distributed internal heat source and wall temperature. *Am. Soc. Mech. Engrs. Paper 58-SA-30*, 1958.
100. Ostrach, S., and Braun, W. H. Natural convection inside a flat rotating container. *NACA Tech. Note 4323*, 1958.
101. Davies, T. V. The flow of a liquid which is heated from below. *Mass. Inst. Technol. Dept. of Meteorol. Sci. Rept. 2*, Apr. 1952.
102. Chandrasekhar, S. The instability of a layer of fluid heated below and subject to Coriolis forces. *Proc. Roy. Soc. London, A217*, 306–327 (1953).
103. Carrier, G. F. Boundary layer problems in applied mechanics. *Advances in Appl. Mech. 3*, 1–19 (1953).
104. Havemann, H. A. Critical cooling. *Mech. Eng., Bangalore, India, 1*, 16–25 (1950).
105. Schmidt, E., Eckert, E. R. G., and Grigull, U. Warmetransport durch Flüssigkeiten in der Nahe des kritischen Zustandes. *Jahrbuch 1939 Deut. Luftfahrtforschung.* (Also *Air Materiel Command Transl. 527.*)
106. Doughty, D. L., and Drake, R. M. Free convection heat transfer from a horizontal right circular cylinder to Freon 12 near the critical state. *Am. Soc. Mech. Engrs. Paper 55-A-100*, Nov. 1955.

107. Illingworth, C. R. Unsteady laminar flow of gas near an infinite flat plate. *Proc. Cambridge Phil. Soc. 46*, 603–613 (1950).
108. Sugawara, S., and Michiyoshi, I. The heat transfer by natural convection in the unsteady state on a vertical flat wall. *Proc. 1st Japan Natl. Congress for Appl. Mech.*, 501–506 (1952).
109. Schuh, H. Calculation of unsteady boundary layers in two-dimensional laminar flow. *KTH Aeronaut. Rapport FL 141, Flygtekniska Laboratoriet, Stockholm, Sweden*, 1953.
110. Siegel, R. Transient free convection from a vertical flat plate. *Trans. Am. Soc. Mech. Engrs. 80*, 347–359 (1958).
111. Courant, R., and Friedrichs, K. O. *Supersonic Flow and Shock Waves.* Interscience Publishers, New York, 1948.
112. Ostroumov, G. A. Unsteady thermal convection about a horizontal cylinder. *Z. Tekhnicheskoi Fiziki* (Russian) *26*, 2720–2730 (1956).
113. Watson, G. N. *A Treatise on the Theory of Bessel Functions.* Cambridge Univ. Press, Cambridge, 1958.
114. Lin, C. C. The stability of two-dimensional parallel flows. Part I—General theory. *Quart. Appl. Math. 3*, 117–142 (1945).
115. Stewartson, K. On the impulsive motion of a flat plate in a viscous fluid. *Quart. Jour. Mech. Appl. Math. 4*, 182 (1951).
116. Izumi, R. Natural heat convection inside the vertical tube. *Proc. 6th Japan Natl. Congress for Appl. Mech.*, 393–396 (1956).
117. Sparrow, E. M., and Kaufman, S. J. Visual study of free convection in a narrow vertical enclosure. *NACA Research Mem. E55L14a*, 1956.
118. Prandtl, L. Einfluss stabilisierener Krafte auf die Turbulenz. *Fortrage Geb. Aerodyn u. nerwandter Gebiete, Aachen, Berlin*, J. Springer, 1–17 (1930).
119. Taylor, G. I. Effect of variation in density on the stability of supported streams of fluid. *Proc. Roy. Soc. London A132*, 499–523 (1931).
120. Goldstein, S. On the stability of supported streams of fluids of different densities. *Proc. Roy. Soc. London A132*, 523 (1931).
121. Schlichting, H. Turbulenz bei Warmeschichtung, *Z. angew. Math. u. Mech. 15*, 313 (1935).
122. Plapp, J. E. *Laminar Boundary-Layer Stability in Free Convection. Ph.D. Thesis*, Calif. Inst. Technol. Dept. of Aero. Eng., 1957.
123. Görtler, H. Uber eine dreidimensionale laminar Grenzschicht an konkaven Wanden. *Nachr. Ges. Wiss. Gott. N.F. 2*, (*1*), 1–26 (1940).
124. Eckert, E. R. G., and Soehngen, E. Interferometric studies on the stability and transition to turbulence of a free-convection boundary layer. *Proc. of General Discussion on Heat Transfer, Inst. Mech. Engrs., London*, 321–323 (1951).
125. Birch, W. D. *On the Stability of Free-Convection Boundary Layers on a Vertical Flat Plate. M.S. Thesis*, U.S. Air Force Inst. Technol., Air Univ., GAE 57-2, 1957.
126. Elsasser, W. M. Hydromagnetism. I.—A review. *Am. J. Phys. 23*, 590–609 (1955).
127. Hartmann, J. Hg-Dynamics I, Theory of the laminar flow of an electrically conductive liquid in a homogeneous magnetic field. *Kgl. Danske Videnskabernes Selskab, Math-fys., Meddeleser XV, 6*, Copenhagen, 1937.
128. Rossow, V. J. On flow of electrically conducting fluids over a flat plate in the presence of a transverse magnetic field. *NACA Rept. 1358*, 1958. (Supersedes *NACA Tech. Note 3971*, 1957.)
129. Rossow, V. J. Magnetohydrodynamic analysis of heat transfer near a stagnation point. *J. Aeronaut. Sci.*, *25*, 334–335 (1958).
130. Neuringer, J. L., and McIlroy, W. Incompressible two-dimensional stagnation-point flow of an electrically conducting viscous fluid in the presence of a magnetic field. *J. Aeronaut. Sci. 25*, 194–198 (1958).
131. Neuringer, J. L., and McIlroy, W. Hydromagnetic effects on stagnation-point heat transfer. *J. Aeronaut. Sci. 25*, 332–334 (1958).
132. Meyer, R. C. On reducing aerodynamic heat-transfer rates by magnetohydrodynamic techniques. *J. Aero/Space Sci. 25*, 561–567 (1958).

133. Kemp, N. H. On hypersonic blunt-body flow with a magnetic field. *AVCO Research Lab., Research Rept. 19,* Feb. 1958.
134. Kemp, N. H., and Petschek, H. E. Two-dimensional incompressible magnetohydrodynamic flow across an elliptical solenoid. *AVCO Research Lab., Research Rept. 26,* Apr. 1958.
135. Resler, E. L., Jr., and Sears, W. R. The prospects for magneto-aerodynamics. *J. Aeronaut. Sci. 25,* 235–245 (1958).
136. Resler, E. L., Jr., and Sears, W. R. Magneto-gas dynamic channel flow. *J. Appl. Math. and Phys. (ZAMP) 9b, Jakob Ackeret Festschrift,* 509–518 (1958).
137. Sears, W. R. Magnetohydrodynamics in aeronautics. *American Rocket Soc. Paper 709-58, presented at 13th Annual Meeting,* Nov. 1958.
138. Shercliffe, J. A. Steady motion of conducting fluids in pipes under transverse magnetic fieids. *Proc. Cambridge Phil. Soc. 49,* 166–168 (1953).
139. Shercliffe, J. A. The flow of conducting fluids in circular pipes under transverse magnetic fields. *J. Fluid Mech. 1,* 644–666 (1956).
140. Panofsky, W. K. H., and Phillips, M. *Classical Electricity and Magnetism.* Addison-Wesley Pub. Co., 1956.
141. Pai, S. I. Energy equation of magneto-gasdynamics. *Phys. Rev. 105,* 1424–1426 (1957).
142. Pai, S. I. *Introduction to the Theory of Compressible Flow.* Van Nostrand, 1959.
143. Rosa, R. *Engineering Magneto-Hydrodynamics, Part 2. Ph.D. Thesis,* Cornell Univ., June 1956.
144. Patrick, R. M. *Magneto-Hydrodynamics of Compressible Fluids. Ph.D. Thesis,* Cornell Univ., 1956.
145. Hess, R. V. Some basic aspects of magnetohydrodynamic boundary-layer flows. *NASA Memo 4-9-59L,* 1959.
146. Rossow, V. J. On magneto-aerodynamic boundary layers. *J. Appl. Math. and Phys. (ZAMP), 9b, Jakob Ackeret Festschrift,* 519–527 (1958).
147. Cowling, T. G. *Magnetohydrodynamics.* Interscience, 1957.
148. Pai, S. I. Magnetohydrodynamics and Magnetogasdynamics. *Univ. Maryland, Tech. Note BN-59,* 1955.

SECTION G

STABILITY OF LAMINAR FLOWS

S. F. SHEN

CHAPTER 1. INTRODUCTION

G,1. The Stability Problem. We shall assume that the fluid flow phenomena obey exactly the fundamental differential equations given previously in Sec. B. The "laminar" solutions can then be found to describe highly idealized special flow problems. For instance, one considers Poiseuille flow as the representation of the flow through a long circular pipe under the action of a pressure drop along the axis. However, there are always departures from the idealized theoretical model in any practical setup. Only if the theoretical solution is insensitive to small deviations of the boundary and initial conditions is it possible to obtain essential agreement between theory and the experiment which may be designed to duplicate the idealized model. In the example of Poiseuille flow, the real pipe might not be strictly axially symmetric, the presence of an entrance section can never be eliminated, the flow must be started through a transient regime, and there is always the possibility of vibrations of the equipment and fluctuations of the environmental conditions. Whether any of these sources of trouble, even though limited to small magnitudes by careful arrangement, might not cause the flow to become radically different from Poiseuille flow is thus a stability problem.

In the theory of laminar stability, it is usually implied that we consider only the stability of a solution with respect to changes of the initial conditions. One begins with an ideal laminar flow and assumes the presence of a disturbance at time $t = 0$. The development with time of the initial disturbance is followed. If, as time goes on, the ideal flow without disturbance tends to reemerge, then it is said to have "stability" (strictly speaking, with respect to the particular disturbance assumed). Conversely, if the ideal flow becomes more and more obliterated by the growth of the disturbance, it is said to have "instability."

The theory of laminar instability is of great practical interest because of its capability in providing qualitative criteria regarding the very important problem of laminar-turbulent transition. The theoretically possible laminar flows are often not observed in practice; turbulent flows prevail instead. One is led to suspect therefore that there might be more than one solution to the equations of fluid motion. The stability theory explains rationally at least the nonoccurrence of the laminar flow under certain combinations of the flow parameters. At present, the achievements of the stability theory are mainly limited to the description of the initial breakdown of the laminar flow due to infinitesimal disturbances. It is still a long way from the regime of the fully developed turbulent flow. On the other hand, the theory predicts the qualitative effectiveness of the various flow parameters in promoting or suppressing stability. These predictions appear to have a direct correlation with the role of the same parameters in promoting or suppressing transition, as evidenced by comparison with available experimental investigations.

Fundamental aspects of the theory of laminar stability have recently been summarized in a monograph by Lin [*1*]. Without going deeply into the subtle mathematical questions, we devote ourselves more to a general understanding of the subject and the various applications which have been made in problems of technical interest. Some of the applications undoubtedly have yet to reach an established status, but the results, however crude, might be of practical significance when nothing better appears available.

G,2. Formulation of the Stability Problem. For the sake of simplicity in clarifying a few of the basic ideas, consider first the Navier-Stokes equations for an incompressible and viscous fluid,

$$\frac{\partial \mathbf{u}}{\partial t} - \mathbf{u} \times \boldsymbol{\omega} = -\nabla\left(\frac{p}{\rho} + \Omega + \frac{1}{2}\mathbf{u}^2\right) - \nu\nabla \times \boldsymbol{\omega} \tag{2-1}$$

$$\nabla \cdot \mathbf{u} = 0 \tag{2-2}$$

where $\mathbf{u}$ = velocity vector
$\boldsymbol{\omega} = \nabla \times \mathbf{u}$ = vorticity vector
Ω = potential for body forces, if any,

and p, ρ, ν are the usual notations for pressure, density, and kinematic viscosity, respectively. Let the laminar solution be $\mathbf{u} = \mathbf{u}_0$, $p = p_0$. We now examine a neighboring solution

$$\mathbf{u} = \mathbf{u}_0 + \mathbf{u}_1, \qquad p = p_0 + p_1$$

in particular, the behavior of $(\mathbf{u}_1, p_1)$ as time goes on, following an arbitrary set of initial values for these variables.

Suppose we restrict ourselves to infinitesimal disturbances and to a steady solution $\mathbf{u}_0$. Then it is reasonable to linearize the equations for $(\mathbf{u}_1, p_1)$:

$$\frac{\partial \mathbf{u}_1}{\partial t} - \mathbf{u}_1 \times \boldsymbol{\omega}_0 - \mathbf{u}_0 \times \boldsymbol{\omega}_1 = -\nabla\left(\frac{p_1}{\rho} + \mathbf{u}_0 \cdot \mathbf{u}_1\right) - \nu\nabla \times \boldsymbol{\omega}_1 \tag{2-3}$$

$$\nabla \cdot \mathbf{u}_1 = 0 \tag{2-4}$$

For the linearized system, the stability characteristics are generally determined by the "frequency response."[1] To describe natural sinusoidal oscillations, we may use the complex notation,

$$\mathbf{u}_1 = \mathbf{u}^{(1)}e^{-i\sigma t}, \qquad p_1 = p^{(1)}e^{-i\sigma t}, \qquad \boldsymbol{\omega}_1 = \boldsymbol{\omega}^{(1)}e^{-i\sigma t} \tag{2-5}$$

and thus obtain from Eq. 2-3 and 2-4,

$$-i\sigma\mathbf{u}^{(1)} - \mathbf{u}^{(1)} \times \boldsymbol{\omega}_0 - \mathbf{u}_0 \times \boldsymbol{\omega}^{(1)} = -\nabla\left(\frac{p^{(1)}}{\rho} + \mathbf{u}_0 \cdot \mathbf{u}^{(1)}\right) - \nu\nabla \times \boldsymbol{\omega}^{(1)} \tag{2-6}$$

$$\nabla \cdot \mathbf{u}^{(1)} = 0 \tag{2-7}$$

Here $\mathbf{u}^{(1)}$, $p^{(1)}$ are functions of the space coordinates only. The solution satisfying the boundary conditions for the disturbances now carries σ as a parameter. Since the original boundary conditions are already satisfied by the steady solution $\mathbf{u}_0$, p_0, the solution $\mathbf{u}^{(1)}$, $p^{(1)}$ must vanish at the boundaries and will be possible only for special values of the real σ. Thus an eigenvalue problem for σ results, and $\mathbf{u}^{(1)}$, $p^{(1)}$ may be referred to as the "mode" of oscillation. Usually, by slightly changing some of the flow parameters, such as the Reynolds number, one finds slightly different eigensolutions for $\mathbf{u}^{(1)}$, $p^{(1)}$ by allowing σ to be complex. If the complex eigenvalue σ has a positive imaginary part, the solution $\mathbf{u}_1$, p_1 is "amplified," because it tends to infinity as $t \to \infty$, and instability is indicated. Conversely, if σ has a negative imaginary part, the solution is "damped" and the steady flow is stable. The transition from stability to instability, in such cases, is marked by the vanishing of the imaginary part of σ.

It may be noted, however, that a degenerate case of the transition from stability to instability occurs when simultaneously the real and imaginary parts of σ vanish. The neutral mode then corresponds to a "secondary flow" rather than a "steady oscillation." In certain problems, it is assumed and sometimes can be shown that the real part of σ always vanishes if the motion becomes unstable. The "secondary flow" simply diverges with time after the stability boundary is crossed. There one locates the stability boundary by finding the conditions for the existence of the steady secondary flow referring to "the principle of exchange of stabilities." (See Art. 22 and 24.)

Because of the mathematical difficulty, it is customary to consider

[1] This is of course well known in dealing with mechanical systems described by ordinary differential equations in the variable t and with constant coefficients. The response to an initial value, or "step-function input," is the Fourier transform of the frequency response, and vice versa. Our fluid problem may be regarded as a system having an infinite number of fluid particles, so to speak, mutually interconnected in the x, y, z space. However, we overlook the question about the completeness of the eigenfunctions, in order to be able to represent an arbitrary initial disturbance, such as that studied by Haupt [2].

simplified cases such that $\mathbf{u}_0 = \mathbf{u}_0(y)$, y being one of the spatial coordinates, and that the other two spatial coordinates, say x and z, extend to infinity in both positive and negative directions. Then it becomes plausible to assume that the solution $\mathbf{u}^{(1)}$, $p^{(1)}$ should be either periodic in x and z or independent of them. Therefore put

$$\left.\begin{aligned} \mathbf{u}^{(1)} &= f_n(y, \sigma)e^{i(\alpha x+\beta z)} \\ p^{(1)} &= \varpi(y, \sigma)e^{i(\alpha x+\beta z)} \end{aligned}\right\} \tag{2-8}$$

with $n = 1, 2, 3$ for the x, y, z components, respectively, and α, β constants. The degenerate cases of $\alpha = 0$, $\beta = 0$, or $\alpha = \beta = 0$ of course correspond to solutions independent of x, or z, or both. Substituting Eq. 2-8 into Eq. 2-6 and 2-7, we get the equations for f_n and ϖ as ordinary differential equations. For every α, β, there may be an infinite set of eigenvalues σ and the corresponding eigensolutions. It is assumed that any disturbance at $t = 0$ could be analyzed into elements of the type of Eq. 2-8. Stability of the given $\mathbf{u}_0$ will thus result when *all* the σ's for *all* the α, β values do not have a positive imaginary part.

While conceptually straightforward, the complete determination of all the eigenvalues is in general a very difficult task. One such study was made by Grohne [*3*], who showed that the lowest mode usually found appears to have the least stability in the simpler cases considered by him. But whether, under complicated circumstances, some of the higher modes might not contribute to instability even though the lowest one should indicate stability, is a question that remains to be answered for the individual case.

G,3. Disturbance Equations in Cartesian Coordinates. Two-Dimensional and Three-Dimensional Disturbances for Parallel Flows over a Flat Surface. So far we have not specified the spatial coordinates x, y, z. In boundary layer-type flows, we shall now regard them as a set of general orthogonal coordinates, with the y axis normal to the solid surface or normal to the main stream direction. The simpler steady flow $\mathbf{u}_0(y)$ then consists of components of the boundary layer velocity vector, their changes in the x and z directions being neglected. In the same approximate manner, we shall further neglect the small y component of the vector $\mathbf{u}(y)$. The steady flow is now a "parallel flow."

In such parallel flows, elementary disturbances are of the type of Eq. 2-8. These "three-dimensional" disturbances, having all of the three components in x, y, z directions, may be interpreted as oblique waves with the wave front direction depending on the ratio of α to β. However, since the directions are arbitrary, it is convenient to choose them such that the x axis coincides with the direction of wave front propagations. By so doing, Lin and Dunn [*4*] were able to arrive immediately and with clear physical insight at the so-called "Squire's Theorem," first derived for incompressible flow over a flat surface by Squire [5], and further extended it to the compressible case.

Let us now write out in detail the linearized differential equations for the disturbances, Eq. 2-6 and 2-7, in Cartesian coordinates. Let $\bar{u}$, 0, $\bar{w}$ be the velocity components of the steady flow $\mathbf{u}_0(y)$, and u', v', w' the components of the disturbance $\mathbf{u}_1$. With a reference velocity $U^* = 1$, density $\rho^* = 1$, and a reference length $L^* = 1$, we define time and pressure in terms of the reference quantities $T^* = L^*/U^*$ and $p^* = \rho^* U^{*2}$, respectively. Then the dimensionless form of the system of Eq. 2-6 and 2-7 becomes

$$\left.\begin{aligned} -i\sigma u' + \bar{u}\frac{\partial u'}{\partial x} + v'\frac{d\bar{u}}{dy} + \bar{w}\frac{\partial u'}{\partial z} &= -\frac{\partial p'}{\partial x} + \frac{1}{Re}\nabla^2 u' \\ -i\sigma v' + \bar{u}\frac{\partial v'}{\partial x} \qquad\quad + \bar{w}\frac{\partial v'}{\partial z} &= -\frac{\partial p'}{\partial y} + \frac{1}{Re}\nabla^2 v' \\ -i\sigma w' + \bar{u}\frac{\partial w'}{\partial x} + v'\frac{d\bar{w}}{dy} + \bar{w}\frac{\partial w'}{\partial z} &= -\frac{\partial p'}{\partial z} + \frac{1}{Re}\nabla^2 w' \end{aligned}\right\} \tag{3-1}$$

$$\frac{\partial u'}{\partial x} + \frac{\partial v'}{\partial y} + \frac{\partial w'}{\partial z} = 0 \tag{3-2}$$

where Re is the Reynolds number U^*L^*/ν and ∇^2 is the Laplacian operator. Introducing Eq. 2-8 we get

$$\left.\begin{aligned} if_1[-\sigma + \alpha\bar{u} + \beta\bar{w}] + f_2\frac{d\bar{u}}{dy} &= -i\alpha\varpi + \frac{1}{Re}\left[\frac{d^2}{dy^2} - \alpha^2 - \beta^2\right]f_1 \\ if_2[-\sigma + \alpha\bar{u} + \beta\bar{w}] \qquad\qquad &= -\frac{d\varpi}{dy} + \frac{1}{Re}\left[\frac{d^2}{dy^2} - \alpha^2 - \beta^2\right]f_2 \\ if_3[-\sigma + \alpha\bar{u} + \beta\bar{w}] + f_2\frac{d\bar{w}}{dy} &= -i\beta\varpi + \frac{1}{Re}\left[\frac{d^2}{dy^2} - \alpha^2 - \beta^2\right]f_3 \end{aligned}\right\} \tag{3-3}$$

$$i\alpha f_1 + \frac{df_2}{dy} + i\beta f_3 = 0 \tag{3-4}$$

Now, the direction of x is chosen to be such that $\beta = 0$. Eq. 3-3 and 3-4 are simplified into

$$\left.\begin{aligned} if_1[-\sigma + \alpha\bar{u}] + f_2\frac{d\bar{u}}{dy} &= -i\alpha\varpi + \frac{1}{Re}\left[\frac{d^2}{dy^2} - \alpha^2\right]f_1 \\ if_2[-\sigma + \alpha\bar{u}] \qquad\qquad &= -\frac{d\varpi}{dy} + \frac{1}{Re}\left[\frac{d^2}{dy^2} - \alpha^2\right]f_2 \\ if_3[-\sigma + \alpha\bar{u}] + f_2\frac{d\bar{w}}{dy} &= \frac{1}{Re}\left[\frac{d^2}{dy^2} - \alpha^2\right]f_3 \end{aligned}\right\} \tag{3-5}$$

$$i\alpha f_1 + \frac{df_2}{dy} = 0 \tag{3-6}$$

We see therefore that the equation for f_3 is decoupled from the rest, and *f_1 and f_2 can be solved independently of f_3 and $\bar{w}$.* That is, for each oblique wave one imagines a cut along the direction of the wave propagation, and calculates stability for the velocity profile $\bar{u}(y)$ in that direction, together with the appropriate boundary conditions for the flow in that same direction, under the action of two-dimensional disturbances u', v' only.

We get "Squire's theorem" in the special case that the steady flow is actually two-dimensional. As the direction of the oblique wave is changed, all velocity profiles $\bar{u}(y)$ are now identical in shape; however the magnitudes are the projections of the actual velocity vector. Thus the wave propagation in the direction of the velocity vector $\mathbf{u}_0(y)$ corresponds to the highest effective velocity—or Reynolds number. Usually a laminar flow is always stable at sufficiently low Reynolds numbers because of viscous dissipation, and a finite "critical Reynolds number" may be reached when at least some self-sustained oscillation becomes possible. In the present case, the conclusion is therefore that the critical Reynolds number for an oblique wave, i.e. the general three-dimensional disturbances, must be higher than that for the wave propagating in the basic flow velocity direction. And, because of the uncoupling, one needs to consider the two-dimensional disturbances only, as if there were no lateral disturbances. This is a very important result, in that it justifies the usual limitations to two-dimensional disturbances only, which otherwise had been based on nothing more than a desire to simplify the mathematics.

It must be further kept in mind, however, that the uncoupling of the lateral oscillations is possible only when the curvature of the surface is negligible, and that the lateral oscillations still exist. The latter point might especially have some bearing on attempts to estimate the magnitude of the energy of amplified disturbances after crossing the stability boundary, since there the resultant must be taken as the resultant of all three components rather than only the u', v' components. In addition, the stability boundary for the oblique wave is obtained by shifting the stability boundary for the two-dimensional disturbance toward higher Re at every wave number α. As a result, however, certain combinations of α and Re may become unstable for the oblique wave although stable for the two-dimensional disturbance [*6,7*].

The extension to compressible flow will be taken up later. It suffices here to remark that a choice of the coordinate axes x, z obviously can only affect the vector velocity distribution but not the scalar quantities, such as density or temperature. Consequently the stability of a compressible flow with respect to oblique waves is not generally derivable from a consideration of the two-dimensional disturbances.

G,4. Stability of Parallel Flows with Respect to Longitudinal Vortex Disturbances and Transverse Wave Disturbances. In the previous article, some useful conclusions were drawn regarding the stability with respect to oblique wave disturbances in a parallel flow, by choosing the x axis to coincide with the direction of wave propagation. The stability problem becomes quasi-two-dimensional, in that we need to consider only the u', v' disturbances on the parallel flow $\bar{u}(y)$. Disregarding the lateral disturbances w', the disturbances u', v' may be visualized as being equivalent to a disturbance *vorticity* distribution, periodic in x and having the vorticity direction in the transverse z direction only. We may refer to this type of disturbance as the "transverse-vortex disturbances." As disturbances of this type are usually found to propagate downstream, they will hereafter be called "transverse wave disturbances." Let us recall that these disturbances are of importance after we set $\beta = 0$ in Eq. 3-3 and 3-4.

On the other hand, suppose we consider the alternative extreme of setting $\alpha = 0$. In Cartesian coordinates, the roles of u and w are simply exchanged, and nothing new follows. However, the disturbances v', w' are now solved independently of u'. Analogous to what was just defined, we may regard this type of disturbance as the "longitudinal vortex disturbance," the vorticity component being parallel to the x axis. Although the longitudinal vortex disturbance is of trivial importance for parallel flows over a flat surface with basic velocity in the x axis direction only,[2] it turns out to be of great importance when there is centrifugal force due to surface curvature in the x direction, or when other body forces such as thermal buoyancy exists.

To illustrate the difference between the stabilities of flows over flat and over curved surfaces, let the equations of motion be written in cylindrical coordinates, the radial direction playing the role of y above. Instead of Eq. 3-1 and 3-2 we now have in dimensionless form and again with parallel flow,

$$-i\sigma u' + v'\frac{d\bar{u}}{dr} + \frac{\bar{u}}{r}\frac{\partial u'}{\partial \phi} + \bar{w}\frac{\partial u'}{\partial z} + \frac{\bar{u}v'}{r} = -\frac{1}{r}\frac{\partial p'}{\partial \phi} + \frac{1}{Re}\left[\nabla^2 u' - \frac{u'}{r^2} + \frac{2}{r^2}\frac{\partial v'}{\partial \phi}\right] \tag{4-1}$$

$$-i\sigma v' + \frac{\bar{u}}{r}\frac{\partial v'}{\partial \phi} + \bar{w}\frac{\partial v'}{\partial z} - 2\frac{\bar{u}u'}{r} = -\frac{\partial p'}{\partial r} + \frac{1}{Re}\left[\nabla^2 v' - \frac{v'}{r^2} - \frac{2}{r^2}\frac{\partial v'}{\partial \phi}\right] \tag{4-2}$$

$$-i\sigma w' + v'\frac{d\bar{w}}{dr} + \frac{\bar{u}}{r}\frac{\partial w'}{\partial \phi} + \bar{w}\frac{\partial w'}{\partial z} = -\frac{\partial p'}{\partial z} + \frac{1}{Re}\nabla^2 w' \tag{4-3}$$

$$\frac{1}{r}\frac{\partial}{\partial r}(rv') + \frac{1}{r}\frac{\partial u'}{\partial \phi} + \frac{\partial w'}{\partial z} = 0 \tag{4-4}$$

[2] With $\bar{w} = 0$, these vortices behave as if in a fluid at rest and the motion is always damped. The solution was explicitly given by Rayleigh [*8*].

where ϕ, r, z are the circumferential, radial, and axial coordinates, u, v, w the respective velocity components, and

$$\nabla^2 \equiv \frac{\partial^2}{\partial r^2} + \frac{1}{r}\frac{\partial}{\partial r} + \frac{1}{r^2}\frac{\partial^2}{\partial \phi^2} + \frac{\partial^2}{\partial z^2}$$

Introducing as in Eq. 2-8,

$$\mathbf{u}^{(1)} = f_n(r, \sigma)e^{i(\alpha\phi+\beta z)}, \qquad p^{(1)} = \varpi(r, \sigma)e^{i(\alpha\phi+\beta z)}$$

we get

$$if_1\left[-\sigma + \frac{\alpha\bar{u}}{r} + \beta\bar{w}\right] + f_2\left[\frac{d\bar{u}}{dr} + \frac{\bar{u}}{r}\right] = -\frac{i\alpha\varpi}{r} + \frac{1}{Re}\left[\left(\nabla_1^2 - \frac{1}{r^2}\right)f_1 + \frac{2i\alpha}{r^2}f_2\right] \tag{4-5}$$

$$if_2\left[-\sigma + \frac{\alpha\bar{u}}{r} + \beta\bar{w}\right] - 2\frac{\bar{u}}{r}f_1 = -\varpi' + \frac{1}{Re}\left[\nabla_1^2 - \frac{1}{r^2} - \frac{2i\alpha}{r^2}\right]f_2 \tag{4-6}$$

$$if_3\left[-\sigma + \frac{\alpha\bar{u}}{r} + \beta\bar{w}\right] + f_2\frac{d\bar{w}}{dr} = -i\beta\varpi + \frac{1}{Re}\nabla_1^2 f_3 \tag{4-7}$$

$$\frac{1}{r}\left[f_2 + r\frac{df_2}{dr}\right] + \frac{i\alpha}{r}f_1 + i\beta f_3 = 0 \tag{4-8}$$

with

$$\nabla_1^2 \equiv \frac{d^2}{dr^2} + \frac{1}{r}\frac{d}{dr} - \frac{\alpha^2}{r^2} - \beta^2 \tag{4-9}$$

By specializing Eq. 4-5 and 4-9 for cases $\alpha = 0$ and $\beta = 0$, it is easy to conclude the following: For $\beta = 0$, the disturbance equation for f_3 is still uncoupled from the rest and $\bar{w}$ does not enter in the equations for f_1, f_2, ϖ; hence we have again a quasi-two-dimensional problem. If the curvature is everywhere small, the solution differs from that over a flat surface previously discussed only through the terms $O(r^{-1})$. For $\alpha = 0$, however, all equations are coupled, the key term being $-2\bar{u}f_1/r$ of Eq. 4-6, which has its origin in the centrifugal force term and vanishes for the flow over a flat surface by letting $r \to \infty$. Furthermore, both $\bar{u}$ and $\bar{w}$ are of importance. The stability characteristics with respect to longitudinal and transverse vortex disturbances are thus expected to behave rather differently.

Historically, the study of stability of laminar flows, such as the classical plane Couette and Poiseuille flows, was started with the transverse vortex disturbances by Rayleigh [*8*], followed notably by Heisenberg [*9*], Tollmien [*10*], and Lin [*11*]. In the literature, these disturbances are sometimes referred to as the "Tollmien waves." The stability of longitudinal vortex disturbances was first demonstrated by Taylor [*12*], on the instability of Couette flow between rotating cylinders. Later Görtler [*13*] showed the extreme instability for flows near a concave wall due to the same type of disturbances. These disturbances are consequently often referred to as the "Taylor-Görtler" vortices. For mathematical simplicity, these two types are often the "modes" assumed in stability investigations. The experiences of cases so far studied appear to indicate that, whenever body forces exist, the stability characteristics would be dictated largely by the action of these body forces. If the body forces tend to excite disturbances of the longitudinal vortex type, the disturbances would as a rule play a more dominant role in causing instability than the transverse vortices.

G,5. The Energy Balance Relation in the Two-Dimensional Problem. Sufficient Condition for Stability. While the growth or decay with time of one component of the velocity disturbances often suffices to reveal the instability or stability, a more significant parameter is probably the growth or decay of the energy of the disturbance. For incompressible flow, the energy consists of the kinetic energy only. The time rate of change of the kinetic energy of the disturbance, at a given *station x, can be obtained from the perturbation equations for the velocity* components, by multiplying with the proper velocity component, integrating throughout the thickness of the basic flow, and summing all three components.

To be specific, we shall consider only the classical case of parallel basic flow with two-dimensional transverse disturbances periodic in x, so that $\mathbf{u}_0$ and $\mathbf{u}_1$ have only x and y components in Eq. 3-1 and 3-2. Integrating throughout the thickness in y and over a wavelength in x, and noting the vanishing of the disturbances at the boundaries for y lead to the following:

$$\frac{\partial E}{\partial t} = \rho M - \mu N \tag{5-1}$$

where

$$\left.\begin{aligned} E &= \iint \frac{1}{2}\rho(u'^2 + v'^2)dxdy \\ \rho M &= -\iint \rho u'v' \frac{d\bar{u}}{dy} dxdy \\ \mu N &= \iint \mu\zeta'^2 dxdy \end{aligned}\right\} \tag{5-2}$$

Thus the time rate of change of E, the kinetic energy of the disturbances, depends on the balance of the two terms on the right-hand side. Eq. 5-1 further suggests a way to estimate the minimum critical Reynolds number. Neutral stability may now be defined to occur when $\partial E/\partial t = 0$, i.e. the two terms on the right-hand side balance each other. In dimensionless form, at neutral stability we get

$$Re_{cr} = \frac{N}{M}$$

The minimum critical Reynolds number then is given by minimizing the quotient N/M with respect to all possible disturbance modes u', v'. Applications of this procedure generally led to minimum critical Reynolds numbers that are too low [*1*, pp. 60, 61]. The reason is presumably that, without sufficient knowledge of the proper distribution of u', v', the trial functions assumed were too inaccurate. This is well illustrated by the recent work of Stuart [*14*], who was able to exploit Eq. 5-1 with success for finite amplitude disturbances. His assumed disturbance modes were calculated from the linear theory, which may be expected to be a close approximation for finite but small amplitudes.

Eq. 5-1 has also been used as the basis for the derivation of sufficient conditions for stability by Synge [*15*]. For convenience, a disturbance stream function ψ' may be introduced,

$$u' = \frac{\partial \psi'}{\partial y}, \qquad v' = -\frac{\partial \psi'}{\partial x}$$

If $\psi' = \phi(y)e^{i\alpha(x-ct)}$, $\alpha > 0$, one obtains from Eq. 5-1 that, if c_i is the imaginary part of c,

$$2\alpha Re\, c_i(I_1^2 + \alpha^2 I_0^2) = -i\alpha Re(Q - \bar{Q}) - 2(I_2^2 + 2\alpha^2 I_1^2 + \alpha^4 I_0^2) \tag{5-3}$$

where
$$I_0^2 = \int_{y_1}^{y_2} |\phi|^2 dy, \qquad I_1^2 = \int_{y_1}^{y_2} |\phi'|^2 dy$$

$$I_2^2 = \int_{y_1}^{y_2} |\phi''|^2 dy, \qquad Q = \int_{y_1}^{y_2} \frac{d\bar{u}}{dy}(\phi'\bar{\phi})dy$$

and $\bar{Q}$, $\bar{\phi}$ are the complex conjugates of Q, ϕ. Now, if $k = |d\bar{u}/dy|_{max}$,

$$\begin{aligned} |Q - \bar{Q}| &\leqq 2\int_{y_1}^{y_2} \left|\frac{d\bar{u}}{dy}\right| |\phi'|\,|\phi| dy \\ &\leqq 2kI_1I_0 \end{aligned}$$

by Schwarz inequality. Thus one gets from Eq. 5-3

$$\alpha Re\, c_i(I_1^2 + \alpha^2 I_0^2) \leqq \alpha Re\, kI_1I_0 - (I_2^2 + 2\alpha^2 I_1^2 + \alpha^4 I_0^2) \tag{5-4}$$

which provides an estimate of c_i, and hence the stability, at given α, Re for any function ϕ satisfying the boundary conditions. When $c_i < 0$ for all ϕ satisfying the boundary conditions, the flow is stable. It is, however, a sufficiency condition since actually only those ϕ's satisfying both the boundary conditions *and* the disturbance equation need be considered.

Synge also gave explicit sufficient conditions for stability without actually substituting specific disturbance modes. For any real constants a and b, one writes

$$\int_{y_1}^{y_2} (\phi + ay\phi' + b\phi'')(\bar{\phi} + ay\bar{\phi}' + b\bar{\phi}'')dy > 0 \tag{5-5}$$

whence

$$b^2I_2^2 > I_1^2(ab - a^2 + 2b) + I_0^2(a - 1) \tag{5-6}$$

Eliminating I_2^2 from Eq. 5-4 by means of Eq. 5-6, we find

$$\alpha Re\, c_i b^2(I_1^2 + \alpha^2 I_0^2) < b^2 k\alpha Re\, I_0 I_1 - I_1^2(2\alpha^2 b^2 + ab - a^2 + 2b) - I_0^2(\alpha^4 b^2 + a - 1) \tag{5-7}$$

Therefore $c_i < 0$ if the following are satisfied:

$$\left.\begin{aligned} (kb^2\alpha Re)^2 &< 4AC \\ A = 2\alpha^2 b^2 + ab - a^2 + 2b &> 0 \\ C = \alpha^4 b^2 + a - 1 &> 0 \end{aligned}\right\} \tag{5-8}$$

As an example, let $a = b = 2$; the sufficient condition for stability reads

$$(k\alpha Re)^2 < (2\alpha^2 + 1)(4\alpha^4 + 1) \tag{5-9}$$

showing that, as $\alpha \to 0$, stability prevails for $\alpha Re < 0(1)$, and that as $\alpha \to \infty$, $Re < 0(\alpha^4)$. The existence of a minimum critical Reynolds number is thus assured. An extension was made by Lessen [*16*] to apply to problems where the fluid region extends to infinity, by simply omitting the weighing factor y in Eq. 5-5. The result, however, is weaker than Synge's when both apply. For example, the one explicit criterion derived by Lessen may be written as

$$kRe < \frac{16}{3\sqrt{3}}\alpha^2 \tag{5-10}$$

which indicates no minimum critical Reynolds number as α varies from zero to infinity.

Returning to Eq. 5-1, we may interpret the terms physically and gain some understanding of the role of viscosity in laminar stability. It is customary to think of viscosity as having a damping action, and such an action is indeed obvious in the term μN—which is always positive. If there is instability in the sense that $\partial E/\partial t > 0$. the other term ρM must be sufficiently positive. Now, the quantity ρM may be interpreted as the transfer of energy from the basic flow to the disturbances through the action of the Reynolds stresses averaged over the wavelength in x, i.e. one may rewrite

$$\rho M = \int \tau' \frac{d\bar{u}}{dy}\, dy$$

$$\tau' = -\int \rho u'v' dx$$

It is then seen that energy is transferred from the basic flow to the disturbances when the Reynolds stress of the disturbances is of the same sign as $d\bar{u}/dy$.

Since the admissable disturbances u', v' are affected by the presence or absence of viscosity, we now see that, besides that of simple damping, the viscosity may have a second role in producing changes of the Reynolds stress τ' above, thereby further modifying the energy transfer balance. This observation, first due to Prandtl, offered the qualitative explanation that the viscosity could be destabilizing. Recently, Lin [*17*] showed explicitly the phase relationship between u' and v' *near the wall.* His result is that the Reynolds stress always acts on the wall in the direction of propagation of the disturbance wave. For a wave propagating in the direction of the basic motion, the Reynolds stress, at least locally near the wall, tends to transfer energy from the basic flow to the disturbance.

CHAPTER 2. THE STABILITY THEORY DEALING WITH TRANSVERSE WAVE DISTURBANCES OF PARALLEL FLOWS

G,6. Differential Equations and the Eigenvalue Problem. While the classical theory has been extensively presented in Lin's book [*1*], it seems desirable to summarize the essential results for readers not too familiar with the general subject, or not particularly interested in the more subtle mathematical questions. For parallel flow with $\bar{u}(y)$ under transverse vortex type disturbances, according to Art. 3, it is only necessary to consider two-dimensional disturbances u', v', and our basic equations of motion are Eq. 3-5 and 3-6. To see the structure of the differential equations clearly, it is convenient to introduce the following variables instead:

$$Z_1 = f_1, \qquad Z_2 = \frac{df_1}{dy}, \qquad Z_3 = f_2, \qquad Z_4 = \varpi \tag{6-1}$$

and, further, let the wave velocity σ/α be denoted by c. In these variables, Eq. 3-5 and 3-6 may be rewritten into a system of first order differential equations:

$$\left.\begin{aligned}
\frac{dZ_1}{dy} &= Z_2 \\
\frac{dZ_2}{dy} &= [i\alpha Re(\bar{u} - c) + \alpha^2]Z_1 + Re\frac{d\bar{u}}{dy}\, Z_3 + i\alpha Re Z_4 \\
\frac{dZ_3}{dy} &= -i\alpha Z_1 \\
\frac{dZ_4}{dy} &= -\frac{i\alpha}{Re} Z_2 - [i\alpha Re(\bar{u} - c) + \alpha^2]\frac{Z_3}{Re}
\end{aligned}\right\} \tag{6-2}$$

Thus it may be concluded that there are four fundamental solutions for the Z_n's ($n = 1, 2, 3, 4$), which are entire functions of the parameters α, c and analytic in the parameter Re in a region excluding the point $Re = 0$. (The point $Re = 0$ is only an apparent singularity which disappears if Z_3 is chosen as the dependent variable, Eq. 6-3.) For fixed y, formal expansion in ascending powers of α or descending powers of Re is therefore justified. However, we shall see later that complications arise for certain values of y.

More often in the literature the system of equations, Eq. 6-2, is reduced to a single differential equation. By successive elimination of Z_4, Z_1, and Z_2, the Orr-Sommerfeld equation is obtained,

$$\left(\frac{d^2}{dy^2} - \alpha^2\right)^2 Z_3 = i\alpha Re\left[(\bar{u} - c)\left(\frac{d^2}{dy^2} - \alpha^2\right) - \frac{d^2\bar{u}}{dy^2}\right] Z_3 \tag{6-3}$$

This form has the advantage of immediately showing the main source of difficulty in the mathematical solution for large αRe; namely, the highest derivative term has a coefficient which tends to zero as αRe tends to infinity. Meanwhile, we are precisely in the range of very large αRe for the stability of basic flows of practical interest.

It is well to note here that Eq. 6-3 may be given a more physical interpretation in terms of the stream functions of the basic and disturbing flows. In a two-dimensional flow the vorticity equation reads

$$\frac{\partial \zeta}{\partial t} + \frac{\partial \psi}{\partial y}\frac{\partial \zeta}{\partial x} - \frac{\partial \psi}{\partial x}\frac{\partial \zeta}{\partial y} = \nu \nabla^2 \zeta$$

Now let $\psi = \psi_0 + \psi_1$, $\zeta = \zeta_0 + \zeta_1$. The linearized disturbance equation becomes

$$\frac{\partial \zeta_1}{\partial t} + \bar{u}\frac{\partial \zeta_1}{\partial x} - \frac{\partial \psi_1}{\partial x}\frac{\partial \zeta_0}{\partial y} = \nu \nabla^2 \zeta_1 \tag{6-4}$$

which is identical to Eq. 6-3 by using the same reference scales and taking $i\alpha\psi_1 = Z_3(y)e^{i\alpha(x-ct)}$.

Leaving aside the problem of how to find the four fundamental solutions, let us assume they are known. For a given basic flow, the boundary conditions are specified. On solid walls, all velocity disturbances must vanish ($Z_1, Z_3 = 0$); if the basic flow extends to infinity, the disturbances must remain bounded at large distances. With such boundary conditions, a secular equation for the determination of the eigenvalue c, say, at given real wave number α and Reynolds number Re results

$$F(\alpha, Re, c) = 0 \tag{6-5}$$

which is generally complex. The eigenvalue c therefore is generally also complex

$$c = c_r + ic_i$$

c_r and c_i denoting the real and imaginary parts, respectively. The complex Eq. 6-5 being equivalent to two real equations, we are in a position to solve

$$c_r = c_r(\alpha, Re), \qquad c_i = c_i(\alpha, Re)$$

Because of the definitions in Eq. 2-5 and $c = \sigma/\alpha$, c_r is seen to be the real wave velocity while αc_i determines the rate of amplification (if $c_i > 0$) and decay (if $c_i < 0$) of the disturbances. The neutral stability thus occurs at $c_i = 0$, defining a locus in the α, Re plane referred to as the "stability boundary" or "neutral curve." Finally, from the neutral curve the *minimum* critical Reynolds number may be found, below which all periodic disturbances are damped.

We shall now take up some of the details.

G,7. General Results from "Inviscid" Considerations. To simplify the notation slightly, we shall use the variable ϕ replacing Z_3/α in Eq. 6-3 which is rewritten in the form

$$\frac{1}{i\alpha Re}\left(\frac{d^2}{dy^2} - \alpha^2\right)^2 \phi = \left[(\bar{u} - c)\left(\frac{d^2}{dy^2} - \alpha^2\right) - \frac{d^2\bar{u}}{dy^2}\right]\phi \tag{7-1}$$

Suppose we formally expand ϕ for $\alpha Re \to \infty$,

$$\phi = \phi^{(0)}(y) + \frac{1}{\alpha Re}\,\phi^{(1)}(y) + \cdots \tag{7-2}$$

Upon substituting Eq. 7-2 into Eq. 7-1 an equation for $\phi^{(0)}$ is obtained:

$$(\bar{u} - c)(\phi^{(0)\prime\prime} - \alpha^2\phi^{(0)}) - \bar{u}''\phi^{(0)} = 0 \tag{7-3}$$

the primed quantities now denoting the derivatives with respect to y. Eq. 7-3 is only of second order instead of fourth. We note that this lowering of the order of the differential equation is similar to what happens when the approximation of potential flow is made in calculating the external flow over bodies in a viscous fluid. An immediate consequence is that slipping along the wall must be allowed, and the boundary condition at a solid wall is only that of no penetration. If the boundaries are solid walls at $y = y_1$ and $y = y_2$, for instance, we prescribe

$$\phi^{(0)} = 0 \quad \text{at} \quad y = y_1 \quad \text{and} \quad y_2 \tag{7-4}$$

or we may consider one half of a symmetrical profile,

$$\phi^{(0)}(y_1) = 0, \qquad \phi^{(0)\prime}(y_2) = 0 \tag{7-5}$$

Although incomplete, Eq. 7-3 provides often useful information regarding the stability at large Reynolds numbers. It is reasonable to expect that, so long as the solution $\phi^{(0)}$ behaves smoothly everywhere, it would be consistent with the omission of the viscous terms for large

αRe—except near the solid wall where the actual no-slipping condition must exert its influence throughout a thin viscous layer. At any rate, the second order equation (Eq. 7-3) is much simpler to analyze and has thus become the subject of study beginning with Rayleigh.

We next list the general results which are obtained without solving the differential equation explicitly. Assume that amplified disturbances exist. Then, the necessary conditions are: (1) $\bar{u}''$ must vanish somewhere between y_1, y_2, i.e. the velocity profile must thus have at least one point of inflection (Rayleigh's theorem); (2) the real wave velocity for such disturbances must lie between the extreme values of $\bar{u}(y)$.

The proof of these results is based upon the same technique. Since only amplified disturbances are considered, c is always complex and Eq. 7-3 may be written

$$\phi'' - \alpha^2\phi - \frac{\bar{u}''}{\bar{u} - c}\phi = 0 \tag{7-6}$$

where the superscript "0" is dropped for simplicity. Multiplying Eq. 7-6 by $\tilde{\phi}$, the complex conjugate of ϕ, and integrating between y_1 and y_2, we get for boundary conditions Eq. 7-4 or Eq. 7-5,

$$\int_{y_1}^{y_2} [|\phi'|^2 + \alpha^2|\phi|^2]dy + \int_{y_1}^{y_2} \frac{\bar{u}''}{\bar{u} - c}|\phi|^2 dy = 0 \tag{7-7}$$

of which the imaginary part is

$$c_i \int_{y_1}^{y_2} \frac{\bar{u}''}{|\bar{u} - c|^2}|\phi|^2 dy = 0 \tag{7-8}$$

As $c_i > 0$ for amplified disturbances, $\bar{u}''$ must change sign between y_1, y_2. Thus (1) is proved.

To prove (2), Eq. 7-3 is recast in another form,

$$\left.\begin{aligned} &\frac{d}{dy}[(\bar{u} - c)\psi'] - \alpha^2(\bar{u} - c)\psi = 0 \\ &\psi = \frac{\phi}{(\bar{u} - c)} \end{aligned}\right\} \tag{7-9}$$

Multiplying the differential equation (Eq. 7-9) now by $\tilde{\psi}$, the complex conjugate of ψ, and integrating between limits, we get

$$\int_{y_1}^{y_2} (\bar{u} - c)^2[|\psi'|^2 + \alpha^2|\psi|^2]dy = 0 \tag{7-10}$$

of which the imaginary part is

$$c_i \int_{y_1}^{y_2} (\bar{u} - c_r)[|\psi'|^2 + \alpha^2|\psi|^2]dy = 0 \tag{7-11}$$

Thus, for $c_i > 0$, $(\bar{u} - c_r)$ must change sign within y_1, y_2, i.e. c_r lies between the extreme values of $\bar{u}$. The only possible exception is when

$$\alpha = 0, \qquad \psi = \text{const}; \qquad \text{i.e. } \phi = \bar{u} - c \tag{7-12}$$

For instance, to satisfy the boundary conditions, Eq. 7-4, one may take $c = 0$. This corresponds to a steady deviation. Its existence is furthermore independent of any restrictions on the profile $\bar{u}$.

The above results will also apply for neutral stability if the latter is regarded as the limiting case by letting $c_i \to 0$ from the positive side. Indeed, if we note that the proofs only require $c_i \neq 0$, it would appear that these results should also hold for damped disturbances. For a velocity profile having no point of inflexion, one would be forced to conclude that no disturbances, amplified or damped, are possible! This difficulty is resolved by the observation that Eq. 7-3 is only meaningful when regarded as the limiting case of $\alpha Re \to \infty$ for the complete Eq. 7-1. Eq. 7-3 remains of the same form when we replace ϕ by $\tilde{\phi}$ and c by $\tilde{c}$ (the respective complex conjugates) so that a damped disturbance goes over into an amplified one satisfying the same differential equation and boundary conditions; but Eq. 7-1 does not. Only by a careful examination of the original Eq. 7-1 can the question be clarified. We shall see later that the correct interpretation is to take $c_i > 0$ as is done here.

We now bring up two more general results of Eq. 7-3 which are in the nature of sufficiency conditions:

(3) If $\bar{u}(y)$ has one point of inflexion at $y = y_0$, and if $\bar{u} > 0$ and $\bar{u}''/[\bar{u} - \bar{u}(y_0)] < 0$ then there exists a neutral solution real $\phi(y)$ by taking $c = \bar{u}(y_0)$; (4) Under the same conditions specified in (3), there also exist amplified solutions at slightly lower values of the wave number α.

Under the conditions of (3) it follows directly from the "oscillation theorem" in the general theory of the Sturm-Liouville equation that the equation

$$\phi'' - \alpha^2\phi - \frac{\bar{u}''}{\bar{u} - \bar{u}(y_0)}\,\phi = 0$$

with boundary conditions Eq. 7-4 or 7-5 has real positive eigenvalues α^2. Therefore we only need to take $c = \bar{u}(y_0)$ to establish the existence of the neutral disturbances for Eq. 7-3. Because of the condition $\bar{u}''/[\bar{u} - \bar{u}(y_0)] < 0$ it may be noted that, of the two types of velocity distributions shown in Fig. G,7, the upper one may not have a neutral oscillation while the lower one will. A similar conclusion for $c_i > 0$ may be drawn if one examines the real part of Eq. 7-7 [*18*]. The proof of (4) may be carried out by a perturbation of the neutral disturbance just mentioned. Multiplying Eq. 7-6 by ϕ and integrating between limits, we get

$$-\int_{y_1}^{y_2} \phi'^2 dy - \alpha^2 \int_{y_1}^{y_2} \phi^2 dy = \int_{y_1}^{y_2} \frac{\bar{u}''}{\bar{u} - c}\,\phi^2 dy \tag{7-13}$$

We now consider Eq. 7-13 for the neutral disturbance $c_N = \bar{u}(y_0)$, $\alpha^2 = \alpha_N^2$,

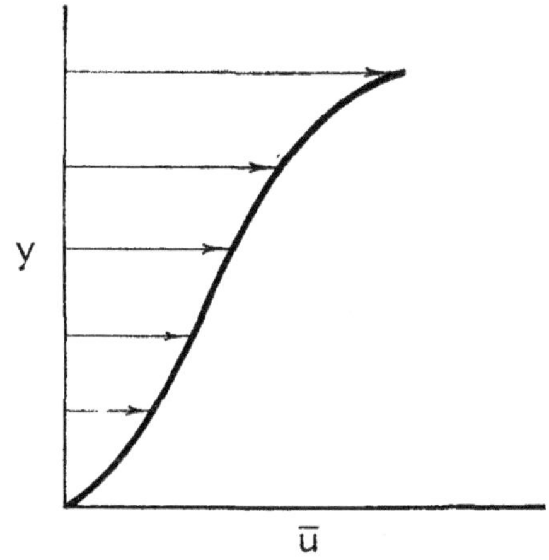

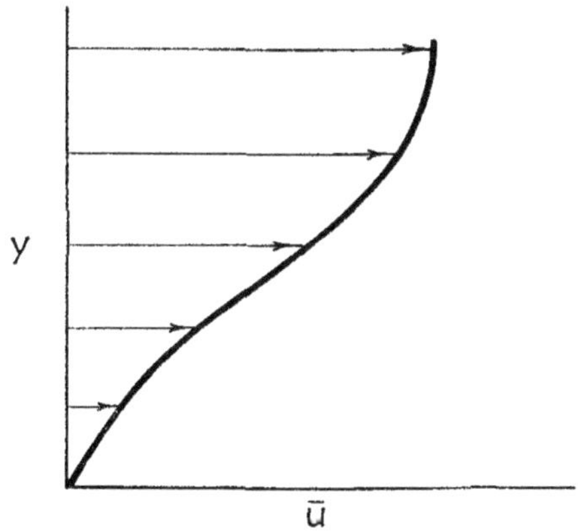

Fig. G,7. Types of velocity profiles.

and a neighboring solution of slightly larger α^2. Differentiating Eq. 7-13 with respect to α^2

$$-2\int_{y_1}^{y_2}\phi'\phi'_{\alpha^2}dy - \int_{y_1}^{y_2}\phi^2dy - 2\alpha^2\int_{y_1}^{y_2}\phi\phi_{\alpha^2}dy + \frac{\partial c}{\partial \alpha^2}\int_{y_1}^{y_2}\frac{\bar{u}''\phi^2}{(\bar{u}-c)^2}dy - 2\int_{y_1}^{y_2}\frac{\bar{u}''\phi\phi_{\alpha^2}}{(\bar{u}-c)^2}dy = 0 \quad (7\text{-}14)$$

where ϕ_{α^2} denotes the derivative of ϕ with respect to α^2. Let $\phi = \phi_N$. Then Eq. 7-14 becomes after simple manipulation,

$$\left(\frac{\partial c}{\partial \alpha^2}\right)_N = \frac{\int_{y_1}^{y_2}\phi_N^2dy}{\lim\limits_{c\to c_N}\int_{y_1}^{y_2}\frac{\bar{u}''}{(\bar{u}-c)^2}\phi_N^2dy} \quad (7\text{-}15)$$

The denominator must be evaluated by the proper limiting process of letting $c_i \to 0$ from the positive side. Then it turns out that

$$\lim_{c \to c_N} \int_{y_1}^{y_2} \frac{\bar{u}''}{(\bar{u} - c)^2} \phi_N^2 dy = A + iB, \qquad B > 0$$

Consequently, splitting Eq. 7-15 into real and imaginary parts for $(\partial c/\partial \alpha^2)_N$, and noting that ϕ_N is real, yields

$$\left(\frac{\partial c_i}{\partial \alpha^2}\right)_N < 0 \tag{7-16}$$

which states that one gets $c_i > 0$ for $\alpha^2 > \alpha_N^2$.

Based upon these results, one has often the so-called "inviscid criterion": For monotonic symmetrical or boundary layer-type profiles, the appearance of a point of inflection ($\bar{u}'' = 0$) is both necessary and sufficient for the existence of a neutral disturbance, of which the wave speed c equals the basic flow velocity at the point of inflection. The disturbance is regular everywhere, and thus is physically possible in the absence of viscous effects.

Aside from the formal mathematical approach summarized above, one should not forget that physically the problem in this section is that of the stability of infinitesimal vorticity disturbances in a given basic flow in the absence of viscosity. In the strictly two-dimensional case, the vorticity sticks with each fluid element. Under disturbances, the fluid elements from different levels of constant vorticity change place with each other, and an apparent excess or deficit of its vorticity relative to the new level results. Lin [*1*, pp. 56–58] was then able to show that, unless the vorticity gradient of the basic flow vanishes, the fluid element will tend to move back to the level to which it belonged. Thus instability will be possible only when the vorticity gradient vanishes, i.e. $\bar{u}'' = 0$, somewhere within the fluid.

The inviscid criterion may also be discussed in terms of the Reynolds stress distribution, which will be brought up in Art. 19.

G,8. Fundamental Solutions of the Orr-Sommerfeld Equation.

THE INVISCID SOLUTIONS. We now return to Eq. 7-3 to discuss its explicit solution, especially in the case of neutral disturbances (real c). Regarded as the limiting case of the complete viscous equation, Eq. 7-3 should yield two fundamental solutions which are the limiting cases of Eq. 7-1 and will be useful for αRe large but not infinite. We therefore consider velocity profiles that do not necessarily have a point of inflection, namely, that may become unstable only after including the viscous effects. But it seems reasonable to expect that the real wave velocity c would still lie between the extreme values of $\bar{u}(y)$. Thus, in the form of Eq. 7-6, the coefficient of the last term has a pole at $\bar{u} = c$.

There are two well-known ways to express the solution of Eq. 7-3. Applying the classical method of Frobenius, Tollmien obtained two solutions in series form:

$$\left.\begin{aligned} \phi_1^{(0)}(y) &= (y - y_c)P_1(y - y_c) \\ \phi_2^{(0)}(y) &= P_2(y - y_c) + \frac{\bar{u}_c''}{\bar{u}_c'}\phi_1^{(0)}(y)\ln(y - y_c) \\ \text{where}\qquad & \\ P_1 = 1 + \frac{\bar{u}_c''}{2\bar{u}_c'}(y - y_c) &+ O(y - y_c)^2, \qquad P_2 = 1 + O(y - y_c)^2 \end{aligned}\right\} \quad (8\text{-}1)$$

(both P_1 and P_2 are regular functions), y_c is the point of singularity defined by $\bar{u}(y_c) = c$, and $\bar{u}_c'$ and $\bar{u}_c''$ are values of $\bar{u}'$ and $\bar{u}''$, respectively, evaluated at y_c. The logarithmic part of $\phi_2^{(0)}$ vanishes only when $\bar{u}_c'' = 0$.

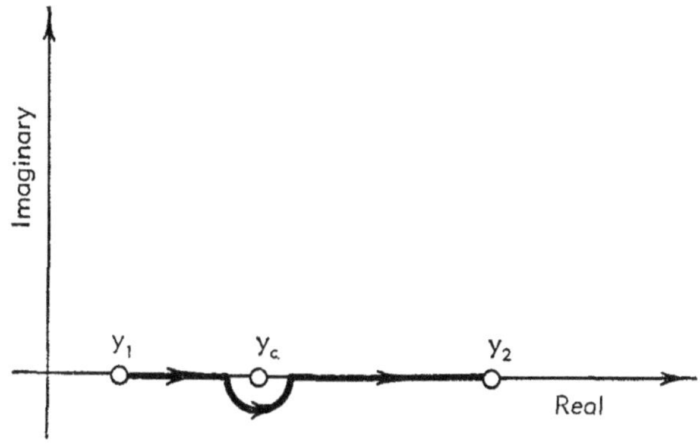

Fig. G,8a. Path of integration in complex y plane.

The chief difficulty here is to decide the proper branch which we should take in dealing with the logarithmic singularity at $y = y_c$. One must realize, however, that the singularity does not appear in the full viscous equation. Its presence here merely reflects the inadequacy of the "inviscid" approach, and the answer must come from a careful examination of the solution of the complete Eq. 7-1. As will be seen later, the proper branch proves to go around the singular point y_c from the under side along the real axis in the complex y plane as shown in Fig. G,8a, when $\bar{u}_c' > 0$; or to go around it from the upper side if $\bar{u}_c' < 0$.

The other method, originated by Heisenberg [*9*] and developed by Lin [*11*], is to expand into a power series in α^2, which converges because of the general nature of the equation. Often the experience is that α^2 is small for neutral disturbances, and the expansion provides certain computational advantages as well as some insight into the numerical problem discussed later on. The result may be written as

$$\phi_{1,2}^{(0)} = (\bar{u} - c)[q_0(y) + \alpha^2 q_1(y) + \cdots + \alpha^{2n}q_n(y) + \cdots] \quad (8\text{-}2)$$

where

$$\left.\begin{aligned} g_0 &= 1 \quad \text{for} \quad \phi_1^{(0)} \\ &= \int_{y_1}^{y} (\bar{u} - c)^{-2} dy \quad \text{for} \quad \phi_2^{(0)} \\ \text{and} \qquad g_{n+1} &= \int_{y_1}^{y} (\bar{u} - c)^{-2} dy \int_{y_1}^{y} (\bar{u} - c)^2 g_n(y) dy \end{aligned}\right\} \tag{8-3}$$

It is recalled that the path of integration should still be the one sketched in Fig. G,8a, along the real axis and bypassing the singularity from the under side when $\bar{u}_c' > 0$.

The Viscous Solutions. Since the formal expansion (Eq. 7-2) led to only two inviscid solutions, we now look toward other methods which enable one to find all four fundamental solutions when $\alpha Re \to \infty$ asymptotically. There have been two usual methods, both originally by Heisenberg [*9*]. The first one is good for fixed y away from the critical point y_c, where $\bar{u} = c$ for real c; the other provides a better approximation in the immediate neighborhood of the critical point y_c, the size of the neighborhood shrinking to zero as $\alpha Re \to \infty$. However, the last mentioned drawback has been removed by a technique due to Tollmien [*19*].

The first method. Suppose we rewrite the solution for Eq. 6-3 as

$$\phi = e^{\int g dy} \tag{8-4}$$

and expand $g(y)$ in descending powers of the parameter $\sqrt{\alpha Re}$:

$$g = \sqrt{\alpha Re}\, g_0 + g_1 + \frac{1}{\sqrt{\alpha Re}} g_2 + \cdots \tag{8-5}$$

Substituting into Eq. 7-1 and equating terms of like powers of $\sqrt{\alpha Re}$, we get a sequence of equations,

$$\left.\begin{aligned} (\bar{u} - c) g_0 &= -i g_0^4, \\ (\bar{u} - c)(g_0' + 2 g_0 g_1) &= -i(4 g_0^3 g_1 + 6 g_0^2 g_0'), \text{ etc.} \end{aligned}\right\} \tag{8-6}$$

The first of Eq. 8-6 admits two possibilities:

$$g_0 = 0 \quad \text{or} \quad g_0 = \pm \sqrt{i(\bar{u} - c)} \tag{8-7}$$

For $g_0 = 0$, the expansion (Eq. 8-5) begins with g_1 and we regain the "inviscid equation" (Eq. 7-3) and its two solutions $\phi_1^{(0)}$ and $\phi_2^{(0)}$, Eq. 8-1 or 8-2. For the other alternative $g_0 = \pm \sqrt{i(\bar{u} - c)}$, we get

$$g_1 = -\frac{5}{2} \frac{g_0'}{g_0} \tag{8-8}$$

Consequently, let suffixes 3 and 4 denote the third and fourth solutions

$$\left.\begin{aligned} \phi_3 &= \exp\left\{-\int_{y_c}^{y} \sqrt{i\alpha Re(\bar{u} - c)}\, dy\right\}\left[(\bar{u} - c)^{-\frac{5}{4}} + O\left(\frac{1}{\sqrt{\alpha Re}}\right)\right] \\ \phi_4 &= \exp\left\{+\int_{y_c}^{y} \sqrt{i\alpha Re(\bar{u} - c)}\, dy\right\}\left[(\bar{u} - c)^{-\frac{5}{4}} + O\left(\frac{1}{\sqrt{eRe}}\right)\right] \end{aligned}\right\} \tag{8-9}$$

the lower limit of the integral having been taken at y_c. The point y_c again shows up as a branch point, and the path of integration must still go under y_c in the complex y plane according to Fig. G,8a. Both ϕ_3 and ϕ_4 oscillate rapidly as y becomes large along the real axis, but ϕ_3 damps while ϕ_4 increases exponentially with y.

The second method. Since the inviscid solution behaves logarithmically near the singular point y_c, one expects rather strong viscous influence in the neighborhood of y_c. To find the solutions which are valid in this region, one magnifies the y direction so that the highest derivative terms on both sides of Eq. 7-1 become of the same order of magnitude. Let

$$\left.\begin{aligned} \zeta &= (y - y_c)/\epsilon, \qquad \epsilon \ll 1 \\ \bar{u} - c &= \bar{u}_c'(y - y_c) + \tfrac{1}{2}\bar{u}_c''(y - y_c)^2 + \cdots \end{aligned}\right\} \tag{8-10}$$

then the procedure leads to

$$\epsilon = (\alpha Re\, \bar{u}_c')^{-\frac{1}{3}} \tag{8-11}$$

The solution is next expanded into an ascending series in ϵ:

$$\phi = \chi^{(0)}(\zeta) + \epsilon\chi^{(1)}(\zeta) + \cdots \tag{8-12}$$

Substituting into Eq. 7-1 and equating terms of like powers of ϵ, we get a sequence of equations:

$$\left.\begin{aligned} &\chi^{(0)''''} - i\zeta\chi^{(0)''} = 0 \\ &\chi^{(1)''''} - i\zeta\chi^{(1)''} = i\frac{\bar{u}_c''}{\bar{u}_c'}\left[\frac{\zeta^2}{2}\chi^{(0)''} - \chi^{(0)}\right] \\ &\cdots \end{aligned}\right\} \tag{8-13}$$

From the first of Eq. 8-13, at once four solutions are found:

$$\left.\begin{aligned} \chi_1^{(0)} &= \zeta \\ \chi_2^{(0)} &= 1 \\ \chi_3^{(0)} &= \int_\infty^\zeta d\zeta \int_\infty^\zeta d\zeta\,\zeta^{\frac{1}{2}} H_{\frac{1}{3}}^{(1)}[\tfrac{2}{3}(i\zeta)^{\frac{3}{2}}] \\ \chi_4^{(0)} &= \int_{-\infty}^\zeta d\zeta \int_{-\infty}^\zeta d\zeta\,\zeta^{\frac{1}{2}} H_{\frac{1}{3}}^{(2)}[\tfrac{2}{3}(i\zeta)^{\frac{3}{2}}] \end{aligned}\right\} \tag{8-14}$$

where $H_{\frac{1}{3}}^{(1),(2)}$ are the Hankel functions of the first and second kind, order $\frac{1}{3}$. The boundary conditions at y_1, y_2, corresponding to negative and positive infinity as $\alpha Re \to \infty$, are on the real axis in the ζ plane. It may be noted that these solutions as functions of ζ do *not* explicitly depend upon the basic flow.

The suffixes 1, 2, 3, and 4 indicate how these should correspond with the ones found by the first method. For finite values of ζ, as $\alpha Re \to \infty$ the distance $y - y_c$ shrinks to zero; so $\chi_1^{(0)}$ and $\chi_2^{(0)}$ clearly correspond to the leading terms of $\phi_1^{(0)}$ and $\phi_2^{(0)}$. The identification of $\chi_3^{(0)}$ and $\chi_4^{(0)}$

with the leading terms of Eq. 8-10 depends on the asymptotic expansion of the Hankel functions, valid as the argument tends to infinity along the real axis. As a result, the identification is possible only when ζ lies within the sector

$$-\tfrac{7}{6}\pi < \arg(\zeta) < \tfrac{1}{6}\pi \tag{8-15}$$

in the complex ζ plane. Since all higher order solutions of Eq. 8-13 contain $\chi_3^{(0)}$ and $\chi_4^{(0)}$, the validity of χ_1 and χ_2, or ϕ_1 and ϕ_2, is restricted by the same condition. *This restriction,* Eq. 8-15, *is the basis for choosing*

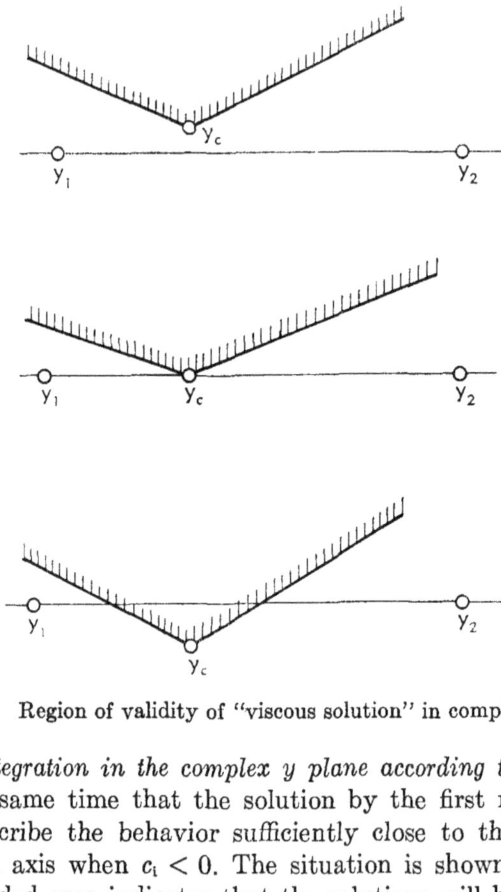

Fig. G,8b. Region of validity of "viscous solution" in complex y plane.

the path of integration in the complex y plane according to Fig. G,8a. It shows at the same time that the solution by the first method will be unable to describe the behavior sufficiently close to the point $y = y_c$ along the real axis when $c_i < 0$. The situation is shown in Fig. G,8b, where the shaded area indicates that the solutions will have to be continued analytically from the given forms in a proper manner to be valid in these regions. The damped disturbances therefore require special handling, although the difficulty will not be of practical importance if only the eigenvalue is desired and the boundary points lie outside of the shaded area ("slightly" damped disturbances). For mathematical discussions, we refer the reader to Lin's book [1].

G,9. Improved Solutions of Tollmien for Real c. As $\alpha Re \to \infty$ the zeroeth order solutions obtained above by either the first or second method may be used for the determination of the eigenvalue problem. The "neutral curve" in the α, Re plane, along which one gets neutral disturbances with $c_i = 0$, can thus be determined. It is, however, realized that both methods suffer from certain drawbacks. The solutions by the first method are valid for fixed y away from the critical point y_c; and they all, except $\phi_1^{(0)}$, exhibit a singular behavior at y_c. Those by the second method behave smoothly near y_c, but their range of validity shrinks to the immediate neighborhood of y_c for large αRe. In the determination of the neutral curve in the α, Re plane, it has been the practice to use the inviscid solutions $\phi_1^{(0)}$ and $\phi_2^{(0)}$ from the first method and the viscous solutions $\chi_3^{(0)}$ and $\chi_4^{(0)}$ from the second method as the set of fundamental solutions. This is based upon the experience that the eigenvalue c for neutral stability is usually quite small for flows with solid walls, such as the Poiseuille flow in a channel or the boundary layer over a flat plate. In other words, the boundary point y_1 lies close to the critical point y_c. Consequently one needs a good representation near y_c to take care of the viscous effects. But toward the other boundary point y_2, $\chi_1^{(0)}$ and $\chi_2^{(0)}$ are obviously too crude, so $\phi_1^{(0)}$ and $\phi_2^{(0)}$ are preferred. So long as an accurate description of the mode is not needed, the singular behavior of $\phi_2^{(0)}$ may be tolerated.

To remove these drawbacks, that is, the restriction to small c and the poor representation of the function itself near y_c, the following improvements were suggested by Tollmien [*19*]. In order to extend the range of validity to more than the immediate neighborhood of y_c, it is first necessary to retain $\bar{u} - c$ instead of expanding it as in Eq. 8-10. Again considering only the leading terms on each side of Eq. 7-1, we have

$$\frac{d^4\phi}{dy^4} - i\alpha Re(\bar{u} - c)\frac{d^2\phi}{dy^2} = 0 \tag{9-1}$$

The idea was to stretch the y axis so that the improved solution preserved more or less the same appearance as that of Eq. 8-13. To do this, Tollmien introduced a new independent variable $Y(y)$ and put

$$\frac{d^2\phi}{dy^2} = \left(\frac{dy}{dY}\right)^{\frac{3}{2}} F(Y) \tag{9-2}$$

Substituting into Eq. 9-1, one gets

$$F''(Y) - \left\{i\alpha Re(\bar{u} - c)\left(\frac{dy}{dY}\right)^2 + \cdot\cdot\cdot\right\} F(Y) = 0 \tag{9-3}$$

where the dotted terms in the bracket are functions of the transformation $Y(y)$ only and may be neglected when compared with the term proportional to αRe. By choosing

$$\alpha Re(\bar{u} - c)\left(\frac{dy}{dY}\right)^2 = Y \tag{9-4}$$

then Eq. 9-3 reduces essentially to

$$F'' - iYF = 0$$

Consequently there is, analogous to $\chi_{3,4}^{(0)''}$,

$$F \cong \sqrt{Y}\, H_{\frac{1}{3}}^{(1),(2)}[\tfrac{2}{3}(iY)^{\frac{3}{2}}] \tag{9-5}$$

But Eq. 9-2 and 9-3 are now valid for arbitrary y, not necessarily in the immediate neighborhood of y_c as $\alpha Re \to \infty$. It is convenient to choose $Y(y_c) = 0$ and to limit ourselves to the case of monotonically increasing $\bar{u}$. Noting that we have made $Y = 0$ at $y = y_c$ and $Y/(\bar{u} - c) > 0$ always, the explicit correspondence between y and Y now follows from Eq. 9-4

$$Y = (\alpha Re)^{\frac{1}{3}}\left[\tfrac{3}{2}\int_{y_c}^{y} \sqrt{\bar{u} - c}\, dy\right]^{\frac{2}{3}} \tag{9-6}$$

so that $\lim_{y \to y_c} Y = \zeta$ and $d^2\phi/dy^2$ as represented by Eq. 9-2 and 9-5 agrees with $\chi_{3,4}^{(0)''}$ near y_c.

To get the viscous solutions corresponding to $\chi_{3,4}^{(0)}$ of Eq. 8-14, Eq. 9-2 remains to be integrated. However, Tollmien proposed to use simply

$$\left.\begin{aligned} \chi_3^{(0)} &= \left(\frac{Y}{\bar{u} - c}\right)^{\frac{1}{2}} \int_{\infty}^{Y} dY \int_{\infty}^{Y} dY \sqrt{Y}\, H_{\frac{1}{3}}^{(1)}[\tfrac{2}{3}(iY)^{\frac{3}{2}}] \\ \chi_4^{(0)} &= \left(\frac{Y}{\bar{u} - c}\right)^{\frac{1}{2}} \int_{-\infty}^{Y} dY \int_{-\infty}^{Y} dY \sqrt{Y}\, H_{\frac{1}{3}}^{(2)}[\tfrac{2}{3}(iY)^{\frac{3}{2}}] \end{aligned}\right\} \tag{9-7}$$

His error estimate showed that Eq. 9-7 represents the solution of the full viscous equation at all y with an error of order ϵ, ϵ being defined by Eq. 8-11. Roughly this may be seen by directly differentiating Eq. 9-7 and neglecting $O(\epsilon)$. Eq. 9-2 is satisfied as long as $Y \sim 0(1)$.

The singularity at y_c for the inviscid solution $\phi_2^{(0)}$ may be smoothed out by including some viscous effects according to $\chi_2^{(1)}$, Eq. 8-13. Since $\chi_2^{(0)} = 1$, Eq. 8-13 leads to

$$\chi_2^{(1)''''} - i\zeta\chi_2^{(1)''} = -i\frac{\bar{u}_c''}{\bar{u}_c'} \tag{9-8}$$

Referring to the series solution $\phi_2^{(0)}$, Eq. 8-2, we see that the particular solution of interest must behave like $(\bar{u}_c''/\bar{u}_c')\zeta \ln \zeta$ for large ζ. This solution has no singularity at $y = y_c$ and has been tabulated by Holstein [20]. Thus we may construct a smoothed "inviscid" solution which agrees with Eq. 8-2 away from $y = y_c$ but is of the form

$$\chi_2 = 1 + \epsilon\chi_2^{(1)}(\zeta)$$

near $y = y_c$ as $\alpha Re \to \infty$. This is given by

$$\phi_2 = P_2(y - y_c) + \epsilon\chi_2^{(1)} \frac{P_1(y - y_c)}{P_1(0)} \tag{9-9}$$

noting that

$$(y - y_c) \ln (y - y_c) \sim \epsilon\zeta[\ln \zeta + \ln \epsilon]$$

so a replacement of the $(\bar{u}_c''/\bar{u}_c')(y - y_c) \ln (y - y_c)$ term in Eq. 8-2 by $\epsilon\chi_{2\phi}^{(1)}$ of Eq. 9-8 is immediately suggested.

The improved solutions (Eq. 9-7 and 9-9) above are for the neutral case $c_i = 0$. More complicated improved solutions of the similar type but for $c_i \neq 0$ have been given by Wasow [*21*], for which the reader is referred to Lin's book [*1*] or Wasow's original paper.

G,10. Secular Equation for the Determination of the Neutral Curve. With the four fundamental solutions $\phi_n^{(0)}$, n = 1, 2, 3, 4, we now examine next in more detail the secular equation in the determination of the neutral curve (the locus in the α, *Re* plane for $c_i = 0$). Let us use the classical Poiseuille flow in a channel as illustration. The basic flow now is symmetrical about the center line of the channel. Any disturbance may therefore be decomposed into a symmetrical and an antisymmetrical part about the same center line. In a linear theory these "modes" may be considered separately. It will be seen later that only the antisymmetrical mode (of u') leads to a stability boundary. Thus the boundary conditions to be satisfied by the solution ϕ are:

$$\phi = \phi' = 0 \quad \text{at wall } y = y_1$$

$$\phi' = \phi''' = 0 \quad \text{at center line } y = y_2$$

since the function ϕ itself must be an even function about the line $y = y_2$. (ϕ represents the magnitude of normal velocity fluctuation, while ϕ' represents the magnitude of the u component by the third equation of Eq. 6-2.) With

$$\phi = \sum_{n=1}^{4} a_n\phi_n^{(0)}, \qquad a_n = \text{const}$$

the secular equation is

$$\begin{vmatrix} \phi_{11} & \phi_{21} & \phi_{31} & \phi_{41} \\ \phi_{11}' & \phi_{21}' & \phi_{31}' & \phi_{41}' \\ \phi_{12}' & \phi_{22}' & \phi_{32}' & \phi_{42}' \\ \phi_{12}''' & \phi_{22}''' & \phi_{32}''' & \phi_{42}''' \end{vmatrix} = 0 \tag{10-1}$$

where the notation is such that

$$\phi_{11} \equiv \phi_1^{(0)}(y_1), \qquad \phi_{12} \equiv \phi_1^{(0)}(y_2), \qquad \phi_{nm} \equiv \phi_n^{(0)}(y_m)$$

and the primes still denote derivatives.

Now, we note that $\phi_3^{(0)}$ and $\phi_4^{(0)}$ may be represented by Eq. 8-9. Because of the exponential factor in these expressions, each differentiation with respect to y increases the order of magnitude by a factor $\sqrt{\alpha Re}$. Consider first the last column of the determinant of Eq. 10-1. If we divide through by ϕ_{42}''', the orders of magnitude of the terms are

$$\frac{\phi_{41}}{\phi_{42}'''} \sim O[(\alpha Re)^{-\frac{3}{2}} e^{-\sqrt{\alpha Re}Q}]$$

$$\frac{\phi_{41}'}{\phi_{42}'''} \sim O[(\alpha Re)^{-1} e^{-\sqrt{\alpha Re}Q}]$$

$$\frac{\phi_{42}'}{\phi_{42}'''} \sim O[(\alpha Re)^{-1}]$$

where $Q \equiv \int_{y_1}^{y_2} \sqrt{i(\bar{u} - c)}\, dy$. Since y_2 is the center line and y_1 the wall, and the path of integration is to go under the singular point y_c, the argument of Q lies within $\pm\pi/4$. Hence, Q has a positive real part. Neglecting errors of order $(\alpha Re)^{-1}$, we may simplify Eq. 10-1 by setting to zero only the principal minor of the determinant,

$$\begin{vmatrix} \phi_{11} & \phi_{21} & \phi_{31} \\ \phi_{11}' & \phi_{21}' & \phi_{31}' \\ \phi_{12}' & \phi_{22}' & \phi_{32}' \end{vmatrix} = 0 \tag{10-2}$$

If further we estimate the orders of magnitude of the last column of the determinant of Eq. 10-2, the same reasoning shows that

$$\phi_{31} : \phi_{31}' : \phi_{32}' \sim 1 : \sqrt{\alpha Re} : \sqrt{\alpha Re}\, e^{-\sqrt{\alpha Re}Q}$$

The dominant term is ϕ_{31}'. When terms of the order $(\alpha Re)^{-\frac{1}{2}}$ are neglected, Eq. 10-2 is reduced to

$$\begin{vmatrix} \phi_{11} & \phi_{21} \\ \phi_{12}' & \phi_{22}' \end{vmatrix} = 0 \tag{10-3}$$

which clearly represents only the "inviscid" problem, requiring $\phi_1^{(0)}$ and $\phi_2^{(0)}$ to satisfy the conditions of zero normal velocity at the wall and zero slope (for even ϕ) at y_2, Eq. 7-5. To include some viscous effects, therefore, we must retain terms of the order $(\alpha Re)^{-\frac{1}{2}}$. The resultant equation becomes

$$\frac{\phi_{31}}{\phi_{31}'} = \frac{\begin{vmatrix} \phi_{11} & \phi_{21} \\ \phi_{12}' & \phi_{22}' \end{vmatrix}}{\begin{vmatrix} \phi_{11}' & \phi_{21}' \\ \phi_{12}' & \phi_{22}' \end{vmatrix}} \tag{10-4}$$

The left-hand side depends only on the viscous solution $\phi_3^{(0)}$, which in fact is better replaced by $\chi_3^{(0)}$ of Eq. 8-14 for $|c| \ll 1$, or by $\chi_3^{(0)}$ of Eq. 9-7 if $c \sim O(1)$. The significant feature of Eq. 10-4 is that the viscous and inviscid solutions are separated. With $\chi_3^{(0)}$ of Eq. 8-14, the left-hand side is a universal function in the variable

$$z = (\alpha Re\, \bar{u}_c')^{\frac{1}{3}}(y - y_c) \tag{10-5}$$

and may be tabulated once and for all. The right-hand side alone must be calculated for each different symmetrical profile.

For boundary layer-type basic flows, the boundary conditions to be satisfied by the solution ϕ are

$$\phi = \phi' = 0 \text{ at wall } y = y_1;$$
$$\phi,\ \phi' \text{ bounded at } y = y_2,\ y_2 \to \infty$$

Now the viscous solution $\phi_3^{(0)}$ decreases exponentially with increasing y, satisfying the second boundary condition automatically; the other solution $\phi_4^{(0)}$ increases exponentially and must be discarded. The boundary condition at infinity thus is actually a requirement on the linear combination of $\phi_1^{(0)}$ and $\phi_2^{(0)}$. The solution ϕ must therefore obey the inviscid equation, Eq. 7-3, as $y \to \infty$. Since $\bar{u}'' \to 0$ as $y \to \infty$, the solution ϕ must behave so that as $y \to \infty$,

$$\phi'' - \alpha^2\phi = 0$$

or $\phi \sim e^{-\alpha y}$ for $\alpha > 0$. In other words, the boundary condition at infinity is reduced to

$$\phi' + \alpha\phi = 0 \quad \text{as} \quad y \to \infty \tag{10-6}$$

The secular equation becomes, in such cases,

$$\begin{vmatrix} \phi_{11} & \phi_{21} & \phi_{31} \\ \phi_{11}' & \phi_{21}' & \phi_{31}' \\ \phi_{12}' + \alpha\phi_{12} & \phi_{22}' + \alpha\phi_{22} & \phi_{32}' + \alpha\phi_{32} \end{vmatrix} = 0 \tag{10-7}$$

Neglecting again the terms of order higher than $(\alpha Re)^{-\frac{1}{3}}$, we have the simplified equation,

$$\frac{\phi_{31}}{\phi_{31}'} = \frac{\begin{vmatrix} \phi_{11} & \phi_{21} \\ \phi_{12}' + \alpha\phi_{12} & \phi_{22}' + \alpha\phi_{22} \end{vmatrix}}{\begin{vmatrix} \phi_{11}' & \phi_2' \\ \phi_{12}' + \alpha\phi_{12} & \phi_{22}' + \alpha\phi_{22} \end{vmatrix}} \tag{10-8}$$

The left-hand side remains the universal function mentioned above.

One other type of basic flow of much interest is that with free boundaries, e.g. the jet or wake flows. Here both $\phi_3^{(0)}$ and $\phi_4^{(0)}$ must be rejected since they grow beyond limit at one or the other boundary. The mechanism of instability differs from that of the channel flows or boundary layer type flows because of the lack of a solid wall, near which the viscous effects play important roles in the latter types. We shall go into this problem separately (Art. 19).

G,11. Numerical Calculation of the Neutral Curve. The Plane Poiseuille and Blasius Flows. If we use $\chi_3^{(0)}$ of Eq. 8-14 as $\phi_3^{(0)}$, the left-hand side of Eq. 10-4 and 10-8 may be evaluated

$$\left.\begin{aligned} \frac{\phi_{31}}{\phi'_{31}} &= (y_1 - y_c)F(z) \\ F(z) &= -\frac{\int_\infty^{-z} d\zeta \int_\infty^{\zeta} \zeta^{\frac{1}{2}} H_{\frac{1}{3}}^{(1)}[\frac{2}{3}(i\zeta)^{\frac{3}{2}}]d\zeta}{z\int_\infty^{-z} \zeta^{\frac{1}{2}} H_{\frac{1}{3}}^{(1)}[\frac{2}{3}(i\zeta)^{\frac{3}{2}}]d\zeta} \\ z &= (\alpha Re\, \bar{u}'_c)^{\frac{1}{3}}(y - y_c) \end{aligned}\right\} \tag{11-1}$$

The function $F(z)$ is the universal function referred to above, first tabulated by Tietjens, and more accurately given by Holstein [*20*]. (The table [*22*] may be used to advantage in computations involving $H_{\frac{1}{3}}^{(1)}$.) Thus one is to solve

$$\left.\begin{aligned} F(z) &= E(\alpha,\ Re,\ c) \\ \text{where}\quad E(\alpha,\ Re,\ c) &= (y_1 - y_c)^{-1} \times \text{R.H.S of Eq. 10-4 or 10-8} \end{aligned}\right\} \tag{11-2}$$

as the case may be. Straightforwardly, the real part of $F(z)$ may be plotted against the imaginary part with z as the running parameter, Fig. G,11a. Then for each α, the inviscid solutions are determined by Eq. 8-1 or Eq. 8-2, or other numerical means (e.g. Timman and Zaat [*23*]), and the real part of the right-hand side of Eq. 11-2 is plotted against the imaginary part on the same diagram, with c as the running parameter. The intersection of the two curves yields a solution (α, c, z), and the Reynolds number is obtained from the definition of z,

$$Re = \frac{\left(\dfrac{z}{y_c - y_1}\right)^3}{\alpha \bar{u}'_c} \tag{11-3}$$

An alternative iterative method has been suggested by Lin [*19*], based upon certain observations of the behaviors of the functions involved. It is noted first that, from experiences in cases where a solid wall is present, both α and c are expected to be small along the neutral curve. Thus the inviscid solutions are given by only a few terms of the α^2 expansion of Eq. 8-2, and the quantity $y_c - y_1$ differs but slightly from $c/\bar{u}'_c$ (y_1 being the solid wall). Now, from Eq. 8-2 and 8-3,

$$\begin{aligned} \phi_{11} &= -c, & \phi_{21} &= 0 \\ \phi'_{11} &= \bar{u}'_1, & \phi'_{21} &= -\frac{1}{c} \end{aligned} \tag{11-4}$$

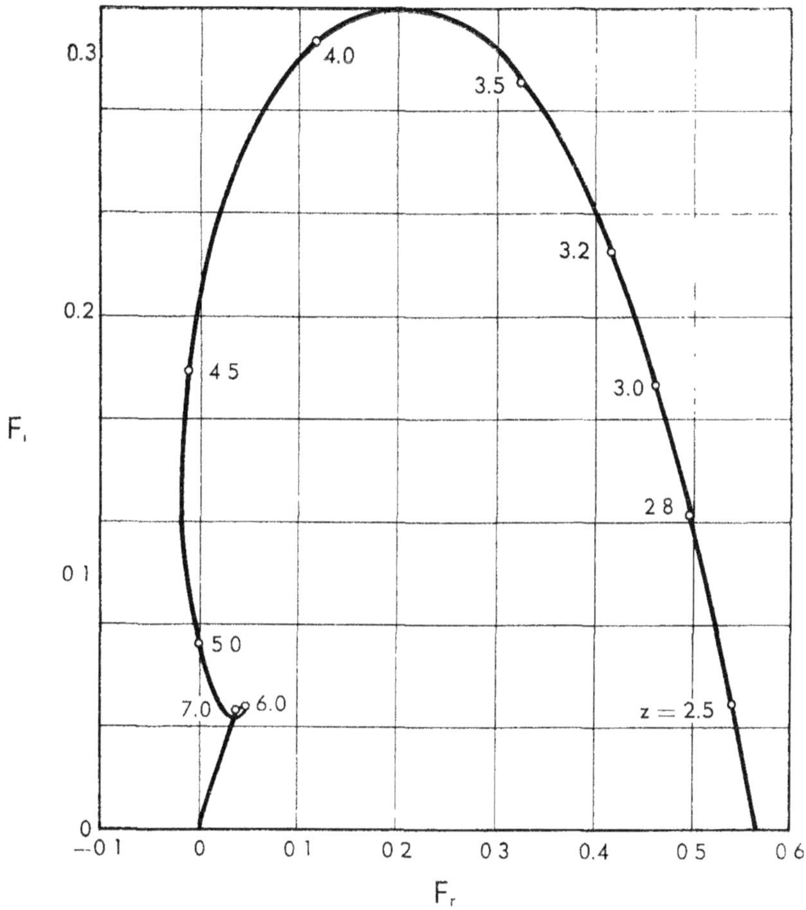

Fig. G,11a. Real and imaginary parts of Tietjen's function.

so Eq. 10-4 is reduced to, after simple manipulation,

$$\left(1 + \frac{\bar{u}_1'}{c}\frac{\phi_{31}}{\phi_{31}'}\right)^{-1} = 1 + \frac{\bar{u}_1' c \phi_{22}'}{\phi_{12}'} \tag{11-5}$$

and, similarly, Eq. 10-8 becomes

$$\left(1 + \frac{\bar{u}_1'}{c}\frac{\phi_{31}}{\phi_{31}'}\right)^{-1} = 1 + \bar{u}'c\,\frac{\phi_{22}' + \alpha\phi_{22}}{\phi_{12}' + \alpha\phi_{12}} \tag{11-6}$$

Defining next a modified Tietjen's function $\mathfrak{F}(z)$

$$\mathfrak{F}(z) \equiv [1 - F(z)]^{-1} \tag{11-7}$$

and a parameter $$\lambda \equiv \left(\frac{\bar{u}_1'}{c}\right)(y_c - y_1) - 1 \tag{11-8}$$

one recasts Eq. 11-5 and 11-6 into the form

$$\mathfrak{F}(z) = \frac{(1 + \lambda)(u + iv)}{1 + \lambda(u + iv)} \tag{11-9}$$

where u and v are the real and imaginary parts of the right-hand side of Eq. 11-5 or 11-6, as the case may be.[3] The function $\mathfrak{F}(z)$ and its first derivative are tabulated as Table G,11 and plotted as Fig. G,11b. The functions u and v are:

(1) for Poiseuille-type flows,

$$u + iv = \bar{u}_1'c\left(K_1 + \frac{1}{\bar{u}_1'c}\right) + \frac{\bar{u}_1'c}{\alpha^2}\,\frac{(1 - \alpha^2 H_2)(1 - \alpha^2 N_2 - \alpha^4 N_4 - \cdots)}{H_1 - \alpha^2 M_3 - \alpha^4 M_5 - \cdots} \tag{11-10}$$

(2) for boundary layer-type flows,

$$u + iv = \bar{u}_1'c\left(K_1 + \frac{1}{\bar{u}_1'c}\right) + \frac{\bar{u}_1'c}{\alpha}\,\frac{(1 - \alpha^2 H_2)[1 - \alpha^2 N_2 - \alpha^4 N_4 - \cdots - (1 - c)^2(\alpha^3 N_3 - \alpha^5 N_5 - \cdots)]}{(1 - c)^2(1 - \alpha^4 M_4 - \cdots) + \alpha(H_1 - \alpha^2 M_3 - \cdots)} \tag{11-11}$$

where

$$\left.\begin{aligned}
K_1(c) &= \int_{y_1}^{y_2} (\bar{u} - c)^{-2}dy \\
K_2(c) &= \int_{y_1}^{y_2} (\bar{u} - c)^{-2}dy \int_{y_1}^{y} (\bar{u} - c)^2 dy \\
K_3(c) &= \int_{y_1}^{y_2} (\bar{u} - c)^{-2}dy \int_{y_1}^{y} (\bar{u} - c)^2 dy \int_{y_1}^{y} (\bar{u} - c)^{-2}dy, \text{etc.} \\
H_1(c) &= \int_{y_1}^{y_2} (\bar{u} - c)^2 dy \\
H_2(c) &= \int_{y_1}^{y_2} (\bar{u} - c)^2 dy \int_{y_1}^{y} (\bar{u} - c)^{-2}dy \\
H_3(c) &= \int_{y_1}^{y_2} (\bar{u} - c)^2 dy \int_{y_1}^{y} (\bar{u} - c)^{-2}dy \int_{y_1}^{y} (\bar{u} - c)^2 dy, \text{ etc.} \\
M_n &= H_2 H_{n-2} - H_n \\
N_n &= K_1 H_{n-1} - K_n
\end{aligned}\right\} \tag{11-12}$$

The imaginary part of Eq. 11-10 or 11-11 is, for small α and c, mainly contained in the term K_1. One has thus *approximately*,

$$v \cong \frac{-\pi \bar{u}_c'' \bar{u}_1' c}{\bar{u}_c'^3} \tag{11-13}$$

a function of c only.

[3] Note that here u and v are not velocity components. We have maintained Lin's notation for easier reference by the readers of the original work.

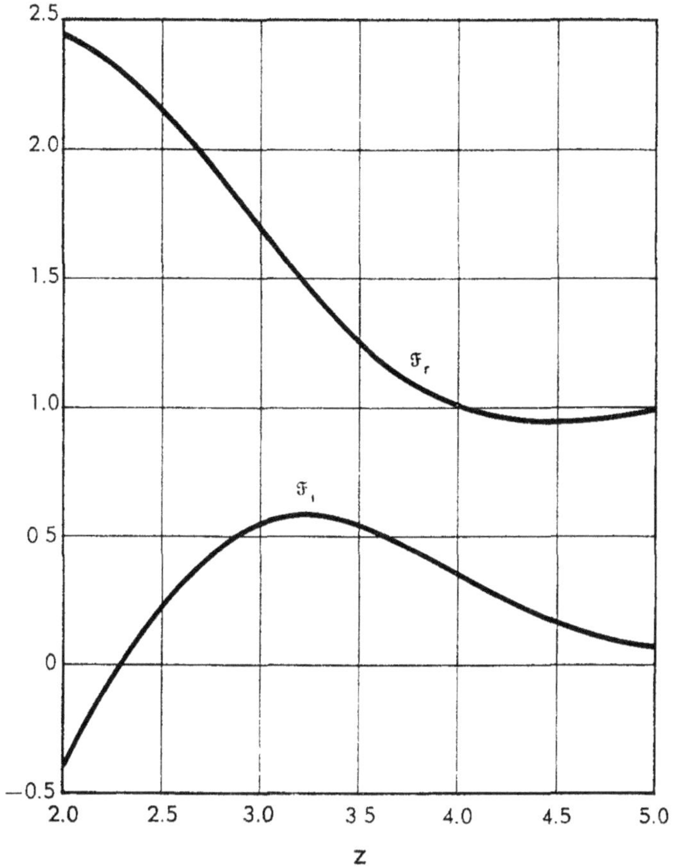

Fig. G,11b. The modified Tietjen's function.

The iterative procedure may now be started by neglecting λ in Eq. 11-9. For a given c, Eq. 11-13 determines $\mathfrak{F}_i(z)$, hence z; the value of z gives $\mathfrak{F}_r(z)$ which is equal to u in this approximation. Now $\lambda(c)$ is also known. We then rewrite Eq. 11-9 as

$$\mathfrak{F}_r + i\mathfrak{F}_i = u + iv + \frac{\lambda(u + iv)(1 - u - iv)}{1 + \lambda(u + iv)} \tag{11-14}$$

The iteration scheme may be carried out by

$$\left.\begin{aligned} \mathfrak{F}_i^{(n+1)} &= v + \text{I.P.}\left[\frac{\lambda(u^{(n)} + iv)(1 - u^{(n)} - iv)}{1 + \lambda(u^{(n)} + iv)}\right] \\ u^{(n+1)} &= \mathfrak{F}_r^{(m+1)} - \text{R.P.}\left[\frac{\lambda(u^{(n)} + iv)(1 - u^{(n)} - iv)}{1 + \lambda(u^{(n)} + iv)}\right] \end{aligned}\right\} \tag{11-15}$$

the superscript denoting the order of approximation. After u and z are obtained to desired accuracy, another iteration must be made to find α

from the definition of u according to Eq. 11-10 or 11-11, always remembering to take advantage of the fact that $\alpha \ll 1$. Finally the Reynolds number is found from Eq. 11-3.

If we note that $\mathfrak{F}_i$ reaches a maximum in Fig. G,11b, the initial approximation above immediately suggests that for each $c \ll 1$ there are generally two solutions of z, hence two branches of the neutral curve, and there is roughly a maximum c corresponding to the maximum of $\mathfrak{F}_i$. Indeed the asymptotic behavior of the two branches as $Re \to \infty$ have thus been demonstrated by Lin [*1*, p. 43]. In addition he also gave an approximate formula for estimating the minimum critical Reynolds number. This was done essentially by using the initial approximation at the maximum of $\mathfrak{F}_i(z)$, which occurs at $z = 3.21$. The maximum of $\mathfrak{F}_i(z)$ corresponds to a constant $\mathfrak{F}_r(z)$ equal to 1.49, which is taken now as $u^{(0)}$. For α, $c \ll 1$, one gets further the approximate formula for α from Eq. 11-10 and 11-11 as

$$\left.\begin{aligned} \alpha^2 &\cong \frac{\bar{u}_1' c}{u^{(0)} H_1(0)} \quad \text{for} \quad \text{Eq. 11-10} \\ \alpha &\cong \frac{\bar{u}_1' c}{u^{(0)}} \quad \text{for} \quad \text{Eq. 11-11} \end{aligned}\right\} \tag{11-16}$$

where $H_1(0)$ is the value of H_1 for $c = 0$. Finally, by Eq. 11-3 and with the approximation $\bar{u}_0' \cong \bar{u}_1'$,

$$Re_{\min} \cong \frac{(3.21)^3 \bar{u}_1'^2}{\alpha c^3}$$

By comparing with his detailed calculations in the Poiseuille and Blasius profiles, Lin adjusted the constants to give

$$\left.\begin{aligned} Re_{\min} &\cong 30\,\frac{\bar{u}_1'}{c^3}\left[\frac{\bar{u}_1'}{c}\int_0^1 \bar{u}^2 dy\right]^{\frac{1}{2}} \text{ for Eq. 11-10} \\ Re_{\min} &\cong \frac{25\bar{u}_1'}{c^4} \text{ for Eq. 11-11} \end{aligned}\right\} \tag{11-17}$$

Here the Reynolds number is calculated on the basis of the artificial boundary layer "thickness" used as the length scale of the problem. The more meaningful Reynolds number based upon the displacement thickness, for instance, must be obtained by multiplying with a factor. The value of c in these formulas is determined by the condition

$$v(c) = \mathfrak{F}_i(z)\Big|_{\max} = 0.58 \tag{11-18}$$

The second of Eq. 11-17 has proved to be quite satisfactory in several other boundary layer-type flows (see Art. 17).

In an attempt to relate laminar-turbulent transition to stability theory, Schlichting [*24*] was the first to calculate the exponential rate of growth, or the "amplification factor" of disturbances. This is given by the product αc_i. With any point in the unstable region of the α, Re plane, there is associated a positive value of c_i. If one restricts consideration to the immediate neighborhood of the neutral curve, it is possible to calculate the c_i's by a perturbation procedure, starting from Eq. 11-2. Thus,

$$\frac{dF}{dz}\left(\frac{\partial z}{\partial \alpha}\Delta\alpha + \frac{\partial z}{\partial Re}\Delta Re + \frac{\partial z}{\partial c}\Delta c\right) = \frac{\partial E}{\partial \alpha}\Delta\alpha + \frac{\partial E}{\partial Re}\Delta Re + \frac{\partial E}{\partial c}\Delta c \quad (11\text{-}19)$$

where $\Delta\alpha$, ΔRe, Δc are the small changes from a point on the neutral curve. Separating Eq. 11-19 into real and imaginary parts, Δc_i and Δc_r may be solved in terms of $\Delta\alpha$ and ΔRe. The values of c_r and c_i at the interior points of the unstable region may then be determined by an interpolation formula satisfying $c_i = 0$ and the rates of change (e.g. $\partial c_r/\partial\alpha$ and $\partial c_i/\partial\alpha$ if the interpolation is taken at constant Re) along the neutral curve.

G,12. Numerical Results for the Plane Poiseuille and Blasius Flows. The results for the two classical cases of the plane Poiseuille

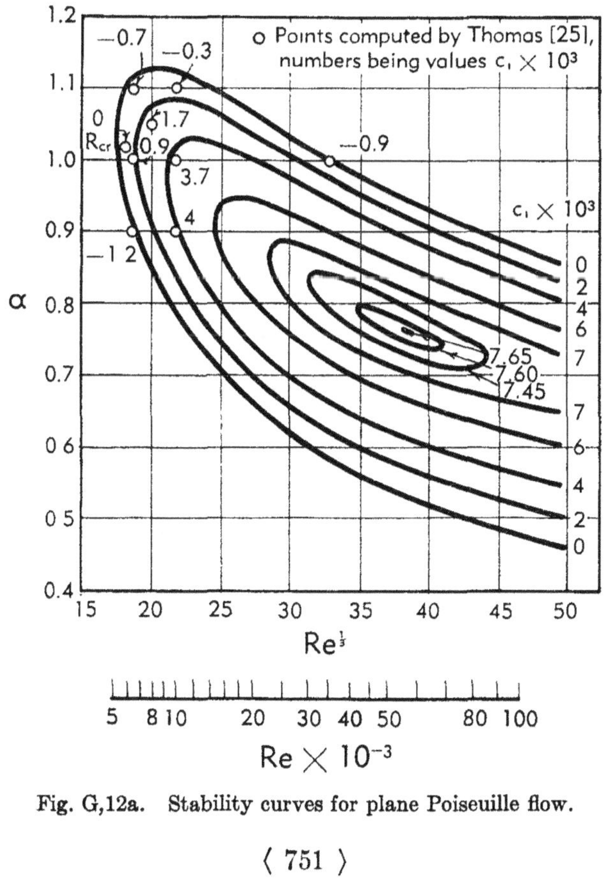

Fig. G,12a. Stability curves for plane Poiseuille flow.

and Blasius flows will now be presented in some detail. We refrain from going into the historical developments and controversies for which the readers are referred to Lin's book [1]. These two cases are of further interest because of the following: In the Poiseuille flow, a direct solution

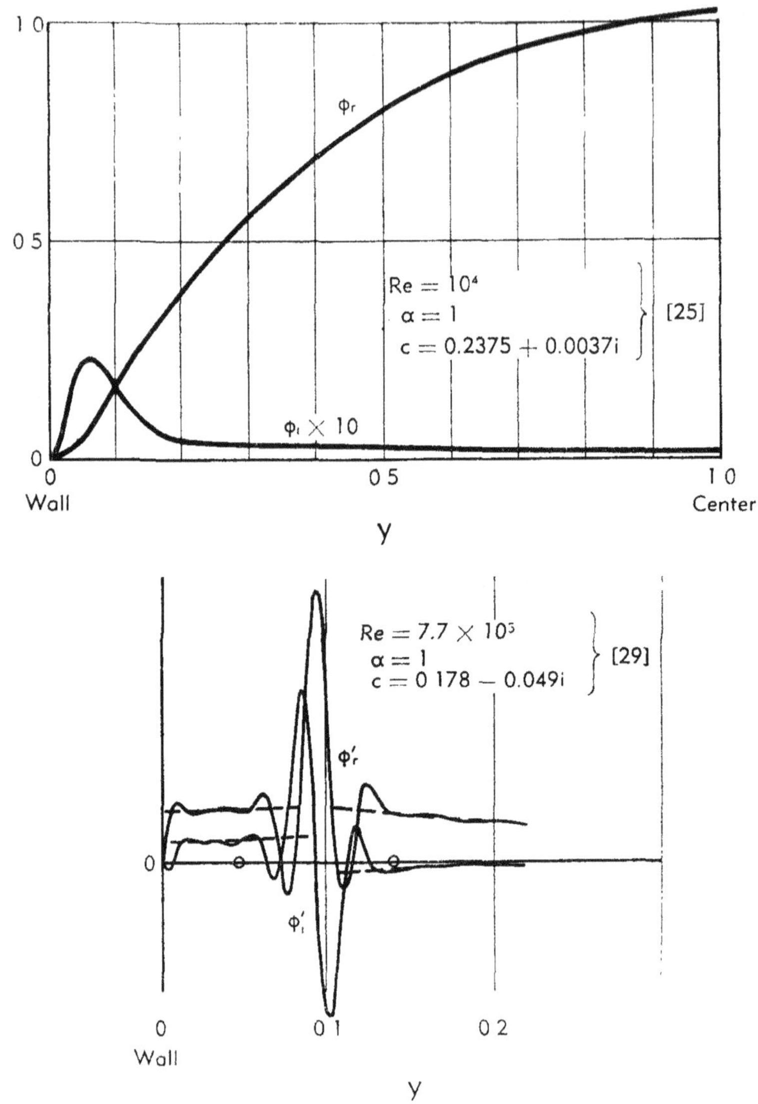

Fig. G,12b. Mode shapes for plane Poiseuille flow.

of the Orr-Sommerfeld equation, Eq. 7-1, by means of an IBM machine has been performed by Thomas [25]. The few complex eigenvalues obtained are estimated to correspond to about 100 years of hand computation. The difficulty lies basically in the rapid changes in the "critical layer." By comparing the results against those obtained from the approximate asymptotic methods described above, one establishes a certain confidence in the accuracy of such approximate methods. In the Blasius case, careful experimental measurements have been made by Schubauer and Skramstad [26], who recorded by the hot wire technique the amplification rates and the disturbance modes downstream from the point where nearly two-dimensional disturbances are generated in the boundary layer over a flat plate. Essentially the same experiment has been studied by Wortmann [27, pp. 460–470] using water as the fluid medium and a special flow visualization technique. Thus the theoretical results may be checked, especially for its applicability to practical cases, where the rather

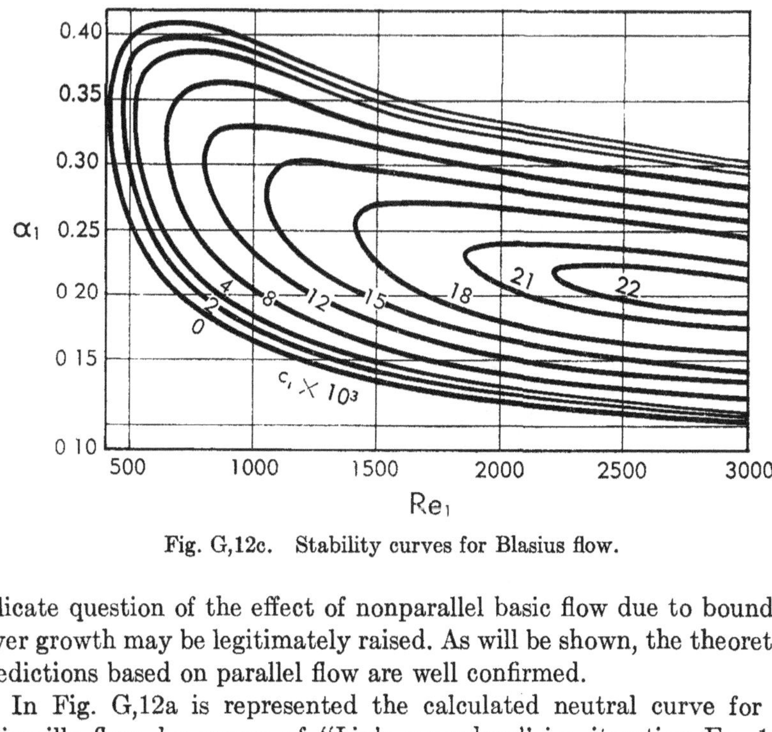

Fig. G,12c. Stability curves for Blasius flow.

delicate question of the effect of nonparallel basic flow due to boundary layer growth may be legitimately raised. As will be shown, the theoretical predictions based on parallel flow are well confirmed.

In Fig. G,12a is represented the calculated neutral curve for the Poiseuille flow, by means of "Lin's procedure" i.e. iterating Eq. 11-9, with the inviscid solutions according to Eq. 8-2 and 8-3 and $\chi_3^{(0)}$ from Eq. 8-14. The lines of constant c_i are obtained by the perturbation and interpolation scheme described above. Superimposed are the points computed by Thomas. Thomas' minimum critical Reynolds number is 5780 at $\alpha = 1.02$. The agreement appears to be generally quite satisfactory.

It is worthwhile to show one mode shape, Fig. G,12b, computed by Thomas for $Re = 10^4$, $\alpha = 1$, $c = 0.2375 + 0.0037i$. Note that ϕ_r is on the whole much greater than ϕ_i. Note also that ϕ_i is concentrated in the "critical" layer region near the wall. The Reynolds stress $-\rho u'v'$, proportional to $(\phi_i\phi_r' - \phi_r\phi_i')$ and responsible for the transfer of energy from the basic flow to the disturbances, is thus concentrated in the same region. When the disturbances are damped, the disturbance would vary even more violently in the critical layer. We include in Fig. G,12b a mode as computed by Grohne following Tollmien's approximate procedure, for $Re = 7.7 \times 10^5$, $\alpha = 1$, $c = 0.178 - 0.049i$, taken from a paper by

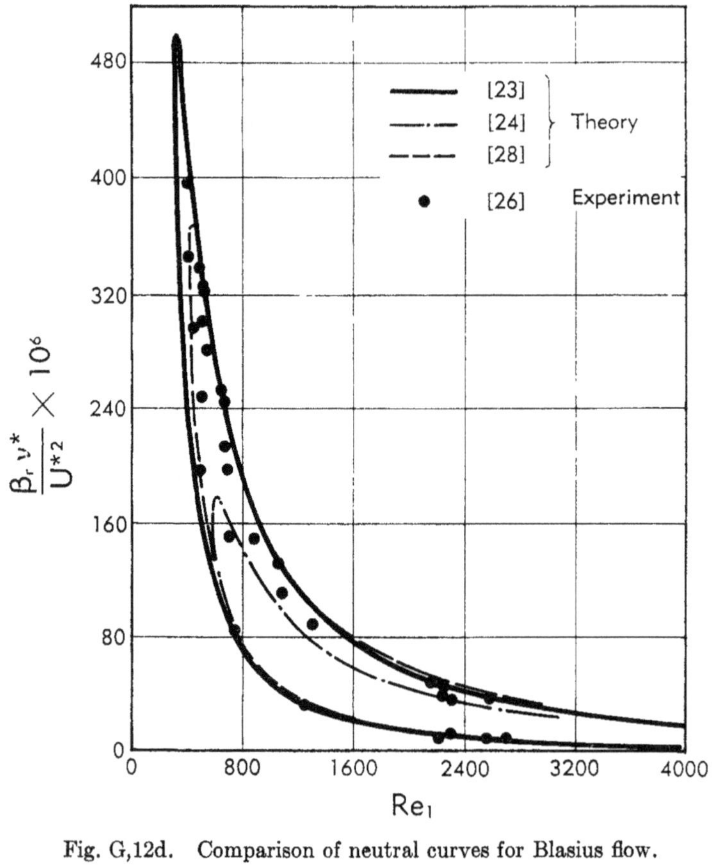

Fig. G,12d. Comparison of neutral curves for Blasius flow.

Tollmien [*29*]. The extent of the critical layer is marked by the two small circles. The rapid oscillation in a small region is truly remarkable and demonstrates the difficulty in direct numerical integration. Tollmien [*29*] conjectured that such "concentration of the vorticity" in a narrow region might have some significance in the "spotty nature" of fully developed turbulence.

For the Blasius flow, in Fig. G,12c are shown Shen's calculated neutral curve and lines of constant c_i by the same procedure [*28*], (α_1, Re_1) now referring to values with the displacement thickness δ_1 as the length scale. In Fig. G,12d, a comparison of the neutral curve with experiment as well as an earlier well-known calculation due to Schlichting [*24*] is shown, in terms of the disturbance frequency $\beta_r = \alpha_1 c_r$ vs. Re_1 to bring out more contrast. The observations of Wortmann define a neutral curve which agrees closely with those of Schubauer and Skramstad. The theoretical result of a more recent calculation by Timman and Zaat [*23*] is also

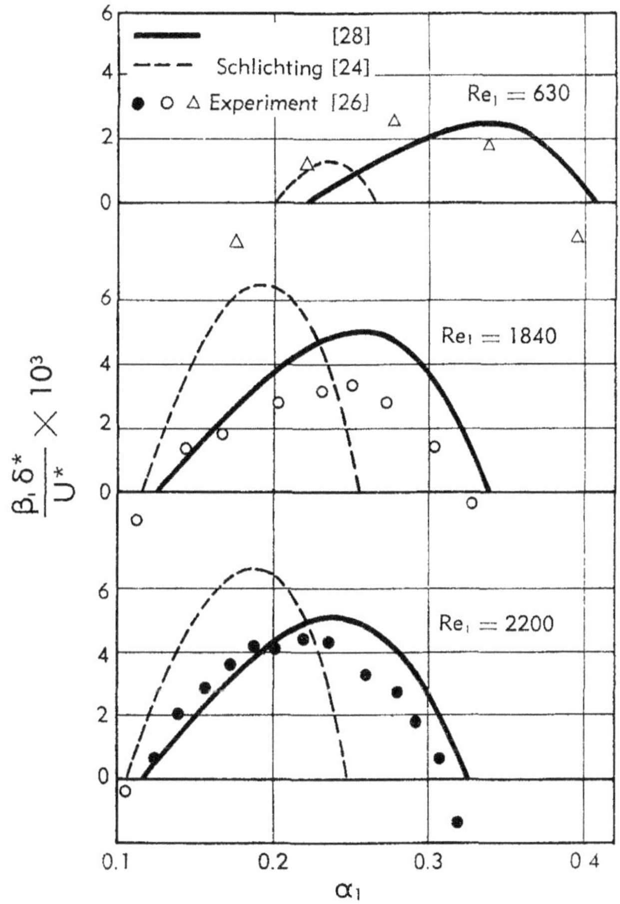

Fig. G,12e. Comparison of amplification rates for Blasius flow.

included, who employed a probably more accurate inviscid solution (by numerical integration of Eq. 7-3). Among the three theoretical results, Timman and Zaat's minimum critical Reynolds number of 320 is the lowest, compared with Lin's 420 and Schlichting's 475. The experimental points, however, seem to agree with Lin's value quite well. The lower branch of all three practically coincide, while Schlichting's upper branch is alone considerably lower than the other two. The experimental points appear to fall closer to the curves of Lin or of Timman and Zaat.

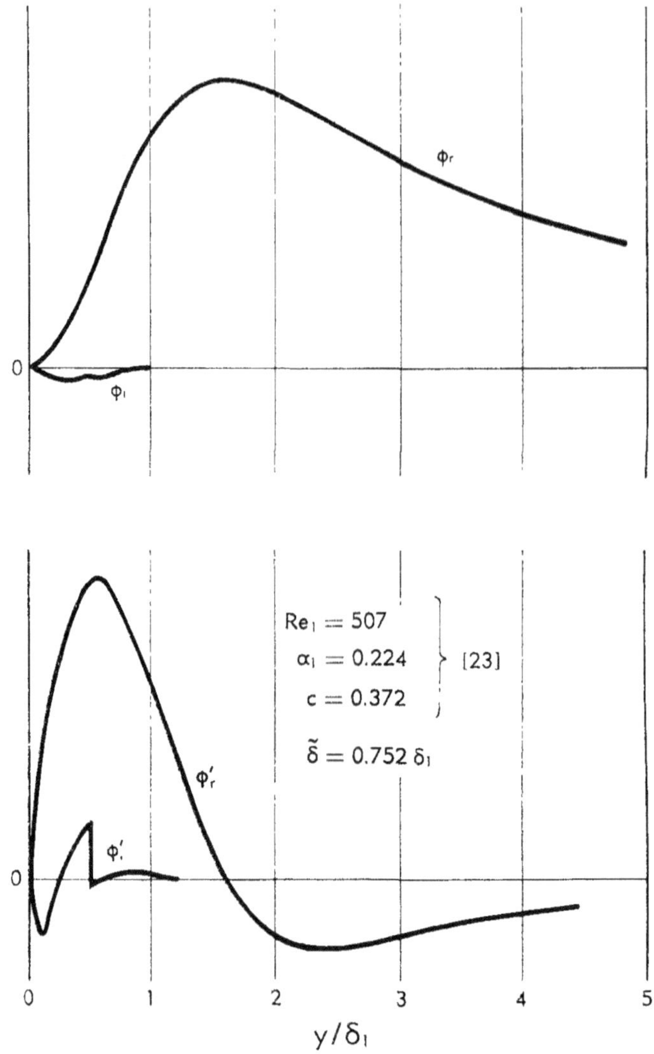

Fig. G,12f. Neutral mode shape for Blasius flow.

The second comparison for the Blasius flow is made in terms of the amplification rate $\beta_i = \alpha_1 c_i$ vs. the wave number α_1 at three different Re_1's, Fig. G,12e. Again the theoretical results of Shen [*28*] and Schlichting [*24*] are both included; the former seems to be quantitatively in much better agreement with the experimental points.

The solution ϕ for one of the points of Timman and Zaat is shown on Fig. G,12f. Notice that ϕ_i and ϕ'_i are generally much smaller than the real part. Thus the absolute values $|u'|$ and $|v'|$ are roughly given by ϕ'_r and ϕ_r. We compare in Fig. G,12g two disturbances modes $|u'|$ corresponding to one on the upper branch of the neutral curve and one on the lower, due to Timman and Zaat [*23*]. The experimental curves are for points slightly off the neutral curve but roughly corresponding. After adjusting the amplitudes to agree at the edge of the boundary layer, the theoretical and experimental distributions are in quite good agreement. The same degree of agreement also existed between Schlichting's mode [*30*] and the experimental one given in Schubauer and Skramstad's original paper [*26*].

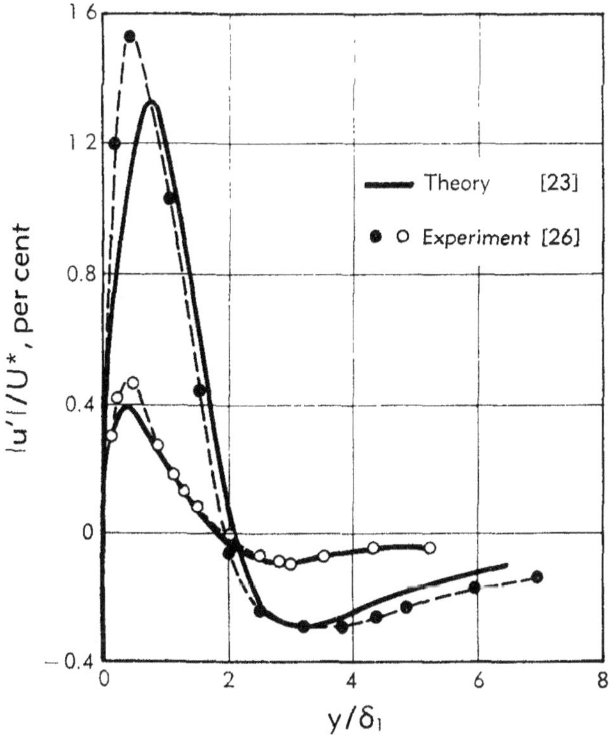

Fig. G,12g. Comparison of neutral modes for Blasius flow.

In Wortman's paper the observed distribution of $|u'|$ was shown to agree with Schlichting's curve to the same degree. In addition, he also evaluated the $|v'|$ and $|w'|$ distributions, mentioning that w' has nearly the same phase and amplitude in the lateral direction. But the comparison of $|v'|$ against Schlichting's ϕ_r showed only a poor qualitative resemblance. Since the observation suggests $\partial w'/\partial z \cong 0$, it is not obvious why his u' and v' fail to obey the continuity relation (Eq. 3-2), which would have assured the agreement between $|v'|$ and ϕ_r as long as $|u'|$ agrees with ϕ_r'!

We may now conclude that the theory would generally give rather reliable results, and the approximate methods are adequate at least in cases where, along the neutral curve, α and c are of the orders of those for the Blasius flow (maximum of α_1 and c about 0.4). Furthermore, the slow boundary layer growth in the Reynolds number range of Schubauer and Skramstad's experiment did not seem to affect appreciably the assumption of "locally parallel" basic flows.

G,13. Calculation of the Eigenvalues for the Higher Modes. The Couette and Poiseuille Flows. We have so far implied that only one eigenvalue of significance for given α, Re is to be found. However, the existence of other eigenvalues for the same α, Re, corresponding to the higher modes, seems to be expected. At the suggestion of Tollmien, these higher eigenvalues of the Orr-Sommerfeld equation have been studied by Grohne [*3*]. The mathematical details will not be reproduced but only some of his results which seem to be of basic interest. Consider first the case studied by Rayleigh [*8*], that the fluid is at rest (or moving at constant velocity). Eq. 7-1 is simplified into

$$\phi'''' - 2\alpha^2\phi'' + \alpha^4\phi + i\alpha ReC(\phi'' - \alpha^2\phi) = 0 \tag{13-1}$$

where the wave velocity is written as C. Let the boundary conditions be

$$\phi(\pm 1) = 0, \qquad \phi'(\pm 1) = 0 \tag{13-2}$$

corresponding to a channel with walls at $y = \pm 1$. This is now an equation with constant coefficients, and all four fundamental solutions are in terms of elementary functions. Introducing $\omega^2 = \alpha^2 - i\alpha ReC$, one finds an infinite set of eigenvalues ω_n, $n = 0, 1, 2, \ldots$ *all imaginary*, which may be arranged so that $0 < \omega_0/i < \omega_1/i < \cdots$. The even indices correspond to the even eigenfunctions, and odd ones to the odd eigenfunctions. From these ω_n's the nth eigenvalue C_n may be estimated,

$$\alpha^2 + \pi^2\left(\frac{n+1}{2}\right)^2 < i\alpha ReC_n < \alpha^2 + \pi^2\left(\frac{n+2}{2}\right)^2 \tag{13-3}$$

At least in this case, it is clear that the magnitude of C_n becomes very large either for $\alpha Re \to 0$ or $n \to \infty$.

Now, consider a variable velocity distribution $\bar{u}(y)$ and the full Orr-Sommerfeld equation. It is reasonable to expect that as $\alpha Re \to 0$ or $n \to \infty$, the new eigenvalues c_n will not differ very much from the C_n's above. We see this, for instance, by *assuming* $c_n \cong C_n$ and noticing that in Eq. 7-1 the terms depending on $\bar{u}(y)$ will then contribute to only minor corrections. Given n and C_n, there is a c_n; in fact, the correction is shown to be

$$c_n - C_n = \frac{1}{2}\int_{-1}^{1} \bar{u}\, dy + O\left(\frac{\alpha Re}{n}\right) \tag{13-4}$$

with fixed αRe, but $n \to \infty$. The result shows at once that the higher order eigenvalues must have their imaginary parts almost equal to that of C_n, thus representing highly damped modes. The other interesting question of how the lower eigenvalues behave at large values of αRe is answered by the formula

$$c_n - \bar{u}_1 \cong \Gamma_n(\alpha Re)^{-\frac{1}{3}} \tag{13-5}$$

for finite n, but $\alpha Re \rightarrow \infty$, where $\bar{u}_1$ is the velocity at the wall ($y = -1$ in this case) and Γ_n is a finite complex number independent of αRe. The

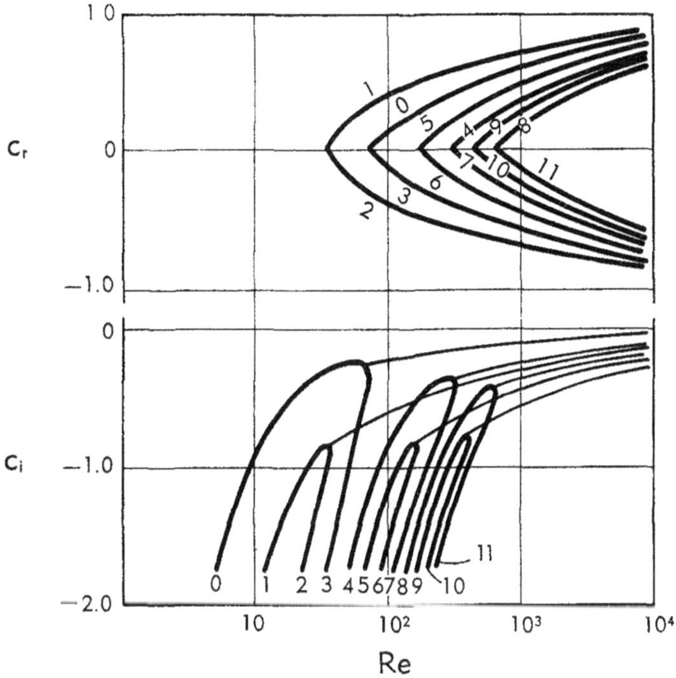

Fig. G,13a. Eigenvalues for plane Couette flow.

result indicates that the imaginary part of c_n always tends to zero as $(\alpha Re)^{-\frac{1}{3}}$. The quantity Γ_n, however, depends on $\bar{u}(y)$.

As numerical examples, the plane Couette and Poiseuille flows were calculated. The classical Couette flow had been treated earlier by several authors. Since the basic flow is characterized by $\bar{u}(y) = y$, $\bar{u}'' = 0$, the solution ϕ may be written down in closed form:

$$\phi = \int_{-1}^{y} \psi(\eta) \frac{\sinh \alpha(y - \eta)}{\alpha} d\eta \tag{13-6}$$

where ψ satisfies

$$\psi'' - [i\alpha Re(y - c) + \alpha^2]\psi = 0 \tag{13-7}$$

which is clearly reducible to the form of the first of Eq. 8-13. The eigenvalues are then to be solved by posing the boundary conditions at

$y = +1,$

$$\int_{-1}^{1} \psi(y) \frac{\sinh \alpha y}{\alpha} dy = 0, \qquad \int_{-1}^{1} \psi(y) \cosh \alpha y \, dy = 0 \tag{13-8}$$

There being no restrictions on αRe, the calculations may be made for the entire range of αRe from zero to infinity.[4] The twelve eigenvalues given by Grohne for $\alpha = 1$ have an interesting topological character,

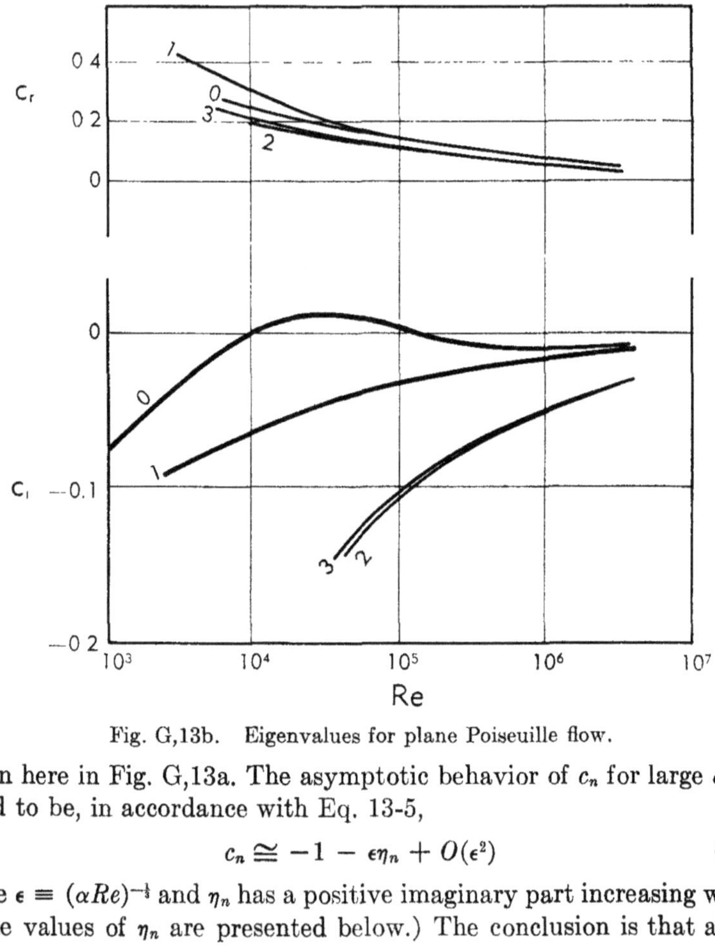

Fig. G,13b. Eigenvalues for plane Poiseuille flow.

shown here in Fig. G,13a. The asymptotic behavior of c_n for large αRe is found to be, in accordance with Eq. 13-5,

$$c_n \cong -1 - \epsilon \eta_n + O(\epsilon^2) \tag{13-9}$$

where $\epsilon \equiv (\alpha Re)^{-\frac{1}{3}}$ and η_n has a positive imaginary part increasing with n. (Some values of η_n are presented below.) The conclusion is that all disturbances of the assumed type must be damped.

The calculation of the Poiseuille case differs from Art. 8 mainly in that Tollmien's improved inviscid solution, Eq. 9-9, was adopted. At $\alpha = 0.87$ the first four eigenvalues were given, Fig. G,13b. The ordering is again that even indices correspond to even eigensolutions and odd

[4] The Couette flow between rotating cylinders with transverse wave-type disturbances is amenable to exact integration in terms of Bessel functions. Stability, again, was reported by Schultz-Grunow [*131*], who also emphasized the considerable effect on stability due to vibration, eccentricity of the cylinders, and other mechanical imperfections in the experimental setup.

indices to odd solutions. An even eigensolution represents an "antisymmetrical mode" of the streamwise fluctuations. Thus the curve $n = 1$ for the lowest symmetrical mode indicates complete stability. The asymptotic behavior for c_n is given as

$$c_n = \bar{u}_1 - \bar{u}'_c \epsilon \eta_n - \frac{\bar{u}''_c}{\bar{u}'_c} k_n \epsilon^2 \ln \epsilon + O(\epsilon^2) \tag{13-10}$$

with the following table to illustrate the values of η_n and k_n:

n	0	1	2	3
η_n	$-4.122 + 1.065i$	$-2.983 + 3.037i$	$-6.8 + 2.5i$	$-5.5 + 4.5i$
k_n	$-1.686 - 1.222i$	$1.902 + 0.851i$	$-2.2 - 1.5i$	$2.4 + 1.14i$

Here, again, the higher eigenvalues move toward more damping.

CHAPTER 3. EXTENSIONS AND APPLICATIONS OF THE STABILITY THEORY OF TRANSVERSE WAVE DISTURBANCES

G,14. Equations and Solutions for Stability of a Compressible Fluid. For higher and higher speed of the basic flows, the effects of compressibility must be considered for the disturbances as well as for the basic flow itself. This very important extension was first made by Lees and Lin [*31*], and later improved by Dunn and Lin [*4*]. One of Lees' theoretical calculations [*32*] predicted that stability could be greatly enhanced by cooling and adversely affected by heating. Experiments of Liepmann and Fila [*33*] and also Higgins and Pappas [*34*] indicate that the transition Reynolds number behaves in a similar fashion. Recent experiments by Laufer [*35*] confirm, in addition, the existence of neutral oscillations, similar to those of Schubauer and Skramsted [*26*], in supersonic boundary layers. (See also [*36,37*].) It appears, therefore, that the theory properly extended again provides a theoretical means to explain the roles of parameters in their qualitative effects on the transition problem.

Let us consider the basic flow as the boundary layer in an ideal gas. Besides the velocity vector $\mathbf{u}$ and pressure p, two more thermodynamic variables must be introduced, the density ρ and the temperature T. The basic differential equations are those given in Sec. B. Suppose the basic flow is again two-dimensional, steady and parallel. Our disturbances now are, instead of Eq. 2-5 and 2-8,

$$\left.\begin{aligned} \mathbf{u}_1 &= f_n e^{i(\alpha x + \beta z - \alpha c t)}, \qquad n = 1, 2, 3; \\ (p_1, \rho_1, T_1) &= (\varpi, r, \theta) e^{i(\alpha x + \beta z - \alpha c t)} \end{aligned}\right\} \tag{14-1}$$

The linearized differential equations satisfied by the disturbances are still too lengthy to write down, complicated by the fact that quantities such as the viscosity coefficient are functions of the thermodynamic variables and thus will have fluctuations, too. A closer examination of the orders of magnitude, however, shows that many terms involving derivatives to the viscosity are ordinarily much smaller than those proportional to the viscosity itself. We note that the wavelength of the disturbances would be $O(\delta)$, δ being the thickness of the boundary layer. But the viscous effects on the disturbances are pronounced only in much thinner layers within the boundary layer, namely near the "critical layer" $y = y_c$ or near the solid wall. Let the thickness of these viscous layers for the disturbances be $O(\delta')$, $\delta' \ll \delta$. To estimate δ', typical terms in the *complete* linearized equation may be examined. For example, in the momentum equation in the x direction,

$$\bar{\rho}\bar{u}\,\frac{\partial u'}{\partial x} + \cdot\cdot\cdot \sim \frac{\partial}{\partial y}\left(\bar{\mu}\,\frac{\partial u'}{\partial y} + \mu'\,\frac{\partial \bar{u}}{\partial y}\right)$$

where $\bar{\rho}$, $\bar{u}$, $\bar{\mu}$ are the density, x component velocity and viscosity coefficient of the basic flow, and u', μ' the disturbances of u and μ. Then it is readily seen that the viscous effects are to be considered only when

$$\frac{\delta'}{\delta} \sim Re_\delta^{-\frac{1}{3}}$$

Re_δ being the Reynolds number based on free stream conditions and the thickness δ. Furthermore, the right-hand side is dominated by $\bar{\mu}\partial^2 u'/\partial y^2$. Since outside δ' the viscous effect is unimportant, a very slight error will be committed if we replace the right-hand side by the dominant term $\bar{\mu}\partial^2 u'/\partial y^2$ *throughout* the thickness δ.

With such considerations, the complete system of linearized equations is greatly simplified: By choosing an arbitrary x direction, one gets

$$\left.\begin{aligned}
\bar{\rho}\left(\frac{\partial u'}{\partial t} + \bar{u}\,\frac{\partial u'}{\partial x} + \bar{w}\,\frac{\partial u'}{\partial z} + v'\,\frac{d\bar{u}}{dy}\right) &= -\,\frac{\partial p'}{\partial x} + \bar{\mu}\,\frac{\partial^2 u'}{\partial y^2}\\
\bar{\rho}\left(\frac{\partial v'}{\partial t} + \bar{u}\,\frac{\partial v'}{\partial x} + \bar{w}\,\frac{\partial v'}{\partial z}\right) &= -\,\frac{\partial p'}{\partial y} + \bar{\mu}\,\frac{\partial^2 v'}{\partial y^2}\\
\bar{\rho}\left(\frac{\partial w'}{\partial t} + \bar{u}\,\frac{\partial w'}{\partial x} + \bar{w}\,\frac{\partial w'}{\partial z} + v'\,\frac{d\bar{w}}{dy}\right) &= -\,\frac{\partial p'}{\partial z} + \bar{\mu}\,\frac{\partial^2 w'}{\partial y^2}\\
\frac{\partial \rho'}{\partial t} + \bar{u}\,\frac{\partial \rho'}{\partial x} + \bar{w}\,\frac{\partial \rho'}{\partial z} + v'\,\frac{d\bar{\rho}}{dy} &= -\bar{\rho}\left(\frac{\partial u'}{\partial x} + \frac{\partial v'}{\partial y} + \frac{\partial w'}{\partial z}\right)\\
\bar{\rho}c_p\left(\frac{\partial T'}{\partial t} + \bar{u}\,\frac{\partial T'}{\partial x} + \bar{w}\,\frac{\partial T'}{\partial z} + v'\,\frac{d\bar{T}}{dy}\right) &= \frac{\partial p'}{\partial t} + \bar{u}\,\frac{\partial p'}{\partial x} + \bar{w}\,\frac{\partial p'}{\partial z} + \bar{k}\,\frac{\partial^2 T'}{\partial y^2}\\
\frac{p'}{\bar{p}} &= \frac{\rho'}{\bar{\rho}} + \frac{T'}{\bar{T}}
\end{aligned}\right\} \tag{14-2}$$

where $\bar{u}$, $\bar{w}$, $\bar{\rho}$, $\bar{T}$, $\bar{\mu}$, and $\bar{k}$ are functions of y only ($\bar{\mu}$ and $\bar{k}$ being the coefficients of viscosity and conductivity in the basic flow, respectively), and the specific heat at constant pressure c_p and the pressure $\bar{p}$ are taken as constants. Now, we may always choose the x direction such that $\beta = 0$ in Eq. 14-1, i.e. along the direction of wave propagation. Then the $\partial/\partial z$ terms drop out from Eq. 14-2, and except for the third equation, Eq. 14-2 no longer contains w' or $\bar{w}$ explicitly. Evidently one may solve u', v', p', ρ', and T' as if, similarly to the incompressible case, the disturbances were two-dimensional, and subsequently, find the w' component from the third equation alone. The "equivalent" two-dimensional basic flow now has the velocity distribution $\bar{u}(y)$ equal to the velocity component in the wave propagation direction (same as for the incompressible case), but the density and temperature, hence also $\bar{\mu}$ and $\bar{k}$, are unaffected by the choice of the x direction. Consequently, the distributions of the flow variables in the "equivalent" two-dimensional basic flow are *not* identical with those of a boundary layer which has the same free stream values of the pressure, density, and temperature, the same wall conditions, but a reduced velocity. The "equivalent" two-dimensional distributions are artificial, useful only for the computation. Consequently, it is not possible to assert whether three-dimensional disturbances are necessarily more stable. The disturbance waves propagating in different directions must be considered separately.

We next nondimensionalize the variables by choosing the free stream values as the respective units, while the length unit is taken to be the boundary layer thickness δ, and the time unit is taken to equal the boundary layer thickness divided by the free stream value of $\bar{u}$. For each wave propagation direction, the equations for the disturbances u', v', ρ', p', and T' are, upon substituting Eq. 14-1 with $\beta = 0$ into Eq. 14-2 and nondimensionalizing,

$$\left.\begin{aligned} \bar{\rho}[i\alpha(\bar{u} - c)f_1 + \bar{u}'f_2] &= -\frac{i\alpha}{\gamma M_\beta^2}\varpi + \frac{\bar{\mu}}{Re}f_1'' \\ \bar{\rho}[i\alpha(\bar{u} - c)f_2] &= -\frac{1}{\gamma M_\beta^2}\varpi' + \frac{\bar{\mu}}{Re}f_2'' \\ i\alpha(\bar{u} - c)r + \bar{\rho}'f_2 + \bar{\rho}[i\alpha f_1 + f_2'] &= 0 \\ \bar{\rho}[i\alpha(\bar{u} - c)\theta + \bar{T}'f_2] &= \frac{i\alpha(\gamma - 1)}{\gamma}(\bar{u} - c)\varpi + \frac{\bar{\mu}\theta''}{RePr} \\ \varpi &= \frac{r}{\bar{\rho}} + \frac{\theta}{\bar{T}} \end{aligned}\right\} \quad (14\text{-}3)$$

where γ is the ratio of specific heats, M_β is the Mach number in the free stream *along the wave propagation direction*, Pr the Prandtl number of the basic flow (assumed constant), and Re the Reynolds number based upon the free stream values and the boundary layer thickness δ. The last of Eq. 14-3 defines r in terms of ϖ and θ, so that we need to deal with a system of only four equations. The eigenvalue problem for c at given α, Re is completely defined when the boundary conditions on f_1, f_2, ϖ and θ are prescribed. We shall come back to the boundary conditions later.

To see the difference between the equations for the compressible and incompressible cases, similarly to Eq. 6-1, we introduce

$$\left.\begin{aligned} &Z_1 = f_1, \quad Z_2 = \frac{df_1}{dy}, \quad Z_3 = \frac{f_2}{\alpha}, \quad Z_4 = \varpi, \\ &Z_5 = \theta, \quad Z_6 = \frac{d\theta}{dy} \end{aligned}\right\} \tag{14-4}$$

In these variables, Eq. 14-3 is reduced to a system of first order equations,

$$\left.\begin{aligned} &\frac{dZ_n}{dy} = G_n(Z_1, Z_2, \ldots, Z_6; \alpha, Re, Pr, M_\beta, \gamma), \\ &\qquad n = 1, 2, 3, \ldots, 6 \\ &G_1 = Z_2 \\ &G_2 = \frac{\alpha Re}{\bar{\mu}}\left[i\bar{\rho}(\bar{u} - c)Z_1 + \bar{\rho}\bar{u}'Z_3 + \frac{i}{\gamma M_\beta^2} Z_4\right] \\ &G_3 = -iZ_1 - \frac{\bar{\rho}'}{\bar{\rho}} Z_3 - i(\bar{u} - c)\left[Z_4 - \frac{1}{\bar{T}} Z_5\right] \\ &G_4 = \left[\frac{1}{M_\beta^2} + \frac{\alpha\gamma\bar{\mu}}{Re} i(\bar{u} - c)\right]^{-1}\left\{-i\alpha^2\gamma\bar{\rho}(\bar{u} - c)Z_3\right. \\ &\qquad + \frac{i\alpha\gamma\bar{\mu}}{Re}\left[\frac{\bar{\rho}'}{\bar{\rho}} Z_1 - Z_2 + i\bar{\rho}\left(\frac{\bar{\rho}'}{\bar{\rho}^2}\right)' Z_3 - \bar{\rho}\left(\frac{\bar{u} - c}{\bar{\rho}}\right)' Z_4\right. \\ &\qquad\qquad \left.\left. + \bar{\rho}\left(\frac{\bar{u} - c}{\bar{\rho}\bar{T}}\right)' Z_5 + \frac{\bar{u} - c}{\bar{T}} Z_6\right]\right\} \\ &G_5 = Z_6 \\ &G_6 = \frac{\alpha Re Pr}{\bar{\mu}}\left[\bar{T}'Z_3 - i\frac{\gamma - 1}{\gamma}(\bar{u} - c) Z_4 + i\bar{\rho}(\bar{u} - c)Z_5\right] \end{aligned}\right\} \tag{14-5}$$

which is reduced to the incompressible case by letting

$$M_\beta \to 0, \; \bar{\rho}' \to 0, \; \bar{\mu} \to 1, \; Z_4 \to \frac{1}{\bar{T}} Z_5 \quad \text{and} \quad Z_6 \to 0$$

in the first four equations. However, for comparison, the α^2 term in the curly bracket of the second and fourth of Eq. 6-2 must be omitted, since it comes from the $\partial^2/\partial x^2$ term of the viscous action, and is therefore absent in Eq. 14-3. In addition Eq. 14-5 is only correct for Re large. Note that in Eq. 14-5, for given y, the functions G_i are now analytic functions of α, c, Re

except when

$$\frac{1}{M_\beta^2} + i\frac{\alpha\gamma\bar{\mu}}{Re}(\bar{u} - c) = 0$$

The structure of the differential equation, Eq. 14-5, being clearly similar to that of Eq. 6-2, it follows that we could attempt to find approximate solutions in much the same manner as in the incompressible case. First, consider the "inviscid" equation by formally dropping all terms of $O(1/\sqrt{\alpha Re})$. After simple manipulation, we obtain

$$\frac{d}{dy}\left[\frac{(\bar{u} - c)\phi^{(0)\prime} - \bar{u}'\phi^{(0)}}{P}\right] - \frac{\alpha^2}{\bar{T}}(\bar{u} - c)\phi^{(0)} = 0 \qquad (14\text{-}6)$$

where $P = \bar{T} - M_\beta^2(\bar{u} - c)^2$, as the generalization of Eq. 7-6. Here again ϕ stands for Z_3, representing the magnitude of the v component fluctuation. The u component fluctuation corresponding to $\phi^{(0)}$ is given by

$$Z_1^{(0)} = i\frac{\bar{T}\phi^{(0)\prime} - M_\beta^2\bar{u}'(\bar{u} - c)\phi^{(0)}}{P} \qquad (14\text{-}7)$$

The other Z_n's ($n = 2, 4, 5, 6$) are likewise linear combinations of $\phi^{(0)}$ and $\phi^{(0)\prime}$. Eq. 14-6 shows that there are still only *two* inviscid solutions. Similar to 8-2 and 8-3, the solution may be written as[5]

$$\phi_{1,2}^{(0)} = (\bar{u} - c)[q_0(y) + \alpha^2 q_1(y) + \cdots + \alpha^{2n}q_n(y) + \cdots] \qquad (14\text{-}8)$$

but with

$$q_0 = 1 \quad \text{for} \quad \phi_1^{(0)}$$

$$q_0 = \int_{y_1}^{y} (\bar{u} - c)^{-2}P\,dy \quad \text{for} \quad \phi_2^{(0)}$$

$$q_{n+1} = \int_{y_1}^{y} (\bar{u} - c)^{-2}P\,dy \int_{y_1}^{y} (\bar{u} - c)^2(\bar{T})^{-1}q_n\,dy$$

The result (2), of Art. 7, that for amplified disturbances $(\bar{u} - c_r)$ must change sign within (y_1, y_2), is equally valid for Eq. 14-6. The proof follows along the same line by taking $\phi = (\bar{u} - c)\psi$.

Thus there is still a singularity at $\bar{u} - c = 0$ for real c. If we choose $\bar{u}_c' > 0$, the path of integration must go below the singular point y_c in the complex y plane as before. The other singularity at $P = 0$, however, can be shown as an apparent one only. This is seen by writing Eq. 14-6 as

$$\phi'' - \frac{P'}{P}\phi' - \left[\frac{(\bar{u}'/P)'}{\bar{u} - c} + \frac{\alpha^2}{\bar{T}}\right]P\phi = 0$$

and there follows $\phi \sim 1$ or $(y - y_P)^2$ near y_P which is defined by $P(y_P) = 0$.

[5] It is of course equally feasible to use the series attack similar to Eq. 8-1 or direct numerical integration as done by Timman and Zaat [23] for the incompressible case. The expansion in α^2 is convenient when $\alpha \ll 1$.

To seek the other four "viscous" solutions, we follow again the "second method" of Art. 8, which leads to solutions accurate near the critical point y_c and furthermore can be improved later by Tollmien's procedure (Art. 9). Introducing again Eq. 8-10 and 8-11 and expanding the Z_n's ($n = 1, 2, \ldots, 6$):

$$\left.\begin{aligned} Z_1 &= \chi_1^{(0)} + \epsilon\chi_1^{(1)}(\zeta) + \cdots \\ \epsilon Z_2 &= \chi_2^{(0)} + \epsilon\chi_2^{(1)}(\zeta) + \cdots \\ \epsilon^{-1}Z_3 &= \chi_3^{(0)} + \epsilon\chi_3^{(1)}(\zeta) + \cdots \\ \epsilon^{-1}Z_4 &= \chi_4^{(0)} + \epsilon\chi_4^{(1)}(\zeta) + \cdots \\ Z_5 &= \chi_5^{(0)} + \epsilon\chi_5^{(1)}(\zeta) + \cdots \\ \epsilon Z_6 &= \chi_6^{(0)} + \epsilon\chi_6^{(1)}(\zeta) + \cdots \end{aligned}\right\} \tag{14-9}$$

where the orders of magnitude of the Z_n's have been estimated from the differential equations, Eq. 14-5. Upon substituting into Eq. 14-5 and equating like powers of ϵ, there follows for $\chi_n^{(0)}$ ($n = 1, 2, \ldots, 6$):

$$\left.\begin{aligned} \frac{d\chi_1^{(0)}}{d\zeta} &= \chi_2^{(0)} \\ \frac{d\chi_2^{(0)}}{d\zeta} &= \frac{\bar{\rho}}{\bar{\mu}}\left[i\left(\frac{\bar{u}-c}{\epsilon\bar{u}_c'}\right)\chi_1^{(0)} + \frac{\bar{u}'}{\bar{u}_c'}\chi_3^{(0)} + \frac{i}{\gamma\bar{p}\bar{u}_c'M_\beta^2}\chi_4^{(0)}\right] \\ \frac{d\chi_3^{(0)}}{d\zeta} &= -i\chi_1^{(0)} \\ \frac{d\chi_4^{(0)}}{d\zeta} &= 0 \\ \frac{d\chi_5^{(0)}}{d\zeta} &= \chi_6^{(0)} \\ \frac{d\chi_6^{(0)}}{d\zeta} &= Pr\,\frac{\bar{\rho}}{\bar{\mu}}\left[\frac{\bar{T}'}{\bar{u}_c'}\chi_3^{(0)} + i\left(\frac{\bar{u}-c}{\epsilon\bar{u}_c'}\right)\chi_5^{(0)}\right] \end{aligned}\right\} \tag{14-10}$$

Since the procedure is valid near $y = y_c$, all coefficients above are to be evaluated at y_c, and $(\bar{u} - c) \sim \bar{u}_c'\epsilon\zeta$. Now Eq. 14-10 are separable into two groups. The first four of Eq. 14-10, the "momentum group," lead to

$$\chi_3^{(0)''''} - \frac{i\zeta}{\bar{\nu}_c}\chi_3^{(0)''} = 0 \tag{14-11}$$

while the last two, the "thermal group," lead to

$$\chi_5^{(0)''} - Pr\,\frac{i\zeta}{\bar{\nu}_c}\chi_5^{(0)'} = Pr\,\frac{\bar{T}_c'}{\bar{\nu}_c\bar{u}_c'}\chi_3^{(0)} \tag{14-12}$$

where $\bar{\nu}_c$ is the kinematic viscosity at $y = y_c$. Both Eq. 14-11 and 14-12 are of the type of Eq. 8-13. The solutions may now be arranged in order, using a double-suffix notation such that $\chi_{nm}^{(0)}$ denotes the mth fundamental solution of the initial approximation for χ_n, as follows:

$$\left.\begin{aligned}
\chi_{31}^{(0)} &= \zeta \\
\chi_{32}^{(0)} &= 1 \\
\chi_{33}^{(0)} &= \int_{+\infty}^{\zeta\bar{\nu}_c^{-\frac{1}{2}}} d\zeta \int_{+\infty}^{\zeta} d\zeta\zeta^{\frac{1}{2}} H_{\frac{1}{3}}^{(1)}[\tfrac{2}{3}(i\zeta)^{\frac{3}{2}}] \\
\chi_{34}^{(0)} &= \int_{-\infty}^{\zeta\bar{\nu}_c^{-\frac{1}{2}}} d\zeta \int_{-\infty}^{\zeta} d\zeta\zeta^{\frac{1}{2}} H_{\frac{1}{3}}^{(2)}[\tfrac{2}{3}(i\zeta)^{\frac{3}{2}}] \\
\chi_{35}^{(0)} &= \chi_{36} = 0
\end{aligned}\right\} \tag{14-13}$$

Since Z_3 corresponds to ϕ of Eq. 8-12, Eq. 14-13 is the generalization of Eq. 8-14. Differentiating Eq. 14-13, we get $\chi_1^{(0)}$ directly. The solution $\chi_4^{(0)}$ may be written as:

$$\chi_{41}^{(0)} = 0, \qquad \chi_{42}^{(0)} = 1, \qquad \chi_{43}^{(0)} = \chi_{44}^{(0)} = \chi_{45}^{(0)} = \chi_{46}^{(0)} = 0 \tag{14-14}$$

The solution $\chi_5^{(0)}$ generally requires quadrature to take care of the inhomogeneous term of Eq. 14-12. But with $\chi_{35}^{(0)} = \chi_{36}^{(0)} = 0$, there are two simpler solutions,

$$\left.\begin{aligned}
\chi_{55}^{(0)} &= \int_{+\infty}^{\zeta(Pr/\bar{\nu}_c)^{\frac{1}{2}}} d\zeta \int_{+\infty}^{\zeta} d\zeta\zeta^{\frac{1}{2}} H_{\frac{1}{3}}^{(1)}[\tfrac{2}{3}(i\zeta)^{\frac{3}{2}}] \\
\chi_{56}^{(0)} &= \int_{-\infty}^{\zeta(Pr/\bar{\nu}_c)^{\frac{1}{2}}} d\zeta \int_{-\infty}^{\zeta} d\zeta\zeta^{\frac{1}{2}} H_{\frac{1}{3}}^{(2)}[\tfrac{2}{3}(i\zeta)^{\frac{3}{2}}]
\end{aligned}\right\} \tag{14-15}$$

The improvement of Eq. 14-13 and 14-15 by the Tollmien technique (Art. 9) is now obvious; it is also necessary because numerical calculations showed that the phase velocity c for neutral disturbances is likely to be larger than in the incompressible case. One obtains first the equations for Z_3 and Z_5 corresponding to Eq. 9-1 by elimination, retaining only the highest derivatives. The result leads to expressions for $(\chi_{33}^{(0)}, \chi_{34}^{(0)})$ or $(\chi_{55}^{(0)}, \chi_{56}^{(0)})$. Using the same transformation (Eq. 9-6), we may still represent the solutions by Eq. 9-7, except that

$$\left.\begin{aligned}
&\text{for } \chi_{33}^{(0)}, \chi_{34}^{(0)}, \text{ replace } (\bar{u} - c) \text{ by } \frac{(\bar{u} - c)}{\bar{\nu}}; \\
&\text{for } \chi_{55}^{(0)}, \chi_{56}^{(0)}, \text{ replace } (\bar{u} - c) \text{ by } \frac{(\bar{u} - c)Pr}{\bar{\nu}}
\end{aligned}\right\} \tag{14-16}$$

Since $Z_1 \cong idZ_3/dy$ in this approximation, we get

$$\left.\begin{aligned}
\chi_{13}^{(0)} &\cong i\sqrt{\alpha Re}\left(\frac{Y\bar{\nu}}{\bar{u} - c}\right)^{\frac{1}{2}} \int_{+\infty}^{Y} \sqrt{Y}\, H_{\frac{1}{3}}^{(1)}[\tfrac{2}{3}(iY)^{\frac{3}{2}}]dY \\
\chi_{14}^{(0)} &\cong i\sqrt{\alpha Re}\left(\frac{Y\bar{\nu}}{\bar{u} - c}\right)^{\frac{1}{2}} \int_{-\infty}^{Y} \sqrt{Y}\, H_{\frac{1}{3}}^{(2)}[\tfrac{2}{3}(iY)^{\frac{3}{2}}]dY
\end{aligned}\right\} \tag{14-17}$$

The solutions naturally are also obtainable by the "first method" of Art. 8, valid for finite y and $\alpha Re \to \infty$. These may be derived by writing, similar to Eq. 8-4,

$$Z_n = \chi_n(y)e^{\int\sqrt{\alpha Re}\, g_0 dy}$$

and expanding

$$\chi_n(y) = \chi_n^{(0)}(y) + \frac{1}{\sqrt{\alpha Re}}\chi_n^{(1)}(y) + \cdots, \qquad n = 1, 3, 4, 5$$

After substitution into Eq. 14-5, g_0 and $\chi_n^{(0)}$ are determined in the same manner as before.

G,15. Some Difficulties of the Stability Theory of Compressible Fluids. By examining the details, we shall see that several difficulties arise in building upon the framework of the previous article to formulate a complete stability theory for the compressible boundary layer. Not all of these difficulties have been completely overcome at this moment.

First, let us take up the question of the boundary conditions. Suppose the basic flow is the boundary layer over a solid surface which may or may not be cooled. The boundary conditions at the wall consist of

$$\left.\begin{array}{lr} \text{(i) the velocity conditions} & (u', v') = 0 \\ \text{(ii) the thermal condition} & a\dfrac{d\theta'}{dy} + b\theta' = 0 \end{array}\right\} \qquad (15\text{-}1)$$

The determination of the constants a and b in Eq. 15-1 at once causes trouble if the results of the theory are to be applied to specific practical cases. As pointed out by Dunn and Lin [4], they are to be solved, strictly speaking, jointly with the unsteady heat transfer problem of the wall under the fluctuating heat input from the assumed disturbances, and hence depend upon the detail design of the vehicle. A general stability theory must therefore carry the ratio a/b as an additional parameter. Three special cases independent of design, however, may be listed:

(a) for insulated wall there must be no heat transfer due to the fluctuation, so, $b = 0$

(b) for very high frequency oscillations, the wall temperature will not respond, so $a = 0$

(c) for very low frequency oscillations, there is no change of the heat transfer at the wall, so (15-2)

$$\frac{a}{b} \cong \left(\frac{k}{\dfrac{dk}{dT}\dfrac{d\bar{T}}{dy}}\right)_{\text{wall}}$$

where k is the coefficient of thermal conductivity, assumed to be a function of temperature only. The last condition results from the analysis of a special cooling model in [4]. The physical explanation put forth is as follows: The thermal fluctuation in the boundary layer will penetrate deeply into the adjacent medium when the frequency is very low. The change of the temperature gradient in the adjacent medium thus becomes small in a distance of the order of the gas boundary layer thickness. While the surface temperature responds to the fluctuations, one may say that the heat transfer rate at the surface is insensitive to these fluctuations. To decide whether the frequency of oscillation is high or low, one must compare against the time constant of the unsteady heat transfer problem of the wall itself. However, Eq. 15-2 is correct in the limit.

Now we turn to the boundary conditions at the outer edge of the boundary layer,

$$u',\ v',\ \theta' \text{ bounded as } y_2 \to \infty$$

Similar to Eq. 10-6, however, we could cast the boundedness requirement into a more definite statement by examining the behavior of the fundamental solutions. Of the four "viscous solutions" ($\chi_{nm}^{(0)}$, $m = 3, 4, 5, 6$) the two involving $H_{\frac{1}{3}}^{(2)}$ grow beyond bounds as $y_2 \to \infty$ and must be rejected; the two involving $H_{\frac{1}{3}}^{(1)}$ decay exponentially with $y_2 \to \infty$, thus satisfying the boundedness requirement automatically. We must therefore consider the asymptotic behaviors of the two "inviscid" solutions of Eq. 14-6. As $y \to \infty$, $\bar{u} \to 1$, $\bar{u}' \to 0$, $\bar{T} \to 1$, Eq. 14-6 reduces to

$$\left.\begin{aligned} \phi'' &= \alpha^2 P_2 \phi = 0 \\ P_2 \equiv P(y_2) &= 1 - M_\beta^2(1-c)^2 \end{aligned}\right\}$$

Two possibilities arise:

(i) $\qquad P_2 > 0, \qquad \phi \sim e^{\pm \alpha \sqrt{P_2} y}$

(ii) $\qquad P_2 < 0, \qquad \phi \sim e^{\pm i\alpha \sqrt{|P_2|} y}$

The first type is at once seen to be the generalization of the incompressible case (where $P_2 = 1$). If we discard the exponentially increasing solution, the disturbance has the anticipated decaying behavior, so that the velocity boundary condition appears to be, as $y \to \infty$,

$$\phi' + \alpha \sqrt{P_2}\, \phi = 0 \tag{15-3}$$

This type of disturbance is named by Lees and Lin [31] as the "subsonic disturbance," since $P_2 > 0$ implies

$$1 - c < \frac{1}{M_\beta} \tag{15-4}$$

which may be interpreted as a statement that the velocity of propagation relative to the free stream velocity (actually the component of free stream velocity in the direction of wave propagation) is subsonic. On the other hand, for the second type, the disturbance simply oscillates faster and faster as y increases but without change of amplitude. Such disturbances furthermore propagate relative to the free stream velocity at a supersonic velocity. These are therefore named "supersonic disturbances" and are regarded as representing acoustic disturbances. After Lees and Lin [*31*], the usual view is that the subsonic disturbances might be the type responsible for laminar stability, leading to discrete eigenvalues and hence definite neutral curves in much the same manner as the incompressible theory. In contrast, the supersonic disturbances would lead to a continuous eigenvalue spectrum, the significance of which has yet to be fully explored.

Before we leave the "inviscid equation," Eq. 14-6, let us search for the "inviscid criterion" for the compressible case, namely, the requirement on the basic flow distributions such that a completely inviscid neutral disturbance follows from Eq. 14-6. It is convenient to rewrite Eq. 14-6 as

$$\left.\begin{aligned} \left(\frac{\phi'}{P}\right)' - \left[\frac{(\bar{u}'/P)'}{\bar{u} - c} + \frac{\alpha^2}{\bar{T}}\right]\phi = 0 \\ P = \bar{T} - M_\beta^2(\bar{u} - c)^2 \end{aligned}\right\} \tag{15-5}$$

which bears a strong resemblance to Eq. 7-6 in the incompressible case $(P = \bar{T} = 1)$,

$$\phi'' - \left[\frac{\bar{u}''}{\bar{u} - c} + \alpha^2\right]\phi = 0$$

We have mentioned already (p. 765) that $(\bar{u} - c_r)$ must still change sign within y_1, y_2 even for Eq. 15-5. If we now seek the eigensolution for a neutral oscillation, the point y_c (where $\bar{u} = c$) will be a logarithmic singularity *unless* $(\bar{u}'/P)'$ vanishes at y_c. A singularity, however, implies rapid variations, and hence the unavoidable viscous effects. Consequently the disturbances can be truly inviscid only when

$$\left(\frac{\bar{u}'}{P}\right)'\bigg|_{y=y_c} = (\bar{\rho}\bar{u}')'\bigg|_{y=y_c} = 0 \tag{15-6}$$

Now, suppose we admit a formal solution with the logarithmic singularity at $\bar{u} - c = 0$. It follows from Eq. 15-5 that, away from $\bar{u} - c = 0$,

$$\frac{d}{dy}\left[\frac{\phi'\tilde{\phi} - \phi\tilde{\phi}'}{P}\right] = 0$$

i.e. $(\phi'\tilde{\phi} - \phi\tilde{\phi}')/P$ is a constant on both sides of $\bar{u} - c = 0$. There is, however, a jump at the critical point equal to $\pi(|\phi_c|^2/\bar{u}'_c)(\bar{u}'/P)'_c$, obtainable by using the series solutions similar to Eq. 8-1 near the critical point. In order to satisfy the boundary conditions on ϕ at the end points (y_1, y_2) the jump must be zero if $P \neq 0$ at y_1 and y_2. There are then two alternatives: $\phi_c = 0$ or $(\bar{u}'/P)'_c = 0$. The former may be ruled out by the argument that $\phi_2^{(0)}$ of Eq. 14-8, being nonzero at y_c, must be excluded from ϕ, while $\phi_1^{(0)}$ certainly cannot go to zero at y_1 in view of its expression in Eq. 14-8 *as long as* $P > 0$. Under this restriction, then, Eq. 15-6 is the necessary condition for neutral disturbances.

To show that indeed there exists a neutral eigensolution when Eq. 15-6 is satisfied, we again consider only the cases with $P > 0$ everywhere. Then by the transformation

$$\eta = \int_{y_1}^{y} P dy \tag{15-7}$$

we reduce Eq. 15-5 to

$$\frac{d^2\phi}{d\eta^2} - \left[\frac{d^2\bar{u}/d\eta^2}{\bar{u} - c} + \frac{\alpha^2}{P\bar{T}}\right]\phi = 0 \tag{15-8}$$

Since $P\bar{T} > 0$, we obtain the same result that $d^2\bar{u}/d\eta^2|_{\eta=\eta_c} = 0$ guarantees real positive eigenvalues α^2 and the real eigensolution, under the same oscillation theorem quoted on p. 734. The condition $d^2\bar{u}/d\eta^2|_{\eta=\eta_c} = 0$ is of course no other than Eq. 15-6 in the original coordinate.[6] The existence of amplified disturbances for wave numbers slightly less than that for the neutral disturbance was also shown by Lees and Lin [*31*].

The necessary and sufficient condition for neutral (subsonic) disturbances in compressible boundary layers is thus $(\bar{\rho}\bar{u}')'_c = 0$ for some $c > 1 - (1/M_\beta)$ within the boundary layer. However, we must point out that the inviscid criterion here is not as strong as its incompressible counterpart because of the restrictive assumption $P > 0$. Reflection on the known behaviors of the temperature distribution in compressible boundary layers immediately suggests that $P > 0$ would be satisfied for most subsonic free streams but is almost certain to be violated, at least in part of the boundary layer with high supersonic M_β and strong cooling of the wall. The inviscid behavior under such circumstances seems to require further investigation.

We next take up the derivation of the secular equation with the solutions of the previous article and the boundary condition equations (Eq. 15-1, 15-2, and 15-3). We have earlier ordered the fundamental

[6] Our proof is different from that of Lees and Lin [*31*], whose argument was based upon a discussion of the Reynolds stress distribution due to the oscillation. The assumption $P > 0$, however, was likewise made by these authors.

solutions as $\chi_{nm}^{(0)}$, $m = 1, 2, 3, \ldots, 6$, while n referring to the index n of Z_n. It has also been seen that the solutions fall into three groups:

1. the "inviscid solutions," $m = 1, 2$
2. the "viscous solution of the momentum group," $m = 3, 4$
3. the "viscous solutions of the thermal group," $m = 5, 6$

In each group, one of the solutions decays exponentially with y and is the proper one to use. The one from the "inviscid solutions" has to be a linear combination of the two given by Eq. 14-8 (for $\chi_{3m}^{(0)}$); let the combination be denoted by $\chi_{nI}^{(0)}$. The ones from the other two groups are clearly $\chi_{n3}^{(0)}$ and $\chi_{n5}^{(0)}$ respectively. It is now only necessary to apply the wall conditions of Eq. 15-1 to get, dropping the superscript $^{(0)}$ and evaluating every term at y_1,

$$a\begin{vmatrix} \chi_{1I} & \chi_{13} & \chi_{15} \\ \chi_{3I} & \chi_{33} & \chi_{35} \\ \chi_{5I}' & \chi_{53}' & \chi_{55}' \end{vmatrix} + b\begin{vmatrix} \chi_{1I} & \chi_{13} & \chi_{15} \\ \chi_{3I} & \chi_{33} & \chi_{35} \\ \chi_{5I} & \chi_{53} & \chi_{55} \end{vmatrix} = 0$$

After a study of the orders of magnitude of the various terms involved, on the basis of the "improved" solutions or those by the first method summarized under Art. 14, Dunn and Lin [4] give the secular equations as

$$\frac{\chi_{3I}}{\chi_{1I}} = \frac{\chi_{33}}{\chi_{13}}\left[1 + \Delta + 0\left(\frac{1}{\sqrt{\alpha Re}}\right)\right] \tag{15-9}$$

where

$$\Delta = \frac{a\chi_{5I}' + b\chi_{5I}}{\chi_{1I}} \frac{\chi_{55}}{a\chi_{55}' + b\chi_{55}} \frac{\chi_{35}}{\chi_{55}} \frac{\chi_{13}}{\chi_{33}} \tag{15-10}$$

They also give the following estimate for Δ:

$$\left.\begin{aligned} &\frac{a}{b} = O(1) \text{ or larger, } \Delta \sim O\left(\frac{1}{\sqrt{\alpha Re}}\right) \\ &\frac{a}{b} = O\left(\frac{1}{\sqrt{\alpha Re}}\right) \text{ or smaller, } \Delta \sim \frac{(\gamma - 1)M_\beta^2 c^2}{\bar{T}_w} O(1) \end{aligned}\right\} \tag{15-11}$$

where $\bar{T}_w$ is the wall-temperature ratio. Thus, referring back to Eq. 15-2, Δ will be negligible if (1) the wall is insulated, (2) we have high frequency oscillation at $M_\beta^2 \ll 1$, or (3) $c^2 \ll 1$, which is approximately true at subsonic M_β but not for even moderate supersonic M_β (Eq. 15-4). In such cases, Eq. 15-9 becomes independent of the thermal boundary conditions. Available numerical computations of the neutral curves in literature (Lees [32] and Dunn and Lin [4]) have all been carried out for $\Delta = 0$.

Now we continue by going into certain details of the calculation procedure. The solution χ_{3I} may be expressed by $\phi_{1,2}$ of Eq. 14-8 as

$$\chi_{3I} = \phi_{11} - C\phi_{21}, \qquad C = \frac{\phi_{12}' + \alpha\sqrt{P_2}\,\phi_{12}}{\phi_{22}' + \alpha\sqrt{P_2}\,\phi_{22}}$$

then by Eq. 14-7,

$$\chi_{1I} = \frac{i}{P_1}[\bar{T}_1\phi_{11}' + M_\beta^2\bar{u}_1'c\phi_{11} - C(\bar{T}_1\phi_{21}' + M_\beta^2\bar{u}_1'c\phi_{21})]$$

where $\phi_{11} \equiv \phi_1(y_1)$, $\phi_{12} \equiv \phi_1(y_2)$, etc., as on p. 743, and $P_1 \equiv P(y_1)$, $\bar{T}_1 \equiv \bar{T}(y_1)$, $\bar{u}_1' \equiv \bar{u}'(y_1)$. Then, similar to Eq. 11-2, we express Eq. 15-9 as

$$F(z^*)(1 + \Delta) = E^*(\alpha, Re, c, M_\beta^2) \tag{15-12}$$

In Eq. 15-12, $F(z^*)$ is the same Tietjen's function defined in Eq. 11-1 except for the argument z^*,

$$z^* = (\alpha Re)^{\frac{1}{3}}\left[\frac{3}{2}\int_{y_1}^{y_c}\sqrt{\frac{c-\bar{u}}{\bar{\nu}}}\,dy\right]^{\frac{2}{3}} \tag{15-13}$$

The function E^* now reads

$$E^* = -\frac{\sqrt{\frac{c}{\bar{\nu}_1}}P_1\left[\frac{3}{2}\int_{y_1}^{y_c}\sqrt{\frac{c-\bar{u}}{\bar{\nu}}}\,dy\right]^{-1}\begin{vmatrix}\phi_{11}\phi_{12}' + \alpha\sqrt{P_2}\,\phi_{12}\\ \phi_{21}\phi_{22}' + \alpha\sqrt{P_2}\,\phi_{22}\end{vmatrix}}{\begin{vmatrix}\bar{T}_1\phi_{11}' + M_\beta^2\bar{u}_1'c\phi_{11}\phi_{12}' + \alpha\sqrt{P_2}\,\phi_{12}\\ \bar{T}_1\phi_{21}' + M_\beta^2\bar{u}_1'c\phi_{21}\phi_{22}' + \alpha\sqrt{P_2}\,\phi_{22}\end{vmatrix}} \tag{15-14}$$

So far numerical calculations have been made only with the Δ term neglected. A procedure similar to the incompressible case may now be naturally formulated. Thus Eq. 11-9 may be replaced by

$$\left.\begin{aligned}\mathfrak{F}(z^*) &= (1 + \lambda^*)\frac{u^* + iv^*}{1 + \lambda^*(u^* + iv^*)}\\ \lambda^* &= \frac{\bar{u}_1'}{c}\sqrt{\frac{\bar{\nu}_1}{c}}\left[\frac{3}{2}\int_{y_1}^{y_c}\sqrt{\frac{c-\bar{u}}{\bar{\nu}}}\,dy\right]^{\frac{2}{3}} - 1\\ u^* + iv^* &= 1 + \frac{\bar{u}_1'c}{\bar{T}_1}\frac{\phi_{22}' + \alpha\sqrt{P_2}\,\phi_{22}}{\phi_{12}' + \alpha\sqrt{P_2}\,\phi_{12}}\end{aligned}\right\} \tag{15-15}$$

where $\mathfrak{F}(z^*)$ is identical to Eq. 11-7 except for the new argument z^* of Eq. 15-13. The definition of λ^* above is the one by Dunn and Lin [*4*], differing from that of Lees [*32*]; it has the advantage of always being rather small. The inviscid solutions are given by, for instance, Eq. 14-8. The calculation, however, is very tedious.

For simplicity, Dunn and Lin [*4*] made calculations for $\alpha \ll 1$.

Omitting terms of α^2 and higher, they get for Eq. 15-15

$$\left.\begin{aligned} u^* &\cong \frac{\bar{u}_1'c}{\bar{T}_1}\left\{K_{1,r}^* + \frac{\bar{T}_1}{\bar{u}_1'c} + \frac{\sqrt{P_2}}{(1-c)^2}\left[\frac{1}{\alpha} - \frac{\sqrt{P_2}}{(1-c)^2}H_1^*\right]\right\} \\ v^* &\cong \frac{\bar{u}_1'c}{\bar{T}_1}K_{1,i}^* \end{aligned}\right\} \quad (15\text{-}16)$$

where

$$\left.\begin{aligned} K_1^* &= K_{1,r}^* + iK_{1,i}^* \equiv \int_{y_1}^{y_2}(\bar{u}-c)^2 P\,dy \\ H_1^* &\equiv \int_{y_1}^{y_2}(\bar{u}-c)^2\bar{T}^{-1}dy \end{aligned}\right\} \quad (15\text{-}17)$$

as compared against K_1 and H_1 of Eq. 11-12. Since v^* and λ^* are both functions of c only, for assigned $c > 1 - (1/M_\beta)$, u^* and z^* may be found in a manner similar to the iteration scheme of Eq. 11-14 and 11-15. Then α is directly solved from u^* by Eq. 15-16; and finally the Reynolds number is obtained from the definition of z^*, Eq. 15-13. Note, however, that the Reynolds number is based on the velocity component of the free stream in the direction of wave propagation, as is the Mach number M_β. It should be emphasized that a complete survey of the "three-dimensional" disturbances in terms of the wave propagation angle is necessary for compressible fluids before the lowest critical Reynolds number based upon the full free stream velocity may be determined.

The assumption of $\alpha \ll 1$ is reasonable for subsonic and slightly supersonic M_β; under extreme cooling of the wall it is also valid for *any* M_β. (See Art. 16.)

Finally, it should be pointed out that all successful calculations to date have been performed for Mach numbers ranging from zero to low supersonic values. Attempts to calculate for somewhat higher Mach numbers, say above about 3, according to the Dunn-Lin theory seem to run into unforeseen difficulties. The fundamental case of the compressible boundary layer over a flat plate has been re-examined by Lees and Reshotko [*136*], retaining more viscous terms on the right hand side of Eq. 14-2 in order to represent the effect of large temperature change at high Mach numbers. In spite of the refinement, however, the Lees-Reshotko theory still has difficulty in getting the neutral curves for the higher Mach numbers of 3 and 5.6. The trouble may have to do with the classical procedure of splitting the solution into "inviscid" and "viscous" parts. In our discussion of the "inviscid equation" on p. 771, the change of behavior of the inviscid solution as P changes sign within the boundary layer was stressed. At Mach numbers higher than about 2.2, this indeed turns out to be responsible for the difficulty in getting a normal neutral curve, according to an unpublished investigation of the same problem, based on the Dunn-Lin theory, by J. Powers of the U.S. Naval Ordnance

Laboratory. No matter how accurately one accounts for the viscous effects, therefore, as long as the inviscid equation is employed in the customary manner, it is perhaps not surprising that the same stumbling block remains. On the other hand, without splitting into "inviscid" and "viscous" solutions, Brown [*137*] attacked the problem by frontal assault, programming the entire set or linearized perturbation equations on a modern computer, and achieved great success. He did not have to neglect *any* viscous terms, and at Mach numbers 5 and 5.8 found neutral curves which show no peculiarity and agree well with Demetriades' data [*37*]. His numerical procedure does not resolve the question of why the analytical methods fail, but obviously is the best current method for getting answers, particularly at high Mach numbers.

G,16. Approximate Calculations for the Compressible Boundary Layer over a Flat Plate. Cooling Required for Complete Stabilization. We describe now certain calculations of practical interest for the compressible boundary layer over a flat plate. These deal with effects on stability of the following parameters:

1. Mach number,
2. wall temperature,
3. disturbance wave propagation angle.

For the flat plate problem, Lees [*32*] gave two sets of neutral curves covering (1) and (2), but all with "two-dimensional" disturbances which propagate in the direction of the free stream velocity. He retained some terms up to α^3 in the inviscid solutions (Eq. 14-8), but used the viscous solution χ_{33} of Eq. 14-13 adequate only for small c. The trend indicated by his results is probably right. Fig. G,16a shows the effect of Mach number for an insulated case, and Fig. G,16b illustrates the effect of surface temperature at a given Mach number. Worth noting are the tendencies toward instability due to increasing Mach number for the insulated case, and due to increased heating of the surface at a given Mach number. The curve for the heated wall also indicates instability at infinite Reynolds number as a result of the "inviscid criterion," (Eq. 15-6).

At a fixed free stream Mach number $M = 1.6$ and the wall temperature ratio $\bar{T}_1 = 1.073$, Dunn and Lin [*4*] gave the neutral curves for three wave propagation angles, shown here as Fig. G,16c. The Reynolds number Re_θ is the free stream Reynolds number based upon the momentum thickness. Of the three wave propagation angles shown, the middle one is the least stable. Based upon Eq. 15-16, the approximate minimum critical Reynolds number as a function of the wave angle may be found directly, after one approximates the corresponding c by solving the second of Eq. 15-11 with[7]

$$v^*(c) \cong 0.58$$

[7] Compare with the discussions on p. 750.

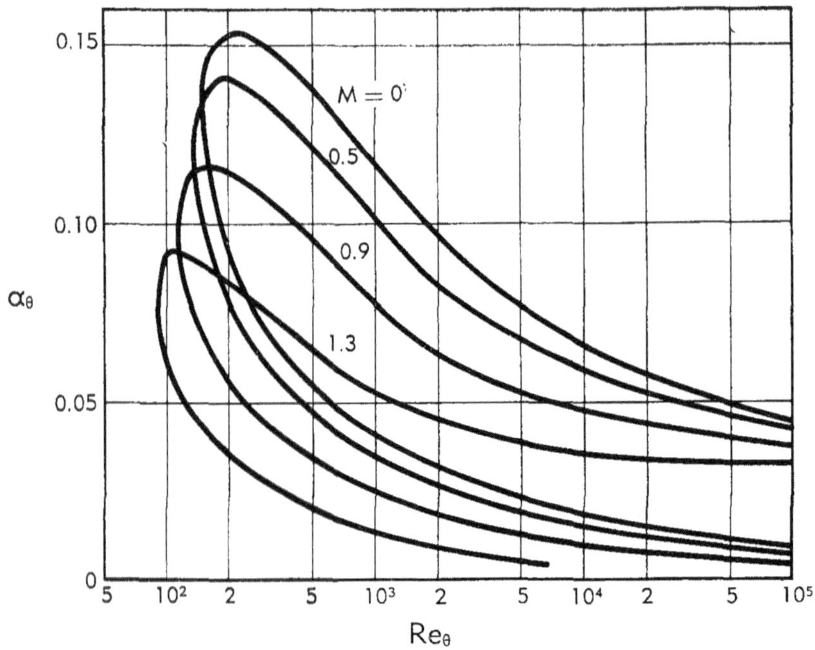

Fig. G,16a. Effect of Mach number on stability of boundary layer over insulated flat plate.

and takes $$u^* \cong 1.49, \qquad z^* \cong 3.21$$

The resultant formula is

$$\left.\begin{aligned} Re_{cr} &\cong \frac{(A + BM_\beta^2)}{M_\beta\sqrt{P_2}} \\ \text{with}\quad A &= K\left\{(1-c)^2\left[\frac{\bar{T}_1}{\bar{u}_1'c}(u^* - L) - M^2\right] + \frac{H_1^*}{(1-c)^2}\right\} \\ B &= K[(1-c)^2 - H_1^*] \\ K &= Mz^{*3}\left[\frac{3}{2}\int_{y_1}^{y_c}\sqrt{\frac{c-\bar{u}}{\bar{\nu}}}\,dy\right]^{-2} \\ L &= \frac{\bar{u}_1'c}{\bar{T}_1}\left[K_{1,r}^* + \frac{\bar{T}_1}{\bar{u}_1'c}\right] \\ M &= \text{free stream Mach number} \\ &K_{1,r}^* \text{ and } H_1^* \text{ from Eq. 15-17} \end{aligned}\right\} \qquad (16\text{-}1)$$

From Eq. 16-1 the lowest minimum critical (free stream) Reynolds number occurs when the wave angle is such that

$$M_\beta = \left[2(1-c)^2 + \frac{B}{A}\right]^{-1} \qquad (16\text{-}2)$$

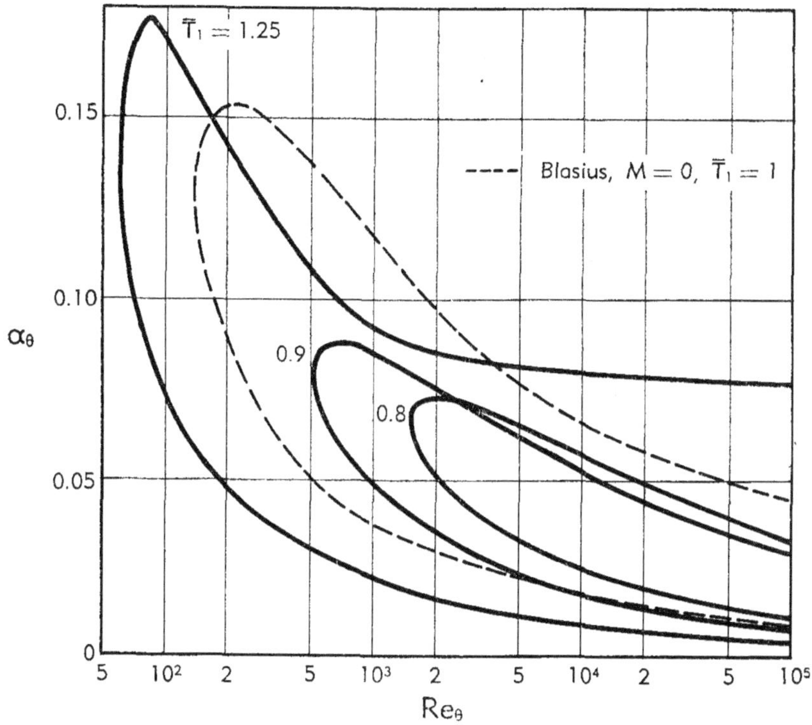

Fig. G,16b. Effect of wall temperature on stability of boundary layer over flat plate at M = 0.7.

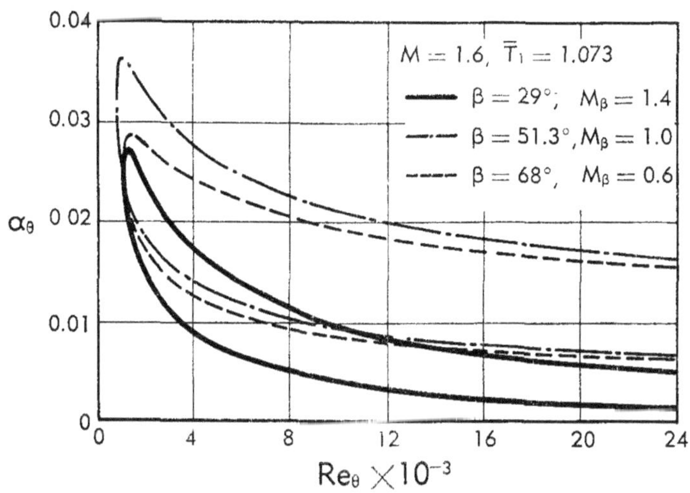

Fig. G,16c. Neutral curves for oblique wave disturbances.

A plot of Eq. 16-1 for the case of $M = 1.6$, $\bar{T}_1 = 1.073$ is given in Fig. G,16d, clearly showing a minimum when $\beta = \cos^{-1}(M_\beta/M) \cong 50°$. The estimated minimum critical Reynolds number at corresponding angles also agree well with the detailed calculation of Fig. G,16c. In this example, the critical Reynolds number for $\beta = 0°$ and $90°$ both tend to infinity. The importance of the oblique disturbance waves is thus demonstrated.

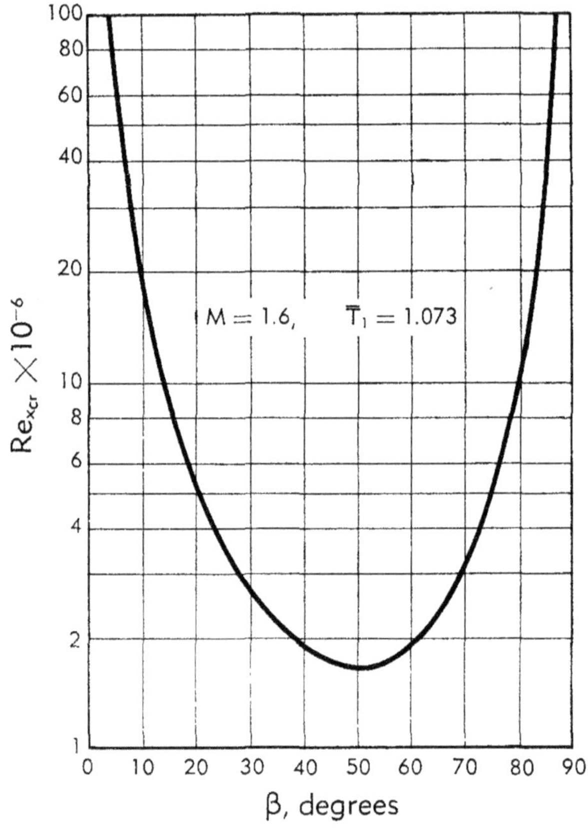

Fig. G,16d. Minimum critical Reynolds number vs. direction of oblique waves.

It may be mentioned at this point that, besides Eq. 16-1, an earlier formula given by Lees [32] is somewhat simpler for estimating the minumum critical Reynolds number, based upon the thickness δ, with "*two-dimensional*" disturbances:

$$Re_{cr} = \frac{25\bar{u}'_1}{c^4} \frac{\bar{\nu}_c}{\sqrt{1 - M^2(1 - c)^2}} \tag{16-3}$$

where c is determined from the condition

$$v^*(c) = -\pi \frac{\bar{u}'_1 c}{\bar{T}_1} \left[\frac{\bar{T}^2}{\bar{u}'^3} \frac{d}{dy} \left(\frac{\bar{u}'}{\bar{T}} \right) \right]_{y=y_c} \cong 0.58 \tag{16-4}$$

These reduce immediately to Eq. 11-17 and 11-18 for the incompressible case. It may likewise be used for three-dimensional disturbances by modifying $\tilde{u}$ and M. However, it is not certain how much inaccuracy might be involved in Lees' formula because of the less adequate viscous solutions used in the original work.

By either formula above, one observes that the critical Reynolds number moves to infinity if the value of c satisfying Eq. 16-4 simultaneously makes $P_2 = 1 - M_\beta^2(1 - c)^2 = 0$; i.e. when the lower limit of admissible c for subsonic disturbances is reached. As the general trend of Fig. G,16b shows, such complete stabilization may be expected with sufficient cooling. But we note that if $M_\beta < 1$, $P_2 > 0$ always, and no cooling is sufficient for complete stabilization. Therefore, the "three-dimensional" disturbances become the cause of instability.

The cooling required, represented by the "critical (wall) temperature ratio" for complete stabilization is just about the simplest and most reliable *calculation* to perform in the stability theory of the compressible boundary layer. It provides some qualitative indication of the stability trend.

The foregoing calculation may be refined by solving v^* in terms of λ^*, $\mathfrak{F}_r$, $\mathfrak{F}_i$ from Eq. 15-15

$$v^* = \frac{(1 + \lambda^*)F_i}{[1 + \lambda^*(1 - \mathfrak{F}_r)]^2 + \lambda^*\mathfrak{F}_i^2}$$

Since c is taken to be $1 - (1/M_\beta)$, λ^* is given and the right-hand side may be maximized with respect to z^*. This maximum value takes the place of 0.58 in Eq. 16-4. It may be noted that, solving the first of Eq. 15-16, one gets $\alpha \sim \sqrt{P_2}$. Thus, near complete stabilization, $\alpha \to 0$ and Eq. 15-16 becomes exact without any truncating error.

In Fig. G,16e are reproduced the curves of the critical temperature ratio required for complete stabilization against two-dimensional disturbances, as calculated by van Driest [*38*] essentially on the basis of Eq. 16-4, with a different combination of viscosity law and Prandtl number. If air closely obeys Sutherland's viscosity law and has a Prandtl number of 0.75, the two other simplifying assumptions would tend to err on the pessimistic side in predicting instability.

Detailed experimental verification of all these theoretical results is not yet conclusive. The neutral curve for an insulated flat plate at Mach number 2.16 was experimentally determined by Laufer and Vrebalovich [*36*], and the same case at Mach number 5.8 was investigated by Demetriades [*37*]. Essentially, artificial disturbances of constant frequency were introduced into the boundary layer, and the minimum (for lower branch) and maximum (for upper branch) responses at fixed values of y/δ were measured by hot wire downstream from the point where the disturbance was initiated. The results were at least consistent with theory. In Fig.

G,16f, a comparison of the experimental neutral curves at the three Mach numbers is shown. Note the shrinking of the unstable region as the Mach number is increased. Recalling that according to theory $c_r > 1 - (1/M)$ for subsonic disturbances, we see that the actual wavelengths of the unstable disturbances would be longer at higher Mach numbers.

There are also, as indirect evidences, a number of observations on the dependence of transition Reynolds number on Mach number and wall temperature. See for instance the reviews of Gazley [*39*] and Dryden [*40*]. The predicted trends seem to be confirmed in most cases, although the sensitivity of the transition Reynolds number to these variables is generally much less pronounced. The experimental results, however, are

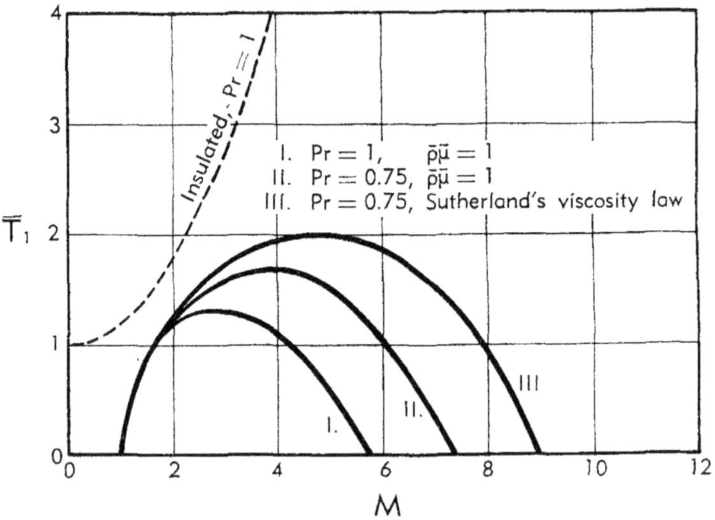

Fig. G,16e. Critical temperature ratio for complete stabilization.

often under the influence of other factors, such as free stream turbulence, roughness, etc., which contribute to the transition phenomenon. We include here Fig. G,16g, due to Dryden [*40*], which shows the likelihood of complete stabilization by cooling if one prolongs the curves A, B, and C to the left. The estimated asymptotes indeed were shown to be in fairly good agreement with the curve for ($Pr = 0.75$, $\mu \sim T$) of Fig. G,16e. A similar conclusion regarding complete stabilization was reached earlier by Gazley [*39*] in analyzing the test results of the nose cone of V-2 rockets. In these transition observations at least, one notes that the three-dimensional disturbances did not appear to have much influence on complete stabilization. It is of course possible that the minimum critical Reynolds numbers, though finite, have moved to values larger than available in the experiments; thus they become *practically* infinite.

As an exception, however, it should be mentioned that a "transition reversal" phenomenon has been recently discovered [*134*]. Under *extreme* cooling, the transition Reynolds number of a group of bodies of revolution, including cones, ogives and blunted cone-cylinders, actually would decrease with decreasing surface temperature. Corresponding to each curve in Fig. G,16f, the experiments suggest a second branch approaching the "critical temperature ratio" as an asymptote from the left. This

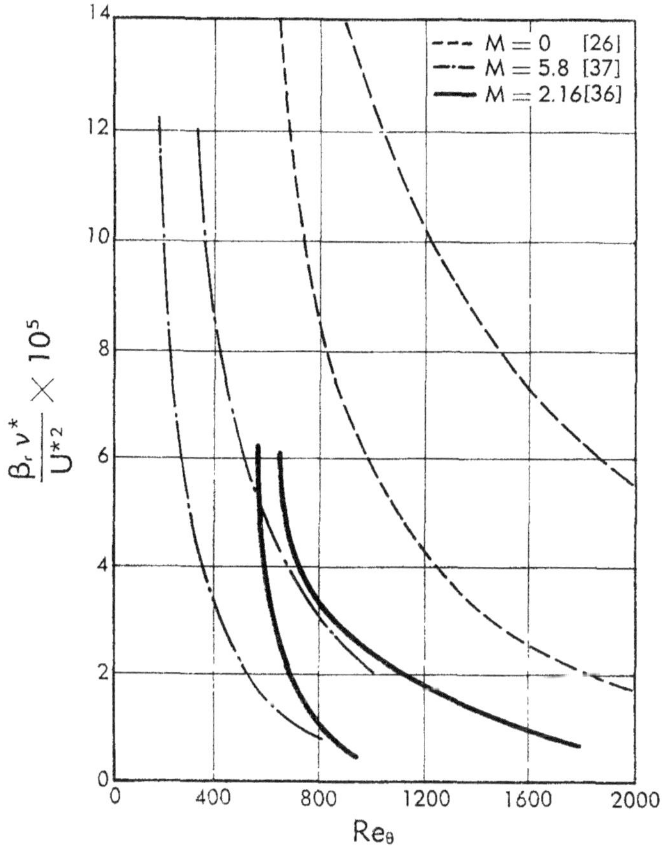

Fig. G,16f. Experimental neutral curves for insulated plate at supersonic and hypersonic Mach numbers.

behavior is thus contradictory to the concept of a "critical temperature ratio" for complete stabilization, according to the usual stability theory. The precise circumstances under which "transition reversal" takes place are yet unsettled. In a recent calculation of the critical temperature ratio for complete stabilization, Reshotko [*138*] found the possibility of a theoretical explanation. Schematically speaking, his results showed that near the point $M = 1$, $T = 0$ in Fig. G,16b, the curve could consist of a second small loop starting and ending at that point, and lying above the (usual) main branch of the curve. Within the small loop, there is again complete stabilization. Thus, over a limited range of supersonic Mach numbers, as the wall temperature ratio is decreased the boundary layer could be first stabilized by entering the small loop through cooling, then destabilized by leaving the small loop through further cooling, and finally again stabilized after crossing the main branch of the curve.

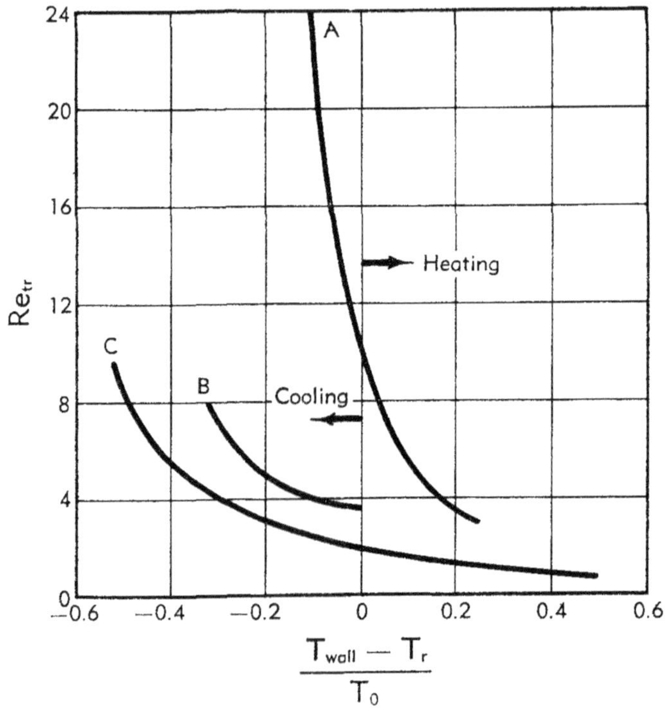

Fig. G,16g. Effect of heating and cooling on transition Reynolds number.

G,17. Boundary Layers with Pressure Gradient, Suction, or Blowing. As the practical purpose of the stability theory is to provide a correlation with the transition phenomenon, there is much interest in examining the stability of boundary layers under influences which are known from experience to play important roles in the maintaining of a laminar boundary layer. If one retains the approximation that the basic flow is locally parallel, the calculations only have to be repeated for the different profiles. The effect of pressure gradient has been studied by many authors. In the incompressible case, Schlichting [*41*] made calculations for the Howarth profiles pertaining to the free stream velocity distribution

$$U = U_0 - ax$$

with various values of the shape parameter $x^* = ax/U_0$. Pretsch [42] considered the Hartree profiles pertaining to

$$U \sim x^m$$

with m ranging between -0.091 and 4, and included a determination of the amplification rates. Schlichting and Ulrich [43] carried out calculations for sixth-degree polynomial approximations of the velocity distribution, the pressure gradient being characterized by the usual Pohlhausen parameter

$$\Lambda = \frac{\delta^2}{\nu}\frac{dU}{dx}$$

Their results are shown as Fig. G,17a. (Similar calculations have been

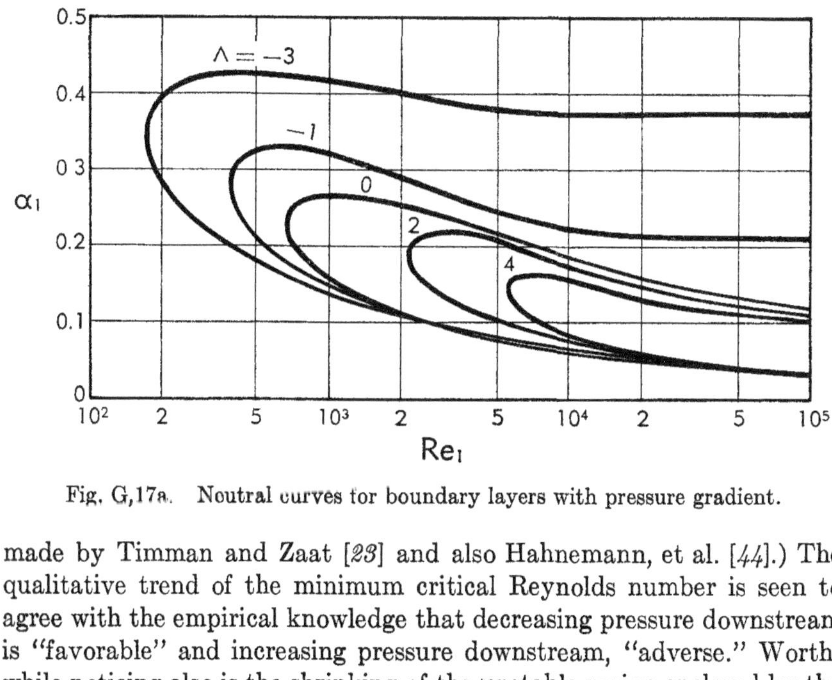

Fig. G,17a. Neutral curves for boundary layers with pressure gradient.

made by Timman and Zaat [23] and also Hahnemann, et al. [44].) The qualitative trend of the minimum critical Reynolds number is seen to agree with the empirical knowledge that decreasing pressure downstream is "favorable" and increasing pressure downstream, "adverse." Worthwhile noticing also is the shrinking of the unstable region enclosed by the neutral curve under favorable pressure gradient, which indicates that the maximum amplification rate would be reduced at the same time.

The compressible case has been calculated by Weil [45], Laurmann [46] and Low [47], all for two-dimensional disturbances only. In general, the effect is qualitatively similar, although at higher Mach numbers it is quantitatively much smaller. (See Fig. G,17b.)

An interesting remark due to Lees [32] is that, in view of the complete stabilization by sufficient cooling of a supersonic boundary layer over a flat plate, there seems to be the possibility of complete stabilization of a supersonic boundary layer, with given thermal conditions, by means of sufficient pressure gradient. Comparing Fig. G,16b and G,17a, we see indeed the qualitative resemblance of the effects of pressure gradient and heating-cooling. Mathematically, the "critical pressure gradient" for

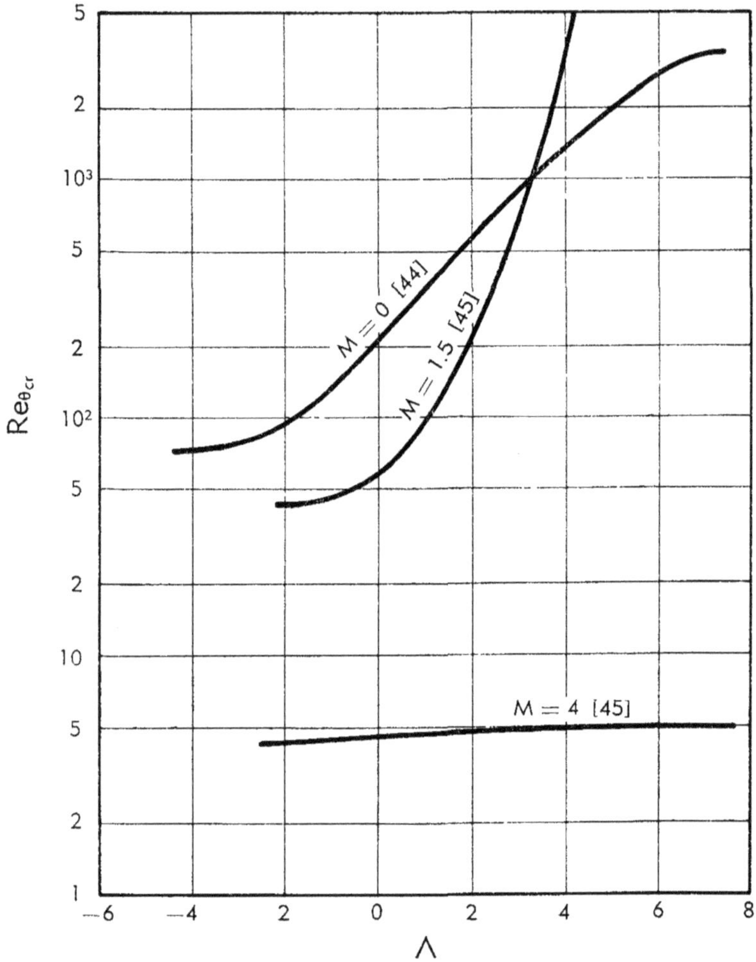

Fig. G,17b. Compressibility effect on minimum critical Reynolds number for boundary layer with pressure gradient.

complete stabilization is to be found by setting $c = 1 - (1/M)$ and satisfying Eq. 16-4 in Lees' approximation. Weil's calculation [*45*] shows that complete stabilization for two-dimensional disturbances by pressure gradient is possible for insulated surfaces in the Mach number range of about 1.3 to 2.4. The behavior at $M = 1.5$ shown in Fig. G,17b reflects this phenomenon.

The effect of suction through the wall likewise has received much attention. The suction parameter in incompressible flow is often defined by the coefficient

$$c_Q = -\frac{v_w}{U_\infty}$$

where v_w is the normal velocity at the wall and U_∞ the free stream velocity. The basic velocity profiles for $c_Q \sim x^{-\frac{1}{2}}$, $c_Q = \text{const}$ and a concentrated slot somewhere downstream, along a semi-infinite plate are all known, and calculations for the neutral curve, or estimates of the minimum critical Reynolds number, have been made for all these cases. (See [*48,49,50,51*].) In the case of constant c_Q, as one moves downstream the so-called "asymptotic profile" will be reached:

$$\bar{u} = 1 - e^{-y/\delta^*}$$

which is particularly convenient for analytical study. The inviscid solution is given by Chiarulli and Freeman [*50*] in terms of a hypergeometric function after introducing a new independent variable $\eta = e^{-y/\delta^*}$. The minimum critical Reynolds number based on δ^* was thus found to be 3.9×10^4,[8] about 100 times higher than that of the flat plate without suction. The approximate formula (Eq. 11-17) gives a value of 4.0×10^4. This close agreement lends some confidence in its accuracy for the intermediate stages between the Blasius profile at $x = 0$ (the leading edge) and the asymptotic profile as $x \to \infty$.

The beneficial effect of suction lies not only in the greatly increased critical Reynolds number, but also in the decreased displacement thickness at a given location. By requiring that at all stations along the plate the local Reynolds number (based upon the displacement thickness) shall be less than the local minimum critical value for stability, the least magnitude of c_Q sufficient for complete stabilization along the entire plate may be found. After Ulrich [*48*], the value is

$$c_{Q_{cr}} \cong 1.2 \times 10^{-4}$$

The inviscid "sink effect" will cost an additional drag coefficient of about twice $c_{Q_{cr}}$, thus a very small price to pay for the reduction of the frictional drag from the turbulent to the laminar value. (See Schlichting [*52*, pp. 233, 234].) However, practical arrangements seem to be unable to realize the extreme increase of the transition Reynolds number. (See discussion by Wuest [*53*].)

To estimate quickly the minimum critical Reynolds number for incompressible boundary layers under pressure gradient and/or suction, Tetervin and Levine [*54*] made a survey based upon the approximate

[8] This number is much lower than Pretsch's value [*49*] of 5.5×10^4 or Bussmann and Münz's value [*51*] of 7×10^4, obtained by following Schlichting's procedure.

formula Eq. 11-17. Replotted against the shape parameter δ^*/θ, this curve is shown as Fig. G,17c. A similar curve was also given in Lin's book [*1*, p. 95]. However, the same parameter may *not* be expected to correlate as well in the compressible case, since it alone will not control effectively the velocity and temperature distributions. An examination of Lees' calculation [*32*] confirms this reasoning.

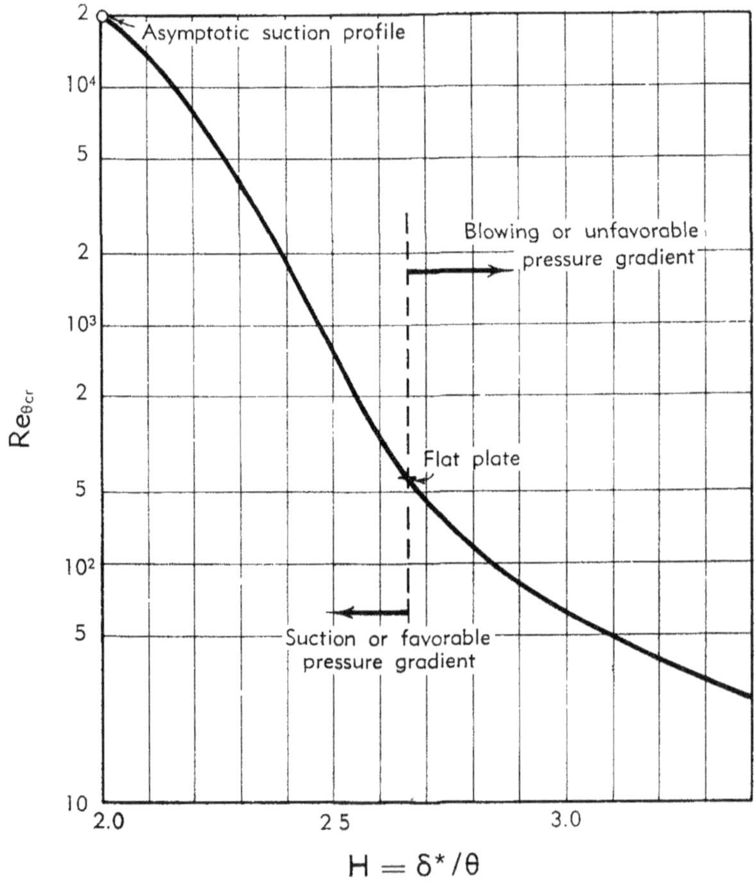

Fig. G,17c. Minimum critical Reynolds number for boundary layer, with suction, blowing, or pressure gradient.

As the opposite of suction, blowing or injection of fluid through the wall is known to be detrimental to the maintenance of a laminar boundary layer. In compressible flow, however, as a means of cooling of the surface the possibility of using injection has been frequently mentioned. Calculations of the critical temperature ratio for complete stabilization, in the case of a flat plate under a mass blowing rate which varies inversely as the square root of the distance from the leading edge, have been made by

Shen and Persh [55] and Low [56]. The free stream and the injected fluid are both air. As to be expected, with more blowing, the curve moves lower and lower from that for the rigid wall presented in Fig. G,16e. The case of stagnation point flow, under cooling by blowing but $M = 0$, has been investigated by Morduchow et al. [57].

G,18. Flows along a Curved Wall and with Axial Symmetry. Poiseuille Flow in a Circular Pipe. The effect of a curved wall on the instability with respect to transverse wave-type disturbances has been studied as a natural consequence to the first success of Tollmien's calculations of the boundary layer flow over a flat plate. The geometry is simplest when the radius of curvature of the wall is taken to be constant. The linearized disturbance equations in the incompressible case then are given by Eq. 4-6, 4-7, 4-8, and 4-9 in cylindrical coordinates.

The usual approach deals with a velocity distribution $\bar{u}(r)$, $\bar{v} = \bar{w} = 0$, and disturbance waves propagating in the ϕ direction parallel to $\bar{u}$, i.e. $\beta = 0$. When the basic flow is within a boundary layer which is of small depth compared to the radius of curvature, the equations may be further reduced by regarding $r \cong r_w$, the radius of curvature of the wall. Thus, for incompressible boundary layers along a curved wall under two-dimensional disturbances, the vorticity equation, Eq. 6-4, may be applied as a first approximation,

$$\frac{\partial \zeta_1'}{\partial t} + \bar{u}\frac{\partial \zeta_1}{\partial x} - \frac{\partial \psi_1}{\partial x}\frac{d\zeta_0}{dy} = \nu\nabla^2\zeta_1$$

with x and y denoting the distance along the curved wall and normal to it, respectively, and ζ_0 the basic flow vorticity. In the limit of a flat plate, the basic flow vorticity gradient $d\zeta_0/dy$ becomes $\bar{u}''$. It is thus obvious that, in the case of curved wall, the parameter $d\zeta_0/dy$ would influence the stability analysis in much the same manner as $\bar{u}''$ in the case of flat plate. The vanishing of the basic flow vorticity gradient defines the "critical layer," and likewise may be taken as the "inviscid criterion" as done by Görtler [58]. In terms of the velocity distribution, the vorticity gradient is of course $\bar{u}'' + (1/r_w)\bar{u}'$. Here r_w must be taken as positive for convex walls and negative for concave walls. For a velocity distribution without a point of inflection, the usual boundary layer has $\bar{u}'' < 0$ and $\bar{u}' > 0$. A convex wall is thus destabilizing and a concave wall stabilizing. However, the concave wall promotes a different kind of instability which will be discussed in the next chapter.

More thorough calculations have been made by Schlichting [59] for the flow inside a rotating cylinder. If there is an inside coaxial cylinder at rest, the basic flow is a Couette flow for which no instability has ever been found. Schlichting made calculations for the velocity profiles which occur in the boundary layer when the outer cylinder is set to a constant speed impulsively. As the profile changes, the displacement thickness δ^* and the minimum critical Reynolds number change at the same time. A minimum critical Reynolds number of all the profiles is found to be

$$\frac{\bar{u}' r_o}{\nu} \cong 66{,}000$$

where $\bar{u}_1$ is the circumferential velocity and r_o the radius of the outer cylinder. The number seems to be of the right order of magnitude when compared against the experimentally observed transition Reynolds numbers which occur with inside cylinders of various sizes. (See, e.g. [*52*, p. 401].) One must of course be aware of the possible effects of basic flow unsteadiness when making such comparisons.

If we still use the equation for the disturbances in cylindrical coordinates, Eq. 4-5 to 4-8, and consider the case,

$$\bar{u} = \bar{v} = 0, \qquad \bar{w} = \bar{w}(r); \qquad \alpha = 0$$

Then the equations describe an axially symmetric basic flow with rotationally symmetric disturbances. Without restricting to a special $\bar{w}(r)$, the equations at once show that the disturbance f_1 is decoupled from the rest. In this case, no generality is lost if f_2 and f_3, together with the eigenvalue c for given β, Re, are calculated as if the disturbances were axially symmetric without f_1. As a matter of fact, Synge [*15*] showed that f_1 will be damped, whatever the values of β, Re, and $\bar{w}$. This may be verified by multiplying Eq. 4-5 by $r\tilde{f}_1$, where $\tilde{f}_1$ is the complex conjugate of f_1, and integrating from $r = 0$ to the wall. The boundary conditions should be f_1 vanishing at the wall, and either $f_1'(0) = 0$ or $f_1(0) = 0$, depending on whether the disturbance is symmetric or antisymmetric. Then one directly gets $\sigma_i < 0$.

After omission of f_1 and elimination of f_3 and ϖ, Eq. 4-6, 4-7, and 4-8 under the above assumptions are reduced to a single equation, comparable to Eq. 6-3,

$$i\beta Re\left[(\bar{w} - c)(L - \beta^2)f_2 - \left(\bar{w}'' - \frac{\bar{w}'}{r}\right)f_2\right] = (L - \beta^2)^2 f_2 \quad (18\text{-}1)$$

where
$$c = \frac{\sigma}{\beta}$$

$$L \equiv \frac{d^2}{dr^2} + \frac{1}{r}\frac{d}{dr} - \frac{1}{r^2}$$

Here we may interpret f_2 as either the v component disturbance or the stream function of the disturbance. The boundary conditions are the vanishing of f_2 and f_2' at the solid wall and the boundedness of the disturbances within the fluid domain.

Eq. 18-1 was transformed by Pretsch [60], using a new independent variable,

$$\rho = 1 - \frac{r^2}{r_c^2} \tag{18-2}$$

where r_c is the critical point where $\bar{u} - c = 0$. The wall corresponds to a negative value of ρ and the center line to $\rho = 0$. The "inviscid equation" now resembles that for the two-dimensional case:

$$(\bar{w} - c)\left[(1-\rho)\frac{d^2 f_2}{d\rho^2} - 4\beta_1^2 f_2\right] - (1-\rho)\left(\frac{d^2\bar{w}}{d\rho^2}\right) f_2 = 0 \tag{18-3}$$

where $\beta_1 = \beta r_c/4$. Since $1 - \rho > 0$ except at one extremity, the only significant critical point is at $\rho = 0$, at which point the singular behavior depends upon $(d^2\bar{w}/d\rho^2)/(\bar{w} - c)$. Consequently, the term $d^2\bar{w}/d\rho^2$ plays the part of $\bar{u}''$ in the two-dimensional case. There will be inviscid instability only if $d^2\bar{w}/d\rho^2$ (which is proportional to $\bar{w}'' - (\bar{w}'/r)$) vanishes somewhere within the fluid. A profile with point of inflection is no longer necessarily unstable, at least with respect to rotationally symmetric disturbances, in the inviscid limit. An example is the axially symmetric jet (p. 796). The inviscid solutions are now slightly different but may be handled as in Art. 8.

For the viscous solutions near the critical point, the same argument leading to Eq. 8-10 and 8-13 may be used. One thus keeps only the highest order derivatives of the disturbance function on both sides of Eq. 18-1 in the lowest approximation,

$$i\beta_1 Re(\bar{w} - c)\left(\frac{dr}{d\rho}\right)_c^2 \frac{d^2 f_2}{d\rho^2} \cong \frac{d^4 f_2}{d\rho^4} \tag{18-4}$$

A change of variable similar to Eq. 8-10 is applicable,

$$\eta = \frac{\rho}{\epsilon}, \qquad \epsilon = \left[r_c\beta_1 Re\left(\frac{d\bar{w}}{d\rho}\right)_c\right]^{-\frac{1}{3}} \tag{18-5}$$

to reduce Eq. 18-4 to the first one of Eq. 8-13. The viscous solutions thus can be immediately written down.

The classical Poiseuille flow in a circular pipe has a velocity distribution $\bar{w} = 1 - r^2$. With Eq. 18-2, the velocity distribution becomes

$$\left.\begin{aligned} \bar{w} &= (1 - r_c^2) + \rho r_c^2 \\ 1 - \frac{1}{r_c^2} &< \rho < 1 \end{aligned}\right\} \tag{18-6}$$

Combining Eq. 18-6 with the "inviscid equation," Eq. 18-3, and the "viscous equation," Eq. 18-4, we have thus essentially a plane Couette flow problem, and stability against the assumed disturbances is expected. Thus the pipe flow should be stable when the parabolic velocity distribution is established. The stability of the profiles in the entrance region of the pipe flow has been studied by Tatsumo [*61*], who obtained a minimum critical Reynolds number for all the profiles prior to the ultimate parabolic distribution.[9] Based upon the average discharge velocity and the pipe radius, this minimum critical Reynolds number is about 10^4, which appears to be of the correct order of magnitude when compared against the experimental *transition* Reynolds number observed by Ekman and Taylor [*62*, p. 321].

Eq. 18-1 of course is also applicable to problems of the boundary layer over a body of revolution under rotationally symmetric disturbances. For "fat bodies" for which the boundary layer is thin compared to the body thickness, the first approximation will be to drop all terms of order $1/r$, thus reducing essentially to the two-dimensional form. In this way, Pretsch [*63*] analyzed the boundary layer over a sphere near the stagnation point by taking Homann's solution as the velocity distribution, obtaining a minimum critical Reynolds number of approximately 4200 (based upon the displacement thickness). It seems questionable, however, that the acceleration along the streamline may be neglected in this problem.

G,19. Flows in Wakes, Jets, and the Mixing Region. The stability problem in wakes, jets, and mixing regions differs from that previously discussed mainly in the absence of any solid boundary. Viscosity therefore can only have its influence felt in the interior of the field of flow. If the approximation of locally parallel basic flow is maintained, the fundamental linearized equations of the stability problem are no different from those used before. Consider first the case of two-dimensional basic flow under periodic disturbances which propagate in the downstream direction only. The disturbances naturally must be bounded anywhere. We still have the Orr-Sommerfeld equation, Eq. 7-1. If we think of the two "viscous solutions" as represented by Eq. 8-9 it becomes evident that, since the flow field now extends to infinity in all directions, both must be discarded. We are left with only the two "inviscid solutions," Eq. 8-1 and 8-2, to satisfy the boundary conditions. When the viscous effects, i.e. the stability at finite Reynolds numbers, is desired, it is necessary to go into higher approximations of the so-called "inviscid equation." The results of such investigations are distinguished from those for basic flows with solid boundaries in several respects. The main effect of viscosity now is *stabilizing* instead of destabilizing at large Reynolds numbers. The "inviscid" case as $\alpha Re \to \infty$ is of course unstable because of the presence of the point of inflection in these flows. The minimum

[9] This is in contrast to the two-dimensional case. For the profiles of the entry region of a channel, Hahnemann et al. [*44*] found that the minimum critical Reynolds number is always higher than that for the ultimate parabolic distribution.

critical Reynolds number is generally very low, so that the method of asymptotic expansion in inverse powers of αRe might be open to question. In addition, the neglect of the change of the velocity profiles in the downstream direction, at such low Reynolds numbers, might also be quite serious when comparison with experimental observations is to be made.

The stability of the mixing region between two parallel streams of different velocities, or the "half-jet," was investigated by Chiarulli [*64*] and Lessen [*65*]. The expansion in inverse powers of αRe, Eq. 7-2, was used and the first term $\phi^{(1)}$ computed. Lessen's numerical results are represented by Fig. G,19a. With only $\phi^{(1)}$, he was not able to find a

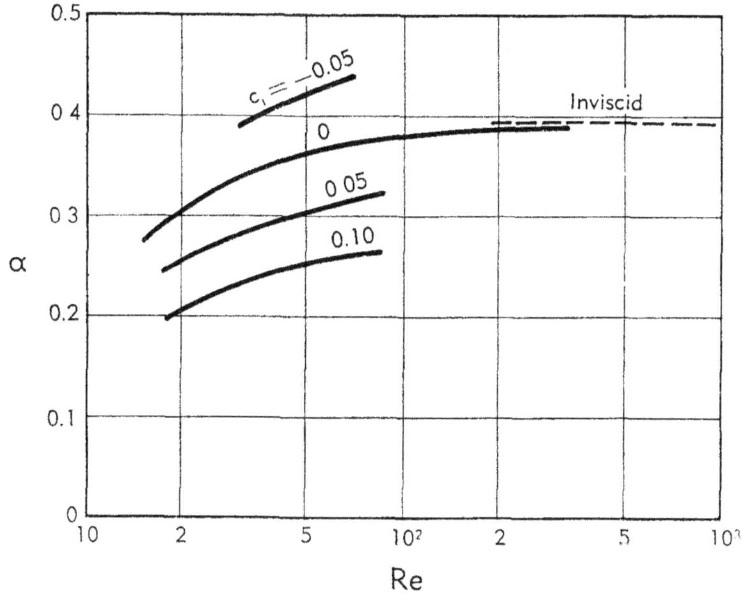

Fig. G,19a. Stability of mixing between two parallel flows.

minimum critical Reynolds number, nor the presence of a lower branch of the neutral curve. The neutral curve moves toward small α as the Reynolds number is decreased. Compared against a typical neutral curve for basic flow with solid walls, this behavior is decidedly different. With solid walls and at sufficiently large Reynolds numbers (i.e. near the upper branch of the neutral curve), a reduction in Reynolds number is stabilizing—a phenomenon sometimes referred to as "Heisenberg's criterion." But here, without solid walls, the viscosity is seen to have a stabilizing influence always.

Lessen also found that the real wave velocity c_r does not change much with either α or Re. This last result is found to agree with Sato's observation of the transition experiment of the half jet [*66*]. The wavelength of the observed oscillations, however, is much smaller than the theoretical value along the neutral curve, presumably corresponding to amplified waves.

If we consider Fig. G,19a as typical of the stability of flows without solid boundaries, further attacks along two different directions suggest themselves. First, at fairly large Reynolds numbers the behavior appears to be essentially no different from the limiting case of $\alpha Re \to \infty$. Only the "inviscid equation" needs to be studied and exploited. Secondly, since the method of asymptotic expansion is probably inadequate for locating the minimum critical Reynolds number as well as the possible existence of a lower branch of the neutral curve, other techniques must be developed which are to be valid for small values of αRe.

The general results of Art. 7 pertaining to the inviscid equation, Eq. 7-3, hold for the half jet, even though the boundaries now extend to infinity in both directions. They are also valid for symmetrical wakes or jets after one breaks down the possible disturbances into symmetrical and antisymmetrical modes, as previously done for the plane Poiseuille flow. Thus we are sure of a neutral oscillation with wave velocity c determined by the condition $\bar{u}''(y_c) = 0$. The eigenvalue problem is now cast for the wave number α. It is not difficult to verify that α may now be determined from a variational integral:

$$[-\alpha^2] = \min \frac{\int_{y_1}^{y_2} \left[\phi'^2 + \frac{\bar{u}''}{\bar{u} - c} \phi^2 \right] dy}{\int_{y_1}^{y_2} \phi^2 dy} \tag{19-1}$$

while keeping $\int_{y_1}^{y_2} \phi^2 dy$ constant. The eigenfunction for these basic flows is generally quite smooth, so that reasonable approximations are possible and Eq. 19-1 appears to be a practical way of getting numerical answers. For instance, in the case of a "full jet" with

$$\bar{u} = \operatorname{sech}^2 by \tag{19-2}$$

the exact solution of the inviscid equation was given by Savic [*67*] for symmetrical oscillations,

$$\phi_1 = \operatorname{sech} by \tanh by, \qquad \alpha_1 = b, \qquad c = \tfrac{2}{3} \tag{19-3}$$

and by Savic and Murphy [*68*] for antisymmetrical oscillations,

$$\phi_2 = \operatorname{sech}^2 by, \qquad \alpha_2 = 2b, \qquad c = \tfrac{2}{3} \tag{19-4}$$

The approximate calculation of Lew and Fanucci [*69*], using one term of the Hermitian polynomial,

$$\phi_k \cong e^{-b^2y^2/2}\mathcal{H}_k(by)$$

($k = 1$ for the symmetrical oscillation and $k = 2$ for the antisymmetrical oscillation) yields $\alpha_1 \cong 0.8b$ and $\alpha_2 \cong 1.96b$.

If the basic flow velocity distribution is not monotonic, or has more than one point of inflection, the inviscid problem falls beyond the scope of Art. 7, and must be given separate consideration. For example, suppose we have an asymmetric jet. Assuming that a neutral oscillation exists, the real wave velocity c can no longer be determined from the condition of $\bar{u}''(y_c) = 0$. An insight into the mechanism in terms of the Reynolds stress associated with the disturbance is provided by Foote and Lin [*70*]. The Reynolds stress is defined, for incompressible flow, by

$$\tau = -\rho\overline{u'v'}$$

averaging over a wavelength in the x direction. In terms of the periodic disturbances, one finds

$$\tau = \tfrac{1}{2}\rho\alpha W e^{2\alpha c_i t}$$

$$W = \frac{i}{2}(\phi\tilde{\phi}' - \tilde{\phi}\phi') \tag{19-5}$$

Consider first $c_i \neq 0$. From Eq. 7-3 there follows

$$\frac{dW}{dy} = \frac{c_i\bar{u}''}{|\bar{u} - c|^2}|\phi|^2 \tag{19-6}$$

Integrating Eq. 19-6 and *then* letting $c_i \to 0$, we get the result that, in neutral oscillation, W is constant except for jumps at $\bar{u} - c = 0$ equal to

$$W(y_c+) - W(y_c-) = \pi\frac{\bar{u}_c''}{\bar{u}_c'}|\phi_c|^2 \tag{19-7}$$

if $\bar{u}$ increases with y. A negative jump occurs if $\bar{u}$ decreases with y.

Now, the boundary conditions at either extremities of the jet require $W = 0$. Therefore the disturbance must be such that either $W = 0$ everywhere, i.e. without jumps, or the several jumps cancel. The former applies when $\bar{u}$ is monotonic, requiring $\bar{u}_c'' = 0$ in previous types of flow. In cases such as the asymmetrical jet, $\bar{u} - c = 0$ may occur at more than one point at which $\bar{u}_c''$ generally does not vanish. The jump-cancelling condition then becomes the necessary condition to be satisfied by the wave velocity c. Since we do not know ϕ to start with, the presence of $|\phi_c|^2$ makes the explicit determination of c impossible. In addition, we also note that the nonvanishing of $\bar{u}_c''$ leads to an inviscid eigenfunction ϕ with a logarithmic singularity at each point $\bar{u} - c = 0$.

When the inviscid neutral oscillation has been found, the behavior for slightly different wave number α may be directly calculated by Eq. 7-16. Thus, Lin [*71*] gave for the half-jet,

$$\frac{\partial c}{\partial \alpha} = 0.093 - 0.287i$$

while Lessen [72] gave for the full jet of Eq. 19-2,

$$\frac{\partial c}{\partial \alpha} = 0.613 - 0.403i \text{ for antisymmetric oscillation}$$

$$\frac{\partial c}{\partial \alpha} = 0.332 - 0.294i \text{ for symmetric oscillation}$$

The rate of change of c with respect to the Reynolds number for large Reynolds numbers may likewise be obtained from the inviscid solution. Essentially, one follows the same steps as in Eq. 7-14, 7-15, and 7-16 except that the full Orr-Sommerfeld equation should be used, and the perturbation taken with respect to the small parameter $1/Re$. Lessen [72, pp. 122–126] proposes to continue on to higher order derivatives, but it appears doubtful whether the rapidly increasing numerical work will justify the end results—for instance in permitting one to say something about the behavior at rather low Reynolds numbers.

If now one assumes that the critical Reynolds number would occur at small values of αRe, it appears natural to develop the solution of the Orr-Sommerfeld equation in ascending powers of αRe instead of the descending powers valid for large αRe.[10] Such a procedure has been carried out by Tatsumi and Kakutani [74] for the full jet, Eq. 19-2, under asymmetric disturbances. Eq. 7-1 may be rewritten as

$$\left.\begin{aligned} (D^2 - \alpha^2)(D^2 - \beta^2)\phi &= i\alpha Re[\bar{u}(D^2 - \alpha^2) - \bar{u}'']\phi \\ D \equiv \frac{d}{dy}, \qquad \beta^2 &\equiv \alpha^2 - i\alpha Rec \end{aligned}\right\} \tag{19-8}$$

with now the boundary conditions

$$\phi'(\infty) = \alpha\phi(\infty) = 0, \qquad \phi'(0) = \phi'''(0) = 0 \tag{19-9}$$

The solution is expressed as

$$\phi(y) = \sum_{n=0}^{\infty} (i\alpha Re)^n \phi^{(n)}(y; \alpha, \beta) \tag{19-10}$$

and a sequence of equations for $\phi^{(n)}$ are obtained by equating like powers of $i\alpha Re$. The four zero-order solutions are

$$\phi_{1,2}^{(0)} = e^{\pm \alpha y}, \qquad \phi_{3,4}^{(0)} = e^{\pm \beta y}$$

while the higher order ones are readily determined by quadrature. The approximate calculation of Tatsumi and Kakutani gave a minimum critical Reynolds number of 4.02 at $\alpha = 0.2$, and the lower branch as $\alpha Re^2 = 1.25$ asymptotically for $\alpha \to 0$. The latter behavior is worth noting because it demonstrates the inadequacy of the usual asymptotic

[10] Yih [73] earlier used the same expansion in the stability problem of flows with a free surface, e.g., that of liquid running down a slope.

expansion in $(\alpha Re)^{-1}$ for determining the lower branch in such problems, even as $Re \to \infty$. The very low critical Reynolds number is mainly of mathematical interest, since the assumption of a locally parallel basic flow would then be seriously questionable.

An alternative approach to come down to rather low values of αRe advanced by McKoen [75] and Curle [76] seems quite interesting. Their crucial step lies in neglecting the highest order derivative $d^4\phi/dy^4$ of the Orr-Sommerfeld equation but keeping all the other viscous terms. The main argument is that, for jet-type flows, as long as the eigenfunction is expected to vary slowly everywhere in the fluid, the contribution of the $d^4\phi/dy^4$ term would be rather small. The remainder of Eq. 7-1 then becomes a second order differential equation containing the viscous parameter αRe. An integral equation is formulated and an approximate solution is effected without the disadvantage of a series expansion in $1/\alpha Re$.

Suppose that the solution of the inviscid equation is known, denoted by the index $_s$,

$$\phi_s'' - \alpha_s^2\phi_s - \frac{\bar{u}''}{\bar{u} - c_s}\phi_s = 0$$

After dropping the $d^4\phi/dy^4$ term, Eq. 7-1 becomes

$$\phi'' - \alpha^2\phi - \frac{\phi}{\bar{u} - c_1}\left(\bar{u}'' + \frac{i\alpha^3}{Re}\right) = 0 \tag{19-11}$$

where $c_1 = c + 2i\alpha/Re$. Subtracting the inviscid equation from Eq. 19-11, we get

$$\left.\begin{aligned} (\phi - \phi_s)'' - \alpha_s^2(\phi - \phi_s) - \frac{\bar{u}''}{\bar{u} - c_s}(\phi - \phi_s) &= g \\ g = \phi\left[\alpha^2 - \alpha_s^2 + \frac{\bar{u}''(c_1 - c_s)}{(\bar{u} - c_1)(\bar{u} - c_s)} + \frac{i\alpha^3}{Re(\bar{u} - c_1)}\right] \end{aligned}\right\} \tag{19-12}$$

essentially a perturbation equation. The solution may be written, by standard procedure, as

$$\phi - \phi_s = \phi_s \int_{y_1}^{y} \phi_s^{-2} \int_{y_1}^{y} g\phi_s dy dy \tag{19-13}$$

with y_1 denoting one extremity of the flow field. The other extremity y_2 may be taken at $+\infty$ without loss of generality, and $\phi_s \to e^{-\alpha_s y}$ as $y \to y_2$.

To satisfy $\phi \to \phi_s$ as $y \to y_2$, the requirement is met by taking

$$\int_{y_1}^{y_2} g\phi_s dy = 0 \tag{19-14}$$

For the full jet, McKoen rewrites Eq. 19-14 explicitly as

$$(\alpha^2 - \alpha_s^2) \int_0^\infty \phi\phi_s dy + (c_1 - c_s) \oint_0^\infty \frac{\bar{u}''\phi\phi_s}{(\bar{u} - c_1)(\bar{u} - c_s)} dy + \frac{i\alpha^3}{Re} \oint_0^\infty \frac{\phi\phi_s}{\bar{u} - c_1} dy = \frac{(\phi\phi_s)_{c_1}}{\bar{u}'_{c_1}} \pi i \left(\bar{u}''_{c_1} + i\frac{\alpha^3}{Re}\right) \quad (19\text{-}15)$$

where subscript c_1 denotes the value at $\bar{u} = c_1$ and $\oint$ means the Cauchy principal value of the integral. The right-hand side comes from going around the singular point ($\bar{u} = c_1$) from *above* in the complex plane, since $\bar{u}'_{c_1}$ is negative in this case. The proper path is in accordance with Art. 9, when the viscous solution in the neighborhood of the singular point is considered. As the viscous region is of size $O(\alpha Re)^{-\frac{1}{3}}$ while c_1 is only $O(\alpha/Re)$ from the real axis, it is still necessary to bypass the viscous region (for large Re at least) as if the singularity lay on the real axis.

By making further simplification, McKoen used Eq. 19-15 to treat the laminar wake profile. Curle applied it also to the jet of Eq. 19-2 under asymmetric disturbances, for which ϕ_s is known in closed form. Noting the other degenerate solution

$$\phi_s = \bar{u} \quad \text{for} \quad \alpha = c = 0$$

he approximated ϕ by linear interpolation

$$\phi \cong \bar{u}(y) + \frac{c_1}{c_s}[\phi_s(y) - \bar{u}(y)] \quad (19\text{-}16)$$

The choice immediately guarantees two branches of the neutral curve as $\alpha Re \to \infty$. His calculated result is shown in Fig. G,19b. (Note that Curle's velocity profile is normalized by taking $b = 1$.) Tatsumi and Kakutani's [*74*] results, in which the upper branch was determined by the customary asymptotic expansion in $(\alpha Re)^{-1}$, are included for comparison in the same figure. The agreement seems quite reasonable. One must be aware, however, of the fact that the integral equation, Eq. 19-14, is developed essentially on the basis of fairly large αRe. Curle's solution, furthermore, is only an approximate one, including the use of an inviscid solution for $\alpha \to 0$, which is strictly speaking incorrect [*74*]. The apparent success probably may be attributed to the smooth nature of the disturbance modes in such highly unstable basic flows.

Galerkin's method has been applied by Lew [*69*] to the complete viscous equation in the cases of the two-dimensional jet under symmetrical disturbances. The one-term approximation leads to a neutral curve with a low minimum critical Reynolds number. The result might be of qualitative value.

The axially symmetric jet issuing from a small hole has the velocity distribution,

$$\bar{w} = (1 + br^2)^{-2}, \qquad b = \frac{3}{64\pi}$$

Its stability with respect to axially symmetric disturbances may be investigated following Art. 18. Since now $\bar{w}'' - (\bar{w}'/r) > 0$, according to the inviscid criterion no instability is possible in the inviscid limit, even though the profile exhibits a point of inflexion.[11] The behavior is therefore quite different from that of the two-dimensional jet. The stabilizing influence of axial symmetry previously has been demonstrated in the case of the channel and pipe flows.

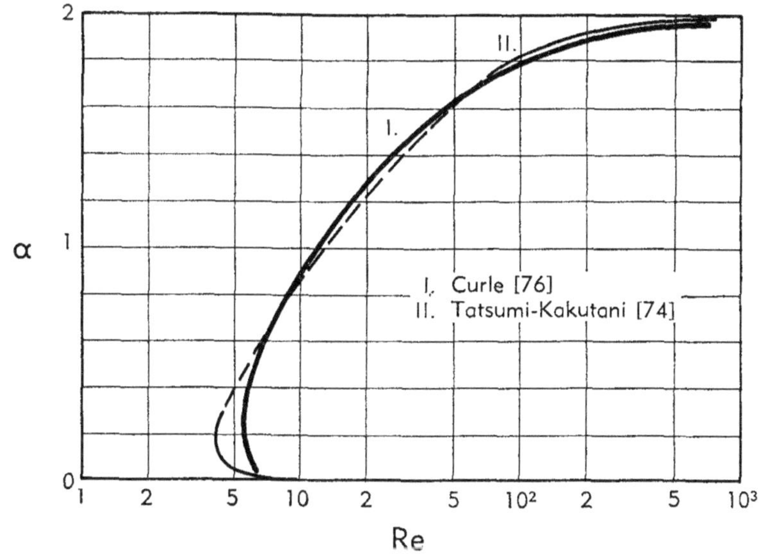

Fig. G,19b. Neutral stability curve for full jet.

For a compressible fluid, the stability of two-dimensional laminar jet flows with respect to two-dimensional disturbances was investigated by Pai [77]. Eq. 14-3 were retained, but again since the viscous solutions increase without bounds at large distances, the higher order solutions of the "inviscid equation" had to be evaluated for finite but large values of αRe. The development essentially parallels that in the incompressible case, with the operator on the left-hand side of Eq. 15-5 replacing that of Eq. 7-6. Detailed calculations, however, are not yet available. As discussed on pp. 769, 770, in a compressible boundary layer one restricts consideration to disturbances which are "subsonic" with respect to the free stream velocity. A similar restriction also applies in jet-type flows.

[11] Lew [*69*] actually calculated a wave number for inviscid neutral stability at $c = 1$ by the one-term Galerkin method. The fallacious result seems to demonstrate the difficulty of applying the Galerkin method to predict complex eigenvalues.

The lack of a solid boundary, however, permits one to superpose a uniform velocity of any magnitude upon the basic flow without affecting the stability problem. It means that the disturbances must be "subsonic" with respect to the prevailing temperature and velocity at *any* point in the jet. That is, we must allow the free stream to have the velocity $\bar{u}$ and

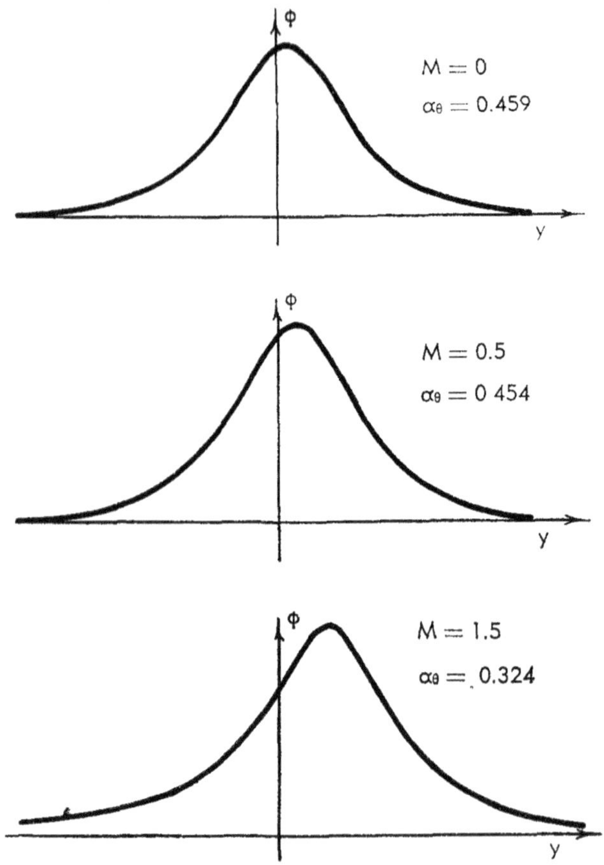

Fig. G,19c. Inviscid neutral mode shapes and wave numbers for laminar mixing region between a moving stream and the other at rest.

temperature $\bar{T}$ that occur at any point in the jet chosen as a reference location. Thus we must have

$$P \equiv \bar{T} - M^2(\bar{u} - c)^2 > 0 \tag{19-17}$$

everywhere, M being a fixed reference Mach number (defined at the jet axis, for example). Thus the *possible* wave velocity c lies between the limits $\bar{u} - \sqrt{\bar{T}}/M$ and $\bar{u} + \sqrt{\bar{T}}/M$. As the reference location is varied, it may turn out that above a certain Mach number the distributions of

$\bar{u}$ and $\bar{T}$ are such that the condition is violated at least somewhere in the fluid. One must then interpret that "subsonic" disturbances cannot exist, and the basic flow is stable in the usual sense (disregarding "supersonic disturbances"), with respect to the two-dimensional disturbances. In the case of constant stagnation enthalpy,

$$\bar{T} + \frac{\gamma - 1}{2} M^2 \bar{u}^2 = 1 + \frac{\gamma - 1}{2} M^2$$

Pai found that no subsonic disturbance is possible if $M > 2.5$ in air.

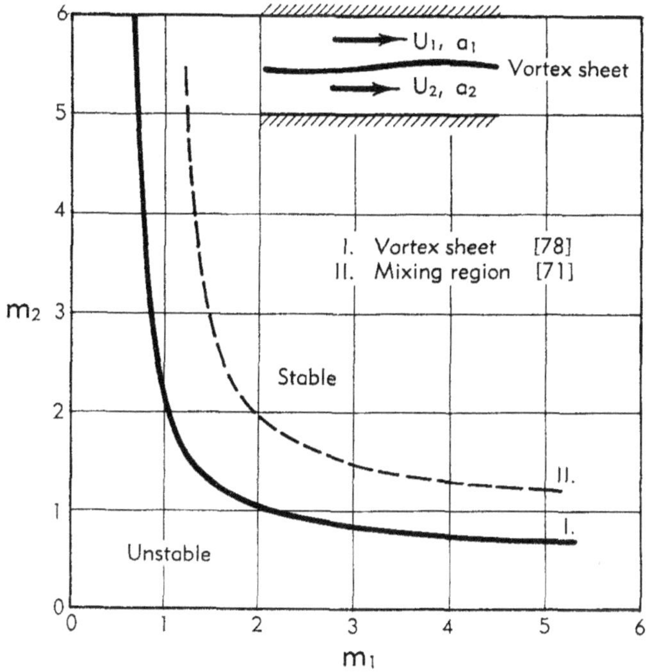

Fig. G,19d. Complete stability of the vortex sheet or mixing region between parallel streams.

The compressible half jet in the mixing of two parallel streams of gas was studied by Lin [*71*]. The existence of subsonic (two-dimensional) disturbances requires that the relative speed of the two streams must not exceed the sum of the speeds of sound in these streams. Some of Lin's calculated results for the inviscid limit are shown in Fig. G,19c.

In the limit when the half jet degenerates to a concentrated vortex sheet separating two parallel uniform streams, the stability with respect to two-dimensional disturbances was solved by Pai [*78*]. The vortex sheet is assumed to distort into a traveling wave and the potential disturbances in the two streams are calculated by the linear theory.

The entire eigenvalue spectrum can be analyzed, and the disturbances may be "subsonic" or "supersonic." Pai's curve is shown as Fig. G,19d, in which Lin's criterion for the half jet is included for comparison.

G,20. Three-Dimensional Boundary Layers. Flow over a Rotating Disk. For three-dimensional boundary layers, as long as we accept the assumption of locally parallel basic flow, i.e. take $\bar{v} = 0$, but assume $\bar{u}$ and $\bar{w}$ to be functions of the normal distance y, the stability problem may be analyzed as a two-dimensional one by choosing the x axis to coincide with the direction of wave propagation (Art. 4). The practical difficulty lies in that, for every direction of wave propagation, a different effective velocity profile must be considered. Only for the flat plate case are the velocity profiles similar in all directions—leading thereby to the famous Squire's theorem that only the two-dimensional disturbance, propagating in the downstream direction, needs to be considered for the minimum critical Reynolds number. In the general case, the resultant velocity vectors at different distances from the wall are not parallel with each other (Fig. G,20a, the x_1 axis being parallel to the resultant velocity at the edge of the boundary layer and the z_1 axis normal to it). The effective $\bar{u}$ profile in the x direction may assume any of the forms of the lower part of Fig. F,20a, depending upon the angle θ between the x and x_1 axes. It may be noted immediately that many of the profiles exhibit a point of inflection I, suggesting a tendency toward instability for disturbances propagating in these directions. Only a detailed calculation can determine the lowest minimum critical Reynolds number among all possible directions of propagation and based upon the *same reference velocity and length.*

However, a closer examination of the role of the inflection point in the inviscid behavior of these profiles is required. One of the assumptions in arriving at the "inviscid criterion" of Art. 7 is the monotonic nature of $\bar{u}(y)$, which does not apply for some of the profiles. Reviewing the arguments, one finds that $\bar{u}'' = 0$ and $\bar{u} - c = 0$ within the boundary layer remain as prerequisites for an amplified disturbance. But the sufficiency proof of a neutral disturbance hinges upon $\bar{u}''/[\bar{u} - \bar{u}(y_c)] < 0$ everywhere, a condition not satisfied by the sketched profiles for $0 < \theta \leqq \pi/2$. Thus it is by no means certain that a neutral disturbance exists with $c = \bar{u}(y_c)$ in such cases. There is, however, a particular profile, designated as profile J, of which the inflexion point moves to $\bar{u}(y_c) = 0$. For this profile the condition $\bar{u}''/[\bar{u} - \bar{u}(y_c)] < 0$ everywhere is again satisfied; likewise for profiles corresponding to even greater values of θ. Only in these cases does the inflexion point regain its usual important role, definitely promoting instability. To determine the minimum critical Reynolds number for all possible oblique disturbances, however, due attention must be given to the proper reference velocity and displacement thickness for each profile when comparison is made against other profiles.

Additional insight may be gained by examining the profiles on the basis of the Reynolds stress distribution, calculated by Eq. 19-5 and 19-7 in the inviscid limit. For neutral disturbances, it is necessary to have zero

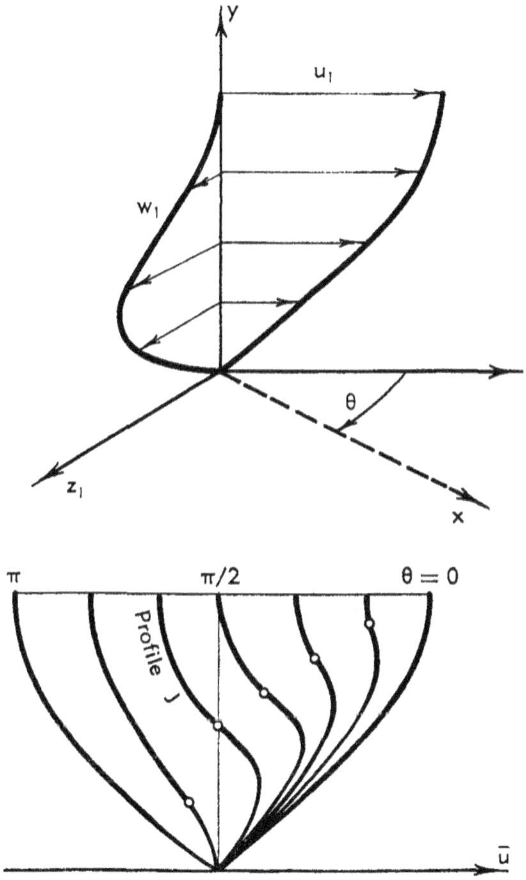

Fig. G,20a. Effective profiles of three-dimensional boundary layers for disturbances propagating in different directions.

Reynolds stress at the wall and at the edge of the boundary layer. Since it is still required that $c < \bar{u}_{max}$, there may be either one or two critical points depending on the magnitude of c and the shape of the profile. For one critical point, it must coincide with the point I to prevent any jump of the Reynolds stress; hence it is only possible for profiles J and those corresponding to larger θ. With two critical points, the jumps of the Reynolds stress must cancel, so that neither critical point can fall on the inflection point I. In fact, consulting Eq. 19-7 and the sentence which follows, one sees that $\bar{u}_c''/\bar{u}_c'$ must have the same sign at the two critical points, otherwise the jumps would be additive. Thus the *possible* value of c is limited to the range between the velocity at the edge of the boundary layer and the velocity at point I. Whether an inviscid neutral distribution exists in such cases cannot be answered without detailed calculation. It is to be noted only that at each critical point a logarithmic singularity *will* result and the solution is to be handled in the same manner as in Art. 8. If one uses Eq. 8-2 and 8-3, the path of integration goes under the singular point y_c if $\bar{u}_c' > 0$, and above if $\bar{u}_c' < 0$. Except for the inviscid solution, the calculation for the minimum critical Reynolds number and the neutral curve is no different from the usual case.

Most of the above considerations were presented by Stuart [*79*] who called attention particularly to the profile J. In the inviscid limit this profile permits a neutral disturbance with $c = 0$, hence actually a secondary flow periodic in the x direction. Its wave number may be approximately determined by Eq. 19-1. The secondary flow may be regarded as a vortex disturbance with an axis transverse to the basic flow direction considered. However, it should not be confused with the "Görtler vortex" of Art. 25, which lies along the basic flow direction.

Calculation of this secondary flow was made by Stuart for the flow over a rotating disk, for which the theoretical basic flow velocity distribution is well-known (Kármán-Corcoran; see, e.g. Schlichting [*52*, pp. 81–84]). Because of the similarity of solutions at various radial distances, the angle between the local vortex axis and the radius vector should remain constant, defining thus a logarithmic spiral. In terms of the angle ϵ as on Fig. G,20b, the theoretical value of ϵ was found to be $13°18'$ by requiring the velocity profile to have an inflexion point at $\bar{u} = 0$ relative to the surface. The wave number calculated was $\alpha = 1.45$, when the reference length was taken to be the displacement thickness based upon the tangential velocity distribution ($\delta^* = 1.271 \sqrt{\nu/\omega}$, ω = angular velocity). The number of spirals over the entire disk is inversely proportional to the wavelength, hence proportional to the wave number, and turns out to be $0.262r \sqrt{\omega/\nu}$ (r being the local radius).

Experimentally these secondary flow vortices were observed by Gregory and Walker [*79*], after coating the rotating disk with china clay, and also by Fales [*80*]. Near the center there was pure laminar flow; near the rim at sufficient speed there was fully turbulent flow. Between these two regions, evenly spaced spiral streaks occurred to indicate the presence of local disturbances more or less stationary relative to the wall. The angle of the spirals agreed very well with Stuart's theory, although the number of spirals in Gregory and Walker's experiment was only about 30 compared to the predicted 113 to 140. Stuart attributed the latter discrepancy to the effects of viscosity. (If we interpret the streaks as due to the amplified disturbances, which would have smaller wave numbers (Art. 7), the agreement would be slightly better. The most amplified disturbance should have a wave number of roughly one half the value at the neutral curve, cutting thereby the predicted number of spirals to about 60.)

In observations with china clay on a swept wing, similar streaks roughly parallel to the local free stream were reported by Anscombe and Illingworth [*81*]. These might again be the secondary flow. Transition to

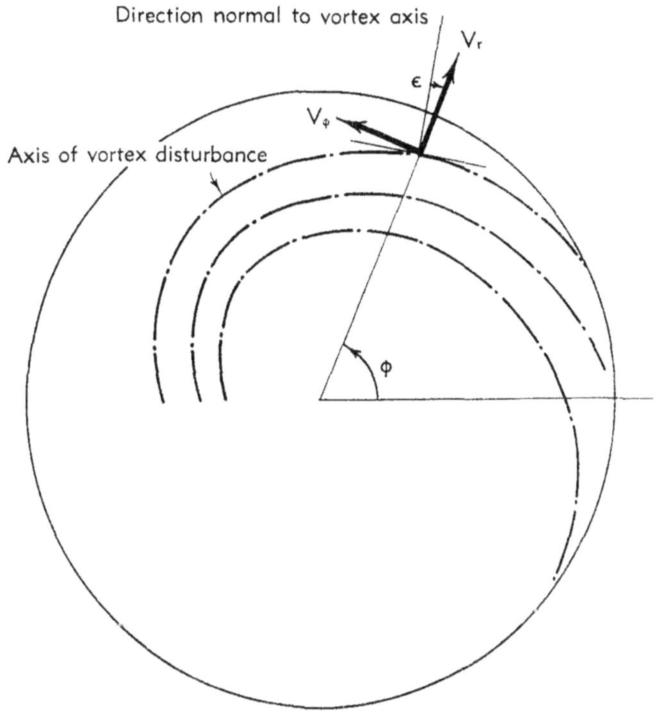

Fig. G,20b. Vortex disturbance on rotating disk.

turbulent boundary layer on the swept wing moved closer to the leading edge compared to the unswept case. If we recall that any three-dimensional boundary layer in general might have all the profiles sketched in the lower part of Fig. G,20a as the direction of disturbance wave propagation is varied, the tendency toward less stability, and hence earlier transition, is readily to be expected. A quantitative correlation with theory, however, is still lacking.

G,21. Flows under the Influence of Additional Physical Mechanisms. It is beyond the scope of this survey to summarize all the applications of the basic laminar stability theory under the influence of various additional physical mechanisms. Problems of importance which will be conspicuously left out include the stability of flows with Coriolis force (of interest in meteorology), thermal instability of fluid under heating, and the stability of the interface between flows of dissimilar fluids. These matters are discussed in Sec. F, Chap. 6. Here we restrict ourselves to a brief account of some investigations in the following directions:

1. the effect of gravity in a density-stratified fluid, or a free convection boundary layer;
2. the effects of a constant magnetic field coplanar or transverse to the boundaries of the plane parallel flow;
3. the effects of diffusion and chemical reaction in a binary gas mixture.

The stability of the flow of a density-stratified fluid has certain meteorological and hydraulic significance. We consider the flow over a surface where the density of the fluid decreases upwards (opposite to the gravity direction, so that there is static stability). As first discussed by Prandtl [*82*, pp. 1–17], using an energy method, the extra dimensionless parameter is the Richardson number.

$$Ri = \frac{-g d\rho/dy}{\rho(d\bar{u}/dy)^2_{\text{wall}}}$$

g being the gravitational acceleration and y pointing upwards. The problem was examined in detail on the basis of infinitesimal disturbances by Taylor [*83*] and Goldstein [*84*], neglecting the viscosity. For small Froude number, the differential equation is of the form

$$(\bar{u} - c)(\phi'' - \alpha^2\phi) - \bar{u}''\phi + Ri\,\frac{\phi}{\bar{u} - c} = 0 \qquad \text{(21-1)}$$

The velocity $\bar{u}$ considered includes both constant and linear distributions, and Ri becomes constant by assuming an exponential density variation with the y coordinate.[12] Both found instability for $Ri < \frac{1}{4}$. The later calculation by Schlichting [*86*] is an application of the classical theory of Chap. 2 to the Blasius profile with density stratification. Thus a continuous variation of the minimum critical Reynolds number against the Richardson number was obtained. The minimum critical Reynolds number starts at 575 for $Ri = 0$, increases slowly until $Ri \cong 0.02$, and afterwards approaches infinity as Ri tends to $\frac{1}{24}$.

The transition from laminar to turbulent of the free convection boundary layer over a vertical plate was made visible by Eckert [*87*]

[12] A continuous velocity distribution extending to infinity at both ends

$$u = \tanh y, \qquad -\infty < y < +\infty$$

has recently been studied by Drazin [*85*].

through the use of an interferometer. It appears that transverse wave disturbances exist before turbulence sets in. The theoretical study extending the classical theory of Chap. 2 to this problem was recently made by Plapp [*88*]. Actually the modification required is not difficult to see. If the working fluid is taken to be a perfect gas and only the two-dimensional case is considered, the governing equations for the disturbances are essentially Eq. 14-2, except for the addition of body force terms due to buoyancy to the right-hand side of the first two equations, and for the simplification that only the velocity and temperature distributions of the basic flow are regarded as variables. The buoyancy force is proportional to the density difference, so that in the disturbance equations it will be proportional to a linear combination of the temperature and pressure disturbances. Now, after nondimensionalizing Eq. 14-2 to arrive at Eq. 14-3 the buoyancy terms become proportional to the inverse of the Froude number $Fr = \bar{u}^2_{max}/g\delta$. But for free convection flow, at a given station along the plate, the Froude number so defined differs from the Reynolds number $\bar{u}_{max}\ \delta/\nu$ only by a constant factor (see, e.g., Schlichting [*52*, p. 304]). Thus the Froude number effect will be felt only in the "viscous" solutions. The "inviscid solutions" (Eq. 14-8) are unchanged, although P reduces to $\bar{T}$ for the very low Mach number in usual free convection flows.

Referring back to the first two equations of Eq. 14-3, we note that the most important viscous term will remain to be f_1'' and f_2'' since the buoyancy terms are proportional to ϖ and θ, and therefore are relatively small in comparison to the second derivative terms. The solutions near the critical point consequently are still those of Art. 14, after putting everything in the basic flow except $\bar{u}$ and $\bar{T}$ as constants. The boundary conditions, too, may be discussed in the same manner. If the Reynolds number of instability is expected to be high, as a first approximation it is only necessary to calculate for the velocity and temperature distributions of the basic free convection flow in the same manner as for the compressible boundary layer of Art. 14. Without the Dunn-Lin Δ correction, one does not need even the temperature disturbance solutions. Such a calculation has been carried out by Plapp for the free convection flow of air adjacent to a hot vertical plate. Curiously Plapp found that the minimum critical Reynolds number is much *higher* than Eckert's experimentally observed transition value (F,30). An explanation of this peculiar result is yet to be found.

The influence of magnetic field on hydrodynamic stability was studied first in papers by Thompson [*89*] and Chandrasekhar [*90*] on thermal instability, and later by Chandrasekhar [*91,92,93*] on the stability of flow between rotating cylinders—all with respect to disturbances not of the type treated in this chapter. The extension to include a magnetic field in the general framework of Chap. 2 was made by Michael [*94*], Stuart

[95], Lock [96], and Rossow [97]. The magnetohydrodynamic equations of motion for an incompressible fluid may be written as [90, Eq. 5, 20, and 21]:

$$\frac{\partial \mathbf{u}}{\partial t} + (\mathbf{u} \cdot \nabla)\mathbf{u} - \frac{\mu_{\mathrm{m}}}{4\pi\rho}(\mathbf{H} \cdot \nabla)\mathbf{H} = -\frac{1}{\rho}\nabla\bar{p} + \nu\nabla^2\mathbf{u} \tag{21-2}$$

$$\frac{\partial \mathbf{H}}{\partial t} + \nabla \times (\mathbf{H} \times \mathbf{u}) = \eta\nabla^2\mathbf{H} \tag{21-3}$$

$$\nabla \cdot \mathbf{u} = 0 \tag{21-4}$$

$$\nabla \cdot \mathbf{H} = 0 \tag{21-5}$$

where $\mathbf{H}$ is the magnetic field vector, μ_{m} is the permeability, $\bar{p}$ is the sum of hydrodynamic and magnetic pressures,

$$\bar{p} = p + \frac{\mu_{\mathrm{m}}|\mathbf{H}|^2}{8\pi}$$

and $\eta = (4\pi\mu_{\mathrm{m}}\sigma)^{-1}$ is the magnetic diffusivity, σ being the electrical conductivity. Note that here the magnetic diffusivity η has the same dimension as the kinematic viscosity ν.

The plane Poiseuille flow is a solution of Eq. 21-2, 21-3, 21-4, and 21-5 if H_x and H_z are constant and $H_y = 0$. This is the basic flow investigated by Stuart [95]. The disturbances are written down in the same manner as before. By similar reasoning as on p. 723, it is apparentt hat any oblique wave disturbance is equivalent to a two-dimensional disturbance on a two-dimensional basic flow, if the latter is taken to consist of the components of the velocity profile and the magnetic field in the direction of wave propagation. After nondimensionalizing and straightforward elimination, the equations corresponding to Eq. 6-3 are

$$(\bar{u} - c)(f_2'' - \alpha^2 f_2) - \bar{u}'' f_2 - S(f_5'' - \alpha^2 f_5) = -\frac{i}{\alpha Re}(f_2'''' - 2\alpha^2 f_2'' + \alpha^4 f_2) \tag{21-6}$$

$$(\bar{u} - c)f_5 - f_2 = -i\frac{T}{\alpha Re}(f_5'' - \alpha^2 f_5) \tag{21-7}$$

where $f_2(y)$ and $f_5(y)$ are the dimensionless disturbance mode functions for the y component of velocity and the magnetic field, respectively; and S and T are two more dimensionless parameters defined by

$$S = \frac{\mu_{\mathrm{m}} H_\beta^2/8\pi}{\frac{1}{2}\rho U_\beta^{*2}}$$

i.e. the ratio of magnetic pressure to the dynamic pressure, and

$$T = \frac{\eta}{\nu}$$

which may be interpreted as the ratio of the viscous Reynolds number to the "magnetic Reynolds number." The suffix $_\beta$ on H and U^* indicates that these are taken along the direction of wave propagation; U^* denotes the basic flow reference velocity. It may now be seen that, if the velocity vector and the magnetic field are in the same direction, both parameters S and T remain constant regardless of the parameter β pertaining to the wave propagation direction. But the Reynolds number will be reduced if the disturbance wave is oblique to the basic flow. The "Squire's Theorem" of p. 724 is thus valid under these conditions.

The boundary conditions at a solid wall of perfect conductor are the vanishing of the velocity components, the normal component of the magnetic field, and the tangential component of the electrical field. (The latter two are referred to as "Fermi's conditions" [*90*].) For Eq. 21-6 and 21-7 and the Poiseuille flow, the proper boundary conditions are that f_2, f_2', and f_5 vanish at both walls.

The solution of Eq. 21-6 and 21-7 is simplified if one considers the special case $T/\alpha Re \gg 1$, i.e. when the product of α and the magnetic Reynolds number is small. Neglecting the f_5 term on the left-hand side of Eq. 21-7, Stuart thus obtained a single equation for f_2 from Eq. 21-6,

$$(\bar{u} - c)f_2'' - (\bar{u}'' + iq)f_2 = \frac{1}{i\alpha Re}(f_2'''' - 2\alpha^2 f_2'' + \alpha^4 f_2) \qquad (21\text{-}8)$$

where $q = \alpha ReS/T$. This is now identical to the Orr-Sommerfeld equation except for the term iq. The eigenvalue problem can be solved in the same way as in Chap. 2, after properly modifying the "inviscid solutions." The neutral curves calculated by Stuart are shown in Fig. G,21a. Note the stabilizing influence of the magnetic field. For $q \cong 0.10$ or above, the flow is completely stable, provided that the parameter $T/\alpha Re$ remains large. In his paper, Stuart pointed out that, since T is fixed for the fluid, the calculation is not valid for sufficiently large αRe. But if we raise q by increasing S (for instance through the magnetic field), then for mercury at room temperature, $T \cong 5 \times 10^6$, and complete stability up to $Re \cong 3 \times 10^5$ should be possible as predicted by his theory.

For a constant transverse magnetic field, $H_x = H_z = 0$, the velocity distribution of the basic flow depends upon the constant H_r. The basic flows corresponding to the plane Poiseuille and Blasius flows were determined, and their stability problems investigated under similar assumptions as above, by Lock [*96*] and Rossow [*97*], respectively. Instead of Eq. 21-8, Lock gave the following equation:

$$(\bar{u} - c)f_2'' - \bar{u}f_2 = \frac{iq}{\alpha^2}f_2'' + \frac{1}{i\alpha Re}(f_2'''' - 2\alpha^2 f_2'' + \alpha^4 f_2) \qquad (21\text{-}9)$$

The "inviscid" and "viscous" solutions by the usual procedure are clearly unaffected by the magnetic field term in Eq. 21-9. Thus one needs to recalculate only for the modified velocity distributions. Lock's calculation for the channel flow also shows a stabilizing influence due to the magnetic field, but one much stronger than that found in the coplanar case as found by Stuart. Rossow made calculations for both coplanar and transverse magnetic fields in the case of flow over a flat plate. According to his results, the coplanar magnetic field still stabilizes; the transverse magnetic field stabilizes even more if it is fixed relative to the fluid, but is slightly destabilizing when fixed to the plate (Fig. G,21b).[13] In a qualitative manner one may visualize the magnetic lines of force as resisting the conducting fluid particles from crossing. In the coplanar case, the

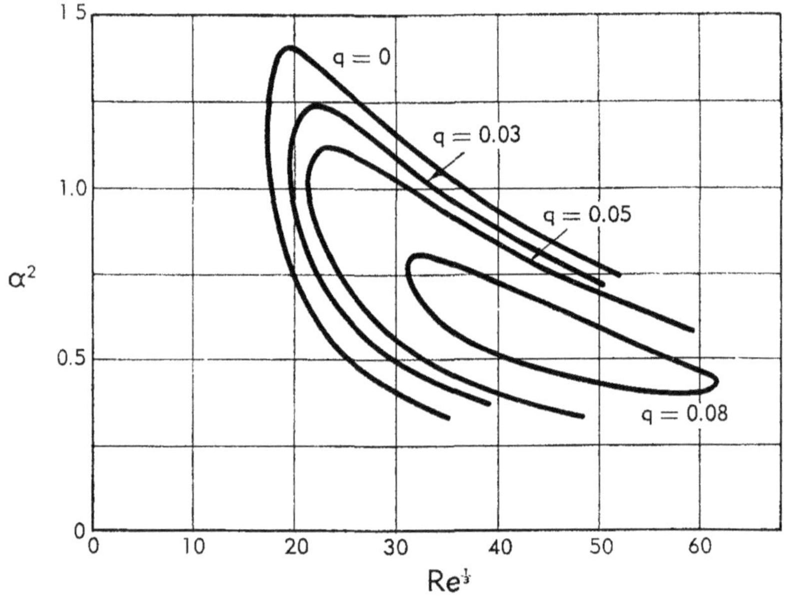

Fig. G,21a. Neutral curves for plane Poiseuille flow under coplanar magnetic field.

velocity distribution of the basic flow is unaffected, but fluctuations tend to move fluid particles across the lines of force and are therefore restrained. In the transverse case, the velocity distribution of the basic flow is modified in a manner depending on whether the lines of force are fixed to the fluid or the plate. The fluid particles tend to remain in the neighborhood of the lines of force. Thus the velocity distribution tends to be fuller when the lines of force are fixed to the fluid, as if under a favorable pressure gradient, and the reverse when the lines of force are fixed to the plate. Movements of the fluid particles in the normal direction (parallel to the lines of force) furthermore are not interfered with by the lines of force. The effects on stability are thus as expected.

[13] For the proper interpretation of Rossow's two cases, see Sears and McCune [*135*].

An experimental investigation of the transition Reynolds number in a pipe flow of mercury under axial magnetic field was reported by Bader and Carlson [*98*]. By the foregoing qualitative argument, one expects the magnetic field to be stabilizing. A slight increase of transition Reynolds number was indeed observed when the Reynolds number was high (above 5000 based upon pipe diameter and mean flow velocity) and initial disturbances were small.

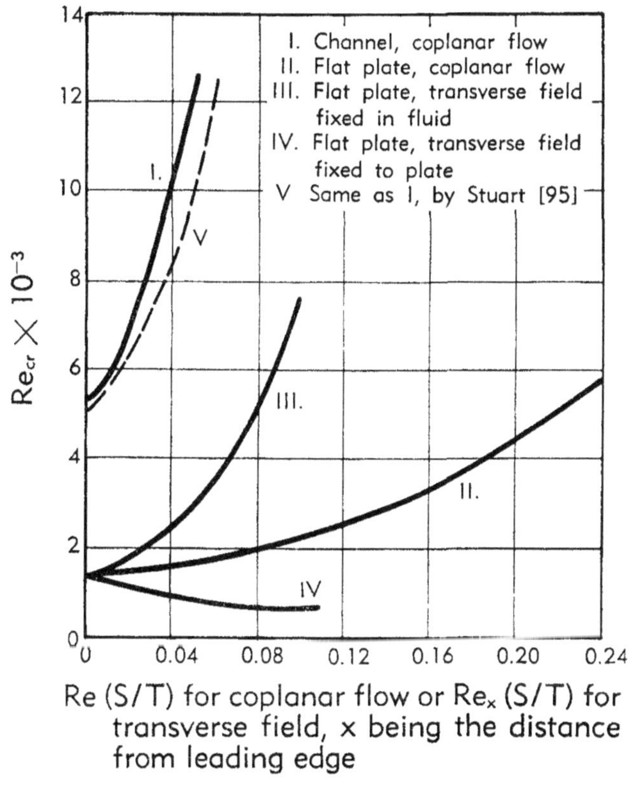

Fig. G,21b. Effect of magnetic field on minimum critical Reynolds number.

Problems involving a binary gas mixture arise because, for efficient cooling of exposed surfaces in high speed flight, it is of interest to consider the possibility of injection of different coolants (see Sec. E, Art. 9). The probable effect on transition is the subject of the stability investigations of Shen [*99*] and, independently, Covert [*100*]. (Covert omitted the variation of gas constant (per unit mass of mixture) with concentration of the injected gas. His results thus are to be taken with reservation unless the molecular weight of the coolant is nearly equal to that of the free stream.) The essential new feature is the appearance of the mass concentration (of the coolant) as a new variable defining the local composition of the gas. The transport properties are functions of the thermodynamic variables and the mass concentration; the distribution of the latter is essentially governed by the diffusion equation. (For the steady laminar boundary layer equation of a binary gas mixture, see e.g. Shen [*101*] or Baron [*102*].)

With the same argument as in Art. 14, it turns out that the following equation may be written down for the disturbances in the two-dimensional case:[14]

$$\left.\begin{aligned}
\bar{\rho}\left(\frac{\partial u'}{\partial t} + \bar{u}\frac{\partial u'}{\partial x} + v'\frac{d\bar{u}}{dy}\right) &= -\frac{\partial p'}{\partial x} + \bar{\mu}\frac{\partial^2 u'}{\partial y^2} \\
\bar{\rho}\left(\frac{\partial v'}{\partial t} + \bar{u}\frac{\partial v'}{\partial x}\right) &= -\frac{\partial p'}{\partial y} + \bar{\mu}\frac{\partial^2 v'}{\partial y^2} \\
\frac{\partial \rho'}{\partial t} + \bar{u}\frac{\partial \rho'}{\partial x} + v'\frac{d\bar{\rho}}{dy} &= -\bar{\rho}\left(\frac{\partial u'}{\partial x} + \frac{\partial v'}{\partial y}\right) \\
\bar{\rho}\left(\frac{\partial h'}{\partial t} + \bar{u}\frac{\partial h'}{\partial x} + v'\frac{d\bar{h}}{dy}\right) &= \frac{\partial p'}{\partial t} + \bar{u}\frac{\partial p'}{\partial x} + \frac{\bar{\mu}}{Pr}\frac{\partial^2 h'}{\partial y^2} \\
&\quad + \frac{\bar{\mu}}{Sc}\frac{\bar{h}}{\bar{c}_p}(c_{p_b} - c_{p_a})\left(1 - \frac{Sc}{Pr}\right)\frac{\partial^2 c'}{\partial y^2} \\
\bar{\rho}\left(\frac{\partial c'}{\partial t} + \bar{u}\frac{\partial c'}{\partial x} + v'\frac{d\bar{c}}{dy}\right) &= \frac{\bar{\mu}}{Pr}\frac{\partial^2 c'}{\partial y^2} \\
\frac{p'}{\bar{p}} &= \frac{\rho'}{\bar{\rho}} + \frac{h'}{\bar{h}} + \frac{c'}{F(\bar{c})}
\end{aligned}\right\} \tag{21-10}$$

In Eq. 21-10, $\bar{h}$ is the enthalpy distribution in the basic flow and $\bar{c}$ the mass concentration defined by

$$\bar{c} = \frac{m_b\bar{n}_b}{m_a\bar{n}_a + m_b\bar{n}_b} \tag{21-11}$$

where m_i = molecular weight of species i

$\bar{n}_i$ = number of molecules of species i per unit volume,

subscript $_a$ = the gas in the free stream

subscript $_b$ = the injected gas

The quantities h' and c' are the disturbances of $\bar{h}$ and $\bar{c}$ respectively. The symbol Sc is the local Schmidt number,

$$Sc \equiv \frac{\mu}{\rho\mathfrak{D}} \tag{21-12}$$

$\mathfrak{D}$ being the binary diffusion coefficient. The specific heat for pure species a or b is assumed constant, while the value for the mixture is approxi-

[14] The role of three-dimensional disturbances may be argued in exactly the same manner as in Art. 14.

mated by a linear interpolation,

$$\bar{c}_p = c_{p_b} + \bar{c}(c_{p_b} - c_{p_a}) \tag{21-13}$$

The mixture gas constant obeys the same linear interpolation,

$$\bar{R} = R_b + \bar{c}(R_b - R_a) \tag{21-14}$$

The quantity $F(\bar{c})$ then represents the effect of the disturbances on the local gas constant and $\bar{c}_p$,

$$\frac{1}{F(\bar{c})} = \frac{R_b - R_a}{R_b + \bar{c}(R_b - R_a)} - \frac{c_{p_b} - c_{p_a}}{c_{p_b} + \bar{c}(c_{p_b} - c_{p_a})} \tag{21-15}$$

The thermal diffusion has been omitted for simplicity. These equations may be compared against Eq. 14-2, which hold for a single species, and the similarity of the mathematical structure is obvious. However, now we have one more equation containing the second derivative. By proceeding in the usual manner under the assumption of periodic disturbances, *eight* fundamental solutions should be found. The "inviscid" equation is of the same form as Eq. 14-6,

$$\left.\begin{aligned} &\frac{d}{dy}\left[\frac{(\bar{u} - c)\phi' - \bar{u}'\phi}{P}\right] - \alpha^2\bar{\rho}(\bar{u} - c)\phi = 0 \\ &\text{but} \\ &P = \frac{1}{\bar{\rho}} - M_\beta^2(\bar{u} - c)^2\left(\gamma - \frac{\gamma - 1}{\bar{\rho}\bar{h}}\right) \end{aligned}\right\} \tag{21-16}$$

The quantities in Eq. 21-16 are defined and nondimensionalized in the same way as in Eq. 14-6. Note that we use $\bar{c}$ and c' for the concentration but c remains the phase velocity as before.

The solution for f_1 is expressible as ϕ and ϕ' a linear combination of

$$f_1 = \frac{i}{P}\left\{\frac{1}{\bar{\rho}}\phi' - M_\beta^2\bar{u}'(\bar{u} - c)\left(\gamma - \frac{\gamma - 1}{\bar{\rho}\bar{h}}\right)\phi\right\} \tag{21-17}$$

and similarly for the other disturbances. The asymptotic behavior of the solutions of Eq. 21-16 as $y \to \infty$ can be exponentially damped only when, again, the disturbances are "subsonic," i.e.

$$c > 1 - \frac{1}{M_\beta}$$

The zeroeth order "viscous" solutions may be handled by considerations identical to those which led to Eq. 14-9, with the understanding that the accuracy for c not very small will not be reliable. The application of Tollmien's technique is of course possible when an improvement is desired. The two extra fundamental solutions involve functions similar to Eq. 14-15 except that the upper limit becomes $-\zeta(S_0/\bar{\nu}_0)^{-\frac{1}{2}}$. On the other hand, the boundary conditions for problems with injection through the wall are difficult to formulate. In addition to those for the thermal disturbances discussed in Art. 15, both the normal velocity disturbance and the concentration disturbance at the wall cannot be stipulated without including the mechanics of the injection device itself. With proper boundary conditions, the Dunn-Lin Δ correction can be worked out and the secular equation (Eq. 15-12) holds. If one accepts the zeroeth order "viscous" solutions, however, the thermal and concentration boundary conditions at the wall will be divorced from the velocity boundary conditions. Within the accuracy of this assumption, Δ may be taken to be zero.

The "inviscid criterion" for a binary gas mixture should still be Eq. 15-6, namely, possible neutral oscillation if

$$\frac{d}{dy}(\bar{\rho}\bar{u}')\bigg|_{y=y_c} = 0, \qquad c > 1 - \frac{1}{M_\beta}$$

since diffusivity is left out in the inviscid considerations. Thus, for subsonic M_β, a density gradient similar to that of a cooled single-species boundary layer will be relatively more stable. The injection of a lighter gas, on such an argument, appears to be less preferable than that of a heavier gas. However, the lighter gas usually has a higher coefficient of heat conductivity per mole. The less amount required for a given cooling problem could have a counteracting influence. In addition, at higher M_β, an examination of the actual distributions is necessary in order to satisfy the requirement of "subsonic" disturbances. Covert [*100*], for instance, concluded that a moderate amount of helium injection would be more stable than air alone at Mach numbers higher than roughly 4.

When chemical reaction takes place between the two species that form the binary gas mixture, a simplified case was briefly examined by Shen [*103*]. The equations of motion are modified by including a "sink" term in the concentration equation (the conversion of species b into species a) and a "source" term in the energy equation (the production of heat due to reaction). If both the sink and the source are taken to be functions of the concentration (of species b) and the mixture temperature, in the disturbance equations (Eq. 21-10), we need only to add a linear combination of the concentration and enthalpy disturbances to the right-hand sides of the fourth and fifth equations. Neglecting all transport terms, the "inviscid" equation may be derived in the usual manner. It turns out that the "inviscid" equation will involve complex coefficients because of the chemical reaction source and sink terms, although remaining a second order differential equation. The immediate conclusion is that the eigenvalue c must in general be complex, so that no "neutral" oscillation is to be expected. The presence of a point of inflection of the velocity distribution or the generalization of such considerations, such as Eq. 15-6, will lose their significance. Indeed, one may make a comparison with Stuart's problem involving magnetic forces, Eq. 21-8. There the "inviscid" equation is likewise complex for $q \neq 0$, and we see that at large enough Reynolds number the flow is always stable in his case. A more thorough investigation of the effect of chemical reaction on stability is still needed.

CHAPTER 4. STABILITY THEORY DEALING WITH LONGITUDINAL VORTEX-TYPE DISTURBANCES

G,22. The Eigenvalue Problem and the Rayleigh Criterion for Stability. Following the discussions of Art. 4, we now consider the disturbance equations in cylindrical coordinates (ϕ, r, z), Eq. 4-5, 4-6, 4-7, 4-8, and 4-9, with the specialization $\alpha = 0$. These disturbances, as mentioned before, may be regarded loosely as "longitudinal vortices," the axes of which lie in the streamline direction of the basic flow. To simplify matters further, let the basic flow be two-dimensional and without an axial component. It then turns out that, after elimination of ϖ and f_3, the following equations result:

$$(L - \beta^2 + i\sigma Re)(L - \beta^2)f_2 = 2\beta^2 \frac{\bar{u}}{r} Re\, f_1 \tag{22-1}$$

$$(L - \beta^2 + i\sigma Re)f_1 = Re \frac{1}{r}\left[\frac{d}{dr}(r\bar{u})\right] f_2 \tag{22-2}$$

where the operator L is defined by

$$L \equiv \frac{d^2}{dr^2} + \frac{1}{r}\frac{d}{dr} - \frac{1}{r^2} \tag{22-3}$$

The boundary conditions must next be specified. Suppose the boundaries coincide with $r = r_1$ and $r = r_2$. On a solid boundary, the no-slip condition requires the vanishing of all disturbance velocities:

$$f_1 = f_2 = 0$$

and $f_3 = 0$ leads to $f_2' = 0$ by Eq. 4-8. If we have boundary layer flow over a curved wall, f_1, f_2, f_2' must be bounded at the edge of the boundary layer. Thus three boundary conditions, homogeneous in nature, are specified at both $r = r_1$ and $r = r_2$. The solutions of Eq. 22-1 and 22-3 are therefore an eigenvalue problem. For given *Re* and wave number β, the eigenvalue $i\sigma$ may be determined. A positive real part of $i\sigma$ is interpreted to mean amplified disturbances.[15]

As an illuminating example of two-dimensional flow over a surface curved in the streamwise direction, we consider the flow in the annulus of two concentric circular cylinders, in the limiting case of an inviscid fluid, i.e. $Re \to \infty$. Formally dropping terms of order $1/Re$, one gets from Eq. 22-1 and 22-2

$$\frac{d}{dr}\left(\frac{1}{r}\frac{dg}{dr}\right) + \beta^2\left[\frac{1}{\sigma^2 r^4}\frac{d}{dr}(r\bar{u})^2 - \frac{1}{r}\right]g = 0 \qquad (22\text{-}4)$$

where $g = rf_2$. The boundary conditions now are $g = 0$ at $r = r_1$ and r_2, the walls of the concentric cylinders. By the theory of the Sturm-Liouville equation, the eigenvalue σ^2 must be real and of the same sign as $(d/dr)(r\bar{u})^2$, if the latter has the same sign throughout the interval (r_1, r_2). Thus we conclude that (1) if $(d/dr)(r\bar{u})^2 > 0$, then $\sigma^2 > 0$, oscillatory motion results; and (2) if $(d/dr)(r\bar{u})^2 < 0$, then $\sigma^2 < 0$, σ is pure imaginary and one of its roots corresponds to divergent motion. In addition, if $(d/dr)(r\bar{u})^2$ changes sign in (r_1, r_2), some σ^2 is negative so that again the basic flow is unstable. The "inviscid criterion" for stability of flows in an annulus is consequently $(d/dr)(r\bar{u})^2 > 0$. This was first due to Rayleigh [*104*], who derived it on an energy consideration; it was interpreted by von Kármán [*105*] on the basis of equilibrium between the centrifugal force and the pressure gradient acting on a displaced fluid element. We see that inviscid instability arises whenever the circulation *squared* begins to decrease with increasing radial distance. In fact the "Rayleigh criterion" $(d/dr)(r\bar{u})^2 > 0$ is shown to be sufficient for stability even with viscosity by Synge [*106*], who used the technique of formulating certain inequalities similar to those in Art. 5.

It may be noted that the "inviscid" equation (Eq. 22-4) does not involve singular points in the interval r_1, r_2, in contrast to the "inviscid" equation (Eq. 7-3), for disturbances of the transverse wave type. The inviscid solution here being everywhere smooth, the delicate question of the role of viscosity in the neighborhood of the "critical layer" does not arise in the present case. It is consequently reasonable to expect that in the present case viscosity will have only its usual damping effect, without being meanwhile a possible agent for promoting instability through the phase-shifting of the velocity disturbances.

Now, the transition from stability to instability in the inviscid case is at $\sigma = 0$. If we assume, from the argument above, that viscosity adds only a damping factor, at finite *Re* the "inviscid" undamped oscillations

[15] For an exhaustive treatment of this type of problem, see the recent book by Chandrasekhar [*139*].

become damped oscillations, the "inviscid" neutral solution becomes damped, and the "inviscid" divergent solutions becomes less divergent. The transition from stability to instability at finite Re thus is still marked by the vanishing of the eigenvalue σ, not by the more general condition that the imaginary part of σ should vanish. This is the so-called "principle of exchange of stabilities." Though often assumed, the rigorous proof of its validity in the general case is difficult. For instance, Pellew and Southwell [*107*] proved it for a thermal instability problem, for which the differential equations are essentially identical to Eq. 22-1 and 22-2. Their technique amounts, in our case, to establishing that σ is pure imaginary, hence that the vanishing of the imaginary part is identical to the vanishing of σ itself. However, such a proof can be carried out only if the basic flow is unstable by the Rayleigh criterion." The limited success is understandable if we recall that damped oscillations, hence complex σ, probably do exist when the basic flow is stable in the inviscid limit.

Accepting the "principle of exchange of stabilities" and that the amplified disturbance would be of the divergent type, we lose no generality if we replace $i\sigma$ by a real parameter $-\lambda$ in Eq. 22-1 and 22-2 for determining the neutral curve and the amplification rates,

$$\left.\begin{aligned}(L - \beta^2 - \lambda Re)(L - \beta^2)f_2 &= 2\beta^2 \frac{\bar{u}}{r}\, Re\, f_1 \\ (L - \beta^2 - \lambda Re)f_1 &= Re\,\frac{1}{r}\left[\frac{d}{dr}(r\bar{u})\right]f_2\end{aligned}\right\} \tag{22-5}$$

The neutral curve will be found by setting $\lambda = 0$ in Eq. 22-5 and a minimum critical Reynolds number may thus be obtained. The neutral disturbances are of the form

$$u' \sim f_1(r)e^{i\beta z}, \qquad v' \sim f_2(r)e^{i\beta z}$$

These represent secondary flows since they do not vary with time, but are periodic in the axial direction. Even in the unstable range, the disturbances tend to diverge but never fluctuate. They are of a nature decidedly different from turbulence. The process by which such instability finally develops into turbulence is still an open question.

G,23. Couette Flow between Rotating Cylinders. One of the classical problems of hydrodynamic stability is that of the Couette flow between rotating cylinders, first exhaustively treated by Taylor [*12*] with beautiful experimental verification of the theoretical results. Let the inner cylinder be of radius r_1, rotating at an angular velocity Ω_1, and the outer cylinder be of radius r_2, at an angular velocity Ω_2. The dimensionless velocity distribution is then given by

$$\left.\begin{aligned} \bar{u} &= Ar + \frac{B}{r} \\ A &= \frac{\left(\frac{r_2}{r_1}\right)^2 \frac{\Omega_2}{\Omega_1} - 1}{\left(\frac{r_2}{r_1}\right)^2 - 1} \\ B &= -\frac{\left(\frac{r_2}{r_1}\right)^2 \left(\frac{\Omega_2}{\Omega_1} - 1\right)}{\left(\frac{r_2}{r_1}\right)^2 - 1} \end{aligned}\right\} \tag{23-1}$$

where the reference length is r_1, and the reference velocity is $r_1\Omega_1$. By taking the time scale as Ω_1^{-1}, Eq. 23-1 is to be substituted into the right-hand side of Eq. 22-5. We thus get

$$\left.\begin{aligned} (L - \beta^2 - \lambda Re)(L - \beta^2)f_2 &= 2\beta^2 \frac{\bar{u}}{r} Re\, f_1 \\ (L - \beta^2 - \lambda Re)f_1 &= 2A Re\, f_2 \\ f_1 = f_2 = f_2' = 0 \quad \text{at} \quad r &= 1 \text{ and } r_2 \end{aligned}\right\} \tag{23-2}$$

Taylor's method of solution is quite elaborate. The functions f_1 and f_2 were expanded in terms of a set of orthonormal functions suggested by the operator L, and the Galerkin method was used for an approximate solution [*15*] or [*1*, pp. 24–26]. The numerical work is somewhat simplified if the clearance is small. The calculated and observed neutral curves are shown in Fig. G,23a. The streamline pattern is shown in Fig. G,23b. A recent experimental investigation is that of Schultz-Grunow and Hein [*108*].

For neutral stability, $\lambda = 0$, and Eq. 23-2 may be reduced to

$$(L - \beta)^3 f_1 = 4A\beta^2 Re^2 \frac{\bar{u}}{r} f_1 \tag{23-3}$$

If the spacing between cylinders is small, further simplification is possible. By setting $r_2 = 1 + \epsilon$, $\epsilon \ll 1$, we write

$$r = 1 + \epsilon y, \qquad \frac{d}{dr} = \frac{1}{\epsilon}\frac{d}{dy}$$

$$\frac{\bar{u}}{r} \cong \Omega_1 \left[1 - \left(1 - \frac{\Omega_2}{\Omega_1}\right) y\right]$$

Then Eq. 23-3 becomes, after dropping the higher order terms on the

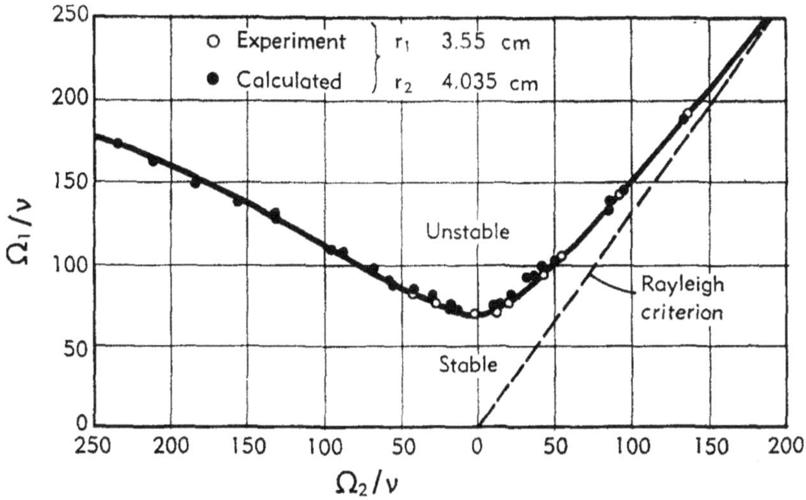

Fig. G,23a. Comparison of theory and experiment for stability of flow between rotating cylinders.

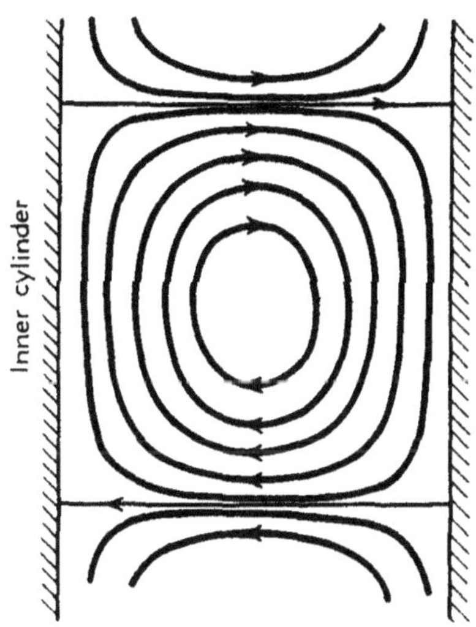

Fig. G,23b. Streamline pattern of vortex disturbance for flow between rotating cylinders.

left-hand side,

$$\left.\begin{aligned} (D^2 - \epsilon^2\beta^2)^3 f_1 &= -\epsilon^2\beta^2\, Ta\left[1 - \left(1 - \frac{\Omega_2}{\Omega_1}\right)y\right]f_1 \\ D &\equiv \frac{d}{dy} \end{aligned}\right\} \tag{23-4}$$

for which the eigenvalue will be $\epsilon\beta$, depending upon the parameters

Ω_2/Ω_1 and the Taylor number Ta,

$$Ta \equiv \frac{-4A\Omega_1\epsilon^4}{\nu^2} \tag{23-5}$$

The Taylor number may be reduced to $Re_1^2[(r_2 - r_1)/r_1]$, similar to the Görtler number ($Re_\theta \sqrt{\theta/R}$, p. 829), where Re_1 is the Reynolds number based on the annulus width.

When the cylinders rotate in the same direction, an average constant may be substituted for the small value y as a further approximation. Thus Eq. 23-5 takes the form

$$\left.\begin{aligned} &(D^2 - \epsilon^2\beta^2)^3 f_1 = -\epsilon^2\beta^2\overline{Ta}\, f_1 \\ \text{with} \qquad &f_1 = D^2 f_1 = D(D^2 - \epsilon^2\beta^2)f_1 = 0 \quad \text{at} \quad y = 0 \text{ and } 1 \end{aligned}\right\} \tag{23-6}$$

The average Taylor number $\overline{Ta}$ may be evaluated at $y = \frac{1}{2}$,

$$\overline{Ta} = \frac{Ta\left(1 + \frac{\Omega_2}{\Omega_1}\right)}{2}$$

For Eq. 23-6 a variational principle is possible (Southwell and Pellew [*107*]),

$$\left.\begin{aligned} &\overline{Ta} = \min \frac{\int_0^1 [g_1''^2 + 2\epsilon^2\beta^2 g_1'^2 + \epsilon^4\beta^4 g_1^2]dy}{\epsilon^2\beta^2 \int_0^1 (f_1'^2 + \epsilon^2\beta^2 f_1^2)dy} \\ \text{with} \qquad &g_1 = (D^2 - \epsilon^2\beta^2)f_1 \end{aligned}\right\} \tag{23-7}$$

The minimum critical Taylor number by this procedure turns out to be

$$\overline{Ta} = 1708 \quad \text{at} \quad \epsilon\beta = 3.13$$

agreeing very well with Taylor's original calculation. The wave number indicates that the spacing of the vortices in the axial direction should be approximately twice the annulus width.

To handle the case of ϵ not too small, the velocity distribution $\bar{u}/r$, eigenfunction f_1, and the eigenvalue $\epsilon\beta$ may all be expanded in ascending powers of ϵ (Witting [*109*]). The lowest order eigenvalue problem is essentially identical to that of Eq. 23-6. The first order correction may then be found by a standard perturbation technique. The calculations were carried out by an integral equation approach, similar to Eq. 25-3 of Art. 25, by Witting. His results are summarized in Fig. G,23c. As $\beta \to 0$, the eigenvalue tends to β^{-2}.

When the cylinders rotate in opposite directions, $\Omega_2/\Omega_1 < 0$, and the linear approximation of $\bar{u}/r$ must be retained. Di Prima [*110*] applied the straight Galerkin method with reasonable success, but the choice

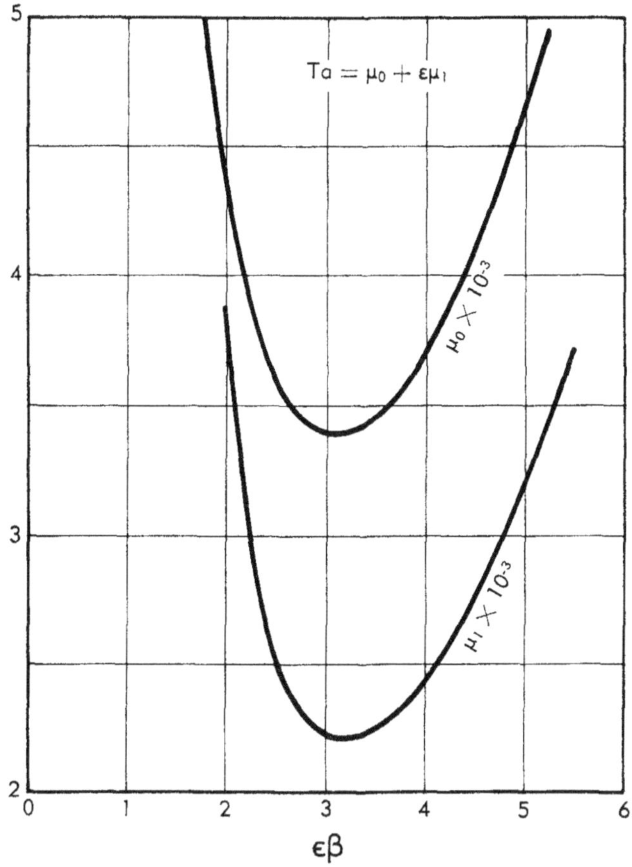

Fig. G,23c. First order correction of radius of curvature on the critical Taylor number.

of the approximate function is quite crucial. The systematic procedure due to Chandrasekhar [*93*] amounts to a refined way of choosing the approximate function, and appears worthy of attention.[16] Rewriting Eq. 23-4 as

$$\frac{1}{1 - \left(1 - \frac{\Omega_2}{\Omega_1}\right) y} (D^2 - \epsilon^2\beta^2)^3 f_1 = -\epsilon^2\beta^2 Ta\, f_1 \tag{23-8}$$

and applying again the operator $(D^2 - \epsilon^2\beta^2)$, we obtain

$$\left.\begin{aligned} (D^2 - \epsilon^2\beta^2)\left[\frac{1}{1 - \left(1 - \frac{\Omega_2}{\Omega_1} y\right)} (D^2 - \epsilon^2\beta^2)^2 W\right] &= -\epsilon^2\beta^2 Ta\, W \\ W = (D^2 - \epsilon^2\beta^2) f_1 \end{aligned}\right\} \tag{23-9}$$

[16] A different method is that of Meksyn [*111*], who expanded the solution asymptotically for large $\epsilon\beta$.

The corresponding boundary conditions are

$$W = DW = (D^2 - \epsilon^2\beta^2)W = 0 \quad \text{at} \quad y = 0 \text{ and } 1 \qquad (23\text{-}10)$$

It is convenient to regard Eq. 23-9 as a pair of equations

$$\left.\begin{aligned} (D^2 - \epsilon^2\beta^2)^2 W &= \left[1 - \left(1 - \frac{\Omega_2}{\Omega_1}\right) y\right] V \\ (D^2 - \epsilon^2\beta^2) V &= -\epsilon^2\beta^2 Ta\, W \\ \text{with} \qquad W = DW &= V = 0 \text{ at } y = 0 \text{ and } 1 \end{aligned}\right\} \qquad (23\text{-}11)$$

From the boundary condition on V, it is natural to put

$$V = \sum_{n=1}^{\infty} C_n \sin n\pi y \qquad (23\text{-}12)$$

The first of Eq. 23-11 may then be exactly solved in terms of the C_n's,

$$\begin{aligned} W = \sum_{m=1}^{\infty} \frac{C_m}{(m^2\pi^2 + \epsilon^2\beta^2)^2} \Bigg\{ &(A_1^{(m)} + A_2^{(m)} y) \cosh \epsilon\beta y \\ &+ (B_1^{(m)} + B_2^{(m)} y) \sinh \epsilon\beta y + \left[1 - \left(1 - \frac{\Omega_2}{\Omega_1}\right) y\right] \sin m\pi y \\ &- \frac{4\left(1 - \frac{\Omega_2}{\Omega_1}\right) m\pi}{m^2\pi^2 + \epsilon^2\beta^2} \cos m\pi y \Bigg\} \end{aligned} \qquad (23\text{-}13)$$

where the constants of integration $A_1^{(m)}$, $A_2^{(m)}$, $B_1^{(m)}$, and $B_2^{(m)}$ are determined by the boundary conditions on W. If Eq. 23-13 is substituted back into the second of Eq. 23-11, a single equation linear in all the C_n's, with coefficients depending on y, will be obtained. The Galerkin procedure may then be used to solve for the C_n's by multiplying the equation by $\sin n\pi y$ and then integrating from $y = 0$ to $y = 1$. Chandrasekhar's numerical results are shown in Fig. G,23d. As $\Omega_2/\Omega_1 \to -\infty$, the asymptotic behavior is

$$Ta_{cr} \to 1180 \left(1 - \frac{\Omega_2}{\Omega_1}\right)^4 \quad \text{at} \quad \epsilon\beta \to 2.03 \left(1 - \frac{\Omega_2}{\Omega_1}\right)$$

The influence of an axial pressure gradient on the stability of the flow between rotating cylinders was studied by Goldstein [*112*]. A significant modification is that the instability will now be in the form of divergent oscillations instead of an exponential growth. The influence of a radial temperature gradient was treated by Chandrasekhar [*113*].

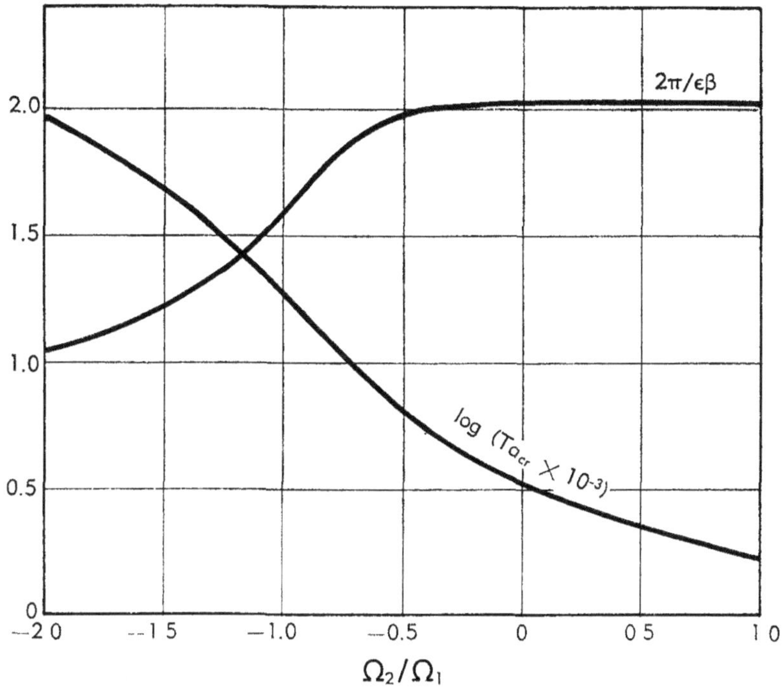

Fig. G,23d. Minimum critical Taylor number and corresponding vortex spacing for flow between rotating cylinders at various speed ratios.

G,24. The Influence of Magnetic Field on Flow between Rotating Cylinders. The results mentioned in Art. 21 indicate that the magnetic field often exerts a stabilizing influence on disturbances of the transverse-wave type. Let us turn now to the question of the behavior of longitudinal vortex disturbances in the presence of a magnetic field. When the basic flow is that between rotating cylinders, the influence of a constant axial magnetic field has been studied by Chandrasekhar [*91*]. For this type of basic flow—two-dimensional and axially symmetric—cylindrical coordinates may again be conveniently used in connection with Eq. 21-2, 21-3, 21-4, and 21-5. Only axially symmetric disturbances are considered for simplicity. The linearized equations of motion for the disturbances before nondimensionalizing are now (cf. Eq. 4-1, 4-2, 4-3, and 4-4):

$$\left.\begin{aligned}
\frac{\partial u'}{\partial t} + \frac{v'}{r}\frac{d}{dr}(r\bar{u}) - \frac{\mu_{\mathrm{m}}H}{4\pi\rho}\frac{\partial h'_\phi}{\partial z} &= \nu\left(\nabla^2 u' - \frac{u'}{r^2}\right)\\
\frac{\partial v'}{\partial t} - \frac{2v'\bar{u}}{r} - \frac{\mu_{\mathrm{m}}H}{4\pi\rho}\frac{\partial h'_r}{\partial z} &= -\frac{1}{\rho}\frac{\partial \tilde{p}'}{\partial r} + \nu\left(\nabla^2 v' - \frac{v'}{r^2}\right)\\
\frac{\partial w'}{\partial t} - \frac{\mu_{\mathrm{m}}H}{4\pi\rho}\frac{\partial h'_z}{\partial z} &= -\frac{1}{\rho}\frac{\partial \tilde{p}'}{\partial z} + \nu\nabla^2 w'\\
\frac{1}{r}\frac{\partial}{\partial r}(rv') + \frac{\partial w'}{\partial z} &= 0
\end{aligned}\right\} \qquad (24\text{-}1)$$

$$\left.\begin{aligned} \frac{\partial h'_\phi}{\partial t} - h'_r r \frac{\partial}{\partial r}\left(\frac{\bar{u}}{r}\right) - H\frac{\partial u'}{\partial z} &= \eta\left(\nabla^2 - \frac{1}{r^2}\right)h'_\phi \\ \frac{\partial h'_r}{\partial t} - H\frac{\partial v'}{\partial z} &= \eta\left(\nabla^2 - \frac{1}{r^2}\right)h'_r \\ \frac{\partial h'_z}{\partial t} - H\frac{\partial w'}{\partial z} &= \eta\nabla^2 h'_z \\ \frac{1}{r}\frac{\partial}{\partial r}(rh'_r) + \frac{\partial h'_z}{\partial z} &= 0 \end{aligned}\right\} \tag{24-2}$$

where $\bar{u}(r)$ is the tangential velocity of the basic flow in the ϕ direction and H the constant magnetic field in the z direction; u', v', w' and h'_ϕ, h'_r, h'_z are the velocity disturbances and the magnetic field disturbances in the ϕ, r, z direction, respectively; $\tilde{p}'$ is the disturbance of $\tilde{p}$; and $\tilde{p}$, η, etc. are all as defined in Art. 21. After nondimensionalizing as in Art. 22 and assuming

$$\left(u', v', w', \frac{h'_\phi}{H}, \frac{h'_r}{H}, \frac{h'_z}{H}\right) \sim (f_1, f_2, \ldots, f_6)e^{-\lambda t + i\beta z} \tag{24-3}$$

Eq. 24-1 and 24-2 may be manipulated to yield

$$(L - \beta^2 - \lambda Re)(L - \beta^2)f_2 = Re\left[2\beta^2\frac{\bar{u}}{r}f_1 - \beta S(L - \beta^2)f_5\right] \tag{24-4}$$

$$(L - \beta^2 - \lambda Re)f_1 = Re\left\{\frac{1}{r}\left[\frac{d}{dr}(r\bar{u})\right]f_2 - \beta S f_4\right\} \tag{24-5}$$

$$\left(L - \beta^2 - \frac{\lambda Re}{T}\right)f_4 = \frac{Re}{T}\left\{\beta f_1 - r\left[\frac{d}{dr}\left(\frac{\bar{u}}{r}\right)\right]f_5\right\} \tag{24-6}$$

$$\left(L - \beta^2 - \frac{\lambda Re}{T}\right)f_5 = \beta\frac{Re}{T}f_2 \tag{24-7}$$

where S is the ratio of magnetic pressure to dynamic pressure and T is the ratio η/ν, both used previously in Art. 21 (p. 806). The boundary conditions for a rigid but perfectly conducting wall are, according to the discussion on p. 807,

$$\left.\begin{aligned} f_1 = f_2 = \frac{d}{dr}(rf_2) = \frac{d}{dr}(rf_4) = f_5 = 0 \\ \text{at} \qquad r = r_1 = 1 \quad \text{and} \quad r = r_2 \end{aligned}\right\} \tag{24-8}$$

Let us again assume the "principle of exchange of stabilities," i.e. the marginal stability still given by $\lambda = 0$. Furthermore, suppose that the cylinders are rotating in the same direction at nearly equal angular speeds $\Omega_1 \cong \Omega_2$ and that the spacing $\epsilon = (r_2/r_1) - 1 \ll 1$. The same change of variable from r to y as in the derivation of Eq. 23-4 from Eq. 23-3 may then be used. For this simplest case, we get, in contrast with

Eq. 23-4,

$$[(D^2 - \epsilon^2\beta^2)^2 + Q\epsilon^2\beta^2]^2 f_4 = -\epsilon^2\beta^2 Ta(D^2 - \epsilon^2\beta^2)f_4 \tag{24-9}$$

where $D \equiv d/dy$, Ta is the Taylor number as defined by Eq. 23-5, and Q is Chandrasekhar's parameter for magnetic interaction,

$$Q \equiv \frac{\mu_m^2 H^2 \sigma}{\rho\nu}(r_2 - r_1)^2 \tag{24-10}$$

i.e. $Q = Re_1^2 S/T$ in our notation, where Re_1 is the Reynolds number based upon the annulus width. (This may be compared with Stuart's q (p. 807), which now equals $Q/\alpha Re$.) The appropriate boundary conditions for Eq. 24-9 are

$$\left.\begin{array}{c} Df_4 = 0, \quad (D^2 - \epsilon^2\beta^2)f_4 = 0, \\ [(D^2 - \epsilon^2\beta^2)^2 + Q\epsilon^2\beta^2]f_4 = 0, \quad D[(D^2 - \epsilon^2\beta^2)^2 + Q\epsilon^2\beta^2]f_4 = 0 \end{array}\right\} \tag{24-11}$$

at the end points $y = 0$ and $y = 1$.

Eq. 24-9 and 24-11 are to be compared with Eq. 23-6. A variational procedure, due to Chandrasekhar [*91*], is

$$Ta = \min \frac{\int_0^1 \{[(D^2 - \epsilon^2\beta^2)P]^2 + Q\epsilon^2\beta^2 P\}dy}{\epsilon^2\beta^2 \int_0^1 \{(DG)^2 + \epsilon^2\beta^2 G^2 + \epsilon^2\beta^2 Q[(Df_4)^2 + \epsilon^2\beta^2 f_4^2]\}dy} \tag{24-12}$$

where

$$\left.\begin{array}{l} P = [(D^2 - \epsilon^2\beta^2)^2 + \epsilon^2\beta^2 Q]f_4 \\ G = (D^2 - \epsilon^2\beta^2)f_4 \end{array}\right\} \tag{24-13}$$

The actual calculation may be carried out in the following manner:

(i) Assume a function DP with arbitrary const A_k. $k = 1, 2, \ldots,$ satisfying $DP = 0$ at the end points;
(ii) Find P by integration, satisfying $P = 0$ at $y = 0$ and $y = 1$;
(iii) Solve the trial f_4 from

$$\left.\begin{array}{c} [(D^2 - \epsilon^2\beta^2)^2 + Q\epsilon^2\beta^2]f_4 = P \\ Df_4 = (D^2 - \epsilon^2\beta^2)f_4 = 0 \quad \text{at} \quad y = 0 \text{ and } y = 1 \end{array}\right\}$$

(iv) Minimize Ta with respect to A_k.

Chandrasekhar used a single A_k and arrived at the results as shown in Fig. G,24. The stabilizing influence seems to be quite pronounced. For nitric acid at room temperature, for instance, $Q = 4 \times 10^{-8}H^2(r_2 - r_1)^2$ with H in Gauss and $r_2 - r_1$ in cm. Thus measurable effects should be present for H of the order of 1000 Gauss and $(r_2 - r_1)$ of the order of 1 cm.

The question whether the "principle of exchange of stabilities" holds under certain circumstances in the presence of the magnetic field is a natural one to be raised. In this connection Chandrasekhar [*90*] has studied the thermal instability problem of a fluid heated from below with a magnetic field parallel to the direction of gravity. The formulation being mathematically quite analogous to that given here. By considering a special case he showed that a sufficient condition for the principle of exchange of stabilities to hold would be $\eta > \kappa$, where η is the magnetic diffusivity and κ the thermal conductivity. (The condition appears to be satisfied by fluids under terrestial conditions, but violated in certain

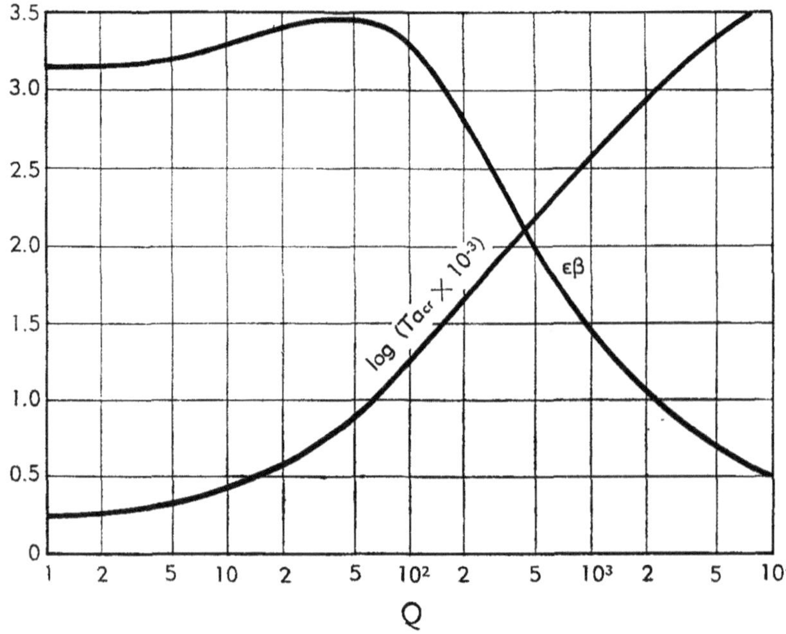

Fig. G,24. Effect of axial magnetic field on minimum critical Taylor number and corresponding vortex spacing for flow between rotating cylinders.

astrophysics problems [*90*].) The same technique probably may also be used here. For instance, without assuming λ real and zero, but specializing for cylinders rotating at nearly equal speeds $\Omega_1 \cong \Omega_2$, Eq. 24-4, 24-5, 24-6, and 24-7 admit the simple solutions.

$$f_1, f_2, f_4, f_5 \sim \begin{cases} \sin \pi y \\ \cos \pi y \end{cases} \tag{24-14}$$

satisfying certain homogeneous conditions at $y = 0$ and 1, but not all of Eq. 24-8. Either of these, on substituting back into the equation, leads to a quartic in λ with constant coefficients which depend on *Re*, *S*, *T*, etc. The principle of exchange of stabilities will then be valid when the roots of the quartic are proved to be either real or complex with negative real parts. Intuitively, it is plausible to expect that, if the simple solution does satisfy the boundary conditions of a certain physical problem, the result is likely to be applicable to problems of the same type. Unfortunately, however, we are unable to interpret the solution, Eq. 24-14, as corresponding to a problem of flow between rotating cylinders.

The "Rayleigh criterion" of Art. 22 may also be generalized to include the effect of the magnetic field. For an inviscid and perfectly conducting fluid, when the basic flow and the basic magnetic field have only tangential components $\bar{u}(r)$ and $H_\phi(r)$ in the cylindrical coordinates ϕ, r, z, such a criterion was derived by Michael [*115*] to show the stabilizing influence. A similar attempt on the problem defined by Eq. 24-4, 24-5, 24-6, 24-7, and 24-8 may be carried through without difficulty. The resulting second order differential equation is

$$(L - \beta^2)f_2 = \frac{\beta^2 r^{-3}}{\lambda^2 + \beta^2 S}\left\{\frac{d}{dr}(r\bar{u})^2 - \frac{4\beta^2 S}{\lambda^2 + \beta^2 S} r\bar{u}^2\right\} f_2 \tag{24-15}$$

Considering $\lambda^2 + \beta^2 S$ as the eigenvalue for given β, we conclude therefore

$$\text{(i)} \qquad \lambda^2 + \beta^2 S > 0 \quad \text{if} \quad \frac{d}{dr}(r\bar{u})^2 - \frac{4\beta^2 S}{\lambda^2 + \beta^2 S} r\bar{u}^2 < 0$$

$$\text{(ii)} \qquad \lambda^2 + \beta^2 S < 0 \quad \text{if} \quad \frac{d}{dr}(r\bar{u})^2 - \frac{4\beta^2 S}{\lambda^2 + \beta^2 S} r\bar{u}^2 > 0$$

The disturbances are certainly stable under (ii). The stabilizing influence of the magnetic field is expressed by the fact that $(d/dr)(r\bar{u})^2$ can be slightly negative without causing instability. However, the assumption of a perfectly conducting fluid $\sigma \rightarrow \infty$ is probably not very realistic in most problems. We see in Art. 21, for instance, that a useful reasonable assumption is $T/\alpha Re \ll 1$, implying a small magnetic Reynolds number, hence *large* rather than small magnetic diffusivity.

G,25. Boundary Layer Flow near a Concave Wall.

In an attempt to search for possible origins of turbulence, Görtler [*58,13*] investigated the stability of boundary layer flows over a curved surface. The curvature of the wall was found to have only minor influence on the instability with respect to transverse wave-type disturbances, but to play an important role in the instability with respect to longitudinal vortex type disturbances similar to those of Art. 22. While a steady secondary flow itself must not be regarded as turbulence, as the Reynolds number is increased more and more of such vortices would appear and become amplified. The basic flow would be greatly distorted, and turbulence would then set in, triggered by some mechanism not completely understood.

Görtler's formulation is equivalent to the solution of Eq. 22-5 when the fluid extends throughout a thin layer over a surface of large radius of curvature. As a first approximation, the radius of curvature may be taken as constant and chosen as the length scale of the dimensionless equation (Eq. 22-5). The boundary conditions on the disturbances at the wall remain the same as before, namely $f_1 = f_2 = f_2' = 0$; at the outer edge of the boundary layer, all disturbances must also eventually damp out. We may thus draw upon the "Rayleigh criterion" of Art. 22 to conclude that instability of the type of disturbances postulated is possible only for boundary layers over a concave surface, since the monotonic nature of the usual velocity distribution cannot have decreasing circulation with increasing radial distance over a convex surface. The simple damping effect of viscosity would not modify this conclusion. Indeed, the boundary layer flow over a concave surface is qualitatively of the same type as the flow between rotating cylinders when only the inner one is rotating; the convex case likewise corresponds to the case when only the outer cylinder is rotating. Referring to Taylor's stability diagram, Fig. G,23a, the possibility of instability only for the concave wall is again confirmed.

Let us consider therefore only concave surfaces and make the simplification that the boundary layer thickness δ is much smaller than the radius of curvature R. In Eq. 22-5, we put

$$r = 1 - \frac{\delta}{R} y$$

where y is the distance from the wall measured with the thickness δ as the length scale. Dropping terms of order δ/R, we get

$$\left.\begin{aligned}(D^2 - \beta_1^2 - \lambda_1 Re_1)(D^2 - \beta_1^2)f_2 &= 2\beta_1 Re_1 \frac{\delta}{R} \bar{u} f_1 \\ (D^2 - \beta_1^2 - \lambda_1 Re_1)f_1 &= -Re_1 \frac{d\bar{u}}{dy} f_2\end{aligned}\right\} \qquad (25\text{-}1)$$

where

$$\beta_1 = \frac{\delta}{R}\beta, \qquad Re_1 = \frac{\delta}{R} Re, \qquad \lambda_1 = \frac{\delta}{R}\lambda$$

That is, the disturbances are now proportional to $e^{\lambda_1 t + i\beta_1 z}$, if δ replaces R in the length and time scales with which z and t are measured. The boundary conditions are

$$\left.\begin{aligned}&f_1 = f_2 = f_2' = 0 \quad \text{at} \quad y = 0 \\ &f_1, f_2, f_2' \rightarrow 0 \qquad \text{as} \quad y \rightarrow \infty\end{aligned}\right\} \qquad (25\text{-}2)$$

The problem was solved by Görtler [*13*] by finding the Green's functions $H(y, y_0)$ and $G(y, y_0)$ respectively, for the first and second of Eq. 25-1, i.e. the solutions of the homogeneous equations with a discontinuity of unity at $y = y_0$. The differential equations may then be cast into two simultaneous integral equations.

$$\left.\begin{aligned} f_1(y) &= -Re \int_0^\infty G(y, y_0) \frac{d\bar{u}}{dy_0} f_2(y_0) dy_0 \\ f_2(y) &= 2\beta_1^2 Re_1 \frac{\delta}{R} \int_0^\infty H(y, y_0) \bar{u}(y_0) f_1(y_0) dy_0 \end{aligned}\right\} \tag{25-3}$$

The solution was finally found by replacing the integral with a finite sum

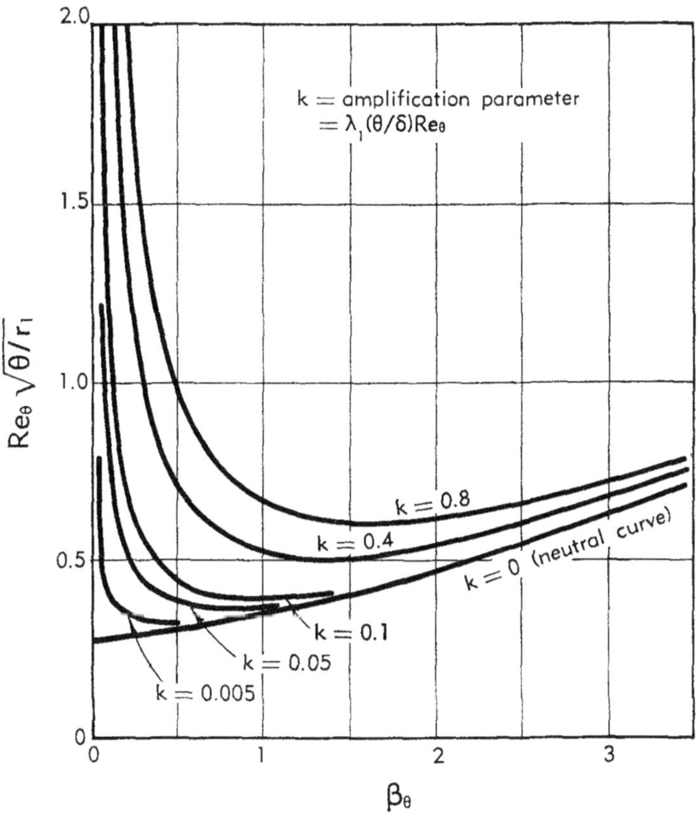

Fig. G,25a. Stability curves for flow over concave wall (Görtler model).

and reducing Eq. 25-3 to a system of algebraic equations. Görtler chose several velocity distributions, and found that the stability behavior is quite insensitive to the precise profile shape, in contrast to that for the transverse wave-type disturbances, if the momentum thickness θ is used as the length scale. An exact iterative solution of Görtler's eigenvalue problem was made by Hämmerlin [*116*], whose curves for neutral stability and constant rates of amplification are shown as Fig. G,25a. It comes out as perhaps surprising that the neutral Reynolds number increases monotonically with the wave number β_1. Actually, Hämmerlin pointed out in [*117*, footnote on p. 307] that the neutral curve near $\beta_1 = 0$ must not be taken seriously, because the neglected terms in arriving at Eq. 25-1 are no longer small compared with β_1 in this region. Witting [*109*] expanded everything in ascending powers of $1/R$, and obtained for the Blasius profile at neutral stability the two lowest order terms

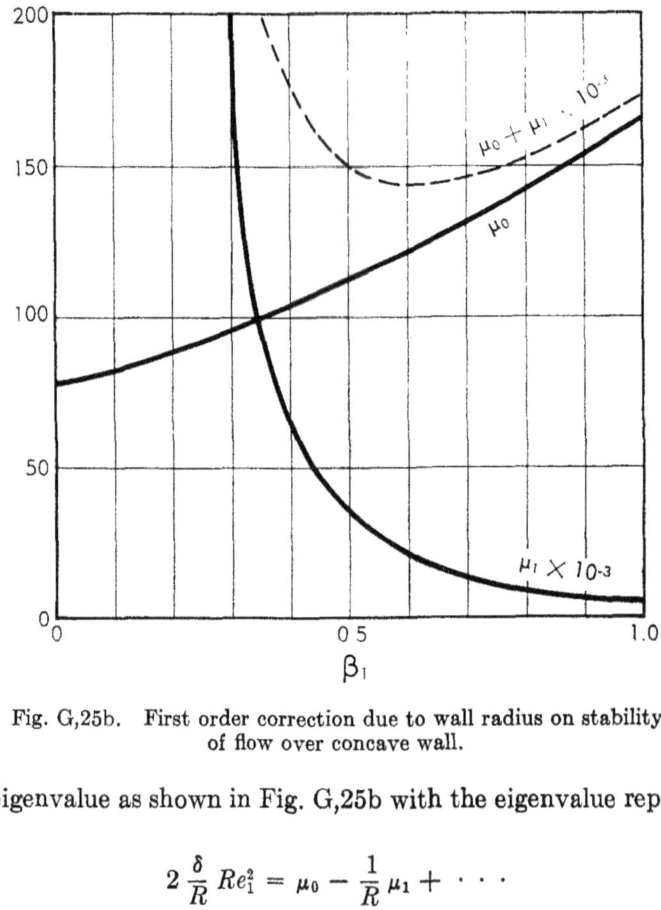

Fig. G,25b. First order correction due to wall radius on stability of flow over concave wall.

of the eigenvalue as shown in Fig. G,25b with the eigenvalue represented by

$$2\frac{\delta}{R}Re_1^2 = \mu_0 - \frac{1}{R}\mu_1 + \cdots$$

The curve for $\delta/R = 10^{-3}$ gives the minimum Reynolds number at $2(\delta/R)Re_1^2 = 143$ and $\beta_1 = 0.62$, or, in terms of momentum thickness ($Re_\theta = Re_1\theta/\delta$, $\beta_\theta = \beta_1\theta/\delta$, since the momentum thickness is now the length scale), $Re_\theta\sqrt{\theta/R}$ about 0.32 and a β_θ of 0.07. These results are in agreement with those of Smith [*118*], who included some curvature and boundary layer growth terms in the differential equation. Interesting to note here is that the least stable longitudinal vortex has a spacing of $2\pi/0.07 \cong 90$ times the momentum thickness, or roughly 10 times the boundary layer thickness for a Blasius profile. These vortices thus extend beyond the boundary layer. (See Fig. G,25c.) The minimum critical Reynolds number may be compared against that for the flow between rotating cylinders of a spacing equal to δ, when the inner cylinder alone is moving. With $\Omega_2 = 0$, $Ta = 1708$ corresponds to

$$Re\sqrt{\frac{\delta}{r_1}} = 41.3$$

or $Re_\theta\sqrt{\theta/r} = 2.81$, since for a linear profile $\theta = \frac{1}{6}\delta$. The absence of a solid surface at the outer edge of the boundary layer is seen to lower the critical Reynolds number.

There seems to be no experimental confirmation as yet of the details of these vortices comparable to Schubauer and Skramstad's investigation

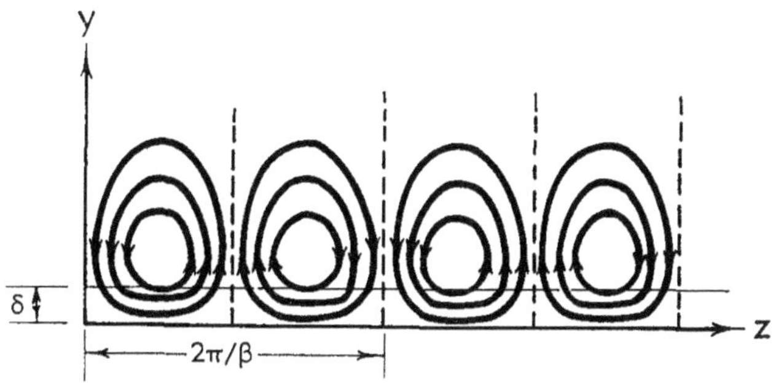

Fig. G,25c. Schematic streamline pattern of vortex disturbance for flow over concave wall.

of the transverse waves, or even the visual observation on the flow between rotating cylinders. Nevertheless, the parameter $Re_\theta\sqrt{\theta/R}$ was indeed found to be the one to use in correlating the transition of flow over a concave surface of constant curvature (Liepmann [*119*]).[17] Gregory and Walker [*120*] noticed longitudinal streaks over the concaved part of an airfoil surface in their transition study using the China clay technique, and interpreted these as indications of the Görtler vortices. The existence of disturbances of this type over concave surfaces prior to transition is generally accepted.

The same problem in a compressible fluid with heat or cooling at the wall was briefly examined by Di Prima and Dunn [*121*]. Extensions of the Rayleigh criterion to the compressible case have been made by Lessen [*122*] and Lees [*123*], both suggesting that the density gradient plays an important role.

[17] For flow over a surface with variable curvature, Smith [*118*] proposes the use of the maximum amplification ratio as computed for constant wavelength disturbances following their paths on Fig. G,25a to correlate transition. The value appears to be about 10, based upon his own chart and comparison with several experiments.

The problem of flow in a channel curved in the streamwise direction under longitudinal vortex-type disturbances was first treated by Dean [*124*]. Only a different velocity distribution needs to be substituted in Eq. 22-5. If the channel width is denoted by $2h$ ($2h = r_2 - r_1$, in our previous notation), the minimum critical Reynolds number based upon mean flow velocity and half width h, is found to be

$$Re\left(\frac{h}{r_1}\right)^{\frac{1}{2}} = 12.72$$

For a very small curvature, say $h/r_1 = 10^{-4}$, the minimum critical Reynolds number is about 1270, which may be compared against the minimum critical Reynolds number of 5300 for stability with respect to transverse wave-type disturbances in a straight channel. Thus even a slight curvature of the wall could be instrumental in setting up longitudinal vortices as secondary flows long before the transverse vortex-type oscillations are excited. Yih and Sangster [*125*] recalculated Dean's problem by Chandrasekhar's method [*93*] (Art. 23) and came up with different numerical results. A recent careful study of Hämmerlin [*126*] confirms the accuracy of Dean's earlier computation.

G,26. Flow near a Stagnation Point. Based upon the Rayleigh criterion of Art. 22, especially if one interprets it physically as the equilibrium condition between the centrifugal force and the pressure gradient, it is expected that instability of the assumed vortices would occur wherever the local streamlines are concaved with respect to the direction of increasing velocity. Such a situation exists near the stagnation point of a two-dimensional body. Experimentally, local velocity fluctuations were indeed found by Piercy and Richardson [*127*], who made hot wire measurements on airfoil-shaped and circular cylinders. The heat transfer experiment of Schuh [*128*] indicated higher rates of heat transfer near the stagnation point than laminar values, again suggesting local turbulence.

A theoretical attack of the instability near a stagnation point was made by Görtler [*117*] and Hämmerlin [*129*, pp. 315–327]. An important feature is that the velocity now increases downstream from the stagnation point, although the boundary layer thickness is constant. A new formulation is therefore necessary. In terms of Cartesian coordinates with origin at the stagnation point, Fig. G,26, the basic flow is well known, given by

$$\left.\begin{aligned} \bar{u}(x, y) &= axF'(\eta) \\ \bar{v}(x, y) &= -\sqrt{\nu a}\,F(\eta) \\ \bar{w}(x, y) &= 0 \end{aligned}\right\} \tag{26-1}$$

where a^{-1} is positive and has the dimension of time, η is dimensionless and

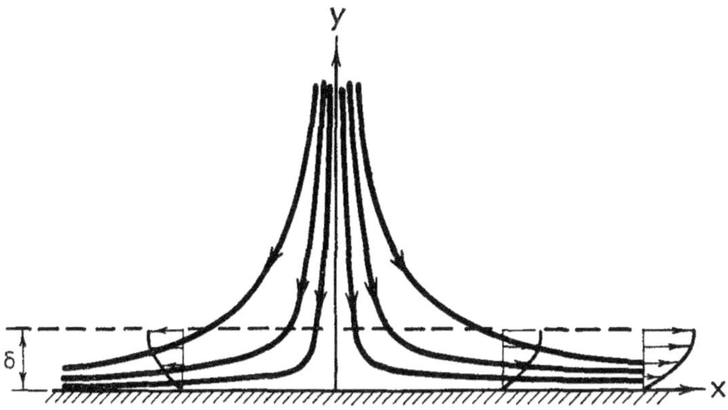

Fig. G,26. Coordinates for flow near stagnation point.

defined by $\eta = \sqrt{a/\nu}y$, and $F(\eta)$ satisfies

$$\left.\begin{aligned} F''' + FF'' - F'^2 + 1 = 0 \\ F(0) = F'(0) = 0 \qquad F'(\infty) = 1 \end{aligned}\right\} \tag{26-2}$$

Since here $\bar{u}$ varies with x, a disturbance is *meaningfully "small"* only when compared with the basic flow at the same station. This suggests, and indeed the hypothesis turns out to be at least workable, that one should try disturbances represented by

$$\left.\begin{aligned} u' &= axf_1(\eta)e^{i\beta z+\lambda t} \\ v' &= -\sqrt{\nu a}\, f_2(\eta)e^{i\beta z+\lambda t} \\ w' &= \beta\sqrt{\nu a}\, f_3(\eta)e^{i\beta z+\lambda t} \\ p' &= \rho\nu a f_4(\eta)e^{i\beta z+\lambda t} \end{aligned}\right\} \tag{26-3}$$

where z and t are made dimensionless by measuring in terms of length $\sqrt{a/\nu}$ and time a^{-1}, respectively. Substituting Eq. 26-3 into the Navier-Stokes equations, we obtain

$$\left.\begin{aligned} f_1'' + Ff_1' - (\lambda + \beta^2 + 2F')f_1 &= -F''f_2 \\ f_2'' + Ff_2' - (\lambda + \beta^2 - F')f_2 &= -f_4' \\ f_3'' + Ff_3' - (\lambda + \beta^2)f_3 &= -f_4 \\ f_1 - f_2' + \beta^2 f_3 &= 0 \end{aligned}\right\} \tag{26-4}$$

After elimination of f_3 and f_4, Eq. 26-4 becomes

$$\left.\begin{aligned} Lf_1 &= 3F'f_1 - F''f_2 \\ L(f_2'' - \beta^2 f_2) &= Lf_1' \\ \text{with} \qquad L &\equiv \frac{d}{d\eta}\left(\frac{d}{d\eta} + F\right) - (\lambda + \beta^2) \end{aligned}\right\} \tag{26-5}$$

The boundary conditions are found to remain

$$f_1 = f_2 = f_2' = 0 \quad \text{at} \quad \eta = 0 \text{ and as } \eta \to \infty$$

According to Hämmerlin, at least a *continuous spectrum* of positive and real eigenvalues β^2 exists, provided that

$$0 < \lambda + \beta^2 < 1$$

Thus the possibility of disturbances of the assumed type is demonstrated, including some with amplification. It may be noted that for the assumed disturbances, no critical Reynolds number appears. The reason may be traced to the fact that no physical length enters into the idealized stagnation point flow. If a wider region is considered such that the radius of curvature of the stagnation point neighborhood is brought into the picture, it is conceivable that a critical Reynolds number may indeed emerge.

CHAPTER 5. BEHAVIOR SUBSEQUENT TO THE ONSET OF INFINITESIMAL DISTURBANCES

G,27. General Remarks. While the calculation of the natural curve for the stability of linear parallel flows with respect to infinitesimal disturbances has now reached a fairly satisfactory status, and while the existence of such disturbances has been confirmed by a few crucial experiments, there remains a wide gap between the onset of these disturbances and the fully developed turbulence. From a theoretical viewpoint, the logical step is to push the theory further toward a prediction or explanation of the events immediately subsequent to the first occurrence of the disturbances. The following mechanism, conceived by Landau [*130*], seems to be quite plausible: Given a basic flow, it may become unstable with respect to certain disturbances as the Reynolds number is increased. The nonlinear effects, however, may keep the disturbance to finite amplitude and thus a distorted laminar flow prevails at slightly higher Reynolds numbers. This distorted flow naturally may fluctuate with time, but should repeat itself periodically. As the Reynolds number continues to increase, it may become unstable with respect to certain other disturbances, in the sense that the flow begins to change its character once more. Again, because of nonlinear effects, equilibrium may be reached between the new disturbances and the first distorted laminar flow. As a result, a second distorted equilibrium laminar flow is created. Additional cycles of instability and equilibrium may follow with higher and higher Reynolds numbers, until finally the flow is so disturbed as to be considered turbulent.

On the experimental side, there exists a certain degree of support.

Taylor's early observation [*111*], and Schultz-Grunow and Hein's recent report [*108*] on the behavior of the flow between rotating cylinders are worthy of notice. The vortices occurred at Reynolds numbers as predicted by theory (Art. 23). At higher Reynolds numbers these vortices merely increased their strengths without much change of the spacing. The resultant flow was truly more and more distorted, but remained laminar for a large range of Reynolds number beyond the critical value.[18] The evidence appears to indicate that there was indeed equilibrium between the basic flow and the amplified disturbances after reaching a certain amplitude, at each of the Reynolds numbers exceeding the minimum critical value. For the boundary layer flow over a flat plate, the methodical investigation by Schubauer and Klebanoff [*131*] shed much light. The details of the growth of the two-dimensional transverse wave-type disturbances after surpassing the minimum critical Reynolds number were carefully examined. At some distances downstream, the presence of secondary disturbances, of much higher frequency than that of the two-dimensional disturbances introduced into the flow, was noticed. In fact, these disturbances were found to be strongest at a distance from the wall where the streamline curvature of the distorted flow *should be* concaved. A periodicity in the transverse direction was also observed. Based on such indications one may be led to the conjecture that perhaps the secondary disturbance in flows of this type could be related to the Görtler-type vortices (Art. 25). This last view has been voiced by Dryden [*40*].

Recent theoretical attempts in going further than the first occurrence of instability of infinitesimal disturbances may be regarded as motivated by viewpoints similar to that of Landau. The behavior of finite disturbances was first treated by Meksyn and Stuart [*132*] in the case of the plane Poiseuille flow. The concept of energy balance, Eq. 5-2, was exploited by Stuart [*14*] for the determination of the equilibrium amplitude of the disturbances when the Reynolds number exceeds the critical value. His calculation of the mean torque required to rotate the inner cylinder of Taylor's experiment at the higher Reynolds numbers agreed very well with the experimental data. Following Görtler, Witting [*109*] examined the possibility of a secondary instability in the form of "Görtler vortices" when the basic flow is taken to be the sum of the steady boundary layer plus a transverse (Tollmien-Schlichting) wave disturbance of given amplitude. The result suggests that such instability will be excited when the latter has an amplitude exceeding $O(10^{-4})$ times the boundary layer displacement thickness. The readiness of the Görtler vortices to be

[18] See the photographs of Schultz-Grunow and Hein [*108*]. In their experiment, the combination of the toroidal vortices and the original basic flow continued to exist at least until the Reynolds number was about 9 times the critical value. Fully developed turbulence was visible when the Reynolds number was about 40 times the critical value.

excited is indeed rather startling. In the recent experimental findings of Klebanoff, Tidstrom and Sargent [*140*] on the incompressible flat plate boundary layer, while secondary longitudinal vortices are indeed observed, they seem to be most pronounced where the distorted streamlines are *convex*, contrary to the prediction of Witting's theory to be described in Art. 30. Explanation of the anomaly is made possible by a different mechanism proposed by Benney and Lin [*141*]. These authors studied the interaction of an oblique transverse wave disturbance and the primary two-dimensional disturbance, for an artificial convenient velocity profile. If the oblique wave has a periodic spanwise variation in amplitude, the interaction is shown to be capable of producing a different type of secondary longitudinal vortex, which can be reconciled with the experimental observations. For an excellent review of this and other advances in nonlinear stability theory, see Stuart [*142*].

G,28. Approximate Formulation of the Stability Problem with Finite Disturbance. For simplicity, suppose we consider incompressible two-dimensional motion only. The dimensionless Navier-Stokes equations in Cartesian coordinates read, after choosing the reference velocity, etc., as on page 723,

$$\left.\begin{aligned} \frac{\partial u}{\partial t} + \frac{\partial}{\partial x} u^2 + \frac{\partial}{\partial y} uv &= -\frac{\partial p}{\partial x} + \frac{1}{Re} \nabla^2 u \\ \frac{\partial v}{\partial t} + \frac{\partial}{\partial x} uv + \frac{\partial}{\partial y} v^2 &= -\frac{\partial p}{\partial y} + \frac{1}{Re} \nabla^2 v \\ \frac{\partial u}{\partial x} + \frac{\partial v}{\partial y} &= 0 \end{aligned}\right\} \tag{28-1}$$

Let the basic flow again be a parallel flow, i.e.

$$\bar{u} = \bar{u}(y, t), \qquad \bar{v} = 0$$

A disturbance, in the form of a disturbance stream function ψ' will be superimposed on the basic flow. When the boundary conditions, too, are independent of x, it is reasonable to assume that

$$\psi'(x, y, t) = \sum_{n=1}^{\infty} \psi^{(n)}(y, t) e^{in\alpha x} \tag{28-2}$$

with

$$u' = Rl[\psi'_y], \qquad v' = -Rl[\psi'_x]$$

as the disturbance velocities. Substituting Eq. 28-2 into Eq. 28-1, one obtains a sequence of simultaneous equations for each of the harmonics

$e^{in\alpha x}$: Independent of x there is

$$\left.\begin{aligned} \frac{\partial \bar{u}}{\partial t} + \frac{\partial}{\partial y}\overline{u'v'} &= -\frac{\partial \bar{p}}{\partial x} + \frac{1}{Re}\nabla^2\bar{u} \\ \frac{\partial}{\partial y}\overline{v'^2} &= -\frac{\partial \bar{p}}{\partial y} \end{aligned}\right\} \tag{28-3}$$

In Eq. 28-3, $\overline{u'v'}$ and $\overline{v'^2}$ are used to designate the zeroeth harmonic terms following direct substitution of Eq. 28-2 into $u'v'$ and v'^2, and may be interpreted as the Reynolds stress terms if the averaging process is taken over a wavelength in the x direction. The term $\bar{p}$ in Eq. 28-3 likewise stands for the pressure averaged over a wavelength in the x direction. Since the Reynolds stress terms $\overline{u'v'}$ and $\overline{v'^2}$ are proportional to the square of the disturbance amplitude, they are of no consequence in a linear theory. On the other hand, the basic flow clearly is modified when the disturbances reach finite amplitude through the transfer of momentum and energy in the presence of the Reynolds stresses.

If we eliminate $\bar{p}$ from these equations, the result is

$$\frac{\partial}{\partial t}\frac{\partial \bar{u}}{\partial y} + \frac{\partial^2}{\partial y^2}\overline{u'v'} = \frac{1}{Re}\frac{\partial^3 \bar{u}}{\partial y^3} \tag{28-4}$$

which, when equilibrium is reached between the basic flow and the disturbances, may be integrated to yield

$$\bar{u} = A_1 + A_2 y + A_3 y^2 + Re\int_0^y \overline{u'v'}dy \tag{28-5}$$

The constants A_1, A_2, and A_3 are to be chosen to satisfy the boundary conditions. The modification due to the Reynolds stress is explicitly given by the last term.

By subtracting Eq. 28-3 from Eq. 28-1, the disturbances satisfy the following:

$$\left.\begin{aligned} \frac{\partial u'}{\partial t} + \bar{u}\frac{\partial u'}{\partial x} + v'\frac{\partial \bar{u}}{\partial y} + G_1 &= -\frac{\partial p'}{\partial x} + \frac{1}{Re}\nabla^2 u' \\ \frac{\partial v'}{\partial t} + \bar{u}\frac{\partial v'}{\partial x} + G_2 &= -\frac{\partial p'}{\partial y} + \frac{1}{Re}\nabla^2 v' \\ \frac{\partial u'}{\partial x} + \frac{\partial v'}{\partial y} &= 0 \end{aligned}\right\} \tag{28-6}$$

where

$$\left.\begin{aligned} G_1 &= \frac{\partial}{\partial x}u'^2 + \frac{\partial}{\partial y}u'v' - \frac{\partial}{\partial x}\overline{u'^2} - \frac{\partial}{\partial y}\overline{u'v'} \\ G_2 &= \frac{\partial}{\partial x}u'v' + \frac{\partial}{\partial y}v'^2 - \frac{\partial}{\partial x}\overline{u'v'} - \frac{\partial}{\partial y}\overline{v'^2} \end{aligned}\right\} \tag{28-7}$$

These equations are exact in the sense that no terms have been dropped.

The terms G_1 and G_2 are neglected in the linear theory. The pressure disturbance p' is also to be represented by a series similar to Eq. 28-2.

Restricting ourselves now to disturbances that have the same wave velocity for all indices n in Eq. 28-2, we write

$$\psi'(x, y, t) = \sum_{n=1}^{\infty} \Phi_n(y) e^{in\alpha(x-ct)} \tag{28-8}$$

whence

$$\left.\begin{aligned} \overline{u'v'} &= \frac{i\alpha}{4} \sum_{n=1}^{\infty} n[\Phi_n' \tilde{\Phi}_n - \Phi_n \tilde{\Phi}_n'] \\ \overline{v'^2} &= -\frac{\alpha^2}{2} \sum_{n=1}^{\infty} n^2 \Phi_n \tilde{\Phi}_n \end{aligned}\right\} \tag{28-9}$$

and the harmonic terms may be explicitly calculated. After elimination of p' in Eq. 28-6 and substitution of Eq. 28-8, the sequence of differential equations for Φ_n obtained by balancing each of the harmonics is

$$\begin{aligned} (\bar{u} - c)(\Phi_1'' - \alpha^2\Phi_1) - \bar{u}''\Phi_1 + \frac{i}{\alpha Re}(\Phi_1'''' - 2\alpha^2\Phi_1'' + \alpha^4\Phi_1) \\ = 2\Phi_2'[\tilde{\Phi}_2'' - \alpha^2\tilde{\Phi}_1] + 4\Phi_2(\tilde{\Phi}_1''' - \alpha^2\tilde{\Phi}_1') - 4\tilde{\Phi}_1'(\Phi_2'' - 4\alpha^2\Phi_2) \\ - 2\tilde{\Phi}_1(\Phi_2''' - 4\alpha^2\Phi_2') + \cdots \end{aligned} \tag{28-10}$$

$$\begin{aligned} (\bar{u} - c)(\Phi_2'' - \alpha^2\Phi_2) - \bar{u}''\Phi_2 + \frac{i}{\alpha Re}(\Phi_2'''' - 8\alpha^2\Phi_2'' + 16\alpha^4\Phi_2) \\ = -(\Phi_1'\Phi_1'' - \Phi_1\Phi_1''') + \cdots \end{aligned} \tag{28-11}$$

etc. The dotted line indicates terms involving higher harmonic Φ_n's. Meksyn and Stuart's approximate treatment is to omit all terms other than Φ_1. Thus only Eq. 28-4 and 28-10 are retained, the latter without its right-hand side. Under equilibrium they are now

$$\left.\begin{aligned} \frac{i\alpha}{4}\frac{d^2}{dy^2}(\Phi_1'\tilde{\Phi}_1 - \Phi_1\tilde{\Phi}_1') &= \frac{1}{Re}\frac{d^3\bar{u}}{dy^3} \\ (\bar{u} - c)(\Phi_1'' - \alpha^2\Phi_1) - \bar{u}''\Phi_1 + \frac{i}{\alpha Re}(\Phi_1'''' - 2\alpha^2\Phi_1'' + \alpha^4\Phi_1) &= 0 \end{aligned}\right\} \tag{28-12}$$

The basic flow $\bar{u}$ and the disturbance Φ_1 must be solved together. If the disturbance has magnitude a, defined by

$$\psi' = a[\Phi_1 e^{i\alpha(x-ct)} + \tilde{\Phi}_1 e^{-i\alpha(x-ct)}] \tag{28-13}$$

then the correction of the Reynolds stress on the basic flow is of the order of a^2Re. One has therefore a one-parameter family of neutral curves, shown schematically in Fig. G,28a.

Not mentioned by these authors is the fact that their procedure is rather identical to that of the "harmonic linearization" of Kryloff and Bogoliuboff [*133*]. In the harmonic linearization of a nonlinear oscillation problem, one assumes that the steady oscillation is dominated by the fundamental harmonic, and considers only the balancing of the fundamental harmonic terms. In certain problems, e.g. when a spring is stiffer for positive displacements and softer for negative displacements, the zeroeth

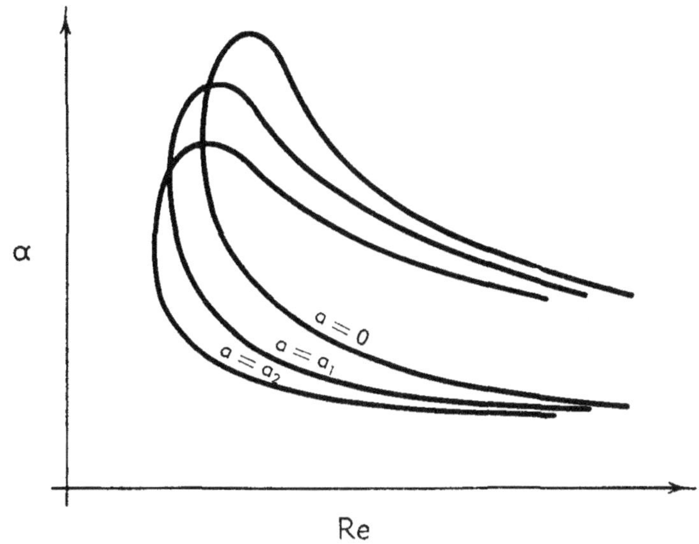

Fig. G,28a. Sketch showing schematic neutral curves at various amplitudes of disturbances.

harmonic (static equilibrium) must also be balanced. A rigorous justification of the procedure, however, is often difficult.

An application to the plane Poiseuille flow was made by Meksyn and Stuart [*132*]. For small a^2Re the basic flow will be only slightly distorted. But the critical Reynolds number is known to be very sensitive to $\bar{u}_c''/\bar{u}_c'$. Essentially they used the same inviscid solutions as those of the parabolic Poiseuille flow but allowed $\bar{u}_c''/\bar{u}_c'$ to vary as a parameter for the distortion. Instead of Eq. 11-12, the corresponding equation is

$$F(z) = E\left(\alpha,\ Re,\ c;\ \frac{\bar{u}_c''}{\bar{u}_c'}\right) \tag{28-14}$$

so that by assigning values of $\bar{u}_c''/\bar{u}_c'$, neutral curves are obtained. For every solution Φ_1, the Reynolds stress term in the modified profile

(Eq. 28-5) may then be evaluated. The distortion is now proportional to a^2Re. In fact one obtains

$$\bar{u} = 1 - \left(1 + Re \int_0^1 \overline{u'v'}dy\right) y^2 + Re \int_0^y \overline{u'v'}dy \tag{28-15}$$

$$\frac{\bar{u}_c''}{\bar{u}_c'} = \frac{2\left(1 + Re \int_0^1 \overline{u'v'}dy\right) - Re\left(\frac{\partial}{\partial y}\overline{u'v'}\right)_c}{2y_c\left(1 + Re \int_0^1 \overline{u'v'}dy\right) - Re(\overline{u'v'})_c} \tag{28-16}$$

The right-hand side of Eq. 28-16 is a function of the assumed $\bar{u}_c''/\bar{u}_c'$ and

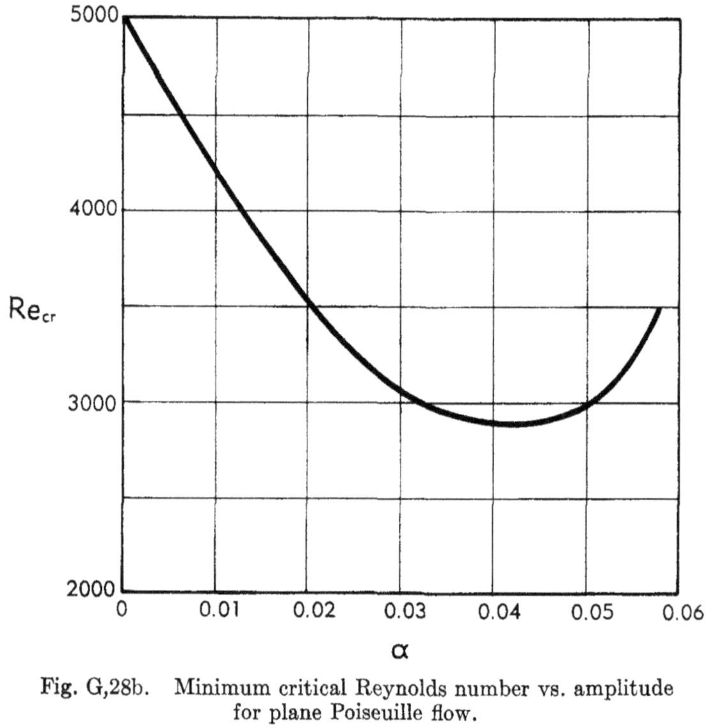

Fig. G,28b. Minimum critical Reynolds number vs. amplitude for plane Poiseuille flow.

the parameter a^2Re, which therefore determines the correct a^2Re for the assumed $\bar{u}_c''/\bar{u}_c'$. The amplitude "a" may finally be computed from a^2Re since the Reynolds number itself along the neutral has already been determined.

Their result [*132*] in the form of the minimum critical Reynolds number vs. amplitude is shown in Fig. G,28b. The minimum critical Reynolds number decreases rather rapidly with amplitude until an absolute minimum of about 2900 is reached for $a \cong 0.042$. (Using Eq. 11-13, the first of Eq. 11-17 and Eq. 11-18 and neglecting the difference between

$\bar{u}_I'$ and $\bar{u}_c'$, one gets $Re_{\min} \sim (\bar{u}_c''/\bar{u}_c')^{3.5}$. With Meksyn and Stuart's solution for a^2Re vs. $\bar{u}_c''/\bar{u}_c'$, about the same Reynolds number at $a = 0.042$ may be obtained.) However, their calculation procedure, based upon a small deviation of the $\bar{u}_c''/\bar{u}_c'$ value from that for the Poiseuille flow, is of but poor accuracy beyond such amplitudes. More fundamentally, the justification of the "harmonic linearization" under such large values of a^2Re may require examination. While it is conceivable that the minimum critical Reynolds number is not likely to decrease indefinitely with increasing amplitude, the number 2900 must be taken with reservation because of the approximate nature of the solution.

We remark finally that, in the same spirit, one might discuss the first order nonlinear effects of three-dimensional disturbances on a two-dimensional basic flow. With first harmonics only, however, Squire's theorem would still hold and Eq. 28-12 remain valid. Three-dimensional disturbances thus are generated probably by a different mechanism, such as described in Art. 30.

G,29. The Energy Method with Finite Disturbances. The previous equations for the energy balance, Eq. 5-1 and 5-2, have been derived without restriction on the type of disturbance or the amplitude. It has not proved fruitful as a means to obtain approximate critical Reynolds numbers by assumed trial disturbance functions, mainly because one has little guidance toward how to assume a reasonably accurate trial function. But suppose the linear stability problem has been solved by some other method. Then the linear disturbance mode ought to be a fair approximation of the disturbance mode for small but finite amplitudes. With this approximation, Stuart [*14*] was able to discuss the growth of an amplifying disturbance as the Reynolds number exceeds the critical value (from linear theory), until final equilibrium between a finite amplitude disturbance and the distorted mean flow is again reached, all based upon the energy balance equation.

If we again assume the finite amplitude disturbance to be of the type represented by Eq. 28-8, Eq. 5-1 becomes now, in dimensionless form,

$$\frac{\partial}{\partial t}\int \frac{1}{2}(\overline{u'^2} + \overline{v'^2})dy = -\int \overline{u'v'}\,\frac{\partial \bar{u}}{\partial y}\,dy - \frac{1}{Re}\int \overline{\zeta'^2}dy \qquad (29\text{-}1)$$

where the barred quantities have been averaged over one wavelength in the x direction. As in the previous section, let us retain only the fundamental harmonic as the dominant term in the disturbance, Eq. 28-13,

$$\left.\begin{aligned} u' &\cong 2a(t)\Phi_1'(y)e^{i\alpha(x-ct)} \\ v' &\cong -2i\alpha a(t)\Phi_1(y)e^{i\alpha(x-ct)} \end{aligned}\right\} \qquad (20\text{-}2)$$

where the growth of the amplitude "a" is taken out explicitly and c

is now real only. As a consequence, the distorted basic flow may be represented as

$$\bar{u} = \bar{u}_0(y) + a^2 Re\, \bar{u}_1(y, t) \tag{29-3}$$

Substituting Eq. 29-3 into Eq. 29-1, we get

$$\gamma_1 \frac{da^2}{dt} = \gamma_2 \alpha a^2 - \gamma_3 \alpha^2 Re\, a^4 - \gamma_4 \frac{a^2}{Re} \tag{29-4}$$

with

$$\left.\begin{aligned} \gamma_1 &= \int [|\Phi_1'|^2 + \alpha^2|\Phi|^2]dy \\ \gamma_2 &= 2\int \frac{d\bar{u}_0}{dy}(\Phi_{1_i}'\Phi_{1r} - \Phi_{1_r}'\Phi_{1i})dy \\ \gamma_3 &= 2\int \frac{1}{\alpha}\frac{\partial \bar{u}_1}{\partial y}(\Phi_{1_i}'\Phi_{1r} - \Phi_{1_r}'\Phi_{1i})dy \\ \gamma_4 &= 2\int [|\Phi_1''|^2 + 2\alpha^2|\Phi_1'|^2 + \alpha^4|\Phi|^2]dy \end{aligned}\right\} \tag{29-5}$$

Now visualize the following situation: Suppose a laminar flow is maintained carefully until the critical Reynolds number from the linear theory is exceeded. At a high Reynolds number a small disturbance takes place. How does it amplify and what final amplitude will be reached? These questions can be answered from Eq. 29-4. As a further approximation, the distortion function $\bar{u}_1$ may be taken as of the same form as that for the final equilibrium state, since its precise form is of little importance when a^2Re at the beginning is small. Thus $\bar{u}_1$ becomes independent of t; and γ_3 becomes a constant as the rest of the γ's. Indeed, Eq. 29-4 becomes identical to a formula first given by Landau [*130*],

$$\frac{dy}{d\tau} = \beta_1 y - \beta_2 y - \beta_3 y^2 \tag{29-6}$$

of which the general solution is

$$y = \frac{Ce^{(\beta_1 - \beta_2)\tau}}{1 + \left(\dfrac{\beta_3}{\beta_1 - \beta_2}\right) Ce^{(\beta_1 - \beta_2)\tau}} \tag{29-7}$$

As $\tau \to -\infty$, one gets an exponential amplification in accordance with linear theory. As $\tau \to +\infty$, a finite limit will be reached regardless of the initial amplitude C. Thus the solution seems to give a qualitatively quite plausible description of the nonlinear effect.

Equilibrium conditions may also be directly deduced from Eq. 29-4. On setting $da^2/dt = 0$, the condition is

$$\gamma_2 \alpha - \frac{\gamma_4}{Re} - \gamma_3 \alpha^2 a^2 Re = 0 \tag{29-8}$$

Corresponding to the linear theory, the a^2Re term may be dropped to give the critical Reynolds number

$$Re_{cr} = \frac{\gamma_4}{\gamma_2\alpha} \tag{29-9}$$

or there may be equilibrium at a finite amplitude a_e given by

$$a_e^2 = \frac{\gamma_2}{\alpha\gamma_3}\frac{Re - Re_{cr}}{Re^2} \tag{29-10}$$

In addition, the amplification rate in the linear theory should correspond to

$$2\alpha c_i = \frac{\gamma_2\alpha}{\gamma_1}\left(\frac{Re - Re_{cr}}{Re}\right) \tag{29-11}$$

To check the assumptions made in arriving at Eq. 29-4, namely, that a single $\Phi_1(y)$ and an equilibrium $\bar{u}_1(y)$ may be used throughout the growth period of the disturbances, Stuart took Φ_1 as Thomas' tabulated ϕ for $\alpha = 1$, $Re = 10^4$, $c = 0.2375 + 0.0037i$ in the case of the plane Poiseuille flow according to the linear theory [*25*]. The values from integrating Eq. 28-5 are

$$\gamma_1 = 2.05146, \ \gamma_2 = 0.040192, \ \gamma_3 = 0.002308, \ \gamma_4 = 247.62$$

A rather crucial check now is to find the critical Reynolds number by Eq. 29-9 with these numbers which are derived from a different mode. It turns out to be $\alpha Re_{cr} \cong 6150$, which is quite close to Thomas' critical value of $Re_{cr} = 5780$ at $\alpha = 1.02$. The calculated c_i at $Re = 10^4$ and $\alpha = 1$, according to Eq. 29-11, is 0.00376 as compared with the true value of 0.0037. The equilibrium amplitude at $Re = 10^4$ and $\alpha = 1$ is calculated to be 0.0256. A comparison of the velocity distributions for laminar Poiseuille flow and the distorted flow is shown in Fig. G,29a, under the condition that the same pressure drop along the channel is maintained.

An application to the Couette flow between rotating cylinders leads to an excellent agreement between calculated and experimental values of the torque required at rotational speeds higher than that corresponding to the critical Taylor number 1708 (Art. 23). The equilibrium amplitude is calculated to be

$$a_e^2 = \frac{21.7 \times 10^4}{Re^2}\left(1 - \frac{1708}{Ta}\right)$$

while the comparison of the torque is shown as Fig. G,29b. In Fig. G,29b, G denotes the torque in gm cm^2 sec^{-2} and N the rotational speed in rev sec^{-1}. The critical Taylor number corresponds to log $(N/\nu) = 1.955$; at ten times the critical Taylor number, log $(N/\nu) = 2.5$, the discrepancy in torque is only 7 per cent.

The excellent agreement between theory and experiment in this particular example perhaps should not be generalized without caution. Indeed, if one notes the different nature of the stability problems of Chap. 2 and 4, it seems highly likely that the simplified "harmonic linearization" (Art. 28) would have a much higher probability of success in stability problems with respect to longitudinal vortices than in those dealing with transverse waves. For the latter, the presence of a critical layer within the fluid suggests high sensitivity toward local distortions.

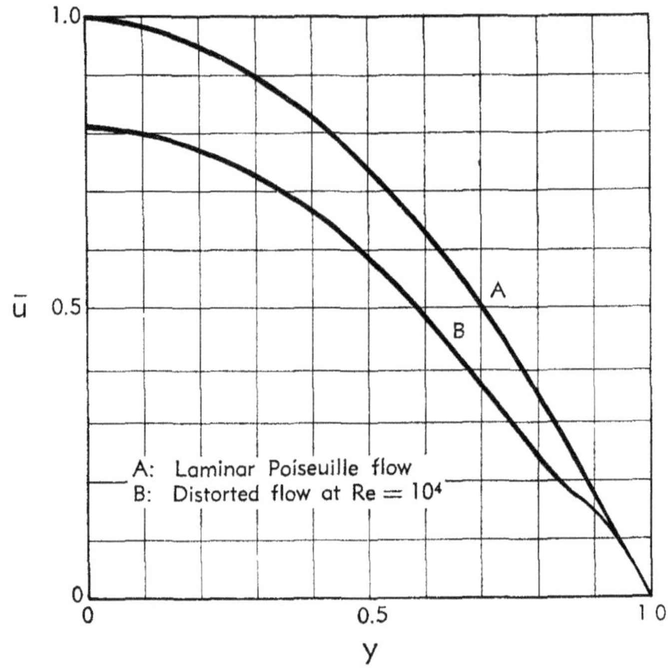

Fig. G,29a. Distorted equilibrium flow in channel at higher than critical Reynolds number but under identical pressure drop.

Those distortions due to the higher harmonics are already beyond the scope of the present approximate theory. Even within the framework of the present theory, the mode function Φ_1 probably would change sufficiently, because of the sensitive critical layer, in different regions of the $\alpha_1 Re$ plane, that treating the γ's of Eq. 29-5 as constants would have only limited validity. For instance, according to Eq. 29-9, for every α one should get two Reynolds numbers for neutral equilibrium, corresponding to the two branches of the usual neutral curve; and the prediction is that both must be higher than the linear critical value given by Eq. 29-9. But very likely only the lower of the two has significance *if* the mode chosen is one near the lower branch from the linear theory, as in Stuart's example. In the application to the Poiseuille flow discussed above, one must use a mode corresponding to the upper branch if the effect of finite amplitude on the critical Reynolds number along the upper branch is desired. When one refers back to Art. 28, the drop of the *minimum* critical Reynolds number with increasing amplitude contradicts Eq. 29-8, unless one accepts the necessity of readjusting the γ's by modifying the mode Φ_1.[19]

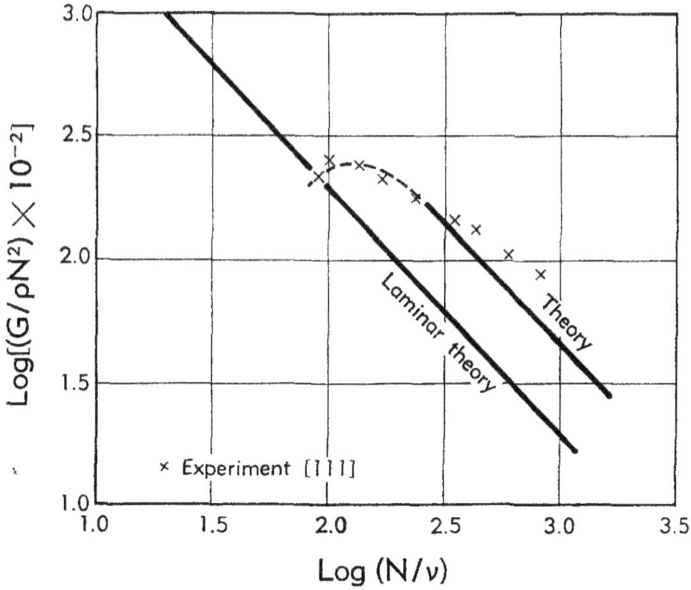

Fig. G,29b. Comparison of theoretical and experimental torque for flow between rotating cylinders.

G,30. Secondary Instability of Boundary Layer Flows. As a step toward the explanation of the origin of three-dimensional disturbances on a two-dimensional basic flow, such as in a boundary layer, the stability with respect to longitudinal vortex-type disturbances when the boundary layer already contains transverse wave-type disturbances was studied by Witting [*109*]. With neutral oscillations of the transverse wave disturbances, the originally straight and parallel streamlines now become curved (see Fig. G,30a). Over the portion which is concaved toward the direction of increasing velocity of the basic flow, a local situation similar to that of Chap. 4 prevails and instability of the same type is dictated at least by the "Rayleigh criterion" of Art. 22. Witting's actual

[19] Eq. 29-8 leads to a solution of αRe, thus a higher α would reduce the Reynolds number. However, Meksyn and Stuart's calculation showed only a 7 per cent increase of α at the lowest Re_{cr} of 2900, not enough to account for the 40 per cent drop of the Reynolds number.

calculation was a perturbation of the results of the linear theory for transverse waves, and showed that a very small amplitude distortion of the streamlines due to transverse wave disturbances, of the order of 10^{-4} times the displacement thickness, would be sufficient to excite the longitudinal vortices. According to the estimates of the previous article, such a small equilibrium amplitude would be obtained when the Reynolds number barely exceeds the critical value. The perturbation scheme is thus justified *a posteriori*.

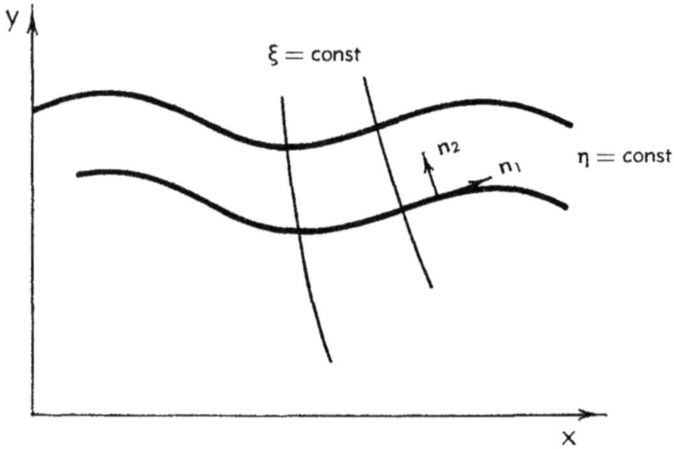

Fig. G,30a. Sketch showing coordinates moving with the wave disturbances.

Consider only the two-dimensional, incompressible problem. Let $\psi(x, y, t)$ be the stream function for the *combined* basic flow and the transverse wave disturbance of small amplitude ϵ,

$$\psi(x, y, t) = \psi_0(y) + \epsilon\psi_T(x, y, t) \tag{30-1}$$

ψ_0 and ψ_T representing the stream functions for the basic flow and the transverse wave disturbances, respectively. Only terms of order ϵ will now be kept in the following derivation. Regarded as the beginning of a Taylor series expansion, Eq. 30-1 may be rewritten as

$$\psi(x, y, t) = \psi_0\left(y + \epsilon\frac{\psi_T(x, y, t)}{\bar{u}}\right) \tag{30-2}$$

where $\bar{u}$ is the basic flow velocity distribution. Thus the streamlines of the combined flow are for constant values of η,

$$\eta = y + \epsilon\bar{u}^{-1}\psi_T(x, y, t)$$

or, since $\psi_T = \phi e^{i\alpha(x-ct)}$, one gets

$$\left.\begin{aligned} \eta &= y + \epsilon Q(y)\cos[\alpha(x - ct) + j(y)] \\ Q &= \frac{\sqrt{|\phi|^2}}{\bar{u}}, \qquad j = \tan^{-1}\left(\frac{\phi_i}{\phi_r}\right) \end{aligned}\right\} \tag{30-3}$$

To simplify the statement of the problem, it seems logical that η should be chosen as one of the new set of coordinates; the other one is chosen as the lines of constant phase angle ξ,

$$\xi = x - ct + \frac{1}{\alpha} j(y) \tag{30-4}$$

The line $\xi = 0$ corresponds to the locus of the "valleys" of the curved streamlines. Clearly, the streamline at $\xi = 0$ has the most curvature,

$$y''_{\xi=0} = \alpha^2 Q(y) \tag{30-5}$$

For longitudinal vortex-type disturbances, we need also the third coordinate in the transverse z direction, which will remain unchanged. The velocity components are decomposed into mutually perpendicular components along $\mathbf{n}_1$, $\mathbf{n}_2$, $\mathbf{n}_3$ directions, with $\mathbf{n}_1$ along the streamlines, $\mathbf{n}_2$ normal to $\mathbf{n}_1$ (but *not* along the ξ-direction), and $\mathbf{n}_3$ in the z direction. In fact, the lengths are slightly stretched (for later convenience) in the $\mathbf{n}_1$, $\mathbf{n}_2$ directions, defined by

$$\begin{aligned} \mathbf{n}_1 &= (1 + \epsilon s)\mathbf{i} + \epsilon r\mathbf{j} \\ \mathbf{n}_2 &= -\epsilon r\mathbf{i} + (1 + \epsilon s)\mathbf{j} \\ \mathbf{n}_3 &= \mathbf{k} \end{aligned} \tag{30-6}$$

where $\mathbf{i}$, $\mathbf{j}$ and $\mathbf{k}$ are the unit vectors of the Cartesian coordinates, and

$$\left.\begin{aligned} r &= \alpha Q \sin \alpha\xi \\ s &= \frac{dQ}{dy} \cos \alpha\xi - Q \frac{dj}{dy} \sin \alpha\xi \end{aligned}\right\} \tag{30-7}$$

Now the velocity vector is likewise decomposed into $\mathbf{n}_1$, $\mathbf{n}_2$, and $\mathbf{n}_3$ components,

$$\mathbf{u} = u_a\mathbf{n}_1 + u_b\mathbf{n}_2 + u_c\mathbf{n}_3 \tag{30-8}$$

so that the *combined* flow, before the longitudinal vortex disturbances are assumed to take place, is described by

$$\bar{u}_a = \bar{u}(\eta), \qquad \bar{u}_b = 0, \qquad \bar{u}_c = 0 \tag{30-9}$$

The equations of motion in these variables (velocity components u_a, u_b, u_c, and pressure p, in terms of ξ, η, z, t) are complicated and the reader is referred to the original work. These are subsequently linearized in the usual manner for the perturbation $\hat{u}_a$, $\hat{u}_b$, $\hat{u}_c$, and $\hat{p}$, which are defined by

$$\begin{aligned} u_a &= \bar{u}_a(\eta) + \hat{u}_a(\xi, \eta, z, t) \\ u_b &= \hat{u}_b(\xi, \eta, z, t) \\ u_c &= \hat{u}_c(\xi, \eta, \xi, t) \\ p &= p_0(\eta) + \hat{p}(\xi, \eta, z, t) \end{aligned}$$

Now, for longitudinal vortices, it may be *postulated* that

$$\left.\begin{aligned} \hat{u}_a &= f_1(\eta) \cos \beta z e^{\lambda t} \\ \hat{u}_b &= f_2(\eta) \cos \beta z e^{\lambda t} \\ \hat{u}_c &= f_3(\eta) \sin \beta z e^{\lambda t} \\ \hat{p} &= \hat{p}_1(\eta) \cos \beta z e^{\lambda t} \end{aligned}\right\} \tag{30-10}$$

and attention is next focused on the least stable part of the streamlines by setting $\xi = 0$. (As Witting stated, the rather drastic simplification $\xi = 0$ is actually closer to reality, even in the case of a fixed curved wall, than Görtler's original model of constant curvature streamlines.)

The lowest approximation finally yields the following differential equations for the neutral case $\lambda = 0$:

$$\left.\begin{aligned} (D^2 - \beta^2)f_1 &= Re \frac{d\bar{u}_a}{d\eta} f_2 \\ (D^2 - \beta^2)^2 f_2 &= -2\epsilon\beta^2 Re\, \alpha^2 Q \left(\bar{u}_a - \frac{c}{2}\right) f_1 \end{aligned}\right\} \tag{30-11}$$

where $D \equiv d/d\eta$, c is the wave speed of the transverse disturbance and the boundary conditions are $f_1 = f_2 = f_2' = 0$ for $\eta = 0$ and as $\eta \to \infty$. Eq. 30-11 may be compared against Eq. 25-1, if we note that in the earlier notation, the radius of curvature for the present case is

$$R = -\frac{1}{\epsilon\alpha^2 Q(y)}$$

The fact that the curvature of streamlines within the fluid is taken into account is an improvement over Eq. 25-1, in problems when both are applicable. (One such application was made by Witting [*109*], considering the flow over a slightly wavy wall.)

The problem now is to find the eigenvalue $\mu_0 = 2Re^2\epsilon\alpha^2 Q_{max}$ with given $\bar{u}_a(\eta)$ and $Q(\eta)$, which was carried out by converting Eq. 30-11 into an integral equation similar to Eq. 25-3 and employing an iterative procedure. For the neutral oscillations calculated by Schlichting [*30*],

(i) $Re = 2620, \quad \alpha = 0.466, \quad c = 0.350$

(ii) $Re = 6080, \quad \alpha = 0.737, \quad c = 0.350$

his results are shown as Fig. G,30b.

Remarkable is the result that the longitudinal vortices would be excited for the extremely small amplitude of the transverse wave disturbances. The model being highly idealized, however, it appears to remain a problem for the experimentalists to investigate in detail just what might be actually happening when such instability is indicated. A limiting amplitude of the secondary flow conceivably would be reached

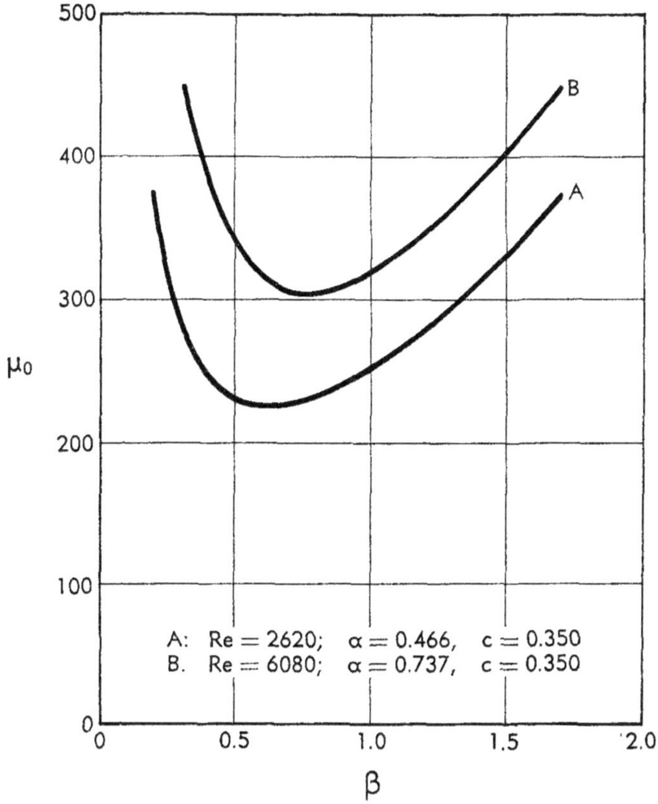

Fig. G,30b. Neutral curves for longitudinal vortex disturbances on basic flow of combined laminar boundary layer and neutral transverse wave.

in a manner not much different from that described in Art. 29. But, further theoretical attack in setting up new stability problems, in the direction outlined in Art. 28, is clearly impossible without first knowing more precisely the equilibrium distorted flow.

G,31. Cited References.

1. Lin, C. C. *The Theory of Hydrodynamic Stability.* Cambridge Univ. Press, 1955.
2. Haupt, O. Uber die Entwicklung einer willkurlichen Funktion nach den Eigenfunktionen des Turbulene problems. *Sitzungsberichte d. K. bayer. Akad. Wiss.*, 289–301 (1912).
3. Grohne, R. Uber das Spektrum bei Engenschwingungen ebener Laminar strömungen. *Z. angew. Math. u. Mech. 35*, 344–357 (1954).
4. Dunn, D. W., and Lin, C. C. On the stability of the laminar boundary layer in a compressible fluid. *J. Aeronaut. Sci. 22*, 455–477 (1955).
5. Squire, H. B. On the stability for three-dimensional disturbances of viscous fluid flow between parallel walls. *Proc. Roy. Soc. London A142*, 621–628 (1933).
6. Jungclaus, G. On the stability of laminar flow with three-dimensional disturbances. *Inst. for Fluid Dynamics and Applied Math., Univ. Maryland, Tech. Note BN-110*, Sept. 1957.

7. Kuethe, A. M. Some aspects of boundary-layer transition and flow separation on cylinders in yaw. *Proc. Midwestern Conference on Fluid Dynamics, Ann Arbor, Michigan*, 44–55 (1950).
8. Rayleigh, Lord (J. W. Strutt). On the question of the stability of the flow of fluids. *Scientific Papers, Cambridge Univ. Press, 3*, 575–584 (1892).
9. Heisenberg, W. Über Stabilität und Turbulenz von Flüssigkeitsströmen. *Ann. Physik. 74*, 577–627 (1924).
10. Tollmien, W. Über die Entstehung der Turbulenz. *Nachr. Ges. Wiss. Göttingen, Math.-phys. Klasse*, 21–44 (1929).
11. Lin, C. C. On the stability of two-dimensional parallel flows, Parts I, II, III. *Quart. Appl. Math. 3*, 117–142, 218–234, 277–301 (1945).
12. Taylor, G. I. Stability of a viscous liquid contained between two rotating cylinders. *Phil. Trans. A223*, 289–343 (1923).
13. Görtler, H. Über eine dreidimensionale laminarer Grenschichten an konkaven Wänden. *Nachr. Ges. Wiss. Göttingen N.F.2*, (*1*), 1–26 (1940).
14. Stuart, J. T. On the non-linear mechanics of hydrodynamic stability. *J. Fluid Mech. 4*, 1–21 (1958).
15. Synge, J. L. Hydrodynamic stability. Semi-centennial publications of the *American Math. Soc. 2* (*Addresses*), 227–269 (1938).
16. Lessen, M. Note on a sufficient condition for the stability of general plane parallel flows. *Quart. Appl. Math. 10*, 184–186 (1952).
17. Lin, C. C. Some physical aspects of the stability of parallel flows. *Proc. Natl. Acad. Sci. 40*, 741–747 (1954).
18. Høiland, E. Two-dimensional perturbation of linear fluid flow. *Geofys. Publ. 18*, (*1*), Oslo, 1953.
19. Tollmien, W. Asymptotische Integration der Störungsdifferentialgleichung ebener laminarer Strömungen bei höhen Reynoldsschen Zahlen. *Z. angew. Math. u. Mech. 25/27*, 33–50, 70–83 (1947).
20. Holstein, H. Über die äussere und innere Reibungsschicht bei Störungen laminarer Strömungen. *Z. angew. Math. u. Mech. 30*, 25–49 (1950).
21. Wasow, W. Asymptotic solution of the differential equation of hydrodynamic stability in a domain containing a transition point. *Annals of Math. 58*, 222–252 (1953).
22. *Tables of the Modified Hankel Functions of Order 1/3.* Harvard Univ., 1945.
23. Timman, R., Zaat, J. A., and Burgerhout, T. J. Stability diagrams for laminar boundary layer flow. *Natl. Aeronaut. Research Inst., Amsterdam, NLL Tech. Note F193*, 1956.
24. Schlichting, H. Zur Entstehung der Turbulenz bei der Plattenströmung. *Nachr. Ges. Wiss. Göttingen, Math.-phys. Klasse*, 181–208 (1933).
25. Thomas, L. H. The stability of plane Poiseuille flow. *Phys. Rev. 91*, 780–783 (1953).
26. Schubauer, G. B., and Skramstad, H. K. Laminar boundary layer oscillations and transition on a flat plate. *NACA Rept. 909*, 1948; *J. Research Natl. Bur. Standards 38*, 251–292 (1947); *J. Aeronaut. Sci. 14*, 69–78 (1947).
27. Wortmann, F. X. Untersuchung instabiler Grenzschichtschwingungen in einem Wasserkanal mit der Tellurmethode. *50 Jahre Grenzschichtforschung.* Vieweg, Braunschweig, 1955.
28. Shen, S. F. Calculated amplified oscillations in plane Poiseuille and Blasius flows. *J. Aeronaut. Sci. 21*, 62–64 (1954).
29. Tollmien, W. Über Schwingungen in Laminaren Strömungen und die Theorie der Turbulenz. *Proc. 8th Intern. Congress Theoretical and Appl. Mech.*, 23–47 (1955).
30. Schlichting, H. Amplitudenverteilung und Energiebilanz der Kleinen Storungen bei der Plattengrenzschicht. *Nachr. Ges. Wiss. Göttingen, Math.-phys. Klasse 1*, 47–78 (1935).
31. Lees, L., and Lin, C. C. Investigation of the stability of the laminar boundary layer in a compressible fluid. *NACA Tech. Note 1115*, 1946.

32. Lees, L. The stability of the laminar boundary layer in a compressible fluid. *NACA Rept. 876*, 1947.
33. Liepmann, H. W., and Fila, G. H. Investigations of effects of surface temperature and single roughness elements on boundary layer transition. *NACA Rept. 890*, 1947.
34. Higgins, R. W., and Pappas, C. C. An experimental investigation of the effect of surface heating on boundary layer transition on a flat plate in supersonic flow. *NACA Tech. Note 2351*, 1951.
35. Laufer, J. Experimental observation of laminar boundary layer oscillations in supersonic flow. *J. Aeronaut. Sci. 23*, 184–185 (1956).
36. Laufer, J. and Vrebalovich, T. Stability of a supersonic boundary layer on a flat plate. *Calif. Inst. Technol. Jet Prop. Lab. Tech. Rept. 20-116*, 1958.
37. Demetriades, A. An experimental investigation of the stability of the hypersonic laminar boundary layer. *Calif. Inst. Technol. Guggenhein Aeronaut. Lab. Hypersonic Res. Proj. Memo. 43*, 1958.
38. Van Driest, E. R. Calculation of the stability of the laminar boundary layer in a compressible fluid on a flat plate with heat transfer. *J. Aeronaut. Sci. 19*, 801–812 (1952).
39. Gazley, C. Boundary-layer stability and transition in subsonic and supersonic flow. *J. Aeronaut. Sci. 20*, 19–28 (1953).
40. Dryden, H. L. Recent investigations of the problems of transition. *Z. Flugwiss. 4*, 89–95 (1956).
41. Schlichting, H. Über die Theoretische Berechnung der Kritischen Reynoldschen Zahl einer Reibungsschicht in Beschleungter und Verzögerter Stromung. *Jahrbuch deut. Luftfahrtforschung I*, 97 (1940).
42. Pretsch, J. Die Stabilität einer ebenen Laminarströmung bei Druckgefälle and Druckanstieg. *Jahrbuch deut. Luftfahrtforschung I*, 58 (1941).
43. Schlichting, H., and Ulrich, A. Zur Berechnung des Umschlages Laminar-Turbulent. *Jahrbuch deut. Luftfahrtforschung I*, 8 (1942).
44. Hahnemann, E., Freeman, J. C., and Finston, M. Stability of boundary layer in the entrance section of channel. *J. Aeronaut. Sci. 15*, 493–496 (1948).
45. Weil, H. Effects of pressure gradient on stability and skin friction in laminar boundary layers in compressible fluids. *J. Aeronaut. Sci. 18*, 311–318 (1951).
46. Laurmann, J. A. Stability of the compressible laminar boundary layer with an external pressure gradient. *College of Aeronautics, Cranfield, Rept. 48*, 1951.
47. Low, G. M. Cooling requirements for stability of laminar boundary layer with small pressure gradient at supersonic speeds. *NACA Tech. Note 3103*, 1954.
48. Ulrich, A. Theoretische Untersuchungen über die Widerstandsersparnis durch Laminar-haltung mit Absaugung. *Schrift. dt. Akad. d. Luftfahrtforschung 8 B*, (*2*), 1944.
49. Pretsch, J. Umschlagbeginn und Absaugung. *Jahrbuch deut. Luftfahrtforschung I*, 1–7 (1942).
50. Chiarulli, P., and Freeman, J. C. Stability of the boundary layer. *Air Materiel Command, Dayton, Tech. Rept. F-TR/1197-1A*, 1948.
51. Bussman, K., and Munz, H. Die Stabilität der Laminaren Reibungsschicht mit Absaugung. *Jahrbuch deut. Luftfahrtforschung I*, 36 (1942).
52. Schlichting, H. *Grenzschicht-Theorie*, 3rd ed. Verlag G. Braun, Karlsruhe, 1958.
53. Wuest, W. Näherungsweise Berechnung und Stabilitätsverhalten von Laminaren Grenschichten mit Absaugung durch Einzelschlitze. *Ing. Arch. 21*, 90–103 (1953).
54. Tetervin, N., and Levine, D. A. A study of the stability of the laminar boundary layer as affected by changes in the boundary-layer thickness in regions of pressure gradient and flow through the surface. *NACA Tech. Note 2752*, 1952.
55. Shen, S. F., and Persh, J. The limiting wall temperature ratios required for complete stabilization of laminar boundary layers with blowing. *J. Aeronaut. Sci. 23*, 286–287 (1956).
56. Low, G. M. The compressible laminar boundary layer with fluid injection. *NACA Tech. Note 3404*, 1955.

57. Morduchow, M., Grape, R. G., and Shaw, R. P. Stability of boundary layer near a stagnation point over an impermeable wall and a wall cooled by normal fluid injection. *NACA Tech. Note 4037*, 1957.
58. Görtler, H. Über den Einfluss der Wandkrümmung auf die Entstehung der Turbulenz. *Z. angew. Math. u. Mech. 20*, 138–147 (1940).
59. Schlichting, H. Über die Entstehung der Turbulenz in einem rotierenden Zylinder. *Nachr. Ges. Wiss., Göttingen, Math.-phys. Klasse*, 160–198 (1932).
60. Pretsch, J. Über die Stabilität einer Laminarströmung in einem geraden Rohr mit Kreisformigen Querschnitt. *Z. angew. Math. u. Mech. 21*, 204–217 (1941).
61. Tatsumi, K. Stability of the laminar inlet-flow prior to the formation of Poiseuille regime. *J. Phys. Soc. Japan 7*, 489–502 (1952).
62. Goldstein, S. *Modern Developments in Fluid Dynamics.* Clarendon Press, Oxford, 1938.
63. Pretsch, J. Über die Stabilität einer Laminarströmung um ein Kugel. *Luftfahrtforschung 18*, 341–344 (1941).
64. Chiarulli, P. Stability of two-dimensional velocity distribution of the half-jet type. *Air Materiel Command, Wright-Patterson Air Force Base, Tech. Rept. F-TS/1228-1A*, 1949.
65. Lessen, M. On the stability of free laminar boundary layer between parallel streams. *NACA Rept. 979*, 1950.
66. Sato, H. Experimental investigation of the transition of a laminar separated layer. *J. Phys. Soc. Japan 11*, 1956.
67. Savic, P. On acoustically effective vortex motion in gaseous jets. *Phil. Mag. 32*, 245–252 (1941).
68. Savic, P., and Murphy, J. V. The symmetrical vortex street in sound-sensitive plane jets. *Phil. Mag. 34*, 139–144 (1943).
69. Lessen, M., Lew, H. G., Pai, S. I., Fanucci, J. B., and Fox, J. A. Hydrodynamic stability. *Penn. State Univ., Dept. of Aeronaut. Eng. Tech. Rept. 2*, 1954.
70. Foote, J. R., and Lin, C. C. Some recent investigations in the theory of hydrodynamic stability. *Quart. Appl. Math. 8*, 265–280 (1950).
71. Lin, C. C. On the stability of the laminar mixing regime between two parallel streams in a gas. *NACA Tech. Note 2887*, 1953.
72. Lessen, M., and Fox, J. A. The stability of boundary layer type flows with infinite boundary conditions. *50 Jahre Grenzschichtforschung.* Vieweg, Braunschweig, 1955.
73. Yih, C. S. Stability of parallel laminar flow with a free surface. *Proc. 2nd U.S. Natl. Congress Appl. Mech.*, 623–628 (1954).
74. Tatsumi, T., and Kakutani, T. The stability of a two-dimensional laminar jet. *J. Fluid Mech. 4*, 261–275 (1958).
75. McKoen, C. H. Stability of laminar wakes. *Brit. Aeronaut. Research Council Current Paper 303*, 1957.
76. Curle, N. On the hydrodynamic stability in unlimited fields of viscous flow. *Proc. Roy. Soc. London A238*, 489–501 (1957).
77. Pai, S. I. On the stability of two-dimensional laminar jet flow of gas. *J. Aeronaut. Sci. 18*, 731–742 (1951).
78. Pai, S. I. On the stability of a vortex sheet in an inviscid compressible fluid. *J. Aeronaut. Sci. 21*, 325–328 (1954).
79. Gregory, N., Stuart, J. T., and Walker, W. S. On the stability of three-dimensional boundary layers with application to the flow due to a rotating disc. *Phil. Trans. A248*, 155–199 (1955).
80. Fales, E. N. A new laboratory technique for the investigation of fluid turbulence. *J. Franklin Inst. 259*, 491–514 (1955).
81. Anscombe, A. and Illingworth, L. N. Wind-tunnel observations of boundary-layer transition on a wing at various angles of sweepback. *Brit. Aeronaut. Research Council Reps. and Mem. 2968*, 1956.
82. Prandtl, L. *Einfluss stabilisierener Krafte auf die Turbulenz.* (Fortrage aus d. Geb. d. Aerodyn. u. nerwandter Gebiete, Aachen.) Springer, Berlin, 1930.

83. Taylor, G. I. Effect of variation in density on the stability of supported streams of fluid. *Proc. Roy. Soc. London A132*, 499–523 (1931).
84. Goldstein, S. On the stability of supposed streams of fluids of different densities. *Proc. Roy. Soc. London A132*, 523 (1931).
85. Drazin, P. G. The stability of a shear layer in an unbounded heterogeneous inviscid fluid. *J. Fluid Mech. 4*, 214–224 (1958).
86. Schlichting, H. Turbulenz bei Warmeschichtung. *Z. angew. Math. u. Mech. 15*, 313 (1935).
87. Eckert, E. R. G. Interferometric studies on the stability and transition to turbulence of a free-convection boundary layer. *Proc. General Discussion on Heat Transfer, Inst. Mech. Eng.*, London, 1951.
88. Plapp, J. E. *Laminar Boundary-Layer Stability in Free-Convection. Ph.D. Thesis*, Dept. of Aeronaut. Eng., Calif. Inst. Technol., 1957.
89. Thompson, W. B. Thermal convection in a magnetic field. *Phil. Mag. 42*, 1417–1432 (1951).
90. Chandrasekhar, S. On the inhibition of convection by a magnetic fluid. *Phil. Mag. 43*, 501–532 (1952).
91. Chandrasekhar, S. The stability of viscous flow between rotating cylinders in the presence of a magnetic fluid. *Proc. Roy. Soc. London A216*, 293–309 (1953).
92. Chandrasekhar, S. The instability of a layer of fluid heated below and subject to the simultaneous action of magnetic field and rotation. *Proc. Roy. Soc. London A225*, 173–184 (1954).
93. Chandrasekhar, S. The stability of viscous flow between rotating cylinders. *Mathematika 1*, 5–13 (1954).
94. Michael, D. H. Stability of plane parallel flows of electrically conducting fluids *Proc. Cambridge Phil. Soc. 49*, 166–168 (1953).
95. Stuart, J. T. On the stability of viscous flow between parallel planes in the presence of a coplanar magnetic field. *Proc. Roy. Soc. London A221*, 189–206 (1954).
96. Lock, R. C. The stability of the flow of an electrically conducting fluid between parallel planes under a transverse magnetic field. *Proc. Roy. Soc. London A233*, 105–125 (1955).
97. Rossow, V. J. Boundary-layer stability diagrams for electrically conducting fluids in the presence of a magnetic field. *NACA Tech. Note 4282*, 1958.
98. Bader, M., and Carlson, W. C. A. Measurement of the effect of an axial magnetic field on the Reynolds number of transition in mercury flowing through a glass tube. *NACA Tech. Note 4274*, 1958.
99. Shen, S. F. The theory of laminar stability of compressible boundary layers with injection of a different gas. *Nav. Ord. Lab. Rept. 6654*, 1958.
100. Covert, E. The stability of binary boundary layers. *Mass. Inst. Technol. Naval Supersonic Lab. Tech. Rept. 217*, 1957.
101. Shen, S. F. The steady laminar boundary layer over a flat plate with injection of a different gas. *Nav. Ord. Lab. Rept. 4235*, 1956.
102. Baron, J. R. The binary-mixture boundary layer associated with mass transfer cooling at high speeds. *Mass. Inst. Technol. Naval Supersonic Lab. Tech. Rept. 60*, 1956.
103. Shen, S. F. Effect of chemical reaction on the inviscid criterion for laminar stability of parallel flows. *Proc. Midwest Conference on Fluid Mech.*, 11–20 (1957).
104. Lord Rayleigh (J. W. Strutt). On the dynamics of revolving fluids. *Scientific Papers, Cambridge Univ. Press, 6*, 447–453 (1916).
105. von Kármán, Th. Some aspects of the turbulence problem. *Proc. 4th Intern. Congress. Appl. Mech.*, 54–91 (1934).
106. Synge, J. L. On the stability of a viscous liquid between two rotating coaxial cylinders. *Proc. Roy. Soc. London A167*, 250–256 (1938).
107. Pellew, A., and Southwell, R. V. On maintained convective motion in a fluid heated from below. *Proc. Roy. Soc. London A176*, 312–343 (1940).
108. Schultz-Grunow, F., and Hein, H. Beitrag zur Couetteströmung. *Z. Flugwiss. 4*, 28–30 (1956).
109. Witting, H. Über den Einfluss der Strömlinienkrümmung auf die Stabilität

laminarer Strömungen. *Habilitations-schrift, Naturwiss-math. Fakultat, Üniv. Freiburg i. Br.*, 1958.

110. Di Prima, R. C. Application of the Galerkin method to the calculation of the stability of curved flows. *Quart. Appl. Math. 13*, 55–62 (1955).
111. Meksyn, D. Stability of viscous flow between rotating cylinders, I, II, III. *Proc. Roy. Soc. London A187*, 115–128, 480–491, 492–504 (1946).
112. Goldstein, S. The stability of viscous flow between rotating cylinders. *Proc. Cambridge Phil. Soc. 33*, 41–61 (1937).
113. Chandrasekhar, S. The stability of viscous flow between rotating cylinders in the presence of a radial temperature gradient. *J. Rat. Mech. and Anal. 3*, 181–207 (1954).
114. Schultz-Grunow, F. Zur Stabilität der Couette Strömung. *Z. angew. Math. u. Mech. 38*, 323 (1958).
115. Michael, D. H. The stability of an incompressible electrically conducting fluid rotating about an axis when the current flows parallel to the axis. *Mathematika 1*, 45–50 (1954).
116. Hämmerlin, G. Über das Eigenwertproblem der dreidimensionalen Instabilität laminarer Grenzschichten an Konkaven Wänden. *J. Rat. Mech. and Anal. 4*, 279–321 (1955).
117. Görtler, H. Dreidimensionale Instabilität der ebenen Staupunktströmung gegenüber wirbelartigen Störungen. *50 Jahre Grenzschichtforschung.* Vieweg, Braunschweig, 1955.
118. Smith, A. M. O. On the growth of Taylor-Görtler vortices along highly concave walls. *Quart. Appl. Math. 13*, 233 (1955).
119. Liepmann, H. W. Investigation of boundary layer transition on concave walls. *NACA Wartime Rept. W-87*, 1945.
120. Gregory, N., and Walker, W. S. The effect on transition of isolated surface excrescences in the boundary layer. *Brit. Aeronaut. Research Council Repts. and Mem. 2772*, 1950.
121. Di Prima, R. C., and Dunn, D. W. The effect of heating and cooling on the stability of the boundary layer flow of a liquid over a curved surface. *J. Aeronaut. Sci. 23*, 913 (1956).
122. Lessen, M. On the hydrodynamic stability of curved laminar flows. *Z. angew. Math. u. Mech. 38*, 95–99 (1958).
123. Lees, L. Note on the stabilizing effect of centrifugal forces on the laminar boundary-layer over convex surfaces. *J. Aeronaut Sci. 25*, 407–408 (1958).
124. Dean, W. R. Fluid motion in a curved channel. *Proc. Roy. Soc. London A121*, 402–420 (1928).
125. Yih, C. S., and Sangster, W. M. Stability of laminar flow in curved channels. *Phil. Mag. 2 (8)*, 305–310 (1957).
126. Hämmerlin, G. Die Stabilität der Strömung in einem gekrümmten Kanal. *Arch. Rat. Mech. and Anal. 1*, 211–224 (1958).
127. Piercy, N. A. V., and Richardson, E. G. The variation of velocity amplitude close to the surface of a cylinder moving through a viscous fluid. *Phil. Mag. 6*, 970–976 (1928).
128. Schuh, H. A new method for calculating the laminar heat transfer on cylinders of arbitrary cross-section and on bodies of revolution at constant and variable wall temperature. *Stockholm KTH Tech. Note 33*, 1953.
129. Hämmerlin, G. Zur Instabilitätstheorie der ebenen Staupunkstströmung. *50 Jahre Grenzschichtforschung.* Vieweg, Braunschweig, 1955.
130. Landau, L. On the problem of turbulence. *Compt. rend. Acad. Sci. U.R.S.S. 44*, 311 (1944).
131. Schubauer, G. B., and Klebanoff, P. S. Contributions on the mechanics of boundary-layer transition. *NACA Rept. 1289*, 1956.
132. Meksyn, D., and Stuart, J. T. Stability of viscous motion between parallel planes for finite disturbances. *Proc. Roy. Soc. London A208*, 517–526 (1951).
133. Kryloff, N., and Bogoliuboff, N. *Introduction to Non-linear Mechanics.* A free translation by S. Lefschetz. Princeton Univ. Press, 1947.

134. Jack, J. R., Wisniewsky, R. J., and Diaconis, N. S. The effect of external surface cooling on boundary layer transition. *NACA Tech. Note 4094*, 1957.
135. Sears, W. R., and McCune, J. E. On the concepts of moving electric and magnetic fields in magnetohydrodynamics. *J. Aeronaut. Sci. 26*, 674–675 (1959).
136. Lees, L., and Reshotko, E. Stability of the compressible laminar boundary layer. *J. Fluid Mech. 12*, 550–590 (1962).
137. Brown, W. B. Exact numerical solution of the complete linearized equations for the stability of compressible boundary layers. Presented at the *1961 Meeting, Fluid Dynamics Division, Amer. Phys. Soc.*, Berkeley, Calif., 1961; *Norair Rept. No. NOR-62-15*, Northrop Corporation, 1962.
138. Reshotko, E. Transition reversal and Tollmien-Schlichting instability. *Phys. Fluid 6*, 335–342 (1963).
139. Chandrasekhar, S. *Hydrodynamic and Hydromagnetic Stability*. Oxford University Press, 1962.
140. Klebanoff, P. S., Tidstrom, K. D., and Sargent, L. M. The three-dimensional nature of boundary-layer instability. *J. Fluid Mech. 12*, 1–34 (1962).
141. Benney, D. J. and Lin, C. C. On the secondary motion induced by oscillations in shear flow. *Phys. Fluids 3*, 656–657 (1960).
142. Stuart, J. T. Non-linear effects in hydrodynamic stability. Proc. *10th Intern. Congress Appl. Mech.*, 63–97, Elsevier Publishing Company, Amsterdam-New York, 1962.

INDEX

GPSR Authorized Representative: Easy Access System Europe - Mustamäe tee 50, 10621 Tallinn, Estonia, gpsr.requests@easproject.com

www.ingramcontent.com/pod-product-compliance
Lightning Source LLC
LaVergne TN
LVHW020957300926
849160LV00007B/178

* 9 7 8 0 6 9 1 6 2 4 7 4 7 *